Enkephalins and Endorphins

Stress and the Immune System

Enkephalins and Endorphins
Stress and the Immune System

Edited by

Nicholas P. Plotnikoff

Oral Roberts University
Tulsa, Oklahoma

Robert E. Faith

University of Houston
Houston, Texas

Anthony J. Murgo

West Virginia University
Morgantown, West Virginia

and

Robert A. Good

University of South Florida
St. Petersburg, Florida

SPRINGER SCIENCE+BUSINESS MEDIA, LLC

Library of Congress Cataloging in Publication Data

Enkephalins and endorphins.

Bibliography: p.
Includes index.
1. Enkephalins—Physiological effect. 2. Endorphins—Physiological effect. 3. Immune response—Regulation. 4. Stress (Physiology) 5. Stress (Psychology) 6. Psychoneuroimmunoendocrinology. I. Plotnikoff, Nicholas P.
QP552.E55E55 1986 616.07′9 85-23234
ISBN 978-1-4899-0559-8 ISBN 978-1-4899-0557-4 (eBook)
DOI 10.1007/978-1-4899-0557-4

Originally published by Plenum Press, New York in 1986
Softcover reprint of the hardcover 1st edition 1986

FOREWORD

Is this a time for a sleeping giant to rise? We have known since study of the lymphocyte and plasma cells really began in earnest in the early 1940's that the pituitary adrenal axis under intimate control of the hypothalamus could influence immunological functions profoundly. We have also known for at least 20 years in my recollection that female sex hormones can maximize certain immunity functions while male sex hormones tend to suppress many immunological reactions. The thyroid hormones accelerate antibody production while at the same time speeding up degradation of antibodies and immunoglobulins and thyroidectomy decreases the rate of antibody production. Further, much evidence has accumulated indicating that the brain, yes even the mind, can influence in significant ways susceptibility to infections, cancers and to development of a variety of autoimmune diseases. More than 20 years ago, my colleagues and I convinced ourselves, if no one else, that hypnosis can exert major influences on the effector limb of the classical atopic allergic reactions. We showed with Aaron Papermaster that the Prausnitz-Kustner reaction may be greatly inhibited, indeed largely controlled, by post-hypnotic suggestion. And it was not even necessary for us to publish our discovery because scientists in John Humphrey's laboratory at Mill Hill Research Center in London had beaten us to the punch. They described hypnotic control of both the PK reaction and delayed allergic reactions to tuberculin by hypnosis. Although he doesn't necessarily attribute the controls to neurological modulation, I have been convinced that the extraordinary rhythms Franz Halberg in Minnesota has elucidated for immunological functions that range from NK cell activity to the amount of antibody produced and even the tempo of allograft rejection reflect neuroendocrinological interactions with the immunological systems.

Halberg's Chronoimmunologic analyses have been underway for at least 20 years and would have begun earlier and progressed even more rapidly than they have had we immunologists been more responsive to his prodding and that of other chronobiologists associated with him.

Why then does the present moment strike me as so propitious a moment for a comprehensive new work considering the interaction of the major body networks?

For me, a new and truly golden age of psychoneuroimmunoendocrinology began when Wybran of Brussels first observed in 1979 that met-enkephalin can talk to lymphocytes directly and its conversation with lymphocytes can be interrupted by naloxone. These rather crude beginnings have now been greatly refined and we know for sure that lymphocytes have receptors for met-enkephalin and have either surface or cytosolic receptors for a number of other hormones and neurohumoral mediators. We see in this volume and in numerous contemporary articles reaching our scientific journals that met-enkephalin, leu enkephalin and endorphins can reproducibly influence cell surface expression and functions of lymphocytes in vivo, antibody production, delayed allergic reactions and development and differentiation of lymphoid cells. We are even witnessing these days the first descriptions of responses of immunoparameters in healthy humans and patients with various kinds of immunodeficiencies including patients with AIDS and cancer. All this is happening right now and this science and this form of immunopharmacology will be rapidly developed in the years ahead.

But this is only one of many many exciting fields in psychoneuroimmunoendocrinology where incredible discoveries are turning up.

As a Visiting Professor at the University of Texas Medical Branch in Galveston last winter, I was introduced to a constellation of related studies by Blalock and his group of young colleagues. These studies established to my satisfaction that lymphocytes, like pituitary cells, can produce a molecule very like ACTH both immunologically and functionally and that like cells of the anterior pituitary, the lymphocytes have cytosolic receptors for cortisol. Through these, cortisol can suppress production of the ACTH-like molecule by these cells. As a classical cellular immunologist such a turn of events could never have entered my mind yet here it was, big as life, and it had been demonstrated by what seemed to me to be commanding and critical scientific methodology.

But these surprises are at least paralleled as surprises presented by Hall and Goldstein's (Washington) discovery that thymosin α_1 now a fully defined molecule (thymic hormone?) exerts functional and electrical influences on certain hypothalamic nuclear cells that can, in turn, exert influences on lymphoid cell function through the thymus. If these descriptions sound fanciful it is because I believe they are. Yet they and much, much more are presented in detail in this volume of collected multi-authored papers that reflect a burgeoning scientific field. Even Robert Ader and Nicholas Cohen's exciting discovery that the immune response can be regulated by taste via prior conditioning in the conditioned -aversion response after simultaneous exposure to a cytotoxic immunosuppressive chemical may before long be explained in precise immunopharmcological terms. I now would bet it will.

This field is developing more rapidly than I thought could ever be the case and I am thankful to the immunopharmacologists like Plotnikoff, Wybran, Hadden, Szentivanyi and others who have urged me to pay close attention.

Robert A. Good, M.D., Ph.D.

CONTENTS

INTRODUCTION: YING-YANG HYPOTHESIS OF IMMUNOMODULATION

N.P. Plotnikoff, A.J. Murgo, R.E. Faith, and R.A. Good

Oral Roberts University School of Medicine
Tulsa, Oklahoma

Originally the inspiration for this book stemmed from a Symposium on Enkephalins; Endorphins: Stress and the Immune System held at the Annual meeting of the American Society of Pharmacology and Experimental Therapeutics in Philadelphia 1983. Since that meeting a significant increase in research has occurred and prompted us to compile this book. Recent research has shown that the immune system is exceedingly sensitive to the effects of enkephalins and endorphins. This sensitivity was illustrated by the immunodepressant effects of morphine on T-cells. In contrast the enkephalins and endorphins have shown to enhance T-cell function.

At the same time a major development in this field was the discovery that the adrenal glands are the source of peripheral enkephalins and endorphins. Two major prohormones have been identified in human adrenal glands, namely, Proenkephalin A as well as proopiomelanacortin hormone.

These prohormones have been shown to release various intermediate peptide fragments(Peptide E and F) as well as the final end products, methionine enkephalin, leucine enkephalin, and beta-endorphin. Since morphine is immunodepressant and is found to have high affinity for the mu and kappa receptors, it was of great interest to find that several of the prohormone peptide fragments also has a high affinity for the same receptors, suggesting that they may also have an immuno-depressant role. In contrast the final end products methionine enkephalin and leucine enkephalin and beta-endorphin have been discovered to be immuno-stimulant and bind preferentially to the delta and epsilon receptors.

This immunomodulation by prohormone peptide fragments and enkephalin-endorphin end products led to the Ying-Yang hypothesis of immunomodulation of T-cell function. We are proposing the working opioid peptide hypothesis that various prohormone fragments are released by different stressors resulting in fluctuations of immunomodulation in concert with the steroid hormones and the catecholamines from the adrenal glands. It is possible that a number of contradictory behavioral studies reported in the literature are a result of this immunomodulation(depression and/or stimulation) as a function of differential prohormone processing and type of duration of stressors.

It is intriguing to consider the possibility that long term stress may result in "depletion or exhaustion phenomenon" of prohormone end products (enkephalins-endorphins) depending upon the state of behavioral coping.

Perhaps 'replacement therapy' with the enkephalins-endorphins would be appropriate in certain selected clinical conditions.

Finally, very recent studies indicate that methionine enkephalin stimulates the production and release of lymphokines(interferons and interleukins) from macrophages and T-cells. Therefore, the clinical effects of methionine enkephalin in enhancing T-cell subsets in Kaposis sarcoma, AIDS, and lung cancer patients may be of therapeutic interest(chapters at the end of the book).

We invite the reader to join us in studying the phenomenon of prohormone fragments released by stress as they impact on the immune system.

REFERENCES

Brown,S.L. and D.E. Van Epps. Beta Endorphin(BE), Met-Enkephalin(ME), and Corticotrophin(ACTH) Modulate the Production of Gamma Interferon(IFN) In Vitro. Fed. Proc. 44,4,949,1985.

Faith, R.E., N.P. Plotnikoff, and A.J. Murgo. Effects of Opiates and Neuropeptides on Immune Functions. pp. 300-311. Ed. C.W. Sharp. Mechanisms of Tolerance and Dependence, NIDA Research Monograph 54,1984.

Herz, A. Multiple Endorphins as Natural Ligands of Multiple Opioid Receptors. 43-52, in Central and Peripheral Endorphins: Basic and Clinical Aspects, ed. by E.E. Muller and A.R. Genazzini. Raven Press, New York, 1984.

Kojima, K.,Kilpatrick, D.L. Stein, A.S. Jones, B.N., Udenfriend, S. 1982. Proenkephalin: a general pathway for enkephalin biosynthesis in animal tissues. Arch. Biochem. Biophys. 215:638-43.

Plotnikoff, N.P., G.C. Miller, J. Wybran, and N.F. Nimeh. Methionine enkephalin T-cell Enhancement in Kaposis's Sarcoma. Serono Symposia, Recent advances in primary and acquired immunodeficiencies(In Press). Copley Plaza Hotel, Boston, April 10,1985.

Plotnikoff, N.P., and A.J. Murgo. Enkephalins:Endorphins: Stress and the Immune System. Fed. Proc. 44,1,1.1985.

Weber, R.J. and C.B. Pert. Peptitergic Modulation of the Immune System. p. 35-42. from Central and Peripheral Endorphins: Basic and Clinical Aspects, ed. by E.E. Muller and A.R. Genazzani. Raven Press, N.Y.,1984.

Youkilis,E.; J. Chapman; E. Woods; and N.P. Plotnikoff. In Vivo immunostimulation and increased in vitro production of interleukin 1(IL-1) Activity by methionine enkephalin. Int. J. Immuno.Pharm. 7,3,79,1985.

CANDIDATE OPIOID PEPTIDES FOR INTERACTION WITH THE IMMUNE SYSTEM

Christopher J. Evans, Elizabeth Erdelyi, and Jack D. Barchas

Nancy Pritzker Laboratory of Behavioral Neurochemistry
Department of Psychiatry and Behavioral Sciences
Stanford University School of Medicine
Stanford, CA 94306

Endogenous opioids are a recently discovered group of peptides which have been implicated as modulators of a number of biological systems. Both opiate receptors and their peptide ligands are widely distributed throughout the CNS and are additionally located in some peripheral sites. The classical effects of opiates as analgesics have perhaps overshadowed many other important biological activities of this group of neuroactive peptides. This book provides strong indications that endogenous opioid peptides play a crucial role in the regulation of the immune system.

Many studies have shown that various stressors have dramatic affects on the functioning of the immune system. The question then becomes one of dissecting out individual components released on the stressful stimuli and determining which may be responsible for the observed changes in immunity. There is a large literature on the effects of glucocorticoids on suppression of the immune function (see Munck et al., 1984 for review). Furthermore, adrenaline has been shown to affect the immune kinetics (Depelchin and Letesson, 1981). A clue that opioids may influence immune responsiveness was indicated in early clinical studies of immuno deficiencies in heroin addicts (Brown et al., 1974). Recent work has shown that opioid peptides can modulate a number of cell types involved in the immune response, and these activities are the subject of other chapters in this book. This chapter will be concerned with the various forms of opioid peptides found in tissues actively secreting into the blood in response to a stressful stimulus, since it is these peptides that will be available for interaction with components of the immune system. Special emphasis will be placed on the opioid peptides that are present in the adrenal gland.

Opioid peptides are generated by enzymic processing of large precursors which themselves are not opiate active. Three opioid precursors have thus far been described: proopiomelanocortin (POMC or pro-ACTH-endorphin); proenkephalin (proenkephalin A), and prodynorphin (proenkephalin B or proneoendorphin/dynorphin). The cDNA representing the messenger RNA encoding the structure of all three opioid precursors has been cloned and the primary amino acid sequence deduced from the cDNA sequence (Nakanishi et al., 1979; Noda et al., 1982; Kakidani et al., 1982; Comb et al., 1982). The endogenous opioid peptides that have been isolated from various mammalian tissues all have at their N-termini an opiate core sequence: Tyr-Gly-Gly-Phe[met or leu]. The primary structure of POMC contains one opioid core sequence, proenkephalin contains seven, and prodynorphin three. The chemical and spatial integrity of certain features of this sequence are crucial for opiate activity. One essential component is the ionizable amino group at the N-terminus of the opioid core sequence in a particular spatial arrangement to the side chain of the tyrosine residue. Removal of the N-terminal positive charge by acylation of the α-amino group completely obliterates opiate-like properties of these peptides (Bradbury et al., 1977). Therefore, if the α-amino group of the opioid core sequence is engaged in a peptide linkage as in the precursor structures, or α-N-acetylated--a modification found in some tissues, the opioid core is completely inactivated. The N-terminal tyrosine residue of every opioid core sequence is preceded in the three precursors by paired basic residues and in all but one core sequence found in bovine proenkephalin this is a lysyl arginine (see Fig. 1). Cleavage of the precursor at the paired basic residues preceding the opioid core is a prerequisite for the generation of peptides with opiate-like properties. Nature has used paired basic residues as precursor processing signals for the generation of many bioactive peptides (see Steiner et al., 1974 for review). In addition, many precursors are cleaved at single arginine residues (Rehfeld, 1981; Roth et al., 1983). If all paired basics and single arginines were precursor processing sites, the opiate-active peptides would be from POMC, β-endorphin(1-27); from prodynorphin, 3 copies of leu-enkephalin; and from proenkephalin, 6 copies of met-enkephalin and one 1 copy of leu-enkephalin. However, in most tissues, complete processing does not occur such that there are whole consortium of peptides with the opioid core at the N-terminus and various C-terminal extensions. It is interesting that the extent and nature of processing can be variable between tissues, an issue addressed in detail later in this chapter. Figure 1 shows the structure of the three opioid precursors and the endogenous biologically active peptides that have been characterized in various mammalian tissues (see Weber et al., 1983b for review).

The importance of the C-terminal tails following the opioid core sequence should not be understated. The tails infer selectivity for various receptor types, provide stability against exopeptidase attack (Austen et al., 1979), and can influence both the on and off rate of receptor/effector activation. These properties may be crucial when considering an opioid peptide as a candidate for interaction with the immune system. It should, however, be remembered that all endogenous opioid peptids are pure agonists for all receptor types and the microenvironment where the peptides are released and act can be a crucial factor with regard to issues relating to the importance of receptor selectivity and ligand stability.

Identification of Precursor Products

The tissue distribution and concentration of various processing products of the three opioid precursors can be studied using specific antisera as probes. Since all endogenous opioids have a common component at the N-terminus--the opioid core sequence--these specific antisera have to be directed to the C-terminal tails of the various products. As nature would have it, rabbits will indeed direct their antigenic response to the C-terminus of short peptides when certain injection protocols are followed (Weber et al., 1982b). Consequently, antisera can be readily raised which differentiate between the processing products. Often it is possible to obtain antisera that require the C-terminal amino acid as part of the antigenic site. A good example is an antisera we raised to dynorphin(1-8) which, when

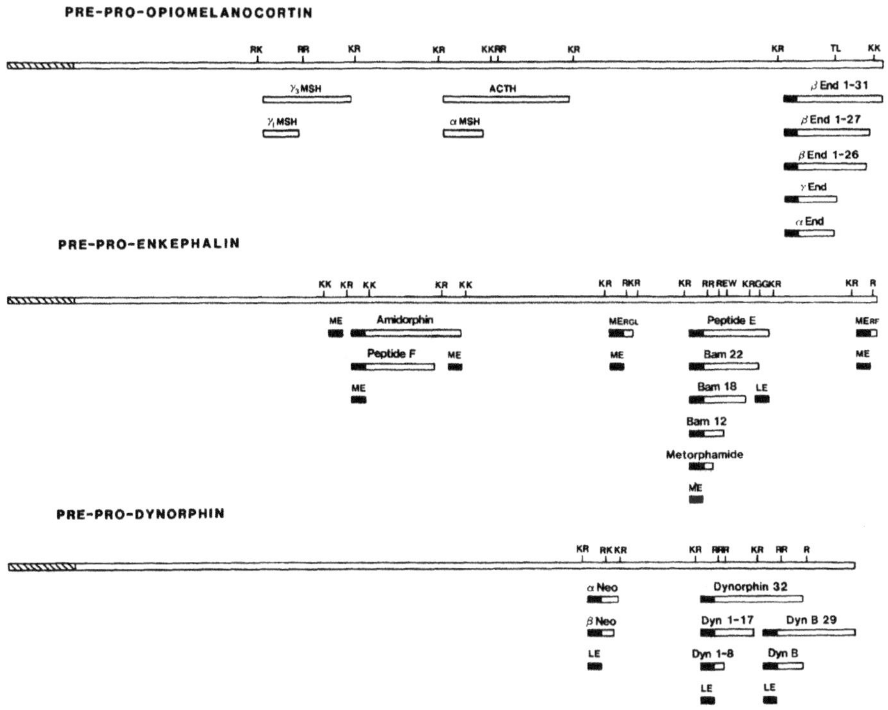

Fig. 1. A large number of endogenous bioactive peptides have been isolated and characterized from POMC, proenkephalin, and prodynorphin. This figure shows the molecular origin of these peptides. The opiate active core sequences are represented by solid bars ( ) and the putative signal regions of the precursors by thatched bars ( ). The following abbreviations have been used: MSH, melanocyte stimulating hormone; ACTH, adrenocorticotrophic hormone; End, endorphin; ME, met-enkephalin; LE, leucine-enkephalin; Neo, neoendorphin; Dyn, dynorphin; R, arginine; K, lysine; T, threonine; L, leucine; E, Glutamic acid; W, tryptophan; G, glycine; and F, phenylalanine. The structures are for bovine pre-pro-opiomelanocortin, bovine pre-pro-enkephalin and porcine pre-pro dynorphin.

used in radioimmunoassay, has less than 0.1% crossreactivity with dynorphin(1-7) or dynorphin(1-17). Radioimmunoassays using such specific antisera have proven to be an invaluable quantitative tool in the study of endogenous opioid peptides. These assays are quick, sensitive and reliable, although extreme caution must be taken when measuring in tissues such as blood and CSF where interfering substances causing displacement can be mistaken for the presence of peptides. It is advisable in these cases that RIAs be validated by chromatographic analysis of the immunoreactive material. Chromatography of immunoreactive peptides can also reveal different forms containing the same antigenic site. A common problem encountered is that most antisera raised to β-endorphin also recognize the opiate inactive precursor β-lipotropin. Gel filtration chromatography rapidly distinguishes between these two immunoreactive forms. The antisera can also be used for immunocytochemistry which provides precise localization of cells producing a particular opiate peptide. The analysis of opioid processing procucts by immunocytochemistry in combination with radioimmunoassays has proved to be very powerful and enabled the construction of detailed maps of the various opioid precursor products in mammalian tissues.

Opioid peptides are found in many peripheral sites that secrete into the blood stream and, consequently, could be available for activation of opiate receptors on cells concerned with the immunoresponse. The presence of β-endorphin-like immunoreactive peptides in the subpopulation of macrophages is of considerable interest in this regard (Lolait et al., 1984). If this β-endorphin-like immunoreactivity proves to be opiate active, an autocrine-like role can be postulated since opiates have been shown to stimulate the IgG mediated antibody-dependent cytotoxicity of macrophages (Foris et al., 1984). Opioid peptides are also present in high concentrations in the pituitary and adrenal glands. Both these endocrine tissues are activated during stress and warrant special attention with regard to their opioid content.

THE PITUITARY

Opioid peptides can be found in all three lobes of the pituitary gland. In the posterior lobe of many species, prodynorphin derived peptides are costored and probably coreleased with vasopressin (Martin and Voigt, 1981). In the oxytocin cells of rat there is immunocytochemical evidence for costorage with met-enkephalin (Martin et al., 1983). In the anterior lobe of the rat pituitary, prodynorphin derived peptides can be detected by radioimmunoassay and there is evidence that these peptides are stored in the gonadotropic hormone secreting cells. More relevant to the theme of this chapter is the β-endorphin-like peptides costored with adrenocorticotropin hormone (ACTH) in the anterior lobe. Proopiomelanocortin has within its structure both the sequence of β-endorphin and ACTH (see Fig. 1). The processing pathway in the anterior lobe of the pituitary is geared for the production of ACTH which is flanked at both the N-terminus and C-terminus by lysyl arginines. The cleavage of POMC at the C-terminus of ACTH leaves β-LPH a fragment of the precursor which contains the β-endorphin sequence yet is not opiate active. This peptide has been shown to be a major product of POMC in the anterior pituitary (Liotta et al., 1978). β-Endorphin (1-31), a very active opioid peptide in many bioassays, is also a product of POMC found in anterior pituitary extracts.

The intermediate lobe of the pituitary contains very high concentrations of POMC-derived peptides. Greater than 90% of the endorphin-like immunoreactive material in rat pituitary is present in the intermediate lobe. It has been demonstrated that certain stressors can stimulate release of the endorphin material not only from the anterior lobe of the pituitary but also the intermediate lobe (Przewlocki et al., 1982). However, the POMC products are very different in the two lobes. More complete processing of POMC occurs in the intermediate lobe than in the anterior lobe such that the major bioactive product of the ACTH portion of POMC is α-MSH--the N-terminal fragment of ACTH, acetylated at the N-terminus and amidated at the C-terminal valine. The endorphin portion of the precursor follows a similar processing pattern. The major products are extensively processed at all paired basic residues and then α-N-acetylated at the N-terminal tyrosine (see Table 1). As previously discussed, the α-N-acetylation is an important modification since it completely obliterates the opiate-like properties of this peptide (for reviews, see Eipper and Mains, 1980; O'Donohue and Dorsa, 1982). The question of why a tissue would synthesize an opioid precursor, then prior to release deactivate the opioid core remains unsolved. Perhaps the MSH portion of the precursor is the important bioactive product or else the acetylated endorphins in the pituitary have an as yet undiscovered nonopioid role.

Using an antisera we raised to α-N-acetyl-β-endorphin that did not crossreact with opiate active β-endorphin, it became possible to measure by radioimmunoassay the concentration of α-N-acetylated endorphins in tissue extracts. This assay, in conjunction with an assay that did not distinguish between the two endorphin forms (i.e., a C-terminal or middle region directed β-endorphin antisera), enabled the ratio of acetylated to nonacetylated endorphin to be assayed in various tissues (Weber et al., 1982b). In our studies, the intermediate lobe of the pituitary was found to be the only tissue where α-N-acetylation of endorphin was a prominent modification.

The human pituitary is different from other species in that it has no classical neurointermediate lobe. It is an interesting and important finding that human pituitary has no acetylated endorphins or acetylated α-MSH (Weber et al., 1982b; Evans et al., 1982). However, processing in some human pituitary cells does proceed to produce β-endorphin and α-MSH sized peptides.

OPIOID PEPTIDES IN ADRENAL GLANDS

Opioid peptides are also present in high concentrations in the adrenal glands of many species. It has been shown by immunocytochemistry and by RIA that the medulla portion of the gland contains the opioid peptides and that they are costored and coreleased with adrenalin (Viveros et al., 1979). Initial studies were performed on bovine adrenals since opiate activity was high and large quantities of tissue are readily available. A number of peptides containing opioid core sequences were isolated and characterized from this source including the BAM (Bovine, Adrenal, Medulla) fragments BAM 12 and BAM 22, peptide E and peptide F, and peptides containing the met-enkephalin-Arg-Phe and met-enkephalin-Arg-Gly-Leu sequence (Stern et al., 1981; Jones et al., 1982). These multiple opioid peptides subsequently were shown to be processing products of a single precursor-proenkephalin (see Fig. 1).

Recently we have identified a novel opioid processing product, BAM-18 from bovine adrenal glands (Evans et al., 1984). It is noteworthy that BAM-18 unlike BAM-12, BAM-22, peptide E and peptide F is in high concentrations in brain tissue as well as adrenal medulla. Additionally, amidorphin, a 26 amino acid opioid peptide with a C-terminal amide, has been isolated from bovine adrenal medulla (Seizinger et al., 1985). Like BAM-18, amidorphin is found in high concentrations in certain areas of brain. In general, processing of proenkephalin in brain is considerably more extensive than in bovine adrenal. For example, met-enkephalin-Arg-Gly-Leu as the heptapeptide is an abundant opioid in brain but in bovine adrenal this sequence is predominantly found as part of a 5.3 kilodalton nonopioid metabolite of proenkephalin (Jones et al., 1982; Lindberg et al., 1984). It is important to note that many of these incompletely processed fragments of proenkephalin are not opiate active, although they contain the opioid core sequence within their structure. The 5.3 and 3.6 kilodalton peptides containing respectively met-enkephalin-Arg-Gly-Leu and met-enkephalin-Arg-Phe at the C-terminus are both examples of proenkephalin processing products found in abundance in the adrenal which although not active at opiate receptors contain an opioid core embedded within their structure.

In our studies we have analyzed adrenals from a number of different species for the processing fragments of the three precursors. Specific radioimmunoassays (RIAs) to processing products have been developed and used for these studies. Table 1 illustrates a summary of the results. As a cautionary note in interpreting these data, it should be remembered

Table 1. Species Variability in Opioid Peptides In Adrenal Glands As Measured by RIA (levels in pmoles/gram)

		Human Medulla	Guinea Pig	Dog	Rabbit	Rat	Sheep
ProDyn	α-Neo-endorphin	BD	216	BD	BD	BD	BD
	Dynorphin 1-17	BD	129	BD	BD	BD	BD
	Dynorphin B	BD	113	BD	BD	BD	BD
	Leu Enkephalin	1132	209	522	85	9	150
ProEnk	Metorphamide	306	BD	33	BD	BD	7
	Met-enkephalin-RGL	957	878	28	8	6	316
	Met-enkephalin-RF	739	484	1541	136	21	200
POMC	β-Endorphin	144	BD	BD	BD	BD	BD
	α-MSH	53	BD	BD	BD	BD	BD
	ACTH	25	BD	BD	BD	BD	BD

Tissues were extracted in 5 wt vol acid/acetone and prepared for RIA using previously procedures (Weber et al., 1983b). Values were obtained from at least 5 adrenal samples for each species. The detection limit was 2-10 pmol/g depending on the assay. In many of the samples no immunoreactive material could be detected (BD). The opioid directed RIAs used above crossreact less than 0.5% with other endogenous opioids with the exception of the leu-enkephalin assay which crossreacts around 7% with met-enkephalin. The β-endorphin RIA recognizes β-LPH, β-endorphin(1-31), β-endorphin(1-27), β-endorphin (1-26) and their respective α-N-acetyl derivatives with equal crossreactivity. The α-MSH assay is C-terminally directed and does not recognize ACTH.

that the structure of the three precursors has not been characterized for all the animals investigated. Although there is considerable structural conservation of the opiate active portions of the precursors that have been sequenced from various species, mutations in certain regions affecting immunoreactivity should not be excluded. In all mammalian adrenals we analyzed, the RIAs detected peptides derived from pro-enkephalin. However, considerable differences were observed in the concentration of the various proenkephalin processing products. For example, metorphamide, an amidated opioid which has been isolated from brain (Weber et al., 1983) and pheochromocytoma (Matsuo et al., 1983) is in high concentrations in human adrenal but is undetectable in guinea pig, rabbit and rat adrenals and is in very low concentrations in dog and sheep adrenals. However, in extracts from rat and guinea pig brain, metorphamide can be detected in many regions (Sonders et al., 1984) indicating that these animals can indeed generate this processing product and that the sequence in this region of proenkephalin is probably conserved. Additional variances between species on the extent of processing have been exposed by sizing the immunoreactive material using column chromatography. Figure 2 shows the gel filtration immunoreactivity profiles of a human adrenal extract assayed with RIAs directed to the proenkephalin processing products metorphamide, met-enkephalin-Arg-Gly-Leu and met-enkephalin-Arg-Phe. All three RIAs detected material eluting at an extrapolated molecular weight of between 0.5 and 1 kilodalton. However, analysis of a guinea pig adrenal extract (Fig. 3) shows no metorphamide and very little low molecular weight met-enkephalin-Arg-Gly-Leu or met-enkephalin-Arg-Phe immunoreactive material. The guinea pig adrenal met-enkephalin-Arg-Phe and met-enkephalin-Arg-Gly-Leu immunoreactivity may correspond to the nonopiate

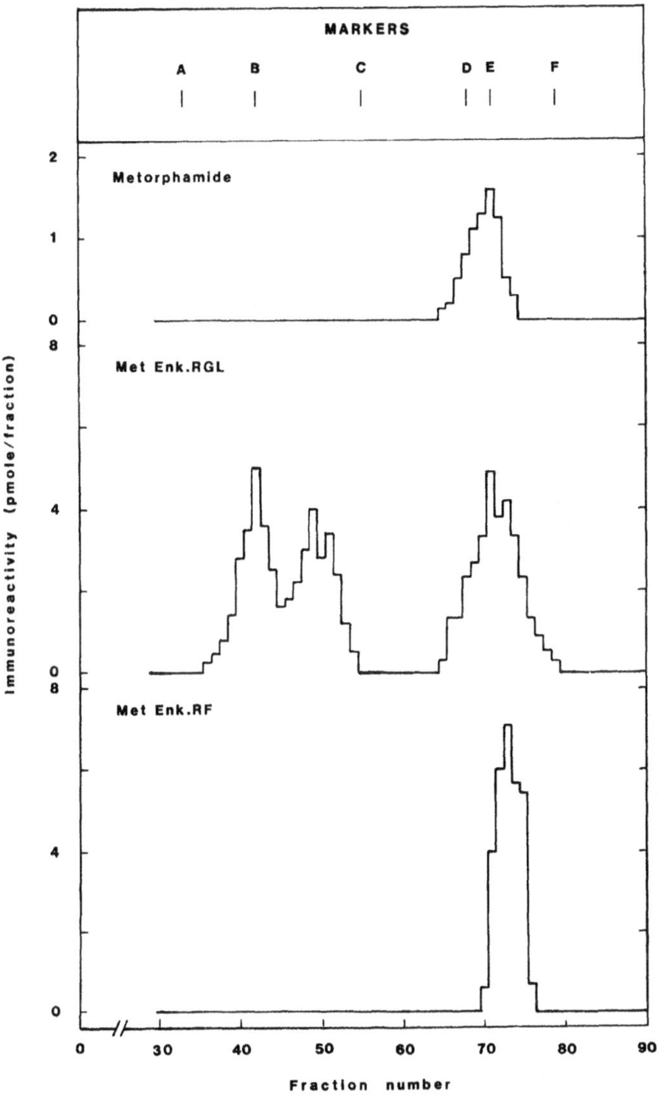

Fig. 2. Gel filtration analysis of the immunoreactive proenkephalin peptides found in human adrenal gland. An acid/acetone extract was loaded onto a 100 x 0.9 CM Sephadex G75 column equilibrated and eluted with 25% acetic acid. Aliquots of the fractions were analyzed using the metorphamide, met-enkephalin-Arg-Phe, and met-enkephalin-Arg-Gly-Leu RIAs. The standard molecular weight markers were A) exclusion, B) ^{125}I-lipotropin, C) ^{125}I-β-endorphin(1-31), D) ^{125}I-α-MSH, E) ^{125}I-α-neo-endorphin, and F) ^{125}I-leucine-enkephalin.

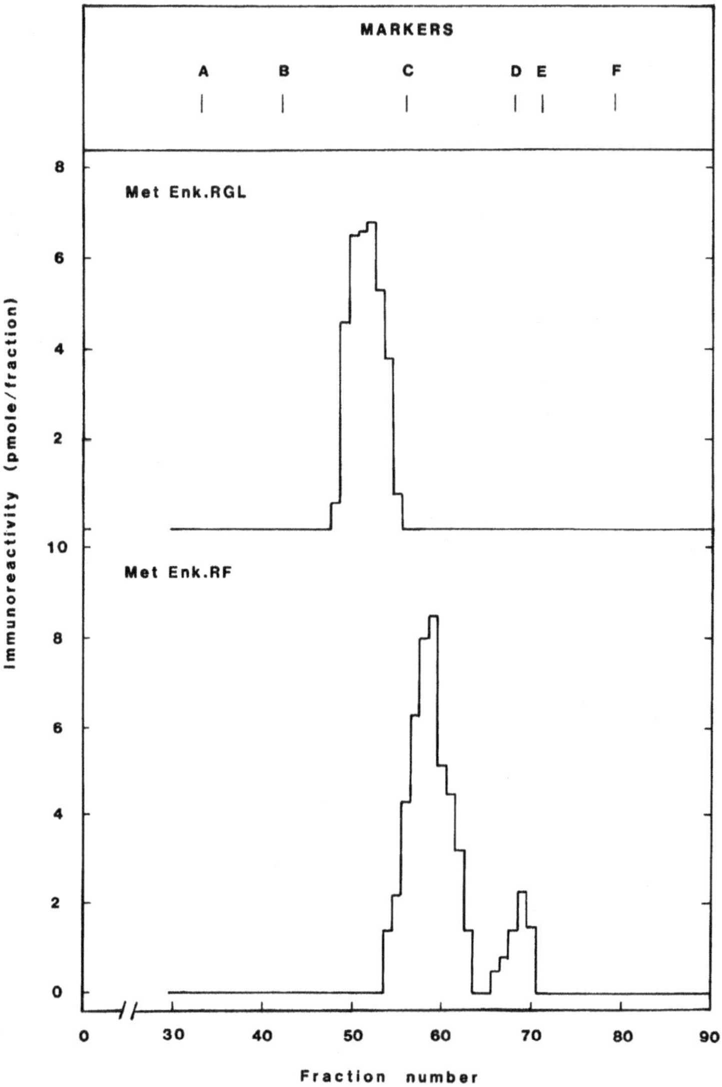

Fig. 3. Gel filtration analysis of the immunoreactive proenkephalin peptides in the guinea pig adrenal gland. The column conditions and molecular weight markers were identical to those used in Figure 3.

3.6 and 5.2 kilodalton fragments found in bovine adrenal (Stern et al., 1981; Jones et al., 1982). We must conclude that between species there is considerable heterogeneity in the stored processing products of adrenal proenkephalin.

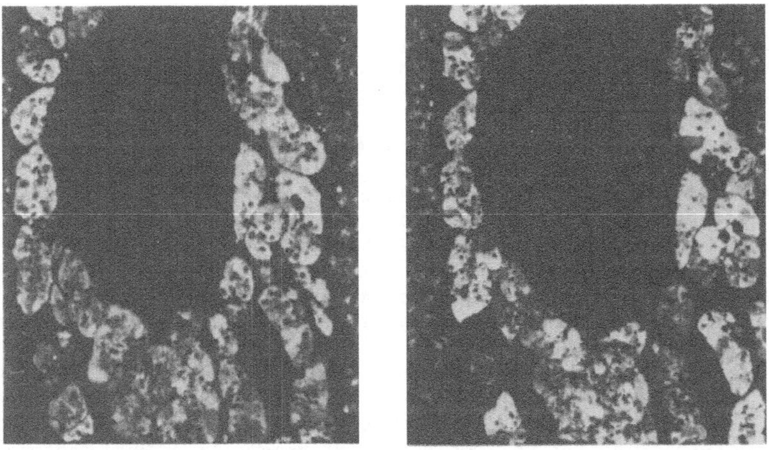

Fig. 4. This figure shows the colocalization of the met-enkephalin-Arg-Gly-Leu (a proenkephalin-derived peptide) and α-neo-endorphin (a prodynorphin-derived peptide) in guinea pig adrenal gland. Sequential 4.5 μ sections through the gland were taken and stained alternatively with α-neo-endorphin or met-enkephalin-Arg-Gly-Leu. The α-neo-endorphin staining could not be blocked by 10 μM dynorphin(1-17), dynorphin B, met-enkephalin-Arg-Phe, met-enkephalin-Arg-Gly-Leu, or leu-enkephalin, but was blocked by α-neo-endorphin. Met-enkephalin-Arg-Gly-Leu staining was not blocked by 10 μM dynorphin(1-17), dynorphin B, met-enkephalin-Arg-Phe, or α-neo-endorphin, but was blocked by met-enkephalin-Arg-Gly-Leu. The blocking controls demonstrated that the staining was specific and the precursor products were indeed present in the same cells.

The guinea pig adrenal glands are somewhat unusual in that prodynorphin-derived peptides can be detected in addition to proenkephalin metabolites. We have analyzed the distribution of these products by immunocytochemistry (see Fig. 4). By staining alternate thin sections with antisera to products of pro-enkephalin and pro-dynorphin we have demonstrated that fragments of both precursors are present in the same cells. Interestingly, in guinea pig adrenal almost every medulla cell is immunoreactive with antisera directed to fragments of the proenkephalin precursor. We do not observe this is all species. Sizing chromatography of the immunoreactive material shows that the processing of prodynorphin and proenkephalin is not extensive in guinea pig adrenal medulla. The dynorphin(1-17) and dynorphin B immunoreactive material is of high molecular weight, ranging from 10 to 12 kilodaltons (Evans et al., 1985). The met-enkephalin-Arg-Gly-Leu and met-enkephalin-Arg-Phe immunoreactive material was also of high molecular weight (see Fig. 3) and consequently may not be opiate active. However, the majority of the α-neo-endorphin immunoreactive material cochromatographed with the authentic peptide on gel filtration. It is at present not clear whether there are discrete prodynorphin-containing or proenkephalin-containing granules--this awaits ongoing electron microscopy localization studies in our laboratory. If the peptides from both precursors are indeed present in the same granules, corelease would be inevitable. However, if the precursor products were stored in separate granules, there is a potential for selective release of products from one or other of the precursors.

The human adrenal medulla is extremely interesting in that processing products derived from POMC as well as proenkephalin can be demonstrated (Nakao et al., 1981; Evans et al., 1983). We have analyzed in considerable detail the processing fragments of these precursors. Since only autopsy material is available, one has to be a little cautious in conclusions drawn from these studies. However, postmortem analysis of neurosecretory peptides has demonstrated that processing patterns in general don't radically alter even a number of hours after death. Results have been consistent in the postmortem tissues we have investigated. Initial RIA shows that there is approximately half the concentration of ACTH as α-MSH. However, gel filtration analysis of the ACTH-like immunoreactivity shows only a small percentage actually chromatographing at the correct molecular weight of ACTH. Therefore ACTH is only a very minor product in these human adrenal medulla cells and α-MSH (identified as des acetyl α-MSH by reverse phase HPLC) is apparently the major product of this region of POMC (Evans et al., 1983). With regard to the endorphin region of POMC, there is processing to form β-endorphin(1-31) which is the major product. Using specific antisera for α-N-acetylated endorphin, we detected no immunoreactive material indicating that there was no α-N-acetylation in this tissue. Interesting, no β-endorphin(1-27) could be detected and less than 10% of the endorphin immunoreactivity could be attributed to lipotropin. This processing pattern is unlike the pituitary in that β-endorphin (1-31) is the major product along with des acetyl α-MSH. It appears from immunocytochemical observations that the POMC producing cells are close to the medulla-cortex interface and are different from the proenkephalin producing cells. The close proximity of POMC producing cells with the adrenal cortex may indicate a direct paracrine-like stimulatory effect on the corticotroph sensitive cells. However, the processing is not geared for ACTH, a considerably more potent corticotroph than α-MSH, and the blood flow is generally accepted to be from the cortex to the medulla. At present it is unclear if the scattered POMC cells contain adrenaline and whether they are indeed activated during stress. With regard to the proenkephalin, the human adrenal unlike many other species fairly extensively processes this precursor to low molecular weight opioids. The met-enkephalin-Arg-Gly-Leu and met-enkephalin-Arg-Phe immunoreactive material is predominantly of low molecular weight (see Fig. 2). Furthermore, the presence of the octapeptide metorphamide would suggest that the human adrenal medulla contains an enzyme capable of cleaving at single arginine residues in addition to an amidating-like enzymic activity.

Interaction of the pituitary and the adrenal gland during a stressful stimuli is extremely relevant to which opiate peptides will be secreted into the blood stream. The release of corticotropic hormone releasing factor (CRF), into median eminance may be accompanied by the release of prodynorphin fragments which have been shown by immunocytochemistry to colocalized in the subpopulation of CRF cells (Hökfelt et al., 1983). The release of CRF in turn will induce release of ACTH and endorphins from the anterior pituitary. The ACTH stimulates the release of steroids from the adrenal cortex. The steroids have multiple actions, including primary effects on the immune system and inhibition of the continuation of ACTH secretion from the anterior pituitary. In a recent study in collaboration with Dr. James Eberwine, we have demonstrated that in guinea pig adrenal glands dexamethasone, a potent steroid agonist, can increase the proenkephalin mRNA approximately 2-fold. Interestingly, the steroid treatment did not affect the co-localized prodynorphin message levels. We found very little change in levels of the proenkephalin peptide processing products. These results could possibly be interpreted as psychological stress (generating the message) preparing the body for a physical stressor (release of opioid peptides with adrenalin). Assuming <u>in vivo</u> opiate peptides are co-released with adrenalin from the chromaffin cells, this could be a major source of circulating opioid peptides released on stress. From our studies on the stored processing products, it would seem that the type and concentration of these opioid peptides from the adrenal would be very dependent on the species under investigation. It is perhaps the human adrenal which should be further investigated with regard to potentially useful opioid peptides that may interact with our immune system. Although in this chapter we have concentrated on the opioids in pituitary and adrenal, it should be remembered that these opioid peptides are only candidates for interaction with opioid receptors in cells involved in the immune response. It may be that other more local sources of opioids are more important factors <u>in vivo</u>.

REFERENCES

Austen, B. M., Evans, C. J., and Smyth, D.G., 1979, Susceptibility of neuroactive peptides to aminopeptidase digestion is related to molecular size, Biochem. Biophys. Res. Commun., 91:1211.

Bradbury, A. F., Smyth, D. G., Snell, C. R., Deakin, W., and Wendlandt, S., 1977, Comparison of the analgesic properties of lipotropin C-fragment and stabilized enkephalins in the rat, Biochem. Biophys. Acta, 74:748.

Brown, S. M., Stimmel, B., Taub, R. N., Kochwa, S., and Rosenfield, R. E., 1974, Immunologic dysfunction in heroin addicts, Arch Intern Med. 134:1001.

Comb, M., Seeburg, P. H., Adelman, J., Eiden, L., and Herbert, E, 1982, Primary structure of the human met- and leu-enkephalin precursor and its mRNA, Nature 295:663.

Depelchin, A., and Letesson, J. J., 1981, Adrenaline influence on the immune response. I. Accelerating or suppressor effects according to the time of application, Immunol. Lett., 3:199.

Eipper, B., and Mains, R., 1980, Structure and biosynthesis of pro-adrenocorticotropin/endorphin and related peptides, Endo. Rev., 1:1.

Evans, C. J., Erdelyi, E., Weber, E., and Barchas, J. D., 1983, Identification of pro-opiomelanocortin derived peptides in the human adrenal medulla, Science, 221:957.

Evans, C. J., Lorenz, R., Weber, E., and Barchas, J. D., 1982, Variants of alpha-melanocyte stimulating hormone in rat brain and pituitary:

evidence that acetylated α-MSH exists only in the intermediate lobe of pituitary, Biochem. Biophys. Res. Commun., 106:910.
Evans, C. J., Erdelyi, E., Makk, G., and Barchas, J. D., 1985, Identification of a novel proenkephalin derived peptide in brain and adrenal, Abstract #69, FASEB Meeting, Anaheim, CA 1985.
Foris, G., Medgyesi, G. A., Gyimesi, E., and Hauck, M., 1984, Met-enkephalin induced alterations of macrophage functions, Mol. Immuno., 21:747.
Geisow, M. J., Deakin, J. F. W., Dostrovsky, J. O., and Smyth, D. G., 1977, Analgesic activity of lipotropin C fragment depends on carboxyl terminal tetrapeptide, Nature, 269:167.
Hazum, E., Chang, K. J., and Cuatrecasas, P., 1979, Specific nonopiate receptors for beta-endorphin, Science, 205:1033.
Hökfelt, T., Fahrenkrug, J., Tatemoto, K., Mutt, V., Werner, S., Hulting, A.-L., Terenius, L., and Chang, K. J., 1983, The PHI (PHI-27)/corticotropin-releasing factor/enkephalin immunoreactive hypothalamic neuron: Possible morphological basis for integrated control of prolactin, corticotropin, and growth hormone secretion. Proc. Natl. Acad. Sci. USA, 80:895.
Jones, B. N., Shively, J. E., Kilpatrick, D. L., Kojima, K., and Udenfriend, S., 1982, Enkephalin biosynthetic pathway: A 5300-dalton adrenal polypeptide that terminates at its COOH end with the sequence [Met]enkephalin-Arg-Gly-Leu-COOH, Proc. Natl. Acad. Sci. USA, 79:1313.
Kakidani, H., Furutani, Y., Takahashi, H., Noda, M., Morimoto, Y., Hirose, T., Asai, M., Inayama, S., Nakanishi, S., and Numa, S., 1982, Cloning and sequence analysis of cDNA for porcine β-neo-endorphin/dynorphin precursor, Nature, 298:245.
Lindberg, I., Yang, H-Y. T., and Costa, E., 1983, A high molecular weight form of $Met^5Arg^6Gly^7Leu^8$ in rat brain and bovine chromaffin granules. Life Sci., 33(Suppl):5.
Liotta, A. S., Suda, T., and Krieger, D. T., 1978, β-Lipotropin is the major opioid-like peptide of human pituitary and rat pars distalis: Lack of significant β-endorphin, Proc. Natl. Acad. Sci. USA, 75:2950.
Lolait, S. J., Lim, A. T., Toh, B. H., and Funder, J. W., 1984, Immunoreactive beta-endorphin in a subpopulation of mouse spleen macrophages, J. Clin. Invest., 73:277.
Martin, R., Geis, R., Holl, R., Schaffer, M., and Voigt, K. H., 1983, Co-existence of unrelated peptides in oxytocin and vasopressin terminals of rat neurohypophyses: Immunoreactive methionine-enkephalin-, leucine-enkephalin- and cholesystokinin-like substances, Neurosci., 8:213.
Martin, R., and Voigt, K. H., 1981, Enkephalins co-exist with oxytocin and vasopressin in nerve terminals of rat neurohypophysis, Nature 289:502.
Matsuo, H., Miyata, A., and Mizuno, K., 1983, Novel C-terminally amidated opioid peptide in human phaeochromocytoma tumour, Nature 305:721.
McCain, H. W., Lamster, I. B., Bozzone, J. M., and Grbic, J. T., 1982, Beta-endorphin modulates human immune activity via non-opiate receptor mechanisms, Life Sci., 31:1619.
Munck, A., Guyre, P. M., and Holbrook, N. J., 1984, Physiological functions of glucocorticoids in stress and their relation to pharmacological actions, Endocrin. Rev., 5:25.
Nakanishi, S., Inoue, A., Kita, T., Nakamura, M., Chang, A. C. Y., Cohen, S. N., and Numa, S., 1979, Nucleotide sequence of cloned cDNA for bovine corticotropin-β-lipotropin precursor, Nature, 278:423.

Nakao, K., Yoshimasa, T., Ohtsuki, H., Oki, S., Tanaka, I., Nakai, Y., and Imura, H., 1981, β-Endorphin, ACTH and γ-MSH in human sympathoadrenal system, Presented at the Eighth International Congress of Pharmacology, Tokyo.

Noda, M., Furatani, Y., Takahashi, H., Toyosato, M., Hirose, T., Inayama, S., Nakanishi, S., and Numa, S., 1982, Cloning and sequence analysis of cDNA for bovine adrenal preproenkephalin, Nature 295:202.

O'Donohue, T. L., and Dorsa, D. M., 1982, The opiomelanotropinergic neuronal and endocrine systems, Peptides, 3:353.

Przewlocki, R., Millan, M. J., Gramsch, G. H., and Millan, M. H., 1982, The influence of selective adeno- and neurointermedio-hypophysectomy upon plasma and brain levels of β-endorphin and their response to stress in rats, Brain Res., 242:107.

Rehfeld, J. F., 1981, Four basic characteristics of the gastrin cholecystokinin system, Am. J. Physiol., 240:6255.

Roth, K. A., Evans, C. J., Lorenz, R. G., Weber, E., Barchas, J. D., and Chang, J.-K., 1983, Identification of gastrin releasing peptide-related substances in guinea pig and rat brain. Biochem. Biophys. Res. Commun., 112:528.

Seizinger, B. R., Liebisch, D. C., Gramsch, C., Herz, A., Weber, E., Evans, C. J., Esch, F. S., and Bohlen, P., 1985, Isolation and structure of a novel C-terminally amidated opioid peptide, amidorphin, from bovine adrenal medulla, Nature, 313:57.

Sonders, M., Barchas, J. D., and Weber, E., 1984, Regional distribution of metorphamide in rat and guinea pig brain, Biochem. Biophys. Res. Commun., 222:892.

Steiner, D. F., Kemmler, W., Tager, H. S., and Peterson, J. D., 1974, Proteolytic processing in the biosynthesis of insulin and other proteins, Fed. Proc., 33:2105.

Stern, A. S., Jones, B. N., Shively, J. E., Stein, S., and Udenfriend, S., 1981, Two adrenal opioid polypeptides: proposed intermediates in the processing of proenkephalin. Proc. Natl. Acad. Sci. USA, 78:1962.

Viveros, O. H., Diliberto, E. J., Hazum, E., and Chang, K-J., 1979, Opiate-like materials in the adrenal medulla: Evidence for storage and secretion with catecholamines, Mol. Pharm., 16:1101.

Weber, E., Evans, C. J., and Barchas, J. D., 1983b, Multiple endogenous ligands for opioid receptors, Trends in Neuroscience, 6:333.

Weber, E., Evans, C. J., and Barchas, J. D., 1982a, Opioid peptide dynorphin: predominance of the aminoterminal octapeptide fragment in rat brain regions, Nature, 299:77.

Weber, E., Evans, C. J., Chang, J-K., and Barchas, J. D., 1982b, Antibodies specific for α-N-acetyl β-endorphins: Radioimmunossays and detection of acetylated β-endorphins in pituitary extracts, J. Neurochem., 38, 436.

Weber, E., Esch, F. S., Bohlen, P., Paterson, S., Corbett, A. D., McKnight, A. T., Kosterlitz, H. W., Barchas, J. D., and Evans, J. C., 1983a, Metorphamide: Isolation, structure and biologic activity of a novel amidated opioid octapeptide from bovine brain, Proc. Natl. Acad. Sci. USA, 80:7362.

CONTROL MECHANISMS IN THE ENZYME HYDROLYSIS OF ADRENAL-RELEASED ENKEPHALINS

L. Giorgio Roda,* Gianna Roscetti,° Roberta Possenti,° Francesca Venturelli,° and Fabrizio Vita°

Laboratorio di Farmacologia, University of Ancona, Ancona, Italy.* Cattedra di Fisiologia Umana, "Tor Vergata" University, Rome, Italy°

A: INTRODUCTION

Opioid peptides are secreted by the adrenal medulla where they are stored within the chromaffin granule (1-3). Consequently, these peptides are released into the blood stream together with the whole soluble content of the granule during the physiological activity of the gland.

Once released, endogenous opioids are carried by the blood stream towards their presumptive target organs. Such a fate is common to any substance released in blood by the endocrine glands. Notwithstanding, the case if adrenal-released opioid peptides seems to be somewhat more complex. Indeed, it is very difficult to demonstrate any pharmacological activity of the smaller opioids when they are artificially administered under in vivo conditions (406, see also 7). This appears--at least in part--to be caused by the rapid inactivation of these peptides by plasma enzymes (8-10, see also 11). In this respect, centrally and peripherally released opioid peptides are in opposite situations: while a rapid inactivation is normal for neurotransmitter-like substances released in the central nervous system, the same phenomenon is hardly expected for hormone-like substances whose activity is supposed to be a relatively long-lasting one.

Thus, there is an apparent discrepancy between blood as carrier of the endogenous opioids and its capacity to degrade the carried substances. Yet, in the last few years a good deal of evidence has been accumulated which clearly indicates the existence of endogenous substances active in preventing a rapid hydrolysis of the opioid peptides released into the blood stream. These results may help to better organize the data available on the stability of the peripheral enkephalins. This can certainly be useful. Indeed, the concept that the immune response can at least partially be influenced by the perpherally-released opioid peptides has been substantiated by two sets of data: the first on the existence of interactions between opioids and immunocompetent cells, and the second on the existence of receptors for opioid peptides on the T-lymphocyte membrane. These data are specifically reviewed elsewhere in this volume. Here, we will simply mention that the problem of the control of the active life of the serum-released opioids is strictly related to the problem of the activity of these peptides on the immunocompetent cells present in the blood stream. The purpose of the following pages is to review the evidence on the hydrolysis of peripheral enkephalins, to add some new data and to extrapolate from them concepts of more general interest.

B: ENKEPHALIN HYDROLYSIS

Unlike the hydrolysis by nervous system related enzymes, for which an extensive literature exists, relatively few works have been performed on the hydrolysis of enkephalins by plasma enzymes. However, this topic has been studied both in vivo and in vitro (8-10, 12) and the enzyme cleavage mechanism has been determined by separation and quantification of the hydrolysis by-products by chromatographic procedures (13).

The results known so far can be summarized as follows: i) enkephalins are hydrolyzed by several enzymes present both in plasma and in brain homogenates; ii) in both cases--at least in man, but not in all mammals--the first hydrolysis by-product is tyrosine (since the remaining tetrapeptide is biólogically inactive, the tyrosine cleavage is also the most relevant step in the hydrolysis sequence; iii) the plasma half-life reported for enkephalins range from a few seconds (9) to 2-3 minutes (8), up to as long as 8-10 minutes (14). The enkephalin-hydrolyzing enzymes present in plasma seem to belong to the same classes, but not to be identical, to the related to the nervous tissue (13).

B1: REQUIREMENTS FOR ENKEPHALIN HYDROLYSIS

It is generally assumed that the rapid enzymatic degradation of the smaller opioids is a direct consequence of their relatively small size--that is, in the case of a molecule as short as a pentapeptide, the specificity of hydrolytic enzymes should necessarily be low. Indeed, no hydrolysis-susceptible bond can be shielded by steric effects, not can the conformation of the molecule hinder the accessibility of the active sites of the enzymes. From these considerations it follows that any proteolytic enzyme able to cleave the peptide bonds present in the molecule should in principle also be able to hydrolyze the enkephalins.

The real situation, however, is probably rather more complex than that. It seems indeed true that the degree of specificity necessary for the proteolytic enzymes to hydrolyze enkephalins is relatively high. Actually, we detected a considerable resistance of enkephalins to non-specific enzyme hydrolysis (submitted). From our data, only pronase is able to completely inactivate leu-enkephalin within minutes (the half-life of leu-enkephalin is 90 seconds in the presence of 1% enzyme w/w at pH 7.0). Of all the many other enzymes assayed, leucine aminopeptidase only was found to hydrolyze leu-enkephalin with measureable kinetics. On the other hand, papain, bommelain, and pepsin, forexample, are practically unable to hydrolyze leu-enkephalin.

These data can be at least partially explained by taking into account the structure of the peptide. Calculations of the minimal energy conformation (15-17), circular dichroism (18-20) and NMR data (19, 21-24), tend to suggest that both enkephalins have a degree of structure which is certainly unusually high for small peptides.

Indeed, a structure in which the C- and N-terminals are held close together has been suggested by several authors (e.g. 22,23). The existing data on enkephalins' structure, taken as a whole, make it highly probable that these peptides are rather rigidly constrained in a folded conformation with a I, I', II, or II'-bend involving the GLY^2, GLY^3, or PHE^4 residues. The band can also be stabilized by non-covalent backbone-backbone interactions and by one (23), two (25) or even three (19) hydrogen bonds. These data have been obtained mainly with the pentapeptides, but they may be extrapolated at least to the hectapeptide, met-enkephalin-ARG-PHE. Indeed, conformational features of the met-enkephalin molecule have been shown to be maintained even in beta-endorphin (26).

It seems therefore that the two sets of data (resistance to enzymatic hydrolysis and the existence of a well defined structure) can be related. Also, it is probably true that the conformation assumed by enkephalins is such as to partially protect the peptide bonds existing within the molecule from enzyme attack.

B2: MECHANISMS OF ENKEPHALIN HYDROLYSIS IN BLOOD

Enkephalin hydrolysis has been primarily studied in the case of nervous tissue-related enzymes. However, a few studies have been published on plasma hydrolysis. Among them, Hambrook et al. studied the kinetics and hydrolysis pattern of enkephalin in rat plasma(8); Hogu-Angeletti and Roda determined the same parameters in rabbit plasma(10); Dupont et al. studied with an in vivo technique the hydrolysis kinetics in rat blood(9); and Possenti et al. investigated enkephalin hydrolysis in human plasma(14). As detailed further on, some of the results of these studies are somewhat conflictual, and quite difficult to fit together in a consistent pattern.

B2a: Hydrolysis in human plasma

Detailed knowledge of the hydrolysis of enkephalins by plasma enzymes is obviously important for understanding protection from hydrolysis itself. Therefore, in this section, we report the experimental procedure which was used in our laboratory to study hydrolysis kinetics and the fragmentation pattern of leu-enkephalin upon cleavage by plasma enzymes.

Leu-enkephalin was incubated with whole plasma and the incubation mixture was fractioned by high pressure steric exclusion chromatography in order to remove the high molecular weight material. The enkephalin-containing low molecular weight fraction separated by the steric exclusion column was then fractioned by reverse phase chromatography, and the hydrolysis by-products were subsequently identified by their retention characteristics and by amino acid analysis(13). Figure 1 reports the kinetics of leuenkephalin degradation and that of the fromation and disappearance of the released fragments. Of the 13 fragments that are theoretically obtainable from pentapeptide, 8 final hydrolysis by-products were actually found: i) all the four possible monomers; ii) 2 dimers out of 4: GLY-GLY and PHE-LEU; iii) 1 trimer out of three: GLY-PHE-LEU; iv) 1 tetramer out of 2: GLY-GLY-PHE-Leu.

All the N-terminal tyrosine peptides are absent. This is due to the high activity of aminopeptidases in human blood. Also lacking are GLY-PHE and GLY-GLY-PHE, probably because of the lack of hydrolysis at the PHE-LEU-bond reported in the next section.

B2b: Enzymes in human plasma

The distribution of the enkephalin-degrading enzymes present in human blood was investigated as follows: whole plasma was fractioned by high pressure steric exclusion or ion exchange chromatography and labelled enkephalins were incubated in the presence of the pools thus obtained. The hydrolysis by-products were separated and identified by thin layer chromatography and quantified by scintillation counting of the thin layer-separated fragments(unpublished). This procedure permits one to identify indirectly and quantitate the enzymes from which the different fragments are generated. Figure 2 shows the pattern of enkephalin-hydrolyzing enzymes in human blood after separation by high pressure steric exclusion chromatography.

In addition, plasma was sequentially fractioned by steric exclusion and ion exchange chromatography. Under these conditions, and allowing for the limitation inherent to the lack of positive identification of the molecules themselves, six enzyme species were apparently separated, relative to the following enzyme activities: three aminopeptidases, one dipeptidylcarboxypeptidases and two dipeptidylaminopeptidases.

It should be noted that the results shown are valid only as far as human plasma is concerned. Indeed, as reported further on, the kinetics of enkephalin hydrolysis and the relative quantities and distribution of the enkephalin-hydrolyzing enzyme activities seems to be markedly dissimilar in man and in other animals.

B3: HYDROLYSIS BY BRAIN HOMOGENATES

The cleavage pattern described in the previous two paragraphs is consistent with inactivation schema followed by enkephalins when hydrolyzed by the soluble fraction of the nervous tissue. In this case too, three enzyme classes are responsible for enkephalin inactivation: aminopeptidase, dipeptidylaminopeptidase and dipeptidylcarboxypeptidase (11). The main difference between plasma and brain enzymes lies in the cleavage of the GLY-GLY bond which, in plasma, is considerably faster than that in the presence of brain soluble homogenates (13).

Although the inactivation of opioid peptides by central nervous system enzymes has been studied in considerable detail, an accurate molecular characterization of the enzyme systems involved is still lacking. Consequently, it is not as yet as easy to decide how similar are the plasma and brain enzyme systems.

Inhibition data may help to clarify this problem. The inhibition of both dipeptidylcarboxypeptidase and dipeptidylaminopeptidase activities by Tiorphan and Captopril in plasma is approximately 10-20% of the inhibition reported for the brain enzymes. Instead, the inhibition of plasma aminopeptidase activities by Bestatin tallies quite well with the values measured for the brain enzymes in the case of the higher molecular weight form. On the contrary, no inhibition at all was observed with the same substances in the case of the lower molecular weight form (unpublished). As a whole, these data tend to disprove the hypothesis of a strict similarity between the enzymes present in the two tissues. These data rather suggest that, while the enzyme activities in both tissues belong to the same class, they nevertheless represent different molecular forms.

B4: PLASMA LEVELS AND ADRENAL RELEASE

Enkephalins circulate as intact pentapeptides in blood, where they can be quantified by radioimmunoassay. In man, under physiologically base conditions, the enkephalin-like material present in the venous district is in the order of $1x10^{-12}$ to $1x10^{-13}$ moles/ml, viz. a total blood content of $3x10^{-7}$ to $3x10^{-6}$g (27,28). The concentration in the adrenal vein is approximately $2x10^{-13}$ moles/ml (27). Since the blood flow through each adrenal vein is approximately 25 ml/min, the net enkephalin addition is $1x10^{-11}$ moles/minute, that is $6x10^{-9}$g released per minute in the blood. Assuming degradation kinetics corresponding to half lives of 8 minutes (see paragraph C3), the inflow required to maintain a steady enkephalin concentration must be $2x10^{-8}$ to $2x10^{-7}$ g/minute.

Assuming that the values thus calculated are correct, the enkephalin released by the adrenal medulla is no more than 30% of the enkephalin needed to maintain a steady peptide concentration. Obviously, either a source for plasma enkephalins other than adrenal medulla can be found, or the degradation kinetics that may be determined under experimental conditions are different from the ones which actually take place under physiological conditions.

C: MECHANISMS PROTECTING PERIPHERAL ENKEPHALINS

As described in the previous section, it is somewhat difficult to accord the data existing on enkephalin blood hydrolysis with the plasma levels of these peptides and with the data concerning adrenal release. The discrepancies percieved induced us to investigate specifically the possible existence of mechanisms acting in vivo, able to protect enkephalins from enzyme proteolysis. This on the assumption that protection mechanisms acting under physiological conditions might not be active on the enkephalins administered under experimental conditions. The results obtained since we started working on this hypothesis have certainly been encouraging, and by now the existence of these mechanisms is certain. The following paragraphs therefore deal with this topic, and primarily report the work carried out in our laboratory over the past few years.

C1: MECHANISMS WITHIN THE CHROMAFFIN GRANULE

The adrenal medulla chromaffin granule stores-within its seemingly uniform matrix- catecholamines(29,30), ions and several proteins(31-33), together with enkephalins and other opioid peptides(34-37), plus several peptides of a non-opioid nature(37-40). Among the proteins, proteolytic enzymes are also present(41,42). These enzymes could be able to hydrolyze other polypeptides present within the granule, notably precursors(43,44) of the opioid peptides(40-42. 45-46, see also 47), as well as enkephalins themselves. The peculiar physio-chemical milieu existing in the granules might alter the enzyme specificities. Yet, there is also a possibility that the smaller polypeptides, in particular enkephalins, are protected from the activity of these enzymes. Actually, we showed that enkephalins bind chromaffin granule components and that bound material is partially protected from enzyme hydrolysis(13-14,48). These data were obtained as described below.

Labelled leu-enkephalin was incubated in the presence of a chromaffin granule soluble lysate and the incubation mixture was fractionated by steric exclusion chromatography. From a column optimized for high molecular weight fractionation(figure 3), two peaks of bound radioactivity, plus one eluted at the free enkephalin volume, were separated. The first one eluted corresponds to a granule-contained lipoprotein, chromolipin(49), the second one was tentatively identified as chromogranin A(50). When the same material was fractioned with a Biogel P2 column(nominal fractionation range up to 2000 Daltons), a considerable amount of the radioactive label was found to be associated with a peptide-containing fraction(figur 4). This material was identified with a peptide family previously discovered in the chromaffin granule(51).

The enkephalins bound to the chromaffin granule-containing material are also partially protected from enzyme hydrolysis, as shown by the following data. Enkephalins - preincubated with chromaffin granule crude lysate - were incubated with whole human plasma and analyzed by ion exchange chromatography. Under these conditons, enkephalin hydrolysis kinetics were considerably altered. Specifically, at medium incubation times, the amount of intact enkephalin was higher than in the non-preincubated controls. After 5 minutes, the difference was 29.1%; after 15 minutes, 31.3%, while after 30 minutes it drops to 15.6%(10).

So far, two enkephalin binding systems - relative proteins and peptides - have been detected within the chromaffin granule. Two major roles may be attributed to these systems. The first one - as briefly outlined at the beginning of this section - is to protect enkephalins while they are still in the chromaffin granule. Indeed, the presence of opioid peptide precursors in the granule(43,44) also require proteolytic enzymes to cleave the parent molecules for the generation of the active peptides. The presence of these enzymes has been demonstrated(41,42), and it seems likely that they may also be able to hydrolyze the active opioid peptides contained in the granule. Thus the presence of enkephalin-binding systems might be instrumental in protecting enkephalins before their release into the blood stream.

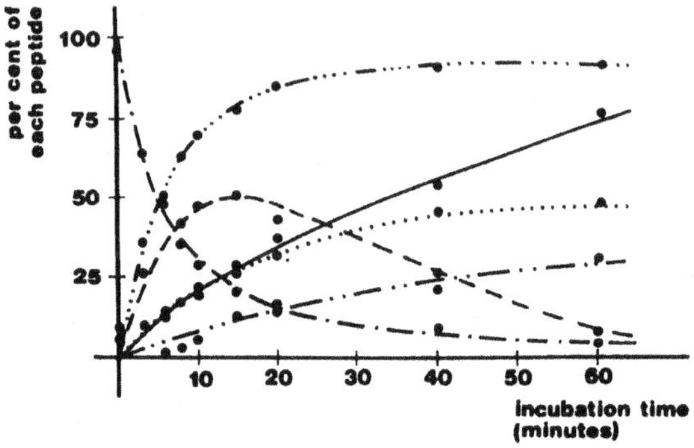

Figure 1. Kinetics of leu-enkephalin hydrolysis and of formation and disappearance of leu-enkephalin fragments by depeptidised plasma. Dash-dotted line: leu-enkephalin; dash-four dots line: TYR; dashed line: GLY-GLY-PHE-LEU and GLY-PHE-LEU; solid line: GLY and GLY-GLY; dotted line: PHE-LEU; dash-two dots line: PHE and LEU. Solid circles indicate experimental points.

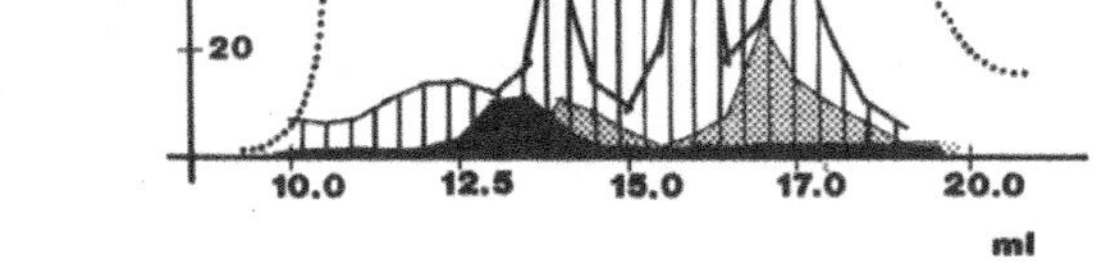

Figure 2. Enkephalin-degrading enzymes in human plasma as fractioned by high pressure steric exclusion chromatography. Enzyme activites were determined as described in the text. Dotted line indicates absorbance at 280 nm. Solid line represents leu-enkephalin; the vertically ruled area represents aminopeptidase; the dotted area represents dipeptidylaminopeptidase and the dark shaded area represents dipeptidylcarboxypeptidase activity.

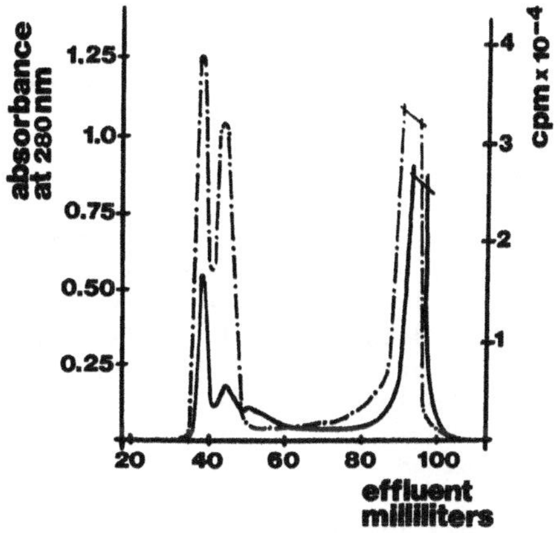

Figure 3: Steric exclusion chromatography(Biogel A 1.5m) of labelled leu-enkephalin incubated in the presence of a chromaffin granule crude lysate. Solid line represents absorbance at 280 nm. Dash-dotted line represents CPM per fraction.

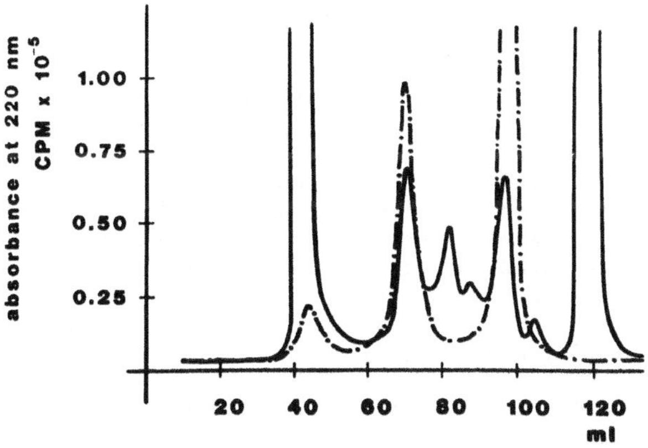

Figure 4: Steric exclusion chromatography of the same incubation mixture as in figure 3.

A second, perhaps more important, role of the chromaffin granule-contained protecting systems may be to contribute to the stability of opioid peptides once they are released into the blood stream. As described in the previous section, several enzymes present in blood are able to hydrolyze enkephalins. On the other hand, on the basis of the aforementioned data, the release of enkephalins is paralleled by the release of molecules able to protect them. If the binding is not destroyed in the different physio-chemical enviroment represented by plasma, this may well be a very effective way to protect the released enkephalins. The relevance of this topic will be discussed in more detail in the next section.

C2: PROTECTION MECHANISMS IN BLOOD

There is also the possibility that opioid peptides bind to plasma components which may act as carriers for these peptides, or that may protect them from enzyme hydrolysis. Indeed, a specific search for enkephalin-binding systems in human plasma revealed the existence of two systems with such an activity(48). The presence of enkephalin-binding substances was evidentiated as described below.

C2a: Binding systems in blood

Whole human plasma was incubated with labelled leu-enkephalin and the incubation mixture was fractioned by steric exclusion chromatography. Under these conditions several radioactive peaks were separated. Figure 5 shows the chromatogram obtained from a column separating high molecular weights. Figure 6 shows the same incubation mixture separated on a column optimized for fractioning low molecular weight material, as it is described further on. As shown by both figures, at least four different radioactive fractions can be separated with this technique. The first-eluted fraction represents serum albumin-bound material, the last-eluted one represents free enkephalin. The material eluted at intermediate volumes - approximately corresponding to molecular weights of 5-6000 and 2000 Daltons(pool 1 and 2 in figure 6) - represents enkephalin-binding material of a peptidic nature.

Enkephalin-binding activity quite naturally suggested the possibility that the bound material was protected from enzyme hydrolysis. Therefore, specific experiments were performed to ascertain this point.

Labelled leu-enkephalin was incubated with whole plasma. Albumin-bound enkephalin was separated from the free peptide by steric exclusion chromatography. The albumin-containing fraction(accounting from approximately 5% of the total radioactivity) was then incubated in 20% ethylene glycol to release bound enkephalin(60% of the albumin-bound enkephalin as released under these conditions after 30 minutes at 30°C). Intact enkephalin and hydrolysis by-products were separated and quantified by reverse phase chromatography. Under these conditions, over 90% of the labelled peptide was found to elute at the position of intact enkephalin, while the unbound enkephalin was almost completely(over 75%) degraded(48).

The same analyses were performed on the enkephalin bound to the peptidic fraction. Enkephalin-binding pools(1 and 2 in figure 6, approximately accounting for 5 and 15% of the labelled material) were incubated for 30 minutes at 37°C in 20%CH_3CN. The incubation mixture was fractioned by reverse phase chromatography on a C-18 column. In this case, incubation and steric exclusion chromatography were found unnecessary since the enkephalin-bound peptides were separated from intact enkephalin by the reverse phase system. (This, in all likelihood, happened on account of the CH_3CN-containing mobile phase used). Even in this case, though, the peptide-bound enkephalin was found to consist of intact pentapeptide for over 95%(14).

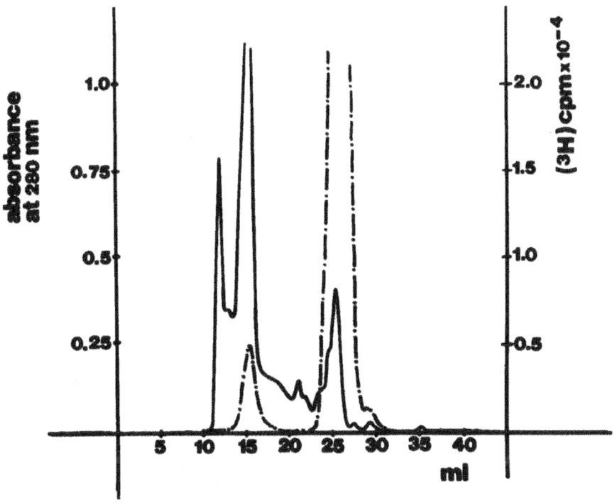

Figure 5. Steric exclusion chromatography(TSK G3000 SW) of labelled leu-enkephalin incubated in the presence of whole human plasma. Solid line represents absorbance at 280 nm; dash-dotted line represents CPM per fraction.

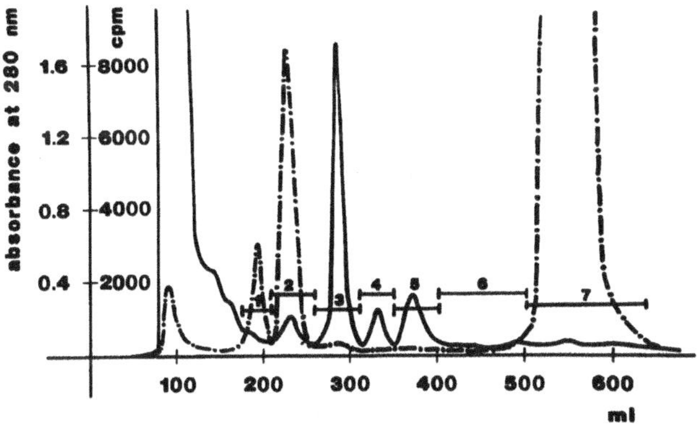

Figure 6. Steric exclusion chromatography of the same incubation mixture as in Figure 5, separated Fractogel HW40. Solid line represents absorbance at 280 nm; Dash-dotted line indicates CPM per fraction.

C2b: In vitro stability of enkephalins

The reliability of the aforementioned hypothesis is also confirmed by direct measurements of the half lives of enkephalins performed in the presence and in the absence of the plasma protection systems(serum albumin and peptides).

The hydrolysis kinetics of leu-enkephalin were determined with whole and de-peptidized human plasma, that is plasma fractioned by steric exclusion chromatography to serparate the peptides from the enzyme containing fractions (13). In the presence of plasma peptides the enkephalin half life is almost exactly three times as long as the one measured in the absence of the peptides (15 minutes instead of 5). In addition after 60 minutes of incubation, the intact enkephalin is approximately 25% in the first case and near to 0(3%) in the second case(13). So as to permit the identification of hydrolysis by-products by column chromatography, the reported values were obtained at enkephalin concentrations several orders of magnitude higher than those of the physiological resting level. Under these conditions, it may be hypothesized that the enkephalin-degrading enzymes are nearly saturated, or else that a high concentration of enkephalin hydrolysis by-products may be competing for the enzyme's active sites, and thus constituting an inhibitor. To circumvent these difficulties, we tried to determine leu-enkephalin plasma half life at a concentration as near to the physiological one as possible. Using a batch of labelled enkephalin of a particularly high specific activity, we were able to measure enkephalin hydrolysis kinetics at a concentration as low as $1x10^{-13}$ moles/ml, which is at the lower end of the physiologically steady-state concentrations of $1x10^{-13}$ to $1x10^{-12}$ moles/ml(27,28). Figure 7 actually shows the degradation kinetics of $1x10^{-10}$ M leu-enkephalin in the presence and in the absence of plasma peptides.

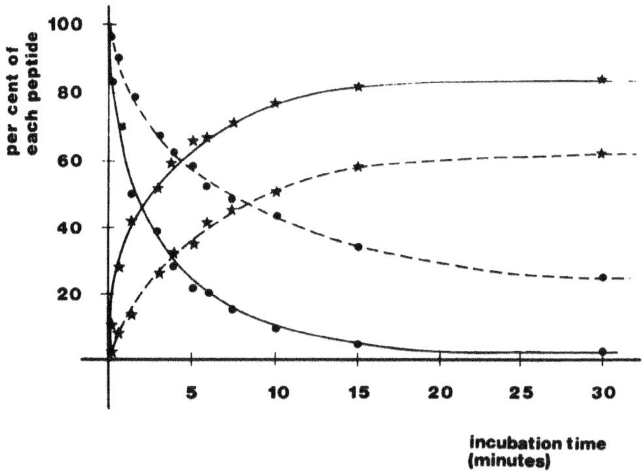

Figure 7. Degradation kinetics of leu-enkephalin in the presence of whole (dashed line) and de-peptidised(dashed line) human plasma. Solid circles represent tyrosine. Incubation and assay conditions as described in text.

Under these conditions, the half-life de-peptidized plasma is 2.5 minutes, while it is 8 minutes in whole plasma. This latter may be considered a reliable value for human plasma. This figure is in perfect accordance with the data previously determined in our laboratory. On the other hand, it strongly contradicts the values reported by other laboratories (8,9) as detailed in section D.

C2c Mechanisms of enzyme inhibition

The mechanisms of aforementioned enzyme inhibition are currently under study in our laboratory. However, some of the data available so far seem worth mentioning.

The peptides with enkephalin-binding and enzyme-degradation protections activities were fractioned as follows: to evidentiate enkephalin binding, whole plasma - preincubated with labelled leu-enkephalin - was fractioned by steric exclusion chromatography. To evidentiate protection from proteolysis, the plasma, not preincubated with enkephalin, was fractioned with the same steric exclusion column. The results of this experiment, combined, are shown in Figure 6. The obtained pools, indicated in Figure 6 by bars and numbers, were assayed for inhibition of the enkephalin-degrading enzymes as follows: the enzymes were fractioned from human plasma by steric exclusion chromatography as described in Section B2b. Labelled leu-enkephalin was incubated with the fractioned enkephalin-degrading enzymes obtained as indicated. The percent of degraded enkephalin was measured by thin layer and reverse phase chromatography. The results of these experiments are reported in Table I.

TABLE I

Inhibition of enkephalin-degrading enzyme activities by plasma components shown in figure 6, defined as percent of the inhibition due to an equimolar amount of leu-enkephalin.

POOL	AP	DAP	DCP
1	0	37	55
2	60	62	*
3	97	28	0
4	*	*	0
5	*	*	0
6	*	*	*
7	*	*	*

* Not significative, less than 10%.
AP = aminopeptidase
DAP = dipeptidyl-aminopeptidase
DCP = dipeptidyl-carboxypeptidase

The results shown indicate that the different peptide groups inhibit different amounts of the three enkephalin-degrading enzyme families. Moreover, the inhibition of the enzyme activities seems not to be due - as we originally supposed - to the binding of the enkephalin to the protecting peptides. Instead, from data available so far, the protection mechanism seems to be due to competition for the active sites of the degrading enzymes. While several of the peptide pools present in human plasma seem to inhibit some of the enkephalin-degrading enzyme systems present in the same tissue, only one of the fractioned pools seems to bind to enkephalin, and to partially inhibit only one(dipeptidylaminopeptidase) of the enzyme activities present in plasma. On the other hand, the data reported in paragraph C2a demonstrate that peptide-bound enkephalins are protected from enzyme hydrolysis. Therefore, two different protection mechanisms seem to be active: competitive inhibition of the enzymes, and a different mechanism acting through binding to the peptide or peptides, as indicated in Table 1.

D: ENKEPHALIN HYDROLYSIS IN DIFFERENT ANIMAL SPECIES

As already pointed out, the blood half-lives of enkephalins, such as they were measured by various authors, differ by a factor of approximately 200. That is, from 2-4 seconds reported by Dupont et al. in rat plasma(9) to the 8 minutes reported by us in human plasma(13). A span of this magnitude can hardly be accounted for by technical reasons alone. In addition, our interpretation of the physiological roles of opioid peptides in plasma can be quite strongly influenced by the degradation kinetics in blood one assumes to be true. Therefore - using the technical procedures listed in Section C3 - we measured enkephalin hydrolysis kinetics in several laboratory animals and in man(52).

Figure 8 shows the hydrolysis kinetics of leu-enkephalin in human, rat and rabbit whole plasma. The corresponding half-lives are 8 minutes, 40 seconds and 90 seconds respectively. Specifically, the half-life measured for rat plasma is still one order of magnitude longer than the one measured by Dupont et al. But, these leatter values were obtained under completely different experimental conditions. Thus, it is possible that the discrepancies observed are mainly caused by technical reasons.

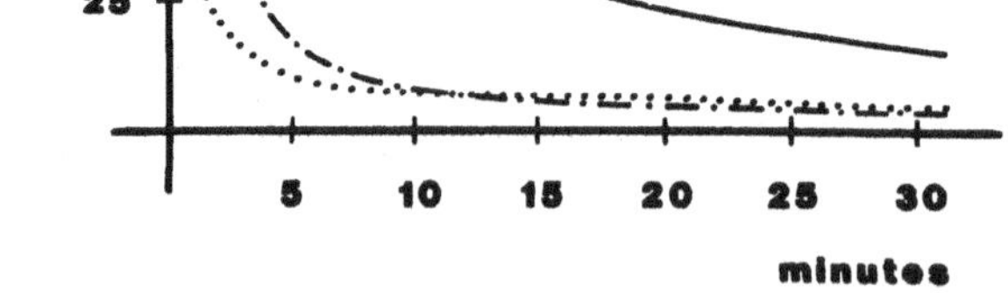

Figure 8. Hydrolysis kinetics of leu-enkephalin in the presence of human (solid-line), rabbit(dash-dotted line) and rat(dashed line) plasma. Incubation and assay conditions as described in the text.

Summing up, it is evident that considerable care should be used in the - frequently unavoidable - procedure of extrapolating data obtained with laboratory animals to the human species. This even more so, since our results indicate that the differences are not only quantitative but qualitative as well. That is, the relative amount of the different enkephalin-hydrolyzing enzymes, as well as the total activity, is considerably different in different mammals. This is evident by comparing Figure 9 and 10, which show the pattern of enkephalin degrading enzymes in rabbit and rat plasma respectively, to Figure 2, showing the distribution of the same enzymes in human plasma.

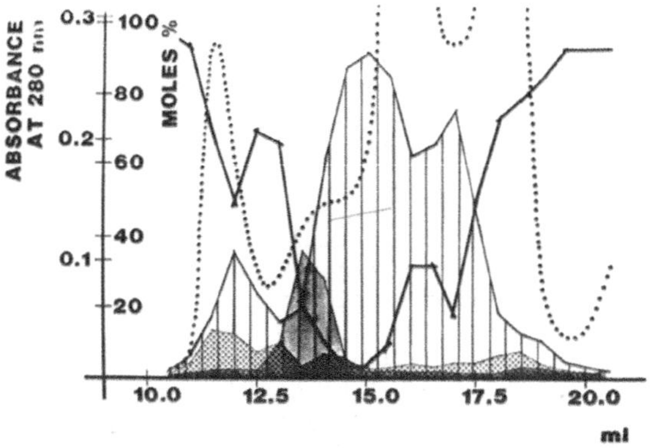

Figure 9. Enkephalin-degrading enzymes in rat plasma as fractioned by high pressure steric exclusion chromatography. Enzyme activities were determined as described in the text. Dotted line:absorbance at 280 nm. Solid line:Leu-enkephalin, Vertically ruled area:aminopeptidase, Dotted area:dipeptidylamino-peptidase. Dark shaded area:dipeptidylcarboxypeptidase activities.

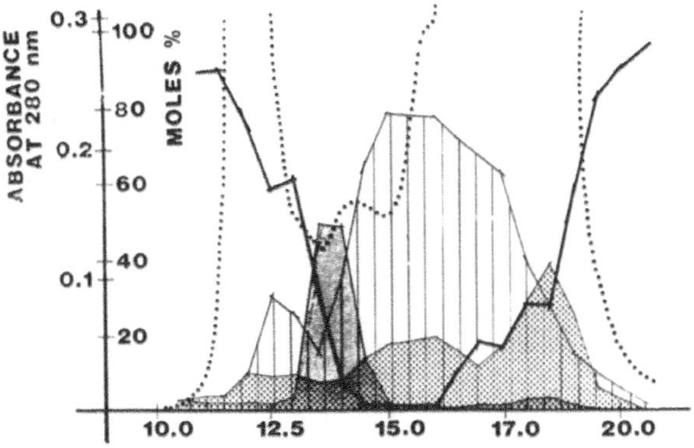

Figure10. Enkephalin-degrading enzymes in rabbit plasma. Symbols as in Fig.9.

E: CONCLUSION

The data reported in the previous sections may be summarized as follows: i) Once in the blood-strea, adrenal-released enkephalins are hydrolyzed by several enzyme groups cleaving TYR-GLY, GLY-GLY and GLY-LEU bonds. ii) The enzyme specificity necessary to hydrolyze enkephalins seems to be relatively high. This seems at least partially due to the unusually high degree of structural organization shown by these peptides. iii) Circulating enkephalins are protected from enzyme hydrolysis by binding to several goups of molecules all of a peptidic nature. Two such systems are present in the chromaffin granule, and two more in the plasma. iv) Taking into account the existence of the protecting systems, the half-lives of the enkephalins in man seem to be long enough to account for medium-lasting physiological effects. v) As far as enkephalin hydrolysis is concerned, the data obtained with a given animal species cannot be extended to other species.

On the whole, release, hydrolysis and protection of enkephalins seem to be a quite complex phenomena. The phylogenetic history of opioid peptides is a relatively long one, since these peptides are found in different, although related, phyla such as Mollusca, Annelida and Arthropoda, as well as in Chordata (53-55). A long molecular evolution can certainly account for the existence of complex interactions as well as for the existance of different physiological roles for the same molecule.

A certain degree of intrinsic naivete nothwithstanding, it seems worthwhile to attempt to interpret the existing data on enkephalins' hydrolysis in terms of the functional significance of the hydrolysis/protection mechanisms discovered so far. In this respect, the hypothesis that best fitls all the available evidence on opioid peptides' release, hydrolysis and protection seems to be the following: once enkephalins are released into the blood-stream. role of the degrading enzymes could be to terminate the physiological activity of these peptides, assuming that no such system exists in the target tissues. Moreover, the exostence in blood of two different systems with opposite functions - degradate as well as to protect enkephalins - makes it appear that a homeostatic regulation of enkephalins' activity is measured through the opposite actions of these systems.

It is clear that the outlined hypothesis is nothing more than a tentative interpretation. Actually, it seems extremely difficult even to devise an experimental procedure to verify it. Yet, it is possible that the addition of more experimental evidence will show us - in the next few years - whether or not the outlined hypothesis can be correct.

ACKNOWLEDGEMENT

The work described in the present article has been carried out within the framework of an with partial support from the "Progetto Finalizzato Chimica e Secondaria" of the CNR (Italian National Council for Scientific Research).

REFERENCES

1. Terenius, L. and Wahlstrom, A. (1974) Inhibitor(s) of narcotic receptor binding in brain extracts and cerebrospinal fluid. Acta Pharmacol. Toxicol. (suppl. 1) 35, 55.
2. Hughes, J. (1975) Isolation of an endogenous compound from the brain with pharmacological properties similar to morphine. Brain Research 88, 295-308.
3. Pasternak, K.W., Goodman, R. and Snyder S.H. (1975) An endogenous morphine-like factor in mammalian brain. Life Sci. 16, 1765-1769.
4. Pert, C.B., Pert, A., Chang, J.K. and Fond, B.T.W. (1976) D-Ala^2-Met-Enkephalinide: a potent, long-lasting synthetic pentapeptide analgesic. Science 194, 330-32.
5. Buscher, H.H., Hill, R.C., Romer, D., Cardinaux, F., Closse, A., Hauser, D. and Pless, J. (1976) Evidence for analgesic activity of enkephalin in the mouse. Nature 261, 423-25.
6. Belluzzi, J.D., Grant, N., Garsky, V., Sarantakis, D., Wise, C.D., and Stein, L. (1976) Analgesia indiced in vivo by central administration of enkephalin in rat. Nature 260, 625-26.
7. Terenius, L., (1978) Endogenous peptides and analgesia. Ann. Rev. Pharmacol. Toxicol. 18, 189-204.
8. Hambrook, J.M., Morgan, B.A., Rance, M.J., and Smith, C.F.C., (1976) Mode of deactivation of the enkephalins by rat and human plasma and rat brain homogenates. Nature 262, 782-783.
9. Dupont, A., Cusan, L., Garon, M., Alvarado-Urbina, G. and Labrie, F. (1977) Extremely rapid degradation of ^{3}H methionine-enkephalin by various rat tissues in vivo and in vitro. Life Sci. 21, 907-914.
10. Hogue-Angeletti, R.A. and Roda, L.G. (1980) In vitro interaction of enkephalin with serum and chromaffin granule components. Experientia 36, 1420-1421.
11. Schwartz, J.C., Malfroy, B. and De La Baume, S. (1981) Biological inactivation of enkephalins and the role of enkephalin-dipeptidyl-carboxypeptidase ("Enkephalinase") as neuropeptidase. Life Sci. 29, 1715-1740.
12. Coletti-Previero, M.A., Matras, H., Descomps. B. and Previero, A. (1981) Purification and substrate characterization of a human enkephalin-degrading aminopeptidase. Biochem. Biophys. Acta 657, 122-127.
13. Roscetti, G., Possenti, R., Bassano, E. and Roda, L.G. (1984) Mechanisms of leu-enkephalin hydrolysis in human blood. Life Sci. in press.
14. Roda, L.G., De Marco, V. and Possenti, R. (1983) Stability of periphral enkephalins in "Degradation of Endogenous Opioid." S. Ehrenpreis and F. Sicuteri eds. pgs. 25-42 Raven Press, New York.
15. Loew, G.H. and Burt, S.K. (1978) Energy conformation study of met-enkephalin and its D-Ala^2 analogue and their resemblance to ridig opiates. Proc. Natl. Acad. Sci. USA 75, 7-11.
16. Momany, F.A.(1977) Conformational analysis of methionine-enkephalin and some analogs. Biochem. Biophys. Res. Comm. 75,1098-1103.
17. Isogai, Y., Nemethy, G. and Scherga, H.A.(1977) Enkephalin-conformational analysis by means of empirical energy calculations. Proc. Natl. Acad. Sci. USA 74,414-418.
18. Spirites, M.A.,Schwartz, R.W., Mattice, W.L. and Coy,D.H.(1978) Circular dichroism and adsorption study of the structure of methionine-enkephalin in solution. Biochem. Biophys. Res. Comm. 81,602-609.
19. Khaled,M.A., Long, M.M., Thompson, W.D., Bradley, R.J., Brown, G.B., and Urry, D.W. (1977) Conformational states of enkephalin in solution. Biochem. Biophys. Res. Comm. 76,224-231.
20. Fillippi, B. Gusti, P., Cima, L. Borin, G., Ricchielli, F. and Marchiori f. (1979) Synthetic enkephalins. Intl. J. Peptide Protein Res. 14,34-40.

21. Levine, B.A.,Rabenstein, D.L., Smith, D. and Williams, R.J.P.(1979) Proton magnetic resonance studies on methionine enkephalin and β-endorphin in aqueous solution. Biochem. Biophys. Acta 579, 279-291.
22. Garbay-Jaureguiberry, C., Marion, D., Fellion, E. and Roques, B.P.(1982) Refinement of conformational preferences of Leu-enkephalin and Tyr-Gly-Gly-Phe by ^{15}N NMR. Int. J. Peptide Protein Res. 20, 443-450.
23. Jones,C.R., Garsky, V. and Gibbons, W.A.(1977) Molecular conformations of met-enkephalin: comparition of the zwitterionic and cationic forms. Biochem. Biophys. Res. Comm. 76,619-625.
24. Garbay-Jaureguberry, C.,Roques, B.P. and Oberlin,R. (1977) ^{1}H and ^{13}C NMR studies of conformational behavior of leu-enkephalin. FEBS Letters 76,93-98.
25. Smith, G.D. and Griffin, J.F.(1978) Conformation of Leu^5 enkephalin from x-ray diffraction: features important for recognition at opiate receptor. Science 199,1214-1216
26. Cabassi, F. and Zetta, L. (1982) Human β-endorphin. Intl. J. Peptide Protein Res. 20,154-158.
27. Clement-Jones,V., Lovry, P.J.,Rees, L.H. and Besser,G.M.(1980) Conformation of Leu^5 enkephalin from x-ray diffraction: features important for recognition at opiate receptor. Nature 283,295-297.
28. Ryder, S.W. and Eng, J. (1982) Radioimmunoassay of leucine-enkephalin like substance in human and canine plasma. J. Clin. Endocrinol. and Metabolism 52,367-369.
29. Blaschko, H. and Welch, A.D.(1953) Localization of adrenaline in cytoplasmic particles of the bovine adrenal medulla. Naunyn Schmiedberg's Arch. Exp. Pathol. Pharmakol. 219, 17-22.
30. Hillarp, N.A., Lagenstedt, S. and Nilson, B. (1953) The isolation of a granular fraction from the suprarenal medulla, containing the sympathomimetic catechol amines. Acta. Phisiol. Scand. 29, 251-263.
31. Winkler, H. (1976) The conposition of adrenal chromaffin granules: an assessment of controversial results. Neuroscience 1, 65-80.
32. Njus, D. and Radda, G.K. (1978). Bioenergetic processes in chromaffin granules: a new prospective on some old problems. Biochem. Biophys. Acta. 463, 219-244.
33. Hogue-Angeletti, R.A., Roda, L.G., Nolan, J.A. and Zarenba, S. Catecholamine storage vesicles in "Protein of the Nervous System" R.A. Bradshaw and D.M. Schneider eds. Raven Press. New York 1980.
34. Schultzberg, M., Lundberg, J.M., Hokfelt, J., Terenius, L., Brandt, J., Elde, R.P. and Goldstein, M. (1978) Enkephalin-like immunoreactivity in gland cells and nerve terminals of the adrenal medulla. Neuroscience 3, 1169-1186.
35. Viveros, O.H., Diliberto, E.J.J., Hazum, E. and Chang, K.J. (1979) Opiate-like materials in the adrenal medulla: evidence for storage and secretion with catecholamines. Mol. Pharmacol. 16, 1101-1108.
36. Di Giulio, A.M., Yang, H.-Y.T., Fratta, W. and Costa, E. (1979). Decreased content of immunoreactive enkephalin-like peptide in peripheral tissues of spontaneously hypertensive rats. Nature 278 646-647.
37. Lewis, R.V., Stern, A.S., Rossier, J., Stein, S. and Udenfriend, S. (1979) Putative enkephalin precursor in bovine adrenal medulla. Biochem Biophys. Res. Comm. 89, 822-829.
38. Stern, A.S., Lewis, R.V., Kimura, S., Rossier, J., Stein, S. and Udenfriend, S. (1979) Isolation of the opioid heptapeptide met-enkephalin-Arg^6-Phe^7 from bovine adrenal medullary granules and striatum. Proc. Natl. Acad. Sci. USA 76, 6680-6683.
39. Lewis, R.V., Stern, A.S., Kimura, S., Rossier, J., Brink, L., Gerber, L.D., Stein, S. and Udenfriend, S. Opioid peptides and precursor in the adrenal medulla in "Neural Peptides and Neural Communication." E. Costa and M. Trabucchi eds. Raven Press, New York 1980.

40. Yang, H.-Y.T., Di Giulio, A.M., Fratta, W., Hong, J.S., Majane, E.A. and Costa, E. (1980) Enkephalin in bovine adrenal gland: multiple molecular forms of Met^5-enkephalin immunoreactive peptides. Neuropharmacology 19,209-215.
41. Troy, C.M. and Musacchio, J.M. (1982) Processing of enkephalin precursors by chromaffin granule enzymes. Life Sci. 31, 1717-1720.
42. Wallace, E.F., Evans, C.J., Jurik, S.M., Mafford, I.N. and Barchas, J.D. (1982) Carboxypeptidase B activity from adrenal medulla. Is it involved in the processing of proenkephalin? Life Sci. 31, 1793-1796.
43. Jones B.N., Shively J.E., Kilpatrick D.L., Stern A.S., Lewis R.V., Kojima K., Udenfriend S., Usa (1979) Adrenal opioid proteins of 8,600 and 12,600 daltons: intermediates in proenkephalin processing. Proc. Natl. Acad. Sci. 2096-2100.
44. Lewis, R.V., Stern, A.S., Kimura, S., Rossier, J., Stein, S. and Udenfriend, S., (1980). An about 50,000-dalton protein in adrenal medulla: a common precursor of Met- and Leu- enkephalin. Science 208, 1459-1461.
45. Fricker, L.D., Supattapone, S. and Snyder, S. (1982) Enkephalin convertase: a specific enkephalin synthesizing carboxypeptidase in adrenal chromaffin granules, brain, and pituitary gland. Life Sci. 31, 1841-1844.
46. Lindberg, I., Yang, H.-Y.T. and Costa, E. (1982) Characterization of a partially pufified trypsin-like enkephalin-generating enzyme in bovine adrenal medulla. Life Sci. 31, 1713-1716.
47. Lewis, R.V. and Stern, A.S. (1983) Biosynthesis of enkephalins and enkephalin-containing polypeptides. Ann. Rev. Pharmacol. Toxicol. 23, 353-372.
48. Possenti, R., De Marco, V., Cherubini, O. and Roda, L.G. (1983) Enkephalin-binding systems in human plasma. Neurochem. Res. 8, 423-432.
49. Hogue-Angeletti, R.A. and Sheetz, PxB. (1978) A soluble lipid protein complex from bovine adrenal medulla chromaffin granules. J. Biol. Chem. 253, 5613-5616.
50. Helle, K.B. (1966) Antibody formation against soluble protein from bovine adrenal medulla chromaffin granules. Biochem. Biophys. Acta 117, 107-110.
51. Roda, L.G. and Hogue-Angeletti, R.A. (1979) Peptides in the adrenal medulla chromaffin granule. FEBS Letters 107, 393-397.
52. Stefano, G.B. and Catapane, E.J. (1979) Enkephalins increase dopamine levels in the CNS of a marine mollusc. Life Sci. 24, 1617-1622.
53. Sundler, F., Hakanson, R., Alumets, J. and Walles, B. (1977) Neuronal localization of pancreatic polypeptide (PP) and vasoactive intestinal peptide (VIP) immunoreactivity in the earthworm (lubricus terrestris). Brain Research Bull. 2, 61-65.
54. Zipser, B. (1980) Identification of specific leech neurones immunoreactive to enkephalin. Nature 283, 857-858.

STUDIES OF THE ENDOGENOUS OPIOID SYSTEM

IN THE HUMAN STRESS RESPONSE

Martin R. Cohen[1], David Pickar[2], Michel Dubois[3], and Robert M. Cohen[4]

1. Ensor Foundation Research Laboratory, William S. Hall Psychiatric Institute, and the Department of Neuropsychiatry and Behavioral Sciences, University of South Carolina.
2. Section on Clinical Studies, Clinical Neuroscience Branch, NIMH, Bethesda, Md.
3. Department of Anesthesiology, Georgetown University School of Medicine, Washington, D.C.
4. Section on Brain Imaging, Laboratory of Psychology and Psychopathology, NIMH, Bethesda, Md.

During the last few years, our group has been actively engaged in clinical studies to understand the role of the endogenous opioid system (EOS) in human behavior. In this chapter, we review those of our studies that are most pertinent to our understanding of human behavior during stress and the implications of results from these studies on future strategies to understand and modify, for therapeutic purposes, the role of the EOS in the human stress response.

PLASMA BETA-ENDORPHIN AND HUMAN STRESS

When we began our studies in 1979, considerable evidence had already been gained from animal experimentation to support an important role for the EOS in the physiological response to stress[1]. Over the years, surgery had been one of the most intensively studied of human stressors. Thus, in our initial studies, we sought to determine whether activation of the EOS occurred with surgery.

The EOS is actually a complex of systems with unknown interrelations, different anatomical locations and different concentrations of endogenous opioids (endorphins)[2]. Nevertheless, due to practical considerations, we limited our study to the evaluation of patients' plasma levels of beta endorphin. This could only be considered a peripheral measurement of EOS activation, but changes in plasma levels of beta-endorphin could probably be attributed to alterations in pituitary secretion, a function known to play an important role in the physiological stress response[3]. In addition, by concomitantly measuring patients' plasma cortisol levels, we could assess the linkage, suggested in animal studies, of the hypothalmic-pituitary-adrenal axis (HPA) and the peripheral EOS during stress[4].

An initial pilot study in patients undergoing limb amputations suggested marked increases in plasma beta-endorphin levels but only in those patients who were administered anesthesia without opiates (exogenous opioids)[5]. We subsequently completed a study on a more uniform cohort[6] as follows:

> Nine patients underwent laparotomy as part of a National Cancer Institute protocol for the assessment of testicular or ovarian malignancy. Patients were free of medical complications other than regional malignancy and were medication free for 72 hours before the surgical procedures. Premedication consisted of 100 mg. of pentobarbital administered intramuscularly. Induction of anesthesia was accomplished with thiopental and succinylcholine, and anesthesia was maintained with enflurane, nitrous oxide, and a muscle relaxant. All patients were intubated and had controlled ventilation. Surgery lasted between 2 and 4 hours. No exogenous opioids were administered before or during surgery.
>
> Blood samples were collected from each patient before, 10 to 15 minutes after induction of anesthesia, 10 minutes after skin incision, at 30-minute intervals during surgery until skin closure and while the patient was conscious following surgery and before post-operative analgesia was initiated. A patient's mean surgery hormone level was calculated as the mean of values derived from samples obtained during surgery at the times just noted. Details of sample care are reported elsewhere.
>
> Plasma cortisol and beta-endorphin levels were determined by radioimmunossay on unextracted samples with reagents supplied by New England Nuclear Corporation. There is in the beta-endorphin assay, 50 per cent crossreactivity to beta-lipotropin, $<$5 percent to ACTH, $<$ 0.01 per cent to alpha-endorphin and alpha-MSH, and $<$ 0.004 per cent to methionine and leucine enkephalin. Thus beta-endorphin (ir) as determined in the study was essentially the sum of circulating beta-endorphin (beta-lipotropin (61-91)) and beta-lipotropin; their ratio was not determined. Details of the assay method may be found in reference 5.

In summary, we found no change in plasma beta-endorphin (ir) following anesthesia and prior to surgical incision, but a marked increase in plasma beta-endorphin (ir) during surgery. Analogous effects of anaesthesia and surgery on plasma cortisol levels were found. In evaluating individual patient's hormone levels, we found across the 3 conditions of before, during and following surgery a strong positive correlation between plasma betaendorphin (ir) and cortisol levels ($r=0.61$, $p<0.01$). Individual hormone responses to surgery are illustrated in figure 1.

Thus we were somewhat comfortable that within the limitations of clinical design, we had been able to demonstrate an activation of the EOS during the human stress response. Similar to the findings in animal experimentation, this activation was concomitant with an activation of the HPA axis.

While we were engaged in our studies of surgical stress, we had the opportunity in collaboration with Drs. Nurnberger and Gershon of the NIMH to monitor the response of plasma beta-endorphin levels to an intravenous

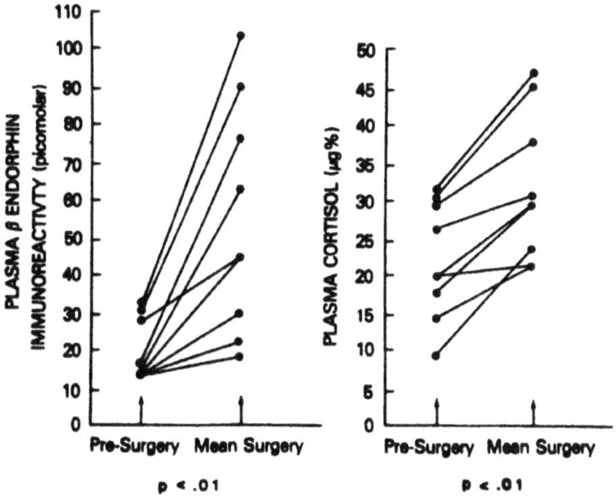

FIGURE 1: Individual patient plasma beta-endorphin (ir) and cortisol responses to surgery. Each line connects an individual's presurgery hormone level to mean-surgery hormone level. Presurgery beta-endorphin levels below the sensitivity of our radioimmunoassay ($<$ 14.4 picomolar) were arbitrarily assigned the value of 14.4 picomolar to minimize any found increases with surgery. The group mean ±SEM were as follows: pre-surgery beta-endorphin 19.9±2.4 picomolar; mean surgery = 54.9±10.2 picomolar; presurgery cortisol = 22.1±2.6 ug per dl: mean surgery cortisol = 32.4±3.2 ug per dl. (From Cohen, M.R., Pickar, D., Dubois, M., et al.: Clinical and experimental studies of stress and the endogenous opioid system. In Verebey, K.: Opioids and Mental Illness. New York, New York Academy of Sciences. 1982)

amphetamine infusion [7]. We were intrigued by this opportunity since in animal models, amphetamine may simulate the effects of stress, and in humans, amphetamine is used in the treatment of behavioral disorders with apparent (narcolepsy) or hypothesized arousal abnormalities (hyperactivity and depression) [8,9].

As in our surgical studies, we found in response to amphetamine a concomitant activation of the EOS and the HPA axis. Although the response to amphetamine was considerably less than the response to surgery, it was demonstrable in a double-blind cross-over study, a design not practical with surgery. However, more recently in collaboration with Dr. C.H. Shatney of the Trauma Center of Baltimore, we have been able to demonstrate in patients coming to a trauma center markedly higher plasma beta-endorphin levels in patients with severe trauma and admission to the center compared to those patients referred elsewhere for treatment because of less severe trauma [10]

Concurrent and subsequent to our studies, many other groups have been able to demonstrate increases of plasma beta-endorphin (ir) in humans during physiological stress[11-13]. However, activation of the EOS only suggests that this system may be involved or may be an indicator of the physiological stress response.

The potential of the EOS to be involved in the behavioral adaptation to a stressor is suggested by the behavioral effects of narcotic alkaloids (exogenous opioids) which are presumed to function by activation of the EOS. In particular, the value of narcotic alkaloids to modify behavior to a continuing painful stimulus is well known[14]. The clinical application of this principle was the mainstay of post-operative analgesia at the NIH. Immediately following surgery, patients were taken to the Clinical Center Surgical Intensive Care Unit (SICU) where they stayed for a minimum period of 24 hours. During their stay, there was a standing routine clinical order for post-operative management of pain. Once patients were conscious (between first and third hour post awakening) they were administered intravenous morphine at 1.5 to 3.5 mg per hour as considered necessary by staff for relief of post operative pain.

We reasoned that if plasma beta-endorphin levels were an accurate indicator of the individual's arousal or stress response, i.e. the ability to physiologically adapt to a stressor, then individuals with high plasma beta-endorphin levels would require less morphine than those with low levels. This would be the clinical extension of the well documented laboratory phenomenon of stress induced analgesia.

A patient's morphine requirement was simply calculated as the total morphine administered to the individual during the first 24 postoperative hours in the SICU. This was obtained from the patient's clinical chart. The postsurgical morphine requirements of our patients ranged from 12 to 56 mg (mean ±SEM, 35.2±16.1). The validity of this simple measurement as a reflection of our patient's physiological need for relief of pain was supported by a positive ($r=0.55$), although not quite significant, correlation with patient weight (range = 47.5-85.5 kg) and a significant negative correlation ($r = -0.69$, p 0.05, 2 tail) with patient age (range = 21 to 60 years). This latter finding is expected from the slowing of morphine metabolism in the aged[15].

We were surprised however to find that not only did those patients with high surgery beta-endorphin (ir) have low morphine requirements (figure 2) but that they were also the patients with high pre-surgery beta-endorphin (ir)[16].

Our findings suggested that elevated plasma beta-endorphin (ir) could be considered a biologic marker of the human stress response much as plasma cortisol levels had previously been used to predict physiological and behavioral adaptation to a stressor. Further, our work suggested that even in the instance of an extreme physical stressor, a strong component of adaptation might lie in the individual's innate responsiveness to stress or his prior psychologic perception of this stress as patients' presurgical stress hormone levels were predictive of their levels during surgery as well as their postsurgery pain perception.

Plasma beta-endorphin has not been directly implicated in analgesia[17] but we might hypothesize that the larger morphine dose was required by some patients to substitute for a relatively poorly stress-activated physiological response. This is supported by the finding of Tamsen et al. of a negative relationship between lumbar CSF endogenous opioid levels

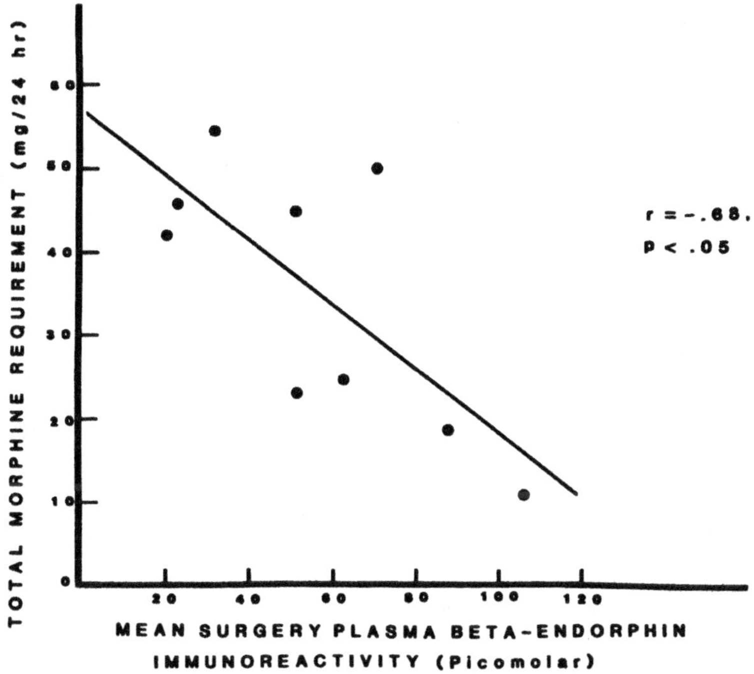

FIGURE 2: Nine patients (+) plotted two-dimensionally for the variables of morphine requirement and mean (surgery) plasma beta-endorphin. Patient morphine requirement can be predicted by the illustrated linear equation: Morphine requirement = 54.9-0.36 x mean (surgery) plasma beta-endorphin. Patient morphine requirement is the total morphine used during the first 24 postoperative hours. Mean (surgery) plasma beta-endorphin is the mean of plasma beta-endorphin levels found in samples taken at 30-minute intervals during surgery. (From Cohen, M.R., Pickar, D., Dubois, M., et al.: Stress induced plasma beta-endorphin immunoreactivity may predict postoperative morphine usage. Psychiatr. Res., 6:7-12, 1982).

measured prior to surgery and calculated 24 hour postsurgery CSF levels of the self-administered opiate pethidine [18]. Further support for a role of the EOS in surgical stress-induced analgesia is provided by the work of Levine et al., demonstrating in the postsurgery period analgesia following 0.2 and 2 mg of naloxone but hyperalgesia following 7.5 and 10 mg doses [19]. A similar biphasic effect of naloxone on analgesia following exercise has been reported [20].

NALOXONE AND HUMAN STRESS

Indeed a common clinical strategy to assess the role of the EOS in human behavior is to evaluate the effect of the administration of naloxone. Naloxone, which is considered a pure opiate receptor antagonist, is assumed to affect an in vivo blockade of the EOS. Thus, if the EOS is involved in an important role in the subject's present mood or behavior, then administration of naloxone should produce demonstrable alterations.

In contrast to the postive findings with stress induced analgesia, the failure in some studies to find modification of the stress response following naloxone administration has led some investigators to exclude the importance of the EOS in these human stress responses[21,22]. However the EOS is a complex system such that parts of the system as a result of location and/or the involvement of different endogenous opioids and receptor subtypes, or the need to be only intermittently active, may be relatively resistant to functional blockade by naloxone. Thus investigators of the stress response may well have generally used too low or restricted a dose range of naloxone to fully assess EOS involvement. Conclusions regarding the role of the EOS in pain perception following surgery or jogging would have differed had not the experimenters increased the dose of naloxone administered to their subjects[19,20].

In recent years, we have been able to demonstrate that bolus doses of naloxone administered in the mg/kg range intravenously to normal volunteers may produce behavioral and physiological responses that mimic responses seen during the human stress response. These effects include an increase in plasma levels of cortisol and growth hormone, an increase in systolic blood pressure and a behavioral syndrome of irritability, anxiety and depression[23,24]. In addition, we have found that hospitalized patients suffering from depression seem to be more sensitive to these naloxone induced dysphoric symptoms[25]. One explanation of the pathogensis of depression is based on physiological adaptation to stress in vulnerable individuals[9]. We believe the previous reports of minimal or inconsistent behavioral effects after naloxone administration to normals are directly attributable to the lower doses of naloxone used in those studies.

The design of our initial study was complex due to the safety precautions that were necessitated by the administration of doses of naloxone not previously administered to humans. The details may be found in reference 24.

> In brief, subjects were without physical illness or a personal history of psychiatric illness or drug abuse. In addition, all subjects were free from acute illness or drug intake for at least three days preceding any individual day of study and free from opiate intake for at least the three preceding weeks.
>
> Subjects participated in three separate days of study which were conducted at least a week apart. At 8:30 and 10:30 a.m. of each study day three-minute IV infusions of increasing doses of naloxone hydrochloride were administered: first day, 0.3 and 1 mg/kg; second, 1 and 2 mg/kg; and third, 2 and 4 mg/kg. Subjects were blind to naloxone doses and sequence. Periodically during the course of the day, each volunteer completed self-rating scales and the non-blind attending psychiatrist the Brief Psychiatric Rating Scale (BPRS).

Results from the rating scales were quite consistent so that only the scores on the self-rating mood adjective checklist are illustrated in Figure 3. In summary, with the administration of increasing doses of naloxone and presumeably increasing functional blockade of the EOS, normals experienced dysphoric symptoms that were also manifest in their behavior and apparent to the experimenters. The most prominent symptoms were anxiety, irritability, depression, a general uncomfortable mental feeling and difficulty with concentration and functioning. Although the effect was found in almost every volunteer, its intensity, duration, and the dose required for the effect was quite individually variable.

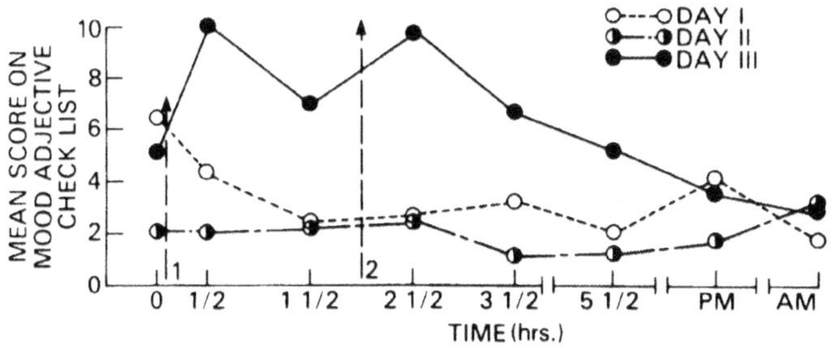

FIGURE 3: Normal volunteers participated in three separate days of study at least one week apart. Bolus doses of naloxone were infused intravenously at 0830 and 1030 hours (Arrows 1 and 2). Doses 1 and 2 were as follows: first day 0.3 and 1 mg/kg, respectively; second day 1 and 2 mg/kg, respectively; and third day, 2 and 4 mg/kg respectively. Total scores on the mood adjective checklist were found to have a main effect for day ($F=3.86$, $df=2,12$, $p = .05$), no significant effect for time ($F=1.25$, $df=7,42$, $p=.3$), but a very significant day x time interaction effect ($F=4.36$, $df=14,84$, $p < 0.0001$). The day x time effect was attributable to significantly increased scores (increased symptoms) on day 3 compared with day 1 after the first infusion and until discharge at 2 pm. Scores before the first infusion were not significantly different between days. Negative scores indicate a decrease in symptoms. The most prominent effects indicated by volunteers were increasing anxiety, irritability and worry, increasing difficulty with concentration and functioning, and an uncomfortable mental feeling.

OPIOID AGONISTS AND HUMAN STRESS

The results from our investigations of opioid levels during surgery and naloxone administration to normals led us to begin to assess the potential therapeutic application of modifying the stress response by activation of the EOS.

In 6 of the surgery patients, we were able to measure plasma beta-endorphin (ir) following surgery, both prior to and following post-operative morphine administration. The marked decrease in all of the patients' hormone levels following morphine administration is illustrated in Figure 4.

We have also been able to study the effect on EOS and HPA activation during surgery of the addition of fentanyl (a synthetic opiate) to anesthesia. Our methods and the results of this study are reported in detail elsewhere[26].

> In brief, patient selection, anesthetic technique, blood collection, and assay methodology were as noted for the previously presented study. However, in this study, 10 patients received, in addition to their premedication and anesthetic drugs, a single medium dose of fentanyl (10 to 20 ug per kg) given intravenously 5 minutes before the surgical incision.

The most important finding is illustrated in Figure 5. A single administration of fentanyl completely blocks the plasma beta-endorphin and cortisol responses early in surgery while partially blocking even the responses late in surgery.

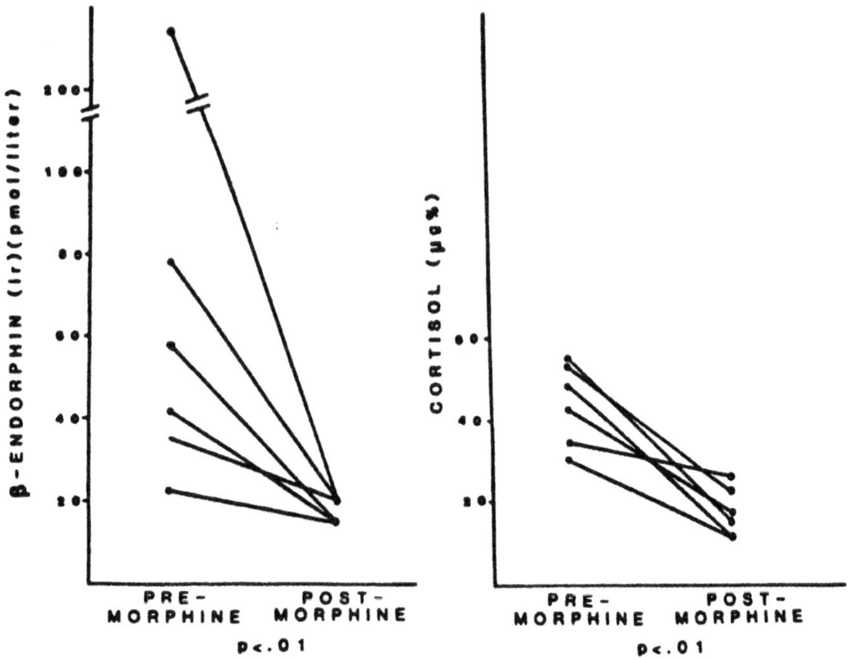

FIGURE 4: Changes in plasma levels of cortisol and beta-endorphin (ir) in six patients following the administration in the postoperative period of standard clinical doses of morphine. Samples were collected while patients were awake in the Surgical Intensive Care Unit. Patients had undergone laparotomy with enflurane as the primary anesthetic and without the aid of opiates. (From Pickar, D., Cohen, M., Naber, D., et al.: Clinical studies of the endogenous opioid system. Biol. Psychiatry., 17:1243-1276, 1981).

We have hypothesized that this suppression of hormone levels represents feedback inhibition from EOS activation. The effects of this activation on behavioral responses following surgery warrants study. What effect would such activation have on the development of phantom limb pain, a not infrequent complication of surgical amputation?

Because of the complexity of the EOS, the effects of its activation with an exogenous opioid do not necessarily demonstrate the potential therapeutic benefit of modifying the stress response by activation of the EOS with an endogenous liqand. Thus we have begun a preliminary study of the effect of intrathecal administration of beta-endorphin on the behavior of patients suffering from the chronic and severe pain of metastatic cancer[27] The intrathecal route was chosen because of the apparent effect of the blood brain barrier to exclude from the central nervous system (CNS) peripherally administered beta-endorphin.

As expected, in the 2 patients who have undergone the study, beta-endorphin administration produced prolonged (>48 hrs.) complete relief from pain. However, one of the patients also experienced a marked generalized behavioral change characterized by confusion, hypomanic/manic behavior and psychosis that also persisted for more than 2 days. In patients undergoing the chronic stress of terminal metastatic cancer and persistent severe pain, it might prove difficult to assess the direct effects of an opioid agonist on behavior and mood from indirect effects resulting from

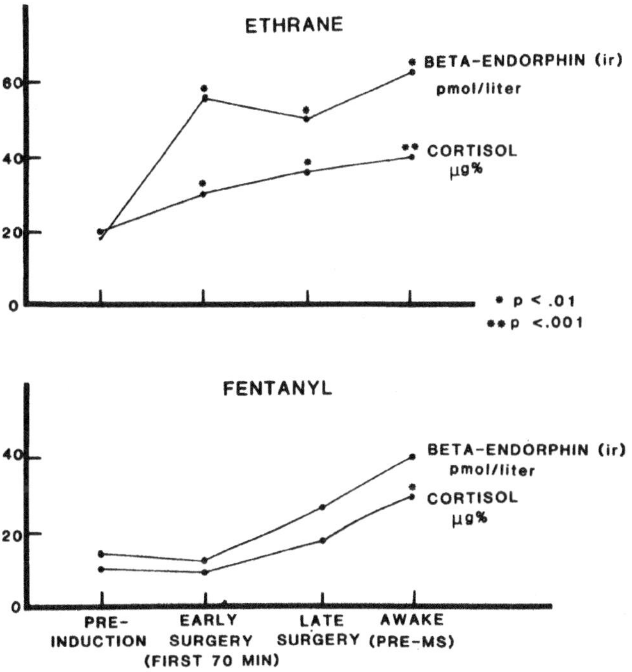

FIGURE 5: Nineteen patients, suffering only from regional malignancy, underwent laparotomy as part of a National Cancer Institute treatment rotocol. All patients received enflurane as the primary maintenance anesthetic. Ten patients received, in addition, 10 to 20 ug of fentanyl intravenously, per kg body weight, five minutes before surgical incision. Surgery lasted between two and four hours. Blood samples for hormone determinations were collected from each patient prior to anesthetic induction (preinduction), 10 minutes after skin incision, at 30-minute intervals during surgery, and when awake in the Surgical Intensive Care Unit, prior to administration of morphine sulfate (MS) for postoperative analgesia. The graphed points represent mean hormone values; p values report the statistical significance of differences from preinduction hormone levels.

pain-relief. However it would be difficult to attribute this patient's psychosis to such an indirect effect.

CONCLUSION

Our work and that of others strongly point to involvement of the EOS in stress induced analgesia. However following a stressful event the human responds with a cluster of behavioral and physiological symptoms. Many of these symptoms may be simulated by the administration of naloxone. If progress in medical science were to be considered the development of an increasingly accurate approximation of how the human body works, we suggest, as a first approximation, that stress disorders are attributable to an exhaustion of the EOS and that activation of the EOS should be helpful in relieving these symptoms. Future studies of this hypothesis must use a broad range of doses of naloxone to ensure functional blockade of the complex EOS. In addition, with the development of our ability to administer or measure various endogenous opioid liqands and to administer opioid antagonists specific to receptor subtypes, we should be able to further modify this first approximation.

REFERENCES

1. Amir, S., Brown, Z.W., and Amit, A., The role of endorphins in stress: Evidence and speculations, Neurosci. Biobehav. Rev. 4:77 (1980).
2. Frederickson, R., and Geary, L., Endogenous opioid peptides: Review of physiological, pharmacological and clinical aspects, Prog. Neurobiol., 19:19-69 (1982).
3. Rossier, J., French, E.D., Rivier, C., Ling, N., Guillemin, R., Bloom, F.E., Foot-shock induced stress increases B-endorphin levels in blood but not brain, Nature 270:618-620 (1977).
4. Guillemin, R., Vargo, T., Rossier, J., Minicck, S., Ling, N., Rivier, C., Vale, W., Bloom, F., Beta-endorphin and adrenocortiocotropin are secreted concomitantly by the pituitary gland, Science 197:1367-1369 (1977).
5. Cohen, M., Pickar, D., Dubois, M., Roth, Y., Naber, D., Bunney, W., Jr., Surgical stress and endorphins, Lancet I 213-214 (1981).
6. Dubois, M., Pickar, D., Cohen, M.R., Roth, Y., MacNamara, T., Bunney, W.E. Jr., Surgical stress in humans in accompanied by an increase in plasma B-endorphin immunoreactivity, Life Sciences 29:1249-1253 (1981).
7. Cohen, M.R., Nurnberger, J., Pickar, D., Gershon, E., Bunney, W., Jr., Dextroamphetamine infusions in normals results in correlated increases of plasma B-endorphin and cortisol immunoreactivity, Life Sciences 29:1243-1247 (1981).
8. Antelman, S., Caggiula, A., Stress-induced behavior: Chemotherapy without drugs in "Psychobiology of Consciousness," Davidson, J. and Davidson, R.M. eds. Plenum Press, New York 65-104 (1980).
9. Cohen, R.M., Campbell, I., Cohen, M.R., Torda, T., Pickar, D., Siever, L., Murphy, D., Pre-synapic noradrenergic regulation during depression and antidepressant drug treatment, Psychiatr. Res. 3:93-105 (1980).
10. Shatney, C.H, Cohen, R.M., Cohen, M.R., Imagawa, D.R. Endogenous opioid activity in clinical hemorrhagic shock. Surgery, Gynecology and Obstetrics (in press).
11. Carr, D., Bullen, B., Skrinar, G., Arnold, M., Rosenblatt, M., Beitins, I., Martin, J.B., McArthur, J., Physical conditioning facilitates the exercise induced secretion of beta-endorphin and beta-lipotropin in women, N. Engl. J. Med., 305:560-563 (1981).
12. Colt, W., Wardlow, S., Frantz, A., The effect of running on plasma beta-endorphin, Life Sciences, 28:1637-1640 (1981).
13. Goland, R., Wardlaw, S., Stark, R., Frantz, A., Human plasma beta-endorphin during pregnancy, labor and delivery, J. Clin. Endocrinol. Metab., 52:74-78 (1981).
14. Jaffe, J.H., and Martin, W.R., Narcotic analgesics and antagonists. In The Pharmacological Basis of Therapeutics, Edition 6 Goodman, A.G., Goodman, A.S., and Gilman, L. (eds), Macmillian New York (1980).
15. Stanski, D., Greenblatt, D., and Lowenstein, E., Kinetics of intravenous and intramuscular morphine. Clin. Pharmacol. Ther., 24:52-59 (1978).
16. Cohen, M.R., Pickar, D., Dubois, M., Bunney, W.E., Jr., Stress induced plasma beta-endorphin immunoreactivity may predict post-operative morphine usage. Psychiatr. Res., 6:7-12 (1982).
17. Foley, K.M., Kourides, I.A., and Inturrisi, C.E., Kaiko, R., Zaroulis, C., Posner, J., Houde, R., Li, C.H., Beta-endorphin: Analgesic and hormonal effects in humans, Proc. Natl. Acad. Sci., 76:5377-5381 (1979).

18. Tamsen, A., Hartvig, P., Dahlstrom, B., Wahlstrom, A., Jerenics,L., Endorphins and on demand pain relief, Lancet, 1:769 (1980).
19 Levine, J., Gordon, N., and Field, H., Naloxone dose dependently produces analgesia and hyperalgesia in post-operative pain, Nature, 278:740-741 (1979).
20. Haier, R., Quaid, K., Mills, J., Naloxone alters pain perception after jogging, Psychiatr. Res., 5:231-232 (1981).
21. Spiler, I.J., and Molitch, M.E., Lack of modulation of pituitary hormone stress response by neural pathways involving opiate receptors, J. Clin. Endocrinol. Metab., 50:516-520 (1980).
22. Enquist, A., Hicquet, J. Saubrey, N., Blichert-Toft, M., Cortisol, glucose and hemodynamic responses to surgery after naloxone administration, Acta Anaesthesiol. Scand., 25:17-20 (1981).
23. Cohen, M.R., Cohen, R.M., Pickar, D., Murphy, D., Bunney, W.E., Jr., Physiological effects of high dose naloxone administration to normal adults, Life Sciences, 30:2025-2031 (1982).
24. Cohen, M.R., Cohen, R.M., Pickar, D., Weingartner, H., Murphy, D.L., High dose naloxone infusions in normals: dose-dependent behavioral, hormonal, and physiological responses, Arch. Gen. Psychiatry, 40:613-619 (1983).
25. Cohen, M.R., Cohen, R.M., Pickar, D., Sunderland, T., Mueller, E. III., Murphy, D.L., High Dose Naloxone in Depression, Biol. Psychiat., 19:6, 825-832 (1984).
26. Dubois, M., Pickar, D., Cohen, M.R., Gadde, P., Macnamara, T.E., Bunney, W.E., Jr., Effects of fentanyl on the response of plasma beta-endorphin immunoreactivity to surgery, Anesthesiology, 57:468-472 (1982).
27. Pickar, D., Dubois, M., and Cohen, M., Behavioral change in a cancer patient following intrathecal B-Endorphin Administration, American Journal of Psychiatry, 141:1, 103-104 (1984).

EXERCISE STRESS AND ENDOGENOUS OPIATES

Peter Farrell* and Anthony Gustafson**

Department of Human Kinetics* and Endocrine Metabolic Section**
University of Wisconsin-Milwaukee*Medical College of Wisconsin**

INTRODUCTION

The use of exercise to reveal physiological responses and adaptations has a long history. Exercise can be quantified and repeated exertion markedly alters body functions. Well established exercise procedures such as the use of an individual's aerobic capacity have allowed exercise stress to be used as an experimental model to investigate the endogenous opiate system. A vast literature had accumulated prior to 1975 concerning the physiological effects of morphine. When the endogenous opiates were discovered[1], researchers already had insights into probable functions of the endogenous opiates. Therefore, it is not surprising that many possible roles for endorphins/enkephalins have been studied using the exercise model.

OPIATES AND MOTOR ACTIVITY

Opiates influence motor activity. Early investigations demonstrated a relationship between spontaneous locomotor activity and exogenous opiates. Goldstein and Sheehan[2] demonstrated the usefulness of opioid-induced "running fit" as a measure of morphine-like drug actions in mice. They showed that levorphanol tartrate (20 mg/kg, i.p.) resulted in increased running activity which was constant for several hours post injection. At higher doses (30 mg/kg) activity decreased. The L-(+)-stereoisomer, dextrophan, had no effect on running activity and the levorphanol effects were antagonized by nalorphine. Tolerance to a repeated constant dose of 2 mg/kg injections of levorphanol over eight hours was evident and this adaptation was shown to have cross-tolerance with morphine.

Babbini and Davis[3] then demonstrated that low doses of morphine sulfate, 1.25 to 5 mg/kg i.p., caused increased locomotor activity in male rats while higher doses, 10 to 40 mg/kg, resulted in an initial depression of activity followed by enhanced activity. However, Goldstein and Sheehan[2] observed no activity change at doses up to 7 mg/kg. This may represent an early indication of species differences.

Babbini and Davis[3] also reported a time-dependent effect of morphine on locomotor activity. Activity was highest approximately 3 hours post injection, 20 mg/kg, i.p., and was followed by return to normal 6 hours post injection. This response became exaggerated after 20 days of daily morphine injection.

Browne and Segal[4] provided two important observations. In rats and mice both exogenous alkaloids (morphine, methadone, etonitazene) and endogenous peptides (Beta-endorphin, as well as D-Ala^2-Met^5-enkephalinamide and D-Met^2-Pro^5-enkephalinamide) produced a biphasic, dose-dependent locomotor activity response. During the first hour following s.c. injection of opiate alkaloids, a decreased locomotor activity was observed while during hours 2-4 a dose dependent (1→10 mg/kg morphine, 0.25→2.5 mg/kg methadone, and 0.5→5.0 µg/mg etonitazene) increase was found. In like manner, intraventricular infusion of Beta-endorphin (1.0→5 µg/10 µl) reduced activity for the first hour followed by an increase over the next 2-4 hrs. Pre-treatment with naloxone (5.0 mg/kg s.c.) antagonized the effect of intraventricular Beta-endorphin. Recently Drust and Crawford[5] found that a low dose of D-Ala^2-Met- enkephalinamide (0.5 µg intracerebroventricular) enhanced locomotor activity in rats. There was no initial depression of activity suggesting that small doses may not result in a biphasic activity pattern as demonstrated by Browne and Segal.[4] This discrepancy may be caused by cannula placement; however, a decrease in spontaneous locomotor activity is not a uniform finding. Oka and Hosoya[6] found that 20 mg/kg morphine sulfate s.c. resulted in an increase in spontaneous locomotor activity in rats lasting several hours.

With the discovery of enkephalins the use of endogenous opiates quickly replaced exogenous opiates in studies examining the relationship between endorphins and activity. Katz et al[7] reported that D-Ala^2-Leu and Met enkephalinamide injections into brain ventricles of mice resulted in a dose-dependent increase in activity which was naloxone (4 or 8 mg/kg) reversible. The stereotyped running behavior produced in the mice was similar to that elicited by morphine injection. Naloxone (8 mg/kg) alone produced a decline of activity in mice who were hyperactive due to initial cage exposure. Amir et al[8] subsequently showed that naltrexone (1 or 2 mg/kg s.c.) reduced activity in rats. While route of administration, use of pharmacological doses, and the species differences tempers our conclusions, these and other studies[9,10,11] demonstrate that blockade of the endogenous opiate system causes decreased locomotor activity.

There is a genetic influence on the interaction between endogenous opiates and activity.[9,10,12,13] Castellano[9] demonstrated that FK 3-824, an enkephalin analogue, (0.25 → 40 mg/kg, i.p.) caused a dose-dependent reduction in locomotor activity in DBA/2J mice while the same enkephalin analogue (1.0 →,40 mg/kg i.p.) produced a dose-dependent hyperactivity in C57BL/6J mice. In both strains naloxone reversibility was demonstrated. Castellano and Oliverio[12] observed that increasing doses of morphine i.p. caused an increased number of cage crossings. Doses ranged from 5 to 20 mg/kg and pawlick latency to hot plate-induced analgesia was also found to increase with increasing doses of morphine. This study also demonstrated strain differences in mice to both activity levels and analgesic responses. For example, DBA/2J mice showed no increase in crossing even at 20 mg/kg while C57BL/6J mice had an approximate 5-fold increase at this dose. However, C57BL/6J mice showed the least hot plate latency (pawlicking as the endpoint) when compared to either BDA/2J or BALB/CJ mice. Pharmacological manipulations do not necessarily imply physiological roles. However, it would appear that opiates are involved in the normal activity patterns of mice and rats.

Endogenous opiates modify immobility following forced exercise. Amir[8] observed that naloxone reduced the immobility time of BALB/CJ and C57BL/6J mice who swam at 22°C for 10 min. Naloxone injections (0.625→40 mg/kg i.p.) resulted in greater reductions in C57BL/6J compared to BALB/CJ mice. Morphine injections (0.15→10 mg/kg, i.p.) caused an increase in immobility during swimming only at the lowest dose (0.15 mg/kg) in both BALB/CJ and C57BL/6J mice. Even high doses of naloxone (40 mg/kg) did not completely block swimming-induced immobility suggesting involvement of non-opiate mechanisms in the survival-induced immobility response. Naloxone decreased cage crossing activity[14,11] but increased survival swimming activity[8] implying that the type or degree of stress influences the degree of endogenous opiate involvement in the physiological response.

Christie et al[15,16,17] examined the relationship between exercise and pain tolerance. Effects of chronic and acute exercise (3 min warm water swim) were compared to chronic and acute exposure to morphine in mice. Exposure to swimming resulted in an increase in tail-flick latency. After a schedule of one 3 minute swim over 48 hours, mice showed less antinociception when compared to naive mice. Cross tolerance with morphine was found since naive mice showed greater tail flick latency at any dose of morphine (1.1 → 8.9 mg/kg) compared to chronically-exercised mice. Christie's group[17] also demonstrated that naloxone-precipitated withdrawal behavior following cessation of chronic warm water swimming after 48 hours is similar to morphine withdrawal behaviors. This observation suggests that swimming elicits a release of endogenous opiates; and, as a result, chronic swimming may result in adaptations of endogenous opiate receptors leading to tolerance similar to that found after chronic exposure to morphine.

Shyu et al[18] provided compelling data to demonstrate that running in rats is associated with an early morning analgesia which persists for at least several hours after running. Rats were given free access to a running session (9 pm - 9 am). Squeak thresholds due to electrical stimulation increased at the end of the running period. Naloxone (1-2 mg/kg i.p.) blocked this exercise-induced analgesia. Non-running control rats did not show elevated pain thresholds. In a small group of rats (n=4) whose running wheels were locked the previous night, squeak thresholds were similar to control periods (mid afternoon). Additionally, the increased squeak threshold in spontaneously hypertensive running rats was positively related ($r = 0.80$, $p < 0.001$) to the distance run on the exercise wheel the preceding night. These data imply that endorphins are involved in exercise-induced analgesia. This analgesia may last for several hours or only for minutes[16] following exercise.

Few studies are available relating changes in endorphins in animals subjected to exercise. Pert and Bowie[19] demonstrated that binding of D-Ala^2-D-Leu-enkephalinamide is reduced in rat brain homogenates after 15 minutes of running or swimming. Further data revealed that the reduced binding was due to enhanced occupancy of opiate receptors by endogenous opiates. However, Metzer and Stein[20] found no change in pituitary, cerebral cortex, and posterior and anterior hypothalamic levels of Beta-endorphin in rats subjected to high intensity running. Significant increases in plasma Beta-endorphin were observed after the acute run. These studies suggest that running may differentially affect central Beta-endorphin and enkephalin systems although investigations using radioreceptor assays[19] cannot be compared directly with studies using radioimmunoassays.[20]

ADRENAL OPIATES

Enkephalins are present in the adrenal medulla.[14,21-28] Quirion et al[24] demonstrated ethyl ketocyclazocine binding sites in the adrenal gland. These sites were concentrated in the adrenal cortex with lower densities found in the medulla. Bindings sites for mu and delta ligands were not demonstrable. Human pheochromocytomas contain both kappa and benzomorphan opiate binding sites[23,29], but intra-tumor variation is large.[23] Yoshimasa et al[29] found enkephalin-like immunoreactivities in sixteen human pheochromocytomas. Epinephrine-producing pheochromocytomas contained high levels of Methionine enkephalin, Leucine enkephalin, and Methionine enkephalin - Arg^6 - Gly^7 - Leu^8 determined by ligand specific radioimmunoassays.

Kumakura et al[30] have provided <u>in vitro</u> evidence for an interaction between adrenal medulla opiates and the release of catecholamines. They observed that nicotine-stimulated catecholamine release was reduced 50% by morphine (10^{-5}M) or Beta-endorphin (10^{-6}M) and this inhibition was blocked by 10^{-5}M naltrexone.

The presence of endogenous opiates in the adrenal gland suggests a possible role in adrenal function. Recently we reported[31] that naltrexone blockade of opiate receptors exaggerated the plasma epinephrine response to isotonic exercise at 70% of each subject's maximal oxygen consumption. Grossman et al[32] demonstrated that naloxone-sensitive endogenous opiates modulate the release of epinephrine during bicycle exercise at 40 and 80% VO_2 max. Plasma epinephrine and norepinephrine were significantly higher with naloxone compared to saline infusion. In like manner, exercise-induced elevations of plasma cortisol, renin activity, aldosterone, luteinizing hormone and prolactin were enhanced by naloxone (12.2 mg total dose). These data along with our study[31] indicate that the endogenous opiate system modulates the hormonal response to exercise. Interestingly, and in agreement with our results, naloxone had no significant effect on exercise heart rate, blood pressure, oxygen consumption or carbon dioxide production. Grossman et al[32] did find a significantly higher exercise ventilation with naloxone. While ventilation was increased in our study[31] by naltrexone (50 mg), the increase was not significant. These investigations reveal a probable relationship between adrenomedullary function and endogenous opiates.

HUMAN STUDIES

The release of hormones into the circulation is directly related to the intensity of aerobic exercise[33]. An important exception is plasma insulin which decreases during exercise. This exercise enhancement of hormone levels is intimately related to metabolic adaptations such as the availability of fuel substrates. The relationship of these metabolic adaptations to the exercise-induced alterations in blood or tissue levels of endorphins/enkephalins awaits elucidation.

Pre and post run plasma levels of Beta-endorphin/Beta-lipotropin-like activity have been reported by many groups and are shown in Table 1. These studies are similar in the following respects: 1) male subjects are in the majority, 2) RIA with antisera crossreacting with Beta-lipotropin from 10 to 50% on a weight basis were utilized, 3) plasma was extracted prior to RIA with the exception of Kelso et al[34], and 4) running protocols were used. However, intensity and duration of the running varied enormously. To provide a further frame of reference for the baseline values in Table 1, resting levels of Beta-endorphin are included which were determined using sequential immune affinity column chromatography prior to RIA which provides for the separation and quantification of Beta-endorphin distinct from Gamma-lipotropin and Beta-lipotropin[42]. Conclusions can be derived from the data in Table 1. All studies reported increases in plasma Beta-endorphin-like material after running. These increases do not appear to be related to intensity of exercise. Lastly, the elevations range from slight to about five-fold.

Immunoreactivity and biological activity are not synonymous for the RIA's employed.[43] Accordingly, we used a radioreceptor assay[44] to examine the effects of exercise on plasma ligands with opiate receptor activity.[45] Venous blood was obtained from 14 volunteers before and after a 10 mile road race. Conditions before and after the run were controlled in terms of diet, warm-up and time elapsed between the end of the race and the post race blood draw. Leucine Enkephalin-like radioreceptor activity (Leu-ERA) was 22.2 $\pm$13.7 pmol/ml (Mean $\pm$ SD) prior to the race and increased non-significantly to 26.1 $\pm$21.5 pmol/ml after the race, $p > 0.05$. However, post run Leu-ERA levels significantly correlated with race pace ($r = +0.52$, $p < 0.05$). No significant relationship was found between changes in mood state assessed by the Profile of Mood State (POMS) and Leu-ERA. These data suggest that no uniform change in Leu-ERA occurs after competitive running. Support for this finding has recently been provided by Grossman et al[32] who found no change in radioimmunoassayable Met-enkephalin after bicycle exercise at 40 and 80% VO_2 max.

Gender may influence the circulating endorphin response to aerobic exercise. In our studies[37,45] no sex difference was observed. However, Gambert et al[41] found a greater plasma Beta-Endorphin response in five men (51.7 $\pm$ 13 pg/ml) compared to four women (9.0 $\pm$ 4.6 pg/ml, $p < 0.05$) after a 20 minute treadmill run. In any event, interesting observations have been reported in studies solely using women.

Carr et al[40] studied seven women before, at mid point, and after an 8 week training program. Plasma Beta-endorphin increased at the end of 60 minutes of exercise at 85 percent of each woman's maximal heart rate. As training progressed, a larger increase in Beta-Endorphin was noted with a 38% increase at 8 weeks versus a 16% increase prior to training. It appears that the peripheral endorphin system may adapt to training in a manner opposite to the stress hormones[38]. Namely circulating opioids increase with training adaptations while the stress hormones diminish in plasma after training at the same absolute or relative workload.[33,46]

If these initial studies are corroborated by future longitudinal training studies, it would provide a strong impetus to training adaptation studies which could benefit from what is known concerning the development of tolerance and dependence on morphine.[48,49] It should be noted, however, that Carr et al[40] do not provide data which demonstrate that a cardiovascular training effect occurred during the training program.

Another investigation involving only female subjects suggests an interesting possibility that endogenous opiates are also involved in the cardiovascular and pulmonary responses to maximal exercise. McMurray et al[50] found no effect from an IV bolus of naloxone (0.4 mg) on submaximal cardiovascular and pulmonary variables. However, maximal ventilation (lower with naloxone), respiratory frequency (lower), end

Table 1
Summary of Human Studies Which Measured Beta-Endorphin/Beta-Lipotropin During Isotonic Exercise. Values Are Means ± SEM.

Reference	Exercise: Duration	Exercise: Intensity	Exercise: Mode	Cross-reactivity of anti-sera with B-LPH (%)	Beta-Endorphin/Beta-Lipotropin (pg/ml): Pre-exercise	Beta-Endorphin/Beta-Lipotropin (pg/ml): Post-exercise	N
Fraioli (38)	8-15 min	exhaustive	run	Sephadex G-75 chomatography prior to RIA = 0%	320±70	1620± 88	8
Colt (39)	6.4-12.8Km	"Comfortable"	run	10.8	11.8±1.8	17.6±3.1	20
Colt (39)	No data	"near maximal"	run	10.8	8.2±1.0	28.0±6.3	15
Gambert (41)	20 min	80% HR max (predicted)	run	20	increased by 51.7±13.0		5M
					increased by 9.0±0.4		4F
Farrell (37)	30 min	60% VO_2 max	run	15	11.2±3.0	58.3±18.0	6
		80% VO_2 max	run		15.4±5.0	37.2±12.3	6
		self-selected (X=75%)			13.8±4.4	39.2±16.6	6
Elliot (36)	no data	exhaustive	run	1:1 molar crossreactivity	24.4±8.1	47.8±10.3	5
Elliot (36)	(45 min intermittent weight training)				20.8±10.3	52.0±27.2	5
Kelso (34)	120	50% VO_2 max	bicycle	50			
conditions	Neutral environment euhydrated				25.1±4.1	32.8±4.8	6
	Hot environment euhydrated				27.8±6.8	32.5±2.7	6
	Hot environment dehydrated				21.7±3.1	40.3±2.8	6
Janal (35)	6.3 miles (43.9 min)	85.3% VO_2 max (estimated)	run	no data	63.±3 (estimated by author from graph in original article)	97.±5	12
Yamaguchi (42)	Resting Reference			0	7.6±0.38		19

tidal PCO_2 (higher), and heart rate (lower) were influenced by naloxone compared to placebo exercise or a control graded stress test. Endogenous opiates may become active regulators of vital physiologic functions with extreme stress levels.

The above in vivo data conflict with a previous in vitro report by Eiden and Ruth.[51] This investigation demonstrated that, using isolated spontaneously-beating rat atria, the maximal chronotropic effect of norepinephrine is greatly reduced by methionine or leucine enkephalin and several enkephalin analogues. This inhibition was abolished by naloxone. Interestingly, submaximal stimulation by norepinephrine was unaffected by either enkephalins or naloxone. These data show that in vitro enkephalins reduce maximal heart rate which is opposite to the effect demonstrated in vivo by McMurray et al.[50] Species differences and the pharmacological nature of the in vitro study may explain these conflicting data.

Most of the investigations examined in Table 2 utilized a thermoneutral environment. Janal et al[35]and Colt et al[39] had their subjects run outdoors; however, neither excessively hot nor cool temperatures were noted. Kelso et al[34] tested their subjects under three different environmental conditions: 1) neutral temperature and humidity, euhydrated (24°C, 50% relative humidity), 2) hot temperature, euhydrated (35°C, 50% relative humidity), and 3) hot temperature, dehydrated (35°C, 50% relative humidity, no fluids, -2.6% body weight) Plasma Beta-endorphin/Beta-lipotropin increased to its highest level during the most environmentally stressful milieu, the hot temperature, dehydrated condition. Again, it would appear that the more stressful the situation, the greater the rise in plasma Beta-endorphin/Beta-lipotropin with exercise.

Haier et at[52] used naloxone infusion and running to determine if exercise-induced analgesia has an opiate basis. They demonstrated that exercise (one mile at a self-selected pace) significantly lengthens the time to the first report of pain due to a three pound weight on a fingertip. Ten mg naloxone completely blocked this exercise-induced analgesia while a 2 mg dose resulted in a significant hyperanalgesia, i.e., it took even longer for the subjects to notice pain after running one mile compared to placebo injection. Elucidation of the relationship between aerobic exercise and the endogenous opiate system will increase rapidly with the further use of opiate receptor blockade.

CONCLUSION

Pharmacological doses of exogenous and endogenous opiates affect locomotor activity in several species of mammals. Exercise stimulates the release of endorphin-like material into the human circulation, but physiological roles of these immunoreactive ligands are unknown. Brain levels of enkephalin-like material are increased after forced exercise in mice and rats, and opiate receptor blockade with naloxone alters post exercise-induced analgesia in both rats and man.

The importance of peripherally-released opiates during exercise may relate to their potential influence on hormonal and metabolic adaptations. Recent evidence suggests a modulating role for endogenous opiates in the adrenal medulla of man. While the metabolic and cardiovascular implications of this observation are unclear at present, elucidation of the physiological actions of the endogenous opioids during exercise should prove to be a fruitful and important area for future investigations.

Acknowledgement This work was supported by NIH grants RO3 MH/OA 36880-01 to PAF and General Clinical Research Center RR-00058; and TOPS Club, Inc., Obesity and Metabolic Research Program, Milwaukee, Wisconsin. Special thanks is extended to Catherine Ladd for her expert assistance in typing this review and to Lynn Cordell Breitlow for her valued assistance in our laboratory.

References

1. Hughes, J., T.W. Smith, H. Kosterlitz, L.A. Forthergill, B.A. Morgan and H.R. Harris. Identification of two related pentapeptides from the brain with potent opiate agonist activity. Nature 258:577-579, 1975.
2. Goldstein, A. and P. Sheenan. Tolerance to opioid narcotics. I. Tolerance to the "Running Fit" caused by levorphanol in the mouse. J. Pharmacol. and Exp. Therapeutics. 169:2, 175-184, 1969.
3. Babbini, M. and W.M. Davis. Time-dose relationships for locomotor activity effects of morphine after acute or repeated treatment. Br. J. Pharmac. 46:213-214, 1972.
4. Browne, R.G. and D.S. Segal. Behavioral activating effects of opiates and opioid peptides. Biol. Psychiat. 15:77-86, 1980.
5. Drust, E.G. and I.L. Crawford. Comparison of the Effects of TRH and Dl-Ala^{2}-Metenkephalinamide on hippocampal electrical activity and behavior in the unanesthetized rat. Peptides. 4:239-243, 1983.
6. Oka, T. and E. Hosoya. Effects of humoral modulators and naloxone on morphine-induced changes in spontaneous locomotor activity of the rat. Psychoparmacol. 47, 243-248, 1976.
7. Katz, R.J., B.J. Carroll and G. Baldrighi. Behavioral activation by enkephalins in mice. Pharmacol. Biochem. Behavior. 8, 493-496, 1978.
8. Amir, S., Z.H. Galina, R. Blair, Z.W. Brown and Z. Amit. Opiate receptors may mediate the suppressive but not the excitatory action of ACTH on motor activity in rats. Eur. J. Pharmacol. 66:307-313, 1980.
9. Castellano, C. Strain dependent effects of the enkephalin analogue FK 33-824 on locomotor activity in mice. Pharmacol. Biochem. Behav. 15:729-734, 1981.
10. Castellano, C. and S. Puglisi-Allegra. Effects of naloxone and naltrexone on locomotor activity in C57Bl/6 and DBA/2 mice. Pharmacol. Biochem. Behav. 16:561-563, 1982.
11. Walker, J.M., G.G. Berntson, T.S. Paulucci and T.C. Champney. Blockade of endogenous opiates reduces activity in the rat. Pharmacol. Biochem. Behav. 14:113-116, 1981.
12. Castellano, C. and A. Oliverio. A genetic analysis of morphine-induced running and analgesia in the mouse. Psychopharmacologica. 41:197-200, 1975.
13. Oliverio, A. and C. Castellano. Genotype-dependent sensitivity and tolerance to morphine and herion dissociation between opiate-induced running and analgesia in the mouse. Psychopharmacologica 39:13-22, 1974.
14. Castanas, E., P. Giraud, Y. Audigier, R. Drissi, F. Boudouresque, B. Conte-Devolx and C. Oliver. Opiate binding sites spectrum on bovine adrenal medullas and six human pheochromocytomas. Life Sciences 33: Supp. 1, 295-298, 1983.
15. Christie, M.J. and G. Chesher. Physical dependence on physiologically released endogenous opiates. Life Sciences, 30:1173-1177, 1982.

16. Christie, J.J., G.B. Chesher and K.D. Bird. The correlation between swim-stress induced antinociception and [^{3}H] Leu-enkephalin binding to brain homogenates in mice. Pharmac. Biochem. Behav. 15:853-857, 1981.
17. Christie, M.J., P. Trisdikoon and G.B. Chesher. Tolerance and cross-tolerance with morphine resulting from physiological release of endogenous opiates. Life Sciences. 31:839-845, 1982.
18. Shyu, B.C., S.A. Anderson and P.Toren. Endorphin-mediated increase in pain threshold induced by long-lasting exercise in rats. Life Sciences 30:833-840, 1982.
19. Pert, C.B. and D.L. Bowie. Behavioral manipulation of rats causes alterations in opiate receptor occupancy. In: E. Usdin, W.E. Bunney and N.S. Kline (Eds.) Endorphins in mental health research, 1979, New York: Oxford University Press, 93-104.
20. Metzer, J.M. and E.A. Stein. β-endorphin and sprint training. Life Sciences 34:1541-1547, 1984.
21. Evans, C.J., E. Erdelyi, E. Weber, J.D. Barchas. Identification of pro-opiomelanocortin-derived peptides in the human adrenal medulla. Science 221:957-960, 1983.
22. Linnoila, R.I., R.P. Diaugustine; A. Hervonen and R.J. Miller. Distribution of [Met^5]- and [Leu^5]-enkephalin-, vasoactive intestinal polypeptide- and substance P-like immunoreactivities in human adrenal glands. Neuroscience 5:2247-2259, 1980.
23. Lundberg, J.M., B. Hamberger, M. Schultzberg, T. Höfelt, P.O. Grandbert, S. Efendic, L. Terenius, M. Goldstein and R. Luft. Enkephalin- and somatostatin-like immunoreactivities in human drenal medulla and pheochromocytoma. Proc. Natl. Acad. Sci. 76:4079-4083, 1979.
24. Quirion, R., M.S. Finkel, F.A.O. Mendelson and N. Zamir. Localization of opiate binding sites in kidney and adrenal gland of the rat. Life Sciences. 33: Supp. 1, 299-302, 1983.
25. Schultzberg, M., T. Höfelt, J.M. Lundberg, L. Terenius, L.G. Elfvin and R. Elde. Enkephalin-like immunoreactivity in nerve terminals in sympathetic ganglia and adrenal medulla and in adrenal medullary gland cells. Acta. Physiol. Scand. 103:475-477, 1978.
26. Viveros, O.H., E.J. Diliberto, E. Hazum and K.J. Chang. Enkephalins as possible adrenomedullary hormones: storage, secretion, and regulation of synthesis. In: Neural peptides and neuronal communication, ed. by E. Costa and M. Trabucchi. Raven Press, N.Y., vol. 22, 191-204, 1980.
27. Viveros, O.H., E.J. Diliberto, E. Hazum and K.J. Chang. Opiate-like materials in the adrenal medulla: evidence for storage and secretion with catecholamines. Mol. Pharmacol. 16:1101-1108, 1979.
28. Viveros, O.H., S.P. Wilson, E.J. Diliberto, E. Hazum. K.J. Chang. Enkephalins in adrenomedullary chromaffin cells and sympathetic nerves. Adv. Physiol. Sci. Vol. 14. Endocrinology, neuroendocrinology, neuropeptides-11, ed. by E. Stark, G.B. Markara, B. Halasz, G. Rappoy, 349-353.
29. Yoshimasa, T., K. Nakao, Y. Ikeda, M. Sakamoto, M. Suda and H. Imura. Methionine-enkephalin, Leucine-enkephalin, methionine-enkephalin-Arg^6-Phe^7 and methionine-enkephalin-Arg^6-Gly^7-Leu^8 in human pheochromocytoma. Life Sciences. 33:85-88, 1983.
30. Kumakura, K., A. Guidotti, H.Y.T. Yang, L. Saiani and E. Costa. A role for the opiate peptides that presumably coexist with acetylcholine in splanchnic nerves. in: Neuronal Peptides and Neuronal Communication, ed. W. Costa and M. Trabucchi, Raven Press, New York, 1980, P. 571-580.

31. Gustafson, A.B., P.A. Farrell, T. Garthwaite and R. Kalkhoff. Endogenous opiates modulate the plasma epinephrine response to submaximal exercise in man. Exerpta Medica, International Congress Series 652:682, abstract 843, 1984.
32. Grossman, A., P. Bouloux, P. Price, P.L. Drury, K.S.L. Lam, T. Turner, J. Thomas, G.M. Besser and J.R. Sutton. Role of opioid peptides in the hormonal responses to acute exercise in man. Clinical Science 67:483-491, 1984.
33. Galbo, H. Hormonal and metabolic adaptation to exercise. Georg. Thieme Verlag, Stuttgart, 1983.
34. Kelso, T.B., W.G. Herbert, F.C. Gwazdauskas, F.L. Goss and J.L. Hess. Exercise-thermoregulatory stress and increased plasma ß-endorphin/ß-lipotropin in humans. J. Appl. Physiol.: Respirat. Environ. Exercise Physiol. 57:444-449, 1984.
35. Janal, M.N., E.W.D. Colt, W.C. Clark and M.Glusman. Pain sensitivity, mood and plasma endocrine levels in man following long-distance running: Effects of naloxone, Pain 19:13-25, 1984.
36. Elliot, D.L., L. Goldberg, W.J. Watts and E. Orwoll. Resistance exercise and plasma beta-endorphin/beta lipotropin immunoreactivity. Life Sciences 34:515-518, 1984.
37. Farrell, P.A., W.K. Gates, W.P. Morgan, and M.G. Maksud. Increases in plasma ß-endorphin/ß-lipotropin immunoreactivity after treadmill running in humans. J. Appl. Physiol.: Respirat. Environ. Exercise Physiol. 52(5): 1245-1249, 1982.
38. Fraioli, F., C. Moretti, D. Paolucci, E. Alicicco, F. Crescenzi, and G. Fortunio. Physical exercise stimulates marked concomitant release of ß-endorphin and adrenocortiocotropic hormone (ACTH) in peripheral blood in man. Experientia 36:987-989, 1980.
39. Colt, W.D., S.L. Wardlow, and A.G. Frantz. The effect of running on plasma ß-endorphin. Life Science 28:1637-1640, 1981.
40. Carr, D.B., B.A. Bullen, G.S. Skrinor, M.A. Arnold, M. Rodenblatt, I.Z. Beitins, J.B. Martin and J.W. McArthur. Physical conditioning facilitates the exercise-induced secretion of Beta-endorphin and Beta-lipotropin in women. N. Eng. J. Med. 305:560-562, 1981.
41. Gambert, S.R., T.L. Garthwaite, C.H. Pontzer, E.E. Cook, F. Tristani, E.H. Duthie, D.R. Martinson, T.C. Hagen and D.J. McCarty. Running elevates plasma ß-endorphin immunoreactivity and ACTH in untrained human subjects. Proc. Soc. Exp. Biol. Med. 168:1-4, 1981.
42. Yamaguchi, H., A.S. Liotla and D.T. Krieger. Simultaneous determination of human plasma immunoreactive ß-Lipotropin, γ-Lipotropin, and ß-Endorphin using immune-affinity chromatography. J. Clin. Endocrinol. Metab. 51:1002-1008, 1980.
43. Li, C.H., A.J. Rao, B.A. Doneen and D. Yamashiro. ß-Endorphin: Lack of correlation between opiate activity and immunoreactivity in radioimmnoassay. Biochem. Biophys. Res. Commun. 75:576-590, 1977.
44. Naber, D., D.P. Pickar, R.A. Dionne, D.L. Bowie, B.A. Ewols, T.W. Moody, M.G. Sable and C.B. Pert. Assay of endogenous opiate receptor ligands in human CSF and plasma. Sub. Alcohol Act./Misuse. 1:113-118, 1980.
45. Farrell, P.A., W.G. Gates, W.P. Morgan and C.B. Pert. Leucine Enkephalin-like radioreceptor activity and tension - anxiety before and after competitive running. In: Biochemistry of Exercise, ed. H.G. Knuttgen, J.A. Vogel and J. Poortmans. Human Kinetics Publishers, Champaign, Illinois, 637-644, 1983.

46. Winder W.W., R.C. Hickson, J.M. Hagberg, A.A. Ehsoni and J.A. McLane. Training-induced changes in hormonal and metabolic responses to submaximal exercise. J. Appl. Physiol. Respirat. Environ. Exercise Physiol. 46:766-771, 1979.
47. Bullen, B.A., G.S. Skrinor, I.Z. Beitins, D.B. Carr, S.M. Reppert, C.O. Dodson, M. deM. Fencl, E.V. Gervino and J.W. McArthur. Endurance training effects on plasma hormonal responsiveness and sex hormone excretion. J. Appl. Physiol.: Respirat. Environ. Exercise Physiol. 56:1453-1463, 1984.
48. Kaymakcalan, S. and L.A. Woods. nalorphine-induced abstinence syndrome in morphine tolerant rats. J. Pharmac. Exp. Ther. 117:112-116, 1956.
49. Martin, W.R., A. Wikler, C.G. Eades and F.T. Pescor. Tolerance to and physical dependence on morphine in rats. Psychoparmacologia 4:247-260, 1963.
50. McMurray, R.G., D.S. Sheps and D.M. Guinan. Effects of naloxone on maximal stress testing in females. J. Appl. Physiol. Respirat. Envion. Exercise Physiol. 56:436-440, 1984.
51. Eiden, L.E. and J.A. Ruth. Enkephalins modulate the responsiveness of rat atria in vitro to norepinephrine. Peptides. 3:475-478, 1982.
52. Haier, R.J., K. Quaid, and J.C. Mills. Naloxone alters pain perception after jogging. Psychiatry Research 5:231-232, 1981.

EFFECTS OF COPING AND NONCOPING WITH STRESSFUL EVENTS ON THE IMMUNE SYSTEM

Wolfgang H. Vogel

Departments of Pharmacology and
Psychiatry and Human Behavior
Jefferson Medical College
Thomas Jefferson University
Philadelphia, PA 19107

INTRODUCTION

All men and animals are continuously exposed to stressful events (stressors) which can affect and change the homeostasis of an organism. A change in homeostasis is usually called stress. Although man must have encountered stressful events and experienced stress from prehistoric times, a scientific concept of stress did not emerge until relatively late in history. Milestones in early stress research include the recognition of the "milieu interior" by C. Bernard as well as the response of the sympathetic-adrenal medullary system (epinephrine) by W.B. Cannon (1) and that of the pituitary-adrenal cortical system (steroids) by H. Selye (2) to stressful events.

Since this time, a large number of investigations have tried to investigate the chemical basis of stress in animals and to a more limited extent in man. In addition to the catecholamines and steroids, a wide variety of other chemicals has been investigated and significant changes have been observed with a number of enzymes, biochemicals and/or electrolytes during and after stress in plasma, brain and other organs of the body.

Stress is usually experienced in response to certain events in our daily environment which are perceived as unpleasant or threatening and which create feelings of tension, anxiety or fear. Experimentally, stress is produced by exposing the organism to a demanding or unpleasant situation. Animals are restrained, exposed to cold temperature, or subjected to an unpleasant footshock. The animal has no control over this situation. In humans, difficult, mental arithmetic is performed, the hand is submerged in ice-cold water or an unpleasant social situation is created. Again, the individual has no control over the experiment but is obliged to follow the instructions of the experimentors. In all these cases, the individual cannot control or cope with the particular situation. However, recent research has shown that not the stressful event per se but the ability or inability to cope with a stressor seems to be the most important determinant in the stress response of an organism. While coping with a stressful event produces few stress-induced changes, the inability to cope with the same event produces marked biochemical and physiological changes, including significant effects on the immune system and the immune response.

DEFINITION AND MEASUREMENT OF COPING

Coping can be defined as gaining and maintaining control over a situation which could otherwise be potentially dangerous, unpleasant or fearful. The ability to control a situation means in many, although not all, circumstances that the typical stress response of uncertainty, apprehension, anxiety or fear and its peripheral consequences such as sweating, tachycardia, rise in blood pressure might be attenuated or might not emerge at all.

The effects of coping and noncoping with a stressful situation can be measured in animals by using a pair of "yoked" animals exposed to a stressful situation. The "yoked" pair to be stressed is placed in a situation in which the pair receives foot or tail shock. However, one animal can press a lever and terminate the shock for both. The other animal has no way to terminate the shock. After onset of a sequence of shocks and rest periods, the first animal learns quickly to use the lever, respond and terminate the shock for both animals. This animal gains control over the unpleasant situation or can cope with it. The other animal can not terminate the shock, has no control and is unable to cope with the situation. Under these conditions, both animals receive the identical number of shocks with identical intensities and durations. Thus, the aversive situation (unpleasantness of footshock) is the same for both animals but the difference rests in the ability or inability to cope with the footshock. These measurements can also be compared with those obtained on a control animal which rests in its home cage. In humans, it is somewhat more difficult to measure coping. In a real life situation, individuals can be asked about their coping strategies and can be judged how successful they are in mastering environmental stressors. Experimentally, two individuals can be given a puzzle and be asked to assemble this "simple" puzzle as many times as they can in a given period of time. However, one individual receives a puzzle which indeed can be assembled easily and repeatedly whereas the other receives a puzzle which is an "unsolvable" puzzle and cannot be assembled. In both cases, a stressful situation exists except that the first individual can cope with this situation whereas the second individual cannot cope.

In experiments like this, a number of biochemical and physiological changes have been observed and a variety of specific pathological consequences have been found which were mostly related to the ability or inability of the individual to control the stressful event and not to the noxious or aversive nature of the stressor per se.

BIOCHEMISTRY OF COPING

Although stress-related chemical changes in noncoping animals have been extensively studied, very few biochemical studies have been performed on animals coping with a stressful event or a stressor. A few crucial studies will be cited because some of the chemicals which were studied may also be directly or indirectly involved in the immune response.

Rats were exposed to a footshock which could be terminated or controlled by one but not the other animal. Again, both animals received the same number of shocks with identical intensities and durations with the only exception that one animal could cope (terminate) whereas the other had no control (noncoping) over the shock. A third animal remained undisturbed in the home cage. Plasma norepinephrine, epinephrine and corticosterone levels rose in both animals as compared to the home cage animal. However, they rose significantly higher in noncoping rats during footshock and remained elevated longer after cessation of the footshock (3). Nor-epinephrine levels fell significantly in the hypothalamus of noncoping animals, while small or no changes were seen in this area in the coping rat during shock (3,4,5,6). In contrast, norepinephrine levels fell in the hippocampus of coping animals while no changes in this catecholamine were seen in noncoping animals (3). Dopamine levels were markedly affected in noncoping rats in the mesocortical and mesolimbic, but not the nigrostriatal, areas whereas no such changes were observed in animals capable of coping with the stressful event; dopamine levels of coping animals were the same as those in resting rats (7,8,9). Serotonin levels in the septum fell below normal levels in noncoping rats but increased above normal levels in coping rats (10). From experiments measuring pain threshold after exposure to controllable and uncontrollable shock, it was found that coping or noncoping can affect the opiate endorphin system differently. Uncontrollable electric shock exposure activates the opiate/endorphin system in the brain markedly and causes significant analgesia whereas controllable shock has only a slight effect on this system and causes only mild analgesia (11,12).

Coping must not necessarily imply gaining or controlling a stressful event directly or successfully. Directing one's attention away from the stressful situation can apparently serve as an alternate coping mechanism. Rats exposed to an uncontrollable footshock show the typical stress induced norepinephrine depletion in the hypothalamus. However, if 2 animals were exposed together to the same footshock and allowed to fight with each other what 2 rats will typically do under these circumstances than no norepinephrine depletion was observed (13) Similarly, wrong coping strategies can be adapted to prevent stress induced changes. Uncontrollable stress causes anxiety and hypothalamic NE depletion in rats. Alcohol can partially prevent this fall and, presumably, reduce stress-induced anxiety (14). Alcohol could now be used as a "wrong" coping strategy to correct temporarily a biochemcial imbalance which, however, eventually might lead to alcoholism and related pathologies.

Thus, chemical changes produced in the brain and periphery do not seem to be the primary results of the aversive nature of the environmental event or situation but seem to depend to a much greater extent on the ability (coping) or inability (noncoping) of the organism to control the aversive nature of the potential stressor. Biochemical changes seem to be most marked in the uncontrollable situation whereas proper, or sometimes improper, coping seems to prevent such changes or, if they occur, seem to be of a different nature. Since evidence is available that some of the above mentioned chemcials are also involved in the regulation of the immune system, the occurrance or absence of such changes as determined by coping must then also exert a crucial role on the immune system.

HEALTH CONSEQUENCES OF COPING BEHAVIOR

Stressful events and the resulting stress have been implicated in the pathogenesis of many diseases such as ulcers, hypertension, myocardial infarctions, depression and, more recently, cancer. Stressful events in man can include separation of spouses or separation of children from parents, death of a beloved person as well as loss of a job or unsurmountable work pressure. However, it is well known that not all individuals who experience such stressful events will become sick. Again, this suggests strongly that not only the stressful event per se but other factors as well must play a role in causing diseases. One factor which becomes more and more important in the etiology of stress and stress-induced illnesses is the particular coping behavior of the individual. And it might just be this mechanism which will turn out to be rather crucial in the response of the body and the potential health consequences to environmental insults. Rats which can escape or control electric shock develop less ulceration than do "yoked" animals which can not escape the identical shock situation (15). As mentioned later on, tumor development is enhanced and tumor rejection is decreased in rats receiving inescapable shock; coping animals show less tumor development and better rejection similar to those found with control animals (16,17). Again, coping must not necessarily mean gaining direct control of the situation. Rats which could not terminate or cope directly with a shock but could fight or chew during a stressful experience developed less ulcers than rats which can not use such alternative coping mechanisms (15). Thus, direct or indirect coping can reduce or abolish the adverse health consequences of a noxious environmental stimulus.

However, the ability to cope reduces detrimental health consequences but must not necessarily mean the absence of pathology; it can mean the occurrance of a different pathology. In "yoked" squirrel monkeys, coping animals show a higher incidence of myofibrillary changes and myocardial lesions whereas noncoping monkeys show more general physical degeneration, bradycardia and ventricular arrest (18). Monkeys which can avoid an electric shock ("executive monkey") for themselves and "yoked" animals develope considerably more gastric pathology than the animals which have no control (19); in the latter case the shock was not terminated but could be avoided which placed the additional stress of attention and preparedness on the coping animal. Lowered cardiovascular activity can be observed in individuals who can easily control an aversive situation with a favorable outcome as compared with individuals who can not (21). However, this beneficial effect seems to depend upon the effort necessary to cope. Increased efforts to cope may enhance and not reduce stress. Individuals who can not easily control an aversive situation and need considerable effort for its control show increased plasma epinephrine in comparison with their noncoping counterparts (21).

Thus, the effects of coping on maintaining health or causing diseases are quite complex. In general, they are beneficial but depend on the particular physiological system studied as well as the intensity of the coping effort.

COPING AND THE IMMUNE SYSTEM

The effects of stress, meaning inescapable or uncontrollable stress, on the immune system have only received a limited amount of attention in spite of the fact that all humans experience more or less stress during their lifes. This might stem partly from the opinion that stress and the brain do not affect the immune system. However, more recent studies have clearly supported anecdotal observations throughout history that stress can influence the body's defense mechanisms and markedly affect the resistance to infections and cancer. In this aspect, stress can increase, decrease or not affect the immune response depending on a number of variables which include the nature of the stressor used and the immunological response studied (22,23,24,25). Unfortunately, only few studies in animals and men have been performed on the effects of coping on the immune system.

Tumor rejection was measured in rats which received 0.2 cm^2 of a tumor (Walker 256 sarcoma) into the left flank (16). This tumor piece induced tumors in 50% of unstressed control rats whereas the other 50% of the animals rejected the implant. Sixty-three percent of the rats which received escapable shock or could cope with the stressor rejected the tumor and did not differ significantly from the unstressed control animals (53%). However, only 27% of the animals which received inescapable shock or could not cope rejected the tumor implants. Thus, only inescapable, but not escapable, shock decreased significantly the ability of an animal to reject a tumor implant.

Mice received by subcutaneous injection 6.25×10^4 viable syngenic P 815 mastocytoma cells which would result in tumor formation in all mice within 7-9 days (17). The animals were then divided into 3 groups. One group received escapable shock, 1 group received inescapable shock and 1 group received no shock. The day of tumor appearance was earliest in the inescapable shock group and delayed, but similar, in the other two groups. Survival time was about 27 days in the group which could not escape the shock but was about 30 days for the control and coping groups. These findings indicate that the inability to cope was responsible for the detrimental effects seen whereas the ability to cope with the same stressful event abolished the detrimental consequences of electric shock.

A direct measure of coping on the immune system was published recently (26). Rats were given a series of identical escapable and inescapable shocks. Lymphocyte proliferation in response to the mitogens phytohemagglutinin and concanavalin A was suppressed only in the inescapable but not the escapable shock or coping group. This correlates well with above mentioned studies.

In man, few measurements of coping on the immune system have been attempted. Using ego strength as a measure of coping ability, subjects received a high or low dose of a vaccine. Antibody titer change was directly related to ego strength and coping ability (27). In another study, undergraduate students were subjected to a psychological evaluation including their coping behavior as well as to influenza immunization. The students were then divided into good and poor coping individuals and natural killer cells and antibody titers were measured. Poor coping individuals showed fewer natural Killer cells as compared to their better coping counterparts (28). These studies were recently summarized as demonstrating that a stressful event itself is not immunosuppressive but suppression of the immune system occurs when the individual is unable to cope effectively (21). In a somewhat reverse study, ego strength or the coping ability of newly diagnosed cancer patient was investigated (30). Patients with better coping abilities manifested less mood disturbances, fewer symptoms, less vulnerability and a better course of the disease.

It is tempting to speculate on the central mechanisms which become operative during coping and noncoping and how they affect the immune system. Although little evidence is available, it becomes more generally accepted that there is a strong interaction between the brain and the immune system; the brain can influence the immune system and the immune system can influence the brain. For instance, it has been shown that stimulation or depression of brain activity, in particular that of the hypothalamus, can modify humoral as well as cell-mediated immune responses (31). Often, a reduction in hypothalamic NE decreases the immune response. And, interestingly, the effets of coping and noncoping are most marked on norepinephrine levels in the hypothalamus. Only noncoping, but not coping, decreases NE levels in the hypothalamus. The decrease in NE could reduce immune activity and explain why noncoping rats develop more tumors and reject less tumor implants. This finding would also explain why depressed patients who have supposedly less coping strategies and lower norepinephrine levels (32), show a higher incidence of cancer (33). On the other hand, the immune system can also influence the brain. A decreased norepinephrine turn-over in the hypothalamus of rats can be observed during the peak of the immune response and a decrease in NE can also be produced by the injection of soluble mediators released by immunological cells (34).

Similar evidence is available for the enkephalins and endorphins. Endorphins have been shown to inhibit the immune response. For instance, splenic natural killer cells activity is markedly suppressed by endorphins (35). Inescapable shock increases endorphin levels which are probably less elevated in escapable shock. Thus, coping animals should have lower endorphin levels and a better functioning immune system than noncoping animals which should have higher levels. Again, these changes correlate well with the tumor experiments. Coping animals have lower endorphin levels and develop lighter tumors and reject more tumor implants whereas noncoping animals with high endorphin levels do worse in both cases. Enkephalins have been shown to stimulate the immune system (36). They also increase during stress and could mediate some of the beneficial effects.

TOWARDS A NEW DEFINITON OF STRESS INVOLVING COPING

The physical and psychological environment challenges organisms constantly with a series of events. In itself, these events are not stressful and are perceived as events only by the brain via sensory afferent nerves. The brain evaluates these events in "screening centers".

If the event has no specific meaning or requires only a minor biochemical or physiological adjustment, then the information will be sent directly to the appropriate brain centers necessary for this adjustment. As an example, pupils will constrict and blood vessels in the skin will dilate when an individual leaves a dark and cold room and enters a sunny, warm street. Other centers of the brain might not be aware of these processes or pay little attention to them. These changes should not be called stress but "adjustments".

If the event is judged important by the "screening centers" with possible consequences to the individual - either good or bad -, information is now channeled into "evaluation" centers. These centers evaluate the nature and intensity of the event. If the event is judged important, these centers will first evaluate if the organism can cope with this event and, second, how much effort is required for successful coping. For instance, an individual arrives at work in the morning and sees his desk loaded with work. Since he is totally indifferent to his work and since he has total job security, this sight is of no importance and does not provoke any stress response. A second person, arriving at his desk loaded with the hypothetically same amount of work, hates work and has experienced trouble to finish jobs in the past; the same sight will now be evaluated as a threat and a task with which he had trouble to cope with in the past and specific biochemical and physiological stress responses will be elicted. A third person who is a "workaholic" will also see the hypothetically identical load; he knows how quickly the job can be done, how easily he can cope with it and how much pleasure he gets out of this work. The event becomes a pleasurable stimuli and no chemical or physiological changes might occur or, if they do, will be different from the second individual. In each case, the event is the same but the response is different depending on its interpretation by the individual judged by its coping potentials.

Thus, the final response to an event will either be no or certain specific chemical and physiological changes which will depend greatly on the ability of the individual to cope with the outside stimuli. In turn, the ability to cope will depend on the individual's genetic background and environmental experiences. Any biochemical and physiological changes which occur during and after evaluation of the event should be called stress. Only events which cause stress should be called stressful events; the stressor is created in the "brain of the beholder".

The exact role of stress in health consequences is unknown. However, stress must not necessarily mean disease. Consequences can be good or bad. Jogging causes cardiac stress responses which are certainly beneficial to the individual. On the other hand, stress can cause ulcers, hypertension and heart attacks. Thus, the actual health consequences of stress must be determined individually before stress can be labelled as "eustress", that is good, or "distress", that is bad, for the individual.

Thus, stress are chemical and physiological changes created in the brain as a result of an evaluation of an environmental stimulus, the ability or inability to cope with this event and the effort required to cope. It is these considerations which make events stressful for certain individuals and which may cause no, negative, or positive health consequences.

CONCLUSION

Stress is usually defined as the chemical or physiological response of an organism to a stressful event. Stressful events and stress are then commonly implicated in the pathogenesis of a large number of diseases. However, stressful events do not produce stress and pathological consequences in all individuals. This controversy is probably due to the fact that stress and stressful events are too loosely defined.

A stricter definition of stressful events and stress must involve the coping mechanisms of the individual. Coping is defined as gaining control over a potentially threatening or unpleasant environmental event. It is a judgemental decision of an individual which derives from genetic factors as well as personal experiences. It is this decision which renders an environmental event either harmless with no or a minor stress response or threatening, uncontrollable and stressful with a major stress response of the organism. Although not all, but most health consequences and diseases seem to arise in the latter situation where the individual can or does not cope with the event.

Stress also affects the immune system. However, stress can inhibit, stimulate or not affect the immune system and its response. Infections and cancer are often less well controlled during stress. However, exposure to stressful events can also produce no or even beneficial effects on the defense system of the body. Again, the answer will probably not be found in the examination of the stressful event per se but in the ability or inability of the individual to cope with environmental challenges. And data are indeed available which suggest that not the aversive nature of the environmental stimulus, but the ability or inability to control the stressor, exerts the major influence on the immune system. It is the inability to cope with a stressor which seems to cause a major reduction in the immune response while the ability to control the same stressor protects the defense mechanisms from its impact.

REFERENCES

1. W.B. Cannon, The emergency function of the adrenal medulla in pain and major emotion, Am. J. Physiol. 33:356 (1914).
2. H. Selye, General adaptation syndrome and diseases of adaptation, J. Clin. Endocr. Metabl. 6:117 (1946).
3. R.M. Swenson and W.H. Vogel, Plasma catecholamine and corticosterone as well as brain catecholamine changes during coping in rats exposed to stressful footshock, Pharmac. Biochem. Behav. 18:689 (1983).
4. J.M. Weiss, H.J. Glazer and L.A. Pohorecky, Coping behavior and neurochemical changes: an alternate explanation for the original "learned helplessness" experiments, in: "Animal Models in Human Psychobiology," G. Serban and A. Kling, ed., Plenum Press, NY (1976).
5. J.M. Weiss, E.A. Stone and N. Harrel, Coping behavior and brain norepinephrine levels in rats, J. Comp. Physiol. Psychol. 72:153 (1970).
6. H. Anisman, A. Pizzino and L.S. Sklar, Coping with stress, norepinephrine depletion and escape performance, Brain Research 191:583 (1980).
7. A.J. Eichler and S.M. Antelman, Sensitization to amphetamine and stress may involve nucleus accumbens and medical frontal cortex, Brain Research 176:412 (1979).
8. A.M. Thierry, J.P. Tassin, G. Blanc and J. Glowinski, Selective activation of the mesocortical DA system by stress, Nature 263:242 (1976).
9. A.J. Maclennan and S.F. Maier, Coping and the stress-induced potentiation of stimulant stereotypy in the rat, Science 219:1091 (1983).
10. F. Petty and A.D. Sherman, A neurochemical differentiation between exposure to stress and the development of learned helplessness, Drug Development Research 2:43 (1982).
11. R.L. Hyson, L.J. Ashcraft, R.C. Drugan, J.W. Grace and S.F. Maier, Extent and Control of shock affects naltrexone sensitivity of stress-induced analgesia and reactivity to morphins, Pharm. Biochem. Behav. 17:1019 (1982).

12. S.F. Maier, R.C. Drugan and J.W. Grace, Controllability, coping behavior and stress-induced analgesia in the rat, Pain 12:47 (1982).
13. J.M. Stolk, R.L. Conner, S. Levine and J.D. Barchas, Brain norepinephrine metabolism and shock-induced fighting behavior in rats: Differential effects of shock and fighting on the neurochemical response to a common footshock stimulus, J. Pharmacol. Expt. Ther. 190:193 (1974).
14. K. DeTurck and W.H. Vogel, Effects of acute ethanol on plasma and brain catecholamine levels in stressed and unstressed rats: evidence for an ethanol-stress interaction, J. Pharm. Expt. Ther. 223:348 (1982).
15. J.M. Weiss, Effects of coping behavior in different warning signal conditions on stress pathology in rats, J. Comp. Physiol. Psychol. 77:1 (1971).
16. M.A. Visintainer, J.R. Volpicelli and M.E.P. Seligman; Tumor rejection in rats after inescapable or escapable shock, Science 216:437 (1982).
17. L.S. Sklar and H. Anisman, Stress and coping factors influence tumor growth, Science 205:513 (1979).
18. K.C. Corley, H.P. Mauck and F.O'M. Shiel, Cardiac responses associated with "yoked-chair" shock avoidance in squirrel monkey, Psychophysiol. 12:439 (1975).
19. J.V. Brady, Ulcers in executive monkeys, Scient. Americ. 199: 95 (1958).
20. D.E. DeGood, Cognitive control factors in vascular stress response, Psychophysiol. 12:399 (1975).
21. R.J. Contrada, D.C. Glass, L.R. Krakoff, D.S. Katz, K. Kehoe, C. Collins and E. Elting, Effects of control over aversive stimulation and type A behavior on cardiovascular and plasma catecholamine responses, Psychophysiol. 19:408 (1982).
22. R.J. Jamasbi and P. Nettesheim, Non-immunological enhancement of tumour transplantability in x-irradiated host animals, British Journal of Cancer 36:723 (1977).
23. R.C. LaBarba, Experiential and environmental factors in cancer, Psychosom. Med. 32:259 (1970).
24. S.N. Pradhan and P. Ray, Effects of stress on growth of transplanted and 7,12-dimethylbenz(a)anthracine-induced tumors and their modification by psychotropic drugs, Journal of the National Cancer Institute 53:1241 (1974).
25. P.R. Goldman and W.H. Vogel, Striatal dopamine-stimulated adenylate cyclase activity reflects susceptibility of rats to DMBA-induced mammary tumors, Carcinogens. (in press).
26. M.L. Laudenslager, S.M. Ryan, R.C. Drugan, R.L. Hyson and S.F. Maier, Coping and immunosuppression: inescapable but not escapable shock suppresses lymphocyte proliferation, Sci. 221:568 (1983).
27. R.J. Roessler, T.R. Cate and J.W. Lester, Ego strength, life change and antibody titers, Presented at the meeting of the American Psychosomatic Society in 1979.
28. S.E. Locke, M.W. Hurst and J.S. Heisel, The influence of stress on the immune response, Paper presented at the meeting of the American Psychosomatic Society in 1978.
29. S.E. Locke, Stress, adaptation and immunity studies in human. Gen. Hosp, Psychiatry 4:49 (1982).
30. J.W. Worden and H.J. Sobel, Ego strength and psychosocial adaptation to cancer, Psychosom. Med. 40:585 (1978).
31. M. Stein, R.C. Schiavi and M. Camerino, Influence of brain and behavior on the immune system, Sci. 191:435 (1970).
32. J. Mendels, S. Stern and A. Frazer, Biochemistry of depression,

Dis. Nerv. Syst. 37:3 (1976).
33. F.G. Surawitz, D.R. Brighwell, W.D. Weitzel and E. Othmer, Cancer, emotions and mental illness, Am. J. Psychiat. 133: 1306 (1976).
34. H. Besedovski, A. DelRey, E. Sorkin, M. DuPrada, R. Burri and C. Honegger, The immune response evokes changes in brain noradrenergic neurons, Sci. 221:564 (1983).
35. Y. Shavit, J.W. Lewis, G.W. Terman, R.P. Gale and J.C. Liebkind, Opioid peptides mediate the suppressive effect of stress on natural Killer cell cytotoxicity, Sci. 223: 188 (1984).
36. N.P. Plotnikoff and G.C. Miller, Enkephalins as immunomodulators, Int. J. Immunopharm. 5:437 (1983).

DEPRESSION, HORMONES AND IMMUNITY

Ziad Kronfol[1] and Janet Schlechte[2]

Departments of Psychiatry[1] and Internal Medicine[2]
University of Iowa College of Medicine
Iowa City, Iowa 52242

Introduction

The last decade or two have witnessed a sharp and steady increase in research and interest in the affective disorders. Clinically, major depressive illness has been delineated from other psychiatric disorders with depressed mood and clinical subtypes of depression have been proposed to reflect the heterogeneity among subgroups of depression. In addition to descriptive and clinical criteria, subtypes of depression have also been differentiated by biological markers such as sleep electroencephalography (EEG) records (1), urinary 3-methoxy-4--hydroxy phenylglycol (MHPG) excretion (2), cerebrospinal fluid 5-hydroxyindol acetic acid (HIAA) concentrations (3) and the dexamethasone suppression test (DST) (4). Particularly relevant to the theme of this book are studies in neuroendocrine and immune regulation in depression. In this chapter we will first summarize major advances in the study of hypothalamic-pituitary-adrenal (HPA) axis activity in the depressive disorders, then review some of the preliminary work in the immunology of depression.

DEPRESSION AND THE HYPOTHALAMIC-PITUITARY-ADRENAL AXIS

A variety of hypothalamic and pituitary hormones have been implicated in the etiology of affective illness but abnormalities of the HPA axis are the most widely studied. Forty to fifty percent of patients with depressive illness have abnormal parameters of cortisol secretion. Failure to suppress serum cortisol after administration of dexamethasone is the most widely investigated abnormality of HPA function and has been extensively reviewed elsewhere (4-6). This review will concentrate on other parameters of cortisol hypersecretion and discuss studies of ACTH secretion and regulation in depression. In addition, we will summarize some of the recent work on the relationship of endogenous opiates to depressive illness.

Cortisol

Abnormal cortisol metabolism in depression was first noted more than a quarter of a century ago when Board and colleagues reported elevated glucocorticoid values in depressed patients (7). Studies since then have largely confirmed the elevation in levels of cortisol or its metabolites in plasma (8), urine (9) and CSF (10) of patients with various forms of depression. Gibbons et al (11) used radioisotope dilution techniques to study the dynamics of cortisol secretion. They showed that both the production rate and metabolic clearance rate of cortisol were elevated during the acute phase of depression, and that following recovery the production rate fell while the metabolic clearance rate remained high. Such techniques are usually difficult to assess, but the elevated cortisol secretion rate corresponds with the elevated plasma cortisol level seen in depressed patients.

The nature of the 24 hour secretory pattern of cortisol in depressed patients was first elucidated when Sachar et al. analyzed plasma cortisol every twenty minutes for twenty-four hours in patients with psychotic depression (12). These investigators found that depressed patients secreted twice as much cortisol per day as normal subjects, and that during illness they had eleven to twelve major cortisol secretory episodes compared to seven to nine episodes seen in normal subjects. After treatment, the number of secretory episodes in the depressed patients returned to normal. They also found that the total time spent in active secretion of cortisol during the day was significantly elevated for the depressives to a mean of about 8 hours and 20 minutes compared to a mean of 6 hours and 20 minutes seen in normal subjects. The average biological half-life of cortisol in plasma both before and after treatment was close to the normal figure of 1 hour and 10 minutes.

Because these cortisol determinations all reflect total cortisol (protein-bound and free) and do not reflect the amount of biologically active cortisol present in the circulation, other parameters of cortisol secretion have been developed which offer an indirect estimate of circulatory free cortisol. One such parameter is urinary free cortisol (UFC) excretion. UFC reflects the effective level of circulating plasma cortisol over time and excreted free cortisol increases in direct proportion to plasma free cortisol (13). Despite the close relationship between urinary and plasma free cortisol, there have been remarkably few studies of this parameter of adrenal function in depressive illness. Using a competitive protein binding assay, Carroll measured UFC levels in 60 depressed patient and 35 non-depressed psychiatric controls (9). He found that depressed patients had mean urinary free cortisol excretion of 90 ug/24 hours vs. 47.6 ug/24 hours in the non-depressed group. Forty-three percent of those patients with elevated UFC had values greater than 100 ug/24 hours. Ferguson et al. (14) utilized a chromatographic technique and found high levels of urinary cortisol that returned to normal after electroconvulsive therapy in six patients with depression. Using a radioimmunoassay technique, we recently compared urinary free cortisol levels in depressed patients with those in normal subjects and in patients with Cushing's disease (15). Only individuals classified as nonsuppressors by the dexamethasone suppression test had urinary free cortisol that exceeded 50 ug/gm of creatinine, and in contrast to Carroll's results, less than 5% of the patients with depressive illness had UFC levels that were in the range seen in patients with Cushing's disease.

The best test of tissue exposure to cortisol is the measurement of plasma free cortisol, but this test of adrenal function has not been widely used in patients with depressive illness because assay methods are not applicable for routine clinical use. Using an equilibrium dialysis technique, Carroll (10) measured free cortisol in nine depressed subjects and five non-depressed psychiatric controls. The mean value for the depressed patients was 0.77 ug/dl compared to 0.31 ug/dl in the non-depressed group. In recent work, we measured plasma free cortisol in patients with depression, Cushing's disease and normal controls (15). Depressed patients who were classified as nonsuppressors after the dexamethasone suppression test had only slightly higher plasma free cortisol than the suppressor group. Although the nonsuppressors had significantly higher free cortisol levels than normal subjects, the levels were still 60% lower than those in patients with Cushing's disease. Furthermore, we found no significant difference in the binding capacity of corticosteroid binding globulin (CBG) in normal subjects and the depressed group. The role of the elevated cortisol values in the etiology of depression remains to be determined.

Adrenocorticotrophic hormone (ACTH)

The concept of hypothalamic-pituitary-adrenal regulation suggests that increases in plasma cortisol cause a decrease of the hypothalamic hormone corticotropin releasing factor (CRF) and a subsequent decrease in pituitary adrenocorticotrophic hormone (ACTH) secretion, thereby bringing circulating cortisol levels back to normal.

It has been assumed that the increased cortisol secretion in patients with depression is due to pituitary ACTH hypersecretion, but until recently plasma ACTH levels have not been studied systematically. High values were obtained by Berson and Yalow (16) in patients with "psycho-depressive disorders" but this was not corroborated by other early investigators (17). More recently, several investigators have begun a more detailed examination of plasma ACTH levels in patients with depression. Fang and co-workers (18) found no significant difference in plasma ACTH levels between depressed patients and control subjects at 8 a.m. before dexamethasone nor at 8 a.m. or 4 p.m. after 1 mg of dexamethasone the night before. Yerevanian and Wolff (19) had corresponding results in a different group of depressed patients and normal controls. Despite significant differences in plasma cortisol, depressed patients and normal controls showed no significant differences in plasma ACTH. Reus (20) and Kalin (21) compared ACTH levels in patients who were classified as suppressors and nonsuppressors after the dexamethasone suppression test. These investigators found significantly higher post-dexamethasone plasma ACTH levels in nonsuppressors. In a more detailed study, Pfohl et al. (22) measured ACTH levels every 20 minutes over a 24 hour period in 25 depressed patients and 21 normal controls. They found that the mean pre-dexamethasone plasma ACTH was higher in the depressives than in the normal subjects at each of the 72 sampling times during the 24 hour period. Pfohl's data support the idea that the increase in plasma cortisol in patients with depression is related to an increase in plasma ACTH. Pfohl et al. also showed that patients who are nonsuppressors after dexamethasone averaged greater elevations in plasma ACTH than suppressors and control subjects (23). While suppressors had cortisol values that were virtually identical to normal controls, they had significant elevations of ACTH compared to normal subjects. A time related analysis of the data revealed that the dexamethasone nonsuppressors also exhibited a blunting in the expected circadian rhythm of ACTH (23). The discrepancies among the studies of ACTH secretion in depressive illness have not been totally explained. While the data by Fang, Yerevanian and their colleagues raise the possibility of adrenal supersensitivity to ACTH, the more frequent sampling technique used by Pfohl and co-workers suggest ACTH hypersecretion in depression.

The whole issue of ACTH secretion can now be studied from a different angle, with the recent discovery of the structure and subsequent synthesis of corticotropin releasing factor (CRF), a hypothalamic neuropeptide thought to be the principal stimulus to ACTH secretion (24). Although work with CRF is still in its very preliminary phase, a recent review by Gold and co-workers (25) suggest that compared to normal controls, patients with depressive illness have a significantly lower ACTH response to CRF. The cortisol response to CRF however was similar in both groups. This decrease in the ratio of ACTH to cortisol response during CRF infusion in depressed patients was interpreted by Gold and colleagues as compatible with the hypothesis that the defect in HPA regulation in depression probably occurs at the hypothalamic level and may involve hypersecretion of endogenous CRF (25). More data are obviously needed to further clarify the relationship between CRF, ACTH and cortisol in depression.

Endorphin

If, as recent literature suggests, ACTH and β-endorphin are secreted concommitantly by the pituitary (26), an elevation in ACTH secretion in depressed patients should be accompanied by a similar increase in β-endorphin release. Previous chapters in this volume have concentrated on the localization and physiological significance of the endogenous opiate peptides and this review will be limited to a discussion of measurements of endorphin levels in patients with depressive disorders and the effects of opiate agonists and antagonists in depression.

Several investigators have attempted to measure circulating β-endorphin levels in patients with psychiatric illness and compare them with values in normal controls. Alexopoulos et al. (27) compared plasma β-endorphin levels in 10 patients with unipolar major depressive disorder and 16 psychiatrically normal controls. They found no difference in the basal levels in the two groups. They did however note a transient elevation in both ACTH and β-endorphin following electroconvulsive therapy in the depressed group. Brambilla and colleagues (28) measured plasma levels of β-endorphin, β-lipotropin and ACTH in patients with schizophrenia, primary and secondary depressions and normal controls. They found that both β-endorphin and β-lipotropin were significantly higher in depressed patients compared to normal controls. Risch (29) also reported higher basal levels of plasma β-endorphin in patients with schizoaffective or major depressive disorder compared to other psychiatric patients and normal controls. He also noted that physostigmine-stimulated release of β-endorphin was significantly greater in the depressed groups. Cohen and associates (30) measured basal plasma cortisol and plasma endorphin levels in patients with "major depression", "nonmajor depression" and normal controls. They noted lower plasma endorphin levels in the nonmajor group and found that a low ratio of plasma β-endorphin to cortisol characterized depressed patients in both groups. Pickar and colleagues (31) reported the case of a patient with manic-depressive illness whose switch from the manic phase to depression was accompanied by a significant decrease in plasma opioid activity.

A different approach to investigate the role of opiate peptides in mental illness is through the study of the behavioral effects of opiate agonists and antagonists in normal subjects and psychiatric patients. Pickar and colleagues (32) administered naloxone, an opiate antagonist, to normal volunteers and noted that at high doses, the behavioral effects consisted of irritability, anxiety, sadness and confusion.

Terenius and others (33) have administered naloxone to patients with primary depression. Results have been varied but the general concensus is that opiate antagonists are ineffective in treating depression. On the other hand, early reports have suggested that administration of β-endorphin may cause improvement in depressive symptoms (34). These reports however still await confirmation in larger samples. Thus, evidence for the involvement of endorphins in affective illness remains confusing and further well-controlled studies are needed to clarify many of the still outstanding issues.

In summary, the bulk of the evidence suggests increased cortisol secretion in many patients with depressive illness. Whether this is due to ACTH supersensitivity at the level of the adrenals or to increased production of pituitary ACTH and/or hypothalamic CRF is now the subject of intense investigation and controversy. The role of various neuropeptides is also being widely studied. While most researchers agree that many patients with depressive illness have high levels of biologically active circulating glucocorticoids, the biological significance of this hypercortisolemia is still unclear. Whether the high cortisol levels are etiologically related to depressive symptoms or whether they are just the by-product of central neurotransmitter disturbances remains to be clearly established. More work is also needed on the effects of hypercortisolemia on various physiological functions in depressed patients. The relationship between hypercortisolemia and different immune parameters in depression will now be discussed.

IMMUNE RESPONSE IN DEPRESSIVE ILLNESS

It is now well known, as evidenced by many chapters in this book, that the central nervous system and the immune system are closely associated. Animal studies have revealed intricate and highly complex interactions between stress, neuroendocrine secretion and immune function. The clinical implications of these phenomena have not been widely appreciated. In particular, the applications of immunological principles to the study of psychiatric illness have been very limited. In this section, we will first review the effects of human stress on a variety of immunological parameters, then discuss the immunological studies conducted so far in depressive illness.

Human stress and immunity

The notion that stress can affect the state of health of a person is not new. In 200 A.D., the Greek scientist Galen noted that melancholic women were more likely to develop cancer of the breast than sanguine women. More recently, several epidemiological studies have noted an association between negative life events, depression and cancer (35). Although work on specific pathophysiological pathways has been somewhat limited in the past, intensive research is presently underway to delineate specific biologic mediators between stress and physical illness. New areas of investigation such as psychoendocrinology and psychoimmunology have emerged to address those specific issues.

The effects of stress on the integrity of the human immune response have recently been the subject of excellent review articles (36,37). Palmblad and associates studied the effects of a prolonged experimental stress paradigm on different immune parameters. They found both a reduction in polymorphonuclear leukocyte phagocytic activity and an elevation in interferon production during the 77-hour vigil as compared to baseline values (38). In a different study,

Palmblad and his group also found a reduction in lymphocyte response to mitogen stimulation following 48 hours of sleep deprivation (39). They concluded that both lymphocyte and granulocyte activity can be altered by experimental stress involving sleep deprivation.

A number of investigators have studied the relationship between academic stress and immunological functioning. Dorian et al. (40) measured in-vitro lymphocyte response to mitogen stimulation in psychiatric residents before and after their Oral Fellowship Examination. They noted a reduction in lymphocyte mitogenic activity the morning before the exam, particularly among candidates who were "highly stressed". Jemmott et al. (41) measured the rate of secretion of salivary immunoglobulin A in dental students at different a priori high-stress and low-stress periods of the students' academic calendar. They found diminished secretion rate of Ig A during the high-stress period, suggesting reduced humoral immunity in relation to stress. Kiecolt-Glaser and associates (42) examined the relationship between academic stress and natural killer cell activity (NKCA). They found a significant decline in NKCA the morning of final exam compared to baseline values taken a month earlier. Locke and colleagues (43) also found lower NKCA among undergraduate students who scored high on both life change scales and psychiatric symptoms scales (poor copers), as opposed to students who scored high on the life change scales but not on the psychiatric symptoms scale (good copers). Baker et al. (44) compared the ratio of helper/suppressor T lymphocytes in medical students under varying degrees of academic stress. First year students had higher anxiety scores, higher plasma cortisol values and higher T_4 (helper T lymphocytes) percentages than second year medical students. The authors concluded that an increase in the percentage of T helper cells is part of the physiological reaction to stress.

Bereavement is another form of stress that has been of interest to psychoimmunologists. Two studies have examined the effects of bereavement on parameters of immune function. In 1977, Bartrop and associates (45) compared in-vitro lymphocyte blastogenesis in a group of bereaved subjects and normal controls. They found reduced lymphocyte mitogenic activity in the bereaved group. The number of T and B lymphocytes, serum immunoglobulin levels and the presence of autoantibodies did not differ in the two groups. These findings were later replicated by Schleifer and associates (46) who studied lymphocyte number and function in the spouses of women with advanced breast cancer. Compared to values obtained before the death of their spouses, lymphocyte responses to mitogen stimulation in the bereaved subjects were significantly reduced one or two months following bereavement. The total number of lymphocytes, T cells and B cells remained unchanged.

Depression and immune function

Indirect evidence for disturbed immunity in depressed patients comes from several sources. First, epidemiologic studies, although still inconclusive, suggest an association between depressive illness and a variety of immune related disorders such as infection, asthma, allergy and cancer. This topic has been reviewed in detail by Kronfol (47). If reports of such an association are confirmed, possible pathophysiological mechanisms may include an impairment of immune responses in depressed patients, thus rendering them more vulnerable to infection, autoimmune disease and cancer (47,48).

Another reason for the growing interest in the immunology of the depressive disorders is the documented disturbance in HPA regulation in many patients with acute depression. As discussed earlier (4-6), a final common pathway of the HPA dysregulation is the excessive secretion of cortisol, an immunosuppressive hormone. The question then is whether this excess in circulating cortisol would affect the integrity of the immune response in depressed patients.

Furthermore, clinicians have often noted the similarities between the reaction to intense stress such as bereavement and clinical depression. Common clinical manifestations include despondency, lack of energy, lack of appetite, insomnia, crying spells, difficulty concentrating, etc. Since acute stress, and particularly bereavement, have been associated with impairment in immunological functioning (45,46), it is therefore reasonable to expect similar changes in depressed patients.

There is also growing direct evidence of impaired immunity in patients with depressive illness. In 1972, vonBrauchitsch (49) reported that about 50% of patients hospitalized with mental depression had positive serum antinuclear factor. These findings have later been replicated by Debert et al. (50), but not by Plantey (51). However, since most of these patients were on psychoactive medications, and since some of these medications, particularly lithium, could produce falsely elevated antinuclear antibodies (52), it is not clear to what extent this positive antinuclear factor is related to depression <u>per se</u> or is an artifact of treatment.

In 1971, Rimon and associates (53) noted an increase in herpes simplex virus (HSV) antibodies in patients with psychotic depression. These findings were replicated by Coppel et al. (54) who also noted an increase in <u>in-vitro</u> lymphocyte response to HSV in psychotically depressed patients. These authors suggested an association between HSV and psychotic depression, and proposed that acute depression may be related to a reactivation of a latent HSV infection with consequent dysfunction in CNS monoamine metabolism leading to depression.

In 1982, Sengar and associates (55) studied the frequency of T and B lymphocytes and lymphocyte mitogenic responses in three groups of subjects: 1) manic-depressive patients in remission, 2) psychiatrically-well siblings and 3) normal controls. They found no significant differences in the distribution of T cells, B cells or lymphocyte mitogenic activity among the three groups. However, since none of the subjects studied was clinically depressed, the question of the integrity of the immune response during the acute phase of depression remained unanswered.

In 1983, Kronfol and associates (56) compared lymphocyte response to mitogen stimulation in 26 drug-free acutely depressed patients and 20 normal controls. They found a generalized and marked reduction in lymphocyte mitogenic activity among the depressive group. These differences could not be explained on the basis of age, sex, diet, diurnal variation or concommittant medical illness. The authors concluded that depression may be associated with an impairment in cell-mediated immunity that could explain the increased risk for immune-related disorders among the depressive population. Kronfol's findings were later replicated by Schleifer and co-workers (57), who also noted a reduced number of both T and B cells in the depressed group, although the percentage of these cells did not differ between depressed patients and normal controls. The ratio of helper/suppressor T lymphocytes was examined by Kronfol and House (58). Depressed patients had lower ratios than normal controls, but the difference was not statistically significant. We are presently studying natural-killer cell activity in depressed patients. Preliminary results from our lab do not suggest major differences in NKCA between depressed and other psychiatric patients (59). Data by Van Dyke and associates (60) however seem to indicate lower NKCA among depressed patients compared to normal controls. It is evident that more data is needed before definite conclusions can be drawn.

Another issue of major interest in this context is the relation of hypercortisolemia to lymphocyte number and function in major depression. In a large retrospective study of 178 untreated depressed patients and 179 untreated schizophrenic controls, Kronfol and associates (61) noted a significant increase in both the percentage and absolute neutrophil count, and significant decreases in both the percentage and absolute lymphocyte and eosinophil counts among the depressed group. These differences, the authors noted, were consistent with the expected hematologic effects of higher levels of circulating glucocorticoids in depressed patients. In a later, prospective study, Kronfol and colleagues (62) were able to confirm much of their earlier findings. In particular, depressed patients had lower concentrations of circulating lymphocytes than a schizophrenic control group. More interestingly, depressed patients with abnormal dexamethasone suppression test (DST) had lower lymphocyte concentrations than depressed patients with normal DST. Furthermore, a significant negative correlation was found between post-dexamethasone plasma cortisol levels and lymphocyte numbers among the depressed group.

The relationship between hypercortisolemia and lymphocyte function in depressed patients is less clear. Kronfol and associates (63) compared lymphocyte response to mitogen stimulation in 3 groups of subjects: 1) depressed patients with hypercortisolemia as indicated by elevated 24-hour UFC, 2) depressed patients without hypercortisolemia and 3) normal controls. Both depressed groups, regardless of UFC excretion, had lower lymphocytic responses than the normal control group. There were however no significant differences in lymphocyte mitogenic activity between the two depressive groups. Furthermore, there was no significant correlation between UFC excretion and lymphocyte mitogenic activity among the depressed group. The authors conclude that depression is associated with an impairment in lymphocyte function that cannot be explained solely on the basis of increased cortisol secretion.

Van Dyke and associates (60) studied two parameters of cellular immunity in depressed patients and patients with Cushing's disease. They found evidence of decreased mitogenesis and reduced NKCA in the depressed patients but not those with Cushing's disease. Within the depressed group, higher cortisol values were more likely to be associated with decreased mitogenesis than reduced NKCA. However, neither lymphocyte mitogenic activity nor NKCA were related to results of the DST. The authors thus concur with Kronfol and associates that immunosuppression in depressed patients is not solely dependent on abnormalities in cortisol secretion.

CONCLUSION

In summary, many patients with depressive illness have documented abnormalities in neuroendocrine and/or immune regulation. The endocrine abnormalities most widely investigated are abnormalities in HPA function. Briefly, these consist of disturbances in the circadian pattern of cortisol secretion, excessive cortisol production and impaired feedback regulation with dexamethasone. The role of ACTH, β-endorphin and CRF has not been adequately investigated. The immune abnormalities are much less know or understood. Lymphopenia, decreased lymphocyte mitogenic activity and a possible reduction of NKCA have been noted. The relation between increased cortisol secretion and immunosuppression needs further investigation. Future research should also address the role of other hormones and/or neurotransmitters in the regulation of the immune response in patients with depressive illness. Special attention should be given to norepinephrine, serotonin, acetylcholine and β-endorphin which have been shown to play a role in the regulation of both mood and immunity. Traditionally, research in the psychobiology of the affective disorders has focused on the genetics, neurochemistry and neurophysiology of depression. More recently, the neuroendocrine approach has been adopted by several investigators. In the meantime, very rapid progress has been achieved in basic immunology and the pharmacology of the opiate peptides. The application of these new discoveries to the study of depression is now just starting.

References

1. D. Kupfer, G. Foster, P. Coble, R. McPartland and R. Ulrich, The application of sleep EEG for the differential diagnosis of affective disorders. Am J Psychiatry, 135:69 (1978).
2. J. Fawcett, J.W. Maas and H. Dekirmemjian, Depression and MHPG excretion: Responses to dextroamphetamine and tricyclic antidepressants. Arch Gen Psychiatry, 26:246 (1972).
3. M. Asberg, P. Thoren, L. Traskman, L. Bertillson and V. Ringberger, Serotonin depression - a biochemical subgroup within the affective disorders? Science, 191:478 (1976).
4. B.J. Carroll, M. Feinberg, J. Greden, J. Tarika, A. Albala, R. Haskett, N. James, Z. Kronfol, N. Lohr, M. Steiner, J.P. deVigne and E. Young, A specific laboratory test for the diagnosis of melancholia: standardization, validation and clinical utility. Arch Gen Psychiatry, 38:15 (1981).
5. B.J. Carroll and J. Mendels, Neuroendocrine regulation in affective disorders. in: "Hormones, Behavior and Psychopathology", E.J. Sachar, ed., Raven Press, New York (1976).
6. B.J. Carroll, The dexamethasone suppression test for melancholia. Br J Psychiatry, 140:292 (1982).
7. F. Board, R. Wadeson and H. Persky, Depressive affect and endocrine functions. Arch Neurol Psychiatry, 78:612 (1957).
8. E.J. Sachar, L. Hellman, D.K. Fukushima and T.F. Gallagher, Cortisol production in depressive illness. Arch Gen Psychiatry, 23:289 (1970).
9. B.J. Carroll, G.C. Curtis, B.M. Davies, J. Mendels and A.A. Sugerman, Urinary free cortisol excretion in depression. Psychol Med, 6:43 (1976).
10. B.J. Carroll, G.C. Curtis and J. Mendels, Cerebrospinal fluid and plasma free cortisol concentrations in depression. Psychol Med, 6:235 (1976).
11. J.L. Gibbons, Cortisol secretion rate in depressive illness. Arch Gen Psychiatry, 10:572 (1964).
12. E.J. Sachar, L. Hellman, H. Roffwarg, F. Halpern, D. Fukushima and T. Gallagher, Disrupted 24-hour patterns of cortisol secretion in psychotic depression. Arch Gen Psychiatry, 28:19 (1973).
13. F.H. Tyler and C.D. West, Laboratory evaluation of disorders of the adrenal cortex. Am J Med, 53:664 (1972).
14. H.C. Ferguson, A.C.G. Bartram, H.C. Fowlie, D.M. Cathro, K. Birchall and F.L. Mitchell, A preliminary investigation of steroid excretion in depressed patients before and after electroconvulsive therapy. Acta Endocrinol, 47:58 (1964).

15. J. Schlechte, B. Sherman and B. Pfohl, A comparison of adrenal corticol function in patients with depression and patients with Cushing's disease. (submitted).
16. S. Berson and R. Yalow, Radioimmunoassay of ACTH in plasma. J Clin Invest, 47:2725 (1968).
17. J.P. Allen, D. Denney, J. Kendall and P.H. Blachly, Corticotropin release during ECT in man. Am J Psychiatry, 131:1225 (1974).
18. V.S. Fang, B.J. Tricou, A. Robertson and H.Y. Meltzer, Plasma ACTH and cortisol levels in depressed patients: Relation to dexamethasone suppression test. Life Sci, 29:931 (1981).
19. B.I. Yerevanian, P.D. Woolf and H.P. Idel, Plasma ACTH levels in depression before and after recovery: Relationship to the dexamethasone suppression test. Psychiatry Res, 9:45 (1983).
20. V.I. Reus, M.S. Joseph and M.F. Dallman, M.F. ACTH levels after the dexamethasone suppression test in depression. N Engl J Med, 306:238 (1982).
21. N.H. Kalin, S.J. Weiler and S.E. Shelton, Plasma ACTH and cortisol concentrations before and after dexamethasone. Psychiatry Res, 7:87 (1982).
22. B. Pfohl, B. Sherman and J. Schlechte, Differences in plasma ACTH and cortisol between depressed patients and normal controls. Biol Psychiatry, in press.
23. B. Pfohl, B, Sherman and J. Schlechte, Pituitary-adrenal axis rhythm disturbances in major depression. Arch Gen Psychiatry, in press.
24. W. Vale, J. Spiess, C. Rivier and J. Rivier, Characterization of a 41-residue ovine hypothalamic peptide that stimulates secretion of corticotropin and beta-endorphin. Science, 213:1394 (1982).
25. P.W. Gold, G. Chrousos, C. Kellner, R. Post, A. Roy, P. Augerinos, H. Schulte, E. Oldfield and L. Lorianx, Psychiatric implications of basic and clinical studies with corticotrophin releasing factor. Am J Psychiatry, 141:619 (1984).
26. R. Guillemin, T. Vargo and J. Rossier, Beta-endorphin and ACTH are secreted concomitantly by the pituitary gland. Science, 197:1367 (1977).
27. G.S. Alexopoulos, C.E. Inturrisi, R. Lipman, R. Frances, J. Haycox, J.H. Dougherty, Plasma immunoreactive beta-endorphin levels in depression. Arch Gen Psychiatry, 40:181 (1983).
28. F. Brambilla, A.R. Genazzani, F. Facchinetti, D. Parrin, F. Petraglia, E. Sacchetti, S. Scarone, A. Guastalla and N. d'Antona, Beta-endorphin and beta-lipotropin plasma levels in chronic schizophrenia, primary affective disorders and secondary affective disorders. Psychoneuroendocrinology, 6:321 (1981).
29. S.C. Risch, Beta-endorphin hypersecretion in depression: possible cholinergic mechanisms. Biol Psychiatry, 17:1071 (1982).
30. M.R. Cohen, D. Pickar, I. Exstein, M. Gold and D. Sweeney, Plasma cortisol and beta-endorphin immunoreactivity in nonmajor and major depression. Am J Psychiatry, 141:628 (1984).
31. D. Pickar, N.R. Cutler, D. Naber, R. Post, C. Pert and W. Bunney, Plasma opioid activity in manic-depressive illness. Lancet, 1:937 (1980).
32. D. Pickar, M.R. Cohen, D. Naber and R.M. Cohen, Clinical studies of the endogenous opioid system. Biol Psychiatry, 17:1243 (1982).
33. L. Terenius, A. Wahlstrom and H. Agren, Naloxone (Narcan) treatment in depression: clinical observations and effects of CSF endorphins and monoamine metabolites. Psychopharmacology, 54:31 (1977).

34. J. Angst, V. Autenrieth, F. Brem, M. Koukkou, H. Meyer, H.H. Stassen and U. Storck, Preliminary results of treatment with beta-endorphin in depression. in: "Endorphins in Mental Health Research", E. Usdin, W. Bunney and N. Kline, eds., Oxford University Press, New York (1979).
35. R.B. Shekelle, W.J. Raynor, A.M. Ostfeld, D.C. Garron, L.A. Bieliauskas, C.L. Shuguey, C. Maliza and P. Ogelsby, Psychological depression and 17-year risk of death from cancer. Psychosom Med, 43:117 (1981).
36. M. Rogers, D. Dubey and P. Reich, The influence of the psyche and the brain on immunity and disease susceptibility: A critical review. Psychosom Med, 41:147 (1979).
37. M. Stein, S. Schleifer and S. Keller, Immune disorders. in: "Comprehensive Textbook of Psychiatry", Third Edition, H. Kaplan, A. Freedman and B. Sadock, eds., Williams and Wilkins, Baltimore, (1980).
38. J. Palmblad, K. Cantell, H. Strander, J. Froberg, C. Karlsson, L. Levi, M. Granstrom and P. Ungar, Stressor exposure and immunological response in man: interferon-producing capacity and phagocytosis. J Psychom Res, 20:193 (1976).
39. J. Palmblad, B. Petrini, J. Wasserman and T. Akerstedt, Lymphocyte and granulocyte reactions during sleep deprivation. Psychosom Med, 41:273 (1979).
40. B. Dorian, D. Garfinkel, G. Brown, A. Shore, D. Gladman and E. Keystone, Aberrations in lymphocyte subpopulations and function during psychological stress. Clin Exp Immunol, 50:132 (1982).
41. J. Jemmott, J. Borysenko, M. Borysenko, D. McClelland, R. Chapman and D. Meyer, Academic stress, power motivation and decrease in secretion rate of salivary secretory immunoglobulin A. Lancet, I:1400 (1983).
42. J. Kiecolt-Glaser, W. Garner, C. Speicher, G. Penn, J. Holliday and R. Glaser, Psychosocial modifiers of immunocompetence in medical students. Psychosom Med, 46:7 (1984).
43. S. Locke, M. Hurst and J. Heisel, J. The influence of stress on the immune response. Annual Meeting, American Psychosomatic Society, Washington, D.C. (1978).
44. G.H. Baker, N. Byrom, M. Irani, D. Brewerton, J.R. Hobbs, R.J. Woods and M. Nagvekar, Stress, cortisol and lymphocyte subpopulations. Lancet, I:574 (1984).
45. R. Bartrop, L. Lazarus, D. Luckhurst and L. Kiloh, Depressed lymphocyte function after bereavement. Lancet, I:834 (1977).
46. S. Schleifer, S. Keller, M. Camerino, J. Thornton and M. Stein, Suppression of lymphocyte stimulation following bereavement. J Am Med Assoc, 250:374 (1983).
47. Z. Kronfol, Depression and immune disease. Psychiatr Med, (in press).
48. Z. Kronfol, Cancer and depression. Br J Psychiatry, 142:309 (1983).
49. H. vonBrauchitsch, Antinuclear factor in psychiatric disorders. Am J Psychiatry, 128:1552 (1972).
50. R. Deberdt, J. VanHooren, M. Biesbrouch and W. Amery, Antinuclear factor-positive mental depression: A single disease entity? Biol Psychiatry, II:69 (1976).
51. F. Plantey, Antinuclear factor in affective disorders. Biol Psychiatry, 13:149 (1978).
52. E. Johnstone and K. Whaley, Antinuclear antibodies in psychiatric illness: their relationship to diagnosis and drug treatment. Br Med J, 2:724 (1975).

53. R. Rimon, P. Halonin, E. Anttinen and E. Evola, Complement fixing antibody to herpes simplex virus in patients with psychotic depression. Dis Nerv Sys, 32:822 (1971).
54. R. Coppel, F. Gregoire, L. Thiry and S. Sprecher, Antibody and cell-mediated immunity to herpes simplex virus in psychotic depression. J Clin Psychiatry, 39:266 (1978).
55. D. Sengar, B. Waters and J. Dunne, J. Lymphocyte subpopulations and mitogenic responses of lymphocytes in manic depressive disorder. Biol Psychiatry, 17:1017 (1982).
56. Z. Kronfol, J. Silva, J. Greden, S. Dembinski, R. Gardner and B.J. Carroll, Impaired lymphocyte function in depressive illness. Life Sci, 33:241 (1983).
57. S. Schleifer, S. Keller, A. Meyerson, M. Raskin, K. Davis and M. Stein, Lymphocyte function in major depressive disorder. Arch Gen Psychiatry, 41:484 (1984).
58. Z. Kronfol and J.D. House, Depression, cortisol and immune function. Lancet, I:1026 (1984).
59. H. Nasrallah, Z. Ballas, Z. Kronfol and S. Chapman, Natural-killer cell activity in major depression. in: "Viruses, Immunity and Mental Diseases", E. Kurstak, ed., Plenum Press, New York, (in press).
60. C. Van Dyke, K. McDaniel, V. Reus, W. Seaman and C. Kaufman, Immune response in depressed and Cushing's patients. American Psychiatric Association Annual Meeting, New Research Abstract #42, Los Angeles, CA (1984).
61. Z. Kronfol, R. Turner, H. Nasrallah and G. Winokur, Leukocyte regulation in depression and schizophrenia. Psychiatry Res, 13:13 (1984).
62. Z. Kronfol, H. Nasrallah, S. Chapman and J.D. House, Depression, cortisol metabolism and lymphocytopenia. J Affect Disord, (in press).
63. Z. Kronfol, J.D. House, J. Greden and B.J. Carroll, Depression, urinary free cortisol excretion and lymphocyte function. Br J Psychiatry, (in press).

ENDOGENOUS OPIOID SYSTEMS, STRESS, AND CANCER

Ian S. Zagon and Patricia J. McLaughlin

Department of Anatomy and Cancer Research Center
The Milton S. Hershey Medical Center
The Pennsylvania State University
Hershey, Pennsylvania 17033

INTRODUCTION

Opium's importance in both medicinal and social contexts dates back thousands of years (Blum, 1970; Terry and Pellens, 1928). Interest in opiates and opiate-like compounds (heretofore collectively referred to as "opioids") had been maintained throughout the years. A major breakthrough in our understanding of opioids came with the discovery of opioid receptors in 1973 by groups led by Simon (Simon et al., 1973), Snyder (Pert and Snyder, 1973), and Terenius (1973), along with the equally elegant finding of endogenous opioids by Hughes, Kosterlitz and colleagues (Hughes, 1975; Hughes et al., 1975) in 1975. Discovery of opioid receptors and the endogenous opioids (collectively referred to as "endogenous opioid systems") stands as a landmark of scientific achievement. Needless to say, this has stimulated intense research activity which has already culminated in important information as to the mechanisms and functions of opioids. One such exciting insight has been the discovery in 1983 that endogenous opioid systems are related to growth of both normal and abnormal cells and tissues (Zagon and McLaughlin, 1983a, b, c, d). This has led to a fascinating story concerning the fundamental nature of growth (Zagon and McLaughlin, 1984a, b, c; 1985a, b). The major purpose of this chapter is to review the evidence relating endogenous opioid systems to cancer. Since some aspects of stress may be associated with the endogenous opioid systems, and ACTH and some endogenous opioids share the same precursor (pro-opiomelanocortin), we have also included relevant information on the topic of stress and cancer in an attempt to provide perspective to the entire field of endogenous opioid systems and tumor biology.

ENDOGENOUS OPIOID SYSTEMS AND CANCER

To fully comprehend the role of endogenous opioid systems in cancer some background information is necessary. Although exogenous opioids such as morphine, methadone, and heroin have well-known analgesic and behavioral effects (Jaffe and Martin, 1980), these compounds have also been reported to alter cell function, particularly in developing neural systems (Zagon and McLaughlin, 1984d; Zagon et al., 1982, 1984). Exogenous opioids are known to be teratogenic in animals (Geber and Schramm, 1975; Harpel and Gautieri, 1968) and a review of the clinical and laboratory literature reveals that these agonists are antimitotic (Andersen, 1966; Simon, 1963), cytotoxic (Ma, 1931; Sinatra et al., 1978) and damaging to chromosomes (Falek and Hollingsworth, 1980; Fischman et al., 1977). Moreover, in vivo and in vitro studies have demonstrated that opioids can retard growth and development in many diverse organisms and cell types, with a suppression of cell division and alterations in polyamine, nucleic acid, and protein synthetic systems thought to be involved (Hui et al., 1976; Slotkin et al., 1980; Zagon and McLaughlin, 1977, 1978).

In view of the growth retarding properties of exogenous opioids in normal tissues undergoing proliferation and development, we were prompted to determine whether these compounds were also effective in inhibiting carcinogenic processes. Since normal neural tissues are especially susceptible to exogenous opioids, we examined the effects of opioids on C1300 mouse neuroblastoma. As a point of information, the C1300 murine neuroblastoma has been widely used as an in vivo model for the study of human neoplasia, particularly in regard to the screening of therapeutic regimens (Arima et al., 1973; Finkelstein et al., 1973). The S20Y neuroblastoma used in these experiments is a well-characterized cholinergic clone (Amano et al., 1972; Zagon and Schengrund, 1978) that contains opioid receptors (Klee and Nirenberg, 1974). We (Zagon and McLaughlin, 1981a) found that chronic administration of heroin (diacetylmorphine) to mice inoculated with 10^6 S20Y cells (a dosage that usually produces 100% tumor take in control mice) inhibited tumor growth and prolonged the lifespan of tumor-inoculated mice. For example, in animals which began receiving heroin at dosages of 3,6,10, or 15 mg/kg starting 2 weeks prior to tumor cell inoculation, a prolongation in mean survival time of 32-39% and median survival time of 8-50%, was recorded. Tumor size of animals receiving heroin also was often smaller than controls but, on the day of death, tumor size of heroin- and saline-treated mice were comparable. In some instances (i.e., dosages of 6 and 15 mg/kg), at least 20% of the animals never developed a tumor. To examine whether the effect of heroin on tumor growth was specific for the opioid receptor, some mice inoculated with neuroblastoma received simultaneous injections of heroin and naloxone (an opioid antagonist). Tumor incidence, survival time, and tumor growth were similar in control and heroin/naloxone animals. The results of this study showed that dosages of heroin (3-15 mg/kg), which were above the analgesic dose of 1 mg/kg (Brands et al., 1975) but well-below the LD_{50} level of 158-190 mg/kg (Brands et al., 1975; Way et al., 1960) in mice, had an extraordinary effect on tumorigenesis that was related to opioid receptor interaction. This antitumor action appeared to be related to heroin's direct influence on tumor cells (presumably by way of the opioid receptors known to be present in these cells) rather than on which reflected a cascade of events from heroin's effects on the entire animal. For example, injections of heroin beginning 2 weeks prior to tumor inoculation, an interval in which tolerance to heroin's actions are known to develop, still altered the course of carcinogenesis. Indeed, an injection schedule beginning 2 weeks prior to tumor inoculation was even more effective than a regimen that was initiated one week after tumor transplantation, although tumor burden may have played a role in this difference.

Additional information which has important bearing on these in vivo experiments with heroin and neuroblastoma, comes from in vitro investigations in which the effects of heroin on neuroblastoma cells in culture were examined (McLaughlin and Zagon, 1984; Zagon and McLaughlin, 1984c). The addition of heroin in concentrations ranging from 10^{-2}M to 10^{-8}M to S20Y cultures beginning 24 hr after seeding (new drug and media was provided daily), produced a dose-dependent inhibition of growth. This growth retarding effect was reversible, since removal of heroin allowed cells to resume normal growth. Experiments with thymidine incorporation, as well as analyses of mitotic figures, revealed that heroin affected cell division. Morphological differentiation also was noted to be decreased in drug-treated cultures, but this activity was not necessarily dependent on heroin dosage. Quite importantly the action of heroin in perturbing cell growth was blocked by concomitant administration of an opioid antagonist, naloxone. Additionally, the sterospecific effects of opioid agonists were investigated by treating logarithmically growing cultures with equimolar concentrations of levorphanol or dextrorphan. Levorphanol is known to be the active isomer. Levorphanol-exposed S20Y neuroblastoma cell cultures were growth-inhibited in a fashion similar to that observed when heroin was applied, whereas dextrorphan had little effect on cell growth. Thus, these data extend in vitro investigations and show that the dose-dependent, sterospecific, and naloxone-reversible effects of opioids on neurotumor cells are the direct effects of opioid agonist action at the opioid receptor level.

Although these observations on the action of exogenous opioids on cancer were revealing, the ethical considerations of utilizing opioid drugs with addictive liability for cancer treatment were apparent. An intriguing observation in our study (Zagon and McLaughlin, 1981a) utilizing heroin was that chronic, once-daily, injections of naloxone (10 mg/kg), a non-addictive opioid antagonist, produced a significant increase (11%) in mean survival time. Additionally, 1 of the 8 mice inoculated with neuroblastoma never developed a tumor, despite a 100% tumor incidence in control animals. We decided to examine further the effects of naloxone on neuroblastoma (Zagon and McLaughlin, 1981a). In utilizing an injection regimen similar to that with heroin, mice inoculated with 10^6 S20Y neuroblastoma received daily subcutaneous injections of either 5, 10, 15, or 20 mg/kg naloxone hydrochloride; injections were initiated either 2 weeks prior to tumor cell inoculation (pre-treatment group) or one week after tumor transplantation (post-treated groups). All mice in the control (saline) and post-treated groups developed tumors within 3 weeks after tumor cell inoculation. In the naloxone pre-treated groups, 4 or 12 mice exposed to 20 mg/kg, 2 of 12 mice exposed to 15 mg/kg, and 1 of 12 mice exposed to 10 mg/kg, did not develop tumors within a 91-day period following tumor inoculation. Moreover, 3 animals in the 20 mg/kg naloxone pre-treated group developed tumors between 43 and 63 days after tumor cell inoculation. Tumor dimensions were often reduced in naloxone-treated mice, although at the time of death tumor sizes of control and naloxone-exposed animals were similar. In general, tumor-bearing mice receiving naloxone lived longer than saline-tumor controls, with animals receiving higher drug dosages surviving for the longest time. In contrast to a mean survival time of 27 days for controls, naloxone pre-treated mice had increases in survival times of 25-61%, whereas naloxone post-treated groups exhibited increases of 20-40%. The median day of death for all mice exposed to naloxone was prolonged by 21-75%, occurring 6-21 days after the 28-day median for saline-tumor controls.

The dosages of naloxone utilized in this study, 5-20 mg/kg, were not exceedingly high, being well-below the LD_{50} level of 252 mg/kg for mice, as well as below the maximum drug dose (i.e., 20 mg/kg) that can elicit evert effects such as ataxia and depression (Giering et al., 1974). The low general toxicity of the dosages employed was also reflected by the fact that tumor-bearing mice did not lose body weight. This observation also sets naloxone apart from many other antitumor agents that cause cachexia.

At this point, we suggested a number of interpretations as to naloxone's antitumor effect. Naloxone had long been regarded as a "pure" antagonist to the many biological actions of opioid substances and as devoid of significant intrinsic activity (Blumberg and Dayton, 1974;

Jasinski et al., 1967; Sawynok et al., 1979). However, an increasing number of reports indicated that naloxone may have a wide variety of pharmacological, physiological, and behavioral effects that are unrelated to opioid receptor blockade (Brown et al., 1980; Geber et al., 1976; Sawynok et al., 1979). Naloxone has been shown to reduce resting serum prolactin and growth hormone levels (Bruni et al., 1977) and to decrease serum titer of interferon (Geber et al., 1976), and these or other alterations in hormonal or immunologic functions may have been responsible for naloxone's antineoplastic activity. In reviewing the studies on heroin and naloxone, it appears that both compounds have a chemotherapeutic influence in mice with neuroblastoma. Yet concomitant administration of heroin and naloxone did not produce a greater (or cumulative) response, but rather simultaneous injections of these agents had little effect on neoplasia. Whether naloxone was capable of acting as an agonist when used alone, and as an antagonist when used in combination with opioid agonists, was unclear. Equally unclear were the mechanisms underlying naloxone's and heroin's ability to retard tumor growth and prolong survival. In this early paper, it was suggested that naloxone and heroin could activate a common mechanism (e.g., similar receptors) to elicit their effects, or these drugs may act through independent channels (e.g., different receptors for each drug). Somehow, in both of these hypotheses concomitant administration of these drugs removed one another's effectiveness. Finally, we suggested that administration of naloxone could provide an over compensating release of endogenous opioids; these endogenous opioids could act like heroin to inhibit cancer.

At this point we reasoned that if naloxone, a short-acting (half-life of 2 hours) opioid antagonist had such marked antitumor properties, we wondered what effect naltrexone, an opioid antagonist that is 8 times as active and 3 times as long-acting as naloxone (Blumberg and Dayton, 1973), would have on tumorigenesis. Once again choosing the neuroblastoma model, we initiated a schedule of chronic daily injections of naltrexone or sterile water. For these experiments, we began drug injections 2 days after inoculation of 10^6 S20Y neuroblastoma cells. Our results (Zagon and McLaughlin, 1983a) indicated that animals receiving the lowest drug dosage (0.1 mg/kg) had a 33% tumor incidence, a 98% delay in the time before the appearance of a palpable tumor, and a 36% increase in survival time of tumor-bearing mice. However, neuroblastoma-inoculated animals receiving the highest drug dosage (10 mg/kg naltrexone) had a 100% tumor incidence, a 27% reduction in the time before tumor appearance, and a 19% decrease in survival time. Inoculation of neuroblastoma cells in control subjects resulted in a 100% tumor incidence within 29 days. It appeared that naltrexone's major action was to modify the interval between tumor inoculation and tumor appearance. Remarkably, these effects occurred with dosages that were 0.01 to 2.0% of the LD_{50} (570 mg/kg; Blumberg and Dayton, 1973) and which did not elicit overt toxicological symptoms such as ataxia, convulsions, or loss of body weight. To examine whether the dosage effects were related to the extent of opioid receptor blockade, mice given 0.1 or 10 mg/kg naltrexone were challenge with an opioid antagonist (morphine). Animals demonstrate analgesia when placed on a hotplate following injection with morphine; when naltrexone is present it prevents the interaction of morphine with opioid receptors and the animals respond in a fashion similar to controls. Our results showed that a dosage of 0.1 mg/kg only blocked the opioid receptors for 4-6 hr per day, while a dosage of 10 mg/kg naltrexone invoked an opioid receptor blockade for at least 24 hr.

To further examine the relationship between tumor response, the pharmacological properties of naltrexone (e.g., ability to block the opioid receptor), and drug dosage a series of experiments were performed (Zagon and McLaughlin, 1984a). Mice were inoculated with S20Y neuroblastoma and injected with eigher (a) 0.1 mg/kg naltrexone given once daily, a dosage previously found to block the receptor for a short ime each day and to have antitumor effects, (b) 0.1 mg/kg naltrexone four times daily, a schedule that utilized a low dosage of drug but blocked the opioid receptor for an entire 24 hr period, or (c) 0.4 mg/kg naltrexone given once daily, a regimen which controlled for the effects of cumulative drug dosages in the four times daily 0.1 mg/kg naltrexone treatment schedule. The results showed that mice receiving 0.1 mg/kg naltrexone four times daily had a 100% tumor incidence, no deviation in time before tumor appearance, and a 17% decrease from control values in total survival time. In contrast, once daily injections of either 0.1 mg/kg naltrexone or 0.4 mg/kg naltrexone resulted in a tumor incidence of 20% and 60% respectively, delays in time prior to tumor appearance of 90% and 65%, respectively, and an increased total survival time of 10% and 24%, respectively, for tumor-bearing mice relative to control levels. Inoculation of neuroblastoma in control mice resulted in 100% tumor appearance within 16 days and a mean survival time of 36 days.

These data beautifully demonstrate the relationship between tumorigenesis and the pharmacological action of opioid antagonists, and show that opioid receptor perturbation is a key element in modulating neuroblastoma. Thus, exacerbation of tumor response is seen when the opioid receptor is blocked for an entire 24 hr period as demonstrated by the 0.1 mg/kg four times daily group in this study or in the study (Zagon and McLaughlin, 1983a) mentioned earlier with 10 mg/kg naltrexone once daily. However, antitumor activity was found at dosages which only blocked the opioid receptor for a short time each day as shown in the 0.1 and 0.4 mg/kg once daily groups and in previously mentioned studies (Zagon and McLaughlin, 1981b) with 0.1 mg/kg once daily or naloxone at dosages of 10-20 mg/kg once daily.

Another way to provide rigorous proof that opioid receptors mediate a particular tumor response, is to explore the stereospecificity of opioid antagonists. Opioid interactions at the receptor level have been shown to be stereospecific (Pert and Snyder, 1973), with isomeric forms showing markedly different affinities for opioid receptors. The (-) isomer of naloxone is known to be 3 or 4 orders of magnitude more active than the (+) isomer in its ability to bind to opioid receptors in rat brain membrane preparations, as well as in antagonizing physiological responses to opioids (Faden and Holaday, 1980; Gayton et al., 1978; Iijima et al., 1978). In fact, in these reports and others (Sawynok et al., 1979), demonstration of stereospecificity of opioid antagonist action has been recommended as a criteria for exclusion of pharmacological effects not due to interaction with opioid receptors. Utilizing the (+) and (-) isomers of naloxone, and providing daily subcutaneous injections of 15 mg/kg beginning 2 days after tumor cell inoculation, we (Zagon and McLaughlin, unpublished observations) found that mice given (-) naloxone had a 71% tumor incidence, a 33% delay in the interval before tumor appearance, and a 28% increase in survival time. Inoculation of neuroblastma cells in control subjects resulted in 100% tumor incidence within 13 days. Mice given the (+) isomer of naloxone were comparable to control subjects in all aspects of neural neoplasia. These data showing that the (-) isomer of naloxone, but not the (+) isomer, alters tumor incidence and survival time is consistent with the enantiomeric specificity of opioids.

An important question that needs to be addressed concerns the type(s) of opioid receptor that is(are) involved in neuro-oncogenesis. At least five different types have been proposed (Martin et al., 1976; Lord et al., 1977; Zukin and Zukin, 1981, 1984). Although the precise physiological function of distinct receptor types is unclear, recently developed selective antagonists (see Takemori and Portoghese, 1985) provide a unique opportunity to elucidate the importance of each receptor type. One such antagonist, β-funaltrexamine (β-FNA), is the fumaramate methyl ester derivative of naltrexone and binds selectively and irreversibly to μ receptors (Portoghese et al., 1980; Takemori et al., 1981; Ward et al., 1982a, b). This compound has been widely used in assessing the relative importance of μ receptors in various opioid actions (Ward and Takemori, 1983; D'Amato and Holaday, 1984). We (Zagon, McLaughlin, Takemori, and Portoghese, unpublished observations) ahve explored the effect of β-FNA in concentrations that antagonize morphine-induced analgesia, in mice inoculated with neuroblastoma. Presumably, tumor response should be altered if μ receptors sensitive to β-FNA were involved in neural neoplasia. We found that chronic injections of 2 mg/kg β-FNA every 48 hr beginning 2 days after tumor inoculation resulted in tumor incidence and survival times that were comparable to controls (i.e., a 100% tumor incidence within 16 days; mean and median survival times of 36 and 35 days, respectively, following tumor inoculation. These results suggest that, in and by themselves, μ receptors selectively antagonized by β-FNA do not play an important role in neuro-oncogenic events. Utilization of agonists that serve as prototypic ligands for particular receptors, as well as exploration of other selective antagonists, are currently being explored.

Another aspect of study in regard to endogenous opioid systems and neural cancer is whether neuro-oncogenesis that includes metastasis is regulated by endogenous opioid systems. In all of the studies described so far, a primary tumor model was used. However, by administering S20Y neuroblastoma cells by the intradermal, rather than subcutaneous, route a metastatic model of disease can be examined. Daily injections of 0.1 mg/kg naltrexone resulted in a 69% tumor take, 70% delay in time prior to tumor appearance, and a 60% increase in median survival time (Zagon and McLaughlin, 1983d). Inoculation of neuroblastoma in control mice resulted in 100% tumor take within 15 days. Mice receiving 10 mg/kg naltrexone were similar to the control group. The absence of detectable acceleration in tumorigenesis as observed earlier with primary tumors may have been due to the fact that a median survival of 50 days was recorded with the primary tumor model but a median survival of 30 days was noted in the metastatic tumor model. Thus, the stimulation of tumor growth by 10 mg/kg naltrexone seen earlier may not have had time to be expressed with the metastatic model. The importance of this study is that in a metastatic system closely resembling the clinical setting, naltrexone markedly influenced neuro-oncogenesis.

In summary, a number of lines of evidence have demonstrated that endogenous opioid systems regulate neural tumors. Exogenous opioid agonists inhibit neoplasia and this action can be completely blocked by concomitant administration of opioid antagonists such as naloxone. Results from in vitro studies reveal that opioid agonists exert a growth inhibitory effect on cells in culture that is dose-dependent, stereospecific, and blocked by naloxone. Moreover, at least 2 opioid antagonists, naloxone and naltrexone, have been shown to affect neural neoplasia. Experiments also demonstrate that the duration of opioid receptor blockade is crucial in determining the course of neuro-oncogenesis. In addition, opioid antagonists appear to alter carcinogenesis in a manner which suggest that opioid receptors and their ligands, rather than the specific antagonists, are the primary influence on cancer. Finally, investigations have demonstrated that opioid antagonists exert a stereo-specific action on neural cancer.

Based on evidence accumulated to this point, it is becoming clear that endogenous opioid systems function in the regulation of neural cancer. The endogenous opioids, as well as their exogenous counterparts, are capable of exerting an inhibitory influence on neoplasia. Data suggest that opioids retard cell proliferation by direct action on opioid receptors associated with neural tumor cells. Continuous occupation of the opioid receptors by opioid antagonists thereby preventing endogenous opioid-opioid receptor interaction, results in an acceleration in tumor development. This would imply that endogenous opioids mediate early events in the establishment of tumors as well as tumor growth. Precisely how opioid antagonists are involved with the growth of cells and tissues needs to be elucidated. It is known that chronic treatment with opioid antagonists results in a striking increase in plasma β-endorphin levels (Recant et al., 1980), an increase in the number of opioid binding receptor sites (Lahti and Collins, 1978; Zukin et al., 1980), and a supersensitivity to opioid agonists (Tang and Collins, 1978). We have conjectured that daily intermittent opioid receptor blockade (e.g., 4-6 hr/day) exerts an anticancer effect by causing the production of more binding sites on tumor cells and a supersensitivity to basal (or elevated) levels of endorphins. The inhibitory action of the endogenous opioids takes place in the 14 to 18 hr interval when the opioid antagonist is no longer present.

Just how the opioid receptor is related to growth processes is an intriguing problem. Opioid receptors have been found to be associated with the amount and intracellular distribution of calmodulin (Simantov et al., 1982), known to be present in developing cells and tissues (Sano and Kitajima, 1983), and to regulate adenylate cyclase in nervous tissue (Bronstrom et al., 1977). In a related study, a calmodulin antagonist has been reported to inhibit tumor growth in mice inoculated with a sarcoma, and the authors suggest modulation of Ca^{++}-calmodulin-dependent processes (Ito and Hidaka, 1983). A link between opioid receptors and adenylate cyclase also has been reported (Sharma et al. 1975a, b). Endogenous (Lampert et al., 1976) and exogenous (Sharma et al., 1975b) opioids cause an inhibition of adenylate cyclase and a decrease in cAMP levels in confluent cultures of mouse neuroblastoma-rat glioma hybrid cells, but have little effect on parental cell lines (Sharma et al., 1975b). Although a connection between opioid receptors and adenylate cyclase would provide the basis of an attractive hypothesis as to how endogenous opioid systems control growth, it should be noted that treatment of neuroblastoma-glioma cultures with 10 μM morphine sulfate over a 9-day period did not alter total protein from control levels (Sharma et al., 1975b). This lack of effect by morphine on the growth of neuroblastoma-glioma cells stands in contrast to numerous reports (Zagon and McLaughlin, 1984c; McLaughlin and Zagon, 1984; Simon, 1971; North-Root et al., 1976) of opioid agonist-induced inhibition of growth that is naloxone-reversible and stereospecific in other cell culture systems (including murine neuroblastoma). Thus, despite the abundance of opioid receptors in neuroblastoma-glioma hybrid cells, the special nature of these hybrid cells (two different cell types, neurons and glia from 2 different species, mice and rat) must be considered.

These results showing that endogenous opioid systems serve to regulate growth of neuroblastoma raise a host of important questions. One of the most important is whether the endogenous opioid systems are involved in the development of other types of tumors. Presently, our laboratory is actively involved in addressing this issue. However, some information in this regard is available from other quarters. Frey and Kebabian (1984) have found opioid receptors in a prolactin-secreting tumor derived from a parent line isolated from the pituitary gland of a female Buffalo rat. In other experiments, Inoue and Hatanaks (1982), working with a rat pheochromocytoma cell line, found very low enkephalin binding in the absence of nerve growth factor but a dramatic increase when cells were cultured in the presence of nerve growth factor. Association of enkephalin with human pheochromocytoma has also been reported by a number of investigators

(Clement-Jones et al., 1980; Sullivan et al., 1978; Osamura et al., 1984).

Two other important studies need to be mentioned in regard to endogenous opioids and cancer. Aylsworth and colleagues (1979) have explored the effects of naloxone and naltrexone on mammary adenocarcinomas induced in rats by injection of 7, 12-dimethylbenz(a)-anthracene (DMBA). Apparently, the motivation for these experiments was that development and growth of mammary tumors in rats is dependent on prolactin and ovarian steroids, as well as growth hormone secretion, and that opioid antagonists are known to cause a decrease in these substances. Presumably, these opioid antagonist-induced changes would in turn influence mammary carcinogenesis. These authors found that both naloxone (0.2 mg/kg, 3 times daily) and naltrexone (1-2 mg/kg, twice daily) suppressed the growth of DMBA-induced mammary cancer, but did not induce regression of tumors. These inhibitory effects were reversible since resumption of growth followed withdrawal of drug treatment. Serum prolactin levels measured in rats receiving 2 mg/kg of naltrexone twice daily, were 50% lower than that of control animals. These investigators concluded by suggesting "that the endogenous opioid peptides may stimulate mammary tumor growth, via their ability to promote prolactin and growth hormone secretion" (Aylsworth et al., 1979, p. 19). In another report, Plotnikoff and Miller (1983) administered subcutaneous daily injections of methionine (met-) or leucine (leu-) enkephalin to mice inoculated with L1210 murine leukemia. Significant increases in the number of survivors were observed. Unfortunately, Plotnikoff and Miller did not try to block enkephalin's antitumor effect with an opioid antagonist in order to elucidate whether these alterations occurred at the receptor level.

ENDOGENOUS OPIOID SYSTEMS, CANCER, AND IMMUNITY

Although _in vivo_ and _in vitro_ studies with the neuroblastoma model argue strongly for a direct influence of opioids on cancer, and one mediated by opioid receptors, little is known about the presence of opioid receptors and responsivity to opioid agonists or antagonists in other tumor types. Additionally, the provocative findings relating endogenous opioid systems to the immune system suggests that immunological mechanisms may play a part in modulating some cancers. Given discussions of endogenous opioids, opioid receptors, and immunity that recur throughout this volume, along with some excellent reviews (Wybran, 1985; Chang, 1984; Fischer, in press; Faith et al., in press; Weber and Pert, 1984), I will not review all aspects of this subject. Rather, I would like to briefly recall some of the highlights in this field.

First, what is known about the effects of exogenous opioids on immunity? Viral hepatitis, endocarditis, bacteremia, cellulitis, injection-site abscesses, pulmonary infections, and other infections are common among individuals using opioids such as heroin (Hussey and Katz, 1950; Cherubin and Brown, 1968), and are frequently attributed to unsterilized needles or injection of contaminated materials. Despite a host of confounding variables that obfuscate interpretation of the results from studies in this area, some workers point towards a link between opioid addiction and susceptibility to infections (Cherubin and Brown, 1968). Brown and colleagues (1974) reported that heroin addicts display impaired _in vitro_ responsiveness in culture studies to at least one of three mitogens and suggested abnormalities occur in both the humoral and cellular immune systems of chronic heroin addicts. Methadone has also been found to have an immunosuppressive effect on the function of immunocompetent cells of human blood. Singh et al. (1980) discovered that methadone, in reaction mixtures of T- and B-cells, reduced T-rosette formation. Moreover, investigations of the function of human lymphocytes, measured _in_ *vitro* by following the mitogen-induced proliferative response of lymphocytes, found that methadone suppressed the lymphocyte stimulation by mitogens. McDonough and co-workers (1980) recorded that opioid addiction produces a significant depression in the absolute number of total T lymphocytes in peripheral blood, an increase in the absolute number of null lymphocytes, but no significant changes in B lymphocytes or total white blood cell count. Naloxone reversed both T cell depression and null cell increase.

A number of animal experiments concerned with exogenous opioids and immunity should be noted. Lefkowitz and Chiang (1975) described the immunosuppressive effect of morphine in mice using a plaque-forming cell assay to evaluate antibody-forming efficiency. Nalorphine, an opioid antagonist, did not diminish this morphine-induced suppression. Lefkowitz and Nemeth (1976) reported that morphine and methadone have a depressive effect in mice in a rosette forming cell test, while Ho and Leung (1979) recorded a reduction in lymphocyte response to concanavalin A-mediated blastogenesis in mice dependent on morphine that was only partially prevented by naloxone. Güngör and colleagues (1980) studied primary immune response in mice chronically receiving morphine and estimated spleen/body weight ratios and serum hemolysis production against sheep red blood cells. They found morphine exerted a dose-dependent inhibitory effect on the immune system which was antagonized by naloxone. In 1983, Tubaro and colleagues (1983) demonstrated that exposure to morphine exacerbates infections in mice. Morphine drastically reduces reticuloendothelial system activity, phagocyte count, phagocytic index, killing properties, and superoxide anion production in polymorphonuclear leukocytes and macrophages. Finally, Shavit et al. (1984) have shown that morphine causes a dose-related suppression of natural killer cell activity in rats; opioid specificity by the use of naloxone/naltrexone antagonism was not performed.

Even though the picture emanating from studies with exogenous opioids generally conclude that these agents depress immunity, investigations into endogenous opioid systems and immunity are less clear. In 1979, Wybran and co-workers showed that normal human blood T lymphocytes have opioid receptors. These investigators found that morphine and dextromoramide inhibited the percentage of active T-rosettes, and that this effect was completely reversed by naloxone. Specificity was further demonstrated by the absence of the effect of levomoramide, the inactive enantiomer upon the rosette system. Interestingly, methionine-enkephalin increased the percentage of T-rosettes and this action was antagonized by naloxone. Evidence for opioid receptors on granulocytes and monocytes (Lopker et al., 1980), mast cells (Yamasaki et al., 1983), the terminal complex of human complement (Schweigerer et al., 1982), as well as lymphocytes (McDonough et al., 1980; Wybran et al., 1979), has been established. Hazum, Chang, and Cuatrecasas (1979) have also shown by a direct radioligand binding technique that transformed cultured human lymphocytes bind β-endorphin in a manner which is not blocked by opioid agonists or antagonists suggesting the existence of a specific, non-opioid type of binding site (receptor) for β-endorphin in these cells.

A number of laboratories have focused on endogenous opioids as immunomodulators. Plotnikoff et al. (1982, 1985) and Plotnikoff and Miller (1983) have found that endogenous opioids stimulate lymphocyte blastogenesis in laboratory animals, and Miller et al. (1983, 1984) have found a similar effect in lymphoma patients as well as normal patients. The enkephalins were also observed to exhibit marked activity in increasing natural killer cell activity in normal human volunteers (Faith et al., 1984). Unfortunately, little is known about the ability to antagonize these effects with opioid antagonists. Gilman et al. (1982)

have reported that β-endorphin enhances rat splenic lymphocyte proliferation in response to concananalin A and phytohemaglutin, although a study my McCain et al. (1982) indicates that β-endorphin suppresses lymphocyte proliferation. Other investigators have shown that β-endorphin suppresses in vitro formation of antibodies (Johnson et al., 1982) and stimulates natural killer cell activity (Mathews et al., 1983; Plotnikoff et al., 1985). Van Epps and Saland (1984) have recently demonstrated that β-endorphin and met-enkephalin interact with human mononuclear cells and stimulate chemotaxis as measured by the in vitro leading front assay for migration; this effect was blocked by naloxone. These investigators also found that injection of β-endorphin into the rat cerebral ventricle resulted in the immigration of macrophage-like cells into the third ventricle.

Another possible connection between opioid peptides and the immune system involves the relationship between opioids and interferon. Administration of opioid agonists such as morphine and methadone to mice results in a dose-related decrease in the level of circulating interferon following injection by polyinosinic:polycytidylic acid (poly I:C), a double-stranded synthetic RNA capable of inducing interferon (Hung et al., 1973; Geber et al., 1975, 1976). Naloxone did not eliminate this effect (Hung et al., 1973). In a subsequent study (Geber et al., 1976), naloxone was found to function as an antagonist to morphine-induced depression of interferon, but acted as an agonist under some situations. Opioid agonists were also tested on cells in vitro using poly I:C as an inducer, and they did not appear to have a major effect on either the induction or the action of interferon in vitro. In regard to endogenous opioids and interferon, Blalock and Smith (1980) have reported that natural interferon has structural and biological similarities to ACTH 1-13 and β-endorphin. Subsequently, human peripheral lymphocytes were shown to be capable of producing ACTH and endorphin-like substances(Smith and Blalock, 1981). In another report (Blalock and Smith, 1981), human leukocyte interferon was shown to bind to opioid receptors in vitro and, when injected intracerebrally into mice, human leukocyte interferon caused potent endorphin-like effects. All of these actions were preventable and reversible by naloxone. Recombinant leukocyte A interferon has also been found to prevent naloxone precipitated abstinence in morphine-dependent rats (Dafny, 1983a), once again suggesting cross-reaction of interferon with opioid systems. Some question as to the relationship of interferon with opioids has been raised by Epstein and co-workers (1982) and Jörnvall and colleagues (1982). Epstein could not detect any functional or structural homology between natural and recombinant inferferon-α with β-endorphin (or ACTH), while Jörnvall excluded structural homologies in amino acid sequence.

ENDOGENOUS OPIOID SYSTEMS, STRESS, AND CANCER

The role of stress and psychological states in the etiology and pathogenesis of neoplasia is an extraordinarily interesting, but complex, subject. I refer the reader to a number of excellent reviews and discussions about this topic (LeShan, 1959; Perrin and Pierce, 1959; Sklar and Anisman, 1980, 1981; Riley, 1981). Certainly there appears to be growing evidence that stress can influence tumor development, as well as immune competence. The results of this research are variable and often contradictory, with some studies reporting an acceleration of oncogenicity (Seifter et al., 1973; Sklar and Anisman, 1979; Visintainer et al., 1982) and others a suppression (Rashkis, 1952; Neuberry et al., 1972; Gershbein et al., 1974; Pradhan and Roy, 1974). A number of factors have been suggested to account for these discrepancies, including the type of tumor model utilized, type of stress applied, timing and duration of stress and whether acute or chronic stress was administered. Exposure to stress can also release opioid peptides, and the purpose of this section is to review our knowledge concerning stress-related opioid activity and tumorigenesis.

Simon and co-workers (1980) have found that electrical stimulation of the midbrain mediates metastatic tumor growth. These investigators discovered that a single session of electrical stimulation of the mesencephalic periaqueductal gray (PAG) region (known to trigger endogenous opioid release) for 7 minutes following intravenous injection of Walker-256 carcinoma cells led to a dramatic increase in the number of "artificial" pulmonary metastases (metastatic-like aggregations which colonize the lung). Intraperitoneal injection of naloxone totally blocked the analgesic effects of the stimulation-produced analgesia (SPA), but did not appreciably block the metastatic effect. In a subsequent study, Simon and colleagues (1984) investigated the influence of the pain suppression system and its associated peptides on tumorigenesis by activating the pain suppressor system directly from the nucleus of the raphe magnus, a non-opioid subsystem. After inducing analgesia by giving a single injection of β-endorphin directly into the nucleus of the raphe magnus, Simon noted an increase in the number of "artificial" pulmonary metastases. This result could be abolished by pretreatment with naloxone. If the nucleus of the raphe magnus was activated by electrical stimulation sufficient to induce analgesia, the metastatic effect was still present but markedly attentuated.

Simon's work showing that SPA of the PAG enhances tumor metastasis, but which was not blocked by naloxone, suggested that the analgesia component was not importantly tied to the increase in neoplasia. However, as noted by Simon, naloxone may not block all opioid receptors. Moreover, opioid receptors on lymphocytes could be mediating some form of tumor enhancement by way of the immune system. To explore this question further, Brechner et al. (1983) designed an experiment in which Fischer 344 rats were implanted with either prolactin-sensitive (13762MT) or prolactin-insensitive (OR-163) tumors and received PAG stimulation. The rationale underlying this experiment was that if immunosuppression is induced by PAG-SPA, both types of tumors would have enhanced growth compared to nonstimulated animals. However, if other mechanisms such as prolactin release (known to be stimulated by opioids) were responsible for rapid tumor growth, then only the growth of prolactin-sensitive tumors would be enhanced compared to controls. Brechner found that the group with the prolactin-sensitive tumors had a statistically higher incidence of enhanced tumor growth than the non-PAG stimulated control group. No differences in tumor growth were observed in rats implanted with the prolactin-insensitive tumor, whether or not they had PAG stimulation. These results suggested that enhanced tumor growth with PAG stimulation may be related to hormonal release rather than immunosuppression.

Another segment of work involving stress-related endogenous opioid activity and cancer comes from studies utilizing the footshock technique. Inescapable foot shock causes a profound analgesia that is antagonized by naloxone (and dexamethasone) when shock is delivered intermittently, but not when it is applied continuously (Lewis et al., 1980). Thus, an equally potent analgesia could be induced by both techniques, but one appears to be mediated by opioid peptides ("opioid stress") and the other does not involve opioids ("nonopioid stress"). In a series of studies (Lewis et al., 1983a, 1983b, Shavit et al., 1983), the relationship between footshock induced stress on tumor growth has been examined. Fischer 344 rats injected with a mammary ascites tumor (13672B) and subjected to inescapable footshock of the "opioid" type manifest an enhanced tumor growth as indicated by decreased survival time and decreased percent survival as monitored over a 30-day post-tumor inoculation period. This effect was prevented by an opioid antagonist, naltrexone. In a preliminary report (Shavit et al., 1983) the non-opioid stress did not significantly affect tumor growth. In the absence of stress, naltrexone (implanted as pellets) did not alter tumorigenicity. In another report, Lewis et al. (1983b) showed that animals receiving morphine administration for 4 days before or after tumor inoculation, or for 14 days after tumor injection, had a decreased survival. The magnitude of morphine's effect was similar to that caused by footshock of the "opioid" type. Interestingly, animals given morphine for 14 days before tumor inoculation did not differ from nonstressed controls, suggesting that tolerance developed to the effect of morphine. In the same study, exposure to footshock of the "opioid" type for 4 or 14 days before or after tumor injection, or for 4 or 14 days after tumor inoculation, was found to produce equivalent and significant decreases in the percent of animals surviving and median survival time. Animals receiving morphine treatment for 14 days before tumor injection and footshock stress ("opioid" type) on the last 4 days of this treatment still exhibited a significant reduction in survival time, indicating that cross-tolerance had not occurred. Thus, footshock stress (of the "opioid" type) can alter tumorigenesis in a manner sharing characteristics of opioid involvement (e.g., responsivity to antagonism by naltrexone) and non-involvement (e.g., no evidence of tolerance or cross-tolerance). In a recent study, Shavit et al. (1984) has found that splenic natural killer cell activity was suppressed by the "opioid", but not the "nonopioid", form of stress and that this suppression was blocked by naltrexone. A similar suppression of natural killer cell activity was induced by high doses of morphine. Shavit suggests that immunological mechanisms may be a common denominator in the alterations of tumorigenicity recorded in footshock stress ("opioid" type) and morphine experiments.

CONCLUSIONS

It is quite clear from this review that endogenous opioid systems play an important role in tumor biology. As with most exciting new directions, we are only at the initial stages of gathering information and for every question answered many more are raised. Are opioid receptors associated with all tumor cells? What is the relationship of endogenous opioids in the plasma and neoplastic tissues of animals with cancer? Is endogenous opioid immunomoduation coincident or vitally important to neoplastic processes? Is "opioid stress" causally or coincidentally related to cancer? Do endogenous opioid systems play a role in human cancer?

At this early junction in this new field, it is often difficult to see an overall pattern to the pieces of information gathered to date. In many ways it would seem that we have seemingly conflicting results that often arrange themselves into two opposing camps: opioids either enhance or retard carcinogenic processes. Alternatively, I would raise the question of whether we are really looking at 2 sides of the same phenomenon? I refer specifically to data gathered in studies with opioid antagonists and neural tumors. Prevention of endogenous opioid binding to opioid receptors exacerbates neural cancer. Promotion of endogenous opioid-opioid receptor interaction inhibits neuro-oncogenicity. Admittedly, most of the evidence gathered so far is drawn from a neural tumor model, yet results with other tumor systems (e.g., L1210 leukemia, mammary carcinoma), as well as *in vitro* experimentation with neural tumor cells, appear to be entirely consistent with this hypothesis. Support for the novel idea that endogenous opioid systems regulate cancer also comes from other quarters. Thompson et al. (1983) has found that gentically obese mice, which have several times the level of β-endorphin compared to their lean littermates, show increased resistance to metastasis of B16 melanoma. In a study by

Lee and Lin (1975), the effects of acupuncture (thought to trigger release of endogenous opioids) on the growth of Erlich ascites tumor was studied in mice. Tumor-inoculated mice receiving acupuncture exhibited slower tumor growth and had extended survival time in comparison to control animals. Moreover, it is now quite clear that endogenous opioid systems control the development of most cells and tissues in the body (Zagon and McLaughlin, 1983b, c, 1984b, 1985a, b), extending the concept from only abnormal tissue regulation.

How does one integrate studies utilizing footshock of the "opioid" type, or electrical stimulation of the brain, or direct injections of β-endorphins into the brain within this scheme? On the surface, the results of these studies appear to contradict the idea that endogenous opioids serve to suppress tumorigenicity. If in fact these situations are causally related to cancer, could it be that these procedures actually elevate endogenous opioids for a short time, but then the body goes through a period of readjustment during which time endogenous opioid levels are low. For example, direct injection of β-endorphin into the brain may lead to a compensatory decrease in endogenous opioids for some time. A decrease in endogenous opioid would permit tumor development to accelerate. Thus, influences on the opioid system, as well as reaction to these influences, must be kept in mind when assessing the role of endogenous opioid systems in cancer. Moreover, the repercussions on immunity by short- and long-term effects resulting from perturbations of endogenous opioid systems also need to be given full consideration.

In conclusion, exploration of endogenous opioid systems in regard to tumor biology is proving to be a fascinating and important new frontier of science. Future studies are certain to reveal fundamental insights into neoplasia, as well as normal cell and tissue development. The exciting information gathered from these studies should have an important bearing in designing strategies for prevention and therapeutic intervention of malignant diseases.

ACKNOWLEDGEMENTS

This research was supported by NIH grants NS 20500 and NS 20623 and a Pennsylvania Research Corporation grant.

REFERENCES

Amano, T. Richelson, E., and Nirenberg, M., 1972, Neurotransmitter synthesis by neuroblastoma clones, Proc. Nat'l Acad. Sci., USA, 69:258.

Andersen, N. B., 1966, The effect of CNS depressants in mitosis, Acta Anaesta Scand. Suppl., 22:7.

Arima, E., Byfield, J. E., Finkelstein, J. Z., and Fonkalsrud, 1973, An experimental model for the therapy of mouse neuroblastoma, J. Pediat. Surg., 8:757.

Aylsworth, C. F., Hodson, C. A., and Meites, J., 1979, Opiate antagonists can inhibit mammary tumor growth in rats, Proc. Soc. Exp. Biol. Med., 161:18.

Blalock, J. E. and Smith, E. M., 1980, Human leukocyte interferons: structural and biological relatedness to adrenocorticotropic hormone and endorphins, Proc. Nat'l Acad. Sci., USA, 77:5972.

Blalock, J. E. and Smith, E. M., 1981, Human leukocyte interferon (HuIFN-α): potent endorphin-like opioid activity, Biochem. Biophys. Res. Comm., 101:472.

Blum, R. H., 1970, A history of opium, in: "Society and Drugs. I.

Social and Cultural Observations," Jossey-Bass, Inc., San Francisco.

Blumberg, H. and Dayton, H. B., 1973, Naloxone and related compounds, in: "Agonist and Antagonist Actions of Narcotic Analgesic Drugs," H. N. Kosterlitz, H. O. J. Collins, and J. E. Villareal, eds., Raven Press, New York.

Blumberg, H. and Dayton, H. B., 1974, Naloxone, naltrexone and related noroxymorphones, in: "Narcotic Antagonists," M. C. Braude, L. S. Harris, E. L. May, J. P. Smith, and J. E. Villareal, Raven Press, New York.

Brands, M., Hirst, M., and Gowdey, C. W., 1975, Duration of analgesia in mice after heroin by two testing methods, Canad. J. Physiol. Pharmacol., 54:381.

Brechner, T., Motyka, D., and Sherman, J., 1983, Growth enhancement of prolactin-sensitive mammary tumor by periaqueductal gray stimulation, Life Sci., 32:525.

Brostrom, C. O., Brostrom, M. A., and Wolff, D. W., 1977, Calcium-dependent adenylate cyclase from rat cerebral cortex, J. Biol. Chem., 252:5677.

Brown, S. M., Stimmel, B., Taub, R. N., Kochwa, S., and Rosenfeld, R. E., 1974, Immunologic dysfunction in heroin addicts, Arch. Internal Med., 134:1001.

Brown, D. R., Blank, M. S., and Holtzman, S. G., 1980, Suppression by naloxone of water intake induced by deprivation and hypertonic saline in intact and hypophysectomized rats, Life Sci., 26:1535.

Bruni, J. F., Van Vugt, D., Marshall, S., and Meites, J., 1977, Effects of naloxone, morphine and methionine enkephalin on serum prolactin, luteinizing hormone, follicle stimulating hormone, thyroid stimulating hormone and growth hormone, Life Sci., 21:461.

Chang, K. J., 1984, Opioid peptides have actions on the immune system, Trends Neurosci., 7:234.

Cherubin, C. E. and Brown, J., 1968, Systemic infections in heroin addicts, Lancet, (i):298.

Clement-Jones, V., Corder, R., and Lowry, P. J., 1980, Isolation of human met-enkephalin and two groups of putative precursors from an adrenal medullary tumour, Biochem. Biophys. Res. Comm., 95:665.

Dafny, N., 1983a, Interferon modifies morphine withdrawal phenomena in rodents, Neuropharmacology, 22:647

Dafny, N., 1983b, Modification of morphine withdrawal by interferon, Life Sci., 32:303.

D'Amato, R. and Holaday, J. W., 1984, Multiple opioid receptors in endotoxic shock: evidence for δ involvement and μ - δ interactions in vitro, Proc. Nat'l Acad. Sci., USA, 81:2898.

Epstein, L. B., Rose, M. E., McManus, N. H., and Li, C. H., 1982, Absence of functional and structural homology of natural and recombinant human leukocyte interferon (IFN-α) with human α-ACTH and β-endorphin, Biochem. Biophys. Res. Comm., 104:341.

Faden, A. I. and Holaday, J. W., 1980, Naloxone treatment of endotoxin shock: stereospecificity of physiologic and pharmacologic effects in the rat, J. Pharmacol. Exp. Ther., 212:441.

Faith, R. E., Plotnikoff, N. P., and Murgo, A. J., Effects of opiates and enkephalins-endorphins on immune functions, NIDA Technical Meeting, Boston, in press.

Faith, R. E., Liang, H. J., Murgo, A. J., and Plotnikoff, N. P., 1984, Neuroimmunomodulation with enkephalins: enhancement of natural killer (NK) cell activity in vitro, Clin. Immunol. Immunopathol., 31:412.

Falek, A. and Hollingsworth, F., 1980, Heroin and chromosome damage, Arch. gen. Psychiat., 37:227.

Finkelstein, J. Z., Arima, E., Byfield, P. E., Byfield, J. E., and Fonkalsrud, E. W., 1973, Murine neuroblastoma: a model of human disease, Canc. Chemother, Rep., 57:405.

Fischer, E. G. and Falke, N. E., 1984, β-endorphin modulates immune

Functions - a review, Proc. of VIIth World Congress Intern. College of Psychosomatic Med.

Fischman, H. K., Roizin, L., Moralshoili, E., Joy, C., and Rainer, J. D., 1977, Effects of prolonged administration of "street heroin" on the chromosomes of Macaca mulatti (Rheses) monkeys, in: "Neurotoxicology", L. Roizin, H. Shirakii and N. Grecevic, Raven Press, New York.

Frey, E. A. and Krebabian, J. W., 1984, A μ-opiate receptor in 7315c tumor tissue mediates inhibition of immunoreceactive prolactin release and adenvlate cyclase activity, Endocrinology, 115:1797.

Gayton, R. J., Lambert, L. A., and Bradley, P. B., 1978, Failure of (+)naloxone to antagonize responses to opioid peptides, Neuropharmacology, 17:549.

Geber, W. F. and Schramm, L. C., 1975, Congenital malformations of the central nervous system produced by narcotic analgesics in the hamster, Am. J. Obstet. Gynec., 123:705.

Geber, W. F., Lefkowitz, S. S., and Hung, C. Y., 1975, Effect of morphine, hydromorphine, methadone, mescaline, trypan blue, vitamin A, sodium salicylate, and caffeine on the serum interferon level in response to viral infection, Arch. Int. Pharmacodyn, 214:322.

Geber, W. F., Lefkowitz, S. S., and Hung, C. Y., 1976, Action of naloxone on the interferon-lowering activity of morphine in the mouse, Pharmacology, 14:322.

Gershbein, L. I., Benuck, I., and Shurrager, P. S., 1974, Influence of stress on lesion growth and on survival of animals bearing parenteral and intracerebral leukemia L1210 and Walker tumors, Oncology, 30:429.

Giering, J. E., Davidson, T. A., Shetty, B. V., and Truant, A. P., 1974, in: "Narcotic Antagonists," M. C. Braude, L. S. Harris, E. L. May, J. P. Smith, and J. E. Villarreal, ed., Raven Press, New York.

Gilman, S. C., Schwartz, J. M., Miller, R. J., Bloom, F. E., and Feldman, J. D., 1982, β-endorphin enhances lymphocyte proliferative responses, Proc. Nat'l Acad. Sci., USA, 79:4226.

Güngör, M., Genc, E., Sugduyu, H., Eroglu, L., and Koyuncuoglu, H., 1980, Effect of chronic administration of morphine on primary immune response in mice, Experientia, 36:1309.

Harpel, H. S. and Gautieri, R. F., 1968, Morphine-induced fetal malformations, J. Pharm. Sci., 57:1590.

Hazum, E., Chang, K.-J. and Cuatrecasas, P., 1979, Specific nonopiate receptors for β-endorphin, Science, 205:1033.

Ho, W.K.K. and Leung, A., 1979, Effects of morphine addiction on concanavalin A-mediated blastogenesis, Pharmacol. Res. Comm., 11:413.

Hughes, J., 1975, Isolation of an endogenous compound from the brain with pharmacological properties similar to morphine, Brain Res., 88:295.

Hughes, J. A., Smith, T. W., Kosterlitz, H. W., Fothergill, L. A., Morgan, B. A., and Morris, H. R., 1975, Identification of two pentapeptides from the brain with potent opiate agonist activity, Nature, 258:577.

Hui, F. W., Krikun, E., Hirsh, E. M., Blaiklock, R. G., and Smith, A. A., 1976, Inhibition of nucleic acid synthesis in the regenerating limb of salamanders treated with dl-methadone or narcotic antagonists, Exp. Neurol., 53:267.

Hung, C. Y., Lefkowitz, S. S., and Geber, W. F., 1973, Interferon inhibition by narcotic analgesics, Proc. Soc. Exp. Biol. Med., 142:106.

Hussey, H. H. and Katz, S., 1950, Infections resulting from narcotic addiction: report of 102 cases, Am. J. Med., 9:186.

Iijima, I., Minamikawa, J., Jacobson, A.E., Brossi, A., and Rice, K.C., 1978, Studies in the (+)-morphinan series .5. Synthesis and biological properties of (+)-naloxone, J. Med. Chem., 21:398.

Inoue, N., and Hatankaka, H., 1982, Nerve growth factor induces specific enkephalin binding sites in a nerve cell line, J. Biol. Chem., 257:9238.

Ito, H. and Hidaka, H., 1983, Antitumor effect of a calmodulin antagonist on the growth of solid sarcoma-180, Cancer Lett., 19:215.

Jaffe, E. H. and Martin, W. R., 1980, Narcotic analgesics and antagonists, in: "The Pharmacologic Basis of Therapeutics," L. S. Goodman and A. Gilman, eds., MacMillan Publ. Co., New York, p. 494.

Jasinski, D. R., Martin, W. R., and Haertzen, C. A., 1967, The human pharmacology and abuse potential of N-allylnoroxymorphone (naloxone), J. Pharmacol. Exp. Ther., 159:420.

Johnson, H. M., Smith, E. M., Torres, B. A., and Blalock, J. E., 1982, Regulation of the in vitro antibody response by neuroendocrine hormones, Proc. Nat'l Acad. Sci., USA 79:4171.

Jörnvall, H ., Persson, M., and Ekman, R., 1982, Structural comparisons of leukocyte interferon and pro-opiomelanocortin correlated with immunological similarities, FEBS Letters, 137:153.

Klee, W. A. and Nirenberg, M., 1974, A neuroblastoma x glioma hybrid cell line with morphine receptors, Proc. Nat'l Acad. Sci., USA, 71:3474.

Lahti, R. A. and Collins, R. J., 1978, Chronic naloxone results in prolonged increases in opiate binding sites in brain, Eur. J. Pharmacol., 51:185.

Lampert, A., Nirenberg, M., and Klee, W. A., 1976, Tolerance and dependence evoked by an endogenous opiate peptide, Proc. Nat'l Acad. Sci., USA, 73:3165.

Lee, S.-C. and Lin, J.-H., 1975, An inhibitory effect of acupuncture on the growth of Ehrlich ascites tumor cells in mice, Chinese Med. J., 22:167.

Lefkowitz, S. S. and Chiang, C. Y., 1975, Effects of certain abused drugs on hemolysin forming cells, Life Sci., 17:1763.

Lefkowitz, S. S. and Nemeth, D., 1976, Immunosuppression of rosette-forming cells, Adv. Exp. Med. Biol., 733:269.

LeShan, L. L., 1959, Psychological states as factors in the development of malignant disease: a critical review, J. Nat. Cancer Inst., 22:1.

Lewis, J. W., Cannon, J. T., and Leibeskind, J. C., 1980, Opioid and nonopioid mechanisma of stress analgesia, Science, 208:623.

Lewis, J. W., Shavit, Y., Terman, G. W., Nelson, L. R., Gale, R. P., and Liebeskind, J. C., 1983a, Apparent involvement of opioid peptides in stress-induced enhancement of tumor growth, Peptides, 4:635.

Lewis, J. W., Shavit, Y., Terman, G. W., Gale, R. P., and Liebeskind, J. C., 1983b, Stress and morphine affect survival of rats challenged with a mammary ascites tumor (MAT 13762B), Nat. Immun. Cell Growth Regul., 3:43.

Lopker, H., Abood, L. G., Hoss, W., and Lionetti, F. J., 1980, Stereoselective muscarinic acetylcholine and opiate receptors in human phagocytic leukocytes, Biochem. Pharm., 29:1361.

Lord, J. A. H., Waterfield, A. A., Huges, J., and Kosterlitz, H. W., 1977, Endogenous opioid peptides: multiple agonists and receptors, Nature, 267:495.

Ma, W. C., 1931, A cytological study of acute and chronic morphinism in albino rat, Clin. J. Physiol., 5:251.

Martin. W. R., Eades, C. G., Thompson, J. A., Huppler, R. E., and Gilbert, P. E., 1976, The effects of morphine and nalorphine-like drugs in the non-dependent and morphine-dependent chronic spinal dog, J. Pharmacol. Exp. Ther., 197:517.

Matthews, P. M., Froelich, C. J., Sibbitt, W. L., and Bankhurst, A. D., 1983, Enhancement of natural cytotoxicity by β-endorphin, J. Immunol., 130:1658.

McCain, H. W., Lamster, I. B., Bozzone, J. M., and Grbic, J. T., 1982, β-endorphin modulates human immune activity via-non-opiate receptor mechanisma, Life Sci., 31:1619.

McDonough, R. J., Madden, J. L., Falek, A., Shafer, D. A., Pline, M., Gordon, D., Bokos, P., Keuhule, J. C., and Mendelson, J., 1980, Alteration of T and null lymphocyte frequencies in the peripheral

blood of human opiate addicts: in vivo evidence for opiate receptor sites on T lymphocytes, J. Immunol., 125:2539.
McLaughlin, P. J. and Zagon, I. S., 1984, Opioid regulation of tumor cell growth in vitro, Soc. Neurosci., 14th Annual Meeting, 10:111.
Miller, G. C., Murgo, A. J., and Plotnikoff, N. P., 1983, Enkephalins-enhancement of active T-cell rosettes from lymphoma patients, Clin. Immunol., Immunol., Immunopathol., 26:446.
Miller, G. C., Murgo, A. J., and Plotnikoff, N. P., 1984, Enkephalins-enhancement of active T-cell rosettes from normal volunteers, Clin. Immunol. Immunopath., 31:132.
Newberry, B. H., Frankie, G., Beatty, P. A., Maloney, B. D., and Gilchrist, J. C., 1972, Shock stress and DMBA-induced mammary tumors, Psychosom. Med., 34:295.
North-Root, H., Martin, D. W., and Toliver, A. P., 1976, Binding of an opiate, levorphanol, to intact neuroblastoma cells in continuous culture, Physiol. Chem. Physics, 8:221.
Osamura, R. Y., Watanabe, K., Yoshimasa, T., Nakao, K., and Imura, H., 1984, Immunohistochemical localization of met-enkephalin, met enkephalin-arg^6-gly^7-leu^8, met-enkephalin-arg^6-phe^7 and leu-enkephalin in human adrenal medulla and pheochromocytomas, Peptides 5:993.
Perrin, G. M. and Pierce, I. R., 1959, Psychosomatic aspects of cancer, Psychosom. Med., 21:397.
Pert, G. B. and Snyder, S. H., 1973, Opiate receptor: demonstration in nervous tissue, Science, 179:1011.
Plotnikoff, N. P. and Miller, G. C., 1983, Enkephalins as immunomodulators, Int. J. Immunopharmacol., 5:437.
Plotnikoff, N. P., Miller, G. C., and Murgo, A. J., 1982, Enkephalins-endorphins: immunomodulators in mice, Int. J. Immunopharmacol., 4:366.
Plotnikoff, N. P., Murgo, A. J., Miller, G. C., Corder, C. N., and Faith, R. E., 1985, Enkephalins: immunomodulators, Fed. Proc., 44:118.
Portoghese, P. S., Larson, D. L., Sayre, L. M., Fries, D. S., and Takemori, A. E., 1980, A novel opiate receptor site directed alkylating agent with irreversible narcotic antagonistic and reversible agonistic activities, J. Med. Chem., 23:233.
Pradhan, S. N. and Ray, P., 1974, Effects of stress on growth of transplanted and 7, 12-dimethylbenz(α)anthracene-induced tumors and their modification by psychotropic drugs, JNCI, 53:1241.
Rashkis, H. A., 1952, Systemic stress as an inhibitor of experimental tumors in Swiss mice, Science, 116:169.
Recant, L., Voyles, N. R., Luciano, M., and Pert, C. B., 1980, Naltrexone reduces weight gain, alters "β-endorphin", and reduces insulin output from pancreatic islets of genetically obese mice, Peptides, 1:309.
Riley, V., 1981, Psychoneuroendocrine influences on immunocompetence and neoplasia, Science, 212:1100.
Sano, M. and Kitajima, S., 1983, Ontogeny of calmodulin and calmodulin-dependent adenylate cyclase in rat brain, Dev. Brain Res., 7:215.
Sawynok, J., Pinsky, C., and LaBella, F. S., 1979, Mini-review on the specificity of naloxone as an opiate antagonist, Life Sci., 25:1621.
Schweigerer, H., Bhakdi, S., and Teschemacher, H., 1982, Specific nonopiate binding sites for human β-endorphin on the terminal complex of human complement, Nature, 296:572.
Seifter, E., Rettura, G., Zisblatt, M., Levenson, S. M., Levine, N., Davidson, A., and Seifter, J., 1973, Enhancement of tumor development in physically stressed mice inoculated with an oncogenic virus, Experientia, 29:1379.
Sharma, S. K., Klee, W. A., and Nirenberg, M., 1975a, Dual regulation of adenylate cyclase accounts for narcotic dependence and tolerance, Proc. Nat'l Acad. Sci., USA, 72:3092.
Sharma, S. K., Klee, W. A., and Nirenberg, M., 1975b, Opiate-dependent

modulation of adenylate cyclase, Proc. Nat'l Acad. Sci., USA, 74:3365.
Shavit, Y., Lewis, J. W., Terman, G. W., Gale, R. P., Liebeskind, J.C., 1983, Endogenous opioids may mediate the effects of stress on tumor growth and immune function, Proc. West. Pharmacol. Soc., 26:53.
Shavit, Y., Lewis, J. W., Terman, G. W., Gall, R. P., and Liebeskind, J. C., 1984, Opioid peptides mediate the suppressive effect of stress on natural killer cell cytotoxicity, Science, 223:188.
Simantov, R., Nadler, H., and Levy, R., 1982, A genetic approach to reveal the action of the opiate receptor in selected neuroblastoma-glioma cells. Interaction with α-adrenoreceptors, calmodulin and calcium ion ATPase, Eur. J. Biochem., 128:461.
Simon, E. J., 1964, Inhibition of bacterial growth by drugs of the morphine series, Science, 144:543.
Simon, E. J., 1971, Single cells, in: "Narcotic Drugs," D. H. Clouet, ed., Plenum Press, New York.
Simon, E. J., Miller, J. M., and Edelman, I., 1973, Stereospecific binding of the potent narcotic analgesic (^{3}H)etorphine to rat brain homogenate, Proc. Nat'l Acad. Sci., USA, 70:1947.
Simon, R. H., Lovett, E. J., Tomaszek, D., and Lundy, J., 1980, Electrical stimulation of the midbrain mediates metastatic tumor growth, Science, 209:1132.
Simon, R. H., Arbo, T. E., and Lundy, J., 1984, β-endorphin injected into the nucleus of the raphe magnus facilitates metastatic tumor growth, Brain Res. Bull., 12:487.
Sinatra, R. A., Milks, L. C., and Ford,D. H., 1978, Structural alterations in normal and axotomized facial nucleus neurons after treatment with morphine, Experientia, 35:1218.
Singh, V. K., Jakubovic, A., and Thomas, D. A., 1980, Suppressive effects of methadone on human blood lymphocytes, Immunol. Lett., 2:177.
Sklar, L. S. and Anisman, H., 1979, Stress and coping factors influence tumor growth, Science, 205:513.
Sklar, L. S. and Anisman, H., 1980, Social stress influences tumor growth, Psychosom. Med., 42:347.
Sklar, L. S. and Anisman, H., 1981, Stress and cancer, Psychol. Bull., 89:369.
lotkin, T. A., Seidler, F. J., and Whitmore, W. L., 1980, Effects of maternal methadone administration on ornithine decarboxylase in brain and heart of the offspring: relationships of enzyme activity to dose and to growth impairment in the rat, Life Sci., 26:861.
Smith, E. M. and Blalock, J. E., 1981, Human lymphocyte production of corticotropin and endorphin-like substances: association with leukocyte interferon, Proc. Nat'l Acad. Sci., USA, 78:7530.
Sullivan, S. N., Bloom, S. R., and Polak, J. M., 1978, Enkephalin in peripheral neuroendocrine tumours, Lancet, 1:986.
Takemori, A. E. and Portoghese, P. S., 1985, Affinity labels for opioid receptors, Ann. Rev. Pharmacol., in press.
Takemori, A. E., Larson, D. L., and Portoghese, P. S., 1981, The irreversible narcotic antagonistic and reversible agonistic properties of the fumarate methyl ester derivative of naltrexone, Eur. J. Pharmacol., 70:445.
Tang, A. H. and Collins, R. J., 1978, Enhanced analgesic effects of morphine after chronic administration of naloxone in the rat, Eur. J. Pharmacol., 47:473.
Terenius, L., 1973, Stereospecific interaction between narcotic analgesics and a synaptic plasma membrane fraction of rat cerebral cortex, Acta Pharmacol. Toxicol., 32:317.
Terry, C. E. and Pellens, M., 1970, Pathology-tolerance-dependence-withdrawal, in: "The Opium Problem," C. E. Terry and M. Pellens, eds., Patterson Smith Publishing Corp., New York.
Thompson, C. I., Kreider, J. W., Black, P. L., Schmidt, T. J., and

Margules, D. L., 1983, Genetically obese mice: resistance to metastasis of B16 melanoma and enhanced T-lymphocyte mitogenic responses, Science, 220:1183.
Tubaro, E., Borelli, G., Croce, C., Cavallo, G., and Santiangeli, C., 1983, Effect of morphine on resistance to infection, J. Infectious Dis., 148:656.
Van Epps, E. E. and Saland, L., 1984, β-endorphin and met-enkephalin stimulate human peripheral blood monomuclear cell chemotaxis, J. Immunol., 132:3046.
Visintainer, M. A., Volpicelli, J. R., Seligman, N. E. P., 1982, Tumor rejection in rats after inescapable or escapable shock, Science, 216:437.
Ward, S. J. and Takemori, A. E., 1983, Relative involvement of mu, kappa, and delta receptor mechanisms in opiate-mediated antinociception in mice, J. Pharmacol. Exp. Ther., 224:525.
Ward, S. J., Portoghese, P. S., and Takemori, A. E., 1982a, Pharmacological profiles of β-funaltrexamine (β-FNA) and β-chlornaltrexamine (β-CNA) on the mouse vas deferens preparation, Eur. J. Pharmacol., 80:377.
Ward, S. J., Portoghese, P. S., and Takemori, A. E., 1982b, Pharmacological characteristics in vivo of the novel opiate β-funaltrexamine, J. Pharmacol. Exp. Ther., 220:494.
Way, E. L., Kemp, J. W., Young, J. M., and Grassetti, D. R., 1960, The pharmacologic effects of heroin in relationship to its rate of biotransformation, J. Pharmacol. Exp. Ther., 129:144.
Weber, R. J. and Pert, C. B., 1984, Opiatergic modulation of the immune system, in: Central and Peripheral Endorphins: Basic and Clinical Aspects, E. E. Muller and A. R. Genazzini, ed., Raven Press, NY, p. 35.
Wybran, J., 1985, Enkephalins, endorphins, substance P and the immune system, in:Neural Modulation of Immunity, R. Guillemin, M. Cohn, and T. Melnechuk, eds., Raven Press, NY, p. 157.
Wybran, J., Appelboom, T., Famaey, J.-P., and Govaerts, A., 1979, Suggestive evidence for receptors for morphine and methionine-enkephalin on normal human blood T lymphocytes, J. Immunol., 123:1068.
Yamasaki, Y., Shimamura, O., Kizu, A., Nakazawa, M., and Ijichi, H., 1983, Interactions of morphine with PGE isoproterenal, dopamine and aminophylline in rat mast cells: their effect on IgE-mediated ^{14}C-serotonin release, Agents and Actions, 13:21.
Zagon, I. S. and McLaughlin, P. J., 1977, The effects of different schedules of methadone treatment on rat brain development, Exp. Neurol., 56:538.
Zagon, I. S. and McLaughlin, P. J., 1978, Perinatal methadone exposure and brain development: a biochemical study, J. Neurochem., 31:49.
Zagon, I. S. and McLaughlin, P. J., 1981a, Heroin prolongs survival time and retards tumor growth in mice with neuroblastoma, Brain Res. Bull., 7:25.
Zagon, I. S. and McLaughlin, P. J., 1981b, Naloxone prolongs survival time or mice treated with neuroblastoma, Life Sci., 28:1095.
Zagon, I. S. and McLaughlin, P. J., 1983a, Naltrexone modulates tumor response in mice with neuroblastoma, Science, 221:671.
Zagon, I. S. and McLaughlin, P. J., 1983b, Increased brain size and cellular content in infant rats treated wtih an opiate antagonist, Science, 221:1179.
Zagon, I. S. and McLaughlin, P. J., 1983c, Naltrexone modulates growth in infant rats, Life Sci., 33:2449.
Zagon, I. S. and McLaughlin, P. J., 1983d, Opioid antagonists inhibit the growth of metastatic murine neuroblastoma, Cancer Lett., 21:89.
Zagon, I. S. and McLaughlin, P. J., 1984a, Duration of opiate receptor blockade determines tumorigenic response in mice with neuroblastoma:

a role for endogenous opioid systems in cancer, Life Sci., 35:409.
Zagon, I. S. and McLaughlin, P. J., 1984b, Naltrexone modulates body and brain development in rats: a role for endogenous opioids in growth, Life Sci., 35:2057.
Zagon, I. S. and McLaughlin, P. J., 1984c, Opiates alter tumor cell growth and differentiation in vitro, NIDA Res. Monog., 49:344.
Zagon, I. S. and McLaughlin, P. J., 1984d, An overview of the neurobehavioral sequelae of perinatal opioid exposure, in: Neurobehavioral Teratology, J. Yanai, ed., Elsevier Science Publishers, Amsterdam.
Zagon, I. S. and McLaughlin, P. J., 1985a, Opiate antagonist induced regulation of organ development, Physiol. Behav., in press.
Zagon, I. S. and McLaughlin, P. J., 1985b, Naltrexone's influence on neurobehavioral development, Pharmacol. Biochem. Behav., in press.
Zagon, I. S., McLaughlin, P. J., Weaver, D. J., and Zagon E., 1982, Opiates, endorphins and the developing organism: a comprehensive bibliography, Neurosci. Biobehav. Rev., 6:439.
Zagon, I. S., McLaughlin, P. J., and Zagon, E., 1984, Opiates, endorphins, and the developing organism: a comprehensive bibliography, 1982-1983, Neurosci. Biobehav. Rev., 8:387.
Zagon, I. S. and Schengrund, C.-L., 1978, Neuronal and non-neuronal properties of neuroblastoma cells, Exp. Cell Res., 114:159.
Zukin, R. S. and Zukin, S. R., 1981, Minireview: multiple opiate receptors: emerging concepts, Life Sci., 29:2681.
Zukin, R. S. and Zukin, S. R., 1984, The case for multiple opiate receptors, Trends Neurosci., 7:160.
Zukin, R. S., Sugarman, J. R., Fitz-Syage, M. L., Gardner, E. L., Zukin S. R., and Gintzler, A. R., 1982, Naltrexone-induced opiate receptor supersensitivity, Brain Res., 245:285.

EFFECTS OF STRESS AND MORPHINE ON NATURAL KILLER CELL ACTIVITY AND ON MAMMARY TUMOR DEVELOPMENT

James W. Lewis, Yehuda Shavit, Fredricka C. Martin, Gregory W. Terman, Robert P. Gale, and John C. Liebeskind

Departments of Psychology and Medicine, University of California, Los Angeles, CA 90024

At the time we began our investigations of the modulation of immune function and tumor development by neural and endocrine opioid systems, several findings suggested that this was a reasonable avenue to explore. For example, focal electrical stimulation of an opioid-rich brain region was found to increase tumor growth (Brechner et al., 1983; Simon et al., 1980), specific brain lesions alter immune functions (Cross et al., 1982; Stein et al., 1982), and immune responses have been correlated with changes in brain neural activity (Besedovsky et al., 1983). Also, evidence was accumulating from *in vitro* studies that opioids influence the functioning of the immune system (for reviews see Plotnikoff and Miller, 1983; Weber and Pert, 1984). It was our intent to use environmental stressors as stimuli for activating endogenous opioid systems and observe their effects on tumor development and immune competence. Examination of the literature pertaining to the effects of stress on these processes, however, presented us with a somewhat confusing picture. This literature contained a host of seemingly contradictory results: stress can either enhance, suppress, or not affect tumor development or immune function (e.g. Jensen, 1968; Newberry and Sengbush, 1979; Nieburgs et al., 1977; Riley, 1981; Sklar and Anisman, 1981).

This seemingly paradoxical situation was reminiscent of our experience with the phenomenon of stress-induced analgesia. Prior to 1980, several investigators had shown that exposure to stress could elicit an analgesic response in laboratory animals although there was no consensus on the involvement of opioid peptides in this analgesia (cf. Akil et al., 1976 and Hayes et al., 1978). In some cases stress analgesia appeared to be opioid-mediated, in others not. This lack of consensus was, in part, due to the heterogeneity of stressors used by different investigators rendering comparison of results difficult. We demonstrated that a single stressor, inescapable footshock, could cause either an opioid or a nonopioid form of stress analgesia depending upon the parameters of its application: Intermittent footshock stress causes analgesia that appears to be mediated by opioid peptides, whereas exposure to the same intensity of shock delivered in a continuous fashion causes an equipotent analgesia that is nonopioid in nature (Lewis et al., 1980, 1981). Moreover, the opioid analgesic response to intermittent footshock stress has been associated with activation of both central and peripheral opioid systems. Thus, multiple endogenous pain-inhibitory systems exist, and it appears that the temporal parameters of stress administration can critically define which system is activated.

Two general principles were illustrated by these stress analgesia experiments. First, stress is not a unitary phenomenon. The same stressor, inescapable footshock, can elicit qualitatively different responses, opioid- vs. nonopioid-mediated, depending upon the parameters of its application. Second, opioid peptides are not involved in all stress responses; but when they are, both central and peripheral opioid systems can be important. It was our goal, therefore, to begin systematic examination of the effects of stress on immune function and tumor development armed with the knowledge of the importance of stressor parameters and with some understanding of the physiological systems activated by these footshock procedures.

EFFECTS OF STRESS AND MORPHINE ON DEVELOPMENT OF A MAMMARY TUMOR

We investigated the effects of footshock stress on the development of a mammary ascites tumor (MAT 13762B) in Fischer 344 female rats. In accordance with our previous work, two footshock stress paradigms were used: intermittent footshock (2.0 mA, on 1 sec every 5 sec for 10 min), a procedure that releases opioids, and continuous footshock (2.0 mA, on continuously for 2 min), a procedure that elicits nonopioid-mediated analgesia. Note that the only apparent difference between these two stressors is the temporal pattern of their administration: In both cases, stimulus intensity and total "shock on" time are the same.

Groups of rats were subjected to either intermittent shock, continuous shock, or no shock, daily for four days. Three hours after the last stress session, all rats were injected intraperitoneally with MAT 13762B tumor cells. Animals were then returned to their home cages and survival was monitored for 30 days. Exposure to intermittent, but not continuous, footshock was found to cause a significant decrease in survival time and percent of rats surviving compared to nonstressed controls (Shavit et al., 1983b).

Considering our previous work on stress analgesia, the finding that development of MAT 13762B is enhanced by intermittent, but not continuous, footshock stress is suggestive of a role for opioid peptides. In order to establish the involvement of opioids in this, or any biological process, however, several criteria should be considered: the effect should be blocked by opiate antagonist drugs; the effect should be reproduced by administration of opiate drugs; and tolerance should develop upon repeated exposure. Subsequent experiments were conducted to address these issues.

Groups of rats were implanted subcutaneously either with pellets that slowly release naltrexone, a potent opiate antagonist drug, or with placebo pellets. Seven days later, half of the rats in each condition were exposed to 4 daily presentations of intermittent footshock and the remaining rats served as non-stressed controls. The tumor-enhancing effect of stress was prevented in rats implanted with naltrexone pellets (Lewis et al., 1983b). Placebo pellets did not alleviate the adverse effect of footshock stress, nor was there any effect of naltrexone on tumor lethality in non-stressed rats. Furthermore, we found that morphine could mimic the tumor-enhancing effect of intermittent footshock. That is, 4 daily injections of a high dose of morphine (50 mg/kg, s.c.) prior to tumor inoculation resulted in the same pattern of decreased survival time and percent survival as did exposure to intermittent footshock (Lewis et al., 1983a). Taken together, these results are compatible with the hypothesized role of opioids in the tumor-enhancing effect of stress.

Our attempts to satisfy the last criterion for opioid involvement, development of tolerance, were less conclusive. In the case of morphine, tolerance was apparent following 14 daily drug injections. Chronic exposure to intermittent footshock (14 sessions), however, did not diminish the impact of stress on mammary tumor development. In this regard, however, it should be noted that while many opioid-mediated behavioral effects of stress (e.g. analgesia, Lewis et al., 1981) do develop tolerance in 14 days, there is recent biochemical evidence indicating that after chronic stress exposure, stress-induced release of pituitary opioids is enhanced, not reduced, compared to control values (Akil et al., 1985).

Overall, these results suggest that opioid peptides are important mediators of stress-induced enhancement of tumor development. Several possible loci and/or mechanisms of action underlying this effect can be suggested. In our studies of stress analgesia we found that intermittent footshock stress caused the release of enkephalin-like peptides from the adrenal medulla (Lewis et al., 1982a, 1982b). It may be that in addition to their role in analgesia, these opioid peptides, circulating peripherally, interact with tumor cells to affect their proliferation. The occurrence of opioid receptors on MAT 13762B cells is, however, unknown. Alternatively, it may be that opioids enhance tumor development indirectly, via hormonal mechanisms. For example, opioids have been shown to modulate the stress-induced release of several hormones, including prolactin (e.g. Ragavan and Frantz, 1981) and corticosterone (Siegel et al., 1982), a hormone with potent immunosuppressant properties (e.g. Riley, 1981). Thus, it may be that enhanced tumor growth in stressed rats is due either to elevated levels of prolactin, a hormone known to facilitate the growth of MAT 13762B (Bogden et al., 1974), or to compromised immune function (Kreider et al., 1978).

EFFECTS OF STRESS AND MORPHINE ON IMMUNE FUNCTION

In parallel with our studies of the effect of footshock stress on tumor development, we have examined the effect of this same stress procedures on immune function. Since natural killer (NK) cells have recently been suggested to play an important role in immune surveillance against tumors (Herberman and Ortaldo, 1981), we elected to use NK cell cytotoxicity as an index of immunocompetence.

In our initial studies, rats were exposed to the same stressors and stress administration regimen as in the tumor experiments. Three hours after the last stress session, spleens were removed and lymphocytes prepared for use in a standard 4 hour chromium release assay (see Shavit et al., 1984a for details). Congruent with the results of the cancer studies, exposure to intermittent, but not continuous, footshock caused a significant suppression of NK cytotoxic activity. Administration of naltrexone prior to intermittent footshock attenuated this immunosuppression (Shavit et al., 1984a). Moreover, daily administration of morphine, at doses which enhanced tumor development, also reduced the cytotoxicity of NK cells. Finally, as before, tolerance developed to the immunosuppressive effects of morphine but not to that of footshock (Shavit et al., 1983a).

In our most recent work, we have continued to use morphine as a pharmacological probe for studying opioid regulation of NK activity. We have found that administration of a single high dose of morphine ($\geq$20 mg/kg, s.c.) induced NK suppression evident at three hours after drug injection, returning to normal values by 24 hours post-injection (Shavit et al., 1984b). This morphine-induced suppression of NK activity was blocked by naltrexone pretreatment. Several pieces of evidence suggest that the site of action of morphine in regulating NK activity is within the central nervous system: First, a very small amount of morphine, 20 μg, injected into the lateral cerebral ventricle causes as great an immunosuppression as is observed following systemic administration of 1000 times this dose (Shavit et al., 1985); Second, systemic administration of N-methylmorphine, an analog of morphine that does not readily penetrate the blood brain barrier, does not affect NK activity; Finally, the immunosuppressive effect of systemically administered morphine is prevented by intracerebroventricular application of naltrexone.

EFFECT OF OPIATES ON NATURAL KILLER CELL FUNCTION:
COMPARISON OF IN VITRO AND IN VIVO STUDIES

Collectively the results of our experiments suggest that exposure to opioids in vivo causes enhanced tumor growth and suppression of NK function. This is in agreement with in vivo studies by others showing facilitatory effects of opioids on tumor development and adverse effects on several indices of immune competence (Cherubin and Brown, 1968; Cherubin and Millian, 1968; Brown et al., 1974; Ho and Leung, 1979; Tubaro et al., 1983; Zagon and McLaughlin, 1981). By contrast, using in vitro methods, others have shown that opioid peptides can either enhance or suppress the responsiveness of T lymphocytes (Chang, 1984; Gilman et al., 1982; McCain et al., 1982; Weber and Pert, 1984), and several opioid peptides appear to enhance the cytotoxic activity of NK cells (Faith et al., 1984; Froelich and Bankhurst, 1984; Kay et al., 1984; Mathews et al., 1983; Wybran, 1985).

In an attempt to resolve or clarify the disparities between in vivo and in vitro actions of opioids, we have carried out the following preliminary studies. Whereas we have reliably observed a suppression of NK cytoxicity 3 hours following injection of morphine in vivo, incubation of NK cells in morphine-containing media (concentrations ranging from 10^{-4} to 10^{-8} M) for 3 hours prior to the NK assay is without effect. These results and those of others (Kay et al., 1984; Mathews et al., 1983) suggest that morphine does not act directly upon NK cells (see however Wybran, 1985). In other work, we have compared the effects of the opiate alkaloid, levorphanol, with that of its opiate-inactive stereoisomer, dextrorphan. In vivo, only levorphanol (10 mg/kg) causes a suppression of NK activity. In vitro, however, it seems that both stereoisomers cause an equivalent, dose-dependent, reduction of NK cytotoxicity (Martin et al., 1985). It appears that the in vivo effects of opioids are mediated via opioid receptors since only morphine and levorphanol are effective. The in vitro effects of these drugs, however, is puzzling and apparently does not rely on opioid receptors since levorphanol and dextrophan, but not morphine, are active under these conditions.

SUMMARY AND CONCLUSIONS

In summary, it is apparent that some, but not all stressors are adequate stimuli for activation of neural and endocrine opioid systems and that opioid peptides released by stress have widespread physiological effects including analgesia, enhancement of tumor development, and immunosuppression. The parallels between the effects of intermittent footshock stress on tumor development and NK activity are striking: Thus far, every manipulation that caused an opioid-mediated effect on tumor development, also did so on measures of NK activity, thereby suggesting that these processes may be causally related.

One theme in psychoneuroimmunological research that has recently gained considerable support is the notion that the impact a particular stressor has on immunity or tumor growth is determined by the degree to which the organism can control the stressor. Originally, it was shown that exposure to inescapable, but not escapable, stressors resulted in a behavioral syndrome called "learned helplessness" (Maier and Seligman, 1976). More recently, work from several laboratories has indicated that exposure to inescapable stress suppressed the proliferative response of T-cells to mitogenic stimulation and facilitated tumor growth, whereas exposure to an equivalent amount of escapable stress is without effect (Laudenslager et al., 1983; Sklar and Anisman, 1979; Visintainer et al., 1982). In a collaborative study with Maier, Ryan, and Laudenslager, we have found that exposure to inescapable, but not escapable, tail-shocks suppresses NK activity (Shavit et al., 1983c). Moreover, it is possible that "perceived controllability" is one of the critical differences between the intermittent and continuous footshock stress procedures employed in our work. Although both stressors are technically inescapable, exposure to intermittent footshock, but not continuous footshock, causes learned helplessness (Maier et al., 1983). Thus, even though the two stressors we have studied have physically equivalent stimulus properties, it is only the stressor that is perceived as uncontrollable that adversely affects the organism.

Several summary messages may be gleaned from the data described in this chapter. For example, continued investigations of stress and immune function and cancer ought to carefully consider the stressors employed noting that a single stressor may cause markedly different effects depending on the particular parameters of its administration. Delineation of the role of opioid peptides in these stress-related phenomena must take into account the existence of multiple opioid systems and the possibility of both central and peripheral sites of action.

REFERENCES

Akil, H., Madden, J., Patrick, R.L., and Barchas, J.D., 1976, Stress-induced increase in endogenous opiate peptides; concurrent analgesia and its partial reversal by naloxone, in: "Opiates and Endogenous Opioid Peptides", H.W. Kosterlitz, ed., Elsevier, Amsterdam.

Akil, H., Shiomi, H., and Matthews, J., 1985, Induction of the intermediate pituitary by stress: Synthesis and release of a nonopioid form of ß-endorphin. Science, 227:424-426.

Besedovsky, H., Del Re, A., Sorkin, E., Da Prada, M., Burri, R., and Honegger, C., 1983, The immune response evokes changes in brain noradrenergic neurons. Science, 221:564-566.

Bogden, A.E., Taylor, D.J., Kuo, E.Y.H., Mason, M.M., and Speropoulos, A., 1974, The effect of perphenazine-induced serum prolactin reponse on estrogen primed mammary tumor-host systems, 13762 and R-35 mammary adenocarcinoma. Cancer Res., 34:3018-3025.

Brechner, T., Motyka, D., and Sherman, J.E., 1983, Growth enhancement of a prolactin-sensitive mammary tumor by periaqueductal gray stimulation. Life Sci., 32:525-530.

Brown, S.M., Stimmel, B., Taub, R.N., Kochwa, S., and Rosenfield, R.E., 1974, Immunologic dysfunction in heroin addicts. Arch. Intern. Med., 134:1001-1006.

Chang, K.-J., 1984, Opioid peptides have actions on the immune system. TINS, 7:234-235.

Cherubin, C.E. and Brown, J., 1968, Systemic infection in heroin addicts. Lancet, i:298-299.

Cherubin, C.E. and Millian, S.J., 1968, Serologic investingations in narcotic addicts I. Ann. Intern. Med., 69:739-742.

Cross, R.J., Brooks, W.H., Roszman, T.L., and Markesbery, W.R., 1982, Hypothalamic-immune interactions: effect of hypophysectomy on neuroimmunomodulation. J. Neurol. Sci., 53:557-566.

Faith, R.E., Liang, H.J., Murgo, A.J., and Plotnikoff, N.P., 1984, Neuroimmunomodulation with enkephalins: Enhancement of human natural killer (NK) cell activity in vitro. Clin. Immun. Immunopath., 31:412-418.

Froelich, C.J., and Bankhurst, A.D., 1984, The effect of ß-endorphin on natural killer cytotoxicity and antibody dependent cellular cytotoxicity. Life Sci., 35:261-265.

Gilman, S.C., Schwartz, J.M., Milner, R.J., Bloom, F.E., and Feldman, J.D., 1982, ß-endorphin enhances lymphocyte proliferative reponses. Proc. Natl. Acad. Sci. (USA), 79:4226-4230.

Hayes, R.L., Bennett, G.J., Newlon, P.G., and Mayer, D.J., 1978, Behavioral and physiological studies of non-narcotic analgesia in the rat elicited by certain environmental stimuli. Brain Res., 155:69-90.

Ho, W.K.K., and Leung, A., 1979, The effect of morphine addiction on concanavalin A mediated blastogenesis. Pharmacol. Res. Commun., 1:413-419.

Herberman, R.B., and Ortaldo, J.R., 1981, Natural killer cells: Their role in defenses against disease. Science, 214:24-30.

Jensen, M.M., 1968, Influence of stress on murine leukemia virus infection. Proc. Soc. exp. Biol. Med., 127:610-614.

Kay, N., Allen, J., and Morley, J.E., 1984, Endorphins stimulate normal human peripheral blood lymphocyte natural killer activity. Life Sci., 35:53-59.

Kreider, J.W., Bartllet, G.L., Purnell, D.M., and Webb, S., 1978, Immunotherapy of an established rat mammary adenocarcinoma (13762A) with intratumor injection of Corynebacterium parvum. Cancer Res., 38:689-692.

Laudenslager, M.L., Ryan, S.M., Drugan, R.C., Hyson, R.L., and Maier, S.F., 1983, Coping and immunosuppression: Inescapable but not escapable shock suppresses lymphocyte proliferation. Science, 221:568-570.

Lewis, J.W., Cannon, J.T., and Liebeskind, J.C., 1980, Opioid and nonopioid mechanisms of stress analgesia. Science, 208:623-625.

Lewis, J.W., Shavit, Y., Terman, G.W., Gale, R.P., and Liebeskind, J.C., 1983a, Stress and morphine affect survial of rats challenged with a mammary ascites tumor (MAT 13762B). Nat. Immun. Cell Growth Regul., 3:43-50.

Lewis, J.W., Shavit, Y., Terman, G.W., Nelson, L.R., Gale, R.P., and Liebeskind, J.C., 1983b, Apparent involvement of Opioid peptides in stress-induced enhancement of tumor growth. Peptides, 4:635-638.

Lewis, J.W., Sherman, J.E., and Liebeskind, J.C., 1981, Opioid and nonopioid stress analgesia: assessment of tolerance and cross-tolerance with morphine, J. Neurosci., 1:358-363.

Lewis, J.W., Tordoff, M.G., Liebeskind, J.C., and Viveros, O.H., 1982a, Evidence for adrenal medullary opioid involvement in stress analgesia. Soc. Neurosci. Abstr., 8:778.

Lewis, J.W., Tordoff, M.G., Sherman, J.E., and Liebeskind, J.C., 1982b, Adrenal enkephalin-like peptides may mediate opioid stress analgesia. Science, 217:557-559.

Maier, S.F. and Seligman, M.E.P., 1976, Learned helplessness: Theory and evidence. J. exp. Psychol. Gen., 105:3-46.

Maier, S.F., Sherman, J.E., Lewis, J.W., Terman, G.W., and Liebeskind, J.C., 1983, The opioid/nonopioid nature of stress-induced analgesia and learned helplessness. J. exp. Psychol. Anim. Behav., 9:80-90.

Martin, F.C., Shavit, Y., Terman, G.W., Pechnick, R.N., Oh, C., and Liebeskind, J.C., 1985, Stereospecificity of opiate immunosuppression in rats. Soc. Neurosci. Abstr., 11.

Mathews, P.M., Froelich, C.J., Sibbit Jr., W.L., and Bankhurst, A.D., 1983, Enhancement of natural cytoxicity by ß-endorphin. J. Immunol., 130:1658-1662.

McCain, H.W., Lamster, I.B., Bozzone, J.M., and Grbic, J.T., 1982, ß-endorphin modulates human immune activity via non-opiate receptor mechanisms. Life Sci., 31:1619-1624.

Newberry, B.H., and Sengbush, L., 1979, Inhibitory effects of stress on experimental mammary tumors. Cancer Detect. Prevent., 2:225-233.

Nieburgs, H.E., Weiss, J., Navarrete, M., Grillone, G., and Siedlecki, B., 1977, Inhibitory and enhancing effects of various stress on experimental mammary tumorigenesis. Cancer Detect. Prevent., 2:463-470.

Plotnikoff, N.P., and Miller, G.C., 1983, Enkephalins as immunomodulators. Int. J. Immunopharmac., 5:437-441.

Ragavan, V., and Frantz, A.G., 1981, Suppression of serum prolactin by naloxone but not anti-ß-endorphin antiserum in stressed and unstressed rats. Life Sci., 28:921-929.

Riley, V., 1981, Psychoneuroendocrine influences on immunocompetence and neoplasia. Science, 212:1100-1109.

Shavit, Y., Depaulis, A., Terman, G.W., Martin, F.C., Gale, R.P., and Liebeskind, J.C., 1985, Evidence for central mediation of morphine's immunosuppressive effects. Fed. Proc., 44:1489.

Shavit, Y., Lewis, J.W., Terman, G.W., Gale, R.P., and Liebeskind, J.C., 1983a, The effects of stress and morphine on immune function in rats. Soc. Neurosci. Abstr., 9:117.

Shavit, Y., Lewis, J.W., Terman, G.W., Gale, R.P., and Liebeskind, J.C., 1983b, Endogenous opioids may mediate the effects of stress on tumor growth and immune function. Proc. West. Pharmacol. Soc., 26:53-56.

Shavit, Y., Lewis, J.W., Terman, G.W., Gale, R.P., and Liebeskind, J.C., 1984, Opioid peptides mediate the suppressive effect of stress on natural killer cell cytotoxicity. Science, 223:188-190.

Shavit, Y., Ryan, S.M., Lewis, J.W., Laudenslager, M.L., Terman, G.W., Maier, S.F., Gale, R.P., and Liebeskind, J.C., 1983c, Inescapable but not escapable stress alters immune function. The Physiologist, 26:A-64.

Shavit, Y., Terman, G.W., Martin, F.C., Gale, R.P., and Liebeskind, J.C., 1984b, Naltrexone-sensitive suppression of the immune system's natural killer cells by morphine. Soc. Neurosci. Abstr., 10:726.

Siegel, R.A., Chowers, I., Conforti, N., Feldman, S., and Weidenfeld, J., 1982, Effects of naloxone on basal and stress-induced ACTH and corticosterone secretion in the male rat - site and mechanism of action. Brain Res., 249:103-109.

Simon, R.H., Lovett, E.J. III, Tomaszek, D., and Lundy, J., 1980, Electrical stimulation of the midbrain mediates metastatic tumor growth. Science, 209:1132-1133.

Sklar, L.S., and Anisman, H., 1979, Stress and coping factors influence tumor growth. Science, 205:513-515.

Sklar, L.S., and Anisman, H., 1981, Stress and cancer. Psychol. Bull, 89:369-406.

Stein, M., Keller, S.E., and Schleiffer, S.J., 1982, The role of brain and the neuroendocrine system in immune regulation - potential links to neoplastic diseases. In: Biological Mediators of Behavior and Disease: Neoplasia, S. Levy, ed., Elsevier, N.Y.

Tubaro, E., Borelli, G., Croce, C., Cavallo, G., and Santiangli, C., 1983, Effect of morphine on resistance to infection. J. Infect. Dis., 148:656-666.

Visintainer, M.A., Volpicelli, J.R., and Seligman, M.E.P., 1982, Tumor rejection in rats after inescapable or escapable shock. Science, 216:437-439.
Weber, R.J. and Pert, C.B., 1984, Opiatergic modulation of the immune system, in: "Central and Peripheral Endorphins: Basic and Clinical Aspects", E.E. Muller and A.R. Genazzani, eds., Raven Press, New York.
Wybran, J., 1985, Enkephalins and endorphins as modifiers of the immune system: Present and future. Fed. Proc., 44:92-94.
Zagon, I.S., and McLaughlin, P.J., 1981, Heroin prolongs survival time and retards tumor growth in mice with neuroblastoma. Brain Res. Bull., 7:25-32.

ENDOCRINE AND IMMUNOLOGICAL RESPONSES TO ACUTE STRESS

F. Berkenbosch, C.J. Heijnen*, G. Croiset*, C. Revers,*
R.E. Ballieux**, R. Binnekade, and F.J.H. Tilders

Dept. Pharmacology, Medical Faculty, Free University,
1007 MC Amsterdam; *Dept. Immunology, University
Hospital for Children and Youth, Utrecht; **Dept.
Clinical Immunology, University Hospital, Utrecht

INTRODUCTION

Although the suggestion that the function of the immune system is regulated by the endocrine system, was postulated some time ago (see Selye, 1980; Besedovsky and Sorkin, 1977), the extent of this interaction is now becoming more obvious by many recent technological advances in the field of endocrinology and immunology. For instance, regulation of the immune response by the hypothalamo-pituitary gonadal axis has been proved to be of great importance (Grossman, 1984) and also growth hormone (Grossman and Roselle, 1983; Michael et al., 1980) prolactin (Bercze et al., 1981), glucocorticoids (Munck et al., 1984), as well as peripheral circulating catecholamines (Johnson et al., 1981; Del Rey et al., 1981) have been shown to have immunomodulatory capacities. Recent studies have presented evidence that lymphocytes possess binding sites for opiates such as morphine and its derivates (Wybran et al., 1979; Gungor et al., 1980; Lopker et al., 1980). The presence of these binding sites would indicate that opiates may be able to affect the immune system directly. In this light, the discovery of the existence of endogenous opiates such as the enkephalins and endorphins motivated many workers to study the role of these peptides on immune function. However, as in other areas of opioid peptide physiology, the effects of opioid peptides in immune regulation are complex and often paradoxical.

Beta-endorphin and related peptides are produced in large quantities in the anterior and intermediate lobe of the pituitary gland as cleavage products from a precursor called pro-opiomelanocortin. It is of interest that stress, which is known to affect immune function (Riley, 1981), causes a rapid release of beta-endorphin and related peptides. In this paper we briefly summarize our studies on the mechanisms involved in stress-induced beta-endorphin secretion. In addition, effects of endorphins on the primary immune response *in vitro*, and effects of acute stress on the proliferative capacity of rat spleen lymphocytes will be presented.

CLEAVAGE PRODUCTS OF POMC IN CORTICOTROPHS AND MELANOTROPHS

The rat pituitary gland contains two cell types being the corticotrophs of the anterior lobe and the melanotrophs of the intermediate lobe that both synthesize a precursor call pro-opiomelanocortin (POMC). After its formation POMC is processed by several enzymes resulting in glycosylation, phosphorylation, cleavage and formation of amidated and acetylated products (see for reviews, Eipper and Mains, 1980; Hope and Lowry, 1981; Roberts et al., 1982). The major POMC-derived peptides produced in the corticotrophs are different from those in the melanotrophs, suggesting cell-specific differences in the enzymatic machinery involved in processing of POMC. In both cell types the processing is started by cleavage of the glycosylated and/or phosphorylated precursor into three fragments being pro-gamma-MSH (16K fragment), ACTH and beta-LPH. The processing of POMC in the corticotrophs terminates at this stage, implicating that pro-gamma-MSH, ACTH and beta-LPH are major secretory products of this cell type, although beta-LPH can be partially processed to gamma-LPH and beta-endorphin.

In the melanotrophs, the processing proceeds to smaller fragments. For instance, ACTH is further cleaved to alpha-MSH and CLIP, whereas beta-LPH serves as an intermediate presursor for gamma-LPH, beta-endorphin-1-31 and smaller peptides being beta-endorphin-1-27, beta-endorphin-1-26, alpha- and gamma-endorphin. In this cell type the processing is completed by N-terminal acetylation of most of the end products (Mains, Eipper, 1981; Serzinger, Hollt, 1980). It is worth noting that the biological activity can be changed completely by N-acetylation. For instance, N-acetylation of beta-endorphin-1-31 causes a complete loss of its opiate activity (Smyth et al., 1979) and increases the melanotrophic activity of ACTH-1-13 approximately 10-fold (Hofman, Vajima, 1961).

SECRETION OF POMC-DERIVED PEPTIDES IN RESPONSE TO ACUTE STRESS

ACTH, alpha-MSH and beta-endorphin assays

Specific radioimmunoassays for the different POMC-derived peptides are required to obtain relevant information on circulating levels of POMC-derived peptides in the rat. Most of the ACTH immunoreactivity and alpha-MSH immunoreactivity as measured by our ACTH and alpha-MSH assays in samples of rat plasma, behaves chromatographically as ACTH-1-39 and alpha-MSH-1-13. In contrast, in our beta-endorphin assays as in most other beta-endorphin assays, acetylated and non-acetylated beta-endorphin-1-31, beta-endorphin-1-27 and beta-endorphin-1-26 and beta-LPH are measured with identical molar efficiencies (Tilders et al., 1984). In view of these considerations, we will refer only in the case of beta-endorphin to beta-endorphin immunoreactivity (beta-ENDi).

Experimental conditions

Experimental procedures such as handling, transport or injections cannot be avoided when rats are subjected to stress-stimuli such as ether, restraint, etc. Such experimental procedures can cause a complete endocrine stress response by itself, thereby complicating the interpretation of the endocrine response to the actual stressor. In order to avoid such interference, the rats used in our experiments are always carefully adapted to experimental procedures by daily handling (transport, injection, etc.) at least 4 days prior to the experiment. Such adapted rats do not show a significant increase in ACTH, alpha-MSH and beta-ENDi in response to unavoidable experimental procedures (Berkenbosch et al., 1983; 1984).

Effects of acute stress

Expoaure of rats, adapted to experimental procedures as described above, to stressful conditions results in a rapid increase of plasma ACTH concentrations and a concomitant rise in plasma beta-ENDi (Berkenbosch et al., 1983; 1984). This observation is in line with those of others (Guillemin et al., 1977; Rossier et al., 1977) and is usually interpreted by assuming that the corticotrophs release ACTH and beta-endorphin and related peptides simultaneously in response to stress. However, it is not generally known that circulating alpha-MSH levels also rapidly increase after a variety of stress stimuli, indicating that stress also activates peptide secretion from the intermediate lobe (Berkenbosch et al., 1984). Therefore, its seems likely that beta-ENDi released during stress can also partially originate from the melanotrophs.

Analysis of the acute response of plasma alpha-MSH, lead us to conclude that stressors with a strong emotional impact (footshocks, passive avoidance, restraint) are good activators of the intermediate lobe, whereas more systemic stressors (ether, formalin or laparotomy under nembutal anesthesia) are rather poor in this respect (Berkenbosch et al., 1984). Studies with neurointermediate lobectomized rats and with dexamethason to selectively suppress corticotroph cell function show that most of circulating beta-ENDi after systemic stress (e.g. formalin) originates from the corticotroph cells, whereas most circulating beta-ENDi after emotional stress (e.g. restraint) originates from tha intermediate lobe (Berkenbosch et al., 1984).

FACTORS MEDIATING STRESS-INDUCED POMC SECRETION

Involvement of corticotropin releasing factor (CRF) and vasopressin (AVP)

Recently, a 41 amino acid has been isolated from bovine and rat hypothalami, that stimulates ACTH release from the anterior pituitary gland *in vivo* as well as *in vitro* (see for review Vale et al., 1983). Subsequently, CRF immunoreactivity has been localized in nerve cells from the paraventricular nucleus and in nerve fibres in the external zone of the median eminence in many species. Two lines of evidence indicate that this CRF peptide plays a key role in both the ACTH and beta-ENDi response to stress. First, anterolateral deafferentation of the hypothalamus, which lead to a loss of CRF-immunostaining in the median eminence (Tilders et al., 1982), completely prevents the ether-induced increase in circulating ACTH and beta-ENDi (Vermes et al., 1981). Second, immunoneutralization of CRF in rats with antisera raised to CRF prevent the increase of ACTH (Linton et al., in press; Tilders et al., 1985) and of beta-ENDi (Tilders, in prep.) to ether, formalin, and laparotomy. In contrast, *in vivo* administration of antisera to CRF partially inhibit peak beta-ENDi levels after emotional stressors, such as restraint stress and footshock, although the ACTH response to such stressors is completely prevented. In addition, no effect of immunoneutralization can be observed on peak levels of alpha-MSH in response to those stressors. Such observations lead us to conclude that CRF does not play a physiological role in the acute response of the intermediate lobe during emotional stress.

Several lines of evidence also indicate that vasopressin is involved in the control of corticotrophic activity (Gillies et al., 1982). Recently, we studied the physiological importance of AVP in stress-induced ACTH release by use of immunoneutralization with an antiserum to AVP (Linton et al., in press; Tilders et al., 1985). Since a marked attenuation is found of the increase in circulating ACTH in response to formalin and restraint, it could be concluded that AVP plays a role in stress-induced ACTH secretion independent of the type of stress stimulus employed.

Involvement of catecholamines

As pointed out above, a special relationship exists between emotional stimuli and reflex activation of the melanotrophs. Emotional stimuli are known to activate the sympatho-adrenomedullary system and lead to a rapid increase in circulating epinephrine (Kvetnansky et al., 1977) to a concentration which by itself can stimulate peptide secretion from the melanotroph (Berkenbosch et al., 1981; Tilders et al., 1982). The action of epinephrine on beta-ENDi and beta-MSH secretion involves beta-adrenoceptors which are known to be present in the intermediate lobe (Cote et al., 1982) and the activation of which leads to an increased peptide secretion (Tilders et al., 1981; Cote et al., 1982). Studies with propranolol to block beta-adrenoceptors lead us to conclude that catecholamines play an important role in stress-induced beta-ENDi and alpha-MSH secretion from the intermediate lobe (Berkenbosch et al., 1983; 1984). Although such studies do not discriminate between involvement of central or peripheral catecholamines, recent studies using rats subjected to extirpation of the adrenal medulla and/or destruction of the sympathetic nerves, suggest involvement of circulating catecholamines in the intermediate lobe response to acute stress(Kvetnansky et al., submitted).

IMMUNOMODULATORY EFFECT OF ENDORPHINS *IN VITRO*

In view of the suggested role of endorphins in immune regulation, we investigated the effect of endorphins on the primary antibody response of B-cells *in vitro*. Isolated human peripheral blood cells consisting of T-lymphocytes, B-lymphocytes and 10-15% monocytes (Boyum, 1968) were cultured with an optimal dose of the antigen ovalbumin (OA; 3 g/ml) (Heijnen et al., 1979) at 37°C in the presence or absence of beta- or alpha-endorphin. Alpha-endorphin represents an enzymatic breakdown product

TABLE 1. Effect of alpha- and beta-endorphin on the number of plaque-forming cells (PFC) in cultures of human peripheral blood.

	CONCENTRATION (M)	PFC/10^6ly*
Control	-	1563 ± 49
alpha-endorphin	0.005	1542 ± 37
	0.05	956 ± 28**
	0.5	656 ± 19**
	5.0	622 ± 25**
beta-endorphin	0.005	1822 ± 41**
	0.05	2978 ± 38**
	0.5	1608 ± 41
	5.0	1269 ± 29

For details of the methodology see Heijnen et al., 1979.

*Number of PFC are given per 10^6 lymphocytes recovered after 6 days of culture. Data represent mean ± SEM (n=3).

**Indicates significant difference in PFC formation between control and endorphin-treated cultures (P 0.01).

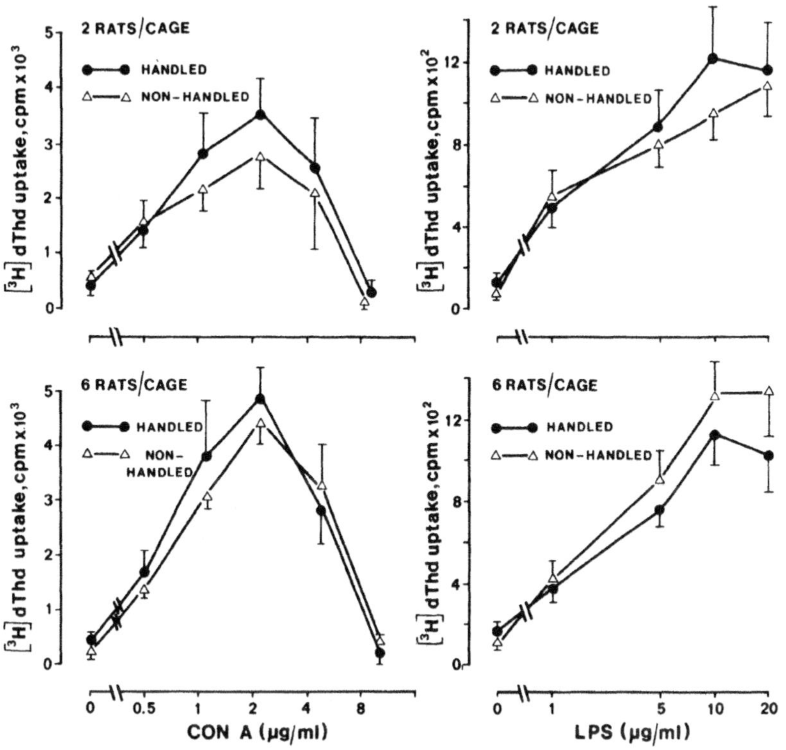

Fig. 1. CON A and LPS proliferation of splenic lymphocytes. Male Wistar rats (140-160 g) were housed two or six per cage in an air-conditioned room at 21-22° C for at least 5 days before the experiment. Food and water were freely available. The room was entered twice daily for adaptation proceures for 4 days before the actual experiment. These adaptation procedures consisted of opening the cage, handling and transfering of animals to an adjacent room. On the day of the experiment, rats were treated identically as described under adaptation procedure except that they were decapitated immediately after entering the adjacent room. Trunkblood was collected in heparinized tubes (0° C) and after centrifugation plasma was aliquoted and stored at -20°C until use or determination of corticosterone and ACTH. The spleen was removed by using sterile scissors and immediately transferred to medium for further processing. Results are expressed as [methyl-^{3}H]-thymidine uptake (cpm) (mean ± SEM; n=4-6).

of beta-endorphin (1-16). When the antibody response of B-cells was studied by enumerating the number of plaque-forming cells after 6 days of culture (PFC) (Heijnen et al., 1979), it could be shown that alpha-endorphin is capable of inhibiting the PFC-response in a concentration of 0.05-0.5 M, whereas beta-endorphin possesses an enhancing effect on the antibody response (see Table I). It is obvious from the results obtained that the influence of alpha- and beta-endorphin on the PFC-response occurs in the presence of relatively high concentrations. This fact does not necessarily implicate that we are dealing with a non-

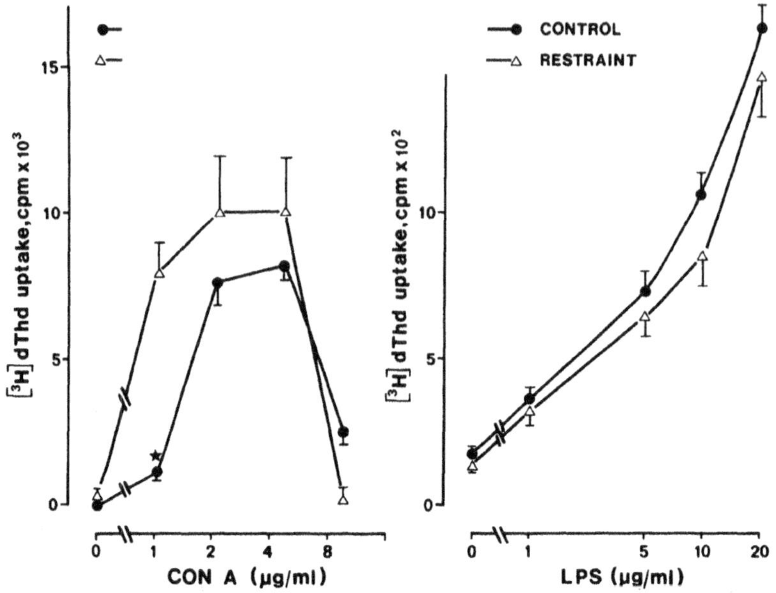

Fig. 2. CON A and LPS induced-proliferation of splenic lymphocytes. Male Wistar rats (140-160 g) were housed six per cage under identical experimental conditions as described in the legend of fig. 1. The adaptation procedure consisted of opening the cage, handling and transfer to an adjacent room. This procedure elsewhere in detail (Berkenbosch et al., 1983) with the exception of a prolonged period of restraining (5 min). Blood and the spleen were obtained in an adjacent room as described in legend 1. Results are expressed as [methyl-^{3}H]-thymidine uptake (cpm) (mean ± SEM; n=6).

physiological effect. The half life of endorphins in medium containing human serum has shown to be short (Burbach et al., 1979), which might explain the need for high concentrations of the peptides in order to exert their regulatory effect. Preliminary studies indicate that one important mechanism by which the endorphins regulate the PFC response is by blocking the production and/or secretion of soluble factors necessary for the differentiation of B-cells into antibody-secreting cells (PFC).

EFFECTS OF ACUTE STRESS ON THE IMMUNE RESPONSE

Experimental conditions

To our knowledge, no systematic studies are available which investigate the reaction of the immune system to short and relatively mild forms of emotional stress, although it can be anticipated that such situations reflect stress stimuli encountered in daily life. Anticipating that the reactivity of the immune system can be influenced by experimental procedures, we first studied the effects of daily handling of rats on the immune response. The reaction of the immune system was tested by measuring the proliferative response of spleen lymphocytes to stimulation *in vitro* with the T-cell mitogen Concanavalin A (CON A) as well as the B-cell mitogen Lipopolysaccharide (LPS) (Vos et al., 1980). As can be concluded from the experiment shown in fig. 1, a 4 day handling procedure does not significantly alter proliferative response to CON A and LPS. In addition, no marked differences are found between groups of rats that were housed 2 or 6 per cage. However, in some experiments handling introduced a slight but significant enhancement of the proliferative response to CON A. Although we cannot explain this observation as yet, it is feasible that manipulation of the non-handled rats sometimes induces a decrease in the proliferative response.

Effects of acute stress

Exposure of handled rats to the adverisve stimulus restraint during 5 minutes results in a marked enhancement of the proliferative response of isolated spleen cells to CON A (fig. 2), indicating that the response of T-cells to CON A is increased immediately in response to the stress stimulus. Since the effect of the stress stimulus was measured immediately after the restraint procedure, it can be deduced that the influence of the stress stimulus on the immune system is almost momentary. In contrast, the B-cell population does not change its reactivity in response to the restraint stress, as can be concluded from the proliferative response to LPS.

Not every type of acute emotional stress evokes an increase in the reactivity of the immune system. For instance, mild emotional stress caused by a retention session of 5 minutes duration, 24 hours after the delivery of an electric footshock in a passive avoidance paradigm induces a decrease in proliferative capacity of T-cells as well as of B-cells (C.J. Heijnen et al., 1985). The suppression is not only apparent in splenic lymphocyte cultures but also in peripheral blood cell cultures. It should be noted, that in control animals receiving only footshocks, a slight but significant increase in T-cell reactivity and no change in B-cell reactivity can be observed, indicating that the suppression of the T- and B-cell reactivity after the retention session is not caused by the footshock delivered 24 hours earlier. In addition, replacement of the retention session by a restraint stress of 5 minutes duration does not result in a decrease but in a dramatic potentiation of proliferating capacity of T-cells without affecting reactivity of the B-cell population. Therefore, the decrease of lymphocyte proliferation after the retention trial is not the result of summation of repeated stimuli but is related to the nature of the retention session itself. From these data the conclusion is warranted that short and mild emotional stress causes immediate changes in the reactivity of the immune system and that the reaction pattern is depending on the nature of the stress stimulus. We hypothesize that the emotional stress experienced in a conflict situation (e.g. passive avoidance) may lead to a prompt reduction of the immune system reactivity , whereas the adversive reaction induced by a sudden novel situation (e.g. restraint, footshock) may lead to a rapid increase in the reactivity of the immune system.

CONCLUDING REMARKS

From data presented in this paper it can be concluded that acute stress clearly affects the proliferative capacity of lymphocytes either originating from the spleen or from the circulation. However, it should be noted that such changes do not necessarily reflect alterations in humoral or cellular immunity. It would be of interest to know whether acute stress affects antibody synthesis or the delayed hypersensitivity to different antigens. Preliminary results in rats indeed indicate that the primary immune response to sheep red blood cells is affected by acute stress (personal observations).

The coincidence between changes in beta-ENDi release and changes in the immune reactivity in response to acute stress may suggest that these phenomena are interconnected. This suggestion finds support in the effects of endorphins on the primary immune response as demonstrated in vitro (Table I). However, it is worth noting that also other substances such as catecholamines, enkephalins and prolactin which are putative modulators of the immune system, are rapidly released in response to acute stress. Therefore, it can be anticipated that the reaction pattern of the immune response to acute stress may be the result of the interaction of many stress-labile substances on the level of the lymphocytes.

REFERENCES

Berezi, I., Nagy, E., Kovacks, K. and Horvath, E., 1981, Regulation of humoral immunity in rats by pituitary hormones, Acta Endocrinol., 98:506.

Berkenbosch, F., Tilders, F.J.H. and Vermes, I., 1983, Beta-adrenoceptor activation mediates stress-induced secretion of beta-endorphin and related peptides from intermediate but not anterior pituitary, Nature, 305:237.

Berkenbosch, F., Vermes, E., Binnkade, R. and Tilders, F.J.H., 1981, Beta-adrenergic stimulation induces an increase of plasma levels of immunoreactive alpha-MSH, beta-endorphin, ACTH and corticosterone, Life Sci., 29:2249.

Berkenbosch, F., Vermes, I. and Tilders, F.J.H., 1984, The beta adrenoceptor blocking drug propranolol prevents secretion of immunoreactive beta-endorphin and alpha-melanocyte stimuli, Endocrinology, 115:1051.

Besedovsky, H. and Sorkin, E., 1977, Network of immunoneuroendocrine interactions, Clin. Exp. Immunol., 27:1.

Boyum, A., 1968, The isolation of lymphocytes with density gradients, Scand. J. Clin. Lab. Invest., suppl. 97:77.

Burbach, H.P.G., Loeber, J.G., Verhoef, J., de Kloet, E.R. and de Wied, D., 1979, Biotransformation of endorphins by a synaptosomal plasma membrane preparation of rat brain and by human serum, Biochem. Biophys. Res. Comm., 86:1296.

Cote, T.E., Eskay, R.L., Frey, E.A., Grewe, C.W., Munemura, M., Stoof, J.C., Tsuruta, K. and Kebabian, J.W., 1982, Biochemical and physiological studies of the beta-adrenoceptors and the D-2 dopamine receptor in the intermediate lobe of the rat pituitary gland: A review, Neuroendocrinol., 35:217.

Del Rey, A., Besedovsky, H.O., Sorkin, E., Da Prada, M. and Arrenbrecht, S., 1981, Immunoregulation mediated by the sympathetic nervous system, II, Cel. Immun., 63:329.

Eipper, R.A., and Mains, R.E., 1980, Structure and biosynthesis of pro- adrenocorticotropin/endorphin and related peptides, Endocrine Rev., 1:1.

Gillies, G.E., Linton, E.A. and Lowry, P.J., 1982, Corticotropin releasing activity of the new CRF is potentiated several times by vasopressin, Nature, 299:355.

Grossman, C.J., 1984, Regulation of the immune system by sex steroids, Endocrine Rev., 5:435.

Grossman, C.J. and Roselle, G.A., 1983, The interrelationship of the HPG-thymic axis and immune system regulation, J. Steroid Biochem., 19 (1B):461.

Guillemin, R.T., Vargo. J., Rossier, J., Minick, S., Ling, M., Rivier, J., Vale and Bloom, F.E., 1977, Beta-endorphin and adrenocorticotropin are secreted concomitantly by the pituitary, Science, 197:1367.
Gungor, M., Genc, E., Sagduyu, H., Eroglu, L. and Koyunoglu, H., 1980, Effect of chronic administration of morphine in primary immune response in mice, Experientia, 36:1309.
Heijnen, C.J., Uytdehaag, F., Gmelig-Meyling, F.J.J., and Ballieux, R.E., 1979,Localization of human antigen-specific helper and suppressor function in distinct T-cell subpopulations, Cellular Immmnol., 43:282.
Heijnen, C.J., Croiset, G., Bevers, C., Veldhuis, H.D., de Wied, D. and Ballieux, R.E., 1985, Interactions between the central nervous system and the immune system. In: New concepts of regulation of autonomic, neuroendocrine and immune systems, H.C. Hendrie, Ed. Martinus Nijhoff, The Hague.
Hofman, K. and Vajima, H., 1961, Studies on polypeptides XX: synthesis and corticotropin activity of a peptide amide corresponding to N-terminal tridecapeptide sequence of the corticotropins, J. Am. Chem. Soc. 83:2289.
Hope, J., Lowry, P.J., 1981, Pro-opiomelanocortin: The ACTH/LPH common precursor molecule, Front. Horm. Res., 8:44.
Johnson, D.L., Ashmore, R.E., and Gordon, M.A., 1981, Effects of beta-adrenergic agents on the murine lymphocyte response to mitogen stimulation, J. Immunopharmac., 3:205.
Kvetnansky, R., Sun, C.L., Lake, C.R., Thoa, N., Torda, R. and Kopin, I.J., 1977, Effect of handling and forced immobilization on rat plasma levels of epinephrine, norepinephrine and dopamine-beta-hydroxylase, Endocrinology, 103:1868.
Kvetnansky, R. Tilders, F.J.H., Van Zoest, I.D., Dobrakovova, M., Berkenbosch, F., Culman, J., Zeman, P. and Smelik, P.G., Sympatho-adrenal activity does not potentiate ACTH secretion but facilitates beta-endorphin and alpha MSH secretion during immobilization stress, submitted.
Linton, E.A., Tilders, F.J.H., Hodgkindon, S., Berkenbosch, F., Vermes, I. and Lowry, P.J., Stress-induced secretion of ACTH in rats is inhibited by administration of antisera to bovine corticotropin releasing factor and vasopressin, Endocrinology, in press.
Lopker, A., Abood, L.G., Moss, W. and Lionetti, F.J., 1980, Stereoselective muscarinic acetylcholine and opiate receptors in human phagocytic leukocytes, Biochem. Pharmacol., 29:1361.
Mains, R.E. and Eipper, R.A., 1981, Differences in posttranslational processing of beta-endorphin in rat anterior and intermediate pituitary, J. Biol. Chem., 256:5683.
Michael, S.D., Taguchi, O. and Nishizuka, Y., 1980, Effect of neonatal thymectomy on ovarian development and plasma LH, FSN, GH and PRL in the mouse, Biol. Reprod., 22:343.
Munck, A., Guyre, N.G. and Holbrook, N.G., 1984, Physiological functions of glucocorticoids in stress and their relation to pharmacological actions, Endocrine Rev., 5:25.
Riley, V., 1981, Psychoneuroendocrine influences on immunocompetence and neoplasia, Science, 212:1100.
Roberts, J.L., Chen, C.C., Eberwine, J.J., Eviner, M.J.A., Herbert, E. and Schachter, R.S., 1982, Glucocorticoid regulation of pro-opiomelanocortin gene expression in rodent pituitary, Rec. Prog. Horm. Res., 38:227.
Rossier, J., Grensch, E.D., Rivier, C., Ling, N., Guillemin, R. and Bloom, F.E., 1977, Footshock induces stress and increases beta-endorphin levels in blood but not in brain, Nature, 270:618.
Seizinger, B.R. and Hollt, V., 1980, In vitro biosynthesis and N-acetylation of beta-endorphin in pars intermedia of the rat pituitary gland, Biochem. Biophys. Res. Comm., 96:535.
Selye, H., 1980, ed. Selye's guide to stress research, Vol. I., Van Nostrand Reinhold Co., New York.

Smyth, D.G., Massey, D.E., Zakarian, S. and Finnie, M.D.A., 1979, Endorphins are stored in biologically active and inactive forms: isolation of alpha-N-acetyl peptides, Nature, 279:252.
Tilders, F.J.H., Berkenbosch, F., and Smelik, P.G., 1982, Adrenergic mechanisms involved in the control of pituitary-adrenal activity in the rat: a beta-adrenergic stimulatory mechanism, Endocrinology, 110:114.
Tilders, F.J.H., Berkenbosch, R. and Vermes, I., 1984, Stress-induced secretion of immunoreactive beta-endorphin in rats: Role of a beta-adrenoceptor mechanism, In: Central and perpheral Endorphins: Basic and clinical effects, E.E. Muller and A.R. Ganazzani, eds., 115.
Tilders, F.J.H., Berkenbosch, F., Vermes, I., Linton, E.A. and Smelik, P.G., 1981, Beta-adrenergic stimulation of the release of ACTH- and LPH-related peptides from the pars intermedia of the rat pituitary gland, Acta Endocrinol., 97:343.
Tilders, F.J.H., Schipper, J., Lowry, P.J. and Vermes, I., 1982, Effect of hypothalamic lesions on the presence of CRF-immunoreactive nerve terminals in themedian eminence and on the pituitary-adrenal response to stress, Reg. Pept., 5:77.
Vale, W., Rivier, C., Brown, M., Spiess, J., Koob, G., Swanson, L., Bilezikjan, L., Bloom, F. and Rivier, J., 1983, Chemical and biological characterization of corticotropin releasing factor. Rec. Progr. Horm. Res., 39:245.
Vermes,I., Berkenbosch, F., Tilders, F.J.H. and Smelik, P.G., 1981, HYpothalamic deafferentation in the rat appears to discriminate between the anterior lobe and intermediate lobe response to stress, Neurosci. Lett., 27:89.
Vos, J.G., Kreeftenberg, J.G., Kruijt, B.C., Kruizinga, W., and Stierenberg, P.A., 1980, The athymic rat. II Immunological characteristics, Clin. Immunol. Immunopathol., 15:229.
Wybran, J., Appelboom, T., Famaey, J.P. and Govaerts, A., 1979, Suggestive evidence for receptors for morphine and methionine-enkephalin on normal human blood T-lynphocytes, J. Immunol., 123:1068.

A COMPLETE REGULATORY LOOP BETWEEN THE IMMUNE AND NEUROENDOCRINE SYSTEMS OPERATES THROUGH COMMON SIGNAL MOLECULES (HORMONES) AND RECEPTORS

Eric M. Smith and J. Edwin Blalock

University of Texas Medical Branch
Galveston, Texas 77550

INTRODUCTION

A link between an individual's psyche and ability to resist disease has been noted for a number of years but was largely based on anecdotal and phenomenological reports. While many mechanisms are possible, a large body of evidence has recently accumulated suggesting a circuit between the nervous, endocrine and immune systems, that operates in both directions. Both structural and functional evidence supports this relationship. Structural studies have shown the thymus, spleen, bone marrow, and possibly other lymphoid organs are innervated with both afferent and efferent nerve fibers (for review see reference 1). Specific perturbations of the central nervous system (CNS) such as brain lesioning (2) or sympathectomy (3) will enhance or suppress immune responses. Also, alterations in noradrenaline turnover and neuron firing rates have been observed in the hypothalamus at the height of an immune response (4,5). In addition to these "hard wired" connections, modulation of immune responses may occur indirectly through neuroendocrine hormones and neurotransmitters such as corticotropin (ACTH), endorphins (END), enkephalins (ENK), vasoactive intestinal peptide (VIP), as well as others (6-9; for review see 10). These hormones and neurotransmitters have been shown to modulate antibody production (7,11), T lymphocyte rosetting (6,12), natural killer (NK) cell activity (13-15), response to mitogens (16), and other examples reviewed in reference 10.

The interaction appears to be a complete circuit in that products of the immune system can also modulate nervous and neuroendocrine system responses. One way this appears to occur is through the production of neuroendocrine hormone-related peptides by stimulated lymphocytes. We recently found that virus-infected lymphocytes produce immunoreactive (ir) ACTH and END that appear identical to their pituitary counterparts on the basis of structure and function (17,18). Lolait et al. (19) found that non-stimulated mouse macrophages also synthesize ir ACTH and END. Furthermore, other products of the immune system such as the thymic hormone, thymosin (α-1) (20), or noncharacterized factors (21) are able to induce an endocrine response such as adrenal gland steroidogenesis. The production of neuroendocrine hormones by leukocytes and the effect of these hormones on primary antibody responses are major interests of our laboratory and thus will be the subject of this review. Two *in vivo* demonstrations of these responses will be described as an attempt to put this phenomena in a physiologic perspective.

Synthesis of Lymphocyte-derived Immunoreactive (ir) ACTH and ENDs

Our prototype and best studied system of lymphocyte-derived hormones involves the production of the proopiomelanocortin (POMC) products ACTH and END in Newcastle disease virus (NDV)-infected lymphocytes (17,18). Initial screening for production of hormones by lymphocytes was accomplished by immunofluorescent staining with a variety of monospecific antisera to neuroendocrine hormones. Six to eight hours post infection with NDV, the lymphocytes began to stain positive for both ACTH and END (17). Interferon-α (IFN-α) was synthesized concomitantly with the ACTH and END (17). Presumably both T and B lymphocytes can synthesize these POMC-related products because, if 100% of the cells are infected and producing NDV antigens, 100% of the cells also synthesize ACTH and END. Lolait et al. (19) found ACTH and END were also produced constituitively by nonstimulated macrophages. If different inducers are used such as foreign cells, which primarily stimulate B lymphocytes (17,22), only 20-30% of the cells stained positive with anti-ACTH antiserum. The B-cell mitogen, lipopolysaccharide (LPS), also induces 20-30% of the lymphocytes to produce ACTH and END . Therefore, the majority of leukocytes appear able to synthesize ACTH and END when properly stimulated. These molecules are synthesized *de novo*, since the ir hormones can be inhibited by induction in the presence of actinomycin D and cycloheximide (23). Further evidence of *de novo* synthesis is that the peptides can be intrinsically labeled by inducing the lymphocytes in the presence of radiolabeled amino acids (18).

Structurally, the lymphocyte-derived ACTH and END are very similar if not identical to their pituitary counterparts. irACTH and pituitary ACTH share: antigenicity, molecular weight, resistance to inactivation by acid, and migration time on reverse phase HPLC (18). Sizing on SDS polyacrylamide gels shows the molecular weight of the ACTH-like molecule is 4.5 kd, which is identical with the full 39 amino acid ACTH molecule (17). The irEND, like β-END, has an apparent molecular weight of 3.6 Kd (24).

Functionally, lymphocyte-derived ACTH is also very similar to pituitary ACTH. The ir ACTH induces both steroidogenesis in mouse adrenal tumor cells and morphologic change (rounding) (17,24). Lymphocyte END also is biologically active, binding to opiate receptors both *in vitro* and *in vivo* (17,25). When analyzed by polyacrylamide gel electrophoresis (PAGE) the ir END had a molecular weight less than 5 kd. *In vivo* intracerebral inoculation of mice with the ir END induced a transient analgesia and catatonia (25). These effects could be immediately reversed or prevented by the opiate antagonist naloxone. Therefore, during an infection lymphocytes may be able to affect pain sensations, temperature regulation, neurotransmission, or other opiate-mediated functions by secretion of END.

Although, initially we thought the ACTH and IFN-α were products of the same gene (26), it is clear now that they are products of different genes which are induced concomitantly. The evidence for this is three-fold. First, the published cDNA sequences of the different IFN-α mRNA species do not contain any ACTH-related sequences. Secondly, the two polypeptides can be dissociated by acid treatment with full retention of IFN's antiviral activity and ACTH's steroidogenic activity (17). Finally, lymphocyte production of ACTH and END can be blocked by the synthetic glucocorticoid, dexamethasone, with no effect on IFN-α synthesis (23). Thus, the two genes can be differentially regulated. It appears that the same POMC gene expressed in pituitary cells is induced in leukocytes. POMC-related RNA can be detected in the BCL_1, B-cell line and in LPS stimulated murine splenocytes (27). This RNA is polyadenylated, but approximtely 200 base pairs longer than mRNA extracted from the AtT-20 pituitary tumor cell line.

Therefore, except for the final, conclusive characterization by sequence analysis, the lymphocyte ACTH and END polypeptides appear identical to their pituitary counterparts and to be products of the same gene.

Hormone production by the immune system is not limited to the POMC derived peptides but also includes immunoreactive (ir) thyrotropin (TSH) (28), vasoactive intestinal peptide (VIP)(29) and somatostatin (30). This laboratory is currently trying to determine the spectrum of peptide hormones which are made by the immune system and the stimuli which elicit their production. Preliminary immunohistochemical studies suggest that leukocytes, in addition to ACTH, END, and TSH also have the potential to produce chorionic gonadotropin (CG) when stimulated in a mixed lymphocyte culture (Table 1).

A Lymphoid Adrenal Axis

If the lymphocyte derived ACTH is biologically relevant, it should be able to replace pituitary ACTH and induce a "stress" response, such as adrenal corticosteroid release. This hypothesis was tested in a deletion type of experiment in which hypophysectomized mice were inoculated with NDV to induce ir ACTH. As a measure of stress the animal's serum was assayed for increases in corticosterone levels. There was a time-dependent increase in corticosterone and interferon which peaked approximately 8 h post inoculation (31). The splenocytes from the NDV infected hypophysectomized mice stained positive for both ACTH and END whereas control mouse splenocytes had no specific staining with either antiserum. Interestingly, if the mice were given dexamethasone *per os* 12h prior to the NDV there was no significant rise in circulating corticosterone and the splenocytes did not stain by IF with antiserum to ACTH. This experiment suggested that stimulated splenocytes can produce the ir ACTH *in vivo* and it is synthesized in biologically significant quantities. Furthermore, the inhibition by dexamethasone suggested that as in the pituitary, ACTH production by lymphocytes is under negative control by glucocorticoids (23,31). These findings strongly suggest the existence of a lymphoid-adrenal axis. An essentially identical time course of corticosterone production in response to NDV was observed in intact mice (data not shown). A similar system has also been observed in chickens (32) and humans. Specifically, we have preliminary data which shows that

Table 1. Leukocytes Produce Different Hormones Depending upon the Stimulus

Hormone	Stimulus	Reference
ACTH, END	Virus (NDV, HSV), Foreign cells, LPS Staph Protein A[a], CRF	17,18,19,24
TSH	T-lymphocyte mitogens (SEA[b])	28
CG	MLC[c]	-

[a]Staphylococcus protein A
[b]Staphylococcus enterotoxin A
[c]Mixed lymphocyte culture

typhoid vaccine induces lymphocyte ir ACTH in both normal and hypopituitary individuals (W.J. Meyer, E.M. Smith, G.E. Richards, A. Cavallo, A.C. Morrill, and J.E. Blalock, manuscript submitted for publication).

Preliminary studies of the effect of immunosuppression on virus induced increases in corticosterone in normal mice were done to determine whether the immune system or the pituitary gland caused the response. Beginning at birth, mice were depleted of B lymphocytes by treatment with rabbit antibody to the μ chain of mouse IgM. Whereas herpes simplex virus (HSV) infection of the eye caused a doubling in corticosterone levels in 5 out of 6 control mice, only 1 out of 6 of the B lymphocyte depleted mice showed an elevation of corticosterone relative to the uninfected controls. Thus, in addition to NDV, HSV can cause an increase in corticosterone levels in normal mice and this response seems to be mediated through lymphocytes rather than the pituitary gland (C. Jordan, E.M. Smith, G.J. Stanton, W.J. Meyer, and J.E. Blalock, unpublished observation).

Lymphocyte Derived END May be a Mediator in Endotoxic Shock

A pathogenic mechanism in which lymphocyte END may be a component has recently come to our attention. It seems that both lymphocytes and ENDs are involved in the pathophysiologic changes that occur during endotoxic shock. This is based on data showing that both naloxone (33), an opiate antagonist, and lymphocyte depletion (34) blocked the physiologic changes induced by LPS (endotoxin) administration. We wondered whether LPS induced lymphocyte END production as was seen with virus infection (17) and if that could be a mediator of endotoxic shock. Therefore, leukocytes were examined for the ability to produce ir END in response to LPS (35). Human peripheral blood lymphocytes and mouse spleen cells were cultured with LPS for 48h in the presence of radiolabeled amino acids. Culture supernatant fluids were passed over an anti-γ-END immunoaffinity column and the bound material sized by gel filtration chromatography. Radiolabeled ir END which eluted from the affinity column co-migrated with α and γ-END molecular weight standards at approximately 1800 daltons. The ir END also competed in a dose dependent fashion with dihydromorphine for opiate receptors in the in vitro receptor assay (25). These results indicated that LPS can stimulate lymphocytes to synthesize ir END. Thus, lymphocytes may be a peripheral source of END and responsible for some of the pathophysiologic changes induced during endotoxic shock. The result that a 1800d END was induced by LPS rather than the 3500d β-END size molecule seen in virus and CRF stimulated lymphocytes and a 2400d ACTH versus a 4500d molecule suggest that different stimuli may induce different processing mechanisms in the lymphocytes.

Immunoregulation by Neuroendocrine Hormones

Our laboratory and others have examined several neuroendocrine hormones for the ability to modulate immune responses. Our reasons for this were two-fold. One was to determine if pituitary hormones have a direct effect on the immune system. The other reason was the possibility that the ir hormones produced by leukocytes were immunoregulatory. Due to the lack of copious quantities of the ir hormones, their impure state, and the close similarity both functionally and structurally, pituitary derived hormones were used for these studies rather than the lymphocyte products. Both ACTH (1-39) and α-endorphin at a concentration of 0.5 μM suppressed over 80% of the in vitro antibody response to the T cell dependent antigen, sheep erythrocytes (SRBC) (7). ACTH also suppressed antibody production to a T-cell independent antigen, dinitrophenyl-Ficoll. Interestingly, only the full (39 amino acid) length ACTH suppresses the in vitro antibody response. Neither ACTH (1-24) or ACTH (1-13) are significantly active in this system (36). Furthermore, only the full ACTH (1-39) and not any shorter fragments will suppress the production of the lymphokine IFN-γ (36). In contrast to α-END, β and γ-END were minimal inhibitors (approximately 20% suppression) at 5-6 μM. However, (Met) and (Leu) enkephalins showed moderate suppression (approximately 60% suppression) at 0.2-2 μM. The interaction appeared to be specifically mediated through opiate receptors since naloxone and β-endorphin blocked the α endorphin mediated suppression of antibody production.

In contrast to ACTH and END, thyrotropin (TSH) enhances the *in vitro* antibody response to SRBC (11). Therefore, the alteration of each immune response is dependent upon the type and length of the hormone.

The immunomodulation appears to be mediated through specific receptors on the lymphocytes. Binding studies with ^{125}I-labeled ACTH suggest that mouse splenocytes possess approximately 3,000 high affinity (Kd 0.1 nM) and 50,000 lower affinity (5 nM) binding sites per average cell (7). This compares favorably with binding studies using isolated purified mouse adrenal cell ACTH receptors which had high and low affinity binding sites with Kds of 0.03 nM and 1 nM respectively (Bost and Blalock, submitted for publication). Binding studies with (^{3}H)-met-enkephalin showed mouse spleen cells possessed approximately 17,000 high affinity (kd=5.9×10^{-10} M) receptors per cell. The receptor binding assay was a membrane preparation of whole mouse spleen so the number of receptor sites per cell is an average of the whole spleen cell population. More than likely future studies of pure cell populations could demonstrate a greater number of receptors on limited cell types. Preliminary structural studies of the lymphocyte ACTH receptor show that it is composed of four subunits, totaling 225 Kd for the intact receptor and is very similar to adrenal cell ACTH receptors (Bost and Blalock, manuscript submitted for publication).

Thus it appears that immunoregulation by neuroendocrine hormones occurs through specific interaction with receptors, and varies with the particular hormone. Also, these results suggest that lymphocyte-derived hormones may play a role in regulating immune responses. The differential effects of various sized ACTH and END species puts a special emphasis on the apparent differential processing of the lymphocyte-derived hormones in response to different inducers. It suggests that differential processing of these molecules by the lymphocyte may be a way for inducers to alter in varied ways immune responsiveness.

Hypothalamic Modulation of the Immune System

During the *in vivo* studies of irACTH production in response to typhoid vaccine, we found that insulin tolerance testing would also induce irACTH synthesis. Since neither insulin or hypoglycemia treatment of lymphocytes was effective, corticotropin releasing factor (CRF) was evaluated as a possible mediator. *In vitro*, synthetic CRF was observed to cause the *de novo* synthesis and release of leukocyte derived ACTH and β-endorphin (24). While it occurred at about 10 fold higher concentrations, arginine vasopressin (AVP) alone was also observed to have intrinsic CRF activity. At concentrations (1-10nM) that are frequently used on cultured pituitary cells, CRF and AVP together acted in an additive fashion to induce these proopiomelanocortin (POMC) derived peptides and such induction was blocked by dexamethasone. Thus, leukocytes seem quite similar to anterior pituitary cells with respect to control of the POMC gene by positive hypothalamic signals (CRF and AVP) and feedback inhibition by a synthetic glucocorticoid hormone.

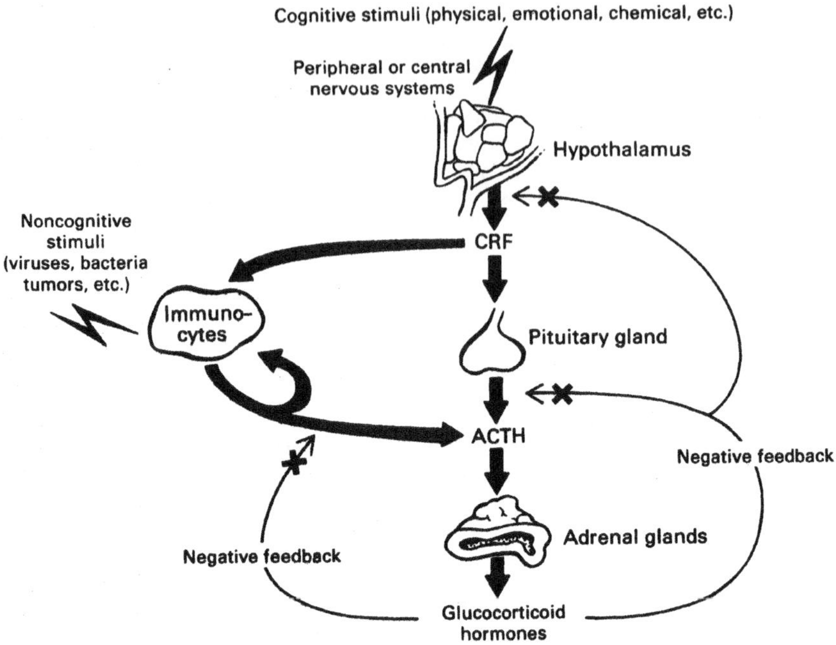

Fig. 1. A regulatory circuit between the immune and central nervous systems mediated by ACTH reproduced with permission from Blalock et al. (37).

A Complete Regulatory Circuit Betweens the Immune and Neuroendocrine Systems

The synthesis of neuroendocrine hormones and possession of high affinity receptors for these same hormones by leukocytes provides the molecules for a bidirectional regulatory circuit between the immune and neuroendocrine systems. The ability of pituitary-derived ACTH, endorphins, enkephalins, and adrenal corticosteroids to modulate immune responses represents one direction of the circuit. Production of ir ACTH and ENDs by lymphocytes, and their contribution to a stress response, represents a pathway in the opposite direction and completes the loop. Figure 1 is a simplified model depicting how a pathway between the immune and neuroendocrine system could fit into the classic hypothalamic-pituitary-adrenal axis. Because the hormones produced by each system can feed back and regulate their own production, intrasystem regulatory loops are included in the model. For example, a stimulus such as a virus infection would induce lymphocyte ACTH and result in adrenal gland steroidogenesis. Glucocorticoids alter host homeostasis as well as suppress immune responses (for review see 10). Furthermore, the lymphocyte ACTH may also suppress the immune response (7,36). Thus, a generalized alteration in host homeostasis and immune response could be affected through activation of the neuroendocrine system. Since different hormones are produced by leukocytes, the particular alteration in homeostasis would be a function of the induction stimulus. Another aspect of this model is that since hypothalamic products will activate leukocytes, there is a mechanism through which the central nervous system can directly modulate the immune system.

An indication that these interactions do occur and are physiologically relevant is that adrenalectomized rats subjected to stress demonstrate a suppressed immune response (38). This suggests that hypothalamic or pituitary hormones are directly immunomodulatory in vivo as well as in vitro (7,10). A naturally occurring case of leukocyte ACTH production in man has recently been reported by Dupont et al. (39). An individual with the clinical symptoms of ectopic ACTH syndrome was found to have normal inflammatory tissue which synthesized ACTH. Removal of the tissue dropped the individual's ACTH levels to normal and immunofluorescent staining showed the tissue to contain leukocytes positive for ACTH. Thus, in at least one instance leukocyte production of ACTH appears to be involved in a disease state.

In conclusion, it seems that the immune and neuroendocrine systems can interact in a regulatory manner by virtue of common ligands and receptors. Through these molecules, the immune system may be able to serve a sensory role, alerting the host of infections and tumors (40). Furthermore, an understanding of these interactions may help explain the pathophysiology of certain diseases and lead to new therapies.

REFERENCES

1. K. Bulloch, Neuroanatomy of lymphoid tissue: A review, in: "Neural Modulation of Immunity," R. Guillemin, M. Cohn, and T. Melnechuk, eds., Raven Press, New York (1983).
2. R.J. Cross, W.R. Markesbery, W.H. Brooks, and T.L. Roszman, Hypothalamic-immune interactions. I. The acute effect of anterior hypothalamic lesions on the immune response, Brain Res. 196:79 (1980).
3. H.O. Besedovsky, A. Del Rey, E. Sorkin, M. Da Prada, and H.H. Keller, Immunoregulation mediated by the sympathetic nervous system, Cell. Immunol. 48:346 (1979).
4. H.O. Besdovsky, A. Del Rey, E. Sorkin, and M. Da Prada, The immune response evokes changes in brain noradrenergic neurons, Science 221:564 (1983).
5. H.O. Besedovsky, E. Sorkin, D. Felix, and H. Haas, Hypothalamic changes during the immune response, Eur. J. Immunol. 7:323 (1977).
6. J. Wybran, T. Appelbroom, J.P. Famaey, and A. Govaerts, Suggestive evidence for receptors for morphine and methionine-enkephalin on normal blood T lymphocytes, J. Immunol. 123:1068 (1979).
7. H.M. Johnson, E.M. Smith, B.A. Torres, and J.E. Blalock, Neuroendocrine hormone regulation of in vitro antibody production, Proc. Natl. Acad. Sci. USA 79:4171 (1982).
8. R.E. Faith, N.P. Plotnikoff, and A.J. Murgo, Effects of opiates and enkephalins-endorphins on immune functions, Proc. Natl. Inst. Drug Abuse Technical Meeting on Mechanisms of Tolerance and Dependence (in press, 1983).
9. E.A. Beed, S. O'Dorisio, T.M. O'Dorisio, and T.S. Gaginella, Demonstration of a functional receptor for vasoactive intestinal polypeptide on Molt 4b T lymphoblasts, Regulatory Peptides, 6:1 (1983).

10. J.E. Blalock, E.M. Smith, and W.J. Meyer, III, The pituitary-adrenocortical axis and the immune system, in: "Clinics in Endocrinology and Metabolism," G.M. Besser and L. Rees, eds., W.B. Saunders Co. Ltd., East Sussex (in press, 1985).
11. J.E. Blalock, H.M. Johnson, E.M. Smith, and B.A. Torres, Enhancement of the in vitro antibody response by thyrotropin, Biochem. Biophys. Res. Commun. 125:30 (1985).
12. G.C. Miller, A. Murgo, and N.P. Plotnikoff, Enkephalins-enhancement of active T-cell rosettes from lymphoma patients, Clin. Immunol. Immunopath. 26:446 (1983).
13. H.W. McCain, I.B. Lamster, J.M. Bozzone, and J.T. Grbic, β-endorphin modulates human immune activity via non-opiate receptor mechanisms, Life Sciences 31:1619 (1982).
14. P.M. Mathews, C.J. Froelich, W.L. Sibbitt, Jr., and A.D. Bankhurst, Enhancement of natural cytotoxicity by β-endorphin, J. Immunol. 130:1658 (1983).
15. N. Kay, J. Allen, and J.E. Morley, Endorphins stimulate normal human peripheral blood lymphocyte natural killer activity, Life Sciences 35:53 (1984).
16. S.C. Gilman, J.M. Schwartz, R.J. Milner, F.E. Bloom, and J.D. Feldman, β-endorphin enhances lymphocyte proliferative responses, Proc. Natl. Acad. Sci. USA 79:4226 (1982).
17. E.M. Smith and J.E. Blalock, Human lymphocyte production of ACTH and endorphin-like substances: Association with leukocyte interferon, Proc. Natl. Acad. Sci. USA 78:7530 (1981).
18. J.E. Blalock and E.M. Smith, A complete regulatory loop between the immune and neuroendocrine systems, Fed. Proc. 44:108 (1985).
19. S.J. Lolait, A.T.W. Lim, B.H. Toh, and J.W. Funder, Immunoreactive β-endorphin in a subpopulation of mouse spleen macrophages, J. Clin. Invest. 73:277 (1984).
20. N.R. Hall, J.P. McGillis, B.L. Spangello, D.I. Healy, G.P. Chrousos, H.M. Schulte, and A.I. Goldstein, Thymic hormone effects on the brain and neuroendocrine circuits, in: "Neuromodulation of Immunity," P. Guillemin, M. Cohn, and T. Melnechuk, eds., Raven Press, New York (1985).
21. H.O. Besedovsky, A. E. Del Rey and E. Sorkin, What do the immune system and the brain know about each other,? Immunol. Today 4:342 (1983).
22. D.A. Weigent, M.P. Langford, E.M. Smith, J.E. Blalock, and G.J. Stanton, Human B lymphocytes produce leukocyte interferon after interaction with foreign cells, Infect. Immun. 32:508 (1981).
23. J.E. Blalock and E.M. Smith, Human lymphocyte production of neuro-endocrine hormone-related substances, in: "Human Lymphokines," A. Khan and N.O. Hill, eds., Academic Press, New York (1982).
24. E.M. Smith, A.C. Morrill, W.J. Meyer, and J.E. Blalock, Corticotropin releasing factor induction of leukocyte derived immunoreactive ACTH and endorphins, Nature (in press, 1985).
25. J.E. Blalock and E.M. Smith, Human leukocyte interferon (HuIFN-α): Potent endorphin-like opioid activity, Biochem. Biophys. Res. Com. 101:472 (1981).
26. J.E. Blalock and E.M. Smith, Human leukocyte interferon: Structural and biological relatedness to adrenocorticotropic hormone and endorphins, Proc. Natl. Acad. Sci. USA 77:5972 (1980).
27. W.A. Kuziel, K.H Brooks, E.S. Vitetta, J.W. Uhr, and P.W. Tucker, Synthesis of pro-opiomelanocortin-related messenger RNA in normal and neoplastic B lymphocytes from the mouse, Science (in press, 1985).
28. E.M. Smith, M. Phan, D. Coppenhaver, T.E. Kruger, and J.E. Blalock, Human lymphocyte production of immunoreactive thyrotropin, Proc. Natl. Acad. Sci. USA 80:6010 (1983).
29. M.S. O'Dorisio, T.M. O'Dorisio, S. Cataland, and S.P. Balcerzak,

Vasoactive intestinal peptide as a biochemical marker for polymorphonuclear leukocytes, J. Lab. Clin. Med. 96:666 (1980).
30. J. Lygren, A. Revhaug, P.G. Burhol, K.E. Giercksky, and T.G. Jenssen, Vasoactive intestinal peptide and somatostatin in leukocytes, Scand. J. Clin. Lab. Invest. 44:347 (1984).
31. E.M. Smith, W.J. Meyer, and J.E. Blalock, Virus-induced increases in corticosterone in hypophysectomized mice: A possible lymphoid-adrenal axis, Science 218:1311 (1982).
32. H.S. Siegel, N.R. Gould, and J.W. Latimer, Splenic leukocytes from chickens injected with *Salmonella pullorum* antigen stimulate production of corticosteroids by isolated adrenal cells, Proc. Soc. Exper. Biol. Med. 178:523 (1985).
33. M.T. Curtis and A.M. Lefer, Protective action of naloxone in hemorrhagic shock, Am. J. Physiol. 239:H416 (1980).
34. L.T. Bohs, J.C. Fish, T.H. Miller, and D.L. Traber, Pulmonary vascular response to endotoxin in normal and lymphocyte depleted sheep, Circ. Shock 6:13 (1979).
35. D.V. McMenamin, E.M. Smith, and J.E. Blalock, Bacterial lipopolysaccharide induction of leukocyte derived ACTH and endorphins, Infect. Immun. (in press, 1985).
36. H.M. Johnson, B.A. Torres, E.M. Smith, L.D. Dion, and J.E. Blalock, Regulation of lymphokine (γ-interferon) production by corticotropin, J. Immunol. 132:246 (1984).
37. J.E. Blalock, D.V. McMenamin, and E.M. Smith, Peptide hormone shared by the neuroendocrine and immunological systems, J. Immunol. (in press, 1985).
38. S.B. Keller, J.M. Weiss, S.J. Schleifer, N.E. Miller, and M. Stein, Stress-induced suppression of immunity in adrenalectomized rats, Science 221:1301 (1983).
39. A.G. Dupont, G. Somers, A.C. Van Steirteghem, F. Warson, and L. Vanhaelst, Ectopic adrenocorticotropin production: Disappearance after removal of inflammatory tissue, J. Clin. Endocrin. Met. 58:654 (1984).
40. Blalock, J.E., The immune system as a sensory organ, J. Immunol. 132:1067 (1984).

ACKNOWLEDGEMENTS

The authors wish to thank Dr. Walter J. Meyer, III, Dr. Howard M. Johnson, and Ms. Audrey C. Morrill for their excellent contributions to these studies. This research was supported in part by Office of Naval Research grant N0014-84-K-0486, NIH research grant 1R01 AM 338-39-01A1. We thank Diane Weigent and Rhonda Peake for their skillful typing of this manuscript.

ENDORPHINS: A LINK BETWEEN PERSONALITY, STRESS, EMOTIONS, IMMUNITY, AND DISEASE ?

George S. Solomon, Neil Kay, and John E, Morley

Geriatric Research, Education and Clinical Center
Sepulveda VA Medical Center (11E), 16111 Plummer Street
Sepulveda, CA 91343 and Minneapolis VA Medical Center
Minneapolis, MN 55417

A growing body of evidence - clinical and experimental, human and animal - that has been accumulating over the last 20 years since Solomon and Moos' then-speculative theoretical integration of emotions, immunity and disease (1) pointed to experiential influences on immune function mediated by the central nervous system and the neuroendocrines, neurotransmitters, and neuropeptides controlled by it. The emerging field of psychoneuroimmunology (2), also referred to as neuroimmunomodulation (3), has rapidly expanded over the last few years due, in part, to the discovery that the endogenous opiods (endorphins), which are released from the anterior pituitary during stress, are potent immune system modulators.

Stress and emotions in the onset and course of immunologically resisted (infectious and neoplastic) and mediated (allergic and autoimmune) disease. George Engel proposed the multifactoral biopsychosocial model of all disease (4). Wise clinicians since antiquity have recognized the role of emotions in disease. Galen is reputed to have said that melancholy women are more prone to cancer than sanguine women and Sir William Osler, the "father of modern medicine," remarked that it is as important to understand what is going on in a man's head as his chest in order to predict the outcome of pulmonary tuberculosis. A number of studies point to specific personality traits, stress and relative failure of psychological defenses and coping as predisposing factors relating to the onset and course of autoimmune diseases, particularly rheumatoid arthritis, but also to others such as systemic lupus erythrematosus and myasthenia gravis (5). [Deficiency of suppressor T-cell function is now felt to be related to autoimmunity (6).] Patients with autoimmune disease tend to be described as self-sacrificing, masochistic, conforming, self-conscious, shy, inhibited, perfectionistic, interested in sports, unexpressive of emotion (particularly anger), compliant, concerned about rejection, and prone to anxiety and depression (5). Poor prognosis relates to degree of anxiety, depression, alienation, and failure to cope (7,8). Similarly, personality studies of patients with cancer tend to report loss of an important relationship prior to development of the tumor with unresolved mourning and consequent depression, inability to express hostile feelings and emotions, unresolved tension about a parent figure and sexual disturbances (9,10). Emotional distress resulting from lack of mature acceptance of reality or from failure of denial defenses is associated with rapidity of death from mestastatic disease (11). [Immune resistance to cancer, an area of great complexity, is related to function both of sensitized cytotoxic T-cells and natural killer cells.] Allergies involve an interplay among genetic, psychological (particularly dependency conflict) and immune factors.

Stress, conditioning, early experience and behavioral effects on immunity in animals (12). Given observations on the relationship of emotional distress to onset and course of the diseases to which immune hypofunction or dysfunction can be related, it seemed rational to study the effects of stress and other experiential manipulations on immune function in animals. In 1969 Solomon first reported that overcrowding stress reduced primary and secondary antibody response to a novel antigen in rats (13). This immunosuppression depended on timing of stress in relation to administration of antigen and to the nature of stress, some stressors either having a lesser or absent effect. Duration of stress is relevant. Under prolonged stress, the immune response may be biphasic. In vitro lymphoproliferative response to T-cell and B-cell mitogens from animals subjected to sound stress is first decreased, followed after continuation of stress for 3 weeks by immunoenhancement (14). It is clear that stress can affect both humoral and cellular immunity (15). Controllability of the stressor appears critical since inescapable, but not the same amount of escapable shock ("controlled" by operantly conditioned lever pushing) leads to suppressed mitogen stimulation and NK cell activity (16,17). (Controllability may be an analogy of "coping" in humans.) Intensity of stress is important--graded stressors producing progressively greater immunosuppression (18). Stress in animals can also modify immunologically mediated disease. Group housing stress (high male/female ratio) significantly increaeses the intensity of adjuvant-induced arthritis and the incidence of spontaneous amyloidosis (19,20). Some separation studies in primates may more closely relate to effects of bereavement and depression on immunity in humans. The "agitation-depression reaction" exhibited by infant monkeys upon maternal or peer separation is associated with decreased lymphocyte stimulation by mitogens (21,22,23).

Russian investigators, in the tradition of Pavlov, claimed as early as 1926 that immunity could be influenced by classical conditioning, which, of course, would clearly implicate the CNS in immune regulation (24). [Editorially, we might comment that modulation of a physiological function, such as immunity, by the CNS is different from its control, as suggested by Eastern Europeans from the 1920s through the 1950s, and is also different from the autonomy of a physiological system, as implied by immunologists until very recent times.]

In elegant work reported in 1975, Ader and Cohen conditioned humoral immunosuppression using the immunosuppressive drug cyclophosphamide as the unconditioned stimulus and saccharin as the conditioned stimulus (saccharin alone reducing antibody response after prior pairing with cyclophosphamide) (25). The work has been confirmed in other laboratories including the utilization of other immunosuppressant agents. Cellular immunity and NK cell activity are also subject to conditioning, and conditioned immunosuppression can prolong the lives of mice with the autoimmune disease systemic lupus erythematosus (26,27,28,29).

There is little work demonstrating an ability experientially to enhance immunity. Rat pups picked up for 3 minutes a day from birth until weaning at 21 days show enhanced primary and secondary antibody response as adults (30). Such early experience seems to be "adaptive" behaviorally and physiologically, for example, leading to a briefer but more vigorous adrenal cortical hormone response to stress (31). [Moral: a good infancy promotes happiness and health!] Inbred female mice that spontaneously develop fighting behavior develop smaller virus-induced tumors, that are immunologically resisted, compared to their more placid "identical twin sisters" (32). [Moral: it is beneficial to express your anger even if you're a mouse!] Socially interactive hamsters develop smaller, less invasive melanomas with greater lymphocyte response than do passive animals (L. Temoshok, personal communication, 1985).

Human studies of stress and immunity. There is a considerable literature on the relationship of recent life changes requiring adaptation and coping and deleterious effects on health if such changes are of sufficient magnitude (33). More recently, life change stress has been correlated with reduction of NK cell activity (34). Social support appears to play a protective role in maintainance of health in the face of stress (35). Likewise, emotional "hardiness" characterized by a strong commitment to self, an attitude of vigorousness toward the environment, a sense of meaningfulness, and an internal locus of control (sense of responsibility for one's own destiny) appears "protective" against stress effects on health (36). (These studies have been done regarding health in general without study of intervening physiological mechanisms, including the immune system.) A large numer of studies point to stress events and/or decompensation of psychological defenses and adaptations as related to the onset of allergic, autoimmune, infectious, and neoplastic diseases (37,5,10).

Palmblad and associates carried out a number of experiments on the effects of food and sleep deprivation on immune mechanisms in humans with generally similar findings of immunosuppression, particularly lymphocyte response to PHA (38). New experiments aboard Spacelab 1 suggest that the decreased stimulability of lymphocytes by PHA from astronauts and cosmonauts after spaceflight is due to effects of weightlessness (39). Hypnosis, long observed to be able to alter the "efferent limb" of the immune response, such as skin response to tuberculin, has recently been claimed to alter mitogen responsiveness of lymphocytes from young, highly hypnotizable subjects (40,41). Bartrop was the first to demonstrate immunosuppression (decreased mitogen response) in bereaved spouses, work that has been confirmed in the laboratory of Schleiffer and Stein, who point out such immunosuppression tends to last up to a year (42,43). (That widowers have higher morbidity and mortality that similar aged married men during the year following their wives' deaths is a well-known observation.) In view of the well-accepted psychodynamic link between mourning and depression, first pointed out by Freud, it is not surprising that hospitalized patients with depressive disorders showed diminished lymphocyte responses to 3 mitogens and lower absolute numbers of T and B cells (44). Anxious college students with poor coping skills have reduced NK cell activity compared to asymptomatic students and those with high ego strength, who respond adaptively to life stress events (S. Heisel, personal communication, 1985).

Mechanisms of psychological and experimental effects on immunity

Neuroendocrines and neurotransmitters. The effect of hormones, which are subject to regulation by the hypothalamus, on immune response has been the subject of many reviews (29). Most clearly implicated, of course, have been adrenal corticosteroids. In vitro, low concentrations of corticosteroids stimulate lymphocyte proliferation and high concentrations inhibit (32). (Various influences on immune response, including stress, may be biphasic.) That stress and other experiential influences on immunity are not merely an adrenally-mediated pheonmenon has been made clear by recent work showing stress-induced suppression of immunity in adrenalectomized rats (33). [Earlier, Amkraut and Solomon had found that stress is more immunosuppressive in a graft vs. host model than is ACTH (15).] A variety of other experientially-influenced neuroendocrines affect immune response including growth, thyroid and sex hormones, insulin, the catecholamines, and histamine. Receptors for a number of CNS-controlled hormones and CNS-active compounds have been (and continue to be) found on lymphocytes or thymocytes by a number of investigators [reviewed by Solomon and Amkraut (45)]. Included are receptors for corticosteroids, insulin, testosterone, estrogens, B-adrenergic agents, histamine, growth hormone, acetylcholine, and B-endorphin as well as sites that seem to be receptors for bezodiazepine and haloperidol neuroleptic drugs (45,46,47). Presumably, the presence of receptor site implies a function for its substrate. Some sites have been identified as playing a role in differentiation of lymphocytes and in controlling their activity (48,49).

Catecholamines, which, of course, are both released into the general circulation and are neurotransmitters, have effects on immunity. Infusions of norepinephrine increase the activity of NK cells (50). Soviet work demonstrates that serotonin and 5-hydroxytryptophan (5-HTP) prolong the latent period of the primary immune response and lower the intensity of primary and secondary response (51). Hypophysectomy and lesions of the pituitary stalk abolish these inhibitory effects. Serotonergic structures of the hypothalamus appear to participate in the production of antibody. (Involvement of such mechanisms is likely in any psychological depression-induced immunosuppression.) Administration of the monoamine oxidase inhibitor antidepressant drug iproniazid, as well as serotonin or 5-HTP, delays involvement of T-cells participating in IgG response (helpers) (52). Similar serotonin-related drug effects are found on delayed hypersensitivity (53). There are claims that administration of serotonin can ameliorate autoimmune disease, both multiple sclerosis and its probable analogue, experimental allergic encephalomyelitis, affecting distribution and quantity of suppressor T-cells (54). Repeated mother-infant separations have persistent efects on serotonin levels (55). Growth hormone secretion is provoked by a variety of stimuli (56). Such responsiveness is diminished in patients with depressive illness, in whom (in contrast to animals) growth hormone can be simulated by thyrotropin-releasing hormone (57). Usually ACTH and growth hormone responses to stress are dissimilar (58). The responsivity of growth hormone to stress may be related to such personality characteristics as "neuroticism" (59) and "coronary-proneness" (60). Prolactin is responsive to stress in man, with a secretion pattern different from that of growth hormone (61). Direct measurement of TSH indicates small, if any, increase following psychological stress (62). However, clinical hyperthyroidism can be precipated by an acute emotional stress (63) (although some question exists whether these reports represent the dating phenomenon) and elevated thyroid hormone levels have been reported in patients with a variety of psychiatric diseases (63a). The thyrotropin response to TRH is blunted in up to one third of patients with bipolar depressive disorders (64). Most studies show a clear decrease in testosterone levels during stress, with poststress recovery to normal levels (65,66).

Important issues are how these multiple hormone responses may be interrelated and what their ultimate metabolic effect may be at the target level--in this case, the immunologically competent cell (67). The ultimate metabolic consequences of a given hormone are influenced strongly by the existing overall "hormonal milieu" (68,69). Endocrine responses to an acute stress seem to be organized in a catabolic-anabolic sequence. The result is enhanced availability of energy substrate (glucose) during the stress as a preparation for "fight or flight," and then poststress restoration of protein and other tissue stores (e.g., glycogen). Such a sequence may account for biphasic immunologic responses to stress. Most hormones are secreted in a pulsatile, episodic fashion. Many hormones have 24-hour secretory rhythms; components of the immune system also show a 24-hour periodicity. For example, stimulation of lymphocytes in vitro by mitogens varies directly with the cortisol rhythm (70). Thus, timing of stress in relation to naturally occurring rhythms may influence its immunologic consequences.

Nervous system and immunity: The hypothalamus, a small area of the diencephalon, rich in neural connections to the limbic system of the brain, that has receptors for humoral influences from blood and cerebrospinal fluid and influences the pituitary through a variety of polypeptide releasing factors (TRH, LHRH, CRF, and somatostatin), is at the interface between the brain and a range of critical peripheral regulatory functions. It is rich in neurohormones and neurotransmitters that may affect immune function and has been implicated in immune function by direct research. Early work demonstrated alterations in globulin levels as a result of the electrical stimulation of the lateral hypothalamus of rats (71). Significant Soviet work revealed that a destructive lesion in a specific portion of the dorsal hypothalamus of rabbits led to complete suppression of primary antibody response, prolonged retention of antigen in the blood, and inability to induce streptoccocal antigen myocarditis, as well as prolonged graft retention (72); whereas, electrical stimulation of the same region enhanced antibody resonse (73). Destruction of the posterior hypothalamus aggravates experimental allergic polyneuritis, which is related to the absence of antibodies to myelin that play a protective role (74). Hypothalamic lesions also affect cell-mediated immunity. Anterior lesions in the guinea pig suppress delayed cutaneous hypersensitivity to picrylchloride and tuberculin (75). Lesions decreased and stimulation enhanced delayed cutaneous hypersensitivity (76). Amkraut and Solomon found that electrolytic lesions of the ventromedial and posterior nuclei of the hypothalamus of hybrid rats impaired the graft versus host reaction in the recipient (unpublished data, 1979). Guinea pigs with anterior hypothalamic lesions had significantly smaller cutaneous tuberculin reactions than nonoperated or sham-operated controls; and there was decreased stimulation in vitro of lymphocytes from animals with hypothalamic lesions (77). Epiphysectmony of the chick embryo leads to failure of development of both thymus and bursa (78).

Most interesting recent French work implicates the cerebral cortex with lateral specificity in immunoregulation. Left cortical lesions decrease T-cells and reduce T-mitogen and NK cell responses (79). B-cells and macrophages are not affected by cortical lesions. Autoimmune and atopic diseases are more common in left-handed individuals (80).

Portions of the immune system itself are significantly innervated (81). The thymus has fibers from the vagus and other sources (82). The bone marrow also has a good neural supply (83). Brain lesions affect marrow function (84). The autonomic nervous system also appears significantly involved in modulation of the immune response. Noradrenaline decreases in the spleen, but not in non-lymphoid organs, during the immune response (85). Further evidence that the hypothalamus is directly involved in regulation of immune response is provided by the work of Besedovsky and Sorkin in Switzerland and of Korneva in the U.S.S.R. (86,73). The firing rates of neurons (as determined by implanted electrodes) in the ventromedial nuclei of the hypothalamus increase following immunization (only in animals responding to antigen) (87).

Thymic hormones, important in T-cell maturation and function, are influenced by the CNS and may be true neuroendocrine modulators. Thymosin B 4 elicits release of lutenizing hormone-releasing factor, and, after intracerebroventricular injection, increases serum levels of luetinizing hormone (88). Thymectomy affects the hypophysis (89). Neuroendocrines influence the production of thymic hormones (90). "Lethargic" mutant mice suffer from a neurologic abnormality that develops before weaning, lasts 30-60 days, and then progressively disappears. The neurological disease is associated with thymic atrophy, which is reversed to normal as soon as the neural disturbances disappear (91). Both thymectomized and "nude" mice (that show thymic atrophy) display a profoundly disturbed neuroendocrine balance (92). Indeed, there is growing evidence of bidirectional interactions between the thymus and the neuroendocrine system. There is a report of the extraction from the anterior pituitary of a low molecular weight peptide with thymocyte-stimulating properties (93). Thymosin factor 5 increases levels of ACTH, cortisol, and β-endorphin (94). Hall suggests that thymic peptides down-regulate glucocorticoidal receptors enabling these to counteract potential suppressive effects of glucosteroids upon lymphocytes. Such an effect may occur in the brain as well, influencing the hypothalamic-pituitary-adrenal cortical feedback circuit including corticotropin-releasing factor (CRF), which, in turn, is influenced by serotonergic, adrenergic and cholinergic mechanisms (95).

Feedback mechanisms in immune regulation appear to act, at least in part, via CNS mediation. Serum cortisol levels are elevated in response to antigenic stimulation or graft rejection (presumably via hypothalamically-controlled ACTH) (86,96). This finding suggests a feedback loop between the immune system and the hypothalamic-endocrine system. Such an endocrine response was postulated as an explanation of the phenomenon of antigenic competition, in which response to one antibody inhibits that to another (97). Non-antigen-stimulated thymocytes are more sensitive to inhibitory effects of steroids than are sensitized T-cells, a probably modulating influence preventing inappropriate cells from being overstimulated by lymphokines released following antigenic stimulation (98). Recent work strongly suggesting such feedback is the finding of decreased noradrenaline turnover in the hypothalami of rats at the peak of immune response (99). This decreased noradrenergic activity was mimicked by injection of soluble mediators released by immunologic cells in vivo.

<u>Stress, endorphins and immunity</u>: Stress activates CRF, in turn leading to the release into the circulation of ACTH and β-lipotropin and β-endorphin from the anterior pituitary (100). In 1974 Brown et al. found that peripheral blood lymphocytes of heroin addicts had reduced proliferative responses to mitogens (101). A major interest in endorphins as physiological modulators of the immune system developed with the report by Hazum et al. (102) that the human lymphocyte cell line RPMI 6237 exhibited specific binding of (^{125}I) (D-Ala^2-1 β-endorphin and the finding of Blalock and Smith (103) that human peripheral blood lymphocytes produce a substance with immunoreactivity similar to γ-endorphin. Various studies established that opiod peptides modulate phytohemagglutin (PHA)-induced proliferation of peripheral blood lymphocytes. These findings were responsible for establishing a role for opioid peptides in the modulation of the immune system (104,105,106).

Recently, dynorphin and β-endorphin have been demonstrated to stimulate superoxide production by human polymorphonuclear leukocytes and macrophages in vitro (127). These findings show that a key metabolic response involved in the microbiodal function of phagocytic cells is triggered by opioids at concentrations in the physiological range.

β-endorphin was found to enhance NK activity (107). The dose response curve had an inverted U-shape. Of particular interest was the fact that this response could only be demonstrated in half the subjects tested. However, in those subjects in whom the response was present, it could be reelicited when lymphocytes from these subjects were tested on multiple other occasions. The β-endorphin effect could be reversed by the opioid antagonist, naloxone. β-lipotropin appeared to be equipotent to β-endorphin as an enhancer of NK activity, and γ-endorphin appeared to be more potent. The γ-endorphin effect was also reversed by naloxone. Morphine, α-endorphin and leucine-enkephalin did not significantly alter NK activity in the system developed by Kay and co-workers. Similar results were reported by Mathews et al., who found that both β-endorphin and methionine-enkephalin produced naloxone reversible enhancement of NK cell activity (108). They, too, found no effect of leucine enkephalin, α-endorphin and morphine. Faith et al. did, however, find effects of leucine-enkephalin as well as methionine-enkephalin on NK activity (109).

A number of Des-Tyr-endorphin fragments (2-16, 2-17, 6-17 and 2-9) have proved to be extremely potent stimulators of NK function with activities at concentrations as low as 10^{-15} M. The fragments (10-16) and (14-16) were inactive. These studies suggest that the ability of endorphins to stimulate NK activity rests predominantly in the (6-9) amino acid fragment. The activity, therefore, lies in the α-helix portion of the molecule (110) (Fig.1). However, the fact that methionine-enkephalin can enhance NK activity--combined with the discrepancy noted above between α- and γ-endorphin--suggests that more than one receptor site may be involved.

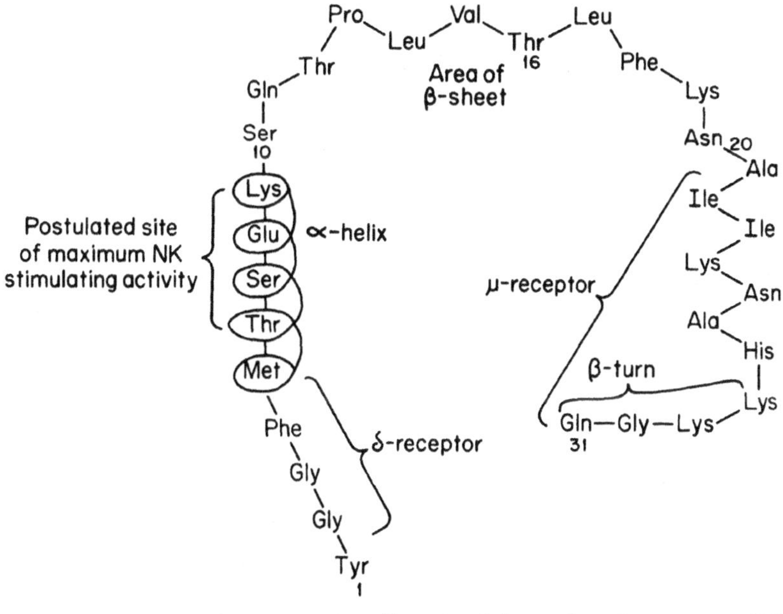

Fig. 1. Postulated sites of the β-endorphin molecule responsible for modulating NK-activity.

Activation of NK activity requires calmodulin (111). Calmodulin is a ubiquitous intracellular calcium binding protein which regulates a variety of intracellular processes and is important in the control of cell shape and cellular secretions. As a number of studies have suggested that opiates exert some of their effects through calmodulin modulation (112,113), Moon and co-workers examined the effect of calmodulin inhibition using the naphthelene sulfonamide W13, on β-endorphin-stimulated NK activity. The β-endorphin enhancement of NK activity was inhibited by W13 suggesting that the opioid stimulation of NK activity may be mediated by calmodulin.

Interferon is a potent stimulator of NK activity. Interferon levels can be altered by experiential manipulation (114). The level of NK enhancement produced by the endorphins is of the same magnitude as that produced by interferon. Interferon has been demonstrated to have a number of opioid-like effects (115). Thus, interferon may produce its enhancement of NK activity through opioid receptors. In support of the hypothesis, it was found that naloxone inhibited the interferon-mediated NK enhancement (107) (Fig. 2).

Interleukin-2 is another NK cell enhancer which has been shown to bind to opioid receptors (116). Preliminary studies suggest that naloxone can partially reverse the effect of interleukin-2 on NK cell activity. Since chronic administration of naltrexone did not affect NK cell activity or stimulability of lymphocytes by PHA, the question of the importance of endorphins in long-term regulation of the immune system remains open (117). Both interferon and interleukin-2 are released after an antigenic exposure or viral infection, making it important to determine if chronic opioid blockade blunts the increase in NK function after such challenges.

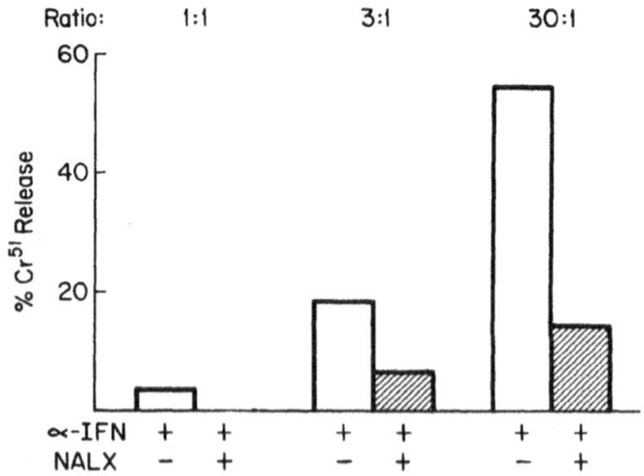

Fig. 2. Effect of naloxone on interferon-mediated NK enhancement.

Recently, the interest in the role of enkaphalins-endorphins in mediating stress effects on immunity has increased considerably. Both ACTH and β-endorphin are part of the precursor molecule pro-opiomelanocortin, and secretion of both is controlled by the hypothalamic corticotropin-releasing factor, CRF (118). In turn, the immune system secretes factors that influence CRF and opiocortin neurons. An anatomical substrate in the CNS exists for participation of peptide systems in autonomic, immune and other physiological homeostatic mechanisms, which tend to be disordered under conditions of stress, in disease and in aging. For example, both ACTH and β-endorphin are released simultaneously from the pituitary by disease or stress (119); the lymphocyte under "stress" (responding to antigen) also releases ACTH (120); the adrenal gland releases enkephalins, catecholamine and steroid hormones during stress (121).

Met-enkephalin-like, morphine-like, and β- and γ-endorphin receptors have been identified on lymphocytes (55). ACTH and β-endorphin can behave like lymphokines (122). As stated by Blalock, "Thus, it appears that the immune and neuroendocrine systems have the ability to signal each other through common or related peptide hormones and receptors." (122).

There is some direct evidence of opioid mediation of stress effects on immunity. Rats subjected to footshock stress develop analgesia; but, depending on its avoidability vs. nonavoidability, such analgesia is or is not reversed by naloxone--thus, is or is not opioid-dependent. Splenic NK cell activity and lymphocyte proliferation by mitogens PHA and ConA are suppressed by the opioid but not the nonopioid form of stress (123,17). Mice injected with enkephalins may have significant increases in thymic size and smaller spleens (125). In addition, BDF_1 mice inoculated with L1210 tumor cells and either Leu- or met-enkephalin have survival advantage over BDF_1 mice with tumor cells alone (126).

Conclusion: Psychoneuroimmunology, the developing field having to do with the effects of emotions and stress on immune function and thus on immunologically resisted and mediated diseases and with the role of the central nervous system (CNS) in modulation of immune function, is rapidly advancing especially because of the unraveling of mechanisms of bidirectional CNS-immune interaction, particularly understanding the role of endogenous opioids as immunomodulators (124).

REFERENCES:

1. Solomon, G.F., Moos, R.H. Emotions, immunity and disease: A speculative theoretical integration. Arch. Gen. Psychiat. 11:556-567, 1964.
2. Ader, R. (ed.). Psychoneuroimmunology, Academic Press, New York, 1981.
3. Guillemin, R., Chone, M., Melnechak, T. (eds.) Neural Modulation of Immunity. Raven Press, New York, 1985.
4. Engel, G.L. A unified concept of health and disease. Perspect. Biol. Med. 3:459-485, 1960.
5. Solomon, G.F. Emotional and personality factors in the onset and course of autoimmune disease, particularly rheumatoid arthritis. In: Ader, R.A. (ed.). Psychoneuroimmunology, Academic Press, New York, 1981, pp. 159-182.
6. Compston, A. Lymphocyte subpopulations in patients with multiple sclerosis. J. Neurol. Neurosurg. Psychiat. 46:105-114, 1983.

7. Moos, R.H. and Solomon, G.F. Personality correlates of the rapidity of progression of rheumatoid arthritis. Ann. Rheum. Dis. 23:145-151, 1969.

8. Solomon, G.F. and Moos, R.H. Psychologic aspects of response to treatment in rheumatoid arthritis. GP 32:113-119, 1965.

9. Le Shan, L.L. and Worthington, R.E. Personality as a factor in the pathogenesis of cancer: Review of the literature. Brit. J. Med. Psychol. 29:49-56, 1956.

10. Bahnson, C.B. Stress and cancer: State of the art. Psychosomatic 21:957-981, 1980.

11. Klopfer, B. Psychological variables in human cancer. J. Prof. Tech. 21:331-340, 1957.

12. Weiner, H. Bronchial Asthma. In: Psychobiology and Human Disease, Elsevier, pp. 223-317, New York, 1977.

13. Solomon, G.F. Stress and antibody response in rats. Int. Arch. Allergy 35:97-104, 1969.

14. Monjan, A.A. and Collector, M.I. Stress-induced modulation of the immune response. Science, 196:307, 1977.

15. Amkraut, A.A., Solomon, G.F., Kasper, P., Purdue, A. Effects of stress and of hormonal intervention on the graft versus host response. Adv. Exper. Med. Biol. 29:667-674, 1972.

16. Laudenslager, M.L., Rayan, S.M., Drugan, R.C., Hyson, R.L., Maier, S.F. Coping and immunosuppression: Inescapable but not escapable shock suppresses lymphocyte proliferation. Science 221:568-570, 1983.

17. Maier, S.F., Laudenslager, M.L., Ryan, S.M. Stressor controllability, immune function and endogenous opiates. Manuscript, 1983.

18. Keller, S., Weiss, J., Schleifer, S. et al. Suppression of immunity by stress: Effect of a graded series of stressors on lymphocyte stimulation in the rat. Science 221:1301-1304, 1981.

19. Amkraut, A., Solomon, G.F., Kraemer, H.C. Stress, early experience and adjuvant-induced arthritis in the rat. Psychosom. Med. 3:203-214, 1971.

20. Ebbesen, P. Spontaneous amyloidosis in differently grouped and treated DBA-2, BALB-C and CMA mice and thymus fibrosis in estrogen-treated BALB-C males. J. Exp. Med. 127:386-396, 1968.

21. Laudenslager, M.L., Reite, M., Harbeck, R.J. Suppressed immune response in infant monkeys associated with maternal separation. Behav. Neurol. Biol. 36:40-48, 1982.

22. Reite, M., Harbeck, R., Hoffman, A. Altered cellular immune response following peer separation. Life Sci 29:1133-1136, 1981.

23. Coe, C.L., Rosenberg, L.T., Levine, S. Immunological Consequences of Maternal Separation in Infant Primates. First international Workshop on Neuroimmunology, Bethesda, MD, November 28, 1984.

24. Metalnikov, S. and Chorin, V. The role of conditioned reflexes in immunity. Ann. Pasteu Inst. 11:1-8, 1926.

25. Ader, R. Behaviorally conditioned immunosuppression. Psychosom. Med. 37:333-340, 1975.

26. Kusnecov, A.W., Sivyer, M.G., King, A.J., Husband, A.W., Cripps, R.L. Behaviorally conditioned suppression of the immune response by antilymphocyte serum. J. Immunol. 189:2117-2120, 1983.

27. Borbjerg, D. and Ader, R. Acquisition and extinction of conditioned suppression of a graft versus host response in the rat. Psychosom. Med. 45:369 (Abstract), 1983.

28. Solvason, H.B., Ghanta, V., Hiramoto, R., Spector, N.H. Natural killer cell activity augmented by classical Pavlovian conditioning. First International Workshop on Neuroimmunology, Bethesda, MD, November 29, 1984.

29. Ader, R. Behaviorally conditioned immunosuppression and murine systemic iupus erythematosus. Science 215:1534-1536, 1982.
30. Solomon, G.F., Levine, S., Kraft, J.K. Early experience and immunity. Nature 220:821-822, 1968.
31. Levine, S. Plasma-free corticosteroid response to electric shock in rats timulated in infancy. Science 135:795-796, 1962.
32. Amkraut, A. and Solomon, G.F. Stress and murine sarcoma virus (Maloney) - Induced tumors. Cancer Res. 32:1428-1433, 1972.
33. Rahe, R.H. Life changes and illness studies: Past history and future direction. J. Hum. Stress 4:3-15, 1978.
34. Locke, S.E., Kraus, L., Leserman, J. Life change stress, psychiatric symptoms and natural killer cell activity. Psychosom. Med. 46:441-453, 1984.
35. Cobb, S. Social support as a moderator of life stress. Psychosom. Med. 38:300-314, 1976.
36. Kobasa, S.C. Stressful life events, personality and health: An inquiry into hardiness. J. Pers. Soc. Psychol. 37:1-11, 1979.
37. Solomon, G.F. and Amkraut, A.A. Emotions, immunity and disease. In: Temoshok, L., Van Dyke, C., Zegans, L.S. (eds.). Emotions in Health and Illness. Theoretical and Research Foundations. New York: Grun & Stratton, pp. 167-186, 1983.
38. Palmblad, J., Petrini, B., Wasserman, J., Akerstedt, T. Lymphocyte and granulocyte reactions during sleep deprivation. Psychosom. Med. 41:273-276, 1979.
39. Cogli, A. and Tschopp, A. Lymphocyte reactivity during spaceflight. Immunol. Today 6:1-4, 1985.
40. Black, S., Humphrey, J.H., Niven, J.S. Inhibition of mantoux reaction by direct suggestion under hypnosis. Br. Med. J. 6:1649-1653, 1963.
41. Hall, H.H. Hypnosis and the immune system: A review with implications for cancer and the psychology of healing. Am. J. Clin. Hypnosis 25:92-103, 1983.
42. Barrop, R.W., Lazarus, L., Luckhurst, E. et al. Depressed lymphocyte function after bereavement. Lancet 1:834-836, 1977.
43. Schleifer, S.J., Keller, S.E., McKegney, F.P., Stein, M. Bereavement and lymphocyte function. Paper delivered at annual meeting of Am. Psychiat. Assn., 1980 and personal communication, 1983.
44. Schleifer, S.J., Keller, S.E., Meyerson, A.T. et al. Lymphocyte function in major depressive disorder. Arch. Gen. Psychiat. 41:484-486, 1984.
45. Solomon, G.F. and Amkraut, A.A. Psychoneuroendocrinologic effects on the immune response. In: Starr, M.P. (ed.) Annual Review of Microbiology 35:155-184, Annual Reviews, Palo Alto, 1981.
46. Weber, R.J., Smith, C.C., Norcross, M.A., Paul, S.M., Pert, C.B. Characterization and Distribution of "Peripheral Type" Benzodiazepine Receptors on Cells of the Immune System. First International Workshop on Neuroimmunology, Bethesda, MD, November 27, 1984.
47. Shaskan, E.G. Probable Lymphoid Cell Sites of Action for Immunomodulation by Haloperidol. First International Workshop on Neuroimmunology, Bethesda, MD, November 27, 1984.
48. Cantor, H. and Gershon, R.K. Immunological circuits. Cellular compositions. Fed. Proc. 39:2058-2064, 1979.
49. Helderman, J.H. and Strom, T.P. Specific binding site on T and B lymphocytes as a marker of cell activation. Nature 274:62-63, 1978.
50. Locke, S., Kraus, L., Kutz, S. et al. Altered Natural Killer Cell Activity During Norepinephrine Infusion in Humans. First International Workshop on Neuroimmunomodulation, Bethesda, MD, November 30, 1984.
51. Devoino, L.V., Eremine, O.F., Ilyutchenok R, Yu, R. The role of

the hypothalamo-pituitary system in the mechanism of action of reserpine and 5-hydroxytryptophan on antibody production. Neuropharmacology 9:67-72, 1970.

52. Devoino, L.V. and Idova, G.V. Influence of some drugs on the immune response. IV. Effect of serotonin, 5-hydroxytryptophan, iproniazid and p-chlorphenylalanin on the synthesis of IgM and IgG antibodies. Eur. J. Pharm. 22:325-331, 1973.

53. Devoino, L.V., Ilyutchenok, R., Yu, . Influence of some drugs on the immune response. II. Effects of serotonin, 5-hydroxytryptophan, reserpine and iproniazid on delayed hypersensitivity. Eur. J. Pharm. 4:449-456, 1968.

54. Abramchik, G.N. Clinical aspects of serotonin treatment for autoimmune diseases. Paper delivered at Soviet Academy of Sciences conference. "Regulation of Immune Homeostasis," Leningrad, 1982 (to be published in USSR, 1984).

55. Mybran, J. Enkephalins and endorphins as modifiers of the immune system: present and future. Fed. Proc. Vol. 44, No. 1, Part 1, pp. 92-94, January 1985.

56. Brown, G.M. and Reichlin, S. Psychologic and neural regulation of growth hormone secretion. Psychosom. Med. 34:45-61, 1972.

57. DeLaFuente, J.R. and Wells, L.A. Human growth hormone in psychiatric disorders. J. Clin. Psych. 42:270-274, 1981.

58. Yalow, R.S., Versano-Sharon, N., Echemendia, E., Berson, S.A. HGH and ACTH secretory responses to stress. Horm. Metab. Res. 1:3, 1969.

59. Miyabo, S., Hisada, T., Asata, T., Mizushima, No, Ueno, K. Growth hormone and cortisol responses to psychological stress. Comparison of normal and neurotic subjects. J. Clin. Endocrinol. Metab. 42:1158, 1976.

60. Friedman, M., Byers, S.O., Roseman, R.H., Newman, R. Coronary-prone individuals (Type A behavior patterns) growth hormone responses. JAMA 217:929, 1971.

61. Mirsky, A. The psychosomatic approach to the etiology of clinical disorders. Psychosom. Med. 19-424-430, 1957.

62. Mason, J.W., Hartley, L.H. Kotchen, T.A., Wherry, F.E., Pennington, L.L., Jones, J.G. Plasma thyroid-stimulating hormone response in anticipation of muscular exercise in humans. J. Clin. Endocrinol. Metab. 37:403, 1975.

63. Mandelbrote, B.M. and Wittkower, E. Emotional factors in Grave's disease. Psychosom. Med. 17:109-117, 1955.

63a. Morley, J.E. and Shafer, R.B. Thyroid function screening in new psychiatric admissions. Arch. Int. Med. 42:591-593, 1982.

64. Linkowski, P., Van Wetere, J.P., Kerhofs, M., Brauman, H., Mendlewicz, J. Thyrotrophin response to thyreostimulin in affectively ill women relationship to suicidal behavior. Brit. J. Psych. 143:401-405, 1983.

65. Kreuz, L.E., Rose, R.M., Jennings, J.R. Suppression of plasma testosterone levels and psychological stress. Arch. Gen. Psych. 26-469, 1972.

66. Hamanaka, Y., Kurachi, K., Aono, T., Mizutani, S., Matsumoto, K. Effects of general anesthesia and severity of surgical streee on serum LH and testosterone in males. Acta. Endocrinol. 65:258, 1975.

67. Mason, J.W. "Over-all" hormonal balance as a key to endocrine organization. Psychosom. Med. 30:791-808, 1968.

68. Mason, J.W. A historical view of the stress field. Hum. Stress 1(1):6, 1(2):22, 1975.

69. Mason, J.W. Levi, L. (ed.) In: Emotions - Their Parameters and Measurement. New York, Raven Press, pp. 143-181, 1975.

70. Tavadia, H.B., Fleming, K.A., Hume, P.D., Simpson, H.W. Circadian rhythmicity of human plasma cortisol and PHA-induced lympho-

cyte transformation. Clinb. Exp. Immunol. 22: 199, 1975.

71. Fessel, W.J. and Forsythe, R.F. Hypothalamic role in control of gamma globulin levels. Arth. Rheum. 6:770 (abstract), 1963.

72. Korneva, E.A. and Khae, L.M. Effects of destruction of hypothalamic areas on immunogenesis. Fiziol ZL SSSR 49:42-46, 1963.

73. Korneva, E.A. The effects of stimulating different mesencephalic structures on protective immune response pattern. Fiziol ZL SSSR 53:42-45, 1967.

74. Konovalov, G.F., Korneva, E.A., Khai, L.M. Effect of destruction of the posterior hypothalamic area on the experimental allergic polyneuritis. Brain Res. 29:283-286, 1971.

75. Macris, N.T., Schiavi, R.C., Camarino, M.S., Stein, M. Effect of hypothalamic lesions on immune processes in the guinea pig. Am. J. Physiol. 210:1205-1209, 1972.

76. Jankovic, B.D. and Isokovic, K. Neuroendocrine correlates of immune response: 1. Effects of brain lesions on antibody production, arthus reactivity, and delayed hypersensitivity in the rat. Int. and Allergy 45:360-372, 1973.

77. Keller, S.E., Shapiro, R., Schleifer, S.J., Stein, M. Hypothalamic influences on anaqhylaxis. Psychosom. Med. 44:302 (abstract), 1982.

78. Jankovic, B.D., Jankovic, D.LJ., Savovski, L. Effect of early epiphysectomy on the immune system of the chick embryo. First International Workshop on Neuroimmunology, Bethesda, MD, November 27, 1984.

79. Behan, P.O. and Geschwind, N. Hemisphere lateralization and immunity. In: Guillemin, R., Cohn, M., Melnechuk, T. Neural Modulation of Immunity. New York, Raven Press, pp. 73-80, 1985.

80. Biziere, K., Guillaumin, J.M., Degenne, D., Bardos, P., Renoux, M., Renoux, G. Lateralized Neocortical Modulation of the T-Cell Lineage. In: Guillemin, R., Cohn, M., Melnechuk, T. Neural Modulation of Immunity. New York, Raven Press, pp. 81-94, 1985.

81. Spector, N.H. Anatomic and physiological connections between the central nervous and immune systems (neuroimmunomodulation). In: Fabris, N., Garaci, E., Hadden, J., Mitchison, N.A. (eds.) Immunoregulation, New York, Plenum Press, 1983.

82. Bulloch, K. and Cullen, M.R. A Comparative Study of the Autonomic Nervous System Innervation of the Thymus in the Mouse and Chicken. At: First International Workshop on Neuroimmunomodulation. Lister Hill Center, NIH, Bethesda, MD, November 27-30, 1984.

83. Calvo, W. The innervation of the bone marrow in laboratory animals. Am. J. Anat. 123:315, 1968.

84. Bacui, I. La regulation nerveuse et humorale di l'erythropoiese. J. Physiol. (Paris) 54:441, 1962.

85. Besdovsky, H.O., Sorkin, E., DaPrada, M., Keller, H.H. Immunoregulation mediated by the sympathetic nervous system. Cell Immunol. 48:346, 1979.

86. Besedocsky, H.O. and Sorkin, E. Network of immune-neuroendocrine interactions. Clin. Exp. Immunol. 27:1-12, 1977.

87. Besedovsky, H.O., DelRey, A., Sorkin, E., DaPrada, M., Keller, H.H. Immunoregulation mediated by the sympathetic nervous system. Cell. Immunol. 48-346-355, 1979.

88. Rebar, R.W. Miyake, A., Low, T.L.K., Goldstein, A.L. Thymosin stimulates secretion of leuterinizing hormone-releasing factor. Science 213:669-671, 1981.

89. Jancovic, B.D. Immunomodulation of neural structures and functions. Paper delivered at Academy of Sciences of USSR conference (to be published in Russia, 1984), Leningrad, 1982.

90. Fabris, N. Endocrine control of thymic factor production in young adult and old mice. Paper prepared for Academy of Science of USSR conference, "Regulation of Immune Homeostasis," Leningrad, 1982.
91. Dung, H.C. Deficiency in the thymus - dependent immunity in "lethargic" mutant mice. Transplantation 23:39, 1977.
92. Fabri, N., Moccegianni, E., Muzzioli, M., Imberti, R. Thymus-neuroendocrine network. In: Fabris, N., Garaci, E., Hadden, J., Mitchison, N.A. (eds.). Immunoregulation, New York, Plenum Press, pp. 341-362, 1983.
93. Saxena, R.K. and Talway, G.P. An anterior pituitary factor stimulates thymide incorporation in isolated thymocytes. Nature 268:57, 1977.
94. Hall, H.R., McGullis, J.P., Spangelo, B.L. et al. Thymic hormone effects on the brain and neuroendocrine circuits. In: Guillemin, R., Cohn, M., Melnechuk, T. (eds.). Neural Modulation of Immunity. Raven Press, New York, pp. 179-193, 1985.
95. Jones, M.T., Hillhouse, E.W., Burden, J. Effects of various putative neurotransmitters on the secretion of corticotropia. J. Endocrinol. 69:1-10, 1975.
96. Besedovsky, H.O., Sorkin, E., Keller, H.H. Changes in the concentration of corticosterone in the blood during skin graft rejection in the rat. J. Endocrinol. 76:175-176, 1978.
97. Desedovsky, H.O., DelRey, A., Sorkin, E., DaPrada, M., Keller, H.H. Immunoregulation mediated by the sympathetic nervous system. Cell Immunol. 48:346-355, 1979.
98. Hall, N.R. and Goldstein, A.L. Role of thymosin and the neuroendocrine system in the regulation of immunity. In: Fabris, N., Garaci, E. Hadden, J., Mitchison, N.A. (eds.). Immunoregulation, New York, Plenum Press, pp. 141-163, 1983.
99. Besedovsky, H.O., DelRey, A., Sorkin, E., DaPrada, M., Burri, R., Honneger, C. The immune response evokes changes in brain noradrenergic neurons. Science 221-564-565, 1983.
100. Morley, J.E. The endocrinology of the opiates and the opioid peptides. Metabolism 30:195-209, 1981.
101. Brown, S.M., Stemmel, B., Taub, R.N., Kochwa, S., Rosenfeld, R.E. Immunologic dysfunction in heroin addicts. Arch. Int. Med. 134:1001-1006, 1974.
102. Hazum, E., Chang, K.J., Cautrecasas, P. Specific non-opiate receptors for B-endorphin. Science 205:1033-1035, 1979.
103. Blalock, J.E. and Smith, E.M. Human leukocyte interferon: Structural and biological relatedness to adrenocorticotropic hormone and endorphins. Proc. Natl. Acad. Sci. USA 77:5972-5974, 1980.
104. Gilman, S.C., Schwartz, J.M., Milner, R.J., Bloodm, F.E., Feldman, J.D. B-endorphin ehances lymphocyte proliferative responses. Proc. Natl. Acad. Sci. USA 79:4226-4230, 1982.
105. McCain, H.W., Lassiter, I.B., Bozzone, J.M., Grbic, J.T. B-endorphin modulates human immune activity via non-opiate receptor mechanisms. Life Sci. 31:1619-1624, 1982.
106. Plotnikoff, N. and Miller, G.C. Enkephalins as immunodilators. Int. J. Immunopharmacol. 5:437-441, 1983.
107. Kay, N.E., Allen, J., Morley, J.E. Endorphins stimulate normal human peripheral blood lymphocyte natural killer activity. Life Sci. 35:53-59, 1984.
108. Mathews, P.M., Froelich, C.J., Sibbit, W.L., Bankhurst, A.D. Enhancement of natural cytotoxicity by B-endorphin. J. Immunol. 130:1658-1662, 1983.
109. Faith, R.E., Liang, H.J., Murgo, A.J., Plotnikoff, N.P. Neuroimmunomodulation with enkephalins: Enhancement of natural killer (NK) cell activity in vitro. Clin. Immunol. Immunopathol.

31:412-481, 1984.
110. Li, C.H., Yamashiro, D., Tseng, L.F., Chang, W.C., Ferrara, P. B-endorphin omission analogs: Dissociation of immunoreactivity from other biological activities. Proc. Natl. Acad. Sci. USA 77:3211-3214, 1980.
111. Moon, T.D., Morley, J.E., Vesella, R.L., Lange, P.H. The role of calmodulin in human NK activity. Scand. J. Immunol. 18:255-258, 1983a.
112. Baram, D. and Simantov, R. Enkephalins and opiate antagonists control calmodulin distribution in neuroblastoma-glima cells. J. Neurochem. 40:55-63, 1983.
113. Clouet, D., Williams, N., Yonehara, N. Is a calmodulin-opiopeptide interaction related to the mechanism of opioid action. Life Sci. 33:727-730, 1983.
114. Solomon, G.F., Merigan, T.C., Levine, S. Variations in adrenal cortical hormones within physiologic ranges, stress and interferon production in mice. Proc. Soc. Exp. Biol. Med. 126:74069, 1967.
115. Dafny, N. Modification of morphine withdrawal by interferon. Life Sci. 32:303-305, 1983.
116. Ahmed, M.S., Llanos, J., Blatties, C.M. Interleukin-1 interacts with opioid binding sites. Proc. Soc. Neurosci. Abstr. 10:1109, 1984.
117. Morley, J.E., Baranetsky, N.G., Wingert, T.D., Carlson, H.E., Heshman, J.M., Melmed, S., Levin, S.R., Januson, K.R., Weitzman, R., Chang, R.J., Varner, A.A. Endocrine effects of naloxone-induced opiate receptor blockade. J. Clin. Endocrinol. Metab. 50:251-257, 1980.
118. Joseph, S.A., Pilcher, W.H., Knigge, K.M. Anatomy of the corticotropin-releasing factor and opiomelanocortin systems of the brain. Fed. Proc. Vol.44, No. 1, Part 1, pp. 100-107, January 1985.
119. Guillemin, R., Vargo, T., Rossier, J., Minick, S., Ling, N., Rivier, C., Vale, W., Bloom, F. B-endorphin and adrenocorticotropin are secreted concomitantly by the pituitary gland. Science 197:1360-1372, 1977.
120. Smith, E.M. and Blalock, J.E. Lymphocyte Production of Neurally Active Pituitary Hormone-like Molecules. First International Workshop on Neuroimmunomodulation, Bethesda, MD, November 27, 1984.
121. Hanbauer, I., Kelly, G.D., Saini, L., Yang, H.Y.T. (Met5)-enkephalin-like peptides of the adrenal medulla: Release by nerve stimulation and functional implications. Peptides 3:469-473, 1982.
122. Blalock, J.E. and Smith, E.M. A complete regulatory 1-op between the immune and neuroendocrine systems. Fed. Proc. Vol. 44, No. 1, Part 1, pp. 108-111, January 1985.
123. Shavit, Y., Lewis, J.W., Terman, G.W. et al. Opioid peptides mediate the suppressive effect of stress on natural killer cell cytotoxicity. Science 223:188-190, 1984.
124. Solomon, G.F. The Emerging Field of Psychoneuroimmunology: Hypotheses, Supporting Evidence and New Directions. Advance (in press).
125. Plotnikoff, N.P., Murgo, A.J., Miller, G.C., Cordev, C.N., Faith, R.E. Enkephalins-Immunomodulation. Fed. Proc. 44:118-122, 1985.
126. Plotnikoff, N.P., Kasten, A.J., Coy, D.M., Christensen, C.W., Schally, A.V., Spirtes, M.A. Neuropharmacological actions of enkephalin after systemic administration. Life Sci. 19:1283-1288, 1976.
127. Sharp, B., Keane, W.F., Suh, H.J., Gekker, G., Peterson, P.K.

Opioid peptides stimulate superoxide production by human polymorphonuclear leukocytes and macrophages. Clin. Res. (in press), 1985.

INVOLVEMENT OF NON-OPIATE PEPTIDES IN PSYCHONEUROIMMUNOLOGICAL MODULATIONS

V. Kluša, R. Muceniece, Š. Svirskis, E. Kukaine, M. Ratkeviča, G. Rosenthal, and G. Afanasyeva

Institute of Organic Synthesis, Latvian SSR Academy of Sciences
Riga, USSR

INTRODUCTION

At present, several teams of investigators have provided evidence that endogenous opiate peptides play an important role in interactions between the nervous, endocrine and immune systems, their dysfunctions being intimately associated with stress (Arrigo-Reina and Ferri, 1980). Hence, the rapid growth of psychoneuroimmunology, a new field of neurobiology, largely relies on investigation of opiatergic modulations of these systems. Non-opiate peptides, however, are equally involved in regulating the integrity of biochemical and physiological processes responsible for homeostasis.

It should be emphasized in this connection that a number of non-opiate analgesic peptides, e.g. substance P, participate in stress adaptation syndromes triggering the release of other peptide hormones, such as vasopressin and oxytocin(Haldar et al.1980), stimulate histamine liberation from mast cells, etc. (Oehme et al., 1981). Vasopressin, which similarly to ACTH is implicated in learning and memory consolidation processes (De Wied, 1979), elicits analgesia (Bertnson and Berson, 1980) and like opiate peptides becomes released during pain-induced stress (Kendler et al., 1978) and serves as a potent regulator of immune homeostasis (Bukharin et al., 1982; Block et al., 1981; Whitefield et al., 1970).

The two posterior pituitary hormones, vasopressin and oxytocin can replace interleukin functions during gamma-interferon (IFN_γ) production by lymphocytes (Johnson, 1983).

Analgesia, influence on immunocompetent cell transmitters, central effects, etc. are also characteristic of other non-opiate peptides - insulin, neurotensin, bradykinin, vasointestinal polypeptide, etc. (Kluša, 1984).

Enhanced release of numerous opiate and non-opiate peptides during stress implies increased biosynthesis of their precursors - protein molecules whose fine processing with proteinases determines the release of oligopeptides of varying specificity (Ashmarin, 1983). Since different protein groups may be related in evolution (Jörnvall et al., 1981) allowing their classification within the same family, several precursors can exist for a single protein. It is interesting to note that hyperglobulinemia (increased IgM and IgG concentration) has been reported in opiate addicts

(Faith and Plotnikoff, in press), while anti-IgE antibody production is affected during stress (Yoshinori et al., 1983). Our knowledge of protein molecules (immunoglobulins, interferons, albumins, enzymes, haemoglobins and so forth) as precursors of physiologically active endogenous protein fragments is very limited. In the past years, special attention has been attracted by IgG fragments: tuftsin (Thr-Lys-Pro-Arg-OH), IgG 289-292 (Najjar and Nishioka, 1970) and rigin (Gly-Gln-Pro-Arg-OH), IgG 341-344 (Chipens et al., 1980) showing immuno- and neuromodulatory properties (see also Kluša, 1984).

Finally, little is known about peptide hormone fragments, i.e. their natural metabolites formed as a result of proteolytic cleavage of intact hormone molecules. Our own findings (Kluša et al., 1982; see also Kluša, 1980) indicate that C-terminal fragments carrying a terminal $-CONH_2$ group (the gastrin, substance P, oxytocin, luliberin fragments) elicit neurotropic effects characteristic of regulatory peptides, possess various immunomodulating properties and can be classified as regulatory peptides with independent functions in the regulation of CNS activities.

We will confine ourselves to the effects exerted by the short non-opiate peptides - protein fragments (tuftsin, rigin), peptide fragments (gastrin, substance P), neuropeptides (TRH and MIF), on neuro- and immunomodulation by comparing their action with the similar effects produced by the short opiate peptides, enkephalins.

The structure-functional organization of peptide hormones is characterized by certain features such as the specific active fragments represented by the common amino acid sequences of the long molecules, e.g. ACTH 4-10, ACTH 4-7 (as well as for α- and β-MSH, β-lipotropin) (De Wied, 1979) or by the common C-terminal sequences in tachykinins, caerulein-like and bombesin-like peptides, etc., which in most cases correspond to the principal metabolites of the full-length hormone molecules (see also Kluša, 1984).

Moreover, apart from the native ACTH 4-10 present in the organism (Ashmarin et al., 1981), high concentrations of some C-terminal hormone fragments considerably exceeding the amounts of the parent hormone in the brain have been demonstrated. For instance, a high cerebral level of the C-terminal tetrapeptide of gastrin was noted in the hog, the fragment being also predominant in the cholecystokinin-producing regions of the gastrointestinal tract (Rehfeld and Golterman, 1979). The C-terminal tripeptide of gastrin was identified as the shortest fragment exhibiting biological activity (Lin, 1972). Furthermore, the possibility of peptide bond breakage in cholecystokinin (gastrin) between 30 and 31 amino acid sequences resembling metalloendopeptidases that cleave enkephalins (Deschodt-Lanckman and Strosberg, 1983):

Tyr^{1}-Gly^{2}-Gly^{3}—Phe^{4}-Met^{5}-OH
-Met^{28}-Gly^{29}-Trp^{30}-Met^{31}-Asp^{32}-Phe^{33}-NH_2
enkephalinase A

The structure of the C-terminal tripeptide (MAF) and met-enkephalin is strikingly similar (Kluša, 1984) as evidenced by the common Met and Phe and equifunctional amino acid residues Gly and Asp (Chipens, 1971). On intracerebral administration to rats, MAF exhibits equal antinociceptive activity in the tail-flick test as the enkephalins (300-400 nmol), but is characterized by a longer duration of action (60-90 min, as compared to 5-7 min in the case of enkephalins. The maximum level of activity is attained 15-30 min following injection. Pentagastrin (BOC-ß-Ala-Trp-Met-Asp-Pre-OH), a protected analogue of the C-terminal fragment, was even more analgesic (150 min, 37.5 nmol) in this test; its peak of activity was attained 30 min following administration.

Unlike the enkephalins that elicit only central analgesia, MAF is also analgesic on peripheral (intraperitoneal) administration in a 5 mg/kg dose. The effect, lasting 15-20 min, is observed 10 min following the injection (Kluša et al., 1984).

Structural similarity has been revealed by us between the non-opiate peptide, substance P, and the immunopeptide tuftsin (Kluša et al., 1984), as the amino acid sequence Lys-Pro-Arg of tuftsin coincides with the retrosequence of the N-terminal tripeptide in substance P. However tuftsin, though evoking a longer duration (15-30 min) of analgesia i. c. v. than the enkephalins, appears less effective in the tail-flick test in rats. Applied in a 300-400 nmol dose it prolongs the latent period of a nociceptive response only by 20%, as compared to controls. The other immunopeptide under study, rigin, resembles tuftsin in its activity. This effect of tuftsin is similar to that of substance P eliciting analgesia in a 5 nmol dose. At the same time, the C-terminal tetra- (Phe-Gly-Leu-Met-NH_2) and tripeptides (Gly-Leu-Met-NH_2) of substance P, sharing common structural elements with the enkephalins, are equally analgesic at 50 and 100 nmol as the full-length substance P molecule at 5 and 100 nmol (Kluša et al., 1981b).

It should be noted that on intracisternal injection to mice, tuftsin, rigin, MAF and the TRH are much more active in the tail-pinch test than the enkephalins (Takagi et al., 1979), but show the same duration (5 min) of action (Kluša et al., 1984) - see Table 1.

Analysis of analgesia mechanisms in the enkephalins and the above peptides reveals both significant differences and similarities in the character of their action. For instance, naloxone (1 mg/kg) given intraperitoneally 5 min prior to peptide injection reversed enkephalin-induced analgesia, but failed to alter the effects of other examined peptides. The radioreceptor assay used for analysis of competitive stereospecific binding of ^{3}H-naloxone and ^{3}H-DADL to synaptic membranes in the rat brain revealed the ability of rigin to bind to opiate receptors (Table 2). However, this effect is attained at very high rigin concentrations considerably exceeding those of Leu-enkephalin.

However, the parallel linearized dose-response curves suggest a comparable intrinsic activity for rigin. Rigin shows binding to δ-opiate recepotrs in concentrations greater by one order of magnitude than those in the case of μ-receptors. Tuftsin fails to bind to opiate receptors in a concentration as high as 5 mmol (Kukaine et al., 1984).

Evidence for rigin binding to opiate receptors in the brain is well-correlated with the findings obtained during a study of its binding to opiate receptors in the smooth muscle (electric stimulation of isolated strips of guinea-pig ileum, GPI, and mouse vas deferens, MVD). In concentrations (IC_{50} = 300 mM) exceeding by 3-4 orders of magnitude the concentration of enkephalins, rigin binds to the GPI μ-receptors and is capable of potentiating by a factor of two the effect of morphine. Neither rigin, nor tuftsin show binding to the MVD δ-receptors (Kukaine et al., 1984). Consequently, despite the low affinity of rigin for opiate receptors, the propensity of rigin for the μ-receptors is in contrast with the prevailing affinity of enkephalins for the δ-receptors. It is difficult to give a straightforward answer to the question as to rigin's identification with opiate peptides.

Table 1. Comparative Analgesic Activity of Peptides on Their Intracisternal Administration in Mice

Peptide	ED_{50} nmol/animal (M ± m)
Met-enkephalin	146 (99.0-215.0)
Leu-enkephalin	223.0 (160.0-310.0)
Tuftsin	46.0 (31.5-66.7)
Rigin	68.0 (46.9-99.3)
MAF	50*
TRH	54.0 (31.8-91.8)

*ED_{50} could not be determined for lack of dose-response effect

An interesting finding is the dualism of action observed by us for rigin, tuftsin and MAF, e.g. the central analgesic and peripheral anti-morphine (naloxone-like) effects (see also Kluša et al., 1984). It follows from Fig. 1 that antagonism to morphine (15 mg/kg, ip) of these peptides (10 and 100 mg/kg ip), as judged from the tail-pinch test in mice is analogous to that of the naloxone-like peptides TRH and MIF described in the literature (Kastin et al., 1980; Bhargava, 1981), tuftsin and rigin being superior to MIF and TRH in their duration of action. The dualism of tuftsin, rigin and MAF action bears some resemblance to the effects of enkephalin. For example, Met-enkephalin given intracerebroventricularly in subanalgesic doses (40 µg) weakens morphine-induced analgesia (Vaught et al., 1982). The non-opiate peptide hormone, bradykinin, when administered centrally elicits analgesia (Ribeiro et al., 1968), whereas on peripheral administration it acts as a pain-producing peptide (Walaszek, 1970) acting like TRH, somatostatin, melanotropin, ACTH and its fragments (Bulaev, 1982), cholecystokinin (Parviz, 1978) as opiate antagonist (see also Bulaev, 1982).

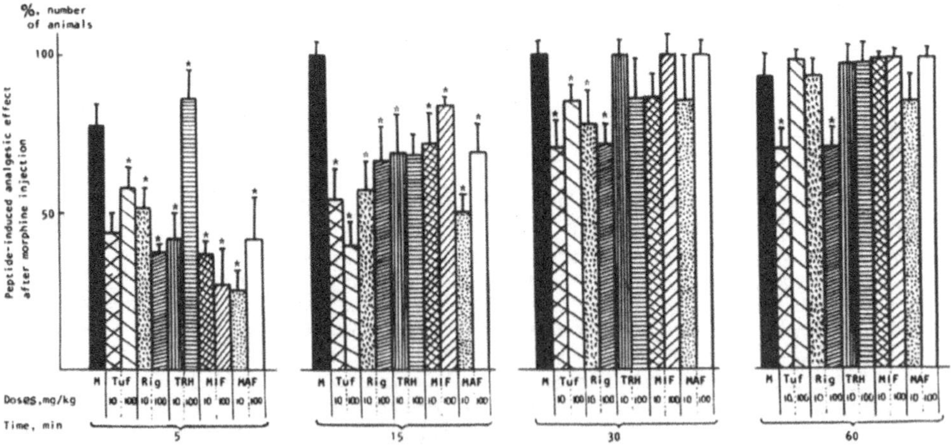

Fig. 1. Anti-morphine action of peptides in mice (ip). M - morphine (control), Tuf - tuftsin, Rig - rigin. (*$p \leq 0.05$ with respect to morphine-induced effects.)

Table 2. Opiate Activity of Peptides Assayed by Radioreceptor Method in the Rat Brain

Peptide	^{3}H-naloxone	^{3}H-DADL
Leu-enkephalin	0.0103 ± 0.0016	0.0117 ± 0.007
Rigin	32.65 ± 6.79	272.0 ± 70.0
Tuftsin	>5000	>5000

Thus, the examined fragments of IgG and gastrin are apparently involved in pain induction and hence can be placed in the family of endogenous factors regulating pain perception. It is conceivable that these peptides elicit conformational changes in opiate receptors, thereby affecting the mechanism of association during receptor-effector coupling and regulating the analgesic effects of opiates (Vaught et al., 1982).

When analyzing the mechanisms of analgesic effects exerted by peptides, the important implications of the monoaminergic system in regulating pain perception must be borne in mind (Sewell and Spencer, 1977). Relying on our own data as to the effect of peptides on the rat brain content of monoamines and their metabolites obtained by using inhibitors of monoamine biosynthesis and postsynaptic monoamine inhibitors (Kluša, 1984) and on the literature data concerning the monoaminergic mechanism of action of TRH (Agarwal et al., 1976), tuftsin (Waldman, 1982) and MIF (Dolzhenko and Komissarov, 1982) we made an attempt to elucidate the principal sites of their involvement in presynaptic monoaminergic transmission. Each peptide in these processes has a unique multidirectional profile as illustrated by Fig. 2. However, these

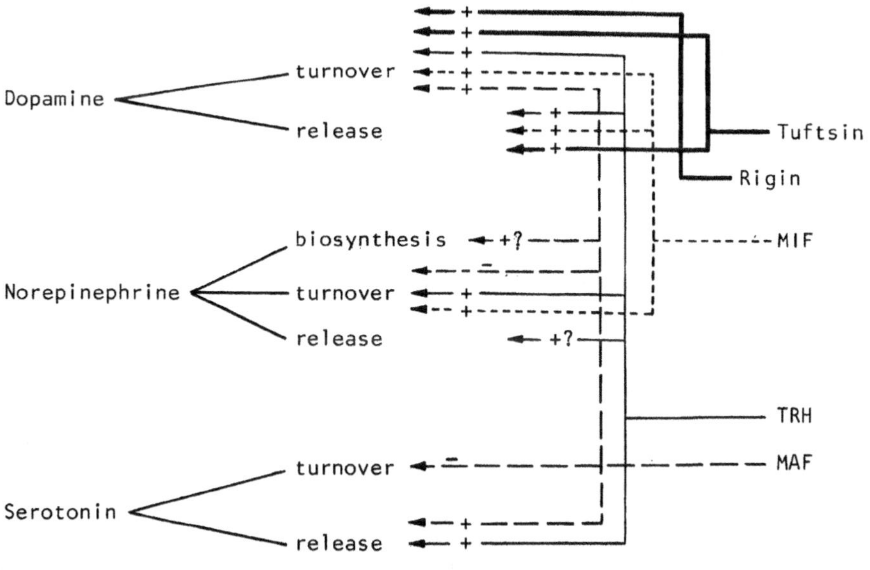

Fig. 2. The mode of action of peptides on presynaptic monoaminergic processes (+ stimulation, - inhibition).

Table 3. Variation in the Analgesic Activity of Compounds (Administered Intracisternally to Mice) in Response to Diethyldithiocarbamate (DDC)

Compound	ED_{50}, nmol/animal	
	Without DDC	In response to DDC
Met-enkephalin	146.0 (99.0-215.0)	18.7 (15.5-21.9)*
Leu-enkephalin	223.0 (160.0-310.0)	20.0 (15.8-24.2)*
Morphine	1.2 (0.6-2.2)	0.015 (0.009-0.024)*
Tuftsin	46.0 (31.5-66.7)	31.2 (27.9-34.5)*
TRH	54.0 (31.8-91.8)	12.5 (9.9-15.1)*

*$p \leqslant 0.05$ with respect to effects without DDC

interactions alone fail to provide a satisfactory explanation of the intricate neurochemical involvements of the peptides and call for more extensive studies. The membrane-affinity (binding to model bilayer phospholipid membranes) of peptides has been recently implicated in their modulating influence (Waldman, 1982): the largest increment in the fluorescence of the probe was induced by MAF, the smallest by tuftsin. Tuftsin was found to enhance GABA-ergic processes in the brain (Nikitina et al., 1983). The monoaminergic mechanism of enkephalins according to our data (Kluša, 1984) is mainly directed towards enhancement of the serotonin turnover. Despite the different neurochemical features of structurally distinct peptides, common pathways do exist by which their analgesic properties are effectuated. A striking example is provided by the fact that preliminary inhibition of dopamine-ß-hydroxylase with diethyldithiocarbamate (600 mg/kg 1.5 h prior to peptide administration) potentiates the central analgesic effects of both the enkephalins, morphine and TRH, tuftsin given intracisternally (Table 3) suggesting a significant role of the catecholaminergic system in the analgesic response to opiate and non-opiate peptides. Monoamines themselves are known to elicit analgesic activity on subarachnoidal administration.

Thus, the results discussed above indicate that the non-opiate peptides under examination affect both the pre- and postsynaptic monoaminergic transmission which allows to qualify them as neuromodulatory peptides. They are also capable of behavioural modulations.

More structurally related peptides (enkephalins, MAF, the C-terminal tetra- and tripeptides of substance P) on intracerebroventricular administration to rats in analgesic doses elicit more uniform locomotor responses: a moderate sedative effect with occasional shakes. Tuftsin enhanced appreciably the locomotor activity, induced grooming and wet-dog shakes resembling the same responses following substance P (25 nmol) and MIF (200 nmol) administration. Increased CNS-activity after rigin administration (marked enhancement of locomotor activity, head shakes, wet-dog shakes and grooming) reminds the effects of TRH (Kluša, 1984).

Our fragments, similarly to the classical neuropeptides, appear to exert both regulatory and modulatory properties (Kluša et al., 1982), which may account for their adaptational role in extremal and stress conditions. Upon peripheral (ip) administration these fragments in a 5 mg/kg dose decreased the immobilization time and increased the swimming time of animals (Table 4) to the same extent as the known neuropeptides when tested under behavioural dispair conditions after the method described by Porsolt et al., 1977. The highest activity was detected for tuftsin, MAF and substance P

Table 4. Comparative Effects of Peptides (5 mg/kg, ip), on Immobilization Time in Mice (Porsolt's Test)

Compound	Immobilization time, %	Difference with respect to control, %Δ
Saline (control)	100.0 ± 5.2	
Tuftsin	50.5 ± 3.9*	-49.5
Rigin	64.5 ± 9.8*	-35.5
MAF	48.4 ± 15.6*	-51.6
MIF	57.3 ± 12.5*	-42.7
TRH	78.6 ± 4.6*	-21.4
Leu-enkephalin	87.26 ± 12.27	-12.74
Met-enkephalin	60.37 ± 4.76*	-39.63
Substance P	48.2 ± 6.1*	-51.8
Substance P C-terminal tripeptide	83.7 ± 6.1*	-16.3

*$p \leqslant 0.05$ with respect to control (100% = 199.2 ± 8.4 s)

(Kluša et al., 1984). Tuftsin in this test was found active also in lower (200-500 µg/kg) doses (Kozlovskaya et al., 1982).

The peptides also show activity in acute emotional stress (Kozlovskaya et al., 1982), thereby behavioural avoidance of a stressful situation under the influence of peptide fragments (Henderson's test, 1970) is dependent on initial emotional reactivity of rats. In other words, the peptides elicit different responses in various subclasses of animals with different emotional backgrounds. For instance, whereas tuftsin given intraperitoneally (200 µg) improves the behaviour of animals in all subclasses, the optimizing effect of MAF (5 mg/kg) is to a lesser extent observed in animals with polar emotional responses (unemotional intact and emotional animals receiving 6-hydroxydopamine). At the same time, the neuropeptide MIF interferes with avoidance behaviour due to enhanced emotional responses of the rats developing in parallel to the high initial level of emotional reactivity.

These data support our assumption that enhanced release of proteins including immunoglobulins and peptide hormones, capable of producing short peptides and fragments active in the regulation of adaptational reactions, occurs in stress reactions. In 1981, we advanced a hypothesis postulating a functional relationship between the neuro-, immuno- and glucoregulating peptides which were shown to involve various aspects of homeostatic regulation (Kluša et al., 1981a). An extensive investigation of immunomodulating action of peptides (Kukaine et al., 1982) has revealed that, similarly to tuftsin and rigin stimulating humoral immunity responses in mice (10-20 mg/kg, ip.), the titre of haemagglutinating antibodies is increased in response to leu-enkephalin (20 mg/kg), MAF (10 mg/kg) and, in particular, to MIF (20 mg/kg). Leu-enkephalin used in a high concentration (100 mg/kg) acts as immunosuppressant. Met-enkephalin in doses below 100 mg/kg fails to affect antibody production (Fig. 3). The in vivo results are somewhat different from in vitro findings (Johnson et al., 1982) where the two enkephalins suppress antibody production in spleen lymphocyte cultures in the presence of sheep erythrocytes.

TRH causes suppression in very high doses (from 100 mg/kg onwards). Generally the immunomodulating properties of peptides are only manifest in doses exceeding those affecting the CNS on peripheral administration (Kluša, 1984). It is interesting to note that TRH inhibits antibody production not only on peripheral administration but also if given centrally (intracisternally, 100-200 µg) when the peptide is not readily accessible to the immunocompetent blood and lymphoid tissue cells but has access to brain neurotransmitters. Neurotransmitters administered in the same manner to animals are also known to affect antibody production by acting either as stimulants (dopamine and serotonin, 10 mg/kg) or inhibitors (noradrenaline, 100 µg/kg) (Kukaine, 1981). Our suggestion that immunopharmacological characteristics of peptides are largely dependent on the activated state of the brain monoaminergic system is supported by the fact that a prior injection of haloperidol, a dopamine receptor inhibitor, (5 mg/kg given ip. 15 min prior to TRH administration) reversed the central immunosuppressive action of TRH (100-400 µg per mouse, intracisternally) (Kukaine et al., 1982).

A study of tuftsin and rigin in T- and B-cell immunity systems (Zālītis et al., 1982) has demonstrated that the two peptides stimulate delayed-type hypersensitivity and are capable of binding the specific antibodies produced against these substances. Rigin (10^{-4} mg/ml <u>in vitro</u> and 20 mg/kg <u>in vivo</u>) is more effective than tuftsin in stimulating rosette-formation in T- and B-lymphocytes from healthy subjects and cirrhotic liver patients; it promotes the adhesive properties of blood monocytes from chronic carriers of hepatitis virus antigen and from mammary tumour patients (Chipens et al., 1982).

As for the immunostimulating role of neuropeptides allowing to predict their effect on resistance to infection and tumours, the first evidence has been obtained of the antitumour activity of MIF in transplantable animal tumours derived from hormone-dependent tissues such as mammary adenocarcinoma Ca-755, cervical carcinoma CCM-5 and melanoma B-16 (Sof'ina et al., 1982).

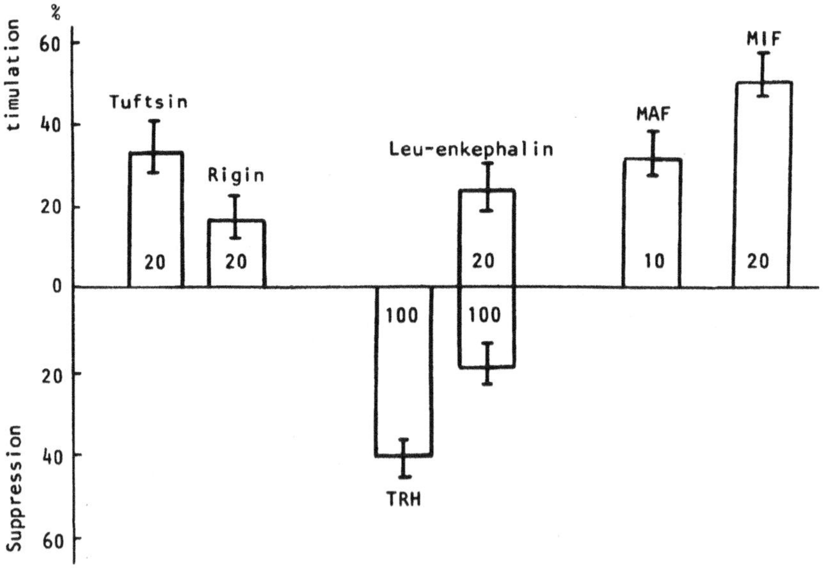

Fig. 3. Effect of peptides on antibody production in mice. Figures in the columns denote the dose of peptides given intraperitoneally. Abscissa corresponds to the control values of haemagglutinating antibody titre (saline = 100%). The peptides were administered 3 days prior to intraperitoneal injection of sheep erythrocytes ($5 \cdot 10^7$ cells per 1 ml of saline).

Experimental evidence accumulated on opiate and non-opiate peptides and the mechanisms of their action is indicative of their universality. A convincing proof is provided by the recent data demonstrating the immunomodulating properties of opiate peptides: enhancement of T-rosettes by enkephalins (Miller et al., 1983), a certain γ-endorphin and interferon similarity, endorphin binding to T-lymphocyte receptors (Weber and Pert, 1984), and so forth. To date, opiate receptors have been located on various immunocompetent cells including the mast cells (DiAugustine et al., 1980), polymorphonuclear leukocytes, and macrophages (Lolait et al., 1984). On the other hand, numerous of receptor populations exist on the membranes of immunocompetent cells (Waksman, 1979) that comprise receptors both for opiate and non-opiate peptides (insulin, ACTH), for prostaglandins, adren- and cholinergic agents, etc. Therefore, it is not surprising that such immunological transmitters as histamine and serotonin are released by immunocompetent cells, e.g. the mast cells, in response to a large number of agents. Morphine was found to release histamine from the mast cells as far back as the late 1940's (Feldberg,1949), in the 1980's the release of serotonin has been demonstrated in response to morphine and endorphins (Yamasaki et al., 1982). Histamine can be also released from the mast cells in response to substance P (Oehme et al., 1981), neurotensin (Rossie and Miller, 1982) and bradykinin. According to our data, tuftsin (10^{-6}-10^{-5}M) released histamine, whereas rigin and enkephalin were fairly ineffective in the test devised by May et al., 1970. So far, the only hormone known to suppress histamine release is the vasointestinal polypeptide (Said, 1984).

At present, histamine is viewed not only as local messenger for immunocompetent cells, but also as neuroendocrine transmitter in the brain (Donoso and Alvarez, 1984), being involved not only in the immunobiological reactions, but also in the regulation of central pain mechanisms (Bulaev, 1982). The universal principles of membrane-associated processes (from changes in adenylate cyclase activity to protein phosphorylation, etc.) mediated by cyclic nucleotides also apply to histamine (Greengard, 1979). Moreover, histamine has been recently demonstrated to affect isolated adrenal chromaffin tissues. Histamine injected intracerebroventricularly enhances ACTH, vasopressin, prolactin and corticosteroid secretion, especially under immobilization stress conditions (see also Donoso and Alvarez, 1984). The above findings suggest that the effects of neuronal histamine are associated with stress.

CONCLUSION

It follows from the aforesaid that both opiate and non-opiate peptides affect all aspects of homeostatic regulation. However, the principles of their action are not identical, although sharing some similarity. Depending on its structural peculiarities each peptide exerts its unique effect on presynaptic neuromediation. The common effects are apparently mediated at the level of postsynaptic receptors, largely monoaminergic, where their interaction takes place with a definite concentration of monoamines being altered by the neuromodulating effects of each particular peptide. Aggregation of various receptor populations on immunocompetent cells plays an equally important role. The effects of peptides on multiple interrelated peptidergic and neurotransmitter interactions mediating universal intracellular processes regulate practically all cellular functions. Variations in the enzyme profiles of tissues, cells and cellular subpopulations account for the prevailing physiological responses.

The immunomodulating properties of non-opiate neuropeptides, on the one hand, and the neurotropic properties on the short IgG and peptide hormone fragments, on the other, as well as their common manifestations, as compared to the short opiate peptides (enkephalins) allow not only classification of natural metabolites as endogenous adaptational agents but also to predict the possible usefulness of their exogenous administration for psychoneuro-immunological homeostasis control, especially in stress-induced disturbances.

REFERENCES

Agarwal, R. A., Rastogi, R. B., and Singhal, R. L., 1976, Changes in brain catecholamines and spontaneous locomotor activity in response to thyrotropin releasing activity, Res. Comm. Chem. Pathol. Pharmacol., 15:743-752.

Ashmarin, I. P., 1983, Neuropeptide complexes, in: Rational Search for New Neurotropic Agents, Zinātne, Riga.

Ashmarin, I. P., Antonova, L. V., Burbaeva, G. Sh., and Kamensky, A. A., 1981, On the possibility of native $ACTH_{4-10}$ being present in the rat, in: Abstr. 4th USSR Symp. on Drug Design, Riga.

Arrigo-Reina, R., and Ferri, S., 1980, Evidence of an involvement of the opioid peptidergic system in the reaction to stressful conditions, Eur. J. Pharmacol., 64:885-888.

Berntson, G. G., and Berson, B. S., 1980, Antinociceptive effects of intraventricular or systemic administration of vasopressin in the rat, Life Sci., 26:255-259.

Bhargava, H. N., 1982, Dissociation of tolerance to the analgesic and hypothermic effects of morphine using thyrotropin releasing hormone, Life Sci., 29:1015-1020.

Block, L. H., Locher, R., and Tenschest, W., 1981, ^{125}J-8-L-Arginine vasopressin binding to human mononuclear phagocytes, J. Clin. Investig., 68:374-381.

Bulaev, V. M., 1982, Opiate receptors and their ligands, in: Advances in Science and Technology, Series: Pharmacology. Chemotherapeutic Drugs. "Opioid Peptides and Their Receptors", I. P. Ashmarin, Ed., Moscow.

Bukharin, O. V., Vasiliev, N. V., and Volodina, E. P., 1982, Oxytocin and vasopressin - regulators of immune homeostasis, in: Abstr. 3rd USSR Symp. on Regulation of Immune Homeostasis, Leningrad, 129-130.

Chipens, G., 1971, Similarity of signature carriers in some physiologically active peptides, in: Chemistry and Biology of Peptides, Zinātne, Riga.

Chipens, G., Blüger, A., Wechsler, H., Timoshenko, Zh., Baumanis, E., Iriste, A., Daija, D., Mergina, G., Poluektova, L., Veretennikova, N., 1982, Comparative study of immunobiochemical effect of tuftsin and rigin, in: Abstr. USSR Symp. on Bioorganic Chemistry and Drug Design, Riga.

Chipens, G. I., Veretennikova, N. I., Zālītis, G. M., Atare, Z. A., Afanasyeva, G. A., and Kukaine, E. M., 1980, New IgG fragments and their analogus possessing immunostimulating and immunoinhibitory activity, in: Abstr. 3rd Intern. Colloquium on Physico-Chemical Information Transfer of Reproduction and Ageing, Varna, Bulgaria, 22-23 Sept.:20.

De Wied, D., 1979, Pituitary and brain peptides and behaviour, in: Brain Peptides: a New Endocrinology, A. M. Gotto, ed., Elsevier/North Holland Biomed. Press, Amsterdam.

Deschodt-Lanckman, M., and Strosberg, A. D., 1983, In vitro degradation of the C-terminal octapeptide of cholecystokinin by enkephalinase, Febs Let., 152:109-183.

DiAugustine, R. P., Lazarus, L. H., Jahnke, G., Khan. M. N., Erisman, M. D., and Linnoilar, R. I., 1980, Corticotropin/ß-endorphin immunoreactivity in rat mast cells. Peptide or protease? Life Sci., 27:2663-2668.

Dolzhenko, A. T., and Komissarov, I. V., 1982, A comparative study of the modulating effect of thyroliberin and melanostatin on neuromediator release by rat brain slices, in: Abstr. USSR Symp. on Bioorganic Chemistry and Drug Design Riga.

Donoso, A. O., and Alvarez, E. O., 1984, Brain histamine as neuroendocrine transmitter, TIPS, March: 98-100.

Faith, R. E., Plotnikoff N. P., Murgo, A. J. (In press), Effects of opiates and enkephalins-endorphins on immune functions.
Feldberg, W., and Paton, W. D. M., 1949, Release of histamine by morphine alkaloids, J. Physiol. (Lond.), 111:19P.
Greengard, P., 1979, Cyclic nucleotides, phosphorylated proteins, and the nervous system, Federat. Proc., 39:2208-2217.
Haldar, J., Hofman, D. L., Nilaver, G., and Zimmerman, E. A., 1980, Oxytocin and vasopressin release by substance P injected into the cerebral ventricles of rats, in: Abstr. 9th Annu. Meet. Atlanta, Georgia, Nov. 2-6, 1979, Bethesda, Soc. Neurosci., 5:447.
Henderson, N. D., 1970, Behavioural reactions of Wistar rats to conditioned fear stimuli, novelty and noxious stimulation, J. Psychol., 75:19-34.
Johnson, H. M., 1983, Neuroendocrine polypeptide hormone regulation of lymphokine production and lymphocyte function, Lymphokine Research, 2:49-56.
Johnson, H. M., Smith, E. M., Torres, B. A., and Blalock, J. E., 1982, Regulation of the in vitro antibody response by neuroendocrine hormones, Proc. Natl. Acad. Sci., 79(13):4171-4174.
Jörnvall, H., Carlström, A., Pettersson, T., Jacobsson, B., Persson, M., and Mutt, V., 1981, Structural homologies between prealbumin gastrointestinal prohormones and other proteins, Nature, 291:261-263.
Kastin, A. J., Olson, R. D., Ehrensing, R. H., Berzas, M. C., Schally, A. V., and Coy, D. H., 1980, MIF-I differential actions as an opiate antagonists, Pharmacol. Biochem. Behav., 11:721-723.
Kendler, K. S., Weitzman, R. E., and Fisher, D. A., 1978, The effect of pain on plasma arginine vasopressin concentration in man, Clin. Endocrinol., 8:89-94.
Kluša, V., 1984, Peptides as Regulators of Brain Functions, Zinātne, Riga.
Kluša, V., Abissova, N., Kukaine, E., Afanasyeva, G., Misiņa, I., Myshliakova, N., Muceniece, R., and Svirskis, Š., 1981a, Possible functional relationship between neuro-, immuno- and glucomodulating peptides: A hypothesis, in: Abstr. 4th USSR Symp. on Drug Design, Riga.
Kluša, V., Abissova, N., Muceniece, R., Svirskis, Š., Bienert, M., and Lipkowski, A., 1981b, Comparative study of substance P and its fragments: analgesic properties, effects on behaviour and monoaminergic processes, Byul. eksp. biol. i med., 12:692-694.
Kluša, V., Muceniece, R., and Svirskis, Š., 1982, Fragments of peptide hormones as possible central regulation factors, in: Neuropharmacology of Peptides, A. V. Waldman, Ed., Moscow.
Kluša, V., Praulīte, G., Svirskis, Š., Muceniece, R., Kukaine, E., and Abissova, N. A., 1984, Comparative study of the central effects of the immunopeptides tuftsin and rigin, LPSR Zin. Akad. vēstis, 12:103-109.
Kozlovskaya, M., Kluša, V., and Bondarenko, N., 1982, A comparison of psychotropic and neurochemical effects of short peptides, in: Neurochemical aspects of psychotropic action, A. V. Waldman, Ed., Moscow.
Kukaine, E., 1981, Effects of neuropeptides and neurotransmitters on the humoral immune system, in: Abstr. 4th USSR Symp. on Drug Design, Riga.
Kukaine, E., Muceniece, R., and Kluša, V., 1982, Comparison of neuro- and immunomodulating properties of low-molecular neuropeptides, Byul. eksp. biol. i med., 8:79-82.
Kukaine, E., Rosenthal, G., and Kluša, V., 1984, Tuftsin and rigin - opiate or non-opiate peptides? in: Abstr. Conf. on Synthesis and Properties of Physiol. Active Agents, Vilnius, 70-71.
Lin, T. M., 1972, Gastrointestinal actions of the C-terminal tripeptide of gastrin, Gastroenterology, 63:922-923.
Lolait, S. J., Lim, A. T. W., Toh, B. H., and Funder, J. M., 1984, Immunoreactive ß-endorphin in a subpopulation of mouse spleen macrophages, J. Clin. Invest., 73(1):277-280.
May, Ch. D., Lyman, M., Alberto, R., and Cheng, J., 1970, Procedures for immunochemical study of histamine release from leukocytes with small

volume of blood, J. Allerg., 46:12-20.
Miller, G. C., Murgo, A. J., and Plotnikoff, N. P., 1983, Enkephalins - enhancement of active T-cell rosettes from lymphoma patients, Clin. Immunol. Immunopathol., 26(3):446-451.
Najjar, V. A., and Nishioka, K., 1970, Tuftsin, a natural phagocytosis stimulatory peptide, Nature, 228:672-673.
Nikitina, Z. S., Batuner, A. S., and Sytinsky, I. A., 1983, Activity of gamma-glutamine transferase in the brain following tuftsin administration, in: Abstr. 4th USSR Conf. on Biochemistry and Nervous System, Yerevan, 216-217.
Oehme, P., Bienert, M., Hecht, K. B., and Bergman, J., 1981, Substance P. Ausgewählte Probleme der Chemie, Biochemie, Pharmacologie, Physiologie und Pathophysiologie, Berlin.
Parviz, M. A., 1978, Psychopharmacology of morphinomimetic peptides in relation to schizophrenia, Gen. Pharmacol., 9:221-222.
Porsolt, R. D., Bertin, A., and Jalfre, M., 1977, Behavioral despair in mice: a primary screening test for antidepressants, Arch. Intern. Pharmacodyn., 229:327-336.
Rehfeld, J. F., and Golterman, N. R., 1979, Immunochemical evidence of cholecystokinin tetrapeptides in hog brain, J. Neurochemistry, 32: 1339-1341.
Ribeiro, S. A., Corrade, A. P., and Graff, F. G., 1968, Analgesia induced by intraventricular administration of bradykinin in the rabbit, Acta Physiol. Latinoam., 18:78-81.
Rossie, S. S., and Miller, R. J., 1982, Regulation of mast cell histamine release by neurotensin, Life Sci., 31:509-516.
Said, S. J., 1984, Vasoactive intestinal polypeptide (VIP): Current Status, Peptides, 5:143-150.
Sewell, R. D. E., and Spencer, P. S. J., 1977, The role of biogenic agents in the actions of centrally-acting analgesics, in: Progress in Medicinal Chemistry, G. P. Ellis, G. B. West, Eds., Elsewier/North Holland Biomedical Press, Amsterdam.
Sof'ina, Z. P., Lagova, N. D., and Kalashnikova, N. M., 1982, Hypothalamic peptide factors as potential antitumour agents, in: Abstr. USSR Symp. on Bioorgan. Chemistry and Drug Design, Riga, 53.
Tagaki, H., Shiomi, H., Ueda, H., and Amano, H., 1979, A novel analgesic dipeptide from bovine brain is a possible Met-enkephalin releaser, Nature, 282:410-413.
Vaught, J. C., Rothman, R. B., and Westfall, T. C., 1982, Mu and delta receptors: their role in analgesia and in differential effects of opioid peptides on analgesia, Life Sci., 30:1443-1455.
Waksman, B. H., 1979, Adjuvants and immune regulation by lymphoid cells, Springer Seminar of Immunopathology, 2:5-33.
Walaszek, E. J., 1970, Effect of bradykinin on the central nervous system, in: Handbook of Experimental Pharmacology, v. 25, Bradykinin, Kallidin and Kallikrein, E. G. Erdös, Ed., Springer Verlag, Berlin - Heidelberg - New York.
Waldman, A. W., 1982, Peptides as modulators of monoaminergic processes, in: Pharmacology of Neropeptides, A. V. Waldman, Ed., Moscow.
Weber, R., and Pert, C., 1984, Opiatergic modulation of the immune system, in: Central and Peripheral Endorphins: Basic and Clinical Aspects, E. E. Müller, A. R. Genazzani, Eds., Raven Press, New York.
Whitefield, J. F., Mac Manus, J. P., and Rixon, R. H., 1970, Potentiation by antidiuretic hormone (vasopressin) of the activity of parathyroid hormone to stimulate the proliferation of rat thymic lymphocytes, Hormone Metabol. Research, 2:235-237.
Yamasaki, Y., Shimamura, O., Kizu, A., Nakagawa, M., and Ijichi, H., 1982, IgE-mediated ^{14}C-serotonin release from rat mast cells modulated by morphine and endorphins, Life Sci., 31:471-478.

Yoshinori, I., Kazunori, M., Yukihiro, A., Tetsuya, N., Michihiro, F., and Showa, U., 1983, Attact stress and IgE antibody production in rats, Pharmacol. Biochem. and Behav., 19:883-886.

Zalitis, G., Afanasyeva, G., G., Golubeva, V., Veretennikova, N., and Chipens, 1981, Immunochemical properties of rigin and tuftsin and their effects on cellular and humoral immunity, in: Abstr. of USSR Symp. Bioorg. Chemistry and Drug Design, Riga, 49.

REGULATION OF HUMAN CELLULAR IMMUNE RESPONSES BY GLUCOCORTICOSTEROIDS

George C. Tsokos and James E. Balow

Arthritis and Rheumatism Branch
National Institute of Arthritis, Diabetes, and Digestive and Kidney Diseases
Building 10, Room 3N114
National Institutes of Health
Bethesda, Maryland 20205

INTRODUCTION

That glucocorticosteroids (GCS) are involved in the regulation of the immune response was implied by Dr. Addison as early as the 1850's, in his observations on disease states which today carry his name. It was not until the 1930's when it was demonstrated that GCS exhibit powerful thymolytic and lympholytic activities in vivo and later on (1940's) in vitro. During the decade of 1940 Hench and his colleagues reported spectacular therapeutic effects of cortisone in patients with rheumatoid arthritis (1). During the next decade GCS were used extensively in the treatment of various immune diseases including systemic lupus erythematosus. During the 1960's several investigators started intensive studies aimed at the dissection of the metabolic effects of GCS on several cells (2-4). The decade of 1970's saw elegant studies which addressed the question of the immunoregulation associated with GCS both in vitro and in vivo (5-8). GCS have been found to modulate every phase of human and animal immune responses. Several described effects of GCS on different phases of the immune response cannot be easily integrated to define the precise mechanism(s) of action of these agents. Studies in animals have helped significantly the understanding of the mode of action of GCS. The ultimate understanding of the modulation of the human immune response has been impeded by the different concentration ranges which obtain optimal effects in vivo and in vitro. In addition the interspecies differences in tissue sensitivity to GCS makes extrapolation of results from animal studies to humans difficult (8,9). The present chapter will update the current knowledge of the effects of GCS on the cellular immune response, i.e. effects on peripheral mononuclear cells, cytotoxic and proliferative responses, suppressor cell function, lymphokine production and B lymphocyte responses.

MECHANISM OF ACTION

The mechanisms of action of GCS at a subcellular level have been previously reviewed (2,3,7) and are elaborated elsewhere in this book.

A main contributor to the observed effects of GCS on cell-mediated immune responses pertains to the shift of the traffic of circulating human leucocytes. These traffic patterns have been dissected in elegant studies and have been reviewed elsewhere (8,10). In brief, the absolute number and the relative proportion of eosinophils, basophils, monocytes and lymphocytes are decreased 4 to 8 hours after the administration of pharmacologic doses of GCS and recover to the previous baseline levels within 24 to 72 hours. In contrast, neutrophils, both in absolute and relative numbers, increase 4-6 hours after administration of GCS and normalize 24 hours later. In particular, among the lymphocytes the thymus (T) derived cells are more sensitive than the bone marrow (B) derived. In regard to the subpopulations of the T lymphocytes, T μ cells (carrying receptors for the Fc portion of the IgM molecule, helper subpopulation) are more sensitive than the T γ cells (carrying receptors for the Fc portion of the IgG molecule, suppressor subpopulation) to GCS (11). Parallel studies of the effects of GCS on T lymphocyte subsets defined by monoclonal antibodies found that OKT4 cells (helper subpopulation) are more sensitive than OKT8 cells (suppressor subpopulation)(12).

EFFECT ON SURFACE MEMBRANE RECEPTORS

The reticuloendothelial system (RES) is currently thought to play a major role in the pathogenesis of immune and autoimmune diseases. Humoral elements interact with the receptors on the RES surface membranes for the Fc portion of the IgG molecule and receptors for breakdown products of the third component of the complement, particularly C3b (13). Corticosteroid treatment of guinea pigs prolonged the circulation of IgG and IgM sensitised autologous erythrocytes probably by decreasing the numbers of Fc and C3b receptors on the surface membranes of the cells of the RES (14). Similarly, GCS treatment of mice delayed the clearance of polymeric immune complexes of human serum albumin and corresponding antibody (15). In vitro studies using human peripheral monocytes showed that hydrocortisone (at concentrations 10^{-3} to 10^{-4} M) treatment for 30 minutes inhibited the binding of IgG sensitized erythrocytes as well as the binding of erythrocytes coated with C3 (16). In potentially related studies involving surface receptors, GCS have been found to inhibit the expression of Fc receptors on a human granulocytic cell line (27).

In vivo administration of GCS in normal individuals and in patients with autoimmune hemolytic anemia resulted in a significant decrease of the number of receptors for the Fc portion of IgG1 on the cell surface membrane of peripheral monocytes. The affinity binding constant (Ka) was not affected by GCS treatment (18). The above findings herald the salutory therapeutic effects of GCS in patients with autoimmune hemolytic anemia. However, the interactions of surface membrane receptor, immune complexes and GCS are far from simple. Thus, in other disease states, such as rheumatoid arthritis (19) and systemic lupus erythematosus (20), where the numbers of the Fc receptors are high, still the clearance of immune complexes is decreased (13) and the role of therapeutic GCS is far from clear.

EFFECTS ON MONOCYTE FUNCTION

GCS have been found to affect multiple functions of the human monocytes, including antigen presentation, lymphokine production, differentiation and phagocytosis. Monocyte chemotaxis is particularly sensitive to pharmacologic doses of glucocorticoids as measured by skin-window tests (21,22).

GCS inhibit the expression of surface Ia antigens by murine peritoneal macrophages both in vitro and in vivo, reduce the production of interleukin-1 (IL-1) and inhibit antigen presentation for T cell proliferation by macrophages (23). Similarly in man, it has been shown that GCS inhibit production of pyrogen (now known as IL-1) (24), interleukin-1 production and antigen (e.g. keyhole limpet hemocyanin) presentation (25), responses of macrophages to migration inhibitory factor (26,27) and macrophage mitogenic factor production (27). The latter in vitro effects of GCS on macrophages have been associated with their in vivo observed anti-inflammatory effects, namely decreased number of mononuclear cells in the inflammatory site and decreased "inflammatory" substances release. In contrast to the observations in mice, GCS increase instead of decreasing Ia antigen expression on the cell surface membrane of monocytes (28). Similar inhibitory effects exerted by GCS were described on the Langerhans cells in the skin (antigen presenting cells) (29). Thus, GCS are clearly involved in the initial phases of the immune response, i.e. antigen presentation.

Upon antigenic challenge monocytes are transformed to macrophages; this differentiation is associated with increased cell size, number of lysosomes, rough endoplasmic reticulum, increased acid phosphatase and 5' nucleotidase content; hydrocortisone inhibits all the above events if present at a concentration of 2.5 μM or higher (30). At the biochemical level, at least in mice, GCS inhibit the incorporation of leucine, glucose uptake and CO_2 production by peritoneal macrophages (31) which might account for the above described effects of GCS on human monocytes and macrophages.

EFFECTS ON THE PROLIFERATIVE RESPONSES OF T-LYMPHOCYTES

GCS inhibit the proliferative response of human T cells to several mitogens and antigens. Proliferative responses to phytohemagglutinin (PHA) are inhibited in vitro, manifested by a rightward shift of the dose-response curve (32). Removal of the phagocytic cells from the cell preparations greatly (100 fold) enhances the sensitivity to GCS (33). Cell cycle studies of human cells stimulated with PHA have indicated that GCS either prevent resting cells from entering the G1 phase (34) or arrest the progression of activated lymphocytes from G1 to S phase (35). In vivo administration of GCS inhibits the proliferative responses of peripheral mononuclear cells to PHA (32,36,37). Pokeweed mitogen (PWM) induced proliferative responses of human T cells (estimated usually by tritiated thymidine uptake on day 3 of the culture) is suppressed by GCS (38,39). GCS also inhibit the proliferative responses to antigens such as PPD (40).

Human allogeneic reactions are markedly decreased by GCS both in vitro (41-44) and after in vivo administration (45,46). The significance of the inhibition of the allogeneic mixed lymphocyte reaction may be relevant to their therapeutic usage in organ transplantation.

The mechanism of inhibition of cell proliferation by GCS can be understood considering the role of interleukin-2 (IL-2) production in T cell proliferation. Mitogenic or antigenic stimulation of T cells leads to the production of IL-2 (previously T cell growth factor) (41,47,48) which is necessary for the proliferation of T cells, sustenance of T cell lines and generation of cytotoxic T cells (49-51). Dexamethasone at concentrations of 1 μM, produces an almost complete inhibition of T cell growth factor production by human peripheral lymphocytes stimulated with PHA (4). GCS also inhibit the production of IL-2 in the human autologous mixed lymphocyte reaction (52). That GCS inhibit the production of T cell growth factor after stimulation with PHA, has also been extensively studied in mice (53,54) and underscores the potential importance of this mechanism in GCS-induced immunosuppression.

EFFECTS ON B LYMPHOCYTE RESPONSES

GCS have various effects on the immunoglobulin production by human B cells. In vivo administration of high doses of methylprednisolone for 3 to 5 days results to 22% decrease in serum IgG, 43% in IgA and 14% in IgM levels. Nadir levels are achieved 3 weeks after the end of the treatment (55). Increased catabolism during the treatment and decreased synthesis were postulated to be responsible for the observed results. GCS significantly depressed the primary IgG and IgM responses in asthmatic children immunised with keyhole limpet hemocyanin, but had no effect on the secondary antibody response (56). Similarly, IgE levels were depressed in asthmatic patients treated with GCS (57).

Mononuclear cells taken from individuals who were given a single dose (58) or 3-5 doses (12) of GCS exhibit diminished immunoglobulin secretion measured by plaque forming cell response after stimulation with pokeweed mitogen (PWM). This is probably due to the redistribution of the helper subpopulation of T lymphocytes caused by GCS treatment, and contrasts with the enhanced spontaneous B cell immunoglobulin production observed in these individuals (12).

In contrast to the suppressive effects of GCS administration in vivo on immunoglobulin production are the effects of GCS in in vitro systems. Addition of prednisolone at concentrations of 10^{-6} M in culture of human mononuclear cells with PWM (59-61) resulted in increased immunoglobulin production in the culture supernatants. This increase was attributed to direct action of GCS on B cells rather than through action on T cells (60). Similarly, in vitro corticosteroids enhance a PWM-driven direct anti-sheep red blood cell response (62), while plaque forming cell responses to trinitrophenyl-polyacrylamide are inhibited (63). The mechanism by which GCS cause enhancement of the PWM-induced immunoglobulin production is not clear. PWM stimulation requires accessory cells, and helper T cells. Recently it was found that nanomolar concentrations of hydrocortisone in vitro induce the generation of immunoglobulin producing cells from human mononuclear cells which peaks after 8-10 days (64). Of note is that these cultures were carried out in fetal calf serum. This immunoglobulin induction is T cell dependent and furthermore it was found that T cells stimulated with dexamethasone produce a lymphokine which has characteristics of T cell replacing factor (65). In short, it appears that GCS have major physiological roles at low concentrations and immunosuppressive effects at pharmacologic levels which is well demonstrated by their effects on B lymphocyte functions.

EFFECTS ON SUPPRESSOR CELL FUNCTION

GCS have been found to affect the suppressor cell limb of the immunoregulatory circuit. Waldmann et al initially described that the suppressor T cells which circulate in patients with common variable immunodeficiency are sensitive to hydrocortisone treatment both in vivo and in vitro. Pretreatment of the T cells from these patients with hydrocortisone abrogated their ability to suppress immunoglobulin production by autologous and allogeneic B cells stimulated with pokeweed mitogen (66,67). Administration of GCS to normal volunteers leads to a temporary loss of suppressor cells generated in vitro after stimulation of the peripheral MNC by PWM (58) but not with concanavalin A (Con A) (68). The difference is apparently due to the fact that Con A is a more potent stimulus of human suppressor cells than PWM. Pharmacologic concentrations of GCS inhibit Con A induced suppressor cells if added prior to initiation of cultures but they cannot modulate the expression of suppressor activity of cells precultured with Con A (69).

The mechanism of expression of suppressive function by Con A stimulated cells seems to be multiple. It has been suggested that Con A stimulated cells can absorb interleukin 2, depriving the culture media of necessary growth factors (70). It is possible that GCS inhibit the expression or the function of interleukin-2 receptors on Con A blasts in a way similar to that by which GCS inhibit the function of Fc and complement receptors on cell surface membranes (16). Another way of expression of suppressor cell function is by secretion of a 70 KD suppressor factor (72) which either directly or indirectly (through monocytes) inhibits immunoglobulin secretion by B cells. Blockage of action of this suppressor factor has been postulated as one mechanism of inhibition of suppressor cell function by GCS (72).

Besides the Con A-associated suppressor cell system, other more specialized suppressor cells are inhibited or abrogated by GCS. Hydrocortisone blocks the expression of suppressor cell activity which is generated by culture in vitro of human mononuclear cells (7,73). Similarly, GCS also abrogate the naturally occurring suppressor cell of the PWM-stimulated anti-sheep red blood cell PFC response (68,69), and the generation of suppressor cells in the human allogeneic mixed lymphocyte reaction (74). Further delineation of the precise nature of the suppressor cell mechanisms and functions is needed prior to understanding their modulation by GCS.

EFFECTS ON AUTOLOGOUS MIXED LYMPHOCYTE REACTION (AMLR)

Human peripheral blood T lymphocytes respond with vigorous proliferation when cultured in vitro with irradiated non-T peripheral mononuclear cells. The presence of hydrocortisone during the culture at concentrations equivalent to those found in normal plasma inhibits this proliferative response (42-45,75). Treatment of adult volunteers with intravenous methylprednisolone led to significant depression of the AMLR (43,46). Interleukin-2 but not interleukin-1 is able to abrogate the suppressive effect of hydrocortisone in these cultures, indicating that suppression is exerted either by inhibition of the production of interleukin-2 or by rendering the interleukin-2 producing T cells unresponsive to interleukin-1 (52). AMLR is an important immune reaction in vitro as it represents the ongoing autoreactivity in vivo. Suppression of AMLR by physiologic concentrations of GCS indicates that the generation of autoreactive cells in vivo is under such control.

EFFECTS ON CELL MEDIATED CYTOTOXICITY

Several forms of cell mediated cytotoxicity responses have been found to be altered by GCS. Allo-specific cytotoxic T cells can be generated in vivo or in vitro. Addition of GCS to cultures during the cytolytic-effector phase has been shown to have little inhibitory effect (76-79). The presence of GCS during the sensitization of cytotoxic cells most consistently reduces the number of cytolytic T cells generated during the culture (77,79), but in some systems also appears to impair generation (76) as well as effector capacity (78) during prolonged incubations. Peripheral blood mononuclear cells taken from normal volunteers infused with hydrocortisone demonstrated enhanced natural killer cell activity on a per cell basis during the nadir of lymphopenia (80,81). The enhancement of natural killer cell activity appears to be due to a selective depletion of non-cytotoxic cells from the blood. Antibody-dependent cell-mediated cytotoxicity is similarly enhanced by in vivo administration of GCS (81,82). As with cell mediated lympholysis, GCS are generally unable to inhibit the cytolytic phase of natural killer and antibody-dependent cellular cytotoxicity in vitro (68). Lectin-mediated cytotoxicity is variably inhibited by GCS (40,83).

CONCLUSION

GCS participate in the human immunoregulation through a variety of mechanisms. The need for elucidation of the mechanisms of action is self-evident given that GCS are normal constituents of living organisms and by the extent GCS are used today in the treatment of various disease states. In Table I we summarise the effects of GCS on several facets of the human cellular immune responses which have been described rather unanimously as inhibitory. In Table II we list effects of GCS on lymphoid cells which have an enhancing character. Despite the fact that an enormous amount of information has been collected so far, the definitive characterization of the mechanism of GCS-induced immunoregulation depends on further knowledge of the regulation of the human immune response in normal individuals and in patients.

Table I: Capitulation of the Inhibitory Effects of GCS on Human Cellular Immune Responses

1. Effects on lymphocyte traffic

   a. Lymphopenia 4-6 hours after GCS treatment.
   b. T μ lymphopenia.

2. Effects on surface membrane receptors

   a. Fc receptor mediated immune clearance.
   b. C3b mediated immune clearance.

3. Effects on the proliferative responses of T cells

   a. Soluble antigens.
   b. Alloantigens and autoantigens.

4. Effects on lymphokine production

   a. IL-2 production in AMLR.
   b. IL-1 production.
   c. T-cell growth factor production after mitogenic stimulation.

5. Effects on monocytes

   a. Ia-antigen presentation.
   b. Interleukin-1 production.
   c. Migration inhibitory factor responsiveness.
   d. Differentiation to macrophages.

6. Autologous mixed lymphocyte reaction

7. Suppressor cells

   a. Spontaneously occurring suppressor cells in normal human peripheral blood.
   b. Spontaneous occurring suppressor cells in diseases (i.e., common variable immunodeficiency).
   c. Spontaneously developing after *in vitro* culture of human MNC.
   d. Generation of suppressor cells after stimulation in vitro with Con A, PWM, autoantigens.

8. Effects on cytotoxic responses

   a. PHA induced cellular cytotoxicity.
   b. Allogeneic cell mediated lympholysis.

9. Effects on B cell responses

   a. Serum immunoglobulins.
   b. Anti-trinitrophenyl plaque forming cell responses.
   c. PWM induced plaque forming cell responses following *in vivo* treatment with GCS.

Table II: Enhancing Effects of GCS on Lymphoid Cell Responses

A. Enhancement of PWM-induced plaque forming cell responses *in vitro*.

B. Immunoglobulin synthesis induction *in vitro* by GCS at nanomolar concentrations.

C. Induction of production of T cell replacing factor.

D. Enhancement of natural killer and antibody-dependent cell-mediated cytotoxic responses *in vivo*.

REFERENCES

1. P. S. Hench, E. D. Kendall, C. H. Slocumb, and H. F. Polley, Effect of a hormone of the adrenal cortex (17-hydroxy-11-dehydrocorticosterone: Compound E) and of pituitary adrenocorticotropic hormone on rheumatoid arthritis. Proc. Staff Meet. Mayo Clin. 24:181 (1949).

2. A. Munck, and K. Leuny, Glucocorticoid receptors and mechanisms of action receptors and mechanism of action of steroid hormones. Part II. Edited by J. R. Pasqualin; New York, Marcel Dekker Inc., p. 311.

3. A. Munck, and D. Young, Corticosteroid and lymphoid tissue, in "Handbook of physiology," Section 7, Vol. VI, Washington, D.C., American Physiological Society, p. 231.

4. G. R. Crabtree, K. A. Smith, and A. Munck, Glucocorticoids and immune responses, Arthritis Rheum. 22:1246 (1979).

5. A. S. Fauci, D. C. Dale and J. E. Balow, Glucocorticosteroid therapy: mechanisms of action and clinical considerations. Ann. Intern. Med. 84:304 (1976).

6. A. M. Dannenberg, The anti-inflammatory effects of glucocorticosteroids. A brief review of the literature. Inflammation 3:329 (1979).

7. J. E. Parrillo, A. S. Fauci, Mechanisms of glococorticoid action on immune processes. Ann. Rev. Pharmacol. Toxicol. 19:133 (1979).

8. T. R. Cupps, and A. S. Fauci, Corticosteroid-mediated immunoregulation in man. Immunological Rev. 65:133 (1982).

9. H. N. Claman, Corticosteroids and lymphoid cells. N. Engl. J. Med. 287:388 (1972).

10. A. S. Fauci, Immunosuppressive and anti-inflammatory effects of glucocorticoids. Monographs in endocrinology. Vol. 12. Glucoroticoid hormone action. Edited by J. D. Baxter and G. G. Rousseau, p. 449-465. Springer-Verlag Berlin Heidelberg.

11. B. F. Haynes, and A. S. Fauci, The differential effects of in vivo hydrocortisone on kinetics of subpopulations of human peripheral blood thymus-derived lymphocytes. J. Clin. Invest. 61:703 (1978).

12. T. F. Cupps, L. C. Edgar, C. A. Thomas, and A. S. Fauci, Multiple mechanisms of B cell immunoregulation in man after administration of in vivo corticosteroids. J. Immunol. 132:170 (1984).

13. M. M. Frank, T. J. Lawley, M. I. Hamburger, and E. J. Brown, Immunoglobulin G Fc receptor-mediated clearance in autoimmune diseases. Ann. Intern. Med. 9o8:206 (1983).

14. J. P. Atkinson, A. D. Schreiber, and M. M. Frank, Effects of corticosteroids and splenectomy on the immune clearance and destruction of erythrocytes. J. Clin. Invest. 52:1509 (1973).

15. A. O. Haakenstad, J. B. Case, and M. Mannik, Effect of cortisone on the disappearance kinetics and tissue localization of soluble immune complexes. J. Immunol. 114:1153 (1975).

16. A. D. Schreiber, J. Parson, P. McDermott, and R. A. Cooper, Effect of corticosteroids on the human monocyte IgG and complement receptors. J. Clin. Invest. 56:1189 (1975).

17. G. R. Crabtree, A. Munck, and K. A. Smith, Glucocorticoids inhibit expression of Fc receptors on the human granulocytic cell line HL-60. Nature 279:338 (1979).

18. L. F. Fries, C. M. Brickman, and M. M. Frank, Monocyte receptors for the Fc portion of IgG increase in number of autoimmune hemolytic anemia and other hemolytic states and are decreased by glucocorticoid therapy. J. Immunol. 131:1240 (1983).

19. S. Katayama, D. Chia, H. Nasau, and D. W. Knutson, Increased Fc receptor activity in monocytes from patients with rheumatoid arthritis: a study of monocyte binding and catabolisms of soluble aggregates of IgG in vitro. J. Immunol. 127:643 (1981).

20. L. F. Fries, W. W. Mullins, K. R. Cho, P. H. Plotz, and M. M. Frank, Monocyte receptors for the Fc portion of IgG are increased in systemic lupus erythematosus. J. Immunol. 132:695 (1984).

21. D. C. Dale, A. S. Fauci, D. Guerry, and S. M. Wolff, Comparison of agents producing a neutrophilic leucocytosis in man. Hydrocortisone, prednisone, endotoxin, and etiocholanolone. J. Clin. Invest. 56:808 (1974).

22. J. J. Rinehart, S. P. Balcerzak, A. L. Sagone, and A. F. Lobuglio, Effects of corticosteroids on human monocyte function. J. Clin. Invest. 54:1337 (1974).

23. D. S. Synder, and E. R. Unanue, Corticosteroids inhibit murine macrophage Ia expression and interleukin-1 production. J. Immunol. 129:1803 (1982).

24. G. M. Dillard, and P. Bodel, Studies on steroid fever. II. Pyrogenic and antipyrogenic activity in vitro of some endogenous steroids of man. J. Clin. Invest. 49:2418 (1970).

25. T. L. Gerrard, T. R. Cupps, C. H. Jergensen, and A. S. Fauci, Hydrocortisone-mediated inhibition of monocyte antigen presentation: Dissociation of inhibitory effect and expression of DR antigens. Cell. Immunol. 85:330 (1984).

26. J. E. Balow, and A. S. Rosenthal, Glucocorticoid suppression of macrophage migration inhibitory factor. J. Exp. Med. 137:1031 (1973).

27. M. R. Duncan, J. R. Saklik, and J. W. Hadden, Glucocorticoid modulation of lymphokine-induced macrophage proliferation. Cell. Immunol. 67:23 (1982).

28. T. L. Gerrard, T. R. Cupps, C. H. Jurgensen, and A. S. Fauci, Increased expression of HLA-DR antigens in hydrocortisone-treated monocytes. Cell. Immunol 84:311 (1984).

29. D. V. Belsito, T. J. Flotte, H. W. Lim, R. L. Baer, G. J. Thorbecke, and I. Gigli, Effect of glucocorticosteroids on epidermal Langerhans cells. J. Exp. Med 155 (1982).

30. J. J. Rinehart, D. Wuest, and G. A. Ackerman, Corticosteroid alteration of human monocyte to macrophage differentiation. J. Immunol. 129:1436 (1982).

31. J. M. Norton, and A. Munck, In vitro actions of glucocorticoids on murine macrophages: Effects on glucose transport and metabolism growth in culture and protein synthesis. J. Immunol. 125:259 (1980).

32. A. S. Fauci, Mechanisms of corticosteroid action on lymphocyte subpopulations. II. Differential effects of in vivo hydrocortisone, prednisone and dexamethasone on in vitro expression of lymphocyte function. Clin. Exp. Immunol. 24:54 (1976).

33. H. Blomgren, Steroid sensitivity of the response of human lymphocytes to phytohemagglutinin and pokeweed mitogen: Role of phagocytic cells. Scand J. Immunol. 3:655 (1974).

34. A. J. Robertson, J. H. Gibbs, R. C. Potts, R. A. Brown, M. C. K. Browning, and J. S. Back, Dose-related depression of PHA-induced stimulation of human lymphocytes by hydrocortisone. Int. J. Immunopharmacol. 3:21 (1981).

35. J. C. Sloman, and PO. A. Bell, Cell cycle-specific effects of glucocorticoids on phytohemagglutinin-stimulated lymphocytes. Clin. Exp. Immunol. 39:503 (1980).

36. J. L. Webel, R. E. Ritts, H. F. Taswell, J. V. Donadio Jr., and J. E. Woods, Cellular immunity after intravenous administration of methylprednisolone. J. Lab. Clin. Med. 83:383 (1974).

37. J. L. Webel, and R. E. Ritts Jr., The effects of corticosteroid concentrations on lymphocyte blastogenesis. Cell. Immunol. 32:287 (1977).

38. D. H. Heilman, Failure of hydrocortisone to inhibit blastogenesis by pokeweed mitogen in human leukocyte cultures. Clin. Exp. Immunol. 11:393 (1972).

39. D. H. Heilman, M. Gambrill, and J. P. Leichner, The effect of hydrocortisone on the incorporation of tritiated thymidine by human blood lymphocytes cultured with phytohemagglutinin and pokeweed mitogen. Clin. Exp. Immunol. 15:203 (1973).

40. A. E. Butterworth, Nonspecific cytotoxic effects of antigen-transformed lymphocytes. II. Inhibition by drugs. Cell. Immunol. 16:60 (1975).

41. J. C. Rosenberg, M. P. Kaplan, K. Lysz, Mechanisms of steroid suppression of immune function: Effect of methylprednisolone on lymphocyte activation and proliferation. Surgery. 88:193 (1980).

42. D. M. Ilfeld, R. S. Krakauer, and R. M. Blaese, Suppression of the human autologous mixed lymphocyte reaction by physiologic concentrations of hydrocortisone. J. Immunol. 119:428 (1977).

43. D. T. Y. Yu, S. J. Ramer, and P. J. Clements, Effect of methylprednisolone on autologous mixed lymphocyte cultures. Transplantation. 163:165 (1978).

44. R. P. MacDermott, and M. C. Stacey, Further characterization of the human autologous mixed leucocyte reaciton (MLR). J. Immunol. 126:729 (1981).

45. P. Katz, and A. S. Fauci, Autologous and allogeneic intercellular interaction. Modulation by adherent cells, irradiation and in vitro and in vivo corticosteroids. J. Immunol. 123:2270 (1979).

46. B. H. Hahn, R. P. MacDermott, Burkholder, S. Jacobs, J. L. Pletscher, and M. G. Beale, Immunosuppressive effects of low doses of glucocorticoids: Effects on autologous and allogeneic mixed lymphocyte reactions. J. Immunol. 124:2812 (1980).

47. D. A. Morgan, F. W. Ruscetti, and R. Gallo, Selective in vitro growth of T-lymphocytes from normal human bone marrows. Science 193:1007 (1976).

48. J. D. Watson, L. A. Aarden, and I. Lefkovitz, The purification and quantitation of helper T-cell replacing factors secreted by murine spleen cells activated by concanavalin A. J. Immunol. 122:204 (1979).

49. S. Gillis, D. A. Smith, Long-term culture of tumor-specific cytotoxic T-cells. Nature 268:154 (1977).

50. W. L. Farrar, H. M. Johnson, and J. J. Farrar, Regulation of the production of immune interferon and cytotoxic T lymphocytes by interleukin-2. J. Immunol. 126:1120 (1981).

51. S. Gillis, and J. Watson, Interleukin-2 induction of hapten-specific cytolytic T cells in nude mice. J. Immunol. 126:1245 (1981).

52. R. Pelacios, and I. Sugawara, Hydrocortisone abrogates proliferation of T cells in autologous mixed lymphocyte reaction by rendering the interleukin-2 producer T cells unresponsive to interleukin-1 and unable to synthesize T cell growth factor. Scand. J. Immunol. 15:25 (1982).

53. E. L. Larsson, Cyclosporin A and dexamethasone suppress T cell responses by selectively acting at distinct sites of the triggering process. J. Immunol. 124:2828 (1980).

54. S. Gillis, G. R. Crabtree, and K. A. Smith, Glucocorticoid induced inhibition of T cell growth factor production. I. The effect on mitogen-induced lymphocyte proliferation. J. Immunol. 123:1624 (1979).

55. W. T. Butler, and R. G. Rossen, Effects of corticosteroids on immunity in man. I. Decreased serum IgG concentration caused by 3 or 5 days of high doses of methylprednisolone. J. Clin. Invest. 52:2629 (1973).

56. M. Tuchinda, R. W. Newcomb, and B. L. De Vald, Effect of prednisone treatment on the human immune response to keyhole limpet hemocyanin. Int. Arch. Allergy 42:533 (1972).

57. W. C. Posey, H. S. Nelson, B. Banch, and D. S. Pearlman, The effects of acute corticosteroid therapy for asthma on serum immunoglobulin levels. J. Allergy Clin. Immunol. 62:340 (1978).

58. A. Saxon, R. H. Stevens, S. J. Ramer, P. J. Clement, and D. T. Y. Yu, Glococorticoids administered in vivo inhibit human suppressor T lymphocyte function and diminish B lymphocyte responsiveness in in vitro immunoglobulin synthesis. J. Clin. Invest. 61:922 (1978).

59. D. A. Cooper, M. Duckett, V. Petts, and R. Penny, Corticosteroid enhancement of immunoglobulin synthesis by pokeweed mitogen-stimulated human lymphocytes. Clin. Exp. Immunol. 37:145 (1979).

60. D. A. Cooper, M. Duckett, P. Hansen, V. Petts, and R. Penny, Glucocorticosteroid enhancement of immunoglobulin synthesis by pokeweed mitogen-stimulated human lymphocytes. Clin. Exp. Immunol. 37:145 (1979).

61. D. A. Cooper, P. Hansen, M. Duckett, J. B. Ziegeler, and R. Penny, Glucocorticosteroid enhancement of immunoglobulin synthesis by pokeweed mitogen-stimulated human lymphocytes. III. Common variable immunodeficiency. Clin. Exp. Immunol. 45:399 (1981).

62. A. S. Fauci, K. R. Pratt, and G. Whalen, Activation of human B lymphocytes. IV. Regulatory effects of corticosteroids on the triggering signal in the plaque-forming cell reponse of human peripheral blood B lymphocytes to polyclonal activation. J. Immunol. 119:598.

63. P. Galanaud, M. C. Crevon, D. Hillion, and J. F. Delfraissy, Hydrocortisone sensitivity of human in vitro antibody response: Different sensitivity of the specific and the nonspecific B cell responses induced by the same agent. Clin. Immunol. Immunopathol. 18:68 (1981).

64. J. Grayson, N. J. Dooley, I. R. Koski, and R. M. Blaese, Immunoglobulin production induced in vitro by glucocorticoid hormones. T cell-dependent stimulation of immunoglobulin production without B cell proliferation in cultures of human peripheral blood lymphocytes. J. Clin. Invest. 68:1539 (1981).

65. F. M. Orson, J. Grayson, S. Pike, V. DeSean, and R. M. Blaese, T cell-replacing factor for glucocorticosteroid-induced immunoglobulin production. A unique steroid-dependent cytokine. J. Exp. Med. 158:1473 (1983).

66. T. A. Waldmann, S. Broder, R. Krakauer, R. P. MacDermott, M. Durm, C. Goldman, and B. Meade, The role of suppressor cells in the pathogenesis of common variable hypogammaglobulinemia and the immunodeficiency associated with myeloma. Fed. Proc. 35:2067 (1976).

67. T. A. Waldmann, R. M. Blaese, S. Broder, and R. S. Krakauer, Disorders of suppressor immunoregulatory cells in the pathogenesis of immunodeficiency and autoimmunity. Ann. Intern. Med. 88:226 (1978).

68. B. F. Haynes, P. Katz, and A. S. Fauci, Mechanisms of corticosteroid action on lymphocyte subpopulations. V. Effects of in vivo hydrocortisone on the circulatory kinetics and function of naturally occurring and mitogen-induced suppressor cell in man. Cell. Immunol. 44:169 (1979).

69. B. F. Haynes, A. S. Fauci, Mechanisms of corticosteroid action on lymphocyte subpospulations. IV. Effects of in vitro hydrocortisone on naturally occurring and mitogen-induced suppressor cells in man. Cell. Immunol. 44:157 (1979).

70. R. Palacios, and G. Moller, T cell growth factor abrogates concanavalin A-induced suppressor cell function. J. Exp. Med. 153:1360 (1981).

71. T. A. Fleisher, W. C. Greene, R. M. Blaese, and T. A. Waldmann, Soluble suppressor supernatants elaborated by concanavalin A-activated human mononuclear cells. II. Characterization of a soluble suppressor B cell immunoglobulin production. J. Immunol. 126:1192 (1981).

72. D. Ilfeld, and R. S. Krakauer, Hydrocortisone reverses the suppression of immunoglobulin synthesis by concanavalin A-activated spleen cell supernatants. Clin. Exp. Immunol. 48:244 (1982).

73. P. E. Lipsky, W. W. Ginsburg, F. D. Finkelman, and M. Ziff, Control of human B lymphocyte responsiveness enhanced suppressor T cell activity after in vitro incubation. J. Immunol. 120:902 (1978).

74. B. M. Frey, F. J. Frey, K. C. Cochrum, and L. Z. Benet, Induction kinetics of human suppressor cells in the mixed lymphocyte reaction and influence of prednisolone on their genesis. Cell. Immunol. 56:225 (1980).

75. G. Opelz, M. Kinchi, M. Takasugi, and P. I. Terasaki, Autologous stimulation of human lymphocyte subpopulations. J. Exp. Med. 142:1327 (1975).

76. J. C. Rosenberg, and K. Lysz, Suppression of the immune response by steroids. Comparative potency of hydrocortisone, methylprednisolone, and dexamethasone. Transplantation 29:425 (1980).

77. J. E. Balow, G. W. Hunninghake, and A. S. Fauci, Corticosteroids in human lymphocyte-mediated cytotoxic reactions. Effects on the kinetics of sensitization on the cytolytic capacity of effector lymphocytes in vitro. Transplantation 23:322 (1977).

78. R. P. Scheimer, A. Jacques, S. H. Shin, L. M. Lichtenstein, and M. Plant, Inhibition of T cell-mediated cytotoxicity by anti-inflammatory steroids. J. Immunol. 132:266 (1984).

79. L. M. Muul, and M. K. Gately, Hydrocortisone suppresses the generation of nonspecific "anomalous" killers but not specific cytolytic T lymphocytes in human mixed lymphocyte-tumor cultures. J. Immunol. 132:1202 (1984).

80. M. Onsrud, and E. Thorsby, Influence of in vivo hydrocortisone on some human blood lymphocyte subpopulations. Scand. J. Immuno. 13:573 (1981).

81. P. Katz, A. M. Zaytoren, and J. H. Lee, The effects of in vivo hydrocortisone on lymphocyte-mediated cytotoxicity. Arthritis Rheum. 27:72 (1984).

82. J. W. Parrillo, and A. S. Fauci, Comparison of the effector cells in human spontaneous cellular cytotoxicity and antibody-dependent cellular cytotoxicity: Differential sensitivity of effector cells to in vivo and and in vitro corticosteroids. Scand. J. Immunol. 8:99 (1978).

83. J. R. Clarke, R. F. Gagnon, F. M. Gotch, M. R. Heyworth, I. M. C. MacLennan, S. C. Truelove, and C. A. Waller, The effect of prednisolone on leucocyte function in man. Clin. Exp. Immunol. 28:292 (1977).

ESTROGEN-MEDIATED IMMUNOMODULATION

Oscar J. Pung, Anne N. Tucker, and Michael I. Luster

National Institute of Environmental Health Sciences
Systemic Toxicology Branch, TRTP, P.O. Box 12233
Research Triangle Park, NC 27709

INTRODUCTION

Early anatomic observations demonstrating that ovariectomy enhanced thymic weights while thymectomy modulated uterine weight provided an impetus for numerous studies which established a relationship between sex hormones and the immune system (see rev. by Ahlquist, 1976; Kalland, 1982; Luster et al., 1985). These studies suggested that pharmacological and to some extent physiological levels of estrogens, as well as inadvertent exposure to estrogenic compounds in the environment, may modulate immune function. Consistent with the immunological alterations, clinical and laboratory studies demonstrated that altered estrogen levels can modulate host resistance to a variety of infectious agents. Illustrative of this are the findings that estrogens precipitate a dramatic increase in *Listeria monocytogenes* susceptibility (Dean et al., 1980; Pung et al., 1984), impair the intestinal expulsion of adult nematode *Trichinella spiralis* (Dean et al., 1980; Luebke et al., 1984) and increase the susceptibility of mice to both transplantable and methylcholanthrene-induced tumors (Dean et al., 1980; Morahan et al., 1984; Kalland and Forsberg, 1981). Estradiol has also been shown to induce more intense chlamydial (Rank et al., 1982) and staphylococcal (Toivanen, 1967) infections. Pharmacological doses of diethylstilbestrol (DES), a nonsteroidal synthetic estrogen, or 17β-estradiol increase the severity of experimental toxoplasmosis in mice (Pung and Luster, submitted). In other instances, estrogen treated mice are less susceptible to bacterial infection, examples being *Pneumococcus* Type I, *Pasteurella* spp. and *Salmonella* spp. (Nicol et al., 1964). Rodents exposed to DES are also less susceptible to the formation of transplantable lung melanoma tumors (Fugmann et al., 1983) as well as to certain plasmodial, babesial and trypanosomal infections (Cottrell et al., 1977; Kierszenbaum et al., 1974; Mankau, 1975).

A number of chemical xenobiotics, including DDT, isomers of polychlorinated biphenyls (PCB's) and chlordecone (Kepone), demonstrate both estrogenic activity, as determined by the rat uterine bioassay (rev. by Katzenellenbogen et al., 1980) and immunotoxicity (rev. by Luster et al., 1985). Naturally occurring environmental estrogens, including Δ^9-tetrahydrocannabinol, the mycotoxin zearalenone and its derivative zearalenol, a commercial anabolic agent, have also been reported to affect the immune system (rev. by Luster et al., 1985). While these xenobiotics are only weakly estrogenic, they have relatively long half-lives and may exert a chronic influence on normal ovulatory cycles (Eroschenko and Palmiter, 1980). DES, one of the most potent synthetic estrogens, has widespread agricultural and therapeutic use, having been employed as a growth promoting agent in the sheep and cattle industries as well as therapeutically in humans in estrogen replacement, as anti-abortive agents and in the treatment of breast and prostate cancer (rev. by McMartin et al., 1978; McLachlan et al., 1976). More importantly, the therapeutic use of DES has been associated with adenosquamous carcinoma of the uterus (Culter et al., 1972) and clear-cell adenocarcinoma of the vagina in young women exposed to DES *in utero* (Herbst et al., 1971).

INFLUENCE OF ESTROGENS ON IMMUNE FUNCTION

Pharmacological Levels

Estrogens induce multiple effects on the immune system of laboratory animals when administered in pharmacological quantities, many of which are listed in Table 1. Suppression of cell-mediated immunity by DES and estradiol has been well documented. In mice, both compounds depress cutaneous delayed hypersensitivity (Kalland and Forsberg, 1978; Rifkind and Frey, 1974; Luster et al., 1980), allograft rejection (Franks et al., 1975; Waltman et al., 1971) and the *in vitro* lymphoproliferative response to mitogens or allogeneic leukocytes (Luster et al., 1980; Kalland et al., 1979; Ways et al., 1980). Immunosuppression following perinatal exposure, i.e. during periods of immune ontogenesis, appears to persist, whereas adult exposure induces more transient responses (Kalland 1982; Luster et al., 1981). Estrogens are reported to decrease the number of splenic T-helper cells (Kalland, 1980) and inhibit the production of interleukin 2 (Henriksen and Frey, 1982) which suggests that they affect T cell maturation. Depressed cellular immunity has been reported in women taking oral contraceptives (Barnes et al., 1974). Studies by Ablin et al. (1979) have also demonstrated depressed cellular immunity in prostate cancer patients receiving estrogen therapy.

Humoral immunity is also modulated in mice administered estrogen, although the alteration does not appear to be at the level of the B cell. Suppression of humoral immunity may occur as a result of stimulation of the RES and subsequent antigen sequestering and/or increased hepatic uptake and clearance of antigen from the blood (Sljivic and Warr, 1974; Bick et al., 1984). Suppression of the antibody response observed in mice exposed perinatally to DES has been reported to occur as a result of altered regulatory cell function, specifically depletion of T-helper cell function (Kalland, 1980a). Several investigators have provided evidence that estrogens enhance antibody synthesis (Kenny et al., 1976), which may be due to a general depletion of T-suppressor cells (Paavonen et al., 1981). It is also evident from clinical observations that the therapeutic use of estrogens can affect humoral immune responses in humans. For example, women taking oral contraceptives exhibit enhanced humoral immunity (Bole et al., 1969).

In contrast to suppression of T-cell functions, pharmacological levels of estrogen stimulate mononuclear phagocyte activity. This is illustrated by enhanced vascular clearance of antigen from the circulation of estrogen-treated animals (Nicol et al., 1964). The rapid clearance is due to hyperphagocytosis and increased division of Kupffers cells (Den Dulk et al., 1979; Kelly et al., 1962). The effects of

Table 1

Altered Host Resistance and Immunity in Laboratory Animals Exposed to Pharmacological Doses of Estrogens

	References
Increased Resistance To:	
Pneumococcus, *Pasteurella*, *Salmonella*	Nicol et al., 1964
lung melanoma tumors	Fugmann et al., 1983; Morahan et al., 1984
plasmodial, babesial and trypanasomal infections	Cottrell et al., 1977; Kierszenbaum et al., 1974
Decreased Resistance To:	
Listeria monocytogenes	Dean et al., 1980; Pung et al., 1984
Trichinella	Dean et al., 1980; Luebke et al., 1984
tumors (transplantable and chemically induced)	Dean et al., 1980; Kalland and Forsberg, 1981
chlamydial infections	Rank et al., 1982
staphylococcal infections	Toivanen, 1967
toxoplasmosis	Kittas and Henry, 1980; Pung and Luster, 1984
Altered Immune Functions:	
enhanced mononuclear phagocytic activity	Boorman et al., 1980; Dean et al., 1984
stimulation of antigen clearance	Loose and DiLuzio, 1976; Nicol et al., 1964; Sljivic et al., 1975
depressed delayed hypersensitivity response	Kalland and Forsberg, 1978; Luster et al., 1980
depressed allograft rejection	Franks et al., 1975; Waltman et al., 1971
depressed mixed leukocyte response	Luster et al., 1980
depressed T cell lymphoproliferative response	Kalland et al., 1979; Luster et al., 1980;
decreased number of splenic T-helper cells	Kalland, 1980a
inhibition of interleukin-2 production	Henriksen and Frey, 1982
decreased numbers of bone marrow stem cells	Boorman et al., 1980; Fried et al., 1974
depressed erythropoiesis	Boorman et al., 1980; Fried et al., 1974
deficient NK activity	Seaman et al., 1979; Kalland, 1980b; Kalland and Forsberg, 1981

estrogens on circulating mononuclear phagocytic cells are less clear. While chronic estrogen exposure decreases the number of circulating monocytes (Loose and DiLuzio, 1976), moderate increases in circulating (Kelly et al., 1962) and resident peritoneal macrophages (Boorman et al., 1980) occur following subchronic exposure, in addition to hyperphagocytosis, increased production of plasminogen activator, and increased inhibition of tumor cell growth, i.e., cytostasis (Boorman et al., 1980; Dean et al., 1984). On the other hand, characteristics of fully activated macrophages such as cytolytic protease secretion and tumor cell binding or lysis are lacking (Dean et al., 1984). Consequently, these cells more closely resemble an inflammatory macrophage population.

Spleen cells from mice treated with estradiol or DES are deficient in natural killer (NK) cell activity (Kalland et al., 1980b; Kalland and Forsberg, 1981; Seaman et al., 1979). DES also impairs natural killer cell activity in humans during treatment for prostatic cancer (Kalland and Haukaas, 1981).

Exposure to pharmacological doses of estrogen depresses hematopoiesis in rodents (Reisher, 1966). Estrogen induced myelotoxicity is characterized by bone marrow hypocellularity and depressed numbers of CFUs (Fried et al., 1974; Boorman et al., 1980). Decreased bone marrow erythropoiesis also occurs although a compensatory increase in splenic erythropoiesis is observed several days following cessation of estrogen administration (Fried et al., 1974). Splenic compensation does not occur for CFU-S or CFU-GM progenitor cells which remain depressed in both the bone marrow and spleen for several weeks (Fried et al., 1974; Boorman et al., 1980; Luster et al., 1984a). Myelotoxicity occurs in the absence of estradiol-induced osteoproliferation, indicating that cell kinetics are not simply due to decreases in marrow volume (Luster et al., 1984a).

Physiological Levels

Pregnancy, age, ovariectomy and the ovulatory cycle influence serum estrogen levels. These quantitative alterations in the serum often correlate with changes in the immune system which may be more regulatory in nature than pathological. However, pregnancy, age, ovariectomy and the ovulatory cycle are also accompanied by fluctuations in a variety of other hormones and serum proteins which are also potential immunoregulators, including human chorionic gonadotrophin (Carter, 1976), corticosteroids (Heslop et al., 1954) and pregnancy-associated α-macroglobin (Stimson, 1976). Thus, in vivo data correlating immunological changes with variations in physiological estrogen levels may not represent a direct relationship. Ovulation, pregnancy and ovariectomy influence the number of peripheral leukocytes; for instance, lymphocyte counts are lowest in women at the midpoint of the menstrual cycle when estradiol levels are maximal (Mathur et al., 1979). The number of peripheral lymphocytes and eosinophils progressively declines during pregnancy as estradiol levels rise and then return to normal following partuition (Sturgis and Bethell, 1943; Valdimarsson et al., 1983). Polymorphonuclear leukocyte (PMN) numbers increase throughout pregnancy whereas peripheral monocyte numbers rise during the first and early second trimester and then return to normal levels (Sturgis and Bethell, 1943; Valdimarsson et al., 1983). The number of endometrial macrophages is decreased by ovariectomy and restored by estrogen treatment (Fluhmann, 1932) and increased during proestrus and metestrus when serum estradiol reaches peak levels (Nicol and Abou-Zikry, 1953).

With regards to the physiological effects of estrogens on immune function, the blastogenic response of murine splenic lymphocytes to B cell and T cell mitogens, as well as the number of IgM plaque forming cells, increases during proestrus and estrus (Krzych et al., 1981). Relative enhancement of antibody production reported in females (Batchelor, 1968; Rifkind and Frey, 1972; Stern and Davidsohn, 1955) may be due to an absence of immunosuppressive androgenic hormones rather than of physiological estrogen. The X chromosome may also play a role in regulation of the humoral response, as suggested originally by Washburn et al. (1965). This idea is supported by familial studies comparing IgM levels among parents and offspring (Grundbacher, 1972). On the other hand, pregnancy depresses the contact allergic reaction to picryl chloride (Carter, 1976). Depression of contact allergic reactions also occurs in ovariectomized pregnant mice reconstituted with estradiol and progesterone (Carter, 1976). Allograft rejection (Heslop, 1954) and the lymphoproliferative response to T-cell mitogens (Purtilo et al., 1972; Valdimarsson, 1983) are depressed during pregnancy, as is IgM antibody production (Suzuki and Tomasi, 1979).

Changes in physiological estrogen levels may also affect the mononuclear phagocyte system. For example, macrophages from female mice are more phagocytic than those from male mice (Fruhman, 1973), and cyclic increases in estradiol during proestrus and metestrus correlate with augmented antigen clearance by macrophages, as well as increased macrophage activity and numbers (Nicol and Vernon-Roberts, 1965; Vernon-Roberts, 1969). Enhanced phagocytosis also correlates with increases in uterine weight during estrus (Vernon-Roberts, 1965), whereas ovariectomy decreases phagocyte activity (Nicol and Vernon-Roberts, 1965).

The effects of cyclic changes in serum estradiol concentrations on resistance to infection appear to be consistent with alterations in the immune system. For example, decreased resistance to viral (Mendelow and Lewis, 1969), bacterial (Kass, 1960; Luft and Remington, 1982) and fungal infections (Vaughan and Ramirez, 1951) have been reported to occur in pregnant women. Furthermore, pregnant mice are more susceptible to *L. monocytogenes* and *Toxoplasma gondii* (Luft and Remington, 1982) and *Plasmodium berghei* (van Zon et al., 1982). Bodel et al. (1972) reported that the bactericidal activity of PMNs is inhibited during pregnancy. In contrast, Mitchell et al. (1966) found PMNs from pregnant women to be hyperphagocytic and highly bactericidal. They also observed (1970) that hexose monophosphate shunt and myeloperoxidase (MPO) enzyme activity was augmented in PMNs from pregnant women. These data suggest that estradiol regulates PMN activity during pregnancy.

Regulation of the immune response by physiological estrogen levels is also suggested by contrasting males and females, although in many studies the sex differences are minimal and require large numbers of experimental animals to achieve statistical significance (rev. by Alquist, 1976). Females are generally considered to have augmented humoral immunity as opposed to males (Terres et al., 1968; Krzych et al., 1981; Kenny and Gray, 1971) although, as mentioned before, this may be due to chromosomal and not hormonal sex differences (Rhodes et al., 1969). In humans, cell-mediated immunity is depressed in females relative to males as evidenced by an increased incidence of autosomal disorders (Inman, 1978) and depressed cell-mediated cytotoxic events (Santoli et al., 1976). The blastogenic response of spleen cells from female mice to B-cell and T-cell mitogens is generally higher than that of male mice (Krzych et al., 1981).

Environmental Estrogens

As mentioned previously, several environmental xenobiotics demonstrate both estrogenic activity, as determined by the rat uterine bioassay, and immunotoxicity (rev. by Luster et al., 1985). Although intriguing, there is no evidence that their immunotoxicity is related to estrogenic activity. In fact, DDT and chlordecone (Kepone), both of which appear to induce persistent uterotrophic activity in the rat (rev. by Katzenellenbogen et al., 1980), induce decreased antibody production and thymic atrophy only at near lethal doses (rev. by Luster et al., 1985). The immunological effects of Δ^9-THC are typical of those seen by other estrogens in that lymphocyte reactivity is depressed and CMI is more affected than HMI, but atypical with respect to the lack of effect on the RES (Munson et al., 1976).

MECHANISMS OF ACTION

Subcellular Mechanisms

Unlike peptide hormones, where effects are often mediated through second messengers such as cyclic AMP, estrogens exert their effects on target cell chromatin to initiate the response. Translocation of estrogen into the nucleus and its association with chromatin is regulated by a specific intracellular receptor (Gorski et al., 1968). While many mammalian tissues contain estrogen receptors, including lymphocytes (Cohen et al., 1983; Luster et al., 1984a) and eosinophils (Tshernitchin, 1983), in general, only certain tissues, usually those that possess relatively high receptor concentrations, have been shown to be responsive. In this respect, accumulating evidence suggests that the thymus, and specifically thymic epithelium, is an estrogen responsive tissue, and some manifestations of estrogen-induced immunotoxicity are clearly related to thymic alterations. Although it was only recently demonstrated that the binding of estrogens to thymic tissue could alter the immune response (Grossman et al., 1982; Luster et al., 1984a,b; Stimson and Hunter, 1980), the existence of an interaction between the thymus and the gonads has been known for many years. For example, both pregnancy (Jolly and Lieure, 1930) and lactation (Bompiani, 1914) are known to decrease thymic weights. Chiodi (1940) observed that age-related thymic involution also occurs in castrated animals suggesting that the phenomenon is not totally due to gonadal involvement. Just as gonadal dysfunction was known to affect the thymus, it was long suspected that the thymus, in turn, could affect the gonads. For example, perinatal thymectomy delays vaginal opening by several days (Besedovsky and Sorkin, 1974). Also, the uterus of athymic nude mice is atrophied by comparison with normal litter mates, a condition which is reversed by thymic tissue transplants (Besedovsky and Sorkin, 1974). These observations, demonstrating a relationship between the gonads and the thymus, were confirmed by Chiodi (1940), who described a dose dependent thymic involution in castrated rats following exposure to either estradiol or testosterone. Estrogen induced thymic atrophy is primarily characterized by a reduction in the number of cortical lymphocytes (Boorman et al., 1980; Scheiff and Haumont, 1979). Castration, on the other hand, induces cellular hypertrophy (Chiodi, 1940). Allen et al. (1984) demonstrated that the injection of thymosin fraction 5 in mice increased circulating levels of serum estradiol and advanced the time to vaginal opening, while administration of estrogen decreased both thymic weight and circulating levels of thymosin $\alpha 1$. Thus, it is probable that both sex hormones and thymic hormones are feedback regulators of thymic and gonadal function.

In 1978, Reichman and Villee demonstrated that rat thymus homogenates contain macromolecular components capable of binding estradiol. Grossman et al. (1979a) later confirmed the presence of estrogen receptors in both bovine and rat thymus and, using crude thymic fractions, suggested that the estrogen binding component was present in cytosols from thymic epithelium. In another study, Grossman et al. (1979b) used autoradiography to demonstrate the binding of tritium-labeled estradiol to the thymic epithelium of exposed rats. Luster et al. (1984b) reported that, although estrogen binding occurred in density gradient enriched mouse thymocytes, estrogen receptors prepared from primary cultures of mouse thymic epithelial cells possessed 10-fold greater concentrations. This suggested that thymic epithelial cells, not thymocytes, are a primary target cell for estrogen.

Several lines of evidence directly support a link between estrogen-induced immunomodulation and the thymus. Stimson and Hunter (1980) observed that serum from estrogen exposed rats inhibited the lymphocyte blastogenic response to PHA and the BCG-induced suppression of leukocyte migration. These effects were not observed when serum from estrogen-exposed, thymectomized rats were added to the cell cultures, suggesting that estrogen exposure induced soluble immunoregulatory products from the thymus. In a related series of experiments, Grossman et al. (1982) observed that serum from castrated rats supported the blastogenic response of thymocytes to Con A and PHA while serum from castrated rats exposed to physiological doses of estradiol inhibited blastogenesis. Thymectomy also abolished the inhibitory effects of serum from estrogen-treated rats on lymphoproliferation (Grossman et al., 1982). Additional evidence supporting a role for the thymus in estrogen mediated immunoregulation was provided in experiments utilizing adult thymectomized mice (Luster et al., 1984a,b). It was found that thymectomy inhibited the ability of estradiol or DES to modulate a number of immune functions. Additionally, adult thymectomy inhibited the effects of estrogens on host resistance to *L. monocytogenes* infection and transplantable tumor cell growth (Pung, *et al.*, 1984). Evidence that estrogen-induced immunomodulation is partially mediated through the binding of estrogens to thymic epithelium, resulting in alterations in the quality and/or quantity of thymic hormones, is also available. Supernatant fluids obtained by culturing thymic epithelial cells in the presence of estradiol, in contrast to cultures not containing estradiol, are deficient in their ability to support T cell maturation (Luster et al., 1984a) and lymphocyte blastogenesis (Stimson and Crilly, 1981). The products in these supernatants have not been characterized and any relationship with known thymic peptide remains to be determined.

Molecular Mechanisms

Certain responses to estrogens, such as those which occur in smooth muscle (Batra and Sjogren, 1983) and selected neoplastic mammary tissues (Gorlich et al., 1981), are mediated by nonspecific events which are independent of the estrogen receptor. In this respect, there is evidence that at least part of immunotoxicity by estrogens, particularly cytotoxic events observed at higher doses, is related to metabolism of the parent compound to reactive quinone intermediates (Pfeifer and Patterson, submitted; Mendelsohn et al., 1977). These metabolites are believed to partition into the lymphocyte membrane and alter cell surface interactions such as response to growth or regulatory factors or cell to cell interaction. In this respect, the catechol estrogen, 2-OH estrone, is present at fairly high levels in plasma, approximating that of estradiol and exceeding that of estriol (Fishman and Martucci, 1980). Recently, it has been shown that catechol estrogen metabolites are more potent compared to other metabolites or the parent compound at suppressing PHA-induced lymphocyte activation _in vitro_ (Pfeifer and Patterson, submitted). A distinctive role for catechol estrogens has previously been described in regulating neuroendocrine feedback response to estrogens (Fishman and Martucci, 1980). Although catechol estrogens demonstrate significant binding to the estrogen receptor, they do not undergo significant nuclear translocation and, therefore, have negligible uterotrophic activity (Fishman and Martucci, 1980). These data suggest a role for these metabolites in modulating the early events of lymphocyte activation at the cell surface, probably independent of a specific estrogen receptor-mediated mechanism. This is further supported by the observation that splenic lymphocytes from mice exposed to DES demonstrate a reduced percentage of cells undergoing lectin-induced capping, suggesting interference with cytoskeletal structures at the cell surface (Kalland, 1982).

Cellular Mechanisms

A number of studies have provided data suggesting that altered regulatory cell functions contribute to estrogen-induced immunoregulation. As described earlier, a relationship has been established between increased phagocytic activity of the RES and antibody responses, suggesting that depressed antibody responses occur as a result of altered antigen redistribution or increased antigen sequestering by the RES (Sljivic and Warr, 1974: Sljivic et al., 1975; Bick et al., 1984). Furthermore, macrophages from estrogen-treated mice are deficient in their ability to support lymphocyte mitogenesis (Luster et al., 1980). Several studies have indicated that production or inhibition of lymphocyte-derived regulatory factors are also influenced by estrogens. Estrogens have been reported to increase T suppressor cell activity in spleen cells of pregnant mice (Suzuki and Tomasi, 1979) while perinatal exposure to DES in mice decreases the number of regulatory cells, specifically T-helper cells (Kalland, 1980a). More recently it has been shown that estrogens reduce the synthesis of interleukin 2 (IL2), a second signal for T-cell proliferation (Henriksen and Frey, 1982). It must be remembered, however, that alterations in growth or regulatory factors may be a manifestation of an earlier lesion in cell maturation or cellular metabolism.

Hormonal Influences

There is evidence demonstrating the presence of estrogen receptors in the brain and the pituitary, particularly in the anterior hypothalamus, the median eminence and the adenohypophysis (rev. by Kato, 1977). The mechanism of action of these receptors is similar to that of the uterine estrogen receptor (Kato, 1977) and, under normal conditions, they serve as sites for the initiation of feedback regulation of reproductive behavior (Pfaff, 1983) and the onset of maturation (Kato, 1977). Estrogen binding to receptors in the hypothalamus or adenohypophysis stimulates the release of gonadotropin releasing factors, pituitary gonadotropins (Kato, 1977; Pfaff, 1983), prolactin (Neill, 1980) and adrenocorticotrophic hormones (ACTH) (Kitay et al., 1965). While there is little evidence that the binding of estrogens to receptors in the brain or the pituitary is responsible for the immunoregulatory effects of estrogenic hormones, certain of the hormones released by these organs are potentially immunosuppressive. Prolactin exposure, for example, reduces resistance to nematode infection (Kelly and Dineen, 1973) and inhibits the ability of lymphocytes to respond to PHA (Karmali et al., 1974). Thus, it is conceivable that estrogens may act as a secondary signal for immunoregulation by stimulating the release of other hormones such as prolactin.

Estrogen binding to receptors in the pituitary and/or adrenals may also induce the release of immunosuppressive glucocorticosteroids. Prior to the discovery of translocatable cytoplasmic estrogen receptors in the pituitary (Kato, 1977) and the adrenals (Muller and Wotuj, 1978; Calandra et al., 1978), it was accepted that estrogens affected various components of the pituitary-adrenocortical axis. Early work demonstrated that ovariectomy caused adrenal atrophy which could be reversed by estrogen treatment (Carter, 1956). Ovariectomy diminishes the secretion of pituitary ACTH and adrenal corticosterone (Kitay, 1963). The levels of both hormones can be restored by estrogen treatment (Kitay, 1963). Estrogens may stimulate glucocorticoid release by acting directly on the adrenals, or indirectly by stimulating ACTH release, or both (Kitay, et al., 1965; Colby and Kitay, 1974). Increased levels of corticosterone occur in ovariectomized, hypophysectomized rats following treatment with both estrogen and ACTH but not estrogen alone, suggesting that estrogens can act directly on the adrenals to alter corticosterone levels only in the presence of pituitary ACTH (Colby and Kitay, 1974). Estrogen binding to adrenal receptors may affect corticosterone levels by inhibiting its metabolism to inactive metabolites (Kitay et al., 1970). However, while physiological amounts of estrogen may induce increased serum levels of glucocorticosteroids, pharmacological amounts of estrogen have the opposite effect (Kitay et al., 1970). Additionally, the effects of pharmacological levels of estrogen on immunity are not altered by adrenalectomy (Luster, 1984a).

SUMMARY AND CONCLUSIONS

Extensive evidence indicates that supraphysiological levels of estrogens alter immune function. Suppression of cell-mediated immunity and stimulation of the RES appear to be the primary effects. The effects of physiological levels of estrogens on the immune system are less clear and probably perform more of a regulatory rather than pathological role. Several mechanisms may be responsible for estrogen-induced immune alterations, particularly at supraphysiological levels (Table 2). In this respect, estrogens are similar to corticosteroids, which also influence immune function by multiple mechanisms (rev. by Cupps and Fauci, 1982). The bulk of evidence presently available indicates that the effects of estrogen on immunity are hormonal in nature and indirectly mediated through the release of immunoregulatory hormones by estrogen responsive lymphoid tissues, particularly the thymic epithelium. Although intracellular estrogen receptors have been described in leukocytes, there is no evidence to suggest that they modulate cell function. Estrogen binding to receptors in the hypothalamus, pituitary and adrenals also influences the release of a variety of potentially immunoregulatory hormones. The metabolism of estrogens to membrane reactive intermediates suggest that under appropriate circumstances estrogens may also have direct impact on cell function.

Table 2

Potential Mechanisms of Estrogen Induced Immunotoxicity

- Cell Membrane Perturbations
- Altered T Cell Maturation (Thymic Microenvironment)
- Alterations in Regulatory Cell Numbers
- Secondary Hormonal Influences
- Inhibition of Interleukin Synthesis

ACKNOWLEDGEMENTS

The authors are indebted to Mrs. S. Wilkins for her assistance in preparing the manuscript.

REFERENCES

Ablin, R.J., Bhatti, R.A., Guinan, P.D., and Khin, W., 1979, Modulatory effects of oestrogen on immunological responsiveness. Clin. Explt. Immunol., 38:83-91.

Ahlquist, J., 1976, Endocrine influences on lymphatic organs, immune responses, inflammation and autoimmunity. Acta Endologica. (Supp. 206), 83:1-136.

Allen, L.S., McClure, J.E., Goldstein, A.L., Barkley, M.S., and Michael, S.D., 1984, Estrogen and thymic hormone interactions in the female mouse. J. Reprod. Immunol., 6:25-37.

Barnes, E.W., Loudon, N.B., MacCuish, A.C., Jordan, J., and Irvine, W.J., 1974, Phytohaemagglutinin-induced lymphocyte transformation and circulating autoantibodies in women taking oral contraceptives. Lancet, May 11, 898-900.

Batchelor, J.R., 1968, Hormonal control of antibody formation, in "Regulation of the Antibody Response", B. Cinader, ed., pp. 276-295, Thomas, Springfield.

Batra, S., and Sjogren, C., 1983, Effect of estrogen treatment on calcium uptake by the rat uterine smooth muscle. Life Sci., 32:315-319.

Besedovsky, H.O., and Sorkin, E., 1974, Thymus involvement in female sexual maturation. Nature, 249:356-358.

Bick, P.H., Tucker, A.N., White, K.L., Jr., and Holsapple, M., 1984, Effects of subchronic exposure to diethylstilbestrol on humoral immune functions in female B6C3F1 mice. Immunopharmacology, 7:27-39.

Bodel, P., Dillard, G.M., Kaplan, S.S., and Malawista, S.E., 1972, Antiinflammatory effects of estradiol on human blood leukocytes. J. Lab. Clin. Med., 80:373-384.

Bole, G.G., Friedlaeuder, M.H., and Smith, C.K., 1969, Rheumatic symptoms and serological abnormalities induced by oral contraceptives. Lancet, 1:323-326.

Bompiani, G., 1914, Der einfluss der sangens auf die restitutions - fahigheit des thymus nach der schwangerschaft. Zbl. Allg. Pathol. u. pathol. Anat., 25:929-935.

Boorman, G.A., Luster, M.I., Dean, J.H., and Wilson, R.E., 1980, The effect of adult exposure to diethylstilbestrol in the mouse on macrophage function and numbers. J. Reticuloendothel. Soc., 28:547-559.

Calandra, R.S., Naess, O., Purvis, K., Attramadal, A., Djoseland, O., and Hansson, V., 1978, Oestrogen receptors in the rat adrenal gland. J. Steroid Biochem., 9:957-962.

Carter, J., 1976, The effect of progesterone, oestradiol and HCG on cell-mediated immunity in pregnant mice. J. Reprod. Fert., 46:211-216.

Carter, S.B., 1956, The influence of sex hormones on the weight of the adrenal gland in the rat. J. Endocrin., 13:150-160.

Chiodi, H., 1940, The relationship between the thymus and the sexual organs. Endocrinology, 26:107-116.

Cohen, J.H.M., Danel, L., Cordier, G., Saly, S., and Revillard, J.P., 1983, Sex steroid receptors in peripheral T cells: absence of androgen receptors and restriction of estrogen receptors to OKT8-positive cells. J. Immunol., 131:2767-2771.

Colby, H.D., and Kitay, J.I., 1974, Interaction of estradiol and ACTH in the regulation of adrenal corticosterone production in the rat. Steroids, 24:527-536.

Cottrell, B.J., Playfair, J.H.L., and de Sousa, B., 1977, Plasmodium yoelii and Plasmodium vinckei: the effects of nonspecific immunostimulation on murine maleria. Exp. Parasitol., 43:45-53.

Cupps, T.R., and Fauci, A.S., 1982, Corticosteroid-mediated immunoregulation in man. Immunol. Rev., 65:133-155.

Cutler, B.S., Forbes, A.P., Ingersoll, F.M., and Scully, R.E., 1972, Endometrial carcinoma after stilbestrol therapy in gonadal dysgenesis. N. Engl. J. Med., 287:628-631.

Dean, J.H., Boorman, G.A., Luster, M.I., Adkins, B., Lauer, L.D., and Adams, D.O., 1984, Effect of agents of environmental concern on macrophage function, in "Mononuclear Phagocyte Biology", A. Volkman, ed., pp. 473-485, Marcell Dekker, Inc.

Dean, J.H., Luster, M.I., Boorman, G.A., Luebke, R.W., and Lauer, L.D., 1980, The effect of adult exposure to diethylstilbestrol in the mouse: alterations in tumor susceptibility and host resistance parameters. J. Reticuloendothel. Soc., 28:571-583.

Den Dulk, M.M.C.D., Crofton, R.W., and Van Furth, R., 1979, Origin and kinetics of Kupffer cells during an acute inflammatory response. Immunology, 37:7-18.

Eroshenko, V.P., and Palmiter, R.D., 1980, Estrogenicity of kepone in birds and mammals, in "Estrogens in the Environment", J.A. McLachlan, ed., pp. 305-325, Elsevier North Holland, New York.

Fishman, J., and Martucci, C., 1980, Dissociation of biological activities in metabolites of estradiol, in "Estrogens in the Environment", J.A. McLachlan, ed., pp. 131-145, Elsevier/North-Holland, New York.

Fluhmann, C.F., 1932, The influence of sex hormones on the reticuloendothelial cells of the uterus and a possible application to the treatment of pelvic inflammatory conditions. Am. J. Obstet. Gynecol., 24:654-655.

Franks, C.R., Perkins, F.T., and Bishop, D., 1975, The effect of sex hormones on the growth of Hela tumor nodules in male and female mice. Brit. J. Cancer, 31:100-110.

Fried, W., Tichlee, T., Dennenberg, I., Bacone, J., and Wang, F., 1974, Effects of estrogens on hematopoietic stem cells and on hematopoiesis of mice. J. Lab. Clin. Med., 83:807-815.

Fruhman, G.J., 1973, Peritoneal macrophages in male and female mice. J. Reticuloendothel. Soc., 14:371-379.

Fugmann, F.A., Aranyi, C., Barbera, P.W., Bradof, J.N., Gibbons, R.D., and Fenterro, J.D., 1983, The effect of diethylstilbestrol as measured by host resistance and tumor susceptibility assays in mice. J. Toxicol. Environ. Health, 11:827-841.

Gorlich, M., Hecker, D., and Heise, E., 1981, Comparison of estradiol receptor investigations and histochemical investigations on enzymes in human mammary cancers. J. Natl. Cancer Inst., 67:521-527.

Grossman, C.J., Sholiton, L.J., and Nathan, P., 1979a, Rat thymic estrogen receptor. I. Preparation, location and physiochemical properties. J. Steroid Biochem., 11:1233-1240.

Grossman, C.J., Sholiton, L.J., Blaha, G.C., and Nathan, P., 1979b, Rat thymic estrogen receptor. II. Physiological properties. J. Steroid Biochem., 11:1241-1246.

Grossman, C.J., Sholiton, L.J., and Rosselle, G.A., 1982, Estradiol regulation of thymic lymphocyte function in the rat: mediation by serum thymc factors. J. Steroid Biochem., 16:683-690.

Grundbacher, F.J., 1972, Human X chromosome carries quantitative genes for immunoglobulin M. Science, 176:311-312.

Henriksen, O., and Frey, J.R., 1982, Control of expression of interleukin-2 activity. Cell. Immunol., 73:106-114.

Herbst, A.L., Ulfelder, H., and Poskanzer, D.C., 1971, Adenocarcinoma of the vagina. Association of maternal stilboestrol therapy with tumor appearance in young women. N. Engl. J. Med., 284:878-881.

Heslop, R.W., Krohn, P.L., and Sparrow, E.M., 1954, The effect of pregnancy on the survival of skin homografts in rabbits. J. Endocrin., 10:325-332.

Inman, R.D., 1978, Immunologic sex differences and the female predominance in systemic lupus erythematosus. Arthritis Rheum., 21:849-852.

Jolly, J., and Lieure, C., 1930, Influence de la gestation sur le thymus. C.R. Soc. Biol., 104:451-454.

Kalland, T., 1980a, Decreased and disproportionate T-cell population in adult mice after neonatal exposure to diethylstilbestrol. Cell. Immunol., 51:55-63.

Kalland, T., 1980b, Reduced natural killer cell activity in female mice after neonatal exposure to diethylstilbestrol. J. Immunol., 124:1297-1321.

Kalland, T., 1982, Long-term effects on the immune system of an early life exposure to diethylstilbestrol, in "Environmental Factors in Human Growth and Development", Banbury Report 11, pp. 217-242.

Kalland, T., and Forsberg, J.G., 1978, Delayed hypersensitivity response to oxazolone in neonatally estrogenized mice. Cancer Lett., 4:141-146.

Kalland, T., and Forsberg, J.G., 1981, Natural killer cell activity and tumor susceptibility in female mice treated neonatally with diethylstilbestrol. Cancer Res., 41:5134-5140.

Kalland, T., and Haukaas, S.A., 1981, Effect of treatment with diethylstilbestrol-polyestradiol phosphate or estramustine phosphate (Estracyt®) on natural killer cell activity in patients with prostatic cancer. Invest. Urol., 18:437-441.

Kalland, T., Strand, O., and Forsberg, J.G., 1979, Long-term effects of neonatal estrogen treatment on mitogen responsiveness of mouse spleen lymphocytes. J. Natl. Cancer Inst., 63:413-421.

Karmali, R.A., Lauder, I., and Horrobin, D.F., 1974, Prolactin and the immune response. Lancet, July (II):106-107.

Kass, E.H., 1960, Bacteriuria and pyelonephritis of pregnancy. Arch. Int. Med., 105:194-198.

Kato, J., 1977, Steroid hormone receptors in brain, hypothalamus, and hypophysis, in "Receptors and Mechanisms of Action of Steroid Hormones", J.R. Pasqualini, ed., pp. 603-671, Marcel Dekker, New York.

Katzenellenbogen, J.A., Katzenellenbogen, B.S., Tatee, T., Robertson, D.W., and Landvatter, S.W., 1980, in "Estrogens in the Environment", J.A. McLachlan, ed., pp. 33-51, Elsevier, North Holland.

Kelly, J.D., and Dineen, J.K., 1973, The suppression of rejection of Nippostrongylus brasilunsis in Lewis strain rats treated with ovine prolactin. Immunology, 24:551-558.

Kelly, L.S., Brown, B.A., and Dobson, E.L., 1962, Cell division and phagocytic activity in liver reticuloendothelial cells. Proc. Soc. Exp. Biol. Med., 110:555-559.

Kenny, J.F., and Gray, J.A., 1971, Sex differences in immunologic response: studies of antibody production by individual spleen cells after stimulus with Escherichia coli antigen. Pediat. Res., 5:246-255.

Kenny, J.F., Pangburn, P.C., and Trail, G., 1976, Effect of estradiol on immune competence: in vivo and in vitro studies. Infect. Immun., 13:448-456.

Kierszenbaum, F., Knecht, E., Budzko, D.B., and Pizzimenti, M.C., 1974, Phagocytosis: a defense mechanism against infection with Trypanosoma cruzi. J. Immunol., 112:1839-1844.

Kitay, J.I., 1963, Pituitary-adrenal function in the rat after gonadectomy and gonadal hormone replacement. Endocrinology, 73:253-260.
Kitay, J.I., Cayne, M.D., Newsom, W., and Nelson, R., 1965, Relation of the ovary to adrenal corticosterone production and adrenal enzyme activity in the rat. Endocrinology, 77:902-908.
Kitay, J.I., Cayne, M.D., and Swygert, N.H., 1970, Influence of gonadectomy and replacement with estradiol or testosterone on formation of 5α-reduced metabolites of corticosterone by the adrenal gland of the rat. Endocrinology, 87:1257-1265.
Krzych, U., Strausser, H.R., Bressler, J.P., and Goldstein, A.L., 1981, Effects of sex hormones on some T and B cell functions, evidenced by differential immune expression between male and female mice and cyclic pattern of immune responsiveness during the estrous cycle in female mice. Am. J. Repro. Immunol., 1:73-77.
Loose, L.D., and DiLuzio, N.R., 1976, Dose related reticuloendothelial system stimulation by diethylstilbestrol. J. Reticuloendothel. Soc., 20:457-460.
Luebke, R.W., Luster, M.I., Dean, J.H., and Hayes, H.T., 1984, Altered host resistance to Trichinella spiralis infection following subchronic exposure to diethylstilbestrol. Int. J. Immunopharmacol., in press.
Luft, B.J., and Remington, J.S., 1982, Effect of pregnancy on resistance to Listeria monocytogenes and Toxoplasma gondii infections in mice. Infect. Immun., 38:1164-1171.
Luster, M.I., Boorman, G.A., Dean, J.H., Lawson, L.D., Wilson, R., and Haseman, J.K., 1981, Immunological alterations in mice following adult exposure to diethylstilbestrol, in "Biological Relevance of Immunosuppression", J.H. Dean and M.L. Padarathsingh, eds., pp. 153-175, Van Nostrand, New York.
Luster, M.I., Boorman, G.A., Dean, J.H., Luebke, R.W., and Lawson, L.D., 1980, The effect of adult exposure to diethylstilbestrol in the mouse: alterations in immunological functions. J. Reticuloendothel. Soc., 28:561-569.
Luster, M.I., Boorman, G.A., Korach, K.S., Dieter, M.P., and Hong, L. , 1984a, Mechanisms of estrogen-induced myelotoxicity: evidence of thymic regulation. Int. J. Immunopharmacol., 6:287-297.
Luster, M.I., Hayes, H.T., Korach, K., Tucker, A.N., Dean, J.H., Greenlee, W.F., and Boorman, G.A., 1984b, Estrogen immunosuppression is regulated through estrogenic responses in the thymus. J. Immunol., 133:110-116.
Luster, M.I., Pfeifer, R.W., and Tucker, A.N., 1985, Influence of sex hormones on immunoregulation with specific reference to natural and synthetic estrogens, in "Endocrine Toxicology", J.A. McLachlan, K. Korach, and J. Thomas, eds., pp. 67-83, Raven Press, New York.
MacMartin, K.E., Kennedy, K.A., Greenspan, P., Alam, S.N., Greiner, P., and Yum, J., 1978, Diethylstilbestrol: a review of its toxicity and use as a growth promotant in food producing animals. J. Environ. Pathol. Toxicol., 1:297-313.
Mankau, S.K., 1975, Host sex and sex hormones as a factor affecting Trypanosoma lewisi population in white rats. Jap. J. Parasitol., 24:379-384.
Mathur, S., Mathur, R.S., Goust, J.M., Williamson, H.O., and Fudenburg, H.H., 1979, Cyclic variations in white cell subpopulations in the human menstrual cycle: Correlations with progesterone and estradiol. Clin. Immunol. Immunopath., 13:246-253.
McLachlan, J.A., and Dixon, R.L., 1976, Transplacental toxicity of diethylstilbestrol: A special problem in safety evaluation, in "Advances in Modern Toxicology", M. Mehlman, and R.E. Shapiro, eds., Vol. 1, pp. 423-448, Hemisphere Publ. Corp., New York.

Mendelow, D.A., and Lewis, G.C., 1969, Varicella pneumonia during pregnancy. Obstet. Gynecol., 33:98-99.
Mendelsohn, J., Multer, M.M., and Bernheim, J.L., 1977, Inhibition of human lymphocyte stimulation by steroid hormones: Cytokinetic mechanisms. Clin. Exp. Immunol., 27:127-134.
Mitchell, G.W., Jacobs, A.A., Haddad, V., Paul, B.B., Strauss, R.R., and Sbarra, A.J., 1970, The role of the phagocyte in host-parasite interactions. XXV. Metabolic and bactericidal activities of leukocytes from pregnant women. Am. J. Obstet. Gynecol., 108:805-813.
Mitchell, G.W., McRipley, R.J., Selvaraj, R.J., and Sbarra, A.J., 1966, The role of the phagocyte in host-parasite interactions. IV. The phagocytic activity of leuocytes in pregnancy and its relationship to urinary tract infections. Am. J. Obstet Gynecol., 96:687-697.
Morahan, P.S., Bradley, S.G., Menson, A.E., Duke, S., Fromtling, R.A., and Marciano-Cabral, F., 1984, Immunotoxic effects of diethylstilbestrol on host resistance: comparison with cyclophosphamide. J. Leukocyte Biol., 35:329-341.
Muller, R.E., and Wotiy, H.H., 1978, Estrogen-binding protein in mouse and rat adrenal glands. J. Biol. Chem., 253:740-745.
Munson, A.E., Levy, J.A., Harris, L.S., and Dewy, W.L., 1976, Effects of Δ^9-tetrahydrocannabinol on the immune system, in "Pharmacology of Marihuana", M.C. Braude, and S. Szara, eds., pp. 187-197, Raven Press, New York.
Neill, J.D., 1980, Neuroendocrine regulation of prolactin secretion, in "Frontiers in Neuroendocrinology", L. Martini, and W.F. Ganong, eds., 6:129-155, Raven Press, New York.
Nicol, T., and Abou-Zikry, A., 1953, Influence of eostradiol benzoate and orchidectomy on the reticuloendothelial system. Brit. Med. J., i:133-134.
Nicol, T., Bilbey, D.L.J., Charles, L.M., Cordingley, J.L., and Vernon-Roberts, B., 1964, Oestrogen: the natural stimulant of body defense. J. Endocrinol., 30:277-291.
Nicol, T., and Vernon-Roberts, B., 1965, The influence of the estrus cycle, pregnancy and ovariectomy on RES activity. J. Reticuloendothel. Soc., 2:15-29.
Paavonen, T., Andersson, L.C., and Aldercreutz, H., 1981, Sex hormone regulation of in vitro immune response. J. Exp. Med., 154:1935-1945.
Pfaff, D.W., 1983, Impact of estrogens on hypothalamic nerve cells: ultrastructural, chemical and electrical effects. Rec. Prog. Hormone Res., 39:127-179.
Pung, O.J., Luster, M.I., Hayes, H.T., and Rader, J., 1984, Influence of steroidal and nonsteroidal sex hormones on host resistance in the mouse: Increased susceptibility to Listeria monocytogenes following exposure to estrogenic hormones. Infect. Immun., 46:301-307.
Purtilo, D.T., Hallgren, H.M., and Yunis, E.J., 1972, Depressed maternal lymphocyte response to phytohaemagglutinin in human pregnancy. Lancet, i:769-771.
Rank, R.G., White, H.J., Hough, A.J., Pasley, J.N., and Barrow, A.L., 1982, Effect of estradiol on chlamydial genital infection of female guinea pigs. Infect. Immun., 38:699-705.
Reichman, M.E., and Villee, C.A., 1978, Estradiol binding by rat thymus cytosol. J. Steroid Biochem., 9:637-641.
Reisher, E.H., 1966, Tissue culture of bone marrow. II. Effect of steroid hormones on hematopoiesis in vitro. Blood, 27:460.
Rhodes, K., Scott, A., Markham, R.L., and Monk-Jones, M.E., 1969, Immunological sex differences. Ann. Rheum. Dis., 28:104-119.
Rifkind, D., and Frey, J., 1972, Sex difference in antibody response of CFW mice to Candida Albicans. Infect. Immun., 5:695-698.

Rifkind, D., and Frey, J.A., 1974, Influence of androgen and estrogen on delayed skin test reactivity to Candida albicans. Infect. Immun., 10:971-974.
Santoli, D., Trichieri, G., Amijewski, C.M., and Koprowski, H., 1976, HLA-related control of spontaneous and antibody-dependent cell-mediated cytotoxicity in humans. J. Immunol., 117:765-770.
Scheiff, J.M., and Haumont, S., 1979, The effect of oestradiol on thymic epithelial cells in the mouse. J. Clin. Lab. Immunol., 2:225-234.
Seaman, W.F., Merigan, T.C., and Talal, N., 1979, Natural killing in estrogen-treated mice responds poorly to poly I.C despite normal stimulation of circulating interferon. J. Immunol., 123:2903-2905.
Sljivic, V.S., and Warr, G.W., 1974, Activity of the reticuloendothelial system and the antibody response. III. The fate of Type III pneumococcal polysaccharide and the antibody response. Immunology, 27:1009-1022.
Sljivic, V.S., Clark, D.W., and Warr, G.W., 1975, Effects of estrogens and pregnancy on the distribution of sheep erythrocytes and the antibody response in mice. Clin. Exp. Immunol., 20:179-186.
Stern, K., and Davidsohn, I., 1955, Effect of estrogen and cortisone on immune hemeoantibodies in mice of inbred strains. J. Immunol., 74:479-486.
Stimson, W.H., 1976, Studies on the immunosuppressive properties of a pregnancy-associated alph-macroglobulin. Clin. Exp. Immunol., 25:199-206.
Stimson, W.H., and Crilly, P.J., 1981, Effects of steroids on the secretion of immunoregulatory factors by thymic epithelial cell cultures. Immunology, 44:401-407.
Stimson, W.H., and Hunter, I.C., 1980, Oestrogen-induced immunoregulation mediated through the thymus. J. Clin. Lab. Immunol., 4:27-33.
Sturgis, C.C., and Bethell, F.H., 1943, Quantitative and qualitative variations in normal leukocytes. Physiol. Rev., 23:279-303.
Suzuki, K., and Tomasi, T.B., 1979, Immune responses during pregnancy. Evidence of suppressor cells for splenic antibody response. J. Exp. Med., 150:898-908.
Tshernitchin, A.N., 1983, Eosinophil-mediated non-genomic parameters of estrogen stimulation - a separate group of responses mediated by an independent mechanism. J. Steroid Biochem., 19:95-100.
Terres, G., Morrison, S.L., and Habicht, G.S., 1968, A quantitative difference in the immune response between male and female mice. Proc. Soc. Exp. Biol. Med., 127:664-667.
Toivanen, P., 1967, Enhancement of staphylococcal infection in mice by estrogens. I. Effect of timing, quantity and quality of the hormone. Ann. Med. Exp. Fenn., 45:138-146.
Valdimarsson, H., Mulholland, C., Fridriksdottir, V., and Coleman, D., 1983, A longitudinal study of leucocyte blood counts and lymphocyte responses in pregnancy: a marked early increase of monocyte-lymphocyte ratio. Clin. Exp. Immunol., 53:437-443.
van Zon, A.A.J.C., Eling, W.M.C., Hermsen, C.C.R., and Koekkoek, A.A.G.M., 1982, Corticosterone regulation of effector function of malarial immunity during pregnancy. Infect. Immun., 36:484-491.
Vaughan, J.E., and Ramirez, H., 1951, Coccidiorodomycosis as a complication of pregnancy. Calif. Med., 74:121-125.
Vernon-Roberts, B., 1969, The effect of steroid hormones on macrophage activity. Int. Rev. Cytol., 25:133-151.
Waltman, S.R., Burde, R.M., and Berrias, J., 1971, Prevention of corneal homograft rejection by estrogens. Transplantation, 11:194-196.
Washburn, T., Medearis, D., and Childs, B., 1965, Sex differences in susceptibility to infections. Pediatrics, 35:57-64.

EFFECTS ON THE IMMUNE SYSTEM OF LESIONING AND STIMULATION OF THE NERVOUS SYSTEM: NEUROIMMUNOMODULATION

Branislav D. Jankovic
Immunology Research Center
Vojvode Stepe 458
11221 Belgrade, Yugoslavia

Novera Herbert Spector
Fundamental Neurosciences Program
National Institutes of Health, NINCDS
Federal Building, Room 916
Bethesda, Maryland 20892, USA

INTRODUCTION

The ancient Egyptians and Greeks, among others, knew that defense against disease involved "mind." Today as monists, many scientists (including the authors of this chapter) believe that "mind" is a function of the nervous system, most particularly the brain. More than a few specialists in science today have developed severe tunnel vision: they see only a small piece of the whole integrated biological entity (the organism). The organism itself does not know that it is supposed to be divided like the departments of a medical school, so it functions as a unit. Thus the ancients were correct on this aspect of life, whereas some educated researchers of today fail to appreciate the connections between the brain and the immune system (Spector, 1980).

Today we are in a much better position than were the ancient philosophers. We need speculate no longer. The evidence for neuroimmunomodulation (NIM) is overwhelming. This entire book details only a part of it. Modern research in this area probably began with Metalnikov and his colleagues at the Pasteur Institute in Paris in the 1920's, and today there is a virtual "information explosion" in solid data on NIM (Spector, 1985).

The data comes from molecular biology, genetics, neurochemistry, cellular immunology, neurophysiology, neuroanatomy, physiology and the behavioral sciences, among others. This chapter will outline that portion of the evidence that derives from experiments using the techniques of partial lesioning (mechanical, chemical, electrical, electromagnetic) of the nervous system and from experiments utilizing stimulation (electrical, mechanical, chemical) of discrete neural loci.

BRAIN LESIONS IN MATURE ANIMALS

Data from brain lesion and brain stimulation experiments (Tables 1 and 5) give us important clues to solving the fascinating puzzle of fitting the details to the "big picture." Based on a number of experimental findings, we could suggest at least seven more detailed possible pathways, all of these always involving the two major routes (see below).

Lesion and stimulation experiments, as well as single-unit recordings from neurons in specific brain nuclei (e.g. Klimenko, 1972; Korneva & Klimenko, 1976; Besedovsky, et al, 1977; Spector and Korneva, 1981) help us to orient ourselves within the myriad pathways of the central nervous system, but the exact "centers" in the hypothalamus and elsewhere and the exact fiber tracts that modulate, regulate or initiate (?) immune responses are still not precisely known. Indeed, depending on the species, the laboratory, the antigen, the time of day, and other variables, experimental results are frequently different (for a fuller and historical discussion of this problem, see Spector, 1979).

It is important also to appreciate both the phylogeny and the ontogeny of these systems (see below for a brief discussion of embryonic brain lesions). Early - in-life lesions do not always have the same effects as lesions in adults (e.g. Paunovic, et al., 1976). Brain lesions in embryos trace the ontogeny of neuroimmunomodulation to even earlier stages .

The timing of a lesion with respect to the immune challenge is likewise critical. We have discussed this in other reviews so let us just mention it here as a reminder to those who wish to design new experimental protocols.

Table one outlines the history of experiments in brain lesioning. Not all papers are listed. Nor do we attempt to list the voluminous literature on lesions of the pituitary (hypophysectomies, partial or total), adrenals or other glands. We will, however, say a few words (below) about epiphysectomies or lesions of the pineal, since this "organ" rests squarely in the brain and is part of the brain itself, both anatomically and physiologically.

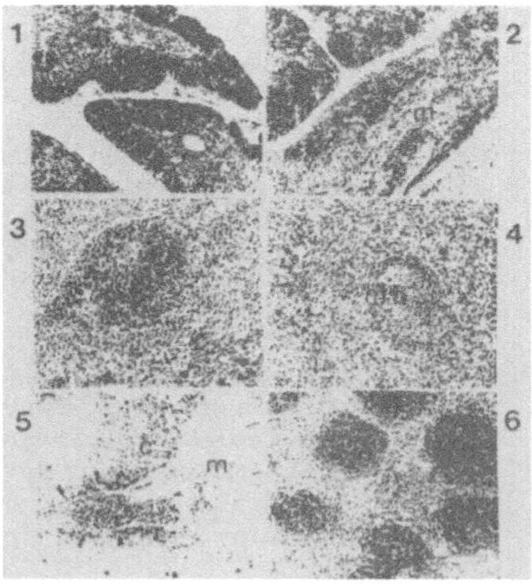

Fig. 1 Normal-looking thymus of a sham-lesioned rat. c, cortex; m, medulla. Haematoxylin and eosin, X 80.
Fig. 2 Profound changes in the thymus from a hypothalamus-lesioned rat. Haematoxylin and eosin, X 80.
3 Normal-looking spleen of a sham-lesioned rat. mb, Malpighian body. Haemotoxylin and eosin, X 100.
Fig. 4 Invòlved Malpighian body (mb) in the spleen of a hypothalamus-lesioned rat. Haematoxylin and eosin, X 100.
Fig. 5 Large area of plasmacellular elements (arrows) in the thymus of a rat with lesioned amygdaloid complex. Methyl green and pyronin, X 80.
Fig. 6 Numerous modules (n) in the lymph node of a rat with lesions in the amygdaloid complex. Methyl green and pyronin, X 80.

Table 1

BRAIN LESIONS (SAMPLING OF EXPERIMENTS)
Hypophysectomy not Included

DATE	AUTHORS	SITE OF LESION	ANIMAL	ANTIGEN	Immune Response: ASSAY	Immune Response: RESULTS
1920s	Speranskii	Rupture of meninges: subarachnoid space	Dog	Many pathogens	Disease symptoms	Alleviated
1930s	Speranskii	Rupture of meninges: subarachnoid space	Human	Many pathogens	Disease symptons	Alleviated
1942	Kopeloff et al.	Cortex	Monkey	Chemical lesions	Production of autoanti-bodies	
1958	Szentivanyi & Filipp	Tuberal region of hypothalamus	Guinea pig	Horse serum	Anaphylaxis	Protection
1958	Szentivanyi & Szekely	Tuberal region of hypothalamus	Guinea pig	Histamine	Histamine shock	Histamine resistance
1958	Freedman & Fenichel	Midbrain rèticulum at superior colliculus	Guinea pig	Egg albumin	Anaphylaxis	Protection
1963	Korneva & Khai	Anterior hypothalamus	Rabbit	Horse serum	Complement binding	Reduction in circulatory antibodies
1964	Khai et al.	Posterior hypothalamus	Rabbit	Horse serum	Complement-fixation reaction	Suppression of antibody production
1964	Luparello et al.	Anterior hypothalamus	Rat	Ovalbumin	Anaphylaxis	Protection
1964	Luparello et al.	Posterior hypothalamus	Guinea pig	Ovalbumin	Anaphylaxis	No effect
1966	Schiavi et al.	Anterior hypothalamus	Guinea pig	Histamine	Histamine shock	Histamine resistance
1968	Bowen	Cortex	Rabbit	Cryogenic lesions	Production of autoantibodies	
1969	Polyak et al.	Posterior hypothalamus		Plague capsule	Antibody formation	Suppression
1969	Thakur & Manchanda	Hypothalamus	Cat		Phagocytosis	Diminution
1969	Thakur & Manchanda	Cingulate gyrus	Cat		Phagocytosis	No effect
1970	Macris et al.	Anterior hypothalamus	Guinea pig	Picryl chloride	Antibody pro-duction, delayed sensitivity	Suppression
1970	Macris et al.	Posterior hypothalamus	Guinea pig	Picryl chloride	Antibody pro-duction, delayed sensitivity	No effect
1971	Thrasher et al.	Hypothalamus, anterior, medial, or posterior	Rat	Egg albumin	Antibody production	No effect
1972	Macris et al.	Anterior hypothalamus	Guinea pig	Ovalbumin	Passive anaphylaxis	Protection
1972	Macris et al.	Posterior hypothalamus	Guinea pig	Ovalbumin	Passsive anaphylaxis	No effect
1972	Tyrey & Nalbandov	Anterior hypothalamus	Rat	Ovalbumin	Antibody production	Decrease
1973	Janković & Isaković	Hypothalamus and reticular formation	Rat	Bovine serum albumin	Antibody pro-duction, Arthus and delayed skin reactions	Decrease
1973	Isaković & Janković	Hypothalamus, reticular formation, thalamus, superior colliculus, caudate nucleus and amygdaloid complex	Rat	Bovine serum allbumin	Thymus	Involution, plasmacytic reaction

(Continued)

Table 1 (Continued)

BRAIN LESIONS (SAMPLING OF EXPERIMENTS)

1973	Isaković & Janković	Hypothalamus	Rat	Bovine serum albumin	Spleen, lymph nodes	Lymphocyte depletion
1974	Spector et al.	Several nuclei of hypothalamus	Rat	Plasmodium berghei	Parasitemia Ab to Plasmodium	Increase
1975	Schiavi et al.	Anterior hypothalamus	Guinea pig	Ovalbumin	Antibody production	No effect
1975	Spector et al.	Anterior hypothalamus	Rat	Ovalbumin	Antibody production	Decrease
1976	Goldstein	Medial hypothalamus	Rabbit	Ovalbumin	Anaphylaxis	Increase
1976	Paunović et al.	Medial and posterior hypothalamus	Rat*	Bovine serum albumin	Antibody production, Arthus and delayed skin reactions	Decrease
1977	Plezitiy	Dorsal hypothalamus	Rabbit	Bovine serum	Arthus reactiön	Decrease
1977	Spector et al.	Several nuclei of hypothalamlus	Mouse	Newcastle disease virus	Interferon, Antibody to NDV	No significant change
1978	Goldstein	Anterior and posterior hypothalamus & thalamus	Rat	Sheep red blood cells	Plaque-forming cells	No effect
1979	Dann et al.	Tuberal hypothalamus	Rat	Tissue	Skin graft	Stimulation of allograft rejection
1979	Dann et al.	Amygdala	Rat	Tissue	Skin graft	No effect
1980	Cross et al.	Anterior hypothalamus	Rat		Total cell number	Decrease thymus & spleen
1980	Keller et al.	Anterior hypothalamus	Guinea pig	Tuberculin	Skin-test	Suppression
1980	Keller et al.	Anterior hypothalamus	Guinea pig	(PHA)	Blastogenesis	Inhibition
1980	Warejcka & Levy	Hypothalamus	Rat	(PHA)	PHA response	Suppression
1982	Banet et al.	Hypothalamic preoptic area (cooling)	Rat	Sheep red blood cells	Antibody production	Increase
1982	Brooks et al.	Frontal cortex, ventromedial hypothalamus, amygdaloid complex, hippocampus anterior hypothalamus			Number of cells	No changes in thymus and spleen
1982	Cross et al.	Anterior hypothalamus	Rat		Thymus	Involution
1982	Cross et al.	Hippocampus	Rat		Thymus	Increased cellularity
1982	Roszman et al.	Preoptic and anterior hypothalamus	Rat	(Con A)	Blastogenesis	Impaired lymphocyte reactivity
1983	Renoux et al.	Cerebral cortex	Mouse		Thy-1 phenotype	Modulation
1984	Baçiu & Ivanov	Anterior hypothalamus	Rat	Salmonella enter. SRBC Myxovirus influ. A	Primary & secondary responses	No changes No changes Reduced
		Lateral hypothalamus	Rat	Salmonella enter. SRBC Myxovirus influ. A	Primary & secondary responses	No changes No changes Reduced
		Tuberal hypothalamus	Rat	Salmonella enter. SRBC Myxovirus influ. A	Primary & secondary responses	No changes No changes Suppressed
		Mammillary areas	Rat	Salmonella enter. SRBC Myxovirus influ. A.	Primary & secondary responses	No sign. change No sign. change Suppressed
1985	Baçiu et. al.	Tuber Cinereum of hypothalamus	Guinea pig	Staphylococcus	Phagacytosis	Decrease

(continued)

Table 1 (Continued)

BRAIN LESIONS (SAMPLING OF EXPERIMENTS)

1985	Biziere et al.	Frontal, parietal or occipital cerebral cortex: unilateral	Mouse	SRBC	PFC(IgG) count	Left: decrease Right: increase
					Weight and number of T-cells in spleen and thymus	Left: decrease Right: increase
				(PHA or Con-A)	Thymidine incorporation	Left: decrease Right: increase
1985	Markesbery et al.	Anterior hypothalamus, preoptic area (AHT)	Rat: young and old	(Con-A)	Thymidine incorporation	Aged: no change Young: decrease
		Hippocampus, frontal cortex			NK activity	Aged: decrease Young: decrease
1985	Ovaida et al.	Anterior hypothalamus	Rat	Rat spinal cord	Clinical signs of EAE	Decrease
		Areas reached via lateral ventricles	Rat treated with reserpine (Norepinephrinr)	" "	" "	"
		" "	Rat treated with 6-OHDA (Catecholamines)	" "	" "	"
		" "	Rat treated with 5,7-DHT (Serotonin)	" "	" "	"
1985	Cross et al.	Brain ventricles	Mouse	SRBC	IgM, IgG, PFC	Decrease
1985	Speplewski & Vogel	Brain ventricles	Rat treated with 5,6-DHT, (Serotonin)		Leucocyte count T-lymphocyte " NK activity	Increase Decrease Reduced
1985	Jankovic et al.	Epiphysis	Chick embryo		Histology	Bursa/Fabricius: size reduced
						Lymphocyte: count down
						Thymus: size reduced, number of pyroninophilic cells reduced
1985	Renoux et al.	Left neocortex Right neocortex	Mouse	SRBC	IgG-PFC	Left lesion-reduced (restored by imuthiol)
						Right lesion-increased
				Chick RBC	Cell-mediated cytotoxicity	No change
				Imuthiol	NK activity	Left lesion-reduction
1985	Wertman et al.	Anterior hypothalamus(AH) Hippocampus	Rat	Rat spinal cord Freund's Adj.	Ab to myelin basic protein (measure of EAE)	(AH) decreased
					Con A reactivity	(AH) increased

*Lesioned as newborns and tested as adults
In parenthesis: mitogens

The immune micromilieu is a multisystem (Janković, 1973, 1979) composed of components of the immune (lymphoepithelial), nervous and endocrine systems. Therefore, the immune response in vivo depends not only on lymphoid cells and their associates (e.g. macrophages, dendritic cells) but also on the function of nervous and endocrine systems. This generalization is especially applicable to the ontogeny of the entire immune complex, and implies that post-embryonic immune responses are rooted in the embryonic phase of dynamic interactions among constituents of the three systems. The nervous system and the immune system function simultaneously in the developing chick embryo (Romanoff, 1960). As for the nervous and endocrine system, the avian brain on the 12th day of incubation already possesses its essential features (Romanoff, 1960) whereas the hypophysis becomes active in the second half of incubation (Fugo, 1940). It has been postulated, therefore, that the brain exerts an early regulatory activity over the anterior hypophysis (Wingstrand, 1951). If the pituitary is the "master gland," then the brain (especially the hypothalamus) is the master of the master gland.

A basic question dealing with the neuroimmune relationship is: are there brain-related structural and functional changes of the immune system during embryogenesis? The answer to this question requires an adequate experimental approach. The surgical removal of the brain or of parts of the brain is a method of choice. Among other reasons: a large number of chemicals which influence the brain of the chick embryo have been found to exert, at the same time, toxic effects on a variety of tissues including the lymphoid ones (Romanoff and Romanoff, 1972). Earlier reports on destruction of the brain (Wolff, 1936) and surgical operations of the head (Schowing, 1959a, 1959b,1968a, 1968b) were concerned mainly with malformations in the development of skeletal elements of neurocranium of the avian embryo. The partial decapitation of the chick embryo which removes the forebrain and hypophysis usually was referred to as "embryonic hypophysectomy" (Fugo, 1940; Betz, 1967; Janković et al., 1978). Retardation in the development of non-lymphoid tissues of the chick embryo due to early hypophysectomy was reported (Fugo, 1940; Betz, 1967). In adult chickens, destruction of the paraventricular area and the anterior hypothalamus produced thymic atrophy (Kanematsu and Mikami, 1970).

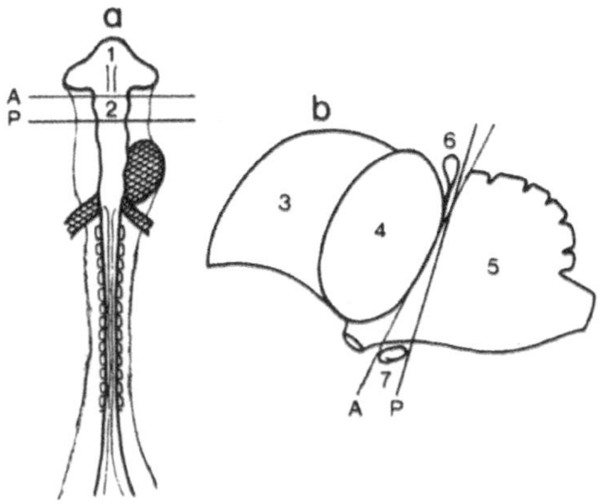

Fig. 7 Diagram of anterior and posterior decapitations performed on 33-hour-old chick embryos. Surgical procedures: A, anterior extirpation line (partial decapitation; Dx embryos), and P, posterior extirpation line (hypophysectomy; Hyx embryos). a. Embryo at the time of operation (about 33 h of incubation 13 somites): 1. prosencephalon, and 2. mesencephalon. b. 17-day-old embryo (lateral view of the brain): 3. forebrain, 4. optic lobe, 5. cerebellum, 6. epiphysis, and 7. hypophysis. Missing parts are 3, 4, and 6 in correctly partially decapitated embryos, and 3,4,6 and 7 in hypohysectomized embryos. (Modified from Janković et al., Develop. Comp. Immunol., 2:479, 1978.)

Experiments examining especially immuno-neuro-endocrine interactions in the developing chick embryo were performed by Janković and associates (Janković et al, 1977, 1978, 1979, 1981a, 1981b, 1982, 1983: Mićić et al., 1983). They developed a new procedure which enabled the removal of the forebrain and optic lobes while leaving the hypophysis intact (Figure 7) thus constituting a sound method for the study of neuroimmune interconnections in the chick embryo. The relationship between the brain, hypophysis and lymphoid organs (thymus and bursa of Fabricius) of the chick embryo is summarized in Table 2. Stereological analysis of the thymus epithelial cells situated in the cortex and medulla revealed an increase in the number of dense bodies and cysts. This implied that the secretory activity of thymic epithelial cells is under the influence of the brain and hypophysis. As for the bursa of Fabricius (a lymphoepithelial organ in birds), there was a pronounced retardation in the development of bursal follicles both in decapitated and hypophysectomized embryos (Figures 8-10). Operated embryos also showed a significant depletion of "light" and "dark" lymphocytes (Isaković et al., 1980) in the bursal follicles.

These studies provided evidence that the brain exerts both stimulatory and inhibitory effects on lymphoid tissues of the chick embryo and that communications between the nervous system and the immune system are functional during embryogenesis. The mechanisms by which brain hormones and neurally-active amines modulate the development and function of embryonic lymphocytes in general and in specific areas of the immune system remain to be elucidated. These mechanisms have been studied more in the adult animal, but only in some respects may be the same in the embryo.

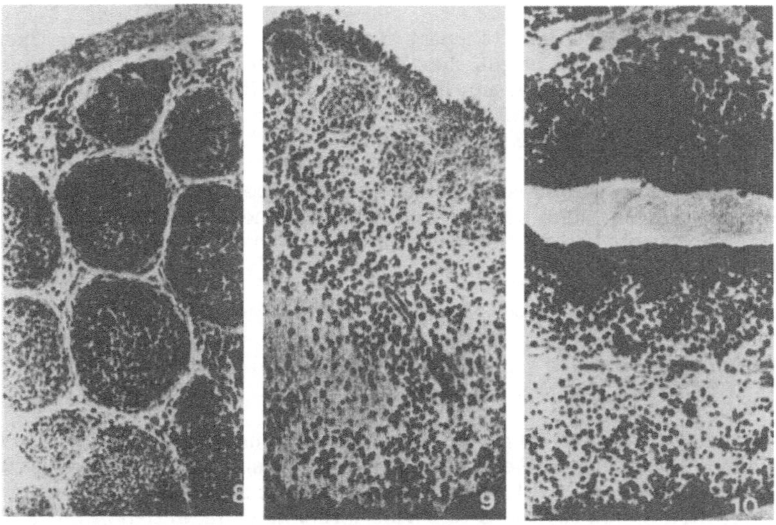

Fig. 8 Bursa of Fabricius of a 19-day-old sham-operated chick embryo Normal looking follicles (f). (X 320)

Fig. 9 Bursa of Fabricius of a 19-day-old decapitated (Dx) embryo (anterior decapitation which leaves intact the hypophysis.) Bursal follicles (f) are underdeveloped and plical mesenchyme (pm) contains large number of embryonic granulocytes. (X 320)

Fig. 10 Bursa of Fabricius of a 19-day-old hypophysectomized (Hyx) embryo (posterior decapitation which removes forebrain, optic lobes and hypophysis). Striking retardation in the development of the organ and follicles (f). (X 320)

All sections were stained with hematoxylin and eosin.

Table 2. Changes in the Thymus and Bursa of Fabricius of]7-Day Old Chick Embryos Surgically Sham-Decapitated (SDx) , Decapitated (Dx) or Hypophysectomized (Hyx) at 33-38 Hours of Incubation

Group	Region	Thymus: Stereological Analysis[a] of Epithelial Cells: Density of Cells[b]	Numerical Density of Dense Bodies	Numerical Density of Cysts	Bursa of Fabricius: Number of Follicles per 1 mm²	Number of Lymphocytes per Follicle: Light Cells	Dark Cells
SDx	Cortex	+	41.6	29.8	26.2	15.4	14.1
	Medulla	+	14.2	16.1			
Dx	Cortex	++	53.7[c]	69.4[c]	17.8[c]	7.9[c]	6.0[c]
	Medulla	++	19.3	32.8[c]			
Hyx	Cortex	+++	73.6[c]	109.2[c]	16.0[c]	5.1[c]	8.4[c]

[a] In analysis of electronmicrographs of the thymus stereological principles for morphometry were applied (Weibel, 1969).

[b] An arbitrary scale.

[c] Statistically significant difference determined by Student's t test by comparing values obtained from Dx and Hyx embryos to those from SDx control embryos.

All figures represent mean values of 14-35 embryos in group.

PINEALECTOMY AND IMMUNE RESPONSE

The epiphysis represents a part of the neuro-endocrine system (Thieblot, 1965). Although of brain origin, the pineal body receives autonomic peripheral innervation (Leeson and Leeson, 1976). In the rat, it seems that the pineal organ is innervated almost exclusively by orthosympathetic fibers originating in the superior cervical ganglia (Kappers, 1960).

The pineal's critical role in circadian and other biorhythms, in part via the superchiastmatic nucleus of the hypothalamus, is now well established, and there is beginning to accumulate a literature on the immunochronobiology of these structures.

The first reports on epiphysectomy are probably those of Exner and Boese (1910) and Foa (1912). The enhancement of transplantable tumor growth in animals without pineals has been repeatedly described (Rodin, 1963; Das Gupta and Terz, 1967; Lapin, 1974).

Janković et al. (1970) reported on the relationship between the pineal body and immune reactions in the rat. Neonatal epiphysectomy performed within 24 hours of birth did not exert any appreciable effect on Arthus reactivity and delayed hypersensitivity skin reactions to bovine serum albumin, anti-BSA antibody production, skin homograft rejection and incidence of experimental allergic encephalomyelitis in rats challenged at the age of 8 weeks. However, in rats pinealectomized at 6 weeks, Arthus and delayed skin reactions were reduced, and allergic encephalomyelitis was less expressed. Absolute lymphocyte and small lymphocyte counts were normal in rats pinealectomized at birth or at 6 weeks. The slight immunosuppressive effect of epiphysectomy in adult life was attributed to the lack of the secretory function of the pineal gland. Similarly, the removal of the epiphysis had little effect on skin homograft rejection, plaque-forming cells, haemagglutinin production and lipopolysaccharide (Rella and Lapin, 1976). The accelerated proliferation of immunocompetent cells in pinealectomized animals was attributed to the loss of the pineal antimitotic activity by the action of melatonin (Banerjee and Margulis, 1973).

It has been suggested that pinealectomy affects the function of T-suppressor lymphocytes and B-lymphocytes by reducing the formation and/or action of prostaglandin E_1 (Dunnane et al., 1979). Studies performed on chick embryos pinealectomized at 6 days of incubation revealed that the thymus and spleen of pineal-less embryos did not differ grossly and histologically from control embryos (Stalberg, 1965). It should be emphasized, however, that regeneration of pineal tissue occurred to some extent, even in completely pinealectomized birds, a fact which makes difficult the interpretation of the results.

Some data suggested the presence of lymphoid cells in the pineal gland of the avian central nervous system (Romieu and Jullien, 1942; Spiroff, 1958; Quay, 1965), whereas other studies failed to demonstrate lymphoid elements in the pineal body (Kappers, 1976). Lymphoid structures are present in the epiphysis of the mouse (Abe, et al., 1971). In the pineal region of the rat, the appearance and gradual increase of T-lymphocytes, beginning from the twentieth day after birth, has been reported (Uede et al., 1981). Studies of the pineal gland of 6-week-old chickens revealed that suspensions of pineal lymphocytes included 42% B-lymphocytes and 51% T-lymphocytes (Cogburn and Glick, 1983). Labeled bursal and thymic lymphocytes migrated to the pineal body, thus suggesting that both the thymus and bursa of Fabricius contributed to pineal lymphoid tissue. Moreover, injection of bovine serum albumin (BSA) into the carotid artery stimulated the production of anti-BSA antibodies in the lymphoid mass of the pineal gland. These studies indicated that the pineal gland is a functional part of the chicken lymphomyeloid system (Cogburn and Glick, 1983). Failure to identify immature forms of lymphocytes in the avian pineal body indicated that pineal lymphocytes are of extrapineal origin (Spiroff, 1958; Boya and Calvo, 1978). The invasion of the chicken epiphysis by lymphocytes was found to be much more massive at 1 month than at 4 months of age (Cogburn and Glick, 1981).

The ontogeny and function of lymphocytes in the pineal body remains unknown at present, and it appears that its T-lymphocytes are derived from T-lymphocyte-forming sites, such as the thymus. It has been reported recently (Janković et al., 1985) that early pinealectomy affects the thymus and bursa of Fabricius of the developing chick embryo, thus suggesting both a structural and a functional relationship between the pineal body and immune system during embryogenesis.

Because lymphocytes accumulate in the pineal capsular tissue, and because the choroid plexus of human, monkey, and rabbit contain receptors for the Fc fragment of immunoglobulin-G and for complement factor C3b (Breathen et al., 1979), it has been hypothesized that lymphocytes from the pineal body and the choroid plexus exert an immune function in the domain of the brain. Round aggregates of mononuclear cells resembling lymphoid follicles and germinal centers were described in the choroid plexus of chickens which developed experimental allergic encephalomyelitis following challenge with chicken or bovine spinal cord in complete Freund's adjuvant (Janković and Isvaneski, 1963).

RECEPTORS

The mechanisms of NIM at the cellular level began to emerge more clearly with the discovery of receptor site on the cells of the immune system for many neurotransmitters and neuromodulators. It seemed that whenever one looked for a receptor for a neurally-active polypeptide or a lymphocyte, it was almost always there! A representative sampling of these is shown on Table 3. Others, especially for endogenous opiates are discussed in more detail in Chapters 81, 263 of this book. There are receptors for a large number of intrinsic and extrinsic ligands on the surface membranes of immunocompetent cells (Cinander, 1977; Schulster and Levitzki, 1980).

Probably the first paper which called attention to the role of catecholamines in the immune response was that of Draškoci and Janković (1964): this described impaired delayed skin hypersensitivity reactions, antibody production and homograft rejection in rats treated with reserpine. Post mortem inspection of lymphoid tissues of reserpine-treated rats revealed pronounced changes in the cellular make-up of the thymus, spleen and lymph nodes. These changes were attributed, inter alia, to the lowering of catecholamine levels in different tissues caused by reserpine. Later investigations provided evidence that catecholamines can modulate cytotoxic activity mediated by T-lymphocytes (Henney et al., 1972), lymphocyte transformation induced by mitogens (Hadden, 1975), and antibody response to thymus-independent antigens (Miles et al., 1981).

We should note also the other side of the coin: production of lymphokines and immune-cell-type-products by cells of the central nervous system (e.g. Fontana, et al, 1985; Ovadia, et al 1985) as well as the production of neuroendocrine-like substances by lymphocytes (e.g. Johnson, et al., 1982; Smith and Blalock, 1985). These findings help us to visualize some of the means by which cells of the immune and nervous systems talk to each other.

Table 3
RECEPTORS FOR NEUROTRANSMITTERS, NEUROPEPTIDES, MEDIATORS AND MODULATORS ON THE LYMPHOCYTE MEMBRANE: DIRECT AND INDIRECT EVIDENCE
(A Partial List)

Receptor (for)	Cells	References
Catecholamines	Lymphocytes	Draškoci & Janković, 1964
Catecholamines	Lymphocytes	Henson et al., 1970
Noradrenaline	Lymphocytes	Hadden et al., 1970
Beta-adrenergic	Lymphocytes	Henney et al., 1972
Nicotinic cholinergic	Lymphocytes	Strom et al., 1972
Adrenaline	Lymphocytes	Sherman et al., 1973
Acetylcholine	Lymphocytes	Illiano et al., 1973
Adrenergic and cholinergic	Lymphocytes	Schreiner & Unanue, 1975
Adrenaline	Lymphocytes	Yu & Clements, 1976
Dopaminergic	Lymphocytes	Mitrova & Mayer, 1976
Beta-adrenergic	Lymphocytes	Williams et al., 1976
Cholinergic	Lymphocytes	Strom et al., 1977
Beta2-adrenergic	Thymocytes	Eskra et al., 1978
Dopaminergic	Lymphocytes	Ferguson et al., 1978
Muscarinic cholinergic	Lymphocytes	Gordon et al., 1978
Muscarinic cholinergic	T-cells	Lane & Strom, 1978
Menthionine-enkephalin	T-cells	Wybran et al., 1979
Morphine	T-cells	Wybran et al., 1979
Nicotinic cholinergic	Lymphocytes	Richman & Arnason, 1979
Beta-endorphin	Lymphocytes	Hazum et al., 1979
Beta-adrenergic	Thymocytes	Singh et al., 1979
Muscarinic cholinergic	T-cells	Shapiro & Strom, 1980
Beta-adrenergic	T and B cells	Bishopric et al., 1980
Dopaminergic	Lymphocytes	Le Fur et al., 1980
Substance P	Neutrophils	Marasco et al., 1981
Dopaminergic (D-1)	Spleen lymphocytes	Spetien et al., 1981
Dopaminergic	B cells	Le Fur et al., 1981
Beta-adrenergic	Lymphocytes	Brodde et al., 1981
Beta-endorphin	Spleen T cells, but not B cells	Gilman et al., 1982
Leucine-enkephalin	Lymphocytes	Plotnikoff, 1982
Adrenaline	Lymphocytes	Krall et al., 1982
Substance P	T-cells	Payan et al., 1983
Methionine-enkephalin	T-cells	Miller et al., 1983
Beta-endorphin	Natural killer cells	Mathews et al., 1983
Menthionine-enkephalin	Natural killer cells	Mathews et al., 1983
Enkephalins (met-and leu-enkephalin)	T-cells	Miller et al., 1984
Opioids	Natural killer cells	Shavit et al., 1984
Enkephalins	Natural killer cells	Faith et al, 1984
Somatostatin	T and B lymphocytes	Payan & Goetzl, 1986

PATHWAYS AND FEEDBACKS

The mechanisms by which brain lesions and brain stimulation alter immune repsonses are still not fully understood. Nonetheless, we do know that the central and peripheral nervous systems have two principal lines of communication with the organs, tissues and cells of the immune system: (1) endocrine-chemical, via the circulatory system, and (2) direct, "hard-wired" via nerve fibers, (principally the autonomic nervous system (Spector 1979, 1980, 1983). Feedback routes that modulate the modulation follow the same general pathways (Spector & Korneva, 1981).

It is now known also that leukocytes can produce certain neurally active polypeptides, which in turn act as lymphokines as well as neuromodulators (Blalock and Smith, 1985; see also page 119 in this volume): these, too, will undoubtedly be shown to be part of feedback cycles. In addition, other hormone-like polypeptides secreted by the thymus (thymosins) may also relay information, not only to the immune system, but to the brain. Hall and Goldstein (1985) have proposed the term "immunotransmitter" to describe the class of circulating substances that relay information back to the central nervous system.

Many circulatory connections are today fairly well recognized (e.g. steroids, catecholamines, neurally-active polypeptides, etc.) and need not be elaborated upon here. Other chapters in this book deal with the effects of opioids. The direct neural connections are less well recognized despite a long documentation of their existence. It is clear, also, that nerve endings exist in the parenchyma of these organs, in direct contact with the blast cells and with mature cells of the immune system. Table 4 provides partial lists, in chronological order of some of the studies describing the innervation of the thymus, the bone marrow, the lymph nodes and the spleen. Other tissues, in various species, related to the immune system are similarly innervated (Bulloch, 1985): we will not list all of them here, but we can mention briefly various elements of the reticuloendothelial system, Peyer's patches, appendix, tonsils, lacrimal glands (e.g. Harderian gland), jugular body and bursa of Fabricius.

Table 4
INNERVATION OF THE IMMUNE SYSTEM
(Partial List in Chronological Order)

I. THYMUS

1899	Bovera	1950	Tcheng
1911	Hallion and Morel	1955	Knoche
1912	Hallion and Morel	1966	Fujiwara
1915	Sjolander and Strandberg	1968	Solov'ev
1918	Crotti	1968	Shavlev
1923	Braeucker	1974	Serge'eva
1928	Terni	1980	Bulloch and Moore
1929	Pines and Majman	1980	Bulloch and Loy
1929	Terni	1980	Bulloch and Moore
1931	Cabanac	1980	Ghali et al.
1931	Terni	1981	Bulloch and Moore
1932	Hammar	1981	Williams and Felten
1933	Cordier	1981	Williams et al.
1933	Terni and Muratori	1982	Bulloch
1934	Koestowiecki	1983	Bulloch and Cullen
1935	Hammar	1983	Bulloch et al.
1938	Kostowiecki	1983	Cullen and Bulloch
1943	Bargmann	1985	Bulloch

(Continued)

Table 4 (Continued)
INNERVATION OF THE IMMUNE SYSTEM
(Partial List in Chronological Order)

II. BONE MARROW

1846	Gros
1880	Variot and Remy
1902	Ottolenghi
1925	Castro
1928	Glaser
1929	Rossi
1929/30	Castro
1936	Takeyama
1945	Kuntz
1952	Fedoroff et al.
1968	Calvo
1970	Calvo
1973	Miller and McCuskey
1981	Calvo
1985	Bulloch

III. SPLEEN

1928	Glaser
1929	Riegele
1944	Harting
1948	Tischendorf
1952	Lentz
1955	Noerthen
1965	Dahlstrom
1966	Gillespie and Kirpekar
1970	Gersbach
1970	Fillenz
1973	Zetterstrom et al.
1976	Reilly, et al.
1979	Reilly et al.
1979	Kudoh et al.
1981	Williams and Felten
1981	Williams et al.
1985	Bulloch

IV. LYMPH NODES

1899	Tonkoff
1965	Volik
1979	Gordon et al.
1980	Giron et al.
1980	Kiss
1985	Bulloch

REVIEW PAPERS: INNERVATION OF IMMUNE SYSTEM

1981	Jankovic et al.
1982	Bulloch
1983	Spector
1985	Bulloch

BRAIN STIMULATION

Brain lesions are useful, but not perfect, tools for studying NIM. A lesion in one area may destroy a "center" (if indeed, one function can ever be relegated to only one center: see Spector, 1979 and 1980 for a discussion of this issue), interrupt a fiber pathway, change a circulatory route, interfere with a secretory process or effect any combination of these. Another difficulty is that no two lesions are ever precisely alike.

Brain stimulation techniques overcome some of these difficulties: if done correctly, a minimum of damage is done, and "normal" "physiologic" processes can be more nearly approximated. However, stimulation experiments have their difficulties also: stimulation spread to other nuclei and to other systems; rebound problems; puncture of the meninges and of brain tissue by the electrodes; the difficulties of finding optimal stimulaton wave form, amplitude and train parameters; and others.

Nonetheless, these experiments have had great value. In some cases (e.g. Korneva and Khai, 1967) stimulation produced the opposite effect from a lesion in the same area, indicating that there was a good chance that we were close to honing in on critical brain areas. In another instance, one of the earliest such experiments, an ingenious one (Baçiu, 1946) remains almost the only physiological (as opposed to anatomic or biochemical evidence) demonstration of a direct, "hard-wired" functional connection between the brain and peripheral hemopoiesis; further experiments by the same group showed a similar neural connection between the brain and the host-defense system; i.e. electrical stimulation of the brain changed the rate of phagocytosis. More recently, Janković et al. (1983b) demonstrated that electrical stimulation of two areas: dorsomedial nucleus of the hypothalamus and of the sensorimotor cortex of the cerebrum in rats resulted in significant changes in Arthus skin reactions. Several investigators have measured changes in other immune reactions following electrical stimulation of various neclei of the hypothalamus. Table 5 outlines briefly, in chronological order, some experiments in NIM utilizing brain stimulation as a tool.

Table 5
BRAIN STIMULATION: (Partial List):

Date, Authors	Site of Stimulation	Animal	Antigen	Immune Response	
				Assay	Results
1949, Benetato et al.	Hypothalamus	Dog		Phagocytosis	Increase
1957, Maslinski & Karszewski	Brain structures			Histamine shock	Reduction
1961, Halvorsen	Hypothalamus	Rabbit		Hematopoiesis	Increase
1961, Petrovski	Hypothalamic tubal region	Rabbit	Bacterial	Antibody production	Increase
1961, Petrovski	Mammillary bodies	Rabbit	Bacterial	Antibody production	Increase
1964, Binder	Anterior hypothalamus	Guinea pig	Horse serum	Anaphylaxis	Decrease
1967, Cipin & Malcev	Posterior hypothalamus	Rabbit		Natural agglutinins	Decrease
1967, Korneva & Khai	Various structures of mesencephalon	Rabbit	Horse serum	Antibody production	Increase
1969, Shiotani & Ban	Hypothalamus (b-sympathetic and c-sympathetic zones)	Rabbit		Thymus	Involution
1973, Abramchik & Shuvalova	Anterior and posterior hypothalamus	?	Spinal cord	Experimental allergic encephalomyelitis	Affected
1977, Plezitiy et al.	Dorsal hippocampus	Rabbit	Bovine serum	Arthus reaction	No effect
1979, Janković et al.	Hypothalamic dorsomedial nucleus	Rat	Bovine serum albumin	Arthus reaction Delayed skin reaction	Increase Increase
1981, Lambert et al.	Hypothalamic ventromedial area	Rat		Phagocytosis	Decrease
1983a, Janković et al.	Hypothalamic dorsomedial nucleus and sensorimotor cortex	Rat	Bovine serum albumin	Arthus and delayed skin hypersensitivity; antibody production	Increase
" " "	"	"	"	Thymus	Moderate increase in lymphoid cells
" " "	Hypothalamic ventromedial nucleus	"	"	Immune responses and thymus	No effect
" " "	Hypothalamic posterior nuclei	"	"	" "	Partial increase

POSSIBLE EXPLANATIONS OF THE EFFECTS EXERTED ON IMMUNE RESPONSES BY LESIONING AND STIMULATION: PATHWAYS AND MECHANISMS

Possible Pathways

Pathway One is: hypothalamus → pituitary → adrenals (glucocorticoids) and gonads (gonadal steroids) → thymus and lymphocytes.

Since the rat thymus contains high affinity androgen (Grossman et al., 1978; Stimson and McCruden, 1981) and estrogen (Grossman et al., 1979) receptors, it is possible that those receptors play a role in the determination of the thymus function. It is well established that the secretion of adrenal corticosteroids is under direct neural control (Mangilli et al., 1966), that there are steroid-sensitive neurons in the hypothalamus (Ruf and Steiner, 1967) and that steroids may play an essential role in the induction of immune response (Ambrose, 1970). The functional heterogeneity of lymhocytes appears to be a reflection of their sensitivity to corticosteroids (Claman, 1972; Fauci, 1976). Specific effects of steroid binding to the thymus (Munk et al., 1971) and variations in the number of glucocorticoid receptors on the lymphocyte membrane (Smith et al., 1977; Crabtree et al., 1980) are consistent with this immunoneuroendocrine pathway.

Pathway Two: hypothalamus → adenohypophysis → somatotropic hormone and/or thyroid hormones → lymphocytes.

Developmental hormones (Pierpaoli et al., 1970; Ahlqvist, 1976) and thyroid hormones (Leger and Masson, 1947; Nilzen, 1954) were shown to influence the immune machinery of the body. Thus, the presence of specific hormone receptors on the lymphocytes (Rimon et al., 1978; Harrison et al., 1979) (see also Table 3) would imply that specific regulation (lymphocytes, antibodies) and hormonally controlled non-specific regulation may work together (Besedovsky and Sorkin, 1977). Cyclic adenosine monophoshate (cAMP) is probably the most common mediator: hormones influence the formation of cAMP in the lymphocyte (MacManus et al., 1971).

Pathway Three: hypothalamus → fasciculus longitudinalis dorsalis (Schutz) postganglionic sympathetic fibres → (norepinephrine and other neurotransmitters) lymphocytes.

This interpretation, suggested in an earlier paper (Janković and Isaković, 1973), concerns the róle of the sympathetic nervous system in the regulation of immune responses. The autonomic innervation of the thymus (Hammar, 1935; Gordon et al., 1979; Ghali et al., 1980; Bulloch, 1985) points to specific neuronal pathways. Indeed, the presence of adrenergic receptors on human lymphocytes (Hadden et al., 1970; see Table 3) and the general involvement of the sympathetic system in immune reactions (Korneva and Khai, 1961; Kozlov, 1973; Williams et al., 1981; Del Rey et al., 1981) have been reported. It is worthwhile mentioning here the presence of both nicotinic (Fuchs, 1979) and muscarinic (Maślinśki et al., 1980) receptors for acetylcholine on the lymphocyte membrane. The increase of cyclic GMP level (Strom et al., 1974) in the lymphocyte induced by acetylcholine would indicate that this neurotransmitter may act as a regulatory ligand (Triggle, 1971) in immunity, suggesting the involvement of cholinergic mechanisms in immune phenomena. Pharmacological influences upon the immune response have been reviewed recently (Maestroni and Pierpaoli, 1981).

Pathway Four: hypothalamus (with hypothetical thymotropic hormone-releasing factor) → adenohypophysis → thymotropic hormone → lymphocytes (Isakovic and Janković, 1973).

Since the first demonstration that the extract from the rat thymus restored immunological potential of rats thymectomized at birth (Janković et al., 1965), the humoral function of the thymus now is well documented (Luckey, 1973; Hall and Goldstein, 1985).

Pathway Five: hormones from different brain structures and centrally acting peptides (Hughes, 1978) ⟶ endocrine glands and lymphoepithelial system.

Several peptide and protein hormones, which are otherwise regarded to be of pituitary origin, are synthesized within the brain (Krieger and Liotta, 1979). These hormones of the brain may regulate not only the secretion of the anterior pituitary (Guillemin, 1978) but also the metabolic processes in the lymphocytes. Thus, it seems that low-molecular polypeptides from the calf cereral cortex were capable of altering the immune response in mice when injected into animals before immunization with sheep erythrocytes (Belokrylov and Sofronov, 1980). Experiments on mice with surgical lesions of the cerebral cortex suggested that the neocortex specifically modulates events mediated by T cells and not by B cells (Renoux, 1980; Renoux et al., 1983). Of particular interest are enkephalins which constitute a major neuroendocrine link between the central nervous system and the immune system (Plotnikoff and Miller, 1983). Their activity in "stress" processes are related to immune responses (Amir et al., 1980; see also pages 221 and 425 in this volume). It has been reported that human blood T-lymphocytes contain receptors for morphine and methionine-enkephalin (Wybran et al., 1979). In mice, met-enkephalin and leu-enkephalin were found to stimulate lymphoblastogenesis in the presence of PHA (Plotnikoff and Miller, 1983). In humans, enkephalins were found to increase significantly the formation of active T-cell rosettes (Miller et al., 1984). It appears that endogenous enkephalins have immunostimulatory activity in vitro (Plotnikoff and Miller, 1983; see also pages 35, 209 and 129, this volume.)

Pathway Six: follows the peripheral divisions of the diffuse neuroendocrine system (Pearse and Takor, 1979).

This system, which probably is the largest organ of the body, and has an excellent strategic position for participation in the neuroimmunomodulation, may exert an influence on lymphocytes and their related cells in the peripheral lymphoid organs.

Pathway Seven: In the case of pathology, as with artificial lesions in experiments animals, other pathways and other factors play a role. For example, any rupture of the meninges alters the blood-brain barriers. For discussion of such "abnormal" and experimental pathways, see Spector, 1979. Suffice it to mention here that in prehistoric times, man probably cured some diseases by trephining the skull (Guiard, 1930) and in "modern" times, i.e. in the 1920's and '30's, A.D. Speranskii (1934) and his colleagues claimed to have cured a host of diseases by "pumping" cerebrospinal fluid until it turned red!

The above-mentioned pathways of neuroimmunomodulation serve ultimately in the transmission of a variety of inhibitory and excitatory information (signals), different in nature and origin, to the intracellular machinery via the ligand-receptor-cAMP/cGMP system (Schulster and Levitzki, 1980) of the lymphoid and nonlymphoid cells of the immune micromilieu (Janković, 1973, 1979). Since the hypothalamus is associated with other brain structures by diffuse fibre systems (Nauta and Haymaker, 1969; Morgane and Panksepp, 1979) and peptidergic neural network (Renaud, 1978), the lesioning and stimulation of the hypothalamus and other brain regions may cause the propagation of impulses (signals, information) via neural fibres to the pituitary gland, adrenergic system, endocrine glands, diffuse neuroendocrine system, epithelial system, and lymphocytes and their associates, and thus modulate the immune response (briefly reviewed by Spector, 1983). The direct "hard-wired" pathways, afferent and efferent, must be considered along with circulatory pathways. In general, brain lesioning and stimulation upset the balance between the inhibitory and facilitatory factors (cellular, subcellular, humoral, molecular and ionic) which control the normal immune response: however, such experiments help greatly to delineate the "normal" physiologic pathways.

CONCLUSIONS AND SUMMARY

In this chapter, we have outlined part of the history of modern research on neuroimmunomodulation (NIM), emphasizing the contributions made to our knowledge by experiments utilizing partial ablations (lesions) of specific brain areas and by other experiments designed to stimulate many of these areas. Such experiments have demonstrated, in many species, the direct influences of the central nervous system upon many immune responses cellular and humoral, and upon many related host-defense mechanisms. While some of the conclusions based on this literature appear to be contradictory, we can say with certainty:

(1) That the nervous, immune and endocrine systems are inextricably linked via numerous feedback loops,

(2) that several nuclei in the hypothalamus, the corpora quadrigemina, the cerebral cortex, the reticular formation of the brainstem and the sympathetic nervous system in general, all play a role in immunity,

(3) that the two-way communicaton is carried out via circulatory and direct neural fiber routes, and

(4) that molecular links among the nervous, endocrine, and immune systems are apparent among the membrane receptor sites on cells of the immune system for neurally-active polypeptides, neurotransmitters and hormones; likewise receptors exist on cells of the nervous system for bioactive substances of the immune system,

(5) that links among the nervous, endocrine and immune systems develop even in the embryo.

Other mechanisms of NIM are explained in detail in many of the chapters in this book.

*POST-SCRIPT

In our introduction and in the table of experiments dealing with brain lesions and immunity, we restricted the list to an outline of "modern" experiments. It should be mentioned, however, that the earliest known experiments were those of Savchenko (1891) and London (1899) which showed that spinal section or removal of the cerebral hemispheres in pigeons made them susceptible to anthrax. Intact pigeons are not susceptible.

Savchenko, I. G. On the unsusceptibility to anthrax (in Russian). Vrach. 5: 132-134, 1891.

London, Ye.S. On the influence of the removal of various parts of the brain of pigeons on their immunity to anthax (in Russian) Arkh. Biol. Nauk 7:177-187, 1899.

The reference list at the end of this chapter should serve as a guide to the literature. In addition, a compilation of over a thousand abstracts covering a five year period in "Mind and Immunity" experiments appeared in 1984 (Locke, et al.), and a collection of important papers going back to the 1920's was published in 1985 (Locke, et al.). Neither of these volumes are comprehensive, but both contain important pieces of the big picture: other recent reviews are included in Ader (1983), Korneva et al. (1985), Guillemin, et al. (1985), Spector, (1979, 1983) and Janković (1985).

For those interested in historic roots, a very early book on NIM, one of great scientific importance and one well worth reading, is that of Metalnikov (1934).

REFERENCES

Abe, K., Matsushima, S. Kachi, T., and Ito, T. Lymphoid tissue in the pineal region of the mouse: a histological and histochemical study, Arch. Histol. Jap., 33:263, 1971.

Abinder, A.A., The effect of electric stimulation of the anterior portion of the hypothalamus on reconstitution of immune reaction of the body, Zh. Mikrobiol. Epidemiol. Immunobiol., 41:47, 1964.

Abramchik, G.V., and Shuvalova N.S. The role of the hypothalamus in the pathogenesis of experimental allergic encephalomyelitis, Zh.Nevropatol.Psikiatr., 73:988, 1973.

Ahlqvist, J. Endocrine Influence on Lymphatic Organs, Immune Responses, Inflammation and Autoimmunity. Almqvist and Wiksell Int. Stockholm, 1976.

Ambrose, C.T., The essential role of corticosteroids in the induction of the immune response in vitro, in Hormones and Immune Response, G.E.W. Wolstenholme and J. Knight, eds. Churchill, London, pp.100, 1970.

Amir, S. Brown, Z.W. and Zelman, A., The role of endorphins in stress: evidence and speculations. Neurosci. Biobehav.Rev., 4:77, 1980.

Baçiu, I., The role of the central nervous system in the inducement of the phagocytic reaction. Doctoral dissertation, Institute of Physiol. and Med. Physics, Univ. of Cluj, 1946. (Romanian: English translation available.)

Baçiu, I., and Ivanov, A. The role of hypothalamic centres in the immune specific response., Physiologie, 21(4):251,1984.

Baçiu, I., Olteanu, A., Prodan, T. Baiescu, M. and Vaida, A. Changes of phagocytic biological rhythm by reproduction of circadian times and by influences upon hypothalamus. Proc., First Int. Workshop on NIM, Bethesda, Md., November 1984, in press, 1985.

Banerjee S., and Margulis, L. Mitotic arrest by melatonin, Exp.Cell.Res., 78:314, 1973.

Banet, S., Brandt, S. and Hensel, H. The effect of continuously cooling the hypothalamic preoptic area on antibody titre in the rat, Experientia, 38:965, 1982.

Bargmann, W., Der Thymus, in Handbuch der Mikroskopischen Anatomie., Springer, Berlin. VI (4): 1-145, 1943.

Barone, R.M., and Das Gupta, T.K. Role of pinealectomy on the Walker 256 carcinoma in rats, J. Surg. Oncol., 2:3132, 1970.

Belokrylov, G.A. and Sofronov., B.N., Effect of thymic and cortical low-molecular polypeptides on different stages of immune response in mice. Immunologia, 4:66, 1980.(in Russian).

Benetato, G., Oprisiu, C. and Baciu, I. Sur le role du systeme nerveux central dans le declenchement de la reaction phagocytaire, Recuil d'Etudes Medicales, Ed. Inst. de Cultura Universala. Bucharest, 1949.

Besedovsky, H.O. and Sorkin, E., Network of immune-neuroendocrine interactions. Clin. Exp. Immunol., 27:1, 1977.

Betz, T.W., The effects of embryonic pars distalis grafts on the development of hypophysectomized chick embryos, Gen.Comp.Endocrinol., 9:172, 1967.

Bishopric, N.H., Cohen, H.J. and Lefkowitz, R.J. Beta-adrenergic receptors in lymphocyte subpopulations, J. Allergy Clin. Immunol., 65:29, 1980.

Bizière, K., Renoux, M. and Renoux, G. Modulation of the T-cell lineage by the cerebral neocortex, Proc., First Intl. Workshop on NIM, Bethesda, MD, Nov. 1984, in press, 1985.

Blalock. J.E., and Smith, E.M. Human leukocyte interferon: structural and biological relatedness to adrenocorticotropic hormone and endorphin, Proc. Natl.Acad. Sci. USA, 77:5972, 1980.

Bovera, A., Sui nervi della ghiandola di timo, Giron Acad. Med. Torino, 62:1, 1899.

Bowen, F.P., Immunological reactions after cortical lesions in rabbits, Arch. Neurol., 19:398, 1968.

Braeucker, W., Die Nerven des Thymus., Z. Anat. Entw., 69:309, 1923.

Breathen, L.R., Forre, O.T. Husby, G., and Williams, R.G., Jr. Evidence for Fc receptors in human choroid plexus, Clin. Immunolo. Immunopathol., 14:284, 1979.

Brodde, O.E., Engel, G. Hoyer, D., Bock, K.D., and Weber, F. The β-adrenergic receptor in human lymphocytes: Subclassification by the use of a new radio-ligand, ()-125 Iodacyanopindolol, Life Sci., 29:2189, 1981.

Brooks, W.H., Cross, R.J. Roszman, T.L., and Markesbery, W.R. Neuroimmunomodulation: neural anatomic basis for impairment and facilitation, Ann. Neurol., 12:56, 1982.

Bulloch, K., Neuroendocrine-immune circuitry pathways included with the induction and persistence of humoral immunity, University Microfilm International, Ann Arbor, MI, 727-1064 (Ph.D. dissertation, 1981, UCSD), 1982.

Bulloch, K., A light and ultrastructural analysis of innervation of the thymus gland during the perinatal period, Neurosc. Abst., 20: 572, 1982.

Bulloch, K., Neuroanatomy of lymphoid tissue: a review, in Neural Modulation of Immunity, Guillemin, R., Cohn,M. and Melnechuk, T.,eds. Raven Press, N.Y., pp. 111- 141, 1985.

Bulloch, K., and Cullen, M.R. An analysis of the thymic CNS relationship in the chick, Immunology, Fifth International Congress of Immunol. (Abst.), 1983.

Bulloch, K., Cullen, M.R. Davis, M.L. and Schwartz, R.H. Neuroimmunology of the thymus gland, Neurology, 33(4) Suppl. #2:194,pp 187 ff., 1983.

Bulloch, K., and Loy, R. The development of innervation in the thymus gland of wildtype and of the neuroimmune mutant Staggerer, Society for Neurosciences, Abstract. 26.5, 1980.

Bulloch, K., and Moore, R.Y. Central nervous system projections to the thymus gland. Possible pathways for the regulation of the immune response, Anat. Rec., 196:25A, 1980a.

Bulloch, K., and Moore, R.Y. Nucleus ambiguous projections to thymus gland -Possible pathways for regulation of the immune response and the neuroendocrine network, Am. Ass. Anat. (Abstr.), 25A, 1980b.

Bulloch, K. and Moore, R.Y. Thymus gland innervation by brainstem and spinal cord in mouse and rat, Am. J. Anat., 162:157, 1981.
Cabanac, J., Les nerfs du thymus, Bull. Assoc. Anat., 25:97, 1931.
Calvo, W., The innervation of the bone marrow in laboratory animals, Am. J.Anat., 123:315, 1968.
Calvo, W., Bone marrow hemopoiesis in the human fetus, in Adv. Physiol. Sci., Vol. 6 Genetics, Structure and Function of Blood Cells, S.R. Hollan, G. Gardos, B. Sardaki, eds., Akademiak Kiado, Budapest, 1981.
Calvo, W., and Forteza-Vila, J., Schwann cells of the bone marrow., Blood, 36: 186, 1970.
Castro, F. de, Technique pour la coloration du systeme nerveux quand il est porvu de ses etuis osseux, Trav. Lab. Rech. Biol. Univ. Madrid., 23:427, 1925.
Castro, F. de, Quelques observations sur l'intervention du système nerveux autonome dans l'ossification. innervation du tissu osseus de la moel osseuse, Trav. Labor. Rech. Biol. Univ. Madrid, 26:215, 1929.
Cinader, B. (ed.), Immunology of Receptors, Marcel Dekker, New York, 1977.
Cipin, A.B., and Malcev. V.N. Effect of hypothalamic stimulation on normal antibodies in the blood, Patol. Fiziol., 11:83 (in Russian), 1967.
Claman, N.H., Corticosteroids and lylmphoid cells. New Engl. J. Med., 287:388, 1972.
Cogburn, L.A., and Glick, B. Lymphopoiesis in the chicken pineal gland, Am.J.Anat., 162:131, 1981.
Cogburn, L.A., and Glick, B. Functional lymphocytes in the chicken pineal gland, J.Immunol., 130:2109, 1983.
Cordier, P. and Coulouma, P. Les nerfs du thymus, Ann. Anat. Path., 1104, 1933.
Crabtree, G.R., Munck, A. and Smith, K.A., Glucocorticoids and lymphocytes. I. Increased glucocorticoid receptor levels in antigen-stimulated lymphocytes. J. Immunol., 124:2430, 1980.
Cross, R.J., Brooks, W.H. and Roszman, T.L. Hypothalamic-immune interactions. I. The acute effect of anterior hypothalamic lesions on the immune response, Brain Res., 196:79, 1980.
Cross, R.J., Brooks, W.H. Roszman, T. L., and Markesbery, W.R. Hypothalamic-immune interactions. Effects of hypophysectomy on neuroimmunomodulation, J. Neurol. Sci., 53:557, 1982.
Cross, R.J., Jackson, J.C. Markesbery W.R., Brooks, W.H. and Roszman, T.L. Modulation of Immune Function by Electrolytic and Chemical Lesions of the Central Nervous System, Proc., First Intl. Workshop on NIM, Bethesda, Md. Nov. 1984, in press, 1985.
Cross, R.J., Markesbery, W.R., Brooks, W.H. and Roszman, T.L. Hypothalamic-immune interactions: Neuromodulation of natural killer activity by lesioning of the anterior hypothalamus. Immunology 51:399.
Crotti, A.,Thyroid and Thymus. Lea & Febiger, Philadelphia and New York, pp 536-559, 1918.

Cullen, M.R., and Bulloch, K. Innervation of thymus transplants in nude mice: An ultrastructural study, Society for Neuroscience, 9(1): 117, Abstract # 3410, 1983.
Cunnane, S.C., Manku, M.S. and Horrobin, D.F. The pineal and regulation of fibrosis: pinealectomy as a model of primary biliary cirrhosis: roles of melatonin and prostaglandins in fibrosis and regulation of T-lymphocytes, Med. Hypotheses, 5:403, 1979.
Dahlström, A.B., and Zetterstrom, B.E.M Noradrenaline stores in nerve terminals of the spleen: changes during hemorrhagic shock, Science, 147:1583, 1965.
Dann, J.A., Wachtel, S.S. and Rubin. A.L. Possible involvement of the central nervous system in graft rejection, Transplantation, 27:223, 1979.
Das Gupta, T.K., and Terz, J. Influence of pineal gland on the growth and spread of melanoma in the hamster, Cancer Res., 27:1306, 1967.
Del Rey, A., Besedovsky, H.O. Sorkin,E., DaPrada, M., and Arrenbrecht, S. Immunoregulation mediated by the sympathetic nervous system, Cell.Immunol., 63:329, 1981.
Draškoci, M., and Janković, B.D. Involution of thymus and suppression of immune response in rats treated with reserpine, Nature, 202:408, 1964.
Eskra, J.D., Stevens, J.S. and Carty, T.J. β2-adrenergic receptors in thymocytes, Fed.Proc., 37:687, 1978.
Exner, A., and Boese, J. Uber experimentelle Extirpation der Glandula pinealis, Dtsch.Z.Chir., 107:182, 1910.
Faith, R.E., Liang, H.J. Murgo, A.J., and Plotnikoff, N.P. Neuroimmunomodulation with enkephalins: enhancement of human natural killer cell activity in vitro, Clin., Immunol. Immunopathol., 31:412, 1984.
Fauci, A.S., Mechanisms of corticosteroid action on lymphocyte subpopulations. Clin. Exp. Immunol., 24:54, 1976.
Federoff, N.A., Terentyeva, E.I. Garfunkel, M.L., Tsesarskaya, T.P. and Rozanova, N.S. The bone marrow after damage to the sacral plexus and the sympathetic innervation, Arch. Pat., 14:25 (in Russian), 1952.
Ferguson, R.M., Schmidtke, J.R. and Simmons, R.L. Effects of psychoactive drugs on in vitro lymphocyte activation, Birth Defects: Original Article Series, 14:379, 1978.
Filipp, G., and Szentivanyi, A. Anaphylaxis and the nervous system, III, Ann. Allergy, 16:306, 1958.
Fillenz, M.,The innervation of the cat spleen, Proc. Roy Soc. B London, 174:459-468, 1970.
Foa, C., Hypertrophie des testicles et de la crête après l'extirpation de la glande pineale chez le coq, Arch. Ital. Biol., 57:233, 1912.
Freedman, D.X., and Fenichel, G. Effect of midbrain lesion on experimental allergy, Arch. Neurol. Psychiat., 79:164, 1958.
Fuchs, S., Immunology of the nicotinic acetylcholine receptor, Curr. Topics Microbiol. Immunol., 85:1, 1979.
Fugo, N.W., Effects of hypophysectomy in the chick embryo, J. Exp.Zool., 85:271, 1940.

Fujiwara, M., Muryobayashi, T. and Shimamoto, K. Histochemical demonstration of monoamines in the thymus of rats, Japan J. Pharmocol., 16:493, 1966.
Gersbach, P., Contribution to the study of the innervation of the spleen. Comparative anatomical study, Arch. Anat. Histol. Embryol. (Strasbourg), 53(5):397, 1970.
Ghali, W.M., Abdel-Rahman, S. Nagib, M. and Mahran, Z.Y. Intrinsic innervation and vasculature of pre-and post-natal human thymus, Acta anat., 108:115, 1980.
Gillespie, J.S., and Kirpekar, S.M. The histological localization of noradrenaline in the cat spleen, J.Physiol, 187:69, 1966.
Gilman, S.C., Schwartz, J.M. Milner, R.J., Bloom, F.E. and Feldman, J.D. Beta-endorphin enhances lymphocyte proliferative responses, Proc. Natl.Acad. Sci. USA, 79:4226, 1982.
Giron, L.T., Crutacher, K.A. and Davis, J.N. Lymph nodes-A possible site for sympathetic neuronal regulation of immune responses, Ann. Neurol., 8:520, 1980.
Glaser, W., Über die motorische Innervation der Blutgefässe der Milz nebst einigen Bemerkungen der intramuralen Nervenversorgung der Blutgefasse Knochenmark, Z. Anat., 87:741, 1928.
Goldstein, M.M., The effect of bilateral destruction of the medial hypothalamic structures on the course of anaphylactic shock, Bull. Exp. Biol. Med., 82:977, 1976.
Goldstein, M.M., Antibody-forming cells of the rat spleen after the injury to the midbrain, Bull. Exp. Biol. Med., 85:185, 1978.
Gordon, D.S., Serge'eva, V.E. and Zelenova, I.G. Functional morphologyof adrenergic innervation and adreno-containing structures in lymphoid organs, Arkh. Anat. Gistol. Embriol., 77:13 (in Russian), 1979.
Gordon, M.A., Cohen, J.J. and Wilson, I. B. Muscarinic cholinergic receptors in murine lymphocytes: Demonstration by direct binding, Proc. Nat.Acad.Sci. USA, 75:2902, 1978.
Gros, M., Note sur les nerfs des os, C.R. Acad. Sci., (Paris), 71:1106, 1846.
Grossman C.J., Nathan, P. and Sholiton, L.J. Specific androgen receptor in the thymus of the castrated male rat, Biol. Reprod., 18:48A, 1978.
Grossman C.J., Sholitan, L.J. Blaja, G.C., and Nathan, P. Rat thymic estrogen receptor: II. Physicochemical properties, J.Steroid. Chem., 11:1241, 1979.
Guiard, E., La Trépanation Cranienne: Chez les Neolithiques et Chez les Primitifs Modernes, Masson, Paris, 1930.
Guillemin,R., The brain as an endocrine organ. Neurosci. Res. Prog. Bull. (Suppl.), 16:1, 1978.
Hadden, J.W., Cyclic nucleotides in lymphocyte function, Ann. N.Y. Acad. Sci., 256:352, 1975.
Hadden, J.W., Hadden, E.M., and Middleton, E., Jr. Lymphocyte blast transformation. 1. Demonstration of adrenergic receptors in human peripheral lymphocytes, Cell. Immunol., 1:583, 1970.
Hall, N. R. McGillis, J.P. Spangelo, B.L., Healy, D.L. and Goldstein, A. Immunoreactive peptides and the central nervous system. Springer Seminar Imunopathol. 8:153, 1985.

Hallion, L., and Morel, L. L'innervation vaso-motrice du thymus, C.R. Soc. Biol. (Paris), 71:382, 1911.
Hallion, L., and Morel, L. L'innervation vaso-motrice du thymus, J. Physiol. Path. Gen., 14:1, 1912.
Halvorsen, S., Plasma erythropoietin levels following hypothalamic stimulation in the rabbit, Scan. J. Clin. Investig., 13:564, 1961.
Hammar, J.A., Glasrekonstruktionen zur Beleuchtung der fruhen embryonalen Enlwicklung der thymus-innervation, Vers. Verh. Anat. Ges., 41:234, 1932.
Hammar, J.A., Innervations-verhältnisse der Krelorgane der Thymus bis in den 4 Fetalmonat, Z.Mikroskanst. Forsch., 8:253, 1935.
Harrison, L.C., Flier, J., Itin, A., Kahn, C.R. and Roth, J., Radioimmunoassay of the insulin receptor: a new probe of receptor structure and function. Science, 203:544-547, 1979.
Harting, K., Vergleichende Untersuchungen über die mikroskopische Innervation der Milz des Menschen und einigen Säugetiere., Erg. Anat., 34:1, 1944.
Hazum, E., Chang, K.-J. and Cuatrecasas, P. Specific nonopiate receptors for beta-endorphin, Science, 205:1033, 1979.
Henney, C.S., Bourne, H.R. and Lichtenstein, L.M. The role of 3',5'-adenosine monophosphate in the specific cytolytic activity of lymphocytes, J.Immunol., 108:1526, 1972.
Henson, E.C., Brunson, J.G., and Everes, C.G. Prevention of the Arthus reaction in rats and mice by combination of epinephrine and a phenothiazine derivate, propiomazine, Int. Arch. Allergy Appl. Immunol., 37:458, 1970.
Hughes, J., Centrally Acting Peptides, MacMillan Press London, 1978.
Illiano, G., Tell, G.P.E., Siegel, M.I. and Cuatrecasas, P. Guanosine 3',5'-cyclic monophosphate and the action of insulin and acetylcholine, Proc. Nat. Acad. Sci. USA, 70:2443, 1973.
Isaković, K., and Janković, B.D. Neuro-endocrine correlates of immune response. II. Changes in the lymphatic organs of brain-lesioned rats, Int. Arch. Allergy Appl. Immunol., 45:373, 1973.
Isaković, K., Janković, B.D., Mićić, M., and Knežević, Z. Thymus-bursa relationship in the developing chick embryo, in Aspects of Developmental and Comparative Immunology, I. J. B. Solomon, ed., pp. 217-220, Pergamon Press, Oxford, 1980.
Janković, B.D., Structural correlates of immune microenvironment, in Microenvironmental Aspects of Immunity, B.D. Janković and K. Isaković, eds., pp.1-4, Plenum Press, New York., 1973.
Jankovic B.D., The immune microenvironment is a multisystem, Immunol.Lett., 1:145, 1979.
Janković, B.D., From immunoneurology to immunopsychiatry: neuromodulating activity of antibrain antibodies, Int. Rev. Neurobiol., 26:249-314, 1985.

Janković, B.D., and Isaković, K. Neuro-endocrine correlates of immune response. I. Effects of brain lesions on antibody production, Arthus reactivity and delayed hypersensitivity in the rat, Int. Arch. Allergy Appl. Immunol., 45:360, 1973.
Janković, B.D., Isaković, K. and Horvat, J. Effect of a lipid fraction from rat thymus on delayed hypersensitivity reactions of neonatally thymectomized rats, Nature, 208:356, 1965.
Janković, B.D., Isaković, K. and Knežević, Z. Ontogeny of the immuno-neuro- endocrine relationship. Changes in lymphoid tissues of chick embryos surgically decapitated at 33-38 hours of incubation, Develop.Comp. Immunol., 2:479, 1978.
Janković, B.D., Isaković, K. and Knežević, Z. Ontogeny of the immuno-neuro- endocrine relationship. Early thymectomy of the chick embryo, Immunol. Lett., 1:7, 1979.
Janković, B.D., Isaković, K. Marković, B.M. and Rajčević, M. Immunological capacity of the chicken embryo. II. Humoral immune responses in embryos and young chickens bursectomized and sham-bursectomized at 52-64 h of incubation, Immunology, 32:689, 1977.
Janković, B.D., Isaković, K., Mićić, M., and Knežević, Z. Thymus-bursa-hypophysis interactions in the developing chick embryo,in, Aspects of Developmental and Comparative Immunology I. J. B. Solomon, ed., pp 529-532, Pergamon Press, Oxford, 1981a.
Janković, B.D., Isaković, K., Mićić, M., and Knežević, Z., The embryonic lympho-neuro-endocrine relationship, Clin. Immunol Immunopathol., 18:108, 1981b.
Janković, B.D., Isaković, K. and Mićić, M. The thymus-hypophysis interaction in the developing chick embryo. Thymic epithelial cells in hypopohysectomized embryos, in In Vivo Immunology, P. Nieuwenhuis, A.A. van den Broek and G. Hanna, Jr., eds., pp. 343-348, Plenum Press, N.Y., 1982.
Janković, B.D., Isaković, K. and Petrović, S. Effect of pinealectomy on immune reactions in the rat, Immunology, 18:1, 1970.
Janković B.D., and Išvaneski, M. Experimental allergic encephalomyelitis in thymectomized, bursectomized and normal chickens, Int. Arch. Allergy Appl. Immunol., 23:188, 1963.
Janković, B.D., Janković, D.Lj. and Savovski, Lj. Effect of early epiphysectomy on the immune system of the chick embryo, Proc., First Int. Workshop on NIM, Bethesda, Md., November, 1984 in press, 1985.
Janković, B.D., Jovanova, K. and Marković, B.M. Effect of hypothalamic stimulation on the immune reactions in the rat, Period. Biol., 81:211, 1979.
Janković, B.D., Mićić, M., Janković, D. Lj. and Isaković, K. The brain-thymus-hypophysis interconnection during embryogenesis, Immunobiology, 165:285, 1983.
Janković, B.D., Nešic, K. and Marković, B.M. Neuroimmunomodulation: electrical stimulations of the hypothalamus and cortex potentiate the immune response, Neurosci. Lett., 14:S180, 1983a.

Johnson, H.M., Smith, E.M., Torres, B.A. and Blalock, J.E. Regulation of the in vitro antibody response by neuroendocrine hormones, Proc. Nat. Acad. Sci USA, 79:4171, 1982.

Kanematsu, S., and Mikami, S.I. Effects of hypothalamic lesions on protein-bound ^{131}Iodine and thyroidal ^{131}I uptake in the chicken, Gen. Comp. Endocrinol., 14:25, 1970.

Kappers, J.A., The development, topographical relations and innervation of the epiphysis cerebri in the albino rat, Z. Zellforsch., 52:163, 1960.

Kappers, J.A., The mammalian pineal gland, a survey, Acta Neurochirurgica, 34:109, 1976.

Keller, S.E., Stein, M., Camerino, M.S., Schleifer, S.J., and Sherman, J. Suppression of lymphocyte stimulation by anterior hypothalamic lesions in the guinea pig, Cell. Immunol, 52:334, 1980.

Khai, L.M., Kovalenkova, M.V. Korneva, E.A., and Seranova, A.E., Further study on the role of the hypothalamic region in the regulation of immunogenesis, Zh. Mikrobiol. Epidemiol. Immunobiol., 41:7, 1964.

Kiss, F., Topographic relationship between the nerve plexuses and lymph nodes of the abdomen, Arch. Surgery, 21:405, 1980.

Klimenko, V.M., The study of some neuronal mechanisms of hypothalamic regulation of immune reactions in rabbits, Avtoref. Kand. diss. Inst. for Exper. Med. Leningrad (Russian), 1972.

Knoche, H., Zur feineren Innervation des Thymus von Menschen, Z.Zellforsch., 41:556, 1955.

Kopeloff, L.M., Barrera, S.E. and Kopeloff, N. Recurrent convulsive seizures in animals produced by immunologic and chemical means, Am.J. Psychiat., 98:881, 1942.

Korneva, E.A., and Khai, L.M. Role of the sympatho-adrenal system in the control of immunogenesis, Fiziol. Zh.SSSR, 47:1298 (in Russian), 1961.

Korneva, E.A. and Khai, L.M. Effect of destruction of areas within the hypothalamic region on the process of immunogenesis, Fiziologicheskii Zh. SSSR. 49(1):42 (in Russian), (English translation in Fed. Proc. Translations Suppl. 23(1): T88, 1964.

Korneva, E.A., and Khai, L.M. Effect of stimulation of various structures of the mesencephalon on the course of immunological reactions, Fiziol. Zh. SSSR I.M. Sechenova, 53:42 (in Russian), 1967.

Korneva, E.A., and Klimenko, V.M., Neuronale hypothalamusakivitat and homoostatische reactionen, Ergeb. exp. Med., 23:373-382, 1976.

Kostowiecki, M., Untersuchungen über Nervenendigungen in den Thymus menschlicher Feten., Vorl. Mitteil. Anat. Anza Bd., 80:231, 1934.

Kostowiecki, M., Uber die Nervenfasern und Nervenendigung in der Thymus wahrend der Fetalperiode, Zool. Pol., 3:23, 1938.

Kozlov, V.A., Anaphylaxis and the vegetative nervous system, Meditzina, Moscow, 1973.

Krall, J.F., Connelly, M. and Tuck, M.L. In vitro desensitization of human lymphocytes by epinephrine, Biochem. Pharmacol., 31:117, 1982.

Krieger, D.T. and Liotta, A.S., Pituitary hormones in brain:where, how and why? Science, 205:366,1979.

Kudoh, G., Hoshi, K. and Murakami,T. Fluorescence microscopic and enzyme histochemical studies of the innervation of the human spleen, Arch. Histol. Jap., 42(2):169-180, 1979.

Kuntz, A. and Richins, C.A. Innervation of the bone marrow, J. Comp. Neurol., 83:213-222, 1945.

Lambert, P.L., Harrell, E.H., and Achterberg,J. Medial hypothalamic stimulation decreases the phagocytic activity of the reticuloendothelial system, Physiol. Psychol, 9:193, 1981.

Lane, M.A., and Strom, T.B. The muscarinic cholinergic receptor: the effect of T cell activation, Fed. Proc., 37:1788, 1978.

Lapin, V., Influence of simultaneous pinealectomy and thymectomy on growth and formation of the Yoshida sarcoma in rats, Exp. Pathol., 9:108, 1974.

Le Fur, G., Phan, T. Canton, T., Tur, C. and Uzan, A. Evidence for a coupling between doipaminergic receptors and phospholipid methylation in mouse B-lymphocytes, Life Sci., 29:2737, 1981.

Le Fur, G., Phan, T. and Uzan, A. Identification of stereospecific (3H) spiroperidol binding sites in mammalian lymphocytes, Life Sci., 26:1139, 1980.

Leeson, C.R., and Leeson, T.S., in Histology, C.R. Leeson and T.S. Leeson, eds., p. 475, W.B. Saunders, Philadelphia, 1976.

Leger, J., and Masson, G. Factors influencing an anaphylactic reaction in the rat, Fed. Proc., 6:150, 1947.

Lentz, H., Die Nervencersorgung der Kanichen-Milz, Z.Zellforsch., 37:494, 1952.

Luckey, T.D., Thymic Hormones, Univ. Park Press, Baltimore, 1973.

Luparello, T.J., Stein, M. and Park, C.D. Effect of hypothalamic lesions on rat anaphylaxis, Am. J. Physiol., 207:911, 1964.

MacManus, J.P. Whitfield, J.F. and Youdale, T., Stimulation by epinephrine of adenylcyclase activity, cyclic AMP formation, DNA synthesis and cell proliferation in populatiions of rat thymic lymphocytes, J. Cell Comp. Physiol., 77:103, 1971.

Macris, N.T., Schiavi, R.C. Camerino, M.S. and Stein, M. Effect of hypothalamic lesions on immune processes in the guinea pig, Am.J. Physiol., 219:1209, 1970.

Macris, N.T., Schiavi, R.C., Camerino, M.S. and Stein, M. Effect of hypothalamic lesions on passive anaphylaxis in the guinea pig, Am. J. Physiol., 222:1054, 1972.

Maestroni, G.J.M. and Pierpaoli, W., Pharmacological control of the hormonally mediated immune response, in Psychoneuroimmunology, pp. 405-428, R. Ader, Ed. Academic press, New York, 1981.

Mangilli, G., Motta, M. and Martini, L. Control of adrenocorticotropic hormone secretion, in Neuroendocrinology,, L. Martini and W.F. Ganong, eds. pp.298-360, Academic Press, N.Y., 1966.

Markesbery, W.R., Cross, R.J., Roszman, T.L. and Brooks, W.H. Aging changes in Neuroimmunomodulation in the Fischer 344 Rat, Proc. First Int. Workshop on NIM, Bethesda, Md. November 1984, in press, 1985.

Maślinski, C., and Karszewski, W. The protective influence of brain stimulation by electric currents on histamine shock in guinea pigs, Bull. Acad. Pol. Sci.1, 5:57, 1957.
Maślinski, W., Grabszewska, E. and Ryzewski, J., Acetylcholine receptors on rat lymphocytes. Biochim. Biophys. Acta, 633:269, 1980.
Mathews, P.M., Froelich, C.J., Sibbitt, W. L., Jr. and Bankhurst, A. D. Enhancement of natural cytotoxicity by beta-endorphin, J. Immunol., 130:1658, 1983.
Mićić, M., Janković, D. Lj., Isaković, K. and Janković, B.D. Forebrain and hypophysis affect development of the bursa of Fabricius in the chick embryo, Period. Biol., 85(Suppl 3):9, 1983.
Miles, K., Quintans, J., Chelmicka-Schorr, E. and Arnason, B.G.W. The sympathetic nervous system modulates antibody response to thymus-independent antigens, J. Neuroimmunol., 1:101, 1981.
Miller, G.C., Murgo, A.J. and Plotnikoff, N.P. Enkephalins-Enhancement of active T-cell rosettes from lymphoma patients, Clin. Immunol. Immunopathol., 26:446, 1983.
Miller, G.C., Murgo, A.J. and Plotnikoff, N.P. Enkephalins-Enhancement of active T-cell rosettes from normal volunteers, Clin. Immunol. Immunopathol., 31:132, 1984.
Miller, M., and McCuskey, R. Innervation of bone marrow in the rabbit., Scand. J. Haemat., 10:17, 1973.
Mitrova E. and Mayer, V. Phenotiatine-induced alterations of immune response in experimental tick-borne encephalitis: morphological model analysis of events, Acta. Virol., 20:479, 1976.
Morgane, P.J., Panksepp, J. Editors. Handbook of the Hypothalamus. Vol I. Anatomy, Dekker, N.Y., 1979.
Munck, A., Young, D.A., Mosher, K.M. and Wira, C.R., Specific metabolic and physiocochemical interactions of glucocorticosteroids with rat thymus cells, in Hormones in Development, pp. 191-201, M. Hamburgh and E.J.W. Barrington, Eds., Appleton Century Crofts, New York, 1971.
Nauta, W.J.H., and Haymaker, W. Hypothalamic nuclei and fiber connections, in: The Hypothalamus, pp. 139-209, W. Haymaker, E. Anderson and W.J.H. Nauta, Eds., Thomas, Springfield, 1969.
Nilzen, A., The influence of the thyroid gland on hypersensitivity reactions in animals. I., Acta. Allerg., 7:231, 1954.
Noerthen, K., Die Nervenversorgung der Katzenmilz., Morph. Jb., 95:55, 1955.
Ottolenghi, D., Sur les nerfs de la moelle des os, Arch. Ital. Biol, 37:73-80, 1902.
Paunović, V.R., Petrović, S. and Janković, B.D. Influence of early postnatal hypothalamic lesions on the immune response in adult rats, Period. Biol.,(Suppl.) 78:50, 1976.
Payan, D.G., Brewster, D.R. and Goetzl, E.J. Specific stimulation of human T- lymphocytes by substance P, J. Immunol., 131:1613, 1983.

Payan, D.G., and Goetzl, E.J. Neuropeptide regulation of immediate and delayed hypersensitivity, in Neuroimmunomodulation II, ed. N.H. Spector et al. Gordon and Breach, N.Y., in press, 1986.
Pearse, A.G.E., and Takor, T.T. Embryology of the diffuse neuroendocrine system and its relationship to the common peptides, Fed. Proc., 38:2288, 1979.
Petrovski, I.N., Effect of stimulation of different brain regions on agglutinin titers, Zh. Mikrobiol. Epidemiol. Immunobiol., 32:103 (in Russian), 1961.
Pierpaoli, W., Fabris, N. and Sorkin , E. Developmental hormones and immunological maturation, in Hormones and the Immune Response, G. E. W. Wolstenholme and J. Knight, eds. pp. 126-143, Churchill, London, 1970.
Pines, L., and Majman, R. The innervation of the thymus, J. Nerv. Dis., 69:361, 1929.
Plezitiy, K.D., Magaeava, S.V. and Evseev, V.A. Effect of lesions and stimulations of dorsal hippocampus on Arthus reaction, in Physiology of Immune Homeostasis,II Symposium, pp. 34-35, Rostov-on-Don, (in Russian), 1977.
Plotnikoff, N.P., and Miller, G.C. Enkephalins as immunomodulators, Int.J. Immunopharmacol., 5:437, 1983.
Polyak, A.I., Rumbeshet, L.M. and Sinichkin, A.A. Antibody synthesis following electrocoagulation of the posterior hypothalamic nucleus, Zh. Mikrobiol. Epidemiol. Immunobiol., 46:52 (in Russian), 1969.
Quay, W.B., Histological structure and cytology of the pineal organ in birds and mammals, Prog. Brain Res., 10:49, 1965.
Reilly, F.D., McCluskey, P.A., Miller, M.L., McCluskey, R. S., and Meineke,H.A. Innervation of the periarteriolar lymphatic sheath of the spleen, Tissue & Cell, 11:121, 1979.
Reilly, F. D., McCluskey, R.S. and Meineke, H.A. Studies of the hematopoietic microenvironment. VIII. Adrenergic and cholinergic innervation of the murine spleen, Anat. Rec., 185:109, 1976.
Rella, W., and Lapin, V. Immunocompetence of pinealectomized and simultaneously pinealectomized and thymectomized rats, Oncology, 33:3, 1976.
Renaud, L.P. Neurophysiological organization of the endocrine hypothalamus, in The Hypothalamus, pp. 269-301, S. Reichlin, R.J. Baldessarini and J.B. Martin, Eds., Raven Press, New York, 1978.
Renoux, G. Differentiation of T-cell lineage by sodium diethyldithiocarbamate (DTC). Influence of the neocortex, in New Trends in Human Immunology and Cancer Immunotherapy, pp. 966-994, Serrou, B. and Rosenfeld, C., Eds, Doin, Paris, 1980.
Renoux, G., Bizière, K., Renoux, M. and Guillaumin, J.M., The production of T-cell inducing factors in mice is controlled by the brain neocortex. Scand. J. Immunol., 17:45, 1983.
Renoux, G., Bizière, K., and Renoux, M. Imuthiol reveals brain cortical asymmetry in the regulation of T-cell activities, Proc., First Int. Workshop on NIM, Bethesda, Md. November, 1984, in press, 1985.

Richman, D.P., and Arnason, B.G.W. Nicotinic acetylcholine receptor: Evidence for a functionally distinct receptor on human lymphocytes, Proc. Natl. Acad. Sci. USA, 76:4632, 1979.
Rimon, G., Hanski, E, Braun, S. and Levitzki, A., Mode of coupling between hormone receptors and adenylate cyclase elucidated by modulation of membrane fluidity, Nature, 276:394, 1978.
Riegele, L., Uber die mikroskopische Innervation der Milz., Z. Zellforsch., 9:511, 1929.
Rodin, A.E., The growth and spread of Walker 256 carcinoma in pinealectomized rats, Cancer Res., 27:1545, 1963.
Romanoff, A.L., The Avian Embryo, MacMillan, New York, 1960.
Romanoff, A.L. and Romanoff, A.J. Pathogenesis of the Avian Embryo, Wiley-Interscience, New York, 1972.
Romieu, M. and Jullien, G. Sur l'existence d'une formation lymphoid dans l'epiphyse de Gallinaces, C.R. Soc. Biol., 136:626, 1942.
Rossi, F., La distribuzione di fibre nervose nell 'uomo e particolarmente nel midollo osseo, studieta con metodi specifici delle neurofibrille., Boll. Soc. Ital. Biol. Sper., 3:863, 1929.
Roszman, T.L., Cross, R.J. Brooks, W.H., and Markesbery, W.R. Hypothalamic-immune interactions. II. The effect of hypothalamic lesioins on the ability of adherent spleen cells to limit lymphocyte blastogenesis, Immunology, 45:737, 1982.
Ruf, K., and Steiner, F.A. Steroid-sensitive neurons in rat hypothalamus and mid-brain:identification by microelectrophoresis, Science, 156:667-669, 1967.
Schiavi, R.C., Adams, J. and Stein, M. Effect of hypothalamic lesions on histamine toxicity in the guinea pig, Am. J. Physiol., 211:1269, 1966.
Schiavi, R.C., Macris, N.T. Camerino, M.S. and Stein, M. Effect of hypothalamic lesions on immediate hypersensitivity, Am. J. Physiol., 228:596, 1975.
Schowing, J., Influence de l'excision du rhombencéphale et du mesencěphale sur la morphogenese du crane chez l'embryon de Poulet, Compt. Rend. Acad. Sci., Paris, 248:2391, 1959a.
Schowing, J., Influence de l'excision du mesencěphale et du prosencephale sur la morphogenese du crâne chez l'embryon du Poulet. Compt. Rend. Acad. Sci., Paris.- 249:170, 1959b.
Schowing, J., Influence inducrice de l'encéphale embryonnaire sur le développement du crâne chez le Poulet. I. Influence de l'excision des territoires nerveux anterieurs sur le dévelopment cranien. J. Embryol. Exp. Morph., 19:9, 1968a.
Schowing, J., Influence inducrice de l'encéphale embryonnaire sur le dévelopment du crâne chez le Poulet. II. Influence de l'excision de la chorde et des territoires encéphaliques moyen et postérieur sur le development cranien. J. Embryol. Exp. Morphol., 19:23, 1968b.
Schreiner, G.F., and Unanue, E.R. The modulation of spontaneous and anti-Ig-stimulated motility of lymphocytes by cyclic nucleotides and adrenergic and cholinergic agents, J.Immunol, 114:802, 1975.

Schulster, D., and Levitzki, A. Cellular Receptors for Hormones and Neurotransmitters, John Wiley & Sons, Chichester, 1980.
Serge'eva, V.E., Histotopography of catecholamines in the mammalian thymus., Bull. Exp. Biol. Med., 77:456, (in Russian), 1974.
Shapiro, H.M., and Strom, T.B. Electophysiology of T-lymphocyte cholinergic receptors, Proc. Nat. Acad. Sci. USA, 77:4317, 1980.
Shavit, Y., Lewis, J.W. Terman, G.W., Gale, R.P., and Liebeskind, J.C., Opoid peptides mediate the suppressive effect of stress on natural killer cell cytotoxicity, Science, 223:188, 1984.
Shavlev, V.N., On the innervation of lymph nodes., Arkh. Anat. Giol. Embriol., 54(2):96 (in Russian), 1968.
Sherman, N.A., Smith, R.S. and Middleton, E. Jr. Effect of adrenergic compounds, aminophylline and hydrocortisone on in vitro immunoglobulin synthesis by normal human peripheral lymphocytes, J. Allergy Clin. Immunol., 52:13, 1973.
Shiotani, Y., and Ban. T. Effect of long-term electrical stimulation of the hypothalamus on pituitary-target gland system in rabbits, Med. J. Osaka Univ., 20:119, 1969.
Singh, U., Millson, D.S. Smith, P.A., and Owen, J.J.T. Identification of beta-adrenoreceptors during thymocyte ontogeny in mice, Eur. J. Immunol., 9:31, 1979.
Sjolander, A., and Strandberg, A. Über die zur Thymusdrüse tretenden Nerven, Upsala, Lak forh Forkh., 20:243, 1915.
Smith, E., and Blalock, J.E. Lymphocyte production of neurally active pituitary hormone-like molecules, Proc., First Int. Workshop on NIM, in press, 1985.
Smith, K.A., Crabtree, G.R., Kennedy, S.J. and Munck, A., Glucocorticoid receptors and glucocorticoid sensitivity of mitogen stimulated and unstimulated human lymphocytes, Nature, 267:523, 1977.
Solov'ev V.N., On the sources of innervation of the thymus gland., Arkh. Anat. Gistol. Embriol., 51:76-82 (in Russian), 1966.
Spector, N.H. Can hypothalamaic lesions change circulating antibody responses to antigens? Current Problems in Experimental and Clinical Allergy (V.I. Pytskii, ed.),pp. 21-37, Moscow, (Russian), 1979.
Spector, N.H. The "central state" of the hypothalamus in health and disease:old and new concepts. Physiology of the Hypothalamus, P, Morgane and J. Panksepp, eds., Dekker, N.Y., pp. 453-517, 1980.
Spector, N.H., Anatomical and physiological connections between the central nervous and the immune systems (neuroimmunomodulation), in Immunoregulation , pp. 231-258, N. Fabris, E. Garaci, J. Hadden, and N.A. Mitchison, eds. Plenum Press, New York, 1983.
Spector, N.H. Information explosions in an old-new research domain. in The Year in Immunology 1984-85, pp. 202-207 J.M. Cruse and R.E. Lewis, Jr. eds. Karger Basel, 1985.

Spector, N.H., Cannon, L.T. Diggs, C.L., Morrison, J.E., and Koob,G.F. Hypothalamic lesions: effects on immunological responses, Physiologist, 18:401, 1975.
Spector, N.H., Koob, G.F. and Baron,S. Hypothalamic influence upon interferon and antibody responses to Newcastle Disease. Virus infection: preliminary report. Proc. Internatl. Union Physiol. Sci. 13:711 (Abstr.)
Spector, N.H. and E. Korneva Neurophysiology and Neuroimmunomodulation in Psychoneuroimmunology R. Ader, ed., Academic Press, N.Y., pp. 449-473, 1981.
Spector, N.H., Martin, L.K., Diggs, C.L. and Koob, G.F. Hypothalamic lesions: effects upon malaria and antibody production in rats. Proceedings of the 26th International Congress, New Delhi, India. Proc. Inter. Union Physiol. Sci (Abstr.), 1974.
Speplewski, C. and Vogel, W. Changes in brain serotonin affect leucocytes, T-cells and natural killer cell activity in rats, in preparation for press, 1985.
Speranskii, A.D., A Basis for the Theory of Medicine (first Russian Ed.) (English translation, International Publisher, N.Y. 1943), 1934.
Spetien, H., Kunert-Radek, J. Karasek, E., and Pawlikowski, M. Dopamine increases cyclic AMP concentration in the rat spleen lymphocytes in vitro, Biochem.Biophys. Res. Commun., 10:1057, 1981.
Spiroff, B.E.N., Embryonic and post-embryonic development of the pineal body of the domestic fowl, Am. J. Anat., 103:375, 1958.
Stalberg, H., Effects of extirpation of the epiphysis cerebri in 6-day chick embryos, Dissertation, Olaf Norlis Forlag, Oslo.,1965.
Stimson, W.H., and McCruden, A.B. Androgen binding cytosol receptors in the rat thymus: physicochemical properties, specificity, and localization, Thymus, 3:105, 1981.
Strom, T.B., Diesseroth, A. Morganroth,J. Carpenter, C.B., and Merrill, J.P. Alteration of the cytotoxic action of sensitized lymphocytes by cholinergic agents and activators of adenylate cyclase, Proc.Natl. Acad. Sci. USA, 69:2995, 1972.
Strom, T.B., Lundin, A.P. and Carpenter, C.B. The role of cyclic nucleotides in lymphocyte activation and function, Prog. Clin. Immunol., 3:115, 1977.
Strom, T.B., Sytkowski, A.J., Carptenter, C.B., and Merrill, J.P., Cholinergic augmentation of lymphocyte-medicated cytotoxicity, A study of the cholinergic receptor of cytotoxic T-lymphocytes. Proc. Natl. Acad. Sci, USA, 71:1330, 1974.
Szentivanyi, A., and Filipp, G. Anaphylaxis and the nervous system, II., Ann. Allergy, 16:143, 1958.
Szentivanyi, A., and Szekely, J. Anaphylaxis and the nervous system, IV., Ann. Allergy, 16:389, 1958.
Takeyama, K., Morphologische Beobachtungen über diesich im Knochenmark verteilenden, peripheren Nerven, Mitt. med. Akad. Kioto, 16:895, 1936.
Tcheng, K.T., Fibres nerveuses momifiées dans les corpuscules de Hassall chez le chat, Bull. Histol. Appl., 27:100, 1950.

Terni, T., Les cellules myoides du thymus des sauropsides et leur innervation., Bull. Ass. Anat., Paris, 3:448, 1928.
Terni, T., Ricerche istologiche sull innervazione del timo dei Sauropsidi, Z. Zellforsch, 9:377, 1929.
Terni, T., L'innervazione del timo, Arch.Zool. Ital., 16:714, 1931.
Terni, T., and Muratori, G. Sulla innervazione del timo e del corporvultim onbranchiale dopo estirpazione del ganglio nodoso del vago, Monit. Zool. Ital., 43: Suppl., 85, 1933.
Thakur, P.K., and Manchanda, S.K. Hypothalamic influence on the activity of reticuloendothelial system in cat, Indian J.Physiol. Pharmacol., 13:11, 1969.
Thiéblot, L., Structure and function of the epiphysis cerebri, Prog. Brain Res., 10:479, 1965.
Thrasher, S.G., Bernardis, L.L. and Cohen,S. The immune response in hypothalamic-lesioned and hypophysectomized rats, Int.Arch. Allergy Appl. Immunol., 41:813, 1971.
Tischendorf, F., Beobachtungen über die feinere Innervation der Milz, Kölner Univ. Verlag, Köln , 1948.
Tonkoff, W., Zur Kenntnis der Neervender Lymphdrüsen, Anat. Anzeiger, 16: 456, 1899.
Triggle, D.J., Neurotransmitter-Receptor Interactions, Academic Press, London, 1971.
Tyrey, L., and Nalbandov. A.V. Influence of anterior hypothalamic lesions on circulating antibody titers in the rat, Am. J. Physiol., 222:179, 1972.
Uede, T., Ishii, Y. Matsuura, A. Shimogawara, I. and Kikuchi, K. Immunohistochemical study of lymphocytes in rat pineal gland: selective accumulation of T-lymphocytes, Anat. Rec., 199:239, 1981.
Variot, P., and Remy, C. Sur les nerfs de la moelle des os., J. Anat. Physiol., 6:273, 1880.
Volik, V.Ia, Development of the neural apparatus of inguinal lymph nodes in man, Arkh. Anat. Gistol. Embriol., 45:(5):34 (in Russian), 1965.
Warejcka, D.J., and Levy, N.L. Central nervous system (CNS) control of the immune response: effect of hypothalamic lesions on PHA responsiveness in rats, Fed. Proc., 39:914, 1980.
Weber, R.J., and Pert, C.B. Opiatergic modulation of the immune system, in Central and Peripheral Endorphins: Basic and Clinical Aspects, E.E. Muller and A.R. Genazzani,eds, pp.35-42, Raven Press, New York, 1984.
Weibel, E.R., Stereological principles for morphometry in electron microscopic cytology, Int. Rev. Cytol., 26:235, 1969.
Wertman, E., Ovadia, H., Feldman, S., and Abramsky, O. Prevention of experimental autoimmune disease by anterior hypothalamus lesion in rat, Proceedings First Int. Workshop on NIM, Bethesda, Maryland, November 1984, in press, 1985.
Williams, J.M., and Felten, D.L. Sympathetic innervation of murine thymus and spleen: A comparative histofluorescence study, Anat. Rec., 199:531, 1981.

Williams J.M., Peterson, R.G. Shea, P.A., Schmedtje, J.F., Bauer, D.C. and Felten, D.L. Sympathetic innervation of murine thymus and spleen: Evidence for a functional link between the nervous and immune systems., *Brain Res. Bull., 6*:83, 1981.

Williams, L.T., Snyderman, R. and Lefkowitz, R.J. Identification of beta-adrenergic receptors in human lymphocytes by (-) [3H] alprenolol binding, *J. Clin. Invest., 57*:49, 1976.

Wingstrand, K.O., *The Structure and Development of the Avian Pituitary,* Gleerup, Lund, 1951.

Wolff, E., Les bases de la teratogenese expérimentale des vertèbres amniotes d'après les resultats des methodes directes, *Arch.Anat. Histol.Embryol. 22*:1, 1936.

Wybran, J., Appelboom, T. Famaly, J.P. and Govaerts, A. Suggestive evidence for receptors for morphine and methionine-enkephalin on normal human blood T-lymphocytes, *J. Immunol., 123*:1068, 1979.

Yu, D.T.Y., and Clements, P.J. Human lymphocyte subpopulations: effect of epinephrine, *Clin. Exp. Immunol., 25*:472, 1976.

Zetterstrom, B.E., Høkfelt, M.T. Norbert, K.A., and Olsson,P. Possibilities of a direct adrenergic influence on blood elements of the dog spleen, *Acta. Chir. Scand., 139*:17, 1973.

ENKEPHALINS: MEDIATORS OF STRESS-INDUCED IMMUNOMODULATION

Anthony J. Murgo[a], Robert E. Faith[b], and
Nicholas P. Plotnikoff[c]

[a]West Virginia University Medical Center, Morgantown, WV;
[b]University of Houston, Houston, TX;
[c]Oral Roberts University School of Medicine, Tulsa, OK

INTRODUCTION

It is well known that environmental stress can influence the immune response and tumor growth (1). Although stress has usually been associated with detrimental effects on the host, under certain experimental conditions of timing and duration of stressful stimuli, immunoenhancement and inhibition of tumor growth can result (1,2,3,4). In addition, stress brings about numerous biochemical changes including the release of neurotransmitters, corticosteroids, and other hormones that can have various effects on the immune system and which may be benificial or detrimental to the host (5,6,7).

Among the potential mediators of stress-induced immunomodulation are the endogenous opioids, including beta-endorphin and the enkephalins. These peptides are found within the central nervous system, pituitary, and other tissues including the adrenals and sympathetic neurons and can be released in response to stressful stimuli (8,9,10,11). This chapter will review in some detail the various effects of the opioids on the immune system and tumor growth.

OPIOID RECEPTORS ON IMMUNE SYSTEM COMPONENTS

The various pharmacological and physiological effects of the opioids on the nervous system such as analgesia, euphoria, and thermoregulation are mediated through interactions with specific membrane receptors (12). Cells vary as to the types and subtypes of opioid receptors they bear and opioids differ in their binding affinities for these receptors. Briefly, [Met]enkephalin and [Leu]enkephalin have relatively high affinities for delta receptors whereas beta-endorphin appears to bind preferentially to epsilon receptors. Morphine has a high affinity for mu and Kappa receptors. Naloxone and naltrexone are pure opioid antagonists which have high affinity binding to mu and Kappa type receptors and lower affinity for delta and epsilon receptors. Since the opioids can directly influence the function of immune cells it is not surprising that these cells possess specific opioid receptors.

Lymphocytes

Indirect evidence: Wybran et al (13) were the first to suggest that lymphocytes possess opioid receptors. This conclusion is based mainly upon the indirect evidence that opioids can influence the binding characteristics of human thymus-derived lymphocytes to sheep erythrocytes. Pheripheral blood lymphocytes treated with 10^{-10} M to 10^{-7} M morphine and 10^{-9} M to 10^{-7}M, dextromoramide inhibit the formation of active (high-affinity) rosettes. T cells forming active rosettes are felt to represent a subpopulation of lymphocytes that may be decreased in certain states of cellular immune deficiency (14). Morphine but not dextromoramide also inhibits total T cell rosettes. In contrast, 10^{-7} M to 10^{-4} M [Met]enkephalin enhances active T cell rosettes and does not affect total T cell rosettes. Similar to mechanisms in the nervous system, the effects of the opioids on the immune system appear to be mediated by interactions with specific receptor types. In the case of active T cell rosette formation specific ligand binding to the mu receptor (morphine) is inhibitory and to the delta receptor ([Met]enkephalin) is stimulatory. Further support for specific opioid receptors on lymphocytes is provided by effects observed with naloxone on the active T cell rosette assay (13). Naloxone, which alone has no effect, very effectively blocks the inhibitory effects of morphine. In contrast, much higher molar concentrations (10 to 100 fold) are needed to block the enhancing effects of [Met]enkephalin.

[Leu]enkephalin as well as [Met]enkephalin can influence the formation of active T cell rosettes. However, in the study by Miller et al (15) only 76 to 77% of the samples tested showed enhancement by the enkephalins and the remainder were either not effected or inhibited. The reason for this variability is not clear but similar inconsistent results with the enkephalins has been noted with other immunological assays (16,17,18). One explanation may be that the binding of opioids may be influenced by endogenous factors and the state of the cell at the time of the assay. For example, enhancement of active T cell rosettes by enkephalin appears to be greater and more consistently found when the lymphocyte samples are obtained from immunosuppressed lymphoma patients (19). In this situation [Met]enkephalin is active over a broader concentration range (10^{-14} to 10^{-2} mg/ml) than [Leu]enkephalin which is only active at a very low concentration (10^{-14} mg/ml) and inactive at higher concentrations (10^{-10} to 10^{-2} mg/ml). This difference in activity between these two endogenous enkephalins may be due to dissimilar binding affinities to the human lymphocyte but this is yet to be determined. Plotnikoff has proposed that the delta [Met]enkephalin receptor and also the delta [Leu]enkephalin receptor are subpopulations of delta (20). Other factors may influence the binding of enkephalins to lymphocytes. Metal ions have been shown to influence the binding of enkephalins to brain receptors (21). Similarly, zinc ions may influence the binding of enkephalin to human T lymphocytes. For example the enhancement of active T cell rosette formation by [Met]enkephalin in the presence of $ZnCl_2$ is greater than the effects of either of these agents alone (22). Furthermore, the enhancement of active T cell rosettes with [Met]enkephalin plus $ZnCl_2$ is not blocked by equimolar concentrations of naloxone but is inhibited by the zinc-chelator 1,10-phenanthroline. Zinc is important for normal lymphocyte function (23) and is essential for the biological activity of serum thymic factor, an immunomodulatory polypeptide (24). Perhaps zinc is an important modulator of enkephalin binding and function in the immune system as well as the nervous system.

There is also indirect *in vivo* evidence supporting the existence of opioid binding to receptors on human lymphocytes (25). The total number of T cell rosettes is markedly decreased and the number of null cells markedly increased in the peripheral blood of morphine addicts (25). These effects are reversed with incubation *in vitro* with naloxone (25).

Direct evidence: Direct evidence for specific opioid receptors on lymphocytes has been obtained by several groups of investigators. Mehrishi and Mills (26) studied the binding characteristics of [^{3}H]naloxone to fresh human peripheral blood lymphocytes and platelets. This radioligand is displaced by morphine to the extent of 43 to 57%. This suggests that at least some, but not all, of the binding sites on lymphocytes are specific opioid receptors of the mu type. Displacement studies using opioids with other specific affinities, such as for delta or epsilon, were not reported. In the presence of Na^+ ions the decrease in binding of [^{3}H]naloxone by morphine is not observed. It appears that opioid receptor sites on lymphocytes and platelets may be similar to those on nerve cell membranes.

Hazum et al (27) studied the binding characteristics of iodinated $beta_H$-[2-D-alanine]endorphin to cultured human lymphocytes. This binding appears to be completely inhibited by beta-endorphin, partially inhibited by [Leu] and [Met]enkephalin but not by opiate agonists, enkephalin analogues, or opioid antagonists. Since alpha-endorphin is not active, the carboxy-terminal region of beta-endorphin is essential for the binding activity. This suggests the existence of a specific beta-endorphin receptor on lymphocytes which is different from beta-endorphin receptors on nerve cells.

Johnson et al (28) reported specific binding sites for ACTH and [Met]enkephalin on mouse spleen cells. Schatchard analysis of ^{125}I-labeled ACTH demonstrated the presence of both high-affinity (Kd=1×10^{-10} M) and low affinity (Kd=4.8×10^9 M) binding sites suggesting the possibility of at least two different receptor sites for this peptide. At least one type of binding site for [^{3}H][Met]enkephalin exists with a Kd=5.9×10^{-10} M.

The function of the various types of opioid receptors on lymphocytes is yet to be determined. The adrenals release many prohormone fragments with preferential binding to different opioid receptors (29). Plotnikoff has proposed (20) the "Ying-Yang" hypothesis which suggests that the final outcome of immunomodulation by stress (suppression versus enhancement) is by virtue of adrenal prohormone fragments binding to mu and kappa receptors (downregulators) and/or delta and epsilon receptors (upregulators).

Phagocytic Leukocytes

[^{3}H]dihydromorphine stereospecifically binds to human granulocytes and monocytes indicating that these bone marrow derived cells possess opioid receptors (30). The calculated Kd for granulocytes is 1×10^{-8} M and for monocytes 8×10^{-9} M. The estimated number of binding sites per cell is 3000 and 4000 for granulocytes and monocytes, respectively.

Complement

Beta-endorphin binds specifically to the terminal (C5b-9) and preterminal (C5b-7 and C5b-8) complexes of complement (31,32). Similar to the binding of beta-endorphin to lymphocytes (27), binding to complement is dependent upon the carboxy-terminus of beta-endorphin rather than the amino-terminus and is not blocked by naloxone (32). This implies that the binding site for beta-endorphin on complement is specific but differs from the brain opioid receptor.

Other Blood Components

As noted above naloxone binds specifically to normal human platelets (26). However, other studies using human platelet lysate preparations failed to demonstrate specific binding of [^{3}H]etorphine (33).

Abood et al (34) studied the binding of dihydromorphine to human erythrocyte membranes. Since naloxone is not competitive in this situation, erythrocyte receptors are distinctly different from brain receptors for dihydromorphine. In addition, it is interesting that the binding is significantly increased when erythrocytes are obtained from heroin addicts (34). This provides further evidence that the *in vivo* mileau can effect the binding characteristics of opioids to cellular elements *in vitro*.

Yamasaki and Way (35) studied the effects of various opioids on the Ca^{++} flux of rat erythrocyte membranes and it appears that erythrocytes bear Kappa type opioid receptors which inhibit the Ca^{++} pump mechanism.

Table 1 outlines some of the data concerning opioid receptors on immunocompetent cells and other blood components. It would be important to extend this information to a larger number of ligands and receptors and to more primitive cells such as those which differentiate in the bone marrow and thymus. Such data would further identify potential target cells which may be influenced by the enkephalins and other opioids.

TABLE 1. Binding Studies To Immune and Blood Components

Target	Ligand	Naloxone Reversible	Possible Receptor(Kd)	Ref.
Transformed Human Lymphocytes	Beta$_H$-[^{125}I] [D-Ala^2]endorphin	No	Beta-endorphin (c-term)specific	27
Human Peripheral Blood Lymphocytes	[^{3}H]naloxone	Partial	Mu and non-mu	26
Mouse Spleen Cells	[^{3}H[Met]enkephalin	NR	Delta (5.9×10^{-10}M)	28
Human Monocytes	[^{3}H]dihydromorphin	NR	Mu(8×10^{-9} M)	30
Human Granulocytes	[^{3}H]dihydromorphin	NR	Mu(1×10^{-8} M)	30
Human Platelets	[^{3}H]naloxone	Partial	Mu and non-mu	26
Human Erythrocytes	[^{3}H]dihydromorphin	No	Non-mu (9×10^{-9}M)	34
Complement	Beta$_H$[^{125}I] endorphin	No	Beta-end(C-term) specific	32

NR = not reported

EFFECTS ON LYMPHOCYTE PROLIFERATION

Murine Studies

Both [Leu]enkephalin and [Met]enkephalin enhance the *in vitro* blastogenic response of mouse spleen cells to phytohemagglutinin (PHA), a predominantly T cell mitogen (36). With suboptimal PHA concentrations [Met]enkephalin is active over a broad concentration range (10^{-12}-10^{-3} M) whereas [Leu]enkephalin is only active at relatively low concentrations (10^{-12}-10^{-8} M). The reason for this apparent difference in activity between these two enkephalins is not clear but is similar to that noted above for the enhancement of active T cell rosette formation (19). At optimal concentrations of PHA both [Met]enkephalin and [Leu]enkephalin are active only at relatively high concentrations (10^{-8}-10^{-3} M).

Gilman et al (16) studied the effects of various opioid peptides on the proliferative response of rat splenic lymphocytes. Beta-endorphin in nM concentrations enhances the proliferative response to PHA and another T cell mitogen, concanavalin A (ConA). It is interesting to note that in this study significant enhancement with beta-endorphin was observed in only 10 of 18 experiments using Lewis rats and 2 of 5 experiments with Brown Norway rats. One explanation for the inconsistency in the results is relative insensitivity of some of the samples (rats) to beta-endorphin rather than due to other variables of the experiments (16). Such inconsistency has been noted above in the active T cell rosette assay with the enkephalins using normal human peripheral blood leukocytes (15). In contrast to beta-endorphin, alpha-endorphin and [D-Ala^2, Met^5]enk do not effect the proliferative response of rat spleen cells to T cell mitogens and none of these agents affect B mitogen-induced or unstimulated blastogenesis (16). The enhanced response to a T cell mitogen with 3.3 mM of beta-endorphin is not inhibited with as much as 10 uM naloxone (16). This would indicate that the enhancement is mediated through beta-endorphin specific but not typical opioid receptors. These receptors may be similar to those identified on human cultured (transformed) lymphocytes by Hazum et al (27) and discussed above.

Human Studies

In contrast to the enhancing effect on rat spleen cell proliferation (16), beta-endorphin appears to be a potent inhibitor of PHA induced human lymphocyte proliferation (37). This suppressive effect, however, appears to be most marked (75% reduction) with a relatively high concentration (10^{-7} M) of beta-endorphin: much higher than that expected to occur *in vivo* (37). A lower concentration of beta-endorphin (10^{-9} M) is also suppressive but to a lesser degree (30% reduction) and only does so at high intensity PHA stimulation (37). These are the only two concentrations of beta-endorphin reported by McCain et al (37) and no indication was given as to how many *different* donors were studied; this would be important since the response to opioid peptides *in vitro* is usually concentration dependent and may vary among the individuals studied as discussed above. Similarly to observations with rat spleen cells (16), the inhibition of human lymphocyte blastogenesis with beta-endorphin is not blocked by naloxone (37). This would indicate that the suppressive effect of beta-endorphin on human lymphocyte proliferation may not be mediated through Mu receptors but possibly through the specific receptors for beta-endorphin previously described by Hazum et al (27) on cultured lymphocytes. Whether this binding is at the epsilon receptor, which is not naloxone sensitive, is yet to be determined.

Wybran (38) studied the effects of morphine and various enkephalins on PHA and pokeweed (a mixed T and B cell mitogen) induced blastogenic responses of human peripheral blood leukocytes. The results were negative with the exception of [Leu]enkephalin, which is active at a very narrow concentration range (10^{-5}M). At suboptimal concentrations of PHA the response is enhanced but with optimal PHA stimulation [Leu]enkephalin suppresses the response. [Leu]enkephalin does not effect the response to pokeweed mitogen (38). In contrast to these results, Bocchini et al. (39) found that both naloxone and morphine effect the PHA response of peripheral blood lymphocytes (i.e. nonadherent mononuclear cells) from 6 healthy donors. Both naloxone (10^{-15} - 10^{-7}M) and morphine (10^{-14} - 10^{-6}M) increased the response whereas a higher concentration of naloxone (10^{-4}M) and morphine (10^{-3}M) inhibited the response in each of the 6 donors tested. Both naloxone (10^{-8}M) and morphine (10^{-7}M) increase the percentage of total T cells and decrease the percentage of null cells (39). It appears that opioids may have opposing effects on different cell types and this may depend upon the specific receptors that they bear. If the _in vitro_ assay system contains several different cell types, the net effect of a particular opioid may be different than that obtained when only one type of target and effector cell is present. This may account for some of the inconsistent results in the literature.

Table 2 summarizes the effects of various opioids on the proliferative response of mitogen stimulated lymphocytes. In general, murine lymphocytes appear to be stimulated by [Met]enkephalin, [Leu]enkephalin, and beta-endorphin but alpha-endorphin and [D-Ala^2, Met^5]enkephalin are inactive. In man the effects appear to be more variable and the results are not definitive. However, beta-endorphin thus far has proven to be depressive. Morphine and naloxone seem to enhance at low concentrations but inhibit at high concentrations.

ENHANCEMENT OF NATURAL KILLER (NK) CELL ACTIVITY

Spontaneous cytotoxicity by NK cells may represent a potential immunosurveillance mechanism against tumors but the role of these cells _in vivo_ remains contraversial (40, 41). Numerous endogenous and exogenous factors may influence NK cell activity (42). Pertinent to this discussion is that murine animal models have shown that these changes in NK cell activity can occur with hypothalamic lesioning (43) and stressful stimuli (44).

Several groups have studied the effects of various opioids on the NK cell activity of human peripheral blood mononuclear cells _in vitro_ and the results are summarized in Table 3. Beta-endorphin and [Met]enkephalin can enhance NK cell activity over a broad concentration range including amounts considered to be physiological (18, 45-47). The increased cytotoxicity can be attributed to both enhanced effector-tumor cell conjugate formation and accelerated kinetics of lysis (45). The enhancing effects of [Met]enkephalin and beta-endorphin can be blocked by naloxone. However Mathews et al., (45) have shown that relatively high concentrations of the antagonist was required for this effect; also the enhancing effect of [Met]enkephalin and beta-endorphin was not blocked by equimolar concentrations of naloxone. Kay et al. (47), however, showed that the enhancing effect of beta-endorphin can be blocked by _high_ concentrations of naloxone. In contrast, Wybran (38) reported that enhancement with 10^6 [Met]enkephalin and beta-endorphin could be completely inhibited with 10^6M naloxone. The reasons for this conflicting data is as yet unclear. It does appear however, that the enhancement of NK cell activity by beta-endorphin and [Met]enkephalin are mediated through opioid receptors but the specific type is yet to be determined. It is not known whether these receptors are similar to those

Table 2. Effect of Opioids in Mitogen-Induced Lymphocyte Prolieration

Ligand	Concentration	Lymphocyte Species	Mitogen	Effect	Reference
[Leu]enk	10^{-12}-10^{-8}M	Mouse	PHA	I	36
	10^{-5}M	Human	Subopt. PHA	I	38
	10^{-5}M	Human	opt. PHA	D	38
	10^{-5}M	Human	Pokeweed	NE	38
[Met]enk	10^{-12}-10^{-3}M	Mouse	PHA	I	36
Beta-end	10^{-9}M	Rat	PHA	I	16
			Con A	I	16
	10^{-9}, 10^{-7}M	Human	PHA	D	37
Alpha-end	10^{-9}M	Rat	PHA	NE	16
			Con A	NE	16
[D-Ala^{2}, Met^{5}]enk	10^{-9}M	Rat	Con A	NE	16
Morphine		Human	PHA	NE	38
			Pokeweed	NE	38
	10^{-14}-10^{-6}M	Human	PHA	I	39
	10^{-3}M	Human	PHA	D	39
Naloxone	10^{-15}-10^{-7}M	Human	PHA	I	39
	10^{-4}M	Human	PHA	D	39

I = enhancement
D = suppression
NE = no effect

described by Hazum et al. (27) on cultured lymphocytes or whether they are delta and/or epsilon which are relatively insensitive to naloxone.

As shown in Table 3, various other opioids can enhance NK activity. These include beta[D-Ala^{2}, Met^{5}]enkephalin (18, 38, 46, 47). Alpha-endorphin has little or no activity (45, 47). Beta-lipotropin, a pituitary hormone which contains beta-endorphin as part of its structure, is also active (47). Interestingly, Wybran (38) has reported that morphine is as active in enhancing NK cell activity as the other opioid peptides studied; however, at least two other groups failed to show an effect with this agent (45, 47).

Peripheral blood mononuclear NK cell activity may vary in the response to opioids depending on the individual donor. Enhancement of NK cell activity may not occur in every subject tested (47). In those who do respond, the degree of enhancement may vary considerably. Mononuclear cells obtained from individuals with low control (baseline) activity appear to be enhanced to a greater degree with enkephalin treatment than those obtained from individuals with relatively high baseline activity; this occurs with both normal individuals and cancer patients (18, 48). In vivo studies in humans treated with [Met]enkephalin have resulted in variable effects on NK cell activity with both increases and decreases (49).

Table 3. Enhancement of Human NK Cell Activity *In Vitro* by Various Opioids

Opioid	Kay et al	Wybran	Faith et al	Mathews et al	Froelich & Bankhurst
Beta-end	15*			30-55	
Beta-[D-Ala^2]end					23
Alpha-end	5			NE	
Gamma-end	25				
Morphine	NE	35		NE	
[Met]enk		31	40	10-45	
[D-Ala^2,Met^5]enk		23			
[Leu]enkephalin	4	29	70	NE	
Beta-lipotropin	16				

* percent increase in activity; NE, no effect

EFFECTS ON PHAGOCYTIC CELL FUNCTIONS

It has previously been shown that morphine and similar opiate alkyloids inhibit phagocytosis and can impair the ability of the host to eradicate infections with bacterial organisms (50, 51). In contrast there is more recent evidence that the endogenous opioids stimulate phagocyte function and, therefore, may increase resistance to infections (52-58).

Van Epps and Saland (54) studied the effects of beta-endorphin and [Met]enkephalin on human peripheral blood leukocyte chemotaxis using various *in vitro* assays including the "leading front technique". Both of these opioid peptides stimulate mononuclear cell chemotaxis and, to a lesser degree, neutrophil chemotaxis. The response with mononuclear cells is approximately 80% of that obtained with formyl-methionyl-leucyl-phenylalanine (f-MLP), a potent chemoattractant. The response with neutrophils, however, is only less than 30% of that of f-MLP. It is interesting that the chemotactic response of mononuclear cells to beta-endorphin and [Met]enkephalin is bimodal with peak activities occurring at 10^{-12}M and 10^{-8}M concentrations. One can speculate that the biomodal chemotactic response of cells to these agents reflects the existence of multiple receptor types on the cell or subpopulations of functionally distinct cells bearing receptors with different binding affinites. It is also interesting that a positive chemotactic response to beta-endorphin or [Met]enkephalin is only obtained in about 70-75% of cases compared to a universal response to f-MLP (53). As discussed above, the lack of universal responsiveness to the opioid peptides *in vitro* has been noted repeatedly with other immunological assays with human (15, 47) and murine (16) cells. Perhaps this variability in the response observed *in vitro* is related to the *in vivo* influence of endogenous opioids, the levels of which fluctuate with various physiological and pathological conditions (52).

The stimulation of chemotaxis by beta-endorphin and [Met]eskephalin is more than 89% inhibited by equimolar concentrations of naloxone indicating that the effect is mediated by opioid receptors (53). The administration of beta-endorphin into the cerebral ventricles of the rat results in the migration of macrophage-like cells which is consistent with the chemoattractant response observed *in vitro* (54).

In addition to its chemoattractant properties, beta-endorphin enhances the migration of f-MLP (55). This effect which is observed with 10^{-9} to 10^{-10}M beta-endorphin, can be blocked by 10^{-6}M naloxone suggesting that the activity is dependent upon the interaction with specific opioid receptors (55).

Opioid peptides may also cause morphological changes in phagocytic leukocytes (56). Treatment with 10^{-9} to 10^{-8}M beta-endorphin, [Met]enkephalin, and opiates, but not with antagonists, causes visible flattening of human neutrophilic granulocytes *in vitro*. This specific change in shape is felt to reflect an adherence phenomenon similar to that which occurs with other known chemoattractants, such as f-MLP. Since the flattening caused by the opioids can be blocked by equimolar concentrations of opioid antagonist, this effect appears to be mediated through specific opioid receptors (56).

Effects of opioids on various macrophage functions *in vitro* have also been noted (57). At relatively low concentrations (10^{-9} to 10^{-7}M) [Met]enkephalin stimulates IgG2a-mediated antibody dependent cytotoxicity (ADCC) of rat peritoneal macrophages and this appears to occur through naloxone-sensitive receptors (57). Similar concentrations of [Met]enkephalin suppress phagocytosis of (IgG2a-coated sheep erythrocytes and induces a significant increase in the generation of luminol dependent chemoluminescence (LDCL) of rat macrophages (57). At relatively high concentrations (10^{-6} to 10^{-5}M) [Met]enkephalin results in an enhancement of phagocytosis and an inhibition of LDCL. The involvement of cyclic nucleotides in [Met]enkephalin induced functional alterations of macrophages is suspected because cGMP accumulation is augmented in the treated cells (57). Also, calmodulin is felt to play a role since the effects of [Met]enkephalin on ADCC is abolished by trifluoroperazine, a calmodulin inhibitor (57).

In regard to LDCL, the response of neutrophils obtained from normal individuals and cancer patients may be influenced by [Leu]enkephalin and [Met]enkephalin *in vitro* (59). The responses obtained may be either increased, decreased, or not significantly changed, depending upon the type of enkephalin, concentration, and the individual being studied. However, an observed increase in activity for LDCL with enkephalin occured only with the cells obtained from cancer patients (59).

As noted in Table 4 the endogenous opioid peptides have numberous effects on phagocyte function. *In vitro* and possibly *in vivo* they behave as chemoattractants and stimulate cellular adhesion and migration of neutrophilic granulocytes. Effects on ADCC, chemoluminesence, and macrophage phagocytosis are more variable, depending on the concentration of opioid and the population studied. Stimulation of cGMP accumulation may be responsible for some of the effects.

Table 4. Phagocyte Functions Effected by Endogenous Opioid Peptides

Function	Effect
Chemotaxis	Stimulated
Adhesion	Stimulated
Antibody Dependent Cytotoxicity	Stimulated
Phagocytosis	Stimulated or Depressed
Chemoluminesence	Stimulated or Depressed
cGMP accumulation	Stimulated

EFFECTS ON HUMORAL IMMUNITY

In Vitro Studies

Johnson et al. (28) studied the effect of various neuropeptides including ACTH, endorphins, and enkephalins on the *in vitro* antibody response of mouse spleen cells to a T cell dependent (sheep erythrocytes) and T cell independent (DNP-Ficoll) antigen. ACTH in uM concentrations markedly inhibits the response to both of these antigens. The inhibitory effect of ACTH does not occur in the presence of the thiol reducing agent 2-mercaptopurine, a property shared with alpha-interferon. Alpha-endorphin (0.5 uM) is also active and causes a 76 to 92% reduction in the response to sheep erythrocytes. [Leu]enkephalin and [Met]enkephalin are somewhat less active than alpha-endorphin resulting in a 64% and 73% reduction, respectively. Beta-endorphin and gamma-endorphin are very weak or inactive as antibody response inhibitors. The suppression in antibody formation caused by 0.5 uM alpha-endorphin is blocked by the addition of 3 uM naloxone and also diminished when combined with 12 uM beta-endorphin. These results imply that the agonistic binding to the endorphin receptor does not necessarily activate immunosuppressive events (28). These results, however, convincingly show that neuroendocrine peptide hormones can regulate the antibody response *in vitro*.

Opioids have also been shown to modulate IgE-mediated ^{14}C-serotonin release from rat mast cells (60). The opioids apparently do not have a direct effect but are capable of reversing the inhibition of serotonin release by PGE_1. Levorphanol appears to be most potent in this regard with 10^{-6}M resulting in a 35% reversal of PGE_1 effects (60). In contrast as much as $3x10^{-5}$M morphine, 10^{-5}M beta-endorphin and 10^{-4}M [Met]enkephalin are required to give an equivalent response. Dextrorphan, an enantiomer of levorphanol, is completely inactive. The effects of the opioids on PGE_1 inhibition of serotonin release are blocked by naloxone (60). These results suggest that opioids can modulate IgE-mediated release from rat mast cells through opioid receptors; however, the types of binding sites on mast cells remains to be defined.

In Vivo Studies

The effect of various neuropeptides including [Met]enkephalin and [Leu]enkephalin on the antibody response of mice to sheep erythrocytes has been studied (61). The antibody response is enhanced by 20 mg/kg of

[Leu]enkephalin when given as a single intravenous dose 3 days prior to antigen. However, the response is inhibited by a higher dose (100 mg/kg) of [Leu]enkephalin (61). [Met]enkephalin has not been shown to effect the *in vivo* antibody response of mice to sheep erythrocytes (61). Similarly, we have treated 4 week old male C57BL/6J mice with 3 mg/kg of [Met]enkephalin subcutaneously daily for seven days and then injected $4x10^6$ or $2x10^8$ sheep erythrocytes subcutaneously on day 8 and found no significant effect on the hemagglutinin response (unpublished data).

In summary, the opioids can modulate the antibody response to T cell dependent and T cell independent antigens and the immunologically mediated release mechanism of mast cells. These effects are dependent upon the specific opioid peptide and its concentration.

EFFECTS ON LYMPHOID TISSUE

Preliminary evidence suggests that the endogenous opioids not only modulate immune function but may also play a role in the growth and development of lymphoid tissue. Rodents injected with enkephalins develop marked changes in thymus and spleen size. The subcutaneous administration of 10 mg/kg of [Met]enkephalin and 30 mg/kg of [Leu]enkephalin daily for 7 days into young adult female BDF_1 mice causes a significant increase in thymus weight (62). Similarly, thymic enlargement occurs in young adult C57BL/6 male mice (Murgo, unpublished observations) and in rats treated with [Met]enkephalin (63). Histological studies in mice treated with enkephalin shows that the thymic cortex is enlarged to a greater degree than the medulla (62). In contrast to the hypertrophy that occurs with a dose of 30 mg/kg of [Leu]enkephalin, 10 mg/kg of this enkephalin results in a significant decrease in thymus weight in mice and both [Met]enkephalin and [Leu]enkephalin result in a decrease in spleen weight (62). In rats, spleen size may actually be increased with [Leu]enkephalin (63).

The mechanism by which the enkephalins can cause changes in lymphoid tissue weight is not known. The fact that endogenous opioids can cause thymic enlargement is particularly interesting since this effect is opposite that reported to occur with stress, corticosteroids, and aging (1). The results with enkephalins provide further support for the role of endogenous opioids as modulators between the neuroendocrine and immune systems. Both of these systems appear to be sources as well as targets for the endogenous opioids. Immunoreactive ACTH and endorphins may be produced concomitantly with alpha-interferon in human leukocytes (64) and in adherent mouse spleen cells (65). In addition, the enkephalins have been identified immunohistochemically within thymic epithelial cells (66). The discovery of the enkephalins in this specific location supports their potential role in T cell growth and differentiation. In addition, preliminary human *in vivo* studies have shown that treatment with [Met]enkephalin increases peripheral blood OKT3, OKT4, and OKT11 subsets (49).

OPIOID MODULATION OF TUMOR GROWTH

It is well known that environmental stress can influence the immune response and neoplastic growth (1). Although there is mounting evidence that a number of substances may play a role in mediating these stress induced changes, early investigations have focused on the adrenal corticosteroids which are generally considered to be immunosuppressive (1). However, under certain experimental conditions of duration and timing of stressful stimuli, immunoenhancement and inhibition of tumor growth can result (1-4). These results support the existence of stress hormones with immunostimulatory and tumor inhibitory properties. The endogenous opioids should be considered in this regard since they are released from the pituitary and the adrenals with stress (8,9,11); however there are conflicting opinions as to whether these neuropeptides are detrimental or beneficial to the tumor-bearing host (20,44,63).

Paradoxically, heroin which is felt to directly contribute to the increased risk of infection and immunological abnormalities that occur in addicts (67, 68) has been found to inhibit tumor growth in mice (69). The administration of heroin at a dose of 3-15 mg/kg daily for 2 weeks prior to or 6 mg/kg daily one week after the inoculation of S20Y neuroblastoma cells results in prolonged survival times and tumor growth retardation (69). Since neuroblastoma cells bear opioid receptors, it is not known whether the effects of heroin are related to direct toxicity against the tumor cells or to immunomodulation.

Opioid antagonists have also been found to influence tumor growth (70-73). Mice treated daily with 5-20 mg/kg of naloxone either 14 days before or 7 days after the inoculation of neuroblastoma cells have a prolonged survival compared to untreated mice; however, the effect is greatest in mice pretreated with naloxone (70). In contrast to the results with naloxone, naltrexone which is a longer acting and more potent opioid antagonist has both stimulatory and inhibitory effects on the growth of neuroblastoma depending upon the dose used (71). Daily injections of 0.1 mg/kg results in a decrease in tumor incidence, delay in tumor appearance, and increase in survival time. In contrast a higher dose (10 mg/kg) shortens the time to tumor appearance and decreases survival. Later studies have implicated the duration of opioid receptor blockade rather than the dosage of the antagonist to be responsible for the variable effects on tumor growth (72).

Opioid antagonists have also been shown to effect the growth of animal tumors other than neuroblastoma. Both naloxone and naltrexone have been shown to inhibit the growth of carcinogen-induced mammary carcinoma in rats (73). The inhibition of mammary tumor growth by these agents may be related to a reduction in prolactin and growth hormone which tend to stimulate mammary tumor growth rather than to immunomodulation (73).

The effects of certain patterns of experimental stress are presumed to be mediated by endogenous opioids (74). Intermittent inescapable footshock stress applied to rats for 4 consecutive days prior to mammary ascites tumor inoculation results in reduced survival compared to nonstressed controls (74). A role for endogenous opioid systems in this process is further implicated since the tumor enhancing effect of this stress is prevented by naltrexone (74). Also, high doses of morphine affects the survival of rats with these tumors but the result depends somewhat upon the schedule of drug administration (75). Animals given 4 daily injections of morphine (50 mg/kg) either before or after tumor inoculation, and those given 14 injections of morphine, but only after tumor inoculation, have a significant decrease in survival comparable to that caused by footshock stress (75). Survival of animals given morphine for 14 days prior to tumor inoculation did not differ from controls, suggesting that tolerance can develop to this morphine effect (75).

The fact that stress can result in the release of numberous hormones including different types of opioid precursers and derivatives makes it difficult to attribute the effects of stress on tumor growth to a specific opioid. This is also true with interpreting the results of naloxone or naltrexone treatment. For example, these agents have a much higher affinity for mu than for delta receptors (12) and not all of the effects of these agents are necessarily related to the antagonism of endogenous opioids (76).

A more direct approach to determining the role endogenous opioids play in tumor growth is to study the effects of these substances when administered to tumor-bearing animals. Systemically administered enkephalins do have activity as tranquilizers, antidepressants and anticonvulsants (77). As discussed above, enkephalins administered subcutaneously to rodents can also cause marked changes in lymphoid tissue weight. In addition, both [Met]enkephalin and [Leu]enkephalin can have beneficial effects on the tumor-bearing host (36,78,79). The daily subcutaneous administration of [Leu]enkephalin (10 mg/kg) increases the survival of BDF, female mice inoculated intraperitoneally with $1x10^2$ L1210 leukemia cells. This protective effect is also seen with [Met]enkephalin at a dose of 30 mg/kg daily in mice inoculated with $1x10^4$ L1210 cells (36).

The enkephalins have also been shown to be beneficial to mice bearing solid tumors. The treatment of C57B1/6J mice with 3 mg/kg of [Met]enkephalin daily for 7 or 14 days beginning the day after the subcutaneous inoculation of 10^5 B16 melanoma cells inhibits local tumor (78,79). This antitumor effect occurs only in mice treated following tumor inoculation, since pretreatment with [Met]enkephalin for 7 days prior to tumor inoculation does not inhibit local tumor growth and may increase the number of metastatic lesions (78).

The mechanism of antitumor effect of the endogenous opioids is not clearly known but immunomodulation is felt to be an important factor (36). Direct cytotoxicity is probably not a major factor since the treatment of various tumor cell lines (including L1210 and B16 melanoma) results in either minimal stimulation, minimal inhibitiion, or no effect on growth in vitro (80;Plotnikoff, unpublished data; Murgo, unpublished data).

Studies with genetically obese (C57BL/6J, ob/ob) mice provides further supportive evidence that the endogenous opioids can be beneficial to the tumor-bearing host. The proliferative response of splenocytes to a T cell mitogen is markedly increased and the frequency of lung metastases from B16 melanoma is greatly reduced in obese mice compared to lean littermates (81,82). Higher levels of beta-endorphin which is known to exist in the obese mice may be responsible for the enhanced level of immunocompetence and relative tumor resistence (82).

There is preliminary evidence that the enkephalins may potentiate the antitumor effects of cytotoxic chemotherapeutic drugs. [Met]enkephalin potentiates the protective effect of cis-platinum in mice inoculated with L1210 leukemia (83). In contrast, morphine administration reduces the survival of tumor-bearing mice treated with cis-platinum (83).

In summary, the opioid peptides are potential mediators of some of the effects of stress on tumor growth. The overall picture, however, is complicated by the fact that stress can result in the release of many different opioid peptides and precursors which differ in their receptor binding affinities and effects on target cells (29). Therefore, the ultimate effect on the immune system and tumor growth may vary depending upon the predominance of a particular ligand and its receptor affinities.

MODULATION OF LYMPHOKINES

The production of endorphin by lymphocytes suggests that the opioids function as lymphokines and in the regulatory loop between the immune and neuroendocrine systems (64). There is also preliminary data which indicates that the endogenous opioids may modulate the production and efects of other lymphokines. Hung et al (84) have reported on the inhibition of interferon by narcotic analgesics. More recently, Brown and Van Epps (85) have shown that beta-endorphin and [Met]enkephalin modulate the production of gamma-interferon in vitro. In addition, the administration of [Met]enkephalin to humans increases interleukin II receptors as measured with monoclonal antibodies (49). The modulation of lymphokines by opioids is discussed in greater detail in other chapters of this book.

CONCLUSION

It is clear that opioids can effect various aspects of the immune system. We feel that part of the function of the endogenous opioids is to participate in immune regulatory pathways, particularly during periods of stress when the levels of these neuropeptides are very high. The apparent divergent activities of the various opioids probably is due to differences in affinities for receptors on a variety of cells. Although the opioids have been associated with both immunoenhancement and immunodepression, the enkephalins appear to enhance cellular immune mechanisms and inhibit tumor growth.

ACKNOWLEDGEMENT

The authors thank Ms. Annorah Cale for her excellent assistance in preparing the manuscript.

REFERENCES

1. V. Riley, Pyschoneuroendocrine influences on immunocompetence and neoplasia, Science 212:1100 (1981).

2. A.A. Monjan and M.I. Collector, Stress induced modulation of the immune response, Science 196:307 (1977).

3. H.A. Rashkis, Systemic stress as an inhibitor of experimental tumors in Swiss mice, Science 116:169 (1952).

4. A. AmKraut and G.F. Solomon, Stress and murine sarcoma virus (Moloney)-induced tumors, Cancer Res. 32:1428 (1972).

5. A.W. Coquelin and R.A. Gorski, Neuroendocrine control, stress, and immunity, in: "Stress, Immunity, and Aging," E.L. Cooper, ed., Dekker, New York (1984).

6. J. Ahlqvist, Hormonal influences on immunologic and related phenomena, in: "Psychoneuroimmunology," R. Ader, ed., Academic Press, New York (1981).

7. N.R. Hall and A.L. Goldstein, Neurotransmitters and the immune system, in, "Psychoneuroimmunology," R. Ader, ed., Academic Press, New York (1981).

8. R. Guillemin, T. Vargo, J. Rossier, S. Minick, N. Ling, C. Rivier, W. Vale, and F. Bloom, Beta-endorphin and adrenocorticotropin are secreted concomitantly by the pituitary gland, Science 197:1367 (1977).

9. S. Amir, Z.W. Brown, and Z. Amit, The role of endorphins in stress: Evidence and speculations, Neurosci. Biobehav. Rev. 4:77 (1980)

10. G.W. Terman, Y. Shavit, J.W. Lewis, J.T. Cannon, and J.C. Liebeskind, Intrinsic mechanisms of pain inhibition: activation by stress,

Science 226:1270 (1984).

11. O.H. Viveros, E.J. Diliberto, E. Hazum, and K-J. Chang, Opiate-like materials in the adrenal medulla: evidence for storage and secretion with catecholamines, Mol. Pharmacol. 16:1101 (1979).

12. W.R. Martin, Pharmacology of Opioids, Pharmacol. Rev. 35:283 (1984).

13. J. Wybran, T. Appelboom, J-P. Famaey, and A. Govaerts, Suggestive evidence for receptors for morphine and methionine-enkephalin on normal human blood T lymphocytes, J. Immunol. 123:1068 (1979).

14. H.H. Fudenberg, J. Wybran, and D. Robbins, T-rosette-forming cells, cellular immunity and cancer, N. Engl. J. Med. 292:475 (1975).

15. G.C. Miller, A.J. Murgo, and N.P. Plotnikoff, Enkephalins:enhancement of active T-cell rosettes from normal volunteers, Clin. Immunol. Immunopathol. 31:132 (1984).

16. S.C. Gilman, J.M. Schwartz, R.J. Milner, F.E. Bloom, and J.D. Feldman, Beta-endorphin enhances lymphocyte proliferative responses, Proc. Natl. Acad. Sci. (USA) 79:4226 (1982).

17. D.E. Van Epps and L. Saland, Beta-endorphin and met-enkephalin stimulate human peripheral blood mononuclear cell chemotaxis, J. Immunol. 132:3046 (1984).

18. R.E. Faith, H.J. Liang, A.J. Murgo, and N.P. Plotnikoff, Neuro-immunomodulation with enkepahlins: enhancement of human natural killer (NK) cell activity in vitro, Clin. Immunol. Immunopathol. 31:412 (1984).

19. G.C. Miller, A.J. Murgo, and N.P. Plotnikoff, Enkephalins: enhancement of active T-cell rosettes from lymphoma patients, Clin. Immunol. Immunopathol. 26:446 (1983).

20. N.P. Plotnikoff. Enkephalins-Endorphins: Emotional stress, depression and immune system, Pyschpharmacol. Bull. in press (1985).

21. K. Stengaard-Pedersen, Inhibition of enkepahlin binding to opiate receptors by zinc ions: possible physiological importance in the brain, Acta. pharmacol. et toxicol. 50:213 (1982).

22. A.J. Murgo, N.P. Plotnikoff and R.E. Faith, Effect of methionine-enkephalin plus $Zn\text{-}Cl_2$ on active T cell rosettes, Neuropeptides 5:367 (1985).

23. R.O. Williams and L.A. Loeb, Zinc requirement for DNA replication in stimulated human lymphocytes, J. Cell Biol. 58:594 (1973).

24. M. Dardenne, J-M, Pleau, B. Nabarra, P. Lefrancier, M. Derrien, J. Choay,and J-F, Bach, Contribution of zinc and other metals to the biological activity of the serum thymic factor, Proc. Natl. Acad. Sci. (USA) 79:5370 (1982).

25. R.J. McDonough, J.J. Madden, A. Falek, D.A. Shafer, M. Pline, D. Gordon, P. Bokos, J.C. Kuehnle, and J. Mendelson, Alteration of T and null lymphocyte frequencies in the peripheral blood of human opiate addicts in vivo evidence for opiate receptor sites on T lymphocytes, J. Immunol. 125:2539 (1980).

26. J.N. Mehrishi and I.H. Mills, Opiate receptors on lymphocytes and platelets in man, Clin. Immunol. Immunopathol. 27:240 (1983).

27. E. Hazum, K-J Chang, and P. Cuatrecasas, Specific nonopiate receptors for beta-endorphin, Science 205:1033 (1979).

28. H.M. Johnson, E.M. Smith, B.A. Torres, and J.E. Blalock, Regulation of the in vivo antibody response by neuroendocrine hormones, Proc. Natl. Acad. Sci. (USA) 79:4171 (1982).

29. A. Herz, Multiple endorphins and natural ligands of multiple opioid receptors, in, "Central and Peripheral endorphins: Basic and Clinical Aspects," E.E. Miller and A.R. Genazzini, ed., Raven Press, New York (1984).

30. A. Lopker, L.G. Abood, W. Hoss, and F.J. Lionetti, Stereoselective muscarinic acetylcholine and opiate receptors in human phagocytic leukocytes, Biochem. Pharmacol. 29:1361 (1980).

31. L. Schweigerer, H. Teschemacher, and S. Bhakdi, Interaction of human beta-endorphin with the terminal SC56-9 and "preterminal" SC56-7 and SC5b-8 complexes of human complement, Life Sci. 31:2275 (1982).

32. L. Schweigerer, S. Bhakdi, and H. Teschemacher, Specific non-opiate binding sites for human beta-endorphin on the terminal complex of human complement, Nature 296:5857 (1982).

33. A. Reches, A. Eldor, Z. Vogel, and Y. Salomon, Do human platelets have opiate receptors?, Nature 288:382 (1980).

34. L.G. Abood, H.G. Atkinson, and M. MacNeil, Stereospecific opiate binding in human erythrocyte membranes and changes in heroin addicts, J. Neurosci. Res. 2:427 (1976).

35. Y. Yamasaki and E.L. Way, Possible inhibition of CA^{++} pump of rat erythrocyte ghosts by opioid K agonists, Life Sci. 33:723 (Sup. I, 1983).

36. N.P. Plotnikoff and G.C. Miller, Enkephalins as immunomodulators, Int. J. Immunopharmacol. 5:437 (1983).

37. H.W. McCain, I.B. Lamster, J.M. Bozzone, and J.T. Grbic, Beta-endorphin modulates human immune activity via non-opiate receptor mechanisms, Life Sci. 31:1619 (1982).

38. J. Wybran, Enkephalins and endorphins as modifiers of the immune system: present and future, Fed. Proc. 44:92 (1985).

39. G. Bocchini, G. Bonanno, and A. Canevari, Influence of morphine and naloxone on human peripheral blood T-lymphocytes, Drug Alcohol Depend. 11:233 (1983).

40. N. Hanna, Role of natural killer cells in host defense against cancer metastasis, in, "Cancer Invasion and Metastatic Biologic and Therapeutic Aspects," G.L. Nicolson and L. Milas, ed., Raven Press, New York (1984).

41. O. Fodstad, C.T. Hansen, G.B. Cannon, C.N. Statham, G.R. Lichtenstein, and N.R. Boyd, Lack of correlation between natural killer activity and tumor growth control in nude mice with different immune defects, Cancer Res. 44:4403 (1984).

42. R.B. Herberman and H.T. Holden, Natural cell-mediated immunity, Adv. Cancer Res. 27:305 (1978).

43. R.J. Cross, W.R. Markesbery, W.H. Brooks, and T.L. Roszman, Hypothalamic-immune interactions: neuromodulation of natural killer activity by lesioning of the anterior hypothalamus, Immunol. 51:399 (1984).

44. Y. Shavit, J.W. Lewis, G.W. Terman, R.P. Gale, and J.C. Liebeskind, Opioid peptides mediate the suppressive effect of stress on natural killer cell cytotoxicity, Science 223:188 (1984).

45. P.M. Mathews, C.J. Froelich, W.L. Sibbitt, and A.D. Bankhurst, Enhancement of natural cytotoxicity by beta-endorphin, J. Immunol. 130:1658 (1983).

46. C.J. Froelich and A.D. Bankhurst, The effect of beta-endorphin on natural cytotoxicity and antibody dependent cellular cytotoxicity, Life Sci. 35:261 (1984).

47. N. Kay, J. Allen, and J.E. Morley, Endorphins stimulate normal human peripheral blood lymphocyte natural killer activity, Life Sci. 35:53 (1984).

48. R.E. Faith, H.J. Liang, N.P. Plotnikoff, A.J. Murgo, and N.F. Nimeh, Neuroimmunomodulation with enkephalins: in vitro enhancement of natural killer (NK) cell activity in peripheral blood lymphocytes from cancer patients, Nat. Imm. Cell Growth Reg. in press.

49. N.P. Plotnikoff, et al, this book.

50. D. Zucker-Franklin, P. Elsbach, and E.J. Simon, The effect of the morphine analog levorphanol on phagocytosing leukocytes: a morphologic study, Lab Invest. 25:415 (1971).

51. E. Tubaro, G. Borelli, C. Croce, G. Cavallo, and C. Santiangeli, Effect of morphine on resistance to infection, J. Infect. Dis. 148: 656 (1983).

52. A.E. Panerai, A. Martini, A. DeRosa, F. Salerno, A.M. DiGiulio, and P. Mantegazza, Plasma beta-endorphin and met-enkephalin in physiological and pathological conditions, in, "Regulatory Peptides: From molecular biology to function," E. Costa and M. Trabucchi, ed., Raven Press, New York (1982).

53. D.E. VanEpps and L. Saland, Beta-endorphin and met-enkephalin stimulate human peripheral blood mononuclear cell chemotaxis, J. Immunol. 132:3046 (1984).

54. L.C. Saland, D.E. VanEpps, E. Ortiz, and A. Samora, Acute injections of opiate peptides into the rat cerebral ventricle: a macrophage-like cellular response, Brain Res. Bull. 10:523 (1983).

55. C.O. Simpkins, C.A. Dickey, and M.P. Fink, Human neutrophil migration is enhanced by beta-endorphin, Life Sci. 34:2251 (1984).

56. E.G. Fischer and N.E. Falke, Beta-endorphin modulates immune functions: a review, Psychother. Psychosom. 42:195 (1984).

57. G. Foris, G.A. Medgyesi, E. Gyimesi, and M. Hauck, Met-enkephalin induced alterations of macrophage functions, Mol. Immunol. 21:747 (1984).

58. M.R. Ruff and C.B. Pert, Neuropeptides as chemoattractants for human macrophages, Proc First International Workshop on Neuroimmunomodulation, in press (1985).

59. G.C. Miller, A.J. Murgo, and N.P. Plotnikoff, The influence of leucine and methionine enkephalin on immune mechanisms, Int. J. Immunopharmacol. 4:367 (1982).

60. Y. Yamasaki, O. Shimamura, A. Kizu, M. Nakagawa, and H. Ijichi, IgE-mediated ^{14}C-serotonin release from rat mast cells modulated by morphine and endorphins, Life Sci. 31:471 (1982).

61. E.M. Kukain, R.K. Muceniece, and V.E. Klusha, Comparison of neuro- and immunomodulator properties of low-molecular-weight neuropeptides, Bull. Exp. Biol. Med. 94:1105 (1982).

62. N.P. Plotnikoff, A.J. Murgo, and R.E. Faith, Neuroimmunomodulation with enkephalins: effects on thymus and spleen weights in mice, Clin. Immunol. Immunopathol. 32:52 (1984).

63. N.P. Plotnikoff, A.J. Murgo, G.C. Miller, C.N. Corder, and R.E. Faith, Enkephalins: immunomodulators, Fed. Proc. 44:118 (1985).

64. J.E. Blalock and E.M. Smith, A complete regulatory loop between the immune and neuroendocrine systems, Fed. Proc. 44:108 (1985).

65. S.J. Lolait, A.T.W. Lim, B.H. Toh, and J.W. Funder, Immunoreactive beta-endorphin in a subpopulation of mouse spleen macrophages, J. Clin. Invest. 73:277 (1984).

66. W. Savino and M. Dardenne, Enkephalin immunohistochemical detection in human thymic epithelial cells, Proc. First Internat. Workshop on Neuroimmunomodulation, in press (1985).

67. S.M. Brown, B. Stimmel, R.N. Taub, S. Kochwa, and R.E. Rosenfield, Immunologic dysfunction in heroin addicts, Arch. Intern. Med. 134: 1001 (1974).

68. R.E. Faith, N.P. Plotnikoff, and A.J. Murgo, Effects of opiates and Neuropeptides on immune functions, in: "Mechanisms of Tolerance and Dependence, C.W. Sharp, ed., NIDA Research Monograph 54, Rockville, MD (1984).

69. I.S. Zagon and P.J. McLaughlin, Heroin prolongs survival time and retards tumor growth in mice with neuroblastoma, Brain Res. Bull. 7:25 (1981).

70. I.S. Zagon and P.J. McLaughlin, Naloxone prolongs the survival time of mice treated with neuroblastoma, Life Sci. 28:1095 (1981).

71. I.S. Zagon and P.J. McLaughlin, Naltrexone modulates tumor response in mice with neuroblastoma, Science 221:671 (1983).

72. I.S. Zagon and P.J. McLaughlin, Duration of opiate receptor blockade determines tumorigenic response in mice with neuroblastoma: a role for endogenous opioid systems in cancer, Life Sci. 35:409 (1984).

73. C.F. Aylsworth, C.A. Hodson, and J. Meites, Opiate antagonists can inhibit mammary tumor in rats, Proc. Soc. Exp. Biol. Med. 161:18 (1979).

74. J.W. Lewis, Y. Sharit, G.W. Terman, L.R. Nelson, R.P. Gale, and J.C. Liebeskind, Apparent involvement of opioid peptides in stress-induced enhancement of tumor growth, Peptides 4:635 (1983).

75. J.W. Lewis, Y. Shavit,G.W. Terman,R.P. Gale, and J.C. Liebeskind, Stress and morphine affect survival of rats challenged with a mammary ascites tumor (MAT 13762B), Nat. Immun. Cell Growth Regul. 3:43 (1983/84).

76. A.A-B Badawy, M. Evans, N.F. Punjani, and C.J. Morgan, Does naloxone always act as an opiate antagonist? Life Sci. 33:739 (Sup. I, 1983).

77. N.P. Plotnikoff, Abba J. Kastin, D.H. Coy, C.W. Christensen, A.V. Schally, and M.A.Spirtes, Neuropharmacological actions of enkephalin after systemic administration, Life Sci. 19:1283 (1976).

78. A.J. Murgo, Modualtion of tumor growth in mice treated with methionine-enkephalin and corticosterone, Proc. First Internat. Workshop on Neuroimmunomodulation, in press (1985).

79. A.J. Murgo, Inhibition of B16 melanoma growth in mice by methionine-enkephalin, J. Natl. Cancer Instit. in press (1985).

80. D.D. Von Hoff and B. Forseth, Modulation of growth of human and murine tumors by human beta-endorphin, Proc. Amer. Assoc. Cancer Res. 23:236 (1982).

81. C.I. Thompson, J.W. Kreider, P.L. Black, T.J. Schmidt, and D.L. Margules, Genetically obese mice: resistance to metastasis of B16 melanoma and enhanced T-lymphocyte mitogenic responses, Science 220: 1183 (1983).

82. P.L. Black, M. Holly, C.I. Thompson, and D.L. Margules, Enhanced tumor resistance and immunocompetence in obese (ob/ob) mice, Life Sci. 33:715 (Sup. I, 1983).

83. N.P. Plotnikoff, R.E. Faith, A.J. Murgo, and G.C. Miller, Enkephalins: immunomodulators and antitumor activities, Proc. First Internat. Workshop on Neuroimmunomodulation in press (1985).

84. C.Y. Hung, S.S. Lefkowitz, and W.F. Geber, Interferon inhibition by narcotic analgesics, Pro. Soc. Exp. Biol. Med. 142:106 (1973).

85. S.L. Brown and D.E. VanEpps, Beta-endorphin, [Met]enkephalin and corticotrophin modulate the production of gamma interferon in vitro, Fed. Proceed. (abst.), 44:949 (1985).

RELATIONSHIP BETWEEN LYMPHOKINE AND OPIATERGIC MODULATION OF LYMPHOCYTE PROLIFERATION

William L. Farrar

Laboratory of Molecular Immunoregulation, Biological Response Modifiers Program, National Cancer Institute-FCRF Frederick, MD 21701 USA

INTRODUCTION

T lymphocytes are primary effector cells of cell-mediated immunity as well as regulators of growth (clonal expansion) and differentiation of the immune system. Until less than a decade ago, their regulatory activity was presumed to be attributed to predominantly cellular interactions mediated by membrane-bound recognitive elements often associated with the major histocompatibility (H2 or HLA) complex. Recently, in the historical context of the development of immunological models, many regulatory functions of T lymphocytes appear to be mediated by molecules secreted by various functionally distinct subpopulations of lymphoid and nonlymphoid cells. Indeed, the principle regulatory cells of the immune system, T lymphocytes, are also under the regulatory influence of specific monocyte/ macrophage derived cytokines referred to as monokines. In addition to regulating lymphocyte activities, there is considerable evidence that lymphocyte-derived molecules, lymphokines, influence the differentiation of hematopoietic cells, mast cells, fibroblasts, and osteoclasts[1-6]. This suggests that the regulatory cytokines produced by specific antigen may also effect hematopoietic homeostasis as well as inflammatory cells involved in immediate hypersensitivities. Lymphokines are considered to be actively synthesized (transcription of DNA) and secreted by either native lymphoid cells or their malignant cell line counterparts (macrophage or lymphocyte). While normal lymphoid tissues or cells must be activated by antigen or polyclonal stimulation to produce lymphokines, some long-term cell lines have been fortuitously shown to constitutively secrete lymphokines. Indeed, such cell lines have been critical for providing the genetic library for the recombinant DNA cloning of several lymphokines. Lymphokines provide nonspecific augmentation to antigen-specific responses. Lymphokines can generally be readily separated by conventional biochemical chromatography, but standard 280 nm absorbance and protein stains have been of little value since they are usually present in culture media in nanomolar concentrations. Finally, with the introduction of monoclonal antibodies, several lymphokines may be neutralized and quantitated in vitro by specific antibody.

Although soluble factors were reported to influence (in a nonspecific fashion) antigen specific immune responses[7], it wasn't until 1978 when chromatographic separation was used to try to discriminate individual lymphokine molecules and their effects on discrete immune responses[8,9].

An apparent enigma from these studies was the observation that biochemically distinct factors derived from cells of different functional phenotypes (i.e., macrophages and T cells) produced identical biological effects. For example, interleukin 1 (IL 1), a soluble factor of 15,000 dalton molecular weight derived from endotoxin-stimulated macrophages, enhanced both antibody producing cell responses to sheep erythrocytes in nude mice, stimulated thymocyte proliferation[9,10] and augmented the production of cytotoxic lymphocytes in macrophage-depleted cultures[11]. The issues were complicated by the observation that three apparently different biological activities mentioned above could be ascribed to a single macrophage-derived molecule (IL 1) and that interleukin 2 (IL 2), derived from T helper lymphocytes (molecular weight 15-21,000 daltons), also exhibited the same three immunobiological activities [9,12,13]. Not only did IL 2 enhance the primary plaque-forming responses of nude mouse spleen cells to sheep erythrocytes and stimulate thymocyte proliferation and cytotoxic T cell responses, but IL 2 could also maintain the selective indefinite growth of cytotoxic T cell clones[12,14]. Although IL 1 and IL 2 were apparently biochemically distinct and were derived from macrophages and T lymphocytes respectively or their corresponding tumor cell lines[15,16], both molecules elicited or enhanced similar immunobiological phenomena. Based on these observations, our laboratory chose to investigate several possible explanations to resolve whether IL 1 and IL 2 acted on the same populations of cells or alternatively were involved in a sequence of factor-cell interactions with a degree of linearity which resulted in the manifestation of similar biological expression (i.e., the development of antibody-forming cells or cytotoxic T cells).

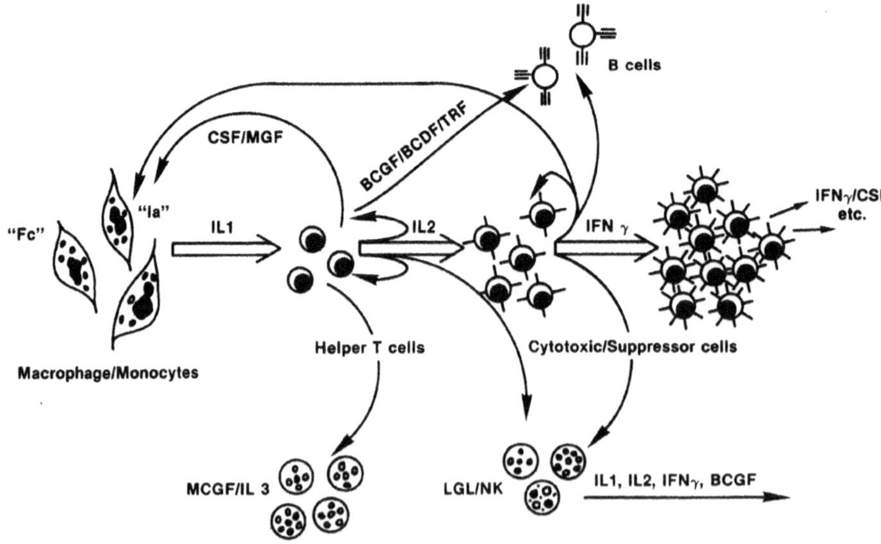

Figure 1: The Lymphokine Regulatory Network

The Lymphokine Regulatory Network

IL 1 Induction of IL 2. Although the similarity of IL 1 and IL 2 bioactivities was striking, data began to accumulate to argue against the possibility that both IL 1 and IL 2 exerted activity on the same target cell populations. The inability of supernatants from activated macrophages to support the growth of long-term T cells[14] and the ability of such supernatants to stimulate production of IL 2 from purified T cells[11] suggested that indeed macrophage-derived soluble mediators did not act on the same target population as did T cell-derived IL 2[17]. Confirmation that IL 1 stimulates T cells to produce IL 2 was demonstrated later by the studies of Smith and colleagues[18] and Larsson *et al.*[19]. The finding that macrophage-derived factors were required for T cells to produce immunoregulatory factors was further supported by the observation of Moore *et al.*[20] who showed that the T cell product, colony stimulating factor (CSF), activated macrophages to produce IL 1, thus completing at least one regulatory circuit (Figure 1). CSF was also shown to enhance the production of prostaglandins and interferon from macrophages[21]. It was apparent from these studies that lymphokines operated in a sequential fashion, with possible feedback loops, which result in a sophisticated system of intercellular communication.

IL 2 Induction of IFN γ. The discovery that IL 2 not only regulated the growth of cytotoxic cells, but also the production of another lymphokine, gamma interferon, (IFN γ) further reinforced the notion that a linear series of factor-cellular interactions was involved in the amplification of cytotoxic T lymphocyte responses (Figure 1)[22]. Because of the "cascading" appearance of these interactions the term "lymphokine cascade" was applied when referring to the interdependent interactions of cells and lymphokines[13]. During the original investigation demonstrating IL 2 regulation of IFN γ, it was shown that antisera directed to gamma interferon inhibited the development of cytotoxic T lymphocytes[22].

In a parallel system, IL 2 has been shown to augment cytotoxic responses of natural killer (NK) cells[23], as well as promote the growth of NK cells[24]. The ability of IL 2 to enhance the cytotoxic responses of NK cells apparently is independent of IL 2 growth promotion and is also neutralized by antisera directed against IFN γ in response to IL 2 and since IFN γ has been described to enhance NK cytolytic activity, it was concluded that IL 2 augments NK activity via the production of IFN [25].

Possible Role of Gamma Interferon in Immunity. Data from a variety of studies have recently demonstrated rather pleotypic differentiating effects of gamma interferon. With regards to the effects observed with NK and cytotoxic T lymphocytes, recent work has suggested that IFN γ may act on T cells by enhancing the expression of IL 2 receptors[26]. Further support for the differentiation signal activity of IFN γ was demonstrated by Steeg *et al.* who showed that IFN γ induced Ia antigen expression on macrophages[27]. Other investigators have shown that IFN γ increased HLA expression on human lymphocytes, Fc receptor expression on macrophages, immunoglobulin synthesis by B cells, and the activation of macrophages and NK cells for tumor cytolysis[28-31]. The evidence remains overwhelming implicating IFN γ as a major differentiating signal required for cellular immunity.

T Cell-Derived Lymphokines and B Cell Responses. The existance of a T cell-derived B cell growth factor (BCGF) was previously implied by the observation that activated mouse B lymphocytes could be grown in vitro by the addition of Con A-conditioned medium or with PMA-induced EL4 cell line supernatant[32]. BCGF has been shown to be produced by helper T lymphocytes and is biochemically separable from IL 2. T cell replacing factor (TRF) is functionally distinct from BCGF in that it is most active in promoting B cell differentiation when added late (day 2 or 3) to cultures of spleen cell cultures undergoing primary immunization in vitro[33]. Although IL 2, BCGF and TRF are biochemically separable, it is not clear whether TRF and B cell differentiation factor (BCDF), a factor which induces B cells to secrete immunoglobulin, are the same or similar molecules.

T Cell Regulation of Other Peripheral Cells. It was mentioned earlier that IL 2 is also known to regulate the growth of NK cells[24]. Subsets of large granulocyte lymphocytes with NK activity have been shown to produce IL 2, IFN and CSF and may also possibly produce IL 1 and BCGF (Scala and Ortaldo, personal communication)[34]. The findings that NK cell subsets produce a variety of immunomodulatory signals may broaden their potential role as central regulatory cells that influence antigen-specific humoral and cellular immunities.

Murine T lymphocytes have been suggested to also produce a unique class of colony stimulating factor(s) referred to as IL 3[35]. Although originally believed to be active on pre-helper T cells, it is now well established that IL 3 does not act on helper T cell precursors and is biochemically identical to burst-promoting activity (BPA), P cell stimulating factor, mast cell growth factor, and multi-colony stimulating factor. The human equivalent has not been well characterized, but it is believed to act as a growth signal for pluripotent stem cells[36].

RELATIONSHIP OF β-ENDORPHIN WITH IL 2 REGULATION OF PROLIFERATION

Among the hormonal consequences of acute stress is the release of corticotropin (ACTH) and β-endorphin by the pituitary. The peripheral effects of β-endorphin and its hormonal relationships to stress remains obscure. Recently immunoenhancing effects of β-endorphin have been reported. β-endorphin has been shown to specifically bind to the human B-cell line RPMI 6237[37], as well as increase the proliferative response of rat splenocytes to T-cell lectins[30]. Additionally, β-endorphin has been shown to increase natural killer (NK) cytotoxic activity[39].

Based on the observations that β-endorphin enhances immunological activities (i.e., lymphocyte proliferation and NK cytotoxicity) that are under the specific regulatory control of interleukin 2 (IL 2), a growth and regulatory peptide of the immune system, our laboratory examined whether β-endorphin may specifically modulate IL 2-dependent proliferative events. Since it was reported by Hazum et al.[37] that cultured transformed RPMI 6237 cells had specific receptors for β-endorphin, we examined whether discrete polulatons of highly purified human T lymphocytes demonstrate any specific binding of ^{3}H-etorphine. The availability of purified lymphocytotropic growth factors such as BCGF and IL 2 have allowed us to selectively expand specific lineages of lymphoid cells. Table 1 shows data obtained from binding studies with subpopulations of lymphocytes expanded in either B-cell growth factor (B cells) or IL 2 (T and LGL). Both T lymphocytes and large granulocyte lymphocytes, which express natural killer cell (NK) activity, demonstrate a similar class of receptors suggesting broad specificity since diprenorphine and β-endorphin effectively displaced ^{3}H-etorphine. B lymphocytes, however, demonstrated competitive displacement of ^{3}H-etorphine, binding of a different pattern

TABLE 1

HUMAN LYMPHOCYTE OPIATE RECEPTOR ANALYSIS

Specificity	Compound [M]		%Inhibition Total Binding* B**	T	LGL (NK)
Broad	Diprenorphine	$[10^{-5}]$	35%	33%	41%
Delta	DADle	$[10^{-6}]$	<1%	<1%	13%
Delta	DSTle	$[10^{-6}]$	<1%	<1%	<1%
Kappa	EKC	$[10^{-6}]$	<1%	22%	20%
Kappa	Bremazocine	$[10^{-6}]$	<1%	14%	13%
Broad	β-endorphin	$[10^{-6}]$	<1%	25%	44%
f-m Specific binding/10^6 cells			67	93	59

*Percentages represent the decrease in total binding of ^{3}H-etorphine at 40 nM.

**Factor-dependent cell lines BD-7 (B cell), TD-1 (T cell) and IL 2 dependent LGL were washed free of conditioned media, resuspended to 20 x 10^6 cells/assay tube. Assay was run at 37°C for 30 min in presence of 50 ul ^{3}H-ligand, 50 ul cold ligand, 50 ul Hanks BSS and 400 ul. Data is result of the mean of triplicate samples. Data courtesy of Dr. Richard Weber, NIH, Bethesda, MD.

suggesting a fundamental difference in the nature of opiate receptor between antibody-producing cell lineage and IL 2 promoted lymphocytes.

Since there is parallel receptor appearance for IL 2 and β-endorphin and similar biological enhancements, we tested whether homogenous IL 2 prepared from the gibbon T lymphoid MLA-144 cell line can possibly compete with β-endorphin bind on brain caudate sections. The results suggest that even when 10^6 M excess concentrations of pure IL 2 are used, absolutely no competition with β-endorphin binding to brain caudate sections occurs (data not shown). Therefore, based on the facts that no primary amino acid sequence homology exists between IL 2 and β-endorphin (or other neuroendocrine hormones; Farrar, unpublished data), and no competition of IL 2 for the opiate receptor was observed, we conclude that β-endorphin enhancement of immune responses does not occur via common structural receptor elements shared with the IL 2 receptor. Additional support for this conclusion comes from our observation that monoclonal anti-TAC antibody, directed against the IL 2 receptor, does not bind to brain sections, further substantiating the possibility that opiate receptors do not immunologically cross-react with IL 2 receptors (Pert, C. and Farrar, W., unpublished data).

<u>β-endorphin Modulation of IL 2 Induced Proliferation.</u> Experiments compared the effects of β-endorphin on the proliferative responses of cloned IL 2 dependent (CT6) murine cytotoxic T cells and peripheral blood lymphocytes (PBL) to PHA. The data in Figure 2 demonstrated that β-endorphin had no modulating activity on the CT6 response to IL 2, however, β-endorphin (10^{-9}M) enhanced the proliferative response of PBL to PHA 30-40%. It was apparent from these experiments that β-endorphin does not have direct modulatory activity on IL 2-mediated events but still maintains a profound influence on the proliferation of mitogen-stimulated T lymphocytes when present in mixed cellular populations. Although both systems represent a proliferative response to IL 2, PBL responses to mitogen are under the exquisite regulatory control of macrophages partially through cellular contact but predominantly through the participation of soluble mediators.

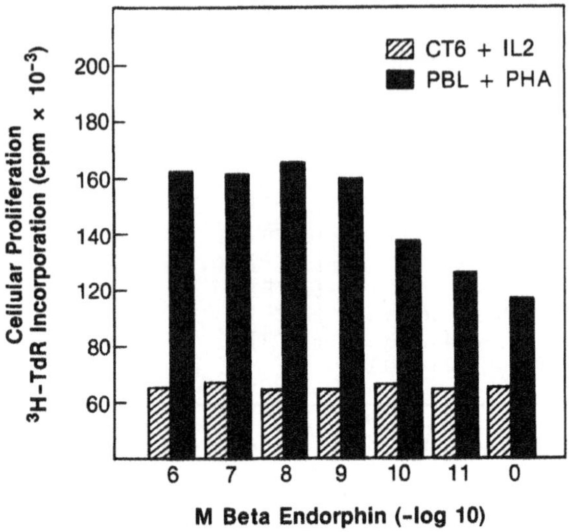

Fig. 2 Effects of β-endorphin on the proliferative responses of IL 2-dependent T cells and PBL mitogen-induced proliferation.

Among the peripheral effects of the endorphins is the observation that β-endorphin can modulate the effects of prostaglandins on mast cells and rat brain cells[40-42]. Since prostaglandins are considered to be produced by macrophages and exert a negative regulatory influence on T-cell proliferaton as well as IL 2 responsiveness[43,44], we test the ability of β-endorphin to influence the effects of prostaglandin E_2 (PGE_2) on CT 6 proliferation to IL 2 as well as purified human T-lymphocyte PHA responses (Figure 3).

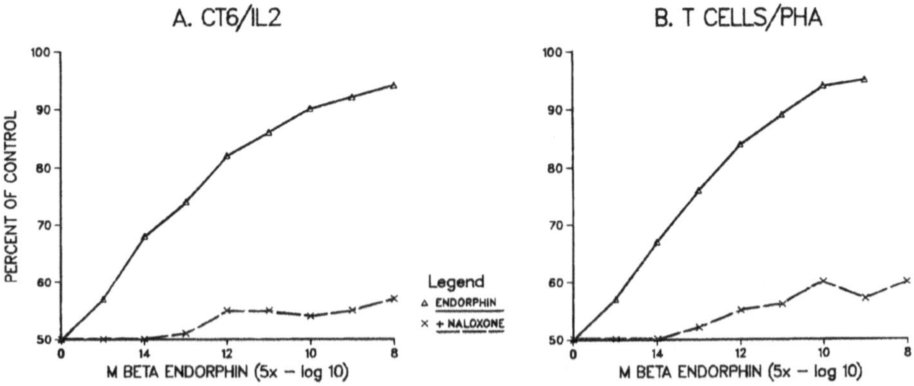

Fig. 3 Reversal of PGE_2 inhibition of proliferative responses by β-endorphin.

From the data, it can be seen that exogenous addition of 10^{-7}M PGE_2 to either CT6 plus 10 units IL 2 or to purified T cells plus PHA results in at least 50% inhibition of proliferation. If either cell population were preincubated with β-endorphin, the inhibitory activity of PGE_2 on proliferative responses is reversed in a dose-dependent manner. This suggests that the enhancement of proliferation of PBL seen in earlier experiments may be due to β-endorphin's ability to reduce the effects of endogenous PGE_2. Similar data are seen when PBL are treated with indomethacin to prevent endogenous production of PGE_2.

When CT6 cells or human T lymphocytes were preincubated with the opiate antagonist naloxone (10^{-6}M), β-endorphin reversal of PGE_2 inhibition of IL 2 responsiveness was not seen, indicating that the effects of β-endorphin observed were the results of specific opiate receptor interaction.

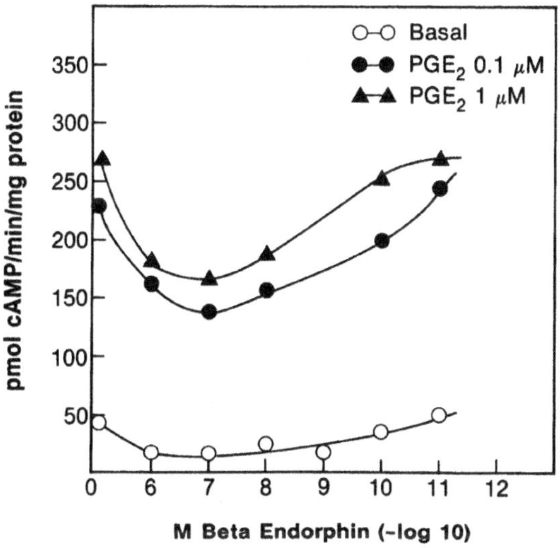

Fig. 4 T lymphocyte adenylate cyclase activity.

<u>Effect of β-endorphin on T Lymphocyte Adenylate Cyclase Activity.</u> Since the data suggested that β-endorphin could modulate the inhibitory effects of PGE_2 on either IL 2 stimulated CT6 proliferation or mitogen induced human T lymphocyte responsiveness, we tested whether β-endorphin had any effects on hormone stimulated membrane-associated adenylate cyclase activity. CT6 cells were pretreated with the indicated doseages of β-endorphin for 10 mins at 37°C. The cells were then centrifuged, lysed and particulate membrane fractions prepared. From the data in Figure 4 it can be seen that β-endorphin pretreatment of CT6 cells decreased the basal activity of adenylate cyclase activity by at least 50%. Unstimulated CT6 membranes exhibited a basal activity of 45 pmoles/min/mg and PGE_2 stimulated particulate membranes increased adenylate cyclase activity to 230 (.1uM) and 266 (1 uM) pmoles/min/mg respectively. The pretreatment of intact CT6 cells with β-endorphin for 10 mins was sufficient to decrease basal adenylate cyclase (AC) activity by at least 50% and PGE_2 induced activity from 30-50% compared to untreated cells. β-endorphin treatment of isolated CT6 membranes reduced basal AC activity by 15-20% and PGE_2 induced AC activity by 22-35% (data not shown). These

experiments provided biochemical support for the hypothesis suggested by the cellular studies (Fig. 3) that β-endorphin could mitigate the negative influence of prostaglandins on the T lymphocyte proliferative response.

Effects of Interleukins on Pituitary Function. Coincident with peak antibody responses in sheep erythrocyte immunized mice are elevations of serum cortisol levels. This observation suggested a casual relationship between the development of immunity and pituitary-adrenal activity[45]. Utilizing a murine pituitary cell line AtT20, we tested whether purified interleukins 1 or 2 had any effects on AtT20 pituitary ACTH or β-endorphin secretion (Table II).

Although AtT20 cells constituatively secrete ACTH and β-endorphin, interleukin 1 and 2 both stimulated 2-3 fold increases in synthesis of the neuropeptides. Gamma interferon had no effect on the secretory activity of AtT20, IL 1 generally stimulated a 60-80% increase in neuropeptide secretion whereas IL 2 stimulated almost 3-fold over background synthesis.

TABLE II

EFFECTS OF LYMPHOKINES ON PITUITARY CELL LINE AtT20 SECRETION OF

β-ENDORPHIN AND ACTH

Stimulant	ng BEP/10^6 cells* 24 hr	48 hr	ng ACTH/10^6 24 hr	48 hr
Media	610	1150	720	1440
Lectin-free** conditioned (5%) media	1240	1650	1490	2360
IL 2 (100 U)***	1870	2115	2670	3318
IFN γ (100 U)	575	1010	698	1384
IL 1 (50 U)	1130	1500	1375	1890

*RIA was performed on duplicate culture supernatants for β-endorphin and ACTH. Supernatants were harvested at 24 and 48 hr filtered and frozen at -70°C until extraction. Data is presented as ng/10^6 cells.

**Lectin-free conditioned media from human PBL was purchased from Cellular Products, Inc.

***Recombinant human IL 2, IFN γ and partially purified human IL 1 was used at the dosages indicated.

DISCUSSION AND SUMMARY

The biochemical resolution of the lymphokines and the recent ability to initiate and maintain antigen-specific lymphoid cells with lymphocytotropic growth factors established the investigative elements for the examination of physiological and molecular events which regulate immunity. Cellular immunologist are no longer burdened with the "black box" of murine spleen cells as the mechanism of investigative research. The clonal expansion and propagation of lineage specific cells allows the development of homogeneous populations of cells which may respond uniformly to a defined growth progession signal, such as IL 2 or a differentiation signal i.e., IFN γ. Utilizing this approach we have been able to expand specific lineages of cells and examine opiate receptor expression and the effects of the opiatergic ligands on the cellular response to the lymphocytotropic hormones of immunity, the interleukins.

Within these studies the data shows an increase in opiate receptors on activated T lymphocytes and large granulocyte lymphocytes, both cell types which have cytolytic activity against tumor cells. B lymphocytes, however, exhibited opiate-like receptors which differ from those found on T lymphocytes (Table 1). Associated with receptor appearance is the ability of β-endorphin to modulate (uncouple) T-lymphocyte responses to prostanglandins (Figure 3,4). The mechanism of β-endorphin activity is mediated through specific naloxone reversible opiate receptors and does not directly involve IL 2 receptors or IL 2 growth-promoting activity. Both T lymphocyte proliferation and natural cytotoxicity are suppressed by the endogenous production or exogenous addition of PGE_2. β-endorphin has been shown to modulate PGE_1 inhibition of serotonin release from mast cells as well as abrogate prostaglandin induced elevation of cyclic AMP in neurocytoma cells[40,41]. The experiments present here (Figure 4) indicate that β-endorphins may also modulate the basal and hormonal induced membrane adenylate cyclase activity of T lymphocytes. Based on these data, it is reasonable to suggest that the enhancement activity reported for IL 2-mediated phenomenon[38,39] by β-endorphin may be attributed to the ability of β-endorphin to mitigate the intrinsic effects of PGE_2 produced in PBL cultures.

This data contributes to a growing body of evidence suggesting a reciprocity of regulatory functions between neuroendocrine hormone systems and peripheral physiologies such as the immune system. Since endorphin release has been associated with acute stress (exercise, sexual activity, etc.), we may be able to extrapolate the notion that endorphins may mitigate some of the effects of chronic inflammation and other deleterious consequences of aberrant prostaglandin production. Of singular significance is the observation of Thomas <u>et al.</u>[46] that the ob/ob obese mouse strain has three- to four-fold greater resistance to metastatic melanoma, heightened T-cell mitogen responsiveness, and elevated serum levels of β-endorphin. This suggests a possible role of β-endorphin in the in vivo maintenance or enhancement of natural immunities.

Following sensitization for T-dependent antigens, rodents exhibit a significant increase in serum corticosterone[45]. This elevation coincides temporarily with peak antibody titers and has been suggested to be under the regulatory control of the immune system . Recent investigators have proposed that either lymphokines or thymic hormones may have activity on the anterior hypothalmus thus regulating corticosterone release[45]. Data from our laboratory using purified lymphokines (Table II) demonstrated a direct effect on ACTH and β-endorphin release by pituitary tumor cells. The findings that IL 2 may have low affinity receptors on pituitary cells and that lymphocytes possess receptors for neuroendocrine hormones further substantiate the possibility of bi-directional peptide-mediated regulation between the chemical regulators of antigen-stimulated immune responses and the pituitary-adrenal axis. The potential effects of β-endorphin interaction with the immune system have been discussed. Corticosterone (CS) production as a consequence of ACTH release, however, may have many negative regulatory effects on the immune system. Corticosterone has been previously shown to inhibit IL 2 production. It is therefore possible that the binding of CS to receptors may be viewed as a possible mechanism for signaling the switching off of lymphokine production which may in turn result in the cessation of the CS response seen during the later stages of development of the immune response (Figure 5).

Hormonal Regulation of the Immune - Pituitary/Adrenal Axis

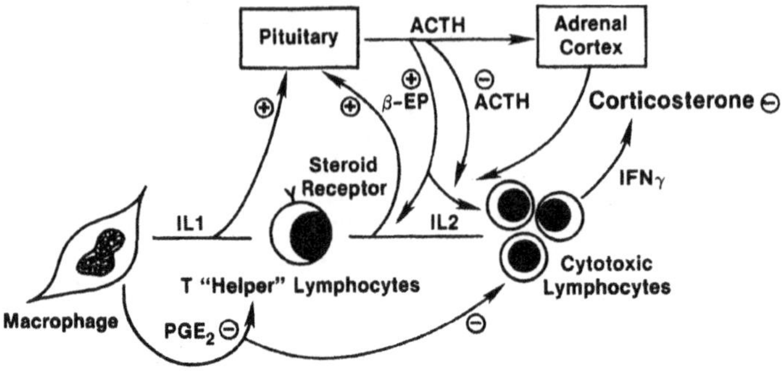

These data suggest that a regulatory circuit of negative-feedback reciprocity between the immune system and the pituitary-adrenal axis exists in response to antigen stimulation and that specific lymphokines such as IL 1 and IL 2 may be involved directly on pituitary function.

REFERENCES

1. R. W. Ruscetti, R. H. Cypess, and P. S. Chervenick, Blood 47:757 (1976).
2. J. J. Oppenheim and I. Gery, Immunology Today 3:113 (1982).
3. K. Tadokoro, B. M. Stadler and A. L. DeWeck, J. Exp. Med. 158:857 (1983).
4. J. W. Adamson, W. J. Dopovic and J. E. Brown, in: "Hematopoitetic Cell Differentiation," D. W. Golde, M. J. Cline, D. Metcalf, C. F. Fox, eds., Academic Press, New York (1978).
5. A. W. Burgess, D. Metcalf, N. A. Nicola and S.H.M. Russel, in: "Hematopoietic Cell Differentiation," D. W. Golde, M. J. Cline, D. Metcalf, C. F. Fox, eds., Academic Press, New York (1978).
6. Y. P. Yung, R. Eger, G. Tertian and M.A.S. Moore, J. Immunol. 127:794 (1981).
7. J. J. Farrar and J. Fuller-Bonar, J. Immunol. 117:274 (1976).
8. J. J. Farrar, W. J. Koopman and J. Fuller-Bonar, J. Immunol. 114:47 (1977).
9. J. J. Farrar, P. L. Simon, W. J. Koopman and J. F. Bonar, J. Immunol. 121:1353 (1978).
10. D. D. Wood, J. Immunol. 123:2400 (1979).
11. W. L. Farrar, S. B. Mizel and J. J. Farrar, J. Immunol. 124:1371 (1980).
12. S. Gillis, P. E. Baker, F. W. Ruscetti and K. A. Smith, J. Exp. Med. 154:983 (1978).
13. J. J. Farrar, W. R. Benjamin, M. L. Hilfker, M. Howard, W. L. Farrar and J. Fuller-Farrar, Immunol. Rev. 63:129 (1982).
14. R. W. Ruscetti, D. A. Morgan and R. C. Gallo, J. Immunol. 122:2527 (1979).

15. S. B. Mizel and D. L. Rosenstreich, J. Immunol. 122:2173 (1979).
16. J. J. Farrar, J. Fuller-Farrar, P. L. Simon, M. L. Hilfiker, B. M. Stadler and W. L. Farrar, J. Immunol. 125:2555 (1980a).
17. J. J. Farrar, S. B. Mizel, J. Fuller-Bonar, M. L. HIlfiker and W. L. Farrar, Microbiol. 36:39 (1980b).
18. K. A. Smith, L. B. Lachman, J. J. Oppenheim and M. F. Favata, J. Exp. Med. 151:1551 (1980).
19. E. L. Larsson, N. N. Iscove and A. Coutinho, Nature 283:664 (1980).
20. R. N. Moore, J. J. Oppenheim, J. J. Farrar, C. S. Carter, A. Waheed and R. K. Shadduck, J. Immunol. 3:119 (1980).
21. R. N. Moore, J. T. Haffeld, J. J. Farrar, S. E. Mergenhagen, J. J. Oppenheim and R. K. Shadduck, Lymphokines 3:119 (1981).
22. W. L. Farrar, H. M. Johnson and J. J. Farrar, J. Immunol. 126:1120 (1981).
23. C. S. Henney, K. Kuribaysashi, D. E. Kern and S. Gillis, Nature 291:335 (1981).
24. W. Domzig, B. M. Stadler and R. B. Herberman, J. Immunol. 130:1970 (1983).
25. D. A. Weignet, G. J. Stanton and H. M. Johnson, Infect. Immun. 41:992 (1983).
26. H. M. Johnson and W. L. Farrar, Cell. Immunol. 75:154 (1983).
27. P. S. Steeg, R. N. Moore, H. M. Johnson and J. J. Oppenheim, J. Exp. Med. 156:1780 (1983).
28. T. Y. Basham and T. C. Merigan, J. Immunol. 130:1492 (1983).
29. B. Harfest, J. R. Huddleston, P. Casali, T. C. Merigun and M.B.A. Olstone, J. Immunol. 127:2146 (1981).
30. D. N. Mannel and W. Falk, Cell. Immunol. 79:396 (1983).
31. J. R. Ortaldo, A. Mantovani, D. Hobbs, M. Rubinstein, S. Pestica and R. B. Herberman, Int. J. Cancer 31:285 (1983).
32. M. C. Howard, J. Farrar, M. Hilfiker, B. Johnson, T. Takatsu and W. L. Paul, J. Exp. Med. 155:914 (1982).
33. A. Schimpl and W. Wicker, Nature 237:15 (1972).
34. T. Kasahara, J. Y. Djeu, S. F. Dougherty and J. J. Oppenheim, J. Immunol. 131:2379 (1983).
35. A. Hapel, J. C. Lee, W. L. Farrar and J. W. Ihle, Cell 25:179 (1981).
36. J. W. Schrader and I. Clark-Lewis, J. Immunol. 129:30 (1982).
37. E. Hazum, K-J. Chang and P. Cuatrecasas, Science 205:1033 (1979).
38. S. C. Gilman, J. M. Schwartz, R. J. Milner, F. E. Bloom and J. D. Feldman, PNAS 79:4226 (1982).
39. P. M. Mathews, C. J. Froelich, W. L. Sibbitt and A. D. Bankhurst, J. Immunol. 130:1658 (1983).
40. Y. Yamasaki, K. Shimamura, A. Kizu, M. Nakagawa and H. Ijichi, Life Sciences 31:471 (1982).
41. H.O.J. Callier and A. C. Roy, Nature 248:24 (1974).
42. R. S. Childers, S. M. Lambert and G. LaRiviere, Life Sciences 33:215 (1983).
43. R. D. Maca, Immunopharmacol. 6:267 (1983).
44. P. E. Baker, J. V. Fahey and A. Munck, J. Exp. Med. 61:52 (1981).
45. P. N. Shek and B. H. Sabiston, Int. J. Immunopharmacol. 5:23 (1983).
46. C. I. Thompson, J. W. Kreider, P. L. Black, T. J. Schmidt and D. L. Margules, Science 220:1183 (1983).

ENKEPHALINS AS MOLECULES OF LYMPHOCYTE ACTIVATION AND MODIFIERS OF THE BIOLOGICAL RESPONSE

Joseph Wybran

Department of Immunology, Hematology and Transfusion
Erasme Hospital, Université Libre de Bruxelles, 1070
Brussels, Belgium

INTRODUCTION

In these recent years, a new stream of investigation has emerged which tightens narrowly the relations existing between the central nervous system, the endocrine system and the immune system through the action of endogenous opioid peptides like enkephalins and endorphins. In the present review, we will attempt to summarize the data related only to the actions of enkephalins upon the immune system.

Enkephalins are small peptides composed of five amino acids whereas endorphins are composed of 16 to 31 amino acids. The first five amino acids of the endorphins are similar and in the same sequence as Methionine-Enkephalin namely Tyr-Gly-Gly-Phe-Met. In contrast, Leucine-Enkephalin is composed by the following sequence Tyr-Gly-Gly-Phe-Leu. A synthetic enkephalin D-Ala^2 Met^5 composed by Tyr-D-Ala-Gly-Phe-Met has been produced. Physiologically the enkephalins are rapidly destroyed by serum enkephalinases.

Enkephalins are synthetized in the adrenals and their main action is upon the nervous system. It is also usually accepted that naloxone is a specific antagonist of the opioid activities of enkephalins. Therefore, any inhibition by naloxone of some immunological function induced by an opioid peptide is accepted as evidence that the immunological activity has been specifically induced by a naloxone sensitive opioid dependent mechanism.

LYMPHOCYTE ENKEPHALIN RECEPTORS

The first demonstration that human T lymphocytes bear a surface receptor for Met-Enk stems from the work of Wybran et al in 1979 (1). Interestingly, at the same time, Hazum et al showed the presence of β endorphin receptors on human lymphoblastoid cell lines (2).

The method for detecting the receptors on human T cells was based on a rosette assay between human T lymphocytes and sheep red blood cells. In this test, termed the active T rosette assay, the technical conditions for performing the rosette assay are not optimal (low ratio of sheep red blood cells to lymphocytes, short incubation time, and temperature of 37 C), so that the active rosette assay is very sensitive to multiple factors (3).

By incubating peripheral blood lymphocytes with morphine or Met-Enk, the percentage of active T rosettes was modified : morphine decreased the percentage whereas Met-Enk increased it. Nonspecificity was ruled out because naloxone restored a normal percentage of active rosettes in the presence of morphine or Met-Enk. On the basis of these results, it was suggested that human blood T cells possess surface receptorlike structures for morphine and Met-Enk.

Finally, and relevant to the receptor problem, Mehrishi et al have recently confirmed the presence of morphine receptors on human blood lymphocytes by direct radioligand techniques. They also showed the presence of such receptors for naloxone and morphine on human blood platelets (4). Also, Johnson et al have shown, by direct labeling, that Met-Enk binds to mice spleen cells (5) allowing the conclusion that enkephalin receptors are present both on human and animal lymphocytes.

The property of Met-Enk to increase,in vitro, active T cell rosettes of normal individuals has been confirmed (6). Furthermore, Leu-Enk can also increase, in vitro, the active T cell rosettes of normal subjects at concentrations lower to the ones which are active for Met-Enk (between 10^{-11} and 10^{-5}M). Interestingly enough, both Met- and Leu-Enk can also, in vitro, enhance the percentage of active T cell rosettes of lymphoma patient (7).

Finally, $ZnCl_2$ appears synergystic with Met-Enk in the active rosette enhancement. This can be blocked by the zinc chelator phenathroline but not by naloxone. These data suggest some type of relationship between zinc may help binding of Met-Enk to the lymphocytes (8).

All the data related to the findings using the active T rosette assay can lead to two possible explanations : one which is purely related to a steric hindrance mechanism which postulates the presence of enkephalin receptor close (or even similar) to the sheep red blood cell (E) receptor on human T lymphocytes and another one related to the functional meaning of the active rosette test (3). Indeed, it has been shown that both specific antigen and drugs (or molecules) able to increase T cell functions (e.g. isoprinosine, thymosin) (9, 10) also enhance, in vitro, the active T cell assay. Therefore, one can also postulate that both Met-Enk and Leu-Enk which increase the active T rosettes assay possess enhancing T cell properties both in normal subjects and in lymphoma patients.

THE TWO-HIT OPIOID PEPTIDE LYMPHOCYTE RECEPTOR HYPOTHESIS

The above findings have allowed the proposal of the two-hit opioid peptide lymphocyte hypothesis in which it is suggested that opioid peptides recognize T and B cells by two different determinants of their molecule, one naloxone sensitive and the other one non naloxone sensitive (indeed, the β endorphin receptor is not inhibited by naloxone). This double recognition of one molecule by two different types of lymphocytes (indeed, human lymphoid cell lines are usually of B cell origin) may provide some type of chemical bridge between T and B cells and therefore allowing T and B cell cooperation (11). Because T and B cell cooperation is of utmost importance in mounting the immune response, the two-hit hypothesis suggests that opioid peptides are a physiological regulator of the immune response and that their modification (in quantity or quality) may influence the immune response.

LYMPHOCYTE PROLIFERATION ASSAYS

The data obtained with enkephalins using lymphocyte stimulation in presence of various mitogens are conflicting. In animals, Gilman et al showed that D-Ala^2 Met^5 Enk does not modify the proliferative response to mitogens (12). In contrast, Plotnikoff et al showed that Met-Enk, and to a lesser extent Leu-Enk, increases the response of mouse spleen cells to phytohemagglutinin. Usually, the results obtained were twice higher than the stimulation performed without enkephalins (13).

In humans, Wybran could usually not show any significant effect of enkephalins using human blood lymphocytes (14). In a series of experiments, the effect of various enkephalins on spontaneous thymidine uptake, the PHA response, and the pokeweed mitogen response was studied (unpublished observations). The results were basically negative. No effect on thymidine uptake was observed except in the case of Leu-Enk at a very narrow concentration range (10^{-5} M). At this concentration, Leu-Enk increased the PHA response when PHA was present at a suboptimal concentration whereas it decreased the optimal PHA response. These results suggest that, in humans, these drugs do not modify nonspecific proliferation assays.

ANTIBODY SYNTHETIS

Johnson et al have studied the in vitro regulation of the antibody response to sheep red blood cells, a T cell dependent antigen (5). They could show that the number of plaque forming cells (antibody producing cells) were moderatly inhibited (around 60 %) by Met-Enk and Leu-Enk.

NATURAL KILLER ACTIVITY

This in vitro assay investigates the spontaneous cytotoxic effect of mononuclear cells (natural killer cells - NK cells) against various target cells which are normally of leukemic or lymphomatous origin. It is thought that NK cells are of importance in controlling the metastasization of tumor cells as well as in various viral and parasitic diseases. Therefore, any agent able to increase NK activity may be considered as potentially useful in the treatment of cancer, viral and parasitic diseases. Interferon, for instance, is known to be able to enhance NK activity and is clinically tried in various malignancies. Mathews et al have published data for humans that show that β-endorphin and Met-Enk could significantly (± 30 %) increase the spontaneous cytotoxicity induced by NK cells against radiolabeled K562 leukemic cells (15). The increased NK activity was dose dependent. Naloxone inhibited the NK augmentation of activity. However, they could not show any modification of NK activity with Leu-Enk.

We have ourselves, in a series of experiments, attempted to see whether morphine and several enkephalins could modify the human NK activity of peripheral blood mononuclear cells against ^{51}Cr radiolabeled K562 cells (14). The effector cells were incubated for 1 h at 37 C, washed, and then mixed at a ratio of 80:1 (effector/target) with K562 cells for 4 h. Several concentrations of drugs (10^{-5}-10^{-8} M) were used. The results can be summarized as follows : morphine, Leu-Enk and Met-Enk increase NK activity by approximately 30 % and usually at a concentration between 10^{-8} and 10^{-5}M. However, D-Ala^2-Met^5 enkephalin increased NK activity less (around 20 %) and in a smaller range of concentrations. Finally, all the NK increments could be abolished by incubation of the effector cells with equimolar naloxone. These results demontrate that NK activity can be enhanced by enkephalins probably acting through opiate-specific (receptors?) mechanisms. Similar results have been obtained with enkephalins by Faith et al (16).

In order to further unravel the possible mechanism by which the NK activity is increased by the opioids, we have investigated whether this effect can be obtained by the supernatant of blood mononuclear cells incubated with opioids (17). The experiments were performed as follows : various concentrations of opioid peptides (10^{-7} M to 10^{-10} M) are incubated at 37 C with blood mononuclear cells of a donor A. After incubation, the cells are washed with culture medium and left overnight at 37 C in culture medium alone. This supernatant is removed by cell centrifugation. Blood mononuclear cells of a donor B are mixed with K562 radiolabeled cells in the presence of the supernatant and NK activity is measured by ^{51}Cr release. A representative experiment is reported hereunder :

NK Cytotoxicity and Supernatants of Enkephalins treated Lymphocytes

Control	Met-Enk (M)				Leu-Enk (M)			
	10^{-7}	10^{-8}	10^{-9}	10^{-10}	10^{-7}	10^{-8}	10^{-9}	10^{-10}
48.2 %	47.3 %	48 %	57.5 %	52.2 %	49.2 %	49 %	57 %	53.3 %

As it can be observed, the supernatant of mononuclear cells incubated with Leu- or Met-enkephalin will increase the NK activity of "naive" blood mononuclear cells of another donor. This suggests that the supernatant contains factor(s) enhancing natural killing. We are currently attempting to identify this (these) factor(s).

Premiminary results suggest that, at least, interleukin 2 might play some role in the enhancement of NK activity by the supernatant.

Finally, we have studied the effect of Met-Enk and Leu-Enk as well as the supernatant of enkephalin treated blood mononuclear cells upon the expression of the Leu 11 phenotype which is associated with NK cells. Clearly, the percentages of Leu 11 cells increase under these conditions which further indicate that enkephalins can enhance natural killer activity (18).

These studies have to be discussed in relation to immunological studies performed in animals during stress. It is generally accepted that stress can influence the immune system either suppressing it or enhancing it, depending mainly on the type of stress as well as the immune mechanisms investigated.

Interestingly, Shavit et al have reported that in rats subjected to one of two inescapable footshock stress paradigms, analgesia can be induced by both, but only one via activation of opioid mechanisms (19). Splenic natural killer cell activity was suppressed by the opioid, but not the nonopioid form of stress. This suppression was blocked by the opioid antagonist naltrexone. These results suggest that some endogenous opioid peptides released during the stress mediate mechanisms regulating the natural killer cytotoxicity of certain forms of stress.

EARLY MARKERS OF ACTIVATION

Lymphocytes in presence of specific antigens or mitogens will undergo a series of events before division. The appearance of some cell surface markers is one of the event of lymphocyte activation (18). In preliminary studies, we have now shown that enkephalins may enhance markers linked to lymphocyte activation like the active T rosette, the OKT10, the Tac and the Ia receptors. Furthermore, naloxone will block this marker enhancement.

These data suggest that enkephalins may be considered as molecules of lymphocyte activation and may play a physiological role in the immune response (e.g., the increased Tac antigen (IL2 receptor) is certainly a regulation mechanism of the immune response).

EFFECT ON TUMOR

In BDF_1 mice challenged with the L1210 murine leukemia line, Plotnikoff and Miller have observed a significant increased survival rate in mice treated with 30 mg/kg s.c. of Met-Enk. A similar improvement in survival was also observed using 10 mg/kg s.c. of Leu-Enk (13). The role of opiates, in general, in the control of tumor growth appears complex. For instance, a high dose of morphine reduces the survival of rats injected with a mammary ascites tumor (20). In contrast, heroin inhibits tumor growth in mice transplanted with neuroblastoma (21). Interestingly, naltrexone has dose-dependent stimulatory or inhibitory effects on neuroblastoma growth in mice (22).

All these experiments are difficult to interpret but would, at least, support the hypothesis according to which opioid peptides and especially enkephalins are able to modulate tumor growth perhaps through immune mechanisms and as such are thus immunomodulators.

HUMAN MONOCYTES AND NEUTROPHILS

Met-Enk has, in vitro, a bimodal stimulating activity on human monocytes chemotaxis with peak activities occuring at 10^{-12} M and 10^{-8} M (23). The distance migrated in response to optimal concentrations of Met-Enk was approximatively 80 % of that obtained with a classical stimulator (f-MLP) and was blocked by 10^{-8} M naloxone.

Human neutrophils showed some migration in response to Met-Enk although the average optimal migration was inferior to 30 % of that observed with f-MLP. All these studies suggest that Met-Enk possesses a chemotactic effect.

Met-Enk, between 10^{-9} M and 10^{-7} M, was found to stimulate IgG2a mediated antibody dependent cytotoxicity of thyroglycollate treated rat peritoneal macrophages through naloxone sensitive mechanisms (24). In contrast to this extracellular phenomenon, phagocytosis of the sensitized sheep red blood cells was suppressed by Met-Enk in the same dose range. Furthermore, cyclic GMP concentration was augmented by Met-Enk treated peritoneal macrophages.

All these data are probably to be related to the finding of receptors for dihydromorphine on human monocytes and neutrophils (25).

HUMAN MAST CELL DEGRANULATION

D-Ala^2-D-Leu^5 enkephalin is able to elicit immediate skin reactivity as judged by a wheal and fare phenomenon. This appears to be related to mast cell degranulation (26). The activity of the synthetic enkephalin appears less potent than other opioids like dymorphin or β-endorphin. The possibility that endogenous opioids may regulate mast cell function can eventually lead to the understanding of nonimmunologic mediator release in normal and diseased states.

PLATELETS

Receptors for naloxone have been identified on human lymphocytes and platelets using as assay the binding of (H^3) naloxone to these cells (4). Since morphine appears able to decrease by half the binding of the radiolabeled naloxone, it appears that some receptors are of the μ type. The possible inhibitory effect of enkephalins upon the binding was not studied.

ENKEPHALINS AND LYMPHOID WEIGHT

The subcutaneous injection of either Met-Enk or Leu-Enk in BDF_1 mice results in increase in thymus weight and decrease in spleen weight (27). These studies are also indirect evidence of the effect of enkephalins upon the immune system.

SURFACE SIMILARITIES AND REGULATION BETWEEN NERVOUS CELLS AND IMMUNE CELLS

One clue to the action of the enkephalins upon the immune cells lies certainly in the presence of specific surface receptors. Therefore, it appears very attractive to know that surface antigens present on immune cells can also be found on nervous cells. For instance, the HNK1 monoclonal antibody detects all the NK cells (through the Leu 7 surface antigen). It has recently been shown that the HNK1 antiserum will also recognize cells of neuroectodermal origin (28). Furthermore, HNK1 also recognizes a common carbohydrate moiety found on neural cell adhesion molecules and myelin-associated glycoprotein (29).

The Thy-1 antigen is expressed on T lymphocytes as well as on the surface of many neurons like in the inner retina. Monoclonal antibodies directed against Thy-1 can selectively enhance the growth of retinal ganglion cells (30).

Using neuroblastoma cell lines, it has also been shown that HLA A, B, C (class I molecules) and β_2 microglobulin can be induced by interferon which, itself, can be a product of immune mechanisms (31). The possibility that neuronal HLA A, B, C expression may be under regulatory control by interferon is interesting in the concept of immunosurveillance.

Conversely, the elimination of limited areas of the cephalic crest in stage 9 or 10 chick embryos markedly reduces the size of the thymus gland or results in its absence (32).

CONCLUDING REMARKS

In this chapter, the author has reviewed the data related to the influence of enkephalins upon the immune system.

Clearly, enkephalins possess immunological properties which are best demontrated by their action upon natural killer activity and in lymphocyte activation. The concentrations necessary to trigger these mechanisms are very low, usually between 10^{-9} M and 10^{-11} M, indicating that these levels can easily be reached in physiological states as well as in, for instance, stress.

The interpretation of these data remains conjectural as long as relevant in vivo studies are not performed.

Nevertheless, one can hypothesize the following physiological pathway : antigenic challenge leads to lymphocyte activation which can be amplified by serum enkephalins (perhaps even, enkephalins are increased during antigenic challenge). The production of lymphokines is enhanced by the enkephalins which also increase the nonspecific recruiting systems of immune mechanisms. One particular system appearing to be an exquisite target of the enkephalins, is the natural killer system. This may explain the enhanced survival observed in mice inoculated with lymphoma cells and treated with enkephalins.

The control of tumor growth may perhaps also be direct since malignant cells, like neuroblastoma, appear also to possess opiate receptors. The relevance of these receptors to tumor growth and metastasization is unclear.

Another hypothesis to be proposed is related to AIDS. Indeed, it has been clinically observed that Kaposi sarcoma is more prevalent in the homosexual group rather than in the addicted group of AIDS patients. We suggest that drug addiction (through opiates) has led to enhanced NK activity and thus provided a better resistance to malignancies like Kaposi sarcoma.

Speculation about the role of enkephalins is also interesting in stress. Indeed, during stress, many hormones are released. Among them, corticosteroids and endogenous opioid peptides. Corticosteroids are known for their immunosuppressive properties whereas enkephalins, as well as endorphins, appear to have immunoenhancing activities. Thus, one can hypothesize that multiple innappropriate stress can lead to some type of immunological unbalance (the global effect of corticosteroids and opioids). This unbalance will disrupt the normal immunosurveillance mechanisms and perhaps lead to somatic diseases like autoimmune diseases (known to be triggered by stress), cancer, or recurrent infections (e.g. tuberculosis is also known to be stress related). This hypothesis can be experimentally tested.

The action of enkephalins appears also to be rather wide since they affect also monocyte and neutrophil chemotaxis as well as mast cell degranulation suggesting a potential role in infections and allergic diseases.

Finally, in view of the in vitro data, it is very tempting to postulate that enkephalins are also modifiers of the biological response.

This hypothesis is under current investigation by injecting enkephalins to normal volunteers as well as patients with various diseases like cancer and AIDS. The preliminary data suggest that enkephalin will also in vivo and in man enhance T cell function and perhaps NK function (33, 34).

In summary, all current evidences indicate that enkephalins are agents of lymphocyte activation as well as modifiers of the biological response. As such, they may play a role in the pathogenesis of stress related diseases and may also become useful agents in the treatment of a variety of disorders associated with immunological deficiencies.

REFERENCES

1. Wybran, J., Appelboom, T., Famaey, J.P. Govaerts, A. (1979). Suggestive evidence for morphine and methionine-enkephalin receptors-like on normal blood T lymphocytes. J. Immunol. 123, 1068-1070.

2. Hazum, E., Chang, K.J., Cautrecasas, P. (1979). Specific nonopiate receptors for β-endorphins. Science 205, 1033-1035.

3. Wybran J., Dupont E. (1982). The active T rosette : an early marker for T-cell activation. Ann. Immunol. (Paris) 133, 211-218.

4. Mehrishi J.N., Mills, I.H. (1983). Opiate receptors on lymphocytes and platelets in man. Clin. Immmunol. Immunopathol. 27, 240-249.

5. Johnson, H.M., Smith, E.M., Torres, B.A., Blalock, J.E. (1983). Regulation of the in vitro antibody response by neuroendocrine hormones. Proceedings of the National Academy of Sciences. 79, 4171-4174.

6. Miller G.C., Murgo, A.J., Plotnikoff, N.P. (1984). Enkephalins : enhancement of active T cell rosettes from normal volunteers. Clin. Immunol. Immunopathol. 31, 132-137.

7. Miller, G.C., Murgo, A.J., Plotnikoff, N.P. (1983). Enkephalins-enhancement of active T-cell rosettes from lymphoma patients. Clin. Immunol. Immunopathol. 26, 446-451.

8. Murgo, A.J., Plotnikoff, N.P., Faith, R.E. (1985). Effect of Methionine-Enkephalin plus $ZnCl_2$ on active T cell rosettes. Neuropeptides, 5, 367-370.

9. Wybran, J., Appelboom, T., Govaerts, A. (1978). Inosiplex, a stimulating agent for normal human T cells and human leucocytes. J. Immunol. 121, 1184-1187.

10. Wybran, J., Levin, A.S., Fudenberg, H.H., Goldstein, A.L. (1975). Thymosin : effects on normal human blood T cells. Ann. N.Y. Acad. Sci. 249, 300-307.

11. Wybran, J., Appelboom, T., Famaey, J.P., Govaerts, A. (1980). Receptors for morphine and methionin-enkephalin on human T lymphocytes: the two hits opioid lymphocyte receptor hypothesis. Serrou, B., Rosenfeld, C., eds. New Trends in Human Immunology and Cancer Immunotherapy. Paris, Doin Publishers, 48-55.

12. Gilman, S.C., Schwartz, J.M., Milner, R.J., Bloom, F.E., Feldman, J.D. (1982). β-Endorphin enhances lymphocyte proliferative responses. Proceedings of the National Academy of Sciences 79, 4226-4230.

13. Plotnikoff, N.P., Miller, G.C. (1983). Enkephalins as immunomodulators. Int. J. Immunopharmacol. 5, 437-441.

14. Wybran, J. (1985). Enkephalins and endorphins as modifiers of the immune system : present and future. Fed. Proc. 44, 92-94.

15. Mathews, P.M., Froelich, C.J., Sibbitt, W.L.,Jr., Bankhurst, A.D. (1983). Enhancement of natural cytotoxicity by β-endorphin. J. Immunol. 130, 1658-1662.

16. Wybran, J. (1985). Enkephalins and endorphins : activation molecules for the immune system and natural killer activity? Neuropeptides. 5, 371-374.

17. Faith, R.E., Liang, J.H., Murgo, A.J., Plotnikoff, N.P. (1984). Neuroimmunomodulator with enkephalins : enhancement of human natural killer (NK) cell activity in vitro. Clin. Immunol. Immunopathol. 31, 412-418.

18. Wybran, J., Schandené, L. : in preparation.

19. Shavit, Y., Lewis, J.W., Terman, G.W., Gale, R.P., Liebeskind, J.C. (1984). Opioid peptides mediate the suppressive effect of stress on natural killer cytotoxicity. Science. 223, 188-190.

20. Lewis, J.W., Shavit, Y., Terman, G.W., Gale, R.P., Liebeskind, J.C. Stress and morphine affect survival of rats challenged with a mammary ascites tumor (MAT 13762B). Nat. Immunol. Cell. Reg. (in press).

21. Zagon, I.S., McLaughin, P.J. (1981). Heroin prolong survival time and retard tumor growth in mice with neuroblastoma. Brain Res. Bull. 7, 25-32.

22. Zagon, I.S., McLaughin, P.J. (1983). Naltrexone modulates tumor response in mice with neuroblastoma. Science. 221, 671-673.

23. Van Epps, D.E., Saland, L. (1984). β-endorphin and Met-Enkephalin stimulate human peripheral blood mononuclear cell chemotaxis. J. Immunol. 132, 3046-3053.

24. Foris, G., Medgyesi, G.A., Gyimesi, E., Hauck M. (1984). Met-Enkephalin induced alterations of macrophage function. Molecul. Immunol. 21, 747-750.

25. Lopker, A.L., Abood, G., Hass, W., Lionetti, F. (1980). Stereoselective muscarine acetylcholine and opiate receptors in human phagocytic leucocytes. Biochem. Pharmacol. 29, 1361-1365.

26. Casale, T.B., Bowman, S., Kaliner, M. (1984). Induction of human cutaneous mast cell degranulation by opiates and endogenous opioid peptides : evidence for opiate and nonopiate receptor participation. J. Allergy Clin. Immunol. 73, 775-781.

27. Plotnikoff, N.P., Murgo, A.J., Faith, R.E. (1984). Neuroimmunomodulation with enkephalins : effects on thymus and spleen weight in mice. Clin. Immunol. Immunopathol. 32, 52-56.

28. Lipinski, M., Braham, K., Cailland, J.M., Carlu, C., Tursz, T. (1983). HNK-1 antibody detects an antigen expressed on neuroectodermal cells. J. Exp. Med. 158, 1775-1780.

29. Kruse, J., Mailhammer, R., Wernecke, H., Faissner, A., Sommer, I., Goridis, C., Schachner, M. (1984). Neural cell adhesion molecules and myelin-associated glycoprotein share a common carbohydrate moiety recognized by monoclonal antibodies L2 and HNK-1. Nature. 311, 153-155.

30. Leifer D., Lipton, S.A., Barnstable, C.J., Masland, R.H. (1984). Monoclonal antibody of Thy-1 enhances regeneration of processes by rat ganglional cells in culture. Science. 224, 303-306.

31. Lampson, L., Fisher, C.A. (1984). Weak HLA and β_2 microglobulin expression of neuronal cell lines can be modulated by interferon. Proceed. Nat. Acad. Sci. 81, 6476-6480.

32. Bockman,D.E., Kirby, M.L. (1984). Dependence of thymus development on derivates of the neural crest. Science. 223, 498-500.

33. Wybran, J., Schandené, L. : in preparation.

34. Plotnikoff, N.P., Miller, G.C., Nimeh, N.F. : in preparation.

THE INFLUENCE OF ENDOGENOUS OPIOID PEPTIDES ON VENOUS GRANULOCYTES

Eike G. Fischer and Nora E. Falke

Sekt.Elektronenmikroskopie
Universität Ulm
D-7900 Ulm , Fed.Rep.Germany

POLYMORPHONUCLEAR LEUKOCYTES (PMNs)

History

Ameboid migrating phagocytes represent a phylogenetically ancient defense system. Amebocytes are found throughout the animal kingdom. All functions of the PMNs (aggregation, adherence, polarization, chemokinesis, chemotaxis, phagocytosis, pinocytosis, endocytosis, exocytosis) are primitive functions, as they occur in most of the protozoa. In mammals they appear in all embryonic cells and in some of the differentiated cells.

In vertebrates the most distinct marker of inflammation is the appearence of granulocytes (PMNs) in tissue. After irritation of the tissue they adhere to the endothelium and after some minutes begin to migrate out of the vessel. This was already observed by Waller(1846) and Cohnheim(1867,1869) in the tongue of the frog. In bacterial inflammations the PMNs and their disintegration products represent the major part of pus. In 1888 Leber interpreted the directed migration towards the source of inflammation as chemotaxis. As a result our present view of PMNs has its origin 100 years ago.

The Cells

In mammals nearly 70 % of the white blood cells are PMNs. In humans PMNs are round-shaped with an average diameter of 8.5 µm . They are specialized in hunting microbes. Granulocytes originate from the bone marrow. In the bloodstream they have a short life span of less than 12 hours and a daily turnover rate of about 100.000.000.000. The name polymorphonuclear leukocytes (PMNs) derives from their lobed nucleus. According to their staining reactions granulocytes are subdivided into neutrophils, basophils and eosinophils. This division is also based on functional criteria. In the following only neutrophils, which represent about 95% of all granulocytes, are taken into account. Neutrophils contain two different populations of granula (vesicles): i) primary granules with lysosomal enzymes and ii) secondary (specific) granules containing lactoferrin (both granula contain of other mediators and enzymes as well).

Substances modulating PMNs behavior

When one speaks about the brain influence on the immune system one should not forget that the endogenous opioids represents only a small voice in a great concert. PMNs respond with chemotactic migration to numerous inflammatory mediators. Some endogenous stimulators like the immune-derived factors (e.g. complement-fragments or the IgG-fragment tuftsin) are always present in the form of inactive precursors. Other stimulators derived from irritated tissues can be released by many cells including PMNs. These factors include denaturated peptides and tissue hormones such as bradykinin, angiotensin, substance P, neurotensin and the nerve growth factor(NGF) and are mainly released by peripheral nerve terminal cells. In physiological concentrations only neurotensin and perhaps NGF seem to be potent enough to stimulate PMNs (Goldman et.al.´83, Gee et.al.´83). Microbial degradation products like N-formylized peptides or oxidized lipids are very potent stimulators. The most potent N-formylized peptide is N-formyl--methionyl-leucyl-phenylalanine (fMLP) and is used as a reference in our testsystems.

Most stimulators depending on their dose successively induce all functions. With low concentrations of stimulators the cells become typically adherent or in suspensions aggregate. In a concentration gradient they migrate towards the (presumptive) center of inflammation and are maximal stimulated to engulf microbes and aberrant autologous cells. During phagocytosis and with high concentrations of stimulators the PMNs release lysosomal enzymes and bacteriocidal oxygen compounds. High concentrations of stimulators also inhibit a further migration.

Other substances inhibit PMN functions. Migrating is impaired by epinephrine and some prostaglandins. The prostaglandin prostacyclin inhibits PMN adherence to endothelial cells.

One finds a considerable variation in the enzymatic contents of PMNs in different species. Therefore, we can expect functional differences and different sensitivities to various stimulators from one species to another. For example, PMNs from human beings, rabbits, rats, guinea pigs and sheep respond well to fMLP, while there is no reaction of PMNs from horses, cows, pigs or dogs. Everyone who is going to investigate PMNs has to consider that blood from different donors can show very different reactions, even if blood from young healthy students is used.

OPIOIDS

The opiate alkaloids like morphine and their effects on pain perception and mood are well known for thousands of years. Since opiate effects can be induced only by active stereoisomers an opiate receptor in the brain was postulated and has been described by several groups in 1973. Consequently an intensive search for an endogenous ligand for this receptor started. Opioid peptides were first isolated from the brain and later isolated in the pituitary, the adrenal medulla and also in the plasma. The opioid peptides include the enkephalins, ß-endorphin (ßEnd) and the dynorphin-related peptides.

High concentrations of opiate alkaloids as found in clinical use and in addicts have inhibitory effects on PMN functions, particulary on migration (Stanley et.al.´76), phagocytosis (Zucker-Franklin et.al.´71) and chemo-luminescence (Gyires et.al.´85). In 1981/82 several groups including ourselves started to investigate the effects of endogenous opioids on immune cells (Fischer and Falke´84). Let us look at the opioid effects in PMNs which we detected in our investigations.

WHAT CAN WE SEE ?

Separate the granulocytes from venous blood, suspend them in Hanks´ buffered salt solution (HBSS). Put a droplet of such a suspension onto a glass coverslip. Within a few minutes all granulocytes (more than 95%) will adhere to the glass, while other cells can be rinsed off with HBSS. Add a droplet of buffer on the coverslip with the adherent cells and turn it upside down onto a slide and surround it with paraffin. In such a preparation most of the PMNs are globular-shaped. Their surface is slightly ruffeled like an untidy sheet. But some cells show spontaneous shape changes and random movements.

If ß-End (10^{-8} to $^{-10}$ M) is added and the cells are observed under a microscope nothing seems to happen within the first 2 minutes. Only after exact measurement of the cell diameters on photographs does it become evident that 1/3 of the cells have an increased diameter of about 20% (Falke and Fischer´85). The scanning electron microscope shows that many cells had lost their rounded shape and were flattened on the substrate, their ruffeled cell membranes were now smoothened, in short, they appeared like fried eggs. While the pseudopodia appearing later are clearly visible under the light microscope the thin layer of hyaloplasma (the ´white of the egg´) could be observed only under the scanning microscope. With PMNs from the rat even under the light microscope the ´fried egg´ phenomenon could be observed in 1/3 of the cells. In venous PMNs from dog, sheep or rabbit this effect did not seem to appear.

The cell spreading which could be observed with ß-End could not be induced by opiate alkaloids in equal concentrations. On the contrary, after addition of levorphanol or the antagonist diprenorphine the cells increased the ruffeling of their membrane and also tended to shrink slightly.

Let us compare the opioid effects with the well known early effects of fMLP on PMNs. If we measure human PMNs treated with fMLP (10^{-8} M) an increase in the cell diameter as occurs with ß-End becomes evident. In PMNs of the rat the 'fried egg' phenomenon could not be observed with fMLP. According to reports of Zigmond and Sullivan'79 and Davis'82 PMNs show a cell spreading within two minutes and a simultaneous ruffeling of the surface. Most of the ruffeling dissapears in the following two minutes. As can be seen, there are fundamental differences in the early reaction to opioids and to fMLP. From this it can be concluded that the stimulation by opioids results in a second messenger pattern which differs from that activated by fMLP.

Now we return to our coverslip with PMNs and ß-End. After 3 min from the start of the preparation more and more cells begin to migrate. At 12 minutes nearly all cells are moving. About one third of the cells are now entirely flattened, one third are nearly rounded showing a thin margin of pseudopodia at the front and a short tail-like uropod. The observer's attention is attracted to the remaining one third of the cells which are extremly elongated more than 3 times normal. This comes as no surprise that the cells react in such a variety of ways. Since these cells have a short life span in the blood they undergo a rapid metamorphoses and thus behave differently.

During the whole incubation time control cells incubated in HBSS had a length over width ratio (a/b) between 1.1 and 1.2 on average for each preparation. The same applies to ß-End within the first 2 minutes. Twelve minutes after the addition of ß-End this ratio reached 1.8 for human and ovine PMNs. In PMNs of the rabbit the elongation is less expressed, and in PMNs of the dog and the rat an elongation is not visible. In human PMNs an elongation of 1.8 could also be induced with the alkaloid levorphanol (10^{-8} M) but not with the dextrorotatory isomer dextrorphan. All effects of ß-End could be inhibited by the opiate antagonist diprenorphine in equimolar concentrations. Thus the effects are stereospectific and antagonizable.

To estimate the influence of ß-End on the cell velocity the same cells were photographed every 20 seconds after 15 min of incubation. For human PMNs it could be shown that the cell velocity of nearly all cells was twofold increased as compared to the control. To our surprise the most elongated cells were hardly moving probably because they could not detatch their retraction fibers. The fastest cells were the relativly rounded ones bearing a short uropod.

If we compare the effect of ß-End on cell elongation and cell velocity up to this point there is no difference than with fMLP. To investigate if endogenous opioids are chemotactic stimulators like fMLP we used the Zigmond-chamber (Zigmond'77). A coverslip with adherent PMNs prepared as described above was placed on a special slide similar to a counting chamber. Between two wells of 2 x 4 mm there is a bridge of 1 mm width. The coverslip leaves a space of 0.01 mm above the bridge. Filling one well with HBSS and the other with a stimulatory substance a gradient is established over

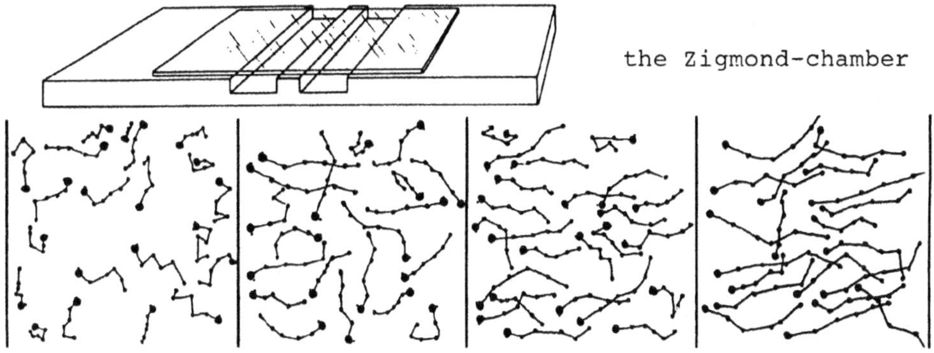

the Zigmond-chamber

a) b) c) d)

Figure 1.

Cells moving in a Zigmond-chamber 20 min after preparation. Points indicate the centers of the moving cells. The start of each cell is marked by a fat point. The lines indicate the distances travelled by single cells within 5 x 20 sec.

a) control cells in buffer only
b) in a gradient of ß-End, which is added 10^{-10} M in the right well
c) in a gradient of fMLP, which is added 10^{-8} M
d) same preparation like c), but with ß-End 10^{-10} M homogenously in the medium

the bridge. Filling one well with HBSS and the other with a stimulatory substance a gradient is established over the bridge. The cells were monitored by time lapse photography and the distances travelled and the degree of orientation in the gradient were measured on a computer tablet.

In a gradient of fMLP nearly all cells were oriented and migrated directly towards the well with fMLP. This does not apply for ß-End. One can see the stimulation of the cell velocity but the cells are moving in all directions. This means they showed chemokinesis but in contrast to fMLP-treated PMNs no chemotactic behavior. Their degree of orientation was the same as in buffer (0.15-0.17 McCutcheon index, McCutcheon´46) To investigate the effects of a combined administration of ß-End and fMLP, ß-End was present homogenously while fMLP was added into only one well. This resulted in an additional increase in cell velocity. This stimulation was significant at a ß-End concentrations of 10^{-11} M and higher. Nearly all cells were stimulated and the maximal effect was achieved with a concentrations 10^{-9} M. With the addition of ß-End the cell orientation in the gradient of fMLP remained unchanged (0.82 - 0.87). The additive effect of ß-End and fMLP was also observed by Simpkins et.al.´84 in a chemotaxis-under-agarose assay. Although a different testsystem was used this group reported equal results and a dose response relationship as was found in our testsystem. An interesting phenomenon is that the broadest interassay variance occurred near the optimal dose of about 10^{-8} M and that with higher doses an inhibitory effect on cell motility was achieved as confirmed by us. This suppression acts parallel to the alkaloid effects. They also showed that the effects of ß-End could be antagonized by naloxone.

fMLP 10^{-8} M gradient

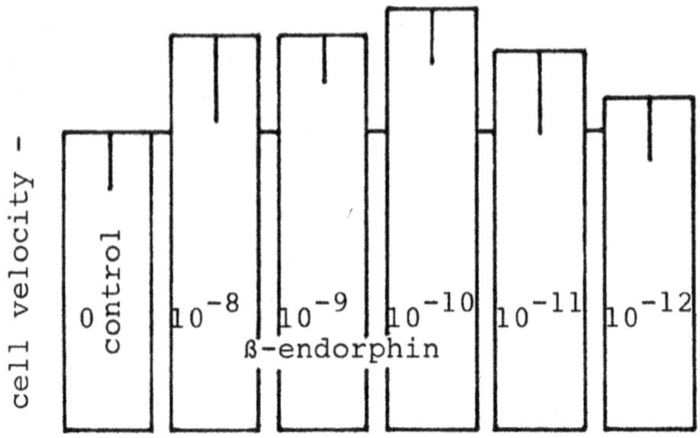

Figure 2.

Zigmond chamber
Additive effect of ß-End on the velocity of cells migrating orientated in a gradient of fMLP. In each of 7 experiments 50 cells were evaluated. Vertical bars indicate deviation between individual experiments.

Having observed an invasion of macrophages into the ventricle after injection of opioid peptides Saland and van Epps´82 tested the chemotactic effects of these peptides in the Boyden-chamber (van Epps et.al.´83 and ´84). The Boyden--chamber consists of an upper compartment containing the cells and a lower compartment containing the stimulatory substance separated by a filter membrane. Using fMLP as a reference this group described a moderate chemotactic effect of the opioids on PMNs, and a stronger effect on monocytes. All findings of this group are consistent with our findings, however, we disagree with their interpretation. We are sure that opioids increase the speed of the cells, but they have no influence on the orientation of a single cell. The Boyden-chamber allows a rapid evaluation but one can not differentiate between a biased random locomotion (called klino-kinesis, which is also found in bacteria) and chemotaxis (the oriented and directed migration)(Keller et.al.´77).

OTHER FUNCTIONS

The opioid effects on cell shape and motility encouraged us to test opioid effects on other granulocyte functions. To investigate if PMNs release the contents of their granula after stimulation by ß-End human PMNs were incubated with the addition of 2 ug/ml cytochalasin B (this addition is necessary for the in vitro test of enzyme release). The release of lactoferrin and the enzymes ß-glucoronidase and myeloperoxidase were measured. Under these conditions fMLP induced a dose dependent release from primary (enzymes) and secondary (lactoferrin) granula, while a release stimulated by ß-End could not be measured. There were no modulatory effects of ß-End on the fMLP-induced release. No release was observed with either ß-End or Met-enkephalin (both at a concentration of 3×10^{-9} M).

To investigate the influence of ß-End (10^{-10} to 10^{-7} M) on phagocytosis PMNs were incubated with heat-inactivated Candida albicans (yeast cells). The exessive uptake of candida cells in presence of serum was not inhibited by ß-End. In incubations without serum there was no stimulatory effect of ß-End and the fMLP-stimulated phagocytosis remained unchanged (fMLP 10^{-9} to 10^{-7} M). There were no effects on oxidization in the NBT-assay (Nitro blue tetrazolium).

RECEPTORS

There were no morphological effects with dextrorphan, an inactive stereoisomer of the opiate alkaloid levorphanol. All effects of ß-End could be inhibited with opiate antagonists. So the effects were stereospecific and antagonizable. These findings suggest that the effects are mediated by an opiate receptor. In contrast to Hazum et.al.´79, who were unable to find any specific binding of iodinated D-Ala2-ß-End in human PMNs, Lopker et.al.´80 described stereospecific opiate binding in human PMNs and monocytes. Our binding studies were performed on living cells incubated under isotonic conditions. No specific binding could be found with the enkephalin analogue DADLE. Specific binding of the opiate antagonists naloxone and diprenorphine occurred in incubations on ice and at 37°C. Binding of the natural opioid peptides Met-enkephalin and dynorphin 1-8 was less at 4°C than at 37°C. These peptides were labeled with tritium at the N-terminal tyrosine. Their rapid degradation accompanied by a tyrosine uptake and the minimal effect of various protease inhibitors made the binding studies rather difficult. Adsorption chromatography was used to differentiate between cell-bound intact peptide and tyrosine. At the beginning of the incubations the major part of the cell bound radioactivity turned out to be specifically bound intact peptide. In summary our studies are evidence for opioid receptors in PMNs (manuscript in preparation).

CONCLUDING REMARKS

Ten years ago it was generally presumed that the various PMN functions are strictly correlated in their stimulation. Only the dose of a mediator was thought to be the discriminating factor. The effects were thought to be mediated by the same structures of the cytoskeleton and by the same second messengers like Ca2+ or cyclic nucleotides (Showell et.al.´76). This view is not unquestioned today (Keller´82) since numerous mediators are known which stimulate certain functional patterns. The stimulatory pattern opioids is well defined among the stimulatory effects of other peptides. To our knowledge the same functional pattern is achieved only with the artifical peptide Gly-His-Gly in the high concentration of 10^{-5} M (Spielberg et.al.´78). In many animal species nearly all functions can be stimulated by fMLP. Up to now there was no blood donor in any species whose cells reacted to opioids but not to fMLP.

In various situations of ´stress´ like sport, sex, labour, and delivery the plasma level of opioid peptides is elevated. Assuming that in these situations there is an increased liability of injury there is a teleological sense in the augmentation of analgesia, wound healing and defense against infections. Other groups working with natural killer cells emphasized an inhibitory effect of stress-induced opioid peptide release (Shavit et.al.´84). According to our investigations there is a short term stimulating effect of opioid peptides between their release and there rapid degradation on the short living PMNs. The results described above indicate that endogenous opioids might play a role in modulating the inital phase of the PMN´s offensive behavior, presumably cell adherence and motility.

The Zigmond-chamber data were partly presented at the Meeting OPIOID PEPTIDES IN PERIPHERY, may 1984, Rome, Italy.

ACKNOWLEDGEMENT

We would like to thank Tony DalleAve for help with the manuscript.

REFERENCES

Cohnheim J., 1867, Ueber Entzündung und Eiterung. Virchow´s Arch.path.Anat.40:1-79

Cohnheim J., 1969, Ueber das Verhalten der fixen Bindegewebskörperchen bei der Entzündung. Virchow´s Arch.path.Anat.45.4:333-350

Davis B.H., Walter R.J., Pearson C.B., Becker E.L. and Oliver J.M., 1982, Membrane actitity and topolography of f-Met-Leu-Phe-treated polymorphonuclear leukocytes. Am.J.Pathol.108:206-216

Falke N.E. and Fischer E.G., 1985, Cell shape of polymorphonuclear leukocytes is influenced by oipids. Immunobiol.169: in press

Fischer E.G. and Falke N.E., 1984, ß-Endorphin modulates immune functions - A review. Psychother.Psychosom.42:195-204

Gee A.P., Boyle M.D.P., Munger K.L., Lawman M.J.P. and Young M., 1983, Nerve growth factor: Stimulation of polymorphonuclear leukocyte chemotaxis in vitro. Proc.Natl.Acad.Sci.USA 80:7215-7218

Goldman R., Bar-Shavit Z. and Romeo D., 1983, Neurotensin modulates human neutrophil locomotion and phagocytic capability. FEBS Lett.159:63-67

Gyires K., Budavarit I., Fürst S. and Molnar I., 1985, Morphine inhibits the carrageenan-induced oedema and the chemoluminescence of leucocytes stimulated by zymosan. J.Pharm.Pharmacol.37:100-104

Hazum E., Chang K.J. and Cuatrecasas P., 1979, Specific nonopiate receptors for ß-endorphin. Science 205:1033-1035

Keller H.U., Wilkinson P.C., Abercrombie M., Becker E.L., Hirsch J.G., Miller M.E., Ramsey W.S. and Zigmond S.H., 1977, A proposal for the definiton of terms related to locomotion of leucocytes and other cells. Clin.exp.Immunol.27:377-380

Keller H.U., Hess M.W. and Cottier H., 1982, Leucocyte activation and the assessment of leucocyte locomotion and chemotaxis. Adv.Exp.Med.Biol.141:9-17

Leber T., 1888, Ueber die Entstehung der Entzündung und die Wirkung der entzündungserregenden Schädlichkeiten. Fortschr.Med.6:460-464

Lopker A., Abood L.G., Hoss W. and Lionetti F.J., 1980, Stereoselective muscarinic acetylcholine and opiate receptors in human phagocytic leukocytes. Biochem.Pharmacol.29:1361-1365

McCutcheon M., 1946, Chemotaxis in leukocytes. Physiol.Rev.26:319-336

Saland L.C., van Epps D.E., Ortiz E. and Samora A., 1983, Acute injection of opiate peptides into the rat cerebral ventricle: A macrophage-like cellular response. Brain Res.Bull.10:523-528

Shavit Y., Lewis J.W., Terman G.W., Gale R.P. and Liebeskind J.C., 1984, Opioid peptides mediate the suppressive effect of stress on natural killer cell cytotoxicity. Science 223:188-190

Showell H.J., Freer R.J., Zigmond S.H., Schiffmann E., Aswanikumar S., Corcoran B. and Becker E.L., 1976, The structure-activity relations of synthetic peptides as chomotactic factors and inducers of lysosomal enzyme secretion for neutrophils. J.Exp.Med.143:1154-1169

Simpkins C.O., Dickey C.A. and Fink M.P., 1984, Human neutrophil migration is enhanced by beta-endorphin. Life-Sci.34:2251-2255

Spielberg I., Mandell B., Mehta J., Sullivan T. and Simchowitz L., 1978, Dissociation of the neutrophil functions of exocytosis and chemotaxis. J.Lab.Clin.Med.92:297-302

Stanley T.H., Hill G.E., Portas M.R., Hogan N.A. and Hill H.R., 1976, Neutrophil chemotaxis during and after general anesthesia and operation. Anesth.Analg., Cleveland 55:668-673

van Epps D.E., Saland L., Taylor C. and Williams R.C., 1983, In vitro and in vivo effects of ß-endorphin and Met-enkephalin on leukocyte locomotion. Prog.Brain Res.59:361-374

van Epps D.E. and Saland L., 1984, ß-Endorphin and Met-enkephalin stimulate human peripheral blood mononuclear cell chemotaxis. J.Immunology 132:3046-3053

Waller A., 1846, Microscopic examination of some of the principal tissues of the animal frame, as observed in the tongue of the living frog, toad, &c. Philosophical Magazine, London 29:271-287 and Microscopic observation on the perforation of the capillaries by the corpuscles of the blood, and on the origin of mucus and pus-globules. :397-405

Zigmond S.H., 1977, Ability of polymorphonuclear leukocytes to orient in gradients of chemotactic factors. J.Cell Biol.75:606-616

Zigmond S.H. and Sullivan S.J., 1979, Sensory adaptation of leukocytes to chemotactic peptides. J.Cell Biol.82:517-527

Zucker-Franklin D., Elsbach P. and Simon E.J., 1971, The effect of the morphine analog levorphanol on phagocytosing leukocytes. Lab.Invest.25:415-421

IMMUNOSUPPRESSIVE EFFECTS OF THE OPIOPEPTINS

Hulon W. McCain,[1,2] Ira B. Lamster,[2,3] and Joanne Bilotta[3]

Departments of Pharmacology,[1] Periodontics and Oral Medicine[2] and The Oral Health Research Center,[3]
Fairleigh Dickinson University School of Dentistry,
Hackensack, New Jersey 07601

INTRODUCTION

The immune system has long been considered an organ system differing significantly from others in the immediacy of its regulation by the central nervous system. Whereas other organs have, to differing degrees, well characterized neural and hormonal afferent as well as efferent regulatory pathways, immune function has traditionally been viewed as an internally regulated effector system essentially devoid of CNS modulation with the possible exception of the CRF-ACTH-corticosteroid link. Such views were reinforced by Jerne's natural antibody selection theory (Jerne, 1955), and the observed ability of immune elements to display significant levels of activity in varied in vitro assay systems (Nowell, 1960). Astute clinicians, however, have for centuries reported antecdotal evidence of increased disease susceptibility associated with diverse physiological and psychological stimuli (Shigami, 1919) including those of stress and emotional distress (Locke, 1982). Patients suffering stressful injury such as burns, and operative or accidental trauma, exhibit decreased immune responsiveness which is often well correlated with post-traumatic susceptibility to bacterial, viral and fungal infections (Constantian et al., 1977; Wolfe et al., 1981; Alexander, 1968; and Goodman et al., 1968). Cancer patients have generally suppressed immunocompetence and, conversely, a large body of experimental data indicates that stress reduces the rate of implanted tumor rejection (Visintaner et al., 1982), and increases tumor growth (Riley, 1981) along with other correlates of invasiveness (Shavit et al., 1983). Recently, intense interest has generated research efforts to elucidate mechanisms by which the central nervous system may modulate normal and stress-induced changes in immune activity, and the physiologic relevance of these putative neuroendocrine-immune system regulatory links. The greatest activity in this regard has been in investigating possible roles of the circulating endorphinergic and enkephalinergic peptides. B-endorphin and several shorter chain products of its enzymatic processing (Akil et al., 1985), as well as numerous high and low molecular weight enkaphalinergic peptides (Evans et al., 1984) are released into peripheral circulation in response to stress, and have as yet undefined biologic relevance. Early studies, however, indicated that cultured human lymphocytes had specific non-opiate binding sites for B-endorphin (Hazum et al., 1979) and that isolated human peripheral blood lymphocytes displayed opiate receptor-like activity with met-enkephalin (Wybran et al., 1979). These cells are key components of immune expression since T-lymphocytes and macrophages are the primary cell types involved in cell-mediated immunity, whereas humoral immunity is characterized by the activity of B-lymphocytes and immunoglobin producing plasma cells. Lymphocytes also play a major role in immunoregulation since when activated they are known to produce regulatory factors influencing the macrophage (i.e., macrophage inhibitory factor), neutrophils (i.e., chemotactic factor), eosinophils (i.e., chemotactic factor), and basophils (i.e., enterleukin-3). They also elaborate compounds controlling activity of other lymphocyte populations (i.e., interleukin-2, suppressing factors and chemotactic factors) (Bellanti and Rocklin, 1985). Ultimate expression of lymphocyte activity may be determined by reciprocal interactions of T-

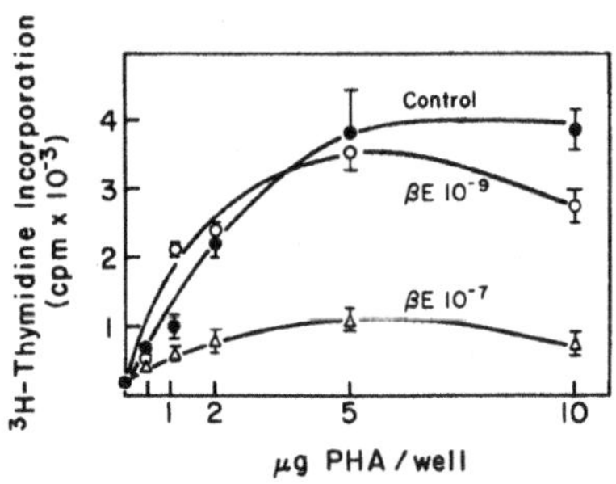

Fig. 1. The effect of B-endorphin on PHA-induced blastogenesis of human T-lymphocytes. Experiments were with lymphocytes freshly isolated from normal healthy volunteers. Lymphocytes were separated from heparinized whole blood by density gradient centrifugation over Lymphoprep. Recovered lymphocytes were washed twice and resuspended in Eagles Minimal Essential Media supplemented with 5% fetal calf serum, glutamine and pen-strep. Cells were cultured in microwells containing $2x10^5$ lymphocytes, PHA and/or B-endorphin. Incubations were at 37°C for 48 hrs in 95% air/5% CO_2. Each well was pulsed with 1uCi ^{3}H-thymidine and incubations continued for 20 hr. ^{3}H-DNA was collected with a cell harvester and quantitated. Viability of cultured cells was 90% as determined by lactic dehydrogenase assay of culture supernatant. PHA and/or BE were simultaneously. Results are reported as mean ± SEM (N = 4).

cell subsets. T-helper cells may be critical to these interactions producing soluble regulators which stimulate activity of relevant cell types including the T-suppressor population, and T-suppressor cells in turn elaborating soluble modulating factors which suppress activity of both B-cells and T-helper populations. This balance of suppressor and helper activities may provide the necessary precise regulation of this system. Specific immunity therefore results from a multi-faceted, integrated response relying on a complex interplay of regulatory modulators. The internal enhancement, regulation and suppression of this system has been characterized through recent investigation. Neuroendocrine/hormonal immunoregulation has only recently come under intense investigation. Although effects of the corticosteroids on immune function have been defined (Parillo and Fauci, 1978), actions of

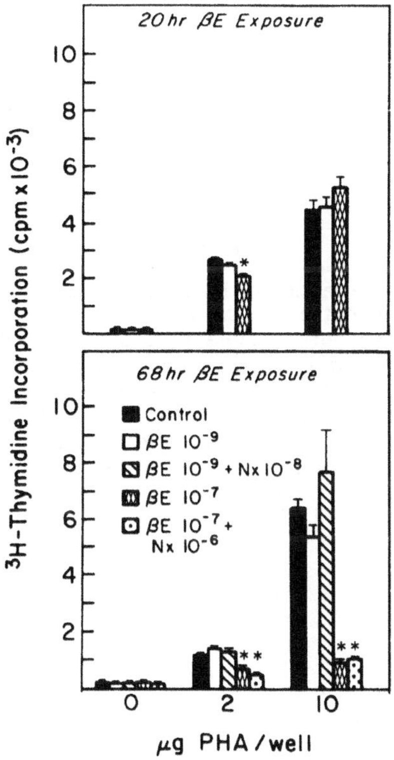

Fig. 2. A. Effect of B-endorphin and naloxone on established blastogenesis of human lymphocytes. BE and/or Nx were added to cultured cells 40 hrs after initiation of mitogenic stimulation. Results are mean ± SEM (N = 4) with significant differences determined by ANOVA. Asterisks indicate $P < 0.01$.

B. Effect of naloxone (Nx) on B-endorphin (BE)-induced suppression of human lymphocyte blastogenesis. Methods were as described in Fig. 1 except when Nx was used, it was added 30 min prior to BE. Results are mean ± SEM (N = 4) with significant differences determined by ANOVA. Asterisks indicate $P < 0.01$.

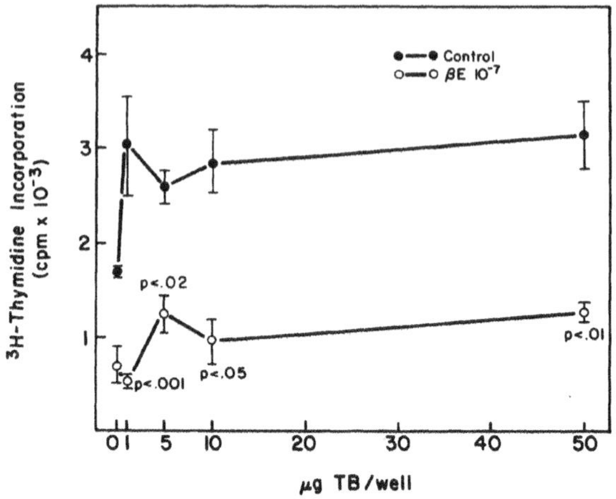

Fig. 3. B-endorphin induced suppression of M. tuberculosis stimulated blastogenesis in sensitized mouse splenocytes. Eight week BALB/c mice were immunized by subcutaneous injection of complete Freund's adjuvant containing M. tuberculosis (TB). Four week later mice were injected with purified protein derivative of tuberculin (PPD). Two weeks after PPD injections, spleens were removed, pressed through a wire mesh, and cells collected, washed and resuspended in RPMI containing fetal calf serum (FCS) and pen-strep. Blastogenesis assay was essentially as described in Fig. 1 except TB was used as antigen. B-endorphin (BE) was added at initiation of cell culture. Results are reported as mean ± SEM (N = 4) with significant differences determined by students T-statistic.

neuroendocrine mediators such as B-endorphin have only recently been studied. This chapter will review our studies of the effects of B-endorphin, its metabolite glycyl-L-glutamine and also isolated adrenal medullary peptides on various aspects of immune functions.

EFFECT OF B-ENDORPHIN AND ITS CARBOXY-TERMINAL DIPEPTIDE ON BLASTOGENESIS OF HUMAN PERIPHERAL BLOOD LYMPHOCYTES AND MOUSE SPLENOCYTES

Our interest in opiopeptin-immunity interactions began following reports by Hazum et al., 1979, of specific non-opiate receptors for B-endorphin on a cultured lymphoid cell line. Our research efforts have since focused on T-helper, T-suppressor and also B-lymphocyte activity assessed by their proliferative and regulatory activities in response to phytohemagglutinin (PHA), concanavalin A (Con-A) Mycobacterium tuberculosis (TB), and partially purified derivative of mycobacterium tuberculosis (PPD). In our human studies we have used normal healthy dental students ranging in age from 20-28 years. Venous blood was collected and used immediately

Table 1. The Effect of GLG and BE on PHA-induced Blastogenesis of Human PBL[a]
Blastogenesis of Human PBL[a]

GLG(10^{-7}M)	BE(10^{-7}M)	(PHA ug/well)	[b]Response (%Control)
+	-	0.5	50
-	+	0.5	51
+	-	1.0	49
-	+	1.0	33
+	-	2.0	57
-	+	2.0	33
+	-	5.0	55
-	+	5.0	21

[a] Glycyl-L-Glutamine (GLG) and B-endorphin (BE) were used in standard blastogenesis assay with human peripheral blood lymphocyte (PBL) as described in Fig. 1.

[b] Response is reported as % of control (without GLG or BE) and is the mean of three separate determinations performed in quadruplicate.

for blastogenesis assays utilizing a microculture system. In our initial studies (McCain et al., 1982) we reported that B-endorphin produced a dose related inhibition of PHA-induced blastogenesis (Fig. 1). Low concentrations of B-endorphin (10^{-9}M) suppressed proliferative responses only with supra-optimal mitogen concentration, whereas, higher concentrations (10^{-7}M) caused significant suppression at all PHA doses. With our experimental design, concentrations of B-endorphin less than 10^{-9}M did not produce reliable suppression of mitogen-induced lymphocyte blast transformation. In order for the opiopeptin to exert its immunosuppressive effect it was necessary that it be present at the time of initial mitogenic stimulation (Fig. 2). When it was added 48 hrs. after culture initiation, minimal effects were observed and then only at the highest concentration of B-endorphin. A similar temporal association has been observed when a low molecular weight ($<$10,000 daltons) immunosuppressive plasma peptide was used in studies of PHA-induced blastogenesis of human lymphocytes (Cooperband, 1976). This suggests that endorphin may interfere with events associated with initiation of the PHA "receptor" transduction-amplification mechanism. Once the mitogenic signal is expressed, it is apparently difficult to repress. The effects of B-endorphin which we observed are not inhibited by the opiate antagonist naloxone, suggesting interaction of the carboxy-terminal portion of the peptide with receptors on the lymphoid cells. A non-opiate receptor for B-endorphin was originally described on a lymphoid cell line (Hazum, 1979). Although once seldom reported, similar naloxone insensitive receptors for opiate drugs and peptides have recently been characterized in rat vas deferens (Wuster el al., 1978), ileum (Nakatsu et al., 1981), adipocytes (Morley et al., 1980), and brain (Johnson et al., 1982). B-endorphin also reportedly binds to a nonpeptide, cytolytic membrane-derived C5b-9(M) complex and also to the serum derived SC5b-9 complex of human complement (Schweigerer et al., 1982). Also, Gilman et al., 1982, observed that B-endorphin effects on PHA-induced lymphocyte transformation were not reversed by treatment with naloxone. It should be noted, however, that there is inconsistency in reports of the actions of B-endorphins on lymphoid proliferation. Gilman et al., (1982) observed that with rat splenocytes 3.2×10^{-8} - 3.3×10^{-9} M B-endorphin produced enhanced responses to PHA, also Weber and Pert, 1984, reported that with

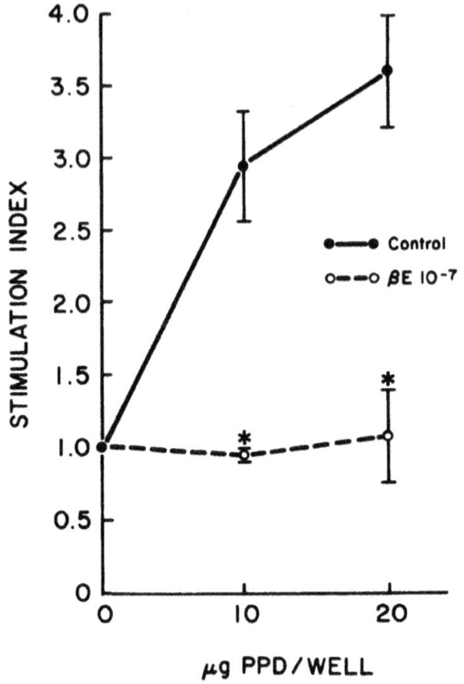

Fig. 4. The effect of B-endorphin on antigen-induced blastogenesis of M. tuberculosis immunized BALB/c mice. Animals were immunized, their splenocytes collected and cultured with or without PPD and/or BE as described in Fig. 3. Results, mean ± SEM (N = 4), are reported as stimulation index (SI).

highly purified, PHA-activated human T-lymphocytes and a mouse derived interleukin-2 dependent T-cell line, B-endorphin enhanced lymphocyte proliferation. Explanation for these discrepancies are at present unclear but may relate to the cellular composition isolated and used in the experiments, since multiple actions (both opiate and non-opiate) for B-endorphin have been demonstrated on various cell types of the immune system (i.e., Wybran, 1979).

Recent studies (Akil et al., 1985) clearly indicate that shorter chained and metabolized forms of B-endorphin may be released from the pituitary differential in chronically stressed rats. Many of these are also devoid of opiate activity and are present in substantial concentrations in the peripheral circulation. Proteolytic cleavage of B-endorphin has also been reported to produce the dipeptide glycyl-L-glutamine within the pituitary (Smyth, 1983). This dipeptide was shown to have widespread inhibitory actions on brainstem neuronal activity which was non-opiate and not related to activation of glycine or glutamate receptive sites. This compound is the carboxy-terminal dipeptide of B-endorphin, and presumably participates in non-opiate interactions of B-endorphin (Schweigerer, 1982). Since the effects of B-endorphin which we observed were non-opiate, we have begun evaluating the effects of this dipeptide on PHA-induced blastogenesis of human lymphocytes (McCain et al., 1985). Glycyl-L-glutamine (GLG) at 10^{-7} M caused a 50% suppression of lymphocyte proliferation over the test range of mitogen concentrations. This was generally less

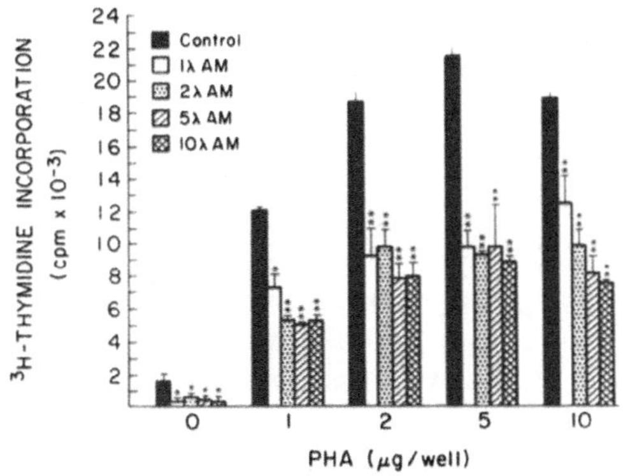

Fig. 5. The effect of adrenal medullary extract on PHA-induced blastogenesis of human peripheral lymphocytes. Adrenals were dissected to provide medallae from which a suspension of chromaffin granules was obtained by differential centrifugation (Smith and Winkler, 1967). Chromaffin granules were lysed, large proteins removed by TCA precipitation and centrifugation, and lipids removed by ether extraction (Stern et al., 1979). Following extensive dialysis and concentration by lyophilization, aliquots were added at initiation of cell culture. Isolation of human lymphocytes and blastogenesis assay were as described in Fig. 1. Results are reported as mean ± SEM (N = 3) with significant differences determined by ANOVA (*P<.05, **P<.01).

efficacious than B-endorphin in these assays. It appears, therefore, that although stress may be associated with increased intermediate pituitary lobe release of N-acetylated endorphinergic compound devoid of opiate activity, non-opiate interactions may still persist if this dipeptide reaches appreciable concentrations in plasma. It should be noted that numerous small peptides have potent immunoregulatory properties. Common examples are muramyl dipeptide and some tripeptide chemotactic factors (Fraser-Smith et al., 1981; Simpkins et al., 1984).

PHA and other lectins provide a potent nonspecific stimulus for lymphocyte transformation, and assays utilizing this or similar mitogens are popular in vitro indices of in vivo immunologic competence. Mitogen stimulated lymphocyte transformation, however, differs from antigen-stimulated activity in that the latter is both specific and generally less intense. In an effort to more closely approximate the in vivo immunologic response, we have examined the effect of B-endorphin on antigen-stimulated blastogenesis of sensitized mouse splenocytes (Fig. 3). BALB/c mice were appropriately immunized with multiple injections of inactivated Mycobacterium tuberculosis (TB) and a partially purified derivative of Mycobacterium tuberculosis (PPD). TB (1-50 ug/well) was used to induce an antigen-specific lympho-proliferative response and did in fact produce a significant two-fold

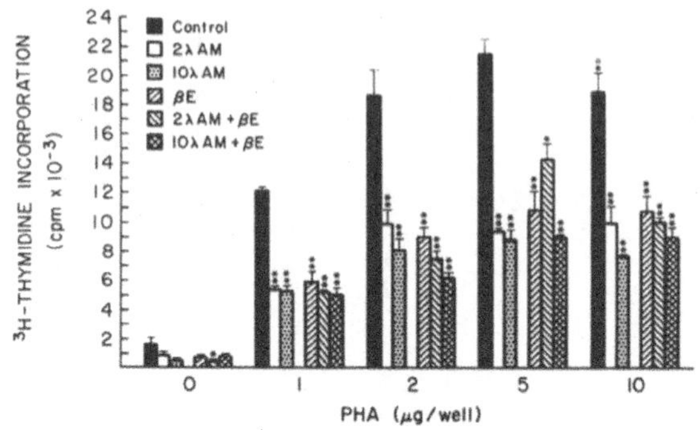

Fig. 6. Interaction of adrenal medullary peptides with B-endorphin in mitogen-induced blastogenesis of human peripheral lymphocytes. Conditions and materials are as described in Fig. 5. Results are reported as means ± SEM (N = 3).

stimulation of ^{3}H-thymidine incorporation with these assays. B-endorphin (10^{-7}M) was added at the initiation of culture, and almost completely prevented the TB dependent blastogenic stimulus. We also used PPD (10-20 ug/well) to induce antigenic stimulation in hopes of eliciting a greater response in the sensitized lymphocytes (Fig. 4). ^{3}H-thymidine incorporation was increased almost four-fold with the higher dose of PPD. B-endorphin (10^{-7}M) completely inhibited antigen-induced proliferation in these assays. While PPD was a more effective antigen than the inactivated bacterial suspension, the immunosuppressive activity of BE was more pronounced for the antigen driven lymphocyte transformation. It should be noted that the degree of stimulation with these antigens was much less than that induced by PHA (stimulation indices of 2-4 as opposed to 10 15 for the lectin). Immunosuppression by B-endorphin was also most prominent with the PPD experiments, confirming our previous observation and also that of others (Gilman et al., 1982) that the effect of this peptide is greater when the stimulus for transformation is mild.

PARTIALLY PURIFIED ADRENAL MEDULLARY PEPTIDES INHIBIT LYMPHOCYTE BLASTOGENESIS

Selye (1956) observed that stress is a complex stimulus resulting in pituitary and adrenal secretions along with other coordinated physiological responses. However, recent studies indicate the truly complex nature of this response and suggest that Selye's original concept may no longer be scientifically acceptable (Terman et al., 1984). Traditional studies have shown stress-associated pituitary release of B-endorphin and ACTH (Rossier et al., 1977) along with concurrent adrenal medullary and cortical secretory activity. The adrenal medulla contains high concentrations of opiate peptides derived from the proenkephalin A and B genes (Udenfriend and Kilpatrick, 1983). Recent reports also suggest that proopiomelanocortin is expressed in human adrenal medulla (Evans et al., 1983). The adrenal enkephalinergic peptides range in size from that of the <1 kilodalton met- and leu-enkephalins to the >50

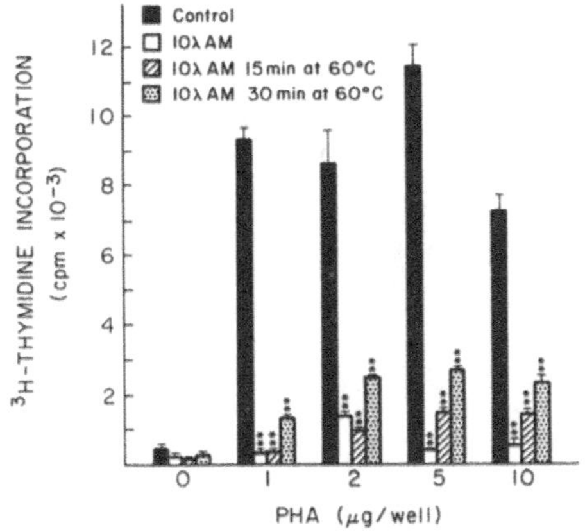

Fig. 7. The effect of heat-treated adrenal medulla extract on PHA-induced blastogenesis of human lymphocytes. Experimental conditions are as described in Fig. 1 except the crude adrenal medulla homogenate was heated at 60°C for 15 or 30 min before isolation of chrommaffin granule proteins.

kilodalton polypeptides (Kilpatrick et al., 1980) and possess widely different opiate receptor binding affinities, as well as biologic activities (Quixion and Weiss, 1983). Although specific proenkephalin converting enzymes have been isolated from adrenal chromaffin granules which produce the free enkephalin pentapeptides, stimulation of the adrenal gland causes release of both high and low molecular weight enkephalinergic peptides (Hanbauer et al., 1982; Chaminade et al., 1983). In the resting condition only the fully processed product of proenkephalin (met-or leu-enkephalin) may be released into peripheral circulation. When more powerful stimuli were utilized, however, greater proportions of higher molecular weight peptides are observed. In recent studies of the pathophysiology of shock, roughly equal quantities of the 31, 8, and 3-5 kilodalton peptides and native enkephalin pentapeptides were found in the plasma of dogs (Evans et al., 1984). Although abundant evidence has been obtained indicating that the enkephalin pentapeptides modify numerous in vitro indices of immune activity (Plotnikoff et al., 1985), no information has been reported regarding possible immunomodulatory activities of the higher molecular weight peptides.

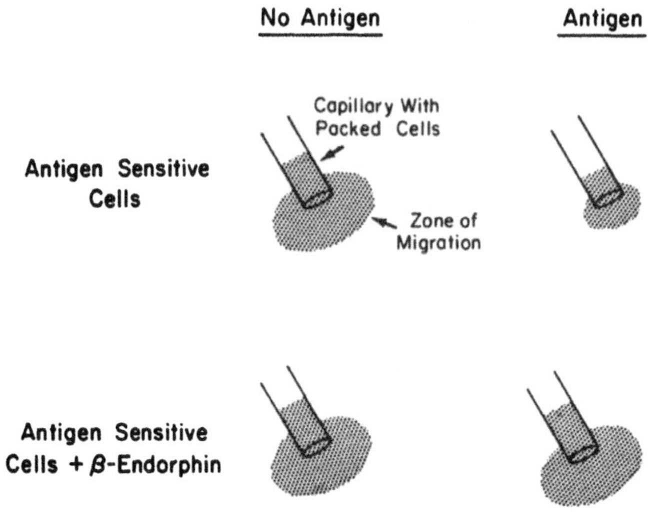

$$\text{Migration Index} = \frac{\text{Migration Area With Antigen}}{\text{Migration Area Without Antigen}} \times 100$$

Fig. 8. Cell migration inhibition assay with PPD-sensitive mouse splenocytes. PPD-sensitive mouse (BALB/c) splenocytes were generated and isolated as described in Fig. 3. Cells were packed into capillary tubes by centrifugation and the tubes cut at the cell-fluid interface (David, 1966). The tube was assembled into a Sykes-Moore Culture Chamber filled with RPMI-1640, 15% fetal calf serum, pen-strep with or without PPD (100 ug/ml) and/or B-endorphin. Chambers were incubated for 24 hrs at 37°C and the "migratory fan" enlarged and quantitated by planimetry.

We have recently begun evaluating the actions of adrenal medullary peptides on in vitro assays of human lymphocyte activity. The method of Stern et al., 1979, was used to isolate a fraction containing enkephalinergic peptides from dissected adrenal medullae. Lyophilized adrenal extract was reconstituted (4 ug dry wt/ul) and added to cultures at their initiation. Adrenal medulla extract (AM) produced a prominent suppression of PHA-induced proliferation at all doses of the mitogen tested (Fig. 5). AM in doses of 1-10 ul (4-40 ug) reduced responses by 40-60% in most cases. We also looked for possible interactions of the adrenal-derived peptides and B-endorphin since presumably both may be released into the bloodstream in a coordinated fashion during times of stress (Fig. 6). B-endorphin (10^{-7} M) and/or AM were added to the cultures at their initiation. B-endorphin reduced the proliferation responses by approximately 50% at all mitogen concentrations as did AM. When used in combination, there was no evidence that these peptides possess any interactive characteristics since B-endorphin and/or AM treatment resulted in an approximate 50% suppression of proliferative responses with PHA. We also examined the heat sensitivity of peptides in this crude adrenal extract (Fig. 7). Heating the adrenal medullae homogenate to 60°C for 15 or 30 min prior to isolation of the peptides did not greatly alter their biologic activities in these assays. However, there were clear indications of reduced actions of the extracted compounds with longer heat treatments. The peptides in this extract apparently have significant heat stable characteristics.

Stress, presumably associated with increased levels of adrenal medullary secretion, is clearly associated with diminished immune activity. It is associated with

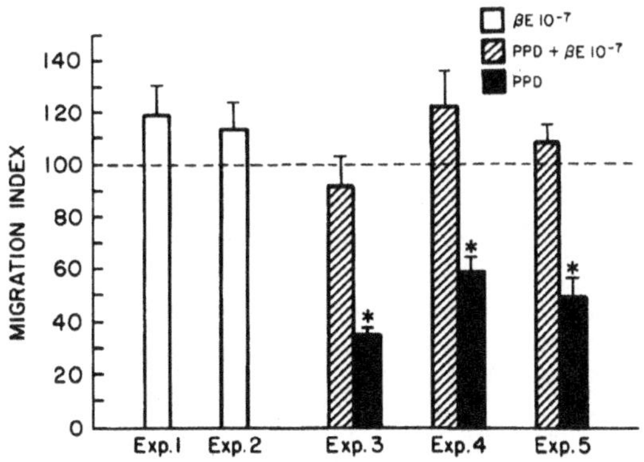

Fig. 9 Effect of B-endorphin on PPD-induced inhibition of cell migration. Experiments 1 & 2 were with normal (not TB immunized) mouse splenocytes. Splenocytes for experiments 3, 4 & 5 were from TB immunized animals. PPD and/or B-endorphin were added at initiation of culture. Results are reported as mean ± SEM (N = 4) with significant differences determined by students T-statistic (* $P < 0.01$).

a markedly reduced lymphoproliferative response to mitogenic stimulation (Shavit et al., 1977; Laudenslager et al., 1983), natural killer cell activities (Shavit et al., 1984) and numerous indices of implanted tumor rejection (Riley, 1981; Sklar and Anisman, 1979). Although the enkephalinergic pentapeptides are generally considered immunopotentiators (Plotnikoff et al., 1985), our studies suggest that adrenal medullary compounds with MW >2000 daltons have significant inhibitory action on human lymphocyte proliferation. *In vivo* and *in vitro* models suggest that intense stimuli (stress) may induce release of these high molecular weight peptides from the adrenal medulla. It should be noted, however, that stress-induced immunosuppression has been observed in adrenalectomized rats (Keller et al., 1983). We are continuing to investigate the biologic activities of this group of compounds.

B-ENDORPHIN INHIBITS ANTIGEN SPECIFIC MIGRATION OF MOUSE SPLENOCYTES

Lymphocytes and macrophages participate in a complex interdependent regulatory scheme by which expression of specific activities of each cell type is modulated by the other. Macrophages liberate interleukin-1, which is necessary for full activation and differentiation of T-helper and B-lymphocytes, as well as numerous other monokines (Bellanti, 1985). T-helper cells elaborate interleukin-2 which together with interleukin-1 promote differentiation of the TDTH (T4+) subpopulation of T-lymphocytes. The TDTH (T4+) cells when fully differentiated can, if activated, release a series of lymphokines which direct activity of the macrophage. These lymphokines include macrophage chemotactic factor and also migration inhibition factor (MIF), which inhibits macrophage migration from sites containing activated lymphocytes. *In vivo* antigenic challenge of sensitized T-cells provides sufficient stimuli to initiate this response. Numerous in vitro tests have been developed for evaluating migratory components of cell-mediated immunity (delayed hypersensitivity), however, one of the oldest and most useful is the cell migration inhibition assay (Likhite and Sehon, 1971). In this assay specific antigen is used to stimulate sensitized lymphocyte production of lymphokines which include MIF (Fig. 8). Antigen-induced in vitro assays of macrophage migration inhibition is in fact an established and sensitive indicator of MIF production by sensitized lymphocytes (Bennett and Bloom, 1968). We have investigated the effects of B-endorphin on cell-mediated immunity using TB sensitized mouse splenocytes in this assay (Fig. 9). Antigen (100 ug/ml PPD) reduced the 24 hr. migration area by 40-60% compared to cultures without antigen. B-endorphin (10^{-7} M) did not significantly alter the migration of nonsensitized splenocytes (Exp 1 & 2). However, when B-endorphin was simultaneously added to cultures of sensitized cells along with PPD, it completely prevented antigen-induced migration inhibition (Exp 3, 4 &5). These results suggest that B-endorphin has no direct effect on nonstimulated migration of mouse splenocytes. The interaction between sensitized (presumably) T-lymphocytes and macrophage may be prevented. Since opiate receptors have been demonstrated on lymphocytes, the initial explanation may be that this opiopeptide inhibits antigen-induced activation of TDTH (T4+) subpopulation which liberates MIF. However, numerous other interactions could produce a similar effect. B-endorphin could interact with macrophage to prevent MIF-induced transduction signal or prevent effective MIF-"receptor" binding. It should be noted that B-endorphin has been reported to act as a neutrophil chemoattractant using the Boyden chamber technique (Van Epps et al., 1983) and to enhance migration of human neutrophils to the chemoattractant peptide N-formyl-methionyl-leucyl-phenylalanine (Simpkins et al., 1984). Regardless of the specific mechanism, it is apparent from our studies that B-endorphin effectively intervenes in this in vitro model of delayed hypersensitivity reaction - a reaction in which subpopulations of T-lymphocytes provide substantial regulatory control.

CONCLUSIONS

It seems clear that the opiopeptins can modulate various components of immune function, however, individual effects as well as physiological relevance of these effects remain uncertain. Our results indicate that B-endorphin and its naturally occurring carboxy terminal dipeptide inhibit mitogen-induced blastogenesis of cultured human peripheral blood lymphocytes. Furthermore, antigen-driven proliferation, and antigen-induced migration inhibition of sensitized mouse splenocytes were potently inhibited by B-endorphin. The immunosuppressive activities of these neuropeptides which we have examined are apparently mediated via non-opiate receptors since the effects were not prevented by naloxone. The adrenal medulla also was shown to contain relatively high molecular weight ($>$ 2000 dalton) compounds which have immunosuppressive action in in vitro blastogenesis assays with human peripheral blood lymphocytes. These compounds were isolated in relatively crude preparations which previously have been shown to contain numerous high molecular weight enkephalinergic peptides.

Stress of varied etiologies induces a myriad of physiological effects including immunosuppression and secretion of "stress hormones" from the pituitary and adrenal glands. The pituitary contains numerous opiatergic peptides including B-endorphin and other products of the proopiomelanocortin gene, many of which contain peptide sequences shared with B-endorphin and the enkephalins. Many of these opiate and non-opiate peptides have been demonstrated in plasma. We have shown that at least two of these compounds, B-endorphin and glycyl-L-glutamine, have potent immunosuppressive activity. However, other products of proopiomelanocortin may be in plasma at higher concentrations than B-endorphin (Akil et al., 1985). Also, numerous non-opiate products of B-endorphin (i.e., N-acetyl-B-endorphin) are released into plasma. None of these acetylated compounds, shorter chained products of B-endorphin metabolism, or larger products of proopiomelanocortin has been evaluated for immuno-modulating activity. Since immunomodulation by these endorphinergic and enkephalinergic neuropeptides apparently occurs via opiate as well as non-opiate interactions with numerous cell types, physiologic effects may well depend upon preferential synthesis and/or release from the pituitary. Recent evidence has suggested that such capacities do exist in pituitary neurosecretory activity (Akil et al., 1985).

The enkephalins are generally reported to potentiate immune activity, and it has been suggested that physiological release of these compounds from the adrenal may serve to counteract immunosuppressive action of the adrenocorticosteroids (Plotnikoff et al., 1985). The adrenal also contains as yet uncharacterized relatively high molecular weight peptides capable of depressing blastogenesis of lymphocytes. Although the fully processed product of proenkephalin (met- or leu-enkephalin) can be released with physiological stimuli, ample evidence has been presented that high molecular weight enkephalinergic compounds may actually predominate in plasma during times of extreme stress such as cardiovascular shock. Also, the adrenal medulla apparently has separate controlling mechanisms for synthesis and processing of its enkephalinergic products (Fleminger et al., 1984). It seems possible therefore that in normal situations the fully processed met- and leu-enkaphalins are released into the peripheral circulation providing positive modulatory influences upon immune expression, perhaps acting in opposition to secreted corticosteroids. However, during particularly stressful situations, secretion of higher molecular weight adrenal peptides may predominate. These compounds may act in unison with the adrenal steroids to induce immunosuppression. Few firm conclusions can be made until systemic investigation determines the contribution of presently uncharacterized adrenal compound(s) to immunologic expression.

Defining the interaction of secretory proopiomelanocortin and proenkephalin derived peptides with the immune system will have a profound impact upon our understanding of immune function and dysfunction. These hormonal regulators may induce multiple immunosuppressive or immunopotentiative effects, as is the situation with the corticosteroids, or more discrete interactions at specific points in the pathways of intrinsic regulatory control. Relatively mild interactions of these putative immunoregulators could induce immunopotentiation or immunodeficiency by shifting the regulatory balance between T-helper and T-suppressor cells. Actions of the endorphins and enkephalins may also summate with ongoing positive and negative regulatory control signals. Nevertheless, the complexity of specific immune expression suggests the need for caution when extrapolating biologic relevance based upon any individual test of immunocompetence.

REFERENCES

Akil, H., Hirohito, S. and Matthews, J., 1985, Induction of the intermediate pituitary by stress: synthesis and release of a nonopioid form of B-endorphin, Sci., 227:424.

Alexander, J.W., 1968, Effect of thermal injury upon the early resistance to infection, J. Surg. Res. 8:128.

Bellanti, J.A., and Rocklin, R.E., 1985, Cell mediated immune reactions, in: "Immunology," J.A. Bellanti, ed., W.B. Saunders, New York.

Bennett, B., and Bloom, B.R., 1968, Reactions in vivo and in vitro produced by a soluble substance associated with delayed-type hypersensitivity, Proc. Natl. Acad. Sci. (USA), 59:756.

Chaminade, M., Foutz, A.S., and Rossier, J., 1983, Co-release of enkephalins and precursors of catecholamines by the perfused rat adrenal in situ, Life Sci., 33:21.

Constantian, M.B., Manzoian, J.O., Nimberg, R.B., Schmid, K., and Mannick, J.A., 1977, Association of a circulating immunosuppressive polypeptide with operative and accidental trauma, J. Surg., 185:73.

Cooperband, S.R., Badger, A.M., and Mannick, J.A., 1976, Non-hormonal serum suppressive factors, in: "Mitogens in Immunobiology," J. Oppeheim and D.L. Rosenstreich, Eds., Acad. Press, New York.

David, J.R., 1966, Delayed hypersensitivity in vitro: its mediation by cell free substances formed by lymphoid cell-antigen interaction, Proc. Natl. Acad. Sci. (USA) 56:72.
Evans, C.J., Erdelyi, E., Weber, E. and Barchas, J.D., 1983, Identification of proopiomelanocortin-derived peptides in the human adrenal medulla, Sci., 221:957.
Evans, S.F., Medbak, S., Hinds, C.J., Tomlin, S.J., Varley, J.F. and Rees, L.H., 1984, Plasma levels and biochemical characterization of circulating met-enkephalin in canine endotoxin shock, Life Sci., 34:1481.
Fleminger, G., Howells, R.D., Kilpatrick, D.L., and Udenfriend, S., 1984, Intact proenkephalin is the major enkephalin-containing peptide produced in rat adrenal glands after denervation, Proc. Natl. Acad. Sci. (USA), 81:7985.
Fraser-Smith, E., and Matthews, T.R., 1981, Protective Effect of muramyl dipeptide analogs against infection of pseudomanas aeruginosa or candida albicans in mice, Infection and Immunity, 34:676.
Gilman, S.C., Schwartz, J.M., Milner, R.J., Bloom, F.E., and Feldman, J.D., 1982, B-endorphin enhances lymphocyte proliferative responses, Proc. Natl. Acad. Sci. (USA), 79:4226.
Goodman, J.S., Schaffner, W., Collins, H.A., Battersby, E.J., and Koenig, M.G., 1968, Infection after cardiovascular surgery: clinical study including examination of antimicrobial prophylaxis, N. Engl. J.Med., 278:117.
Hanbauer, I., Kelly, G.D., Saiani, L., and Yang, H.Y.T., 1982, (Met)-enkephalin-like peptides of the adrenal medulla: release by nerve stimulation and functional implications, Peptides, 3: 469.
Hazum, E., Chang, K., and Cuatrecasas, P., 1979, Specific non-opiate receptors for B-endorphin, Sci., 205:1033.
Ishigami, T., 1919, The influence of psychic acts on the progres of pulmonary tuberculosis, Am. Rev. Tuberculosis, 2:470.
Jerne, N.K., 1955, The national selection theory of antibody formation, Proc. Natl. Acad. Sci. (USA), 41:849.
Johnson, N., Houghten, R., and Pasternack, G., 1982, Binding of ^{3}H-B-endorphin in rat brain, Life Sci., 31:1381.
Keller, S.E., Weiss, J.M., Schleifer, S.J., Miller, N.E., and Stein, M., 1983, Stress-induced suppression of immunity in adrenalectomized rats, Sci., 221:1301.
Kilpatrick, D.L., Lewis, R.V., Stein, S., and Udenfriend, S., 1980, Release of enkephalins and enkephalin-containing peptides from perfused adrenal gland, Proc. Natl. Acad. Sci. (USA), 77:7473.
Laudenslager, M.L., Ryan, S.M., Drugan, R.C., Hyson, R.L., and Mailer, S.F., 1983, Coping and immunosuppression: inescapable but not escapable shock suppresses lymphocyte proliferation, Sci., 221:568.
Likhite, V., and Sehon, A., 1971, Migration inhibition and cell-mediated immunity: a review, Rev. Can. Biol., 30:135.
Lock, S.E., 1982, Stress, adaptation and immunity: studies in humans, Gen. Hospital Psychiatry, 4:49.
McCain, H.W., Bilotta, J.M., and Lamster, I.B., 1985, Life Sci. (submitted for publication).
McCain, H.W., Lamster, I.B., Bozzone, J.M., and Grbic, J.T., 1982, B-endorphin modulates human immune activity via non-opiate receptor mechanisms, Life Sci., 31:1619.
Morely, J., and Levine, A., 1980, Stress-induced eating is mediated through endogenous opiates, Sci., 209:1259.
Nakatsu, K., Goldenburg, E., Penning, D., and Jhamandas, K., 1981, Enkephalin-induced inhibition of the rat ileum is not blocked by naloxone, Can. J. Physiol. Pharmacol., 59:901.
Johnson, N., Houghten, R., and Pasternack, G.W., 1982, Binding of ^{3}H-B-endorphin in rat brain, Life Sci., 31:1381.
Nowell, P.C., 1960, Phytohemagglutinin: indication of mitosis in cutures of normal human leukocytes, Cancer Res., 20:462.

Parillo, J.E. and Fauci, A.S., 1978, Comparison of the effector cells in human spontaneous cellular cytotoxicity and antibody dependent cytotoxicity: Differential sensitivity of effector cells to in vivo and in vitro corticosteroids, Scand. J. Immunol., 8:99.
Plotnickoff, N.P., Murgo, A.J., Miller, G.C., Corder, C.N., and Faith, R.C., 1985, Enkephalins: immunomodulators, Federation Proc., 44:118.
Quirion, R., and Weiss, A., 1983, Peptide E and other proenkephalin-derived peptides are potent kappa opiate receptor agonists, Peptides, 4:445.
Riley, V., 1981, Psychoneuroendocrine influences on immunocompetence and neoplasia, Sci., 212:1100.
Rossier, J., French, E., Rivier, C., Ling, N., Guillemin, R., and Bloom, F.E., 1977, Foot-shock induced stress increases beta-endorphin levels in blood but not brain, Nature, 270:618.
Schweigerer, L., Bhakdi, S., and Teschemacher, H., 1982, Specific non-opiate binding sites for human beta-endorphin on the terminal complex of human complement, Nature, 296:572.
Selye, H., 1956, " The stress of life," McGraw-Hill, New York.
Shavit, Y. Lewis, J.W., Terman, G.W., Gale, R.P., and Liebeskind, J.C., 1984, Opioid peptides mediate the suppressive effect of stress on natural killer cell cytotoxicity, Sci., 223:188.
Shavit, Y., Lewis, J.W., Terman, G.W., Gale, R.P., Liebeskind, J.C., 1983, Endogenous opiates may mediate the effects of stress on tumor growth and immune function, Proc. West. Pharmacol. Soc., 26:53.
Simpkins, C.O., Dickey, C.A., and Fink, M.P., 1984, Human neutrophil migration is enhanced by B-endorphin, Life Sci., 34:2251.
Sklar, L., and Anisman, H., 1979, Stress and coping factors influence tumor growth, Sci. 205:513.
Smith, A.D., and Winkler, H., 1967, A simple method for the isolation of adrenal chromaffin granules on a large scale, Biochem. J., 103:480.
Smyth, D.G., Parish, D.C., Normanton, J.R., and Wolstencroft, J.H., 1983, The C-terminal dipeptide of beta-endorphin: a neuropeptide with inhibitory activity, Life Sci., 33:575.
Stern, A.S., Lewis, R.V., Kimura, S., Rossier, J., Gerber, L.D., Brink, L., Stern, S., and Udenfriend, S., 1979, Isolation of the opioid hepta-peptide met-enkephalin-arg-6, Phe-7 from bovine adrenal medullary granules and striatum, Proc. Natl. Acad. Sci. (USA), 76:6680.
Terman, G.W., Shavit, Y., Lewis, J.W., Cannon, J.T., and Liebeskind, J.C. 1984, Intrinsic mechanims of pain inhibition: activation by stress, Sci., 226:1270.
Udenfriend, S., and Kilpatrick, D.L., 1983, Biochemistry of the enkephalins andenkephalin-containing peptides, Arch. Biochem. Biophys., 221:309.
Van Epps, D., Saland, L., Taylor, C., and Williams, R., 1983, In vitro and in vivo effects of beta-endorphin and met-enkephalin on leukocyte locomation, Prog. Brain Res.. 59:361.
Visintaner, M.A., Volpicelli, T.R., and Seligman, M.E.P., 1982, Tumor rejection in rats after inescapable or escapable shock, Sci., 216:437.
Weber, R.J., and Pert, C.B., 1984, Opiatergic Modulation of the Immune System, in: "Central and peripheral endorphins: Basic and clinical aspects," E.E. Muller and A.R. Genazzani, eds., Ravin Press, New York.
Wolfe, J.H.N., Saporoschetz, I., Young, A.E., O'Connor, N.E., and Mannick, J.A., 1981, Suppressive serum, suppressor lymphocytes, and death from burns, Ann. Sug., 193:513.
Wuster, M., Schulz, R., Herz, A., 1979, Specificity of opioids towards the mu-, delta- and epsilon-opiate receptors. Neurosci. Lett., 15:193.
Wybran, J., Appelboom, T. Famaey, J., and Gavaerts, A.M. 1979, Suggestive evidence for receptors for morphine and methionine-enkephalin on normal human blood T-lymphocytes, J. Immunol., 123(3):1068.

T-Cell E-RECEPTOR MODULATION BY OPIATES, OPIOIDS AND OTHER BEHAVIORALLY ACTIVE SUBSTANCES

Robert Donahoe, John Madden, Felicia Hollingsworth,
David Shafer, and Arthur Falek

Department of Psychiatry, Emory University
and Georgia Mental Health Institute
1256 Briarcliff Road
Atlanta, Georgia 30306

INTRODUCTION

Homeostasis has been defined as a state of metastable equilibrium wherein the constancy of the internal milieu is forever being challenged [1,2] Using this definition, any antihomestatic stimulus may be regarded as a stressor[2]. As reviewed eleswhere [3,4], the neural and immune systems share many basic physiologic processes in response to environmental stressors. The ultimate purpose of the neuroimmune system is maintenance of homeostasis. Neurons and lymphocytes are pluripotent sensory cells[5] capable of eliciting and responding to various hormone and hormonelike peptides that have systemic and local effects characteristic of endocrine, exocrine, and paracrine functions. The primary physiologic basis for the adaptability of the neuroimmue system resides in a complex network of receptor-response systems. This review focuses on the means by which opiates and other behaviorally active substances, as examples of environmental stressors of known behavioral modifying capacity, can modulate receptor activity on T-cell lymphocytes. This capacity for receptor modulation appears to be a primary means by which opiates and like factors evoke both neurobiological and immunobiological changes in host physiology and responsive behavior.

FIRST EVIDENCE OF T-CELL E-RECEPTOR MODULATION BY OPIATES

About 5 years ago, Wybran et al.[6] reported that in vitro treatment of T-cells with morphine depressed their ability to form active and total E-rosettes with sheep erythrocytes (E). Similar treatment with methionine enkephalin enhanced active E-rosette formation. They demonstrated that these effects were reversible by the opiate receptor antagonist, naloxone, which confirmed their probable mediation through opiate receptors. Contemporary with these observations, McDonough et al.[7] from our laboratory demonstrated that T-cells of heroin addicts were similarly depressed in regard to total T-cell E-rosette formation and that this effect was reversible in vitro with naloxone. Table 1 shows recent data which indicates that active T-cell E-rosettes are also depressed in heroin addicts. In addition, Miller et al.[8] have shown that treatment of peripheral blood lymphocytes from cancer patients with methionine-enkephalin significantly enhanced active E-rosette formation. These observations have established that opiates and their endogenous congeners alter T-cell function by affecting expression of the E-receptor. We have since explored these phenomena through kinetic analyses of E-rosette formation.

Table 1. Effect of heroin addiction on active T-cell E-rosette formation[a].

Experimental group	% Active E-rosettes
Addicts n = 21	18.6 ($\pm$ 10.2) ($p \leq 0.05$)
Controls n = 11	30.1 ($\pm$ 10.4)

[a] Active E-rosettes were formed in accordance with procedures of Wybran and Fudenberg[11] with the following modifications: 25/1 E/T-ratio; centrifugation at 600 x g for 3 min; assessment of rosettes 2 min after centrifugation.

DESIGN OF KINETIC ANALYSES OF E-ROSETTE FORMATION

Various investigators[9,10] have reported methods for kinetic analyses of E-rosetting based on variations of the active E-rosette protocol of Wybran and Fudenburg[11]. However, these kinetic assessments did not account for events occurring in the earliest stages of binding of E to T-cells. Consequently, we developed assays to assess early interactions of E with T-cells. Our first approach to this end involved mixing of E with T-cells in a 25:1 E/T ratio along with 50% fetal bovine serum (heat inactivated and pre-adsorbed with E); incubating the mixture

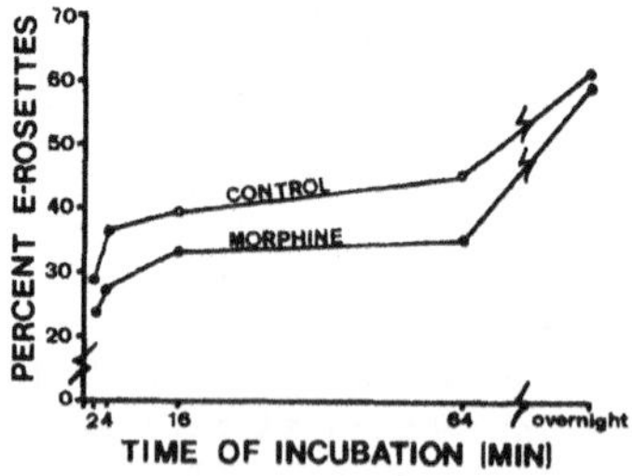

Fig. 1. Effect of *in vitro* treatment of leukocytes with 10^{-9}M morphine on the rate of E-rosette formation. The kinetic active E-rosette assay was used as described in the text.

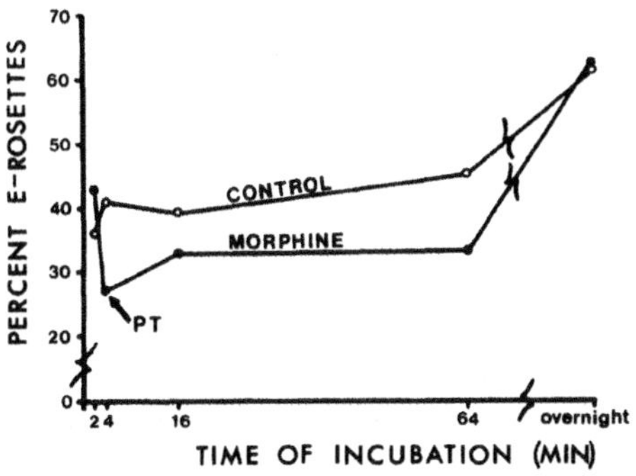

Fig. 2. Effect of in vitro treatment of leukocytes with 10^{-9}M morphine on the rate of E-rosette formation. The kinetic active assay, adapted by first incubating cells on wet ice for 10 min, was used.

for 5 minutes at 37°C; and then centrifuging at 2,000 x g for 20 seconds. By centrifuging rapidly for brief duration, assessment of the kinetics of E-rosette formation was possible soon after E and T-cells made initial contact. A typical result with this kinetic active assay, when used to determine the effects of in vitro treatment of leukocytes with morphine (10^{-9}M, 1 hr, 23°C before mixing with E), is shown in Fig. 1. Morphine inhibited the extent and rate of active E-rosette formation; but, unlike Wybran et al.[6], not the formation of total T-cells.

We then examined the effects of varying assay temperatures on rosetting kinetics in the presence and absence of morphine. Using the kinetic active assay as defined above, altered so that the reaction mixture of E and T-cells was chilled on wet ice for 10 minutes before initiating the 37°C incubation, we found that morphine actually enhanced E-rosette number relative to the control (Fig. 2) at the 2-minute assessment time in the kinetic procedure. However, this E-rosette enhancement shifted to depression by 4 minutes which continued thereafter throughout the portion of the kinetic curve representative of active types of E-rosettes (up to 64 minutes). Again, no effect of morphine on total T-cell E-rosettes was seen.

To understand how morphine could cause phase transition kinetics of E-rosette formation by simply shifting pre-test incubation conditions as in the preceding experiment (Fig. 2), we examined effects of thermal changes on E-rosetting kinetics. In the kinetic active assay, the time before the 37°C incubation step was regarded as the pre-incubation stage. During this stage, cells are normally kept at room temperature (~23°C). We changed these thermal conditions by holding cells on wet ice until the start of centrifugation while eliminating incubations at both 37°C and 23°C before centrifugation. For this procedure, centrifugation was conducted at 4°C and at 600 x g for 3 minutes. In

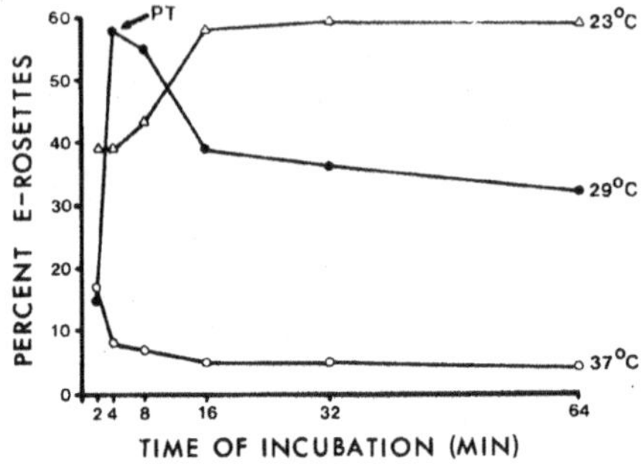

Fig. 3. Kinetic rates of E-rosetting measured with a 0-29°C kinetic assay as described in the text. The different temperatures represent the varied incubation conditions used during the kinetic assessment phase.

this way, the cells were kept chilled throughout the assay up to the point when kinetic assessments were begun. Then E-rosetting kinetics were examined for purposes of determining the effects of variation in incubation temperatures during the kinetic assessment stage, at either 23°C, 29°C, or 37°C. As shown in Fig. 3, kinetics of E-rosetting at 23°C were biphasic and devoid of phase transitions. At 29°C, kinetics of E-rosette formation were multiphasic with obvious phase transitions cycling between increases and decreases in E-rosette percentages. At 37°C, biphasic kinetics of the loss of E-rosettes through capping events was apparent.

This experiment defined conditions for a new type of kinetic assay (the 0-29°C kinetic assay) that accentuates phase transition kinetics. When this assay was employed along with variation in conditions of centrifugation of E with T-cells, it was found that the magnitude (amplitude) and duration (frequency) of phase transitions differed with respect to the intensity of contact between E and T-cells. These observations indicated that E modulate expression of their own T-cell receptor as a function of thermal and physical conditions that affect the ability of E and T-cells to initiate and maintain contact with each other. This is a notable observation because self modulation of receptor activity is a common feature in cell surface receptor-ligand systems[12]. Thus, we found that centrifugation at 600 x g for 1 1/2 minutes in the 0-29°C kinetic assay accentuated phase transitional kinetics of E-rosette formation and that, under these conditions, morphine (10^{-9}M) caused extension of the transitional phases of E-rosette formation and ultimate depression of active E-rosetting (Fig. 4). This slowing or extension of the phase transition effect identified with the way morphine affected E-rosette formation when cells were first chilled on wet ice before the kinetic active assay as in Fig. 2.

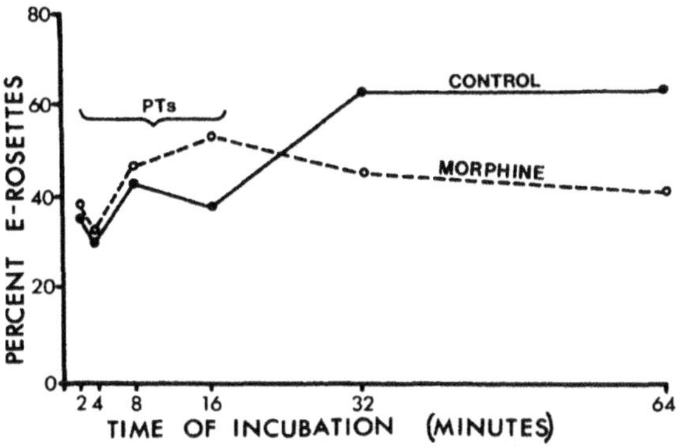

Fig. 4. Effect of in vitro treatment of leukocytes with 10^{-9}M on the rate of E-rosette formation with the 0-29°C kinetic assay adapted by centrifuging for 90 sec at 600 x g (see text).

AN EXPLANATION FOR THE PHASE TRANSITION PHENOMENON

As shown above, observation of phase transitions in the kinetics of E-rosette formation was owing primarily to judicious application of assay thermal and centrifugation conditions. Phase transitions in kinetics of E-rosette formation with untreated lymphocytes were only seen when active-assay thermal and centrifugation conditions were adjusted to accentuate this effect as in Figs. 3 and 4. However, in the presence of morphine, phase transitions were even more easily demonstrated if initial thermal conditions were cold, as in Figs. 2 and 4. Because of these thermal restrictions on expression of phase transitions, it was concluded that the portion of phase transitory kinetic curves wherein the percentage of E-rosettes was in decline was related to capping as described by Yu[9].

Capping involves the lateral mobilization of patched receptor-ligand complexes (E bound to the T-cell E-receptor) through the cell membrane into a cap at the polar end of the cell[9,13,14]. Patching and capping are both temperature dependent requiring physiologically warm conditions for their efficient expression, but only capping is energy dependent[14]. Following capping, E-rosettes are typically stripped from the cell as receptor-ligand complexes undergo endocytosis. T-cell E-rosettes are particularly susceptible to the processes of patching, capping and endocytolytic stripping because of their innate ability to serve as a receptor crosslinker.

Because of this probable involvement of capping in the phase transition phenomenon, we examined the kinetics of E-rosette capping. Yu[9] has described similar studies but did not document the very early events of this phenomenon. We studied kinetics related to 2-fold geometrically progressive increases in E-rosette assessment times beginning at 0.5 minutes. Total T-cell E-rosettes were generated by standard procedures[7] and placed in a water bath at 37°C. Then E-rosettes were evaluated at the appropriate intervals thereafter.

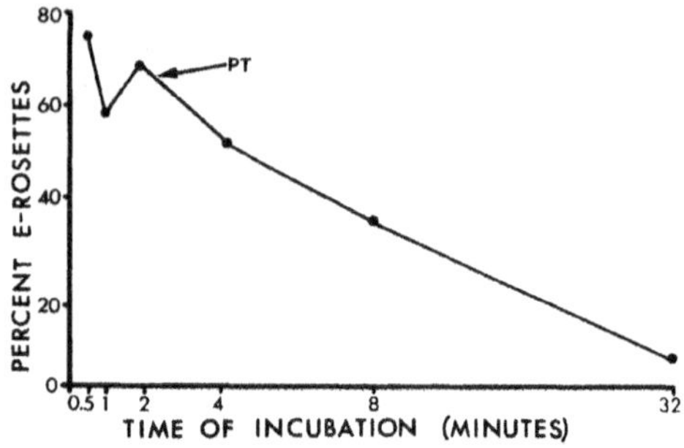

Fig. 5. Derosetting response generated by incubating total T-cell E-rosettes at 37°C before pellet disruption and E-rosette ascertainment.

Surprisingly, even during derosetting, phase transition kinetics were observed (Fig. 5).

Together, the preceding findings appeared somewhat dichotomous. When E-rosettes formed in the 0-29°C kinetic assay, there was a steady increase of E-rosettes with periods of intermittent losses of E-rosettes. On the other hand, when total T-cell E-rosettes were derosetted at 37°C, there was a steady decline in E-rosettes with intermittent periods where gain in E-rosette number occurred.

Explanation for these findings has been constructed in a model (Fig. 6) based on the assumption that E-receptor expression by T-cells is controlled by cytological processes of E-receptor microdisplacement (as originally defined by Bernard et al.[15], see later discussion) together with patching, capping and endocytolytic stripping of E-receptor-ligand complexes. It is assumed that a normal sample of peripheral blood lymphocytes is comprised of a variety of T-cells in various states of differentiation related in part to the level of expression of E-receptors[16,17], both in terms of affinity[18] and density[16]. Thus, lymphocytes initially exist in a receptor-level gradient. Within this gradient, some T-cells have a high density of high-affinity E-receptors. Others have lesser numbers of E-receptors on their surface and/or E-receptors of lower affinity. Under satisfactory thermal conditions, when E and T-cells make adequate contact for receptor-ligand binding, receptor microdisplacement processes that change the E-binding capacity of T-cells are activated so that more avid binding occurs across the T-cell gradient. T-cells lower on the E-receptor gradient would need time to reach optimal expression of ligand receptivity so that stable E-rosettes could form. This stability, however, would be dictated by thermal conditions. At temperatures of 23°C or lower (to 0°C) capping of E-rosettes would essentially cease and efficient microdisplacement of dormant E-receptors would result in accumulation of E-rosettes over time

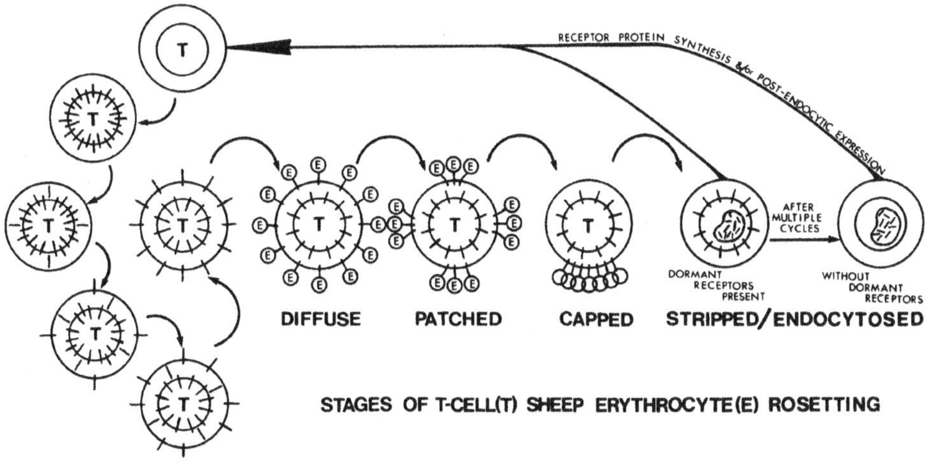

Fig. 6. Gradient of T-cell receptor expression. Lines attached to the inner circles of T-cells represent E-receptors that are lying dormant within the cell membrane probably attached to a transducer-effector complex. Those protruding from the cell represent ligand-receptive receptors capable of binding E. Others are floating within the lipid bilayer in the process of being microdisplaced. This gradient correlates with phase transition kinetics in that T-cells accumulate rosettes after sufficient expression of E-receptors through microdisplacement (the gain portion of phase transitions); then the loss portion of the phase transitions will occur when thermal conditions are conducive to capping and endocytolytic stripping of E-rosettes. As long as these stripped T-cells have dormant E-receptors available for microdisplacement, the processes described for gaining and losing E-rosettes will occur cyclically. When dormant E-receptors are expended as the result of these processes new receptors will have to be synthesized and/or re-expressed from endocytic vesicles before new E-rosettes can be formed, and the gradient and receptor cycling re-activated.

(see Fig. 1-3). At 37°C, capping of E-rosettes would be so efficient that microdisplacement of E-receptors would occur too rapidly for detection, as in Fig. 3 where assessment of active types of E-rosettes was involved, or only be barely perceptible if total T-cell E-rosettes were examined through kinetic analysis of derosetting (Fig. 5). However, at temperatures intermediate between 23°C and 37°C (in our studies 29°C), the rates of E-receptor microdisplacement and E-rosette capping would be nearly equal allowing both processes to be observable in kinetic analyses of E-rosette formation as cyclical phase transition kinetics (Figs. 2-4). Thus, gain portions of the phase transition kinetics occur when dormant E-receptors are recruited into the rosetting process via microdisplacement processes. Losses occur when sufficient ligand-receptive receptors are constituitively expressed or have become expressed via microdisplacement of dormant receptors so that when these receptors bind E they will initiate patching, capping and endocytolytic stripping if thermal conditions are warm enough for the successful activation and maintenance of these post-binding processes.

As mentioned previously, capping of E-rosettes is a well documented phenomenon[9,13,14]. However, microdisplacement of E-receptors has not been studied intensively since originally described by Bernard et al.[15]. By using monoclonal antibodies that bind the 9.6 and D66 E-receptor epitopes, they[15] showed that E-receptors lying dormant within the sugar layer of the cell membrane assume a new molecular conformation after capping of E-receptors which makes them highly E-ligand receptive. They[15] also showed that microdisplacement and capping events can be cyclical, as we have seen by observations of phase-transition kinetics of E-rosette formation and loss. Our kinetic E-rosetting data further show that the interdependency of microdisplacement and capping processes is related to the metabolic and physiologic state of the T-cell. Contrary to the need for active metabolic conditions requiring energy sources to drive capping events[14], we have seen that microdisplacement of E-receptors occurs efficiently at all temperatures between 4°C and 37°C; and, thus, that receptor microdisplacement appears not to be strictly reliant on capping. In fact, microdisplacement occurs most efficiently at cooler temperatures, as does E-rosetting. This circumstance probably relates to increases in membrane fluidity because of relaxation of metabolic constraints upon the lipid bilayer. Indeed, the division of T-cell biotypes into early (active, high affinity) E-rosette formers and late formers[6,18,19] is most likely a reflection of variable metabolic and physiologic constraints on E-receptor expression on one T-cell type versus another. These considerations also help explain why total T-cells, as determined by E-rosetting, require relatively long durations of incubation of E with T-cells and are best expressed when assayed at relatively cool temperatures (4°C usually, and up to 23°C). That is, given sufficient time and relaxation of metabolic and physiological constraints on E-receptor expression, enough E-receptors will become microdisplaced on all T-cells to allow for total T-cell E-rosette formation.

INTERPRETATION OF THE EFFECTS OF OPIATES ON E-ROSETTE FORMATION

In accordance with the E-receptor gradient model in Fig. 6, both the initial state of T-cells within this receptor gradient and their state during and after assay, as regards E-receptor expression, would be a reflection of the influence of various factors that affect homeostasis. Such factors as listed below would interdependently modulate the metabolic and physiologic set-points of receptor expression: 1. assay thermal conditions; 2. the intensity of the E-ligand and T-cell E-receptor interaction as dictated by the relative concentrations of each reactant and the centrifugal forces used to bring them into contact; 3. metabolites, ions and other nutrients in the assay medium; 4. presence of, or previous exposure of T-cells to drugs and the like which influence membrane functions either nonspecifically or through specific receptor-ligand interaction; 5. genetic restriction of receptor expression on the T-cell surface.

Opiates represent a type of factor as listed above in the fourth category. By examining the data in Figs. 1-5, it can be surmised that opiates affect E-rosette formation by altering the configurational expression of E-receptors through interference with their microdisplacement. This explanation accounts for the ability of opiates to both slow phase transition kinetics (Figs. 2 & 4) and the rate of E-rosetting (Figs. 1, 2, 4); and, also, for their ability to reduce the absolute number of active E-rosettes.

There are several observations about the physiological and metabolic changes that accompany binding of morphine to its receptor that support the possibility that morphine inhibits E-receptor microdisplacement.

Morphine receptors couple with adenylate cyclase through a guanine nucleotide-sensitive regulatory protein[20]. Much evidence supports the notion that coupling of membrane receptors and transducer-effector molecules of a variety of receptor systems occurs in a shared manner[21,22], which indicates that activation of coupling processes by morphine may result in coordinate coupling of both morphine receptors and E-receptors within the cell membrane. This coupling event could inhibit the ability of the E-receptor to undergo microdisplacement. An essential element in efficient coupling of receptors to their transducer-effector systems, at least with opiate receptors and others like them that are inhibitory for adenylate cyclase, is that Na^+ and GTP are required as cofactors[20]. This leads to the possibility that Na^+ fluxes are critical in determining the outcome of E-rosetting, especially under the influence of morphine. In fact, E-rosettes are poorly formed when the assay medium is deficient in Na^+ (unpublished data). The potential role of sodium is further suggested by the fact that receptor affinity for agonists is influenced by Na^+ concentrations[23]. Such effects could be mediated through receptor microdisplacement. Also, Na^+ has been reported to alter the reactivity of sulfhydryl groups at or near the binding site for opiates[24]. This finding, along with the fact that crosslinking of receptor-ligand complexes occurs through joining of sulfhydryl bonds[25], indicates that Na^+ may also be a controlling element in the E-receptor patching phenomenon, as also suggested by Yu[19]; and that such effects are relevant to the phase transition process. The notion of coupling of E-receptors and the potential role of Na^+ are represented schematically in Fig. 7.

AN EXPLANATION FOR THE NATURE OF ACTIVE E-ROSETTING ASSAYS

The nature of the active E-rosetting process is better understood because of the preceding considerations. For generation of the data for Fig. 1, thermal conditions of assay were identical to those used in 'classical' active E-rosetting assays, while, for the data in Fig. 2, cells were first chilled on wet ice (not held at room temperature, as was done for the data in Fig. 1) before active assay conditions were pursued (i.e., incubation at 37°C for 5 minutes before centrifugation). These thermal differences account for the ability of morphine to cause phase transitions in kinetics of E-rosette formation in the second experiment, but not in the first. Thus, it appears that the cooler conditions of the second experiment combined with an ability of morphine to slow E-receptor microdisplacement processes led to sufficient slowing of the phase transition processes to make them observable by kinetic assessment of E-rosette formation. This implies that in active assays, in situ (during centrifugation of E with T-cells), E-receptors normally cycle between dormancy, full expression consequent to microdisplacement, and endocytolytic stripping after formation, patching and capping of E-rosettes. However, this cycling is dependent on the initial warm thermal conditions promoted by incubating the cells at 37°C immediately before centrifugation. In this way, the cycling will stop during a late stage of centrifugation because the thermal conditions of centrifugation at that time will have reached ambient temperature of approximately 23°C which results in the essential discontinuation of capping processes. The consistency of results in active assays, therefore, appears to be dependent on the relative efficiency of the E-receptor cycling processes. Indeed, the findings of West et al.[18] that active assays are more reproducible if pelleted E and T-cells are incubated after centrifugation for 1 hour at 29°C (their high-affinity E-rosette assay procedure) are probably a consequence of the fact that these procedures promote stability of the E-receptor cycling processes under discussion here. Because of these interpretations, one can view active assays as a measure of the metabolic and physiologic equilibrium between microdisplacement and

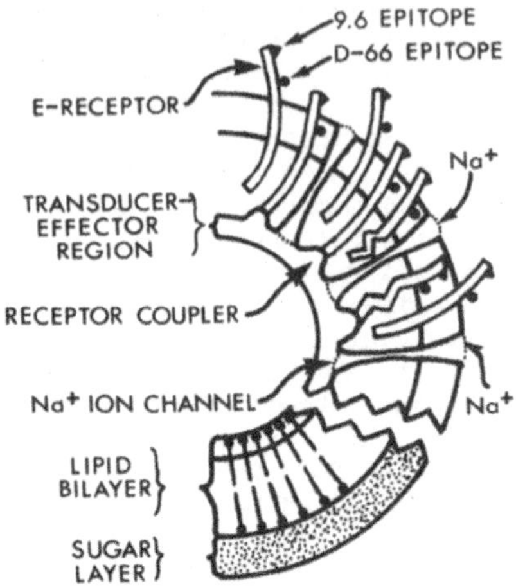

Fig. 7. Illustration of hypothetical receptor conformations in the T-cell membrane based on the observations of Bernard et al.[15]. This illustration is a magnified projection of what happens to E-receptors during their microdisplacement. Some E-receptors are linked to a transducer-effector complex, some are floating freely in various states of expression, and some are fully expressed. As shown by Bernard et al.[15], partial expression indicates that only the 9.6 E-receptor epitope is exposed at the cell surface while full expression requires that both the 9.6 and D66 epitopes be expressed. The latter fully expressed receptors bind E with high affinity. It is possible (see text) that binding of ligands such as E itself or morphine (not shown) to their respective T-cell receptors would activate physiological changes via opening of Na^+ channels, thereby modulating E-receptor coupling to a transducer-effector complex within the cell membrane and thus influencing the ability of the receptor to undergo microdisplacement.

capping processes. In this way, factors, like morphine, that inhibit active E-rosetting[6] can be envisioned as inducing metabolic and physiologic changes that inhibit microdisplacement while those, like enkephalins, which enhance active E-rosetting[6] enhance this receptor modulating process.

FURTHER EVIDENCE FOR RECEPTOR MODULATION BY BEHAVIORALLY ACTIVE DRUGS

In a recent study we showed that alcohol use by pregnant women and treatment of normal leukocytes <u>in vitro</u> with alcohol associate with enhancement in the rate of E-rosette formation[26]. We have also found that heroin use by addicts associates with enhanced expression of T^4 markers on T-helper lymphocytes[27]; and that, if addicts also use cocaine regularly, depression in E-rosetting rate that is otherwise apparent can be reversed[28]. In conjunction with the information presented in this report, these observations suggest that modulation of receptors on the surface of T-cells is a characteristic property of behaviorally active drugs to which they are sensitive. Such conclusions identify well with the fact that enkephalins modulate E-receptor expression[6,8] while endogenous opioids and other neurotransmitters and neurohormones are known to serve as modulators of neuronal receptors[3,4,29]. Likewise, these conclusions are compatible with the fact that alcohol modulates both E-receptor[26] and neuronal receptor expression[30,31] through an apparent enhancement of the fluidity of cell membranes[30,31]. Indeed, many receptor-ligand systems are autoregulatory[12,21], as are E-receptors in response to E; and many are regulated by secondary receptor interaction[21,22], as are E-receptors in conjunction with opiate receptors. These observations suggest that receptor modulation is a fundamental property of behaviorally active substances regardless of their tissue tropisms. These conceptualizations fit well with the notion that lymphocytes like neurons may be regarded as receptor-responsive sensory cells[5].

CONCLUSIONS

The foregoing discussion has indicated how study of the kinetics of T-cell E-rosette formation and loss can lead to better understanding of the cytological processes involved in receptor modulation. A new phenomenon of phase transition kinetics has been described which shows that E-receptors are regulated both independently and co-jointly with other receptors by processes of receptor microdisplacement along with patching, capping and endocytolytic stripping of receptor ligand complexes. It is possible that such findings identify microdisplacement processes with the concept of interconversion of receptor specificity and affinity as proposed to occur with mu and delta opiate receptors[32]. Notably, fibroblast receptors for insulin and epidermal growth factor have also been shown to undergo up-down regulation through capping processes[33] which further attests to the probability that our findings about E-receptor regulation can be generalized for most, if not all, cell surface receptor systems. Indeed, the link between up-down regulation of receptors and capping and endocytosis of receptor ligand complexes has been well established [22,34]. This information indicates that homeostatic maintenance relates in large measure to the ability of an organism to alter its receptor expression in response to changes in its environment and consequently that the ability of stressors to cause aberrant changes in receptor expression might lead to homeostatic dysfunction and disease. In as much as the findings reported here show that modulation of receptor expression is often tantamount to modulation of the differentiation status of cellular elements of a given organism, the concepts presented in this report seem as relevant to the study of cell differentiation as they are for investigation of immunomodulation and behavioral expression.

Because the endogenous opioid system is activated during stressful situations along with adrenocorticotropin and cortisol production[35], the present findings are of particular relevance to understanding how stress events might precipitate immunological dysfunction. That is, one might suspect that chronically stress-induced opioids might alter T-cell functioning in much the same way as chronic heroin abuse has been shown to alter the differential expression of E-receptors[7] and T4 antigenic markers[27] with the possible effect of making heroin addicts more susceptible to opportunistic infections and cancers [27,36].

E-receptors are physiological regulators of T-cell function which emphasizes their relevance in homeostatic regulation of immune response systems. They serve as negative feedback receptors for T-cell division[37] and are associated with the differential expression of thymocytes, and of T-suppressor/cytotoxic and T-helper/inducer lymphocytes[16,38]. E-receptors are also functionally linked to expression of the T-cell antigen receptor[39], Fc receptors[18,40], Ia antigens[40], gamma-interferon receptors[41] and interleukin-2 receptors[41,42]. Consequently, one can envision numerous ways by which E-receptor modulation might influence the normal functioning of T-cells as well as immune dysfunction. However, the complexities of T-cell function extend well beyond the inter-receptor linkages outlined here. T-cells have receptors for a variety of neurotransmitters and neuroendocrine hormones[3,4,6,43] as well as foreign antigens and numerous immunohormones and immunotransmitters. Therefore, it is likely that all these receptor-response systems interact to maintain homeostatic balance within the neuroimmune system and that the cytological mechanisms described in this report are broadly relevant in moderating these inter-receptor processes.

Recently Nadi et al.[44] showed that the familial association of affective disorder was genetically linked with enhanced expression of acetylcholine receptors. Such information leads to an appreciation of the potential importance of genetic restriction of receptor expression as a variable in determining the means by which individuals mount receptor-mediated responses to changes in their environment. This information in conjunction with the data and other conclusions in the present report indicate that future investigation into the mechanisms of homeostatic maintenance can profit considerably by detailed examination of the processes involved in receptor and inter-receptor function within the neuroimmune system.

REFERENCES

1. W. B. Cannon, Organization for physiological homeostasis, Physiol. Rev. 9:399 (1929).
2. T. L. Sourkes, Neurotransmitters and central regulation of adrenal functions, Biol. Psychiatry 20:182 (1985).
3. R. Ader, "Psychoneuroimmunology," Academic Press, N.Y. (1981).
4. R. Guillemin, M. Cohn, and T. Melnechuk, "Neural modulation of immunity," Raven Press, N. Y. (1984).
5. J. E. Blalock, The immune system as a sensory organ, J. Immunol. 132:1067 (1985).
6. J. Wybran, T. Appelboom, J.-P. Famaey, and A. Govaerts, Suggestive evidence for receptors for morphine and methionine-enkephalin on normal human T-lymphocytes, J. Immunol. 123:1068 (1980).
7. R. J. McDonough, J. J. Madden, A. Falek, D. A. Shafer, M. Pline, D. Gordon, P. Bokos, J. C. Kuehnle, and J. Mendelson, Alteration of T and null lymphocyte frequencies in the peripheral blood of human opiate addicts: In vivo evidence for opiate receptor sites on T lymphocytes, J. Immunol. 125:2539 (1980).
8. G. C. Miller, A. J. Murgo, and N. P. Plotnikoff, Enkephalin-enhancement of active T-cell rosettes from lymphoma patients, Clin. Immunol. Immunopathol. 26:446 (1983).
9. D. T. Y. Yu, Human lymphocyte receptor movement induced by sheep erythrocyte binding: Effect of temperature and neuraminidase treatment, Cell Immunol. 14:313 (1974).
10. D. T. Y. Yu, J. B. Peter, H. E. Paulus, and K. M. Nies, Human lymphocyte subpopulations. Study of T and B cells and density distribution, Clin. Immunol. Immunopathol. 2:333 (1974).
11. J. Wybran and H. H. Fudenberg, Thymus-derived rosette-forming cells in various human disease states: Cancer, lymphoma, bacterial and viral infections, and other diseases, J. Clin. Investig. 52:1026 (1972).

12. E. Loumaye and K. J. Catt, Homologous regulation of gonadotropin-releasing hormone receptors in cultured pituitary cells, Science 215:983 (1982).
13. E. R. Unanue, W. D. Perkins, and M. J. Karnovsky, Ligand-induced movement of lymphocyte membrane macromolecules. I. Analysis by immunofluorescence and ultrastructural radioautography, J. Exp. Med. 136:885 (1972).
14. G. B. Ryan, E. R. Unanue, and M. J. Karnovsky, Inhibition of surface capping of macromolecules by local anaesthetics and tranquilizers, Nature (London) 250:56 (1974).
15. A. Bernard, C. Gelin, B. Raynal, D. Pham, C. Gosse, and L. Boumsell, Phenomenon of human T-cell rosetting with sheep erythrocytes analyzed with monoclonal antibodies. "Modulation" of a partially hidden epitope determining the conditions of interaction between T-cells and erythrocytes, J. Exp. Med. 155:1317 (1982).
16. F. D. Howard, J. A. Ledbetter, J. Wong, C. P. Bieber, E. B. Stinson, and L. A. Herzenberg, A human T-lymphocyte differentiation marker defined by monoclonal antibodies that block E-rosette formation, J. Immunol. 126:2117 (1981).
17. R. B. Ashman, M. F. L. LaVia, S. Sawanabori, and A. J. Nahmias, Studies on the mechanism of E-rosette formation and inhibition, Clin. Immunol. Immunopathol. 16:39 (1980).
18. W. H. West, S. M. Payne, J. L. Weese, and R. B. Herberman, Human T-lymphocyte subpopulations: Correlation between E-rosette-forming affinity and expression of the Fc receptor, J. Immunol. 119:548 (1977).
19. D. T. Y. Yu, Human lymphocyte subpopulation: Early and late rosettes, J. Immunol. 115:91 (1975).
20. G. Koshki, R. Streaty, and W. A. Klee, Modulation of sodium-sensitive GTPase by partial opiate agonists. An explanation for the dual requirement for Na^+ and GTP in inhibitory regulation of adenylate cyclase, J. Biol. Chem. 257:14035 (1982).
21. C. R. Kahn, Membrane receptors for hormones and neurotransmitters, J. Cell Biol. 70:261 (1976).
22. M. Bornens, E. Karsenti, and S. Avrameas, Receptor mobility and its cooperative restriction by ligands, in: "Immunology of Receptors," B. Cinader, ed., Immunol. Series 6:41 (1977).
23. C. B. Pert, G. Pasternak, and S. H. Snyder, Opiate agonists and antagonists discriminated by receptor binding in brain, Science 182:1359 (1973).
24. E. J. Simon and J. Groth, Kinetics of opiate receptor inactivation by sulfhydryl reagents: Evidence for conformational change in presence of sodium ions, Proc. Natl. Acad. Sci. USA 72:2402 (1975).
25. E. Hazum, K.-J. Chang and P. Cuatrecasas, Role of disulphide and sulfhydryl groups in clustering of enkephalin receptors in neuroblastoma cells. Nature (London) 282:626 (1975).
26. J. J. Madden, R. M. Donahoe, I. E. Smith, D. E. Martinson, S. Moss-Wells, L. Klein, and A. Falek, Increased rate of E-rosette formation by T-lymphocytes of pregnant women who drink alcohol, Clin. Immunol. Immunopathol. 33:67 (1984).
27. R. M. Donahoe, J. K. A. Nicholson, J. J. Madden, F. Hollingsworth, A. Falek, D. A. Shafer, D. Gordon, and P. Bokos, Drug abuse, leukocyte antigenic markers, and AIDS, (submitted 1985).
28. R. M. Donahoe, J. J. Madden, F. Hollingsworth, D. A. Shafer, A. Falek, J. K. A. Nicholson, and P. Bokos, Immunomodulation by behaviorally active drugs provides a paradigm for connecting AIDS-like processes and psychoneuroimmunology, Proc. 1st Int. Workshop Neuroimmunol. (in press 1985).
29. W. Shain and D. O. Carpenter, Mechanisms of synaptic modulation, Int. Rev. Neurobiol. 22:205 (1981).

30. J. M. Hiller, L. M. Angel, and E. J. Simon, Multiple opiate receptors: Alcohol selectively inhibits binding to delta receptors, Science 214:468 (1981).
31. M. E. Charness, A. S. Gordon, and I. Diamond, Ethanol modulation of opiate receptors in cultured neural cells, Science 222:1246 (1983).
32. W. D. Bowen, S. Gentleman, M. Herkenham, and C. B. Pert, Interconverting μ and δ forms of the opiate receptor in rat striatal patches, Proc. Natl. Acad. Sci. USA 78:4818 (1981).
33. J. Schlessinger, Y. Shechter, M. C. Willingham, and I. Pastan, Direct visualization of binding, aggregation and internalization of insulin and epidermal growth factor on living fibroblast cells. Proc. Natl. Acad. Sci. USA 75:2659 (1978).
34. G. M. Edelman, Surface modulation in cell recognition and cell growth, Science 192:218 (1976).
35. R. Guillemin, T. Vargo, J. Rossier, S. Minnick, N. Ling, C. Rivier, C. Vale, and F. Bloom, β-Endorphin and adrenocorticotropin are secreted concomitantly by the pituitary gland, Science 197:1367 (1977).
36. J. D. Sapira, J. C. Ball, and H. Penn, Causes of death among institutionalized narcotic addicts, J. Chron. Dis. 22:733 (1970).
37. R. Palacios and O. Martinez-Maza, Is the E-receptor on human T-lymphocytes a "negative signal receptor"?, J. Immunol. 129:2479 (1982).
38. R. Mandle, R. E. Birch, S. H. Polmar, G. M. Kammer, and S. A. Rudolph, Abnormal adenosine-induced immunosuppression and cAMP metabolism in T-lymphocytes of patients with systemic lupus erythematosus, J. Immunol. 79:7542 (1982).
39. S. C. Meuer, R. E. Hussey, M. Fabbi, M. Fox, D. Acuto, O. Fitzgerald, K. A. Hodgdon, J. C. Protentis, S. F. Schlossman, and E. L. Reinherz, An alternate pathway of T-cell activation: A functional role for the 50 KD T11 sheep erythrocyte receptor protein, Cell 36:897 (1984).
40. G. Semenzota, G. Basso, V. Fagiolo, A. Pezzutto, C. Agostini, M. G. Cocito, and G. Gasparotto, Active and later rosette-forming cells: Immunological and cytochemical characterization, Cell. Immunol. 64:227 (1981).
41. M. Wilkinson and A. Morris, The E-receptor regulates interferon-gamma production: Four receptor model for human lymphocyte activation, Eur. J. Immunol. 14:708 (1984).
42. G. H. Reem and N. H. Yeh, Interleukin 2 regulates expression of its receptor and synthesis of gamma interferon by human T-lymphocytes, Science 225:429 (1984).
43. D. G. Payan, J. D. Levine, and E. J. Goetzl, Modulation of immunity and hypersensitivity by sensory neuropeptides, J. Immunol. 132:1601 (1984).
44. N. S. Nadi, J. I. Nurnberger, Jr., and E. S. Gershon, Muscarinic cholinergic receptors on skin fibroblasts in familial affective disorder, New Engl. J. Med. 311:225 (1984).

SIGNIFICANT ROLE OF RECEPTOR COUPLING IN THE

NEUROPEPTIDE-INDUCED ALTERATIONS OF MACROPHAGE CYTOTOXICITY

Gabriella Foris*, George A. Medgyesi**, and
Jozsef I. Szekely***

First Department of Medicine, University Medical School
Debrecen*, Institute of Heaematology and Blood Transfusion
Budapest** and Institute for Drug Research, Budapest***
Hungary

INTRODUCTION

Although in 1975 (Hughes et al., 1975) the year of discovery of the endogenous opiates, great expectations accompanied the research of these peptides, our knowledge regarding the central and peripheric effects of these compounds remain incomplete and superficial. Presumably the unexpected difficulties are due partly to the fact that the "secret" of the complex function of the opiate receptors has not been revealed, and on the other hand, some smaller peptide hormones derived from the large precursor molecules and displaying various biological effects could always confuse the interpretation of the essential processes.

In spite of the great amount of data regarding the function of the peptidergic system and published during the recent years, the exact role opioid peptides in the endocrine regulation cannot be defined unequivocally.

Obviously, nobody could predict 10 years ago that just the discovery of the opioid peptides would lead to the recognition of the fact that the peptidergic regulation of the neuroendocrine and the immune systems are closely related to each other. The serial discoveries of the recent years clarifying the immunomodulator functions of the opioid peptides on one hand, and the evidence showing that stimulated lymphocytes produce neuropeptides or neuropeptide-like factors on the other hand, may further stimulate both the endocrinological and immunological research activities.

This chapter will report on the results obtained when studying the effects of Met-enkephalin and its analogue D-Met^2-Pro^5 enkephalinamide on the peritoneal macrophages of rats. This type of study was also performed by using 2 other neuropeptides, angiotensin II and somatostatin, and the conclusion was reached that all these peptides affect the macrophages in a strongly similar way, and this effect resembles partly also that of the chemotactic peptides. The most important reason for such a similarity of their effect is probably the fact that all these peptides act through receptors being negatively coupled to the adenylate-cyclase system, and this coupling determines unequivocally their effects on the phagocytic cells.

Obviously,it was possible to reach a correct interpretation of the similar effects of neuropeptides and chemotactic peptides only when considering the results of the recent years, which clarify the intracellular regulatory mechanisms of phagocytes.

INTRACELLULAR REGULATING MECHANISM OF PHAGOCYTIC CELLS (PC)

The two major types of PC dependent effector functions are the extracellular and intracellular degradation of antigen (target) following its adherence through specific receptors. The first step of this process is the binding of target (either directly or by the mediation of antibody and/or complement) to one or more specific receptors of PCs. In the next step, signals induced by receptors upon target binding trigger processes leading to the destruction of target. Depending on the cell type, the size and physicochemical nature of target, two different kinds of reaction -cascade may follow: 1. The incorporation of target gets started involving concerted changes in microfilaments (MF) and microtubules (MT) or, 2. The attached target remains fixed on the other surface of PC. In the former process membrane ruffling is followed by invagination and phagosome formation, whereas in the later case the lysosomes cumulate at the inner site of PC membrane. The participation of cytoskeletal elements in these early steps of effector functions are well documented. Cytochalasin B (CB) known to disrupt MF inhibits markedly all kinds of incorporation, rosette formation, however, is diminished only slightly by relatively high concentration of this drug. On the other hand, CB stimulates the extracellular cytotoxicity indicating that lysosome-migration is not MF-dependent process. Colchicin and Vinblastin which inhibit MT assembly are able to augment rosette formation even in low concentration with a subsequent inhibition of incorporation, whereas the ADCC (antibody dependent cellular cytotoxicity) of PCs are only slightly affected by these drugs (Oliver, 1978; Foris et al., 1981; Medgyesi et al., 1980, 1984). The accumulation of lysosomes in the proximity of phagosome seems also to be independent on cytoskeleton, therefore the later steps of effector functions are not affected by drugs which damage cytoskeleton (Lowrie et al., 1980).

What kinds of regulation operate in PCs, which affect surface receptors through the directed assembly/disassembly of cytoskeletal elements? During yeast cell phagocytosis in PCs an early enhancement of intracellular cAMP level was recorded (Stolz, 1981; Fülöp et al., 1984) and furthermore all drugs which stimulate adenylate cyclase as well as the lipophilic cAMP analogues are able to stimulate the rosette formation and the early phase of incorporation (Vogel et al., 1981; Oliver and Zurier, 1976; Ogmündsdottir and Weir, 1980; Hartwig et al., 1980). On the other hand, lysosomalenzyme release during the extracellular cytotoxicity and the fusion of lysosomes with the phagosomes during the phagocytosis are associated with a significant elevation of intracellular cGMP (Fülöp et al., 1984; Ogmundsdottir and Weir, 1980). Drugs, which stimulate guanylate cyclase and lipophilic cGMP analogues augment both extracellular cytotoxicity and the intracellular killing activity of PCs (Lowrie et al., 1980; Oliver and Zurier, 1976)

It should be pointed out that enhancement of effector function by stimuli elevating cGMP level is only partly accounted for the mechanism described above. Important messenger functions of cyclic nucleotides are directed to the oxidative processes, in so far as cGMP stimulates and cAMP inhibits the activation of NADPH oxidase (McPhah et al., 1983), an initial step in the respiratory burst. However, during activation of Fcγ receptors by immune complexes, Ca^{2+} influx into the cells ensues, although the phagocytosis does not depend on the extracellular Ca^{2+} content of medium, as the intracellular level of free Ca^{2+} may also be increased using intracellular pools (Onazaki et al., 1983).

Summarizing the data available on the intracellular regulating mechanism of PCs, specific receptors mediating effector functions appear to be coupled to adenylate cyclase resulting in a transient stimulation, which is followed by a switch of receptors to the inositol triphosphate -diacylglycerol system (Berridge, 1984). Therefore, agents recognized by specific receptors on the surface of PCs and consequently trigger intracellular messenger production, are able to modulate the effector functions of PCs.

ALTERATIONS OF PC FUNCTIONS BY NATURAL CHEMOTACTIC PEPTIDES (CP)

A wide variety of natural and synthetic peptides affect PCs through specific receptors in a similar way: they stimulate the phagocytosis through Fc and C3 receptors (Najjar and Nishioka, 1970; Foris et al., 1984a), induce respiratory burst, aggregation and spreading inhibition (Holian and Daniele, 1979; Fehr and Huber, 1984; Keller et al., 1976), positive chemotaxis and lysosomal enzyme release. Binding to their specific receptors these peptides cause Na^+ influx, K^+ and Ca^{2+} efflux from cells, and an elevation in intracellular free Ca^{2+} level by mobilization from the intracellular Ca^{2+} pools (Holian and Daniele, 1982; Gennaro et al., 1984). All these specific receptors are coupled positively to inositol triphosphate-diacylglycerol system resulting in increase of intracellular cGMP level (Berridge, 1984; Gilman, 1984). These CPs are of various origin, they may be cleaved from large molecules during the immue response such as C5a from complement component C5 (Damerau, 1978) tuftskin and other fragments from immunoglobulins (Najjar and Nishioka, 1970; Kolb et al., 1982) and CP s from lymphokine components (Honda and Havashi, 1982; Coleman et al., 1984; Foris et al., 1983c, 1983b). Lymphokine derived oligopeptides (isolated from rat lymphokine by purification on Sephadex G 150 column) in higher than 10 μg/ml concentrations stimulated phagocytosis by elicited macrophages through Fcγ and C3b receptors, whereas in lower than 10 μg/ml concentrations they inhibited the Fcγ receptor mediated engulfment of ^{51}Cr-labeled sheep erthrocytes (Foris et al., 1983c).

In the last decade, a large number of formyl and muramyl di and tripeptides were isolated from bacterial cell walls (Wahl et al., 1979; Wilkinson, 1979) and in addition, their synthetic analogues were also prepared. For experimental studies the synthetic f-Met-Leu-He (FMLP) is applied the most frequently, since it has a pronounced influence on phagocytic cells under in vitro circumstances (Keller et al., 1981). FMLP exerts its effect on PCs through specific receptors with "high" and "low" affinity for the ligand, however, the former type of receptors are in strong and even morphologically recognizable association with the cytoskeletal structure of cells (Jesaits et al., 1984). The FMLP-triggered processes in PCs are similar or even identical with those which are induced by the natural CPs mentioned above.

The question arose about the physiological role of natural CPs released during the immune response either in circulation or in tissues? It is evident, that in consequence of their chemotactic effect they are able to accumulate PCs at the site of inflammation. Furthermore, their stimulating effect on the activity of both Fc and C3b receptors, as well as on the respiratory burst of PCs attach a great importance to these immune mediators. These peptides released from lymphokines, complement components and from immunoglobulins represent a significant amplifying mechanism that may be involved in the fine regulation of the effector cell functions during the immune response. It is reasonable to assume that they exert their effect more intensively at the site of their formation and that they are inactivated immediately or shortly after their generation by peptidases. Consequently, some of the most effective amino acid sequences might not have been determined so far.

As suggested by the earlier reviews of David and Remold (1976) it is possible that both "activating" and "inhibiting" peptides are released from lymphokines at the same time and site during the macrophage-lymphokine interaction. The same authors found the sensitivity of guinea pig macrophages to be increased in the presence of protease inhibitors such as soybean trypsin inhibitor, which acts only extracellularly. The assumption that peptides which are split from larger lymphokine components by macrophage proteases exert inhibitory effect on lymphokine induced activation, seems plausible.

Information on the role of these natural CPs beyond the immune system is scarce. As it is well documented, fragments released during complement activation (C3a, C5a, earlier called "anaphylatoxins"), may exert vasoconstriction at the periphery (Hugli et al., 1981). Furthermore, CPs are able to stimulate NA^+ leak-influx and consequently, the ATPase dependent Na^+ pumping mechanism (Korchak and Weissmann, 1980). Indeed, Na, K-ATPase stimulating factor or factors were demonstrated in sera of patients suffering from essential hypertension with low renin activity (Horky et al., 1984) and factor(s) was detected in supernatant of Concanavalin A-stimulated blood lymphocytes of the patients (unpublished data). Therefore, CPs or pepetides released together with CPs can also play some role in the development of essential hypertension. On the other hand, not only gamma interferon (Blalock and Smith, 1981) but also tuftsin and muramyl peptides exert morphin-like effects after intracerebral administration in mice (Herman et al., 1981; Krueger et al., 1984).

Summarizing the available knowledge about natural CPs, they are significant amplifying components of the immune response, but they may also have some undefined hormone-like effects outside the immune system.

GENERAL ASPECTS OF IMMUNOMODULATION BY OPIOID PEPTIDES (OP)

The immunodeficiency of morphine addicts has been known for a long time. However, because of the living conditions of these subjects, this observation could not be considered as an evidence of the immunosuppressive effect of morphine (Sheagran and Tuazon, 1977). On the other hand, both the immunosuppression caused in animals with large doses of morphine (Hung et al., 1973) and the existence of specific morphine receptors in leukocytes (Merishi and Mills, 1983) represents the most important proofs for a direct immunosuppressive effect of morphine. Moreover, morphine administration decreased the mean survival time of tumor bearing mice. Therefore, one may conclude that the immune system is suppressed unambigously by morphine (Fischer and Falke, 1985). In addition, OPs administered to animals and elevation of the OP level induced by stress, caused immunosuppression and an increase of tumor growth in most of the cases (Lewis et al., 1984). On the other hand, the survival time of mice treated with low doses of heroin increased after transplantation of neuroblastoma cells. It should be noted, however, that neuroblastoma cells possess also morphine receptors, therefore a direct effect of morphine on tumor growth may not be excluded (Zagon and McLaughlin, 1981). A significant increase in number of survivors was observed by Plotnikoff and Miller (1983), if BDF_1 mice inoculated with L 1210 cells were treated with enkephalins. It has been found that the subcutaneous injection of either methionine-enkephalin or leucine-enkephalin in healthy BDF_1 mice resulted in a significant increase in thymus weight and a significant decrease in spleen weight (Plotnikoff et al., 1984). According to the in situ model experiment of Moore (1984) carried out in sheep popliteal lymph node, the output of lympocytes increased under the effect of acute infusion of serotonin, substance P, bombesin, metenkephalin, isoprenaline and phenylephrine: all these compounds elevate cGMP level both in monocytes and the nervous tissue. Casale et al. (1984) demonstrated that the immediate skin response of healthy volunteers to endogenous OPs-occured as a consequence of mast cell degranulation. Similarly, in a number of in vitro experiments, it was found that opioid peptides stimulate leukocytes: lymphocyte proliferation (Gilman et al., 1982) and the number of active T cell rosettes obtained from healthy volunteers and lymphoma patients (Miller et al., 1984, 1983) were enhanced. In addition, OPs appeared to augment the natural killer activity of human lymphocytes in vitro (Faith et al., 1984; Mathews et al., 1983). Fischer and Falke (1985), have found a stimulated random migration of human granulocytes in the presence of beta endorphin, whereas Epps and Saland (1984), reported on the chemotactic effect of this peptide. Recently, Foris et al. (1984b), reported that Met-enkephalin stimulated the macrophage cytotoxicity by a calmodulin dependent process. Experiments with the opioid receptor antagonist naloxone suggest that both opioid and non-opioid receptors may be involved in this effect. Some of the multiple effects of OPs may be abolished by their antag-onist naloxone, however, specific opioid and also non-opioid receptors for beta endorphin were detected on human phagocytic leukocytes (Lopkor et al., 1980).

Despite the above contradictory findings, it can definitely be stated that OPs possess a significant immunomodulating effect (Weber and Pert, 1984; Faith et al., 1985). However, the immunomodulation may manifest itself either as a stimulation or a suppression, possibly dependent on the applied concentrations.

The main starting point of our study was that the OPs and also some other NPs have specific leukocyte receptors with positive coupling to inositol triphosphate-diacylglycerol system and possess a common mechanism of action. In lower concentration-range they have a stimulating effect on the immune system like the CPs, whereas the suppressive effect of the higher concentrations of OPs may represent a special kind of down regulation of their receptors.

COMPARATIVE STUDIES WITH THREE NEUROPEPTIDES (NP) ON THE EFFECTOR FUNCTIONS OF ELICITED RAT PERITONEAL MACROPHAGES (PM)

The starting point of our work was an earlier finding in our laboratory, namely that oligopeptides obtained from a rat lymphokine (Foris et al., 1984c) had an effect on the Fcγ receptor mediated processes of rat PMs like angiotensin II (At II) (Dezső and Foris, 1981; Foris et al., 1983a). In addition, both peptides have chemotactic activity (Honda and Hayashi, 1982; Goetzl et al., 1980). Experiments were carried out applying two further peptides which bind to specific receptor on the surface of leukocytes (Thomas and Hoffman, 1984; Bhathena et al., 1981) and which are well known immunomodulators (Weinstock and Kassab, 1984; Payan et al., 1984; Foris et al., 1985). Like At II, somatostatin (SS) and Met-enkaphalin (ME) exert their effect on target tissues through specific receptors which are coupled to the adenylate cyclase system in an inhibiting way, since all of the applied NPs affect their targets elevating the intracellular level of cGMP (Gilbert and Richelson, 1983; Wodcock and Johnston, 1982; Spada et al., 1984).

The IgG mediated phagocytosis and antibody dependent cellular cytotoxicity (ADCC) of PMs are their most significant function in the defense against pathogens and tumor cells and they are regulated intracellularly by secondary messengers with help of cytoskeleton. We assumed that NPs like CPs may influence these effector functions altering cation transport and/or triggering enzyme activities in phagocytic cells. In Fig. 1. the concentra-tion dependent effect of At II, SS and ME is demonstrated on the Fcγ receptor mediated phagocytosis and ADCC activity of rat PMs elicited by thioglycolate administration. Firstly, it should be noted that all of the applied NPs affected PMs in a similar way: in the 10^{-9}-10^{-7}M range of concentrations they stimulated the extracellular lysis of IgG2a coated ^{51}Cr-sheep erythrocytes, whereas their incorporation was diminished in the

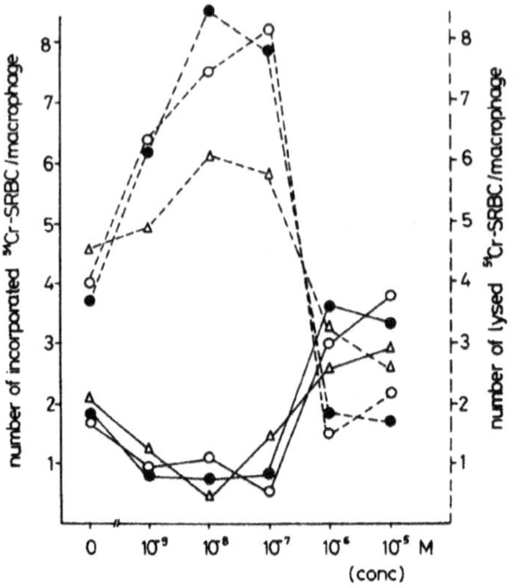

Fig. 1. The NP-induced alterations in Fcγ receptor mediated phagocytosis and ADDC of PMs after 60 min preinculbation. Experiments were carried out on PM monolayers applying IgG2a coated ^{51}Cr-sheep red blood cells (SRBC) as earlier described (Foris et al., 1983c). The target: effector ratio was 16:1. Each point represents the mean of six experiments. At II Δ, SS ● and ME o.

same range of concentrations. By contrast, the ADCC activity of PMs was significantly decreased by higher concentrations of NPs (10^{-6}-10^{-5}M), and the incorporation of SRBC was augmented.

Furthermore, the intracellular killing capability of PMs was also found altered upon a 60 min pretreatment with NPs (Fig. 2.). Namely, At II, SS and ME in higher, phagocytosis stimulating concentrations (10^{-6}-10^{-5}M) markedly diminished the killing of ^{51}chromium labeled Candida albicians, whereas they had no effect on killing capability of PMs when applied in lower concentrations. The ability of oligopeptides to stimulate phagocytosis accompanied by a decreased intracellular degradation, may play an important role in the prolonged survival of pathogens during chronic infections as well as in the pathomechanism of different storage disorders. In Fig. 3. the total number of incorporated +lysed + adhered erythrocytes is demonstrated as a function of the hormone concentration applied. Data show that all of the applied NPs were able to increase the number of "affected" SRBCs only in the lower (10^{-9}-10^{-7}M) concentration-range, whereas when the higher concentrations were applied the number of these target cells was similar to the control levels or even slightly lower. The question arose therefore whether NPs induce an increase in the Fcγ receptor expression. As the capacity of PM monolayers to bind ^{125}I-IgG2a was not changed after treatment with the peptides (Table 1), NPs do not appear to affect the rosette formation, phagocytosis or ADCC by enhancing the number and/or affinity of Fcγ receptors, but by modulating the functional activity of these receptors possibly through the cytoskeleton. Although the ADCC activity enhancing effect of NPs applied in the 10^{-9}-10^{-7}M concentration range could have been explained by a functionally enhanced Fcγ receptor activity one should take into account that beyond receptor activity, ADCC depends on the production of biologically active oxygen species (BAOS) which play an essential role in both the intracellular and extracellular degradation of targets (Babior, 1984). In addition, various types of natural and synthetic CPs are known stimulants of BAOS production (Tanabe et al., 1983). Therefore, a study

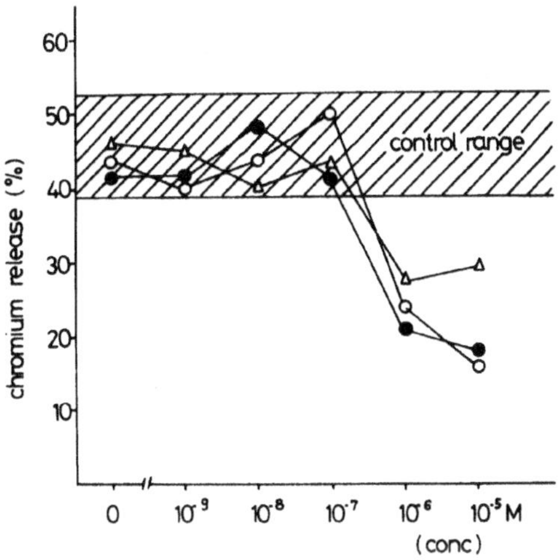

Fig. 2. The alteration of intracellular killing activity of PMs after 60 min preincubation with NPs. Experiments were carried out according to the method of Yamamura et al. (1976) in the presence of 5% fresh rat serum. ^{51}Cr release from Candida albicans was calculated as percentage of total releasable radioactivity. At II Δ, SS ● and ME o. Each value represents the mean of five experiments.

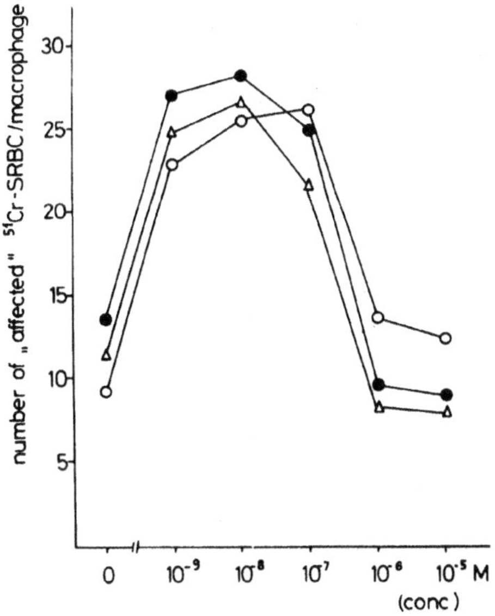

Fig. 3. The elevated number of "affected" ^{51}Cr-erythrocytes through Fcγ receptors under the effect of NPs. The number of adhered + lysed + engulfed erythrocytes was calculated as "affected" cells. Each point represents the mean of five experiment. At II o, SS ● and ME Δ. Experimental conditions see on Fig. 2.

Table I. The effect of NPs in different concentrations on the ^{125}I-labeled IgG2 a binding capacity of PM monolayers

^{125}I-labeled rat IgG2a^{x} bound (% of control)

conc M	At II	SS	ME
none	100.0	100.0	100.0
10^{-9}	96.3	108.6	89.6
10^{-8}	109.0	102.7	104.3
10^{-7}	100.5	95.3	96.5
10^{-6}	91.2	101.1	111.2
10^{-5}	98.8	93.8	94.8

^{x}IgG2 a was separated from rat immune sera according to the method of Medgyesi et al. (1983). Each value represents the mean of four experiments.

Summarizing the available data on the modulating effect of NPs on phagocyte function, NPs stimulate macrophage ADCC in the 10^{-9}–10^{-7}M range of concentration. Both an enhancement of Fcγ receptor activity and stimulation of respiratory burst may underlie this effect. In higher concentrations both ADCC and intracellular killing activity were inhibited by these hormones in parallel with a suppression of superoxide anion production.

of the effect of NPs on superoxide generation looked warrented (Fig. 4.). The effect of NPS was assessed on the phorbol myristate acetate (PMA)-stimulated superoxide anion production of PMs, since the low level of the spontaneously generated superoxide anion was not sufficient to study an inhibiting effect. Indeed, all of the applied NPs enhanced markedly the superoxide generation, i.e. the NADPH oxidase activity in rat PMs, but only in the concentration range in which an ADCC stimulating effect was observed, whereas the amount of superoxide anion produced was inhibited by higher concentrations. These data are not easy to interpret as the most well characterized CPs such as C5a and FMLP did not suppress BAOS production in the same concentration range. On the other hand, some synthetic substrates for peptidases such as N-benzoyl-L-tyrosine ethyl ester and L-1-tosylamide -2-phenylethyl chloromethyl ketone exert a respiratory burst suppressing effect on PCs, as a consequence of an increase in the H^+ concentration in the extracellular space upon the cleavage of these substrates (Dri et al., 1981; Kitagawa et al., 1979). The lowered pH at the outer site of membrane inhibits the H^+ excretion from cells resulting in an enhancement of intracellular H^+ concentration (Lynn and Mohapatra, 1980). The activity of glutathione peroxidase (GSH•Px) which is involved in the elimination of the most significant BAOS, i.e. H_2O_2 (Baker and Cohen, 1983) was slightly diminished by the higher (inhibiting ADCC and intracellular killing) concentrations of NPs. The present experimental data do not allow an answer to the question, whether intracellular elevation of H^+ concentration exerts its effect through inhibition of GSH•Px activity or affects NADPH oxidase by a direct way. Baker and Cohen (1983) reviewed the problem of a feed back mechanism between GSH•Px and NADPH oxidase as it was demonstrated in granulocytes of rats fed by a selenium deficient diet.

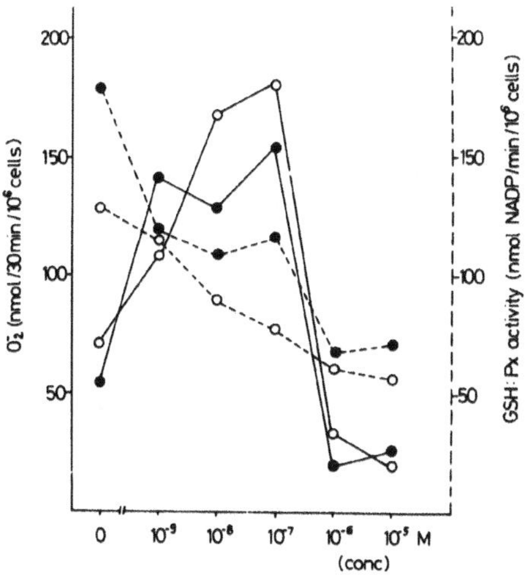

Fig. 4. The SS(•) and ME (o) induced changes in PM suspension O_2^- generation (——) and glutathione peroxidase (GSH·Px) activity (- - -). The GSH·Px activity was measured in the 30th min of incubation according to the method of Holmes et al. (1970).

MECHANISMS WHICH ARE INVOLVED IN THE DUAL EFFECT OF OPs

Finding a correct explanation for the effects of the higher ME concentrations is a real challenge. On one hand, ME may be subject of enkephalinase present in the membrane of PMs and a decrease of BAOS production is induced either by inhibition of H^+ excretion from cells or by releasing biologically active break down products of ME. On the other hand, it was reported by Gilbert and Richelson (1983) that delta opiate receptors which are coupled inhibiting to adenylate cyclase in cultured murine neuroblastoma clones and their glioma hybrids, after incubation with the high concentrations of ME may form clusters with a positive coupling to the adenylate cyclase system. This change in the receptor coupling serves as a form of down regulation at the post-receptor level without any reduction in the number of affinity of delta receptors. As it was observed by the same authors, cluster formation depends on the membrane fluidity of cells. Prompted by these data, we determined the cyclic nucleotide level in PMs during incubation with low (10^{-8}M) and high (10^{-6}M) concentrations of ME (Fig.5.). The intracellular cAMP level was increased only in the presence of 10^{-6}M) of ME with a peak in the 30th min of incubation. The cGMP level rose progressively during the 120 min incubation period in the presence of 10^{-8}M of ME. This finding is in good agreement with the observations of the above mentioned authors, that ME is able to stimulate adenylate cyclase in higher concentrations through the guanylate cyclase stimulating delta opiate receptors. (In Fig. 6.) the intracellular cAMP levels are presented after 30 min incubation with different concentrations of ME and it was evident that ME in 10^{-6}-10^{-5}M concentrations increased the adenylate cyclase activity of PMs. In our system the altered receptor coupling was abolished with the acetylated form of low density lipoprotein (LDL) isolated from sera of healthy human volunteers. acLDL after binding to scavenger receptors is endocytosed and degraded intracellularly with a subsequent increase of membrane cholesterol, i.e. membrane rigidity (Zechner et al., 1984). Cytochalasin B (CB) which disrupts the contractile elements of cytoskeleton and blocked in many cells all forms of receptor movement in the plane of membrane lipid bilayer also abolished the cAMP elevation induced by the higher concentrations of ME. To elucidate the role of enkephalinase in the ME-triggered adenylate cyclase activation, the effect of the known enkephalinase inhibitor puromycin (Hersch, 1982) was also investigated. Puromycin was found to abolish the cAMP increasing effect of ME in 10^{-6}M concentration. In addition, a potent ME analogue (D-Met^{2}, Pro^{5}) enkephalinamide (MP-EA) with high resistence against proteolysis (Bajusz et al., 1977) did not induce adenylate cyclase activation even in higher concentrations. Therefore, the change in the receptor coupling does not depend only on the membrane fluidity but also on ME processing by enkephalinase. Undoubtedly, ME appears to affect PMs through two different ways depending on the concentration applied, however, at this point of our experiments the mechanism of action induced by the higher concentrations, i.e. this special form of down regulation remains a puzzle.

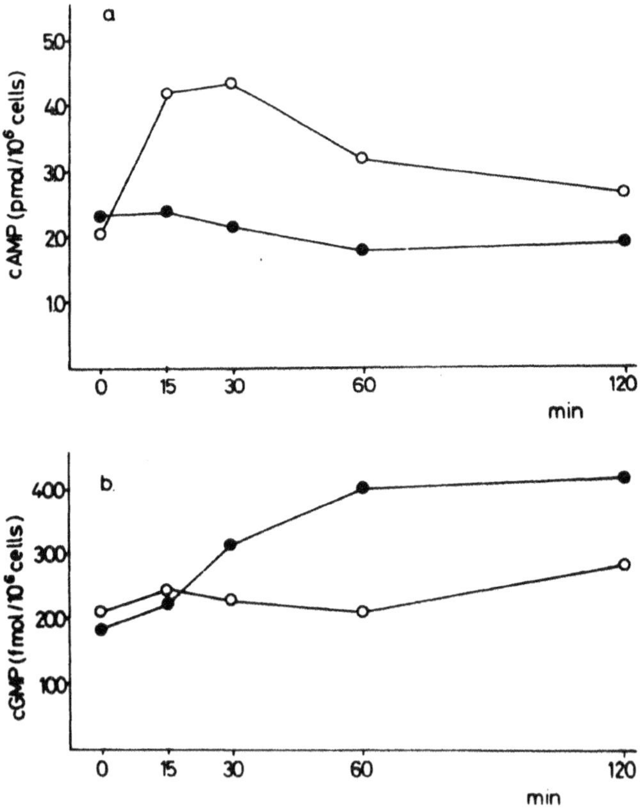

Fig. 5. The alteration of cAMP (a) and cGMP (b) level in PMs under the effect of 10^{-8}M (-●-) and 10^{-6}M (-o-) of ME. Each point represents the mean of four experiments. Experiments were carried out according to the method of Stabinsky et al. (1980).

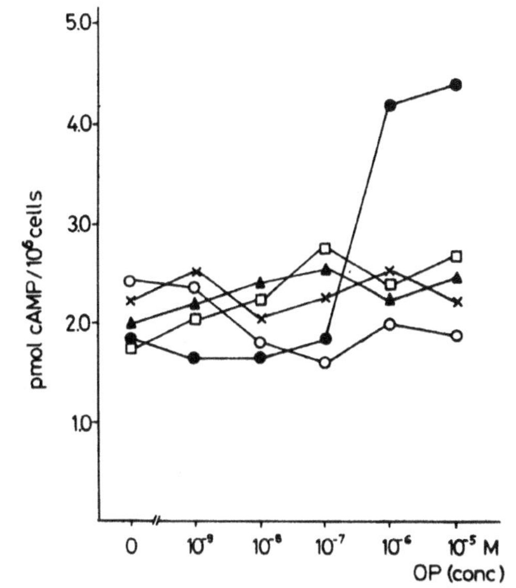

Fig. 6. The inhibition of ME induced cAMP elevation in PMs measured in the 30th min of incubation. ME alone ●, (D-Met^{2}, Pro^{5}) EA alone □, ME 50 μg/ml acLDL x, ME + 5 μg/ml CB ▲, ME + 10^{-5}M puromycin o. Each value represents the mean of four experiments.

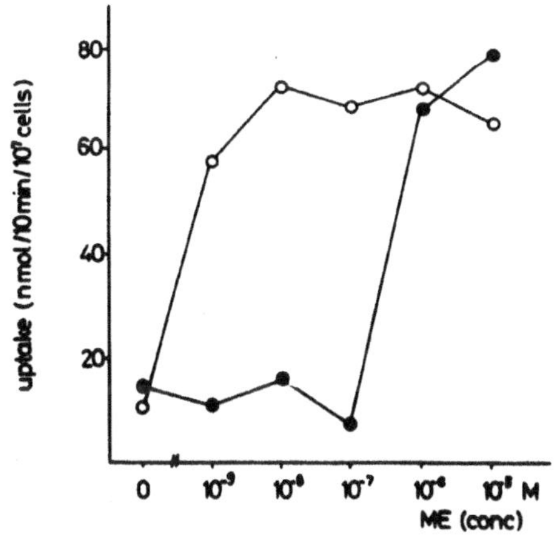

Fig. 7. The $^{22}Na^{+}$ (o) and $^{45}Ca^{2+}$ (●) uptake by PM monolayers from the medium containing 140 Na^{+}, 5 K^{+}, 1.4 Ca^{2+}, and 1.2 Mg^{2+} Each value represents the mean of four experiments.

Examining the early cation uptake, ME was found to enhance the $^{22}Na^{+}$ influx into cells when applied in low concentrations (found to activate guanylate cyclase). No. Ca^{2+} influx was found at the same ME concentrations (Fig.7.). However, a pronounced $^{45}Ca^{2+}$ influx was measured upon exposure of PMs to 10^{-6}-10^{-5}M ME. One can conclude therefore, that opioid receptor clustering triggered by the higher concentrations of ME may induce Ca channel opening. Alternatively, binding a large amount of extracellular Ca^{2+} by the Ca^{2+}-dependent enkephalinase in the presence of high ME concentrations can be assumed (Pant et al., 1982). To clarify the role of enkephalinase and of membrane fluidity dependent receptor clustering in the cAMP level elevating effect of the higher concentrations of ME, the enkephalinase activity of living PMs was measured applying fluorogenic substrate L-alanine-beta naphthylamide (ABNA). Enzyme which cleaves ABNA is the common peptidase for both ME and At II and in the presence of its natural substrates the cleavage of the synthetic substrate is markedly diminished (Fig. 8.). Thus, enhancing the concentration of ME in the system the enkephalinase activity of living PMs by the inhibition of cleaved beta naphtylamide may be determined. ME was able to inhibit the cleavage of ABNA only in the higher (10^{-9}-10^{-5}M) concentrations and this inhibition was not merely augmented by the addition of enkephalinase-inhibitor puromycin or EGTA (Hersch, 1982). The cleavage of ABNA was not inhibited by MP-EA, corroborating the earlier observed resistence of this analogue against peptidases (Bajusz et al., 1977). However, At II in 10^{-5}M concentration resulted a similar inhibition of ABNA-cleavage, the inhibition induced by ME and At II was not additive. It should be noted, that ABNA specific peptidase which was not inhibited by ME, At II or by puromycin was located in the lysosome fraction of PMs and it was inhibited by epsilon amino caproic acid (unpublished data). We assume that the crucial point of our observations is the abolition of Me induced inhibition of ABNA cleavage by both LDL and CB. In other words, both the normal condition of membrane fluidity and an intact cytoskeletal structure are required not only for cAMP elevation, but for the sufficient ME processing by enkephalinase. Based upon earlier results (Fig. 6.) that in the presence of 10^{-6}-10^{-5}M of ME the increase in cAMP level was abolished by puromycin as well as our further observations about the role of membrane fluidity in the ME processing by enkephalinase indicate that both delta opioid receptor clustering and enkephalinase activity are involved in the modified signallization triggered by high concentrations of ME.

It has been pointed out, that enkephalinase of living PMs is able to cleave ME if its natural ligand is processed by clustered receptors. Consequently, if no receptor clustering has occurred, enkephalinase is not able to split ME. Ca^{2+} uptake resulted either by cluster formation of by enkephalinase activation may serve as trigger for adenylate cyclase stimulation. This hypothesis is corroborated by data of Fig. 9. in so far as the ME induced enhancement of ADCC was abolished by the opiate antagonist naloxone applied in 10^{-5}M concentration, whereas the inhibition of ADCC resulting from the higher concentrations of ME was not abolished by this drug. On the other hand puromycin, CB and acLDL inhibited markedly the concentration dependent decrease of ADCC activity, however, the nalaxone sensitive stimulation of ADCC remained unaltered by these agents. In addition, the enkephalinase resistent analogue MP-EA increased the ADCC even if it was added to the PMs in higher than 10^{-7}M concentration. Lastly, the calmodulin inhibitor trifluoperazine (Smith et al., 1981) abolished both suppressing and inhibiting effects of ME indicating the role of calmodulin in the ME-triggered alterations of effector functions.

Concluding our considerations, the unique form of receptor down regulation with change of receptor coupling appear by a good explanation for the dual effect of ME on phagocyte function which is observed both in vivo and in vitro.

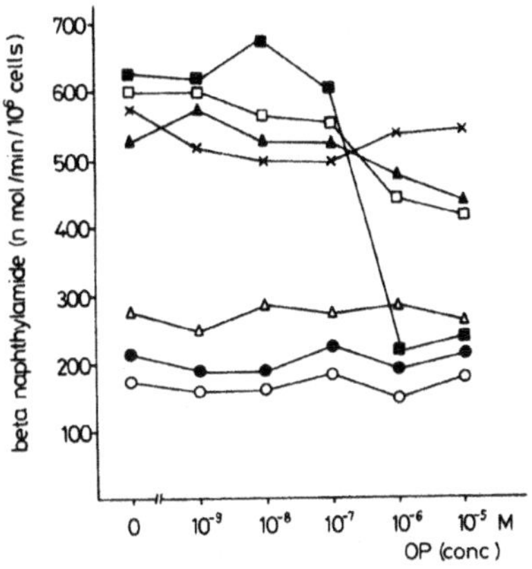

Fig. 8. The effect of various drugs on the enkephalinase activity of living PM suspension measured by cleavage of fluorogenic substrate L-alanine-beta naphthylamide (ABNA in the presence of different concentrations of ME. The assay was performed in Hitachi-NK spectrofluorimeter at 480 nm ME alone ■, ME + 10^{-6}M At II Δ ME + 10^{-5}M Puromycin ●, ME+5 mM EGTA o, ME+5μg/ml CB ▲, ME+50 μg/ml acLDL x (D-Met,Pro/EA alone □. Each value represents the mean of four experiments.

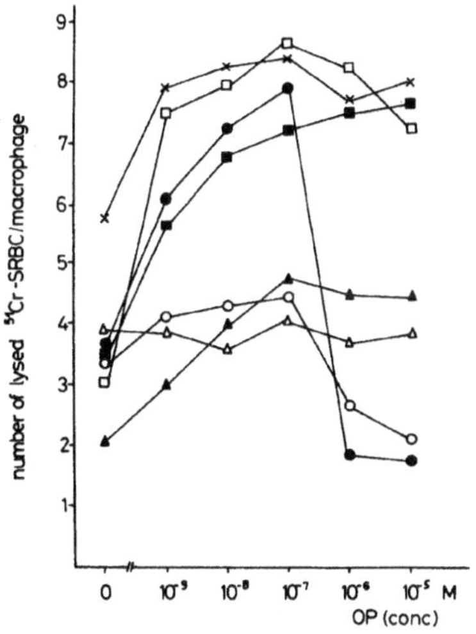

Fig. 9. The effect of various agents on the ME-induced alterations of ADCC activity of PM-monolayers. Each value represents the mean of five experiments. The monolayers were preincubated with different agents for 30 min before the 60 min incubation with different concentration of ME.

ME alone ●, 10^{-5}M naloxone+ ME o, 10^{-6}M trifluoperazine+ ME Δ, 50 μg/ml acLDL + ME ▲, 10^{-5} μ puromycin + ME □, 5 μg/ml CB + ME x, (D-Met^2, Pro^5) EA alone ■.

Our investigations on human neutrophil granulocytes obtained from aged, arteriosclerotic and diabetic patients as well as from patients suffering from Alzheimer disease corroborated our assumption that one of the most vulnerable points in the regulation of phagocytic function is the post-receptor signallization (unpublished data). We suggest that this is also the case with opioid receptors. It must be kept in mind, that the final effect brought about by OPs is not only determined by their actual dose, but also individual changes of cellular sensitivity defined by the membrane fluidity and/or by the effective peptidase activity.

CONCLUSION

The in vitro effect of three neuropeptides (At II, SS and ME) was investigated on rat PMs elicited by thioglycolate provocation. All of the applied peptides equally affected PMs. In lower concentrations (10^{-9}-10^{-7}M) they markedly inhibited the phagocytosis through Fc γ receptors whereas the ADCC activity of PMs was stimulated by both the increased Fc γ receptor activity and the enhancement of oxidative processes. On the other hand, these NPs in higher concentrations (10^{-6}-10^{-5}M) inhibited both ADCC and intracellular killing capability of PMs possibly as a consequence of diminished respiratory burst of PMs.

In the case of ME the mechanism of this dual effect appear to be based on the presence of naloxone sensitive opioid receptors on PMs with negative coupling to adenylate cyclase. These receptors are able to form clusters and positive coupling to the adenylate cyclase in the presence of high concentrations of ligand. The incubation with higher concentrations (10^{-6}-10^{-5}M) of ME caused an elevation of the intracellular cAMP level which was abolished by acLDL, CB as well as by the enkephalinase inhibitor puromycin. Therefore, we concluded that the adenylate cyclase stimulating (and consequently ADCC suppressing) effects of ME are caused by changes in the post-receptorial coupling. It should be noted that this type of coupling is strongly dependent on the actual state of the cell membrane fluidity and membrane-cytoskeleton interaction. In addition, the enkephalinase activity of PMs may play a significant role in this special form of down regulation, because (D-Met, Pro) EA (an ME analogue with high resistence to enkephalinase) exerts only stimulating effect on PM cytotoxicity. Furthermore, puromycin as a potent inhibitor of enkephalinase, abolished the cAMP elevation and ADCC suppression induced by the higher concentrations of ME.

REFERENCES

Babior, B.M., 1984, The respiratory burst of phagocytes, J. Clin. Invest., 73:599.

Bajusz, S., Ronai, A.Z., Szekely, J.I., Graf, L., Dunai-Kovacs, Z., and Berzetei, I., 1977, A superactive antinociceptive pentapeptide, (D-Met^2, Pro^5) enkephalinamide, Febs Letters, 76:91.

Baker, S.S.,and Cohen, H.J., 1983, Altered metabolism in selenium deficient rat granulocytes, J. Immunol., 130:2856.

Berridge, MJ., 1984, Inositol triphosphate and diaclyglycerol as second messengers, Biochem. J., 220:345.

Bhathena, S.J., Loule, J., Scheitter, G.P., Redman, R.S., Wahl, I., and Recant, L., 1981, Identification of human mononuclear leukocytes bearing receptors for somatostatin, and glucagon, Diabetes, 30:127.

Blalock, J.E., and Smith, E.M., 1981, Human leukocyte interferon: potent endorphin-like opioid activity, Bioch. Biophys. Res. Commun., 101:472.

Casale, T.B. Bowman, S., and Kaliner, M., 1984, Induction of human cutaneous mast cell degranulation by opiates and endogenous opioid peptides: evidence for opiate and non opiate receptor participation, J. Allergy Clin. Immunol., 73:775.

Coleman, D.L., Culver, K.E., and Ryan, J.L., 1984, Enhancement of macrophage immune and non immune receptor mediated phagocytosis by a low molecular weight soluble factor from resident thymocytes, J. Immunol., 133:3121.

Damerau, B., Grünefeld, E., and Vogt, W., 1978, Chemotactic effect of complement-derived peptides C3ai, C5a (classical anaphylatoxin) in rabbit and guinea pig polymorphonuclear leukocytes, Naunyn-Schemeideberg's Pharm., 305:181.

David, J.R., and Remold, H.G., 1976, Macrophage activation by lymphocyte mediators and studies on the interaction of macrophage inhibitory factor (MIF) with its target cells, In: "Immunobiology of the Macrophage," D.S. Nelson, ed., Academic Press, New York, San Francisco.

Dezso, B., and Foris, G., 1981, Effect of angiotensin II on the Fc receptor activity of rat macrophages, Immunology, 42:277.

Dri, P., Berton, G., and Patriarca, P., 1981, A twofold effect of L-tosylamide-2-phenylethyl chloromethyl ketone on the oxidative metabolism of guinea pig phagocytes, Inflammation, 5:223.

Epps, D.E., and Saland, L., 1984, β endorphin and Met-enkephalin stimulate human peripheral blood mononuclear cell chemotaxis, J. Immunol., 132:3046.

Faith, R.E., Liang, H.J., Murgo, A.J., and Plotnikoff, N.P., 1984, Neuroimmunomodulation with enkephalins: enhancement of human natural killer (NK) cell activity in vitro, Clin. Immunol. Immunopathol., 31:412.

Faith, R.E., Plotnikoff, N.P., and Murgo, A.J., 1985, Effects of opiates and enkephalins-endorphins on immune functions, In: "NIDA Technical Meeting on Mechanisms of Tolerance and Dependence"(In Press).

Fischer, E.G., and Falke, N.E., 1985, β endorphin modulates immune functions In:"Psychotherapy and Psychosomatics, II," World Congress of the International College of Psychosomatic Medicine, (In Press).

Foris, G., Dezso, B., Medgyesi, G.A., and Bazin, H., 1981, Role of cytoskeleton in the Fc receptor activity of rat peritoneal macrophages, Int. Archs. Allergy appl. Immun., 65:138.

Foris, G., Dezso, B., Medgyesi, G.A., and Füst, G., 1982a, Effect of angiotensin II on macrophage functions, Immunology, 48:529.

Foris, G., Füst, G., and Medgyesi, G.A., 1983b, The effect of oligopeptides on the C3b receptor-mediated functions of rat macrophages, Immunol. Letters, 6:7

Foris, G., Hauck, M., Dezso, B., Medgyesi, G.A., and Füst, G., 1983c, Effect of low-molecular weight lymphokine components on the Fc and C3b receptor mediated macrophage functions, Cell. Immunol., 78:276.

Foris, G., Medgyesi, G.A., and Fust, G., 1984a, Regulating role of natural peptides in the Fc and C3b receptor mediated functions of resident and provoked macrophages, In: "Tissue Culture and RES," P. Rohlich and E. Baczy, eds., Publishing House of the Hung. Acad. Sci.

Foris, G., Medgyesi, G.A., Gyimesi, E., and Hauck, M., 1984b, Met-enkephalin induced alterations of macrophage functions, Mol. Immunol., 21:747.

Foris, G., Gyimesi, E., and Komaromi, I., 1985, The mechanism of antibody dependent cellular cytotoxicity stimulation by somatostatin in rat peritoneal macrophages, Cell. Immunol., 90:217.

Fülöp, T., Foris, G., and Leövey, A., 1984, Age related changes in cAMP and cGMP levels during phagocytosis in human polymorphonuclear leukocytes, Mech. Ageing Develop., 27:233.

Gennaro, R., Pozzan, T., and Romeo, D., 1984, Monitoring of cytosolic free Ca^{2+} in C5a-stimulated neutrophils: Loss of receptor modulated Ca^{2+} uptake in granule-free cytoplasts, Proc. Natl. Acad. Sci. USA, 81:1416.

Gilbert, J.A., and Richelson, E., 1983, Function of delta opioid receptors in cultured cells, Mol. Cell. Biochem., 55:83.

Gilman, S.C., Schwartz, J.M., Milner, R.J., Bloom, F.E., and Feldman, J.D., 1982, endorphin enhanced lymphocyte proliferative responses, Proc. Natl. Acad. Sci. USA, 79:4226.
Gilman, A.G., 1984, Guanine nucleotide binding regulatory proteins and dual control of adenylate cyclase, J. Clin. Invest., 73:1.
Goetzl, E.J., Klickstein, L.B., Watt, K.W.K., and Wintroub, B.U., 1980, The preferential human mononuclear chemotactic activity of the substituent tetrapeptides of angiotensin II. Bioch. Biophys. Res. Commun., 97:1097.
Hartwig, J.H., Yin, L., and Stossel, T.P., 1980, Contractile proteins and mechanism of phagocytosis in macrophages, In: "Mononuclear Phagocytes," R. van Furth, ed., Martinus Nijhoff, The Hague, Boston, London.
Herman, Z.S., Stachure, Z., Opielka, L., Siemion, I.Z., and Nawrocka, E., 1981, Tuftsin and D-Arg^{3}-tuftsin possess analgesic action, Experientia, 37:76.
Hersch, L.B., 1982, Degradation of enkephalins: the search for an enkephalinase, Mol. Cell. Biochem., 47:35.
Holian, A., and Daniele, R.P., 1979, Stimulation of oxygen consumption and superoxide anion production in pulmonary macrophages by N-formyl methionyl peptides, Febs Letters, 108:47.
Holian, A., and Daniele, R.P., 1982, Formyl peptide stimulation of superoxide anion release from lung macrophages: sodium and potassium involvement, J. Cell. Physiol., 113:413.
Holmes, B., Prk, D.H., Malavista, S.E., Qui, P.G., Nelson, D.L., and Good, R.A., 1970, Chronic granulomatous disease in females. A deficiency of glutathione peroxidase, N. Engl. J. Med., 283:217.
Honda, M., and Hayashi, H., 1982, Characterization of three macrophage chemotactic factors from PPD-induced delayed hypersensitivity reaction sites in guinea pigs, with special reference to a chemotactic lymphokine, Am. J. Pathol., 108:171.
Horky, K., Schreiber, V., and Pribyl, T., 1984, Radioactive ^{86}Rubidium influx into red blood cells in essential hypertension in relation to the plasma renin activity, Horm. Metab. Res., 16:41.
Hughes, J., Smith, T.W., Kosterlitz, H.W., Fothergill, L.A., Morgan, B.A., and Morris, H.R., 1975, Identification of two related pentapeptides from the brain with potent opiate agonist activity, Nature, 258:577.
Hugli, T.E., Gerard, C., Kawahara, M., Scheetz, M.E., Barton, R., Briggs, S., Koppel, G., and Russel, S., 1981, Isolation of three separate anaphylatoxins from complement activated human serum, Moll. Cell. Biochem., 41:59.
Hung, C.Y., Lefkowitz, S.S., and Geber, W.F., 1973, Interferon inhibition by narcotic analgestics, Proc. Soc. Exp. Biol. Med., 142:106.
Jesaits, A.J., Naemura, J.R., Sklar, L.A., Cohrane, C.G., and Painter, R.G., 1984, Rapid modulation of N-formyl chemotactic peptide receptors on the surface of human granulocytes: formation of high-affinity ligand-receptor complexes in transient association with cytoskeleton, J. Cell. Biol., 98:1378.
Keller, H.U., Gerber, H., Hess, M.W., and Cottier, H., 1976, Studies on the regulation of neuthrophil chemotactic response using a rapid and releable method for measuring random migration and chemotaxis of neutorphil granulocytes, Agents and Actions, 6:326.
Keller, H.U., Wissler, J.H., and Damarau, B., 1981, Diverging effects of chemotactic serum peptides and synthetic f-Met-Leu-Phe on neutrophil locomotion and adhesion, Immunology, 42:379.
Kitagawa, S., Takaku, F., and Sakamoto, S., 1979, Possible involvement of proteases in superoxide production by human polymorphonuclear leukocytes, Febs Letters, 99:275.
Kolb, G., Köppler, H., Gramse, M., and Haveman, I., 1982, Cleavage of IgG by elastase-like protease (ELP) of human polymorphonuclear leukocytes (PMNL): isolation and characterization of Fab and Fc frag-

ments and low molecular weight peptides stimulation of granulocyte function by ELP-derived Fab and Fc fragments, Immunobiol., 161:507.

Korchak, H.M., and Weissmann, G., 1980, Stimulus response coupling in the human neutrophil. Transmembrane potential and the role of extracellular Na^+, Biochem. Biophys. Acta, 601:180.

Krueger, J.M., Walter, J., Karnovsky, M.L., Chediak, L., Choay, J.P., Lefrancier, P., and Lederer, E., 1984, Muramyl peptides. Vatiation of somnogenic activity with structur, J. Exp. Med., 159:68.

Lewis, J.W. Shavit, Y., Terman, G.W., and Gale, R.P., 1984, Stress and morphine affect survival of rats challenged with a mamary ascites tumor (MAT 31762B), Nat. Immun. Cell. Growth Regul., 3:43.

Lopkor, A., Abood, L.G., Hass, W., and Lionetti, F.J., 1980, Stereoselective muscarinic acethylcholine and opiate receptors on human phagocytic leukocytes, Biochem.Pharmacol., 29:1361

Lowrie, D.B., Jackett, P.S., Aber, V.K., and Carol, M.E.W., 1980, Cyclic nucleotides and phagosome-lysosome fusion in mouse peritoneal macrophages, In: "Mononuclear Phagocytes," R. van Furth, ed., Martinus Nijhoff, The Hague, Boston, London.

Lynn, W.S., and Mohapatra, N., 1980, Control of leukocyte functions. Role of internal H^+ concentration and a membrane-bound esterase, Inflammation, 4:329.

Mathews, P.M., Froelich, C.J., Sibbit, W.L., and Bankhurst, A.D., 1983, Enhancement of natural cytotoxicity by β endorphin, J. Immunol., 130:1658.

McPah, L.C., and Snyderman, R., 1983, Activation of the respiratory burst enzyme in human polymorphonuclear leukocytes by chemoattractants and other soluble stimuli, J. Clin. Invest., 72:192.

Medgyesi, G.A., Foris, G., Dezso, B., Gergely, J., and Bazin, H., 1980, Fc receptors of rat peritoneal macrophages: immunoglobulin class specificity and sensitivity to drugs affecting the microfilament and microtubule system, Immunology, 40:317.

Medgyesi, G.A., Miklos, K., Kulich, J., Fust, G., Gergely, J., and Bazin, H., 1981, Classes and subclasses of rat antibodies: reaction with the antigen and interaction of the complex with the complement system, Immunology, 43:171.

Medgyesi, G.A., Foris, G., Fust, G., and Bazin, H., 1984, Regulation of Fc receptor mediated functions of resident and provoked peritoneal macrophages, Immunobiol., 167:293.

Merishi, J.N., and Mills, I.H., 1983, Opiate receptors on lymphocytes and platelets in man, Clin. Immunol. Immunopathol., 27:240.

Miller, G.C., Murgo, A.J., and Plotnikoff, N.P., 1983, Enkephalins - enhancement of active T-cell rosettes from lymphoma patients, Clin. Immunol. Immunophathol., 26:446.

Miller, G.C., Murgo, A.J., and Plotnikoff, N.P., 1984, Enkephalins - enhancement of active T-cell rosettes from normal volunteers, Clin. Immunol. Immunopathol. 31:132.

Moore, T.C., 1984, Modification of lymphocyte traffic by vasoactive neurotransmitter substances, Immunology, 52:511.

Najjar, V.A., and Nishioka, K., 1970, A natural phagocytosis stimulating peptide, Nature, 228:672.

Oliver, J.M., 1978, Cell biology of leukocyte abnormalities-membrane and cytoskeletal function in normal and defective cells, Am. J. Pathol., 93:221.

Oliver, J.M., and Zurier, R.B., 1976, Correction of characteristic abnormalities of microtubule function and granule morphology in Chediak-Higashi syndrome with cholinerg agonist, J. Clin. Invest., 57:1239.

Onazaki, K., Takenawa, T., Homma, Y., and Hashimoto, T., 1983, The mechanism of macrophage activation induced by Ca^{2+} ionophore, Cell Immunol., 72:242.

Ogmundsdottir, H.M., and Weir, D.M., 1980, Mechanism of macrophage activation, Clin. Exp. Immunol., 40:223.

Pant, H.C., Gallant, P.E., Gould, R., and Gainer, H., 1982, Distribution of calcium-activated protease activity and endogenous substrates in the squid nervous system, J. Neurosci., 2:1587.

Payan, D.G., Hes, C.A., and Goetzl, E.J., 1984, Inhibition by somatostain of the proliferation of T-lymphocytes and Molt-r lymphoblasts, Cell. Immunol., 84:433.

Plotnikoff, N.P., and Miller, G.C., 1983, Enkephalins as immunomodulators, Int. J. Immunopharmac., 5:437.

Plotnikoff, N.P., Murgo, A.J., and Faith, R.E., 1984, Neuroimmunomoduation with enkephalins: effects on thymus and spleen weights in mice, Clin. Immunol. Immunopathol., 32:52.

Sheagran, J.N., and Tuazon, C.U., 1977, Immunological aspects, In: "Drug Abuse. Clinical and Basic Aspects," S.N. Pradhan and S.N. Dutta, eds., C.V. Mosby Co., St. Louis.

Smith, R.J., Bowman, B.J., and Iden, S.S., 1981, Effects of trifluoperazine on human neutrophil function, Immunology, 44:677.

Spada, A., Vallar, L., and Giannatasio, G., 1984, Presence of an adenylate cyclase dually regulated by somatostain and human pancreatic growth hormone GH-releasing factor in GH-secreting cells, Endocrinology, 115:1203.

Stabinsky, Y., Bar-Shavit, Z., Fridkin, M., and Goldman, R., 1980, On the mechanism of action of the phagocytosis-stimulating peptide, tuftsin, Mol. Cell. Biochem., 30:71.

Stolz, V., 1981, Stimulatory effect of latex and zymosan particles on cAMP content in human granulocytes, Mol. Immunol., 18:773.

Tanabe, A., Kobayashi, Y., and Usui, T., 1983, Enhancement of human neutrophil oxygen consumption by chemotactic factors, Experientia, 39:604.

Thomas, D.W., and Hoffman, M.D., 1984, Identification of macrophage receptors for angiotensin: a potential role in antigen uptake for T-lymphocyte responses, J. Immunol., 132:2807.

Vogel, S.N., Weedon, L.L., Oppenheim, J.J., and Rosenstreich, D.Z., 1981, Defecive Fc-mediated phagocytosis in C3HeJ macrophages II. Correction by cAMP agonists, J. Immunol., 126:441.

Wahl, S.M., Wahl, L.M., McCarthy, J.B., Chedid, L., and Mergehagen, S.E., 1979, Macrophage activation by mycobacterial water soluble compounds and synthetic muramyl dipeptides, J. Immunol., 122:2226.

Weber, R.J., and Pert, C.B., 1984, Opiaterg modulation of the immune system, In: "Central and peripheral endorphins: basic and clinical aspects," E.E. Müller and A.R. Genezzani, eds., Raven Press, New York.

Weinstock, J.V., and Kassab, J. T., 1984, Angiotensin II stimulation of granuloma macrophage phagocytosis and actin polymerization in murine Schistosomiasis mansoni, Cell Immunol., 89:46.

Wilkinson, P.C., 1979, Synthetic peptide chemotactic factors for neutrophils: the range of active peptides, their efficacy, inhibitory and susceptibility of the cellular response to enzymes and bacterial toxins, Immunology, 36:579.

Woodcock, E.A., and Johnston, C.I., 1982, Inhibition of adenylate cyclase by angiotensin II in rat renal cortex, Endocrinology, 111:1687.

Yamamura, M., Boler, J., and Valdimarsson, H., 1976, A ^{51} chromium release assay for phagocytic killing of Candida albicans, J. Immunol. Meths., 13:227.

Zagon, I.S., and Mc Laughlin, P.J., 1981, Heroin prolongs survival time and retards tumor growth in mice with neuroblastoma, Brain Res. Bull., 7:25.

Zechner, R., Dieplinger, H., Roscher, A., and Kostner, G.M., 1984, The low density lipoprote in pathway of native and chemically modified low density lipoprotein isolated from plasma incubated in vitro, Biochem. J., 224:569.

HORMONAL MODULATION OF THE HUMAN AUTOLOGOUS MIXED LYMPHOCYTE REACTION

Georges J.M. Maestroni

Laboratory of Cellular Pathology
Istituto cantonale di patologia
6604 Locarno, Switzerland

INTRODUCTION

"Mens sana in corpore sano". This roman proverb expresses beautifully the ancient recognition of the organism unity and it implies awareness of mutual influences between mind and body. As a consequence, it should not be surprising to find, in our analytical era, specific connections between the major homestatic device of the mind - the neuroendocrine system- and the major homestatic device of the body - the immune system. In fact, several years details of these connections are increasingly piled to give a complicated and still confusing picture. A rather exhaustive review of our knowledge of the field may be found in the recently published book "Psychoneuroimmunology" (R. Ader, ed., 1981). Most recently however, the field of neuroimmunomodulation has become fashionable and the list of neurohormones known to have important immunomodulatory properties has increased (Johnson et al., 1982; Mathews et al., 1983; Plotnikoff and Miller, 1983; Miller et al., 1984; Faith et al and Murgo et al., 1984). On the other hand, the already proposed view of lymphocytes producing messengers for the neuroendocrine system (Pierpaoli and Maestroni, 1977; Maestroni and Pierpaoli, 1981) has been further confirmed and detailed (Smith and Blalock, 1981; Smith et al., 1983). The mechanisms by which both the neuroendocrine and the immune system synergizes in order to preserve the organism from environment challenges are thus becoming slowly clearer. The so called psychosomatic integrity appears as a finely tuned balance of neuroendocrine and immune interactions. It becomes also more conceivable that a deranged or exaggerated psychoneuroendocrine response to environmental stimuli or a continuous exposure to stress events may lead to pathological alterations of the immune system.

Concrete examples of such situations might be those autoimmune diseases whose etiology clearly involves hormonal factors such as rheumatoid arthritis (RA) and systemic lupus erythematosis (SLE). As a matter of fact, most connective tissue autoimmune diseases are far more frequent in females than in males (Theophilopoulos and Dixon, 1982). For example, the incidence of SLE in sexually mature women is nine times that in men (Dubois, 1974).

The reasons for such differences are not clear but experimental and clinical evidence incriminate, at least in part, female hormones (Yocum et al., 1975; Lahita et al., 1970; Steinberg et al., 1979). Pregnancy and oral contraceptives are known to affect onset and development of autoim-

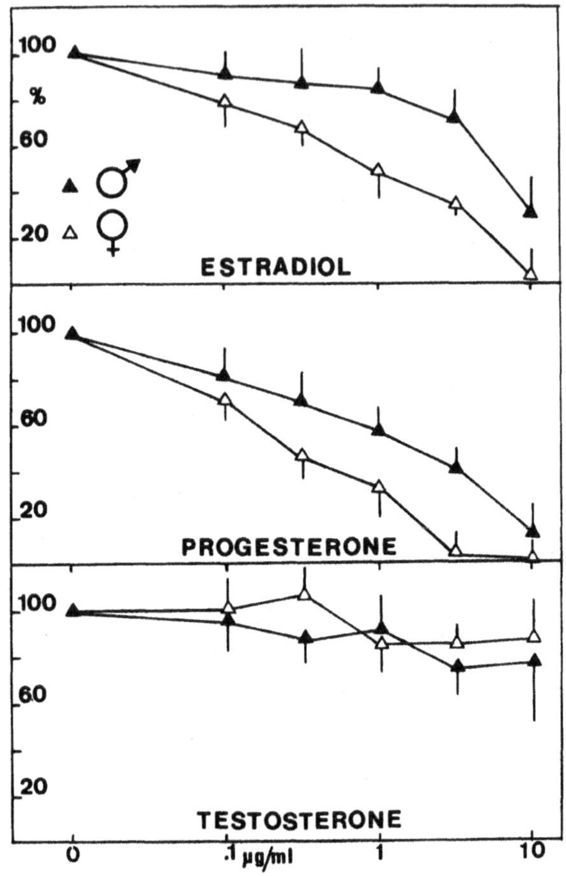

Fig. 1. Sex steroid effect on the human AMLR- Human blood cells were obtained from buffy coats of normal, healthy male and female blood donors. Mononuclear cells were separated by Ficoll centrifugation and subsequently layered on discontinuous Percoll gradients for the separation of T cells from non-T cells according to the method of Gutierrez et al. (1979). Two separate fractions of T cells are obtained by this method. The cultures were set up by mixing the non-T cell fraction with the second T cell fraction.

One hundred μl of medium containing 10^5 non-T cells were added to the same volume and to the same number of T lymphocytes in round bottomed microplates. In order to be closer to a possible in vivo situation, non-T cells were not irradiated. On the other hand, irradiation was found not to affect the AMLR as reported also by Katz and Fauci (1979). Culture medium was RPMI 1640 supplemented with 10% pooled serum from untransfused healthy men, 2mM L-glutamine and 50 μl/ml gentamycin. Each culture was set up in quintuplicate. The cells were incubated for 6 days at 37 C in a 5% CO_2 atmosphere before pulsing with 1 μCi/well. Twenty hours later the cells were harvested and the activity measured by liquid scintillation. The steroid hormones were dissolved in 100% ethanol and then diluted in culture medium. The volume added was 20 μl/well for any final concentration tested. At the concentrations used, ethanol alone did not affect the AMLR. The values represent the mean ± standard mune diseases. For instance, RA is often ameliorated by pregnancy and oral contraceptives seem to lower significantly its incidence (Persellin, 1977; Vandenbroucke, 1982). Moreover, inability to cope with stressful stimuli has been listed amongst personality features of RA patients (Solomon, 1981). A sex specific difference in the neuroendocrine response to stress has also been shown (Frankenhauser et al., 1978).

On the other hand, amongst the various immunological abnormalities found in patients suffering from autoimmune diseases, a most interesting alteration of the in vitro T cell reaction in response to autologous antigens has been found to correlate with active phases of the disease. This in vitro T lymphocyte reaction is called autologous mixed lymphocyte reaction (AMLR) and has been found decreased in most autoimmune diseases as well as in other pathological situations including cancer (Weksler et al., 1981; Smith and deHoratius, 1982). The AMLR is the in vitro proliferation of T lymphocytes in response to autologous B lymphocytes and monocytes. It has been found in both experimental animals and man. The stimulation is mediated by HLA-DR or Ia glycoproteins and the response displays memory and specificity (Opelz et al., 1975; Weksler et al., 1981). Helper, suppressor and cytotoxic functions may be generated during the AMLR (Opelz et al., 1975; Miller and Kaplan, 1978; Smith and Knowlton, 1979; Weksler et al., 1981; Kotani et al., 1984). Thus the AMLR may represent basic immunoregulatory circuits taking place also in vivo. Helper and suppressor mechanisms may regulate humoral responses while cytotoxicity may serve a surveillance mechanism.

In this context we investigated the effect of steroid, pituitary and hypothalamic hormones on the human AMLR.

EFFECT OF SEX STEROID HORMONES

All the autologous cultures (AMLR) used for the experiments described in this chapter were set up by mixing peripheral blood T lymphocytes with autologous non-T cells isolated from buffy-coats of normal healthy female and male blood donors (see legend to fig. 1).

Here, different concentrations of progesterone, beta-estradiol and testosterone (Sigma Co., St. Louis, USA) were added to the cultures. The results are shown in fig. 1. Pharmacological concentrations of female sex steroid hormones (> 1 µg/ml) produced a significant ($P<0.01$) depression of the AMLR of both male and female cells. Similar effects have been already described in a variety of immunological in vitro models (Mendelsohn et al., 1977; Wyle and Kent, 1977; Herr, 1982). However, here it has to be noted that both progesterone and beta-estradiol depressed significantly ($p < 0.1$ and $p < 0.02$) more the AMLR of female than that of male cells. The inhibition produced by testosterone was weaker and no significant sex difference was apparent.

---

deviation of 3 experiments and are expressed as percentage of the ^{3}H-thymidine incorporation given by the relative controls. The depression of female AMLRs was significantly different from that of male cultures for estradiol ($p < 0.02$) and for progesterone ($p < 0.01$).

EFFECT OF HUMAN POLYPEPTIDE HORMONES

The human polypeptide hormones used in this study were a generous gift from the National Institute of Arthritis, Diabetes and Digestive and Kidney Diseases (NIADDK), Bethesda, Ma, U.S.A., except vasopressin (VP), oxytocin (OT) and somatostatin that were purchased from Sigma, Co., St. Louis, MO. USA. The human pituitary hormones growth hormone (GH), prolactin (PRL), follicle stimulating hormone (FSH), luteinizing hormone (LH), thyroid stimulating hormone (TSH), adrenocorticotropic hormone (ACTH), beta-lipotropin (β -LPH), alpha-endorphin (α -EN), VP, OT, estrogen stimulated neurophysin (ESN), nicotine stimulated neurophysin (NSN) and the hypothalamic tetradecapeptide somatostatin were tested at different concentrations for their effect on the human AMLR.
Table 1 illustrates the results of three experiments. The majority of the hormones tested did not affect the human AMLR, only ESN and NSN showed a decided inhibition. However, in contrast with the other hormones human neurophysins were sent by NIADDK already dissolved and containing 1 % NaN_3 as preservative. Therefore, in order to establish whether the observed inhibition was due to NaN_3 we performed experiments where similar dilutions of a NaN_3 solution were added to the AMLR as controls. NaN_3 is known to be an energy poison and therefore its profound inhibition of the human AMLR was not surprising. On the contrary, the surprising thing was that at certain dilutions the NaN_3-mediated inhibition was counteracted by the presence of ESN but not of NSN (table 2). These puzzling findings suggested continued studies in two directions. The investigation of a possible effect of NaN_3-free ESN was the first one. In order to get rid of NaN_3, ESN was exhaustively dialyzed against sterile phosphate buffer at 4 C and the dialyzate tested in the AMLR. The second approach was to go more deeply investigate the interesting NaN_3-ESN competition. The next paragraphs illustrates the results of these studies.

Human neurophysins: a brief description

Neurophysins are specific proteins of about 100 aminoacids synthetized in association with the well known neurohypophyseal nonapeptides arginine-vasopressin (VP) and oxytocin (OT). These hormone- bound proteins are packed with the hormones in neurosecretory granules and transported along axonal pathways to the site of secretion. Neurophysins are released into the blood stream and it is postulated that their secretion always accompanies hormone secretion. Two human neurophysins have been identified, nicotine-stimulated-neurophysin (NSN) and estrogen - stimulated-neurophysin (ESN). NSN secretion is associated to VP release while ESN is postulated to appear in plasma at times when would expect OT release (Robinson, 1978 a, b). Stress seems to inhibit ESN and stimulates NSN secretion (Robinson, 1978 a,b). The concentration of plasma ESN increases as response to pharmacological doses of estrogen and to physiological changes in estrogen. In fact, ESN is elevated in plasma of women on oral contraceptives (Robinson, 1974) and during pregnancy (Robinson et al., 1973). In spite of these interesting relationships of ESN with estrogen and of NSN with stressfull stimuli, no physiological function of circulating neurophysins is known.

Table 1. Effect of human polypeptide hormones on the AMLR

HORMONES	CONCENTRATION ng/ml (% of ^{3}H-thymidine incorporation of controls ± SD)											
	10^{-3}	$5x10^{-3}$	10^{-2}	10^{-1}	0.5	1	5	10	50	10^{2}	$5x10^{2}$	10^{3}
GH				103±6	-	90±16	112±7	115±12	95±8			
PRL						106±19	-	101±9	100±4	93±6	120±7	
LH						98±11	-	110±9	87±9	89±11	104±16	
FSH								95±12	106±8	74±20	107±6	83±4
TSH								73±27	118±12	95±8	95±7	114±8
ACTH						68±32	-	78±25	90±36	89±30	78±12	
β-LPH						76±7	-	125±12	81±11	68±20	74±22	
α-EN						94±33	-	76±33	108±12	73±25	89±31	
VP	112±20	92±8	89±16	123±30	98±10							
OT	78±18	102±9	91±4	107±8	102±11							
ESN				96±12	82±20	39±10	8±5	4±5				
NSN				35±7	19±3	5±6	6±2	5±3				
ST						79±20	102±13	97±23	87±20	85±29		

Cultures were set up with cells obtained from healthy blood donors. Hormone concentrations were chosen in order to include and exceed physiological plasma levels. The data represent the mean ± standard deviation (SD) of three experiments and are expressed as percentage of the values of ^{3}H-thymidine incorporation given after 6 days of incubation (see legend of fig. 1) by control cultures in which no hormone was added.

Table 2. Human, estrogen stimulated neurophysin prevents the NaN_3-mediated inhibition of the AMLR

Hormone concentration ng/ml	ESN + NaN_3 % of controls	NSN + NaN_3 % of controls	NaN_3 % of controls	NaN_3 concentration µg/ml
1	116+4	8+2	8+3	100
0.1	118+18	20+9	11+4	10
0.05	174+87	30+13	27+16	5
0.01	103+13	107+6	75+51	1
0.005	94+10	85+17	106+23	0.5

The hormones and NaN_3 were diluted in supplemented medium and the volume added was 20 µl/well for any final concentration tested. The values represent the mean ± standard deviation of three experiments and are expressed as percentage of ^{3}H-thymidine incorporation given by controls.

Evidence that some batches of NIADDK human ESN enhance the AMLR

Concentrations of NaN_3-free ESN higher than that shown in table 2 were tested in the human AMLR. In the series of experiments performed with the first two neurophysins batches sent by the NIADDK we obtained encouraging results showing that ESN alone may affect the AMLR as reported in fig. 2. Physiological concentrations of human ESN significantly ($p < 0.005$) increased T cell proliferation during the AMLR.

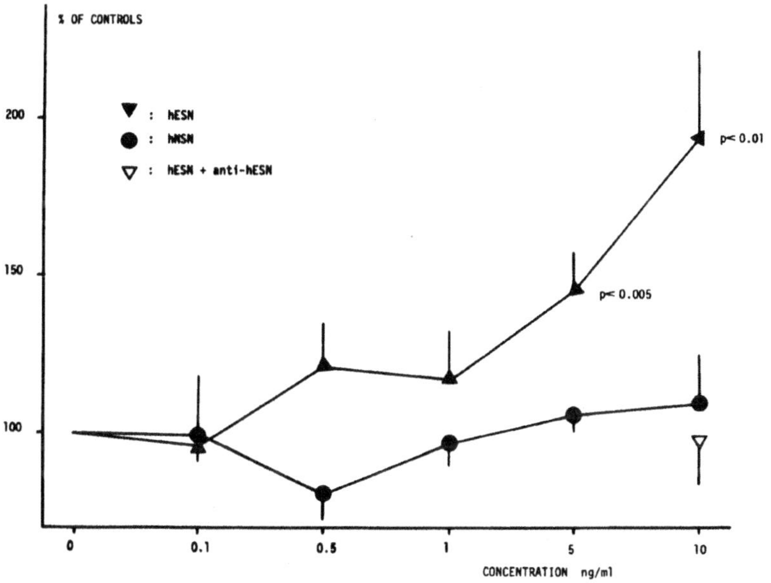

Fig. 2. Effect of ESN and NSN on the human AMLR - The values represent the mean ± standard deviation of 5 experiments and are expressed as percentage of the incorporation of ^{3}H-thymidine given by the relative controls. Cultures were set up with cells obtained from healthy donors. The AMLR was evaluated as described. The final dilution of the anti-ESN antiserum was 1:10,000.

IS THE MEMBRANE ENZYME ATPase (E.C.3.6.1.3) THE TARGET OF HUMAN ESN ?

At concentrations in the range of 10^{-1}-10^{-4}M, NaN_3 has been reported to interfere with a variety of membrane associated phenomena including inhibition of ATPase activity (Kobayashi, 1977) antigenic modulation (Schlesinger and Chaovat, 1972) and capping of surface immunoglobulins (Stackpole et al., 1978). ATPase activity is associated with energy utilization and ion transport. In particular the ATPase mojety showing sensitivity to the glycoside ouabain is known as Na^+, K^+ pump. The activity of this enzyme appears therefore fundamental for most cellular events. From the immunological point of view it seems that membrane receptors and ion transport mechanisms have important roles in lymphocyte activation leading to T-lymphocyte proliferation (DeCoursey et al., 1984). Increased production of interleukin 2 (IL 2) by stimulated normal human lymphocytes has been reported to occur in presence of 0.01 % NaN_3 (Feldman et al., 1983). Ouabain has been reported to inhibit mitogen or alloantigen activated T cells, production and utilization of IL 2 and binding of cytotoxic T cells to their target (Stoek et al., 1983).

By these considerations, the observed ESN-azide competition appears extremely interesting. In order to confirm the phenomenon, normal human peripheral blood mononuclear cells (PBMC) have been stimulated by suitable concentrations of pokeweed mitogen (PwM) in presence of various concentrations of NaN_3 or NaN_3+ESN and at three different concentrations of fetal calf serum (FCS). The results of this experiment are reported in table 3 and show that ESN is capable of reversing the NaN_3-mediated inhibition of B cell polyclonal stimulation. We have no explanation for the apparent correlation between FCS concentration and degree of the NaN_3-mediated inhibition (table 3). This result confirms that human ESN competes with azide at concentrations of the poison known to inhibits ATPase activity, therefore the enzyme ATPase might be, if not the cellular target of ESN, at least closely connected with it. This possibility was investigated by measuring the ouabain-sensitive and insensitive ATPase activity of normal human PBMC after addition of ESN and/or an equal amount of phosphate saline as control. Table 4 illustrates the results of these experiments and shows that ESN seems to interfere with the ouabain-mediated inhibition of the (Na,K) ATPase activity. Alternatively one might infer that ESN selectively stimulates the ouabain insensitive (Mg)ATPase activity while keeping the level of total ATPase constant.

ARE HORMONE-ASSOCIATED HUMAN NEUROPHYSINS THE ACTIVE AGENTS IN NIADDK PREPARATIONS ?

The experiments described above were performed using the first two batches of human neurophysins sent by NIADDK. The investigation was obviously continued in order to establish the modalities of the observed ESN effects on the AMLR. But with our great disappointment, the successive batches of NIADDK human neurophysins were found completely inactive. ESN apparently lost any activity in the AMLR and as azide inhibitor. At this point we reasoned that the activity previously observed might be due to contaminants contained in variable amounts in human neurophysins preparations. In fact, NIADDK human neurophysins are extracted from human neurohypophysis. Furthermore, neurophysins are synthetized in association with VP or OT. Hence, nobody may exclude that small and variable amounts of OT and/or VP "contaminate " some neurophysins preparations and that the activity observed was indeed due to these hormone-associated forms of ESN rather than to ESN alone. Both neurophysins bind OT and VP in vitro (Breslow, 1984). Therefore, different concentrations of OT and/or VP were mixed with a rather high concentration of ESN and NSN because of the low binding affinity at 37 C and physiological pH (Breslow, 1984). The mixtures were added to the AMLRs. The results are shown in fig. 3 and indicate that ESN-OT had the most significant although contrasting effects. OT-associated -ESN showed the ability of modulating human AMLRs. In fact, out of 10 AMLRs, 3 cultures were enhanced, 4 depressed and 3 unaffected by the presence of ESN-OT. Also, 2 cultures were enhanced and 2 depressed by ESN-VP. The ESN-OT dependent positive and negative alterations of the AMLRs were significant with a $p < 0.05$ (100 mg neurophysins / ml) if confronted with the values given by the other hormone combinations (fig. 3).
OT and ESN alone had no effect while anti-ESN antiserum completely abolished the ESN-OT action as shown in fig. 4. Also OT associated NSN never showed significant effects as well as the specific anti-NSN antiserum (Fig. 4). Therefore it seems that the immunomodulatory properties of OT-ESN mainly depended on the neurophysin, or perhaps on the precursor molecule of both OT and ESN. We cannot possibly know whether the stimulation given by the first batches of ESN depended on the presence of this precursor protein.

Table 3. ESN reverses the NaN_3-mediated inhibition of PwM-induced proliferation of peripheral blood mononuclear cells.

	FCS 1%		FCS 5%		FCS 10%		
Hormone concent. ng/ml	ESN+NaN_3 CPM±SD (% of control)	NaN_3	ESN+NaN_3 CPM±SD (% of control)	NaN_3	ESN+NaN_3 CPM±SD (% of control)	NaN_3	NaN_3 concent. µg/ml
0.5	73476±10522 (83%)	65056±2329 (72%)	57561±4073 (58%)	47061±6097 (47%)	49498±5559 (49%)	45449±4754 (45%)	50
0.1	99576±11141 (110%)	5910&±3235 (65%)	69444±10610 (69%)	48303±5535 (48%)	64189±6218 (64%)	36293±2653 (36%)	10
0.05	100495±10085 (11%)	63216±5280 (70%)	89050±5328 (89%)	49266±4449 (49%)	89987±6525 (89%)	41724±7148 (41%)	5
0 (control)	90281±4758		9952±11591		100729±8804		0 (control)

The experiment shown is one of three experiments yielding similar results. FCS: fetal calf serum, CPM: counts per minute, SD: standard deviation.

Table 4. Human, estrogen-stimulated-neurophysin prevents the ouabain-mediated inhibition of the (Na,K) ATPase activity of peripheral blood mononuclear cells

	(Na,K,Mg)ATPase total ATPase	(Mg)ATPase ouabain-insensitive	(Na,K)ATPase ouabain-sensitive
		(nmole / hr / 10^6 cells)	
PBMC + 10 ng/ml ESN	77.79 ± 7.1	76.47 ± 8.9	1.33 ± 1.8
PBMC	72.41 ± 8.6	41.80 ± 1.9	30.71 ± 7.6

(Na,K,Mg) adenosine-triphosphatase (EC.3.6.1.3) and (Mg) adenosine triphosphatase were assayed according to the method described by Losa et al. (1982). The difference between the total and the ouabain-insensitive ATPase activity was considered as the activity of the ouabain-dependent (Na,K)ATPase. A preincubation was performed in duplicate at 37 C for 10 min in a total volume of 0.5 ml. $2x10^6$ human PBMC suspended in 0.2 ml of 0.9% NaCl were added to a mixture containing 1mM EDTA, 25 mM Tris-HCl buffer at pH 7.4 and 0.02% of Na-deoxycholate at this concentration allowed the maximal activity to be reached. A further incubation at 37 C for 30 min was performed after adding to the preincubation mixture 0.5 ml of an aqueous solution containing 10 mM $MgCl_2$, 6mM ATP-Tris, 100 mM NaCl, 40 mM KCl, 2 mM EDTA and 100 mM Tris, adjusted to pH 7.5. The reaction was stopped with 1 ml of ice-cold TCA 10%. After centrifugation, the inorganic phosphate was determined in an aliquot of supernatant by the method of Chen et al. (1956). Enzyme activity was expressed as nanomole of substrate liberated per hour and per 10^6 peripheral blood mononuclear cells (PBMC). The values represent the mean ± SD of three experiments. The hormone diluted in phosphate buffer pH 7.4 was added to cell suspensions 1 hr before initiating the assay. The same volume of phosphate buffer (10 µl) was added to the controls.

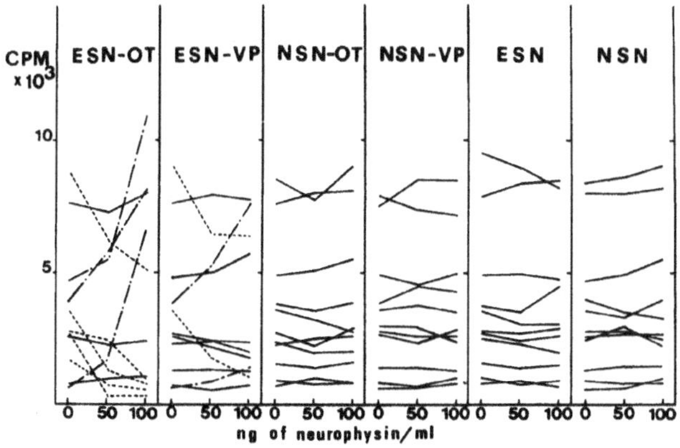

Fig. 3. Oxytocin-associated ESN modulates the human AMLR - Each curve represent the data from one single AMLR performed in presence of different concentrations of neurophysins and of OT and/or VP at a given final concentration of 100 pg/ml. The values are expressed in counts per minute (CPM) relative to the ^{3}H-thymidine incorporation after 6 days of culture. ———— : no change, ·——·——: significant ($p < 0.05$, 100 ng of neurophysin / ml) increase, ---------: significant ($p < 0.02$, 100 ng of neurophysin / ml) decrease.

As a matter of fact, those previous preparations never inhibited AMLRs, but this might depend on the relative small number of experiments performed. Anyhow, OT-associated ESN seems to possess powerful immunomodulatory properties whose sense changes according to unknown individual variables. In the attempt to investigate the nature of these variables, experiments were devised in which human AMLRs were performed with a monocyte depleted non-T cell fraction. Monocytes are known to have various influence on the AMLR. The presence of monocytes has been reported to lower T cell proliferation in the AMLR (Smolen et al., 1981) but, on the other hand, the Ia+ monocytes subpopulation is known to stimulate the AMLR (Weksler et al., 1981). We reasoned that the ESN-OT action might be mediated by monocytes, implying that monocytes would have been the cellular targets of ESN-OT. However, the results of such experiments were negative (data not shown). The presence or absence of monocytes did not change the sense and the extent of the ESN-OT effect on the AMLR.
A recent paper by Kotani et al. (1984) reports that T lymphocytes that had been activated for three days in AMLR helped PwM stimulated immunoglobulin synthesis by autologous B cells. However, cells activated for 6 days preferentially exerted strong suppressor activity. It is thus conceivable that intrinsic individual differences in functions activated in the AMLR may account for the different sense of the ESN-OT mediated modulation. Inhibition of suppressor cell activation and stimulation of helper cell formation would have contrasting effect on proliferation (and thus on ^{3}H-thymidine incorporation as measured in the AMLR) but an identical final

immunoregulatory effect. Experiments to investigate this important point are in progress. The first and simplest attempt consisted in adding the hormone-associated neurophysins to Concanavalin A stimulated PBMC. No effect was apparent in this model (data not shown). However, mitogen stimulated PBMC might be sensitive to higher concentrations of hormone-associated neurophysins. A similar situation is already known: physiological concentrations of corticosteroids almost abolish the AMLR while effect on alloantigen or mitogen stimulated lymphocytes (Ilfeld et al., 1977).

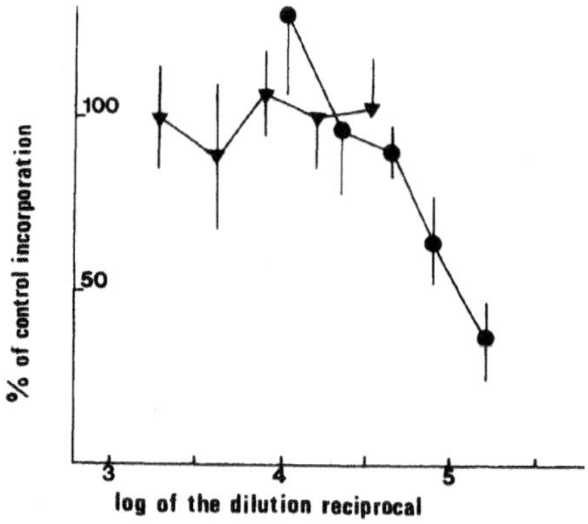

Fig. 4. Anti-ESN antiserum abolish the ESN-OT mediated effect on the AMLR The curves represent the results of one AMLR in which different dilutions of anti-NSN and anti-ESN specific antisera were added to a given concentration of NSN+OT (100 ng/ml + 100 pg/ml) and of ESN+OT (100 ng/ml + 100 pg/ml). The values are expressed as percentage of the control cultures without antiserum and are the mean of the quintuplicate of each dilution tested ± standard deviation. Similar experiments gave identical results. The antisera were a gift from the NIADDK, Bethesda, Ma, U.S.A.
▲: (NSN+OT) + anti-NSN, ●: (ESN+OT) + anti-ESN.

CONCLUSIONS

Effect of steroid hormones

The most interesting observation is the sex difference in the steroid-mediated depression of the AMLR. Although the effect was apparent at rather high concentrations this sex difference seems to reflect a different cellular sensitivity to female steroid hormones. Estrogen receptors have been recently found on human lymphocytes (Danel et al., 1983), however it seems unlikely that hormone receptors are involved in our case because of the high level of the effective steroid concentrations. The major depression of the AMLR might thus express an intrinsic and receptor independent major weakness of female cells toward sex steroid hormones. It has been reported that human monocytes-macrophages produce a factor that protects lymphocytes from the direct toxic effect of corticoid hormones (Blomgren, 1974). As the composition of male and female AMLRs was very similar with respect to the concentration of monocytes, it follows that female monocytes might be less able to produce this factor. A deficient production of this factor would render lymphocytes much more sensitive to the effect of sex steroid hormones. The possibility that such a mechanism might play a role in autoimmunity deserves attention.

In general, depression of the AMLR might mean impaired regulation of immune responses and thus major susceptibility to disturbances such as autoimmunity. As a matter of fact it has been recently reported that 6 days long AMLRs as those performed in this study, preferentially express suppressor functions (Kotani et al., 1984). Depression of suppressor functions means enhanced immune responses and possibly autoimmunity. In this respect, the weaker depression of the AMLR given by testosterone may assume a particular relevance.

Effect of polypeptide hormones

Although interesting, the results of this section appears rather puzzling. First of all, human pituitary hormones previously reported to affect *in vitro* and *in vivo* immune responses as for example, GH, ACTH and endorphins (Ahlqvist, 1981; Solomon and Amkraut, 1981; Johnson et al., 1982; Mathews et al., 1983) had no effect on human AMLR. Then, the variable effects of ESN and of ESN-OT bring several questions to the fore.

We do not know whether the active batches of ESN contained OT or were "contaminated" by other substances or forms of precursor hormone that actually would be the covalent association of ESN with OT. However, ESN maintains a weak *in vitro* affinity for OT (Breslow, 1984). It is so conceivable that ESN and OT might act on their targets in the associated form.

Anyhow, the ESN or ESN-OT mediated activity seems important for a series of reasons:

a) Plasma concentration of ESN is elevated ($>$ 10 ng/ml) in situations such as pregnancy and use of oral contraceptives that are associated with changes in onset and development of autoimmune diseases (Persellin, 1977; Vandenbroucke et al., 1982).

b) Secretion of ESN and/or OT is inhibited by stress. Stress factors are clearly implicated in the etiopathogenesis of autoimmune diseases (Solomon, 1981; Solomon and Amkraut, 1981).

c) A poorly explained sex difference in normal immune responses is well recognized. Generally human and animal females respond significantly better than males to a given antigen.

COMMENTS

The individual variability of the effects shown by ESN-OT might reflect two situations: a basic inconsistency of the model used or real immunomodulatory properties. Further *in vitro* and *in vivo* investigations are clearly needed to answer this question. The ESN-OT affected immune functions, the cellular targets and the suggested ESN-ATPase connection have to be better defined and confirmed.

The confirmation of ESN dependent biological effects might mean more than a mere addition of a polypeptide to the list of neurohormones affecting in some way the immune system. The modulation of a putative immunoregulatory reaction (AMLR) by estrogen stimulated polypeptides might represent a neuroendocrine buffer against variations of the immune performance induced by other factors just as it may happen during the female hormonal cycle.

Chronic disturbance of this buffer system such as those that, for example, stress events may induce, might lead to basic alterations of the immune reactivity including autoimmunity.

ACKNOWLEDGEMENTS

I am grateful to Mrs. E. Hertens for her skillful technical assistance and to Miss T. Bacciarini for typing the manuscript. I thank Dr. A. Conti for the ATPase determination and Dr. D. Castelli with the technical staff of the Centro Trasfusione Sangue della Croce Rossa Svizzera for providing the buffy coats used in this study.

REFERENCES

Ahlqvist, J., 1981, Hormonal influences on immunologic and related phenomena, in: "Psychoneuroimmunology," R. Ader, ed., Academic Press, New York.

Blomgren, H., 1974, Steroid sensitivity of the response of human lymphocytes to phytohemagglutinin and pokeweed mitogen: role of phagocytic cells, Scand. J. Immunol., 3:655.

Breslow, E., 1984, Neurophysin: biology and chemistry of its interactions, in: "Cell biology of the secretory process," M. Cantin, ed., S. Karger, Basel.

Chen, P. S., Toribara, T. Y., and Warner, H., 1956, Microdetermination of phosphorus, Anal. Chem., 28:1756.

Danel, L., Souweine, G., Monier, J.C., and Saez, S., 1983, Specific estrogen binding sites in human lymphoid cells and thymic cells, J. Steroid Biochem., 18:559.

DeCoursey, T. E., Chandy, K. G., Gupta, S., and Cahalan, M. D., 1984, Voltagegated K^+ channels in human T lymphocytes: a role in mitogenesis ?, Nature, 307:465.

Dubois, E. L., 1974, The clinical picture of systemic lupus erythematosus, in: "Lupus erythematosus," 2nd edition, E. L. Dubois, ed., University of Southern California Press, Los Angeles.

Feldman, S. P., Mertelsman, R., Venuta, S., Andreed, M., Welte, K., and Moore, A. S. M., 1983, Sodium azide enhancement of interleukin-2 production, Blood, 61:815.

Frankenhauser, M., Rauste-von Wright, M., Collins, A., von Wright, J., Sedvall, G., and Swahn, C. G., 1978, Sex differences in psychoneuroendocrine reactions to examination stress, Psychosom. Med., 40:334.

Gutierrez, C., Bernabe, R. R., Vega, J., and Kreisler, M., 1979, Purification of human T and B cells by a discontinuous density gradient of percoll, J. Immunol. Meth., 29:57.

Herr, H. W., 1982, Effect of estrogen on the mixed lymphocyte reaction in normal individuals and prostatic cancer patients, The Prostate, 3:17.

Ilfeld, D. N., Krakauer, R. S., and Blaese, M. R., 1977, Suppression of the human autologous mixed lymphocyte reaction by physiologic concentrations of hydrocortisone, J. Immunol., 119:428.

Johnson, H. M., Smith, E. M., Torres, B. A., and Blalock, J. E., 1982, Regulation of the in vitro antibody response by neuroendocrine hormones, Proc. Natl. Acad. Sci. U.S.A., 79:4171.

Katz, P., and Fauci, A. S., 1979, Autologous and allogeneic intercellular interactions: modulation by adherent cells, irradiation, and in vitro and in vivo corticosteroids, J. Immunol., 123:2270.

Kobayashi, H., Maeda, M., and Anraku, Y., 1977, Membrane-bound adenosine triphosphatase of Echerichia Coli. IV effect of sodium azide on the enzyme functions, J. Biochem., 81:1071.

Kotani, H., Takada, S., Ueda, Y., Murakawa, Y., Suzuki, N., and Sakane, T., 1984, Activation of immunoregulatory circuits among OKT4+ cells by autologous mixed lymphocyte reactions, Clin. Exp. Immunol., 56:390.

Lahita, R. G., Bradlow, H. L., Kunkel, H. G., and Fishman, J., 1979, Alterations of estrogen metabolism in systemic lupus erythematosus, Arthritis Rheum., 22:1195.

Losa, G., Morell, A., Barandun, S., 1982, Correlations between enzymatic and immunologic properties of human peripheral blood mononuclear cells, Am. J. Pathol., 107:191.

Maestroni, G. J. M., and Pierpaoli, W., 1981, Pharmacological control of the hormonally mediated immune response, in: "Psychoneuroimmunology," Ader, ed., Academic Press, New York.

Mathews, P. M., Froelich, G. J., Sibbit, W. L., and Bankhurst, D., 1983, Enhancement of natural cytotoxicity by β-endorphin, J. Immunol., 130:1658.

Mendelsohn, J., Multer, M. M., and Bernheim, J. L., 1977, Inhibition of human lymphocyte stimulation by steroid hormones: cytokinetic mechanisms, Clin. Exp. Immunol., 27:127.

Miller, R. A., and Kaplan, H. S., 1978, Generation of cytotoxic lymphocytes in the autologous mixed lymphocyte culture, J. Immunol., 121:2165.

Miller, G. C., Murgo, A. J., and Plotnikoff, N. P., 1984, Enkephalins-Enhancement of active T-cells rosettes from normal volunteers, Clin. Immunol. Immunopath., 31:132.

Opelz, G., Kiuchi, M., Takasugi, M., and Terasaki, P. I., 1975, Autologous stimulation of human lymphocytes subpopulations, J. Exp. Med., 142: 1327.

Persellin, R. H., 1977, The effect of pregnancy on rheumatoid arthritis, Bull. Rheum. Dis., 27:922.

Pierpaoli, W., and Maestroni, G. J. M., 1977, Pharmacological control of the immune response by blockade of the early hormonal changes following antigen injection, Cell. Immunol., 31:355.

Plotnikoff, N. P., and Miller, G. C., 1983, Enkephalins as immunomodulators, Int. J. Immunopharmac., 5:437.

"Psychoneuroimmunology," 1981, R. Ader, ed., Academic Press, New York.

Robinson, A. G., Archer, D. F., and Tolstoi, L. F., 1973, Neurophysin in women during oxytocin-related events, J. Clin. Endocrinol. Met., 37:645.

Robinson, A. G., 1974, Elevation of plasma neurophysin in women on oral contraceptives, J. Clin. Invest., 54:209.

Robinson, A. G., 1978a,Neurophysins, an aid to understanding the structure and function of the neurohypophysis, in: "Frontiers in Neuroendocrinology," Vol. 5, W. F. Ganong and L. Martini, eds., Raven Press, New York.

Robinson, A. G., 1978b, Neurophysins, in: "Clinical Neuroendocrinology," L. Martini, and G. M. Besser, eds., Academic Press, New York.

Schlesinger, M., and Chaovat, M., 1972, Modulation of the H-2 antigenicity on the surface of murine peritoneal cells, Tissue Antigens, 2:427.

Smith, J. B., and Knowlton, R. P., 1979, Activation of suppressor T cells in human autologous mixed-lymphocyte culture, J. Immunol., 123:419.

Smith, E. M., and Blalock, J. E., 1981, Human Lymphocyte production of ACTH and endorphin-like substances: association with leukocytes interferon, Proc. Natl. Acad. Sci. U.S.A., 78:7530.

Smith, J. B., and DeHoratius, R. J., 1982, Deficient autologous mixed lymphocyte reactions correlate with disease activity in systemic lupus erythematosus and rheumatoid arthritis, Clin. Exp. Immunol., 48:155.

Smith, E. M., Phan, M., Coppenhaver, D., Kruger, T. E., and Blalock, J. E., 1983, Human lymphocyte production of immunoreactive thyrotropin, Proc. Natl. Acad. Sci. U.S.A., 80:6010.

Smolen, J. S., Sharrow, S. O., Roeves, J. P., Boegel, W. A., and Steinberg, A. D., 1981, The human autologous mixed lymphocyte reaction I suppression by macrophages and T cells, J. Immunol., 127:1987.

Solomon, G. F., 1981, Emotional and personality factors in the onset and course of autoimmune disease, particularly rheumatoid arthritis, in: "Psychoneuroimmunology," R. Ader, ed., Academic Press, New York.

Solomon, G. F., and Amkraut, A. A., 1981, Psychoneuroimmunological effects on the immune response, Ann. Rev. Microbiol., 35:155.

Stackpole, C. W., Jacobson, J. B., and Smith, M. P., 1978, Two distinct types of capping of surface receptors on mouse lymphoid cells, Nature, 248:232.

Steinberg, A. D., Melez, K. A., Raveche, E. S., Reeves, J. P., Boegel, W. A., Smathers, P. A., Taurog, J. D., Weinlein, L., and Duvic, M., 1979, Approach to the study of the role of sex hormones in autoimmunity, Arthritis Rheum., 22:1170.

Stoeck, M., Northoff, H., and Resh, K., 1983, Inhibition of mitogen-induced lymphocyte proliferation by ouabain interference with interleukin-2 production and interleukin-2 action, J. Immunol., 131:1433.

Theophilopoulos, A. N., and Dixon, F. J., 1982, Autoimmune diseases. Immunopathology and etiopathogenesis, Am. J. Pathol., 108:321.

Vandenbroucke, J. P., Valkenburg, H. A., Boersma, J. W., Cats, A., Festen, J. J. M., Huber-Bruning, O., and Rasker, J. J., 1982, Oral contraceptives and rheumatoid arthritis; further evidence for a preventive effect, The Lancet, ii:839.

Weksler, M. E., Moody, Jr. C. E., and Kozak, R. W., 1981, The autologous mixed-lymphocyte reaction, Adv. Immunol., 31:271.

Wyle, F. A., and Kent, J. R., 1977. Immunosuppression by sex steroid hormones I the effect upon PHA- and PPD-stimulated lymphocytes, Clin. Exp. Immunol., 27:407.

Yocum, M.W., Grossman, J., Waterhouse, C., Abraham, G. N., May, A. G., and Condemi, J. J., 1975, Monozygotic twins discordant for systemic lupus erythematosus, Arthritis Rheum., 18:193.

IMPLICATIONS INVOLVING AUTOIMMUNE MECHANISMS TO THE PCP RECEPTOR IN SCHIZOPHRENIA

H.D. Whitten, K.Y. Tsang, P. Arnaud, N.K. Khansari, and H.H. Fudenberg
Department of Basic and Clinical Immunology, Medical University of South Carolina, 171 Ashley Avenue
Charleston, S.C.

INTRODUCTION

Substantial evidence indicates that autoimmune mechanisms may either accompany or even precipitate some forms of psychoses. Such findings include: 1. Up to 50% of systemic lupus erythematosus (SLE) patients demonstrate some kind of neuropsychiatric disease; the most commom symptom (psychotic alteration in behavior or thinking) is not due to any known pre-existing psychiatric illness.[1] 2. The deposition of immune complexes in the choroid plexus is common in such cases and often has been shown to culminate ultimately in both DNA-anti-DNA antibody complexes in the cerebrospinal fluid (CSF)[2] and lymphocytic perivascular cuffing of small blood vessels in the frontal cortex and hippocampus.[1] Interestingly, although CNS involvement usually develops late in the course of SLE, it often occurs early and can appear as a sporadic disturbance of mental function.[3] Thus, evidence exists supporting the concept of altered blood-brain barrier permeability accompanying immune complex disease.[1,4] In fact, Rudin has suggested that every SLE patient would develop psychosis if other lethal complications were avoided.[5] 3. Anti-brain antibodies are commonly detected in SLE and schizophrenia. Bluestein[6] reported that 78% of SLE sera contain IgM antineuronal antibody and 17% had IgG antineuronal antibody. Antibrain antibodies have been detected in the sera and sometimes the CSF of 48% of schizophrenics by Pandey and co-workers.[7] Others have found 17%, 25%, 21%, 29%, 86%, and 63%[8-14] depending on the sensitivity of the test utilized for antibody detection. One of the most interesting studies came from Luria and Domashneva[15] in which sera (1:32 dilution) of schizophrenics demonstrated a 100% positive immunofluorescence toward mouse thymocytes in contrast to no reactivity from healthy donors. This finding was later confirmed and extended by Goldstein et al.[16]

Thus, it is apparent that the CNS choroid plexus blood-brain barrier can at times be transgressed. Whether it be an emphatic assault with overt lesion as in SLE or of a more subtle nature as in schizophrenia, the "target" antigen to which the immune response is directed may be of paramount importance. The most obvious candidates for immune attack in the psychoses would appear to involve one or more of the neurotransmitter receptors. In this chapter, we chose to emphasize the PCP(phencyclidine,"angel dust")receptor that may be akin or synonymous with the sigma(σ) receptor for several reasons. First, it has been proposed that PCP and adenosine may compete for a subpopulation of receptor sites in that adenosine receptor agonists have been shown capable of blocking the discriminative properties of PCP.[17] Interestingly, adenosine receptors are apparently localized to the axon terminals of the only excitatory type neuron in the cerebellum - the granule cell. The remaining four neuronal types are inhibitory in nature.[18] Second, the PCP receptor itself is a pharmacologically excitatory receptor (vide infra). Third, PCP reportedly induces diverse effects on several neurotransmitter systems.[19] And finally, a small but certain percent of PCP abusers are ultimately diagnosed as chronic schizophrenics. These individuals, although a very circumscribed patient group, may well comprise a unique model system in that symptomology follows the interaction of PCP with specific receptors on cells. This cell membrane sigma receptor site has been well characterized by binding, saturation, and inhibition studies[20-22] and is purportedly the prototype excitatory ligand among the five major pharamacological classes of opiate receptor.[23] The sigma receptor may well be a class of entities instead of a single receptor site at which chemically diverse psychoactive drugs can produce various effects; these include some benzomorphans as well as PCP and some of its analogs. Conceptual support for this resides in the fact that PCP-binding materials include; 1) cell membrane sigma σ receptor sites on brain slices and various neurotumor cell lines, 2) the alpha and delta subunits of the acetylcholine receptor, 3) a ninety-five thousand dalton synaptosome protein involved in potassium channeling, 4) a forty-three thousand dalton protein in the synaptosome of uncertain function and 5) low and high affinity binding sites within the synaptosome complex.[24,25] PCP-binding to any of these receptors might well result in an anti-receptor process, especially in genetically predisposed individuals. Interesting, the development of anti-nuclear antibodies in psychiatric patients chronically treated with chlorpromazine was strikingly increased in the HLA-BW44 type compared to non-BW44 controls.[26] A similar genetic susceptibility to PCP-induced psychoses is an intriguing possibility.

Most contemporary hypotheses concerning neurotransmission dysfunction in schizophrenia emphasize dopaminergic hyperactivity.[27,28] Schizophrenics who respond to anti-psychotic drugs have recently been speculated to possess "autoantibodies" which interact with and stimulate dopamine receptors.[29] However, dopamine receptors appear to be heterogeneous and the increased dopamine (D-2) receptors observed in the schizophrenic brain may be due to neuroleptic drug treatment. This has been experimentally confirmed in rats.[28] Yet, it is probable that these receptors are somehow linked anatomically or physiologically to the cell membrane sigma receptor since chronic PCP administration in vivo leads to a down regulation or diminution of dopamine and acetylcholine as well as PCP receptors.[30]

We have postulated also that some subsets of schizophrenia have an immune basis[31] in that multiple forms of schizophrenia may have deficiencies in receptor number and/or function in common. Human diabetes is the relevant analogy; one type is associated with high levels of antibody against insulin receptors; a second (insulin dependent diabetes) is associated with antibodies to the pancreatic islet cells producing insulin that arises in a genetically predisposed group after cell damage by cytomegalovirus infection; a third, adult-onset noninsulin-dependent diabetes, appears due to deficiency in numbers and/or affinity of insulin receptors. The second type is associated with pancreatic islet cell damage following coxsackievirus infection suggesting that both genetic prediposition (high incidence of the gentically determined histocompatability antigen, HLA-B8), and an environmental triggering agent (the virus) are prerequisites for development of diabetes.

In PCP-induced chronic schizophrenia, PCP might block a receptor which is normally present, so that a naturally occurring PCP-like "orphin" cannot bind to the PCP receptor. Candidates for such an endogenous ligand have been detected in preliminary experiments and tentatively termed "angeldustin".[20,31] Additional forms of chronic schizophrenia may result from other mechanisms including viruses whose "epitopes" (antigenic determinants) are partially similar to the receptor, analogous to the situation in which a selective affinity of rabies virus exists for the acetylcholine receptor.[32]

Other schizophrenia subsets might be analogous to autoantibodies to the receptor in myasthenia gravis (in which patients have antibodies to acetylcholine receptors), and some forms of "autoimmune" thyroiditis, in which patients have antibodies to the receptors for the thyroid stimulating hormone. Thus, we postulate that immunological parameters can be productively applied also to an analysis of the schizophrenic syndrome.

Antigenically distinct immunocytes regulate the immune response by simultaneously helping and suppressing and some share antigens with neural cells but no other body cells. In autoimmune disease a number or functional decrease of suppressor T cells is often accompanied by increased autoantibodies and immune complexes, i.e., agents interfering with suppressor cell physiology. Removal or inactivation of suppressor cells may induce the hypergammaglobulinemia observed in systemic lupus erythematosus (SLE). Suppression can be exerted by T cell and/or monocyte subsets; for instance, suppressor T cells are defective in human SLE[33-36] and certain monocyte subsets appear to suppress B cell function.[37] The decrease in suppressor T cell subsets and/or suppressor monocytes may culminate in increased autoantibodies.

The murine theta antigen is present on both T lymphocytes and brain cells (indeed on synaptosomes); also a monoclonal antibody against human suppressor T cells also binds specifically to oligodendrocytes[38] indicating the sharing of some brain cell surface structures with those of lymphocytes. Receptors for morphine and met-enkephalin have also been detected recently on a human T lymphocyte subset.[39] Also, the neurotransmitter "Substance P"[40] and beta-endorphin[41] have been shown to bind and stimulate proliferation of T lymphocytes and to greatly enhance natural killer cell activity, respectively. The specificity of binding of these ligands to these receptors was implied by the reversibility of binding in the presence of suitable concentrations of naloxone, an antagonist of the ligand drugs. These competition studies suggest that at least some of the drug binding sites on human lymphocytes are specific, identical or cross-reactive with opiate receptors and lend credence to the hypothesis that opioid receptors on lymphoid cells are likely to be similar in their physicochemistry to the brain receptors. Resolution of the enigmas of immunogenetic subgroups in some subsets of schizophrenia, of sigma and/or PCP receptor structure, and of susceptibility of some genetically defined groups to mental disorders may well be clarified upon scrutiny of the impact that these opioids have on particular immunocyte subpopulations.

Pharmacological studies have incriminated PCP as an indirect dopamine agonist in that PCP has been shown to release a pool of intracellular dopamine. Thus, PCP-induced chronic psychoses in humans and stereotypy in animals have been stated to respond to administration of dopamine receptor blockers.[27,29] Therefore, an alteration in the PCP receptor through direct stimulation or blocking effects by antibody binding might lead to symptoms similar or identical to that seen in schizophrenia. In addition, since PCP has been shown to bind and block the K^+ channels of synaptosomes, it may be that autoantibodies to this receptor might also prevent the naturally occurring endogenous ligand from binding to these sites. The knowledge gained from a scrutiny of the human PCP-induced schizophrenia model might be applied to an examination of endogenous ligand-induced autoimmune receptor alteration. Indeed, several investigators have emphasized that of those currently available the PCP-model most closely approximates human schizophrenia.[42,43]

In the transfer of information between nerve cells in schizophrenia, aberrations may well result from the selective blockage of certain potassium (K^+) channels in the brain cells or in the synaptosomes.[24,25] Recent startling and provocative studies indicate that PCP binds to such channels and can be used to photoaffinity label them. This binding of PCP to particular brain ion channels appears to increase the nerve impulse duration, leading to elevated amounts of released neurotransmitter. It is obvious that autoantibodies directed against such PCP-binding ligands could cause topographical rearrangements resulting in clustering and an imbalance in the local molecular environment leading to neurotransmitter release dysfunction.

Plotz[44] recently suggested that autoantibodies are in reality anti-idiotypic antibodies to antiviral antibodies. He suggested that a cell membrane receptor (designed for some particular cell function) is recruited by a virus for its own use and that to the extent that the viral structure is different from the host, it is a potential immunogen. The antiviral antibodies elicited will resemble in structure the host element with which the viral immunogen interacts. If these antiviral antibodies elicit anti-idiotypic antibodies (via heightened helper T cell or defunct suppressor T cell activity) some may be able to interact with the initial host receptor. It is a simple extension to envision that in addition to interacting with the neuronal sigma receptor and/or synaptosome K^+ channels, PCP could possibly have a similar effect on immunocyte populations if they possessed sigma-like receptors. It is interesting in this regard to note that purified B lymphocytes were the lymphocyte subpopulation most heavily labelled with [^{3}H]PCP (Table 2).

To determine if PCP (sigma) receptors were present on human peripheral blood lymphocytes (PBL), the NCB-20 mouse neuroblastoma-chinese hamster brain clonal hybrid cell line known to have sigma receptors was used as a prototype control sigma receptor positive cell line. The saturability of the receptor and the reversibility of binding (cold inhibition) were measured both in the NCB-20 cell line and in peripheral blood lymphocytes. The results of these experiments are shown in Figures 1 and 2. Interestingly, partial inhibition of PCP binding to lymphocytes was accomplished by lymphocyte preincubation with dopamine; this may indicate either the proximity of the dopamine-sigma receptor, partial identity, or an allosteric effect. Saturability and reversibility of PCP binding to the NCB-20 cell sigma receptor is similar to that of peripheral blood lymphocytes, indicating that sigma-like receptors are present on both cells. Therefore, it can be contended that human PBL possess sigma receptors approximating those on NCB-20 cells. Thus, the major difficulty of tissue availability in conducting research within the neuropsychiatric field can be circumvented by looking for opiate receptors on lymphoid tissues. In addition, adenosine receptors have been detected on human peripheral blood T cell and perhaps null cell subpopulations[45]. The activities of these receptors (PCP, dopamine and adenosine) are easily quantifiable with immunological parameters.

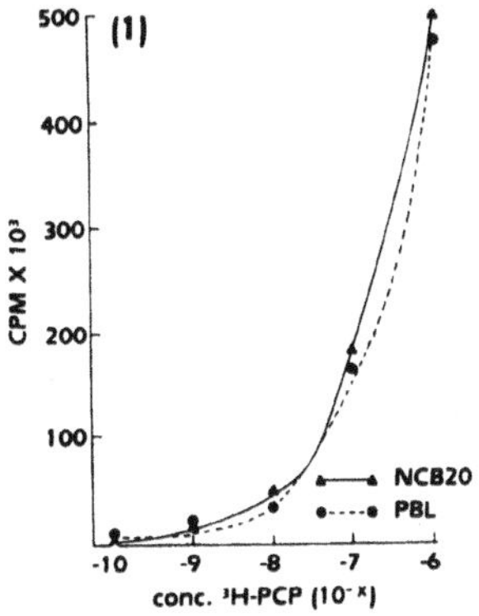

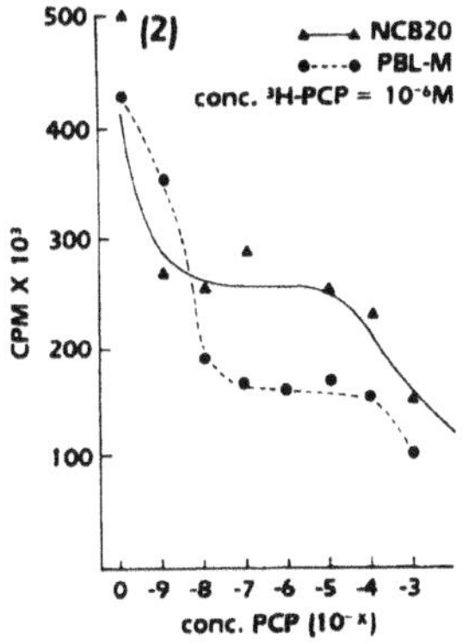

Fig. 1. Binding curve of radiolabelled PCP to peripheral blood lymphocytes and NCB20 neurotumor cells.

Fig. 2. Inhibition of binding of radiolabelled PCP by various concentrations of unlabelled PCP.

In other experiments, preliminary indications hint the existence of PCP-binding materials in both NCB-20 supernatants and normal human serum. When concentrated (100X) and incubated with [^{3}H] PCP, the following percentages (Table 1) of radioactive counts were found to be trichloroacetic acid (TCA) precipitable:

Table 1

Source (100X)	Incubation time (hrs)	Temperature (°C)	% counts ppt.
NCB-20 supernate	6	4	23
NCB-20 supernate	18	37	56
Human serum	6	4	35
Human serum	18	37	40
RPMI-1640 Media Control	6 and 18	4 and 37	0.2%

These above findings indicate that the sigma receptor positive NCB-20 cell line may be releasing their membrane PCP-receptors into the supernatant and can be compared with the results of Bailey and Zettner[46] using equilibrium dialysis; these investigators found that PCP can bind to some undefined component in normal human serum (more than 50% bound at 4°C). This binding apparently was not to α-globulins, γ-globulins (or albumin?) at the physiological concentrations of these proteins. These PCP-binding entities may have been circulating PCP receptors per <u>se</u> and the fact that γ-globulin containing immunoglobulin showed no binding does not negate the possibility that anti-PCP antibodies would be detected had the human serum been obtained from drug abusers and/or those with chronic PCP-induced psychosis. Further investigations in this area are underway in our laboratory.

Of further interest is our preliminary findings indicating that [^{3}H] PCP binds selectively to certain subpopulations of human T lymphocytes (non-autologous rosette forming cells, non-ARFC) and to a greater extent to B cells (Table 2). The binding profiles of these "autologous" rosette forming T and B cells suggest that PCP can indeed interact with unknown consequences with certain immunocompetent cell subpopulations.

Table 2

$[^3H]$ PCP Binding* to Human Lymphocytes

Cells	CPM
ARFC-T	5,040 cpm
non-ARFC-T	34,731 cpm
B cells	88,671 cpm

*250 X 10^3 cpm total added

[See Khansari, N., Whitten, H.D. and Fudenberg, H.H.: Phencyclidine (PCP)-induced immunodepression. Science 225: 76-78, 1984].

The small, but significant binding of ARFC-T is particularly interesting in that this T cell subpopulation appears to exert suppressor activity.[49] Although comprising a small percentage of peripheral blood lymphocytes, their numbers are greatly increased by Concanavalin A (ConA) activation. Hence the ARFC have been regarded as the ConA-inducible suppressor cell.[47,48] PCP-induced anomalies in their immunoregulatory capacities could lead to autoantibody formation. Possible consequences of PCP-immunocyte interactions are shown in Figure 3.

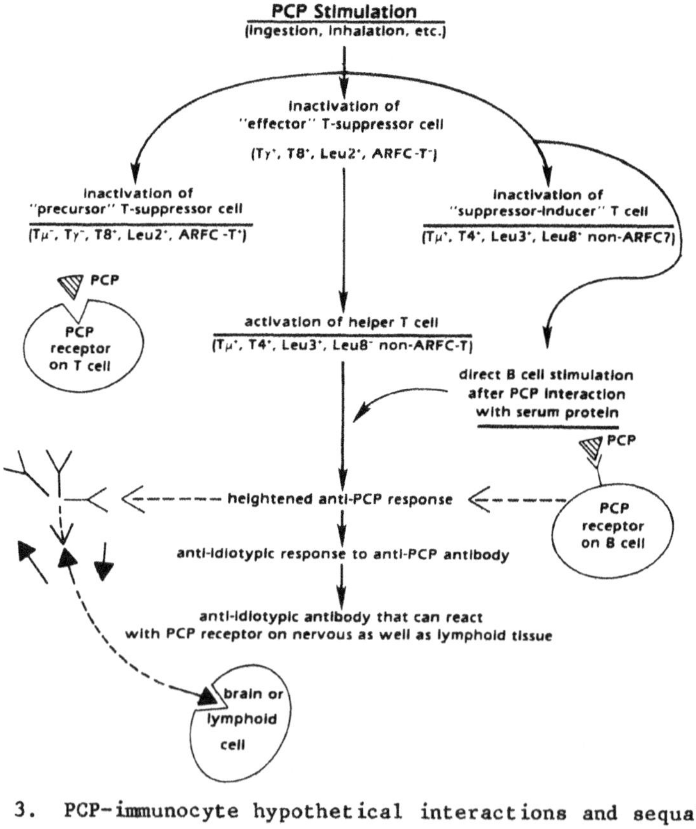

Fig. 3. PCP-immunocyte hypothetical interactions and sequalae.

Also an accelerated $^{86}Rb^{+}$ (K^{+}) release has been noted to be the first physiologic event associated with the lethal hit on target cells delivered by cytotoxic T lymphocytes.[50] If PCP should block lymphocyte K^{+} channels in an analogous fashion to that of synaptosome K^{+} channels (with resultant increased neurotransmitter production), then helper T lymphocytes may well be induced by PCP to form excessive helper factors. This too is a potential mechanism of autoantibody production. Currently studies are underway to clarify these interactions and to investigate the mitogenic activity of PCP on different lymphocyte subpopulations.

REFERENCES

1. D.A.S. Compston, "Neurological Disease", in: Immunology in Medicine Medicine, E.J. Holborow and W. Reeves, ed., Grune & Stratton, New York, N.Y. (1983).

2. R.J. Harbeck, A.A. Hoffman, S.A. Hoffman and D.W. Shucarel, CSF and and the choroid plexus during acute immune complex disease, Clin. Immunol. Immunopathol. 13:413 (1979).

3. R.T. Johnson and E.P. Richardson, The neurological manifestations of SLE, Medicine 47:337 (1968).

4. L.D. Petz, G.C. Sharp, N.R. Cooper and W.S. Irwin, Serum and CSF complement and serum autoantibodies in SLE, Medicine 50:259 (1971).

5. D.O. Rudin, The choroid plexus and system disease in mental illness. II. Systemic lupus erythematosus: a combined transport dysfunction model for schizophrenia, Biol. Psychiat. 16:373 (1981).

6. H.G. Bluestein, Neurocytotoxic antibodies in serum of patients with with systemic lupus erythematosus, Proc. Natl. Acad. Sci. (USA) 75:3965 (1978).

7. R.S. Pandey, A.K. Gupta and U.C. Chaturvedi, Autoimmune model of schizophrenia with special reference to antibrain antibodies, Biol. Psychiat. 16:1123 (1981).

8. R.T. Rubin, Investigation of precipitins to human brain in sera of psychotic patients, Brit. J. Psychiat. 110:84 (1965).

9. N.I. Kuznetoza and S.F. Semenov, Detection of antibrain antibodies in the sera of patients with neuropsychiatric disorders, Zh. Neuropatol. Psikhiatr. 61:869 (1961).

10. M. Yokoyama, E.G. Trams and R.O. Brady, Sphingolipid antibodies in animals and patients with central nervous system lesions, Proc. Soc. Exptl. Med. 111:350 (1962).

11. R.G. Heath and I.M. Krupp, Schizophrenia as an immunologic disorder. I. Demonstration of antibrain globulins by fluorescent antibody techniques, Arch. Gen. Psychiat. 16:1 (1967).

12. M. Stern, R. Anavi and J.P. Witz, Tissue-binding factor in schizophrenic sera: A clinical and genetic study. Biol. Psychiat. 12:199 (1977).

13. R.F. Willikens, R.R. Whitaker, R.V. Anderson and D. Berven, Significance of antinuclear factors in older persons. Ann. Rheumatic Diseases 26:306 (1967).

14. N.I. Kuznetzova, "Some data concerning the immunological properties of brain mitochondria", _in_: "Biological Research in Schizophrenia", M. Vartanian, ed., Ordina Lennia, Moscow (1967).

15. E.A. Luria and I.V. Domashneva, Antibodies to thymocytes in sera of patients with schizophrenia. _Proc. Natl. Acad. Sci. (USA)_ 71:235 (1974).

16. A.L. Goldstein, J. Rossio, G.I. Kolyeskina, L.E. Emory, J.E., Overall, G.B. Thurman and J. Hatcha, "Immunological components in schizophrenia", _in_: "Perspectives in Schizophrenia Research, C. Baxter and T. Melnechuk, eds., Raven Press, N.Y. (1980).

17. R.G. Browne and W.M. Welch, Stereoselective antagonism of phencyclidine's discriminative properties by adenosine receptor agonists, _Science_ 217:1157 (1982).

18. R.R. Goodman, M.J. Kuhar, L. Hester and S.H. Synder, Adenosine receptors: autoradiographic evidence for their location on axon terminals of excitatory neurons, 220:967 (1983).

19. L.L. Hsu, R.C. Smith, C. Rolstein and D.E. Leelavathei, Effects of acute and chronic PCP on neurotransmitter enzyme in rat brain, _Biochem. Pharmacol._ 29:2524 (1980).

20. J.P. Vincent, J. Vignon, B. Kartalouski and C.B. Pert, "Receptor sites for phencyclidine in mammalian brain and peripheral organs", _in_: "PCP: Historical and Current Perspectives", E.F. Domino, ed., Ann Arbor, Michigan (1981).

21. R. Quirion, R.P. Hammer, M. Horkenham and C.B. Pert, Phencyclidine (angel dust/sigma "opiate" receptor): Visualization by tritium-sensitive film, _Proc. Natl. Acad. Sci. U.S.A._ 78:5881 (1981).

22. R.S. Zukin and S.R. Zukin, Multiple opiate receptors: Emerging concepts, _Life Sciences_ 29:2681 (1981).

23. A.R Gintzler, R.S. Zukin and S.R. Zukin, Effects of phencyclidine and its derviatives on enteric neurons, _Brit. J. Pharmacol._ 75:261 (1982).

24. R. Oswald and J.P. Changeux, Ultraviolet light-induced labeling by noncompetitive blocking of the acetylcholine receptor from _Torpedo marmorata_, _Proc. Natl. Acad. Sci. U.S.A._ 78:3925 (1981).

25. E.X. Albuquerque, L.G. Aguayo, J.E. Wannick et al., The behavioral effects of phencyclidines may be due to their blockade of potassium channels, _Proc. Natl. Acad. Sci. U.S.A._ 78:7792 (1981).

26. R.T. Canoso, M.E. Lewis and E.J. Yunis, Association of HLA-Bw44 with chlorpromazine-induced autoantibodies, _Clin. Immunol. Immunopathol._ 25:278 (1982).

27. T.J. Crow, Molecular pathology of schizophrenia: More than one disease process? _Brit. Med. J._ 280:66 (1980).

28. R. Rodnight, Schizophrenia: Some current neurochemical approaches, _J. Neurochem._ 41:12 (1983).

29. J.G. Knight, Dopamine-receptor-stimulating autoantibodies: A possible cause of schizophrenia, _The Lancet_ ii:1073 (1982).

30. R. Quirion, M.A. Bayorh, R.L. Zorlie and C.B. Pert, Chronic phencyclidine treatment decreases phencyclidine and dopamine receptors in rat brain, Pharmacol. Biochem. Behav. 17:699 (1982).

31. H.H. Fudenberg, H.D. Whitten, E. Merler and O. Farmati, Is schizophrenia an anti-receptor disorder? Medical Hypothesis 12:85 (1983).

32. T.L. Lentz, T.E. Burrage, A.L. Smith, J. Crick and G.H. Tignor, Is the acetylcholine receptor a rabies-virus receptor? Science 215:182 (1982).

33. M.F. Hamilton and J.B. Winfield, T gamma cells in systemic lupus erythematosus, Arth. Rheumatol. 22:1 (1979).

34. C. Morimoto, Loss of suppressor T-lymphocyte function in patients with systemic lupus erythematosus, Clin. Exp. Immunol. 32:125 (1978).

35. A.S. Fauci, A.D. Steinberg, B.F. Haynes and G. Whalen, Immunoregulatory aberrations in systemic lupus erythematosus, J. Immunol. 121:1473 (1978).

36. A. Sagawa and N.T. Abdou, Suppressor-cell dysfunction in systemic lupus erythematosus, J. Clin. Invest. 62:789 (1978).

37. F. Gmelig-Meyling, and T.A. Waldman, Human B cell activation *in vitro*: Augmentation and suppression by monocytes of the immunoglobulin production induced by various B cell stimulants, J. Immunol. 126:529 (1981).

38. J. Oger, J. Szudhet, J. Antel, B.G.W. Arnason, A monoclonal antibody against human T suppressor lymphocytes binds specifically to the surface of cultured oligoendrocytes, Nature 295:66 (1982).

39. J. Wybran, T. Appelboom, J.P. Famasy and A. Govaerts, Suggestive evidence for receptors for morphine or methionine-enkephalin in normal human blood lymphocytes, J. Immunol. 123:1068 (1983).

40. D.G. Payan, D.R. Brewster and E.F. Goetzl, Specific stimulation of human T lymphocytes by Substance P, J. Immunol. 131:1613 (1983).

41. P.M. Matthews, C.J. Froehlick, W.L. Sibbitt and A.D. Edmonds, Enhancement of natural cytotoxicity by beta-endorphin, J. Immunol. 131:1658 (1983).

42. E.D. Luby, Phencyclidine revisited: PCP research and functional psychoses, Psychopharmacol. Bull. 16:85 (1980).

43. S.H. Synder, Phencyclidine, Nature 285:355 (1980).

44. P.H. Plotz, Autoantibodies are anti-idiotype antibodies to antiviral antibodies, Lancet 290:824 (1983).

45. C. Moroz, H. Bessler, M. Djaldetti and R.H. Stevens, Human adenosine receptor bearing lymphocytes: enumeration, characterization and distribution in peripheral blood lymphocytes, Clin. Immunol. Immunopathol. 18:47 (1981).

46. D.N. Bailey and A. Zettner, PCP binding in human serum, Res. Commun. Subst. Abuse 3:237 (1982).

47. R. Palacios, Role of autologous rosette-forming T cells in the concanavalin A-induced supppressor cell functions, Cell. Immunol. 61:273 (1981).

48. T. Sakane, M. Honda, Y. Taniguchi and H. Kotani, Separation of concanavalin A-induced human suppressor and helper T cells by the autologous erythrocyte rosette technique, J. Clin. Invest. 68:447 (1981).

49. N. Khansari, M. Petrini, F. Ambrogi et al., Role of autorosette forming cells in antibody synthesis in vitro: suppressive activity of ARFC in humoral immune response, Immunobiology 166:1 (1984).

50. J.H. Russell and C.B. Dobos, Accelerated $^{86}Rb^+$ (K^+) release from the cytotoxic T lymphocyte is a physiologic event associated with delivery of the lethal hit, J. Immunol. 131:1138 (1983).

EFFECTS OF MORPHINE OPIATES ON IMMUNE FUNCTION

Hikmet Koyuncuoğlu and Mehmet Güngör

Department of Pharmacology and Clinical Pharmacology
Istanbul Medical Faculty
Çapa, Istanbul

The general anesthesia and surgery on the morphine opiate addicts, in which total blood volume and blood cell mass had been found to be less than normal, were more problematic and complicated than on the patients and prisoners in similar health conditions (Williams and Oberst, 1946; Isbell, 1947; Eiseman et al., 1964). There is substantial evidence that opiate addicts suffer from a mortality rate considerably higher that what would be expected in relation to their age. The follow-up studies of opiate addicted persons discharged from the treatment institutions reveal that a marked proportion died shortly after discharge although they were mostly young adults. Among the causes of death found in postmortem examinations the major medical complications of opiate addiction mostly related to the use and administration routes such as abscesses, cellulitis, endocarditis, hepatitis, pneumonia, septic pulmonary embolism, tetanus, thrombophlebitis etc. were called up (Tolentino et al., 1961; Cherubin, 1967; Louria et al., 1967; Cherubin et al., 1968; Espiritu and Medina, 1980; Tarr, 1980; Sequeira et al., 1982) ignoring the fact that opiate addicts are highly prone to infectious diseases and death (Cherubin, 1967).

It has long been known (Clouet and Ratner, 1967; 1968) that either a single dose of morphine or its long-term administration decrease the rate of removal of the labeled leucine from the free amino acid pool of brain and the rate of incorporation of labeled leucine into protein. In HeLa cells, morphine congeners, levorphanol and levallorphan, also inhibit protein synthesis (Noteboom and Müller, 1966 a; b). The mechanism underlying this inhibition was then reported to be the inhibitory effects of morphine and its surrogates on uridine and thymidine incorporating abilities, on RNA polymerase activity and on aminoacyl-tRNA synthethase (Datta and Antopol, 1972; 1973).

Postpartum studies of the hamster offspring that were exposed to morphine opiates <u>in utero</u> indicate deficiencies in growth rates and possible increased susceptibility to infections (Geber and Schramm, 1974). Additionally morphine and related opiates, administered to the pregnant hamster during the critical period of organogenesis, have been reported to be capable of altering the embryonic developmental pathways to such an extent as to produce a variety of important anatomical malformations especially in the central nervous system. When they were given to the weanling hamster offspring they have been demonstrated to be powerful inhibitors of the ability of the tissue cells to rise the serum titer of interferon, an integral component of the host-defense mechanism (Geber and Schramm, 1975). Morphine opiates exert the effects similar to those of immunosuppressive agents not only during the gestation (Harpel and Gautieri, 1968; Dawis and Lin,1972; Buchenauer, 1974) but before mating as well. Treatment of male or female rats with morphine opiates prior to mating reduces neonatal viability and increases the number of stillborns (Friedler and Cochin, 1972; Smith and Joffe,1975). The antagonism by opiate antagonists of growth retardation brought about by morphine opiates (Smith et al., 1977) and the interesting experimental findings of Zagon and McLaughlin, who have previously showed the inhibitory effect of morphine on growth rate and the increasing effect of morphine on the percentage of stillborns and infant mortality (Zagon and McLaughlin, 1977 a; b) regarding the increase in brain size and brain cellular content in infant rats treated with naltrexone, an opiate antagonist (Zagon and McLaughlin, 1983) can be considered as experimental results supporting the idea that the administration and/or free intake of morphine opiates, and the normal or excessive release of their endogenous ligands may play a substantially important role in the development of immunity and immune system, and in the acquisition of immune response.

In spite of the suggestive experimental findings and clinical observations some of which have been mentioned above, very few experimental and clinical studies concerning the effects of morphine opiates on the immune system have been reported. In one of them (Güngör et al., 1980) chronic morphine administration has been demonstrated to inhibit the primary immune response in mice, and an impaired cellular immunity, a markedly decreased ratio of T helper/inducer cells to T suppressor/cytotoxic cells and associated viral infections in drug abusers have been reported in the other (Small et al., 1983). It has also been shown that morphine markedly inhibits the in vitro mitogenic response of human lymphocytes unstimulated as well as stimulated by phytohaemagglutinin (Brown et al., 1974; Zagon and McLaughlin, 1977).

Furthermore, opiate addicts were found to have a considerable reduction in the total number of T-lymphocytes as investigated by the ability of the lymphocytes to form rosettes with sheep erythrocytes (McDonough et al., 1980). The results of the experiments showing that in vitro treatment of peripheral blood lymphocytes obtained from normal human subjects with morphine decreases the ability of the lymphocytes to form rosettes with sheep erythrocytes probably through the presence of opioid receptors on lymphocytes (Wybran et al., 1979), the inhibition of natural killer cytotoxicity (Shavit et al., 1984) and the decrease of antibody dependent cellular cytotoxicity (Morgan et al., 1984) provide additional support for the idea that opiates alter immune system.

Even though none of the productive parts of the mammalian immune system are located in the central nervous system the immune system and immune response are greatly regulated by the central nervous system via the secretion of different hormones and/or the release of the neurotransmitters from the autonomic nervous system innervating the organs and tissues which provide the essential humoral and cellular elements of the immunity against microorganisms and degenerated cells giving access to tumor development. Thus, morphine opiates which cause many changes in the functions and hormonal status of the brain cannot be thought ineffective on the immune system and the preparation of immune response. Because of the very limited number of the experimental works concerning the direct or indirect effects of morphine opiates on the immune system we only have to overview the nearly well-established central actions of morphine opiates which can influence the immune system and response to give informations about the possible actions of morphine opiates on the immune system and immune response. Nevertheless, some direct effects of morphine opiates, unrelated to the functioning of the cent. ner. sys. which may be regarded to be rather peripheral, will be added to others. On the other hand, the actions of the chronic administration or intake of morphine opiates will be emphasized even though their single dose also can, rather for a very short time, alter the immune system and immune response through the effect on the release of acetylcholine, dopamine, noradrenaline, adrenaline, 5-hydroxytryptamine and the secretion of some hormones (Vogt, 1954; Beleslin and Polak, 1965; Akera and Brody, 1968; Inturrisi and Fujimoto, 1968; Jhamandas et al., 1971; Fennessy and Lee, 1972; Fukui and Takagi, 1972). The small percentage of morphine opiates used as a single dose especially for medical purposes in the huge amount of opiates consumed largely for non-medical aims is the other reason to take this decision.

The brain regulation and modulation of the immune system and immune response are possibly mediated by direct secretion of brain peptides into the general circulation and/or by the release of the neurotransmitters of the autonomic nervous system. Not only the relationship between the secretion of the peptides and the peripheral release of the neurotransmitters has to be taken into consideration but the modulation of the secretion of the peptides by the neurotransmitters within the central nervous system and viceversa has to be emphatically pointed out. Among the neurotransmitters and peptides by which the brain might modulate immune functions and processes acetylcholine, dopamine, noradrenaline, adrenaline, 5-hydroxytrytamine growth hormone, vasopressin, corticotropin, thyrotropin stimulating hormone, follicule stimulating hormone and luteinizing hormone have mostly been subject to the studies of the effects of morphine opiates.

In addition to the decreasing effect of single dose morphine opiates on the release of acetylcholine from brain neurons (Beleslin and Polak, 1965; Jhamandas et al., 1971) Maynert (1967) observed an increase in brain levels of acetylcholine, presumably due to the inhibition of acetylcholine release, in chronically morphinized rats. It has similarly been found that striatal acetylcholine and acetylcholinesterase were increased after chronic administration of morphine (Merali et al., 1974). On the basis of these central effects of morphine opiates and their effects on gastrointestinal tract it has almost generally been accepted that morphine opiates decrease the release of acetylcholine from the cholinergic nerve endings. The decrease of acetylcholine release may influence immune system as acetylcholine stimulates cyclic guanosine monophosphate which in turn has been demonstrated to activate B-cell as well as T-cell responses (Bourne et al., 1974; Goldberg et al., 1974). Furthermore, several investigators have reported the existence of acetylcholine receptors in thymus tissue (Eremina and Devoino, 1973; Lindstrom et al., 1976). Singh (1979) showed that thymocytes also contain cholinergic receptors and cholinergic stimulation can increase the cell yield from thymus tissue (Sing, 1979). Additionally it has been revealed that bone marrow stem cells bear cholinergic receptors which might be involved in the activation of these precursor cells (Byron, 1975; 1976) and cholinergic innervation might influence the production and/or release of thymic hormones via the stimulation of cholinergic receptors on epithelial cells in the thymus which produce thymic hormones as a consequence of the increase cyclic guanosine monophosphate content (Oosterom et al., 1979; Oosterom and Kater, 1980; Kater et al., 1980).

From the reported changes that occur in the immune system and immune response in the patients with neurologic disorders involving the dopaminergic pathways and dopamine metabolism, as well as experimental manipulation and determination of dopamine in animals it has been concluded that the central dopaminergic system has an evident stimulatory effect on various components of the immune system. The total number of T-lymphocytes as well as their ability to respond to mitogenic stimulation and in T-dependent skin tests were shown to be reduced in the patients with Parkinson's disease in which neuropathology is associated with dopaminergic neurons in the Substantia nigra (Hoffman et al., 1980). The production of immunoglobulins in response to the antigen of Salmonella-Nashua strain, which never causes an infection in human, was found to be much slower and lower level in the chronic and defective schizophrenic patients than in the controls, and in the psychotic patients treated with antipsychotics the level of immunoglobulins reached the control level after a letancy period (Özek et al., 1971). Moreover the administration of the antigens to animals which causes an increase in dopamine-stimulated adenyl cyclase activity in Nucleus caudatus was considered as an experimental finding supporting the fact that the decrease in brain dopaminergic activity increases the mortality in the mice bearing mammary tumors (Cotzias and Tang, 1977). As morphine opiates inhibits dopamine release, augments brain dopamine content and brain dopamine turnover, and blocks dopamine receptors (Clouet and Ratner, 1970; Kuschinsky and Hornykiewicz, 1972; Perez-Cruet et al., 1972; Puri and Lal, 1973; Lal, 1975; Lal and Numan, 1976; Koyuncuoğlu et al., 1977; Puri et al., 1978) their single or multiple doses is expected to exert an inhibitory action on immune system and immune response.

Noradrenaline and adrenaline have long been regarded as the neurotransmitters modulating immune system and immune response rather periferically. Adrenergic fibers innervating the interlobular septa of the rat thymus have been demonstrated (Fujiwara et al., 1966). The surfaces of T-cells, B-cells and macrophages contain a considerable amount of beta-adrenergic receptors (Bourne et al., 1974). Kasahara et al. (1977) and Hall et al. (1979) reported that chemical sympatectomy by 6-hydroxydopamine resulted in a significant depression of antibody production lasting till the return to normal of splenic noradrenaline content. On the other hand, Anderson and Slotkin (1975) found that the administration of morphine to rats produces a biphasic change in catecholamine content of adrenal medulla; in acute phase there is a reflex stimulation-induced depletion and during chronic administration the depletion is followed by an increase in the catecholamine content indicating an inhibition of catecholamine release (Anderson and Slotkin, 1975). Recently, it has been shown that endogenous opiates diminish the release of catecholamines from sympathetic nerve endings and the adrenal medulla by acting on the opiate receptors present in the membranes of the nerves and adrenal chromaffin cells (Weiner, 1979; Kumakura et al., 1980). On the basis of these experimental results the inhibition of immune system and immune response would be occurred during chronic administration or intake of morphine opiates by their blocking effect on catecholamine release from sympathetic nerve endings and adrenal medulla, in addition to others.

The experimental findings, resulted from the inverse relationship between antibody production and brain 5-hydroxytryptamine contents, suggest that 5-hydroxytryptamine exerts a clear inhibitory influence over immunogenesis. After administration of 5-hydroxytryptamine and its immediate precursor 5-hydroxytryptophan an increase in the latent period of antibody formation and a decrease in the intensity of both the primary and secondary response have been observed (Devoino et al., 1970; Idova and Devoino, 1972). In contrast to the inhibition by increased brain 5-hydroxytryptamine contents of the immunogenesis the reduction of brain 5-hydroxytryptamine levels appeared to have a stimulatory action on immune system. The destruction of the nucleus of the midbrain raphe, which has 5-hydroxytryptamine rich neurons, stimulates antibody production to bovine serum albumin (Eremina and Devoino, 1973). The extirpation of pineal gland which contains very high levels of 5-hydroxytryptamine is able to enhance the growth of certain tumors (Rodin, 1963; Lapin 1974; 1976). Although the ability of 5-hydroxytryptamine to alter immune system and immune response has saticfactorily been shown no conclusive conciliation has come out yet from the experimental results to say whether single dose or long-term use of morphine opiates can cause increases or decreases in the brain level of 5-hydroxytryptamine or its metabolism (Cochin and Axelrod, 1959; Maynert and Klingman, 1962; Gunne, 1963; Sloan et al., 1963; Algeri and Costa, 1971; Koyuncuoğlu et al., 1977). Therefore it is not possible, at the moment, to say in which direction, stimulation or inhibition, chronic use of morphine opiates may influence immune system and/or immune response by means of the changes in central 5-hydroxytryptamine metabolism.

Experimental findings have revealed that the intensity of immunologic responses such as antibody production, anaphylactic shock, skin reaction of the delayed type etc., are dependent on the action of neurohumoral factors. It has been shown that the intensity of the process of immunogenesis is affected by the actions of the hypothalamus, limbic system, pituitary body, adrenal glands, thymus and thyroid gland (Kanda, 1959; Luparello et al., 1964; Tsypin and Maltzev, 1967; Stein et al., 1969; Spector et al., 1975; Stein et al., 1976; Filipp, 1973; Amkraut and Solomon, 1974). Hormonal alterations generally follow antigenic challenge and they provide basic conditions for the pharmacologic control of the immunogenesis. Indeed, it appears logical to assume that a drug-mediated interference with neuroendocrinological events related to the central nervous system presumably promoting the hormonal variations observed would affect the immunogenesis and immune response to the antigen that triggers this very complex chain of reactions. Morphine opiates as the agents influencing directly the hypothalamus as well as the limbic system, to a lesser extent, indirectly the peripheral hormonal systems may play an important role in the orientation of the immune response. Without going in details of the interactions of the hormonal systems with the chemical neurotransmitters which take part in their release and/or metabolism, and the relationships between the neurotransmitters and hormonal systems and *viceversa*, the probable actions of morphine opiates on immune system and immune response *via* their effects on hormonal systems will shortly be reviewed.

It has been reported that growth hormone increases the weight of the thymus after hypophysectomy in neonatal Snell pituitary dwarf mice (Okouchi, 1976) Growth hormone has been found to stimulate rat thymocyte mitosis *in vitro* (Ahlqvist, 1976). Gisler (1974) showed that growth hormone reverses the immunosuppressive effects of corticosteroids in the plaque-forming cell assay. Moreover, growth hormone enhances graft rejection (Comsa et al., 1975). In spite of these experimental findings favouring the stimulatory effect of growth hormone on immune system there have been other contradictory reports (Dougherty, 1952; Ahlqvist, 1976; 1981). Growth hormone has long been regarded as a prophylogistic hormone enhancing a large number of different experimental vasculitic, arthritic, carditic and other inflammatory reactions, prevented tuberculosis in cortisone-treated rats, and turning monilial *foci* into proper granulomas (Ahlqvist, 1981). Growth hormone increases the fibrinogen level and erythrocyte sedimentation rate and it induces a neutrophilic reaction in dogs and it stimulates hepatic fibrinogen synthesis in rats (Ahlqvist, 1976). Even though it cannot surely be said that growth hormone inhibits the immune system, on the basis of the information mentioned above, it may be concluded that the direct evidence seems to be in favour of an inhibition. Therefore, the long-term administration of morphine opiates may be considered to have an inhibitory effect or at least contributory effect to the other inhibitory factors on immune system through the increased growth hormone secretion since the release of growth hormone is stimulated by exogenous as well as endogenous opioids and since opiate antagonists abolish the stimulatory effects of opioids on growth hormone release (Brown and Vale, 1975; Bruni et al., 1977; Koenig et al., 1980; Okajima et al., 1980; Sarne et al., 1981).

Antidiuresis, which follows the administration of morphine opiates, has been attributed to the direct action of these opiates on the hypothalamoneurohypophyseal axis, causing the release of vasopressin (DeBodo, 1944; Inturrisi and Fujimoto, 1968; Meites et al., 1979). Recently it has been shown that morphine, D-aspartic acid and PLG (prolyl-leucyl-glycinamide), all of which are the inhibitors of the brain L-asparaginase activity (Koyuncuoğlu et al., in press), elevate the urine osmolality (Koyuncuoğlu et al., 1984 c) and cause the release of vasopressin also in homozygous Brattleboro rats suffering from severe diabetes insipidus due to the lack of vasopressin. It has been concluded that the mechanism underlying the release of vasopressin is the inhibition of the brain L-asparaginase activity which was claimed to be the mechanism of physical dependence on morphine (Koyuncuoğlu, 1983; Koyuncuoğlu et al., 1984 b) . Vasopressin has not only antidiuretic effect at the level of kidneys but it has many other effects and functions inside and/or outside of the central nervous system. Vasopressin is one of the main factors causing the release of corticotropin from the anterior part of the pituitary gland (Hahmeier et al., 1980; Lutz-Bucher et al., 1980; Knepel et al., 1982).. It also can alter the turnover of the catecholamines in various brain regions (Meisenberg and Simmons, 1983). Vasopressin stimulates Na^+,K^+-ATPase and adenylate cyclase activity (Roy, 1979; Radominska-Pyrek et al., 1982; Courtney and Raskind, 1983) and it induces the increase of cAMP level. Moreover, vasopressin affects directly and/or indirectly carbohydrate, protein and lipid metabolism (Caisova et al., 1980 a, b; Edwards et al., 1981; Wood, 1981; Kneer and Lardy, 1983; Knudtzon,1983; Rofe and Williamson, 1983). So, from the information given above it is possible to reach the conclusion that morphine opiates may negatively affect immune system also <u>via</u> the release of vasopressin which increases the release of corticotropin and exerts direct metabolic effects.

As well known, hypothalamic lesions have a protective effect on anaphylaxis presumably related to changes in endocrine activity. It has been reported that the administration of thyroxine partially restores the sensitivity to anaphylaxis of actively immunized guinea pigs with lesions in the tuberal area of the hypothalamus (Filipp and Mess, 1969). When thyroxine in combination with metopirone, an inhibitor of adrenocorticol hormone synthesis, is administered to the sensitized guinea pigs with lesions in the tuberal region the protective action of the lesions from anaphylactic state is completely abolished, suggesting that the antianaphylactic effect of hypothalamic damage is connected with the combined effect of decreased thyroid function and increased adrenocortical activity (Filipp and Mess, 1969). It has also been demonstrated that thyroid enhances immune processes in the rat, mouse, and guinea pig. The resistance to anaphylaxis is increased in thyroidectomized rats, mice and guinea pigs. Inhibition of thyroid activity suppresses local and systemic anaphylaxis, abolishes circulating precipitins, and decreases the susceptibility of the animals to histamine administration (Leger and Masson, 1947; Nilzen, 1955; Csaba et al., 1977; Denckla, 1978). Thyroxine and triiodothyronine increase the phagocytic activity (Luri, 1960). Even though the direct effects of thyrotropin on tissues, other than the thyroid gland, such as the stimulation lipolysis in fat, and the inflammation in orbital muscle and connective tissue have been demonstrated, the effects of thyrotropin on the immunogenesis, a part from the effects mediated by thyroid hormones, have not been shown yet (Maclean and Reichlin, 1980). Administration of morphine opiates results in decreased thyrotropin secretion (Kim and Michael, 1975; Krulich et al., 1977; Tache et al., 1977). The activation by morphine of an inhibitory hypothalamic system under basal conditions has been claimed to be the mechanism of the inhibitory effect of morphine on thyroid functions (Lomax and George, 1966). While morphine prevents the cold stimulated rise in TSH it significantly increases thyrotropin releasing hormone content of the medial-basal hypothalamus reflecting morphine suppression of TRH release (Krulich et al., 1977). These findings are most consistent with a hypothalamic site for morphine action on thyrotropin secretion is mediated through decreased portal thyrotropin releasin hormone. No tolerance develops to the inhibitory effect of morphine on thyrotropin release (Morley et al., 1979). Chronic administration of morphine inhibits ^{131}I uptake and release of ^{131}I-labeled hormone by the thyroid (George and Lomax, 1965). Morphine addiction decreases thyroxine levels measured by radioimmunoassay (Morley et al., 1979).

The acute involution of the rodent thymus following stressful events is generally attributed to the increase in glucocorticoid levels (Dieckhoff et al., 1965). Rat thymus size is inversely correlated with plasma levels of unbound corticosterone, the main glucocorticoid in rodents (Westphal, 1971). Elevations of plasma corticosterone levels cause thymus involution, lymphocytopenia, alterations in amino acid metabolism and may produce various effects upon tumor growth, latent period and incidence (Spackman and Riley, 1974; Spackman et al., 1974; Fauci, 1975; Monjan and Collector, 1977). One of the most important effects of corticosteroids is their influence on amino acid metabolism and on protein synthesis. Elevated blood levels of these hormones significantly alter the levels of amino acids in physiological fluids and tissues, and result in an increased excretion of urinary nitrogen and a negative nitrogen balance presumably due to both an inhibition of protein synthesis and increased protein catabolism causing the elevated concentrations of plasma amino acids, as well as increased gluconeogenesis (White et al., 1968; Turner and Hagnara, 1971). These alterations in protein and amino acid metabolism have potential influences upon disease processes, <u>via</u> their adverse effects on the synthesis of enzymes and cellular proteins needed for normal immune surveillance. In may experiments it has been shown that the basic cellular elements forming the immunological system, involving macrophages, T cells and B cells, are all subject to modification, impairment and/or destruction by glucocorticoids (Santisteban and Dougherty, 1954; Gisler and Schenkel-Hullinger, 1971; Monjan and Collector, 1977). The lymphocytopenia resulting from elevated plasma glucocorticoids undoubtedly indicates destruction of T cells and probably also B cells (Monjan and Collector, 1977; Solomon and Amkraut, 1979). On the other hand, adrenalectomy increases the size of thymus (Dougherty, 1952), and seems to stimulate the mitotic index in both epithelial cells and thymocytes after X-ray-induced involution (Gregoire and Duchateau, 1956). Adrenalectomy may also enhance antibody formation and susceptibility to anaphylaxis and Arthus reaction (Ahlqvist, 1976). When glucocorticoids are administered before antigen challenge they suppress antibody formation (Ahlqvist, 1976). Although there have been reports claiming that exogenous and endogenous opioids have an inhibitory effect on the pituitary-adrenal axis (Blankstein et al., 1980; Grossman and Besser, 1982) it has generally been ascribed that acute administration of morphine stimulates corticosterone release (Lotti et al., 1969; George, 1971; George and Kokka, 1976). The corticotropin releasing action of morphine is mainly mediated <u>via</u> the hypothalamus; lesions of the median eminence blocks the adrenal cortical response to morphine (George and Way, 1959; Kokka and George, 1974). Recently morphine has been demonstrated to potentiate the adrenal steroidogenic response to corticotropin (Heybach and Vernikos, 1981). However, the release by morphine opiates of vasopressin may also play an important role in the secretion of corticotropin (DeBodo, 1944; George and Way, 1959; Inturrisi and Fujimoto, 1968; Meites et al., 1979; Koyuncuoğlu et al., 1984 b). Even though tolerance to the corticosterone releasing effect of morphine opiates develops and higher doses are required to elevate corticosterone levels (Kokka and George, 1974) the parallel increase in compulsory intake by addicts because of the development of tolerance may not cause any significant change in the elevations of plasma glucocorticoid levels.

Although gonadectomy in experimental animals of both sexes has been reported to increase the size of the thymus and to counteract the thymic involution induced by stress and fasting (Dougherty, 1952; Ahlqvist, 1976) quite a large number of experimental findings is in favour of the enhancing effects of sex hormones especially androgens on antibody formation. The "anti-infective" action of 4-chloro testosterone in experimental infections was first demonstrated by Ghione (1958). The androgen prolonged the survival time of the experimentally infected animals and it provided a more rapid healing of infectious lesions. Using metyrapone, an 11-beta-hydroxylase blocker, which decreases the synthesis of cortisol and increases the synthesis of adrenal androgens, it has been shown that androgens stimulate the antibody formation (Tolentino et al., 1961). Then the confirming data obtained from the experiments performed on experimental animals as well as on human beings utilizing different antigens and different anabolic agents helped to reach the almost generally accepted conclusion that sex hormones stimulate immune system especially by means of their anabolic action (Sawyer et al., 1955; Coraggio et al., 1962; Heboyan and Messeri, 1962; Jannuzzi and Bassi, 1962; 1964; Terragna and Januzzi, 1963; Dieckhoff et al., 1965; Januzzi and Gemme, 1965; Wakabayashi, 1966). On the other hand morphine and opioid peptides consistently produce naloxone reversible suppression of luteinizing hormone and to a lesser extent of follicle stimulating hormone (Cicero et al., 1975; Zimmerman and Pang, 1976; Bruni et al., 1977). Chronic morphine administration causes regressive changes in Leydig cells with minimal alterations in testicular weight (Hohlweg et al., 1961; Cicero et al., 1975). Morphine and methadone both suppress the growth of sex organs (Cicero et al., 1975; 1976). No tolerance to the inhibitory effects of morphine on luteinizing hormone, follicle stimulating hormone and testosterone levels was demonstrated 14 days after the continuous morphine administration (Cicero and Badger, 1977; Morley et al., 1979).

Malnourished animals and humans appear to be more susceptible to all forms of infection and recovery is often delayed. Infections occur more often and may be more severe in chronically undernourished populations. Based on this often observed coincidence it has been suggested that undernutrition predisposes to infections and malnutrition has generally been accepted as one of the main factors causing secondary immunodeficiency. As most studies have been on malnourished children they have been found particularly prone to septicaemia with Gram negative organisms or disseminated herpes simplex infections (Harmon et al., 1963; Scrimshaw et al., 1968; McFarlane, 1976; Hayward, 1977; Suskind, 1977; Murray and Murray, 1979; Gross and Newberne, 1980; Chandra, 1983). Furthermore it has been shown that caloric restriction can increase tumour incidence and growth rates (Visshee et al., 1942; Rusch, 1944; Tannenbaum and Silverstone, 1953). In spite of many studies the nature of increased infection, and of the high tumour incidence and growth rate in malnutrition is not quite clear. Whereas the majority of malnourished individuals have been reported to have normal concentrations of the serum immunoglobulins (Smythe et al., 1971) the deprivation of certain amino acids in the diet has been shown to cause an impaired synthesis of antibodies (Jose and Good, 1973). In addition various alterations of host defense in anorectic patients such as blood polymorphonuclear granulocyte glucose oxidation, adherence and bactericidal capacity (Gotch et al., 1975; Kjösen et al., 1975; Palmblad et al., 1977) and depression of factors C1, C2, C3, C1s inactivators, B and total hemolytic complement (Kim and Michael, 1975) have been described, and the decrease in numbers of lymphoid cells in spleen and thymus of rats with malnutrition has been suggested as an impaired cell-mediated immune response (McFarlane and Hamid, 1973). Malnutrition is almost always accompanied by deficiencies of vitamins some of which have an important function in immune response. Deficiencies of the B group vitamins have long been associated with reduction in antibody production to antigenic stimuli. Vitamin B6 (pyridoxine) deficiency causes a much more complete inhibition of the immune response (Axelrod and Trakatellis, 1964). The effects of Vitamin B6 deficiency on tumour suppression have also been shown (Mihich, 1962; Mihich and Nichol, 1965). Vitamin B12 also has been reported as affecting the immune response (Ludovici and Axelrod, 1951). Recently some more experimental findings regarding the stimulation by Vitamin C supplements of the immune system have been added to previous ones (Feigen et al., 1982; Kennes et al., 1983).

Long term use of morphine opiates by subjects, physically dependent or not, is usually detrimental to the users for many important reasons also in terms of nutrition. Although the malnutrition seems to be related, to some extent, with financial, social and familial problems the most prevalent factor is the effect of morphine opiates on appetite. Even though it has been hypothesized that opioid systems take part in the regulation of behaviour in association with acquisition of nutrients and in the conservation and expenditure of energy (Margules, 1979) and morphine like agents have, in favour of the hypothesis, been reported to increase food and fluid intake (Grandison and Guidotti, 1977; Maickel et al., 1977; Belluzzi and Stein, 1978) several other studies have generally found a decrease (Kumar et al., 1971; Chance and Rosecrans, 1977; Mark-Kaufman and Kanarek, 1980; Koyuncuoğlu et al., 1984 a). This is why the users of morphine opiates generally lose appetite, and the majority of them are thin even cachectic, and malnourished.

As known, interferon is a small protein which takes part in the general cellular defense mechanisms against chemically induced carcinogenesis, solid tumour growth, both cancer and noncancer-inducing virus invasion in animals. It can be effective in the therapy of respiratory and other virus infections in humans. Animal experimental work showed that a single administration of morphine is capable of depressing or inhibiting the circulating serum interferon level for 3-9 days which can render the body more susceptible to virus-induced pathology (Hung et al., 1973; Geber et al., 1976 a,b).

As most actions, including vasopressin release, of the acute and chronic administration of morphine opiates have been attributed to the inhibitory effect of these substances on the activity of L-asparaginase (Koyuncuoğlu et al., 1979; 1982 a,b; 1984 b; Koyuncuoğlu 1983; Koyuncuoğlu et al., in press) which has long been used in the treatment of lymphomas because of the catalytic effect on the deamidation of asparagine and, to a lesser extent, glutamine resulting in the depletion of these amino acids required for peptides and proteins synthesis, it seems to be logical to think that the inhibitory effect of morphine opiates on L-asparaginase activity can be one of the main factors in their modulating action on immune function. The promising experimental studies regarding the discovery of the physical dependence development on morphine opiates via their effects on L-asparaginase can be taken into consideration as the basis of the additional innovative experimentations in connection with immunity in all senses of the word, and the resistance to experimental pyelonephritis of homozygous Brattleboro rats suffering from a genetic severe hypothalamic diabetes insipidus (Güngör et al., in press) and having a higher L-asparaginase activity than normal rats (Koyuncuoğlu et al., 1984 b) can be regarded as one of the primary experimental findings in favour of this suggestion.

REFERENCES

Ahlqvist, J., 1976, Endocrine influences on lymphatic organs, immune responses, inflammation and autoimmunity, Acta Endocrinol. (Copenhagen), Suppl., 206.

Ahlqvist, J., 1981, Hormonal Influences on Immunologic and Related Phenomena, in: "Psychoneuroimmunology", R. Ader and R.A. Good, Eds., Academic Press, New York.

Akera, T., and Brody, T.M., 1968, The addiction cycle to narcotics in the rat and its relation to catecholamines, Biochem. Pharmacol., 17:675.

Algeri, S., and Costa, E., 1971, Physical dependence on morphine fails to

increase serotonin turnover in rat brain, Biochem.Pharmacol., 20:877.
Amkraut, A., and Solomon, G.F., 1974, From the symbolic stimulus to the pathophysiologic response: Immune mechanisms, Int.J.Psychiatry Med., 5:542.
Anderson, T.R., and Slotkin, T.A., 1975, Maturation of the adrenal medulla -IV. Effects of morphine, Biochem.Pharmac. 24:1469.
Axelrod, A.E., and Trakatellis, A.C., 1964, Relationship of pyrodoxine to immunological phenomena, Vitamins Hormones 22:591.
Beleslin, D., and Polak, R.L., 1965, Depression by morphine and chloralose of acetylcholine release from the cat's brain, J.Physiol.(Lond.), 177:411.
Belluzzi, J.D., and Stein, L., 1978, Do enkephalin systems mediate drive reduction? Soc.Neurosci.Abstr., 8:405.
Blankstein, J., Reyes, F.I., Winter, J.S.D., and Faiman, C., 1980, Effects of naloxone upon prolactin and cortisol in normal women, Proc. Soc.Exp.Biol.Med., 164:363.
Bourne, H.R., Lichtenstein, L.M., Melmon, K.L., Henney, C.S., Weinstein, Y., and Shearer, G.M., 1974, Modulation of inflammation and immunity by cyclic AMP, Science, 184:19.
Brown, M., and Vale, W., 1975, Growth hormone release in the rat: Effects of somatostatin and thyrotropin-releasing factor, Endocrinology, 96: 1333.
Brown, S.M., Stimmel, B., Taub, R.N., Kochwa, S., and Rosenfield, R.E., 1974, Immunologic dysfunction in heroin addicts, Arch.Intern.Med., 134:1001.
Bruni, J.F., Van Vugt, D., Marshall, S., and Meites, J. 1977, Effects of naloxone, morphine and methionine enkephalin on serum prolactin, luteinizing hormone, follicle stimulating hormone, thyroid stimulating hormone and growth hormone, Life Sci., 21:461.
Buchenauer, D., Turnbow, M., and Peters, M.A., 1974, Effects of chronic methadone administration of pregnant rats and their offspring, J.Pharmac.Exp.Ther., 189:66.
Byron, J.W., 1975, Manipulation of the cell cycle of the hamopcietic stem cell, Exp.Hematol., 3:44.
Byron, J.W., 1976, Cyclic nucleotides and the cell cycle of the hematopoietic stem cell. in: "Cyclic Nucleotides and the Regulation of Cell Growth", M.Abou-Sade, ed., Dowden, Hutchinson and Ross, Pennsylvania.
Caisova, D., Stajner, A., and Suva, J., 1980 a, Modification of fat and carbohydrate metabolism by neurohypophyseal hormones: I.Effects of lysine-vasopressin on non-esterified fatty acid, glucose, triglyceride and cholesterol levels in the serum of female rats, Endokrinologie, 76:315.
Caisova, D., Suva, J. and Stajner, A., 1980 b, Modification of fat and carbohydrate metabolism by neurohypophyseal hormones: II. Effect of lysine-vasopressin on non-esterified fatty acid, glucose triglyceride and cholesterol levels in the serum of male rats, Endokrinologie, 76:326.
Chance, W.T., and Rosecrans, J.A., 1977, Inhibition of drinking by intrahypothalamic administration of morphine, Nature, 270:167
Chandra, R.K., 1983, Nutrition and immune responses, Can.J.Physiol.Pharmacol. 61:290.
Cherubin, C.E., Baden, M., Kavaler, F., Lerner, S., and Cline, W., 1968, Infective endocarditis in narcotic addicts, Ann.Int.Med. 69:1091.
Cherubin, C.E., 1967, The medical sequelae of narcotic addiction, Ann.Int. Med., 67:23.
Cicero, T.J., and Badger, T.M., 1977, A comparative analysis of the effects of narcotics, alcohol and the barbiturates on the hypothalamic-pituitary-gonadalaxis, Adv.Exp.Med.Biol., 85:95.
Cicero, T.J., Meyer, E.R., and Bell, R.D., 1976, Effects of morphine on serum testosterone and luteinizing hormone levels and on secondary

sex organs of the male rat, Endocrinology, 98:367.
Cicero, T.J., Meyer, E.R., and Wiest, W.A., 1975, Effects of chronic morphine administration on the reproductive system of the male rat, J.Pharmacol.Exp.Ther., 192:542.
Clouet, D.H., and Ratner, M., 1967, The effect of the administration of morphine on the incorporation of ^{14}C-leucine into the proteins of rat brain in vivo, Brain Res., 4:33.
Clouet, D.H., and Ratner, M., 1968, The effect of morphine administration on the incorporation of ^{14}C leucine into protein in cell-free system from rat liver and brain, J.Neurochem., 15:17.
Clouet, D.H., and Ratner, M., 1970, Catecholamine biosynthesis in brains of rats treated with morphine, Science, 168:854.
Cochin, J., and Axelrod, J., 1959, Biochemical and pharmacological changes in the rat following chronic administration of morphine, nalorphine and normorphine, J.Pharmacol.Exp.Ther., 125:105.
Comsa, J., Leonhardt, H., and Schwarz, J.A., 1975, Influence of the thymus -corticotropin-growth hormone interaction on the rejection of skin allografts in the rat, Ann.N.Y.Acad.Sci., 249:387.
Coraggio, F., Coto, V., Oriente, P., De'Longis, G., and Galeota, C.A., 1962, Influenza del 19-norandrostenolone sulla produzione di anticorpi antimorbillosi nel coniglio, Boll.Soc.Ital.Biol.Sper., 38:1316.
Cotzias, G.C., and Tang, L.C., 1977, An adenylate cyclase of brain reflects propensity for breast cancer in mice, Science, 197:1094.
Courtney, N., and Raskind, M., 1983, Vasopressin affects adenylate cyclase activity in rat brain: A possible neuromodulator, Life Sci., 32: 591.
Csaba, G., Sudar, F., and Dobozy, O., 1977, Triiodothyronine receptors in lymphocytes of newborn and adult rats, Horm.Metab.Res. 9:499.
Datta, R.K., and Antopol, W., 1972, Inhibitory effect of chronic administration of morphine on uridine and thymidine incorporating abilities of mouse liver and brain subcellular fractions, Toxicol.appl.Pharmacol., 23:75.
Datta, R.K., and Antopol, W., 1973, Effect of chronic administration of morphine on mouse brain aminoacyl-tRNA synthetase and tRNA-amino acid binding, Brain Res., 53:373.
Dawis, W.M., and Lin, C.H., 1972, Prenatal morphine effects of survival and behaviour of rat offspring, Res.Commun.Chem.Pathol.Pharmacol., 3:205.
DeBodo, R.C., 1944, The antidiuretic action of morphine, and its mechanism, J.Pharmacol.Exp.Ther., 82:74.
Dencklа, W.D., 1978, Interactions between age and neuroendocrine and immune systems, Fed.Proc.Fed.Am.Soc.Exp.Biol., 37:1263.
Devoino, L.V., Eremina, O.F.N., and Ilyutchenok, R.Yu., 1970., The role of the hypothalamopituitary system in the mechanism of action of reserpine and 5-hydroxytryptophan on antibody production, Neuropharmacology, 9:67.
Dieckhoff, J., Schneeweiss, B., Schicke, R., and Hübschmann, K., 1965, Tierexperimentelle Untersuchungen über die Antikörperbildung unter gleichzeitiger, Anabol.Applikation,Mschr.Kinderheilk., 113:468.
Dougherty, T.F., 1952, Effects of hormones on lymphatic tissue, Physiol. Rev., 32:379.
Edwards, M.W., Brooks, S.L., Gove, C.D., Hems, D.A., and Cawthorne, M.A., 1981, Effects of vasopressin on lipogenesis in obese mice, FEBS Letters, 127:25.
Eiseman, B., Lam, R.C., and Bush, B., 1964, Surgery on the narcotic addict, Ann.Surg., 159:748.
Engel, W.K., Trotter, J.L., McFarlin, D.E., and McIntosh, C.L., 1977, Thymic epithelial cell contains acetylcholine receptor, Lancet, 1:1310.
Eremina, O.F., and Devoino, L.V., 1973, Production of humoral antibodies in rabbits with destruction of the nucleus of the midbrain raphe, Bull.Exp.Biol.Med.(Engl Transl.), 75:149.

Espiritu, M.B., and Medina, J.E., 1980, Complications of heroin injections of the neck, Laryngoscope, 90:1111.

Fauci, A.S., 1975, Mechanisms of corticosteroid action on lymphocyte subpopulations. I.Redistribution of circulating T and B lymphocytes to the bone marrow, Immunology, 28:669.

Feigen, G.A., Smith, B.H., Dix, C.E., Flynn, C.J., Peterson, N.S., Rosenberg, L.T., Pavlovic, S., and Leibovitz, B., 1982, Enhancement of antibody production against systemic anaphylaxis by large doses of vitamin C, Res.Comm.Chem.Path.Pharmac., 38:313.

Fennessy, M.R., and Lee, J.R. 1972, Comparison of the dose-response effects of morphine on brain amines, analgesia and activity in mice, Brit.J.Pharmacol., 45:240.

Filipp, G., 1973, Mechanism of suppressing anaphylaxis thorough electrolytic lesion of the tuberal region of the hypothalamus, Ann.Allergy, 31: 272.

Filipp, G., and Mess, B., 1969, Role of the adrenocortical system in suppressing anaphylaxis after hypothalamic lesion, Ann.Allergy, 27:607.

Friedler, C., and Cochin, J., 1972, Growth retardation of offspring of female rats treated with morphine prior to conception, Science, 175:654.

Fujiwara, M., Muryobayashi, T., and Shimamoto, K., 1966, Histochemical demonstration of monoamines in the thymus of rats, Jpn.J.Pharmacol., 16:493.

Fukui, K., and Takagi, H., 1972, Effect of morphine on the cerebral contents of metabolites of dopamine in normal and tolerans mice: its possible relation to analgesic action, Brit.J.Pharmacol., 44:45.

Geber, W.F., Lefkowitz, S.S., and Hung, C.Y., 1976 a, Action of naloxone on the interferone lowering activity of morphine in the mouse, Pharmacology, 14:322.

Geber, W.F., Lefkowitz, S.S., and Hung, C.Y., 1976 b, Role of spleen in the interferon-lowering action of morphine, Gen.Pharmac., 7:255.

Geber, W.F., and Schramm, L.C., 1974, Postpartum weight alteration in hamster offspring from females injected during pregnancy with either heroin, methadone, a composite drug mixture, or mescaline, Am.J. Obstet.Gynecol., 120:1105.

Geber, W.F., and Schramm, L.C., 1975, Congenital malformation of the central nervous system produced by narcotic analgesics in the hamster, Am.J. Obstet.Gynecol. 123:705.

George, R., 1971, Hypothalamus: Anterior Pituitary Gland. in: "Narcotic Drugs, Biochemical Pharmacology", D.M.Clouet, ed., Plenum Press, NewYork.

George, R., and Lomax, P., 1965, The effects of morphine, chlorpromazine and reserpine on pituitary-thyroid activity in rats, J.Pharmacol. Exp.Ther., 150:129.

George, R., and Way, E.L., 1959, The role of the hypothalamus in pituitary -adrenal activation and antidiuresis by morphine, J.Pharmacol.Exp. Ther., 125:111.

Ghione, M., 1958, Anti-infective action of an anabolic steroid, Proc.Soc. Exp.Biol.Med., 97:773.

Gisler, R.H., 1974, Stress and the hormonal regulation of the immune response in mice, Psychother.Psychosom., 23:197.

Gisler, R.H., and Schenkel-Hullinger, L., 1971, Hormonal regulation of the immune response. II.Influence of pituitary and adrenal activity on immune responsiveness in vitro, Cell. Immunol., 2:646.

Goldberg, N.D., Haddox, M.K., Estensen, R., White, J.G., Lopez, C., and Hadden, J.W., 1974, Evidence of a dualism between cyclic GMP and cyclic AMP in the regulation of cell proliferation and other cellular processes, in: "Cyclic AMP, Cell Growth, and the Immune Response", W.Braun, L.M.Lichtenstein and C.W.Parker, eds., Springer -Verlag, NewYork.

George, R., and Kokka, N., 1976, The Effects of Narcotics on Growth Hormone,

ACTH and TSH Secretion, in: "Tissue Responses to Addictive Drugs", D.M.Ford and D.H.Clouet, eds., Spectrum Publications, NewYork.
Gotch, F.M., Spry, C.J.F., Mowat, A.G., Beeson, P.M., and MacLennan, I.C.M., 1975, Reversible granulocyte killing defect in anorexia nervosa, Clin.Exp.Immunol., 21:244.
Grandison, S., and Guidotti, A., 1977, Stimulation of food intake by muscimol and beta-endorphin, Neuropharmacology, 16:533.
Gregoire, C., and Duchateau, G., 1956, Study on lympho-epithelial symbosis in thymus. Reactions of the lymphatic tissue to extracts and to implants of epithelial components of thymus, Arch.Biol. 67:269.
Gross, R.L., and Newberne, P.M., 1980, Role of nutrition in immunologic function, Physiol.Rev., 1:188.
Grossman, A., and Besser, G.M., 1982, Opiates control ACTH through a noradrenergic mechanism, Clin.Endocrinology, 17:287.
Gunne, L.-M., 1963, Catecholamines and 5-hydroxytryptamine in morphine tolerance and withdrawal, Acta physiol.scand. 58:Suppl., 204.
Güngör, M., Genç, E., Sagduyu, H., Eroglu, L. and Koyuncuoğlu, H., 1980, Effect of chronic administration of morphine on primary immune response in mice, Experentia, 36:1309.
Güngör, M., Ang, Ö., Uysal, V., Sagduyu, H., Inanç, D., Ağ, M., and Koyuncuoglu, H., Comparison of experimental pyelonephritis in homozygous Brattleboro diabetes insipidus, heterozygous control and normal Wistar rats, Infection (in press).
Hahmeier, W., Fenske, M., Pitzel, L., Holtz, W., and König, A., 1980, Corticosteroids and testosterone in adult male pig, Acta Endocrinologica, 95:518.
Hall, N.R., McClure, J.E., Hu, S.-K., Tick, N.T., Seales, C.M., and Goldstein, A.L., 1979, Effects of chemical sympathectomy upon thymus dependent immune responses, Soc.Neurosci.Abstr., 26.4.
Harmon, B.G., Miller, E.R., Hoeffer, J.A., Ullrey, D.E., and Luecke, R.W., 1963, Relationship of specific nutrient deficiencies to antibody production in swine. I.Vitamin A, J.Nutr., 79:263.
Harpel, H.S., and Gautieri, R.F., 1968, Morphine-induced malformations, J. Pharm.Sci., 57:1590.
Hayward, A.R., 1977, "Immunodeficiency", Edward Arnold Ltd, London.
Heboyan, M., and Messeri, E., 1962, Variazioni immunitarie in ratti tenuti a dicta normale e a dicta di Handler, vaccinati con Salmonella typhi e trattati con 4-idrossi-19-nortestosterone-17-ciclopentilpropionato, Rass.Ital.Gastroenter., 11:590.
Heybach, J.P., and Vernikos, J., 1981, Naloxone inhibits and morphine potentiates the adrenal steroidogenic response to ACTH, Eur.J.Pharmacol., 75:1.
Hoffman, P.M., Robbins, D.S., Nolte, M.T., Gibbs, jr, C.S., and Gajdusek, D.C., 1980, Immunity and immunogenetics in Guamanians with amyotrophic lateral aclerosis (ALS) and Parkinsonism-dementia, J. Supramol.Struct., 8:Suppl. 2.
Hohlweg, W., Knappe, G., and Domer, G., 1961, Tierexperimentelle Untersuchungen ueber den Einfluss von Morphine auf die gonadotrope und thyrectrope Hypophysenfunktion, Endokrinologie 40:152.
Hung,C.Y., Lefkowitz, S.S. and Geber, W.F., 1973, Interferon inhibition by tion of morphine in the rat, Eur.J.Pharmacol. 2:301.
Idova, C.V., and Devoino, L.V., 1972, Dynamics of formation of M- and G-antibodies in mice after administration of serotonin and its precursor 5-hydroxtryptophan, Bull.Exp.Biol.Med.(Engl.Transl.), 73:294.
Inturrisi, C.E., and Fujimoto, J.M., 1968, Studies on the antidiuretic acnarcotic analgesics, Proc.Soc.Exp.Biol.Med., 142:106.
Isbell, H., 1947, The effect of morphine addiction on blood plasma and extracellular fluid volumes in man, Public Health Reports, 62:1499.
Jannuzzi, C., and Bassi, A., 1962, Ormoni steroidei ed anticorpopoiesi. Nota III. Potenziamento della vaccinazione antidifterica nel bambino da parte di ormoni anabolizzanti, Boll.Ist.Sieroter.Mil., 41:221.

Jannuzzi, C., and Bassi, A., 1964, Vaccinazione antitetanica e ormoni anabolizanti, G.Mal.Inf., 16:748.
Jannuzzi, C., and Gemme, G., 1965, Tentativi di immunizzazione attiva del neonato verso i ceppi enteritogeni di E.coli, G.Mal.Inf., 17:74.
Jhamandas, K., Phillis, J.W., and Pinsky, C., 1971, Effects of narcotic analgesics and antagonists on the in vivo release of acetylcholine from the cerebral cortex of the cat, Brit.J.Pharmacol., 43:53.
Jose, D.G. and Good, R.A., 1973, Quantitive effects of nutritionally essential amino acid deficiencies upon immune responses to tumors in mice, J.Exp.Med., 137:1.
Kanda, R., 1959, Studies of the regulation centre on promotion of antibody II. On the migration and relation of normal precipitin antibody and leucocyte in the peripheral blood by electric stimuli in the hypothalamus of rabbit, Jpn.J.Bacteriol., 14:542.
Kater, L., Oosterom, R., McClure, J., and Goldstein, A.L., 1980, Presence of thymosin-like factors in human thymic epithelial conditioned medium, Int.J.Immunopharmacol. 1:273.
Kasahara, K., Tanaka, S., Ito, T., and Hamashima, Y., 1977, Suppression of the primary immune response by chemical sympathectomy, Res.Commun. Chem.Pathol.Pharmacol. 16:687.
Kennos, B., Dumont, I., Brohee, D., Hubert, C., and Neve, P., 1983, Effect of Vitamin C supplements on cell-mediated immunity in old people, Gerontology 29:305.
Kim, Y., and Michael, A.E., 1975, Hypocomplementemia in anorexia nervosa, J.Pediatr., 87:582.
Kjösen, B., Bassoe, H.H., and Myking, O., 1975, The glucose oxidation in isolated leukocytes from female patients suffering from overweight or anorexia nervosa, Scand.J.Clin.Lab.Invest., 35:447.
Kneer, N.M., and Lardy, H.A., 1983, Regulation of gluconeogenesis by norepinephrine, vasopressin, and angiotensin II: A comparative study in the absence and presence of extracellular Ca^{2+}, Arch.Biochem. Biophys. 225:187.
Knepel, W., Benner, K., and Hertling, C., 1982, Role of vasopressin in the ACTH response to isoprenaline, Eur.J.Pharmacol., 81:645.
Knudtzon, J., 1983, Acute effects of oxytocin and vasopressin on plasma levels of glucogon, insulin and glucose in rabbits, Horm.metabol. Res., 15:103
Krulich, L., Giachetti, A., Marchlewska-Koj., Hefco, E., and Jameson, H.E. 1977, On the role on the central noradrenergic and dopaminergic systems in the regulation of THS secretion in the rat, Endocrinology, 100:496.
Koenig, J., Mayfield, M.A., Coppings, R.J., McCann, S.M., and Krulich, L., 1980, Role of central nervous system neurotransmitters in mediating the effects of morphine on growth hormone-and prolactin-secretion in the rat, Brain Res., 197:453.
Kokka, N. and George, R., 1974, Effects of Narcotic Analgesics, Anesthetic and Hypothalamic Lesions on Growth Hormone and Adrenocorticotrophic Secretion in Rats, in:"Narcotics and the Hypothalamus", E.Zimmerman and R.George, eds., Raven Press, NewYork.
Koyuncuoğlu, H., 1983, The treatment with L-aspartic acid on persons addicted to opiates, Bull.Narcotics (United Nation Publication), 35: 11.
Koyuncuoğlu, H., Berkman, K., Wildmann, J., and Matthaei, H., Antagonistic effect of L-aspartic acid on decrease in body weight, food and fluid intake, and naloxone reversible rectal temperature caused by D-aspartic acid, Pol.J.Pharmacol.Pharm., 34:333.
Koyuncuoğlu, H., Berkman, K., and Sabuncu, H., 1984 a, Feeding, drinking, urine osmolality in DI Brattleboro rats: Changes by morphine, naloxone, D-amino acids, prolyl-leucyl-glycinamide (PLG),

Pharmacol.Biochem.Behav., 20:29.
Koyuncuoğlu, H., Berkman, K., Hatipoğlu, I., and Sabuncu, H., 1984 b, Vasopressin release by D-aspartic acid, morphine and prolyl-leucyl-glycinamide (PLG) in DI Brattleboro rats, Pharmacol.Biochem.Behav., 20:519.
Koyuncuoğlu, H., Berkman, K., and Matthaei, H., Effects of morphine, D-aspartic acid, d_phenylalanine, D-leucine and PLG on L-asparaginase activity in rats, Med.Bull.Ist.Med.Fac. (in press).
Koyuncuoğlu, H., Berkman, K., and Sabuncu, H., 1984 c, Feeding, drinking, urine osmolality in DI Brattleboro rats: Changes by morphine, naloxone, D-amino acids, prolyl-leucyl-glycinamide (PLG), Pharmacol.Biochem.Behav., 20:29.
Koyuncuoglu, H., Genç, E., Güngör, M., Eroglu, L., and Sağduyu, H., 1977, The antagonizing effect of aspartic acid on the brain levels of monoamines and free amino acids during the development of tolerance to and physical dependence on morphine, Psychopharmacology, 54:187.
Koyuncuoğlu, H., Keyer-Uysal, M., Berkman, K., Güngör, M., and Genç, E., 1979, The relationship between morphine, aspartic acid and L-asparaginase, Eur.J.Pharmacol., 60:369.
Koyuncuoğlu, H., Wildmann, J., Berkman, K., and Matthaei, H., 1982 b, The effects of D- and/or L-aspartic acid on the total weight of body and the weights of certain organs, and their protein, triglyceride and glycogen contents, Drug Res., 32:738.
Kumakura, K., Karoum, F., Guidotti, A., and Costa, E., 1980, Modulation of nicotinic reseptors by opiate receptor agonists in cultured adrenal chromaffin cells, Nature, 283:489.
Kumar, R., Mitchell, E., and Stolerman, I.P., 1971, Disturbed patterns of behavior in morphine tolerant and abstinent rats, Br.J.Pharmacol. 42:473.
Kuschinsky, K., and Hornykiewicz, O., 1972, Morphine catalepsy in the rat: Relation to striatal dopamine metabolism, Eur.J.Pharmacol. 19: 119.
Lal, H., 1975, Narcotic dependence, narcotic action and dopamine receptors, Life Sci. 17:483.
Lal, H., and Numan, R., 1976, Blockade of morphine-withdrawal body shakes by haloperidol, Life Sci., 18:163.
Lapin, V., 1974, Influence of simultaneous pinealectomy and thymectomy on the growth and formation of metastases of the Yoshida sarcoma in rats, Exp.Pathol., 9:108.
Lapin, V., 1976, Pineal gland and malignancy, Österr.Z.Onkol., 3:51.
Leger, J., and Masson, G., 1947, Factors influencing an anaphylactoid reaction in the rat, Fed.Proc.Fed.Am.Soc.Exp.Biol., 6:150.
Lindstrom, J.M., Lennon, V.A., Seybold, M.E., and Whittingham, S., 1976, Experimental autoimmune myasthenia gravis and myasthenia gravis: Biochemical and immunochemical aspects, Ann.N.Y.Acad.Sci., 274: 254.
Lomax, P., and George, R., 1966, Thyroid activity following administration of morphine in rats with hypothalamic lesions, Brain Res., 2:361.
Lotti, V.J., Kokka, N., and George, R., 1969, Pituitary-adrenal activation following intrahypothalamic microinjection of morphine, Neuroendocrinology, 4:326.
Louria, D.B., Hensle, T., and Rose, J., 1967, The major medical complications of heroin addiction, Ann.Int.Med., 67:1.
Ludovici, P.P., and Axelrod, A.E., 1951, Circulating antibodies in vitamin -deficiency states: Pteroylglutamic acid, niacin-tryptophan, vitamin B12 and D deficiencies, Proc.Soc.Exp.Biol.Med., 77:526.
Luparello, T.J., Stein, M., and Park, C.D., 1964, Effect of hypothalamic lesions on rat anaphylaxis, Am.J.Physiol., 207:911.
Luri, M.B., 1960, The reticuloendothelial system, cortisone, and thyroid function: Their relation to native resistance to infection, Ann.N.Y.Acad.Sci., 88:83.

Lutz-Bucher, B., Koch, B., Mialhe, C., and Briaud, B., 1980, Involvement of vasopressin in corticotropin-releasing effect of hypothalamic median eminence extract, Neuroendocrinology, 30:178.
Maclean, D., and Reichlin, S., 1980, Neuroendocrinology and the Immune process, in: "Psychoneuroimmunology" R.Ader and R.A.Good, eds., Academic Press, NewYork.
Maickel, R.P., Braude, M.C., and Zabit, J.E., 1977, The effects of various narcotic agonists and antagonists on deprivation-induced fluid consumption, Neuropharmacology, 16:863.
Maravelias, C.P., and Coutselinis, A.S., 1984, Suppressive effects of morphine on human blood lymphocytes: an in vitro study, IRCS Med. Sci., 12:106.
Margules, D.L., 1979, Beta-endorphin and endoloxone. Hormones of the autonomic nervous system for the conservation or expenditure of bodily resources and energy in anticipation of famine or feast, Neurosci. Biobehav.Rev., 3:155.
Mark-Kaufman, R., and Kanarek, R.B., 1980, Morphine selectively influences macronutrient intake in the rat, Pharmacol.Biochem.Behav., 12:427.
Maynert, E.W., 1967, Effect of morphine on acetylcholine and certain other neurotransmitters, Arch.Biol.Med.exp.(Santiago), 4:36.
Maynert, E.W., and Klingman, G.I., 1962, Tolerance to morphine. I.Effects on catecholamines in the brain and adrenal glands, J.Pharmacol.Exp. Ther., 135:285.
McFarlane, H., 1976, Malnutrition and impaired response to infection, Proc. Nutr.Soc., 35:263.
McFarlane, H., and Hamid, J., 1973, Cell-mediated immune response in malnutrition, Clin.Exp.Immunol. 13:153.
McFarlane, H., Reddy, S., Adcock, K.J., Adeshina, H., Cocke, A.E., and Akene, J., 1970, Immunity, transferrin and survival in kwashiorkor, Brit.Med.J., 4:268.
McDonough, R.J., Madden, J.J., Falek, A., Shafer, D.A., Pline, M., Gordon, D., Bokos, P., Kuehnle, J.C., and Mendelson, J., 1980, Alteration of T and null lymphocyte frequencies in the peripheral blood of human opiate addicts: In vivo evidence for opiate receptor sites on T lymphocytes, J.Immunol., 125:2539.
Meisenberg, G., and Simmons, W.H., 1983, Centrally mediated effects of neurohypophyseal hormones, Neurosci.Biobehav.Rev., 7:263.
Meites, J., Bruni, J.F., VanVugt, D.A., 1979, Relation of endogenous opioid peptides and morphine to neuroendocrine functions, Life Sci., 24: 1325.
Merali, Z., Chosh, P.K., Hrdina, P.D., Singhal, R.L., and Ling, G.M., 1974, Alterations in striatal acetylcholine esterase, and dopamine after methadone replacement in morphine-dependent rats, Eur.J.Pharmacol., 26:375.
Mihich, E., 1962, Host defense mechanisms in the regression of sarcoma 180 in pyridoxine deficient mice, Cancer Res., 22:218.
Mihich, E., and Nichol, C.A., 1965, Differences in the selective antitumor effects of 4-deoxypyridoxine and dietary pyridoxine deficiency, Cancer Res. 25:153.
Monjan, A.A., and Collector, M.I., 1977, Stress-induced modulation of the immune response, Science, 196:307.
Morgan, D.R., DuPont, H.L., Wood, L.V., and Kohl, S., 1984, Cytotoxicity of leucocytes from normal and Shigella-susceptible (opium-treated) Guinea pigs against Shigella sonnei, Infect.Immun., 46:22.
Morley, J.E., 1981, The endocrinology of the opiates and opioid peptides, Metabolism, 30:195.
Morley, J.E., Yamada, T., and Shulkes, A., 1979, Effects of morphine addiction and withdrawal on thyrotropin releasing hormone (TRH), somatostatin (SLI) and vasoactive intestinal peptide (VIP), Clin.Res., 27:75A.

Murray, M.J., and Murray, A.B., 1979, Anorexia of infection as a mechanism of host defense, Am.J.Clin.Nutr. 32:593.
Needham, W.P., Shuster, L., Kanel, C.C., and Thompson, M.L., 1981, Liver damage from narcotics in mice, Toxicol.Appl.Pharmacol. 58:157.
Nilzen, A., 1955, The influence of the thyroid gland on hypersensitivity reactions in animals, Acta Allergol., 7:231.
Noteboom, W.D., and Müller, G.C., 1966 a, Effect of levallorphan on RNA and protein in Hela cells, Fed.Proc., 25:646.
Noteboom, W.D., and Müller, G.C., 1966 b, Inhibition of protein and RNA synthesis in Hela cells by levallorphan and levorphanol, Molec.Pharmacol., 2:534.
Okajima, T., Motomatsu, T., Kato, K., and Ibayashi, H., 1980, Naloxone inhibits prolactin and growth hormone release induced by intracellular glucopenia in the rats, Life Sci., 27:755.
Okouchi, E., 1976, Thymus, peripheral tissue and immunological responsiveness of the pituitary dwarf mice, J.Physiol.Soc.Jpn., 38:325.
Oosterom, R., and Kater, L., 1980, Target cell subpopulations for human thymus, Clin.Immunol.Immunopathol., 17:183.
Oosterom, R., Kater, L., and Oosterom, J., 1979, Effects of human thymic epithelial conditioned medium on mitogen responsiveness of human and mouse lymphocytes, Clin.Immunol.Immunopathol., 12:460.
Özek, M., Töreci, K., Akkök, I., and Güvener, Z., 1971, Die Wirkung der Neuroleptica-Behandlung auf die Antikörperbildung, Psychopharmacologia (Berl.), 21:401.
Palmblad, J., Fohlin, L., and Lundström, M., 1977, Anorexia nervosa and polymorphonuclear (PMN) granulocyte reactions, Scand.J.Haematol., 19:334.
PerezCruet, J., Chiara, G.D., and Gessa, G.L., 1972, Accelerated synthesis of dopamine in the rat brain after methadone, Experientia, 28: 926.
Puri, S.K., and Lal,H., 1973, Effect of dopaminergic stimulation or blockade on morphine-withdrawal aggression, Psychopharmacologia, 32:113.
Puri, S.K., Spaulding, T.C., and Mantione, C.R., 1978, Dopamine antagonist binding: A significant decrease with morphine dependence in the rat striatum, Life Sci., 23:637.
Radominska-Pyrek, A., Kraus-Friedmann, N., Lester, R., Little, J., and Denkins, Y., 1982, Rapid stimulation of Na^+,K^+-ATPase by glucagon, epinephrine, vasopressin and cAMP in perfused rat liver, FEBS Letters 141:56.
Rodin, A.E., 1963, The growth and spread of Walker 256 carcinoma in pinealectomized rats, Cancer Res., 23:1545.
Rofe, A.M., and Williamson, D.H., 1983, Metabolic effects of vasopressin infusion in the starved rat: Reversal of ketonaemia, Biochem.J., 212: 231.
Roy, C., 1979, Vasopressin-sensitive adenylate cyclase: Reversibility of hormonal activation, Biochim.Biophys.Acta, 587:433.
Rusch, H.P., 1944, Extrinsic factors that influence carcinogenesis, Physiol. Rev., 24:177.
Santisteban, G.A., and Dougherty, T.F., 1954, Comparison of the influences of adrenocortical hormones on the growth and involution of lymphatic organs, Endocrinology, 54:130.
Sarne, Y., Gilt-Ad, I., and Laron, Z., 1981, Regulation of hypophysial secretion by endogenous opiates: Humoral endorphin stimulates the release of growth hormone, Life Sci., 28:681.
Sawyer, C.H., Critchlow, B.V., and Barrclough, C.A., 1955, Mechanism of blockage of pituitary activation in the rat by morphine, atropine and barbiturates, Endocrinology, 57:345.
Scrimshaw, N.S., Taylor, C.E., and Gordon, J.E., 1968, "Interactions of nutrition and infection", World Health Organization, Geneva.
Sequeira, W., Jones, E., Siegel, M.E., Lorenz, M., and Kallick, C., 1982,

Pyogenic infections of the pubis symphysis, Ann.Int.Med., 96:604.
Sing, U., 1979, Effect of catecholamines on lymphopoiesis in fetal mouse thymic explants, J.Anat., 129:279.
Shavit, Y., Lewis, J.W., Terman, G.W., Gale, R.P., and Liebeskind, J.C., 1984, Opioid peptides mediate with suppressive effects of stress on natural killer cell cytotoxicity, Science, 223:188.
Sloan, J.W., Brooks, J.W., Eisenman, A.J., and Martin, W.R., 1963, The effect of addiction to and abstinence from morphine on rat tissue catecholamine and serotonin levels, Psychopharmacologia(Berl.), 4:261.
Small, C.B., Klein, R.S., Friendland, G.H., Moll, B., Emerson, E.E., and Spigland, I., 1983, Community-acquired opportunistic infectious and defective cellular immunity in heterosexual drug abusers and hemosexual men, Am.J.Med., 74:433.
Smith, A.A., Hui, F.W., and Crofford, M.J., 1977, Inhibition of growth in young mice treated with d,l-methadone, Eur.J.Pharmacol., 43:307.
Smith, D.J., and Joffe, J.M., 1975, Increased neonatal mortality of offspring of male rats treated with methadone or morphine before mating, Nature, 253:202.
Smythe, P.M., Shonland, M., Brereton-Stiles, G.G., Coovadia, H.M., Grace, H.J., Loening, W.E.K., Mafoyame, A., and Pavent, M.A., 1971, Thymolymphatic deficiency and depression of cell-mediated immunity in protein-calorie malnutrition, Lancet, II:939.
Spackman, D.H., and Riley, V., 1974, Increased corticosterone, a factor in LDH-virus induced alterations of immunological responses in mice, Proc.Am.Assoc.Cancer Res., 15:143.
Spackman, D.H., Riley, V., Santisbetan, G.A., Kirk, W., and Bredburg, L., 1974, The role of stress in producing elevated corticosterone levels and thymus involution in mice, Abstr.Int.Cancer.Congr., 11th, 3:382.
Spector, N.H., Cannon, L.T., Diggs, C.L., Morrison, J.E., and Koob, G.F., 1975, Hypothalamic lesions: Effects on immunological responses, Physiologist, 18:401.
Solomon, G.F., and Amkraut, A.A., 1979, Neuroendocrine aspects of the immune response and their implications for stress effects on tumor immunity, Cancer Detect.Prev., 2:197.
Stein, M., Schiavi, R.C., and Camerino, M.S., 1976, Influencer of brain and behavior on the immune system, Science, 191:435.
Stein, M., Schiavi, R.C., and Luparello, T.J., 1969, The hypothalamus and immune process, Ann.N.Y.Acad.Sci., 164:465.
Suskind, R., 1977, in: "Malnutrition and the Immune Response", RavenPress, NewYork.
Tache, Y., Lis, M., and Collu, R., 1977, Effects of thyrotropin-releasing on behavioral and hormonal changes induced by β-endorphin, Life Sci., 21:841.
Tannenbaum, A., and Silverstone, H., 1953, Nutrition in relation to cancer, Adv.Cancer Res., 1:451.
Tarr, K.H., 1980, Candida endophthalmitis and drug abuse, Aust.J.Ophthalmol. 8:303.
Terragna, A., and Jannuzzi, C., 1963, Ormoni steroidei e anticorpopoiesi. Nota II. Valutazione comparativa di vari ormoni anabolizzanti, G.Mal.Inf., 15:360.
Tolentino, P., Terragna, A., and Jannuzzi, C., 1961, Ormoni stereoidei e anticorpopoiesi. Nota I. Blocco enzimatico della 1. idrossilasi surrenalica. Effetto della somministrazione di androgeni, G.Mal. Inf., 13:561.
Tsypin, A.B., and Maltzev, V.N., 1967, The effects of irritation of the hypothalamus on the concentration of normal antibodies in blood, Patol.Fiziol., 11:83.
Turner, C.D., and Hagnara, J.T., 1971, in: "General Endocrinology" 5th ed., Saunders, Philadelphia.

Visscher, B., Ball, Z., Barnes, R.H., and Silvertsen, I., 1942, The influence of caloric restriction upon the incidence of spontaneous mammary carcinoma in mice, Surgery, 11:48.
Vogt, M., 1954, The concentration of sympathin in different parts of the central nervous system under normal conditions and after the administration of drugs, J.Physiol.(Lond.), 123:451.
Wakabayashi, M., 1966, On the effect of hormones on antibody production, J. Jap.Obstet.Gynaec.Soc., 13:209.
Weiner, N., 1979, Multiple factors regulating the release of norepinephrine consequent to nerve stimulation, Fed.Proc., 38:2193.
Westphal, U., 1971, in: "Steroid-Protein Interactions", Springer-Verlag, NewYork.
White, A., Handler, P., and Smith, E.L., 1968, in: "Principles of Biochemistry" 4th ed. McGraw-Hill, NewYork.
Williams, E.G., and Oberst, F.W., 1946, A cycle of morphine addictions, I: Biological investigations, Public Health Reports, 61:1.
Wood, C.L., Babcock, C.J., and Blum, J.J., 1981, Effects of vasopressin on fructose and glycogen metabolism in hepatocytes from fed and fasted rats, Arch.Biochem.Biophys., 212:43.
Wybran, J., Appleboom, T., Famaey, J.P., and Govaerts, A., 1979, Suggestive evidence for receptors for morphine and methionine-enkephalin on normal blood T lymphocytes, J.Immunol., 123:1068.
Zagon, I.S., and McLaughlin, P.J., 1977, Morphine and brain growth retardation in the rat, Pharmacology 15:276.
Zagon, I.S., and McLaughlin, P.C., 1977 b, Effects of chronic morphine administration on pregnant rats and their offspring, Pharmacology, 15: 302.
Zagon, I.S., and McLaughlin, 1983, Increased brain size and cellular content in infant rats treated with an opiate antagonist, Science, 221:1179.
Zimmerman, E., and Pang, C.N., 1976, Acute Effects of Opiate Administration on Pituitary Gonadotrophin and Prolactin Release, in: "Tissue Response to Addictive Drugs", D.M.Ford and D.H.Clouvet, eds., Spectrum Publications, NewYork.

OPIOID PEPTIDE EFFECTS ON LEUKOCYTE MIGRATION

S. Lori Brown*, Sei Tokuda,** Linda C. Saland,*** and Dennis E. Van Epps*

*Departments of Pathology and Medicine
**Department of Microbiology
***Department of Anatomy

The University of New Mexico School of Medicine
Albuquerque, New Mexico 87131

INTRODUCTION

The recognition that morphine bound stereospecifically to receptors in the brain to exert its effects led to a concerted search for an endogenous ligand for the opiate receptor (Terenius and Wahlstrom, 1975). In 1975, peptides which bound to the opiate receptor were isolated from porcine brain extracts and characterized (Hughes et al., 1975; Hughes et al., 1976). The first opioid peptides characterized were the pentapeptides methionine-and-leucine enkephalin (met- and leu-enkephalin). Simultaneously, a larger (31 amino acid), more active peptide containing the amino acid sequence of met-enkephalin as its first 5 N-terminal amino acids was isolated from porcine and camel pituitary and was named beta-endorphin (a contraction for "endogenous morphine") (Teschemacher et al., 1975; Li and Chung, 1976). Beta-endorphin and alpha-endorphin (amino acids 1-16 of beta-endorphin), gamma-endorphin (amino acids 1-17 of beta-endorphin), (Ling et al., 1976), as well as other neuropeptides, may arise from a 31 Kd glycoprotein, the proopiomelanocortin molecule (reviewed by Frederickson and Geary, 1982). The discovery that beta-endorphin is released from the pituitary along with the steroidogenic hormone, adrenocorticotropin (ACTH) in response to acute stress suggests that in addition to its numerous actions in the central nervous system, beta-endorphin, like ACTH, also has peripheral hormonal effects (Guillemin et al., 1977; Rossier et al., 1977). The release of both ACTH and beta endorphin may be under the control of the hypothalamic peptide, corticotropin releasing factor (CRF) in some species (Hook et al, 1982 and Knepel et al, 1984).

Endogenous ligands for the opiate receptor are traditionally defined as ligands which compete for binding at the opiate receptor with the agonist morphine or with naloxone, the opiate receptor antagonist. Leukocytes have both classical receptors as defined by morphine and naloxone binding (Lopker et al., 1980; Mehrishi and Mills, 1983) and a specific non-opiate receptor for beta-endorphin (Hazum et al., 1979). In addition, leu-enkephalin binds to Jurkat cells, a human lymphoid line, and binding is not prevented by morphine or naloxone suggesting the existence of other lymphocyte receptors which are specific for opioid peptides but unreceptive to exogenous ligands (Ausiello and Roda, 1984).

That leukocytes possess receptors for opiates and opioid peptides suggests that these cells may be targets for peripheral hormonal effects of opioid peptides. A multitude of recent studies have suggested that plasma beta-endorphin levels rise in response to stressful stimuli in both humans and experimental animals (Genazzani et al., 1981; Puolakka et al., 1982; Wilkes et al., 1980; Cosontos et al., 1979; Goland et al., 1981; Mueller, 1981; Wardlaw et al., 1981; Nakao et al., 1981; Fraioli et al., 1980; Colt et al., 1981; Dubois et al., 1981). In humans, the "stressed" level of circulating beta-endorphin has been reported to be as high as 10^{-10} M. Like beta-endorphin, met-enkephalin is also found in the plasma (Clement-Jones et al., 1980; Shanks et al., 1981). Though its relationship to stress is not known, met-enkephalin has recently been detected in bovine adrenal glands (Kilpatrick et al.,1980; Livett et al., 1982; Pelto-Huikko et al., 1982). Because the adrenal glands are activated in response to stressful stimuli, one may cautiously posit that the adrenal glands may be a source of stress-provoked circulating met-enkephalin. Leu-enkephalin is also found circulating in plasma, and its concentration increases in response to stress (Ryder and Eng, 1981).

Another possible source of circulating opioids are leukocytes themselves. Smith and Blalock (1981) have discovered that in response to viral infection human peripheral blood lymphocytes produce a protein which is immunologically indistinguishable from gamma endorphin. Utilizing immunofluorescence and radioimmunoassay, Lolait et al (1984) have shown that mice spleen cells produced a beta endorphin like peptide.

Evidence for a Role for Opioid Peptides in Immunomodulation

Since the discovery of the opioid peptides in 1975, numerous studies have been devoted to determining the location of endogenous opioids and to understanding their role in the central nervous system (reviewed by Smith and Simon, 1981; Hughes et al., 1980; Frederickson and Geary, 1982; Olson et al., 1984). In addition to their analgesic properties centrally administered endogenous opioids mediate numerous behavioral effects (Barchas et al., 1978). Because beta-endorphin and enkephalin have been shown to be released into the peripheral circulation, this suggests that in addition to their numerous roles in the central nervous system, these opioid peptides also act as peripheral hormones (Amir et al., 1980). In the past few years, accumulating evidence indicates that endogenous opioids modulate numerous measurable immune functions (reviewed by Fischer and Falke, 1984; Weber and Pert, 1984). To date, receptors for opiates or opioid peptides have been found on both mononuclear and polymorphonuclear leukocytes (Ausiello and Roda, 1984; Hazum et al., 1979; Lopker et al., 1980; Mehrishi and Mills, 1983). Likewise, opiates and opioid peptides have been found to exert effects on each of these cell types. A summary of *in vitro* studies is presented in Table 1 and briefly in the following text.

The earliest report that an opioid peptide could bind to and modulate leukocyte function was by Wybran et al (1979). They demonstrated that both the exogenous opiate morphine and the endogenous opioid, met-enkephalin, affected the formation of active T-cell rosettes with sheep red blood cells. Though these drugs had opposing effects on active rosette formation, the effects of either were prevented by naloxone, the opiate receptor antagonist. Miller et al (1984) have confirmed the finding that met-enkephalin enhances active T-cell rosette

TABLE 1: OPIOID PEPTIDES AND OPIATE EFFECTS ON LEUKOCYTES IN VITRO

IMMUNE PARAMETER	β	α	γ	M	L	MO	N	REF
Active T-Rosette Formation				↑*		↓*		Wybran et al, 1979
Active T-Rosette Formation				↑	↑			Miller et al, 1984
Total T-Rosette Formation				→		↓*		Wybran et al, 1979
Human Peripheral Blood Lymphocyte Proliferation	↓						→	McCain et al, 1982
Human Peripheral Blood Lymphocyte Proliferation						↑	↑	Bocchini et al, 1983
Rat Spleen Lymphocyte Proliferation	↑	→		→			↓±	Gilman et al, 1982
Mouse Spleen Lymphocyte Proliferation				↑	↑			Plotnikoff & Miller, 1983
Mouse Spleen-Specific Antibody Production	↓	↓*	↓	↓	↓		↓±	Johnson et al, 1982
Human Antibody-Dependent Cellular Cytotoxicity	→							Froelich & Bankhurst, 1984
Human Natural Killer-Mediated Cytotoxicity	↑*	→		↑*	→	→	→	Mathews et al, 1983
Human Natural Killer-Mediated Cytotoxicity				↑	↑			Faith et al, 1984
Human Natural Killer-Mediated Cytotoxicity	↑*	→	↑*	↑	↑		→	Kay et al, 1984
Rat Mast Cell Serotonin Release	↑*			↑*		↑*	↓	Yamasaki et al, 1982
Human Leukocyte Chemotaxis								
Mononuclear Cells	↑*			↑*			→	Van Epps & Saland, 1984
Neutrophils	↑			↑				Van Epps et al, 1983
Response to Chemotactic Factor After Preincubation in Opioid:								
Mononuclear Cells	↓							Van Epps & Saland, 1984
Neutrophils	→							Van Epps & Saland, 1984
Neutrophils	↑*						↑±	Simpkins et al, 1984
Production of Lymphocyte Chemotactic Factor	↓	→		↓			↓	Brown & Van Epps, 1985
Leukocyte Adherence	↑			↑				Van Epps (unpublished)

Key:

α = Alpha-Endorphin
β = Beta-Endorphin
γ = Gamma-Endorphin

M = Met-Enkephalin
L = Leu-Enkephalin
MO = Morphine
N = Naloxone

↑ Enhanced
↓ Inhibited
→ No Effect
± Insignificant Effect

*Effect prevented by naloxone

formation and have shown that leu-enkephalin has similar effects in this assay. Met- and leu-enkephalin not only increase the percentage of T-lymphocytes forming active rosettes from normal human donors but have similar effects on T cells from lymphoma patients (Miller et al., 1983).

Opioid peptides and opiates also have effects on in vitro T-lymphocyte proliferation in response to the mitogenic lectins, phytohemmaglutinin (PHA) and concanavalin A (Con A). McCain et al (1982) have shown that 10^{-7} M beta-endorphin suppressed the human peripheral blood lymphocyte (PBL) blastogenic response to PHA. The suppressive effects of beta-endorphin were not prevented by 10^{-6} M naloxone which, alone, had no effect on PHA mitogenesis. Bocchini et al (1983) have shown that naloxone and morphine both enhance the proliferative response of human PBL to PHA. Phencyclidine, a drug which may bind to opiate receptors (Vincent et al, 1978), has suppressive effects on mitogen-induced proliferation of human lymphocytes (Khansari et al., 1984). Though human PBL proliferation was suppressed by some concentrations of beta-endorphin, rat spleen lymphocyte (RSL) proliferation appears to be enhanced (Gilman et al., 1982). In these experiments, naloxone, which slightly suppressed RSL proliferation itself, did not prevent the effects of beta-endorphin. Mouse spleen lymphocyte proliferation has also been shown to be enhanced by leu- and met-enkephalin (Plotnikoff and Miller, 1983).

The response to the B-cell mitogen, lipopolysaccharide/dextran sulfate, by RSL was not affected by beta-endorphin (Gilman et al., 1982). However, production of antibody by B cells in response to a T-cell-dependent antigen was depressed by alpha-endorphin, met- and leu-enkephalin and to a lesser degree by beta- and gamma-endorphin (Johnson et al., 1982). Naloxone prevented the suppressive effect of alpha-endorphin on antibody production by B cells in this system.

Beta-endorphin has been shown to have no effect on antibody-dependent cellular cytotoxicity (Froelich and Bankhurst, 1984), but beta-endorphin, met-enkephalin, leu-enkephalin and gamma-endorphin enhance human natural killer (NK) activity against the NK-sensitive K562 cell line (Mathews et al., 1983; Faith et al., 1984; Kay et al., 1984). The enhancing effects of beta-endorphin, met-enkephalin and gamma-endorphin were prevented by naloxone. Others, however, have reported that opioid peptides depress human NK activity as well as killing by cytotoxic T-lymphocytes (Prete and Levin, 1985).

In addition to the effects of opioids on lymphocytes, opioids may also bind to mast cells and cause degranulation. Yamasaki et al (1982) have shown that beta-endorphin, met-enkephalin, and morphine reverse or prevent the PGE_1-mediated inhibition of serotonin release from rat mast cells. Others have shown that opiates and opioid peptides induce wheal and flare reactions in skin testing and enhance histamine-induced itch and flare (Casale et al., 1984; Fjellner and Hagermark, 1982).

All of these findings indicate that leukocytes have functional opiate receptors or receptors for opioid peptides which when bound by ligand may modulate the function of these cells. Furthermore, taken together, these studies indicate that opioid peptides may potentially affect both the afferent and efferent limbs of the immune response. One aspect of the expression of immunity which may be influenced by the opioid peptides is leukocyte chemotaxis. The directional locomotion of cells in response to a gradient of chemotactic factor is a functional characteristic of polymorphonuclear leukocytes (PMN), monocytes and lymphocytes and will be the subject of the remainder of this review.

LEUKOCYTE CHEMOTAXIS AND OPIOID PEPTIDES

Chemotaxis is the directional movement of cells in response to a chemical gradient. The ability of leukocytes to sense and respond to chemotactic factors may, in part determine the outcome of immune or inflammatory responses. In vivo, this mobilization of leukocytes to a source of chemotactic factor is believed to be the major mechanism of accumulating PMN, monocytes or lymphocytes at an inflammatory focus or site of antigenic challenge. The chemotactic factors may be generated as a result of complement activation and the consequent generation of C5a, activation of arachidonic acid metabolism to produce leukotrienes, or by antigen or mitogen triggering of lymphocytes to release chemotactic factors for neutrophils, monocytes or lymphocytes.

The elaboration of chemotactic lymphokines by inflammatory or immune cells may be important in stimulating the migration of immunocytes to sites of inflammation or of antigen deposition. The type of cell accumulating at the inflammatory site is probably dictated by a number of factors, one of which is the cell specificity of the chemotactic factor generated at that site. For example, antigen-antibody complex deposition and the release of C5a derived from the activation of the complement system stimulates predominantly a neutrophil influx. A monocyte-specific chemotactic factor can also be derived from antigen- or mitogen-challenged lymphocytes and may be responsible for the influx of monocytes and macrophages of cellular immune reactions. More recently, the release of a T-cell-specific chemotactic factor from antigen- or mitogen-challenged lymphocytes has been described which may account for the concentration of T cells at sites involving cellular immune reactions, such as delayed cutaneous hypersensitivity reactions, allograft rejections, or tumor rejections. It is apparent that any agents that may alter the migration of leukocytes to an inflammatory site may potentially result in a compromised or, alternatively, an enhanced expression of immunity in the host.

Several recent studies have shown that opioid peptides can directly stimulate leukocyte chemotaxis. In addition, more recent studies from this laboratory indicate that opioids can also modulate the production of a T-lymphocyte chemotactic factor by human mononuclear cells.

Beta-endorphin and Met-enkephalin Stimulate Leukocyte Locomotion In Vivo

Acute injection of the opioid peptides beta endorphin and met enkephalin into the cerebral ventricle of cannulated rats, results in the rapid influx of macrophage-like cells (Saland et al, 1983 and Van Epps et al, 1983). Within an hour following infusion of these opioid peptides into the ventricle, scanning and transmission electron microscopy revealed numerous cells exhibiting a typical macrophage morphology (high cytoplasmic to nuclear ratio, multiple granules and vacuoles, pseudopod-like projections, and ruffled membrane) atop the ependyma. In these previous studies, beta endorphin was much more effective than met enkephalin at stimulating migration of these cells into the ventricle. Alpha endorphin infused animals were indistinguishible from saline infused controls who displayed only a minimal cellular infiltration. The origin of these macrophage-like cells is not known but they could conceivably arise from peripheral blood monocytes (Baldwin, 1980) or from a pool of macrophage-like microglial cells residing in close proximity to the ventricle (Bleier and Marsh, 1978).

More recent studies (Saland et al, 1984) demonstrate that chronic infusion of beta endorphin into the rat cerebral ventricle, using osmotic pumps over 24- or 48-hours, induces an inflammatory-like cellular

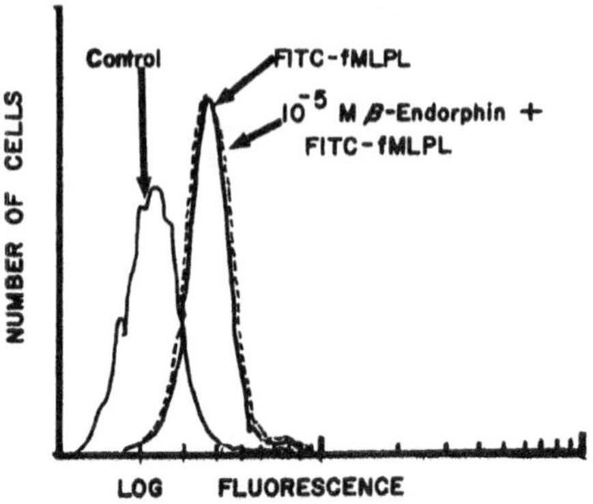

Figure 1. Beta-endorphin does not bind to the f-MLP receptor on neutrophils. Control neutrophils were incubated in media alone and represent autofluorescence of neutrophils. Neutrophils treated with FITC-fMLPL bind the fluorescent probe (solid line). Those neutrophils pretreated for 15 minutes with beta-endorphin (broken line) bind the f-MLPL equally well indicating that the endorphin is not binding at the f-MLP receptor.

response, including the appearance of neutrophils, lymphocyte-like cells, and macrophages. Infusion of other related compounds (naloxone, alpha endorphin, met enkephalin) or saline produced little or no cellular response.

That these opioids stimulated a cell infiltration indicated that they could be chemotactic for leukocytes and suggested that macrophages have a mechanism for recognizing and responding to opioid peptides. Although it is difficult to visualize how a gradient of opioid peptide may form spontaneously _in vivo_, the effects of opioid peptides on leukocytes may be particularly relevent to the recent clinical use of endogenous opiates. Several reports have described the use of opioid peptides as an analgesic by infusion into cerebral spinal fluid (CSF) during labor and delivery (Oyama et al, 1980) or into individuals with chronic pain (Hosobuchi and Li, 1978). The potential also exists for interaction of opiate drugs, such as morphine or its derivatives, with cells of the immune system within the CSF during administration to patients via a lumbar puncture route (Stoelting, 1980 and Hughes et al, 1984).

Opioid Peptides Stimulate Chemotaxis of Human Leukocyte In Vitro

Numerous methods for studying chemotaxis _in vitro_ are currently used (Gee, 1984). The response of human peripheral blood leukocytes can be measured _in vitro_ using a modification of the Boyden assay (Boyden, 1962). In this assay, factors being assessed for their ability to

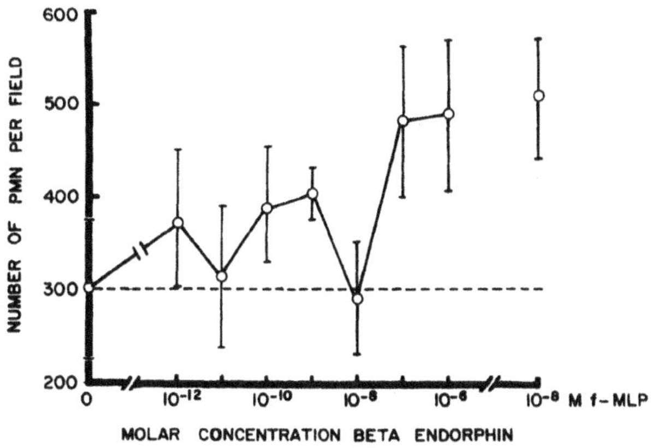

Figure 2. Beta-endorphin enhances neutrophil adherance. Data are expressed as the number of neutrophils adhering per high power field. The adherence of neutrophils in response to 10^{-8} M f-MLP is shown for comparison. Each point represents the Mean±SD of 3-experiments.

stimulate locomotion are placed in the lower chamber of a blind well chemotaxis chamber. A membrane of known pore size is laid over the test solution, and separates leukocytes, in the upper compartment, from the test solution. The distance that leukocytes migrate into the nitrocellulose membrane in response to the test solution can be assessed by using a microscope with micrometer. These techniques have been applied to evaluate the response of human neutrophils, lymphocytes and monocytes to opioid peptides *in vitro*, and results of these studies are summarized below.

Previous studies from this laboratory have shown that both beta endorphin and met enkephalin stimulated migration of human peripheral blood mononuclear cells *in vitro* (Van Epps et al, 1983C and Van Epps and Saland, 1984). It was determined that the effect of these opioid peptides on cell locomotion was due to an enhanced directional locomotion (chemotaxis) and not due to enhanced random migration (chemokinesis). This response was dose-dependent and attributable to the glass adherent population in the mononuclear cell preparation. When beta endorphin or met-enkephalin were tested in concentrations ranging from 10^{-14} to 10^{-7} M an interesting phenomenon was observed. Both opioids stimulated bimodal peaks of mononuclear cell locomotion. The significance of a bimodal peak is not clear but could be due to, 1.) multiple opiate receptor types on mononuclear cells, 2.) differential binding of opioid peptides by mononuclear cell subsets, or 3.) a combination of 1 and 2. The observed migration of freshly isolated mononuclear cells to 10^{-8} M beta endorphin in these studies was approximately 80% of that seen in response to 10^{-8} M formyl-methionyl-leucyl-phenylalanine (f-MLP). F-MLP is a synthetic peptide which is a relatively potent chemotactic factor for monocytes (Cianciolo and Snyderman, 1980) and for neutrophils (Showell et al, 1976). Although neutrophils have been shown to have more receptors for dihydromorphine than monocytes (Lopker et al, 1980), similar measurements of neutrophil migration in responses to beta endorphin showed variable response averaging only 30% of that observed utilizing f-MLP as a chemotactic stimulus. This may reflect the differences in PMN and monocyte responsiveness to f-MLP since monocytes do not migrate as far in response to fMLP as neutrophils. In the same study, monocytes also showed enhanced migration in response to met-enkephalin, but the magnitude of the response was less than that seen in response to a gradient of beta endorphin. These studies imply that monocytes recognize and respond more effectively to opioid peptides than neutrophils. It is feasible that the lower responsiveness of neutrophils may be due to their demonstrated ability to oxidize met-enkephalin to its sulfoxide derivative (Turkall et al, 1982) and render it inactive. This may also hold true for beta-endorphin which also contains methionine.

In our previous studies experiments were performed to determine whether the migration of mononuclear cells to beta-endorphin and met-enkephalin was opiate receptor mediated. Monocytes were preincubated in naloxone and then placed in the upper compartment of chemotaxis chambers containing beta endorphin or met enkephalin in the lower compartment. Naloxone prevented the migration of mononuclear cells to both of the opioid peptides. Naloxone also showed some slight inhibition of the mononuclear cell response to f-MLP, although not approaching that seen with the opioid peptide-mediated chemotaxis.

Because of the presence of some similar amino acids in met-enkephalin (Tyr-gly-gly-phe-met), beta-endorphin, and f-met-leu-phe, we wished to determine whether opioids were binding at the f-MLP receptor. Studies were performed to determine if either met enkephalin or beta endorphin could block the binding of the f-MLP analog, FITC-f-met-leu-phe-lysine (Van Epps and Saland, 1984). By flow cytometry, 30% of the mononuclear cells bound the fluoresceinated conjugate of f-MLPL which binds to the f-MLP receptor. This binding is attributable to monocytes of which approximately 70-80% bind f-MLP. Neither beta endorphin nor met enkephalin blocked the binding of FITC-f-MLPL indicating that these opioid peptides were not interacting with the f-MLP receptor on monocytes. Similar studies were conducted using human neutrophils. Isolated neutrophils were preincubated for 15 minutes in media alone or with 10^{-5}M beta-endorphin. FITC-fMLPL was then added to the neutrophils for 15 minutes at 25°C, washed and fixed with paraformaldehyde, and binding of the fluorescent fMLP analog to neutrophils was analyzed by flow cytometry. As can be seen in Figure 1, the f-MLPL binding to neutrophils was not blocked by beta-endorphin, indicating that endorphin was not binding to f-MLP receptors on neutrophils. Similar experiments performed with 10^{-5}M met-enkephalin indicated that it did not reduce the binding of f-MLPL to its receptor.

In vivo, one can imagine that leukocytes may be exposed to circulating opioids prior to their migrating from the blood to an inflammatory focus. This type of pre-exposure was simulated in an *in vitro* model in which leukocytes were preincubated in 10^{-8}M beta endorphin for 15 minutes and their chemotactic response to a second stimulus, such as f-MLP or the chemotactic factor casein, is tested (Van Epps et al, 1983c). When mononuclear cells or isolated lymphocytes were tested in this fashion, there was an inhibition of the migration in response to f-MLP or casein. This phenomenon of decreased responsiveness to a second chemotactic stimuli after exposure to a first chemotactic stimuli is known as down regulation or desensitization and has been previously reported (Nelson et al, 1978). In the modified Boyden assay, beta-endorphin did not down regulate the response of neutrophils to a second chemotactic stimulus (Van Epps et al, 1983c). However, others have reported an enhanced response to f-MLP by neutrophils preincubated in beta endorphin when chemotaxis was measured by the chemotaxis under agarose technique (Simpkins et al, 1984). In their studies the migration response of neutrophils to 10^{-7} M f-MLP was significantly enhanced when PMN were preincubated in beta endorphin. It is interesting that although naloxone alone tended to enhance the chemotactic response of neutrophils, it antagonized the effects of beta endorphin on neutrophils. Is it possible that naloxone alone exerts some agonistic effects but in the presence of an agonist for the opiate receptor, such as beta endorphin, it acts as an antagonist? An alternative explanation for these seemingly incongruent effects of naloxone may be that additive effects of the two factors result in a suppression of the response. Similar mixed effects of naloxone have been observed in other systems. In studies by Johnson et al (1982), naloxone mildly inhibited specific antibody production, while it prevented the inhibitory effects of alpha endorphin in this same system. Likewise, naloxone, like the opiate agonist morphine, enhanced the proliferative response of human lymphocytes (Bocchini et al, 1983). It may be that naloxone is not a pure antagonist for the opiate receptor.

One critical requirement for leukocyte chemotaxis is the necessity for adherence. It is well known that chemotactic factors enhance neutrophil adherence (O'Flaherty et al., 1978; Hoover et al., 1980) and that variation in the adherence of neutrophils to a substratum will influence their ability to migrate (Wilkinson et al., 1982). In view of the fact that our previous studies of PMN and monocyte responses to opioid peptides and those of Simpkins et al, (1984) indicate that the chemotactic response of these cells may be influenced by opioid peptides, experiments were designed to determine the effects of both beta-endorphin and met-enkephalin on human neutrophil adherence. In these studies, 1.5 X 10^5 isolated human PMN were placed in 1 X 2 cm compartments on 5% fetal calf serum-coated tissue culture slides and incubated for 20 minutes at 37°C with or without opioid peptide. Slides were then washed gently, fixed with paraformaldehyde, and the number of cells per 100X field quantified. As shown in Figure 2, when 10^{-12} to 10^{-6} M beta-endorphin is added to neutrophils, a marked increase in the number of cells adhering to glass slides following a 20-minute incubation was observed. This response was dose dependent, and the titration curve was suggestive of a bimodal response like that seen previously in the chemotaxis assays. The adherence of neutrophils in the presence of 10^{-8} M f-MLP is shown for comparison. Met-enkephalin, too, enhanced the adherence of neutrophils to the coated slides. The observed enhancing effects of opioid peptides on neutrophil adherence may account in part for some of the effect on PMN locomotion observed by Simpkins et al where enhancement of neutrophil migration to f-MLP was observed.

The demonstration that opioid peptides directly influence leukocyte migration and that these effects are naloxone reversible is additional evidence for the existence of functional opioid receptors on leukocytes. It can be imagined that _in vivo_ exposure of leukocytes to circulating beta endorphin or met enkephalin may result in an altered response to chemotactic factors which might affect the host's immune status.

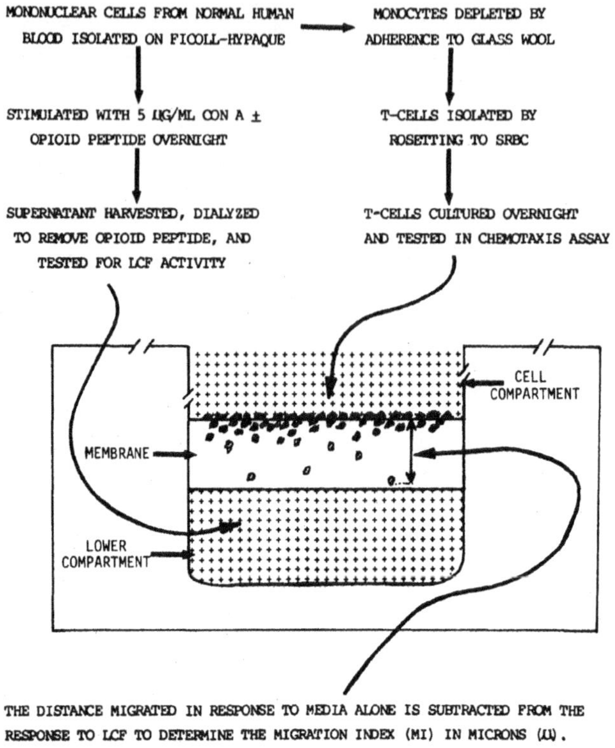

Figure 3 The production and measurement of T-lymphocyte chemotactic factor (LCF).

EFFECTS OF OPIOID PEPTIDES ON T-LYMPHOCYTE CHEMOTACTIC FACTOR PRODUCTION

In recent years, the central role of lymphokines in regulating the immune response has become apparent. The T-lymphocyte chemotactic factor (LCF) is a lymphokine produced by Con A or antigen stimulated mononuclear cells (Van Epps, 1982, Van Epps et al, 1983A, Van Epps et al, 1983B, Center and Cruikshank, 1982, Cruikshank and Center, 1982) This lymphokine is produced by lymphocytes and is chemotactic for T-lymphocytes. Because only a small number of the lymphocytes that are attracted to a site of antigen deposition are antigen specific (McCluskey et al, 1963), it may be hypothesized that this chemotactic lymphokine is important in recruiting unsensitized lymphocytes to a site of antigenic challenge. The fact that this or a similar lymphokine is produced in the human mixed lymphocyte reaction (El-Naggar et al, 1982) suggests that recruiting lymphocytes to the site of an allograft may be a critical step in allograft rejection.

Our recent studies show that the production or secretion of this lymphokine is inhibited by both beta endorphin and met enkephalin (Brown and Van Epps, 1985). LCF is produced by Con A-stimulated human mononuclear cells and can be measured utilizing the leading front T-cell chemotaxis assay (Van Epps et al, 1983a). Figure 3 outlines the procedure for generating LCF containing supernatants and measuring its activity in the T-cell chemotaxis assay. Both beta endorphin and met enkephalin suppressed the generation of this lymphokine by mononuclear cells. On the contrary, alpha-endorphin did not significantly alter the production of this lymphokine. The effect of high concentrations

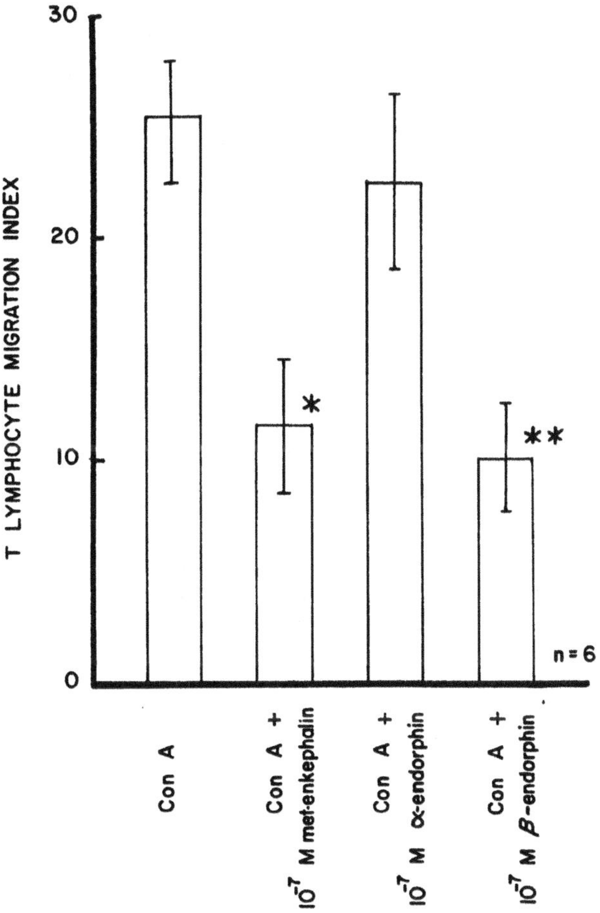

Figure 4. LCF production in the presence and absence of opioid peptides. The production of LCF by opioid containing cultures is compared to the control (Con A) supernatant. *$p \leq .05$, **$p \leq .001$. Supernatants generated with alpha-endorphin do not differ significantly from the control.

(10^{-7}M) of beta endorphin, met enkephalin, and alpha endorphin on LCF production are shown in Figure 4. These data are expressed as the migration index: the difference in the distance in microns (u) migrated by T lymphocytes in response to LCF containing supernatants and to media alone. As can be seen, the distance that T cells migrated in response to supernatants generated with 10^{-7}M beta-endorphin or met-enkephalin present is significantly decreased, reflecting a decrease in the generation of LCF in these supernatants. Alpha-endorphin, however, had no significant effect on the production of LCF (Brown and Van Epps, 1985). As can be seen in Figure 5, a titration of beta endorphin resulted in bimodal peaks of inhibition of LCF production. In the two representative experiments shown, it should be noted that peak inhibition varied slightly, but the bimodal curve was a consistent finding and the peaks of inhibition were between 10^{-6} - 10^{-8}M and 10^{-10} to 10^{-12}M. This variability probably represents differences in the sensitivity of mononuclear cells from different donors to opioid peptides as well as differences in responding T-cell preparations. The same effect was seen with met enkephalin. Again, the phenomenon of a bimodal

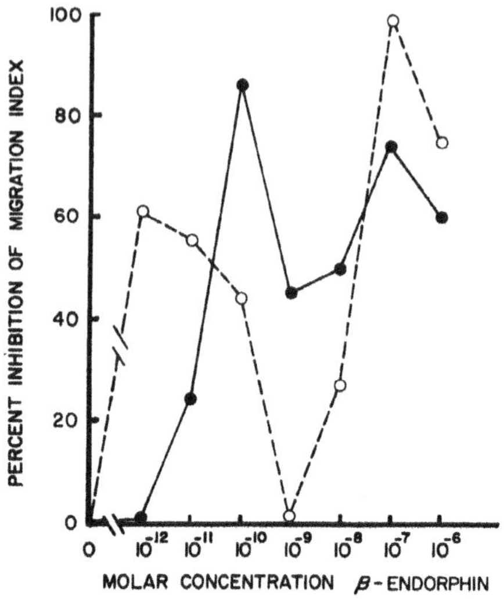

Figure 5. Inhibition of LCF production is bimodal. Experiment 1 and experiment 2 demonstrate that although the peak inhibition may vary from donor to donor, it is a consistent finding.

peak may be due to multiple types of receptors for opioids on these cells or to different receptors on mononuclear cell subpopulations which in turn may influence LCF production.

Experiments to determine whether the effect of these peptides on LCF production were mediated by the opiate receptor showed that naloxone, like the opioid peptides, inhibited the production of LCF. However, unlike the opioid peptides, mononuclear cells titrated with naloxone exhibited only a single peak of inhibition at 10^{-9}M. It was surprising then, that although naloxone exhibited some agonistic qualities, it was a partial antagonist with respect to the effects of both beta endorphin and met enkephalin as is shown in Figure 6. For instance, the migration index in response to the control LCF supernatant (Con A alone) was 35 ± 6u. When 10^{-8}M naloxone was tested, the migration index was decreased to 13 ± 7u. LCF supernatants generated in the presence of 10^{-8}M beta-endorphin stimulated a response of 17 ± 6u. However, when both 10^{-8}M naloxone and 10^{-8}M beta-endorphin were included in culture, mononuclear cells produced slightly more LCF activity than was seen in response to beta-endorphin alone, but not as much as was seen in control supernatants. This finding was consistent with several concentrations of naloxone and opioid peptide suggesting that it was not an artifact. Again, this speaks strongly for a re-evaluation of naloxone's status as a pure antagonist of the opiate receptor. Indeed it has been suggested that naloxone may exert effects independent of the opiate receptor (Badawy et al, 1983) or in some circumstances act as an agonist (Ferreira and Nakamura, 1979).

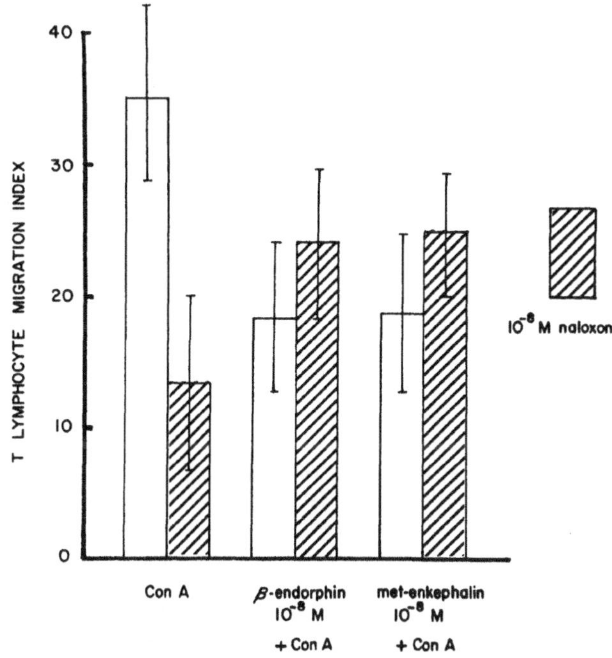

Figure 6. Naloxone alone inhibits the production of LCF but in the presence of opioid peptide acts as a partial antagonist. These data are the mean ± S.E. of 3 experiments.

When glass adherant monocytes are depleted from mononuclear cells resulting in a monocyte-depleted lymphocyte preparation (MDL), there was a pronounced loss of the inhibitory effects of both beta endorphin and met enkephalin on LCF production. When the production of LCF by MDL is compared to that by mononuclear cells (MNC) in the presence of opioid peptide ranging in concentration from 10^{-12} to 10^{-6}M, the suppression of LCF production is diminished. For instance, it was shown that with 10^{-7}M beta endorphin, MNC production of LCF was inhibited by 57±13% (n=6). In response to LCF generated by MDL preparations from the same donors, the migration index of T-cells was only inhibited by 2±1%. The efficacy of met enkephalin, too, was diminished when monocytes were depleted from mononuclear cell preparations. These data suggested that the inhibitory effects of opioids on LCF production were mediated by monocytes.

Like monocyte depletion, inclusion of the cyclooxygenase inhibitor, indomethacin, also prevented the inhibitory effects of opioids on LCF production. As is shown in Table 2, indomethacin prevented the effects of both beta endorphin and met enkephalin in a dose dependent manner when used in concentrations between 10^{-6}M and $5X10^{-6}$M. A major product of monocyte cyclooxygenase activity is PGE_2 (Bonney et al 1979 and Kurland et al , 1978). PGE_2, but not other prostaglandins (PGA, PGF_2) inhibit the production of LCF (Van Epps et al, 1985).

The PGE_2 mediated suppression of LCF production is dose dependent between 10^{-12} - 10^{-6}M with lower doses of PGE_2 being less effective. In the presence of inodomethacin and either beta-endorphin or met-enkepahlin subinhibitory concentrations of PGE_2 were inhibitory as is shown in Table 3. These data suggested an opioid peptide mediated enhancement of mononuclear cell sensitivity to the suppressive effects of

TABLE 2: INDOMETHACIN PREVENTS OPIOID PEPTIDE SUPPRESSION OF LCF PRODUCTION*

INDOMETHACIN CONCENTRATION	BETA-ENDORPHIN 0	BETA-ENDORPHIN 10^{-7}M	MET-ENKEPHALIN 0	MET-ENKEPHALIN 10^{-7}M
0	22±3	11±4	25±5	13±5
$1X10^{-6}$M	23±4	19±4	31±4	20±4
$5X10^{-6}$M	23±4	20±8	28±5	26±4

*Expressed as the migration index (u). Beta endorphin data represent the Mean±SE of 5 experiments and the met enkephalin data of 3 experiments.

TABLE 3: SUBINHIBITORY DOSES OF PGE_2 RECONSTITUTE THE SUPPRESSIVE EFFECT OF OPIOID PEPTIDES IN THE PRESENCE OF INDOMETHACIN*

	PGE_2 CONCENTRATION#		
OPIOID PEPTIDE	0	10^{-12}M	10^{-10}M
Control	0±0%	15±3%	19±8%
10^{-8}M Beta Endorphin	11±4%	53±8%	49±8%
10^{-8}M Met Enkephalin	15±6%	42±13%	40±8%

* All LCF supernatants were generated in the presence of $5X10^{-6}$M indomethacin.

\# The data is expressed as the percent inhibition of the migration index.

** Each point represents the Mean ± SE from 6-experiments.

PGE_2 on LCF production. It was not determined whether this was a direct effect of PGE_2 on LCF producing cells or an indirect effect mediated by another cell type. Enhanced sensitivity to PGE_2 mediated effects on mononuclear cells has been reported by others to occur in a stressed and an aged population (Goodwin et al, 1981 and Goodwin and Messner, 1979).

CONCLUSIONS

These studies indicate that opioid peptides may modulate the immune response under some circumstances. Opioid peptides are found circulating in normal individuals and at higher concentrations in stressed individuals. Their binding to leukocytes may modulate the expression of immunity in these individuals.

The continued study of these interactions will enhance our understanding of the effects of neuroendocrine hormones on the immune system in stressed as well as in healthy individuals and will define the link between the neuroendocrine and immune systems.

ACKNOWLEDGEMENTS

We gratefully acknowledge Jeff Potter and Susan Griewe for their expert technical assistance. We thank Patricia Eckhardt for her skill and patience in preparing this manuscript.

REFERENCES

Amir, S., Brown, Z. W., and Amit, Z., 1980, The role of endorphins in stress: Evidence and speculations, Neurosci. Biobehav. Rev., 4:77.

Ausiello, C. M., and Roda, L. G., 1984, Leu-enkephalin binding to cultured human T lymphocytes, Cell Biol. Intern. Rep., 8:97.

Badawy, A. A., Evans, M., Punjani, N. F., and Morgan, C. J., 1983, Does naloxone always act as an opiate antagonist? Life Sci., 33 (suppl.):739.

Baldwin, F., The reticuloendothelial system. A comprehensive treatise, in: Vol. 1 "Morphology, Microglia and Brain Macrophages," I. Carr and W. T. Daems, eds., Plenum Press, New York (1980).

Barchas, J. D., Akil, H., Elliottt, G. R., Holman, R. B., and Watson, S. J., 1978, Behavioral neurochemistry: Neuroregulators and behavioral states, Science, 200:964.

Bleier, R., and Albrecht, R., 1980, Supraependymal macrophages of third ventricle of hamster: Morphological, functional and histochemical characterization in situ and in culture, J. Comp. Neurol., 192:489.

Bocchini, G., Bonanno, G., and Canevari, A., 1983, Influence of morphine and naloxone on human peripheral blood T lymphocytes, Drug and Alchol. Depend., 11:233.

Bonney, R. J., Naruns, P., Davies, P., and Humes, J. L., 1979, Antigen-antibody complexes stimulate the synthesis and release of prostaglandins by mouse peritoneal macrophages, Prostaglandins, 18:605.

Boyden, S. V., 1962, The chemotactic effect of mixtures of antibody and antigen on polymorphonuclear leukocytes., J. Exp. Med., 115:453.

Brown, S. L., and Van Epps, D. E., 1985, Suppression of T-lymphocyte chemotactic factor production by the opioid peptides beta endorphin and met enkephalin, J. Immunol., 134:3384.

Casale, T. B., Bowman, S., and Kaliner, M., 1984, Induction of human cutaneous mast cell degranulation by opiates and endogenous opioid peptides: Evidence for opiate and non-opiate receptor participation, J. Allergy Clin. Immunol., 73:775.

Center, D. M., and Cruikshank, W., 1982, Modulation of lymphocyte migration by human lymphokines. Identification and characterization of chemoattractant activity for lymphocytes from mitogen-stimulated mononuclear cells, J. Immunol., 128:2563.

Cianciolo, G. J., and Snyderman, R., 1981, Monocyte responsiveness to chemotactic stimuli is a property of a subpopulation of cells that can respond to multiple attractants, J. Clin. Invest., 67:60.

Clement-Jones, V., Lowry, P. J., Rees, L. H., and Besser, G. M., 1980, Met enkephalin circulates in human plasma, Nature, 283:295.

Cohen, M., Pickard, D., Dubois, M., Roth, Y. F., Naber, D., and Bunney, W. E. Jr., 1981, Surgical stress and endorphins, Lancet, 1:213.

Colt, E. W., Wardlaw, S. L., and Frantz, A. G., 1981, The effects of running on plasma beta endorphin, Life Sci., 28:1637.

Cosontos, K., Rust, M., Hollt, V., Mahr, W., Kromer, W., and Teschemacher, H.J., 1979, Elevated plasma beta endorphin levels in pregnant women and their neonates, Life Sci., 25:835.

Cruikshank, W., and Center, D. M., 1982, Modulation of lymphocyte migration by human lymphokines. II. Purification of a lymphotactic factor, J. Immunol., 128:2569.

Dubois, M., Pickar, D., Cohen, M. R., Roth, Y. F., Macnamara, T., and Bunney, W. E. Jr., 1981, Surgical stress in humans is accompanied by an increase in plasma beta endorphin immunoreactivity, Life Sci., 29:1249.

El-Naggar, A., Van Epps, D. E., and Williams, R. C. Jr., 1982, A human lymphocyte chemotactic factor produced by the mixed lymphocyte reaction, J. Lab. Clin. Med., 100:4.

Faith, R. E., Liang, H. J., Murgo, A. J., and Plotnikoff, N. P., 1984, Neuroimmunomodulation with enkephalins: Enhancement of human natural killer (NK) cell activity in vitro, Clin. Immunol. and Immunopath., 31:412.

Ferreira, S. H., and Nakamura, M., 1979, Prostaglandin hyperalgesia: The peripheral analgesic activity of morphine, enkephalins, and opioid antagonists, Prostagland., 18:191.

Fischer, E. G., and Falke, N. E., 1984, Beta-endorphin modulates immune function, Psychother. Psychosom., 42:195.

Fjellner, B., and Hagermark, O., 1982, Potentiation of histamine-induced itch and flare responses in human skin by the enkephalin analogue FK 33-824, beta-endorphin, and morphine, Arch. Derm. Res., 274:29.

Fraioli, F., Moretti, C., Paolucci, D., Alicicco, E., Crescenzi, F., and Fortunio, G., 1980, Physical exercise stimulates marked concomitant release of beta endorphin and adrenocorticotropic hormone (ACTH) in peripheral blood in man, Experientia, 36:987.

Frederickson, R. C. A., and Geary, L. E., 1982, Endogenous opioid peptides: Review of physiological, pharmacological and clinical aspects, Progress in Neurobiol., 19:19.

Froelich, C. J., and Bankhurst, A. D., 1984, The effect of beta endorphin on natural cytotoxicity and antibody dependent cellular cytotoxicity, Life Sci., 35:261.

Gee, A. P., 1984, Advantages and limitations of methods for measuring cellular chemotaxis and chemokinesis, Mol. Cell. Biochem., 62:5.

Genazzani, A. R., Facchinetti, F., and Parrini, D., 1981, beta-lipotrophin and beta endorphin plasma levels during pregnancy, Clin. Endocrin., 14:409.

Giagnoni, G., Santagostino, A., Senini, R., Fumagalli, P., and Gori, E., 1983, Cold stress in the rat induces parallel changes in plasma and pituitary levels of endorphin and ACTH, Pharm. Res. Comm., 15:15.

Gilman, S. C., Schwartz, J. M., Millner, R. J., Bloom, F. E., and Feldman, J.D., 1982, Beta endorphin enhances lymphocyte proliferative responses, Proc. Natl. Acad. Sci., 79:4226.

Goland, R. S., Wardlaw, S. L., Stark, R. I., and Frantz, A. G., 1981, Human plasma beta endorphin during pregnancy, labor and delivery, J. Clin. Endocrin. Metab., 52:74.

Goodwin, J. S., Bromberg, J., Staszak, C., Kaszubowski, P., Messner, R. P., and Neal, J. F., 1981, Effect of physical stress on sensitivity of lymphocytes to inhibition of prostaglandin E_2, J. Immunol., 132:246.

Goodwin, J.S., and Messner, R. P., 1979, Sensitivity of lymphocytes to prostaglandin E_2 increases in subjects over age 70, J. Clin. Invest., 64:434.

Guillemin, R., Vargo, T., Rossier, J., Minick, S., Ling, N., Rivier, C., Vale, W., and Bloom, F., 1977, Beta endorphin and adrenocorticotropin are secreted concomitantly by the pituitary gland, Science, 197:1367.

Hazum, E., Chang, K., and Cuatrecasas, P., 1979, Specific nonopiate receptors for beta endorphin, Science, 205:1033.

Hook, V. Y. H., Heisler, S., Sabol, S. L., and Axelrod, J., 1982, Corticotropin releasing factor stimulates adrenocorticotropin and beta-endorphin release from AtT-20 mouse pituitary tumor cells, Biochem. Biophys. Res. Comm., 106:1364.

Hoover, R. L., Folger, R., Haering, W. A., Ware, B. R., and Karnovsky, M. J., 1980, Adhesion of leukocytes to endothelium: Roles of divalent cations, surface changes, chemotactic agents and substrate, J. Cell. Science, 45:73.

Hosobuchi, Y., and Li, C. H., 1978, The analgesic activity of human Beta-endorphin in man, Communications in Psychopharmacology, 2:33-37.

Hughes, J., Beaumont, A., Fuentes, J. A., Malfroy, B., and Unsworth, C., 1980, Opioid peptides: Aspects of their origin, release and metabolism, J. Exp. Biol., 89:239.

Hughes, J., Smith, T. W., and Kosterlitz, H. W., 1975, Identification of two related pentapeptides from the brain with potent opiate agonist activity, Nature, 258:577.

Hughes, J., Smith, T., Morgan, B., and Fothergill, L., 1975, Purification and properties of enkephalin—the possible endogenous ligand for the morphine receptor, Life Sci., 16:1753.

Hughes, S. C., Rosen, M. A., Shnider, S. M., Abboud, T. K., Stefani, S. J., and Norton, M., 1984, Maternal and neonatal effects of epidural morphine for labor and delivery, Anesthesia and Analgesia, 63:319-324.

Johnson, H. M., Smith, E. M., Torres, B. A., and Blalock, J. E., 1982, Regulation of the in vitro antibody response by neuroendocrine hormones, Proc. Natl. Acad. Sci., 79:4171.

Kay, N., Allen, J., and Morley, J. E., 1984, Endorphins stimulate normal human peripheral blood lymphocyte natural killer activity, Life Sci., 35:53.

Khansari, N., Whitten, H. D., and Fudenberg, H. H., 1984, Phencyclidine-induced immunodepression, Science, 225:76.

Kilpatrick, D.L., Lewis, R.V., Stein, S., and Udenfriend, S., 1980, Release of enkephalins and enkephalin containing polypeptides from perfused beef adrenal glands, Proc. Natl. Acad. Sci., 77:7473.

Knepel, W., Homolka, L., Vlaskovska, M., and Nutto, D., 1984, Stimulation of adrenocorticotropin/beta-endorphin release by synthetic ovine corticotropin-releasing factor in vitro. Enhancement by various vasopressin analogs, Neuroendocrin., 38:344.

Kurland, J. I., and Bockman, R., 1978, Prostaglandin E production by human blood monocytes and mouse peritoneal macrophages, J. Exp. Med., 147:952.

Li, C. H., and Chung, D., 1976, Isolation and structure of an untriakontapeptide with opiate activity from camel pituitary glands, Proc. Natl. Acad. Sci. (USA), 73:1145.

Ling, N., Burgus, R., and Guillemin, R., 1976, Isolation, primary structure, and synthesis of alpha-endorphin and gamma-endorphin, two peptides of hypothalmic-lypophysial origin with morphomimetic activity, Proc. Natl. Acad. Sci., 73:3942.

Livett, B. G., Day, R., Elde, R. P., and Howe, P. R. C., 1982, Co-storage of enkephalins and adrenaline in the bovine adrenal medulla, Neuroscience, 7:1323.

Lolait, S. J., Lim, A. T., Toh, B. H., and Funder, J.W., 1984, Immunoreactive beta-endorphin in a subpopulation of mouse spleen macrophages, J. Clin. Invest., 73:277.

Lopker, A., Abood, L. G., Hoss, W., and Lionetti, F. J., 1980, Stereoselective muscarinic, acetylcholine and opiate receptors in human phagocytic leukocytes, Biochem. Pharmacol., 29:1361.

Mathews, P. M., Froelich, C. J., Sibbitt, W. L., Jr., and Bankhurst, A. D., 1983, Enhancement of natural cytotoxicity by beta endorphin, J. Immunol., 130:1658.

McCain, H. W., Lamster, I. B., Bozzone, J. M., and Grbic, J. T., 1982, Beta endorphin modulates human immune activity via non-opiate receptor mechanisms, Life Sci., 31:1619.

McCluskey, R. T., Benacerraf, B., and McCluskey, J. W., 1963, Studies on the specificity of the cellular infiltrate in delayed hypersensitivity reactions, J. Immunol., 90:466.
Mehrishi, J. N., and Mills, I. H., 1983, Opiate receptors on lymphocytes and platelets in man, Clin. Immunol. and Immunopath., 27:240.
Miller, G.C., Murgo, A. J., and Plotnikoff, N. P., 1983, Enkephalins—enhancement of active T-cell rosettes from lymphoma patients, Clin. Immunol. Immunopath., 26:446.
Miller, G. C., Murgo, A. J., and Plotnikoff, N. P., 1984, Enkephalins—enhancement of active T-cell rosettes from normal volunteers, Clin. Immun. and Immunopath., 31:132.
Mueller, G.P., 1981, Beta endorphin immunoreactivity in rat plasma: Variations in response to different physical stimuli, Life Sci., 29:1669.
Nakao, K., Nakai, Y., Jingami, H., Oki, S., Fukata, J., Imura, H., 1979, Substantial rise of plasma beta endorphin levels after insulin induced hypoglycemia in human subjects, J. Clin. Endocrin. Metab., 49:838.
Nelson, R. D., McCormack, R. T., Fiegel, V. D., and Simmons, R. L., 1978, Chemotactic deactivation of human neutrophils: Evidence for non-specific and specific components, Infect. Immun., 22:441.
O'Flaherty, J. T., Showell, H. J., Becker, E. L., and Ward, P. A., 1978, Substances which aggregate neutrophils, Am. J. Pathol., 92:155.
Olson, G. A., Olson, R. D., and Kastin, A. J., 1984, Endogenous opiates: 1983, Peptides, 5:975.
Oyama, T., Matsuki, A., Taneichi, T., Ling, N., and Guillemen, R., 1980, Beta-endorphin in obstetric analgesia, Am. J. Obstet. Gynecol., 137:613-617.
Pelto-Huikko, M., Salminen, T., and Hervonen, A., 1982, Enkephalin-like immunoreactivity is restricted to the adrenaline cells in the hamster adrenal medulla, Histochemistry, 73:493.
Plotnikoff, N. P., and Miller, G. C., 1983, Enkephalins as immunomodulators, Int. J. Immunopharm., 5:437.
Prete, P., and Levin, E., 1985, In vitro effect of neuropeptides on human natural killer and cytotoxic T-cell function and T-cell subsets, Clin. Res., (abstr.) 33:54A.
Puolakka, J., Kauppila, A., Leppaluoto, J., and Vuolteenaho, O., 1982, Elevated beta endorphin immunoreactivity in umbilical cord blood after complicated delivery, Acta Obstet. Gynecol. Scand., 61:513.
Rossier, J., French, E. D., Rivier, C., Ling, N., Guillemin, R., and Bloom, F. E., 1977, Footshock induced stress increases beta endorphin levels in blood but not brain, Nature, 270:618.
Ryder, S. W., and Eng., J., 1981, Radioimmunoassay of leucine enkephalin-like substance in human and canine plasma, J. Clin. Endocrin. Metab., 52:367.
Saland, L. C., Van Epps, D. E., Ortiz, E., and Samora, A., 1983, Acute injections of opiate peptides into the rat cerebral ventricle; a macrophage-like cellular response, Brain Res. Bull., 10:523.
Saland, L. C., Ortiz, E., and Samora, A. 1984. Chronic infusion of opiate peptides to rat cerebrospinal fluid with osmotic pumps, Anat. Rec., 210:115.
Shanks, M.M., Clement-Jones, V., Linsell, C.J., Mullen, P.E., Rees, L.H., and Besser, G.M., 1981, A study of 24 hour profiles of plasma met enkephalin in man, Brain Res., 212:403.
Shavit, Y., Lewis, J. W., Terman, G. W., Gale, R. P., and Liebeskind, J. C., 1984, Opioid peptides mediate the suppressive effect of stress on natural killer cell cytotoxicity, Science, 223:188.

Showell, H. J., Freer, R. J., Zigmond, S. M., Schiffman, E., Aswanikumar, S., Corcoran, B., and Becker, E. L., 1976. The structure activity relationship of synthetic peptides as chemotactic factors and inducers of lysosomal secretion for neutrophils, J. Exp. Med., 143:1154.

Simpkins, C. O., Dickey, C. A., and Fink, M. P., 1984, Human neutrophil migration is enhanced by beta-endorphin, Life Sci., 34:2251.

Smith, E.M., and Blalock, J.E., 1981, Human lymphocyte production of corticotropin and endorphin-like substances: Association with leukocyte interferon, Proc. Natl. Acad. Sci., 78:7530.

Smith, J.R., and Simon, E.J., 1981, Endorphins, opiate receptors, and their evolving biology, Pathobiol. Ann., 11:87.

Stoelting, R. K., 1980, Opiate receptors and endorphins: Their role in anesthesiology, Anesthesia and Analgesia, 59:874-880.

Terenius, L., and Wahlstrom, A., 1975, Search for an endogenous ligand for the opiate receptor, Acta Physiol. Scand., 94:74.

Teschemacher, H., Opheim, K. E., Cox, B. M., and Goldstein, A., 1975, A peptide like substance from the pituitary that acts like morphine, Life Sci., 16:1771.

Tubaro, E., Borelli, G., Croce, C., Cavallo, G., and Santiangelli, C., 1983, Effect of morphine on resistance to infection, J. Infect. Dis., 148:656.

Turkall, R. M., Denison, R. C., and Tsan, M., 1982, Degradation and oxidation of methionine enkephalin by human neutrophils, J. Lab. Clin. Med., 99:418.

Van Epps, D. E., 1982, Mediators and modulators of human lymphocyte chemotaxis, Agents and Actions, 12 (suppl.):217.

Van Epps, D. E., Durant, D. A., and Potter, J. W., 1983a, Migration of human helper/inducer T cells in response to supernatants from Con A-stimulated suppressor/cytotoxic T cells, J. Immunol., 131:697.

Van Epps, D. E., Potter, J., and Brown, S. L., 1985, Production and regulation of human T-lymphocyte Chemotactic factor (LCF), in: "Mediators of Inflammation," G. A. Higgs and T. J. Williams, eds., MacMillan, London (in press).

Van Epps, D. E., Potter, J. W., and Durant, D. A., 1983b, Production of human lymphocyte chemotactic factor by T-cell subpopulations, J. Immunol., 130:2727.

Van Epps, D. E., and Saland, L., 1984, Beta endorphin and met enkephalin stimulate human peripheral blood mononuclear cell chemotaxis, J. Immunol., 132:3046.

Van Epps, D. E., Saland, L., Taylor, C., and Williams, R. C. Jr., 1983c, In vitro and in vivo effects of beta endorphin and met enkephalin on leukocyte locomotion, Prog. Brain Res., 59:361.

Vincent, J. P., Cavey, D., Kamenka, J. M., Geneste, P., and Lazdunski, M., 1978, Interaction of phencyclidines with the muscarinic and opiate receptors in the central nervous system, Brain Res., 152:176.

Wardlaw, S. L., Stark, R. I., Daniel, S., Frantz, A. G., 1981, Effects of hypoxia on beta endorphin and beta lipotropin release in fetal, newborn and maternal sheep, Endocrin., 108:1710.

Weber, R. J., and Pert, C. B., Opiatergic modulation of immune system, in: "Central and Peripheral Endorphins: Basic and Clinical Aspects," E. E. Muller and A. R. Genazzini, eds., Raven Press, (New York) (1984).

Wilkes, M. M., Stewart, R. D., Bruni, J. F., Quigley, M. E., Yen, S. S. C., Ling, N., and Chretien, M., 1980, A specific homologous radioimmunoassay for human beta endorphin: Diect measurement in biological fluids, J. Clin. Endocrin. Metab., 50:309.

Wilkinson, P. C., Haston, W. S., and Shields, J. M., 1982. Some determinants of the locomotor behavior of phagocytes and lymphocytes *in vitro*, *Clin. Exp. Immuno.*, 50:461.
Wybran, J., Appelboom, T., Famaey, J. P., and Govaerts, A., 1979, Suggestive evidence for receptors for morphine and met enkephalin on normal human blood T lymphocytes, *J. Immunol.*, 123:1068.
Yamasaki, Y., Shimamura, O., Kizu, A., Nakagawa, M., and Ijichi, H., 1982, IgE-mediated ^{14}C-serotonin release from rat mast cells modulated by morphine and endorphins, *Life Sci.*, 31:471.

NEUROPEPTIDES ARE CHEMOATTRACTANTS FOR HUMAN MONOCYTES AND TUMOR CELLS: A BASIS FOR MIND-BODY COMMUNICATION

Michael R. Ruff and Candace B. Pert

Cellular Immunology Section, Laboratory of Microbiology and Immunology, National Institute of Dental Research, NIH and Section on Brain Biochemistry, Clinical Neuroscience Branch, NIMH, Bethesda, Maryland 20892

INTRODUCTION

Monocytes, or their non-circulating counterparts, macrophages, are a heterogeneous population of cells which regulate many aspects of immune functioning including antigen processing, T cell activation, lymphokine and antibody production, as well as defense against neoplastic cells[1]. Recently, several other cell types, histologically and morphologically different from classical macrophages, have come to be included into this family of cells. They include the several minor cell populations present in skin (called Langerhans cells), brain (microglia), and dendritic cells sparsely distributed throughout basal epithelial layers of numerous organs and endocrine tissues[2], as well as osteoclasts in joints. The functions of these macrophages are less clear although a role of the Langerhans cell in local immune reactions has been inferred.

All of these cells derive from bone marrow stem cell precursors and a particularly interesting aspect of their physiology is their ability to egress from bone marrow into the circulation and migrate to various body sites where they undergo final differentiation. These cells can also be actively recruited to specific locations during pathological states by chemotactic signals where, in a programmed sequence, they combat infection and participate in the repair of damaged tissue during wound healing[3]. Many stimuli have been shown to induce migration of macrophages including bacterial products, peptides generated during clotting and complement activation, proteolytic fragments of structural matrix proteins such as collagen, elastin, and fibronectin, as well as lymphocyte and fibroblast products[1]. The signals which regulate the normal trafficking and ultimate localization of these related but discrete macrophage populations is, however, largely unknown. Recently, we have begun to document the receptor-mediated abilities of human monocytes to migrate toward very low concentrations of several neuropeptides including the opiates[4,5], bombesin[6], substance P[7], and the benzodiazepines[8] (whose endogenous ligand is a neuropeptide only recently sequenced). These agents are not currently considered to be products of inflammatory reactions (with the possible exception of substance P)[9], and thus it seems that although neuropeptides may have roles in the inflammatory, wound healing aspects of macrophage function they likely also have broader function in the day-to-day maintenance of health and homeostasis[10] throughout the organism by virtue of their action on the disseminated members of the macrophage population. Because neuropeptides have been postulated to be the biochemical mediators of behavior and emotion[11] it seems plausible that these same mediators of various mood states in brain also communicate to macrophages and other cells involved in healing and thus offer a potential mechanism for the effect of emotional states[11] on the course and outcome of illness.

METHODS

Mononuclear cells, obtained from heparinized blood from healthy human volunteers, were isolated by sedimentation over Ficoll-Paque (Pharmacia). This population of cells was comprised of approximately 30% monocytes and 70% lymphocytes as determined by specific antibody and esterase staining. Lymphocytes do not migrate in this assay. Migration was studied using blind-well chemotaxis chambers (Neuro Probe, Bethesda, MD) in which the upper and lower compartments were separated by a 5 μM pore size nucleopore polycarbonate membrane[7]. Mononuclear cells (5.5×10^4) in Geys balanced salt solution (GBSS) (pH 7.4) containing 2% bovine serum albumin were placed in the upper wells and test attractants in the lower wells of the chambers. All drugs were diluted into GBSS containing 2% BSA and were prepared from a 10^{-4} stock solution stored in 0.02 M acetic acid at -20°C and diluted just prior to assay. Monocyte migration was assessed after a 90 min incubation at 37°C by fixing and staining the distal membrane-adherent cells. These migrating cells were enumerated by counting three fields in triplicate utilizing an optical image analyzer (Optomax Inc., Hollis, NH). Data are expressed as the migration index, which is the ratio between migration toward test attractants and buffer alone. The number of migrating cells in the buffer alone controls generally ranged from 20-60 cells/field. The values shown represent the mean and S.E.M. of the migration indices of multiple individual experiments, as indicated. Some experiments report results as actual number of migrating cells. All experiments have been repeated three to twelve times with similar results.

RESULTS

Opiate Receptor. Utilizing the modified Boyden chamber method it was possible to demonstrate that opiate peptides and drugs are chemoattractants for human monocytes (Fig. 1). Results show that the naturally occurring opiate peptides β-endorphin and dynorphin were most potent in this assay with EC_{50} (concentration of half-maximal response) values of 5×10^{-13}M. The peptide met-enkephalin was only slightly less active, EC_{50} of 10^{-12}M, although this comparison may reflect the greater lability of the enkephalins relative to endorphins. Stereospecificity of opiate chemotaxis is demonstrated by the results which show that levallorphan (a synthetic opiate) is greater than 10,000 times more active than its enantiomer dextrallorphan, which was largely inactive in this assay. Bremazocine, another synthetic opiate, showed intermediate activity, EC_{50} of 5×10^{-11}M. Interestingly, morphine, the oldest and perhaps best known opiate was largely inactive in promoting chemotaxis. This apparent selectivity in opiate pharmacologic response (e.g., endorphins vs. morphine) has been used to operationally define subtypes of receptor specificity[11] although the relationship of these varied functional responses to the actual receptor structure is as yet unknown. The responses to β-endorphin, dynoprhin [1-13], and D-ala^2-D-leu^5-enkephalin (DADL) are stereospecifically reversed by the opiate antagonist naloxone (Fig. 2) thereby indicating that the responses measured are mediated through an opiate receptor. The inactive enantiomer of naloxone did not reverse opiate monocyte chemotaxis.

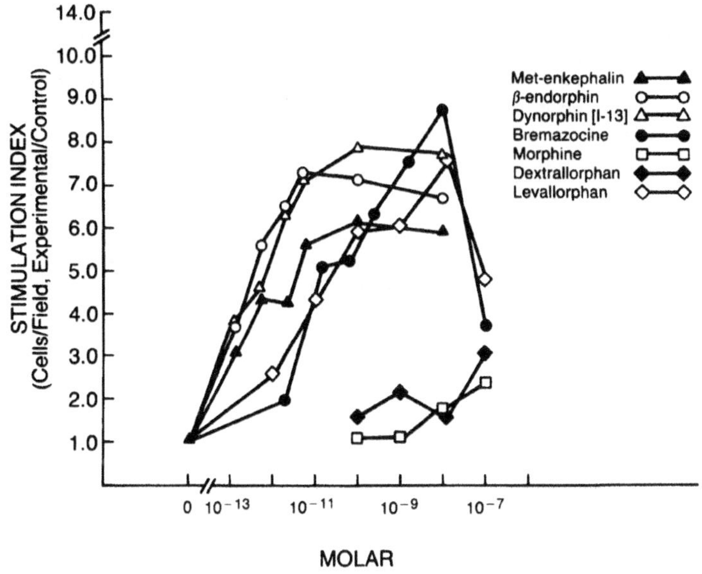

Fig. 1. Chemotactic response of human monocytes to opiates and opiate peptides[5]. Chemotaxis was performed as described. Data is expressed as a stimulation index. The maximal response for all analogs occurred at 10^{-8}M, although the peptides β-endorphin, D-Ala-D-Leu enkephalin, and dynorphin [1-13] were similarly active at 10^{-10}M. The EC_{50} for these agents was $5x10^{-12}$M. Stereospecificity is demonstrated in this figure by levallorphan, which shows intermediate activity compared to opiate peptides, while its enantiomer, dextrallorphan, is inert. The stimulation index for the chemotactic peptide f-Met-Leu-Phe, 10^{-7}M, was 14.2.

Substance P Receptor. Substance P, an undecapeptide ligand first sequenced by Leeman and colleagues[13] and recently proposed to stimulate hypersensitivity and inflammation[9] is also active in promoting monocyte chemotaxis. This result, therefore, supports a role for substance P in the development of experimental arthritis and provides an additional mechanism to explain the monocyte inflammatory infiltrate of arthritic lesions. In order to establish the receptor mediated nature of substance P responses a series of closely related analogs was evaluated for their effect on chemotaxis (Fig. 3). Substance P showed a biphasic response, also observed with N-formyl peptides[1] as well as other neuropeptide-induced chemotaxis[4-8], with maximal activity occurring in the range between $5x10^{-11}$M and 10^{-12}M. As was shown for opiate peptides[4,5], chemotaxis is an extremely sensitive bioassay and substance P showed a significant enhancement of chemotactic activity even at $5x10^{-14}$M, the lowest concentration tested. Lack of effect at high concentrations has been attributed to receptor desensitization or gradient breakdown[1]. Thus, substance P at concentrations higher than 10^{-10}M elicited reduced chemotactic responsiveness. Other analogs (e.g., SP[3-11]) showed similar or reduced potency in this assay; peak responses are shown for all compounds and no increase in activity over the range [10^{-14}-10^{-6}M] of tested concentrations was observed (data not shown) for any of the indicated analogs. Intact substance P had greatest potency in this assay and the results suggest that the C-terminal amino acid residues are required for maximal biological activity. Thus, increasingly N-terminus deleted analogs showed increasing loss of activity; e.g., SP[3-11] > SP[9-11], and C-terminal-deleted peptides were largely inactive, e.g., SP[1-9] or SP, free acid. Interestingly, the ability of these analogs to induce monocyte chemotaxis is consistent with previous studies demonstrating specificity of binding to other tissues or cells. Thus, the rank potency orders of substance P analogs in chemotaxis induction were comparable to their potencies in displacing substance P binding to brain membranes[14], and slices[15]; SP > SP[3-11] > SP[8-11] ≅ SP[9-11] >> SP[1-9], SP, free acid. Although these exact analogs were not used, a recent report on substance P binding to a human T lymphocyte line also yielded comparable results[16]; SP > SP[4-11] >> SP[1-4] for rank potency orders in binding displacement.

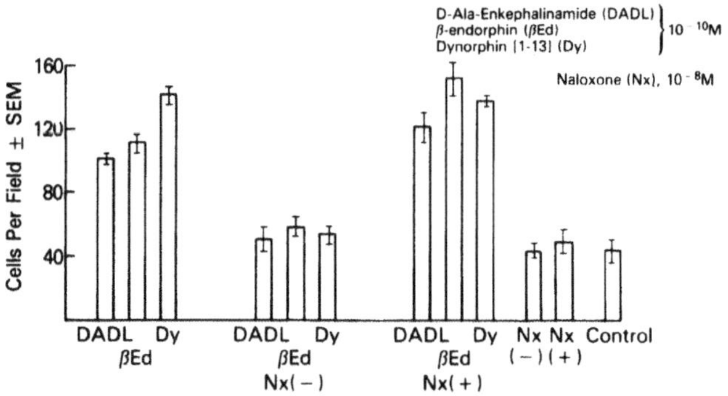

Fig. 2. Stereospecific reversibility of opiate chemotaxis by naloxone. Peptides (10^{-10}M) were co-incubated with active (-) or inactive (+) enantiomers of the antagonist naloxone (10^{-8}M). Controls contained diluent or naloxone only.

Benzodiazepine chemotaxis. Our observations of opiate and substance P mediated chemotaxis suggested the possibility that other behavior and mood modifying drugs which work through neuropeptide receptors might also have effects on macrophage and immune function. The benzodiazepines are a group of drugs widely used for their anxiolytic, hypnotic, and anticonvulsant properties - effects which are mediated through high affinity receptors present in the central nervous system[17]. Recently, an 18 amino acid endogenous peptide ligand has been proposed for the benzodiazepine receptor[18]. We tested a number of benzodiazepines for their effects on monocyte chemotaxis (Fig. 4) and we were able to show responses for this family of drugs as well[8]. Thus, the benzodiazepine receptor ligands diazepam (valium) and Ro5-4864 had potent activity, while clonazepam was largely inert. These results suggested that the monocyte benzodiazepine receptor was of the "peripheral" type (operationally the peripheral benzodiazepine receptor is defined as having selectivity for Ro5-4864 compared to clonazepam). A suitable antagonist exists for the peripheral receptor, the compound PK-11195, and it was possible to show antagonism of both Ro5-4864 and diazepam chemotaxis by this drug. Thus, (Fig. 5) equimolar mixtures of the antagonist with either Ro5-4864 or diazepam resulted in complete inhibition of chemotaxis. This inhibition is specific for benzodiazepine ligands since PK-11195 did not inhibit chemotaxis to the N-formyl peptide receptor. In this case we have been able to directly demonstrate high affinity benzodiazepine receptors on human monocytes utilizing receptor binding methods (Fig. 4 inset)[8]. The

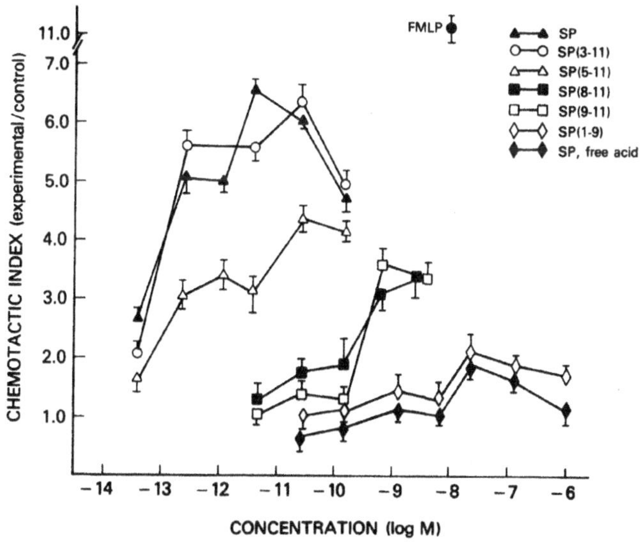

Fig. 3. Human monocyte chemotaxis to substance P and its analogs. Indicated test attractants were evaluated for their ability to stimulate monocyte migration over a broad concentration range. Data are expressed as a stimulation index, which is the ratio of the number of cells/field in experimental compared to control wells, ± S.E.M. Determinations were made as described. Data for substance P (SP) was averaged from three replicate experiments, SP[3-11], SP[5-11], SP[8-11] and SP[9-11] were averaged from two replicate experiments, while data for SP[1-9] and SP, free acid were from a single experiment which was repeated with similar results. The background cell migration, in the absence of specific attractants, ranged from 20-40 cells/field for these experiments. f-Met-Leu-Phe (FMLP) was also tested at 10^{-7}M.

potencies in binding were in agreement with chemotactic activity shown here. Thus the binding activity of Ro5-4864 > diazepam >> clonazepam confirm that chemotaxis is mediated through a peripheral type benzodiazepine receptor.

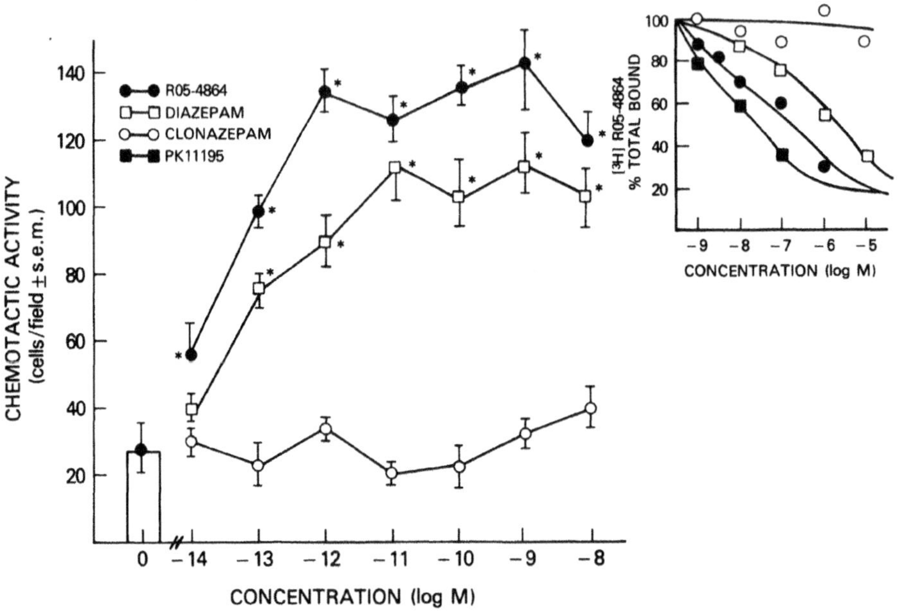

Fig. 4. The effects of various benzodiazepines on the chemotaxis of human peripheral blood mononuclear cells. The data represent the mean ± S.E.M. of the cell migration of an individual experiment that was replicated with similar results. The responses with Ro5-4864 and diazepam were significantly different from control (*) ($p<0.01$, t test). (Inset). Displacement of specific ^{3}H-labeled Ro5-4864 bound to purified monocytes by various drugs. Displacement potencies of Ro5-4864, PK-11195, diazepam, and clonazepam were determined by incubating 4×10^6 monocytes in 0.05M tris and 0.15M NaCl, pH 7.4 (final volume, 0.5 ml) with ^{3}H-labeled Ro5-4864 (4 nM; specific activity, 82.7 Ci/mmol) and various concentrations of unlabeled drug as described. After a 45 minute incubation at 0°C, the reactions were terminated by rapid vacuum filtration of the solution over GF/B glass fiber filters (Whatman) and then by washing the filters twice with 5.0 ml of ice-cold buffer. Specific binding, defined as the total binding minus that observed in the presence of 10^{-5}M unlabeled Ro5-4864 (nonspecific binding), was approximately 75% at a ligand concentration of 4 nM. Displacement curves for each drug were computed as a percentage of the maximal specific ^{3}H-labeled Ro5-4864 binding in the presence of each drug over the indicated concentration ranges. IC_{50} values were estimated from displacement curves such as those shown here. Values are from a typical experiment repeated three times with similar results.

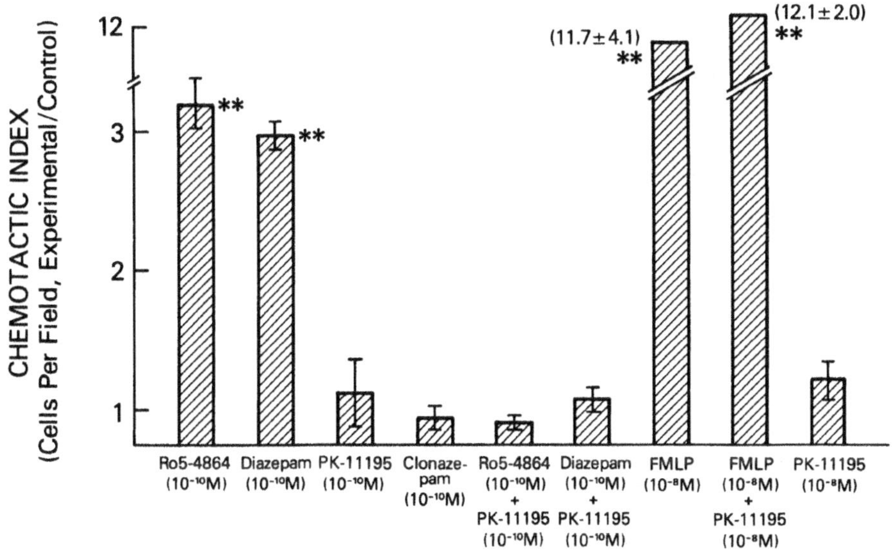

Fig. 5. Benzodiazepine receptor-mediated chemotaxis of human monocytes. Chemotaxis assays were performed by using modified Boyden chambers as described. Data are expressed as a migration index. The number of migrating cells in the buffer alone controls was 40 cells/field (200X). Values shown represent the mean and S.E.M. of four experiments. Ro5-4864, (4-chlorodiazepam), and diazepam yielded statistically significant responses compared to control (**, $p<0.001$). Clonazepam, a benzodiazepine analog with low activity on non-neural cells, was not active in this assay. These receptor-mediated events were blocked by the appropriate antagonist, PK-11195, when co-mixed with the agonist Ro5-4864 or diazepam. The response of the tripeptide, f-Met-Leu-Phe (FMLP) was not inhibited when co-mixed with equimolar (10^{-8}M) PK-11195.

Tumor cell chemotaxis. In an attempt to gain some insight into the factors which may affect metastatic neoplastic diseases we have also examined the role of neuropeptides as chemoattractants for human tumor cells[6]. In these experiments a modification of the monocyte chemotactic methodology was employed[6]. Chemotaxis time was extended to four hours and filters were coated with collagen IV to promote cell adherence. Small cell lung cancer (SCLC) cells were used in this study. SCLC accounts for almost 25% of all human lung cancers and is a rapidly progressing disease with widespread and early metastases observed. These cells and tumors have been proposed to be derived from macrophage precursors[19]. Results (Fig. 6) revealed that SCLC cells express chemotactic receptors for several neuropeptides. Among the active peptides were bombesin[6], β-endorphin, substance P, and arg-vasopressin. Not all peptides were active in this assay. For example, oxytocin, which is closely related to vasopressin, did not appear to be active. Other peptides such as neurotensin and melanocyte stimulating hormone were not active (data not shown) further indicating the specificity of the responses and supporting the concept that unique neuropeptide receptors are responsible for these results. Interestingly, these tumor cells also seem to express chemotactic receptors for N-formyl peptides since f-Met-Leu-Phe was active.

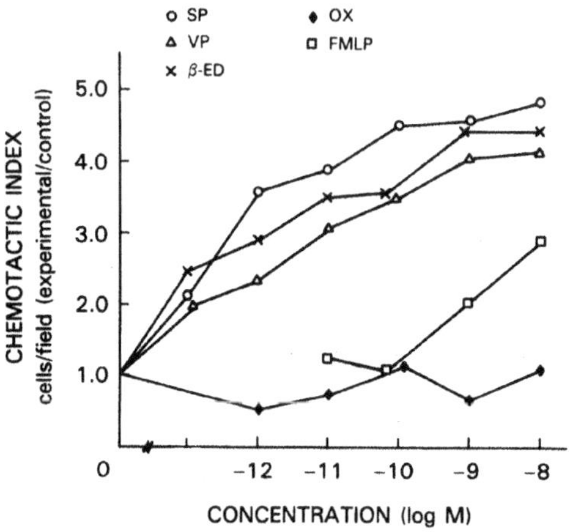

Fig. 6. Chemotaxis of small cell lung cancer (SCLC) cells to various neuropeptides. Results are expressed as a stimulation index, the ratio of experimental to control migration. SP; substance P, VP; arg-vasopressin, β-ED; β-endorphin, OX; oxytocin, FMLP; f-Met-Leu-Phe. Backround migration in this experiment was 6±2 cells/field. Chemotaxis was performed for four hrs on collagen IV coated filters[6].

DISCUSSION

We have previously emphasized that careful structure-activity analysis of the correlation between binding and physiological activity is a useful strategy for demonstrating receptor mediated events[20]. Thus, it has been previously shown that N-formyl peptides, which are potent chemoattractants, bind specifically to rabbit neutrophils with a structure-activity correlation between binding and chemotaxis of $p<.001$[21]. In this present review we have utilized this principle to infer the receptor-mediated effects of several neuropeptides or drugs on chemotaxis of human monocytes. In at least one case (benzodiazepines) it has been possible to directly demonstrate binding sites on monocytes, the pharmacology of which is in agreement with the functional chemotactic effects[8]. By these arguments and methods we have been able to show that human monocytes express receptors for multiple and distinct neuropeptides. Some tumor cells, such as SCLC, also express functional neuropeptide receptors. The existence of discrete neuropeptide receptors on monocytes is further suggested by observations of blockade by selective antagonists for a given neuropeptide. Rather than speculate on the role of the individual ligands on monocyte function, which at this time is largely unknown, we will focus this discussion on what we perceive to be the major conceptual themes of this work.

Neuropeptides, short chains of amino acids present in brain as well as non-neural tissues, are primarily known for their role as neurotransmitters[22]. In the central nervous systems of the recently evolved mammals, neuropeptides are regionally distributed being largely localized in those limbic and cortical regions primarily responsible for sensory processing, memory, behavior and mood[10]. As such, neuropeptides are candidates for the biochemical mediators of these events. It is becoming clear that these compounds, far from acting solely within the nervous system, are localized throughout the brain and body, and have pleiotropic effects serving as transmitters, growth hormones, and signals for many body systems. These peptides and their receptors have been extensively conserved structurally and are widely distributed across phyla[23,24] where they presumably also function as components in a network for intercellular communication. The presence of varied, distinct neuropeptide receptors on macrophages and other immune system cells, coupled with reports of neuropeptide synthesis by immune cells, suggests that immune cells are in communication with, potentially, themselves as well as other neuropeptide secreting and responding cells. Neuropeptides and their receptors thus join the nervous system, glands, and immune system in a network of information exchange which exists throughout the body, as well as the brain. We have called this a "psychoimmunoendocrine" network and the purpose of such a network may be to prioritize the body's defense and repair mechanisms within the context of the whole internal milieu. Because neuropeptide-rich loci in the central nervous system have been inferred to mediate numerous behavioral and mood states we have proposed that this neuropeptide net may form the biochemical basis for observations that emotional states can significantly alter the course and outcome of biological illnesses previously considered to exist strictly in the somatic realm. By this rationale the same mediators which activate mood states in brain also communicate to other cells involved in healing and homeostatic processes. Although speculative it may well be that a particular mood state is associated, biochemically, with a particular neuropeptide tone. Opiates, for example, clearly mediate a state of intense pleasure, while substance P is associate with pain, and benzodiazepines with anxiety. If mood is associated with release of various neuropeptides then the hypothesis of a psychosomatic network implies that all the cells which participate in the network contribute to creation of a neuropeptide tone (mood) which then is perceived throughout the body. All illness and disease may result in alterations in the neuropeptide tone of the mind/body and such changes may have repercussions for health and well-being.

The physiological correlate of *in vitro* macrophage chemotaxis by neuropeptides is currently unknown although a localized tissue site is implicated as the source of chemoattractant. Macrophages are, as discussed in the introduction, a very heterogeneous population of cells that subserve important roles in immune activation, as well as healing and cancer defense. Chemoattractants often alter other aspects of macrophage physiology and the effects we report here may not be solely limited to enhancement of migration. With respect to the concept that release of neuropeptides may have impact on health, in addition to the effects mediated through the immune and endocrine systems, to cite one example, it may also have direct effects on establishment or maintenance of certain cancers.

Small cell lung cancer (SCLC), which accounts for 25% of all lung cancer deaths, is a tumor whose cells store and secrete numerous neuropeptides. These cells and their normal, non-cancerous progenitors, are therefore components of the psychoimmunoendocrine network and would be expected to be under its tonic influence. Indeed, a most exciting recent paper documents that the characteristic neuropeptide secreted by these cells, the peptide bombesin[25], is also a potent growth factor for these cells[26]. Antibodies to bombesin inhibited cell growth _in vitro_ and tumor formation _in vivo_. Other work we have been involved in suggests that these cells are derived from macrophages[19] and shows that both macrophages and SCLC tumor cells will chemotax to bombesin[6] and other neuropeptides (Fig. 6)[6]. Neuropeptides released in the brain, as well as the body, due to cognitive, emotional, or other stimuli, may there by have effects on tumor and/or macrophage growth and tissue localization. These observations may be relevant to the metastatic process _per se_ in that cells which have dissociated from the primary tumor mass invade secondary sites by following gradients of neuropeptide signals[6]. Disseminated neoplastic diseases, such as SCLC, breast, or hemopoietic cancers, may, to some extent, develop as a result of dysfunctions in the neuropeptide network which then modifies the course and progression of disease.

One additional prediction of this formulation is that the network and its composite cells can be accessed via any of its members. Thus, to continue with the model for SCLC, an approach which seeks to linearly interrupt the flow of information to the tumor by blocking the action of one peptide or growth factor, such as bombesin, does not accommodate the possibility that, at least on the organismal level, other neuropeptide signals acting directly, or relayed through other cells, may also have effects on tumor growth or the general state of well-being. An understanding of these principles may prove essential to the ultimate goal, which is to re-establish growth homeostasis and health throughout the individual.

REFERENCES

1. E. Schiffmann and J. I. Gallin, Biochemistry of phagocyte chemotaxis, _Curr. Topics in Cell Reg._ 15:203 (1979).

2. V. H. Perry, D. A. Hume and S. Gordon, Immunohistochemical localization of macrohages and microglia in the adult and developing mouse brain, _Neuroscience_ 15:313 (1985).

3. S. M. Wahl and L. M. Wahl, Modulation of fibroblast growth and function by monokines and lymphokines, _Lymphokines_ 2:179 (1981).

4. D. Van Epps and L. Saland, β-endorphin and met-enkephalin stimulate human peripheral blood mononuclear cell chemotaxis, _J. Immunology_ 132:3046 (1984).

5. M. R. Ruff, S. M. Wahl, S. Mergenhagen and C. B. Pert, Opiate receptor-mediated chemotaxis of human monocytes, _Neuropeptides_ (Europe) 5:363 (1985).

6. M. R. Ruff, E. Schiffmann, V. Terranova and C. B. Pert, Neuropeptides are chemoattractants for human macrophages and tumor cells: a mechanism for metastasis, _Clinical Immunol. Immunopath._, in press.

7. M. R. Ruff, S. M. Wahl and C. B. Pert, Substance P receptor-mediated chemotaxis of human monocytes, _Peptides_, in press.

8. M. R. Ruff, C. B. Pert, R. J. Weber, L. M. Wahl, S. M. Wahl and S. M. Paul, Benzodiazepine receptor-mediated chemotaxis for human monocytes, Science 229:1281 (1985).

9. J. D. Levine, R. Clark, M. Devor, C. Helms, M. A. Moskowitz and A. I. Basbaum, Intraneuronal substance P contributes to the severity of experimental arthritis, Science 226:547 (1984).

10. D. G. Payan, J. D. Levine, E. J. Goetzl, Modulation of immunity and hypersensitivity by sensory neuropeptides, J. Immunol. 132:1601 (1984).

11. C. B. Pert, M. R. Ruff, R. J. Weber and M. Herkenham, Neuropeptides and their receptors: a psychosomatic network, J. Immunol. 135:820S (1985).

12. S. J. Paterson, L. E. Robson and H. W. Kosterlitz, Classification of opioid receptors, Br. Med. Bull. 39:31 (1983).

13. M. M. Chang, S. E. Leeman and H. D. Niall, Amino-acid sequence of substance P, Nature (N.B.) 232:86 (1971).

14. M. A. Cascieri and T. Liang, Characterization of the substance P receptor in rat brain cortex membranes and the inhibition of radioligand binding by guanine nucleotides, J. Biol. Chem. 258: 5158 (1983).

15. R. B. Rothman, M. Herkenham, C. B. Pert, T. Liang and M. A. Cascieri, Visualization of rat brain receptors for the neuropeptide, substance P, Brain Res. 309:47 (1984).

16. D. G. Payan, D. R. Brewster and E. J. Goetzl, Stereospecific receptors for substance P on cultured human IM-9 lymphocytes, J. Immunol. 133:3260 (1984).

17. S. Garattini, E. Mussini, L. O. Randall, "The Benzodiazepines," Raven, New York (1973).

18. P. Ferrero, A. Guidotti, B. Conti-Tronconi and E. Costa, A brain octadecaneuropeptide generated by tryptic digestion of DBI (diazepam binding inhibitor) functions as a proconflict ligand of benzodiazepine recognition sites, Neuropharmacology 23:1359 (1984).

19. M. R. Ruff and C. B. Pert, Small cell carcinoma of the lung: macrophage-specific antigens suggest hemopoietic stem cell origin, Science 255:1034 (1984).

20. C. B. Pert and M. Herkenham, From receptors to brain circuitry, in: "Biochemistry of Taste and Olfaction," R. Cagan, ed., Academic Press, New York (1981).

21. S. Aswanikumar, B. A. Corcoran, E. Schiffmann, A. R. Day, R. J. Freer, H. J. Showell, E. L. Becker and C. B. Pert, Demonstration of a receptor on rabbit neutrophils for chemotactic peptides, Biochem. Biophys. Res. Comm. 74:81 (1977).

22. D. Krieger, Brain peptides: What, where, why?, Science 222:975 (1983).

23. D. LeRoith, J. Shiloach and J. Roth, Is there an earlier phylogenetic precursor that is common to both the nervous and endocrine systems?, Peptides 3:211 (1982).

24. D. LeRoith, A. S. Liotta, J. Roth, J. Shiloach, M. E. Lewis, C. B. Pert and D. T. Krieger, Corticotropin and β-endorphin-like materials are native to unicellular organisms, Proc. Natl. Acad. Sci. USA 79:2086 (1982).

25. T. W. Moody, C. B. Pert, A. F. Gazdar, D. N. Carney and J. D. Minna, High levels of intracellular bombesin characterize human small cell lung carcinoma, Science 214:1246 (1981).

26. F. Cuttitta, D. Carney, J. Mulshine, T. Moody, J. Fedorko, A. Fischler and J. Minna, Bombesin-like peptides can function as autocrine growth factors in human small cell lung cancer, Nature 316:823 (1985).

METHIONINE ENKEPHALIN:

IMMUNOMODULATOR IN NORMAL VOLUNTEERS (*IN VIVO*)

N.P. Plotnikoff*, G.C. Miller**, S. Solomon**, R.E. Faith*, L. Edwards***, and A.J. Murgo*

Departments of Pharmacology* and Medicine***
School of Medicine
Oral Roberts University and Immuno-Diagnostics Lab**
Tulsa, Oklahoma

INTRODUCTION

The enkephalin, methionine enkephalin (met-enkephalin) was originally discovered to be one of the endogenous ligands for morphine receptors (Hughes et al., 1975; Simontov and Snyder, 1976). However, Plotnikoff et al., (1976) discovered anti-depressant, anti-anxiety, and anti-convulsant effects of the enkephalins, identifying a much broader spectrum of pharmacological effects than originally anticipated.

Earlier studies by several researchers have identified enkephalin receptors on human T-cells, granulocytes, complement, platelets and B cells (Wybran et al., 1979; Lopker et al., 1980; Schweigerer et al., 1982; Hazum et al., 1979). Plotnikoff et al., (1982, 1983) explored the possibility that methionine enkephalin might have an immunomodulatory role. Other investigators studied similar effects with the endorphins (Blalock and Smith, 1980; Gilman et al., 1982; Mathews et al., 1983). In mice, Plotnikoff and Miller (1983) identified enhanced lymphocyte blastogenesis with phytohemagglutinin and also increased size of thymus (Plotnikoff et al., 1985). In rats, Plotnikoff et al., 1984 showed similar effects on the thymus gland.

Studies in humans (*in vitro*) identified enchanced active T-cell rosettes and NK cell activity in normal volunteers as well as cancer patients (Faith et al., 1984).

The present study is an evaluation of the *in vivo* effects of methionine enkephalin on levels and activities of T-cells and NK cells in human volunteers.

PATIENTS AND METHODS

Male subjects age 21-39 were recruited and informed consent forms signed. They had complete abstinence from all drugs, including alcohol and tobacco at least three weeks prior to the study. All subjects were within twenty percent of ideal body weight and free of acute and chronic cardiovascular, renal, metabolic and neurological diseases. Those subjects satisfying admission criteria were admitted to the hospital the evening prior to the study, given the drug infusion in the morning of the second day, and discharged on the third day.

TABLE 1: ACTIVE T-CELL ROSETTES

Volunteer	One microgram/kg #1	#2	Ten micrograms/kg #3	#4	Fifty micrograms/kg #5	#6
Baselines	17/200	31/200	30/200	34/200	33/200	6/200
2 Hours	29/200 (71%↑)	77/200 (148%↑)	85/200 (183%↑)	77/200 (126%↑)	82/200 (148%↑)	14/200 (133%↑)
24 Hours	26/200 (53%↑)	17/200 (45%↓)	61/200 (103%↑)	43/200 (26%↑)	15/200 (54%↓)	44/200 (633%↑)

$p<0.05$ - t-test paired difference ($\frac{10 \text{ volunteers}}{12 \text{ samples}}$)

significant positive increase over controls

METHIONINE ENKEPHALIN FORMATION

A sterile solution of met-enkephalin was prepared. Met-enkephalin (Sigma-Aldrich Chem. Co., St. Louis) was dissolved in sterile saline and filter sterilized by the City of Faith pharmacy. Met-enkephalin has the following characteristics: M.W. 574, water soluble and amino acid sequence Tyr-Gly-Gly-Phe-Met-OH. It was dissolved in 60 ml. of saline at a concentration sufficient to infuse at 2 ml/minute for thirty minutes to give a delivered dose of one, ten, or fifty micrograms/kilogram body weight.

IMMUNOLOGIC TESTS

TOTAL AND ACTIVE T LYMPHOCYTE ROSETTE DETERMINATION

Peripheral blood was collected by antecubital venipuncture, diluted with RPMI 1640 medium, layered over a Ficoll-Paque gradient and centrifuged for thirty minutes at 400 x g. The mononuclear band was collected, washed twice, and resuspended in RPMI 1640. Total E rosettes were quantitated by mixing one million lymphocytes with sheep red blood cells in RPMI supplemented with 10% fetal calf serum at $4^{\circ}C$ overnight. The cells were gently resuspended and the rosettes quantified. The active T-cells were determined by mixing 0.1 ml volume of 1×10^7 cells/ml and an equal volume of fetal calf serum and incubated at $37^{\circ}C$ for 60 minutes. Sheep red blood cells (2×10^7) were added, the cell suspension centrifuged at 200 x g for five minutes, gently resuspended, and rosette forming cells quantitated with a hemocytometer.

NK (NATURAL-KILLER) CELLS

Target cells (K562) were adjusted to 2×10^6 cells in 0.2 ml RPMI 1640 and labeled with $Na^{51}CrO_4$ (1mCi) for 45 minutes at $37^{\circ}C$ in 5% Co2. The cells were washed twice, resuspended in 2 ml RPMI 1640, and adjusted to 5×10^4 cells/ml.

The effector cells were collected on Ficoll-Paque gradients as described above. The cells were collected, washed and adjusted to 5×10^6 cells/ml and three serial three fold dilutions were made.

Equal volume of effector cells and target cells were added to wells in microtiter plates. The cell mixtures were incubated at $37^{\circ}C$ in 5% CO_2 for four hours. One-tenth milliliter supernatant was harvested from each well, and quantified in a liquid scintillation counter. The killing capability of the NK cells was calculated as follows:

$$\text{NK ratio} = 1 - \frac{\text{Experimental counts} - \text{SR}}{\text{Total counts} - \text{SR}} \times 1000$$

where SR = spontaneous release of ^{51}Cr.

RESULTS

ACTIVE T-CELL ROSETTES

Significant increases in the levels of active T-cell rosettes were observed in all six volunteers at all dose levels (1, 10, and 50μg/kg). The range of increases were 26% to 633% over control levels (Table 1).

TABLE 2: TOTAL T-CELL ROSETTES

	One microgram/kg		Ten micrograms/kg		Fifty micrograms/kg	
Volunteer	#1	#2	#3	#4	#5	#6
Baseline	100/200	116/200	62/200	113/200	95/200	110/200
2 Hours	106/200 (6%↑)	124/200 (4%↑)	123/200 (98%↑)	117/200 (4%↑)	128/200 (35%↑)	102/200 (7%↑)
24 Hours	108/200 (8%↑)	110/200 (3%↑)	113/200 (82%↑)	123/200 (9%↑)	117/200 (23%↑)	134/200 (21%↑)

Statistically significant difference from controls

$p<0.05 - \frac{10 \text{ volunteers}}{12 \text{ samples}}$ showing positive increase over controls

TOTAL T-CELL ROSETTES

Only a slight increase in total T-cell rosettes was seen in five of the volunteers. However, volunteer #3 demonstrated a very marked increase of 98% and 82% over control levels, (Table 2). A significant increase in 10/12 (5/6 at 2 hours and 5/6 at 24 hours) volunteer samples was obtained with ranges of 4% to 35%.

NK CELL ACTIVITY

11:1 Effector/Target Cell Ratio

Two hours after infusion, volunteers 1 and 5 had increases of 15% and 72% respectively at doses of 1 and 50μg/kg while the other four had decreases of 33%, 59%, and 10%. Volunteer #6 demonstrated no change.

Three of the volunteers administered doses of 1, 10, and 50μg/kg were found to have significant increases in NK activity, 24 hours after infusion. This increase of activity was 30%, 33% and 128% over control levels. Three of the volunteers at doses of 1, 10 and 50μg/kg showed decreases (2%, 39% and 66%) (Table 3).

Similar changes were seen at effector to target cell ratios of 33;1 and 100:1 (Tables 4&5). Overall analyses of all samples indicated that 16/36 showed a positive increase (average 28.2% over controls) ($p<0.05$). At the same time 20/60 samples showed a significant decrease (average 25.3%).

33:1 Effector/Target Cell Ratio

Two hours after infusion of met-enkephalin, at doses of 1 and 10μg/kg, four volunteers showed a slight decrease (6% to 30%). Two volunteers (50μg/kg) showed increases of 19% and 61%.

Twenty-four hours after infusion, three volunteers at doses of 1 and 50μg/kg had increases of 17%, 28% and 106%. Three volunteers (1 and 10μg/kg) had decreases of 16%, 34% and 56%.

TABLE 3: NK CELL ACTIVITY
11:1 Effector/Target Cell Ratio

Volunteer	One microgram/kg #1	#2	Ten micrograms/kg #3	#4	Fifty micrograms/kg #5	#6
Baseline	20.4%	14.3%	32.08%	18.6%	17.48%	7%
2 Hours	23.5% (15%↑)*	9.6% (33%↓)	13.27% (59%↓)	16.8% (10%↓)	30.13% (72%↑)*	7% (0%)
24 Hours	26.6% (13%↑)*	8.7% (39%↓)	11.06% (66%↓)	24.8% (33%↑)	17.06% (2%↓)	16% (128%↑)*

Significant difference from controls
5/12 positive increase $p<0.05$
7/12 negative decrease n.s.

Overall Analyses (11:1, 33:1, 100:1)
16/36 samples positive increase (average 28.2%) $p<0.05$
20/36 samples negative decrease (average 25.3%) $p<0.05$

TABLE 4: NK CELL ACTIVITY
33:1 Effector/Target Cell Ratio

Volunteer	One microgram/kg #1	#2	Ten micrograms/kg #3	#4	Fifty micrograms/kg #5	#6
Baseline	39.1%	31.4%	51.0%	37.7%	34.5%	16%
2 Hours	35.9% (8%↓)	22.1% (30%↓)	39.0% (24%↓)	35.5% (6%↓)	55.6% (61%↑)*	19% (19%↑)*
24 Hours	50.1% (28%↑)	20.7% (34%↓)	22.6% (56%↓)	31.8% (16%↓)	40.2% (17%↑)	33% (106%↑)*

Significant difference from controls
5/12 positive increase $p<0.05$
7/12 negative decrease $p<0.05$

Overall Analyses (11:1, 33:1, 100:1)
16/36 samples positive increase $p<0.05$
20/36 samples negative decrease $p<0.05$

100:1 Effector/Target Cell Ratio

Two hours after infusion, met-enkephalin at all doses showed a slight increase in three volunteers (0.4%, 2% and 7%) and three volunteers (0.04%, 2% and 7%) and three volunteers showed a slight decrease (11%, 17% and 7%).

Twenty-four hours after infusion, three volunteers showed an increase (10%, 5% and 127%) and three showed a decrease (25%, 35% and 17%).

DISCUSSION

The present in vivo study of met-enkephalin at doses of 1, 10 and 50μg/kg confirms and extends our original in vivo studies (Plotnikoff et al.,

1981, 1982, 1983). In fact the minute doses of met-enkephalin employed (1, 10 and 50 μg/kg) corresponds to 10^{-5} mg/ml blood which was a high dose range of out *in vitro* studies. Marked activity on T-cells and NK cells was seen at concentrations of 10^{-6} to 10^{-14} mg/ml blood.

TABLE 5: NK CELL ACTIVITY
100: 1 Effector/Target Cell Ratio

	One microgram/kg		Ten micrograms/kg		Fifty micrograms/kg	
Volunteer	#1	#2	#3	#4	#5	#6
Baseline	58.8%	58.8%	70.9%	57.8%	58.3%	30%
2 Hours	59.8% (2%↑)*	48.9% (17%↓)	62.9% (11%↓)	58.1% (0.4%↑)*	68.6% (18%↑)*	28% (7%↓)
24 Hours	64.9% (10%↑)*	44.2% (25%↓)	46.3% (35%↓)	48.3% (17%↓)	61.5% (5%↑)*	68% (127%↑)*

Statistical difference from controls
6/12 positive increase n.s.
6/12 negative decrease n.s.

Overall Analyses
16/36 positive increase $p<0.05$
20/36 negative decrease $p<0.05$

Previous *in vitro* studies have shown that volunteers with low NK cell baseline values had significant increases of NK activity with met-enkephalin treatment. Low baseline conditions appear to be most sensitive to the stimulant effects of met-enkephalin in elevating NK cell activity have less response following enkephalin treatment (Faith et al., 1984). Similar effects on NK cell activity have been reported with the interferons (Lotzova et al., 1983).

Such increases in active T-cell rosettes as well as NK cells may have implications in the immunomodulatory control of infectious and neoplastic diseases, (Wybran and Fudenberg, 1973; Herberman, 1982). Certainly the elevations of blood levels of enkephalin as well as NK cell activity in times of stress have been now documented by several researchers (Udenfriend and Kilpatrick, 1983, Gilman et al., 1982; Hanbauer et al., 1982; Viveros et al., 1980; Farrell et al., 1982). It is possible that the enkephalins have a reciprocal relationship with the steroid hormones in modulating immune function in times of stress (Riley, 1981; Ader, 1982; Besedovsky et al., 1983).

It is hoped that the administration of met-enkephalin to patients with infections as well as cancerous tumor growth will be found useful in modulating the immune system, as measured by its restorative effects on T-cells and NK cells.

REFERENCES

Ader R., 1982, in: "Psychoneuroimmunology", Academic Press, New York.

Besedovsky, H. O., del Rey, A. E., and Sorkin, E., 1983, What do the immune system and brain know about each other? *Immunol. Today*, 4:342.

Blalock, J. E., and Smith, E. M., 1980, Human leukocyte interferon: structural and biological relatedness to adrenocorticotropic hormone and endorphins, *Proc. Nat'l. Acad. Sci.*, 77:5972.

Faith, R. E., Liang, H. J., Murgo, A. J., and Plotnikoff, N. P., 1984, Neuroimmunomodulation with enkephalins: enhancement of human natural killer (NK) cell activity *in vitro*, *Clin. Immunol. Immunopathol.*, 31:412.

Farrell, P. A., Getes, W. K., Maksud, M. G., and Morgan, W. P., 1982, Increases in plasma beta-endorphin/beta-lipotropin immunoreactivity after treadmill running in humans, *J. Appl. Physiol.*, 52:1245.

Gilman, S. C., Schwartz, J. M., Milner, R. J., Bloom, F. E., and Feldman, J. D., 1982, Beta-endorphin enhances lymphocyte proliferative responses, Proc. Nat'l. Acad. Sci., 79: 4226.
Hanbauer, J., Kelly, G. D., Saiani, L., and Yang, H. Y. T., 1982, Met-enkephalin like peptides of adrenal medulla: release by nerve stimulation and functional implications, Peptides, 3: 469.
Hazum, E., Chang, K. J., and Cuatrecasas, P., 1979, Specific non-opiate receptors for beta-endorphins, Science, 205: 1033.
Herberman, R. B., 1982, in: "NK Cells And Other Natural Effector Cells", Academic Press, New York.
Hughes, I., Smith, T. W., Kosterlitz, H. W., Fotergill, L. A., Morhan, B. A., and Morriss, H. T., 1975, Identification of two related pentapeptides from the brain with potent opiate agonist activity, Nature, 258: 577.
Laszlo, J., Muang, A. T., Brenchman, W. D., Jeffs, C., Koren, H., Cianciolo, G., Metzgar, R., Cashdollar, W., Cox, E., Buckley III, C. E., Tso, C. Y., and Lucas, V. S., 1983, Phase I study of pharmacological and immunological effects of human lymphoid blasts interferons given to patients with cancer, J. Cancer Research, 43: 4458.
Lopker, A., Abood, L. G., Hoss, W., and Lionetti, F. J., 1980, Stereo-selective muscarinic, acetylcholine and opiate receptors in human phagocytic leukocytes, Biochem. Pharmacol., 29: 1361.
Lotzova, E., 1983, Function of natural killer cells in various biological phenomena, Surv. Synth. Path. Res., 2:41.
Mathews, P. M., Froelich, C. J., Sibbitt, Jr., W. L., and Bankhurst, A. D., 1983, Enhancement of natural cytotoxicity of beta-endorphin, J. Immunol., 130: 1658.
Miller, G. C., Murgo, A. J., and Plotnikoff, N. P., 1984, Enkephalins-Enhancement of active T-cells rosettes from normal volunteers, Clin. Immunol. and Immunopathol., 31: 132.
Plotnikoff, N. P., 1982, The central nervous system control of the immune-system-Enkephalins: Anti-tumor activities, Psychopharm. Bulletin, 18: 148.
Plotnikoff, N. P., Kastin, A. J., Coy, D. H., Christensen, C. W., Schally, A. V., and Spirtes, M. A., 1976, Neuropharmacological actions of enkephalin after systemic administration, Life Science, 19: 1283.
Plotnikoff, N. P., and Miller, G. C., 1983, Enkephalins as immunomodulators, Int. J. Immunopharmac., 5: 437
Plotnikoff, N. P., and Murgo, A. J., Faith, R. E., 1984, Neuroimmunomodulation with enkephalins: Effects on thymus and spleen weights in mice, Clinical Immunology and Immunopathology, 32: 52.
Plotnikoff, N. P., Murgo, A. J., Miller, G. C., Corder, C. N., and Faith, R. E., 1985, Enkephalins: immunomodulators, Fed. Proc. 44: 118.
Prange, A. J. Jr., Wilson, J. C., Zara, P. P., Alltop, L. B., and Breese, G. R., 1972, Effects of thyrotropin-releasing hormone in depression, The Lancet, 2: 999.
Riley, V., 1981, Psychoneuroendocrine influences on immunocompetence and neoplasia, Science, 212: 1100.
Robbins, D. S., Donnan, G. G., Fudenberg, H. H., and Strelkauskas, A. J., 1981, Functional subsets of human T cells defined by 'active' rosette formation, Cell. Immunol., 59: 205.
Rossier, J., Dean, D. M., Livett, B. G., and Udenfriend, S., 1981, Enkephalin congeners and precursors are synthesized and released by primary cultures of adrenal chromaffin cells, Life Sci., 28: 781.
Schweigerer, L., Bhakdi, S., and Teschemacher, H., 1982, Specific non-opiate binding sites for human beta-endorphin on the terminal complex of human complement, Nature, 296: 572.
Simantov, R., and Snyder, S., 1976, Morphine-like peptides in mammalian brain: isolation, structure elucidation, and interactions with the opiate receptor, Proc. Nat'l Acad. Sci., 73: 2515.

Targan, S., Britvan, L., and Dorey, F., 1981, Activation of human NKCC by moderate exercise: increased frequency of NK cells with enhanced capability of effector-target lytic interactions, Clin. Exp. Immunol., 45: 352.

Udenfriend, S., and Kirkpatrick, D. L. 1983, Biochemistry of the enkephalins and enkephalin-containing peptides, Arch. Biochem. Biophys., 221: 309.

Vaught, J. L., and Takemori, A. E., 1979, Differential effects of leucine and methionine enkephalin on morphine-induced analgesia, acute tolerance and dependence, J. Pharmacol. Exp. Ther., 208: 86.

Viveros, O. H., Diliberto, E. J., Hazum, E., and Chang, K. J., 1980, Enkephalins as possible adrenomedullary hormones: storage, secretion, and regulation of synthesis, Adv. Biochem. Psychopharmacol., 22: 191.

Wybran, J., Appelboom, T., Famaey, J. P., Govaerts, A., 1979, Suggestive evidence for receptors for morphine and methionine-enkephalin on normal human blood T lymphocytes, J. Immunol., 123: 1068.

Wybran, J., and Fudenberg, H. H., 1973, Thymus-derived rosette-forming cells in various human disease states: cancer, lymphoma, bacterial and viral infections and other diseases, J. Clin. Invest., 52: 1026.

Wybran, J., Appelboom, T., Famaey, J. P., and Goverts, A., 1980, in: "International Symposium on New Trends in Human Immunology and Cancer Immunotherapy", B. Serrou and C. Rosenfeld, ed., Doin Press, Paris.

ACKNOWLDEGEMENTS

These studies were generously supported by a grant from TNI Pharmaceuticals, Inc., Tulsa, Oklahoma and Travenol Laboratories, Morton Grove, Illinois.

METHIONINE ENKEPHALIN: CLINICAL PHARMACOLOGY

N.P. Plotnikoff, S. Solomon, J.L. Valentine, M. Fesen, R.E. Faith, L. Edwards, R. Richter, A.J. Murgo, and G.C. Miller*
Departments of Pharmacology and Medicine
School of Medicine
Oral Roberts University
Immuno-Diagnostics Laboratory*
Tulsa, OK 74171

INTRODUCTION

Earlier studies indicated that methionine enkephalin infusions (1,000 micrograms) in normal volunteers induced facial flushing, general vasodilation and noisy stomach (borborgymi). Blood pressure and pulse rate were normal in all experiments. Plasma levels of pancreatic polypeptide, somatostatin, T_3 and T_4, TSH, FSH, PRL, ACTH, and GH were unchanged. However, there was some reduction of LH (Rolandi et al., 1980, Goldstein et al., 1981). The present study is an evaluation of the effects of methionine enkephalin (over a broad dose range) on all of the vital signs as well as measurements of blood level.

PATIENTS AND METHODS

Twelve male subjects age 21-39 years were recruited for these studies and given appropriate informed consent documents approved by Oral Roberts University Institutional Review Board. They also provided a signed statement of complete abstinence from drugs, including alcohol and tobacco, at least three weeks prior to the study. All subjects were within 20% of ideal body weight and free of acute and chronic cardiovascular, renal, metabolic and neurological diseases. Those subjects satisfying admission criteria were admitted to the hospital the evening prior to the study, given the drug infusion in the morning of the second day and discharged on the third day.

LABORATORY/PHYSICAL EXAMINATION--PREHOSPITALIZATION CLINIC

A routine physical examination was done which included medical and social history, habits, medications and comprehensive physical examination. The examination included supine and erect heart rate, blood pressure, and examination of occult blood.

LABORATORY (CERTIFIED CLINICAL LABORATORY)

The laboratory work-up included: 12 lead electrocardiograms, chest x-rays, PA and lateral. A standard blood chemistry profile (SMAC 26) was done which routinely includes sodium, potassium, chloride, carbonate, glucose, BUN, creatinine, calcium, osmolality, phosphate, total protein, albumin, globulin, total bilirubin, SGOT, AST, ALK., phosphatase, LDH, cholesterol, triglyceride, uric acid and iron. Also included in the laboratory testing were serum amylase, VDRL, ANA, HbSAg, clinical blood count (CBC), which includes WBC, RBC, Hg., HCT., Diff., platelet and RBC indices. Urinalysis was also done including pH, Sp. Gr., protein and microscopic. All patients had to have normal data for No. 1 and 2 before entering the hospital phase of the study.

METHIONINE ENKEPHALIN FORMULATION

An investigational new drug (IND) sterile solution of methionine enkephalin was prepared at 10 mg. per ml. Methionine enkephalin (Sigma-Aldrich Chem. Co., St. Louis, Mo.) was dissolved in sterile saline (0.9 gm NaCl per 100 ml) and passed through a 0.22 micron millipore filter. Sigma-Aldrich Chem. Co. provided the Master file for the FDA. The ultra-filtration through the millipore filter was done in the City of Faith Hospital pharmacy. Methionine enkephalin has the following characteristics: M.W. 574, water soluble, Amino acid sequence Tyr-Gly-Gly Phe-Met-OH. It was dissolved in 60 ml NS (0.9 gm NaCl per 100 ml) at a concentration sufficient to infuse at 2 ml per minute for 30 minutes to give a total delivered dose of 1, 10, 50, 100, 150 and 200 micrograms per kilogram body weight.

HOSPITALIZATION

Patients were admitted to the hospital at approximately 1600 on Day 1. A brief physical examination was done and recorded in their chart, and the consent form was signed. The protocol was as follows: Day 1, the patients were admitted, daily body weight taken. Breakfast on Day 2 was a clear liquid diet; thereafter, a regular diet was served. Activity was restricted to room. No visitors were allowed. Supine, standing heart rate and blood pressure were taken at 0800, 1400, 2000 and other times as indicated. Day 2 at 0800: a No. 18 gauge intracath was placed in the antecubital vein with open NS, 1 L. EKG leads and blood pressure cuff were attached. Baseline EKG, supine, erect blood pressure and heart rate were recorded.

SPECIAL PRECAUTIONS

Seizure precautions were taken from 0800 Day 2 to discharge. Usual cardiovascular emergency precautions were taken, including the availability of Naloxone injectibles. Patients used bed pans and urinals for voiding on Day 2 from 0830 to 1200 so that the patient would remain supine.

BEHAVIORAL TEST

100 mm. Mood Line Test. Clinical behavioral mood states were determined by the use of 100 mm Line Test. In this test, a 100 mm line was drawn on a sheet of plain paper and one end was identified "as well as I could be", the other end "as depressed as I could be". The patient marked where he/she stood at the moment of testing. The score, the length in millimeters from the "well" end of the line, was based on the premise that length is analagous to the severity of depression (Prange et al., 1972).

RIA-METHIONINE ENKEPHALIN METHODS

The Immuno Nuclear radioimmunoassay (Immuno Nuclear Corp., Minn.) for methionine enkephalin employs simultaneous addition of sample, rabbit anti-met-enkephalin antibody, and ^{125}I met-enkephalin, followed by an overnight incubation at $4^{o}C$. Phase separation is accomplished by the addition of an equal volume of saturated ammonium sulfate in the presence of carrier gamma globulin. Met-enkephalin is extracted from plasma on ODS-silica, eluted, dried, and reconstituted for assay. Standard were prepared from 1-2000 ng/ml.

PROLACTIN DETERMINATION

The [^{125}I] Prolactin RIA kit by Diagnostic Products Corp., Los Angelos, CA was used for prolactin determinations. This kit requires no preliminary sample extraction. Furthermore, the calibration curve prepared in the presence of serum proteins has a range equivalent to circulating prolactin of 5-200 ng/ml.

RESULTS

Cardiovascular, Respiratory and Temperature Parameters

No significant changes in any parameter measured were recorded. Only transient increases or decreases were seen in heart rate and blood pressure. Representative findings are illustrated in Tables 1-6, showing effects of 1, 10, 50, 100, 150 and 200 micrograms/kg.

TABLE 1
VOLUNTEER #3
MET-ENKEPHALIN (0.001 mg/kg)

11/20/83	BLOOD PRESSURE Supine/Erect		HEART RATE(S/E)	RESP. RATE	TEMPERATURE (°F)
0800	110/80	110/80	52/60	12	98
0830	110/80	110/80	56/60	12	98
0855	110/80	110/80	52/60	12	98
0900	ENKEPHALIN INFUSION				
0910	110/80		55	12	98
0920	110/80		55	12	98
0930	110/80		50	12	98
1000	110/80		54	10	98
1100	110/80	120/80	56/62	12	98
1300	110/80	110/80	55/58	12	98
1600	110/80	110/80	55/58	12	98
11/21/83					
0855	110/80	120/80	54/60	12	98

TABLE 2
VOLUNTEER #1
MET-ENKEPHALIN (0.010 mg/kg)

	BLOOD PRESSURE		HEART RATE(S/E)	RESP. RATE	TEMPERATURE (°F)
11/12/83					
0800	100/70		44/45	12	98
0855	100/70		50/52	13	98.8
0900	ENKEPHALIN INFUSION				
0910	100/70		60	14	98.8
0920	100/70		60	12	98.8
0930	100/70		60	12	98.8
1000	110/70		64	12	98.8
1100	100/70		60	14	98.6
1300	100/70	100/70	45/60	12	98.4
1600	100/70	100/70	44/56	13	98.4
11/13/83					
0855	100/70	100/70	45/60	12	98.4

TABLE 3
VOLUNTEER #5
MET-ENKEPHALIN (0.050 mg/kg)

	BLOOD PRESSURE Supine/Erect		HEART RATE(S/E)	RESP RATE	TEMPERATURE (^{o}F)
12/19/83					
0800	110/70	110/70	71/80	12	97.6
0855	110/70	110/70	68/76	12	97.6
0900	MET-ENKEPHALIN INFUSION				
0910	110/70		75	11	97.8
0920	110/70		81	12	97.4
0930	110/70		68	12	97.6
1000	110/70		65	11	97.6
1100	110/70	100/70	68/90	12	97.6
1300	110/70	100/70	64/90	12	98
1600	110/70	110/70	74/88	12	98
12/20/83					
0855	110/70	110/70	68/76	12	97

TABLE 4
VOLUNTEER #7
MET-ENKEPHALIN (0.100 mg/kg)

	BLOOD PRESSURE	HEART RATE(S/E)	RESP. RATE	TEMPERATURE (^{o}F)
2/5/84	Supine			
0800	130/80	60/74	14	97.6
0855	120/80	57/64	12	98
0900	MET-ENKEPHALIN INFUSION			
0910	120/80	56	12	98
0920	120/80	54	12	98
0930	120/80	54	13	98
1000	120/80	63	12	98
1100	120/80	54/64	14	98
1300	120/80	64/72	14	98
1600	120/80	72/76	13	98
2/6/84				
0855	120/80	70/76	14	97.4

TABLE 5
VOLUNTEER #9
MET-ENKEPHALIN (0.150 mg/kg)

	BLOOD PRESSURE	HEART RATE(S/E)	RESP. RATE	TEMPERATURE (^{o}F)
3/11/84	Supine			
0800	120/80	62/76	14	97.4
0855	120/80	76/80	13	98.2
0900	MET-ENKEPHALIN INFUSION			
0910	120/80	90	13	98
0920	120/80	81	12	98
0930	120/80	73	12	98
1000	120/80	74/102	14	98
1100	120/80	87/100	12	98
1300	120/80	74/80	13	98
1600	120/80	76/80	12	98
3/12/84				
0855	120/80	72/80	13	98

TABLE 6 - VOLUNTEER #11

MET-ENKEPHALIN (01.200 mg/kg)

	BLOOD PRESSURE	HEART RATE(S/E)	RESP RATE	TEMPERATURE (°F)
6/17/84				
0800	120/80	80/94	12	97.2
0855	120/80	74/76	12	97.2
0900	MET-ENKEPHALIN INFUSION			
0910	120/80	76	13	97.2
0920	120/80	88	14	97.2
0930	120/80	81	12	97.2
1000	120/80	71	12	97.4
1100	120/80	75/78	13	98
1200	120/80	70/74	14	98
1300	120/80	68/76	13	98
1600	120/80	70/74	12	98
6/18/84				
0830	120/80	61/68	12	97

TABLE 7 - HEART RATE

Maximum Increase Over Baseline Control-8:55 a.m.	Peak Time Increase	Maximum % Increase Over Baseline
#1 50 - 64 =	10:00 a.m.	<u>28%</u>
#2 70 - 76 =	10:00 a.m.	9%
#3 52 - 56 =	11:00 a.m.	8%
#4 60 - 74 =	10:00 a.m.	<u>23%</u>
#5 68 - 81 =	9:20 a.m.	<u>19%</u>
#6 76 - 82 =	9:30 a.m.	8%
#7 57 - 72 =	4:00 p.m.	<u>26%</u>
#8 70 - 76 =	9:30 a.m.	9%
#9 76 - 90 =	9:10 a.m.	<u>18%</u>
#10 55 - 62 =	10:00 a.m.	<u>12%</u>
#11 74 - 88 =	9:20 a.m.	<u>19%</u>
#12 66 - 75 =	9:30 a.m.	<u>14%</u>

8/12 10% Increase

7/8 a.m. 1/8 p.m.

<u>Heart Rate</u>

Eight subjects exhibited an increase in heart rate following the infusion of methionine enkephalin. However, no dose response was apparent and the increase in heart rate may have been due to anxiety(Table 7).

<u>100 mm Test</u>

Eight subjects indicated a shift to the "sad scale". (6/8)- Six of the eight subjects indicated this shift in the morning following infusion suggesting "sedation." Three subjects appeared to have transient mystagmus (attributabel to "sedative" effect).

TABLE 8
MET-ENKEPHALIN
100 mm LINE TEST

Maximum Change Over Controls 8:55 a.m.	Post-Infusion Peak Time	%Decrease Over Controls
#1 59mm - 38 =	10:00 a.m.	36%
#2 46mm - 44 =	9:30 a.m.	4%
#3 37mm - 28 =	10:00 a.m.	24%
#4 29mm - 26 =	10:00 a.m.	10%
#5 42mm - 30 =	4:00 p.m.	28%
#6 5.5mm - 3.0 =	4:00 p.m.	45%
#7 26mm - 26 =	11:00 p.m.	0%
#8 37mm - 12 =	4:00 p.m.	67%
#9 50mm - 17 =	1:00 p.m.	66%
#10 33mm - 21 =	4:00 p.m.	36%
#11 34mm - 24 =	1:00 p.m.	29%
#12 39mm - 31 =	4:00 p.m.	21%

$\frac{10}{12}$ 10% Decrease $p < 0.05$ Paired Difference Analyses

$\frac{3}{10}$ a.m. $\frac{7}{10}$ p.m.

TABLE 9
MET-ENKEPHALIN
100 mm LINE TEST

Maximum Change Over Controls 8:55 a.m.	Peak Time	%Increase Over Controls
#1 59 - no increase		0%
#2 46 - 50 =	11:00 a.m.	8%
#3 37 - 41 =	9:30 a.m.	10%
#4 29 - 38 =	4:00 p.m.	24%
#5 42 - 53 =	10:00 a.m.	26%
#6 5.5 - 17.0 =	1:00 p.m.	209%
#7 26 - 29 =	10:00 a.m.	11%
#8 37 - 61 =	11:00 a.m.	64%
#9 50 - no increase		0%
#10 33 - 38 =	9:30 a.m.	13%
#11 34 - 37 =	9:30 a.m.	8%
#12 39 - 54 =	10:00 a.m.	38%

8/12 10 % Increase $p < 0.05$ Paired Difference Analyses

6/8 a.m. 2/8 p.m.

Ten subjects indicated a shift to the "happy side of the scale". (7/10)- Seven of the ten subjects reported a shift to the "happy side of the scale" in the afternoon. See Table 9.

Prolactin

No significant alteration in plasma prolactin levels were observed over the course of the study except for two subjects who showed an increase in blood levels of prolactin after recieving a dose of ten micrograms/kilogram.

Blood Levels-Methionine Enkephalin

The pre-infusion levels of methionine enkephalin were quite high. This was probably due to the stress of pre-infusion procedures(indwelling catheter, EKG, as well as anticipation of the infusion). Volunteers maintained high levels throughout the study.

Levels elevated above baseline were seen in volunteers administered 50 micrograms/kilogram, 100 micrograms/kilogram, 150 micrograms/kilogram, and 200 micrograms/kilogram. No apparent dose response was seen due to administered methionine enkephalin.

TABLE 10:PROLACTIN BLOOD LEVELS(ng/ml)

DOSE	1 microgram/kg		10 micrograms/kg		50 micrograms/kg
Volunteer Number	1	2	3	4	5
0855 Day One	a	4.9	10.6	3.1	20.1
0910	0.2	3.8	7.7	3.3	2.6
0920	1.2	4.4	6.5	8.5	22.3
0930	0.9	2.1	10.6	29.1	14.0
1000	a	0.8	15.3	16.9	12.4
1100	a	6.4	25.1	42.4	9.4
0855 Day Two	0.3	6.3	14.3	24.0	18.7

[a]Below level of detection for the assay

TABLE 11: BLOOD LEVELS-MET ENKEPHALIN (pg/ml)

DOSE	1 microgram/kg	10 micrograms/kg	50 micrograms/kg
0855 Day One	763.5	377.9	130.3
0910	618.1	216.4	221.1
0920	557.2	162.6	179.9
0930	540.0	140.3	218.2
1000	706.7	138.8	860.9
1600	890.2	--	240.8
0855 Day Two	608.9	--	--

TABLE 12: BLOOD LEVELS-MET ENKEPHALIN (pg/ml)

DOSE	100 micrograms/kg	150 micrograms/kg	200 micrograms/kg
Control 0855	157.1	652.7	1315.4
0910	--	735.9	660.7
0920	252.4	759.6	1679.3
0930	196.4	1311.9	779.9
1000	--	785.8	264.5
1600	180.6	664.4	269.2
0855 Day Two	115.5	559.2	237.0

SIDE EFFECTS

No consistent changes or any dose response were observed.

Initial facial flushing following infusion; sedation, "stomach noises", skin tingling, were reported similar to other human enkephalin studies (Table 13).

SMAC 26 and Urinalysis

No consistent changes or dose response changes were observed.

The principal side effects observed were vaso-flushing(5/12) and gastro-intestinal disturbances(4/12). See Table 13.

TABLE 13: MET-ENKEPHALIN - PHASE I SUMMARY(12 Volunteers)

Side Effects	Frequency
Flushing	
Warm feeling during infusion	1 (50 μg/kg)
Tingling and flushing scalp	1 (150 μg/kg)
Burning sensation -- head	1 (200 μg/kg)
Flushing face	1 (150 μg/kg)
(hands-head chest-body)	1 (200 μg/kg)
	TOTAL...5/12
Gastro-Intestinal	
Crampy abdominal pain	1 (200 μg/kg)
Increased bowel sounds	1 (150 μg/kg)
Intra-umbilical abdominal cramps	1 (150 μg/kg)
Slightly nauseated	1 (100 μg/kg)
	TOTAL...4/12
Headache	2/12 (1 & 10 μg/kg)
Euphoric and light-headed	1/12 (1 μg/kg)
Angry for a couple of minutes	1/12 (50 μg/kg)
Drowsy-sleeping	1/12 (200 μg/kg)

DISCUSSION

The present study demonstrates that methionine enkephalin can be administered safely to humans over a wide dose range. No significant alterations in vital signs were observed. Biphasic behavioral changes were recorded on the 100 mm line test ("sedated" effects following the infusion in the morning and a "stimulant" effect in the afternoon). Only slight effects were seen in elevating prolactin levels (two volunteers). Blood levels of methionine enkephalin recorded were highly variable and appeared to be independent of any dose response relation. The side effects recorded were minimal and consisted predominantly of "vaso-dilation-flushing" (5/12) and gastro-intestinal effects (4/12). In general these effects were transient and seen at high doses.

In the earlier study reported by Golstein et al. (1981) methionine enkephalin at a dose of 1000 μg induced side effects of facial flushing, vasodilation and borborgymi, while blood pressure and pulse rate were normal. Hormone levels remained unchanged.

Other clinical studies with derivatives of methionine enkephalin indicated that various degrees of analgesia were observed because of stability as well as more side effects. Drowsiness, sensation of heaviness in extremeties, burning at injection site, and nasal congestion were the side effects reported (Calimlin et al., 1982).

FK 33-824, another enkephalin derivative apparently resulted in a larger rate of side effects in man; oppression of chest, red eyes, whole body flushing, as well as a flare reaction with intradermal injection (Von Grafferied et al., 1978). In addition, FK 33-834 increased PRL and GH and reduced cortisol (del Pozo et al., 1980).

The present study supports and extends the original findings by Golstein et al., (1981) that methionine enkephalin can be administered safely to man over a wide range.

The blood levels of methionine enkephalin observed would support the findings of Roda et al. (this volume) that there is protective binding to plasma proteins resulting in sustained blood levels. Also the difference in side effects compared to the synthetic derivatives suggest that methionine enkephalin may have great receptor specificity in the delta family of receptors.

REFERENCES

Calimlin, J. F., Sriwatanakul, K., Mardell, W. M., Lasagna, L., and Cox C., 1982, Analgesic efficacy of parenteral met-enkephamide acetate in treatment of post-operative pain (LY 127623), The Lancet, 1(2):1374.

del Pozo, E., Kleinstein, J., Brun del Re R., Derrer F., and Martin-Perez, J., 1980, Failure of oxytocin and lysine-vasopressin to stimulate prolactin release in humans, Horm. Metab. Res., 12:26.

Golstein, J., Cantraine, F., Copinschi, G., L'Hermite, M., Pipeleers, D., Robyn, C., Velkeniens, B., and Vanhaelst, L., 1981, Effects of enkephalin infusion on hormonal levels in man: Inhibition of the episodic secretion of LH by methionine enkephalin, IRCS Med. Sci., 9:218.

Plotnikoff, N. P., Murgo, A. J., Miller, G. C., Corder, C. N., and Faith, R. E., 1985, Enkephalins: Immunomodulators, Fed. Proc., 44:118.

Prange, A. J. Jr., Wilson, J. C., Zara, P. P., Alltop, L. B., and Breese, G. R., 1972, Effects of thyrotropin-releasing hormone in depression, The Lancet, 2(1):999.

Rolandi, E., Pescatore, D., Milesi, G. M., Gilberti, C., Sannia, A., and Barreca, T., 1980, Evaluation of LH, FSH, TSH, PRL, and GH secretion in patients suffering from prostatic neoplasms, Acta Endocrinol. (Copenh), 95:23

Von Graffenried, B., Del Pozo, E., Roubicek, J., Krebs, E., Poldinger, W., Burmeister, P., and Kerp, L., 1978, The effects of the synthetic enkephalin analogue FK 33-824 in man, Nature, 272:729.

METHIONINE ENKEPHALIN: T-CELL ENHANCEMENT IN NORMAL VOLUNTEERS (*IN VIVO*)

N.P. Plotnikoff, G.C. Miller*, S. Solomon
R.E. Faith, L. Edwards, and A.J. Murgo

Departments of Pharmocology and Medicine
Oral Roberts University, School of Medicine
Immuno-Diagnostics Laboratory*
Tulsa, Oklahoma 74171

INTRODUCTION

Earlier studies by our group indicated that methionine enkephalin enhanced PHA induced lymphocyte proliferation in mice, active rosette forming cells and NK cells in the cells of normal volunteer and cancer patients,(Plotnikoff et. al.,1985). More recently, several groups have reported that methionine enkephalin stimulates macrophage function as well as chemotaxis(Foris, 1984;Weber, 1984). Finally several groups have reported increased lymphokine production by methionine enkephalin(Brown, et al.,1985;Youkilis et al.,1985).

The present study demonstrates that methionine enkephalin increases circulating lymphocyte numbers, exhibited by increased numbers of various T-lymphocyte subpopulations, and enhances lymphoproliferation of PBL's stimulated with PHA.

METHODS

PATIENTS

Seven normal volunteers (males were studied clinically and immunologically prior to and following methionine enkephalin treatment.

METHIONINE ENKEPHALIN

Methionine enkephalin(Sigma-Aldrich Chem. Co., St. Louis, MO) was dissolved in sterile saline and filter sterilized by the hospital pharmacy. It was dissolved in 60 ml of saline at a concentration sufficient to infuse at 2 ml per minute for 30 minutes to give a total delivered dose of 10 to 25 micrograms per kilogram per body weight.

CELL PREPARATIONS

Heparinized peripheral blood was collected, incubated with 25 microliters of a 1:100 dilution of latex bead preparation/10^6 cells for 30 minutes at 37C, diluted with phosphate buffered saline(PBS), layered over a Ficoll-Paque(Pharmacia) gradient and centrifuged for 30 minutes at 400 x g. The mononuclear cell band was collected, washed twice with PBS, resuspended in RPMI 1640 and quantified.

ACTIVE T-CELLS

An equal volume of 1×10^6 lymphocytes and heat inactivated fetal calf serum was incubated at 37°C for 60 minutes. The fetal calf serum had been absorbed against both sheep and human AB erythrocytes. The lymphocytes were then gently mixed with an equal volume of sheep red blood cells (SRBC) at 4×10^7/ ml suspended in saline, centrifuged at 200x g at 4°C for 5 minutes gently resuspended and rosettes quantified using a hemacytometer. An active rosette forming cell is defined as a lymphocyte which has bound three or more SRBC.

T LYMPHOCYTES AND T SUBSETS

10^6 cells in 50 microliters PBS with 0.1% sodium azide was incubated for 30 minutes at 4°C with monoclonal antibodies (MoAb-s(10 micrograms/ml)). The cells were washed twice, incubated with FITC labeled goat anti-mouse serum at 4° for 30 minutes, washed twice again with PBS with sodium azide and gently resuspended in 1 drop of PBS with the sodium azide and quantified under a Nikon epi-illumination fluorescent microscope. Three hundred cells were counted: the monoclonal antibodies utilized were OKT 3, OKT 4, OKT 8, and OKT 11 (Ortho Pharmaceu., N.J.)

LYMPHOCYTE PROLIFERATION ASSAY

Lymphocyte proliferation, stimulated by Phytohemagglutinin (PHA), pokeweed mitogen (PWM), concanavalin A (con A) and Staphylcoccal Protein A was performed in triplicate. The mitogens, in varying concentrations, were plated with 1×10^5 cells per well in 96 microculture plates and were incubated for 96 hours. Sixteen hours before harvesting, 1 microCurie of ^3H thymidine (2 Ci/mmole; Amersham, Arlington Heights, Bethseda, MD) suction-filter apparatus and the activity of tritium bound to acid insoluble material was quantified in a liquid scintillation spectrometer.

NATURAL KILLER (NK) CELL ASSAY

K562 target cells, adjusted to 2×10^6 cells in 0.2 ml RPMI 1640, were labeled with $Na^{51}CrO_4$ (1mCi) for 45 minutes at 37°C in 5% CO_2. The cells were washed twice, resuspended in 2 ml RPMI 1640 and adjusted to 5×10^4 cells per ml. Peripheral blood lymphocytes were collected on a Ficoll-Paque gradient, washed twice and adjusted to 5×10^6 cells/ml and 3 serial 3 fold dilutions made. An equal volume of effector cells and target cells were added to wells in microtiter plates. These cell mixtures were incubated at 37°C in 5% CO_2 for 4 hours. A 0.1 ml aliquot was collected from each of the wells and quantified. The killing capabilities of the NK cells was calculated as follows:

$$\text{NK ratio} = 1 - \frac{\text{Experimental counts} - \text{SR}}{\text{Total counts} - \text{SR}} \times 100$$

where SR = spontaneous release of ^{51}Cr.

INTERLEUKIN 2 RECEPTOR

Lymphocytes collected from a Ficoll-Paque gradient were incubated with Phytohemagglutinin for 96 hours, the cell harvested, washed twice and incubated with anti-human IL-2 receptor (Becton-Dickinson, Mtn. View, Calif.) for 30 minutes at 4°C. The cells were washed twice in PBS with sodium azide, incubated with FITC goat anti-mouse antibody, washed twice with PBS, resuspended in 1 drop of PBS with sodium azide and examined under the fluorescent microscope. Three hundred cells were counted.

RESULTS

TOTAL LYMPHOCYTES - There was a significant increase in numbers of circulating lymphocytes two hours after infusion in six volunteers (Table 1).

B LYMPHOCYTES - At two hours, there was a slight decrease in three volunteers. At 24 hours there was a significant increase in five subjects (Table 1).

ACTIVE ROSETTE FORMING CELLS - Significant increases of E rosettes were found at both the 2 hour(six subjects) and 24 hour period(seven subjects) (Table 1).

T-LYMPHOCYTES (OKT 11) - Significant increases in number of OKT 11 cells were seen at both 2 hours in six volunteers and 24 hours in all seven volunteers(Table II).

T-HELPER LYMPHOCYTES (OKT 4)- Six subjects recieving methionine enkephalin showed a significant increase in OKT 4 labeled cells 2 hours after infusion. All seven subjects had a significant increase at 24 hours(Table II).

T-SUPPRESSOR LYMPHOCYTES (OKT 8)- Five subjects exhibited an increase in OKT 8 labeled cells 2 and 24 hours after infusion. Two subjects had a slight decrease. There was no significant alteration of the ratio OKT 4 and OKT 8 labeled cells in those subjects at baseline and 1.6 at 2 and 24 hours (Table II).

ACTIVE T-LYMPHOCYTES (ROSETTES) - Significant increases in number of rosettes was seen in seven subjects at 2 hours and in six subjects at 24 hours (Table II).

BLASTOGENESIS ConA - Significant increases in proliferative rates were seen at 2 hours in six subjects (Table III).

BLASTOGENESIS ConA - Four subjects were found to have significant increases at 24 hours while three subjects had significant decreases (Table III).

BLASTOGENESIS-POKEWEED - Five subjects were found to have a significant increase at 2 hours (Table III).

BLASTOGENESIS-STAPH A - No significant changes were observed in four subjects studied(Table III).

NK CELL ASSAY - At 2 hours, five of the volunteers were recorded to have a significant decrease in NK activity at effector to target ratios of 100:1 and 11:1. At 24 hours, six volunteers registered a significant decrease at a ratio of 100:1 while four subjects expressed a decrease at a ratio of 11:1 (Table IV).

NK CELLS (LEU 7) - No significant changes recorded in the four subjects studied (Table IV).

DISCUSSION

The present study in normal volunteers demonstrates that methionine enkephalin significantly elevates numbers of lymphocytes, active rosette forming cells, and T lymphocyte subsets. In addition there is a significant increase in PHA and ConA stimulated blastogenesis in some but not all individuals. Earlier studies indicated that *in vitro* , methionine enkephalin increased numbers of rosette forming cells in normal volunteer cells(Plotnikoff et al.,1985). In this regard, significant increases in T lymphocyte subsets have now been

TABLE I - LYMPHOCYTES FROM VOLUNTEERS
ADMINISTERED METHIONINE ENKEPHALIN

	0 hours	2 hours	24 hours
Lymphocytes			
	1864±90	2375±185* (6/7 +)	--
	0 hours 1739±129		2212±223 (7/7 +)
B Lymphocytes			
	0 hours	2 hours	24 hours
	212±25 (4/7 +)	295±27 (4/7 +)	--
	247±33 (3/7 +)	225±40* (4/7 +)	--
	226±24 (5/7 +)		*323±39 (5/7 +)
T Lymphocytes(OKT 11 Rosettes)			
	0 hours	2 hours	24 hours
	1004±74 (6/7 +)	1495±133* (6/7 +)	
	972±70 (7/7 +)		1184±215* (7/7 +)

* Significant difference from controls p<0.05 paired difference analysis

TABLE II - T LYMPHOCYTES FROM VOLUNTEERS

ADMINISTERED METHIONINE ENKEPHALIN

	0 hours	2 hours	24 hours
T Lymphocytes (OKT 11)			
	1353±108 (6/7 +)	1791±138* (6/7 +)	
	1282±115 (7/7 +)		1774±171* (7/7 +)
T Helper Lymphocytes (OKT 4)			
	719±89 (6/7 +)	983±55* (6/7 +)	
	690±81 (7/7 +)		979±110* (7/7 +)
T Supressor Lymphocytes (OKT 8)			
	419±39 (5/7 +)	635±89* (5/7 +)	
	374±82 (5/7 +)		632±163* (5/7 +)
	Ratio 1.7	Ratio 1.6	Ratio 1.6
Active T Lymphocytes			
	31±6 (7/7 +)	66±13* (7/7 +)	
	34±6 (6/7 +)		87±17* (6/7 +)

* Significant difference from controls p<0.05 paired difference analysis

TABLE III-BLASTOGENESIS FROM VOLUNTEERS ADMINISTERED METHIONINE ENKEPHALIN

Blastogenesis PHA

0 hours	2 hours	24 hours
219±77 (6/7 +)	468±130* (6/7 +)	
246±88 (5/7 +)		357±81 (5/7 +)

Blastogenesis ConA

0 hours	2 hours	24 hours
173±54 (4/7 +)	238±87 (4/7 +)	
211±69 (4/7 +)		375±50* (4/7 +)
192±15 (3/7 -)		136±24* (3/7 -)

Blastogenesis –Pokeweed

0 hours	2 hours	24 hours
33±10 (5/7 +)	53±16 * (5/7 +)	
37±10 (5/7 +)		63±11 (5/7 +)

Blastogenesis-Staph A

0 hours	2 hours	24 hours
12±7 $(\frac{2}{4} + \frac{2}{4} -)$	15±11 $(\frac{2}{4} + \frac{2}{4} -)$	
11±7 $(\frac{2}{4} + \frac{2}{4} -)$		13±8 $(\frac{2}{4} + \frac{2}{4} -)$

*Significant difference from controls $p<0.05$

found in a patient with Kaposis' sarcoma(AIDS) as well as a lung cancer patient(Plotnikoff et al.,1985). Therefore, it may be of great interest to expand the clinical stúdies into patients with T-cell deficiencies. However, in the present study no significant increase in NK cell activity was observed in contrast to earlier in vitro studies, perhaps because of high baseline control levels. The highly significant increase in T lymphocytes and T-cell subsets and PHA induced blastogenesis is supported by our earlier findings that methionine enkephalin may be stimulating in the maturation of T-cells in the thymus gland. This effect on T-cells is separate from other natural hormone immunomodulators such as interferons, and interleukins(Einhorn et al.,1983).

TABLE IV - NK CELLS FROM VOLUNTEERS

ADMINISTERED METHIONINE ENKEPHALIN

	0 hours	2 hours	24 hours
NK CELL ASSAY			
100:1 Effector/Target Ratio			
	54± 6 (5/7 -)	51±6* (5/7 -)	
	56±4 (6/7 -)		47±4* (6/7 -)
33:1 Ratio			
	27±6 (4/7 +)	29±5 (4/7 +)	
	37±3 (3/7 -)	35±2 (3/7 -)	
	34±4 (6/7 -)		29±4 (6/7 -)
11:1 Ratio			
	20±2 (5/7 -)	15±3* (5/7 -)	
	19±4 (4/7 -)		16±4* (4/7 -)
NK CELLS(LEU 7)			
	252±68 (2/4 + 2/4 -)	374±95	
	297±66 (2/4 ± 2/4 -)		288±86

* Significant difference from controls p less than 0.05

REFERENCES

Brown, S. L. and Van Epps, E. E., 1985, Beta endorphin (BE), met-enkephalin (ME), and corticotrophin (ACTH) modulate the production of gamma interferon (INF) in vitro, Fed. Proc., 44:949.

Einhorn, S., Blomgren, H., Stander, H., and Troye, M., 1983, Activity following treatment with Interferon In Vivo, in: "Mediation of Cellular Immunity in Cancer by Immune Modifiers", M. A. Chirigos, ed., Raven Press, New York.

DeMaeyer, G., Galasso, G., and Schellenkens, H., eds. 1983, in: "The biology of the interferon system", 347-352, Elsevier/North Holland Biomedical Press, Amsterdam.

Foris, G., Medgyesi, G. A., Gyimesi, E., and Hauck, M., 1984, Met-enkephalin induced alterations of macrophage functions, Mol. Immun., 21:747.

Plotnikoff, N. P., Murgo, A. J., Miller, G. C., Corder, C. N., and Faith, R. E., 1985, Enkephalins: Immunomodulators, Fed. Proc. 44:118.

Plotnikoff, N. P., Miller, G. C., Wybran, J., and Nimeh, N. F., April 10, 1985, Methionine enkephalin T-cell enhancement in Kaposis' sarcoma, Serono Symposium, Copley Plaza Hotel, Boston, Recent Advances in Primary and Acquired Immunodeficiencies (in Press).

Weber, R. J., and Pert, C. B., 1984, Opiatergic modulation of the immune system, in: "Central and Peripheral Endorphins: Basic and Clinical Aspects", E. E. Muller and A. R. Genezzani, eds., Raven Press, New York.

Youkilis, E., Chapman, J., Woods, E., and Plotnikoff, N., 1985, In Vivo immunostimulation and increased in vitro production of interleukin 1 (IL-1) activity by met-enkephalin, Int. J. Immunopharm., 7:79.

METHIONINE ENKEPHALIN: ENHANCEMENT OF T-CELLS IN

PATIENTS WITH KAPOSIS' SARCOMA, AIDS, AND LUNG CANCER

N.P. Plotnikoff[*], J. Wybran[**], N.F. Nimeh[***], and G.C. Miller[****]

Departments of Pharmacology[*] and Medicine[***]
Oral Roberts University School of Medicine,
Immuno-Diagnostics Laboratory, Inc.[****]
Tulsa, Oklahoma; Hospital Erasme, Universite Libre de
Bruxelles, Brussels, Belgium[**]

INTRODUCTION

The enkephalins were originally described as endogenous opiate peptides modulating perception of pain (1). Other behavioral functions for the peptides were identified including anti-depressant and anti-anxiety actions (2).

More recently we reported that the enkephalins were immunomodulators, resulting in very significant activation of the T-cells in mice and men (3,4,5,6,7,8). Based on the latter studies we started a series of in vivo studies in normal volunteers as well as cancer patients.

Patient one, age 27, was admitted to our hospital with swollen lymph nodes, fever, and suspected Kaposis' sarcoma and Pneumocystitis. Initially he was placed on Pentamidine therapy for suspected Pneumocystis carinii pneumonia. After the initial treatments with Pentamidine, the patient was placed on methionine enkephalin treatment. The schedule employed was daily infusion the first week for five days and twice a week thereafter. The dose employed was ten micrograms per kilogram prepared in a normal saline solution and infused over a thirty minute time course. Twenty four hours after infusion, blood samples were withdrawn and analyzed for T-cells (E rosettes), OKT 3 T lymphocytes, mature thymocytes, OKT 11 T lymphocytes-early development stage thymocytes, T helper cells (OKT 4), T suppressor cells (OKT 8), blastogenesis or mitogen stimulated proliferation, phytohemagglutinin (T helper cell mitogen), concanavalin A (a T helper and T suppressor cell mitogen), pokeweed (a T dependent B cell mitogen), and Staph A (a B cell mitogen). NK cell assay (natural killer cells) was conducted against the K-562 tumor cell line labeled with ^{51}Cr (7).

This patient received infusions of methionine enkephalin (10 micrograms per kilogram) each day of the first week (five days, Monday thru Friday). Blood samples were taken each day before infusion of methionine enkephalin. The data in Table 1 indicates that absolute number of T helper cells increased (57 to 94/mm^3) and the absolute number of suppressor cells (OKT 8) decreased from 382 to 264 per mm^3 on the fifth day. There was a slight increase in OKT 3 and OKT 11 cells. The T helper/T suppressor ratio rose from 0.149 to 0.354. The total lymphocyte count dropped from 516 on day 1 to 368 on day 4 and rose again to 550 on day 5. Active T-cells rose significantly from 201 on day 1 to 396/mm^3 on day 5.

No methionine enkephalin was infused on day 6 or day 7. Day 8 blood samples were withdrawn followed by methionine enkephalin infusion (dose of 25 micrograms per kilogram) corresponding to one-half of the first week total dose. The T-cells were still elevated on day 8 compared to day 5. T helper cells were 119 compared to 94; T suppressor cells increased to 409 from 264; OKT 3 increased from 341 to 491/mm^3; and OKT 11 increased from 429 on day 5 to 558 on day 8. The active T-cells increased to 565/mm^3 on day 8 compared to 396/mm^3 on day 5. Futhermore, the total lymphocytes increased to 774/mm^3 on day 8 compared to 550/mm^3 on day 5. The NK cell assay on day 8 was slightly decreased compared to day 1 values.

On day 11 blood samples were taken once again followed by methionine enkephalin infusion. All of the T-cell subsets increased except active T-cells. The total lymphocyte count diminished slightly.

The patient was discharged from the hospital with signs of subjective improvement. Of great interest was that the original lesion of the sarcoma including biopsy hole, was crusting and healing. Lymph nodes in the axillary and groin area were palpably smaller.

AIDS PATIENT STUDY BY DR. J. WYBRAN (BRUSSELS, BELGIUM)

Patient 2

Methionine enkephalin was given to a 26 year old male caucasian AIDS patient from Zaire, Africa whose clinical manifestations included resected cerebral lymphoma and opportunistic infections (Candida albicans and Mycobacterium bovis). Methionine enkephalin (UCB, Brussels) was administered at a dose of 20 micrograms/kilogram in saline perfusion over a period of 30 minutes. Lymphocyte population were studied using monoclonal antibodies (Ortho, Raritan, New Jersey and Becton-Dickinson, Mtn. View, California). After incubation with the monoclonal antibodies, the cells were further incubated with goat anti-mouse coupled particles of colloidal gold (Janssen Life Science Products, Beerse, Belgium) and the smears were put in a silver developer and counter-stained with May-Grunwald-Giemsa (9). This smear method allows morphology and is about twenty times more sensitive than immunofluorescence.

RESULTS

Lymphocytes	Before	After 2 hours	After 24 hours
OKT 3	30 %	26 %	27 %
OKT 10	45 %	53 %	49 %
OKT 11	31 %	34 %	31 %

Blood T-cells increased in number slightly as judged by the OKT 10 and OKT 11 antisera. Of special interest is the increase in OKT 10 which is similar to <u>in vitro</u> data obtained by incubating human lymphocytes with methionine enkephalin (10).

LUNG CANCER PATIENTS (W.A.)

Patient 3

This woman was found refractory to chemotherapy and had a positive X-ray diagnosis of lung cancer. The dose schedule was the same as our Kaposis' sarcoma patient (10 micrograms/kilogram per day the first week and 25 micrograms per kilogram two to three times a week thereafter).

Patient 1	Day	1	2	3	4	5	8	11
ASSAYS	Rx Conc	10 μg/K/da	10 μg/K/da	10 μg/K/da	10 μg/K/da	10 μg/K/da	25 μg/K/2x wk	25 μg/K/2x wk
WBC		$4300/mm^3$	$2600/mm^3$	$2300/mm^3$	$2300/mm^3$	$2500/mm^3$	$3100/mm^3$	$2950/mm^3$
Lymphocytes		$516/mm^3$	$442/mm^3$	$391/mm^3$	$368/mm^3$	$550/mm^3$	$774/mm^3$	$738/mm^3$
Active T cells								
Percent		39%	83%	79%	72%	72%	76%	26%
Absolute		$201/mm^3$	$367/mm^3$	$309/mm^3$	$265/mm^3$	$396/mm^3$	$565/mm^3$	$192/mm^3$
T LYMPHOCYTES & SUBSETS								
T lymphocytes (OKT 3)								
Percent		60%	80%	82%	82%	62%	66%	76%
Absolute		$310/mm^3$	$354/mm^3$	$321/mm^3$	$302/mm^3$	$341/mm^3$	$491/mm^3$	$561/mm^3$
T lymphocytes (OKT 11)								
Percent		79%	88%	80%	80%	78%	75%	81%
Absolute		$408/mm^3$	$389/mm^3$	$313/mm^3$	$294/mm^3$	$429/mm^3$	$558/mm^3$	$598/mm^3$
T lymphocytes (OKT 4)								
Percent		11%	13%	14%	17%	17%	16%	17%
Absolute		$57/mm^3$	$58/mm^3$	$55/mm^3$	$63/mm^3$	$94/mm^3$	$119/mm^3$	$125/mm^3$
T lymphocytes (OKT 8)								
Percent		74%	63%	53%	53%	48%	55%	61%
Absolute		$382/mm^3$	$278/mm^3$	$207/mm^3$	$195/mm^3$	$264/mm^3$	$409/mm^3$	$450/mm^3$
T helper/T suppressor ratio		0.149	0.209	0.264	0.321	0.354	0.291	0.278
BLASTOGENESIS								
Phytohemagglutinin		20X	---	---	---	---	134X	---
Concanavalin A		11X	---	---	---	---	49X	---
Pokeweed		12X	---	---	---	---	18X	---
Staph A		1X	---	---	---	---	3X	---
Interleukin 2 Receptor		14%	---	---	---	---	48%	---
NK cell Assay								
100:1		26%	---	---	---	---	16%	---
33:1		11%	---	---	---	---	8%	---
11:1		5%	---	---	---	---	3%	---
Slope of killing		0.237	---	---	---	---	0.156	---

Patient three

ASSAYS	1	2	3	4	5	8	11	15	18	22	25	29	32	36
	DAYS													
Lymphocytes(per mm^3)	1360	1088	1952	1488	2002	2040	1908	1716	1974	N.D.	N.D.	1830	3159	1452
Active T-cells														
Percent	52%	54%	34%	43%	23%	58%	30%	40%	16%	36%	74%	40%	23%	52%
Absolute(per mm^3)	707	578	664	639	460	1183	572	686	316	685	N.D.	732	727	755
T LYMPHOCYTES & SUBSETS														
T lymphocytes (OKT 3)														
Percent	64%	-	-	-	-	47%	-	-	-	-	-	67%	-	-
Absolute (per mm^3)	870	-	-	-	-	959	-	-	-	-	-	985	-	-
T lymphocytes (OKT 11)														
Percent	77%	-	-	-	-	50%	-	-	-	-	-	74%	-	-
Absolute(per mm^3)	1047	-	-	-	-	1020	-	-	-	-	-	1088	-	-
T lymphocytes (OKT 4)														
Percent	36%	-	-	-	-	28%	-	-	-	-	-	44%	-	-
Absolute(per mm^3)	489	-	-	-	-	571	-	-	-	-	-	647	-	-
T suppressor cells (OKT 8)														
Percent	20%	-	-	-	-	17%	-	-	-	-	-	22%	-	-
Absolute(per mm^3)	272	-	-	-	-	347	-	-	-	-	-	323	-	-
T helper/suppressor														
Ratio	1.8	-	-	-	-	1.65	-	-	-	-	-	2.0	-	-
BLASTOGENESIS														
Phytohemagglutinin	723X	-	-	-	-	432X	-	-	-	-	-	864X	-	-
Concanavalin A	704X	-	-	-	-	576X	-	-	-	-	-	711X	-	-
Pokeweed	83X	-	-	-	-	78X	-	-	-	-	-	73X	-	-
Staph A	21X	-	-	-	-	84X	-	-	-	-	-	44X	-	-
Interleukin-2 Receptor	52%	-	-	-	-	60%	-	-	-	-	-	80%	-	-
NK Cell Assay														
100:1	50%	-	-	-	-	28%	-	-	-	-	-	30.6%	-	-
33:1	23%	-	-	-	-	11%	-	-	-	-	-	19.3%	-	-
11:1	11%	-	-	-	-	6%	-	-	-	-	-	8.8%	-	-
Slope of killing	.4369	-	-	-	-	0.2515	-	-	-	-	-	.227	-	-

The accompanying table illustrates the finding that OKT 3, OKT 4, OKT 8, increase significantly over the 29 days. The degree of blastogenesis with phytohemagglutinin also increased. However, one of the most striking increases was seen in the percent of interleukin-2 receptors. A slight decrease in NK cell activity was observed. Finally, there was a significant increase in the absolute lymphocyte count.

DISCUSSION

The patient with Kaposis' sarcoma was diagnosed as an AIDS patient on the basis of defective T helper cell levels compared to T suppressor cell levels. Seligmann et al. have indicated that AIDS is characterized by a decreased OKT 4 lymphocyte subset (11).

The present study with methionine enkephalin is based on our earlier findings that methionine enkephalin stimulates PHA-induced blastogenesis (a T helper cell antigen) in mice and also markedly stimulates OKT 4 (T helper cell) production in normal volunteers (Plotnikoff et al.). Similar elevation of T helper cells by methionine enkephalin was seen in the present AIDS patient with Kaposis' sarcoma. In addition there is a marked elevation of interleukin-2 receptors to 48% of the lymphocytes compared to 14% on day one. In addition, the primary sarcoma lesion appears to be healing. Similar findings on T-cell subsets were seen both in the AIDS patient studied by J. Wybran as well as the lung cancer patient studied by our group.

REFERENCES

1. J. Hughes, T. W. Smith, H. W. Kosterlitz, Z. A. Fothergill, B. A. Morgan, and H. T. Morris, Identification of two related pentapeptides from the brain with potent opiate agonist activity, Nature, 258:577 (1975).
2. N. P. Plotnikoff, A. J. Kastin, D. H. Coy, C. W. Christensen, S. V. Schally, and M. A. Spirtes, Neuropharmacological action of enkephalins after systemic administration, Life Science, 19:1283 (1976).
3. N. P. Plotnikoff and G. C. Miller, Enkephalins as immunomodulators, Int. J. Immunopharm., 5:437 (1983).
4. N. P. Plotnikoff, A. J. Murgo, and R. E. Faith, Neuroimmunomodulation with enkephalins: Effects on thymus and spleen weights in mice, Clin. Immunol. Immunopathol., 26:446 (1983).
5. G. C. Miller, A. J. Murgo, and N. P. Plotnikoff, Enkephalins: Enhancement of active T-cell rosettes from lymphoma patients, Clin. Immunol. Immunopathol., 26:446 (1983).
6. G. C. Miller, A. J. Murgo, and N. P. Plotnikoff, Enkephalins: Enhancement of active T-cell rosettes from normal volunteers, Clin. Immunol. Immunopathol., 31:132 (1984).
7. R. E. Faith, H. J. Liang, A. J. Murgo, and N. P. Plotnikoff, Neuroimmodulation with enkephalins: Enhancement of human natural killer (NK) cell activity in vitro, Clin. Immunol. Immunopathol., 31:412 (1984).
8. J. Wybran, Enkephalins and endorphins as modifiers of the immune system: Present and future, Fed. Proc., 44:92 (1985).
9. F. Romasco, J. Rosenberg, and J. Wybran, An immunogold silver staining method for the light microscopical analysis of blood lymphocyte subsets with monoclonal antibodies, Amer. J. Clin. Pathol., (in press).
10. J. Wybran and L. Schandene (in preparation).
11. M. Seligmann, L. Chess, J. L. Fahey, A. A. Fauci, P. Lachmann, J. L'Age-Stehr, J. Ngu, A. J. Pinching, F. S. Rosen, T. J. Spira, and J. Wybran, AIDS: An immunologic re-evaluation, N. Eng. J. Med., 311:1286 (1984).

EPILOGUE

Robert E. Faith[a], Anthony J. Murgo[b], and Nicholas P. Plotnikoff[c]

[a]University of Houston, Houston, TX; [b]West Virginia University Medical Center, Morgantown, WV; [c]Oral Roberts University School of Medicine, Tulsa, OK

When immunology was a new science the immune system and immune response were viewed in rather simple terms. The immune system was viewed to be rather autonomous in function and to be a primairly internally regulated system. This view was supported by the fact that various elements of the immune system are capable of functioning in _in vitro_ systems and by the natural antibody selection theory (1). As our knowledge of immunity and the immune system has increased it has become increasingly clear that the immune system is quite complex, both in its make up and its functions.

In the early 1960's it was demonstrated that there is a division of labor in the immune response and that lymphocytes could be divided into thymus-derived lymphocytes (T-cells) and bursal or bursal-equivalent derived lymphocytes (B-cells). Subsequent studies have shown that the two divisions are further subdivided with each having several functionally different subpopulations of cells. The two divisions have been shown to interact in various ways providing both help and suppression.

As the complexity of the immune system has become increasingly clearer it has also become apparent that the immune system interacts with other bodily systems, especially the endocrine system and the central nervous system. In fact, it has recently been suggested that perhaps the immune system should be viewed as an internal sensory organ recognizing non-cognitive stimuli such as bacteria, viruses, antigens, etc and relaying information to the neuroendocrine system by lymphocyte derived hormones (2).

The recurrent theme in this volume is that there exists an intimate interrelationship between the immune, endocrine and central nervous systems. Indications of this interrelationship discussed in this volume are briefly reviewed here. An early indication of the interaction of the endocrine and immune systems was provided by the now well known effects of adrenal glucocorticosteroids on immune function. Other, more recent findings include the observation that the estrogens suppress immune functions, primarily through interactions with thymic epithelial cells. In contrast, androgens have been shown to be immunoenhancing. Thyroxine has been shown to increase phagocytic activity and the inhibition of thyroid activity to suppress immune responses.

The endogenous opioid peptides, which have now been shown to be produced and stored in and secreted by the adrenals, as well as the central nervous system, now appear to play a role in regulation of immune function. These peptides are apparently protected from hydrolysis by enzymes in circulation which allows for half-lives sufficiently long for the peptides to be physiologically active. Lymphocytes, monocytes, granulocytes and mast cells have been shown to have receptors for the opioid peptides. The enkephalins and beta-endorphin have been shown to modulate a number of immune functions *in vitro*, generally enhancing T cell functions and natural killer (NK) cell activity. These peptides also modulate the functions of neutrophils and macrophages. It appears that these peptides may be shown to be critically involved in the regulation of immnue responses.

In addition to the effects of the neuropeptides on immune function there are other manifestations of central nervous system effects on immune function. Undoubtedly some of these central nervous system effects will be observed to be mediated via the neuropeptides while others will be shown to be mediated in other ways. It has now been demonstrated that stress may or may not induce suppression of immune response depending upon whether the stressed individual is able to "cope" with his stress or not. Indeed some forms of stress actually appear to be healthful while other forms of stress can lead to disease conditions. Other interactions of the central nervous system with the immune system include the observations that depressive illness may lead to abnormalities in immune regulation and in neuroendocrine regulation. Physical lesioning of various areas in the hypothalmus lead to modulation of immune responses. Lesions in some areas lead to suppression while lesions in other areas lead to enhancement. One of the most intriguing observations is that immune responses can be conditioned in a Pavlovian manner just as other physiological responses.

The interaction of the central nervous and endocrine systems with the immune system is not a one way street. There is a growing body of evidence that the cells of the immune system produce molecules which allow communication with the other two systems. It has now been well documented that lymphocytes produce active neuropeptide hormones (ACTH, endorphins, etc.) and also produce thyrotropin, vasoactive intestinal peptide and somatostatin. Further evidence for feedback regulation from the immune system to the central nervous system is provided by the observation that there are receptors for lymphokines in the central nervous system and lymphokines such as interleukin-1 and interleukin-2 may directly influence pituitary fuction.

Other observations of interest are those which show endogenous opioid levels to be increased both by exercise and trauma. These observations lead to speculation that the increased levels of enkephalins/endorphins may be a partial explanation as to why well conditioned individuals are apparently healthier and more resistant to disease than sedentary individuals. Perhaps the increased levels of endogenous opioids enhance immune function in well conditioned individuals leading to increased resistance to disease. By the same token increased levels of endogenous opioids following trauma would function to enhance the body's ability to resist infection at the site of the injury as well as provide an endogenous source of analgesia.

We feel that further investigation will show the enkephalins/endorphins to be significant endogenous regulators of immune function. As such we feel they will become important tools for immunotherapy of various disease states, especially certain immunodeficiency states and neoplastic disease. Indeed the final chapters of this volume provide the initial observations indicating that this is the case.

REFERENCES

1. N. K. Jerne, 1955, The natural selection theory of antibody formation. Proc. Natl. Acad. Sci. (U.S.A.), 41: 849.
2. J. E. Blalock, 1984, The immune system as a sensory organ. J. Immunol., 132: 1067.

CONTRIBUTORS

INDEX

GPSR Compliance

The European Union's (EU) General Product Safety Regulation (GPSR) is a set of rules that requires consumer products to be safe and our obligations to ensure this.

If you have any concerns about our products, you can contact us on ProductSafety@springernature.com

In case Publisher is established outside the EU, the EU authorized representative is:

Springer Nature Customer Service Center GmbH
Europaplatz 3
69115 Heidelberg, Germany

Batch number: 10761599

Printed by Printforce, the Netherlands

MICROBIOLOGY OF THE EYE

MICROBIOLOGY OF THE EYE

DEBORAH LOCATCHER-KHORAZO, M.D.

Formerly Director of the Laboratory of Microbiology,
Edward S. Harkness Eye Institute of Presbyterian Hospital, New York;
Assistant Professor of Microbiology,
College of Physicians and Surgeons, Columbia University,
New York, New York

BEATRICE CARRIER SEEGAL, M.D.

Professor Emeritus of Microbiology,
College of Physicians and Surgeons, Columbia University,
New York, New York

With 258 illustrations, including 4 color plates

(1972)

THE C. V. MOSBY COMPANY

Saint Louis 1972

Printed in the United States of America

International Standard Book Number 0-8016-3024-X
Library of Congress Catalog Card Number 70-181594

Distributed in Great Britain by Henry Kimpton, London

Contributors

HENRY F. ALLEN, M.D.

Henry Willard Williams Clinical Professor and Head, Department of Ophthalmology, Harvard Medical School; Chief of Ophthalmology, Massachusetts Eye and Ear Infirmary, Boston, Massachusetts

JAMES H. ALLEN, M.D.

Professor of Ophthalmology, Associate Dean and Medical Director, Tulane Clinics, Tulane University School of Medicine, New Orleans; Consultant in Ophthalmology, U.S. Public Health Service Hospital, Carville, Louisiana

MARGARET BRAND, M.B., B.S. (London)

Ophthalmologist, U.S. Public Health Service Hospital, Carville, Louisiana; Associate Clinical Professor of Ophthalmology, Tulane University School of Medicine, New Orleans, Louisiana; formerly Ophthalmic Consultant, Schieffelin Leprosy Research Institute, Karagiri, S. India

DAVID H. BROWN, M.D.

Assistant Resident in Ophthalmology, Edward S. Harkness Eye Institute of Presbyterian Hospital, New York, New York

HAROLD W. BROWN, M.D., Sc.D., D.P.H., LL.D.

Professor Emeritus of Parasitology, College of Physicians and Surgeons of Columbia University; Parasitologist, Presbyterian Hospital, New York, New York

RICHARD W. DARRELL, M.D.

Instructor in Ophthalmology, Columbia University; Assistant Ophthalmologist, Edward S. Harkness Eye Institute of Presbyterian Hospital, New York, New York

ARTHUR GERARD DeVOE, M.D.

Edward S. Harkness Professor of Ophthalmology, Columbia University; Director, Edward S. Harkness Eye Institute of Presbyterian Hospital, New York, New York

ELIZABETH H. GUTIERREZ, B.A.

Registered Microbiologist of The American Academy of Microbiology; Supervisor of Microbiology, Edward S. Harkness Eye Institute of Presbyterian Hospital, New York, New York

SEYMOUR P. HALBERT, M.D.

Professor of Pediatrics, National Children's Cardiac Hospital, University of Miami, Miami, Florida

G. RICHARD O'CONNOR, M.D.

Associate Professor in Ophthalmology (in Residence) and Director, the Francis I. Proctor Foundation, University of California, San Francisco Medical Center, San Francisco, California

MARGARITA SILVA-HUTNER, Ph.D.

Associate Professor of Dermatology, College of Physicians and Surgeons of Columbia University, New York, New York

FREDERICK H. THEODORE, M.D., F.A.C.S.

Clinical Professor of Ophthalmology, Mount Sinai School of Medicine; Associate Clinical Professor of Ophthalmology, New York University Medical Center; Attending Ophthalmic Surgeon and Director, External Disease and Infections Clinic, Manhattan Eye, Ear and Throat Hospital; Attending Ophthalmologist, New York Hospital; Consultant Ophthalmic Surgeon, The Mount Sinai Hospital, New York, New York

RANDOLPH WHITFIELD, Jr., M.D.

Assistant Resident in Ophthalmology, Edward S. Harkness Eye Institute of Presbyterian Hospital, New York, New York

EMIL WIROSTKO, M.D.

Instructor in Ophthalmology, Director of Microbiology Laboratory, Assistant Ophthalmologist, Edward S. Harkness Eye Institute of Presbyterian Hospital, New York, New York

It is with pleasure that this volume is dedicated to Dr. Frederick Parker Gay, Professor and Chairman of the Department of Bacteriology, College of Physicians and Surgeons, Columbia University, from 1923 to 1939.

During forty years devoted to bacteriology Dr. Gay's researches covered many aspects of this subject. He began his experiences with infectious diseases while still a medical student at Johns Hopkins University, when he accompanied Dr. Simon Flexner on a world tour that included a study of cholera and dysentery in the Philippines. Other investigations included the relative roles of humoral and cellular mechanisms in immunity, the characterization of anaphylactic shock, the carrier state and immunity in typhoid fever, the pathogenesis and chemotherapy of infections caused by the hemolytic streptococcus, and the pathogenesis of herpesvirus infections. He was a member of many scientific societies, including the National Academy of Sciences and the Association of American Physicians.

Dr. Gay's vision of the scope and importance of infectious diseases of the eye led him to play an important role in the development of the Laboratory of Bacteriology at the Institute of Ophthalmology, Columbia-Presbyterian Medical Center. When Dr. John Wheeler became Director of the Institute of Ophthalmology he wished to establish in the Institute a laboratory for the diagnosis of eye infections and the investigation of the source, nature, prevention, and treatment of all types of such ocular disease. For this reason he sought the advice and cooperation of Dr. Gay, who was in charge of diagnostic bacteriology at Presbyterian Hospital. Dr. Gay enthusiastically supported the idea of a separate laboratory for the Institute of Ophthalmology. He recognized the special problems incident to infection of the eyes, the need for personnel trained in this field, and the great importance of rapid, accurate diagnosis and therapy. Dr. Gay also agreed with Dr. Wheeler's desire to place the laboratory in charge of a director who would participate in investigations of problems relevant to bacteriology of the eye. On this basis of cooperation the present Laboratory of Microbiology was established in 1933.

Foreword

Ophthalmology must freely acknowledge its great debt to laboratory scientists, who, though devoted to basic disciplines, have nevertheless contributed their skills to the identification of causal agents of ocular disease and, in no small measure, to their treatment. Nowhere has this been more evident than in the close relationship between microbiology and ophthalmology, a clinical specialty that was the first to diverge from the mainstream of medicine and surgery. As the specialty developed, serious students realized the necessity of firmer knowledge in certain basic areas. The first of these were anatomy and histology, followed shortly by pathology and microbiology. For many years, these were the only nonclinical fields intensively studied by the ophthalmologist in training, and it was necessary for him to travel abroad to get this knowledge. In the past few decades, although we have witnessed an explosion of information in such divergent fields as electronics, immunology, biochemistry, electron microscopy, and pharmacology, microbiology remains one of the major disciplines upon which the practicing clinician heavily depends. In good measure through the efforts of our colleagues in this field, serious complications of bacterial disease of the eye have been largely eliminated. Some progress has occurred in the treatment of mycotic invasion, and a start has been made against viral disorders. It is only a matter of time before these, too, will be controlled.

We are most fortunate that two such dedicated scientists as Dr. Locatcher-Khorazo and Dr. Seegal have devoted a lifetime to the study of microbiology, with particular reference to ophthalmic disease. It is most appropriate that the authors now present the distillation of their lifelong efforts, not only for the practical guidance of the clinician struggling with a diseased eye, but for the stimulation and, hopefully, the inducement of young clinicians to extend their interests in the fundamentals of disease.

Arthur Gerard DeVoe, M.D.

Preface

The eye, because of its location and opportunity for direct and indirect microbial involvement, offers unique possibilities for ophthalmologists concerned with clinical and experimental infection. That they have taken advantage of this opportunity is evident from the data presented by the contributors to this volume. In the text they describe many common and rare examples of infection of the eye and various prophylactic, diagnostic, and therapeutic procedures.

This book contains an analysis of the data gathered mostly in the Laboratory of Microbiology at the Edward S. Harkness Eye Institute, Columbia-Presbyterian Medical Center. In 1933 this laboratory was established under the direction of Dr. Richard Thompson, who in 1941 accepted the chairmanship of the Department of Microbiology at the University of Denver. Dr. Deborah Locatcher-Khorazo became associated with the laboratory in 1934 as Dr. Thompson's assistant and directed the laboratory upon Dr. Thompson's departure. She held this position until her retirement in 1968. In the period 1938 to 1968 over 50,000 patients and normal controls were examined in the microbiology laboratory. Thus the patterns of ocular infections were determined from the study of over 100,000 eyes. The data have been analyzed, tabulated, and compared with reports from other laboratories throughout the world.

During the laboratory's existence many members of the staff of the Eye Institute became interested in various infections of the eye and contributed valuable information regarding such diseases as trachoma, toxoplasmosis, toxocariasis, and fungal and viral infections of the eye. The composition and activity of the normal ocular flora were studied. These experiences are supplemented by three chapters—Aseptic Technique in Ophthalmology, Ocular Leprosy, and Allergy in Relation to Ophthalmology—from contributors not associated with the Columbia-Presbyterian Medical Center. This collected material, not readily available elsewhere, should prove valuable to the ophthalmologist and some microbiologists as a ready reference; it offers a potentially useful background for effective diagnosis and management of the infected eye.

The preparation of this manuscript was aided by many friends and associates both through financial assistance and in myriad other ways during the years in which the data were gathered.

Financial aid was received from Dr. Byung W. Kong, Seoul, Korea; Mr. Max Kade, of the Max Kade Foundation; Dr. Anthony Donn; Mr. and Mrs. Samuel

Kastle; Dr. Raymond L. Pfeiffer; and Dr. Algernon B. Reese.

We particularly wish to thank the past and present directors of the Institute of Ophthalmology—the late Dr. John M. Wheeler, Dr. Phillips Thygeson, and Dr. John H. Dunnington—and the present director, Dr. Arthur G. DeVoe, who made it possible to gather the data presented in this book. Dr. Harry M. Rose, Chairman of the Department of Microbiology, gave valuable assistance. The professional staff, both doctors and nurses, have offered fine cooperation in the laboratory and clinic throughout the study. Without their support the objectives could not have been realized. Dr. Konrad C. Hsu, Dr. David Seegal, and Dr. Takeo Iwamoto provided assistance in many aspects related to the preparation of the manuscript.

We are grateful to Mr. Emil G. Bethke, Mr. James Lafayette, and Mr. Edward Gonzalez for their skills in the preparation of drawings, photographs, and photomicrographs, respectively. Miss Margaret Kane and Miss Marian Johnson have provided generous administrative assistance. The staff of the record room of the Presbyterian Hospital and the libraries of the Institute of Ophthalmology and the College of Physicians and Surgeons facilitated the use of these services. Miss Elizabeth Chaffee and Mrs. Katherine Lewis, who were from the Volunteer Department, contributed secretarial help. Mrs. Nora McElligott, Mrs. Thoma Milani, and Mr. Heinz Rosskothen gave technical assistance, and Miss Regina Loewenstein and Mrs. Anita MacDonnell gave statistical assistance.

Dr. Ross M. Grey, Chairman of the Institute of Comparative Medicine and Associate Professor of Pathology, provided both space and encouragement, with the ever cheerful aid of Miss Adeline M. Elia, Administrative Assistant, and Mrs. M. E. Matthies, Senior Secretary of the Institute of Comparative Medicine.

Deborah Locatcher-Khorazo, M.D.
Beatrice Carrier Seegal, M.D.

Contents

Color plates

MICROBIOLOGY OF THE EYE

1
Historical perspectives

Ophthalmologists have made outstanding contributions to the early acquisition of knowledge on infectious diseases of the eye. These contributions can be appreciated better if the few landmarks they had to guide them are enumerated first. As early as 1688, van Leeuwenhoek (1632-1723), who had developed a combination of lenses to produce a microscope giving a magnification of 150×, observed spherical granules in beer yeast; and in a series of letters to the Royal Society of London he carefully drew and described recognizable infusoria, protozoa, and bacteria found in saliva and various excretions. In 1762 von Plenciz, aware of van Leeuwenhoek's findings and seeing yeasts himself, sponsored a sweeping theory that disease was caused by "contagium animatum," with a specific agent for each disease. Unfortunately, his concept was much ahead of the technical means to verify it, and little or no attention was paid to it. In 1845 von Langenbeck saw in an infected human bone what we now know to be actinomyces, and in 1850 Pollender saw large bacilli (anthrax) in blood from a sheep with splenic fever; these bacilli were grown by Koch in 1876. From this point on, a search was made for living agents producing disease; and a great number were found, though the hunt is not yet over.

Before the discovery of agents that cause infectious diseases, observant ophthalmologists supported the theory that certain eye diseases were contagious; and rules for prevention, anticipating many of today's routine precautions, were laid down. Table 1-1 includes information about some of these early observations. Charles Saint-Yves (1667-1736),[38] a much respected French ophthalmologist, had a young male patient who made a practice of washing his eyes with his urine. After this patient had contracted gonorrhea, he developed a purulent conjunctivitis. Saint-Yves had no difficulty in diagnosing the eye disease, and he is credited with having given the first exact description of ophthalmia neonatorum. He concluded that the patient had transmitted gonorrhea to his eyes. Other contributions made by Saint-Yves include the first record of what was apparently an ophthalmitis caused by herpes zoster and descriptions of ocular lesions in smallpox. He is said to have used silver nitrate "plentifully" in treating abscesses or ulcers of the eye and "probably" in treating ophthalmia neonatorum.

Carris[3] credits Quelmalz, a German physician, with having pointed out around 1785 the possible connection between "babies' sore eyes" and the discharge present in the mother's birth canal. After many years Dr. Gibson,[11] an Englishman, in 1808 confirmed the observation of Quelmalz and added that proper cleansing of the mother's vaginal canal and removing mucus from the baby's eyes might prevent the occurrence of this infection.

Joseph F. Pieringer (1800-1879),[9,17] shown in Fig. 1-1, was among the most notable of ophthalmologists. He performed empirical experiments, particularly in his

Table 1-1. Some early observations on agents responsible for ocular infections

Present designation	Clinical description	Transmission, prophylaxis, treatment	Causative agent
Gonococcal conjunctivitis	*Galen*, 132 AD, introduced term gonorrhea, "flow of seed"[42]* *Severus*, 3rd century AD, described disease of newborn with lids swollen because of abundance of matter[38]	*Saint-Yves*, 1722, described case of a young man, accustomed to wash his eyes with his urine, who transmitted his gonorrhea to his eyes[38] *Quelmalz*, 1785, noted connection between babies' sore eyes and discharge in mothers' vaginal canals[3,17] *Gibson*, 1808, recommended proper cleansing of mother's birth canal and baby's eyes to lessen chances for blindness[3,36] *Pieringer*, 1841, proved pus derived from urethral discharge produced infection when applied to conjunctiva of blind volunteers[9,17] *Credé*, 1881, introduced the extensive use of silver nitrate in eyes of newborn[3,31]	*Neisser*, 1879, described a diplococcus in urethral and conjunctival exudate[32] *Leistikow*, 1880, first cultivated the gonococcus[26] *Krause*, 1882, cultured the gonococcus and produced disease with pure culture[24]
Mycotic keratitis	*Leber*, 1879, saw corneal ulcer, greyish turbid base, in 5 days membrane from upper border lifted, had mycelial elements, hypopyon, progressive course[25]	*Leber*, 1879, contaminated corneal wounds of rabbits with culture from patient, produced similar lesions[25]	*Leber*, 1879, grew *Aspergillus glaucus* from corneal ulcer[25]
Koch-Weeks conjunctivitis	*Weeks*, 1886, reported acute epidemic mucopurulent conjunctivitis[43]	*Weeks*, 1886, demonstrated pure cultures of the bacillus-produced characteristic conjunctivitis in his own eye and those of human volunteers.[43]	*Hirschberg and Krause*, 1881, observed delicate rods in some cases of trachoma[18] *Koch*, 1883, described similar organism in some cases of Egyptian ophthalmia[23] *Weeks*, 1886, isolated the Koch-Weeks bacillus (Hemophilus aegyptius)[43]
Morax-Axenfeld angular blepharoconjunctivitis	*Morax*, 1896, observed subacute conjunctivitis, redness at angles of lid margin, mild, bilateral[28]	*Morax*, 1896, transmitted disease to human eye; zinc sulphate used therapeutically[28] *Axenfeld*, 1897, reported finding and cultivating same organism independently[1]	*Morax*, 1896,[28] *Axenfeld*, 1897,[1] independently isolated Morax-Axenfeld diplobacillus (*Moraxella lacunata*)
Keratitis of Petit	*Petit*, 1899, described superficial serpiginous hypopyon keratitis[34]	*Petit*, 1899, showed organism not pathogenic for animals[34]	*Petit*, 1899, isolated diplobacillus of Petit (*Moraxella liquefaciens*)[34]

*Superscript numbers indicate references at end of chapter.

It might be guessed from his pioneering adventures in microbiology that Dr. Weeks had qualities of the explorer, including excellence of performance and hardihood. Dr. Knapp[22] said of him that his "modesty and superb courage were remarkable," and when asked where he would advise a young ophthalmologist to settle, he replied "wherever the competition is the keenest."[22] Certainly one can only settle for the best in such surroundings.

Harold Gifford (1858-1929) was another American ophthalmologist who brought the viewpoint of the bacteriologist to the practice of ophthalmology at a very early date. In many respects his career started in a manner similar to that of Dr. Weeks. Born in Milwaukee, Dr. Gifford received his medical degree from the University of Michigan. He went abroad for postgraduate studies in Germany and Switzerland, where he carried out early research in bacteriology of the eye. Like Dr. Weeks, he later became associated with Dr. Knapp.[19]

Dr. Gifford's studies in ophthalmic bacteriology included his important observations on the flora of the normal eye. He found potential pathogens among organisms present in the healthy eye and indicated that these might be carried into the eye following the trauma of operation. This indeed proved true. His studies on attempts at sterilization of the conjunctiva demonstrated that a sterile state could not be maintained, since the bacteria lie under the epithelium, in ducts of the glands, and in the eyelash roots.[13] Dr. Gifford's other investigations in ocular bacteriology covered a wide field. He was the first to publish a description in English (1896) of acute pneumococcus conjunctivitis and to demonstrate that the infection was transmissible by inoculation of secretions containing the pneumococci.[12] For this experiment he used the conjunctivas of his own eye and those of his assistants. He studied infections associated with the diplobacillus of Morax-Axenfeld, demonstrating this organism not only in angular conjunctivitis but also in corneal ulcers.[14] In 1910 he reported bulbar conjunctivitis caused by *Sporotrichum schenckii,* which was the first time a publication on ocular infection by this organism had appeared in English.[15] Dr. Gifford also carried out early research in experimental bacterial sympathetic ophthalmia and followed the spread of the

Fig. 1-7. Victor Morax (1866-1935), born in Morges on Lake Geneva, was the son of a physician who had interned at the Hôpitaux de Paris. His father had great enthusiasm for his years in Paris, and the young Victor looked forward to living in this city. In the course of his studies he acquired a passion for chemistry and a great curiosity about the pathogenic microorganisms and powerful toxins they produced. After his training at the University of Lausanne and Fribourg-en-Brisgau, a visit to Indochina focused Morax's attention on the bacteriology of the eye, which led, among other things, to his isolation of the Morax-Axenfeld diplobacillus, the causative agent of one type of conjunctivitis. His very productive scientific career was spent in his adopted country, France, and in 1900 he was nominated Ophthalmologist of the Hôpitaux de Paris. (From Duke-Elder, S.: System of ophthalmology. Diseases of the outer eye, vol. 8, part 1, St. Louis, 1965, The C. V. Mosby Co., p. 48.)

A

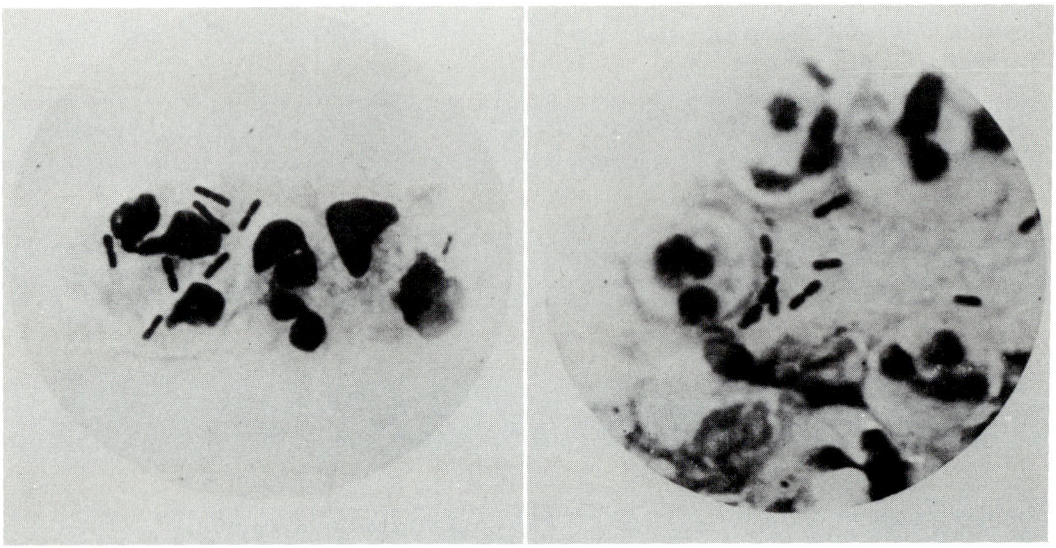

B

Fig. 1-8. A, Diplobacilli included in a polymorphonuclear leukocyte from conjunctival secretions in subacute conjunctivitis; **B,** diplobacilli free in the conjunctival secretions. The bacillus is now classified *Moraxella lacunata.* (From Morax, V.: Ann. Inst. Pasteur **10**:337, 1896.)

microorganisms from the inoculated eye through lymph paths into the orbit and from there to the nerve and the suprachoroidal spaces of the second eye.[19]

Victor Morax[27] (1886-1935), shown in Fig. 1-7, was born in Switzerland but received his medical education in France, where he continued to live and to contribute extensively to ophthalmology as related to bacteriology and to medicine and surgery. In 1896, while at the Pasteur Institute, independently and simultaneously with Dr. Axenfeld, Dr. Morax[28] discovered a gram-negative diplobacillus in angular conjunctivitis, which became known as the Morax-Axenfeld diplobacillus. A member of the Hemophilus group, its official name now is *Moraxella lacunata.* Fig. 1-8 is an illustration from Dr. Morax's original paper. Dr. Morax also carried out studies in cases of acute conjunctivitis, from which he isolated the Koch-Weeks bacillus. In 1901, while residing in Egypt in order to study trachoma, he observed during the acute stage of the disease multiple bacterial infections, which he considered were responsible for much of the pathology.[29,30] He continued to be closely associated with problems of trachoma and related folliculosis and established *La Revue du Trachome.*[27] He also was concerned with sympathetic ophthalmia and recommended enucleation as a precautionary measure against complications.

Theodor Axenfeld[41] (Fig. 1-9), of German origin, studied medicine at the Universities of Marburg and Berlin and bacteriology with Rubner and Fraenkel. He went from a Professorship of Ophthalmology at the University of Rostok to the University of Freiburg in 1901, where he remained for the succeeding 29 years of his life. There his Ophthalmological Institute became a mecca for students from many lands. He held the belief, revolutionary in that day, that the head of a clinic should be free from private practice in order to devote his time to teaching and research. Professor Axenfeld's research covered almost every branch in ophthalmology, but his name is kept especially in mind because of his independent discovery of the Morax-Axenfeld diplobacillus. At Freiburg Dr. Axenfeld built up the *Klin-*

Fig. 1-9. Theodor Axenfeld (1867-1930), though born in Smyrna, obtained his early education in Bonn and his medical training at the Universities of Marburg and Berlin. The following years were spent in training in ophthalmology, pathology, and bacteriology. In 1897 he became Director of the Eye Clinic in Rostock, but moved in 1901 to Freiburg, where he remained for the rest of his life. In 1904 he published a book, *Bakteriologie in der Augenheilkunde,* which became the standard work in this field for many years. It was translated into English by MacNab in 1908. Axenfeld's presentation of his experiences during a long life devoted to the subject are still a delight to workers in this field. (From von Szily, A.: Klin. Mbl. Augenheilk. **85**(4):i-xvi, 1930.)

Fig. 1-10. Paul J. Petit (1875-1950), an ophthalmologist and bacteriologist, was concerned with the importance of using the bacteriologic laboratory to assist in the diagnosis of ocular disease. He was associated with Morax and with Nicolle in these endeavors. One of Petit's main interests was the cause and management of trachoma; some of his studies on this subject were carried out in Tunisia. He was Secretary General of the Ligue Contre le Trachome from 1924 to 1939. Another major interest of Petit was the identification of the possible etiologic agents of keratitis. This work led to his discovery in 1899 of a bacterium, the diplobacillus of Petit (now called *Moraxella liquefaciens*), responsible for a somewhat clinically distinctive type of keratitis. Petit was much beloved, as was gracefully said by Bailliart: "C'est un savant très modeste mais de grande valeur." (From Petit, P. J.: Rev. Int. Trachome **25**:194, 1948; legend material from Baillairt, P.: Ann. Oculist. **183**:1007, 1950.)

ische Monatsblätter für Angenheilkunde, still a leading ophthalmologic journal. The diversity of his interest in problems of eye infection was great. In 1894 he received the Graefe Prize for his "classical research" on suppurative metastatic ophthalmia. This work could still be used as a model for a complete experimental bacteriologic and pathologic examination of a disease.[41]

Mention should be made of other areas in which Professor Axenfeld contributed to bacteriology. For example, he concerned himself with studies of the etiology of serpiginous corneal ulcer, demonstrating the important role the pneumococcus plays in this disease. In 1907 he wrote a book on the bacteriology of the eye, which was translated into English in 1908 by Angus MacNab.[2] This book, which includes a consideration of mycotic and parasitic infections of the eyes, laid the foundation for the principles and techniques of ophthalmic bacteriology. Although it is more than 50 years old, the book can still be read with profit. Later monographs on trachoma and vernal conjunctivitis were written. Dr. Axenfeld kept pace with the developments in his field, and his writings are filled with records of the contributions of a great number of investigators in the same and allied fields. His own work was very stimulating to the productive research of others. Ophthalmologists are greatly indebted to Dr. Axenfeld for clarifying the relations of bacteriology to ophthalmology.

In 1899 another small gram-negative diplobacillus was isolated by Petit (Fig. 1-10) from patients with keratitis with hypopyon.[34] This organism, long known as the diplobacillus of Petit, is now classified as *Moraxella liquefaciens*. Dr. Petit was a French ophthalmologist who, as an associate of Charles Nicolle, devoted his energy to the investigation of ocular infections. He contributed much to the fight against trachoma in Tunisia[35] and collaborated with Victor Morax in a book on this subject.[30]

Trachoma has been of greatest concern to ophthalmologists and other physicians for most of recorded history. In China it dates back to about 2700 BC, and it has been described in Mesopotamia since the twentieth century BC. Celsus, in 14 AD, noted, roughness of the eyes, and Aetius of Amido, in 550 AD, described the four stages of the disease. In 1823 the ophthalmologist Werneck[17,21] transmitted the disease with follicular material to volunteers. The significance of Werneck's observation was not advanced for almost a 100 years because of the lack of techniques for carrying the studies further. Modern scientific investigation into the cause of trachoma may be said to have started with the classic observations made by Halberstaeder and von Prowazek[8,16] in 1907. When on a joint expedition to Java to study syphilis, they became interested in the conjunctivitis of trachoma, which was common there. Scrapings of the trachomatous conjunctivas revealed small granular intracellular bodies, which stained red with the Giemsa stain and lay in a blue matrix. They believed the small granular bodies to be the cause of trachoma. They were successful in producing infection typical of trachoma in the eyes of orangutans, using scrapings from trachomatous eyes. The cells in the exudates from the animals' eyes also contained foreign bodies similar to those seen in man. The illustrations that they published seemed to show both elementary bodies and inclusion bodies.

In 1912 Nicolle, Blaisot, and Cuénod[33] found that it was possible to transmit trachoma to the *Macaca inuus* and the chimpanzee with material that had been rendered bacteria free by filtration through a Berkefeld filter. In 1935 Thygeson and Proctor[40] reproduced the disease in a human volunteer with material from pooled exudates from trachomatous children. The material had been filtered through a 0.6 μg gradocol membrane filter. They showed that the filtrate was bacteria free but contained elementary bodies about 0.25 μ in diameter. Thus, at this time there was no reason to doubt that the agent that caused trachoma was filtrable but large enough to be seen at high magnification. Although fairly sophisticated methods were being developed for the cultivation of viruses and rickettsiae, the agent of trachoma escaped capture until 1957 when it was grown in China in the chick embryo by T'ang, Chang, Huang, and Wang.[39] Previous attempts to use this method had apparently failed because of the injection of penicillin with the trachoma washings into the eggs. The use of penicillin was standard procedure to eliminate bacterial contamination in viral studies, but it also eliminated the agent causing trachoma. When T'ang and his associates employed streptomycin as a bacteriostatic agent, they obtained growth of the agent causing trachoma, a microorganism belonging to the psittacosis, ornithosis, lymphogranuloma venereum group of organisms, now named *Chlamydia trachomatis* (Chapter 14). This organism has now been cultured by many investigators throughout the world.

Much is owed Dr. Ernest Fuchs (1851-1930) for his pioneer studies on keratoconjunctivitis. An Austrian ophthalmologist, Dr. Fuchs was a guide and teacher of ophthalmologists throughout the world, largely through his book *Treatise on Diseases of the Eye,* which appeared in numerous editions and translations. Dr. Fuchs[10] in 1889 gave the classical description of the clinical findings characterizing superficial punctate keratitis. Although the etiology of this condition, now called epidemic keratoconjunctivitis, was not known until the mid-1950's, Dr. Fuchs' careful delineation of the disease maintained it as a clinical entity until the specific causative agent was isolated. In 1930 Wright,[45] a professor of ophthalmology in Madras, reported on the transmissibility of this disease to man with filtered material from conjunctival scrapings, thus establishing the probable viral etiology of the disease. The agent was not identified, however, until 1955 when Jawetz, Kimura, Nicholas, Thygeson, and Hanna[20] isolated a new type of adenovirus—type 8—from patients with epidemic keratoconjunctivitis and proved this virus to be the etiologic agent of the disease (Chapter 17).

REFERENCES

1. Axenfeld, T.: Beiträge zur Aetiologie der Bindehautentzündungen, Bericht Ophthalmologischen Gesellschaft, Munich, 1897, J. F. Bergmann, p. 140.
2. Axenfeld, T.: The bacteriology of the eye, translated by Angus MacNab, New York, 1908, William Wood and Co.
3. Carris, L. H.: Preventing blindness through social hygiene co-operation, Sightsav. Rev. **4**:28, 1934.
4. Delbanco, E.: Leo Leistikow, Derm. Wschr. **66**:153, 1918.
5. Derby, G. S.: Lucien Howe: 1848-1928, Arch. Ophthal. **1**:241, 1929.
6. Duke-Elder, S.: System of ophthalmology. Diseases of the outer eye, vol. 8, part 1, St. Louis, 1965, The C. V. Mosby Co., p. 260.
7. Dunnington, J. H.: John Elmer Weeks, Amer. J. Ophthal. **32**:595, 1949.
8. Finzi, N. S.: Ludvig Halberstaedter, Brit. J. Radiol. **22**:347, 1949.
9. Friedenwald, H.: Joseph Friederich Pieringer: his methods and investigations, Johns Hopkins Hosp. Bull. **8**:191, 1897.
10. Fuchs, E.: Keratitis punctata superficialis, Wien. Klin. Wschr. **2**:837, 1889.
11. Garrison, F. H.: History of medicine, ed. 4, Philadelphia, 1929, W. B. Saunders Co., p. 578.
12. Gifford, H.: Pneumococcus of Fraenkel as a frequent cause of acute catarrhal conjunctivitis, Arch. Ophthal. **25**:314, 1896.
13. Gifford, H.: Notes on ophthalmic bacteriology, partly with reference to asepsis, Arch. Ophthal. **27**:616, 1898.
14. Gifford, H.: On the clinical importance of the diplobacillus of Morax and Axenfeld, Ophthal. Rec. **14**:511, 1905.
15. Gifford, H.: Sporotrichosis of the eyeball and eyelids, Ophthal. Rec. **19**:573, 1910.
16. Halberstaedter, L., and von Prowazek, S.: Zur Aetiologie des Trachoms, Deutsch. Med. Wschr. **33**:1285, 1907.
17. Hirschberg, J.: Charles St. Yves, S. T. Quelmalz, W. Werneck, J. Pieringer, Graefe-Saemisch Handbuch der Gesamten Augenheilkunde **14**(1):9, 203, 561, 577, 1911.
18. Hirschberg, J., and Krause, F.: Zur Pathologie der ansteckenden Augenkrankheiten, Zbl. Augenheilk. **5**:278, 1881.
19. Jackson, E.: Harold Gifford, Arch. Ophthal. **3**:217, 1930.
20. Jawetz, E., Kimura, S., Nicholas, A. N., Thygeson, P., and Hanna, L.: New type of APC virus from epidemic keratoconjunctivitis, Science **122**:1190, 1955.
21. Julianelle, L. A.: The etiology of trachoma, New York, 1938, Oxford University Press, p. 43.
22. Knapp, A.: John Elmer Weeks, Arch. Ophthal. **41**:752, 1949.
23. Koch, R.: Thätigkeit der deutschen Cholerakommission in Aegypten und Ostindien, Wien. Med. Wschr. **52**:1548, 1883.
24. Krause, F.: Die Mikkrokokken der Blennorrhoea neonatorum, Zbl. Praktische Augenheilk. **6**:134, 1882.
25. Leber, T.: Keratomycosis aspergillina als Ursache von Hypopyonkeratitis, Graefes Arch. Ophth. **25**:285, 1879.
26. Leistikow, L.: Ueber Bacterien bei den venerischen Krankheiten, Charité-Annalen, **7**:750, 1880.
27. Magitot, A.: Victor Morax, Ann. Oculist. **172**:529, 1935.
28. Morax, V.: Note sur un diplobacille pathogène pour la conjonctivite humaine, Ann. Inst. Pasteur **10**:337, 1896.
29. Morax, V., and Lakah, M.: Recherches cliniques sur l'étiologie du trachome en Égypte, Ann. Oculist. **126**:353, 1901.
30. Morax, V., and Petit, P. J.: Le trachome, Paris, 1929, Jean Morax, Éditeur.
31. Narayanrao, U. B.: Lives of great men: Carl Siegmund Franz Credé (1819-1892), Indian J. Venereal Dis. **2**:161, 1936.

32. Neisser, A.: Ueber eine der Gonorrhoe eigentümliche Micrococcusform, Zbl. Med. Wiss. **17**:497, 1879.
33. Nicolle, C., Blaisot, L., and Cuénod, A.: Le magot animal réactif du trachome: filtrabilite du virus; pouvoir infectant des larmes, C. R. Acad. Sci. **155**:241, 1912.
34. Petit, P. J.: Sur une forme particuliere d'infection cornéenne a type serpigineux, Ann. Oculist. **121**:166, 1899.
35. Petit, P. J.: Souvenirs d'une mission en Tunisie, Rev. Int. Trachome, **25**:194, 1948.
36. Posey, W. C.: The romance in the movement for the prevention of blindness, Sight-sav. Rev. **4**:83, 1934.
37. Pusey, W. A.: Albert Neisser. In The history of dermatology, Springfield, Ill., 1933, Charles C Thomas, Publisher, p. 155.
38. Shastid, T. H.: Charles Saint-Yves, Amer. Encyc. Dict. Ophthal. **15**:11496, 1919.
39. T'ang, F., Chang, H., Huang, Y., and Wang, K.: Studies of the etiology of trachoma with special reference to isolation of the virus in chick embryo, Chin. Med. J. **75**:429, 1957.
40. Thygeson, P., and Proctor, F. I.: The filtrability of trachoma virus, Arch. Ophthal. **13**:1018, 1935.
41. von Szily, A.: Theodor Axenfeld, Klin. Mbl. Augenheilk. **85**:1, 1930.
42. Vertue, H. St. H.: An enquiry into venereal disease in Greece and Rome, Guy. Hosp. Rep. **102**:277, 1953.
43. Weeks, J. E.: The bacillus of acute conjunctival catarrh, or "pink eye," Arch. Ophthal. **15**:441, 1886.
44. Weeks, J. E.: A treatise on diseases of the eye, Philadelphia, 1910, Lea and Febiger, p. 255.
45. Wright, R. E.: Superficial punctate keratitis, Brit. J. Ophthal. **14**:257, 1930.

2

The bacterial flora of the healthy eye

THE NORMAL FLORA OF THE EYE

The eye harbors bacteria from the time of birth throughout life. During his passage through the birth canal, the infant acquires some of the microorganisms in the mother's vagina. Although the dramatic example of this is *Neisseria gonorrhoeae*, causing gonorrheal conjunctivitis, the infant routinely picks up bacteria carried in the healthy vagina. Grünberger and Kofler[12] carried out a simultaneous study of the vaginal flora of the mother and the eye flora of the newly born infant. In Table 2-1, adapted from their table, it is seen that *Staphylococcus epidermidis, Staphylococcus aureus*, streptococci, and sometimes *Escherichia coli* were transmitted to infants' eyes during birth. The authors point out the failure of silver nitrate to influence parasitism, particularly by *S. aureus*. Franklin and Loeb[10] examined the bacterial flora of 200 infants, covering a period from 4 to 40 hours after birth. The cultures were taken following the instillation of penicillin in the eyes of 100 of the children and silver nitrate in the eyes of the others. Nevertheless, cultures revealed the presence of a variety of staphylococci, streptococci, diphtheroids, and gram-negative bacilli in 41% and 45% of the children, respectively.

Several studies of the ocular flora of healthy eyes of children as compared with adults have been reported. Khorazo and Thompson[14] in 1935 examined the normal eyes of 1,084 individuals ranging in age from birth to over 70 years. The results obtained in the various age groups are shown in Table 2-2. The bacteria commonly found in the eyes at all ages were the staphylococci and diphtheroids; streptococci

Table 2-1. Incidence of individual bacteria in birth canal of 53 mothers and in eyes of the newly born infants

Bacteria	Mothers' vaginal canal	Infants' eyes: Before treatment	Infants' eyes: After silver nitrate	
		At birth	2 days	5 days
S. aureus	26	26	28	30
*S. albus**	22	20	16	18
Streptococci	12	7	3	1
B. coli†	21	4	0	2

Data from Grünberger, V., and Kofler, E.: Uber den Keimgehalt der kindlichen conjunctiva nach der Geburt, Gynaecologia **137**:331, 1954.

**Straphylococcus epidermidis.*

†*Escherichia coli.*

Table 2-2. Bacteria recovered from normal eyes in individuals of various ages

	Age group					
	0-5	6-14	15-29	30-49	50-69	70+
Total cases in each age group	64	106	143	220	412	139
Staphylococcus epidermidis†	41%	39%	43%	39%	42%	45%
Staphylococcus aureus	25%	24%	24%	21%	25%	23%
Streptococcus hemolyticus	(1)*	(1)	3%	(1)	2%	(1)
Streptococcus viridans	8%	4%	2%	3%	4%	1%
Pneumococci	13%	2%	2%	3%	2%	1%
Diphtheroids	25%	24%	28%	38%	42%	45%
Mixed cultures	26%	16%	18%	19%	30%	34%
No growth	20%	23%	20%	20%	14%	12%

Data from Khorazo, D., and Thompson, R.: The bacterial flora of the normal conjunctiva, Amer. J. Ophthal. **18:**1115, 1935.
*Parentheses indicate actual numbers, not percent.
†*Staphylococcus albus.*

Table 2-3. Incidence of microorganisms cultured from conjunctivas and eyelid margins*

	1-18 years		20-35 years		40-90 years	
	Percent positive (range)†		Percent positive (range)		Percent positive (range)	
Microorganisms	Conjunctivas	Eyelids	Conjunctivas	Eyelids	Conjunctivas	Eyelids
Staphylococcus epidermidis	56–73	62–81	61–78	63–76	54–83	62–82
Staphylococcus aureus	35–38	32–45	30–41	31–44	31–45	31–47
Diphtheroids	21–23	20–24	32–39	33–43	21–34	28–43
Streptococcus viridans	1.1–2.1	0.7–2.5	0.3–0.7	0.3–0.8	0.5–5.0	0.5–5.0
Diplococcus pneumoniae	5.0–9.0	5.0–9.0	0.9–2.7	0.5–1.3	0.4–0.8	0.2–1.2
Escherichia coli	0.1–0.2	0.2–0.4	0.2–2.7	0.3–1.5	0.4–3.0	0.5–4.0
Klebsiella pneumoniae	0.6–3.0	0.6–3.0	0.3–3.0	0.3–1.3	0.2–1.2	0.2–2.2
Klebsiella ozenae	—	—	—	—	0.2–0.6	0.2–0.6
Proteus vulgaris	0.1–0.2	0.1–0.2	0.1–0.2	0.1–0.3	0.6–3.0	0.9–4.0
Proteus morgani	0.1–0.2	0.1–0.2	0.1–0.2	0.1–0.6	0.3–3.0	0.3–3.0
Proteus mirabilis	0.1–0.2	0.1–0.2	0.1–0.4	0.2–0.9	0.7–2.4	0.7–2.4
Proteus rettgeri	—	—	—	—	0.5–0.9	0.5–0.9
Neisseria catarrhalis	0.9–3.0	1.2–0.6	0.1–1.3	0.2–1.5	0.2–1.2	0.2–1.2
Neisseria sicca	0.5–0.9	0.9–2.0	0.2–2.0	0.2–3.0	0.2–1.1	0.5–1.1
Neisseria flava	—	—	0.1–0.7	0.2–0.9	0.3–1.2	0.3–1.2
Bacillus subtilis	0.6–1.3	0.5–1.3	0.7–2.3	0.8–4.0	0.2–1.8	0.6–2.6
Bacillus cereus	—	—	—	—	0.2–1.2	0.2–1.5
Bacillus megaterium	—	—	—	—	0.4–1.1	0.3–1.5
Sarcina	0.8–1.2	0.9–1.2	0.5–1.0	0.5–1.5	0.2–1.8	0.2–2.4
Micrococcus tetragenus	0.7–1.2	0.6–1.2	0.5–2.0	0.7–3.2	0.4–4.0	0.8–4.1

*From 1,024 young people 1-18 years old, 1,786 adults 20-35 years old, and 7,461 patients awaiting ocular surgery, 1952-1968. None had infected eyes. Ages 1-18 drawn from children awaiting surgery, those accompanying parents to clinic, and visiting high school students; ages 20-35 years drawn from graduate students and from some of the personnel at the Columbia-Presbyterian Medical Center, 1957-1962.
†Figures given are from the years showing the lowest and the highest incidence for each microorganism.

and pneumococci were present infrequently. The bacterial flora was not modified by the age of the subjects except for a rise in the incidence of diphtheroids in those over 50. From 1952 to 1968 Locatcher-Khorazo and Gutierrez[16] determined the bacterial flora of noninfected eyes of 10,271 individuals 1 to 90 years of age. Table 2-3 shows data from these studies in which the conjunctivas and eyelid margins of both eyes were cultured. The percent of persons carrying the individual isolated bacteria was determined for each year. Since there was considerable fluctuation in the carriage rate from year to year, the lowest percentage and the highest percentage of positive cultures obtained during the years of study are recorded. Three age groups, 1 to 18 years, 20 to 35 years, and 40 to 90 years are represented. The data confirm the preponderance of staphylococci and diphtheroids in the normal eye, an observation that is supported in reports from many other laboratories.[1,7a,15,18] Other bacteria were found in a few cases. The incidence of diphtheroids was somewhat less in the age group 1 to 18 than in the other age groups. There was no significant difference in the floras recovered from the eyelid margins and the conjunctivas.

Further analysis of the data from which Table 2-3 was constructed has given the following information: (1) there was no difference in bacterial flora between the healthy eyes of males and females; (2) the data examined month by month showed no seasonal changes in the carriage rate; (3) a scrutiny of the yearly fluctuations observed failed to indicate any significant trend in the yearly variations in the number of staphylococci and diphtheroids recovered; (4) the flora obtained from the right and left eyes was almost always identical; there was a difference in only 244 individuals out of the 7,000 tested; (5) in no instance were the cultures sterile.

The routine techniques used in the studies recorded in Table 2-3 served to isolate only aerobic bacteria. Anaerobic cultures were done, however, on 2,000 eyes. Facultative anaerobes were rarely found, and true anaerobes were not seen. Others have reported anaerobes in the normal eye.[18a] Fungi were present in a small percentage of eyes, as is described in Chapter 13.

In Table 2-4 further information is given regarding the bacterial flora cultured from the eyes of the 7,461 individuals awaiting ocular surgery. The data have been presented in such a way as to show the frequency with which the eye carries only a single organism as compared with two or more organisms. In about half of these patients, only single bacteria were isolated.

Table 2-4. Bacterial flora in eyes of 7,461 patients prior to ocular surgery, 1952-1968

Organisms isolated	Conjunctivas		Eyelids	
	Number	Percent	Number	Percent
Staphylococcus epidermidis	2,804	36.2	2,813	37.8
Staphylococcus aureus	1,295	17.3	1,310	17.6
Diphtheroids	158	2.1	69	0.9
Staphylococcus epidermidis and *S. aureus*	645	8.6	688	9.2
Staphylococcus epidermidis and diphtheroids	862	11.5	864	11.6
Staphylococcus aureus and diphtheroids	534	7.4	562	7.5
Staphylococcus epidermidis, *S. aureus*, and diphtheroids	503	6.7	500	6.7
Miscellaneous organisms in various combinations with above*	638	8.5	653	8.7

**Bacillus subtilis*, coliform bacilli, *Neisseria catarrhalis*, *N. sicca*, pneumococci, *Streptococcus viridans*, *Proteus* group, and *Sarcina*.

Table 2-5. Conjunctival flora of normal eyes from various parts of the world (expressed in percentages)

Location of study	New York, Birmingham, United States		London, England	Copenhagen, Denmark	Jerusalem, Israel	Freiburg, Germany	Peking, China	Gifu-shi, Japan	Algiers, Algeria	Freiburg, Germany
Study	Locatcher-Khorazo and Gutierrez[16]*	Cason and Winkler[6]	Smith[22]	Barfoed[4]	Bachrach and associates[3]	Doden[8]	Chang[7]	Yamazaki[24]	Orfila and Courden[20]	Axenfeld[2]
Microorganisms isolated	**7,461**	**1,604**	**5,000**	**501†**	**277**	**915†**	**100**	**564†**	**120**	**61**
Staphylococcus epidermidis	68.5	68.0	33.8	35.7	58.6	75.7	62.0	40.2	50.0	65.0
Staphylococcus aureus	42.0	7.0	7.9	7.6	6.9	6.1	2.0	5.5	25.8	11.0
Diphtheroids	28.7	33.0	31.5	32.1	39.2	76.4	18.0	27.7	32.0	83.0
Streptococcus viridans	2.5	<1.0	0.9	0.6	5.4		12.0	3.7	1.1	3.0
Streptococcus hemolyticus	—	<1.0	0.3	—	—	—	1.0	—	1.7	—
Nonhemolytic streptococci	—	—	—	—	—	—	—	—	1.7	—
Streptococcus faecalis	—	—	—	—	0.54	—	—	—	—	—
Diplococcus pneumoniae	0.3	<1.0	3.2	0.8	1.08	2.3	1.0	2.8	0.8	5.0
Gram-negative bacilli‡	5.0	<4.0	2.9	0.2	2.1	0.5	2.0	6.3	11.6	—
Gram-negative diplococci§	0.7	<1.0	2.3	—	2.16	—	—	4.5	10.8	—
Gram-positive bacilli‖	3.2	<1.0	—	0.2	2.1	—	3.0	3.5	—	—
Micrococcus tetragenus and *Sarcina*	2.6	1.0	—	—	—	—	9.0	—	0.8	1.0
No organisms found	0	23.0	47.0	44.5	32.9	8.7	9.0	13.3	3.4	—

*Superscript numbers indicate references at end of chapter.
†Eyes. Other figures indicate number of individuals examined.
‡Gram-negative bacilli: *Escherichia coli, Klebsiella pneumoniae, K. ozenae, Proteus vulgaris, P. morgani, P. mirabilis, P. rettgeri, Aerobacter aerogenes, Alcaligenes faecalis, Hemophilus influenzae*, Koch-Weeks bacillus, *Moraxella*, and unclassified gram-negative bacilli.
§Gram-negative diplococci: *Neisseria catarrhalis, N. sicca*, and *N. flava*.
‖Gram-positive bacilli: *Bacillus subtilis, B. cereus*, and *B. megaterium*.

The two species of staphylococci were more often found alone or in combination with each other; the diphtheroids usually were in association with one or another of the staphylococci. It is apparent from the figures that the distribution of the bacterial flora in the eyelids was essentially the same as that found in the conjunctivas.

Several studies of the flora of the healthy conjunctiva have been reported from various countries. In Table 2-5 some of these data from cities around the world are tabulated. The organisms most often present in the eye throughout these studies were those already reported, namely, *Staphylococcus epidermidis, S. aureus,* and diphtheroids. *Streptococcus viridans* and *Diplococcus pneumoniae* also were found in small numbers in all the reports. The remaining groups of bacteria listed were found by some but not all of the investigators. There were variations in the reported incidence of the three major groups, particularly of *Staphylococcus aureus;* but considering the large differences in the sizes of the sampling, in the geographic location, and in the period in time when the cultures were obtained, the uniformity of the results is quite striking. Eight of the workers have reported from 3.4% to 47% of persons with no bacteria in the eyes. The exceptions were the early studies of Axenfeld and the report of Locatcher-Khorazo and Gutierrez. This later report, covering the years 1952 to 1968, is at variance with the earlier paper by Khorazo and Thompson (Table 2-2), where 12% to 23% of the subjects appeared not to have bacteria in their conjunctivas. Experience with the techniques of eye culture has led to the adoption of a routine procedure, described later, which always gives growth of some organisms, except in patients who have just received a preoperative course of local antibiotic treatment.

SOURCES OF BACTERIAL OCULAR FLORA

The bacteria cultured from the eye usually are similar to those found on the skin and in the upper respiratory tract. Occasionally, gram-negative bacilli, more common in the gastrointestinal tract, are isolated from the eye. In contrast, bacteria commonly found in the air are rarely recovered from the eye. This may be in part because of the high content of lysozyme present in tears. This agent, first described by Fleming[9] in 1922, is an enzyme, a glucosaminidase that brings about the lysis of *Micrococcus lysodeikticus, Sarcina lutea,* and some staphylococci by acting on a hexosamine bound in a mucopolysaccharide fraction incorporated in the cell wall.[19] Some ocular bacteria themselves produce antibiotic agents that also may serve to reduce the number of certain organisms in the eye (Chapter 3).

Somerville[23] has reported cultures taken from the forehead of normal men at various ages. Staphylococci, diphtheroids, streptococci, and a miscellaneous group of less frequently seen bacteria duplicate the flora reported for the normal eye. At the Institute of Ophthalmology, Columbia-Presbyterian Medical Center (IOCPMC), cultures usually have been obtained simultaneously from the eye, nose, face, and hands. Experience has shown that when one attempts to free the eye of bacteria previous to operation, it is important to give attention to neighboring areas. The incidence of bacteria on the face has averaged *Staphylococcus epidermidis* 50%, *Staphylococcus aureus* 22%, diphtheroids 37%, *Bacillus subtilis* 3%, *Streptococcus viridans* 1.5%, *Proteus* group 2%, gram-negative diplococci 5%, and *Klebsiella pneumoniae* 1%. Although many of these organisms are found in the nose and on the hands, the highest incidence and greatest variety usually have been on the skin of the face.

Staphylococcus aureus has been the most common cause of postoperative infections; therefore, the presence of this organism on areas neighboring the eye has been tested. Table 2-6 summarizes some data gathered from 1966 to 1968. Forty-three percent of 2,379 healthy individuals who were found to carry S. *aureus* in the eye also had S.

Table 2-6. Incidence of *Staphylococcus aureus* in neighboring areas in relation to incidence in noninfected eyes

Noninfected eyes	Incidence of *Staphylococcus aureus*					
	Skin of face		Hands		Nose	
	Number positive	Percent	Number positive	Percent	Number positive	Percent
Carrying *Staphylococcus aureus* in conjunctivas Number: 2,379	1,026	43	949	40	921	38
Not carrying *Staphylococcus aureus* in conjunctivas Number: 2,104	398	19	735	35	714	34

aureus on the skin of the face, whereas only 19% of 2,104 individuals who did not show this bacterium in the eye carried it on the skin. The incidence of *S. aureus* on the hands or in the nose was only slightly less in the group that did not carry this bacterium in the eye. It may be that the skin of the face is an important source of contamination of the eye. *S. epidermidis* and the diphtheroids may equally well enter the eye from the skin. The nose and hands, however, also are potential areas from which the eye may acquire bacteria.

Gowen[11] made a study of the possible role the eyelids play in maintaining the flora of the eye. He noted the identity of the flora of the eye and eyelid margins and further pointed out that when the eyes are closed, the upper eyelids overlap the lower lids; thus, the upper eyelids are exposed to bacteria of the skin. Gowen believed that the eyelashes play a major role in maintaining the skin flora in the conjunctival sac and that the eyes are never devoid of staphylococci, because the source of contamination is always present.

FLORA OF CADAVERIC EYES

The bacterial flora of cadaveric eyes used for grafting has been studied by Polack and his associates[21] and by others[5,13] in order to determine changes in ocular flora of eyes after death. Methods to promote sterilization of the donor eye were tested. It may be seen from Table 2-7 that the flora is essentially the same in both cadaveric and living eyes. However, a few organisms, notably hemolytic streptococci, *Escherichia coli, Klebsiella pneumoniae, Pseudomonas aeruginosa,* and *Flavobacterium,* are present more frequently in cadaveric eyes. The density of the growth of these bacteria is also greater in the cadaveric eyes. Attempts to utilize disinfecting agents indicated the difficulty of producing sterility without injury to the tissue.

RELATIONSHIP OF NORMAL OCULAR FLORA TO INFECTIONS OF THE EYE

It is apparent from the presence of certain microorganisms in the normal conjunctiva that man often carries with him potentially pathogenic organisms in his eyes just as he may be a carrier of such bacteria in his upper respiratory tract. *Staphylococcus aureus* is the potential pathogen most frequently found. In Chapter 6 evidence is given to indicate that this organism is the commonest cause of postoperative infection and that this infection may arise from the patient's own eye.[17] *Hemophilus influenzae* and some of the gram-negative bacilli found in normal eyes, namely, *Pseudomonas aeruginosa, Escherichia coli, Proteus vulgaris,* and *Klebsiella pneumoniae,* may produce grave infections, such as corneal ulcers or postoperative complications (Chapters 5 and 6). The importance of all these organisms

Table 2-7. A comparison of the ocular flora of normal living eyes with that of postmortem donor eyes

Microorganisms isolated	Percent positive in 566 normal living eyes	Percent positive in 240 postmortem donor eyes
Staphylococcus epidermidis	69.7	71.7
Staphylococcus aureus	30.2	36.3
Diphtheroids	73.1	60.8
Pneumococci	—	0.8
Streptococcus viridans	3.7	8.3
β-Hemolytic streptococci, presumptive group A (bacitracin sensitive)	—	8.3
Neisseria catarrhalis	0.9	0.4
Escherichia coli	1.1	10.0
Proteus mirabilis	1.9	3.3
Proteus morgani	1.6	0.8
Klebsiella pneumoniae	1.4	18.8
Klebsiella ozenae	—	1.7
Pseudomonas aeruginosa	—	4.2
Bacillus megaterium	0.5	—
Bacillus cereus	—	1.7
Sarcina	0.7	2.5
Flavobacterium	—	10.8

From Polack, F. M., Locatcher-Khorazo, D., and Gutierrez, E.: Bacteriologic study of "donor" eyes, Arch. Ophthal. **78**:221, 1967.

is such that, when they are present in the eye of a patient awaiting surgery, prophylactic preoperative treatment with antibiotics has been initiated at the IOCPMC (Chapter 6).

TECHNIQUE FOR OBTAINING CULTURES FROM CONJUNCTIVAS AND EYELIDS

It seems appropriate to include the techniques employed to obtain material for culture, since this must often be performed by the ophthalmologist. The basic media used for the initial cultivation of organisms from the normal external eye is nutrient agar enriched with blood. Fig. 2-1 shows the kit that is used to obtain the material for culture. The wax pencil in the foreground is for labelling the plates in the basket. Zephiran is for cleansing the skin around the eye. The sterile forceps and the cotton on the left and the sterile saline bottle, rear right, are used to wash out any previously applied ointments or makeup. Moist, sterile swabs in test tubes containing a few drops of nutrient broth (see basket) are used to obtain cultures of the conjunctiva and eyelid, because the moistened swab lessens trauma to the patient and ensures optimal yield of bacteria. Table 2-8 presents data that show the great superiority of the moistened swab over the dry swab for cultivation of the bacteria. One of these moistened swabs is passed back and forth twice over the greater part of the lower conjunctiva, carefully avoiding the eyelid border and the eyelashes (Fig. 2-2). Fig. 2-3 shows the inoculation of the blood agar plate. The plate should be inoculated at the patient's bedside, or the material should be sent immediately to the laboratory and the plate promptly inoculated there. The swab is rubbed over the entire surface of the plate; the bottom of the plate is labelled with the individual's name, the date, and the source of the culture (such as the conjunctiva or lid margins) and is incubated overnight. The top of the swab is broken to ensure that it will not be reused. Fig. 2-4, *A* shows the ap-

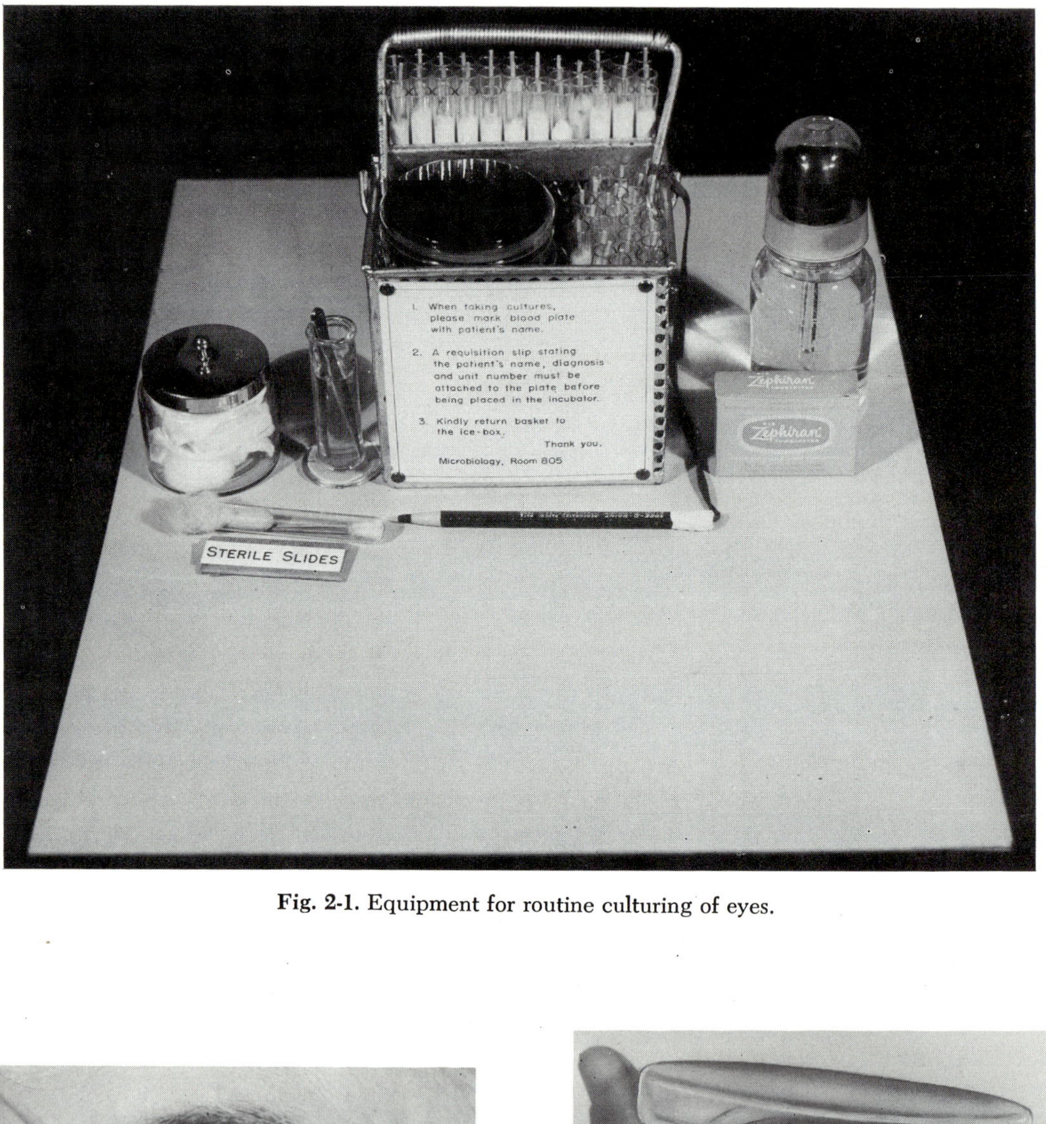

Fig. 2-1. Equipment for routine culturing of eyes.

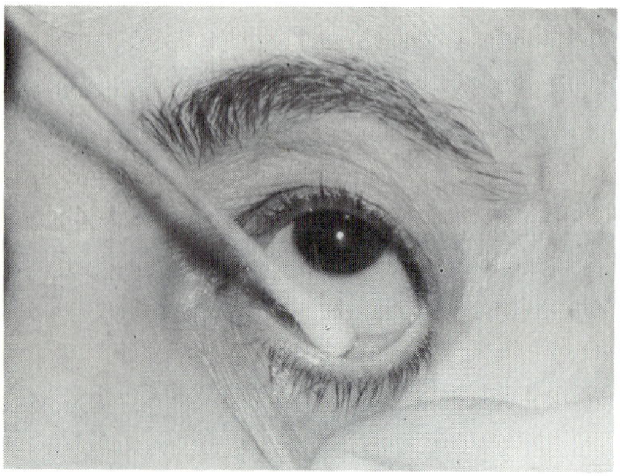

Fig. 2-2. Obtaining culture from lower conjunctiva.

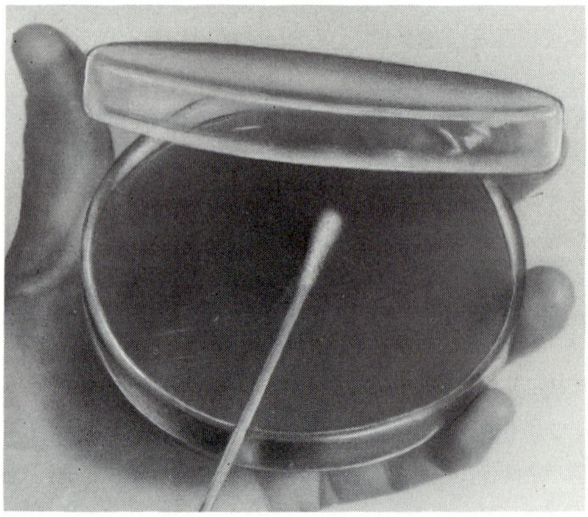

Fig. 2-3. Inoculation of blood agar plate.

Table 2-8. Comparison of cultures of conjunctiva obtained with dry and wet swabs

Case no.	Diagnosis	Dry swab	Wet swab
1	Cataract	No growth	14 colonies *Staphylococcus epidermidis* 6 colonies *Staphylococcus aureus*
2	Cataract	No growth	16 colonies *Corynebacterium hoffmannii* 14 colonies *Staphylococcus epidermidis*
3	Cataract	43 colonies *C. hoffmannii* 18 colonies pneumococci	125 colonies *C. hoffmannii* 63 colonies pneumococci 3 colonies *Staphylococcus aureus*
4	Cataract	1 colony pneumococci	23 colonies pneumococci 24 colonies *Staphylococcus aureus*
5	Cataract	110 colonies *C. hoffmannii*	120 colonies *C. hoffmannii* 32 colonies *Staphylococcus aureus*
6	Normal eye	No growth	Numerous diphtheroids 12 colonies *Staphylococcus aureus*
7	Dacryocystitis	3 colonies *C. hoffmannii*	14 colonies *C. hoffmannii* 2 colonies *Staphylococcus aureus* 6 colonies pneumococci
8	Acute conjunctivitis	6 colonies *Staphylococcus aureus*	Numerous colonies *Staphylococcus aureus* and *Hemophilus influenzae*
9	Chronic conjunctivitis	14 colonies *Staphylococcus epidermidis* 6 colonies *Staphylococcus aureus*	Numerous colonies *Staphylococcus epidermidis* 34 colonies *Staphylococcus aureus* 2 colonies *Escherichia coli*
10	Corneal ulcer	No growth	6 colonies *Staphylococcus aureus*

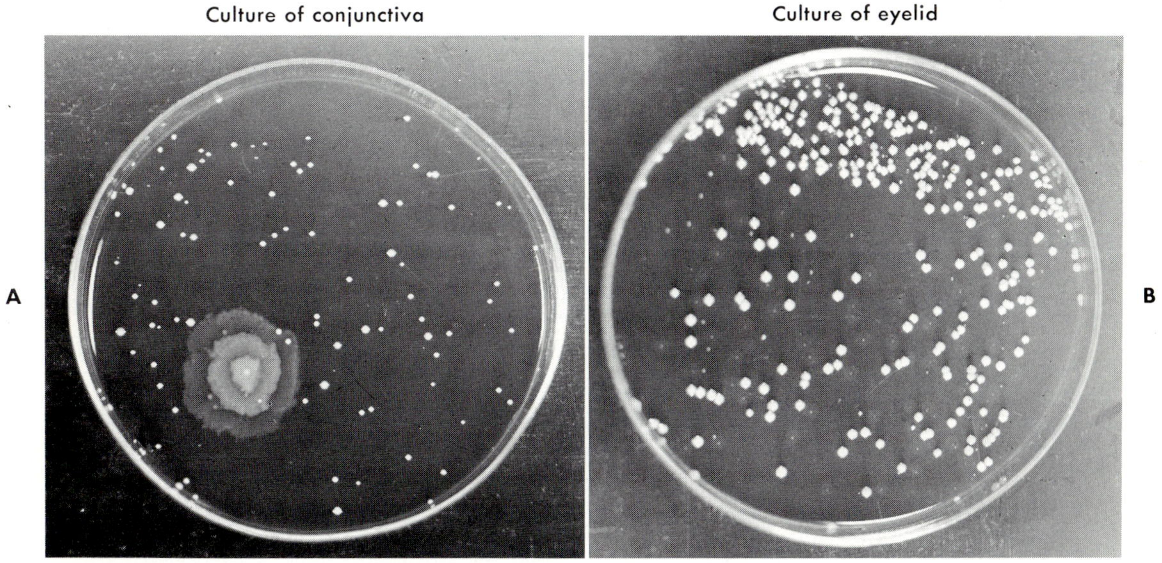

Fig. 2-4. Cultures of conjunctiva, **A,** and eyelid margin, **B,** from the same normal eye. Note difference in number of colonies. Because the eyelid is rich in bacteria, the lid swab was first passed over a portion of the upper part of the plate; then a sterile moist swab was used to complete the inoculation of the plate.

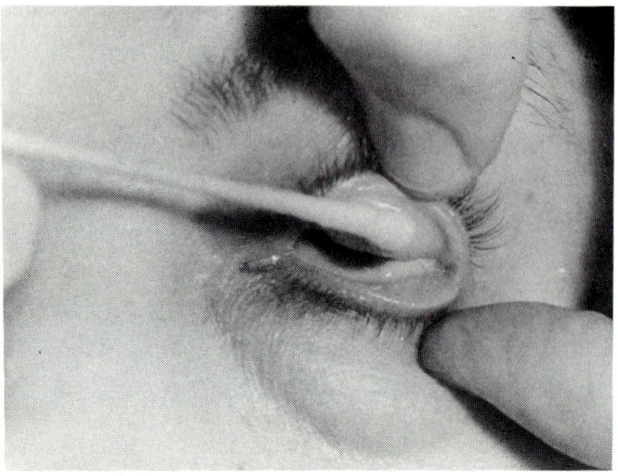

Fig. 2-5. Obtaining culture from upper conjunctiva.

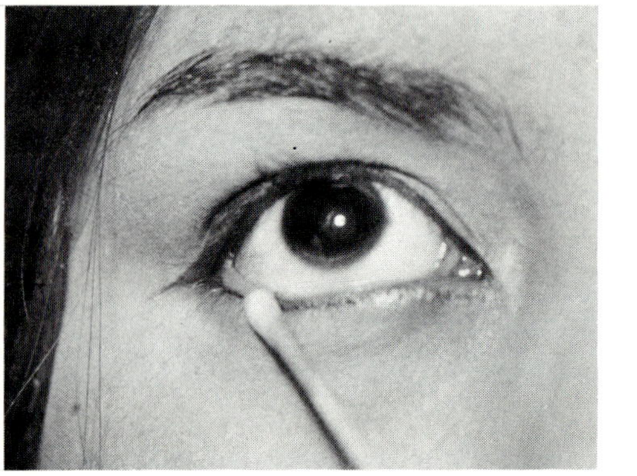

Fig. 2-6. Obtaining culture from eyelid margin.

pearance of growth on such a plate of material from a normal conjunctiva after 24 hours. Note the increased number of bacteria recovered from the eyelid of the same individual (Fig. 2-4, *B*).

Cultures from the upper conjunctiva are not usually necessary except when there is a local area of inflammation. However, the number of bacteria recovered are often more than are found in the lower conjunctiva. If cultures are negative or growth is sparse, a culture of material from the upper conjunctiva is indicated. The technique for obtaining cultures from the upper conjunctiva is illustrated in Fig. 2-5. The material is handled in the same manner as for the lower conjunctiva.

For eyelid cultures another moistened swab is applied to the outer margin of each lid (Fig. 2-6) as the eye is held open, and the material from each lid is swabbed on another blood agar plate. Growth of bacteria is usually heavier from the lid, as is seen in Fig. 2-4, *B*. When an infection is present the swab is rubbed over only one third of the plate, and a fresh sterile swab is used to spread the heavy inoculum further (Chapter 5). The entire process must be repeated for the second eye if there is any question of infection in that eye. It may also be important to obtain cultures from both normal eyes in order to determine whether there is a difference in the flora of the two eyes. Direct smears taken from normal eyes for histologic examination are not helpful, because the organisms are relatively few in number.

SUMMARY

Bacterial flora of the conjunctiva of the normal eye is remarkably constant around the world. Cultures taken from persons of all ages in many parts of the world have shown essentially the same findings, namely, that *Staphylococcus epidermidis,* diphtheroids, and *Staphylococcus aureus* are the predominant organisms. This also was true over 65 years ago, according to Axenfeld.[2] A number of other organisms, some of them potential pathogens, occur very much less frequently but nevertheless must be recognized because of their potential importance.

The source of bacteria in the eye may be the skin of the face, the upper respiratory tract, and the skin of the hands.

Eyes obtained for the eye bank carry the same flora found in life in the normal eye with a few exceptions. All bacteria are more numerous than in the living eye.

Directions are included for obtaining cultures from healthy eyes. These stress the use of moist swabs and the direct inoculation of the media at the bedside or promptly in the laboratory. Blood-enriched nutrient agar is the basic medium for primary isolation.

REFERENCES

1. Allansmith, M. R., Ostler, H. B., and Butterworth, M.: Concomitance of bacteria in various areas of the eye, Arch. Ophthal. **82**:37, 1969.
2. Axenfeld, T.: The bacteriology of the eye, translated by Angus MacNab, New York, 1908, William Wood and Co.
3. Bachrach, U., Gurevitch, J., Landau, J., and Birnbaum, D.: The flora of the normal conjunctiva of healthy people in Israel, Acta Med. Orient. **12**:10, 1953.
4. Barfoed, P.: Conjunctivas Bakterieflora. Thesis presented at Ejnar Munksgaard, Copenhagen, 1953.
5. Boberg-Ans, J., Badsberg, E., and Rasmussen, J.: Frequency of infection in donor eyes post mortem: a method of obtaining sterile eyes for corneal grafting, Brit. J. Ophthal. **46**:365, 1962.
6. Cason, L., and Winkler, C. H.: Bacteriology of the eye: normal flora, Arch. Ophthal. **51**:196, 1954.
7. Chang, H. L.: Bacterial flora of the normal conjunctiva, Chin. Med. J. **75**:233, 1957.

7a. De Ocampo, G., Salceda, S., and de Leon, A.: Bacterial flora of the healthy conjunctiva among Filipinos, Philippine J. Surg. Surgical Spec. **20**:94, 1965.

8. Doden, W.: Vergleischende Untersuchungen über die Wirkung verschiedener Augensalben auf die Keimflora des gesunden Bindehautsackes, Klin. Mbl. Augenheilk. **131**:237, 1957.
9. Fleming, A.: On a remarkable bacteriolytic element found in tissues and secretions, Proc. Roy. Soc. London (Series B) **93**:306, 1922.
10. Franklin, H. C., and Loeb, L. N.: Bacterial flora in infants encountered at time of delivery: a study of one hundred cultures from the eyelids, Amer. J. Obstet. Gynec. **56**:738, 1948.
11. Gowen, G. H.: Source of Staphylococci on normal conjunctiva of human eye, Amer. J. Ophthal. **17**:36, 1934.
12. Grünberger, V., and Kofler, E.: Uber den Keimgehalt der kindlichen Conjunctiva nach der Geburt, Gynaecologia **137**:329, 1954.
13. Kanski, J. J.: Bacterial flora of the human conjunctiva after death, Brit. J. Ophthal. **49**:445, 1965.
14. Khorazo, D., and Thompson, R.: The bacterial flora of the normal conjunctiva, Amer. J. Ophthal. **18**:1114, 1935.
15. Khristov, G.: Bacterial flora of the normal conjunctiva and in conjunctivitis, Izv. Microbiol. Inst. (Sofiia) **7**:107, 1956.
16. Locatcher-Khorazo, D., and Gutierrez, E.: Ocular flora of 1,024 children (1-18 years old), 1,786 young adults (20-35), and 7,461 patients awaiting ocular surgery, with no known infection. Unpublished data from 1952 to 1968.
17. Locatcher-Khorazo, D., and Gutierrez, E.: Eye infections following cataract extraction with special reference to the role of Staphylococcus aureus, Amer. J. Ophthal. **41**:981, 1956.
18. Lucic, H.: Bacteriology of the normal conjunctival sac, Amer. J. Ophthal. **10**:829, 1927.

18a. Matuura, H.: Anaerobes in the bacterial flora of the conjunctival sac, Jap. J. Ophthal. **15**:116, 1971.

19. Meyer, K.: Lysozyme, Bull. N. Y. Acad. Med. **30**:995, 1954.
20. Orfila, J., and Courden, B.: Contribution a l'etude de la flore conjonctivale normale, Ann. Oculist. **194**:892, 1961.
21. Polack, F. M., Locatcher-Khorazo, D., and Gutierrez, E.: Bacteriologic study of "donor" eyes, Arch. Ophthal. **78**:219, 1967.
22. Smith, C. H.: Bacteriology of the healthy conjunctiva, Brit. J. Ophthal. **38**:719, 1954.
23. Somerville, D. A.: The effect of age on normal bacterial flora of the skin, Brit. J. Derm. **81**(Suppl. 1):14, 1969.
24. Yamazaki, T.: Bacterial examination of conjunctival sac by thioglycolate medium and human blood agar, Acta Soc. Ophthal. Jap. **62**:1291, 1958.

3

Inhibitory properties of the ocular flora

SEYMOUR P. HALBERT

GENERAL FACTORS IN RESISTANCE TO SUPERFICIAL OCULAR INFECTION

The mechanisms involved in the resistance of the conjunctiva to superficial bacterial infections are not clearly established, but several factors appear to play significant roles. The mechanical trapping action of the eyelashes to catch fine particles of dust almost certainly has a protective function, since it is fairly well established that many airborne microorganisms are carried on such dust motes. The blinking of the eyelids and the flushing action of the tear flow also have a cleansing and protective effect. Experimental studies in which the nasolacrimal duct was tied off so that the normal flow of tears was interrupted clearly demonstrate the increased susceptibility of the conjunctiva to infection. There seems to be little question that disturbance of the nasolacrimal duct clinically results in increased incidence of superficial ocular infection.

Various chemical factors in the tear film may have protective significance. The relative importance of lysozyme as a resistance factor is not yet unequivocally established. It is conceivable that the saprophytic microorganisms that are highly susceptible to this enzyme are devoid of pathogenicity largely because of lysozyme. It is also possible that even the known pathogenic bacterial species that reveal relatively slight susceptibility to this substance might be inhibited adequately to prevent active infection. Since local antibody concentration seems to be a significant factor in certain infections, such as influenza, a knowledge of the antibody content of tears would be of some importance. However, no such data seems to be available.

POTENTIAL PROTECTIVE ROLE OF INHIBITORY OCULAR FLORA

The bulk of this chapter will be devoted to the possible role of the normal ocular flora as a protective factor in resistance to superficial conjunctival bacterial infections. As was pointed out in Chapter 2, the organisms that make up the bulk of the ocular flora are staphylococci and diphtheroids. The precise species involved and their apparent fluctuation have been discussed in Chapter 2. It is clear that they are always present in variable numbers and that they become implanted shortly after birth. Pathogenic microorganisms must multiply and invade tissues in spite of the presence of normal flora. Considerable data has accumulated demonstrating the inhibitory effects of this normal flora on pathogenic bacteria.[1-10]

In occasional instances, primary cultures of the conjunctival flora reveal inhibition of one organism by another. An example of this is shown in Fig. 3-1, which represents the growth on a blood agar plate from the

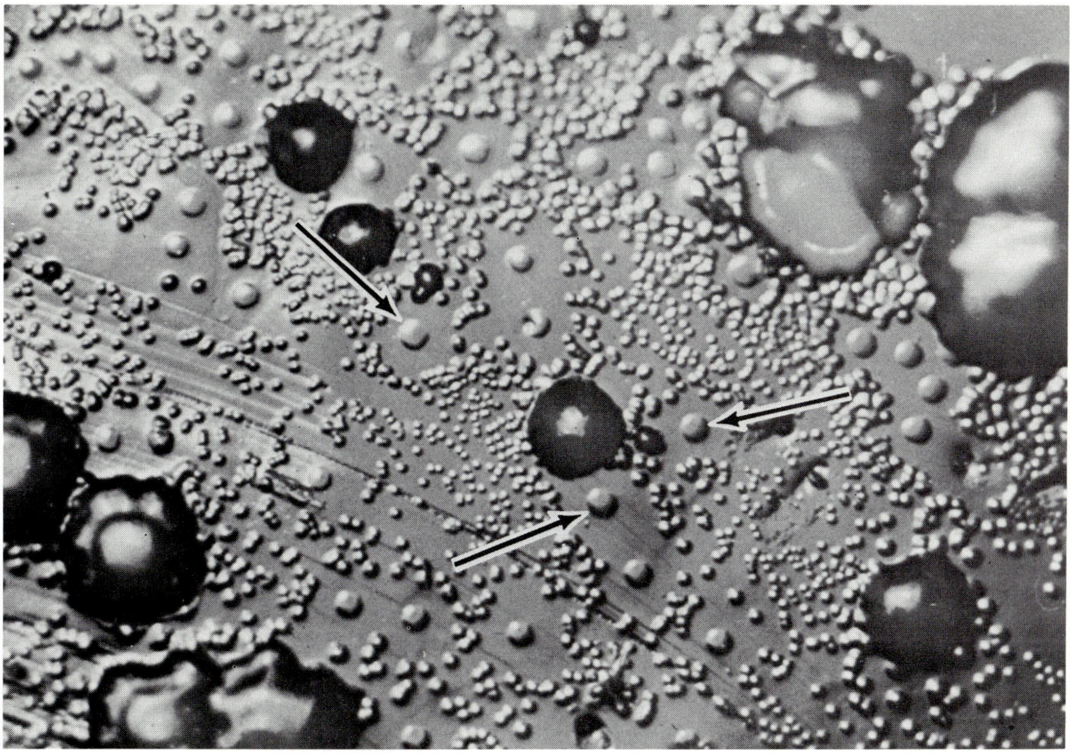

Fig. 3-1. "Spontaneous" appearance of inhibiting activity (arrows) in the original streak from the eye swab of a patient. (From Halbert, S. P., and Swick, L. S.: Amer. J. Ophthal. **35**:78, 1952.)

conjunctival swab of material from a patient with conjunctivitis. It is clear that the slightly larger translucent colonies interfered with the growth of the more numerous minute colonies of microorganisms in their immediate vicinity.

Inhibiting strains of ocular organisms can be more readily detected by other methods. When the conjunctival culture is allowed to grow for 24 hours and then the entire surface is seeded with an indicator strain by a spray technique, variable numbers of colonies can be demonstrated to have secreted inhibitors, as seen in Fig. 3-2. In this instance, the indicator organism is a strain of *Staphylococcus albus* of ocular origin, and the medium is blood agar.

Still another way of demonstrating the inhibitory effect of strains of the ocular flora consists of first growing the primary conjunctival culture and then transferring individual colonies to the surface of an agar plate already uniformly seeded with an indicator strain. Examples of results obtained with 20 colonies, each selected at random from single individuals, are shown in Fig. 3-3. In *A*, *B*, and *C* the indicator organism is a highly susceptible *Corynebacterium pseudodiphtheriticum*. It is of interest that the inhibitory activities of these strains from the same person were so variable, in spite of the fact that many of the individual colonies were otherwise indistinguishable. The striking differences in the susceptibility of different indicator strains is also shown in Fig. 3-3, *D*, where identical colonies were tested against an *S. albus* culture as well as against a *C. pseudodiphtheriticum* culture.

That the inhibitions noted resulted from secreted antibiotic substances in the case of highly active strains is revealed by the readily demonstrated activity of crude culture filtrates. The crisp inhibition zones

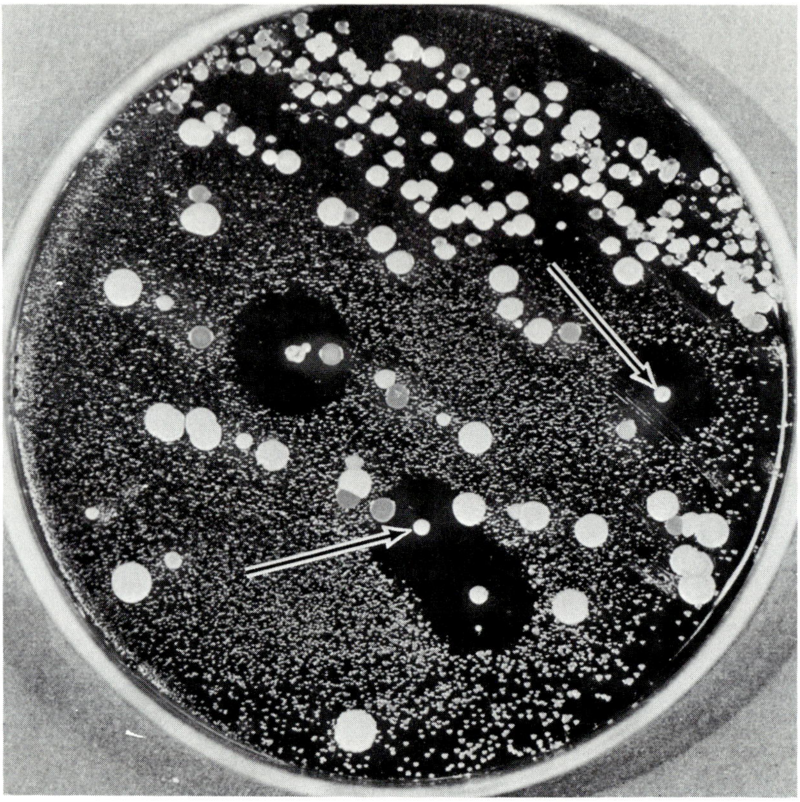

Fig. 3-2. *Staphylococcus albus* 7-6 spray-seeded onto blood agar after 48-hour incubation of original inoculum from one patient. The arrows point to highly active antibiotic-producing organisms. (From Halbert, S. P., and Swick, L. S.: Amer. J. Ophthal. 35:74, 1952.)

produced by such filtrates from three potent organisms, 224-4, 166-4, and 303-3, are shown in Fig. 3-4. When a series of both highly and slightly inhibitory strains of ocular staphylococci were similarly tested for the secretion of antibiotics, results with the highly potent strains were always positive, although only a portion of the slightly active strains could be readily shown to secrete inhibitors. However, heat lability of the antibiotics and other factors could account for the failure to demonstrate antibiotics in a higher proportion of these latter strains. Nonactive organisms never revealed inhibitors in the crude filtrates. Some of these results are summarized in Table 3-1, where it may also be noted that the crude filtrates of many slightly active organisms produced zones of decreased opacity of the indicator organisms. When similar tests were carried out using the more sensitive indicator, *C. pseudodiphtheriticum,* about one half of the weakly inhibitory strains revealed antibiotic substances in the culture supernates. These data are shown in Fig. 3-5, which also shows the high degree of variability in the potency of the antibiotic crude filtrates.

CHARACTERISTICS OF INHIBITORY OCULAR FLORA AND THE ANTIBIOTICS THEY PRODUCE

In this study most of the inhibitory organisms of the ocular flora proved to be strains of staphylococci, although occasional active streptococci and diphtheroids were detected. A number of the staphylococcal cultures were subjected to more detailed identification tests; the results are shown in Table 3-2. It may be seen that

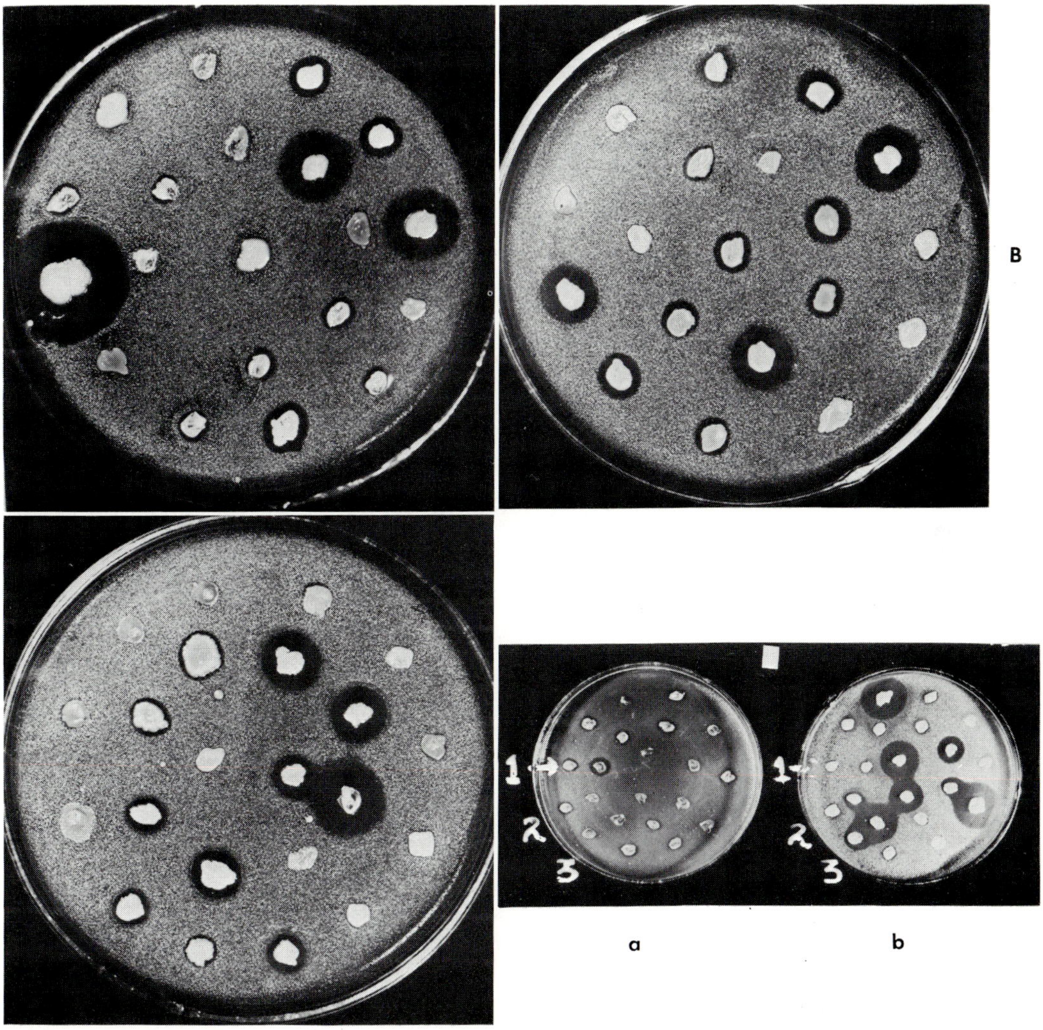

Fig. 3-3. Anitbiotic activity of sample colonies of ocular microorganisms from single individuals. **A, B,** and **C,** Tested against *C. pseudodiphtheriticum;* **D,** tested against both *S. albus* 7-6 (*a*) and *C. pseudodiphtheriticum* (*b*). In these plates the same colonies were tested against both indicators. (From Halbert, S. P., Locatcher-Khorazo, D., Sonn-Kazar, C., and Swick, L. S.: Arch. Ophthal. **58**:68, 1957.)

most of the strains were typical or atypical *Staphylococcus albus,* with occasional organisms of other *Micrococcus* species detected. It is thus clear that the inhibitory bacteria so frequently encountered in the ocular flora are derived from the most abundant species residing on the superficial eye tissues. It is also clear that the inhibitory effects are caused by the secretion of antibiotic substances.

It was found that occasional resistant variants of the indicator strains occurred with regularity upon exposure to minimal concentrations of a given antibiotic. When these resistant variants were subcultured, it was found that they were not affected by the antibiotic of the strain to which they had been exposed but that they might still be susceptible to antibiotics of other strains. In addition, investigation of the chemical and physical properties of the antibiotics produced by 22 strains of highly inhibitory ocular microorganisms demonstrated clear differences between many of them. Nine distinct groups of antibiotics could be differentiated. Some of these data are summarized in Table 3-3. Since many of the antibiotics were destroyed by the proteolytic enzymes trypsin and chymotrypsin and were precipitable by high concentrations of ammonium sulfate, it is probable that many were of polypeptide nature. Some were rather stable to heat (100° C), whereas others were destroyed by such temperatures. The fact that some were freely or partially dialyzable through Visk-

Table 3-1. Crude antibiotic production in relation to the inhibitory activity of living ocular staphylococcal cultures, tested against other staphylococci as indicators

Strains	Number of strains	Number producing active crudes (percent)	Potency of crudes: mm width*	Potency of crudes: Number of strains	Median potency (mm width)
Highly active	28	28 (100%)	0.0–3.0	3	5.0
			3.1–5.0	11	
			5.1–7.0	8	
			7.1+	6	
Slightly active	36	8 (22%)	0.0	9	—
			Zones of decreased opacity†	19	
			1.0‡–2.8	8	
Nonactive	22	0 (0%)	—	—	0

From Halbert, S. P., Swick, L. S., and Sonn, C.: Characteristics of antibiotic-producing strains of the ocular flora, J. Immun. **70**:402, 1953. © 1953, The Williams and Wilkins Co.

*Refers to the distance from the rim of the well to the edge of normal growth. The figures represent the maximum inhibition against any of the three *Micrococcus* indicator strains, 7–6, 15–1, or 13–2.

†Although no distinct inhibition was noted, the colonies surrounding these wells were of less opacity than the normal growth.

‡The edges of the inhibition zones of most of these were poorly demarcated, both with the crudes and with the living systems.

Table 3-2. Identification of a series of ocular staphylococci in relation to inhibitory activities

Biochemical pattern					Number of strains			
Coagulase production	Mannitol fermentation	Gelatin liquefaction	NO_3 reduction	(NH_4)-H_2PO_4 utilization	Highly active	Slightly active	Nonactive	Tentative identification*
0	0	+	+	0	9	30	25	"Atypical" *M. pyogenes* var. *albus*
0	+	+	+	0	8	1	2	*M. pyogenes* var. *albus*
0	+	0	+	0	1	0	2	*M. aurantiacus*
0	0	0	+	0	1	1	0	*M. epidermidis*
0	0	+	0	0	3	0	0	"Atypical" *M. candidus*
0	0	0	0	0	0	1	1	*M. candidus*
+	+	+	+	0	0	0	1	*M. pyogenes* var. *aureus*

From Halbert, S. P., Swick, L. S., and Sonn, C.: Characteristics of antibiotic-producing strains of the ocular flora, J. Immun. **70**:401, 1953. © 1953, The Williams and Wilkins Co.

*According to *Bergey's Manual of Determinative Bacteriology*, 6th edition.

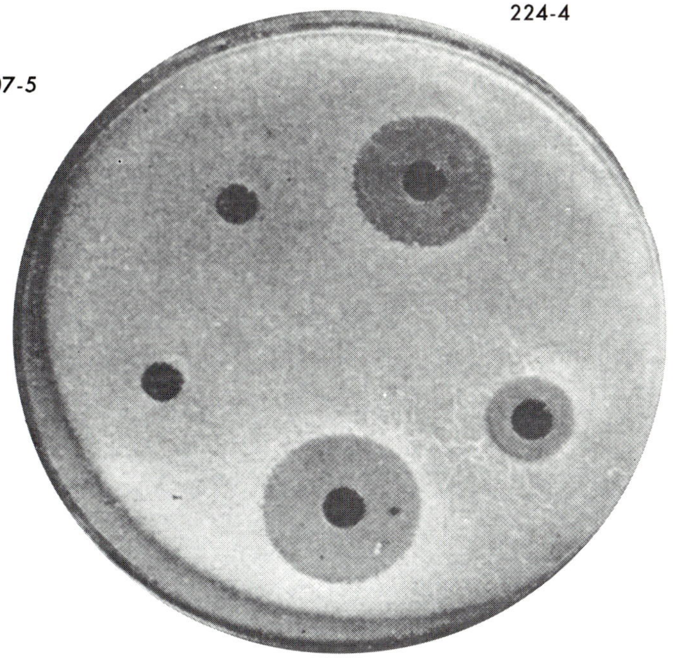

Fig. 3-4. Antibiotic activity of some crude supernates from several active staphylococci tested against strain 15-1. (From Halbert, S. P., and Swick, L. S.: Amer. J. Ophthal. 35:77, 1952.)

Table 3-3. Differentiation of the antibiotics of 22 strains of ocular microorganisms into 9 distinct groups, on the basis of tests with resistant variants of an indicator *Staphylococcus* (15-1), and chemical properties

Activity vs. resistant variants of strain 15-1	Group	Number of strains	Properties				Enzyme digestibility	
			Acteone precipita-bility	$(NH_4)_2SO_4$ precipita-bility	Heat stability	Dialyza-bility	Trypsin	Chymo-trypsin
Inactive vs. R_1 and R_2,	I	7	0	+†	S‡	±*	0	0
active vs. R_3 and R_4	II	1	0	+	L	±	0	0
Inactive vs. R_3 and R_4,	III	4	±	+	S	+	+†	0
active vs. R_1 and R_2	IV	2	±	+	L	+	+	0
	V	3	0	+	S	+	+	0
Active vs. R_1, R_2, R_3,	VI	2	±	0	L	0	+	+
and R_4	VII	1	±	+	L	±	+	0
	VIII	1	0	0	L	0	+	+
	IX	1	0	+	L	±	+	+

From Halbert, S. P., Swick, L. S., and Sonn, C.: Characteristics of antibiotic-producing strains of the ocular flora, J. Immun. **70**:404, 1953. © 1953, The Williams and Wilkins Co.

*+ represents free dialyzability, ± partial dialyzability, and 0 nondialyzability.

†+ represents precipitability with ammonium sulfate or acetone, or destruction by the enzyme. 0 represents failure to precipitate with $(NH_4)_2SO_4$ or acetone, or resistance to the enzyme.

‡S represents heat stability, and L represents heat lability, especially at alkaline pH levels.

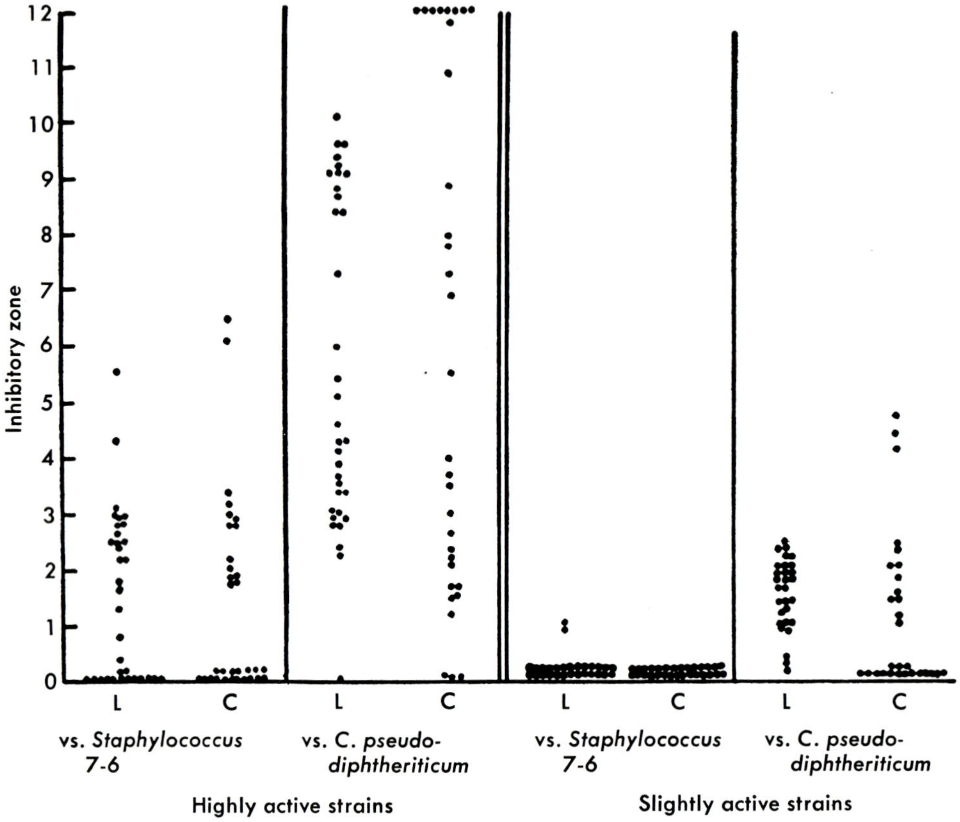

Fig. 3-5. Production of crude antibiotic solutions by sample strains of inhibitory ocular organisms. Each point represents the value obtained with one organism. *L,* Inhibition zones (in millimeters) in tests with the living antibiotic-producing strain; *C,* inhibition zones in tests with crude antibiotics from these strains. (From Halbert, S. P., Locatcher-Khorazo, D., Sonn-Kazar, C., and Swick, L. S.: Arch. Ophthal. **58**:69, 1957.)

ing cellophane membranes suggests that these antibiotics were of rather small molecular sizes.

ANTIBACTERIAL SPECTRA OF INHIBITORY OCULAR FLORA

The antibacterial spectra of these antibiotics were found to be largely limited to the gram-positive species and the neisseria. The gram-negative rods were quite resistant. Even among the gram-positive organisms, however, there was considerable variability in the susceptibility of different strains of the same species. This is clearly shown in Fig. 3-6, which shows the susceptibility of a number of bacterial species to 10 antibiotic-producing ocular microorganisms. The figure clearly indicates that differing percentages of test strains were affected by any given antibiotic and that, in general, most other staphylococci were susceptible. In the case of *Corynebacterium,* most strains were highly sensitive to the antibiotic of *Staphylococcus* 283-8, but most were resistant to strain 306-6. In addition to the organisms listed in Fig. 3-6, it was found that numerous strains of pathogenic and nonpathogenic *Mycobacterium, Neisseria,* and *Clostridium* were frequently and highly susceptible to such antibiotics. Strains of *Salmonella, Shigella, Proteus, Pseudomonas, Escherichia, Klebsiella,* and *Vibrio* were included in the Enterobac-

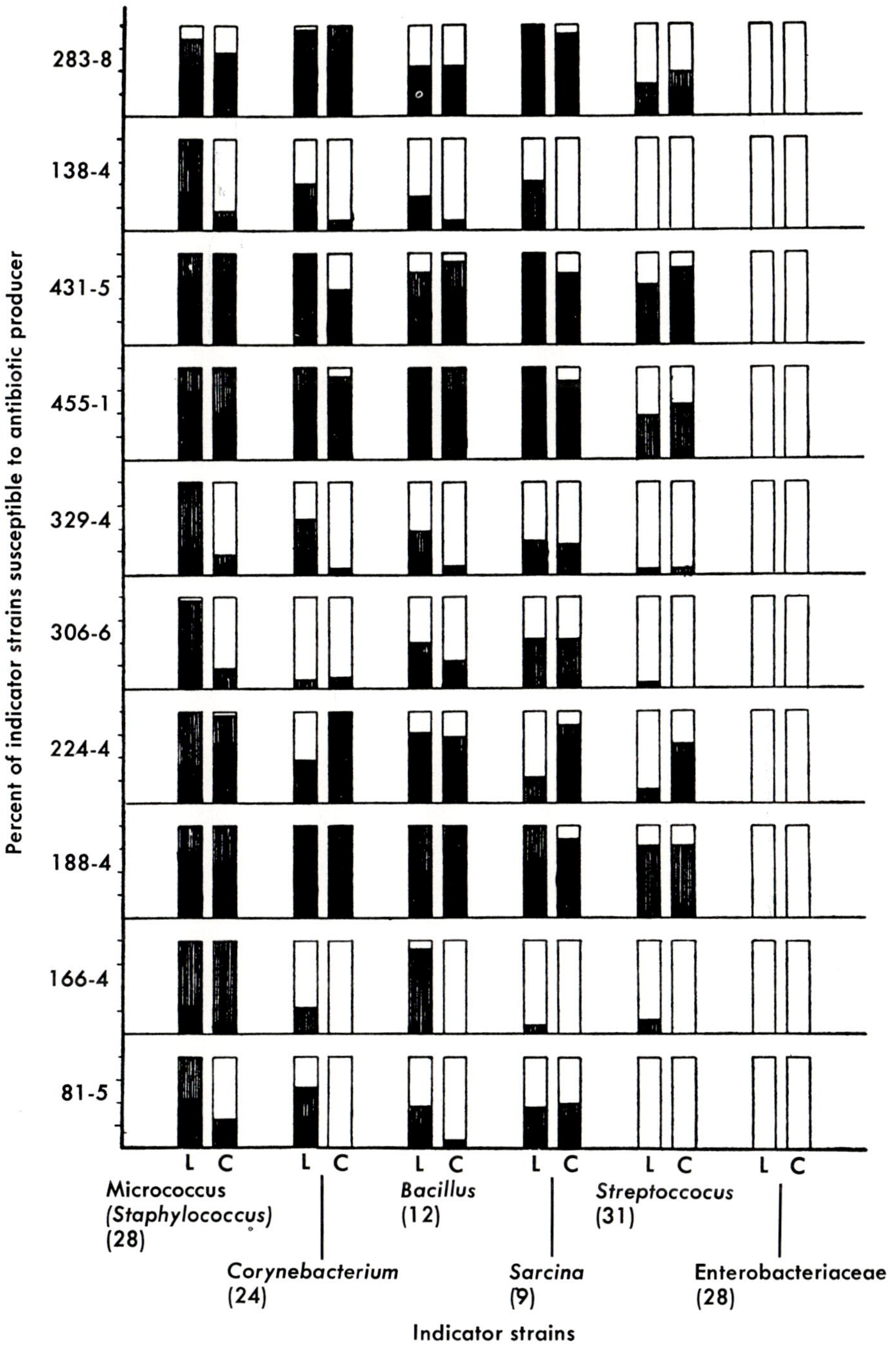

Fig. 3-6. Susceptibility of some bacterial species to a group of antibiotic-producing ocular micrococci. *L,* Tests with living microorganisms; *C,* tests with crude antibiotic derived from that strain; *( ),* number of strains examined.

■ Represents percent of highly susceptible strains; that is, those being inhibited at least one half the size of the most susceptible stock indicator strains. Many showed zones much higher.

▥ Represents percent of slightly susceptible strains; that is, those being inhibited to a width of less than one half the size of the most suscpetible stock indicator strain.

□ Represents percent of resistant strains; that is, no significant inhibition noted.

(From Halbert, S. P., Swick, L. S., and Sonn, C.: J. Immun. **70**:407, 1953.)

teriaceae that were tested; these were found to be quite resistant.

CLINICAL OBSERVATIONS RELATED TO INHIBITORY OCULAR MICROORGANISMS

In a number of individuals, simultaneously sampled strains of staphylococci from the eye, nose, and skin of different patients were tested for antibiotic activities. In about one fourth of the cases the inhibitory patterns of the flora of the various tissues were strikingly different from each other. Table 3-4 shows some examples of cultures from the conjunctiva, anterior nares, and skin of the ventral surface of the hand in a series of such individuals. In patient 2 all twenty staphylococcal strains from the ocular flora produced inhibitory zones of 1.8 mm, whereas 10 such colonies from the nose and 10 from the skin had no activity whatsoever. In the remaining three fourths of patients so tested, organisms from the various tissues showed quite similar inhibitory patterns.

A series of observations were made to see if the antibiotic-producing organisms of the ocular flora could be related to superficial conjunctival infection or to inflammation. In such a study it was of some importance to characterize the microorganisms isolated from the patients with conjunctivitis and to compare them with the control group with normal eyes. When this was done in one such study, as shown in Table 3-5, it could be readily seen that no striking differences were observed between the two groups. However, only two patients with conjunctivitis revealed *Hemophilus* strains, and another two harbored diplobacilli; none of these microorganisms were found in the control eyes. These findings suggest that bacterial infections may not have been involved in the etiologies of most of the conjunctival disturbances or that the strains isolated from patients with disease differed in some subtle ways from those isolated from normal individuals.

When the patients with conjunctivitis were compared with the normal individuals with respect to antibiotic producers, no significant differences were found between the two when a staphylococcal indicator strain was used (Table 3-6). In a similar study on a different group of patients, a highly sensitive corynebacterial indicator was used. When the number of strains active against this organism from the patients with conjunctivitis were compared with the group who had normal eyes, again no differences were seen. These data are summarized in Fig. 3-7, which shows that striking fluctuations were observed in the incidence of such strains at various periods throughout the year. During March 1952, for example, about 70% of all strains tested

Table 3-4. Comparison of antibiotic activities of organisms (mostly staphylococci) of the ocular, nasal, and skin flora of single individuals

Patient	Antibiotic activity against *Corynebacterium pseudodiphtheriticum* of strains from		
	Eye	Nose	Skin (hand)
1	20 (1.2)* 0 (0)	12 (0)	4 (0.6) 6 (0)
2	20 (1.8) 0 (0)	10 (0)	10 (0)
3	15 (1.7) 5 (1.0) 0 (0)	10 (0)	2 (1.0) 8 (0)
4	20 (0)	20 (0)	7 (1.7) 2 (0)
5	11 (1.0) 9 (0)	7 (4.0) 1 (1.5) 3 (0)	10 (0)
6	20 (4.5) 0 (0)	10 (0)	—
7	20 (0)	10 (1.3) 1 (0)	—
8	12 (1.3) 8 (0)	—	15 (0)
9	20 (0)	10 (0)	10 (0)
10	20 (6.0)	10 (6.0)	11 (5.5)

From Halbert, S. P., Locatcher-Khorazo, D., Sonn-Kazar, C., and Swick, L. S.: Further studies on the incidence of antibiotic-producing micro-organisms of the ocular flora, Arch. Ophthal. **58**:73, 1957.
*The first figure represents the number of colonies; the figure in parentheses represents the zone of inhibition in millimeters.

Table 3-5. Comparison of species in the bacterial flora from patients with normal eyes and those with conjunctivitis

Group	Normal						Conjunctivitis					
	*Staphylococcus aureus**	Streptococci*	Hemophilus*	Diphtheroids*	Diplobacilli	No "pathogens"	*Staphylococcus aureus**	Streptococci*	Hemophilus*	Diphtheroids*	Diplobaccili	No "pathogens"
H	3†	0	0	0	0	20	5	0	1	1	0	10
Iota	3	0	0	0	0	17	4	1	0	(1)‡	0	15
Q	9	0	0	0	0	25	3	0	0	0	0	3
S	3	0	0	0	0	18	1	2	0	0	0	16
Y	6	4	0	1	0	23	2	(1)	0	0	0	4
Delta	1	0	0	1	0	20	5	(2)	0	0	1	12
Theta, 6th visit	0	3	0	1	0	13	3	2	1	1	1	14
Theta, 7th visit	0	4	0	1	0	14	3	5	0	4	0	9
Totals	25	11	0	4	0	150	26	10	2	6	2	83
			190						129			

From Halbert, S. P., Swick, L. S., Sonn, C., and Locatcher-Khorazo, D.: Ocular antibiotic-producing bacteria in normal eyes and in conjunctivitis, Arch. Ophthal. **51**:10, 1954.
*Only cultures showing more than 50 colonies of these microorganisms are recorded here.
†Figures represent number of patients.
‡Patients in parentheses showed two of the species indicated, usually with *S. aureus*.

Table 3-6. Comparison of patients with conjunctivitis and normal eyes, with respect to antibiotic-producing organisms of the ocular flora

Cottage	Number of patients with					
	Normal conjunctivas			Conjunctivitis		
	Highly active*	Slightly active†	Total number of patients	Highly active	Slightly active	Total number of patients
Q	1	14	34	0	3	6
Y	2	6	33	0	3	7
S	1	9	21	0	9	19
Iota	0	6	20	0	4	20
H	2	11	23	1	8	17
Delta	3	17	2	1	14	18
Theta (last 3 visits)	4	28	48	18	19	69
Totals	13	91	201	20	60	156
	(6.5%)	(45.3%)		(12.8%)	(38.5%)	
	51.8%			51.3%		

From Halbert, S. P., Swick, L. S., Sonn, C., and Locatcher-Khorazo, D.: Ocular antibiotic-producing strains of the ocular flora, Arch. Ophthal. **51**:10, 1954.
*Patients revealing any highly active strains; some had both highly and slightly active organisms.
†Patients revealing any slightly active strains; all also revealed nonactive strains.

showed some activity, whereas about 35% revealed inhibition zones of 2.0 mm or greater. During December 1952, however, only about 30% of all strains showed any inhibitory properties, and merely 3% were rather highly active. Despite the fluctuations with time, no clear differences could be observed between the group with conjunctivitis and the group with normal eyes. This lack of difference was also apparent when the number of individuals revealing *any* inhibitory strains were compared.

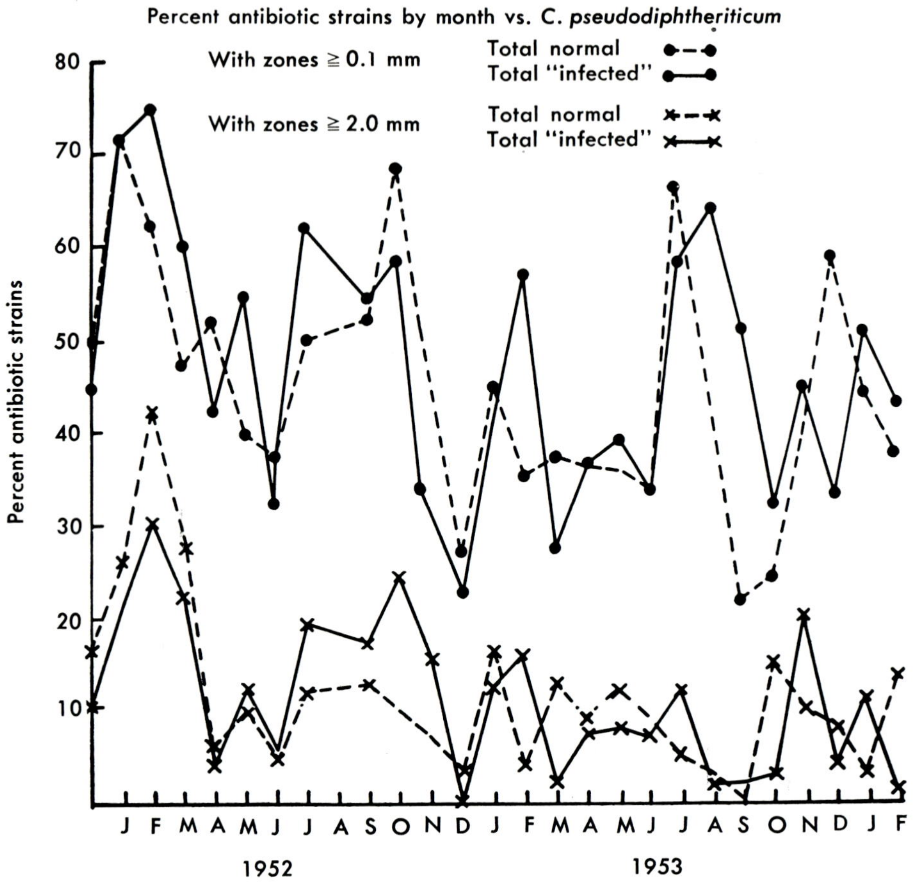

Fig. 3-7. Monthly incidence of antibiotic strains from the ocular flora during the period of observation. Each point represents the percentage of the total number of strains examined during the month. (From Halbert, S. P., Locatcher-Khorazo, D., Sonn-Kazar, C., and Swick, L. S.: Arch. Ophthal. **58**:72, 1957.)

IN VIVO SIGNIFICANCE OF INHIBITORY OCULAR MICROORGANISMS IN EXPERIMENTAL INFECTIONS

In an attempt to obtain evidence regarding the significance of these inhibitory strains in ocular resistance mechanisms, simplified experimental models were sought. Mixed infections were carried out in animals in which one member of the bacterial pair was lethally pathogenic and the other member was an ocular staphylococcus. Ocular staphylococci were chosen to represent strains both antibiotically active and nonactive against the pathogen. The virulent pathogens included a gas gangrene-producing organism (*Clostridium septicum*) and the diphtheria-producing *Corynebacterium diphtheriae*. Since superficial ocular infections would have been complicated by the presence of the normal flora of the animal itself, other routes of infection were employed.

In the case of mixed infections with *Clostridium septicum*, it was soon clear that the antibiotic activities do have in vivo biologic significance. This may be seen in Table 3-7, which summarizes the results of intramuscular infection of mice with this pathogen and mixtures of ocular staphylococci of varying antibiotic potencies. The protection against fatal outcome was

Table 3-7. Mixed infections of *Clostridium septicum* and ocular staphylococci of varying antibiotic activities in experimental infections of mice by the intramuscular route

Antibiotic group	*Micrococcus* strain*	Inhibition of *Clostridium septicum* (mm)†	Results of *Clostridium septicum* challenge with					
			10⁻⁴‡			10⁻²		
Highly active	117-3	7.1	S	S	S	S	S	S
	K1328-1	7.0	D4§	S	S	D2	S	S
	96-6	6.8	S	S	S	D3	S	S
	K88-8	5.9	S	S	S	S	S	S
	188-4	5.5	S	S	S	D2	D2	S
	K21-4	5.5	D2	S	S	D2	S	S
	K610-18	5.2	D1	D1	D1	D1	D1	D1
	K618-6	4.6	S	S	S	D2	D4	S
	455-1	4.3	D4	S	S	D4	D4	S
	K102-11	4.0	S	S	S	D1	D2	S
	K609-16	3.9	D2	S	S	D2	D2	D2
	K423-11	2.7	D2	S	S	D2	D2	D4
Slightly active	K1745-5	2.4	D1	D1	D1	D1	D1	D1
	K462-2	1.5	D2	S	S	D2	D2	S
	K36-2	1.4	D1	D2	S	D1	D4	S
	K1750-5	1.9	D1	D1	D2	D1	D1	D1
	K1771-11	0.7	D1	D2	S	D1	D1	D1
	100-5	2.9‖	D1	D1	D1	D1	D1	D2
Nonactive	191-4	0	D1	D1	D1	D1	D1	D1
	431-5	0	D1	D1	D1	D1	D1	D1
	K887-9	0	D1	D1	D1	D1	D1	D1
	K1115-19	0	D1	D1	D1	D1	D1	S
	81-5	0	D1	D1	D1	D1	D1	D1
	166-4	0	D1	D1	D1	D1	D1	D1
	303-3	0	D1	D1	D1	D1	D1	D1
	127-1	0	D1	D1	D1	D1	D1	D1
	K619-5	0	D1	D1	D1	D1	D1	D1
	K-1101-6	0	D1	D1	D1	D1	D1	D1
	K420-2	0	D1	D1	D1	D1	D1	D1
	K582-7	0	D1	D1	D1	D1	D1	D1
Controls	—	—	D1	D1	D1	D1	D1	D1
			D1	D1	D1	D1	D1	D1
			D1	D1	D1	D1	D1	D1

From Halbert, S. P., Sonn, C., and Swick, L.: Mixed bacterial infections in relation to antibiotic activities: *Clostridium septicum*–micrococcus infections, J. Immun. **73:**172, 1954. © 1954, The Williams and Wilkins Co.
*The micrococcal inocula averaged about 300 million organisms (as judged by colony counts), or 0.3 mg of bacterial cell solids. All suspensions were prepared to roughly similar concentrations.
†Millimeter width of inhibition zone of *Clostridium septicum* produced by the living micrococcal strains, assayed at the time of infection.
‡These dilutions of *Clostridium septicum* spores represented inocula of 10–100 LD_{50} and 1,000–10,000 LD_{50}, equivalent to 1,000 and 100,000 spores respectively.
§S = survived 7 days; D1 = died first day, etc.
‖This zone was very poorly demarcated.

clearly apparent when the latter organism was highly active antibiotically against the *Clostridium* strain. The complete lack of protective effect of the noninhibitory strains was also clear, as was the small increase in survival when the ocular staphylococci were slightly active.

It was also evident that this protective property was much more effective when the staphylococcal partner in these mixed

Table 3-8. Effect of heat-killed and living ocular staphylococci in mixed intramuscular infections with *Clostridium septicum* in mice

Group	Dilution of *Clostridium septicum* spore suspension											
	Experiment 1				Experiment 2				Experiment 3			
	10^{-3}	10^{-4}	10^{-5}	10^{-6}	10^{-3}	10^{-4}	10^{-5}	10^{-6}	10^{-3}	10^{-4}	10^{-5}	10^{-6}
Clostridium controls	D1	D1	D1	D1	D1	D1	D1	D1	D1	D1	D1	D1
	D1	D1	D1	D1	D1	D1	D1	D1	D1	D1	D1	D1
	D1	D1	D1	D2	D1	D1	D1	D1	D1	D1	D1	D1
	D1	D1	D1	S	D1	D1	D2	D2	D1	D1	D1	D1
	D1	D1	D1	S	D1	D1	D2	S	D1	D1	D1	D1
	D1	D1	S	S	D1	D1	D2	S	D1	D1	D1	S
	D6	S	S	S	D1	S	S	S	D1	D4	S	S
Clostridium + *Micrococcus* K88-8 (living)	S	S	S	S	D2	S	S	S	D2	D2	D4	S
	S	S	S	S	S	S	S	S	D4	D4	S	S
	S	S	S	S	S	S	S	S	S	D4	S	S
	S	S	S	S	S	S	S	S	S	D7	S	S
	S	S	S	S	S	S	S	S	S	S	S	S
	S	S	S	S	S	S	S	S	S	S	S	S
	S	S	S	S	S	S	S	S	S	S	S	S
Clostridium + *Micrococcus* K88-8 (heat-killed)	D1	D1	D2	S	D1	D3	S	D6	D1	D1	D2	D1
	D1	D1	S	S	D1	S	S	S	D1	D2	D2	D2
	D1	D1	S	S	D2	S	S	S	D1	D2	D4	S
	D2	D2	S	S	D4	S	S	S	D1	D2	D4	S
	D2	D2	S	S	S	S	S	S	D1	D2	S	S
	D2	S	S	S	S	S	S	S	D1	D4	S	S
	D2	S	S	S	S	S	S	S	S	S	S	S

From Halbert, S. P., Sonn, C., and Swick, L.: Mixed bacterial infections in relation to antibiotic activities: *Clostridium septicum*–micrococcus infections, J. Immun. **73**:175, 1954.

infections was living. This is shown in Table 3-8, which indicates that heat-killed staphylococci were much less protective than were the living organisms. Such data thus indicated that the staphylococcus must have been secreting an antibiotic in the in vivo environment and that this secretion inhibited the growth of the clostridia in the tissues.

That this was actually the case was demonstrated when an antibiotic from one of these ocular staphylococcal strains was partially purified and tested for its protective capacity against the *Clostridium septicum* infection. The high potency of this agent is illustrated in Table 3-9, where it may be seen that a dose of as little as 0.1 mg given intramuscularly at the infection site completely protected against as much as 1,000 LD_{50} doses of the virulent organism. A dose of the antibiotic of 1 mg given intraperitoneally also protected most of the animals against this size of challenge infection. In view of this data it seems clear that the active protective agent in the mixed infections is the staphylococcal antibiotic produced in vivo. If similar production of such antibiotics occurs in vivo on the surface of the ocular tissues, as seems reasonable, it is fair to suspect that such substances and activities are of considerable significance in superficial resistance mechanisms of the eye.

Even when the ocular staphylococcal antibiotic was administered at varying times following the infection dose, definite therapeutic effects could still be observed. The data from such an experiment are summarized in Table 3-10. It may be recalled that this infection system is usually fatal within 18 to 24 hours (Table 3-7), and a 6-hour delay in starting treatment repre-

Table 3-9. Prophylaxis of *Clostridium septicum* infections of mice by antibiotic concentrate derived from an ocular staphylococcus

Group	Antibiotic dose (mg)	Dilution of *Clostridium septicum* spores			
		10^{-3}	10^{-4}	10^{-5}	10^{-6}
Controls	—	0/18*	1/18	2/27	18/22
Staphylocococcal antibiotic,	1.0	8/8	9/11	8/8	7/8
locally intramuscular	0.5	8/8	11/11	11/11	8/8
at infection site	0.2	8/8	8/8	8/8	8/8
	0.1	8/8	7/8	8/8	8/8
Staphylococcal antibiotic,	8.0	8/8	11/11	8/8	7/8
intraperitoneal	4.0	7/8	9/11	11/11	8/8
	2.0	6/8	8/11	9/11	8/8
	1.0	6/8	4/8	7/8	8/8

From Halbert, S. P., Sonn-Kazar, C., and Swick, L.: Mixed bacterial infections in relation to antibiotic activities: the role of staphyloccal antibiotic in a *Clostridium septicum–Staphylococcus albus* infection, Antibiot. Chemother. **7**:237, 1957.
*The figures represent the number of survivors/total number injected.

Table 3-10. "Therapeutic" activity of staphylococcal antibiotic in *Clostridium septicum* infection

Group	*Clostridium septicum* spore dilution	Time (hours) following infection when antibiotic given			
		0	2	4	6
Staphylococcal antibiotic, intraperitoneal:	10^{-4}	14/15*	14/15	13/15	11/15
8 mg/0.5 ml	10^{-5}	14/15	15/15	15/15	13/15
Staphylococcal antibiotic, intramuscular,	$10\ ^{4}$	15/15	6/15	7/15	4/15
locally at infection site: 1 mg/0.05 ml	10^{-}	15/15	11/15	12/15	11/15
Controls	10^{-3}	0/15			
	10^{-4}	0/15			
	10^{-5}	3/12			
	10^{-6}	7/10			

From Halbert, S. P., Sonn-Kazar, C., and Swick, L.: Mixed bacterial infections in relation to antibiotic activities: the role of staphyloccal antibiotic in a *Clostridium septicum–Staphylococcus albus* infection, Antibiot. Chemother. **7**:237, 1957.
*The figure represents the number of survivors/total number injected.

sents a relatively long incubation period. In spite of this, large numbers of animals were cured of roughly 10 or 100 LD_{50} when the antibiotic was given locally or intraperitoneally 4 or 6 hours following infection.

The surprising potency of this type of antibiotic secreted by an ocular staphylococcus was apparent from comparative tests with clinically valuable antibiotics, such as penicillin and oxytetracycline. Some in vitro tests are summarized in Table 3-11, which shows that the staphylococcal antibiotic is somewhat more potent than either penicillin or oxytetracycline in agar plate assays, producing inhibition zones at concentrations of 0.1 μg/ml against the four pathogen indicators utilized.

When such a comparison was made in vivo on a limited scale, it was found that the ocular staphylococcal antibiotic revealed protective properties in vivo of the same order of magnitude as oxytetracycline or penicillin, using *Clostridium septicum*

Table 3-11. Comparative in vitro potency of the partially purified antibiotic of an ocular staphylococcus, and several clinically valuable antibiotics

Indicator strain	Antibiotic	Inhibition zone (μg/ml)						
		1000	100	10	1	0.1	0.01	0.001
Clostridium septicum	*Staphylococcus* K88-8	9.7†	9.5	8.3	3.1	3.0*	0	0
	Penicillin G	17.5	15.8	12.2	5.9	0	0	0
	Oxytetracycline	18.3	15.7	11.7	7.4	0	0	0
	Streptomycin	3.3	0	0	0	0	0	0
Clostridium perfringens (III)	*Staphylococcus* K88-8	8.9	7.0	6.4	4.9	3.8	3.0*	0
	Penicillin	26.5	24.9	16.9	5.3	0	0	0
	Oxytetracycline	19.3	17.1	14.6	7.4	0	0	0
	Streptomycin	4.1	2.3*	0	0	0	0	0
Corynebacterium diphtheriae (*mitis*)	*Staphylococcus* K88-8	10.0	9.2	8.5	5.5	2.5	0	0
	Penicillin	14.3	13.0	10.4	6.4	0	0	0
	Oxytetracycline	15.0	11.5	6.1	tr	0	0	0
	Streptomycin	11.3	8.2	2.2*	0	0	0	0
Corynebacterium pseudodiphtheriticum (A6485)	*Staphylococcus* K88-8	7.9	7.0	6.6	5.0	2.5	0	0
	Penicillin	20.1	15.0	10.8	3.9*	0	0	0
	Oxytetracycline	10.4	9.5	5.2	2.4*	0	0	0
	Streptomycin	10.8	7.5	2.5	0	0	0	0

From Halbert, S. P., Sonn-Kazar, C., and Swick, L.: Mixed bacterial infections in relation to antibiotic activities: the role of staphylococcal antibiotic in a *Clostridium septicum–Staphylococcus albus* infection, Antibiot. Chemother. **7**:238, 1957.
*These inhibition zones were poorly demarcated.
†The figures represent the width of the inhibition zone from the edge.

Table 3-12. Comparison of the prophylactic effects in mouse *Clostridium septicum* infections of the ocular staphylococcal antibiotic with penicillin, oxytetracycline, and streptomycin

Clostridium septicum **challenge, intramuscular**	**Control infections**	**Antibiotic**	**Antibiotic dose/mouse (mg intraperitoneal)**					
			4	**2**	**1**	**0.5**	**0.25**	**0.13**
10^{-3}	1/14* (7%)	*Staphylococcus* K88-8	15/17 (90%)	16/22 (75%)	9/13 (70%)	2/5	—	—
		Penicillin	—	6/11 (55%)	3/14 (20%)	3/7	1/5	—
		Oxytetracycline	—	6/6	11/14 (80%)	7/8	3/5	3/5
		Streptomycin	—	0/8	2/8	—	—	—
10^{-4}	2/14 (14%)	*Staphylococcus* K88-8	18/20 (90%)	17/25 (70%)	6/13 (45%)	2/5	—	—
		Penicillin	—	7/11 (65%)	7/14 (50%)	5/8	2/5	—
		Oxytetracycline	—	5/6	13/14 (95%)	6/8	4/5	5/5
		Streptomycin	—	0/8	2/8	—	—	—
10^{-5}	2/14 (14%)							
10^{-6}	8/14 (57%)							

From Halbert, S. P., Sonn-Kazar, C., and Swick, L.: Mixed bacterial infections in relation to antibiotic activities: the role of staphyloccal antibiotic in a *Clostridium septicum–Staphylococcus albus* infection, Antibiot. Chemother. **7**:239, 1957.
*Number of survivals/number of mice injected (percent survivors).

infection as the challenge system. These data are shown in Table 3-12, which indicates that the staphylococcal preparation appeared to be somewhat more efficient than penicillin and slightly less so than oxytetracycline. The comparability of all three is quite clear, however, even with the relatively small number of animals utilized.

Similar results were obtained in mixed infections using diphtheria inoculations of guinea pigs as the test system. When highly active ocular staphylococci were mixed with the subcutaneous challenge suspension of *Corynebacterium diphtheriae*, striking protection was usually obtained. All of the animals given nonantibiotic ocular staphylococci in the mixture succumbed

Table 3-13. Mixed *Staphylococcus*-diphtheria infection in guinea pigs (subcutaneous)

Ocular *Staphylococcus* component	Antibiotic activity of *Staphylococcus* vs. *Corynebacterium diphtheriae*	Antibiotic group	Results of mixed infections		*Staphylococcus* controls, survivors/total injected
			Survivors*/total injected	Percent	
None, diphtheria alone	—	—	0/20	0	—
None, diphtheria antitoxin	—	—	18/20	90	—
K88-8	8.4†	—	11/11	100	2/2
K609-16	8.3	—	7/8	90	4/4
K102-11	8.1	—	7/7	100	2/2
117-3	8.1	I	16/16	100	8/8
K423-11	7.4	—	7/7	100	2/2
K618-6	7.2	—	7/8	90	3/4
K17-5	7.0	I	10/12	85	6/6
K610-18	6.9	—	8/8	100	4/4
431-5	6.5	V	0/8	0	4/4
188-4	6.4	I	8/8	100	3/4
100-5	6.0	I	7/8	90	4/4
455-1	4.4	II	9/12	75	6/6
290-7	4.1	IX	0/8	0	3/4
191-4	2.4	VII	0/16	0	10/10
224-4	2.1	III	0/8	0	4/4
212-1	2.0	—	0/4	0	2/2
81-5	1.0	VII	0/8	0	4/4
127-1	0	—	0/8	0	3/4
280-3	0	—	0/8	0	4/4
223-1	0	—	0/8	0	4/4
217-1	0	—	0/8	0	3/4
K887-5	0	—	0/3	0	—
K582-7	0	—	0/3	0	—
K595-1	0	—	0/3	0	—
K1758-2	0	—	0/4	0	2/2
K1084-17	0	—	0/3	0	—
K1796-1	0	—	0/4	0	2/2
K661-1	0	—	0/3	0	—
K609-1	0	—	0/4	0	2/2

From Halbert, S. P., Sonn-Kazar, C., and Swick, L.: Mixed bacterial infections in relation to antibiotic activities: staphylococcal-diphtheria infections of guinea pigs, Arch. Ophthal. **57**:718, 1957.
*Survival time used was 14 days.
†The figures represent the zone of inhibition of the mitis strain used for infection. They are average values of a number of assays.

Table 3-14. Mixed diphtheria—ocular staphylococcal infections by the intratracheal route

Infective mixture	Results of mixed infections		Results of control staphylococcal infections	
	Experiment 1	Experiment 2	Experiment 1	Experiment 2
Diphtheria and buffer	D1 D2 D2 D2 D4	D1 D2 D2 D2	—	—
Diphtheria and antitoxin	D2 D2 S* S S	D13 S S S	—	—
Diphtheria and antibiotic *Staphylococcus* 117-3	D1 D2 D3 S S	D6 S S S	S S S S S	D2 S S S
Diphtheria and nonantibiotic *Staphylococcus* 127-1	D1 D1 D1 D1 D2	D1 D2 D2 D3	S S S S S	S S S S

From Halbert, S. P., Sonn-Kazar, C., and Swick, L.: Mixed bacterial infections in relation to antibiotic activities: staphylococcal-diphtheria infections of guinea pigs, Arch. Ophthal. **57**:720, 1957.
*S = survived one month.

like the controls. These data are shown in Table 3-13, which also shows that the slightly active organisms were without protective effect with this challenge diphtheria dose.

In an attempt to obtain results somewhat more analogous to the situation that might occur in the natural habitat of the organisms involved, mixed infections were carried out by the intratracheal route on a limited scale in guinea pigs. These data are summarized in Table 3-14, where it may be seen that mixture of the challenge dose of diphtheria organisms with antibiotic-producing staphylococci resulted in significant protection against fatal outcome, roughly comparable to that found with antitoxin treatment.

In summary, the accumulated data are strongly suggestive that the antibiotic properties of numerous organisms of the ocular flora have significance in the resistance of the eye to superficial bacterial infections and are another protective factor of this tissue against disease.

REFERENCES

1. Dujardin-Beaumetz, E.: Action antibiotique d'une variete staphylocoque a l'egard des bacilles gram-positifs et acido-resistants, Compt. Rend. Soc. Biol. **110**:1210, 1932.
2. Halbert, S. P., Locatcher-Khorazo, D., Sonn-Kazar, C., and Swick, L. S.: Further studies on the incidence of antibiotic-producing micro-organisms of the ocular flora, Arch. Ophthal. **58**:66, 1957.
3. Halbert, S. P., Sonn, C., and Swick, L. S.: Mixed bacterial infections in relation to antibiotic activities: *Clostridium septicum*–micrococcus infections, J. Immun. **73**:169, 1954.
4. Halbert, S. P., Sonn-Kazar, C., and Swick, L. S.: Mixed bacterial infections in relation to antibiotic activities: the role of staphylococcal antibiotic in a Clostridium septicum–Staphylococcus albus infection, Antibiot. Chemother. **7**:235, 1957.
5. Halbert, S. P., Sonn-Kazar, C., and Swick, L. S.: Mixed bacterial infections in relation to antibiotic activities: staphylococcal-diphtheria infections of guinea pigs, Arch. Ophthal. **57**:716, 1957.
6. Halbert, S. P., and Swick, L. S.: Antibiotic-producing bacteria of the ocular flora, Amer. J. Ophthal. **35**:73, 1952.
7. Halbert, S. P., Swick, L. S., and Sonn, C.: Characteristics of antibiotic-producing strains of the ocular flora, J. Immun. **70**:400, 1953.
8. Halbert, S. P., Swick, L. S., Sonn, C., and Locatcher-Khorazo, D.: Ocular antibiotic-producing bacteria in normal eyes and in conjunctivitis, Arch. Ophthal. **51**:7, 1954.
9. Scrivener, C. A., Myers, H. I., Moore, N. A., and Warner, B. W.: Some antagonistic activity of bacteria from the human oral cavity, J. Dent. Res. **29**:784, 1950.
10. Su, T. L.: Micrococcin: an antibacterial substance formed by a strain of micrococcus, Brit. J. Exp. Path. **29**:473, 1948.

4

Cytology of exudates and scrapings from normal and inflamed eyes

HISTORICAL NOTES

Cellular response reflects some degree of inflammation in the eye and often reveals the presence of infection. Cytologic examinations of smears from exudates and scrapings obtained from the conjunctivas, corneas, or eyelids may contribute to differential diagnosis in various ocular infections. Before bacteria were known as causative agents of disease, interest in the cells of ocular exudates existed, and Werneck[12,25] in 1839 wrote letters about the microscopic examination of exudates from normal and inflamed eyes to Dr. Florent Cunier, who later published these after Werneck's death.

In the early records on eye infections one finds that ocular smears and scrapings have been studied extensively. Leber[16] in 1879 wrote about a patient who had developed an ulceration of the cornea, and hypopyon, following an injury received from a small chaff of oats while working with a threshing machine. Microscopic examination of a smear from a light gray membrane over the ulcer, lifted out from the upper border, revealed packed masses of mycelial elements of a fungus that proved to be *Aspergillus glaucus*. Hirschberg and Krause[11] in 1881 obtained smears from the conjunctiva and urethra of an adult with gonorrhea, and, as may be observed in Fig. 4-1, the micrococci were present in or on both epithelial cells and polymorphonuclear cells. The Koch-Weeks bacillus was first described by Koch, who saw the organism in an exudate from a patient with trachoma. Kamen[15] in 1899 reported finding Koch-Weeks bacilli on or in epithelial cells (Fig. 4-2) from a patient with conjunctivitis caused by this organism. Schaudinn and Hoffmann[23] in 1905 identified *Treponema pallidum* as the etiologic agent of syphilis, and Aubineau[2] in 1907 described a chancre of the lower lid. Ulcer scrapings spread on a cover glass and stained with gentian violet showed many spirochetes (Fig. 4-3). Long before the etiologic agent of trachoma was isolated, Halberstaedter and von Prowazek[8] described small masses in the cytoplasm of epithelial cells found in conjunctival smears from patients with trachoma; these masses stained blue or violet with Giemsa stain. An example of one such inclusion is seen in Fig. 4-4. Axenfeld,[3] who contributed the definitive text on ocular microbiology in the first decade of this century, stated that "the mere examination of a film gives a great deal of information, and it is absolutely necessary, in dealing with external and internal disease, that the material should be collected at suitable stages in the disease." He also advised the exercise of greater caution in obtaining material from the cornea or infiltrated tissues, since scrapings can loosen the tissue, resulting in further spreading of the infection.

An early extensive study comparing the

presumptive diagnosis of bacterial infections of the eye, made from examinations of exudates and scrapings, with the diagnosis made by cultivation of the bacteria, is that reported in 1912 by Browning.[5a] One thousand consecutive cases of conjunctival infection were examined simultaneously by direct smear and by culture. The presumptive diagnosis made from smears was particularly accurate in infections caused by

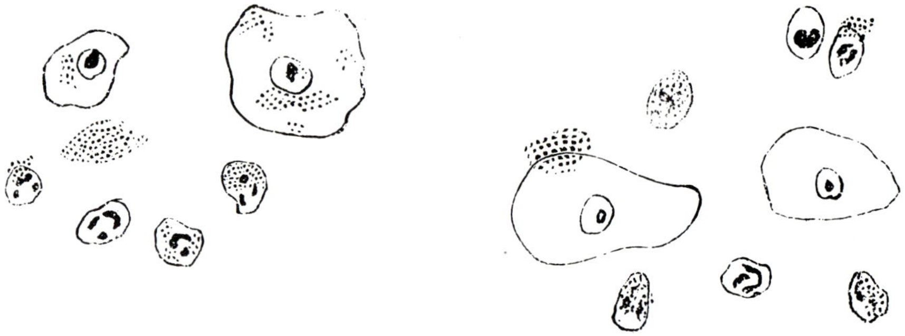

Fig. 4-1. Smears from a case of gonorrhea with eye infection in a male. The micrococci are seen in or on the epithelial and polymorphonuclear cells as well as lying free in smears from both the urethra (left) and eye (right). (From Hirschberg, J., and Krause, F.: Zbl. Augenheilk. 5:272, 1881.)

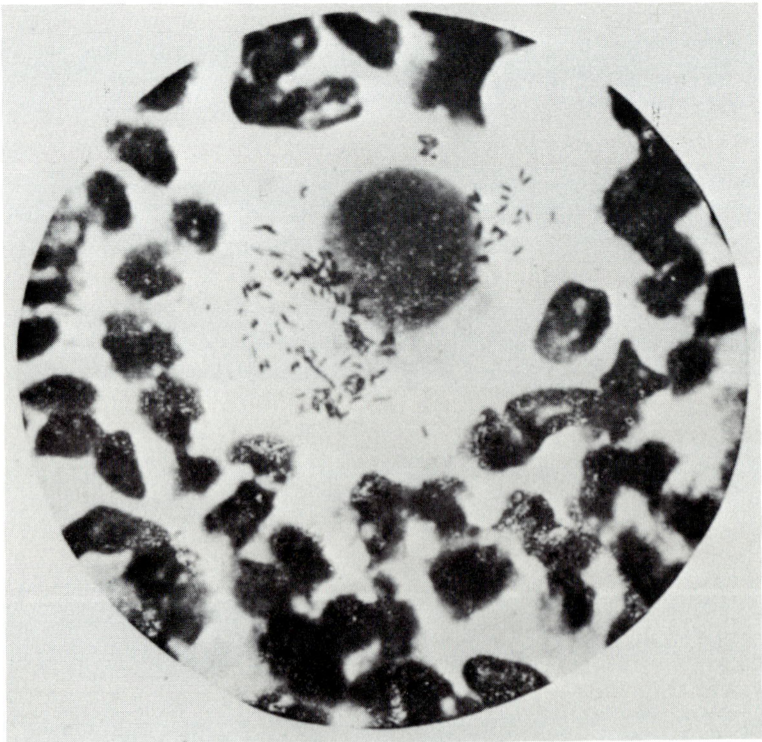

Fig. 4-2. Smear from exudate from a patient with conjunctivitis caused by the Koch-Weeks bacillus, showing numerous organisms on or in an epithelial cell. (From Kamen, L.: Zbl. Bakt. Parasit. Infekt. [Jena] 25:449, 1899.)

gonococci, pneumococci, streptococci, the bacillus of diphtheria, and *Diplobacillus pneumoniae* (*Klebsiella pneumoniae*). McKee[19] in the same year also reported on the usefulness of scrapings in cases of ophthalmia neonatorum. He found that although cultivation of pus from the conjunctiva of suspected gonococcal infections might give negative results, when a scraping was prepared in such a way that the

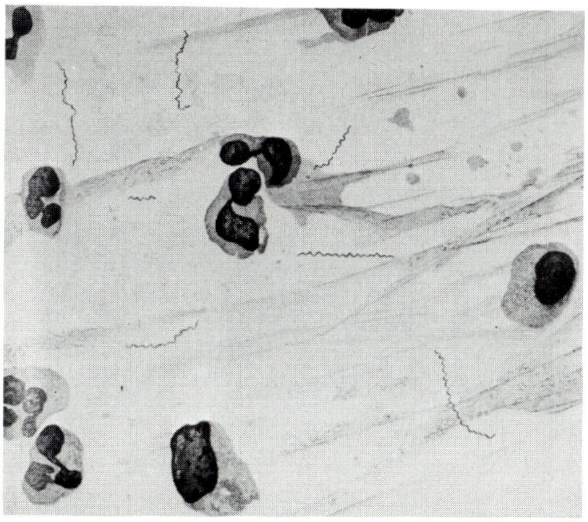

Fig. 4-3. Spirochetes and polymorphonuclear neutrophils in a smear from a scraping of a chancre of the lower lid. (From Aubineau, E.: Ann. Oculist. 138:16, 1907.)

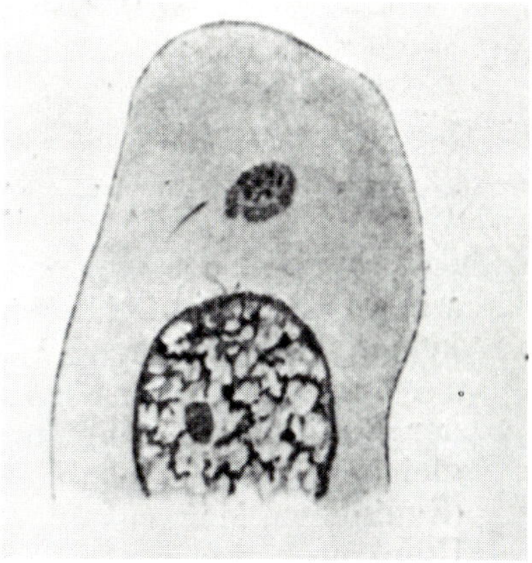

Fig. 4-4. Trachoma bodies seen in the cytoplasm of an epithelial cell in the first stage of the disease. (From Halberstaedter, L., and von Prowazek, S.: Deutsch. Med. Wschr. 33:1285, 1907.)

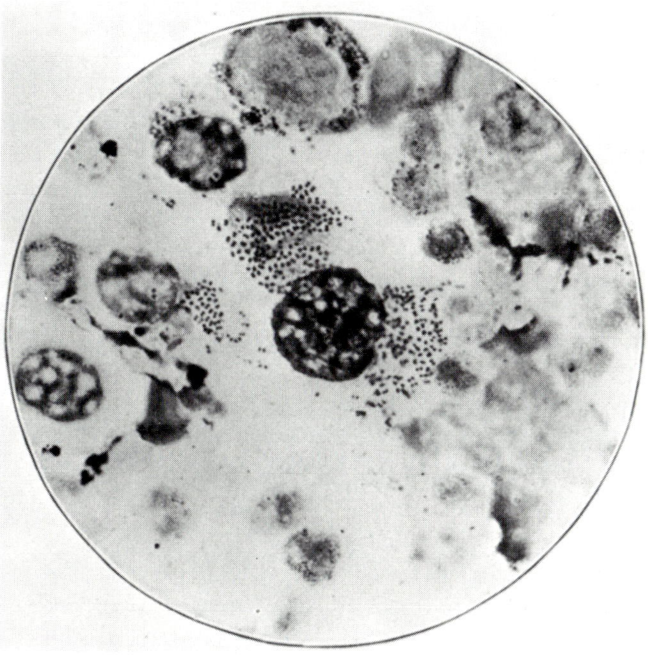

Fig. 4-5. A smear obtained from a case of ophthalmia neonatorum shows numerous gonococci in or on epithelial cells. (From McKee, H. A.: Ophthal. Rec. 21:5, 1912.)

epithelial cells were examined, masses of bacteria similar in morphology to gonococci were found. With the aid of an ear curette the palpebral conjunctiva was gently stroked, and the material was spread carefully over a glass slide and stained with Giemsa stain. Epithelial cells filled with biscuit-shaped diplococci, as shown in Fig. 4-5, were seen. The scraping containing the epithelial cells was cultured and yielded a profuse growth of gonococci. McKee concluded that the method for preparing films showing epithelial cells would demonstrate the presence of gonococci in many cases where smears of exudates yielded negative results; he did not commit himself to the relation of the microorganisms to the epithelial cells. McKee's illustrations indicate the possibility that the gonococci were colonizing in the epithelial cells.

In 1921 Lindner[17] reported that certain microorganisms directly attacked epithelial cells and in early stages of infection might be confined within the cells and be absent or sparse in the secretions until the infection was well established. Lindner considered microorganisms that thrive and multiply on living epithelial cells and confine themselves mainly to the superficial epithelium as epithelial parasites. He recognized only six such organisms: *Neisseria gonorrhoeae, Neisseria meningitidis, Diplococcus pneumoniae, Hemophilus Koch-Weeks, Hemophilus influenzae,* and *Corynebacterium diphtheriae.* Lindner believed the steps of conjunctival infection were as follows: the microorganisms localize on the normal conjunctival epithelial cells where they multiply and adhere. From this point the microorganisms spread over the conjunctiva as a result of the movements of the eyelids, establishing new foci of infection. The toxins that they produce diffuse into the tissue, and by virtue of the dissecting effect of edema these toxins may penetrate even further. In the second stage the epithelial layer is elevated and exfoliated, but the microorganisms are still on the surface. Fig. 4-6 illustrates a smear of an epithelial scraping from the bulbar conjunctiva in one of Lindner's patients who had pneumococcal conjunctivitis.

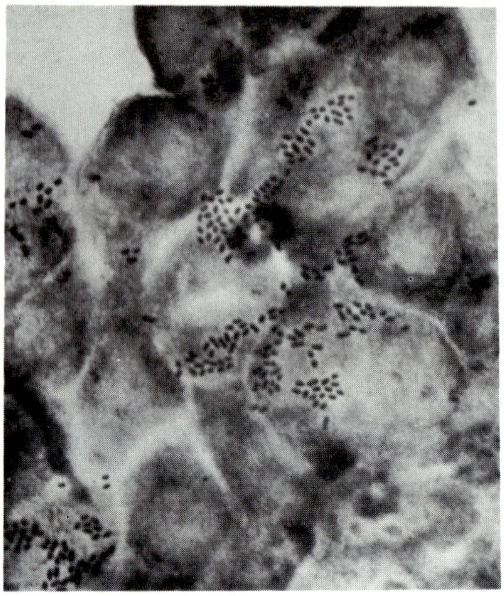

Fig. 4-6. Pneumococcal conjunctivitis, early stage. Epithelial scraping from the conjunctiva of the lower lid on the second day of infection in a 19-year-old girl. The scraping demonstrates the organisms on the epithelial cells. (From Lindner, K.: Graefes Arch. Ophth. **105**:758, 1921.)

Howard,[14] working in Lindner's laboratory, confirmed Lindner's observations; and while studying gonococcal, pneumococcal, and Koch-Weeks infections, he noted that the conjunctival epithelial cells had phagocytic characteristics (Fig. 4-7). The conjunctival epithelial cells not only phagocytized the bacteria but often immediately disgested them in the same manner characteristic of the leukocytes. He observed that the epithelial cells were not affected during phagocytosis and believed the phagocytic action of the young epithelial cells was probably one of the most important defense mechanisms against those pathogens that actively invade epithelium.

CYTOLOGY OF NORMAL OCULAR SCRAPINGS

The first recorded observations of the appearance of cells scraped from normal conjunctivas may have been those of Werneck.[25] He found in adult subjects epithelial cells in various stages of normal aging.

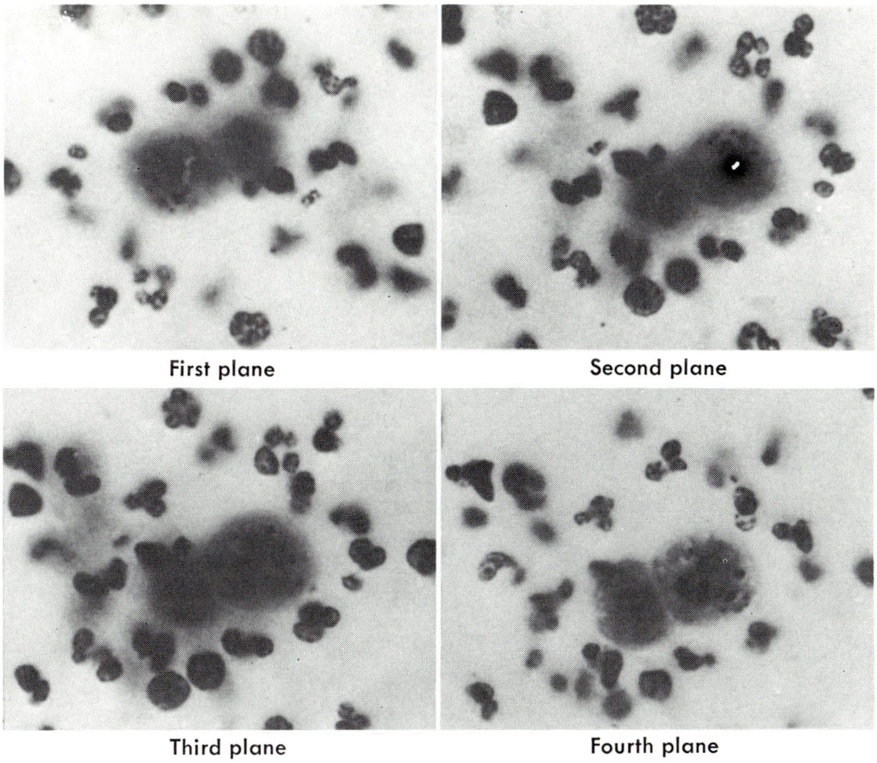

Fig. 4-7. Gonorrheal conjunctivitis in newborn, ninth day. Scrapings from palpebral conjunctiva. Phagocytosis by epithelial cell of 25 to 30 gonococci. Photographs taken in different planes show some microorganisms at different levels in the cell. (From Howard, H.: Amer. J. Ophthal. 7:909, 1924.)

Leukocytes were seen only in scrapings from inflamed eyes. Norn[21] recently has studied the cytology of conjunctival fluid in the newborn as well as in children and adults. Lymphocytes were not normally present in the newborn but were demonstrable in all children from 3 to 10 years of age; after this age the number of lymphocytes decreased gradually, and in the eye of the normal adult it was rare to see other than epithelial cells. Agarwal and Malhoutra[1] also have compared the appearance of conjunctival scrapings from normal individuals of various ages and have pointed out that the epithelial cells may be polyhedral as well as round. They found a small percentage of goblet cells that we, however, have not encountered in the normal eye. These authors also describe variations in the appearance of the epithelial cells associated with the menstrual cycle. De Azevedo[7] has made a comparison of the character of the epithelium of the normal eye as seen in histologic preparation with that seen in smears. Illustrations of the histology of the eye in health and disease are to be found in the textbook by Hogan and Zimmerman.[13] A description of the cells of the normal as well as diseased conjunctiva has been given by Liotet and Iris.[18]

In our experience with adults the normal epithelium from the conjunctiva is composed of cells that, when stained with Giemsa stain, show a single violet-colored nucleus surrounded by a light blue zone of cytoplasm (Plate 1, *A*). There is some variation in the size and shape of the epithelial cells, which may be in part because of the depth of the scraping. In addition to the normal epithelial cells, many scrapings con-

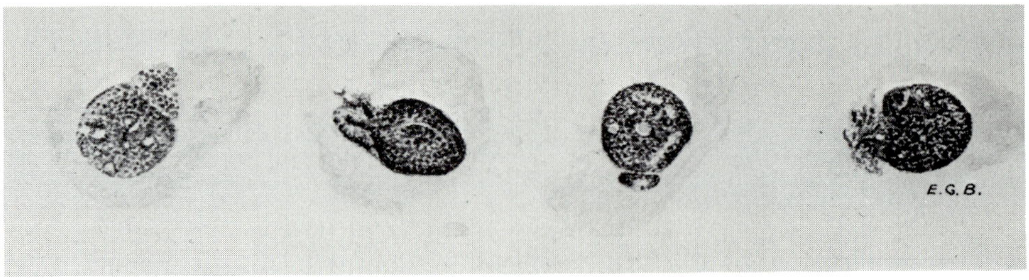

Fig. 4-8. Four epithelial cells showing a variety of so-called pseudoinclusion bodies, which are actually disrupted nuclear material sometimes seen in smears of scrapings from normal individuals.

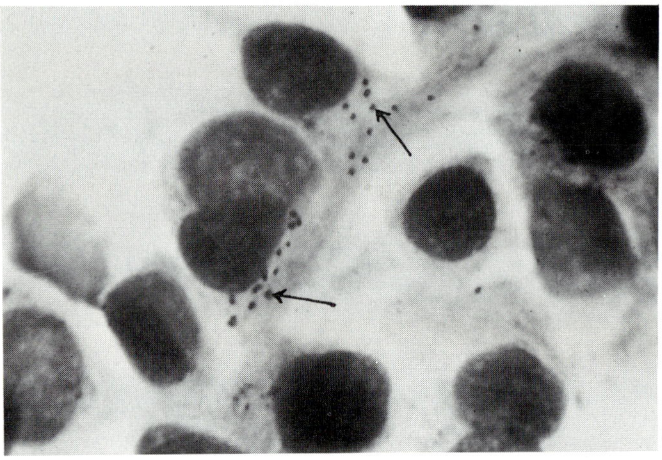

Fig. 4-9. Normal epithelial cells, some with "green granules" (arrows). (Giemsa stain.)

tain a few cells that show an abnormality caused by a fragmenting nucleus (Fig. 4-8). A superficial similarity to some viral inclusions has led to the designation pseudoinclusions, but this term does not indicate disease. Another variation of the normal epithelial cell is illustrated in Fig. 4-9, which shows intracellular granules that stain green with Giemsa stain. These occur most frequently in Orientals and people with dark complexions and are not associated with disease. Similar intracellular granules staining dark blue may be observed in normal epithelial cells. They sometimes are mistaken for viral cytoplasmic inclusions but are nonspecific and of no diagnostic value.

Goblet cells, though normal to the conjunctiva of the eye, are rarely seen in scrapings from the normal eye, since they occur mainly in the fornices. Plate 1, *B* illustrates these cells, which increase in numbers in keratoconjunctivitis sicca.

It is important in preparing scrapings of the conjunctiva to avoid obtaining material from the lid margin. Keratinized epithelial cells, which are abnormal when occurring in the conjunctiva, may be found normally on the lid margins. In addition, scrapings from lid margins may contain *Pityrosporum orbiculare,* staphylococci, and diphtheroids. The latter two organisms are present in the normal conjunctiva but usually not in sufficient numbers to be seen readily in conjunctival scrapings.

Structures identified as the female sex chromatin may be present in different ocular tissues. These structures were origi-

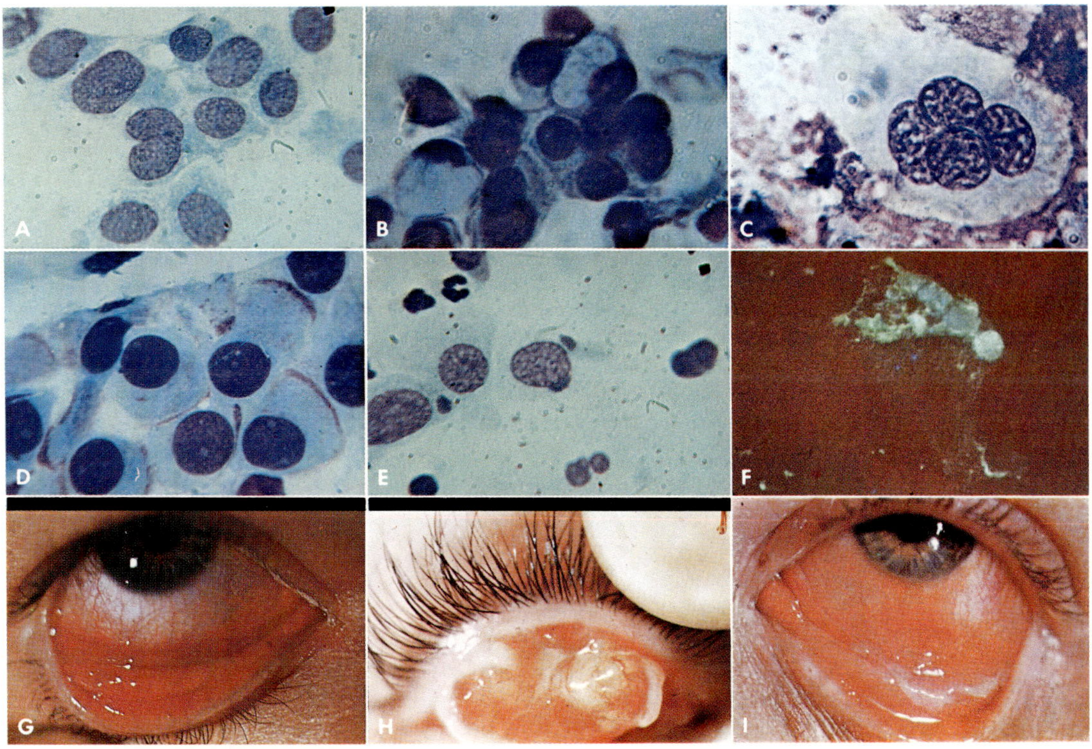

Plate 1. A, Normal epithelial cells. (Giemsa stain.) **B,** Two goblet cells with clear cytoplasm seen among epithelial cells in the early stage of keratoconjunctivitis sicca. (Giemsa stain.) **C,** Multinucleated epithelial cell from a scraping of a corneal ulcer caused by herpes simplex infection. (Giemsa stain.) **D,** Lavender keratin granules in epithelial cells in chronic conjunctivitis. (Giemsa stain.) **E,** Scraping of conjunctiva from a patient with acute trachoma. Two cytoplasmic inclusions in center epithelial cell and one in cell on left. (Giemsa stain.) **F,** Corneal scraping stained with fluorescein-labeled antibody to herpes simplex. The viral antigen is stained green. **G,** Acute bulbar conjunctivitis caused by *Staphylococcus aureus.* **H,** Membranous conjunctivitis caused by *Streptococcus hemolyticus.* **I,** Membranous conjunctivitis caused by *Corynebacterium diphtheriae.*

nally detected by Barr and Bertram[4] while studying accelerated nucleoprotein synthesis in the neurons of cats. They observed that these peculiar structures were intranuclear and lay adjacent to the large nucleoli in nerve cells of the cat. It has been shown that this chromatin mass was related to sex and could be demonstrated in the nuclei of most tissue cells from females, both human and animal, when stained with basic dyes or preferably with the Feulgen stain. Thus sections of tissues of unknown origin may be identified as to sex. Moore and Barr[20] made scrapings from the epithelial oral mucosa and detected the sex chromatin in the epithelial cell nucleus of females, whereas a similar chromatin mass was absent in males. Pedler and Ashton[22] have demonstrated that structures identified as the female sex chromatin are found in nuclei of conjunctival, ciliary, and corneal epithelium, in stromal cells of the iris and ciliary body, in the layers of the retina, and in the ganglion cells where the chromatin masses are well defined. They concluded that it is evident that sex may be determined as accurately from the epithelium of the cornea as from that of the skin. Fig. 4-10 shows sex chromatin masses in nuclei of corneal cells.

When trying to apply the Papanicolaou method of studying cytology to ophthalmology, de Azevedo[6] also found the presence of characteristic sex chromatin in the exfoliated epithelial cells of the conjunctiva. Conjunctival epithelial scrapings were made with a sterile de Wecker's iris spatula, and the material obtained was stained by Papanicolaou's method.

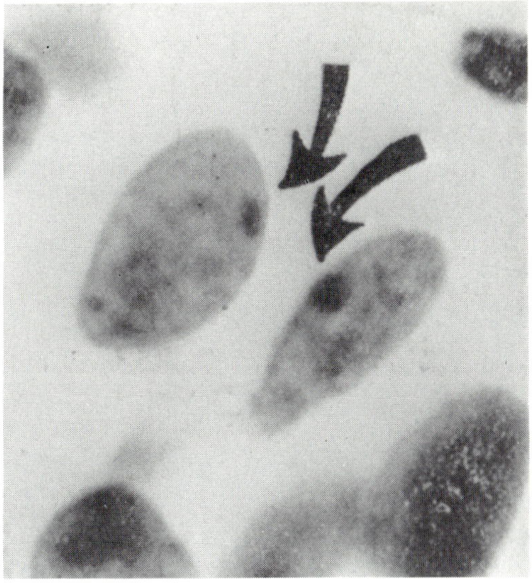

Fig. 4-10. Female sex chromatin mass (arrows) in nuclei of epithelial cells from a corneal scraping. (From Pedler, C., and Ashton, N.: Brit. J. Ophthal. **39**:364, 1955.)

Table 4-1 summaries some of the variations seen in epithelial cells of normal adult conjunctivas.

CELLS FOUND IN SMEARS AND SCRAPINGS IN INFLAMED CONJUNCTIVAS AND CORNEAS

In inflamed eyes the epithelial cells seen in scrapings from normal eyes are replaced in whole or in part by various polymorphonuclear leukocytes, lymphocytes, large mononuclear cells, plasma cells, and ab-

Table 4-1. Various intracellular granules found in epithelial cells seen in scrapings from normal eyes

Pseudoinclusion bodies	Intracellular free green pigment	Intracellular free blue granules	Sex chromatin
Extrusion of nuclear material into cytoplasm through a break in nuclear membrane	Present in cytoplasm in individuals with dark complexion	Present in cytoplasm in about 12% of smears from normal individuals; not to be confused with viral inclusions	Present in the nuclei of exfoliated epithelial cells; in females only

normal epithelial cells, a process giving expression to pathologic changes.

Polymorphonuclear neutrophils, because of their phagocytic ability, play an important role in resistance to infection and present the first line of defense. These cells are concerned primarily with phagocytosis of bacteria in the acute phase of infection. The ingestion of bacteria by polymorphonuclear cells is variable, however, since some virulent strains, such as group A *Streptococcus hemolyticus,* may be totally resistant to phagocytosis. The polymorphonuclear neutrophil exhibiting no phagocytic activity against spirochetes was well illustrated by Aubineau[2] in 1907 (Fig. 4-3); whereas Fig. 4-11 shows active phagocytosis by neutrophils in a scraping from a case of gonococcal conjunctivitis.

Even where there is rapid or complete phagocytosis by the polymorphonuclear cells, the infecting parasite may multiply within the leukocyte. Dead phagocytized bacteria lose their shape and staining properties.

Allergic reactions in the eye are notable for an exudate that usually contains *eosinophils* and sometimes *basophils* in addition to the usual polymorphonuclear neutrophils. For example, according to Thygeson,[23a] basophilic cells may be found in considerable numbers in one kind of allergic reaction, vernal catarrh, if the numbers of eosinophils are small. A few basophils also may be seen in some mild subacute ocular infections. The eye of a patient with follicular swelling caused by vernal catarrh is illustrated in Fig. 4-12.

Small *lymphocytes* are about the size of an erythrocyte or slightly larger with a single, sharply defined, round, dense violet nucleus surrounded by a varying amount of delicate, pale blue cytoplasm when stained with Giemsa stain. The cytoplasm, however, may form only a fine rim around the nucleus. Larger lymphocytes usually contain a more abundant zone of cytoplasm. Unlike the polymorphonuclear neutrophils, the lymphocytes do not exhibit phagocytosis.

Some acute infections of presumed viral origin are characterized by the presence of

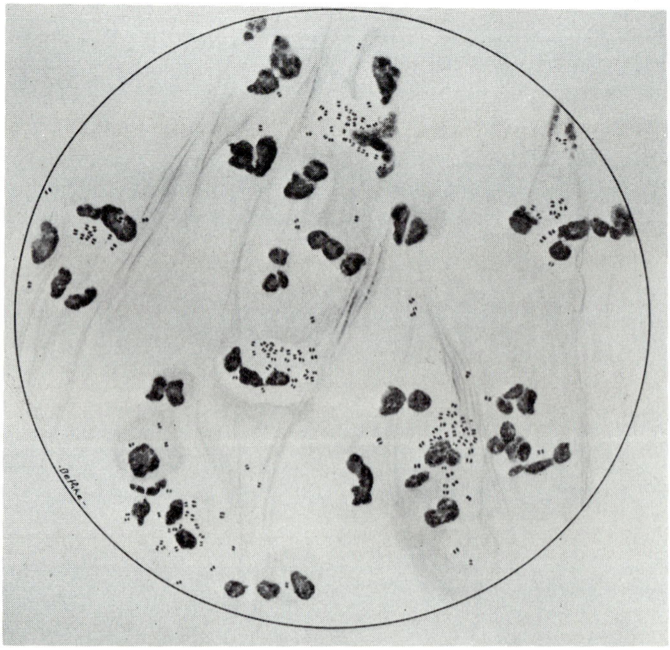

Fig. 4-11. Gonococcal conjunctivitis. Phagocytosis by polymorphonuclear cells.

lymphocytes (Fig. 4-13). Lymphocytes, usually in small number, are found also in the conjunctival scrapings from patients with various chronic ocular infections.

The large *monocyte* (macrophage) is the largest cell of the leukocytic series seen in ocular scrapings. It contains a single nucleus, which stains less deeply than the lymphocyte, and which may be round or irregularly oval. The zone of cytoplasm surrounding the nucleus of the monocyte is wider than that of the small lymphocyte. Mononuclear cells are phagocytic and engulf not only bacteria but also debris and whole cells. Monocytes are an important part of cellular reaction in ocular infections, and they may be the predominating cell in certain infections, such as pharyngoconjunctival fever. They are also found in the later stages of inflammation or infection.

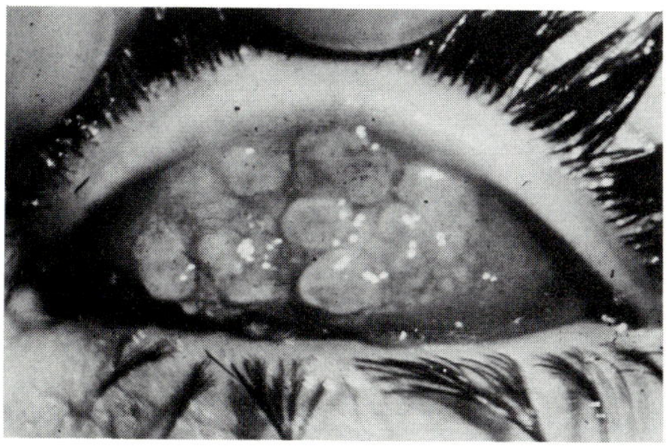

Fig. 4-12. Vernal conjunctivitis. Note follicular swelling.

Plasma cells, somewhat larger than lymphocytes, have a characteristic eccentric nucleus with distinctive distribution of chromatin, as shown in Fig. 4-14, *1.* The cytoplasm is opaque, abundant, and stains blue. These cells are found in chronic infections of the eye, notably in trachoma, where they occur in large numbers. Thygeson[24] reports that they are present in 40% of all scrapings from trachomatous eyes, whereas they are seen in only 0.5% of scrapings from nontrachomatous cases. So far as is known, the function of the plasma cells is the production of antibodies. Fig. 4-14 also shows an atypical large mono-

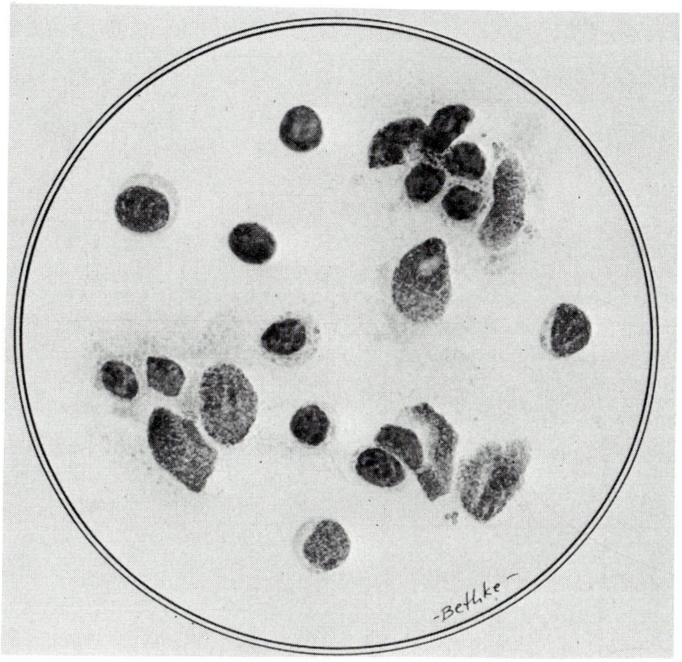

Fig. 4-13. Lymphocytes in conjunctival scrapings from a patient suspected of having a viral infection.

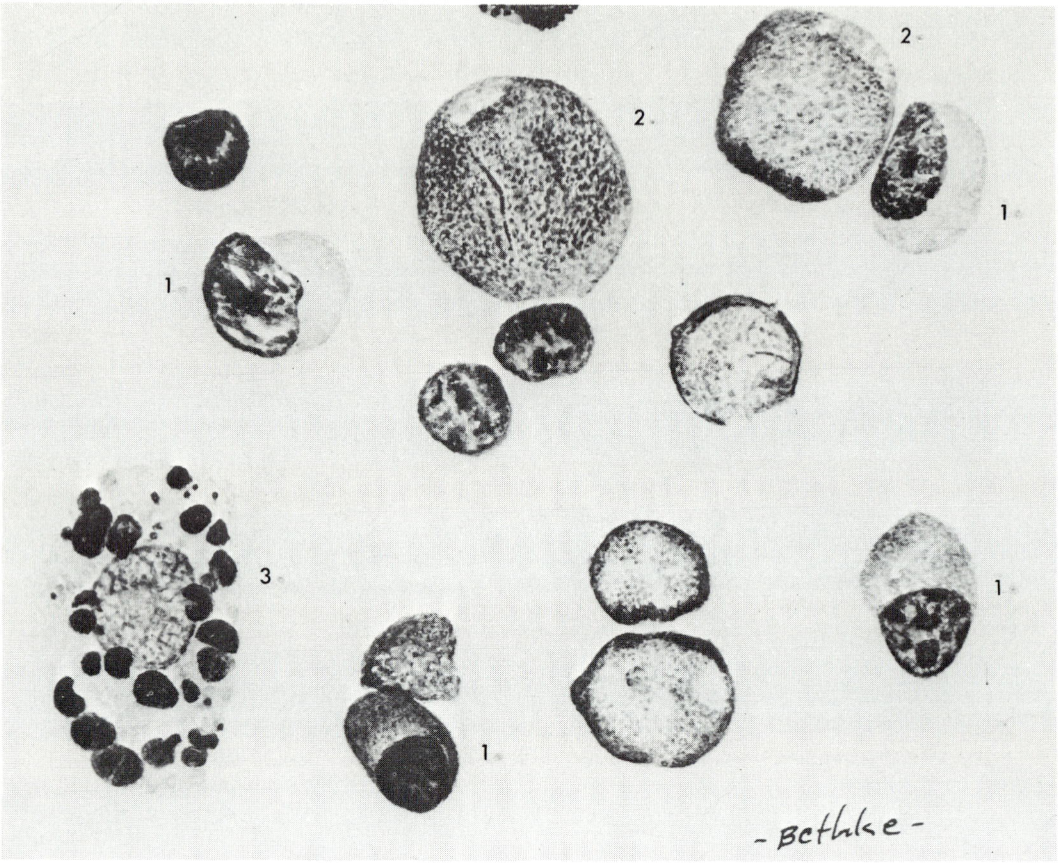

Fig. 4-14. Illustration of cells seen in a follicular expression from the conjunctiva in a case of trachoma. *1,* Plasma cells; *2,* large mononuclear cells; *3,* phagocytic macrophage.

nuclear cell (*2*) and a macrophage involved in phagocytosis (*3*).

The *multinucleated epithelial cell* is found in scrapings from eyes with certain viral infections, notably in eyes with herpetic corneal ulcers. Plate 1, *C* shows one such cell found in a corneal scraping from a patient with herpetic keratitis. Damaged epithelial cells may be found in conjunctival scrapings from cases of chronic inflammations of the eye, which are so varied as to include inflammation caused by vitamin A deficiency, keratoconjunctivitis sicca, and some chronic diseases of bacterial or viral origin. One such cell abnormality is that seen in keratoconjunctivitis sicca, where the cytoplasm contains keratin granules. Vacuoles in these epithelial cells may be few or numerous, and since they do not

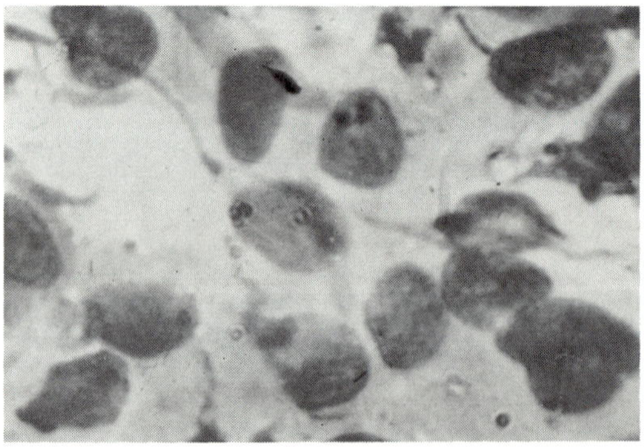

Fig. 4-15. Vacuolated epithelial cells seen in a smear of a conjunctival scraping from a patient with keratoconjunctivitis sicca. There are also strands of mucus.

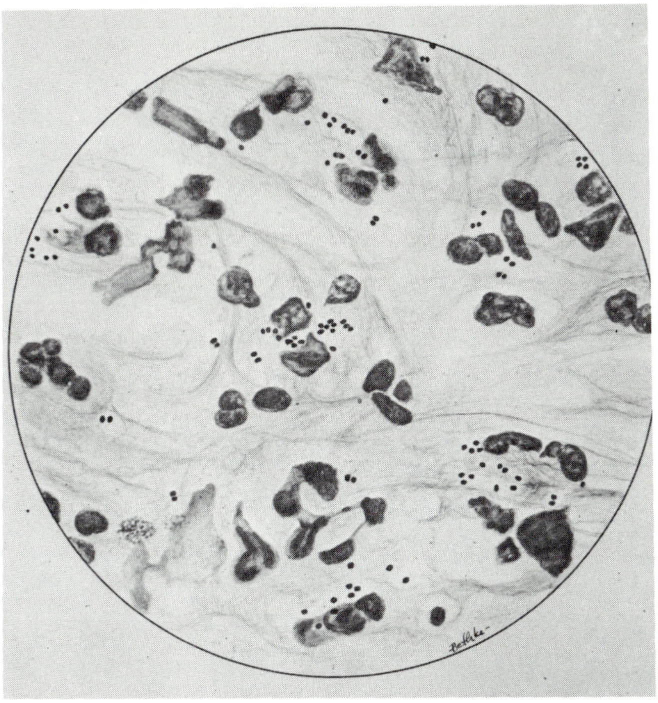

Fig. 4-16. Conjunctival scraping from a patient with conjunctivitis caused by *Staphylococcus aureus.* Note absence of phagocytosis. (Gram stain.)

stain, they appear as clear round spaces (Fig. 4-15). Cells containing keratin granules are also found in some long-standing cases of chronic conjunctivitis (Plate 1, *D*).

ACUTE BACTERIAL INFECTIONS OF THE CONJUNCTIVA AND CORNEA

Before the era of antimicrobic agents, acute bacterial infections ran a relatively protracted course, and much was to be learned from observations of cytologic changes during the disease. In the present era so many acute infections are readily controlled that few run their natural course, and some that were common years ago now are rarely encountered. Today one seldom sees a patient with ocular infection who has not already received antibiotic treatment prior to the laboratory visit. At this point cytologic examination of smears of exudates or scrapings may no longer be helpful. To a limited extent, however, cytology is still useful in the diagnosis of some acute infections of the conjunctiva and cornea.

Smears taken during acute bacterial infection are likely to show numerous polymorphonuclear cells, together with the microorganism involved. If the organism possesses a characteristic morphology, it can be provisionally diagnosed, and time may be saved by using an optimal antibiotic before cultures provide the final identification. In cases where special media are indicated, such as in tuberculosis, the *Moraxella* group, or *Corynebacterium diphtheriae,* morphologic findings may serve to provide the laboratory with a clue toward the selection of special media for cultures. Fig. 4-16 illustrates a gram-positive coccus that was seen in a conjunctival scraping from a patient who proved to have conjunctivitis caused by *Staphylococcus aureus.* There was no evidence of phagocytosis. Such a scraping as this is helpful, since it almost certainly indicates a *Staphylococcus* infection and thus permits immediate

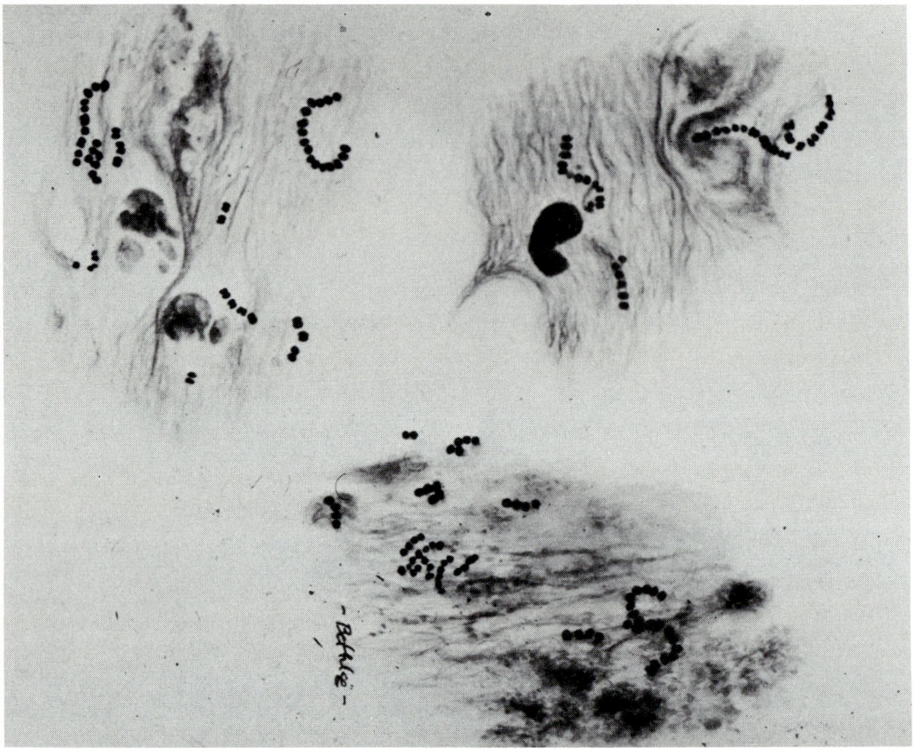

Fig. 4-17. Scraping from a membranous conjunctivitis caused by *Streptococcus hemolyticus*. (Gram stain.)

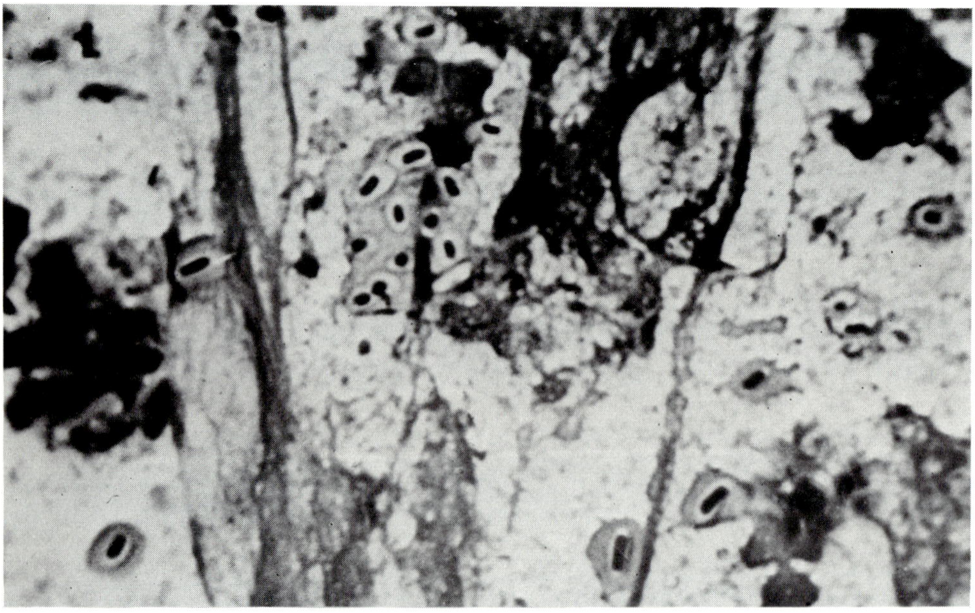

Fig. 4-18. Scraping from corneal ulcer. The encapsulated gram-negative rods proved to be *Klebsiella pneumoniae*. (Gram stain.)

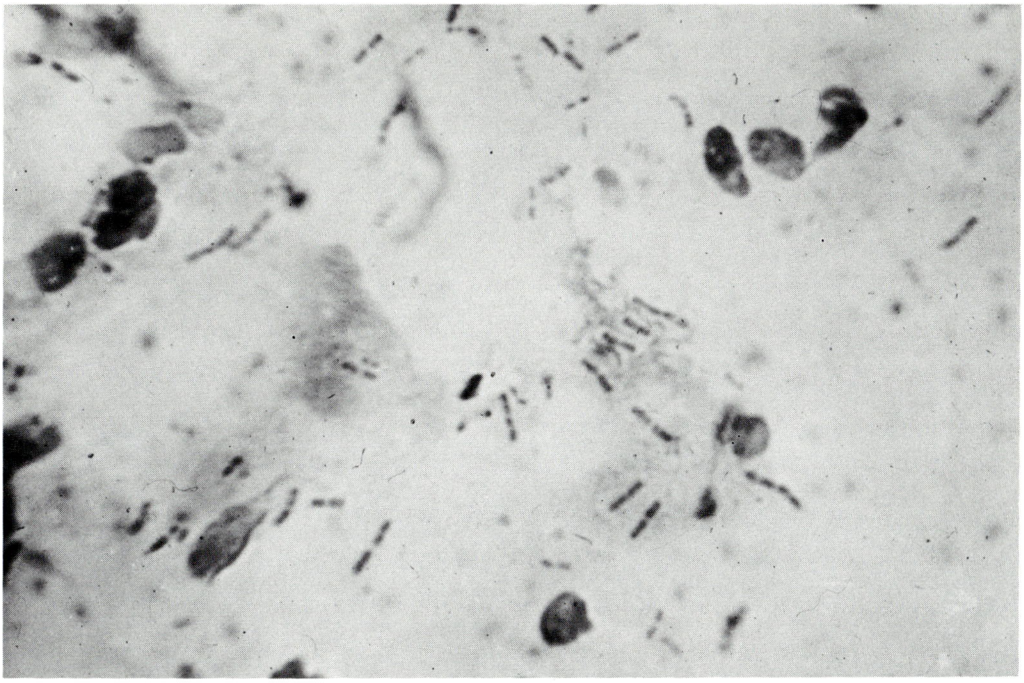

Fig. 4-19. Conjunctival scraping from a patient with angular conjunctivitis caused by the Morax-Axenfeld bacillus. (Gram stain.)

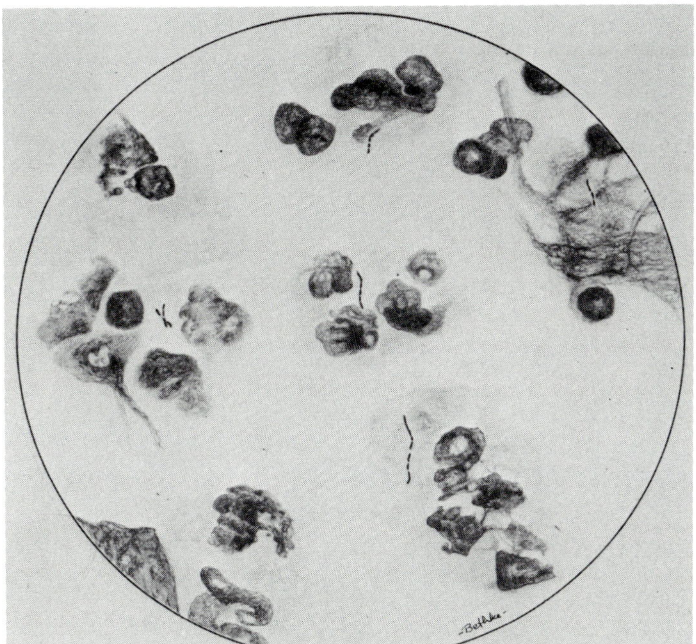

Fig. 4-20. Acid-fast bacilli in primary ocular tuberculosis. (Ziehl-Neelsen stain.) (From Bruce, G. M., and Locatcher-Khorazo, D.: Arch. Ophthal. **37**:375, 1947.)

treatment. Similarly, the presence of chains of gram-positive cocci suggests the probability of *Streptococcus hemolyticus* infection. Fig. 4-17 is a scraping from a membranous conjunctivitis caused by *S. hemolyticus* (see also Plate 1, *H* showing the eye from this patient). Similarly, the presence of gram-negative rods may suggest infection with an enteric pathogen. Fig. 4-18, a scraping from a corneal ulcer resulting from the encapsulated *Klebsiella pneumoniae,* is such an example. Antibiotic therapy should be started immediately, as gram-negative infections usually present the most serious therapeutic problems.

Not all gram-negative bacilli pathogenic for the eye are of enteric origin, for the diplobacillus of Petit (*Moraxella liquefaciens*), the Koch-Weeks bacillus (*Hemophilus aegypticus*), and the Morax-Axenfeld bacillus (*Moraxella lacunata*) are specific to the eye. Fig. 4-19 is a scraping from a patient with angular conjunctivitis caused by *Moraxella lacunata,* which shows the appearance of this gram-negative diplobacillus.

Primary tuberculosis of the eye is rare. Fig. 4-20 illustrates a smear of exudate from a tuberculous conjunctival ulcer; the smear has been stained by the Ziehl-Neelsen technique and shows scattered acid-fast bacilli (see Plate 2, *G,* showing the conjunctival ulcer from which the scraping was taken).

The examples of smears or scrapings from patients with acute bacterial infecions illustrate the variety of cellular response seen. It is evident that the cells are preponderantly polymorphonuclear cells but that some epithelial cells may be seen in the scrapings and occasionally a large mononuclear cell may be seen. The presence of the bacteria within or outside leukocytes can be of diagnostic value.

CHRONIC BACTERIAL INFECTIONS OF THE CONJUNCTIVA AND CORNEA

Chronic conjunctivitis is most frequently caused by staphylococci, members of the enteric groups of gram-negative bacilli, certain fungi, and occasionally a Vincent's infection. The cytology of chronic infections is characterized by a decrease in polymorphonuclear neutrophils and an increase in lymphocytes and large mononuclear cells. This cytology is not in itself pathognomonic of chronic infection nor does it make specific etiologic diagnosis, which depends upon the cultivation of the organism. However, there are a few important situations

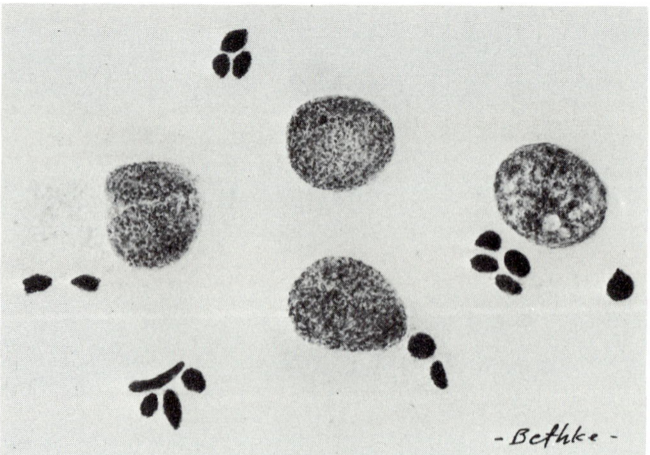

Fig. 4-21. A conjunctival scraping from a case of chronic conjunctivitis and meibomitis showing large gram-positive, oval, yeastlike organisms and a fragment of mycelium. This proved to be *Candida parapsilosis.*

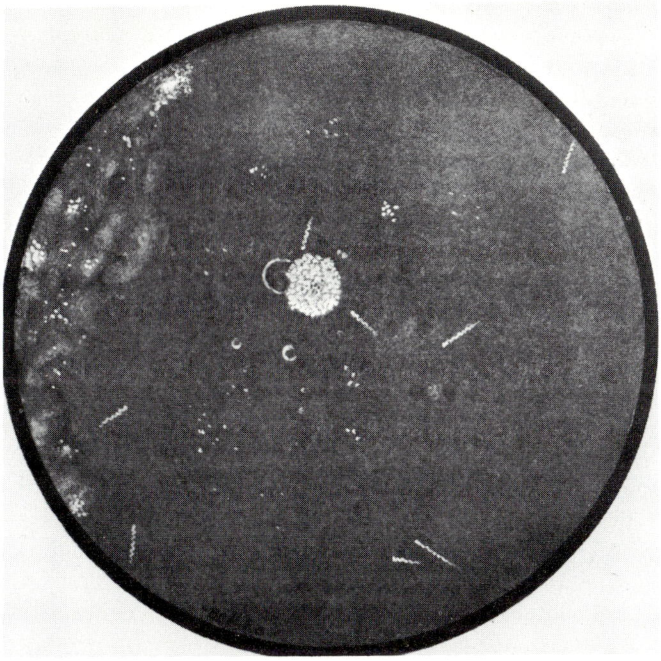

Fig. 4-22. A darkfield preparation of exudate from a patient with unilateral palpebral conjunctivitis with canalicular involvement caused by Vincent's infection. This field shows only the spirochetes. (From Burns, R. P., Macnie, J. P., Pfeiffer, R. L., and Locather-Khorazo, D.: Arch. Ophthal. **59**:236, 1958.)

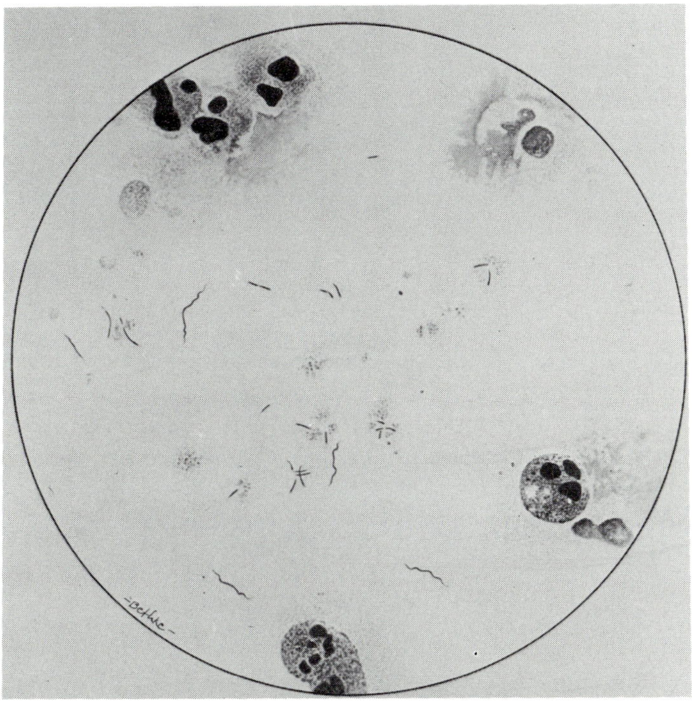

Fig. 4-23. A gentian violet stain of a smear from the same eye seen in Fig. 4-22. Fusiform bacilli as well as spirochetes can be observed. (From Burns, R. P., Macnie, J. P., Pfeiffer, R. L., and Locatcher-Khorazo, D.: Arch. Ophthal. **59**:236, 1958.)

where a provisional specific diagnosis is possible from the examination of smears. Fig. 4-21 shows a conjunctival scraping from a case of chronic conjunctivitis and meibomitis of 2 years duration. It proved to be caused by *Candida parapsilosis.* Large gram-positive yeastlike organisms lie free, and a fragment of mycelium is also present. These findings are diagnostic of a *Candida* infection. The same fungus was found in expressions of the meibomian glands. Smears for the recognition of mycotic infections are in general of great value. Certain rare chronic diseases of the eye may be diagnosed by such smears from scrapings. Fig. 4-22 is a picture of a darkfield fresh preparation showing spirochetes in exudate from a patient with unilateral follicular conjunctivitis and canaliculitis of 6 years duration, which proved to be caused by Vincent's infection. In Fig. 4-23, a gentian violet stain of a smear from the same case, fusiform bacilli as well as spirochetes may be identified. Since neither of these agents, which combine to produce Vincent's infection, are readily cultivated, the smear or fresh preparations examined by darkfield microscopy and the stained smear are of great diagnostic value.

CHLAMYDIA

Although acute and chronic chlamydial diseases can be identified best by cultural means, the presence of these agents causes the formation of cytoplasmic inclusion bodies, which may be recognized in scrapings. Plate 1, *E* shows a smear of a scraping from a case of acute trachoma in which two epithelial cells contain one and two small dark blue inclusions, respectively. The distribution of inclusion granules in epithelial cells follows a varied pattern that may be characterized by only a few scattered granules, or small clumps of granules, to massive deposits in the cytoplasm. These inclusion bodies are similar to those seen in inclusion conjunctivitis. They are difficult to find in chronic disease. Their reappearance indicates acute exacerbation. In contradistinction to viral infections, infections with *Chlamydia* produce an exudate rich in neutrophils and containing plasma cells (Fig. 4-14).[9,24]

VIRAL INFECTIONS

Conjunctival scrapings from patients with viral conjunctivitis show a variety of small and large mononuclear cells and lymphocytes. When no bacteria can be isolated to explain the conjunctivitis, it is assumed that it may be a chronic viral disease. Isolation of the virus is rare under such circumstances. The value of smears is largely that the character of the cells suggests the results of a previous viral infection. Fig. 4-13 illustrates the variety of cells that may be seen in chronic conjunctivitis free of bacteria. This scraping was obtained from a patient with chronic conjunctivitis whose past history and present bacteriologic status provided no explanation for the conjunctivitis, which therefore may have been of viral origin. No multinucleated cells were present, although such cells have been seen in about 170 cases of herpes simplex keratitis (Plate 1, *C*).

It is quite likely that the staining of scrapings with fluorescein-labeled antibodies specific to viruses known to cause viral infection may prove a useful adjunct in the diagnosis of some ocular viral diseases. Plate 1, *F* shows a scraping from the cornea of an 11-month-old child with acute keratoconjunctivitis; the scraping has been stained with fluorescein-labeled antibody to herpes simplex virus and was viewed in ultraviolet light. The viral antigen in the cell has bound the labeled antibody; and where this has occurred, blue-green color is seen and identifies the presence of the virus, thus confirming the presumptive clinical diagnosis.

NONINFECTIOUS INFLAMMATION OF THE EYE

In 1907 in India Herbert[10] examined ocular exudates in patients with spring catarrh; his diagnosis was largely based on the presence of eosinophils in the secretion. The greater the allergic reaction the higher the proportion of eosinophils found. It has since been learned that eosinophils may be observed in conjunctival epithelial scrapings as well as in smears of secretions. However, allergic inflammation of the eye, as in other areas of the body, can only be diagnosed with certainty when the specific allergen is found. This problem is discussed in Chapter 10.

Vernal conjunctivitis is the conjunctivitis attributed to an allergic response that was seen most frequently at the IOCPMC. This disease is characterized by a cobblestone type of follicular swelling, as is illustrated in Fig. 4-12. The scrapings contain epithelial cells, polymorphonuclear neutrophils, and eosinophils. This laboratory has examined ocular scrapings from 2,068 patients with allergic inflammation; 1,755 cases were diagnosed as vernal catarrh, 263 as hay fever, and 50 as allergic response to a variety of drugs or cosmetics. Neutrophils were present in all the eyes, eosinophils in 1,903, and basophils in only 74. In all 50 scrapings from patients with drug allergies no neutrophils were seen, and eosinophils were present only in the 16 cases caused by atropine. Other conditions in which eosinophils are sometimes found in the eye include ocular pemphigus, chronic trachoma, and chronic staphylococcus conjunctivitis.

Several factors may be responsible for keratinization of the epithelial cells of the eye: Sjögren's syndrome (keratoconjunctivitis sicca); nutritional disturbances, particularly vitamin A deficiency; exposure to radiation; age; or deficiency of lacrimal secretion, to mention some. The epithelium of both conjunctiva and cornea may be involved. The presence of keratinized cells should be considered abnormal when found in conjunctival or corneal scrapings, provided there is no contamination from the lid margins or caruncle. Numbers of goblet cells are found in keratoconjunctivitis sicca.

In mild degrees of vitamin A deficiency, keratinization of the epithelial cells is slight. A microscopic examination may indicate the possibility of deficiency in the early stages of vitamin A depletion. Clinical diagnosis includes three points: (1) wrinkling of the conjunctiva, (2) lack of luster, and (3) in some cases the development of Bitot's spots just outside the limbus. Bitot[5] in 1863 described these conjunctival spots, which are associated with night blindness (Fig. 4-24). These spots appear at the temporal limbus and are white, sharply defined, triangular patches; they are grossly visible and covered with a silver-gray foamlike substance, which is readily scraped off and which re-forms and may appear like

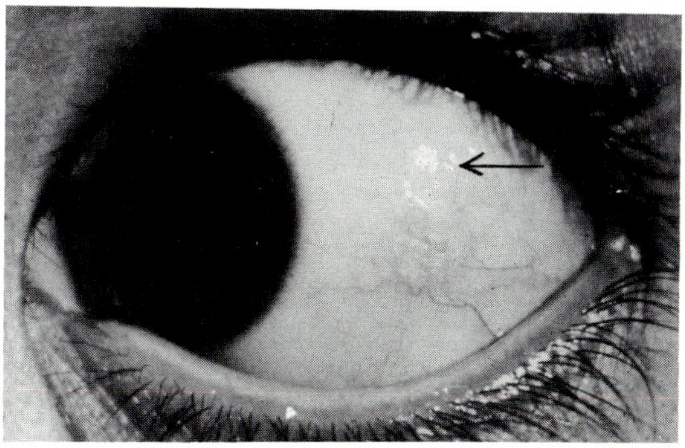

Fig. 4-24. Patient showing Bitot spots (arrow), small areas of keratinization. (Courtesy P. Thygeson.)

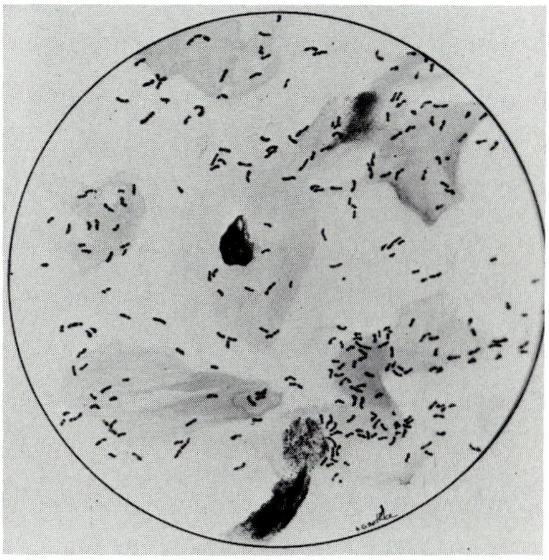

Fig. 4-25. Epithelial scrapling of conjunctiva of a patient with vitamin A deficiency. Note diphtheroids apparently growing on the dead cells. (From Thygeson, P.: Amer, J. Ophthal. **29**:1504, 1946.)

Table 4-2. Cytology observed in smears of exudates and scrapings in various ocular diseases

Disease	Cytology
None	Normal epithelial cells. Some may contain pseudoinclusion bodies or granules. A rare goblet cell.
Acute bacterial infections	Watery or mucopurulent discharge containing predominantly polymorphonuclear neutrophils. Bacteria may be identified.
Chlamydia infection	Cytoplasmic inclusion bodies in epithelial cells. Neutrophils, lymphocytes, plasma cells, macrophages.
Acute viral infection	Lymphocytes and large mononuclear cells. Some multinucleated epithelial cells.
Chronic bacterial, mycotic, and mycobacterial infections	Decrease in polymorphonuclear neutrophils and the appearance of some lymphocytes and large mononuclear cells. Fungi may be identified, also acid-fast bacilli.
Chronic infection of known or possible viral origin	Large mononuclear cells. Lymphocytes. Some multinucleated epithelial cells.
Allergies	Notable for numerous eosinophils and sometimes basophils.
Nutritional, physical, and chemical injuries, keratoconjunctivitis sicca, pemphigus	Squamous and/or keratinized epithelial cells, cicatrization. In vitamin A deficiency there may be numerous diphtheroids; in keratoconjunctivitis sicca, goblet cells and mucus; in pemphigus, occasional eosinophils.

plaques. They are more prevalent in children but may be present in adults. Fig. 4-25 shows scrapings from the bulbar conjunctiva of a patient with vitamin A deficiency. Note the epithelial cells covered with masses of diphtheroids. The relationship of Bitot's spots to vitamin A deficiency is under scrutiny.[15a,21a]

In several forms of trauma to the conjunctiva, such as allergic conjunctivitis, keratoconjunctivitis sicca, and vitamin A deficiency, the normal bacterial flora of the eye increases in volume, with the result that numbers of staphylococci and diphtheroids are seen in the scrapings.

A summary of the cellular changes found in ocular smears and scrapings in various types of inflammation is given in Table 4-2.

TECHNIQUE FOR PREPARING SMEARS FROM EXUDATES AND SCRAPINGS

Fig. 4-26 illustrates the tray of equipment needed for obtaining smears of exudates and scrapings from the conjunctiva, eyelid margin, and cornea. This tray is equipped with everything necessary to obtain the material at the bedside. If the patient is ambulatory, the scrapings are taken in the microbiology laboratory.

The three types of platinum spatulas employed for scrapings are standing to the left in front of the alcohol lamp and matches. The lamp is used to sterilize the spatula. Sterile gauze is in the jar in the shadow to the left. The spatula most frequently used is on the left; the "spade" in the center is sometimes helpful in recovering material from fungal infections and is preferred by some over the rounded spatula. The plastic bottle contains ophthaine 0.5%, for local anesthesia, and the glass holds the slips necessary for the laboratory records. Slides thoroughly cleaned with soap and water and alcohol are ready for use. Next to them in a glass test tube are small sterile swabs on toothpicks for recovering exudates. A wax pencil for writing on glass and a diamond glass-marking pencil, in the foreground next to two test tubes holding sterile cotton swabs in case of abundant amounts of exudate, are avail-

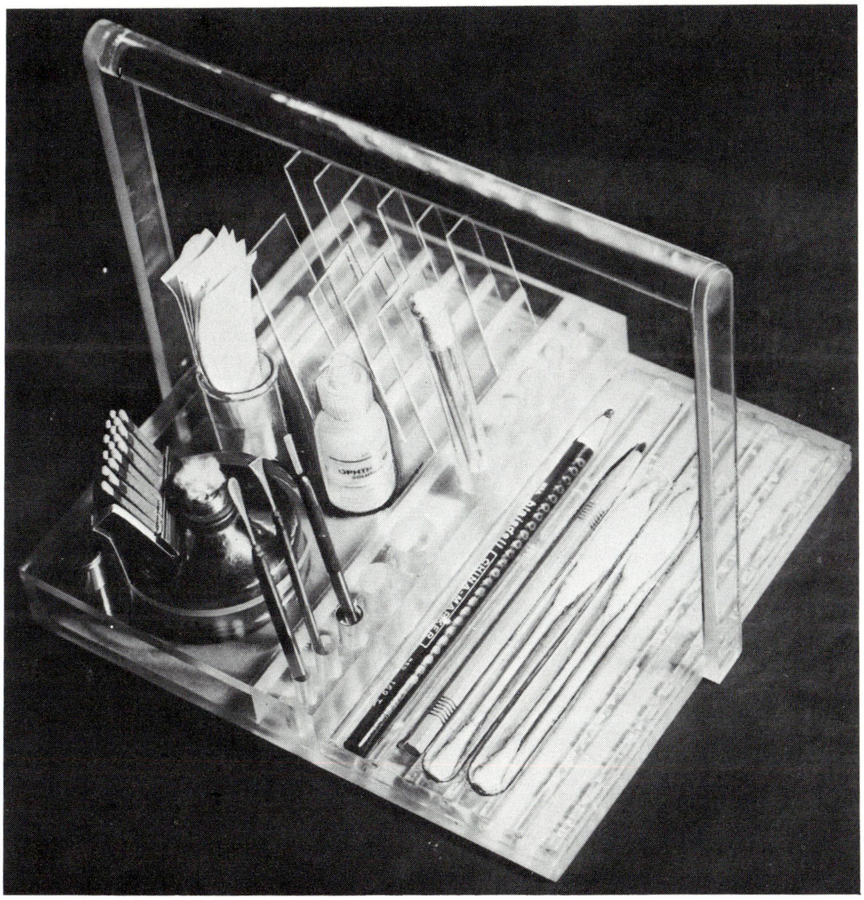

Fig. 4-26. Tray equipped for taking smears of exudates and scrapings from the eye.

able. The slides are marked with the diamond glass-marking pencil and carry the patient's name, the source of the material, such as OD, OS, C (for conjunctiva), K (for cornea), KU (for corneal ulcer), L (for lid), or any other information needed.

Fig. 4-27, *A* through *D*, illustrates the approach to taking ocular scrapings. The application of the material to the slide is most important. As indicated in Fig. 4-28, the spatula is gently rubbed over a small area of the slide. At least two slides are prepared. If the specimen is dry, a very small bit of water may be added to the slide to help spread the film more evenly. Much of the success for finding favorable fields for examination depends on obtaining a good distribution of the material without crushing the cells. In some instances, especially in suspected tuberculosis, after the smears are made the spatula may be placed in a few drops of broth. When the adhering material is loosened, the tube is put aside for centrifugation in case no acid-fast bacteria are found in the smears. The concentrated sediment from the centrifugate may be found positive for acid-fast bacilli. The slides containing the smears are returned to the rack to dry.

Usually a Gram stain is done immediately so that antibiotic therapy may be started if a presumptive diagnosis can be made.[7a] The Giemsa stain, a process that requires about 1¼ hours, is the stain of choice for histologic examination; and this stain, or some special stain indicated by the presumptive diagnosis, is used on the second

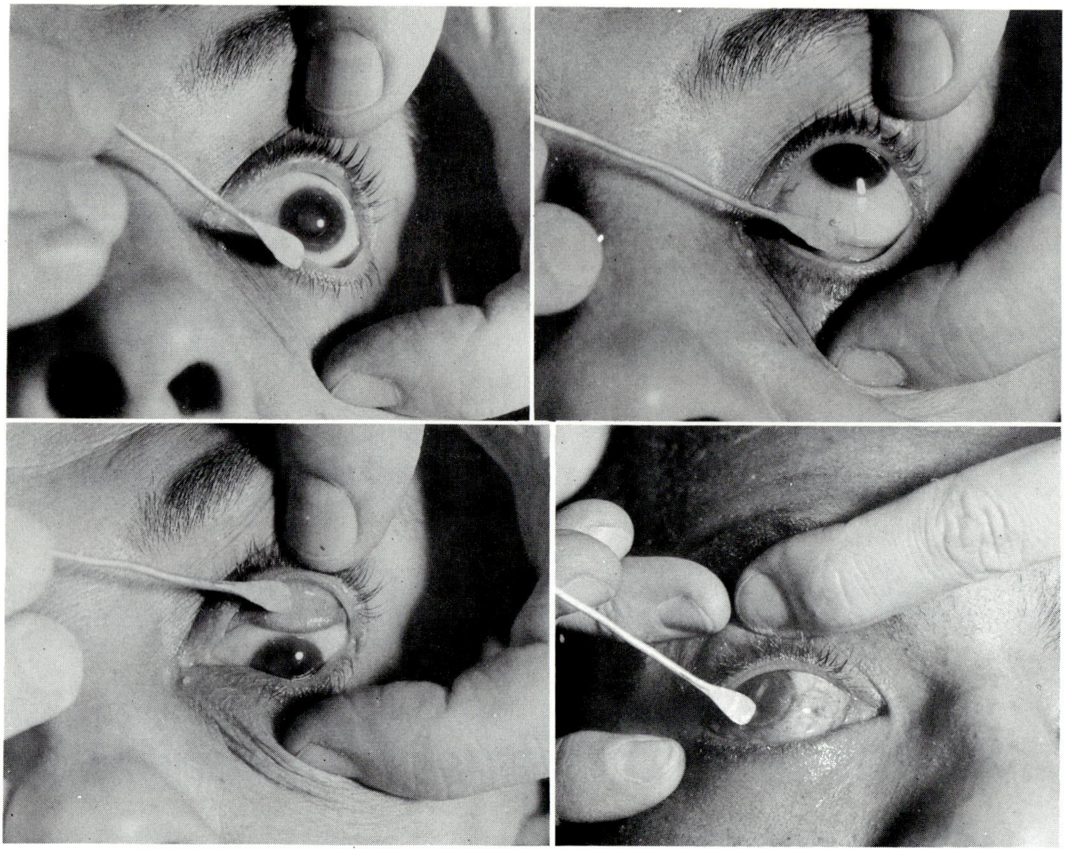

Fig. 4-27. A, Lid margin scraping; **B,** scrapings of the lower conjunctiva; **C,** scrapings of the upper conjunctiva; **D,** corneal scraping.

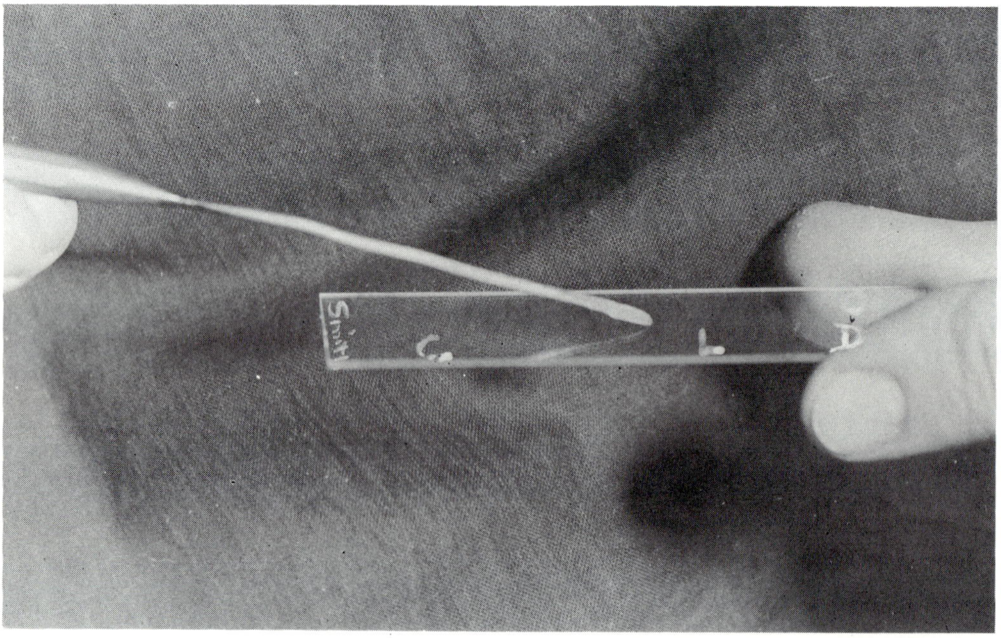

Fig. 4-28. The material obtained is spread on the slide.

slide. The Giemsa stain is made by fixing the slide in absolute methyl alcohol in a coplin jar for 5 minutes. The slide is then transferred to a jar containing freshly prepared Giemsa stain and placed in the incubator at 36° C for ½ to 1 hour. After passing through two changes of 95% alcohol, the slide is dried in the air and covered with emersion oil and a cover slip. The bacteria stain blue; the inclusions of trachoma or inclusion conjunctivitis appear deep blue; the eosinophils and their granules are pink; nuclear material and other miscellaneous granules are blue or green. In some laboratories Wright's stain is preferred.[7a]

SUMMARY

The examination of smears from exudates and scrapings offer the following diagnostic and therapeutic aids:

1. In the healthy eye only epithelial cells are found.
2. Acute infections of bacterial origin are characterized by numerous polymorphonuclear cells and often by the presence of the causative agent.
3. Although identification of the specific bacterium depends on its cultivation, this may be facilitated by preliminary inspection of the smear, after which special media may be selected.
4. Tentative recognition of the organism may indicate initiation of antibiotic therapy.
5. Trachoma or inclusion conjunctivitis can be diagnosed from the presence of epithelial cytoplasmic inclusion bodies.
6. The value of the smear from scrapings in acute viral infection lies in observing the lack of bacteria and the presence of lymphocytes.
7. The smears of scrapings obtained from cases of chronic trachoma or viral infection are often similar and show decreasing polymorphonuclear cells and increasing lymphocytes and monocytes.
8. Identification of fungal infection is possible.
9. Acid-fast rods, mycobacteria, can alert one to the possibility of tuberculosis.
10. Eosinophils are most frequently associated with an allergic disease.
11. Smears showing keratinized epithelial cells arise from a variety of physical and chemical injuries, malnutrition, cicatrization, and keratoconjunctivitis sicca.

It thus may be seen that the routine examination of smears from exudates and scrapings is well worth the extra time and care required, since diagnosis is often facilitated and therapy accelerated because of the information that such an examination yields.

REFERENCES

1. Agarwal, L. P., and Malhoutra, R. L.: Conjunctival smear cytology in xerosis, Ophthalmologica **130**:378, 1955.
2. Aubineau, E.: Chancre induré de la conjonctive, Ann. Oculist. **138**:16, 1907.
3. Axenfeld, T.: The bacteriology of the eye, translated by Angus MacNab, New York, 1908, William Wood and Co.
4. Barr, M. L., and Bertram, E. G.: A morphological distinction between neurons of the male and female, and the behavior of the nuclear satellite during accelerated nucleoprotein synthesis, Nature **163**:676, 1949.
5. Bitôt, P.: Sur une lesion conjonctivale non encore decrite coincidant avec l'hemeralopie, Gaz. Hebd. Med. Chir. **10**:284, 1863.

5a. Browning, S. H.: The value of the direct smear in the bacteriology of conjunctivitis, with analysis of a thousand cases, Ophthal. Rev. **31**:97, 1912.

6. De Azevedo, M. L.: Sex chromatin detection in epithelial cells of the conjunctiva, Ophthalmologica **142**:425, 1961.
7. De Azevedo, M. L.: Citologia ocular. Doctoral dissertation presented to the Faculty of Medicine at the University of São Paulo, São Paulo, 1962.

7a. Fedukowicz, H. B.: External infections of the eye, Des Moines, 1963, Meredith Corporation, p. 206.

8. Halberstaedter, L., and von Prowazek, S.: Zur Aetiologie des Trachoms, Deutsch. Med. Wschr. **33**:1285, 1907.
9. Hardy, D., Surman, P. G., and Howarth, W. H.: Cytologic survey of conjunctival smears

from aboriginal school children at Yalata, South Australia, Brit. J. Ophthal. **51**:54, 1967.

10. Herbert, H.: Clinical observations on spring catarrh, Brit. Med. J. **2**:1232, 1907.
11. Hirschberg, J., and Krause, F.: Zur Pathologie der ansteckenden Augenkrankheiten, Zbl. Augenheilk. **5**:270, 1881.
12. Hirschberg, J.: Wilhelm Werneck, Graefe-Saemisch Handbuch der Augenheilkunde **14**: 561, 1911.
13. Hogan, M. J., and Zimmerman, L. E.: Ophthalmic pathology: an atlas and textbook, ed. 2, Philadelphia, 1962, W. B. Saunders Co.
14. Howard, H.: Role of the epithelial cell in conjunctival and corneal infections, Amer. J. Ophthal. **7**:909, 1924.
15. Kamen, L.: Zur Aetiologie der epidemischen Bindehautentzündung, Zbl. Bakt. Parasit. Infekt. (Jena) **25**:449, 1899.

15a. Kimura, S. J., and Thygeson, P.: The cytology of external ocular disease, Amer. J. Ophthal. **39**:137, 1955.

16. Leber, T.: Keratomycosis aspergillina als Ursache von Hypopyonkeratitis, Graefes Arch. Ophth. **25**:285, 1879.
17. Lindner, K.: Über die Topographie der parasitären Bindehautkeime, Graefes Arch. Ophth. **105**:726, 1921.
18. Liotet, S., and Iris, L.: Étude analytique de la cytologie conjonctivale normale et pathologique, Arch. Ophtal. (Paris) **23**:143, 1963.
19. McKee, H. A.: A new method of making film preparations to demonstrate the presence of the gonococcus, Ophthal. Rec. **21**:4, 1912.
20. Moore, K. L., and Barr, M. L.: Smears from oral mucosa in the detection of chromosomal sex, Lancet **2**:57, 1955.
21. Norn, M. S.: Age variation in the frequency of lymphocytosis in the conjunctival fluid, Acta Ophthal. (Copenhagen) **40**:69, 1962.

21a. Paton, D., and McLaren, D. S.: Bitot spots, Amer. J. Ophthal. **50**:568, 1960.

22. Pedler, C., and Ashton, N.: Sex of nuclei in ocular tissues, Brit. J. Ophthal. **39**:362, 1955.
23. Schaudinn, F., and Hoffmann, E.: Vorläufiger Bericht über das Vorkommen von Spirochaeten in syphilitischen Krankheitsprodukten und bei Papillomen, Arbeiten, Aus Dem Kaiserlichen Gesundheitsamte **22**:527, 1905.

23a. Thygeson, P.: The cytology of conjunctival exudates, Amer. J. Ophthal. **29**:1499, 1946.

24. Thygeson, P.: The cytologic diagnosis of trachoma, Rev. Int. Trachome **32**:421, 1955.
25. Werneck, W.: Lettres au Docteur Florent Cunier: sur la structure de la conjonctive, Ann. Oculist. **14**:145, 241, 1845.

5

Bacterial infections of the eye

with **ELIZABETH H. GUTIERREZ**

During the period in which one of us (D. L.-K.) was in charge of the microbiology laboratory at the Institute of Ophthalmology, Columbia-Presbyterian Medical Center (IOCPMC), cultures from the eyes of some 40,000 patients were obtained. Records of findings have been kept in large ledgers, which, though somewhat reminiscent of a Dickens accounting house, have proved very valuable in that the clinical diagnosis and chart unit number accompany the detailed account of organisms seen in smears, isolated, identified, and checked for sensitivity to antibiotics. Through the years these readily available data have been assembled in secondary ledgers in various ways according to incidence of ocular disease, incidence of etiologic agents, relation to season, relation to known trauma or chemical treatment, incidence of infection per month and per year, and changing sensitivities of the microorganisms. Periodic cultures of normal eyes have been carried out in order to establish the ocular flora for comparison with that determined in disease and in relation to postoperative infections. Data for the normal flora are given in Chapter 2. Both eyes of every patient were cultured even when only one was involved. In this chapter, bacterial infections of the eye, as seen at the IOCPMC over a 30-year period, are discussed.

Neisser in 1879 was the first to describe a bacterial ocular pathogen, the gonococcus. A few years later, while studying the etiology of cholera in Egypt, Koch observed the second organism considered to be a primary pathogen of the eye. This organism was later isolated by Weeks and came to be known as the Koch-Weeks bacillus (*Hemophilus aegypticus*). In 1897 Morax isolated a diplobacillus from a case of subacute angular conjunctivitis. This bacterium, identified independently by Axenfeld, is now familiar as the Morax-Axenfeld diplobacillus, or *Moraxella lacunata.* A similar diplobacillus, *Moraxella liquefaciens,* described by Petit in 1898, causes serpiginous corneal ulcers. The experimental proof of the role of these organisms as agents of infection of the eye could not be established by transmission to animals, since they proved pathogenic only for man; but their etiologic relationship to ocular disease was demonstrated by experimental infection of the eyes of volunteers (Chapter 1). The flora of the normal eye contain bacteria that also may be found at times in ocular infections. These include *Staphylococcus aureus, Hemophilus influenzae,* and *Diplococcus pneumoniae,* among several others.

ACUTE CONJUNCTIVITIS

Acute conjunctivitis is the commonest type of infection seen at the IOCPMC. Table 5-1, covering the period 1938 to

Table 5-1. Etiologic agents in 8,068 patients with acute bacterial conjunctivitis: 1938-1968*

Etiologic agents	Premature babies (38)	Term babies (1,223)	Children to 18 years (3,062)	Adults (3,745)
Diplococcus pneumoniae	7	324	1,135	1,337
Hemophilus influenzae	3	136	1,032	1,148
Staphylococcus aureus	13	341	401	776
Neisseria gonorrhoeae	—	306	4	215
Streptococcus viridans	1	23	408	136
Hemophilus aegypticus	—	—	16	82
Hemolytic streptococci	2	7	12	20
Neisseria catarrhalis	—	5	18	16
Anhemolytic streptococci	1	—	—	6
Pseudomonas aeruginosa	7	—	2	—
Neisseria meningitidis	—	—	1	—
Proteus vulgaris	—	16	—	—
Escherichia coli	2	14	—	—
Diplococcus pneumoniae and *H. influenzae*	—	—	—	6
Diplococcus pneumoniae and *S. aureus*	—	25	—	—
Hemophilus influenzae and *S. aureus*	2	13	33	—
Streptococcus viridans and *S. aureus*	—	13	—	—
Hemolytic streptococci and *D. pneumoniae*	—	—	—	3

*Patients were premature or term babies, young children, and adults.

1968, presents data on 8,068 patients with acute bacterial conjunctivitis. These patients included premature babies, term babies, children to 18 years, and adults. In the 38 premature babies, *Staphylococcus aureus, Diplococcus pneumoniae,* and *Pseudomonas aeruginosa* accounted for 27 of the infections. The infections caused by *P. aeruginosa* were the most severe, and 4 of the children died of sepsis. This organism has proved to be difficult to control in patients of all ages. The pneumococci, *H. influenzae, S. aureus,* and *Neisseria gonorrhoeae* topped the list of agents causing acute conjunctivitis in the newborn. In 735 of these 1,223 babies the conjunctivitis developed before they left the hospital. Pneumococci, *H. influenzae,* and *S. aureus* also were agents of acute conjunctivitis in children to 18 years of age. *Streptococcus viridans* as well played a major role in the conjunctivitis of this group. *D. pneumoniae, H. influenzae,* and *S. aureus* each proved to be of major importance in acute conjunctivitis of the adult. Plate 1, *G* illustrates an acute bulbar conjunctivitis in an adult; the conjunctivitis was caused by *S. aureus.* Although *S. aureus* is a common inhabitant of the eye, the other two bacteria are only rarely found in the normal eye. There were 1,077 cases of acute conjunctivitis in adults in which no pathogenic bacterium was found.

Two hundred and seventeen cases of membranous or pseudomembranous conjunctivitis not included in Table 5-1 are presented in Table 5-2. Since such cases present a distinctive lesion, it was of interest to compare the etiologic agents producing this type of conjunctivitis with those of other cases. The single organisms producing membranous conjunctivitis most often were *Staphylococcus aureus* and *Streptococcus hemolyticus,* neither of which headed the list of organisms causing conjunctivitis reported in Table 5-1. Plate 1, *H* illustrates a streptococcal membranous conjunctivitis; Plate 1, *I* shows membranous conjunctivitis caused by *Corynebacterium diphtheriae.*

Table 5-3 indicates the sites of conjunctival lesions caused by some of the more important eye pathogens, and the nature of the exudates. From this table it may be

Table 5-2. Etiologic agents in 217 patients of all ages with membranous or pseudomembranous conjunctivitis: 1938-1968

Etiologic agents	Membranous conjunctivitis (163)	Pseudomembranous conjunctivitis (54)
Diplococcus pneumoniae	9	6
Staphylococcus aureus	47	27
Streptococcus viridans	6	2
Hemolytic streptococci	39	4
Escherichia coli	4	1
Hemophilus influenzae	8	2
Corynebacterium diphtheriae	4	3
Diplococcus pneumoniae and *S. aureus*	11	2
Hemolytic streptococci and *S. aureus*	12	1
Streptococcus viridans and *S. aureus*	5	3
Hemophilus influenzae and *S. aureus*	1	1
Escherichia coli and *S. aureus*	1	1
Staphylococcus aureus and *Neisseria catarrhalis*	10	1
Staphylococcus aureus and diphtheroids	6	—

seen that some bacteria are epithelial parasites and that in at least two instances (infection with the gonococcus or the pneumococcus) epithelial scrapings may give an early diagnosis.

CHRONIC CONJUNCTIVITIS

The diagnosis of chronic conjunctivitis, usually a low-grade infection of long duration, is made primarily by the attending physician. The determination of the etiologic agent is difficult because of several factors; perhaps the most important is that the patient has often been treated with antibiotics before the bacteriologic examination has been made. Other conditions, such as allergic reactions, eyestrain, or exposure to irritants, may introduce a nonbacterial factor that causes the appearance of infection. Furthermore, the fact that S. *aureus* and sometimes other potential pathogens are present in the normal conjunctiva makes difficult the evaluation of the relationship of these organisms to chronic infection. In spite of these hazards, the findings in the bacteriology laboratory were felt to justify the identification of the etiologic agent of chronic conjunctivitis in 562 children and 9,147 adults. The criteria applied were: (1) the organism must be present alone or in combination with another pathogen; (2) the organism must be present in large numbers; and (3) the use of antibiotics to free the eye of the pathogens must be associated with improvement in the conjunctivitis. In 4,857 instances the etiologic agent was undetermined. Cultures were sterile in 43 cases.

The bacterial agents causing chronic conjunctivitis in children and in adults are presented in Table 5-4. It may be seen that the mixed infections predominate in children. S. *aureus* alone or in association with *H. influenzae, D. pneumoniae,* or *S. viridans* accounted for 82% of the infections. *Escherichia coli* or paracolon bacilli caused another 10% of the cases. The majority (6,273) of cases of chronic conjunctivitis in adults were the result of infection with S. *aureus* only. The following combinations of organisms—*D. pneumoniae* with *H. influenzae* or S. *aureus;* S. *viridans* with *H. influenzae,* S. *aureus,* or *E. coli; Proteus vulgaris* with gram-positive cocci; and *E. coli* in association with paracolon bacilli and S. *aureus*—accounted for another 2,256

Table 5-3. Causative agents, sites of lesions, and nature of exudate in some types of conjunctivitis

Bacterium	Relationship to conjunctival epithelium	Site of conjunctival involvement	Character of exudate
Neisseria gonorrhoeae	Ephithelial parasite; found early in epithelial scraping	Bulbar conjunctiva; may involve entire conjunctiva	Early stage, serosanguineous discharge; later, profuse mucopurulent discharge with intracellular cocci
Neisseria catarrhalis	Perhaps an epithelial parasite	Affects whole conjunctiva	Sparse to profuse mucopurulent discharge with intracellular cocci
Hemophilus aegypticus	Epithelial parasite	Bulbar conjunctiva	Profuse mucopurulent discharge with numerous bacilli
Hemophilus influenzae	Epithelial parasite but may penetrate the epithelium to slight extent only	Bulbar conjunctiva	Thin, watery, or flocculent discharge with numerous bacilli
Diplobacillus of Petit (*Moraxella liquefaciens*)	Not an epithelial parasite	Usually invades whole conjunctiva	Fibrinous; absence of polymorphonuclear neutrophils
Diplobacillus of Morax-Axenfeld (*Moraxella lacunata*)	Secretes proteinase destructive of epithelium; tears protect most of eyes, however	Invades transitional epithelium near inner and outer angles of eyelid	Mild purulent discharge containing diplobacilli, adheres to lashes, especially near angles
Gram-negative bacilli*	Not epithelial parasites	Usually invade whole conjunctiva	Mild purulent discharge containing bacilli
Diplococcus pneumoniae	Epithelial parasite	Bulbar conjunctiva	Mucopurulent or serous; numerous organisms in epithelial scraping
Streptococci	Not epithelial parasites	Affect whole conjunctiva; exotoxin from virulent strains may rupture epithelium, give rise to membranous conjunctivitis	Catarrhal, rarely purulent
Staphylococcus aureus	Not an epithelial parasite	Proliferates on the surface of ocular membrane; inflammation results from liberation and diffusion of toxic products into the tissues	Thin, occasionally mucopurulent; generally scanty number of cocci
Corynebacterium diphtheriae	Epithelial parasite	Bulbar conjunctiva invaded only in severe cases; usually membranous; tarsal and retrotarsal folds affected	Yellow-glairy sticks to eyelashes; bacilli usually present
Francisella tularensis	Epithelium thickened; granulomatous reaction	Usually unilateral; small discrete ulcers more frequent on the lower than upper lid; bulbar conjunctiva rarely involved	Little if any exudate; diagnosis by culture difficult; animal inoculation and serologic tests used

**Escherichia coli, Proteus* group, *Aerobacter aerogenes*, *Klebsiella pneumoniae*, paracolon bacilli, and *Pseudomonas aeruginosa*.

Table 5-4. Etiologic agents in chronic bacterial conjunctivitis in 9,709 patients: 1938-1968

Etiologic agents	Children to 18 years (562)	Adults (9,147)
Staphylococcus aureus	60	6,273
Pseudomonas aeruginosa	2	28
Proteus group*	21	124
Escherichia coli or paracolon bacilli	60	90
Diplobacillus of Morax-Axenfeld (*Moraxella lacunata*)	9	76
Diplobacillus of Petit (*Moraxella liquefaciens*)	8	13
Klebsiella group†	5	30
Alcaligenes faecalis	—	4
Diplococcus pneumoniae with *Hemophilus influenzae* or *Staphylococcus aureus*	187	450
Streptococcus hemolyticus with *Hemophilus influenzae*, *Streptococcus viridans*, or *Staphylococcus aureus*	7	15
Anhemolytic streptococci with *Staphylococcus aureus*	—	3
Streptococcus viridans with *Hemophilus influenzae*, *Staphylococcus aureus*, or *Escherichia coli*	161	234
Klebsiella group with *Pseudomonas aeruginosa*, *Staphylococcus aureus*, *Proteus* group, or *Streptococcus viridans*	16	62
Proteus vulgaris and gram-positive cocci‡	—	913
Escherichia coli and paracolon bacilli with *Staphylococcus aureus*	—	659
Alcaligenes faecalis and *Staphylococcus aureus*	—	6
Pseudomonas aeruginosa with *Streptococcus viridans* or *Staphylococcus aureus*	—	20
Neisseria catarrhalis or *Herellea* species with *Staphylococcus aureus*	23	136
Nocardia species	—	7
Fusiform bacilli, *Treponema macrodentium*, and *Treponema microdentium*	—	4
Mycobacterium boris	3	—

**Proteus vularis, P. morgani, P. mirabilis,* and *P. rettgeri.*
†*Klebsiella pneumoniae* and *K. ozenae.*
‡The spreading *Proteus* organisms obliterate the colonial form of the cocci, preventing more precise diagnosis.

infections. Thus, when the etiologic agents in chronic conjunctivitis in children are compared with those seen in adults, considerable variation in the incidence of the various pathogens is apparent.

When the findings in acute conjunctivitis (Table 5-1) are compared with those in chronic conjunctivitis (Table 5-4), some marked differences between the incidence of the bacteria causing these two diseases are evident. *D. pneumoniae, H. influenzae,* and *S. viridans* may be as prominent as the staphylococci in initiating acute conjunctivitis.

Table 5-5 presents some of the findings in the four cases of chronic follicular conjunctivitis resulting from fusospirochetal or fusiform bacilli infection.[3] Figs. 4-22 and 4-23 illustrate the appearance of a darkfield preparation and a gentian violet stain of a scraping from the lesion of the second patient. The follicular conjunctivitis seen at the time the patient came to the IOCPMC is shown in Fig. 5-1.

Bacteriologic examination of the 1,077 cases of acute conjunctivitis and the 4,857 cases of chronic conjunctivitis in which no bacterial agent was thought responsible for

Table 5-5. Four cases of fusospirochetal or fusiform bacillus follicular conjunctivitis

Patient no.	Age	Sex	Clinical findings	Direct examination of purulent exudate	Isolation
1 (1935)	20	F	"Cold" OD for 2 weeks; purulent discharge; congestion and thickening of palpebral conjunctiva; few enlarged follicles in lower cul-de-sac; cornea clear; lacrimal sac not involved; left eye normal	Darkfield examination of secretion; few spirochetes; fusiform bacilli	Fusiform bacilli
2 (1938)	44	M	6 years' duration; follicular conjunctivitis of right eye and stasis of lacrimal canal; expression and irrigation of lacrimal canaliculi	Darkfield examination of secretion; spirochetes and fusiform bacilli	Fusiform bacilli Spirochetes
3 (1944)	51	F	History of tearing, redness, and stickiness of left eye for 1 year; severe papillary purulent conjunctivitis; no involvement of cornea; canaliculi dilated, curetted, and irrigated; rapid improvement	Darkfield examination; only fusiform bacilli seen	Fusiform bacilli
4 (1956)	28	F	16 months; redness, tearing, discharge of right eye; irrigation of painful canaliculi; follicular conjunctivitis	Darkfield examination; only fusiform bacilli seen	Fusiform bacilli

Data from Burns, R. P., Macnie, J. P., Pfeiffer, R. L., and Locatcher-Khorazo, D.: Unilateral conjunctivitis and canaliculitis due to fusospirochetal infection, Arch. Ophthal. **59**:235, 1958; and Dunnington, J. H., and Khorazo, D.: Conjunctivitis due to fusospirochetal infection, Arch. Ophthal. **16**:252, 1936.

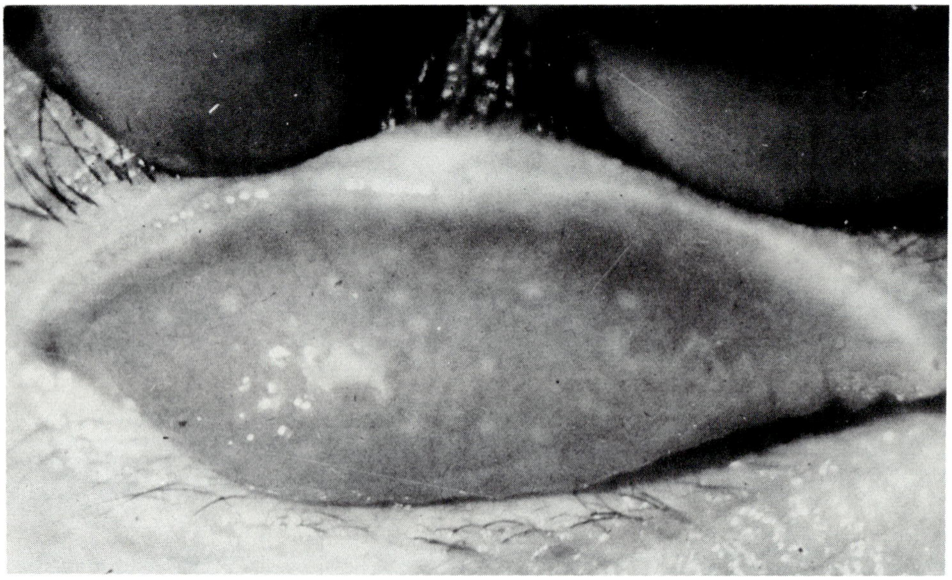

Fig. 5-1. Chronic unilateral follicular conjunctivitis caused by fusospirochetal infection. (From Burns, R. P., Macnie, J. P., Pfeiffer, R. L., and Locatcher-Khorazo, D.: Arch. Ophthal. **59**:236, 1958.)

Table 5-6. Etiologic agents in corneal ulcers in 3,535 patients: 1938-1968

Etiologic agents	Number
Diplococcus pneumoniae	1,233
Staphylococcus aureus	1,755*
Streptococcus viridans	15
Hemolytic streptococci	4
Pseudomonas aeruginosa	194
Proteus vulgaris	30
Escherichia coli	18
Paracolon bacilli	33
Diplobacillus of Petit (*Moraxella liquefaciens*)	77
Klebsiella pneumoniae	11
Nocardia species	1
Nocardia asteroides	1
Streptomyces species	1
Diplococcus pneumoniae and *S. aureus*	72
Hemolytic streptococci and *S. aureus*	17
Hemophilus influenzae and *S. aureus*	6
Pseudomonas aeruginosa and *S. aureus*	5
Pseudomonas aeruginosa and *K. pneumoniae*	1
Escherichia coli and *S. aureus*	61

*212 of these were marginal corneal ulcers.

the conjunctivitis yielded only *S. albus* or *S. aureus* in combination with diphtheroids. Forty-three of the cases had sterile cultures, a finding that is attributed to prior treatment with antibiotics. Many of these cases, on the basis of cytology, were probably of viral origin; some were allergic in nature. However, about half gave no evidence of these conditions. (See Chapter 4 on evaluation of scrapings.)

CORNEAL ULCERS

The bacterial agents isolated in 3,535 cases of corneal infections are listed in Table 5-6, which covers the period 1938 to 1968. The data indicate that *S. aureus* and *D. pneumoniae* are by far the most frequent causes of such infections, and if combined they account for 86% of the total. Plate 2, *A* illustrates a serpiginous corneal ulcer caused by the pneumococcus. *P. aeruginosa* occupies third place in these infections and accounts for 194 cases. Since *P. aeruginosa* usually is not pathogenic or when it does cause infection is often associated with trauma, the charts of the 194

Table 5-7. Type of injury causing 296 cases of bacterial infection of the cornea in adults: 1938-1968

Etiologic agents	Type of lesion	Number	Types of corneal injury
Pseudomonas aeruginosa	Corneal epithelium absent centrally	1	Struck by a piece of sandstone
	Corneal abrasion	1	Scratch with a newspaper
	Corneal ulcer	85	Incomplete removal of foreign bodies
	Hypopyon ulcer	1	While drilling struck by particles
	Corneal abscess	1	Instillation of contaminated drops
	Corneal ulcer	10	Instillation of contaminated drops
	Central corneal ulcer	1	Explosion of refrigerator, ammonium splashed on face
	Corneal exposure	1	Dryness, exophthalmos
Staphylococcus aureus	Corneal abrasion	57	Struck with shrubbery, finger, fingernails, fork; cut with glass, foreign particles; rust embedded in cornea; burns with lime and alkali
	Hypopyon ulcer	2	Injury with venetian blind; cinder removed by druggist
	Corneal ulcer	61	Injury with foreign particles
	Corneal aerosion	12	Foreign particles
Diplococcus pneumoniae	Corneal ulcer	1	Patient attacked and hit
		62	Injury with foreign particles

patients with corneal ulcers caused by pseudomonas were examined for history of trauma. The same procedure was carried out on records of some corneal ulcers caused by other bacteria. The results of this study are recorded in Table 5-7. This table shows that 102 patients with *P. aeruginosa* infections had a history of preceding trauma. Incomplete removal of foreign bodies preceded 85 of the infections.

P. aeruginosa may produce a most devastating infection of the eye and is also a difficult organism to treat with antibiotics. Within 24 to 48 hours it can produce a rapidly spreading ulcer with central necrosis. The infection should be considered of primary importance; and if the organism is suspected on the basis of the clinical picture and the presence of gram-negative bacilli, treatment should be initiated immediately. Plate 2, *B* illustrates such a rapidly progressing infection following removal of a foreign body.

Traumas of similar origin were present in the histories of 132 patients with corneal ulcers caused by *S. aureus* and in 63 patients with pneumococcal ulcers. Plate 2, *C* illustrates a marginal staphylococcal ulcer. Other types of predisposing injury in corneal ulcers have been described by Allen.[1] Although *Moraxella liquefaciens* (Diplobacillus of Petit) was among the first specific pathogens to be identified, it is one of the rarer causes of corneal ulcer seen in this country. Plate 2, *D* shows a corneal ulcer caused by *M. liquefaciens.*

The etiologic agents in 922 cases of corneal ulcers were not determined. One hundred and nine of these cases had sterile cultures, presumably because of antibiotic treatment prior to admission to the hospital. From the remaining eyes only staphylococci and diphtheroids were grown, and these organisms did not seem to be related to the cause of the lesion. It may be assumed that in some instances a virus was the etiologic agent, whereas in others prior antibiotic treatment had already removed the causative agent.

An attempt was made to determine if the strains of *S. aureus* causing either acute or chronic conjunctivitis or corneal ulcer were different from those strains carried by the patient in uninfected areas of the body. In order to carry out these studies, phage typing of staphylococci was performed. Many staphylococci are susceptible to lysis by one or more of 24 phages (viruses) that infect this bacterium. Such lysis is readily determined; and because of the multiple combinations of phages causing lysis of a specific type of staphylococcus, the organism is marked, making it possible to achieve accurate identification of a given strain. Cultures were taken from neighboring tissues as well as from the lesion in order to determine if the same staphylococcus was present in those noninfected tissues. Twenty-eight strains from the eyes of patients with acute conjunctivitis were phage typed; and strains isolated from the nose, throat, or hands, or in one case from an infected leg of the same patient, were also typed. The pattern of phage lysis (type) was the same for the staphylococci obtained from the infected and noninfected areas. Further studies on the normal disposition of typed *S. aureus* have shown that individuals usually carry the same type for many months.

Studies were carried out in the case of 35 patients who had chronic conjunctivitis. The results obtained were entirely comparable to those observed in cases of acute conjunctivitis. In addition, 31 cases of corneal ulcer yielded staphylococci that fell among the same types as were seen in the preceding two studies.[5,6]

Two other organisms, the pneumococcus and *H. influenzae,* causing conjunctivitis or corneal ulcers, were also typed in an effort to determine if a unique member of the group was responsible for these ocular infections. One hundred and twelve strains of *D. pneumoniae* isolated from acute conjunctivitis in children were typable; 62 belonged to type 1. In 236 adults with typable pneumococcal conjunctivitis, types 1, 2, and 3 accounted for 155 of the infections. Other types were found scattered among

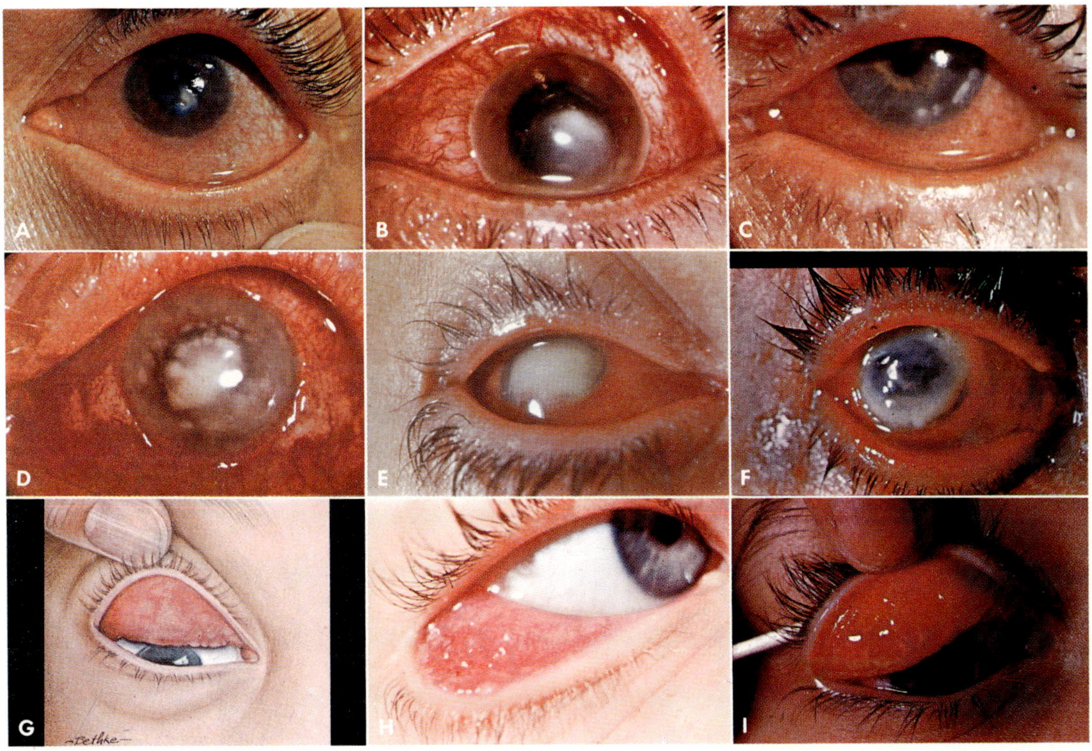

Plate 2. A, Serpiginous corneal ulcer caused by *Diplococcus pneumoniae.* **B,** Corneal ulcer caused by *Pseudomonas aeruginosa.* **C,** Marginal corneal ulcer with blepharitis caused by *Staphylococcus aureus.* **D,** Corneal ulcer caused by *Moraxella liquefaciens.* **E,** Endophthalmitis caused by *Staphylococcus aureus.* **F,** Endophthalmitis caused by *Proteus vulgaris.* **G,** Primary tuberculosis of the tarsal conjunctiva in a 9-year-old child. Note at about the middle of the upper tarsal conjunctiva a shallow, indolent ulcer, 4 × 5 mm in size with fairly well-defined reddish margin and a gray sloughing base. (See also Fig. 4-20.) **H,** Primary tuberculoma of conjunctiva in a 10-year-old boy. **I,** Papillary hypertrophy of the right upper lid with small areas of ulceration in an 11½-year-old girl with primary tuberculosis of the conjunctiva. (**B** and **D** courtesy Dr. R. W. Darrell; **G** from Bruce, G. M., and Locatcher-Khorazo, D.: Arch. Ophthal. **37**:376, 1947; **H** from DeVoe, A. G., and Locatcher-Khorazo, D.: Trans. Amer. Ophthal. Soc. **62**:205, 1964; **I** from Chandler, A. C., Jr., and Locatcher-Khorazo, D.: Arch. Ophthal. **71**:203, 1964.)

the remaining cases. Five hundred and eighty-six strains of *D. pneumoniae* isolated from corneal ulcers were also typed. Types 1, 2, and 3 accounted for 473 of the infections, and various types were found in the remainder. The distribution of types in the infected eye, exclusive of types 1 and 2, was not strikingly different from that which might be found in the normal throat.

Virulent *H. influenzae* may also be typed serologically, but the organisms isolated from conjunctivitis and corneal ulcers caused by this bacterium were with one exception not typable. It is concluded from the attempt to determine specific strains in infection that these three microorganisms, the staphylococcus, *D. pneumoniae*, and *H. influenzae*, show no specific strain causing conjunctivitis or corneal ulcer.

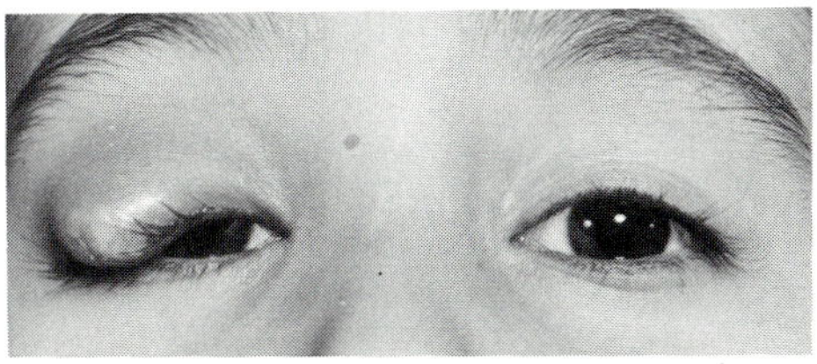

Fig. 5-2. Abscess, caused by *Staphylococcus aureus*, of right upper lid.

INFECTIONS OF THE EYELID

In the period of our studies (1938 to 1968) 8,486 patients with various bacterial infections of the eyelids were seen. The findings from these eyes are listed in Table 5-8 under six categories, which represent the clinical diagnoses. *Staphylococcus aureus* (Fig. 5-2) alone or in combination with other individual pathogens accounted for over 81% of the infections seen in all the groups. It was somewhat unexpected to find that almost all of the other infections

Table 5-8. Bacterial agents in 8,486 patients with various infections of the eyelids: 1938-1968

Microorganisms isolated	Eyelid abscess (228)	Hordeolum (230)	Meibomitis (325)	Acute and chronic chalazion (384)	Blepharitis (1,637)	Blepharoconjunctivitis (5,682)
Staphylococcus aureus	193	224	146	292	1,029	2,614
Streptococcus viridans	7	—	11	13	12	231
Diplobacillus of Morax-Axenfeld (*Moraxella lacunata*)	—	—	4	—	29	224
Escherichia coli	3	1	14	6	26	223
Klebsiella pneumoniae	1	—	44	—	29	213
Proteus vulgaris	—	—	11	5	16	160
Pseudomonas aeruginosa	—	2	8	1	15	146
Diplobacillus of Petit (*Moraxella liquefaciens*)	—	—	5	3	12	36
Diplococcus pneumoniae	7	—	6	11	29	26
Hemophilus influenzae	—	—	—	—	3	1
Hemolytic streptococci	3	—	4	3	3	—
Diplococcus pneumoniae and hemolytic *S. aureus*	7	2	16	9	20	275
Streptococcus viridans and *S. aureus*	3	—	7	10	226	215
Klebsiella pneumoniae and *S. aureus*	—	1	14	10	58	324
Proteus vulgaris and *S. aureus*	1	—	16	6	25	467
Escherichia coli or paracolon bacilli and *S. aureus*	—	—	6	6	49	363
Pseudomonas aeruginosa and *S. aureus*	—	—	11	4	46	136
Hemophilus influenzae and *S. aureus*	—	—	2	5	7	12
Hemolytic streptococci and *S. aureus*	3	—	—	—	3	16

were caused by one or another of 10 organisms, which, with the exception of *Streptococcus viridans,* are rarely found on the skin adjacent to the eye, or in the normal eye. In an additional 357 patients the etiologic agent was not determined.

INFECTIONS OF THE LACRIMAL SAC AND CANALICULI

Dacryocystitis is a common acute or chronic disease, most often unilateral, which is almost always secondary to obstruction of the nasal lacrimal ducts. Table 5-9 describes the results of cultures of 891 cases of dacryocystitis or lacrimal duct obstruction resulting from bacterial infection. The 551 cases carrying the diagnosis of acute and chronic dacryocystitis showed a preponderance of infections by *S. aureus,* with a scattering of cases involving the other 12 pathogens. Fig. 5-3 illustrates a staphylococcal dacryocystitis with draining abscess.

The three categories of lacrimal canaliculus disease showed a scattering of infectious agents, except for the 287 cases diagnosed lacrimal obstruction, 216 of which were caused by *Nocardia* species (Fig. 5-4). Three of the four cases of fusospirochetal or fusiform bacillus infection listed with the chronic conjunctivitis cases in Table 5-4 also had involvement of the lacrimal canals and might have been placed in Table 5-9 under the heading "lacrimal conjunctivitis." The findings from these patients are briefly summarized in Table 5-5.

Table 5-9. Etiologic agents in 891 patients with bacterial infections of the lacrimal apparatus: 1938-1968

Microorganisms isolated	Acute and chronic dacryocystitis (551)	Lacrimal obstruction (287)	Lacrimal abscess (25)	Lacrimal conjunctivitis (28)
D. pneumoniae (types 1, 3, 4, 6, 10)	43	12	3	1
Staphylococcus aureus (with or without diphtheroids)	282	17	3	12
Streptococcus viridans	4	1	1	1
Hemolytic streptococci	22	2	1	1
Pseudomonas aeruginosa	9	2	1	1
Proteus vulgaris	9	14	1	—
Escherichia coli	5	2	1	1
Klebsiella pneumoniae	15	4	4	1
Diplobacillus of Morax-Axenfeld (*Moraxella lacunata*)	5	1	—	—
Hemophilus influenzae	3	—	—	—
Corynebacterium diphtheriae (non–toxin producing)	—	1	—	—
Nocardia species	—	216	—	7
Streptomyces species	—	4	—	—
Diplococcus pneumoniae and *S. aureus*	68	—	—	—
Hemolytic steptococci and *S. aureus*	24	1	1	—
Streptococcus viridans and *S. aureus*	18	2	2	1
Klebsiella pneumoniae and *S. aureus*	25	2	1	1
Escherichia coli and *S. aureus*	13	3	3	1
Escherichia coli and pneumococci	1	1	1	—
Pseudomonas aeruginosa, Staphylococcus aureus, and *Streptococcus viridans*	3	1	1	—
Diplobacillus of Morax-Axenfeld and *S. aureus*	2	1	1	—

There were no failures to obtain a definitive etiologic diagnosis in the lacrimal area of the eye.

INFECTIONS OF THE SOCKET

Infection of the socket was seen in 987 patients; in all but two cases the causative agent was a bacterium. The two exceptions were infections caused by fungi (Chapter 13). *S. aureus* accounted for more than one third of the cases, with *Pseudomonas* and *Proteus* groups holding second and third place (Table 5-10). The bacterial flora in infections of the socket is similar to that found elsewhere in the eyes.

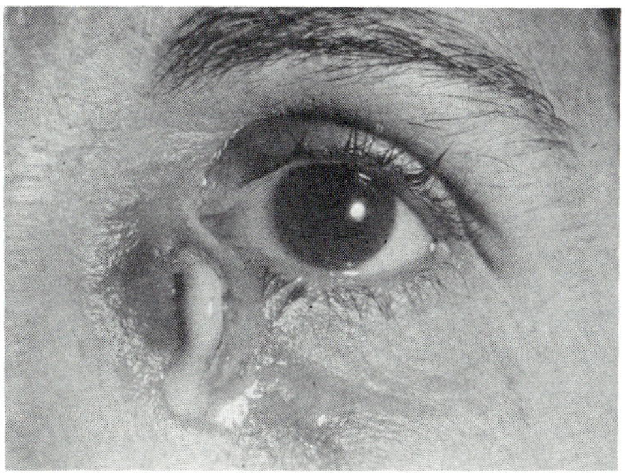

Fig. 5-3. Staphylococcal dacryocystitis with draining abscess.

ORBITAL CELLULITIS

From 1944 to 1968 cultures were obtained from 1,004 patients with a diagnosis of orbital cellulitis (Table 5-11). In 950 instances S. *aureus* was the causative agent, and in another 34 patients the staphylococcus in association with streptococci or pneumococci were isolated. Only infections of the eyelid show a somewhat comparable preponderance of staphylococcal infection.

ENDOPHTHALMITIS

Table 5-12 presents data from cases of endophthalmitis in which the etiologic agent was determined in the years 1945 to 1968. This represents only a portion of the

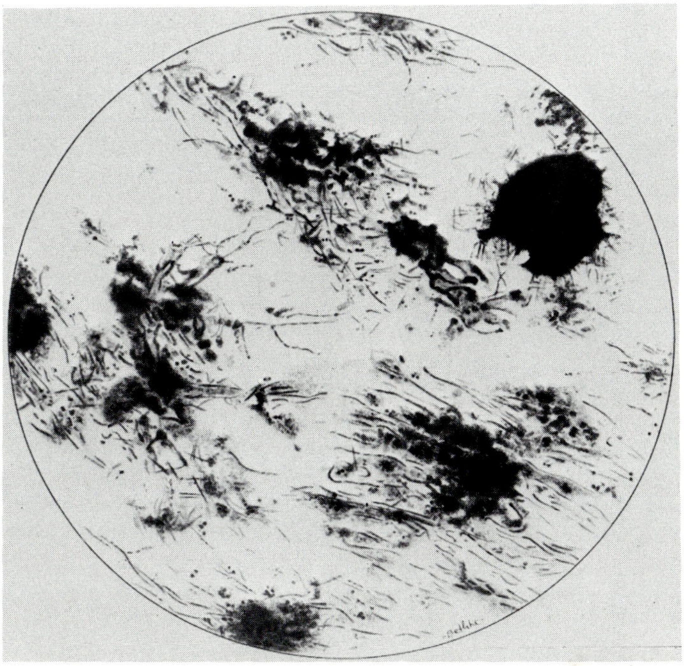

Fig. 5-4. Smear of expression from canaliculus of patient with lacrimal obstruction caused by *Nocardia* species. This bacterium was formerly classified as a fungus because of the branching filamentouslike appearance seen in the illustration.

Table 5-10. Bacterial infection of the socket in 987 patients: 1938-1968

Microorganisms isolated	Number	Percent positive
Staphylococcus aureus	351	35.5
Pseudomonas aeruginosa	112	11.3
Proteus group*	83	8.4
Escherichia coli	40	4.5
Diplococcus pneumoniae	37	3.7
Streptococcus viridans	25	2.5
Klebsiella pneumoniae	16	1.6
Hemolytic streptococci	5	0.51
Diplobacillus of Petit (*Moraxella liquefaciens*)	4	0.45
Diplococcus pneumoniae and *S. aureus*	69	6.9
Proteus vulgaris and gram-positive cocci†	52	5.2
Paracolon bacilli and *S. aureus*	44	4.4
Klebsiella pneumoniae and *S. aureus*	34	3.4
Pseudomonas aeruginosa and *S. aureus*	34	3.4
Streptococcus viridans and *S. aureus*	28	2.8
Hemophilus influenzae, *S. aureus*, and *D. pneumoniae*	16	1.6
Hemolytic streptococci and *S. aureus*	16	1.6
Beta-hemolytic streptococci and *D. pneumoniae*	10	1.1
S. aureus and *Neisseria catarrhalis*	7	0.7
S. aureus and *N. sicca*	2	0.22
Diplobacillus of Petit and *S. aureus*	2	0.22

**Proteus vulgaris*—62; *Proteus mirabilis*—21.

†The spreading *Proteus vulgaris* obliterates the colonial form of the organism, preventing more precise diagnosis.

patients with this clinical diagnosis, since bacteriologic examination was not made in some instances. The largest series of cultures were from eyes with postoperative infections, and here the staphylococcus was the most important pathogen.

SUCCESSFUL CULTIVATION OF OCULAR FLORA

The eye offers two special problems to challenge the skill of the microbiologist. The first of these is the extremely small amount of material available in many instances for culture. This amount may limit the number of inoculations of different, specialized media.[2] The small amount of material also puts pressure on the ophthalmologist or technician who takes the culture to return it to the laboratory promptly to prevent drying. The ideal condition is to have the patient come to the laboratory. The second problem that faces the microbiologist is related to the various flora that may cause infection of the eye. When a specimen comes to the routine microbiology laboratory in a general hospital, the accompanying slip states the source of the culture—sputum, throat, gastric lavage, stool, or abscess, for example. The selection of media for primary isolation is dictated by this information. By contrast, one may find in the eye almost any one of the upper respiratory tract, enteral canal, or skin inhabitants in addition to some bac-

Table 5-11. Bacterial agents in orbital cellulitis in 1,004 patients of all ages: 1944-1968

Microorganisms isolated	Number
Staphylococcus aureus	950
Diplococcus pneumoniae	8
Hemolytic streptococci	7
Pseudomonas aeruginosa	2
Proteus vulgaris	2
Anhemolytic streptococci	1
Streptococcus viridans and *S. aureus*	16
Hemolytic streptococci and *S. aureus*	13
Diplococcus pneumoniae and *S. aureus*	5

Table 5-12. Endophthalmitis—causative agents and preceeding injury in 105 patients: 1945-1968

Microorganisms isolated	Corneal ulcer	Surgery*	Trauma	Endogenous
Neisseria gonorrhoeae	1	—	—	—
Clostridium perfringens	—	—	1	—
Staphylococcus aureus	12	51†	4	2
Pseudomonas aeruginosa	5	1	—	3
Proteus vulgaris	2	3	1	—
Diplococcus pneumoniae (types 3, 8, 9, 14, 17, 21)	2	4	—	—
Streptococcus viridans	—	3	—	1
Hemolytic streptococci	—	1	—	—
Escherichia coli	—	1	3	1
Pseudomonas aeruginosa and *S. aureus*	—	1	—	—
Bacillus subtilis	—	—	1	—
Alcaligenes faecalis	—	—	—	1

*Thirty-five of these infections were "late" infections.
†In 13 of these cases the aqueous humor provided material for culture. This source was frequently the best for obtaining the etiologic agent in endophthalmitis.

teria peculiar to the eye. Judgment as to the handling of the culture has to be made promptly on a minimum amount of data. The microbiologist may receive considerable help from examination of smears of scrapings if these are available. Often the clinical judgment of the ophthalmologist is of great assistance. Each case is a "special" problem, and the success achieved in the laboratory often depends on cooperation from the ophthalmologist.

The prompt determination of the sensitivity of a pathogen to antibiotics is one of the most important functions of the laboratory. To isolate individual bacteria and determine their sensitivity in pure culture is routinely done. With a fast-moving infection, however, the minimum of 48 hours that this process requires may be highly damaging to the eye. Therefore, when severe inflammation is present, two extra plates are streaked at once and the entire spectrum of antibiotics is tested. Although the plates contain a mixture of organisms, it is rare to meet a situation in which at least one of the antibiotics has failed to inhibit all growth. This information, obtained overnight, is sometimes very helpful to the handling of the infection.

The sensitivity of S. *aureus* to the various antibiotics in common use at the IOCPMC was tabulated by month and by year. The strains of staphylococci found in this patient population showed a surprisingly constant seasonal and yearly pattern of sensitivity. The Institute has been free of problems associated with the spread of resistant, virulent organisms. Spot data analyzed from the sensitivity test on the other eye pathogens showed a similar stability among these organisms. The stability may be because of the fact that most of the hospital population are well except for the specific eye problem. Furthermore, the period of hospitalization is usually only a few days. Many of the cultures received came from the physicians' private offices and the clinic, where the infection could be handled on an outpatient basis. The variation in sensitivities of various strains within a species, however, makes the determination of the sensitivities of each strain still essential. If the patient fails to respond to treatment, frequent reculturing of the lesion and retesting of sensitivity is important.

Sensitivity of Staph

REFERENCES

1. Allen, H. F.: Current status of prevention, diagnosis and management of bacterial corneal ulcers, Ann. Ophthal. 3:235, 1971.

2. Burns, R. P.: Laboratory methods in diagnosis of eye infection. In Infectious diseases of the conjunctiva and cornea, Symposium of the New Orleans Academy of Ophthalmology, St. Louis, 1963, The C. V. Mosby Co.
3. Burns, R. P., Macnie, J. P., Pfeiffer, R. L., and Locatcher-Khorazo, D.: Unilateral conjunctivitis and canaliculitis due to fusospirochetal infection, Arch. Ophthal. **59**:235, 1958.
4. Dunnington, J. H., and Khorazo, D.: Conjunctivitis due to fusospirochetal infection, Arch. Ophthal. **16**:252, 1936.
5. Locatcher-Khorazo, D., and Gutierrez, E.: Bacteriophage typing of Staphylococcus aureus: a study of normal, infected eyes and environment, Arch. Ophthal. **63**:774, 1960.
6. Locatcher-Khorazo, D., Sullivan, N., and Gutierrez, E.: Staphylococcus aureus isolated from normal and infected eyes, phage types and sensitivity to antibacterial agents, Arch. Ophthal. **77**:370, 1967.

6

Postoperative infections of the eye

with **ELIZABETH H. GUTIERREZ**

Ocular surgery usually can be done at a selected time when the eye shows no evidence of infection. The operations considered in this chapter were for cataract extraction, muscle correction, repair of detached retinas, and treatment of glaucoma. The literature contains many references to infections that follow such surgery.[2,4,5,9,20] It is practically impossible to maintain sterility in the eye for any length of time, open as it is to the air and to the contiguous nose and skin. *Staphylococcus aureus,* a potential pathogen, is frequently found in the healthy eye (Chapter 2); and other, perhaps more virulent, organisms are sometimes present.[1a,19a] In the preantibiotic era, a semblance of sterility was achieved by the instillation of protein silver; with the advent of antibiotics the eye may be rendered apparently bacteria free for short periods of time.[6,12,18] Nevertheless, the area is one where some postoperative infections may be unavoidable.

Infections following cataract surgery at the Institute of Ophthalmology, Columbia-Presbyterian Medical Center (IOCPMC) from 1938 through 1955 have been reported. From 1938 to 1944, 2,508 cataract operations were performed on patients whose eyes had been cultured prior to surgery.[6] There were 11 cases of endophthalmitis caused by S. *aureus,* all occurring in the group of 529 patients who were carrying this organism. This was a 2% incidence of endophthalmitis in patients harboring this pathogen. There were no infections in the 1,706 patients who were not carrying S. *aureus.* In the last few years of this study 104 patients carrying S. *aureus* were given preoperative treatment with sodium sulfathiazole or penicillin ointments until the conjunctivas were free of the pathogens. Since no infection developed, treatment consisting of 5 local applications of antibiotics given over a 24-hour period prior to operation were performed routinely, starting with the year 1944. The antibiotics employed during the succeeding years have varied with the development of new agents and with the results of routine sensitivity tests.[13]

From 1945 to 1955[13] there were a total of 7,662 cataract extractions from patients who had preoperative antibiotic treatment, although not all patients had preoperative cultures taken (Table 6-1). In 567 patients, operation was deferred for supplementary treatment, since it was found that pathogens were not always eliminated by the routine preoperative application of antibiotics for 24 hours. There was a total of 6 cases of endophthalmitis, 5 of which occurred in a group of 3,854 patients in whom no preoperative cultures were taken. No postoperative infections were present in 2,263 patients known to be free of S. *aureus.* There was one infection among a group of 978 patients known to carry the

Table 6-1. Postoperative endophthalmitis following 7,662 operations for cataract extraction:* 1945-1955

Number of operations	Preoperative cultures	Postoperative endophthalmitis	Causative organisms
3,854	Not available	4 1	*Staphylococcus aureus* *Escherichia coli*
2,263	*Staphylococcus aureus* absent	None	
978	*Staphylococcus aureus* present	1	*Staphylococcus aureus* and *Pseudomonas aeruginosa*
567†	*Staphylococcus aureus* (507) *Escherichia coli* and *Proteus vulgaris* (60)	None	

*All patients received 24-hour preoperative antibiotic treatment.
†Operations deferred for supplementary local antibiotic treatment until the eye was free of potential pathogens.

Table 6-2. Incidence of postoperative infections in 15,278 patients with 24 hours of routine applications of antibiotics preceding cataract surgery: 1955-1968

Number of operations	Preoperative culture	Postoperative infection	Causative agent
13,319	Not done	Endophthalmitis (15) Conjunctivitis (2) Abscessed conjunctiva (1) Chalazion (1)	*Staphylococcus aureus*
		Endophthalmitis (2) Conjunctivitis (1)	*Streptococcus viridans* *Streptococcus viridans* and *S. aureus*
1,065	*Staphylococcus albus*	None	
726	*Staphylococcus aureus*	Endophthalmitis (1) Conjunctivitis (1)	*Staphylococcus aureus*
28	*Proteus* group	Endophthalmitis (1)	*Proteus vulgaris*
4	*Escherichia coli*	None	
136	*Diplococcus pneumoniae* *Streptococcus viridans* *Klebsiella pneumoniae* *Neisseria catarrhalis* with *S. aureus*	None	

staphylococcus; the infection was mixed, caused by *S. aureus* and *Pseudomonas aeruginosa.* No infection followed surgery among the 567 patients carrying pathogenic organisms whose operations were deferred until the eyes were free of pathogens.

Table 6-2 summarizes the incidence of infections following cataract surgery in the period 1955 to 1968 in 15,278 patients receiving 24 hours of routine prophylactic treatment with antibiotics (4 to 5 local applications in a 24-hour period). In 13,319 patients who had no preoperative cultures

Table 6-3. Pathogens found prior to local antibiotic treatment in 2,419 patients awaiting cataract extraction: 1943-1968*

Number of patients	Microorganisms isolated preoperatively	Treatment used
34	*Staphylococcus aureus*	Sulfathiazole
4	*Diplococcus pneumoniae*	
2	*Streptococcus hemolyticus*	
72	*Staphylococcus aureus*	Penicillin
147	*Staphylococcus aureus*	Combination of penicillin and streptomycin
2	*Hemophilus influenzae* and *S. aureus*	
57	*Staphylococcus aureus*	Terramycin
912	*Staphylococcus aureus*	Chloramphenicol
18	*Diplococcus pneumoniae*	
6	*Streptococcus viridans*	
3	*Streptococcus hemolyticus*	
20	*Diplococcus pneumoniae* and *S. aureus*	
4	*Hemophilus influenzae* and *S. aureus*	
683	*Staphylococcus aureus*	Erythromycin
3	*Pseudomonas aeruginosa*	Polymyxin
149	*Proteus* group	Chloramphenicol
108	*Escherichia coli*	
32	*Klebsiella pneumoniae*	Colymycin
113	*Proteus* group and *S. aureus*	Furacin
38	*Escherichia coli* and *S. aureus*	Neomycin
12	*Klebsiella pneumoniae* and *S. aureus*	Tetracyclines

*Operation deferred until pathogens were eliminated; no postoperative infections occurred.

taken, there were 15 cases of endophthalmitis caused by S. *aureus* and 2 caused by *Streptococcus viridans*. In the 1,065 patients who had preoperative cultures taken and who were found to carry only S. *albus*, no infections occurred. In the 726 patients whose cultures were positive for S. *aureus* there was one case of endophthalmitis. One case of endophthalmitis caused by *Proteus vulgaris* was present in 1 patient out of 28 who had carried *Proteus* organisms at preoperative culture time. There were another 136 patients carrying other potential pathogens, but no postoperative infection was noted.

Table 6-3 is a summary of the results obtained in a total of 2,419 patients who were seen from 1943 to 1968. On the basis of preoperative culture, these patients were found to carry potential pathogens, and they were treated with the various antibiotics listed until the eyes gave negative cultures (in 2 to 7 days). There were no postoperative infections in this group. These figures support the concept that the bacterial flora of the preoperative eye may be an important source of postoperative infection. The importance of S. *aureus* that many patients carried is emphasized by the observation that in the period 1938 to 1968, 5,034 patients who did not carry S. *aureus* in their preoperative cultures also did not have any postoperative infection. Adding the 5,034 patients free of S. *aureus* to the 2,419 in whom antibiotics were used to eliminate S. *aureus* gives a total of 7,453 cataract extractions performed in the absence of S. *aureus* and without a single postoperative infection.

The patient can be shown how to carry out preoperative antibiotic treatment at home, thus saving unnecessary days of stay in the hospital. S. *aureus* is usually eliminated in 2 to 3 days; but the gram-negative organisms—the *Proteus* group, *Klebsiella pneumoniae, Escherichia coli,* and *Pseudomonas aeruginosa*—may persist for as long as 7 days. The period of time required is determined in part by the care the patient takes. Since adjacent areas usually carry the same type of S. *aureus,* the patient is given instructions to wash his hands and face with a good disinfectant soap.

Other less serious infections have occurred following intraocular surgery. Conjunctivitis complicated recovery from cataract surgery in 32 patients in the period 1938 to 1968. In 28 instances the causative agent was S. *aureus,* in 4 *Proteus vulgaris.* S. *aureus* was also responsible for one case each of corneal ulcer, abscess of the conjunctiva, stitch abscess, infected bleb on the conjunctiva, and chalazion.

Surgery involving eye muscle, detached retina, or glaucoma was followed by postoperative infection in some instances. Table 6-4 presents the findings in 20,966 patients operated on at the IOCPMC from 1950 to 1968. Although preoperative cultures were very seldom taken, preoperative local anti-

Table 6-4. Incidence of postoperative infections in 20,966 patients following ocular surgery: 1950-1968

Number of operations	Types of operations	Postoperative complications	Causative agents*
12,263	Muscle surgery	Orbital cellulitis (11) Conjunctivitis (38) Sloughing ulcer on conjunctiva (1) Stitch abscess (1) Subconjunctival abscess (1) Suppurative wound infection along with fascia lata suspension LUL (1)	*Staphylococcus aureus*
		Corneal ulcer (ptosis; Motais procedure) (1)	*Klebsiella pneumoniae*
		Conjunctivitis (1)	*Diplococcus pneumoniae* and *S. aureus*
		Conjunctivitis (1)	*Hemophilus influenzae* and *S. aureus*
4,912	Retinal detachment	Endophthalmitis (1)	*Pseudomonas aeruginosa*
		Conjunctivitis (3) Orbital cellulitis (2) Endophthalmitis (4) Corneal ring abscess (1) Infected surgical wounds (22) Infected silicone sponge (2) Infected suture site (3) Cloudy vitreous (1)	*Staphylococcus aureus*
		Infected surgical wound (1)	*Escherichia coli*
3,791	Glaucoma	Endophthalmitis (2) Corneal infiltration (1) Infected blebs (2)	*Staphylococcus aureus*
		Endophthalmitis (1)	*Proteus vulgaris*
		Endophthalmitis (1)	*Diplococcus pneumoniae* Type 17

*The nonoperated eye, nose, face, and hand carried the same microorganisms as the infected eye.

biotic treatment was usually employed in cases of detached retina and in a few cases of glaucoma. Among the 12,263 patients undergoing muscle surgery there were 38 instances of conjunctivitis, 1 conjunctival ulcer, 1 abscess, 11 cases of cellulitis, and 1 suppurative wound infection, all caused by *S. aureus.* A corneal ulcer was the result of

Table 6-5. Susceptibility to bacteriophage lysis of strains of *Staphylococcus aureus* causing infection following cataract extraction compared with susceptibility of staphylococci isolated prior to surgery: 1957-1962

Case number	Type of infection	Lysed by phage types			
		Preoperative cultures	Postoperative cultures		
		Both eyes	Operated eye	Nonoperated eye	Other sources
1	Endophthalmitis	7, 53, 42B, 81	7, 53, 42B, 81	7, 53, 42B, 81*	
2	Endophthalmitis	53, 75, VA4	53, 75, VA4,	53, 75, VA4*	
3	Endophthalmitis	3B, 3C, 55	3B, 3C, 55	3B, 3C, 55*	Nose, 3B, 3C, 55*
4	Endophthalmitis	52, 80, 81	52, 80, 81	52, 80, 81*	Nose, 52, 80, 81*
5	Endophthalmitis	52, 80, 42B, 81	52, 80, 42B, 81	52, 80, 42B, 81*	Nose, face, skin, 52, 80, 42B, 81*
6	Endophthalmitis	52A, 79	52A, 79	52A, 79*	Nose, face, skin, 52A, 79*
7	Endophthalmitis	Nontypable	Nontypable	Nontypable*	Nose, nontypable*
8	Endophthalmitis	No culture	Nontypable	Nontypable*	
9	Endophthalmitis	No culture	52A, 79	52A, 79	Nose, 52A, 79
10	Endophthalmitis	No culture	52, 52A, 80	52, 52A, 80	Nose, 52, 52A, 80
11	Endophthalmitis	No culture	80, 53	80, 53	Nose, skin, 80, 53
12	Endophthalmitis	No culture	80, 53, VA4	80, 53, VA4	Nose, 80, 53, VA4
13	Endophthalmitis	No culture	52, 52A, 80, 42B, 81	52, 52A, 80, 42B, 81	Nose, 52, 52A, 80, 42B, 81
14-19	Endophthalmitis	No culture	Nontypable	Nontypable	Nose, nontypable
20	Endophthalmitis	No culture	Nontypable	29	Nose, 29; hand, . skin, nontypable
21	Endophthalmitis	No culture	29, 52A, VA4	29, 52A, VA4	Nose, 29, 52A, VA4; hand, nontypable
22	Endophthalmitis	No culture	52, 52A, 80	52, 52A, 80	Nose, 52, 52A, 80
23	Conjunctivitis	52, 80	52, 80	52, 80*	Nose, 52, 80*
24	Abscess of conjunctiva	No culture	52, 80, 42B, 81	52, 80, 42B, 81	Nose, skin, air of patient's room, 52, 80, 42B, 81
25	Chalazion	No culture	47, 54, VA4	47, 54, VA4	Nose, 47, 54, VA4
26	Bleb on conjunctiva	No culture	80, 81	80, 81	Nose, 80, 81

*These cultures from the face and other sources were taken after operation.

K. pneumoniae infection, and *S. aureus* with the pneumococcus or *H. influenzae* caused conjunctivitis in two instances. Of 4,912 cases of detached retina only 40 were followed by infection. One case of endophthalmitis was caused by *Pseudonomas aeruginosa,* whereas 3 cases of conjunctivitis, a ring abscess of the cornea, 2 cases of cellulitis, and 4 cases of endopthalmitis were caused by *S. aureus.* Twenty-two infections of the wound were caused by *S. aureus* and one by *Escherichia coli.* Seven infections followed the 3,791 glaucoma operations; 2 cases of endophthalmitis, 1 corneal infiltration, and 2 infected blebs were caused by *S. aureus.* One case of endophthalmitis was produced by *Proteus vulgaris* and another by *Diplococcus pneumoniae* type 17.

Twenty-six strains of *S. aureus* isolated from cases of endophthalmitis or other infections following cataract surgery were phage typed. In 8 of the patients, preoperative isolates of the staphylococcus were also typed, as described in Chapter 5, p. 70. The purpose of this study was to determine if the patient carried the same type of staphylococcus in the eye prior to surgery as the type causing the infection. Table 6-5 shows that in 9 instances where preoperative strains were typed, the same strain was found to cause the postoperative infection. The same strain was present in the noninfected eye in 25 of the 26 patients, and the same strain causing the infection was also present in the nose and on the skin of the hands and face.

Plate 2, *E* and *F* illustrates postoperative endophthalmitis caused by *S. aureus* and *P. vulgaris,* respectively.

Others have reported postoperative infections, some or all of which may have been caused by exogenous sources of contamination. Winkler and Cason[20] have described a series of 11 cases of endophthalmitis that occurred following cataract extractions, 10 of which were caused by gram-negative bacilli (either *Aerobacter aerogenes, Pseudomonas aeruginosa,* or *Proteus vulgaris*) and one caused by *S. aureus.* In 4 of these 11 cases the infection was considered to be endogenous, since the causative organism was cultured preoperatively. In two other cases the possibility of an endogenous infection existed. Three cases probably were related to contamination of cotton pledgets used on the day of the operation. Custodis[4a] has reported a similar epidemic of infections involving 9 cases of iritis with hypopyon and 3 with less severe infection of the iris, all occurring within 3 days of the operation and traced to a solution of pilocarpine contaminated with *E. coli.*

It is important to examine factors that may contribute to the invasion by the bacteria present in the normal eye. One series of observations may have some bearing on this problem. The temperature of the eye is less than that of the internal organs of the body[1,19] and hence is lower than the optimal temperature for growth of the potential pathogens normally present in the eye. In 1954 cultures from the eyes of 15 patients were taken prior to a variety of operations and again 24 hours following operation. The results are presented in Table 6-6. During the 24-hour period the eye had been bandaged and probably had maintained a temperature nearer to that optimal for growth of the potentially pathogenic bacteria that were present. From Table 6-6 it can be seen that the colonies of the various microorganisms found preoperatively were considerably fewer than those found after 24 hours. Furthermore, in 12 instances an additional species of microorganisms was seen on the postoperative culture plate. This suggests that a bandage promotes conditions favorable for the multiplication of bacteria; and though this does not prove that bandaging would contribute to the development of infection, it suggests this possibility. This point of view may also gain some support from the latent period preceding the onset of postoperative infection; this period has varied from 2 to 7 days, with the peak incidence of infection occurring on the fourth day. Linhart[11] has reported an increase in bacterial colony

Table 6-6. Effect of surgical dressing on ocular flora in conjunctiva

Case number	Clinical diagnosis	Preoperative culture	Second culture 24 hours after surgery
1	Absolute glaucoma	15 colonies *S. aureus*, 12 colonies pneumococcus type 3, diphtheroids	Numerous *S. aureus*, pneumococci type 3, diphtheroids
2	Glaucoma	81 colonies diphtheroids	18 colonies *S. aureus*, diphtheroids
3	Pterygium	1 colony *S. aureus*, diphtheroids	Numerous *S. aureus*, diphtheroids
4	Extirpation of sac	23 colonies *S. albus*, diphtheroids	Numerous *S. albus*, diphtheroids, 13 colonies *S. aureus*
5	Esotropia	36 colonies *S. albus*	Numerous *S. albus*, *S. aureus*
6	Esotropia	3 colonies *S. aureus*	Numerous *S. albus*, *S. aureus*
7	Esotropia	7 colonies *S. aureus*	Numerous *S. aureus*, 6 colonies *S. albus*
8	Esotropia	1 colony *S. aureus*, diphtheroids	Numerous *S. aureus*, 14 colonies *S. albus*, diphtheroids
9	Esotropia	1 colony *S. aureus*	Numerous *S. aureus*, *S. albus*, diphtheroids
10	Congenital ptosis	45 colonies *S. albus*	Numerous *S. albus*, 27 colonies *S. aureus*
11	Paralysis of lateral rectus	42 colonies *S. albus*	Numerous *S. albus*, 10 colonies *S. aureus*
12	Esotropia	16 colonies *S. albus*	Numerous *S. albus*, *S. aureus*
13	Exotropia	22 colonies *S. albus*, diphtheroids	Numerous *S. albus*, 12 colonies *S. aureus*, diphtheroids
14	Exotropia	17 colonies *S. albus*	Numerous *S. albus*, 5 colonies *S. aureus*, diphtheroids
15	Congenital ptosis	2 colonies *S. albus*	Numerous *S. albus*, 44 colonies *S. aureus*

counts in patients with normal conjunctivas who wore an eye patch for 2 days.

Another factor that may influence the occurrence of infections following cataract extraction is the nature of the operation. At the IOCPMC since 1938 there have been 13,524 operations involving intracapsular extraction of the lens, 16 of which were followed by endophthalmitis, an incidence of 0.12%. In the same period there were 3,119 cases of extracapsular lens extraction, followed in 11 instances by endophthalmitis, an incidence of 0.35%. It would seem from these figures that endophthalmitis may be less likely to occur following intracapsular extractions. The added hazard of extracapsular extraction is also suggested by some other reported results.[2,5]

Culturing of the air from the operating room[14] and the routine monitoring of the sterility of instruments, dressings, solutions, and other objects that might come in contact with the patient's eye during surgery have failed to incriminate these objects as possible exogenous sources for the infections seen at the IOCPMC. In this respect the Institute of Ophthalmology at the Columbia-Presbyterian Medical Center has so far been fortunate.

Hughes and Owens[8] have stated that in a series of 2,086 operations performed without prophylactic antibiotic treatment, purulent endophthalmitis occurred in 21 patients. Staphylococci, streptococci, pneumococci, or *E. coli* were isolated from cultures of secretions in the lower cul de sac. These investigators believed that the sources of infection were "(1) obstruction of the lacrimal duct with resulting low grade chronic dacryocystitis; (2) inadequately treated chronic conjunctivitis; (3) sudden onset of an acute infection of the upper respiratory tract . . . ; (4) abscess around corneal scleral sutures . . . , and (5) undetermined causes." After they employed prophylactic measures with prolonged treatment, only 2 cases of postoperative endophthalmitis resulted out of a total of 1,200 cataract extractions. Maumenee and Michler[15] have emphasized the possible role of microorganisms carried on the skin as a source of postoperative infection. They recommend the topical application of antiseptics. Boyd, Maumenee, and McLean[3] recommend thorough bacteriologic studies of the lid and conjunctiva preoperatively, if the eye is inflamed, followed by adequate treatment prior to surgery. They use a subconjunctival injection of penicillin following surgery. Pearlman[16] also reports favorably on this type of prophylactic antibiotic treatment. Callahan[4] in a series of 1,653 cataract extractions lost 5 eyes with endophthalmitis. He believed that delaying the operation for adequate antibiotic treatment had reduced the number of infections in his series and might prevent future cases of endophthalmitis as more specific antibiotics were developed. Liehn and Schlagenhauff[10] reviewed 1,000 cases of cataract extraction in which 2 eyes were lost, one because of a mixed infection with pneumococci and a paracolon bacillus, the other from *E. coli* infection. They employed penicillin prophylactically.

Fine and Zimmerman[7] have collected the literature on mycotic infections of the eye and have reported the results of examination of eyes sent to the Registry of Ophthalmic Pathology. They employed stains appropriate for identifying fungi in the deep tissues. They found histologic evidence of fungal infection in 13 cases, 4 with a history of trauma and 9 following cataract extraction. These authors probably were in the position of receiving highly selected material in which the etiologic agent had not been identified. Fungi causing disease have not been cultured from postoperative infections in our experience, although expert assistance has been provided from the mycology laboratory.

Leopold and Apt,[9] and Theodore,[16a] and Theodore, Littman, and Almeda[17] have summarized the literature on fungal infections of the eye and have discussed the therapy in ocular infections in general. Chapter 13 describes the role of fungi in ocular infections.

REFERENCES

1. Adler, F. H.: Physiology of the eye: clinical application, ed. 3, St. Louis, 1959, The C. V. Mosby Co., p. 45.

1a. Allansmith, M. R., Anderson, R. P., and Butterworth, M.: The meaning of preoperative cultures in ophthalmology, Trans. Amer. Acad. Ophthal. Otolaryng. 73:683, 1969.

2. Allen, H. F., and Mangiaracine, A. B.: Bacterial endophthalmitis after cataract extraction, Arch. Ophthal. **72**:454, 1964.

3. Boyd, B. F., Maumenee, A. E., and McLean, J. M.: Advances in cataract surgery, Highlights Ophthal. 3:243, 1960.

4. Callahan, A.: Effect of sulfonamides and antibiotics on panophthalmitis complicating cataract extraction, Arch. Ophthal. **49**:212, 1953.

4a. Custodis, E.: Beitrag zur intraokularen Infektion nach Staroperationen, Klin. Mbl. Augenheilk. **133**:632, 1958.

5. Dellaporta, A., Riemer, F., and Harmuth, E.: Eine vergleichende Untersuchung über die Infektionshäufigkeit nach Kataraktoperation in den Jahren 1936, 1941, 1946 und 1948 mit Berucksichtigung der Penicillinbehandlung, Klin. Mbl. Augenheilk. **115**:369, 1949.

6. Dunnington, J. H., and Locatcher-Khorazo, D.: Value of cultures before operation for cataract, Arch. Ophthal. **34**:215, 1945.

7. Fine, B. S., and Zimmerman, L. E.: Exogenous intraocular fungus infections with particular reference to complications of intraocular surgery, Amer. J. Ophthal. **48**:151, 1959.

8. Hughes, W. F., and Owens, W. C.: Postoperative complications of cataract extraction, Arch. Ophthal. **38**:577, 1947.

9. Leopold, I. H., and Apt, L.: Postoperative intraocular infections, Amer. J. Ophthal. **50**: 1225, 1960.

10. Liehn, R., and Schlagenhauff, K.: 1000 intrakapsuläre Starextraktionen, ihre Ergehnisse und Komplikationen, Wien. Klin. Wschr. **64**: 988, 1952.

11. Linhart, R. W.: The effect of the eye patch on organisms of the conjunctival sac, Amer. J. Ophthal. **33**:1280, 1950.

12. Locatcher-Khorazo, D.: The effect on the ocular bacterial flora of local treatment with Chloromycetin (chloramphenical), Terramycin, or penicillin-streptomycin ophthalmic ointment in preoperative cataract cases and miscellaneous infections, Amer. J. Ophthal. **36**:475, 1953.
13. Locatcher-Khorazo, D., and Gutierrez, E.: Eye infections following cataract extraction with special reference to the role of Staphylococcus aureus, Amer. J. Ophthal. **41**:981, 1956.
14. Locatcher-Khorazo, D., and Gutierrez, E.: Bacteriophage typing of Staphylococcus aureus: a study of normal, infected eyes and environment, Arch. Ophthal. **63**:774, 1960.
15. Maumenee, A. E., and Michler, R. C.: Sterility of the operative field after ocular surgery, Trans. Pacif. Coast Otoophthal. Soc. **32**:172, 1951.
16. Pearlman, M. D.: Prophylactic subconjunctival penicillin and streptomycin after cataract extraction, Arch. Ophthal. **55**:516, 1956.
16a. Theodore, F. H.: Bacterial endophthalmitis after cataract surgery, Intern. Ophthal. Clin. **4**:839, 1964.
17. Theodore, F. H., Littman, M. L., and Almeda, E.: The diagnosis and management of fungus endophthalmitis following cataract extraction, Arch. Ophthal. **66**:163, 1961.
18. Thompson, R., Isaacs, M. L., and Khorazo, D.: A laboratory study of some antiseptics with reference to ocular application, Amer. J. Ophthal. **20**:1087, 1937.
19. Thygeson, P.: Ocular bacteriology, The Lancaster Course in Ophthalmology, The Ophthalmological Study Council, Division of Summer and Special Programs, Colby College, Waterville, Maine, p. 3.
19a. Whiston, G. J.: Review of postoperative endophalmitis, Montreal General Hospital—1955-1964, Canad. J. Ophthal. **2**:63, 1967.
20. Winkler, C. H., Jr., and Cason, L.: Bacteriology of the eye, role of gram-negative bacilli in infections following cataract extraction, Arch. Ophthal. **51**:200, 1954.

7

Aseptic technique in ophthalmology

HENRY F. ALLEN

Aseptic technique in hospital and office practice is a systematized, continuing effort to prevent infection by eliminating pathogenic microorganisms from the patient's environment. It is directed by the chief of service through his representative and must be supported by every member of the staff. It is of direct concern to the hospital administrator. It is above all a team effort, demanding the intelligent participation of all hospital and clinic personnel.

Aseptic technique can be no stronger than its weakest component. An error on the part of one individual can break the chain of asepsis. The duty of the ophthalmologist is to educate, to organize, and to inspire with the desire for perfection the members of his team. Only when discipline becomes second nature and any break in technique is detected instantly can it be said that the system is functioning as it should. Only by striving for the ideal of no infections can the incidence of infection be reduced to a minimum.

Foundations of asepsis

The basis for aseptic procedure rests on fundamentals relating to the communicability, virulence, and destruction of microbes and the resistance or susceptibility of tissues. These foundations are established in microbiology, epidemiology, and general surgery; specialized considerations determine their application to ophthalmology. Bacteriologic concepts include thermal death times of various groups and species of organisms, the relative sterilizing efficiency of moist and dry heat, and the ability of certain species to multiply in unpreserved solutions. Epidemiologic methods reveal modes of transfer and vectors of disease. Surgical principles stress operating room procedure and discipline, hemostasis, and respect for tissues. When infections occur, violation of basic principles should always be suspected. When claims for a new method or compound are at variance with existing knowledge of similar procedures and substances, acceptance should follow only after careful experimental and clinical evaluation of these claims.

Susceptibility of ocular tissues

Whereas other tissues can usually deal successfully with smaller or larger numbers of pathogenic bacteria, the cornea, lens, and vitreous being avascular are by nature especially vulnerable to attack by pathogens and even by organisms that are relatively nonpathogenic for vascular structures. The latter, though invaded, may emerge functionally unimpaired after elimination of infection by normal defense mechanisms with or without the help of specific therapy. But since transparency is

Modified from Allen, H. F.: Aseptic technique in ophthalmology, Trans. Amer. Ophthal. Soc. **57**:377-472, 1959.

synonymous with function in the cornea, lens, and vitreous, and since repair of damaged neural elements can occur to no significant degree in the retina, the development of an infectious process in the eye may lead to irreversible disability in spite of prompt and ultimately effective treatment. In intraocular infections there is literally no substitute for prevention.

Misconceptions and pitfalls

All ophthalmologists wish to prevent infection and presumably believe that they are taking all possible steps to do so. However, available evidence shows the need for wider understanding of the principles and practices on which aseptic eye surgery must be based. This evidence consists of: (1) continued occurrence of postoperative infections, both reported and unreported; (2) breaks in technique apparent to the casual observer in operating rooms and clinics; (3) use of preparations and devices, which not only are inefficient antiseptics, disinfectants, and sterilizers but may actually serve as vehicles of the very infections they are intended to prevent; and (4) the false sense of security generated in the minds of ophthalmic surgeons by the absence of infection from a series of intraocular operations. In some cases the absence of infection must be attributed more to the fact that infectious agents do not readily establish themselves in the eye than to the perfection of the technique employed. Only the busiest ophthalmic surgeons can cite a series large enough to have statistical significance. When postoperative infection strikes, it does so without warning. It may involve the eyes of several patients at once, and it usually destroys useful vision in the eyes it touches. For these reasons every avenue of exogenous infection must be considered and every means employed to block these routes and to eliminate organisms at their source.

Prophylactic antibiotics

Antibiotics may supplement but can never substitute for careful aseptic technique. In the brief history of these drugs, the development of bacterial resistance and of human hypersensitivity have refuted their infallibility against organisms and their innocuousness to patients. Antibiotic-resistant staphylococci and such refractory genera as *Proteus* and *Pseudomonas* forbid complacency. No antibiotic administered by any route can prevent infection resulting from irrigation of the anterior chamber with a pure culture of organisms in an aqueous solution. Anaphylactic deaths and other severe reactions make one consider the risk of sensitizing patients to future injections or of provoking a reaction in a previously sensitized patient. Before administering the antibiotics for prophylactic purposes, the surgeon must weigh their advantages against their hazards for each individual patient.

Responsibility for asepsis

The responsibility for aseptic technique, as for other aspects of patient care and protection, belongs to the physician who treats the patient. The surgeon who operates in ignorance of conditions and procedures in his hospital cannot shift responsibility to higher authority or impute negligence to subordinate personnel. Personal responsibility carries both the right and the obligation to investigate and to correct or insist in writing upon correction of deficiencies. Rigid self-discipline must precede criticism of infractions on the part of others.

In hospitals where ophthalmology is not a separate service, the ophthalmic chief must be prepared to educate the chief of surgery in the specific requirements of aseptic eye surgery. The hospital administrator, the chief of the nursing service, and the hospital pharmacist must be oriented in the same direction.

The initial institution of seemingly adequate precautions is not sufficient in itself. Many factors operate to inactivate or to infiltrate the barriers erected against bacteria. Periodic checks must, therefore, be carried out to ensure the continued effectiveness of current practices. Cultures of air, instruments, and supplies from the autoclave should be performed at regular intervals. Methods of conducting sterility surveys are outlined in this chapter.

THE DEVELOPMENT OF ANTISEPTIC AND ASEPTIC TECHNIQUE IN GENERAL AND OPHTHALMIC SURGERY

Concepts of infection before 1850

The history of aseptic technique began with the discovery of microorganisms and of methods for their destruction. Before the modern era, the value of dry heat (stoving) for rendering innocuous the fomites of scarlatina and smallpox was demonstrated by William Henry in 1831. It remained for Koch and Wolfhuegel[28] fifty years later to work out thermal death times of anhydrous spores. A device designed by a French physicist in 1679 and known as "Papin's Digester" was the precursor of the modern autoclave. Pasteur employed steam under pressure at 120° C for sterilization. Redard[41] demonstrated that this was an effective means of killing spores.

During the first half of the nineteenth century, postoperative suppuration and methods for its prevention received but scant mention in the ophthalmic literature. Infection was considered to be endogenous and therefore unavoidable in a certain percentage of cases. Travers[47] believed that the degree of inflammation following cataract extraction depended upon "bodily habit." He recommended cupping, purging, and an abstemious life; but his treatise contains no reference to cleanliness or to the care of instruments.

Ophthalmic surgery in the preantiseptic era, 1850-1875

The invention of the ophthalmoscope by von Helmholtz at midcentury marked the beginning of modern ophthalmology. A few years later the techniques of Albrecht von Graefe gave to intraocular surgery its first major advance since the introduction of cataract extraction through a corneal incision over 100 years before.

All of this preceded proof of the germ theory of disease by several years. The members of the First International Congress of Ophthalmology, meeting at Brussels in 1857, could scarcely have dreamed that 3 years later Pasteur would awaken the world to the science of bacteriology with his *Mémoire sur la fermentation alcoolique*. The rest of the third quarter-century would pass before Lister's contributions to antiseptic surgery would find their modified expression in the ophthalmic field.

Describing conditions as they were in the preantiseptic period, Ernst Fuchs[19] noted that when he started as an aspirant at the clinic of the University of Vienna in 1873, von Arlt "taught us to wipe Daviel's spoon between our lips before putting it into the eye, in order that it be moistened and made slippery." It is surprising that the incidence of cataract wound infection was only 8% to 10% under those circumstances. Following iridectomy, infections of the wound were exceptional. Fuchs attributed this difference to the role of free lens material in providing a good culture medium for bacteria. This it undoubtedly does. Kinoshita has observed *Pseudomonas* contamination of lens homogenates at 4° C.

The following year Fuchs was sent on detached service to Billroth's general surgical clinic. By a stroke of good fortune, Lister himself came to demonstrate his methods to Billroth over a period of several weeks, and personally assisted Fuchs during his stay. When Fuchs returned in 1876 to become an assistant in von Arlt's clinic, he applied antiseptic techniques to preparation of the operative field and to sterilization of instruments and supplies. By this time others were beginning to follow the new trend, and it was possible for the first time to visualize the beginning of the end of septic complications in intraocular surgery.

The antiseptic era, 1876-1886

The arrival of the new era was signalized in the ophthalmic literature in 1878 by Alfred Graefe's paper, "Antiseptic Wound-Treatment in Cataract Extraction."[20] Before his adoption of antiseptic measures, Graefe reported a minimum loss of 4% of eyes operated for cataract extraction, with an upper limit of 10% and an average loss of 5% to 6%. He succeeded in reducing these figures to the loss of 2 eyes in 114 consecutive operations by adoption of a modified Listerian routine.

Graefe's figures, both in regard to the preantiseptic incidence of intraocular infection and to reduction of this incidence by antiseptic techniques, were confirmed by a larger series reported by Horner[23] 3 years later. Both Graefe and Fuchs mentioned specifically that the carbolic acid spray, used as an integral part of the Listerian regimen during operations elsewhere in the body, was too irritating to patient and surgeon to permit its use in ocular surgery.

Beginnings of asepsis

In ophthalmology also, the trend changed from carbolic acid to bichloride of mercury. Landolt and Hirschberg were operators meticulous in their attention to details of asepsis and antisepsis. The former cleansed the lids and irrigated the conjunctival sac before and during operation with bichloride of mercury 1:5,000. The latter, with the Prussian passion for discipline, made every visitor wash his hands with soap, soak them in bichloride of mercury, and don a white robe before witnessing an operation. Hirschberg also subjected all his dressings, bandages, solutions, and instruments to high degrees of heat in containers specially made for the purpose. Like Landolt, he insisted on the most careful cleansing of the eye and the ocular adnexa before operation. Millikin[35] noted that pain could be avoided by making up bichloride of mercury with 4% cocaine solution, but stated that hyperemia of the conjunctiva was considerable as a result of the irrigation.

By the turn of the century all the basic principles of asepsis as we know them today had been clearly set forth, and few if any significant advances have been made since that time in this area. That infections have occurred and continue to occur can usually be explained by the failure to observe one or more of these principles.

THE PROBLEM OF OCULAR INFECTION

Etiology of ocular infections

Bacterial infections that follow elective intraocular surgery are almost always caused by a member species of one of two groups of vegetative organisms. The gram-positive cocci are most often represented by *Staphylococcus aureus,* less often by pneumococci, and more rarely by hemolytic streptococci. Among the gram-negative rods, *Pseudomonas aeruginosa,* the bacillus pyocyaneus, is the chief offender, followed by *Aerobacter aerogenes, Proteus* species, and coliform bacilli. Although spore-forming organisms are seldom identified in postoperative infections, they are often found in operating rooms and must always be taken into account. Mixed infection by *Proteus* organisms and *Bacillus subtilis* has followed retinal detachment surgery.

Postoperative intraocular fungus infections are not so rare as was formerly believed. Fine and Zimmerman[18] in 1959 reported 13 new exogenous cases of which 9 followed intraocular surgery. In 8 of these the operation was a cataract extraction. A latent period of days, weeks, or months after surgery preceded development of the clinical picture of plastic iridocyclitis with anterior endophthalmitis.

The globe is resistant to inward extension of external infection from the conjunctiva and orbital tissues. Trephine blebs and filtering cicatrices form an exception to this rule. They are particularly susceptible to pneumococcal infection.

In some cases of presumed bacterial endophthalmitis, the offending organism is never recovered. This failure may result from failure to tap the anterior chamber or from faulty culture techniques. Unless a blood culture is positive, a surgeon has no justification for attributing the cause of infection to endogenous sources. The recovery of a pathogenic organism from the conjunctival sac by no means establishes its origin. By the time endophthalmitis has developed, many of the possible sources of infection are no longer available for analysis. Since treatment is usually ineffective, the best hope lies in prevention of infection. The keys to prevention are anticipation and vigilance.

Epidemiology of ocular infections

History records many mechanisms for transmission of ocular disease. These range from the use of urine as an eyewash to the role of insects in carrying bacteria from eye to eye. Mother-to-child transmission is observed prenatally in gonorrheal ophthalmia and inclusion blennorrhea; postnatally, in herpes simplex and trachoma. Insect vectors are responsible for ocular filariasis and onchocerciasis. Swimming pools have been indicted in outbreaks of adenovirus infection and inclusion blennorrhea of adults. Animal hosts have been identified in cat scratch disease, tularemia, and others. Ingestion of infected meat may lead to trichinosis; of raw milk, to brucellosis and tuberculosis.

The most ironic methods of transfer are those mediated by medical personnel. Outbreaks of *Pseudomonas* infection from contaminated solutions and transmission of epidemic keratoconjunctivitis by infected tonometers, ophthalmoscopes, and unwashed fingers furnish the most common examples of iatrogenic spread of ocular disease. Ophthalmologists, among other physicians, can transmit serum hepatitis by breaking the skin with a knife or needle bearing this virus. Strict precautions are necessary if the eye physician is to avoid becoming a vector of disease.

Education of the public and of professional colleagues in the prevention of ocular infection is essential. In many cases, socioeconomic factors place the prevention of a given disease in the province of the public health officer. Where personal hygiene and individual precautions play a part, it is the ophthalmologist's duty to instruct his patient in the steps necessary for the protection of himself and others.

Incidence of postoperative infection

During the modern era of ophthalmic surgery, comprising approximately the past 100 years, statistics of the incidence of surgical infections have been derived almost entirely from cataract operations. The incidence of infection in intraocular procedures of all types is probably lower than that reported for cataract surgery, since it is generally admitted that the hazard of infection is less in simple iridectomy and other operations for glaucoma.

Figures such as those cited in Table 7-1 are necessarily derived from larger clinics. A series of fewer than a thousand operations has little statistical significance. Apart from large series, numerous cases of infection occur, which are never reported in the literature.

Incidence of infection is influenced by operative technique. Prolonged operating time, multiple entries into the anterior chamber, and excessive manipulation of the wound invite infection. Blood clots, traumatized tissues, and lens cortex are all good culture media. The infection rate is greater after extracapsular extraction than when the lens is delivered intact. There may also be a greater risk of postoperative infection in eyes of which the fellow eye has been operated during the same hospital admission. Patients may unconsciously rub the unoperated eye during the period intervening between the two operations. Topical antibiotics should be used in both eyes after operating on the first eye.

Table 7-1. Endophthalmitis after cataract extraction

Period	Rate per 1,000
Before 1875	40-100
1875-1900	10-20
1900-1950	11-12
1950-1970	1-2

Sources of ocular infection

Since it is not always possible to trace the origin of a given infection, all potential sources of contamination must be considered. Close attention must be paid to solutions and instruments that come into contact with intraocular tissues. Serious outbreaks of infection have been traced to preparations employed for their supposed antiseptic properties.

The main categories or sources of infection that must be considered in planning a system of aseptic technique are:

1. Solutions and medications
   - Saline for irrigation
   - Collyria
   - Ointments
   - Instrument disinfectants
   - Skin antiseptics
   - Injectables
2. Tissues
   - Skin of hands
   - Skin of operative field
   - Lid margins and lashes
   - Conjunctival sac
   - Lacrimal sac
   - Nasal mucosa
   - Corneal grafts
   - Vitreous implants
   - Fellow eyes
3. Objects and materials
   - Surgical instruments
   - Optical instruments
   - Tonometers
   - Magnets
   - Hypodermic needles, blood lancets
   - Drapes, dressings, masks and gowns
   - Rubber gloves, bulbs, droppers
   - Glass syringes, bottles, irrigating tips
   - Plastic tubing, sheeting, retractors
   - Contact lenses
   - Orbital implants
   - Intraocular lenses
   - Sutures
4. Airborne contaminants
   - Respiratory origin
   - Surface origin—skin, clothing
   - Ventilating system

These categories are separately considered in subsequent sections of this chapter.

DEFINITION OF TERMS

The following definitions reflect current usage.

STERILIZATION. Sterilization is the use of any process, physical or chemical, that destroys all forms of life. The concept of sterilization is absolute and permits no exceptions. It implies the destruction of all types of microorganisms. Since boiling water and most antiseptics will not destroy resistant bacterial spores, to refer to such agents as sterilizers is incorrect. Under the terms of this definition, sterilization can be achieved for practical purposes only with inanimate objects and restricted spaces. Living tissues may or may not be sterile; if unsterile, they can seldom be sterilized without being themselves destroyed, as is true for skin. A completely sterile environment can be achieved only under experimental conditions. In surgery, the effort to achieve sterility of environment can be considered successful if all potential pathogens have been eliminated from the operative field.

The suffix "-cide" is used to indicate an agent that will destroy that form of microorganism to the name of which it is appended. A sporicide is likely to be a complete sterilizer, since certain bacterial spores are the most resistant of organisms. On the other hand, germicides and bactericides are not necessarily sporicidal or virucidal and will therefore not be sterilizers if spores or viruses are present. The use of the term "cold sterilization" as commonly employed is a misnomer, since the processes usually designated by this term are not lethal to resistant spores.

DISINFECTION. Disinfection refers to killing of pathogenic agents by chemical or physical means directly applied. In the accepted sense, disinfection does not necessarily include the killing of resistant spores or the destruction of viruses. Certain disinfectants may have a sporicidal or virucidal effect; this is usually specified on the label. The term disinfectant is commonly used with substances applied to inanimate objects. Degerming the skin is accomplished with the help of antiseptics rather than disinfectants.

ANTISEPTIC. Antiseptics are substances that prevent or arrest growth or action of microorganisms, either by inhibiting their activity or by destroying them, especially agents applied to living tissue. Although we commonly think of antiseptics as germicides, the term originally implied only the ability to arrest putrefaction. In the modern sense of the word, antiseptic implies merely a bacteriostatic effect. Ability to kill spores or viruses is not implied by the term antiseptic. Bactericidal properties may rest with one or another antiseptic substance, depending on the conditions under which it is used. The first antiseptic in general surgery was carbolic acid, used by Lister as a 5% or 2½% solution.

SANITIZATION. Sanitization refers to reduction of the number of bacterial contaminants to safe levels as judged by public health requirements. This term is applied chiefly to the cleansing of utensils for eating and drinking. The standards implied are neither precise enough nor rigid enough for application to surgical procedures, especially in ophthalmology.

CORRECT USE OF TERMS. Certain of the above terms may have more than one meaning, and these meanings may overlap. Ambiguity can be avoided by specifying the circumstances and the microbiologic agents against which a given process is to be used and by limiting the term sterilization to those circumstances under which all forms of life are destroyed.

RESISTANCE OF ORGANISMS

Natural versus acquired or induced resistance

The ability of microorganisms to withstand exposure to adverse environmental influences is a subject of great complexity and of infinite variations. Its interest, both theoretical and practical, has been heightened by the development of chemotherapeutic and antibiotic agents. The appearance of acquired resistance, through processes of selection and of induced resistance (through mutation) is of obvious significance. However, the range in which such adaptation can occur is so limited as to be of little or no practical importance for purposes of surgical sterilization. The problem, therefore, lies not so much in the discovery of new means of killing organisms as in selecting and adapting agents of proved effectiveness on the basis of their suitability to the practical problems at hand. For this reason, only a brief résumé of the natural resistance of groups and species is pertinent to the present discussion.

Environmental factors

The principal physical agencies to which organisms may be subjected include heat, light, ionizing radiations, dessication, and supersonic vibrations. The range of chemical compounds employed against bacteria is almost unlimited. In every case the resistance of a given species or strain is influenced by the environmental conditions prevailing at the time of exposure to the particular agency. The most important of these conditions are the presence of moisture, the presence of organic matter, and the pH of solutions.

Within the range of ordinary environmental factors, microorganisms vary greatly in their ability to survive exposure to the various physical and chemical agents that may be applied to them. The most significant difference exists between the resistance of bacterial spores and that of all other microbial forms.

Spore-forming bacteria

Two bacterial genera, the aerobic bacilli and the anaerobic clostridia, have the property of forming endospores. Spores presumably represent a resting stage in the bacterial life cycle. Formation of spores is *sporulation; germination* signifies their return to the vegetative phase upon incubation in a nutrient medium.

Bacterial spores are the most resistant form of life, and it is believed that any process capable of killing resistant spores will result in the destruction of all other organisms as well. The resistance of spores extends to most types of physical and chemical agents and forms the basis of differentiation between absolute and relative methods of sterilization. Most spores are known to withstand boiling, ultraviolet light, and a wide range of chemicals that are lethal to other forms of microorganisms. Spores can survive indefinitely in solutions of disinfectants that kill vegetative bacteria in a matter of minutes. The differential between spores and vegetative forms of the same species is much smaller for gaseous sterilants.

Spore-forming species in the vegetative phase have no greater resistance than species that exist only in that phase.

Failure to use inocula or test strips rich in spore content can lead to false conclusions in evaluating sterilizing procedures.

Vegetative bacteria and fungi

The differences in resistance among non–spore-forming species are of less practical importance. The spores of fungi, though far less resistant than those of bacteria, are somewhat more resistant than vegetative forms. Bacteria and fungi in the vegetative phase are destroyed in 1 hour or less by boiling, by heating in the anhydrous state, and by many disinfectants. Certain species, such as the staphylococci, are more resistant than others, such as the neisseriae. For example, *N. gonorrhoeae,* an unusually delicate species, succumbs to 41.5° C in a few minutes; many strains of staphylococci withstand 60° C for 1 hour.

Most vegetative bacteria die out when dried, but tubercle bacilli have been found viable in dry dust after many weeks. Gram-negative organisms are seldom found in dust, although they occur occasionally in samples of room air.

Other differences between gram-positive and gram-negative species include the greater susceptibility of the former to penicillin and the cationic detergents.

Viruses

Although viruses can multiply only in the presence of living cells, most viruses can survive for a while in their absence. This property can readily be tested with viruses that can be propagated in tissue culture or in animal hosts. One of the more delicate viruses is that of herpes simplex. This agent is readily inactivated by heat (56° C for ½ hour) and by storage at room temperature. The latter result is the effect partly of oxidation and partly of phosphatase activity. Protection against both mechanisms is afforded by reducing agents such as cysteine and glutathione.

The resistance of adenoviruses to ether suggests that cleaning of tonometers with ether alone may not prevent transmission of epidemic keratoconjunctivitis. Although there is some evidence that swimming pools may be an epidemiologic factor in the spread of pharyngoconjunctival fever, it is probable that adenovirus type 3 in purified suspension is destroyed by 0.4-0.6 ppm residual chlorine.

The ability of inclusion blennorrhea virus (*Chlamydia oculogenitale*) to survive in swimming pools and to resist Credé prophylaxis with 1% silver nitrate suggests that the chlamydiae are either resistant to chemicals or protected from chemical action, possibly by the carbohydrate matrix that they possess.

Viruses that proliferate only in a human host present obvious difficulties to anyone wishing to test their resistance to heat and chemicals. In the case of serum hepatitis virus, studies with human volunteers have shown that this agent is killed by boiling for 15 minutes and by steeping at 60° C for 10 hours, but may survive for 1 hour or more at 60° C. The same virus has apparently remained infective after prolonged exposure to 0.2% tricresol. The practical importance of these data is apparent in relation to the sterilization of needles and sharp instruments.

PRINCIPLES OF STERILIZATION

Methods that kill all forms of microorganisms, including resistant spores, are always to be preferred to methods that merely kill certain species under certain conditions. Spore-forming species are constantly present in the environment and can never be discounted as a potential hazard. The sporicidal requirement provides a margin of safety over and above the minimum circumstances necessary for the destruction of vegetative forms. Because the eye may respond to infection in all-or-none fashion, acceptance of methods less than infallible is equivalent to playing "ophthalmic roulette."

The theory and practice of sterilizing processes are fully described in standard texts. The following paragraphs are intended only to provide a convenient summary. Ophthalmologists desiring fuller treatment of the subject are referred to source material.[4,32,38,50]

Requirements common to all methods

PROCESSING. The first requirement for sterilization is that the material must actually be subjected to the sterilizing process. Obvious though this may seem, instances of infection have occurred through the use of objects erroneously believed to have been processed. This accident can be avoided. Unsterile packages should never be stored in the sterilizing room or in a place where they can be picked up in the mistaken belief that they have already been sterilized. Sterile packages must be clearly labelled with heat-sensitive adhesive tape. Other types of thermal controls that indicate that a certain temperature has been maintained for a given period of time, may be used. None of these thermal controls is infallible, but they are useful adjuncts.

INCOMPLETE STERILIZATION. Next in disastrous effect to complete omission of the sterilizing procedure, and of far more frequent occurrence, is faulty or incomplete execution of the prescribed routine. Haste, impatience, ignorance, fatigue, and carelessness are the components of the human element, a factor that must be considered by every surgeon who does not, as indeed he cannot, personally supervise every step in the sterilization of every object used in the course of an operation. Indoctrination, training, and frequent checks on sterility are the only means of reducing such errors to a minimum.

TIME FACTOR. Time is essential to any sterilizing process. The critical time interval is that during which the object to be sterilized is actually exposed to the sterilizing agent, whether this be a certain degree of heat, a certain concentration of vapor, or a chemical in solution. Additional time must be allowed for penetration of the package and permeation of all parts of the object or material. Large packages and instruments require a longer time than small ones. The time required for objects to reach the sterilizing temperature is in addition to that required for sterilization.

CLEANING. Mechanical cleaning is necessary, not merely for esthetic reasons, but because the sterilizing action of heat, moisture, and chemicals is prevented or seriously delayed by the presence of organic residues. Few eye surgeons insist upon cleaning their own instruments with the aid of special illumination and of magnification. This practice should recommend itself to any surgeon who has found bits of tissue adherent to the teeth of forceps or to the points of diathermy electrodes. Whenever possible, instruments should be taken apart for cleaning. Joints should be worked back and forth in a detergent solution, and a stiff brush is essential for getting into crevices. Drying is important to prevent corrosion and because bacteria can proliferate wherever moisture is present. Ultrasonic cleaning is an important advance.

PACKAGING. The manner of wrapping must permit delivery without contamination of the contents. The material used for packaging must be permeable by the sterilizing agent. Cloth wrappings are suitable for steam sterilization, as are certain types of paper. Plastics are usually waterproof but may be permeable by certain gases. However, if they are permeable by the sterilizing gas but impermeable by air, the gas concentration attained within the package may be below that required for sterilization, for the simple reason that air connot be exhausted from the package. Glass and metal containers are suitable only for aqueous solutions or for materials that are to be subjected to dry heat.

Packages that are not to be used immediately after sterilization must be stored in such a way as to prevent contamination and should be resterilized after a long period of storage under certain conditions. Moisture or wetting is to be avoided at all costs, and any package that is not absolutely waterproof is considered to be contaminated if it has been in contact with liquid. Dry, imperviously wrapped packages may remain sterile for indefinitely long periods.

STERILIZING METHODS

Steam under pressure

Correctly performed, exposure of objects to saturated steam under pressure in the autoclave is the most rapid and dependable means of complete sterilization for practical use. For sterilization to occur, all objects in the autoclave must be exposed in their entirety to the action of saturated steam at the sterilizing temperature (121° C or higher). Moisture, heat, and time are essential; pressures achieved in the autoclave have no sterilizing action per se.

Saturated steam is water vapor at the maximum pressure and density for a given temperature. Because of Dalton's law of partial pressures, this situation cannot exist if another gas is present. If air has not been completely removed, steam will not be saturated, no matter what the pressure. At a given pressure, the maximum temperature for steam cannot be reached if air is present. Since air and steam do not mix readily, objects in contact with air are insulated from both heat and moisture. A film of oil or grease is also an effective means of shielding bacteria from the action of moisture. Since air is heavier than steam, modern autoclaves are evacuated through an opening at the bottom of the chamber, and the thermometer element is located at this point. In this way the sterilizing temperature is not recorded until all of the air has been exhausted from the chamber. Loading of packages, containers, and of the autoclave itself must be such that no air can be trapped by gravity at any point in the autoclave.

In the presence of saturated steam, temperature at equilibrium is a function of pressure. Ordinary autoclaves operate at 15 pounds per square inch (psi) and 121° C; high-speed autoclaves achieve 132° C at 27 psi. These temperatures kill resistant spores in minutes by moist heat; when moisture is absent, the same temperatures require almost the same number of hours to kill the same type of spores. Once the sterilizing temperature has been reached, exposure for 10 min at 121° C is considered sufficient to kill the spores of pathogenic bacteria. Solutions in containers are no exception to the time requirement. It takes longer to sterilize 1,500 ml of saline in a flask than 500 ml, just as it takes longer to boil a large volume of water than a smaller one.

Personnel entrusted with the responsibility of sterilizing instruments and other supplies must be thoroughly trained in the principles and use of the autoclave. The foregoing brief discussion is intended only as a review of principles with which all surgeons should be thoroughly familiar. Technical details of operation of the autoclave and its maintenance are available in manuals provided by the manufacturer. Periodic checks by professional maintenance men are advisable.

Dry heat (hot air)

Exposure to hot air in a gas or electric oven is the time-honored method of sterilizing glassware for bacteriologic purposes. This rigid test in daily use is sufficient indication of the efficacy of the process, but conditions of time and temperature must be met in all cases. Although dry heat is a dependable means of killing spores, it is far less efficient for this purpose than is saturated steam. Therefore, both a higher temperature and a much longer period of exposure are required for the sterilizing temperature to be reached at the interior of a package; 20 to 30 minutes are required for a cataract knife in a manufacturer's cardboard box to reach 140° C in a preheated oven. Much more time is needed for dry heat to penetrate a bulky package wrapped in several layers of insulating material.

When time is available, dry heat has clear advantages for sterilization of objects that are damaged by or impermeable by moisture and even of those that are not. The time can be shortened by the use of higher temperatures for glassware or metals.

Because of the poor heat-conducting properties of air, close attention must be paid to ensure that free circulation of air can occur through all parts of the loaded oven. Overloading or crowding is almost certain to mean that gross differences in temperature will exist in different parts of the oven.

The simplest hot air sterilizer is the electrically heated gravity-type oven, which depends upon circulation of air by thermal convection. Several such ovens are on the market for sterilization of small instruments used in ocular, aural, and dental surgery. Since these ovens are small and carry loads that normally permit free circulation of air through the entire chamber, they are usually reliable. Nevertheless, their accuracy should always be checked with the thermometer and their efficiency verified by the use of packaged dried spores as controls.

Larger ovens may be of the gravity-convection type or may include a turboblower to provide forced circulation of air. The latter type is greatly to be preferred when the load includes numerous packaged instruments, as it usually does in the surgical unit of a busy hospital.

Gaseous sterilization

A field of expanding potential is the use of gaseous agents for sterilization of materials and objects that would be damaged by ordinary methods of heat sterilization.

An ideal gaseous sterilant would be: (1) noncorrosive, (2) harmless to materials, (3) penetrative, (4) easily removed by aeration, (5) rapid in action, (6) of low toxicity to humans and animals, (7) nonflammable and nonexplosive, (8) convenient to store, (9) easy to handle, (10) readily available commercially, (11) inexpensive, and (12) bactericidal, sporicidal, virucidal, and fungicidal at ordinary atmospheric conditions.

Formaldehyde vapor has limited usefulness because of its corrosiveness to aluminum and other metals, its tendency to polymerize on surfaces, and the resulting difficulty of removing the compound following treatment of objects and spaces. Like formalin solution, it finds little application in ophthalmic sterilization.

Ethylene oxide combines sporicidal effect with penetration into packages, some liquids, and certain plastics. Although flammable, it can be handled in small quantities as the pure liquid and in larger volumes in combination with relatively inert gases, such as carbon dioxide and freon. Its effect is greatest in the range of moderate humidity, rather than at extremes of dryness and moisture. In hospital and surgical use its greatest application lies in sterilization of blankets and of plastics. It can also be used for sterilization of ophthalmic instruments, ampoules, plastic prosthesis and implants, and other items that would be damaged by heat or excessive moisture. Experience with ethylene oxide indicates the need for continual spore controls.

Flaming

Organisms subjected to the open flame are killed by charring. In bacteriology, flaming the mouths of test tubes containing culture media is a time-honored procedure. In ophthalmology, platinum wire loops and spatulas are sterilized by flaming before cultures or scrapings are taken from the conjunctiva and cornea. Flaming is an effective method of sterilizing stainless steel tonometers, but electronic tonometers cannot be flamed without incurring permanent damage.

Ionizing radiations

Although x-rays, beta rays, and gamma radiation have sterilizing properties, beams of high-voltage accelerated electrons (cathode rays) have proved to be the most practical for processing applications. Bacterial spores, yeasts, molds, and vegetative spores, in order of decreasing resistance, can be destroyed under conditions similar to those that may be used for practical sterilization of objects of limited thickness.

Miscellaneous methods

Under certain circumstances numerous agencies other than those listed above may prove destructive to living matter. Ultraviolet light, ultrasonic vibrations, strong acids and alkalies, boiling water, formaldehyde, and various other physical and chemical means may prove lethal to organisms if they are allowed to act under optimal conditions for a long enough time. Most of them, unfortunately, have limitations that render them unsuitable for surgical sterilization, or at least impractical by comparison with the more reliable methods just described.

DISINFECTANTS

Although we have noted that sterilization is, properly speaking, an absolute concept, it is possible, under certain circumstances, to achieve sterility by agencies other than those discussed above. The conditions and circumstances under which these methods may be effective sterilizers must be clearly defined. Since these circumstances usually exclude the presence of bacterial spores, the application of these methods for surgical purposes is limited. Only when it is impracticable to apply one of the absolute methods can the use of a method of relative effectiveness be justified. So rare are these occasions in ophthalmology that there is hardly any indication for using one of the latter methods. Nevertheless, antiseptics and disinfectants have a place in environmental sanitation and may be valuable when they are used with full realization of their limitations.

With the exception of strong denaturants unsuitable for clinical purposes, most chemicals are not sporicidal. Many, however, are bactericidal and qualify as disinfectants. Newer standards for the evaluation of germicides show a trend away from the historic but limited concept of the "phenol coefficient" and stress the importance of "use-dilution" tests and of evaluating compounds under circumstances paralleling as closely as possible those under which they are to be used.

Quaternary ammonium compounds

The quaternary ammonium derivatives, often referred to as cationic wetting agents or cationic detergents, have been extensively employed as instrument disinfectants, skin antiseptics, and preservatives for ophthalmic solutions. The quaternaries, or quats, as they are sometimes called, include several hundred compounds. The terms benzalkonium chloride and benzethonium chloride are often applied to some of the mixed alkyl quaternary ammonium derivatives. Ophthalmologists will more readily recognize such familiar trade names as Zephiran, Ceepryn, Phemerol, and Roccal.

Although cationic detergents are effective germicides in vitro against certain vegetative organisms, especially gram-positive cocci, their germicidal effectiveness is greatly reduced in the presence of organic matter. Moreover, they suffer from other serious limitations that restrict their usefulness: (1) lack of uniform sporicidal effect, (2) lack of uniform bactericidal effect, (3) possibility of contamination by pseudomonas, (4) reduction of efficiency by inorganic cations that may be present in tap water, (5) inactivation by soap and other anions, (6) incompatibility with a large number of other compounds, (7) toxicity to corneal endothelium and epithelium, (8) ability to bind bacteria to the skin, and (9) adsorption to skin, glass, rubber, and gauze.

The fact that the quaternaries in aqueous solution may carry pseudomonas renders them unsuitable for routine use in ophthalmology. Not only are the quaternaries less effective than other methods for the purposes for which they are employed in ophthalmic practice, but also their use has been attended by deleterious and even disastrous results.

Synthetic phenolic compounds

Environmental sanitation is aided by the residual surface disinfectant action of such compounds as orthophenylphenol incorporated in solution with other appropriate constituents. The low volatility of these compounds makes it possible to heat them to temperatures close to boiling without any objectionable odor. At such temperatures they kill resistant spores that are otherwise destroyed neither by boiling in plain water for several hours nor by the same compounds acting at room temperature. Although no direct application of this effect to ophthalmic surgery is immediately apparent, the usefulness of these compounds in hospital hygiene is considered in a later section.

Summary

Disinfectants must be evaluated under the conditions for which they are to be used. They are never to be used for sterilization when more reliable methods are practicable.

SOLUTIONS AND MEDICATIONS

Role of medications in ocular infections

The ability of tap water, distilled water, and aqueous solutions of many different substances to support bacterial multiplication has often led to ocular infections.

Through fear of admitting legal liability, hospitals and individual physicians are reluctant to publish details of infections traceable to contaminated solutions that were or may have been used in affected eyes. Most ophthalmologists have either first- or second-hand knowledge of one or more eyes that have been lost in this way. Instances of contaminated solutions, reported[5] and unreported, are too numerous to mention. Such incidents, in addition to the occasional cases reported in the literature, make it clear that a hazard exists whenever an aqueous solution of any description is instilled into an eye. Even solutions of antibiotics may become contaminated with organisms that lie outside their range of effectiveness.

The following examples are cited to prove this point:

1. Postoperative infection occurred in five eyes subjected to intraocular surgery during a single morning on a large eye service. All five eyes were lost from endophthalmitis. *Pseudomonas* organisms were recovered in pure culture from the five eyes and from undiluted, previously unopened stock bottles of benzalkonium chloride 1:1,000 and from the more dilute solutions that were used for preoperative preparation of the skin of these patients.
2. Postoperative infection occurred in six eyes subjected to intraocular surgery on two successive days in a large eye hospital. All six eyes were lost from endophthalmitis. *Pseudomonas aeruginosa* was recovered in pure culture from the six eyes and from a large demijohn used to fill 1,500 ml flasks with saline. After autoclaving, one flask was used to fill trays and basins for three operations. Among these were the containers for anterior chamber irrigant. Although none of the flasks available for culturing at the time infection was discovered was found to be contaminated, it was postulated that two or more flasks used for the six operations had escaped processing in the autoclave.
3. Postoperative infection occurred in three eyes subjected to intraocular surgery in a large general hospital. All these eyes were lost from endophthalmitis. *Pseudomonas* organisms were recovered from these eyes and from a solution of a cationic detergent in which cataract knives used on these cases were immersed. It was said that bronchoscopic instruments had been immersed in the same solution.
4. Following removal of embedded corneal foreign bodies in a large industrial eye dispensary, pseudomonas corneal ulcers developed in five eyes. Four eyes lost useful vision. *Pseudomonas* organisms in pure culture were recovered from previously unopened stock bottles of a proprietary sulfonamide solution that had been instilled into these eyes for an intended prophylactic effect.

In analyzing this series of clinical disasters we note that in three of the four instances the contaminated solution had been employed because of its supposed antibacterial effect. In one case the offending substance was used as a skin antiseptic; in another, as an instrument disinfectant; and in the third, as a chemoprophylactic agent. The causative agent of all these infections was the organism that is at once the most common contaminant of aqueous solutions and the most lethal in its effect upon the eye, *Pseudomonas* or *Bacillus pyocyaneus*. This omnipresent organism grows within a temperature range of 4° to 42° C and can reproduce in the presence of little or no nutritional substrate. It thrives in the anterior chamber and spreads rapidly through the corneal stroma, where it often forms a series of concentric ring abscesses. Other ocular pathogens often found in solutions are S. *aureus* and members of the coliform group.

Credit for arousing interest in this problem and for bringing it to the attention of the authorities belongs chiefly to Theodore and Feinstein.[45] Since 1953 the Federal Food, Drug and Cosmetic Act has been interpreted as requiring manufacturers of eye medicaments to prepare sterile solutions fortified with adequate preservatives. This requirement does not yet apply to eye ointments or to ophthalmic medications compounded and dispensed by local pharmacies or hospital pharmacies. It therefore behooves every ophthalmologist to familiarize himself with the principles of asepsis as they apply to ophthalmic medications.

Saline solutions

Aqueous solutions of sodium chloride are used in ophthalmology for irrigation of the anterior chamber, corneoconjunctival irrigation, wound irrigation, filling of gonioscopy lenses, and for intravenous infusion. Saline for any of these purposes is to be considered as a pharmaceutical in the strictest sense. Commercial saline is suitable for all.

SALINE FOR INTRAOCULAR SURGERY. The most dangerous solutions in ophthalmology are those of saline used to irrigate the anterior chamber at surgery or the conjunctival sac in the presence of an epithelial defect in the cornea. Contamination of such a solution means introduction of a pure or mixed bacterial culture into the interior of the eye or into the defenseless corneal tissues. In either event, loss of the eye is the most probable result. For this reason the principle of sterilization in the final container must be rigidly applied, and the container may be opened when, and only when, the operation is ready to begin.

Immediately before operation, the circulating nurse removes the cap from a 500 ml flask of saline, taking care not to contaminate the sterile lip of the container. The pernicious practice of pouring away a small quantity should be avoided, because the few drops of solution that run down the outside of the flask can run back again after being contaminated. The requisite number of sterile receptacles placed close together can be filled at a single pouring without repeated inversion of the flask. The flask is then emptied and permanently discarded. Under no circumstances should any fluid remaining in the bottle be used in another operation.

SALINE FOR CORNEOCONJUNCTIVAL IRRIGATION. Although the intact corneal epithelium is normally resistant to pseudomonas, broken epithelium and the stroma are not. Strict asepsis must be observed in all cases of corneal foreign body or abrasion. Many such eyes have been lost from unnecessary instillation of contaminated drops when otherwise healing would have been uneventful if no medication had been used. Examples of this have been cited above.

The same rigid criteria governing the use of saline in the anterior chamber must be applied when the corneal epithelium is broken. Even in the case of preserved solutions there is no substitute for the unit-dose preparation that has been sterilized in the final container. Every patient who requires irrigation of the conjunctival sac is entitled to a sterile solution. If an epithelial break is suspected, a previously unopened unit should always be used.

Individual disposable 15 ml bottles of 1.4% sterile sodium chloride are used in the clinics and wards of the Massachusetts Eye and Ear Infirmary. These bottles are handled aseptically and are used to moisten sterile cotton balls or fluorescein strips. Opened units remaining at the end of a clinic session are discarded. Bedside units are recapped after each use and are discarded when empty or when the patient is discharged.

Silicone bulb syringes used for irrigation are prepared in the central supply room as follows. The unsterile bulb is rinsed with distilled water, of which a little is purposely left in the syringe. It is then autoclaved and is ready for use. Failure to leave some fluid in the interior may mean that only the outside will be sterilized by steam at 121° C, the inside receiving only a substerilizing exposure to dry heat.

SALINE FOR GONIOSCOPY AND CONTACT LENSES. Since corneal abrasions can occur during gonioscopy and in fitting contact lenses, the principles governing the use of solutions for intraocular procedures should always apply.

SALINE FOR POSTOPERATIVE DRESSINGS. Saline is often used for swabbing the encrusted lid margins of operated eyes each time a dressing is changed. Wet swabbing increases the patient's comfort by relief of itching. Individual units are employed and a single container should never be used for more than one patient.

Collyria

The following principles apply to the manufacture and use of solutions of drugs commonly used in ophthalmic practice: (1) preparation under sterile conditions, (2) sterilization in the final container, (3) preservation against bacteria and fungi, (4) handling with aseptic technique, and (5) testing for sterility at intervals when open.

PREPARATION OF STERILE COLLYRIA. The compounding of collyria, formerly done by the local pharmacist, has fallen into the province of the wholesale manufacturer. This means that it has come under the jurisdiction of the Food and Drug Administration. Manufacturers have acquired the knowledge and the equipment necessary to produce sterile medications. Sterility is usually achieved by bacterial filtration of prepared solutions, which are then aseptically distributed into sterile containers. Filtration will not remove most viruses, but the likelihood of viral contamination is very slight.

The ophthalmologist is responsible for the sterility of the solutions he uses. For solutions to be used by patients it is better to specify commercially manufactured preparations or to dispense medications of known sterility than to run the risk that a contaminated solution might be dispensed by a local pharmacy. Collyria used in the operating room should be freshly opened.

STERILIZATION IN THE FINAL CONTAINER. As long as a given solution is sterile, the manner in which its sterility was achieved is immaterial. Sterilization in the final container offers a guarantee of sterility and should be employed whenever there is doubt about the manner of preparation. It usually requires autoclaving in sealed glass or metal containers and the observation of certain other requirements.

PRESERVATION OF COLLYRIA. Once the seal is broken, both the screwcap vial and the plastic dispensing bottle are subject to contamination by bacteria, fungi, and viruses. Bacteria and fungi can be inhibited by a suitable preservative.

A suitable preservative must be bacteristatic, fungistatic, effective in the presence of other substances, innocuous to the corneal endothelium and other tissues, compatible with other ingredients, chemically stable, and unlikely to sensitize patients when used over long periods of time. All of these properties must be maintained under conditions of actual use and especially within the pH range of maximum stability of alkaloids. The antibacterial powers of any preservative must be demonstrated against a representative number of strains of *Pseudomonas.*

Although many substances have been recommended and used as preservatives for collyria, only a few have been submitted to rigid testing of the above properties. These include benzalkonium chloride (Zephiran), chlorobutanol, phenylmercuric nitrate, phenylethyl alcohol, and a combination of polymyxin B sulfate with benzalkonium chloride.

The use of bactericidal preservatives must await the appearance of a single substance that is uniformly effective and that combines with its bactericidal effect the other properties desired in a preservative. Until such a drug becomes available, chlorobutanol 0.5% most nearly approaches the requirements for the ideal preservative, provided that its use is combined with careful handling.

Like other preservatives, chlorobutanol has certain disadvantages. It evaporates slowly from aqueous solutions. This effect can be minimized by the use of small, tightly stoppered containers. These are desirable in any event to prevent concentration of active ingredients by evaporation of water. Sealers inserted in plastic screwcaps should always be of rubber, never of cardboard.

Although chlorobutanol does not have a rapidly bactericidal effect, experimentally contaminated solutions containing chlorobutanol 0.5% almost invariably are free of vegetative forms at the end of 24 hours. In comparative studies the bacteristatic effect of chlorobutanol is superior to that of phenylethyl alcohol, benzalkonium chloride, and phenylmercuric nitrate in the strengths commonly used for preservation.

Finally, the case of fluorescein solutions is special. All authors agree upon the ineffectiveness of attempts to preserve this medication against contamination by *Pseudomonas* organisms. The use of sterile fluorescein papers, first suggested by Kimura,[26] is the standard method for staining the cornea in the presence of an epithelial break.

Handling of collyria. Contamination of sterile collyria can be avoided in the first place by careful technique in handling. Although viruses do not multiply in solutions, viral contamination of collyria can occur and virus particles can survive in them. In one instance 42 cases of epidemic keratoconjunctivitis were attributed to a single bottle of tetracaine hydrochloride. Another accident that has often occurred is contamination of one drug by another introduced by a dropper. Atropine contamination can produce acute angle-closure glaucoma through accidental dilatation of a pupil. For these reasons every solution that may be administered to more than one patient should be protected from microbial and pharmacologic contamination by the interposition of a sterile intermediary dropper, which enters the solution once and only once before being discarded.

In the Eye Clinic of the Massachusetts Eye and Ear Infirmary, autoclaved glass droppers are used once and discarded. Droppers do not need to be individually wrapped provided they are carefully handled and used in solutions containing an effective preservative.

Polyethylene "squeeze" bottles are convenient for use by patients and are less subject to contamination of their contents than the old-style dropper bottles. The contents must be preserved against bacterial contamination, which can be introduced by aspiration of the last drop after contamination of the dropper tip. Because ordinary polyethylene is pervious to chlorobutanol, some other preservative must be used. In the hands of the ophthalmologist and of properly trained nurses, contamination of the delivery tip by contact with conjunctival or skin surfaces can be avoided. If such contact occurs, the vial should be discarded. In the hands of careless or untrained personnel, contamination of the tips is almost certain to occur and to be ignored or overlooked. The contaminated tip can then convey infection to other patients. Sterile disposable droppers provide a greater safeguard in clinic practice where close supervision of personnel administering drops to large numbers of patients is not feasible.

Recent progress has been made in the development of unit-dose containers. They are, however, expensive.

Sterility testing of collyria. Since no preservative offers a complete guarantee of sterility after the seal is broken, periodic surveys of collyria in use should be carried out. Details of this procedure will be discussed in the section on sterility testing.

Ophthalmic ointments

Preparation and sterilization of ointments. The problem of ensuring sterility of ointments has not yet been satisfactorily solved in all cases. With certain exceptions, ointments are not marketed with the implication that they are sterile or have been sterilized. Contaminants have been found in eye ointments when they have been looked for. Fortunately, these are usually nonpathogenic bacteria and fungi.[9,10,48]

Heat sterilization of ointments in the tube is not practical for several reasons. Most commercial ointments are prepared by grinding the dry alkaloid powder into a petrolatum base in a milling machine. An alternative method is to dissolve the free alkaloid base in a fatty acid ester (such as polyethylene glycol 200 monostearate) that is freely miscible with petrolatum. In either case, heating the resulting anhydrous mixture to the required temperature of 140° C for 4 hours would result in layering of the suspension with consequent loss of uniform distribution of the active ingredient.

A compromise is reached by sterilizing the petrolatum base separately and adding the alkaloid to it. Heat sterilization of alkaloids is not always possible; for example, atropine base melts at 114° C. However, physostigmine salicylate and the sulfonamides will withstand 140° C.

Handling of ointments. As in the case of collyria, a single tube of ointment should not pass from patient to patient without the interposition of a sterile intermediary. This may be a sterile glass rod or a sterile cotton swab.

Selection of vehicle. The relative merits of prescribing a drug in aqueous solution or in an ointment base must always be considered. In general, ointments are indicated for treatment of conditions of the lids and lid margins. Certain disadvantages may attend their use in the conjunctival sac, namely: (1) greater risk of sensitization, (2) inhibition of epithelial regeneration by the ointment base, (3) retention of fat globules as foreign bodies under corneal epithelium, under the conjunctiva, or in the anterior chamber, (4) uncertain sterility, and (5) blurring of the cornea for ophthalmoscopy.

On the other hand, ointments prepared under sterile conditions offer less opportunity for bacterial proliferation following accidental contamination and adhere well to the lid margins.

Instrument disinfectants

Solutions of many chemicals used to disinfect instruments are dangerous for two reasons. First, they create a sense of security unwarranted by their antibacterial properties. Not only are most disinfectants not sporicidal, but their ability to kill vegetative bacteria is often contingent upon the absence of organic matter. In the second place, solutions of certain substances, especially the quaternary ammonium compounds, may themselves be contaminated with gram-negative bacteria.

The following solution is satisfactory for maintaining the sterility of instruments:

Ingredient	*Parts per 1,000 ml*
Isopropanol 99%	700.00
Roccal 50%	2.00
Triethanolamine	10.00
Potassium nitrite	1.00
Safranin O solution 2%	0.10
Water ad	1,000.00

This formula will kill most vegetative bacteria upon contact of 30 minutes or longer, but it is not sporicidal.

Skin antiseptics. Like disinfectants, sulfonamides, and antibiotics, some skin antiseptics may be contaminated by the very bacteria they are intended to inhibit. A disastrous episode of this nature has been cited. Isopropanol 70% is satisfactory.

Substances for injection. Any of the injectables used in general medicine and surgery may be administered to ophthalmic patients. Some of the drugs used most commonly in ophthalmology are insulin, epinephrine, procaine, meperidine hydrochloride (Demerol), and aqueous or procaine penicillin. Most of these are prepared commercially and are fortified with one of several preservatives. Instances of contamination are rare.

Epinephrine for subconjunctival injection or other parenteral use can be obtained from a single dose unit such as the familiar 1 ml ampoule.

CONTACT LENS SOLUTIONS. Contact lens solutions have been found contaminated with *Pseudomonas* organisms. If a corneal abrasion occurs in the presence of a contaminated solution, the results can be extremely serious.

Case history. A 27-year-old language teacher had worn contact lenses for 2 years. One day she noted blurring of vision in the left eye. Examination showed a corneal ulcer from which *Pseudomonas* organisms was recovered. The same strain was present in the cassette equipped with sponge-rubber clips in which her contact lenses were stored in the presence of wetting solution. The ulcer was superficial and the strain was not highly virulent, but useful vision in the eye was lost.[19b]

Summary

The role of contaminated medications and irrigants in producing ocular infections of the most serious nature has been demonstrated. Because bacteria can multiply in unpreserved aqueous solutions, and because such solutions may contact the intraocular tissues or broken corneal epithelium, saline irrigants should be sterilized in the final container, dispensed on an individual basis, and discarded after a single use.

Collyria must be sterile, fortified with a suitable preservative and handled with aseptic precautions. Medications, whether collyria or ointments, which are to be used for more than one patient, should be administered by means of a sterile intermediary dropper or applicator.

Medications in use should be dated and checked for sterility at intervals. The use of unit-dose containers for operating rooms and accident rooms is recommended.

TISSUE CONTAMINANTS

Skin of hands

CUTANEOUS FLORA. The bacterial flora of normal human skin consists of transient and resident organisms. The transient population includes any organisms with which the individual has recently come in contact. They are the first to be removed by scrubbing. The resident bacteria maintain themselves by continuous reproduction on the skin, the birth rate being equal to the death rate unless it is altered by external influences. The superficial resident flora exists in the folds and crypts, the deep flora in the sweat glands, oil glands, and follicles of the skin. Nonpathogenic staphylococci form the bulk of the bacterial population, but pathogenic staphylococci and other pathogenic species may be found. Removal of the deep flora is never accomplished by scrubbing.

When an upper respiratory infection exists, the number of transient organisms and the relative proportion of pathogens on the skin are both markedly increased. Contamination of the skin with pathogenic transient organisms may result in the adaptation of these organisms to the skin, where they become part of the resident flora and may remain for weeks. For this reason the ophthalmic surgeon should protect his hands from contamination with pathogenic bacteria insofar as possible. The surgeon who operates on an infected patient should wear gloves not only for protection of the patient but also to prevent contamination of his own hands.

Because the normal skin is never sterile and cannot be made sterile, the wearing of sterile rubber gloves is recommended for all operators. But because rubber gloves may be torn or punctured during an operation, removal of dirt, grease, and bacteria must be carried out before an operation. The process of reducing the cutaneous flora has been designated *degerming* the skin. The most efficacious methods of accomplishing this object have been evaluated by surgical and skin bacteriologists.

DEGERMING THE SKIN. Rapid reduction of the cutaneous flora is accomplished in two stages: (1) mechanical removal of bacteria by scrubbing all accessible areas with a detergent and (2) killing the bacteria that remain on the skin by a skin antiseptic. A third process, (3) inhibition of the resident flora by a bacteriostatic substance deposited on the skin, is effective when it is allowed to act over a period of days. The principles underlying these processes should be understood by ophthalmic surgeons.

The effectiveness of mechanical cleansing of the skin depends upon (1) scrubbing in the presence of a surface-active compound, (2) friction, (3) time, and (4) hydrogen ion concentration (pH). The first two factors are self-explanatory. The time required is a function of these and of pH.

Surface-active agents include cationic, anionic, and nonionic compounds. There are hundreds of these compounds on the market, with more appearing yearly. The higher the alkalinity of any scrubbing process, the higher will be the negative charge on the cutaneous surface and on the bacteria.[6] Mechanical removal of bacteria is thus promoted by the mutual repulsion of similarly charged bodies. The alkaline detergents are therefore more efficient than the acid detergents.

Anionic detergents include soaps, sulfated alcohols and fatty acid esters, and sulfonated aliphatic and aromatic compounds. Soaps having a pH of 9.5 to 10.5 are the most alkaline and therefore the most efficient degerming agents in water of ordinary softness. The newer detergents, unlike soaps, are not precipitated by either acids or calcium and magnesium salts and therefore retain their cleansing properties in either acid solutions or hard waters.

Cationic surface-active agents are not effective for degerming the skin. They alter the electrostatic charge on the skin from negative to positive while leaving bacteria negatively charged. Bacteria thus temporarily bound to the skin by electrostatic attraction can be released by a subsequent alkaline rinse. These agents also lose some of their bactericidal power when in contact with the cutaneous surface. There have been instances of pseudomonas contamination of these agents resulting in loss of eyes. Thus, the quaternary ammonium compounds in aqueous solution are not only ineffective for purposes of degerming the skin but may even carry infection to the skin.

Killing of bacteria, other than spores, which remain on the skin after mechanical cleansing, is best accomplished by alcohol allowed to act for a sufficient time. Ethyl alcohol 70% is effective but is expensive and is subject to alteration in its concentration. Isopropyl alcohol is cheaper and is germicidal over a wide range of concentration. Iodine and its derivatives are not superior to isopropyl alcohol as skin antiseptics. The combination of mechanical scrubbing with soap and water followed by a 2-minute soak in 70% isopropyl alcohol renders the skin more nearly sterile than any other procedure for *rapid* degerming of the cutaneous surface.

Hexachlorophene leaves a bacteristatic film on the skin and lowers the bacterial population when it is used continually for washing. Added to soap it offers no advantage for *rapid* degerming of the skin.[7] Its value consists entirely in its cumulative effect over a period of time. For this reason, its addition to solutions of alcohol for soaking is recommended.

Fats are removed from the skin by emulsification during scrubbing with soap and water and by solution in alcohol. The skin feels dry when fat is thus removed. Cetyl alcohol added to the soaking solution provides a satisfactory substitute that leaves the skin soft and smooth without producing a greasy sensation. This emollient has the additional advantage of enabling the surgeon to don rubber gloves without the use of powder, thus eliminating objectionable talcs and starches from the operating room.

The above principles are applied in the following routine that has been used at the Massachusetts General Hospital and the Massachusetts Eye and Ear Infirmary since 1952.[8]

SURGICAL SCRUB ROUTINE

1. Wet hands and arms under running water and work up a lather with soap and brush.
2. Scrub each surface (fingers, hands, nails, arms to elbows) using three double strokes of the brush on each

surface. Rinse under running water (time—about 1 minute).

3. Clean nails with an orangewood stick.
4. Scrub each surface again with three double strokes of the brush (time—about 1 minute).
5. Rinse off soap completely and dry hands and arms with sterile towel.
6. Wet hands and arms thoroughly with the following solution:

Isopropyl alcohol	70.0 parts per 1,000 ml
(kills bacteria with 2 minutes of rubbing)	
Cetyl alcohol	0.5 parts per 1,000 ml
(substitutes for fats and oils of skin)	
Hexachlorophene	0.1 parts per 1,000 ml
(reduces resident flora when used daily)	
Water q.s. ad	100.0 parts per 1,000 ml

7. Rub hands and arms for a full 2 minutes, being sure that the skin is kept wet with alcohol solution the entire time. (This may be accomplished by continuous immersion, repeated immersion, or by repeated delivery of the solution from a dispenser.)
8. At the end of this 2-minute period, allow the hands and arms to dry by evaporation.

Recontamination of the hands can be avoided by wearing a separate pair of sterile gloves for preparation of the patient and injection of the local anesthetic. Alternatively, the surgeon may scrub his hands and the patient's skin with soap and water, administer the local anesthetic, and return to the alcohol soak for 1 minute before putting on his gown and gloves. Accidental contamination of the hands does not require repetition of the entire surgical scrub. The last bacteria to reach the skin are the first to be removed by soap or killed by an antiseptic.

The operative field

In ophthalmology the cutaneous operative field comprises the brow, the eyelids, and the anterior part of the temple and the cheek. Two stands of hair grow here, the eyebrow and the eyelashes.

The bacterial population of this area is comparable to the flora of other exposed areas of the body; that is, there is a preponderance of nonpathogenic staphylococci, but there may be pathogenic organisms among the transient and resident flora. Because skin pathogens adhering to instruments or sutures can be introduced into the eye, pathogens must be eliminated or reduced to a minimum.

Preparation of the operative field must be preceded by inspection to make sure that there are no traces of lid infection such as styes, furuncles, cellulitis, or impetigo. The bases of the lashes and the orifices of the meibomian glands should receive careful scrutiny for the presence of crusting and erythema.

Since it is impossible to study quantitatively the rate of removal of bacteria from a facial area, the principles of degerming derived from studies of hand scrubs must be applied. The scrub described for the hands is adapted, with minor modifications, to the skin of the operative field; that is, a scrub with sterile green soap solution is followed by a 70% isopropyl alcohol rub that should last a full 2 minutes. Cetyl alcohol is omitted; a few drops of eosin in the alcohol serve to indicate the area that has been prepared in this manner. The lids and lid margins are swabbed with alcohol on toothpick applicators, care being taken that the alcohol solution does not enter the conjunctival sac.

Proprietary cutaneous detergents are used in many hospitals for preparing the operative field. Because of their lower alkalinity, these are probably less effective in removing bacteria from the skin than are soaps. More serious are reports of severe keratitis resulting from prolonged contact of pHisoHex with the cornea. Browning and Lippas,[11] reporting two human cases and the results of animal experiments, identified the toxic agent as sodium octylphenoxyethoxyethyl ether sulfonate (Entsufon), an ingredient of pHisoderm and pHisoHex. Pure hexachlorophene 3%, present in pHisoHex, produced only mild stippling in a small percentage of animal eyes, whereas the sulfonated ether compound reproduced the epithelial denudation and deep stromal edema that characterized the clinical cases reported. Although both patients recovered full visual acuity after several weeks, the discomfort, disability, and the inevitable psychic trauma to patients should dictate extreme caution in the use of these agents in the vicinity of the eye.

Although shaving of hair is basic to skin preparation elsewhere on the body, shaving the eyebrow is not necessary provided scrub solutions are vigorously worked into the eyebrow and provided the brow is carefully walled off from the palpebral fissure during the operation itself.

Clipping of the lashes before intraocular surgery may do more harm than good. It is doubtful whether this practice is helpful in preventing infection. A solid-bladed speculum (Metscher) walls off the lid margins better than an open-bladed one.

The conjunctival sac

Pathogenic bacteria in the conjunctival secretions pose an obvious threat of infection at the time of intraocular surgery. Surveys of patients admitted for ocular surgery show that clinically healthy conjunctivas often harbor pathogens. Smith[42a] found pathogens in the eyes of 15% of 5,000 patients admitted to two London hospitals; only 3.5% of the 5,000 patients were classified as clinically "dirty." Cason and Winkler,[13] reporting conjunctival cultures from patients who showed no clinical evidence of an infectious process in either eye, found pathogens in 12%, including gram-negative rods in 5%. Khorazo and Thompson[25a] found S. *aureus* in 23% of 1,122 patients and other pathogens in an additional 11%. Dunnington and Locatcher-Khorazo[17] found pathogenic staphylococci preoperatively in 26% of 2,508 patients, alone or in combination with other pathogens. Postoperative endophthalmitis developed in 11 of the 529 patients in whom pathogenic staphylococci were found, but in none of 74 patients in whom other potential pathogens were cultured. Species that are pathogenic when introduced into the cornea or anterior chamber may thus be present in the conjunctival sac without producing symptoms or signs of conjunctivitis.

There is no infallible method for sterilizing the conjunctival sac preoperatively. Assuming that the lacrimal passages are patent, a suitable regimen is the use of drops containing chloramphenicol and polymyxin 4 times daily for 2 days before surgery, combined with erythromycin ointment to the conjunctiva and lids at bedtime.[3a]

The continued use of antibacterial medication in the conjunctival sac is desirable for at least the first week after operation, in view of the well-known effect of patching in creating an environment favorable for the proliferation of conjunctival bacteria.

The lacrimal sac

The lacrimal sac is of the greatest importance as a source of virulent organisms. Every well-trained eye surgeon is thoroughly imbued with the principle that reflux of mucopurulent material on pressure over the sac requires deferment of intraocular surgery until this focus of infection has been eradicated. In the absence of epiphora or of a mucocele, many surgeons prefer to proceed without irrigation or probing of the lacrimal passages than to take the chance of causing infection or edema by such manipulation. If epiphora is present with or without signs of infection, cultures should invariably be taken to determine the presence of pathogenic organisms before operating.

AIR HYGIENE FOR HOSPITALS

The role of airborne organisms in postoperative infections, like that of other environmental factors, is difficult to evaluate. Circulation of air in the operating room and loss of viability of most respiratory pathogens in air render futile retrospective attempts to recover an etiologic agent from the air by the time signs of endophthalmitis have appeared.

Coagulase-positive staphylococci are the most frequent cause of postoperative endophthalmitis and the pathogen most frequently recovered from air samples taken during eye operations. They are often recovered from the inside and outside surfaces of cloth masks worn by operating room personnel. Pneumococci and coliform bacilli, next in importance to staphylococci and pseudomonas in postoperative infections, may also be derived from the upper respiratory tract. Among rarer agents isolated from infected eyes, *Neisseria catarrhalis* has been cited by Givner[19a] as having a probable respiratory origin.

Early concepts and modern methods

The problem of organisms falling from the air upon open wounds became a concern of surgeons after Pasteur's demonstration of the role of airborne microorganisms in fermentation. Lister's techniques were directed primarily toward destroying airborne bacteria and isolating open wounds from them.

The introduction of the surgical mask by von Mikulicz in 1896 was perhaps the first attempt to check the dissemination of respiratory contaminants at their source. Twelve years later, at least one American eye surgeon felt that the use of the mask as a mouth protector was extreme and that its principal use in eye surgery was to control the surgeon's beard!

During the 1930's there was a reawakening of interest in the infection-producing role of airborne bacteria in general surgery. Renewed attention was directed toward their control by improved masking, ultraviolet irradiation of circulating air, and the use of germicidal aerosols and mists.

In recent years, mechanical ventilation by air of controlled temperature and humidity has been installed in most operating suites. The source of ventilation and the method of humidity control can have a profound influence on the number and species of organisms discharged from supply ducts. Aerosols of pathogenic bacteria can be generated in concentrations that could never be reached in ordinary atmospheric air and blown into spaces, such as operating rooms, where they can do the greatest damage. At the same time, great advances have been made in freeing air of particulate matter, chiefly through the development of filtering media of high efficiency against atmospheric dust and even smaller particles. Familiarity with the properties of these media can avoid inefficient, unnecessarily elaborate, and expensive installations.

Origin, nature, and fate of airborne contaminants

Microorganisms enter rooms in two principal ways. Some are carried in with atmospheric dust by currents of air. They include bacteria, spores, and molds in the dry state.

Other contaminants are carried in by personnel in their respiratory secretions, on the skin, on their clothing and shoes, or in bedclothes. They are liberated from these sources by activity on the part of the individuals concerned. The concentration of air contaminants derived from human sources is proportional to the following factors: (1) duration of occupancy, (2) number of occupants, (3) activity of occupants, that is, movements and forced expirations, (4) carriers of pathogenic organisms, and (5) intensity of growth from respiratory tracts of carriers. Of all these factors, it appears that the activity of individuals is the most important. Relatively few bacteria are liberated by people sitting quietly without speaking, whereas talking, laughing, coughing, and bustling about result in the liberation of many organisms from the mouth, skin, clothing, and bedding.

These organisms are distributed in air in three forms: (1) attached to dust particles, (2) in droplets, and (3) as droplet nuclei. A single particle may contain one or more bacteria. Dust particles vary greatly in size. Those larger than 10 μ settle at the rate of 1 to 2 ft per minute in still air; those smaller than 1 μ remain suspended for indefinite periods. Droplets evaporate at a rate proportional to their size and to the temperature and relative humidity of the environmental air. Large droplets are usually projected for several feet and fall to the ground before evaporating. Droplets of 10 μ diameter are propelled for shorter distances and evaporate almost instantaneously to form droplet nuclei. These lie almost entirely within the range of 1.34 to 2.68 μ diameter and settle at an average rate of 0.04 ft per minute in still air. Under ordinary conditions they remain in continuous suspension because of thermal and mechanical air currents.

The ultimate fate of airborne organisms depends upon environmental conditions. Spores remain viable indefinitely. Molds and fungi of various species are common airborne contaminants. On the other hand, *Pseudomonas* and other gram-negative species are seldom found in air samples.

Pathogenic bacteria do not multiply in ordinary room air or upon dry surfaces. However, certain bacteria can and do reproduce on moist surfaces, particularly if nutrient substances happen to be present. The rate of bacterial multiplication depends upon the growth requirements of the species involved and the extent to which these requirements are met under existing circumstances.

When the palpebral fissure is open, airborne particles impinge upon the cornea and the exposed conjunctiva just as they do upon other areas of the body's surface. Nonpathogenic gram-positive bacteria are usually destroyed by the lysozyme of the tears. Other organisms are mechanically removed by blinking and are washed away by the normal flow of tears. More rarely, airborne pathogens in droplets or spatters may establish themselves in the conjunctival sac. For example, the conjunctiva may be directly infected by meningococci expelled from the nasopharynx of a carrier or a case of meningitis. Adenovirus infections of the eye may be transmitted in the same manner.

Collection of data for periodic checks

Relevant data. Three items of information are required for periodic checks of conditions relevant to air contamination: (1) the temperature and relative humidity of the air in the operating rooms, (2) the direction and rate of flow of air currents in the surgical suite and the individual operating rooms, and (3) the bacterial count per unit volume of air and its breakdown into pathogenic and nonpathogenic species. The last of these is the most important, since it provides an index of total contamination in the environmental air.

Temperature and relative humidity. The temperature and relative humidity are determined from readings of thermometers, psychrometers, and hygrometers. Variation of either of these factors within the limits found in operating rooms in the temperate zone probably has little effect upon bacterial contamination of room air or of surfaces in the operating room. Different species show marked differences in their rate of viability loss in relation to drying. Decay rates of gram-negative bacilli are faster than those of micrococci, corynebacteria, and mycobacteria.

Air current tracing. A simple, inexpensive indicator of air flow is the smoke from a cigarette. When a lighted cigarette is placed near the bottom of a closed door, it becomes apparent at once whether air is flowing into or out of the room at that point, and how fast it is moving. When it is held near the ceiling, the direction of air currents from a supply duct can be traced in the same manner. These currents indicate whether a positive or a negative pressure exists in the operating room relative to the corridor and to adjoining spaces, such as scrub rooms and anterooms for anesthesia. The same test can be applied to the operating suite as a whole by testing air currents at either end of the main corridor.

Quantitative measurement of air flow from duct outlets requires the use of flowmeters and anemometers. Such determinations are easily performed by air-conditioning engineers, who should make periodic checks and furnish pertinent data to the hospital.

Collection of airborne organisms. The simplest method for collection of organisms in air is the exposure of open petri dishes containing nutrient agar at strategic points in the operating room. The disadvantages of this method are: (1) only particles heavy enough to settle out by gravity will be collected, and (2) the volume of air sampled is unknown, a fact that prevents quantitative comparison of air samples. Nevertheless, the open plate technique, by permitting qualitative detection of pathogenic species, is better than nothing at all.

The problem of quantitative air sampling has stimulated the development of numerous ingenious devices too expensive for routine use.

Counts of air contaminants. In recording counts of organisms collected in the hospital it is essential to break down the total count into nonpathogens and pathogens. Under the former heading we include nonhemolytic staphylococci, species of sarcina, hay bacilli, and diphtheroids. These are usually of no interest, but their total count may be important as an index of air turbulence and human activity. When pathogens are found, the species should always be recorded. These often include *Staphylococcus aureus* and more rarely *Streptococcus hemolyticus, Diplococcus pneumoniae, Pseudomonas aeruginosa,* and other gram-negative rods. Because *Bacillus subtilis* and species of aspergilli may be pathogenic for the eye, notice should be taken if these occur with undue frequency in the operating room air.

Ventilation and air conditioning

Air entering the operating room suite should be free from dust and bacteria, controlled as to temperature and humidity, and supplied under slight positive pressure. Control of temperature and humidity is important for human comfort and efficiency as well as for minimizing explosion hazard from static electricity. Positive pressure is necessary in order to prevent entrance of unsterile air from uncontrolled sources. Under no circumstances should air be drawn into the operating suite from other parts of the hospital.

Achievement of these objectives is the problem of the ventilation engineer and the hospital architect. In existing hospitals the supply of air may or may not meet requirements. In hospitals under construction an opportunity for providing ideal conditions is present. The administration and the medical staff should understand the means of obtaining a properly cleaned and conditioned supply of air.

The intake for outside air should be located as far as possible above street level, preferably on the roof and always far removed from any exhaust outlets of other ventilating systems. Cold air must be warmed and sometimes humidified, while warm air must be cooled and sometimes dehumidified before delivery to the operating room. When the outside air is below 70° F, dehumidification is almost never necessary, because the relative humidity of heated air is automatically lowered by expansion.

Thomas,[45a] reviewing practical aspects of air conditioning in hospitals, recommends a central system of air conditioning to include the following features for operating and delivery rooms:

1. Winter humidification to reduce fire and explosion dangers of anesthetic gases
2. Summer cooling with some dehumidification to reduce fatigue and sweating
3. Eight to twelve changes of air per hour
4. A controlled temperature range of 72° to 80° F
5. A controlled relative humidity ranging from 50% to 60%
6. Proper filtering devices to remove particulate matter

Thomas takes issue with the Guide published by the American Society of Heating and Air Conditioning Engineers for requiring 100% outside air. He cites evidence suggesting that the use of 50% recirculated air presents no greater explosion hazard and is much more economical.

Air cleaning

Clean air is needed to maintain dilution of airborne contaminants derived from sources within the operating room; contaminated ventilating air would obviously compound the chances of infection. To ensure satisfactory performance over long periods, devices for delivering clean air should be simple, trouble free, and should require a minimum of maintenance. Recent developments in the field of air filtration permit realization of these objectives at reasonable installation and operating costs.[1]

Dry or fibrous filters. Dry filters are fabriclike media of wool felt, cotton batting, or thick mats composed of very fine glass fibers that may be bonded together by phenolic resins. By means of pleating, a great increase in effective filtering surface can be achieved. The more efficient of these filters may arrest up to 95% of ordinary atmospheric dust particles at a high rate of air flow and with a pressure drop across the filter bank usually less than 0.5 inch water gauge for a new filter. Examples of deep-bed glass fiber media are the American and Cambridge types.* Specifications of these media are available in manufacturers' brochures. They are composed of glass fibers averaging 0.8 to 3.0 μ diameter. Individual differences exist in the disposition of the medium (whether pocket, pleated, or expansion type) and in the manner of adaptation of the filter frame to the duct. Units presenting 4 sq ft superficial area provide an air flow of 1,000 cfm (superficial face velocity 250 fpm).

Efficiency ratings of air filters. Although surgeons do not ordinarily carry out performance tests on air filters, an understanding of the basis for efficiency rating is important. A filter that is highly efficient against particles of 10 μ diameter may be extremely inefficient against particles of 1 μ diameter.

Deep-bed media are rated by the National Bureau of Standards (NBS) Discoloration Test using atmospheric dust as the test material; that is, against particles averaging 0.5 μ diameter. "Stain" efficiencies increase with the useful life of the filter as particles build up on the filter face. Efficiencies from 35% to 95% over the life of the different filters are achieved at the rated face velocity. The pressure drop for newly installed filters varies from 0.14 to 0.48 inches of water, depending upon the efficiency of the filter medium. Thus they combine high retention of particles with the advantages of lower initial and operating cost, easier and less expensive maintenance, and absolute freedom from shock and sparking hazards.

In order to conserve the efficiency of deep-bed media for fine (smaller than 5 μ diameter) particles, a viscous panel filter should be incorporated upstream from the former to screen out large particles and to postpone buildup of maximum allowable (dirty) pressure drop. The water gauge manometer should be inspected at regular intervals to prevent exceeding the maximum dirty drop.

Because no filter is better than the efficiency of its installation, this should be inspected for possible bypassing of the filter bed. A performance test should be carried out by the consultant, the manufacturer's representative, or the installing contractor to ascertain whether the completed assembly is functioning in a satisfactory manner. Faulty adaptation of the filter frame to the duct can reduce filtration by 50%. Under these circumstances, low manometer readings will give the erroneous impression that the pressure drop across the filter is below the maximum level permissible when actually the reverse may be the case.

A correctly installed, performance-tested bank of fibrous filters requires a minimum of maintenance. Its efficiency for fine particles increases with buildup of arrested dust at the filter face. A panel-protected filter bank may operate below the critical pressure drop for many months. Replacement of filter cartridges should be accom-

*American Air Filter Co., Louisville, Ky.; Cambridge Filter Corp., Syracuse, N. Y.

panied by careful vacuum cleaning of the duct.

Organisms liberated in the operating room

SOURCES OF CONTAMINATION. Most bacteria in operating rooms enter by ways other than ventilating air. Such bacteria are carried by: (1) particles containing bacteria and viruses freshly emitted from the respiratory tracts of the occupants as droplets or droplet nuclei; (2) particles of human origin arising from the skin, hair, and clothes of the occupants; (3) particles of nonhuman origin, including soil bacteria and spores, associated with clothes and bedding, and others similarly dispersed.

Particles in all three categories are continuously produced by activity of the occupants, and their concentration is continuously reduced by an adequate supply of bacteria-free ventilating air. Bacteria in the first category lose viability by decay. The resultant of the three forces of liberation, dilution, and decay determines the moment-to-moment concentration of airborne contaminants. Since the last two factors should be constant in suitably ventilated rooms, the main variable must be the rate of liberation of bacteria by the occupants. This factor is also the most difficult to control.

RESPIRATORY CONTAMINANTS. Most individuals expel very few bacteria in the course of quiet respiration. Forced expiration, as in talking, laughing, coughing, or sneezing, increases this number by expulsion at high speeds of droplets laden with bacteria. In the presence of an upper respiratory infection, the concentration of bacteria in secretions, the proportion of pathogens, and the virulence of these pathogens are all increased.

The two most obvious means of controlling respiratory contamination of the air are the avoidance of talking and the exclusion of individuals with coughs and colds from the operating room. In most operating rooms these means are seldom utilized.

MASKING TECHNIQUE. Cloth masks are inadequate and obsolete. Modern synthetic fibrous masks are effective filtering devices.

Air that escapes around a mask is unfiltered and carries its bacterial burden into the room air.

Masks should be worn in the operating room even when an operation is not in progress. Since bacteria can remain airborne in a viable state for many hours, failure to observe this precaution can liberate organisms that may eventually find their way into open wounds.

The cooperation of anesthesia personnel is essential in preventing unnecessary dissemination of organisms. Induction of general anesthesia in the operating room is attended by a rise in airborne bacteria recovered in samples. By the same token, extubation of a patient on the table after operation may give rise to a spasm of coughing that is productive of atomized tracheobronchial secretions.

BACTERIA FROM SKIN AND CLOTHING. Particles in these categories can be partially controlled by requiring a change to laundered cotton clothing of all individuals entering the operating room, by minimizing the movements of such individuals, and by applying grease or cream to exposed skin surfaces of the face and neck.

BLANKETS AND BEDDING. One of the richest sources of bacteria, including many pathogens, and one of the easiest to control, is the blanket that accompanies the patient from the ward to the operating room. Blankets are easily sampled by passing an inverted blood agar plate rapidly and firmly across a section of blanket resting on a hard flat surface. The action of the nap in springing back throws up many bacteria and provides an adequate sample as an index of contamination. Large numbers of pathogenic staphylococci and other organisms are found in bed blankets.

Because of the danger from static electricity, wool blankets are never sent with patients to the operating room. Instead, a sterile cotton "ether blanket" should be placed over the patient after he has been transferred to the litter on which he is to be wheeled to surgery. The number of bacteria that settle upon this blanket during transportation of the patient will be few in comparison with those that collect in wool blankets in use for but a short time. Blankets should never be stripped off a patient in the operating room either before or after an operation, as large numbers of bacteria are liberated by these procedures. Instead, a sterile sheet should be draped over the patient and allowed to hang down over the sides of the table as soon as it has been wheeled in.

Practical studies

Bacterial counts taken in empty operating rooms ventilated under positive pressure by air filtered through deep-bed glass fiber media at the rate of 8 to 12 air changes per hour show negligible initial contamination and rapid disappearance of organisms introduced as droplet nuclei by nebulization. During operations the concentration of airborne bacteria-bearing particles, as calculated from colony counts, may vary from 0.66 to 5.2 colonies per cubic foot of air. The pathogen most frequently recovered is S. *aureus*. When this species has been found in the air, it has usually been demonstrated on the inside or the outside surface of the mask of a member of the surgical team. When S. *aureus* has been isolated from the air, it has often been found on the instruments at the end of an operation. On the other hand, instruments have been found to carry this organism when it was not recovered from the air or from the surgeons' masks.

Summary

Air samples taken in operating rooms always contain nonpathogenic bacteria and occasionally pathogenic species. Wounds can become infected by airborne organisms. Periodic checks of the concentration and species of airborne bacteria should be made.

Inadequately cleaned ventilating air can carry bacteria into operating rooms. Suitable fibrous filters clean air supplied to operating rooms.

The principal sources of organisms in the operating room air are the skin, clothing, and upper respiratory tracts of the occupants. Simple precautions, together with adequate ventilation, are more important than elaborate efforts to destroy bacteria liberated into the air. Unnecessary talking and activity are to be avoided. Masking technique and discipline should be strict.

STERILIZATION OF SPECIFIC ITEMS

Stainless steel items

Stainless steel instruments of all types, including scissors, are most rapidly and efficiently sterilized in the autoclave at 121° C. The time-honored formula of "15 min at 15 lb" offers an adequate margin of safety provided timing is begun only after the sterilizing temperature has been registered by the thermometer in the exhaust line. Unless air has been entirely evacuated from the chamber, this temperature will never be reached.

A complete set of instruments for a cataract operation can be folded into a towel and sterilized in a shallow, rectangular stainless steel basin wrapped in two more towels. The entire sterile package is delivered to the sterile nurse or to the operative assistant after he has donned mask, gown, and gloves.

Individual instruments needed during an operation can be sterilized in 5 minutes at 132° in a small, high-pressure autoclave.

Stainless steel instruments can also be sterilized in the final package by dry heat when time permits. In addition to instruments, stainless steel basins, trays, and other components of the operating kit can be sterilized in the autoclave. An item frequently overlooked are the steel trays that fit on the Mayo stands used for skin preparation and for instruments. Even though these are covered with sterile towels, they almost always get wet and can become contaminated from an underlying unsterile tray. For this reason all trays used in our operating rooms are autoclaved in paper wrappers sealed with Steritape.

Cutting-edge instruments

Cataract knives, glaucoma knives, keratomes, trephine blades, and other keen-bladed instruments cannot be autoclaved without corrosion of their cutting edges. Because most disinfectants lack sporicidal properties, dry heat offers the best method for sterilizing these objects.

The following method has proved satisfactory. The carefully cleaned instrument is packed in the original manufacturer's cardboard box with a silica gel packet* to absorb incidental moisture. The box is inserted into a syringe envelope, which is open at one end. Over this is passed a larger envelope, of which the open end is folded down to the level of the closed end of the inner sleeve and sealed with a wire strip. The sealed package is labelled with the surgeon's name, the name of the instrument, and the date.

A load of instruments packaged in this manner is then carefully packed in the sterilizing oven, which is equipped with a turboblower and provided with a thermometer and a timing device. In the event of power failure, the timer will not record a complete cycle. To provide an ample margin of safety, a cycle of 6 hours at 140° C is employed. Spore controls are included weekly. Instruments are sterilized after use and are carefully stored so that they are ready for reuse whenever they are needed.

Several complete cataract sets can be processed overnight and thus can be ready for an early morning operation. No damage to cutting edges results from multiple exposures to sterilizing cycles at 140° C. However, the blades should always be wiped with water (not with saline) and dried after wiping. Rinsing in saline may cause pitting of blades in a short time.

Not only eye instruments but other sharp instruments used in ear and dental surgery can be sterilized in the oven. Hypodermic needles, syringes, and many other heat-stable items are suitable for sterilization by dry heat. A well-insulated oven with provision for forced circulation of air is a worthwhile investment for any hospital.

*Dri-Pax Silica Gel Air Dryer, Davison Chemical Co., Baltimore, Md.

Dry heat sterilizes instruments without introducing toxic substances or contaminants by way of disinfectants carried on knife blades. Christy[15] found a high incidence of striate keratitis after cataract operations in which knife blades soaked in Zephiran were used. When this practice was discontinued, marked reduction in the incidence of striate keratitis was observed.

Rubber bulbs and droppers

Silicone bulb syringes and eye droppers are suitable for sterilization by autoclaving. Sterilization will be readily effected if these items are partly filled with water before autoclaving. Sterilization is not complete unless some water is aspirated into the bulb, because trapped air prevents the entrance of steam from the outside.

Bulb syringes are conveniently sterilized in individual paper envelopes. If glass droppers with rubber bulbs are to be sterilized in a screwcap jar, the jar should be laid flat and the cover left off during sterilization. Otherwise air retained in the jar will prevent attainment of sterilizing conditions.

Droppers and irrigating bulbs that are to be used in intraocular surgery should be filled with fluid and autoclaved together with instruments for the same operation. Because of the intimate nature of their contact with the intraocular tissues, these precautions cannot be too strongly emphasized.

Heat-sensitive items

Optical instruments, electric cords, polyethylene tubing, diathermy points, enucleation implants, plastic prostheses, intraocular acrylic lenses, contact lenses, and many other objects that are subject to damage by sterilizing temperatures must be sterilized by means other than moist or dry heat. Ethylene oxide is recommended.

Table 7-2. Relative merits of steam and gas sterilization

	Steam	Ethylene oxide
Action	Protein denaturation	Alkylation of SH group
Penetration	Rapid	Very rapid
Efficiency	Complete	Complete
Time required	15 min or less	4-12 hours
Equipment	Autoclave	Large or small chamber
Technical snags	Air removal	Prevacuum, correct RH, temperature over 70° F
Danger	Very little	Explosion hazard of pure ETO
Hazard	Inadequate if air not all removed	Toxic residual effect if aeration inadequate
Indications	Steel, glass, cloth	Plastics, delicate instruments
Contraindications	Rustable edges, heat-sensitive items	Previously irradiated PVC

Adapted from Rendell-Baker, L., and Roberts, R. B.: Medical and Surgical Rev. fourth quarter, 1969.

Tonometers

Causation of bacterial keratitis from contact with a contaminated tonometer is a rare possibility that could be eliminated by sterilization of the instrument between patients. Tonometers have transmitted infections by the viruses of inclusion blennorrhea and epidemic keratoconjunctivitis. In mass screening surveys for glaucoma, an infected tonometer could give rise to an epidemic of adenovirus keratoconjunctivitis. To cause an epidemic disease through an attempt to diagnose an endemic one would be ironic indeed.

The principal difficulty in sterilizing tonometers of the Schiøtz type is that a film of lacrimal fluid is drawn up by capillarity between the barrel and the plunger of the instrument each time it is used. This film of moisture, which is inaccessible to ultraviolet radiations, preserves bacteria and viruses in a viable state, so that succeeding patients are infected during displacement of the plunger.

Flaming of stainless steel tonometers is harmless to the instrument and offers the best guarantee of true sterility. To avoid coagulation of protein in the bore, the bore should be swabbed after removal of the plunger. The plunger and footplate should then be flamed separately. Tonofilms may be used instead of flaming.

Electronic tonometers present a delicate problem in that they must not be subjected to heat or to certain chemicals. The minimum acceptable routine for sanitization of electronic tonometers should include removal of the plunger and thorough swabbing of it and the barrel with sterile distilled water, using sterile cotton swabs and sterile pipe cleaners held in well-washed fingers. Applanation tonometers can be swabbed with sterile distilled water.

Sterilization versus disinfection

There are few items used in ophthalmic practice that cannot be sterilized. With a little forethought, presterilized, properly protected instruments and materials can be kept on hand to meet ordinary requirements. Disposable blood lancets and hypodermic needles help to prevent serum hepatitis. Office autoclaves and hot air ovens are available for instruments that are not damaged by heat. For those that are, hospitals need a device for gaseous sterilization. When means of sterilization are at hand, mere sanitization cannot be justified.

HOSPITAL HYGIENE

Staff attitudes

If aseptic technique is to succeed, all personnel in the hospital must be convinced of the necessity for energetic participation in the practices and procedures necessary for reducing the numbers of bacteria in the environment and for preventing their transmission from patient to patient. For all personnel to become convinced of this, the surgical staff must believe in the validity of such elementary precautions as washing the hands between visits to patients, wearing masks over nose and mouth, minimizing conversation and refraining from laughter in the operating room, and carrying out approved practices. If surgeons flout aseptic technique, nurses and corpsmen will find little incentive to observe it themselves or to enforce discipline upon orderlies and attendants. Because of the law of averages, single infractions may not show any effect upon the infection rate, but when breaks in technique are multiplied by many instances and individuals, it is only a question of time before disaster strikes.

Role of the laboratory

The hospital bacteriologic laboratory that limits its activity to isolation, identification, and sensitivity testing of strains recovered from patients is fulfilling only part of its proper function. A physician member of the committee on hospital infections should be given responsibility for supervision of spot checks, bacteriologic controls, and sterilizing practices in the operating and central supply rooms. The extent to which such an individual participates in inspections, sterility surveys, and in the taking and reading of cultures will largely determine his effectiveness. In general hospitals, a member of the ophthalmic staff should concern himself with both the specialized and the broader aspects of aseptic technique and hospital hygiene. By so doing he can provide liaison between the laboratory and the professional staffs, gain experience as an infections control officer, and be instrumental in preventing infections that might otherwise occur.

Committee on hospital infections

Because of accreditation requirements, almost all hospitals have a committee on hospital infections. The hospital bacteriologist and pharmacist should be ex officio members of this committee, which should also have a representative from each of the special services, including ophthalmology. The functions of such committees have been described elsewhere.

Routine controls

A schedule should be set up for routine sterility control of all lots processed in the pharmacy and central supply room. Spore samples processed with each lot should be sent for sterility testing, and the entire lot should be released only upon receipt of the bacteriologist's report of negative cultures, which becomes part of the permanent record of that lot. Carriers bearing 500,000 or more spores and suitably protected by envelopes can be included with objects to be sterilized, and their contents should be cultured aerobically and anaerobically. A log or ledger of all such routine checks should be kept.

Sterility surveys

A comprehensive survey of all the items used during a single operation should be carried out at intervals. Even though the results cannot affect that particular operation, possible sources of contamination may be revealed, and a body of information will be accumulated that may point to some potential source of infection. At the same time, the resident and nursing staffs become more conscious of infection and acquire some knowledge of sterility surveys.

Such surveys are best performed by residents and nurses, working as a team. About 30 to 40 cultures are taken each time. The following materials are furnished by the bacteriologic laboratory: culture tubes of fluid medium in racks, sterile swabs, and blood agar plates.

The manner of conducting a survey is as follows. At the beginning of an operation bacterial sterility controls (resistant spores) are removed from the autoclave, and settling plates are set out. At the end of the operation all instruments that have been used are swabbed with sterile swabs, which are immediately inoculated into tubes of fluid medium. Miscellaneous small items, such as sponges, swabs, sutures, and glass droppers, are dropped into culture tubes without swabbing. Samples of all solutions used in the preparation of the patient and during the operation itself are inoculated into individual tubes of broth. These may include liquid soap, isopropanol, procaine, epinephrine, saline (extraocular and intraocular), and a random sampling of collyria and ointments, whether or not these were used during the operation. The inside surfaces of gloves and masks used by the operating team are cultured by pressing them flat against the surface of blood agar in petri dishes. The cultures are incubated at 37° C and read at the end of 24 and 48 hours. Disruption of operating room routine is minimal; if all preparations have been made beforehand, an entire survey can be taken in 10 to 15 minutes.

Bedding precautions

Every patient entering the hospital is entitled to a sterile blanket and an impermeable pillowcase and mattress cover. These precautions are as elementary as the accepted routine of clean sheets daily. Blankets, pillows, and mattresses are often a formidable reservoir of staphylococci, other bacteria, and fungi. Because of the rapid turnover of patients on busy surgical services, a large, automatic ethylene oxide sterilizer is essential for larger hospitals.

Environmental sanitation and dust control

The high bacterial content of dust removed from modern vacuum cleaners is evidence of their effectiveness in reducing environmental contamination. Used in combination, adequate ventilation, blanket sterilization, vacuum cleaning, and residual surface disinfection with orthophenylphenol offer a four-way approach to reducing the number of bacteria throughout a hospital.

SUMMARY

Although the incidence of preventable infection in ophthalmology has been greatly reduced, major outbreaks (each involving the loss of several eyes) and an undetermined number of isolated cases of postoperative endophthalmitis continue to occur. There are also localized epidemics of adenoviral keratoconjunctivitis. Most, if not all, of the surgical infections and many of the cases of virus infection are preventable by stricter precautions on the part of opthalmologists.

Unfortunately, the standards observed in clinics in this country and in others offer no guarantee that similar surgical disasters and medically mediated epidemics will not continue. Community droppers and dispensing bottles pass from eye to eye in many clinics. Stagnant solutions of unpreserved irrigants can be seen in private offices as well as in hospital wards and clinics. Although unopened collyria and stock solutions are now required to be sterile, the possibility that opened units may become contaminated during careless handling is ever present. Collyria and irrigants, antiseptics and disinfectants, and antibiotics and sulfonamides must all be regarded with the suspicion that they may be contaminated by *Pseudomonas* organisms. When that suspicion arises, it should be confirmed or dismissed by cultural methods. Only a new bottle should be used for an injured eye.

For operating rooms, we must continue to stress all details of aseptic technique, with particular emphasis upon the elimination of unnecessary conversation before, during, and after operation. Adequate ventilation by adequately cleaned air is essential. An environmental survey of 1 case every 6 months is recommended as the minimum essential in spot checking for incidental contaminants.

Probably the most important source of postoperative endophthalmitis is the indigenous flora of the patient's conjunctival sac and lids.[3,31] Elimination or suppression of this flora, according to our latest figures, is best accomplished by the preoperative instillation of drops containing chloramphenicol 0.5% and polymyxin B sulfate 0.1% four times daily, combined with the nightly application of erythromycin ointment 0.5% to the conjunctival sac and the lid margins.[3a] This regimen should be instituted 36 to 48 hours before scheduled surgery.

The three As of asepsis are:
Aseptic technique
Atraumatic surgery
Antibiotic prophylaxis

Assiduous application of these principles will minimize postoperative infection in ophthalmic surgery.

REFERENCES

1. Allen, H. F.: Air hygiene for hospitals: efficiency of fibrous filters against staphylococcic droplet nuclei and bacteria-bearing dust, J.A.M.A. **170**:261, 1959.
2. Allen, H. F.: Aseptic technique in ophthalmology, Trans. Amer. Ophthal. Soc. **57**:377, 1959.
3. Allen, H. F., and Mangiaracine, A. B.: Bacterial endophthalmitis after cataract extraction, Arch. Ophthal. **72**:454, 1964.
3a. Allen, H. F., and Mangiaracine, A. B.: Bacterial endophthalmitis after cataract extraction. II. Incidence in 36,000 consecutive operations with special reference to preoperative topical antibiotics, Trans. Amer. Acad. Ophthal. Otolarying. To be published.
4. Aseptic methods in the operating suite: a report to the Medical Research Council, Lancet **1**:709, April 13, 1968.
5. Ayliffe, G. A. J., Barry, D. R., Lowbury, E. J. L., Roper-Hall, M. J., and Walker, W. M.: Postoperative infections with Pseudomonas aeruginosa in an eye hospital, Lancet **1**:1113, 1966.
6. Blank, I. H., and Coolidge, M. H.: Degerming the cutaneous surface: I. Quarternary ammonium compounds, J. Invest. Derm. **15**:249, 1950.
7. Blank, I. H., and Coolidge, M. H.: Degerming the cutaneous surface: II. Hexachlorophene (G-11), J. Invest. Derm. **15**:257, 1950.
8. Blank, I. H., Coolidge, M. H., Soutter, L., and Rodkey, G. V.: A study of the surgical scrub, Surg. Gynec. Obstet. **91**:577, 1950.
9. Bowman, F. W.: Microbial contamination of antibiotic ophthalmic ointments, J. Pharm. Sci. **58**:277, 1969.
10. Bowman, F. W., and Holdowsky, S.: Sterility of antibiotic ophthalmic ointment, J. Amer. Pharm. Ass. **48**:95, 1959.
11. Browning, C. W., and Lippas, J.: pHisoHex keratitis, Arch. Ophthal. **53**:816, 1955.
12. Burns, R. P.: Postoperative infections in an ophthalmologic hospital, Amer. J. Ophthal. **48**:519, 1959.
13. Cason, L., and Winkler, C. H.: Bacteriology of the normal eye: normal flora, Arch. Ophthal. **51**:196, 1954.
14. Chalkley, T. H. F., and Schoch, D.: An evaluation of prophylactic subconjunctival antibiotic injection in cataract surgery, Amer. J. Ophthal. **64**:1084, 1967.
15. Christy, N. E.: Chemically caused striate keratitis, Amer. J. Ophthal. **45**:5, 1958.
16. Dreisler, K.: Sterilization of tonometers, Arch. Ophthal. **53**:860, 1955.
17. Dunnington, J., and Locatcher-Khorazo, D.: Value of cultures before operation for cataract, Arch. Ophthal. **34**:577, 1947.
18. Fine, B. S., and Zimmerman, L. E.: Exogenous intraocular fungus infections with particular reference to complications of intraocular surgery, Amer. J. Ophthal. **48**:151, 1959.
19. Fuchs, E.: Memories of Vienna clinics, Amer. J. Ophthal. **9**:605, 1926.
19a. Givner, I.: Neisseria catarrhalis endophthalmitis, Amer. J. Ophthal. **32**:5, 1949.
19b. Golden, B., Fingerman, L. H., and Allen, H. F.: Pseudomonas corneal ulcers in contact lens wearers, Arch. Ophthal. **85**:543, 1971.
20. Graefe, A.: Die antiseptische Wundbehandlung bei Star Extraction, Graefes Arch. Ophth. **24**:223, 1878.
21. Guyton, H. G., and Decker, H. M.: Respiratory protection provided by five new contagion masks, Appl. Microbiol. **11**:66, 1963.
22. Guyton, J. S., and Woods, A. C.: Oral use of sulfadiazine for cataract extractions, Amer. J. Ophthal. **26**:1278, 1943.
23. Horner, M.: La methode antiseptique dans la chirurgie oculaire, Ann. Oculist. **86**:56, 1881.
24. Hughes, W. F., Jr., and Owens, W. C.: Postoperative complications of cataract extraction, Arch. Ophthal. **38**:577, 1945.
25. Jennison, M. W., and Edgerton, H. E.: Droplet infection of air, Proc. Soc. Exp. Biol. Med. **43**:455, 1940.
25a. Khorazo, D., and Thompson, R.: The bacterial flora of the normal conjunctiva, Amer. J. Ophthal. **18**:1114, 1935.
26. Kimura, S. J.: Fluorescein paper: a simple means of insuring the use of sterile fluorescein, Amer. J. Ophthal. **34**:466, 1951.
27. King, J. H.: Contamination of eye medications: practical methods of prevention, Amer. J. Ophthal. **36**:1389, 1953.

28. Koch, R., and Wolfhuegel, G.: Disinfection by hot air, Mitt. a.d. Kgl. Gesundheitsamt **1**:301, 1881, translated by W. W. Cheyne in Recent essays on bacteria in relation to disease, London, 1886, New Syndenham Society.
29. Lawrence, C. A.: Surface-active quaternary ammonium compounds, New York, 1950, Academic Press, Inc.
30. Locatcher-Khorazo, D., and Gutierrez, E.: Eye infections following cataract extraction, Amer. J. Ophthal. **41**:981, 1956.
31. Locatcher-Khorazo, D., and Gutierrez, E.: Bacteriophage typing of Staphylococcus aureus, Arch. Ophthal. **63**:774, 1960.
32. Lowbury, E. J. L.: Contamination of cetrimide and other fluids by Ps. pyocyanea. Brit. J. Industr. Med. **8**:22, 1958.
33. Lowbury, E. J. L., and others: Control of infection in ophthalmic surgery, Proc. Roy. Soc. Med. **60**:349, 1967.
34. Maumenee, A. E., and Michler, R. C.: Sterility of the operative field after ocular surgery, Trans. Pacif. Coast Otoophthal. Soc. **32**:172, 1951.
35. Millikin, B. L.: Cleanliness in eye surgery, Amer. J. Ophthal. **7**:175, 1890.
36. Murphy, J. T., Allen, H. F., and Mangiaracine, A. B.: Preparation, sterilization and preservation of ophthalmic solutions, Arch. Ophthal. **53**:63, 1955.
37. Parker, W. R.: Senile cataract extraction: comparative study of results in 1,421 operations, Tr. Sect. Ophth. A.M.A. **26**:202, 1921.
38. Perkins, J. J.: Principles and methods of sterilization, Springfield, Ill., 1956, Charles C Thomas, Publisher.
39. Price, P. B.: Bacteriology of normal skin: a new quantitative test applied to the study of the bacterial flora and the disinfectant action of mechanical cleansing, J. Infect. Dis. **63**:301, 1938.
40. Proskauer, C.: Development of the rubber glove in surgery and gynecology, J. Hist. Med. **13**:373, 1958.
41. Redard, P.: Desinfection des instruments chirurgicaux et des objets de pansement, Rev. Chir. **8**:360, 1888.
42. Reddish, G. F., editor: Antiseptics, disinfectants, fungicides, and chemical and physical sterilization, ed. 2, Philadelphia, 1957, Lea and Febiger.
42a. Rendell-Baker, L., and Roberts, R. B.: Gas versus steam sterilization: when to use which, Medical and Surgical Rev. fourth quarter, 1969.
42b. Smith, C. H.: Bacteriology of the healthy conjunctiva, Brit. J. Ophthal. **38**:719, 1954.
43. Swan, K. C.: Reactivity of the ocular tissues to wetting agents, Amer. J. Ophthal. **27**:1118, 1940.
44. Theodore, F. H.: Contamination of eye solutions, Amer. J. Ophthal. **34**:1764, 1951.
45. Theodore, F. H., and Feinstein, R. R.: Practical suggestions for the preparation and maintenance of ophthalmic solutions, Amer. J. Ophthal. **35**:656, 1952.
45a. Thomas, G.: Air conditioning in hospitals, Anesth. Analg. (Cleveland) **36**:32, 1957.
46. Thygeson, P.: The epidemiology of epidemic keratoconjunctivitis, Amer. J. Ophthal. **32**: 951, 1949.
47. Travers, B.: A synopsis of diseases of the eye, ed. 3, London, 1824, Longman, Hurst, Rees, Orme, Brown, and Green.
48. van der Wyk, R. W., and Granston, A. E.: A bacteriological study of ophthalmic ointments, J. Amer. Pharm. Ass. **47**:193, 1958.
49. Vaughan, D. G., Jr.: Contamination of fluorescein solutions with special reference to Pseudomonas aeruginosa, Amer. J. Ophthal. **39**:55, 1955.
50. Walter, C. W.: The aseptic treatment of wounds, New York, 1948, The Macmillan Co.

8

Ocular tuberculosis

Man probably has been subject to tuberculosis for as long as or longer than history is recorded. Fig. 8-1 shows the skeleton of an Egyptian priest of Amen dating back to 2500 BC; the skeleton exhibits evidence of Pott's disease.[33] While excavating in Grennelle, France, Pales discovered two neolithic skeletons, one showing a hip lesion probably caused by tuberculosis and the other showing tubercularlike lesions of the spine.[15] Bartels stated that a neolithic skeleton from Heidelberg exhibited Pott's disease. Gordon further records the observations of Moodie, who found Pott's disease in a Peruvian skeleton from prehistoric times. Moodie was of the opinion that tuberculosis was widespread among Peruvian Indians. Gordon[15] credits Antoine Maitre-Jan with being the first to describe what probably was a tuberculous lesion of the eye. As quoted by Veith,[36a] Pott stated that his disease was characterized in part by "tedious obstinate ophthalmy." Lloyd[23] in 1930 presented a chronologic bibliography covering the published references to tuberculosis of the eye to that date.

Only clinical data and histologic examination were available for making the diagnosis of tuberculosis until the latter part of the nineteenth century. Furthermore, until the discovery of the ophthalmoscope by von Helmholtz in 1851, observations on tuberculosis of the eye were limited to external examination. Jaeger in 1851 was the first to observe tubercles in the choroid.[15] The tuberculous origin of these lesions was established microscopically by Manz,[25] who reported the case in 1858 and two additional cases in 1866. Cohnheim demonstrated that the choroidal nodules were histologically identical with the tubercles elsewhere in the body.[15]

DISCOVERY OF MYCOBACTERIA

In 1882 Robert Koch[19] stained a bacillus in tuberculous tissue by means of alkaline methylene blue, using vesuvine as a counterstain. He was able to demonstrate the consistent presence of these bacilli in such lesions. He found that cultivation was difficult and that special media were required to grow the organism. By successful animal inoculation he showed that the organism was the causative agent of tuberculosis. It was Ehrlich who discovered the fundamental acid-fast property of mycobacteria by using solutions of nitric acid for decoloration. Both Koch and Klebs recognized bovine tuberculosis as infectious for man; and both believed that the two strains, the human and the bovine, were the same.[35] In 1898 Theobald Smith[34] differentiated *Mycobacterium hominis* from *Mycobacterium bovis*. This was a second important step in the control of tuberculosis, because *M. bovis*, which infects cattle and which is shed in the milk, was soon found to be an important pathogen for man and more particularly for children. Though the figures differ from country to country, through the 1930s *M. bovis* accounted for 50% to 100% of lupus vulgaris infections in children to the age of 16 in Great Britain, 74% to 85% of cervical gland infections, and about 80%

Fig. 8-1. Mummy of a priest of Amen (Twenty-first Dynasty) afflicted with Pott's disease. (From Gay, F.: Agents of disease and host resistance, Springfield, Ill., 1935, Charles C Thomas, Publisher, p. 958.)

of abdominal infections. The percentage of tuberculosis of the bone caused by *M. bovis* was variously reported as 18% to 60%.[35]

Today in many countries bovine tuberculosis is rare, as dairy herds are carefully screened for the disease and milk is pasteurized. The figures on the incidence of *M. bovis* in tubercular infections of the eye are unfortunately full of gaps because of the failure of many investigators to type the mycobacterium isolated from the eye. But from the data to be presented it appears that this pathogen has perhaps played a more important role in primary ocular infections than has *M. hominis*.

PRIMARY TUBERCULOSIS OF THE EYE

The occurrence of primary tuberculous infection of the eye is rare. At the Institute of Ophthalmology, Columbia-Presbyterian Medical Center (IOCPMC) from 1933 to 1970, among 163,948 patients examined only 3 cases of tuberculous infection of the conjunctiva that could be considered primary lesions were seen. In Table 8-1 reports

Table 8-1. Incidence of primary tuberculosis of the conjunctiva in various parts of the world*

Site of observations	Number of patients observed, incidence, or years covered	Number positive	Reported by
Baltimore Eye, Ear and Throat Charity Hospital, Baltimore, Maryland, USA	41,730	None	Bordley, 1902
Same	1870–1902	48	Bordley, 1902 (review of literature)
Guy's Hospital, London, England	1 in 3,200	29	Eyre and Durh, 1912
Finsen Institut, Copenhagen, Denmark	1909–1932	40	Blegvad, 1933; Lundsgaard, 1915
Seraphimer Hospital, Stockholm, Sweden	181,000 (1915–1934)	3	Samuelson, 1936
Oran, Algeria	60,000 (1936–1937)	2	Benalioua, 1938
Moscow Eye Clinic, Moscow, USSR	94,000	3	Ivanov, 1939
New York Eye and Ear Infirmary, New York, USA	30,000 per year for 20 years	None	Burchell, cited by Sitchevska and Sedan, 1947
Institute of Ophthalmology, Columbia-Presbyterian Medical Center, New York, USA	163,948 1933-1968	3	Bruce and Locatcher-Khorazo, 1947; Chandler and Locatcher-Khorazo, 1964; DeVoe and Locatcher-Khorazo, 1964

*This table presents only those studies containing information on frequency of disease. Individual case histories of primary tuberculosis of the eye are not included, except for Bordley's review.

from some other institutes of ophthalmology show equally small numbers of cases.

Primary tuberculous conjunctivitis occurs more often in children than in young adults and more frequently in females than in males. Such infection presumably is introduced into the eye by contaminated hands, fomites, or by exposure to dust or sputum particles containing tubercle bacilli. The patient sometimes mentions associated trauma. There is usually some fever with malaise; and the regional preauricular, submaxillary, and cervical lymph nodes become enlarged. This enlargement is in contrast to secondary infections of the eye, for in patients with some degree of immunity the adjacent lymphatic system is generally not grossly involved.

The first illustration of corneal tuberculosis is credited to Roy and Alvarez[28] (Fig. 8-2). The types of lesions produced in conjunctival tuberculosis are varied and are essentially the same as those listed by Sattler[30a,31] in 1874 and in 1891 and recorded by Duke-Elder.[12] Basically they are ulcerative or hyperplastic (nodular, hypertrophic, or polypoid). The *ulcerative* type is characterized by one or more small miliary ulcers, which may coalesce. They occur more often on the palpebral than on the bulbar conjunctiva. Tubercle bacilli may be recovered from the base of the ulcer. In severe cases the infection may spread to the cornea with pannus formation. Such lesions also may burrow through the thickness of the lid or penetrate the sclera. The *nodular* type shows the presence of small yellow or gray subconjunctival nodules, somewhat resem-

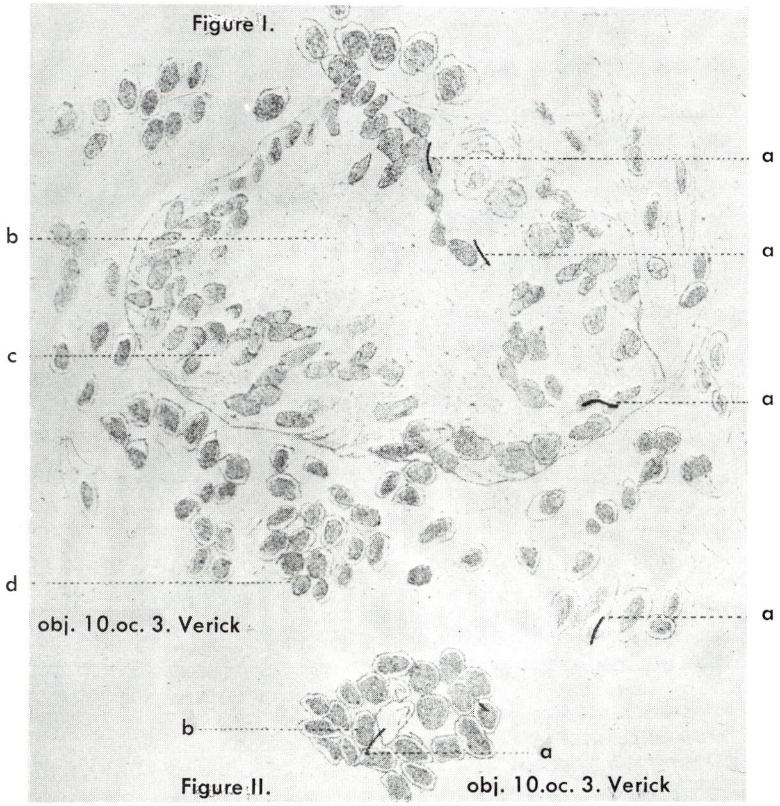

Fig. 8-2. A case of ulcerative tuberculosis of the cornea in a 4½-year-old girl. The corneal biopsy shown was taken about 8 months after the onset of symptoms. *a,* Tubercle bacilli; *b,* giant cell; *c,* numerous nuclei in a giant cell; *d,* round cells from infiltrated cornea. (From Roy, L., and Alvarez: Rev. Clin. Ocul. [Paris] **5**:187, 1885.)

bling granules of trachoma.[12] They tend to increase in size and become surrounded by follicles and granulation tissue. On histologic examination nodules prove to be miliary tubercles. The *hypertrophic* papillary type of conjunctival lesion is characterized by growth of granular tissue, usually developing from the fornices and associated with edema of the lids. In the fornices the granulations may become large, pedunculated, jellylike excrescences and may develop superficial ulcers. The *polypoid* type simulates a fibroma growing from the tarsal conjunctiva. On histologic examination a framework of connective tissue is found enclosing round cell infiltration with scattered giant cells containing tubercle bacilli. Bordley[4] reviewed the literature up to 1902 and assigned the cases to the categories described by Sattler.[31]

Table 8-2. Three cases of primary tuberculosis of the eye*

Special features of ocular disease	Other involvement	Course
Patient A: 9 years old, male September 1937: Painless swelling, right upper lid, for 9 days; shallow, indolent ulcer, upper tarsal conjunctiva; defined reddish margins; gray, sloughing base; surrounding follicles ulcerated; scraping from ulcer showed acid-fast bacilli, *M. bovis* cultured; positive tuberculin test; no history of previous exposure; x-ray films of chest, abdomen, and long bones negative[5]† (Plate 2, *G*)	Two enlarged preauricular lymph nodes, one ruptured, one incised at 5 months; tonsils removed in 4 years showed area of confluent tubercles	Treatment—irrigation with bland fluid; epithelium covered conjunctival lesion in 10 weeks; slow repair of lymph tissue; 9 years free of symptoms
Patient B: 10 years old, male March 1945: Painless swelling, lid of right eye, with mucous discharge; admitted in 2 months to another hospital; x-ray films of chest and abdomen negative; positive tuberculin test; biopsy of involved conjunctiva showed tuberculous lesions, acid-fast bacilli cultured; May 1945: admitted to the IOCPMC; yellowish granules on posterior surface of red and swollen right lid, also on bulbar conjunctiva near outer canthus; scraping and biopsy of swollen conjunctiva gave culture of *M. bovis*; histology typical of tuberculosis; intensive study gave no evidence of tuberculosis elsewhere except for local nodes[10] (Plate 2, *H*)	Concurrent with conjunctivitis, preauricular mass noted; submandibular gland enlarged 2 months later; preauricular node excised; tuberculous lesion	Conjunctivitis subsided over 6 month period; after 1½ years all lesions healed
Patient C: 11½ years old, female November 1961: Painless swelling, right upper lid; tearing "for a few days"; membranous tarsal conjunctivitis; scrapings negative for acid-fast organisms; in 2 weeks admitted to the IOCPMC; cobblestone papillae and small areas of ulceration; x-ray film of chest negative; positive tuberculin test; no known exposure to tuberculosis; 1 month later, scrapings and biopsy of conjunctiva showed acid-fast bacilli; *M. bovis* cultured; histologic examination of biopsy—tubercles, some with caseation, all with plasma cells, epithelioid cells, and Langhans' giant cells[6] (Plate 2, *I* and Fig. 8-3)	Concurrent with conjunctivitis, enlarged right preauricular node; subsequently mandibular node and right lateral glands of neck enlarged	PAS therapy started after biopsy; February 1962: conjunctival lesions healed; May 1962: submandibular and neck glands excised; 1963: patient doing well

*Patients seen at the IOCPMC.
†Superscript numbers refer to references at end of chapter.

Differential diagnosis of mycobacterial infection of the eye is usually difficult, since the lesion is variable and may simulate diseases such as Parinaud's syndrome, ophthalmia nodosa, trachoma, and others listed by Anhalt and his associates.[1] The difficult diagnostic situation is not helped by the infrequency with which tuberculous conjunctivitis is seen. The definitive diagnosis is made by identifying the mycobacterium.

In Table 8-2 three cases of primary tuberculosis of the conjunctiva seen at the IOCPMC are tabulated. Though the series is small, it represents several of the clinical and diagnostic features of this type of infection. All the patients were children under 12 years of age. The onset was insidious, since it was relatively painless. The patients did not present themselves for treatment until there was visible edema of the eyelid, and in each case the opinion was expressed that the condition had been present only a few days. But by the time the child had reached the ophthalmologist there was quite extensive involvement of the eye. Ulceration had taken place, and adjacent lymph nodes were enlarged. The mycobacterium was isolated, cultured, and typed in each instance. All three strains proved to be *M. bovis*. Even in the preantibiotic era the course of the disease under supportive treatment and surgical handling of the infected lymph nodes were satisfactory. The third patient received the usually highly effective general treatment with streptomycin and para-aminosalicylic acid (PAS).

It was quite unexpected to find that all three cases of primary tuberculosis of the eye seen at the IOCPMC were caused by *M. bovis*. This organism is most readily distinguished by its high pathogenicity for the rabbit, which is resistant to the human strain. A review of the literature has revealed a great sparsity of information about the type of tubercle bacillus found in primary lesions of the conjunctiva. Diagnosis was sometimes made only from acid-fast stains and histology. If the organism was grown and animal inoculation made, the

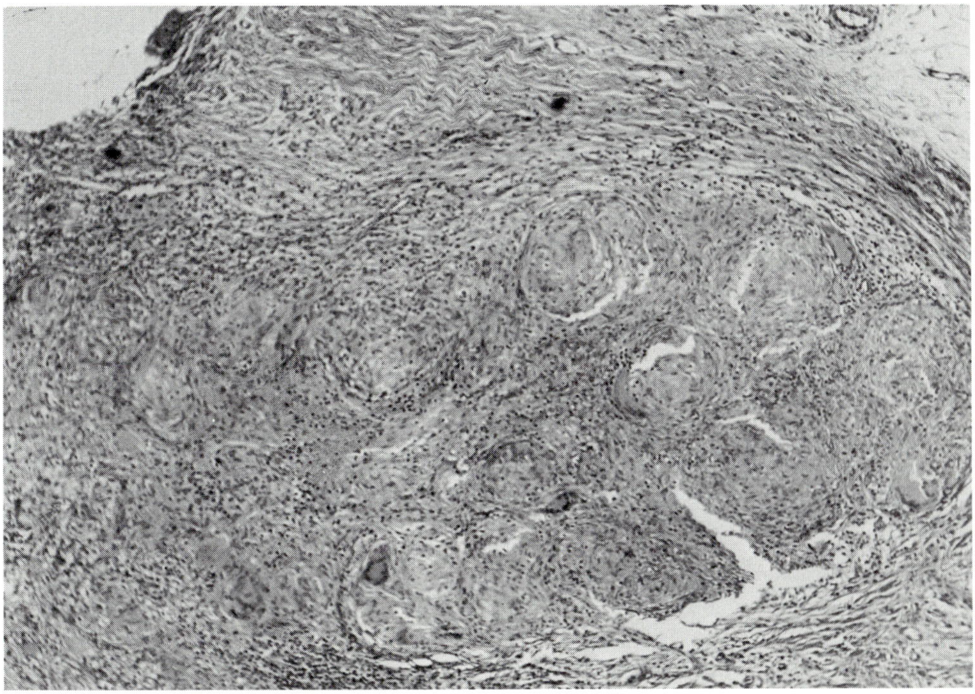

Fig. 8-3. Tubercles in conjunctival biopsy of lesion seen in Plate 2, *I*. Giant cells are evident. (From Chandler, A. C., Jr., and Locatcher-Khorazo, D.: Arch. Ophthal. 71:203, 1964.)

guinea pig was often the only animal tested. The guinea pig is equally susceptible to *M. hominis* and *M. bovis;* thus, such a test does not identify either. The following figures were obtained, however. Blegvad[3] reported 11 cases of primary tubercular conjunctivitis from the Finsen Institut; the mycobacteria were typed, and all were *M. bovis.* Samuelson[30] presented 3 cases of primary tuberculosis, one of which he was able to type. The infection was caused by *M. hominis.* Healy[16a] reported on 2 cases of primary tuberculosis of the conjunctiva, one of which was caused by *M. hominis* and one by *M. bovis.* Eadie[12a] reported 1 case caused by *M. hominis.* With the 3 cases from the IOCPMC, the totals are 15 cases caused by *M. bovis* and 3 by *M. hominis.* This data suggests that the eye of the child may be rather selectively susceptible to the bovine strain. More data on this point is needed.

SECONDARY TUBERCULOUS LESIONS OF THE EYE

Involvement of the eye, secondary to a primary tuberculous lesion elsewhere, may result from the manual introduction of the organism into the eye by the patient, extension of the disease from an adjoining area such as the skin or nasal passages in lupus vulgaris or the meninges in tuberculous meningitis, or from the dispersion of the bacteria via the bloodstream. In the latter situation any part of the eye may be infected (Table 8-3). The incidence of ocular involvement in a tuberculosis sanatorium, as reported by Donahue,[11] was 154 cases out of 10,524, or 1.4%, seen in the years 1940 to 1966.

Lupus vulgaris of the skin is thought to arise by direct inoculation of organisms into an abraded area. A subcutaneous tubercle is formed, which pushes to the surface. This is followed by proliferation of surrounding tubercles, leading to the formation of a lupus patch. When this occurs near the eye, the lids and the conjunctiva may become involved. The same type of evolution takes place in the mucous membranes of the conjunctiva. As in primary infections of the eye, the lesion is painless. The associated lupus tuberculosus of the skin indicates the correct diagnosis.

Miliary tuberculosis resulting from the spread of organisms through the bloodstream may seed tubercles in any part of the eye. Massaro and his co-workers[26] have

Table 8-3. Areas of the eye involved in tubercular infection

Orbit	Orbital abscess—extension from lacrimal gland or sinuses Periostitis—commonly involves tissues and bones of lateral orbital rim; painless red swelling
Lacrimal gland	Chronic dacryoadenitis
Eyelids	Primary lesions rare, but may be involved by extension from the skin
Conjunctiva	Conjunctivitis primary, or secondary to lesions elsewhere in the body
Cornea	Interstitial keratitis; deep (sclerosing) keratitis; corneal ulcer or abscess
Sclera	Proliferative ulcerative lesion; allergic exudative response
Iris	Iritis, a feature of tuberculous panuveitis
Ciliary body	Involved with iris and choroid in tuberculous panuveitis
Choroid	Infection most likely to occur with tuberculous meningitis or miliary tuberculosis, but may occur as only clinical manifestation of tuberculosis
Retina	Involved by contiguity with choroidal lesions; periphlebitis or exudative retinitis
Vitreous	Involved by retinal lesions (new blood vessels proliferate into vitreous with repeated hemorrhages)
Optic nerve	With tuberculous meningitis, optic neuritis not uncommon; also secondary to uveal and orbital lesions

recommended periodic mydriatic examination of the fundus of the eyes of patients with pulmonary tuberculosis. They describe 10 cases in which new choroidal tubercles were found, leading to the institution of therapy. According to Chapman and Whorton,[7] three fourths of 310 cases of acute generalized miliary tuberculosis, discovered at autopsy, had been missed clinically. Presumably a careful examination of the fundus would have corrected some of the missed diagnoses. This point has also been stressed by Olazábal.[27]

At the IOCPMC 2 cases of secondary ocular tuberculosis have been reported.[9,10] The first was in a 74-year-old man who in July 1962 became aware of pain and redness of the left eye. An "ulcer" of the sclera was cauterized elsewhere. In February 1963 the increasing ulceration of the sclera proximal to the limbus led to admission to another hospital where active pulmonary tuberculosis was diagnosed. Scrapings of the eye lesion did not show tubercle bacilli, nor were mycobacteria grown. Local treatment with steroids and atropine was given. In June 1963 the patient was admitted to the IOCPMC. At this time he was receiving PAS and isoniazid. A chest consultant considered the pulmonary condition clinically inactive. There was moderate edema of the lids and conjunctival injection of the left eye. The most striking finding was what appeared to be a large pigmented mass, 6 × 4 × 3 mm, herniating through a dehiscence in the sclera, which was absent except for a few remaining tags of tissue (Plate 3, *A*). Although long-term prognosis was poor, it appeared that an attempt should be made to repair the dehiscence. With the patient under general anesthesia, the conjunctiva was dissected free around the prolapsed uveal tissue. A large lamellar graft 15 × 20 mm was sutured into position. There were no unusual complications during the postoperative course. Acid-fast stains, cultures, and animal inoculation were made from conjunctival scrapings. Acid-fast bacilli were seen in the scrapings; cultures were positive; and guinea pig and rabbit inoculation identified the organism as *M. hominis.* The eye one year after the operation is shown in Fig. 8-4.

The second case was one of acute tuberculous panophthalmitis.[9] It is of particular interest because the primary lesion was discovered 40 years earlier in this 73-year-old man. The patient, first seen on February 4, 1966, had had pain and decreasing vision in the left eye during the previous 6 weeks. A diagnosis of granulomatous uveitis was made. Reactions to skin tests, including those using intermediate strength PPD,

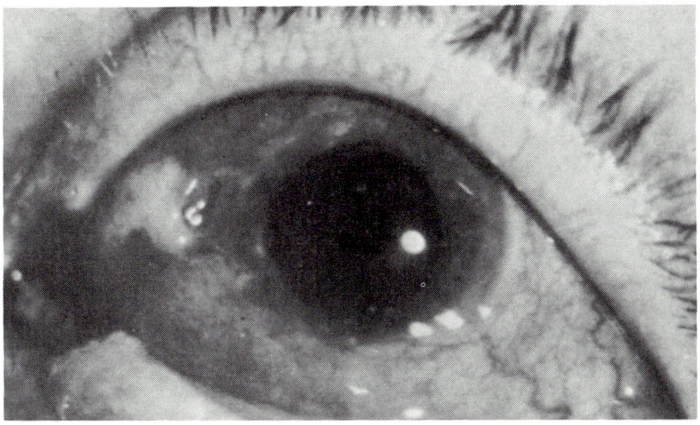

Fig. 8-4. Eye seen in Plate 3, *A* one year after operation. (From DeVoe, A. G., and Locatcher-Khorazo, D.: Trans. Amer. Ophthal. Soc. **62**:209, 1964.)

Table 8-4. Ocular infections caused by atypical pathogenic mycobacteria

Lesion	History	Mycobacterium	Course	Reported by
Left eye, injection of cornea and sclera. Central corneal ulcer, 3 mm.	Piece of steel removed 5 months previously. Corneal ulcer treated with antibiotics and steroids for 3 months.	Scrapings negative for fungi. Sabouraud's agar, blood agar, thioglycollate broth, and Petragnani media inoculated. *M. fortuitum* isolated.	First treated with Chloromycetin, then amphotericin B until mycobacterium identified. Healed on streptomycin, followed by other antibiotics, in 3 months.	Turner and Stinson, 1965.
Right eye markedly injected and painful. Corneal ulcer, 2-3 mm, with margin of infiltrate two-thirds through cornea.	Piece of steel removed 6 months previously. Treatment with various antibiotics and steroids.	Cultures of scrapings for fungi and viruses. Diphtheroids cultured. Patient received lamellar corneal graft; from the removed corneal button *M. fortuitum* seen and identified.	Graft remained clear; one year later, right eye 20/50 + 2 vision, was painless, not injected.	Levenson and Harrison, 1966.
Right eye, limbal flush. Dense stromal corneal opacity 1 × 2 mm. Anterior chamber, trace of flare, 1+ cells.	Five weeks previously, struck by telephone cable in right eye. Minimal discomfort. No treatment.	Scrapings negative for fungi. Unclassified *Mycobacterium* group IV seen and isolated from corneal tissue on 21st hospital day.	Despite negative test reaction for fungi, antifungal treatment given, and lesion became worse. After organism identified, isoniazid, erythromycin, streptomycin therapy, later corticosteroids administered. Eye healed in 3 months.	Lauring, Wergeland, and Sack, 1969.
Necrotizing ulcer involving almost entire cornea, with hypopyon. (Patient seen in 1963.)	Eye believed to be end stage mycotic infection. Smear thought to show *Nocardia*. In 3 months eye had to be enucleated.	Culture of recovered organism identified as *M. fortuitum*. Thought to be nonpathogenic contaminant.	Enucleated eye diagnosed as having chronic keratitis. Six years later, after 2nd case, review of eye tissue showed many acid-fast bacilli.	Zimmerman, Turner, and McTigue, 1969.
Three weeks after keratoplasty, painful, inflamed eye. Antibiotics and steroids unsuccessful. Full-thickness graft after 4 months.	Continued inflammation. Patient transferred to Mt. Sinai Hospital, New York City.	Cultures taken prior to transfer identified as *M. fortuitum*. Keratectomy specimen, many acid-fast bacilli.	At Mt. Sinai Hospital, diagnosis of *M. fortuitum* confirmed. Mixed antibiotic treatment given. In 4 weeks lamellar graft sloughed, removed. No further deterioration.	Zimmerman, Turner, and McTigue, 1969. Wunsh, Boyle, Leopold, and Littman, 1969.

Table 8-4. Ocular infections caused by atypical pathogenic mycobacteria—cont'd

Lesion	History	Mycobacterium	Course	Reported by
Edema, scleral conjunctiva. Lower nasal quadrant of cornea, 4-5 mm epithelial defect.	Hit by piece of brick. Foreign body removed. After 6 weeks no improvement of persistent keratitis.	Gram-positive organism isolated, finally identified as *M. fortuitum*.	Initially treated as *Nocardia* infection. No improvement. When *M. fortuitum* recognized, PAS, isoniazid, and topical streptomycin treatment given.	Sexton, 1970.
Localized keratitis, outer one-third of left eye. Circle fibrin precipitate.	Foreign body acquired when using a grinder, removed some time previously from left eye.	Organism grown, *M. fortuitum*.	Same treatment as above successful.	

toxoplasmin, histoplasmin, blastomycin, and coccidioidin, were negative. The reaction to the test using brucellergen was "slightly positive." X-ray films of the chest showed fibrotic and fibrocalcific disease. Two weeks later exophthalmos of the left eye was noted. When the patient was admitted on February 27 for enucleation, the reaction to the tuberculin test was positive. The iris had become necrotic, and the corneal epithelium showed bullous elevation. Examination of the globe revealed a chorioretinal granuloma. There was necrosis of the overlying retina and inflammatory reaction in the vitreous. Areas of caseous necrosis were surrounded by lymphocytes, plasma cells, and giant cells. A large number of acid-fast bacilli were identified in the necrotic retina. The diagnosis was uveitis, scleritis, and orbital cellulitis specific to tuberculosis.

These two experiences with secondary tuberculosis of the eye seen at the IOCPMC give the extremes of ocular involvement; on the one hand, a localized and chronic lesion of conjunctiva and sclera; on the other, acute, rapidly progressing panophthalmitis. They illustrate the protein nature of pathology caused by *M. tuberculosis* and perhaps reinforce the plea of Massaro, Katz, and Sachs[26] for routine periodic examination of the fundus in cases of known tuberculosis.

ROLE OF ATYPICAL ACID-FAST BACTERIA IN OCULAR DISEASE

Several acid-fast bacteria have long been recognized as atypical mycobacteria.[14,18,20,29] They were formerly considered harmless saprophytes for man, although some were pathogenic for animals. Recent bacteriologic and clinical studies have shown that some of these mycobacteria may produce disease in man. The medical significance of these organisms and their classification may not yet be complete, but a tentative grouping offered by Runyon[29] in 1959 is currently widely accepted. Three of his four groups—I, III, and IV—contain bacteria that have caused human disease. One general infection caused by the Battey strain, group III has been reported to include infection of the orbit.[22a] Members of group IV have been reported in ocular disease. Table 8-4 summarizes data obtained from seven patients, six of whom had corneal ulcers from which *M. fortuitum*, a member of group IV, was isolated. An unclassified bacterium belonging to group IV was cultured from the corneal ulcer of the seventh patient.

The histories in each instance had some

common characteristics. The first of these was preceding trauma. Four patients had had foreign bodies removed from the cornea; one had been struck in the eye by a cable 3 to 5 months previously. Dr. Zimmerman's[38] first patient had come to him 6 years prior to the final infection with a lesion that resembled a crescentic laceration. Five years later the patient had an active dendritic ulcer. Two flare-ups with ulceration occurred at 4 month intervals. Eventually enucleation was necessary. Dr. Zimmerman's last patient developed intransigent inflammation 3 weeks after penetrating keratoplasty.

The second common characteristic of these infections was the failure to isolate immediately a recognized pathogenic organism. This led to treatment with a very wide spectrum of antibiotics and steroids. Four of the lesions were at one time suspected of being of fungal origin. Some of the difficulties of making an earlier bacteriologic identification of *M. fortuitum* in the patients reported in Table 8-4 probably were caused in some measure by the temperature for incubation, 36° C, and the use of Sabouraud's media, which is designed to favor only the growth of fungi. The cultural characteristics of *M. fortuitum* include rapid growth (2 to 5 days) on ordinary media and an optimum temperature for growth of about 25° C. The first culture obtained from Dr. Turner's[36] patient took 3 weeks to grow on Sabouraud's media, and the organisms were first identified as *Nocardia.* The first cultures from the patient of Dr. Levenson[22] (media not specified) grew diphtheroids. When the piece of cornea was cultured in thioglycollate broth and on blood agar, there was growth of a pleomorphic gram-positive rod resembling a diphtheroid. The Communicable Disease Center in Atlanta identified the organisms as *M. fortuitum.* Mycobacteria may resemble corynebacteria on Gram stain.

Dr. Lauring[21] had a similar experience. Organisms that grew on Sabouraud's media in 21 days were initially identified as *Nocardia.* Again the Communicable Disease Center, together with the National Jewish Hospital and Fitzsimons General Hospital in Denver, made the diagnosis of *Mycobacterium* group IV serotype II. Cultures from Dr. Zimmerman's first case were originally identified as *Nocardia* or *Actinomyces.* Again the Communicable Disease Center made the correct identification. *Nocardia* was also the original identification in Dr. Zimmerman's second case.

The histology of the lesions produced were similar and were reported as follows. Dr. Levenson reported "ulceration into the cornea stroma with a cellular infiltrate consisting of polymorphonuclear leukocytes and round cells in almost equal numbers and many slender pleomorphic bacilli with rounded ends . . . deep within the substantia propria." Dr. Lauring reported numerous bacilli in the corneal tissue and a stromal keratitis without abscess formation. The corneal tissue from Dr. Zimmerman's first patient showed a large area of corneal necrosis and inflammation with many typical beaded acid-fast bacilli present in the exudate. The iris temporally was densely infiltrated by lymphocytes and plasma cells. The keratectomy specimen from the second patient contained a deep ulcerated area with many polymorphonuclear leukocytes containing large numbers of acid-fast bacilli. In other areas of the eye additional similar microabscesses were seen.

The case histories of these 7 patients with infection caused by one of the group IV mycobacteria have been given in some detail because they pose an important differential diagnostic problem. These organisms are not sensitive to the drugs that are effective for the human and bovine strains of mycobacteria. Treatment of such cases requires knowledge of their specific sensitivities.

The question of the source of these bacteria is one that may also have some bearing on the origin of tuberculous infections of the eye. Chapman and Speight[8] have reported that of 458 samples of pasteurized milk sold in cartons, 2.8% contained mycobacteria, including *M. fortuitum* and others from group III. Harrington and Karlson[16] have raised the question as to whether the atypical mycobacteria as well as *M. tuberculosis* and *M. bovis* are always killed by pasteurization. They have shown that some of the atypical strains do survive this treatment. The 2 strains of *M. tuberculosis* and the 10 strains of *M. bovis* that they tested were killed by the treatment. This point becomes of special interest when the high incidence of bovine strains among the cases of primary tuberculosis of the eye is considered. Useful information would be obtained if possible sources of infection for all cases of primary ocular tuberculosis caused by the bovine strain were followed up.

Another practical diagnostic problem between infection by *M. hominis* or *M. bovis* and the atypical mycobacteria is the observation that there is crossing between tuberculins produced by *M. tuberculosis* and those derived from the atypical organisms.[9a] It may be that in some areas a large number of positive reactions to PPD are actually cross-reactions resulting from antibody response to atypical organisms. In evaluating the significance of a positive reaction to the tuberculin test one must thus realize that the positive reaction may not rule out one of the atypical mycobacteria as the causative agent of eye infection. Therefore such a reaction to the tuberculin test must not prevent looking for *M. fortuitum.*

SUMMARY

Primary tuberculosis of the eye that involves the cornea and conjunctiva is a rare disease today. It usually is unilateral only; the onset is insidious and painless; and, in contradistinction to mycotic infection, preceding trauma is often not mentioned. Lymph nodes contiguous to the involved eye become enlarged. Scrapings from the lesions usually will show acid-fast bacilli. In the 18 cases found where typing was carried out, 15 resulted from infection by *M. bovis.* The histology of the lesions is characteristic of typical tuberculous infection with infiltration of epithelioid cells and giant multinucleated cells, round cells, and scattered bacilli.

Secondary tuberculosis of the eye can be more difficult to diagnose when the lesions are not susceptible to direct culture and the primary disease is not recognized. Any or all parts of the eye may be involved. Sometimes the ophthalmologist may be the first to make the diagnosis of tuberculosis by seeing choroidal tubercles.

Differential diagnosis of infection caused by one of the atypical mycobacteria, notably *M. fortuitum,* may present an important problem because these latter organisms are not sensitive to the chemotherapeutic agents used for tuberculosis. The few infections with the atypical organisms so far reported have been associated with previous trauma and at one time in the course of determining the etiologic agent were considered as possibly caused by fungi. Biopsies of the lesions have proved most useful both for obtaining sufficient material to permit isolation and identification of the bacterium and for histologic evaluation. The lesions have shown many acid-fast organisms in fields filled with many polymorphonuclear neutrophils as well as mononuclear cells and no epithelioid or multinucleated giant cells.

REFERENCES

1. Anhalt, E. F., Zavell, S., Chang, G., and Bryon, M. H.: Conjunctival tuberculosis, Amer. J. Ophthal. **50**:265, 1960.
2. Benalioua: La primo-infection tuberculeuse de la conjonctive, Ann. Oculist. **175**:746, 1938.
3. Blegvad, O.: Fortgesetzte Untersuchungen über die Conjunctivaltuberkulose, Acta Ophthal. (Copenhagen) **14**:200, 1936.
4. Bordley, J.: Tuberculosis of the conjunctiva, Ophthal. Rec. **11**:368, 1902.
5. Bruce, G. M., and Locatcher-Khorazo, D.: Primary tuberculosis of the conjunctiva, Arch. Ophthal. **37**:375, 1947.
6. Chandler, A. C., Jr., and Locatcher-Khorazo, D.: Primary tuberculosis of the conjunctiva, Arch. Ophthal. **71**:202, 1964.
7. Chapman, C. B., and Whorton, C. M.: Acute generalized miliary tuberculosis in adults, New Eng. J. Med. **235**:239, 1946.
8. Chapman, J. S., and Speight, M.: Isolation of atypical mycobacteria from pasteurized milk, Amer. Rev. Resp. Dis. **98**:1052, 1968.

9. Darrell, R. W.: Acute tuberculous panophthalmitis, Arch. Ophthal. **78**:51, 1967.
9a. Davis, B. D., Dulbecco, R., Eisen, H. N., Ginsberg, H. S., and Wood, W. B., Jr.: Microbiology, New York, 1967, Harper and Row, Publishers, p. 863.
10. DeVoe, A. G., and Locatcher-Khorazo, D.: The external manifestations of ocular tuberculosis, Trans. Amer. Ophthal. Soc. **62**:203, 1964.
11. Donahue, H. C.: Ophthalmologic experience in a tuberculosis sanatorium, Amer. J. Ophthal. **64**:742, 1967.
12. Duke-Elder, S.: System of ophthalmology. Diseases of the outer eye, vol. 8, part 1, St. Louis, 1965, The C. V. Mosby Co., pp. 212-225.
12a. Eadie, S.: Primary tuberculosis of the conjunctiva treated with streptomycin and PAS, Brit. J. Ophthal. **38**:568, 1954.
13. Eyre, J. W. H., and Durh, M. S.: Tuberculosis of the conjunctiva, its etiology, pathology and diagnosis, Lancet **2**:1319, 1912.
14. Furguson, S. H., Prather, E. C., and Mayfield, C. L.: Clinically significant Mycobacteria, Southern Med. J. **63**:318, 1970.
15. Gordon, B. L.: Ocular and general tuberculosis, Hebrew Med. J. **2**:322, 1963.
16. Harrington, R., Jr., and Karlson, A. G.: Destruction of various kinds of Mycobacteria in milk by pasteurization, Appl. Microbiol. **13**:494, 1965.
16a. Healy, J. J.: Two cases of exogenous tuberculosis conjunctivitis displaying Parinaud's conjunctivo-adenitis syndrome, Trans. Ophthal. Soc. U. K. **66**:455, 1946.
17. Ivanov, N. K.: Tuberculosis of the conjunctiva, Vestn. Oftal. **14**(5):24, 1939.
18. Jawetz, E., Melnick, J. L., and Adelberg, E. A.: Review of medical microbiology, ed. 9, Los Altos, Calif., 1970, Lange Medical Publications, p. 186.
19. Koch, R.: Die Aetiologie der Tuberculose, Berl. Klin. Wschr. **19**:221, 1882.
20. Kubica, G. P., and Dye, W. E.: Laboratory methods for clinical and public health mycobacteriology, U. S. Department of Health, Education and Welfare Public Service, 1967.
21. Lauring, L. M., Wergeland, F. L., and Sack, G. E.: Anonymous mycobacterium keratitis, Amer. J. Ophthal. **67**:130, 1969.
22. Levenson, D. S., and Harrison, C. H.: Mycobacterium fortuitum corneal ulcer, Arch. Ophthal. **75**:189, 1966.
22a. Levine, R. A.: Infection of the orbit by an atypical mycobacterium, Arch. Ophthal. **82**:608, 1969.
23. Lloyd, R. I.: Tuberculosis of the conjunctiva, Amer. J. Ophthal. **13**:753, 1930.
24. Lundsgaard, K. K. K.: Die sogenanate primäre Konjunktivaltuberkulose und die Konjunktivaltuberkulose bei Lûpuspatienten, Klin. Mbl. Augenheilk. **55**:97, 1915.
25. Manz, W.: Tuberculose des Choriodea, Graefes Arch. Ophth. **4**:120, 1858.
26. Massaro, D., Katz, S., and Sachs, M.: Choroidal tubercles: a clue to hematogenous tuberculosis, Ann. Intern. Med. **60**:231, 1964.
27. Olazábal, F., Jr.: Choroidal tubercles: a neglected sign, J.A.M.A. **200**:374, 1967.
28. Roy, L., and Alvarez: Observation chinique de bacille de la tuberculose, dans la cornée, Rev. Clin. Ocul. (Paris) **5**:185, 1885.
29. Runyon, E. H.: Anonymous Mycobacteria in pulmonary disease, Med. Clin. N. Amer. **43**:273, 1959.
30. Samuelson, A.: Primary tuberculosis of the conjunctiva, Arch. Ophthal. **15**:975, 1936.
30a. Sattler, H.: Tubercular ulceration of the conjunctiva, J. Med. Sci. **68**:276, 1874.
31. Sattler, H.: Ueber die Behandlung de verschiedenen Formen der Conjunctivaltuberculose mit Tuberculin nebst experimentellen Untersuchungen über die Wirkung desselben, Ber. (21 Versammlung der) Ophth. Ges. **27**:33, 1891.
31a. Sexton, R. R.: Mycobacterium fortuitum infection of the cornea. In Polack, F. M.: Corneal and external diseases of the eye, Springfield, Ill., 1970, Charles C Thomas, Publisher, p. 25.
32. Sitchevska, O., and Sedam, M.: Primary tuberculosis of the conjunctiva, Arch. Ophthal. **30**:196, 1943.
33. Smith, G. E., and Dawson, W. R.: Egyptian mummies, New York, 1924, The Dial Press.
34. Smith, T.: A comparative study of bovine tubercle bacilli and of human bacilli from sputum, J. Exp. Med. **3**:451, 1898.
35. Steinbach, M. M.: Mycobacteria: the tubercle bacilli and tuberculosis. In Gay, F. P.: Agents of disease and host resistance, Springfield, Ill., 1935, Charles C Thomas, Publisher.
36. Turner, L., and Stinson, I.: Mycobacterium fortuitum as a cause of corneal ulcer, Amer. J. Ophthal. **60**:329, 1965.
36a. Veith, I.: This long disease, my life—postscript to "HE," Perspect. Biol. Med. **15**:110, 1971.
37. Wunsh, S. E., Boyle, G. L., Leopold, I., and Littman, M. L.: Mycobacterium fortuitum infection of corneal graft, Arch. Ophthal. **82**:602, 1969.
38. Zimmerman, L. E., Turner, L., and McTigue, J. W.: Mycobacterium fortuitum infection of the cornea: a report of two cases, Arch. Ophthal. **82**:596, 1969.

9

Ocular leprosy

JAMES H. ALLEN AND MARGARET BRAND

Leprosy is the oldest recorded disease of mankind.[26] In ancient cultures the visible features of the disease, its chronicity, and its incurability, coupled with lack of exact information on its mode of transmission, resulted in fear and superstition. From time to time many afflictions have been thought to be retribution for moral trangression, but leprosy has been unique in the intensity, ubiquity, and persistence of this superstition. In ancient India it was thought to be a curse of the gods. In China it was spoken of as "T'ien Ying," meaning "Retribution from Heaven."[29] The disease remains a great challenge to every branch of medicine, for in spite of the progress that medicine has made in general, leprosy presents many unsolved problems.

HISTORY

There has been much controversy regarding the origin of leprosy. Disease presumed to be leprosy was described in the Bible,[4,20] in the Susruth Samhita (India, approximately 600 BC), and in Chinese medical writings between 500 and 300 BC.[30] These writings infer much older awareness of the disease, although it is now agreed that a number of skin diseases and other afflictions were included. Nevertheless, fairly clear concepts of the disease seem to have emerged in the Middle East and the Orient by the second or third century before Christ. Pascha, quoted by Cochrane,[10] felt that the cradle of leprosy was ancient India, for sacred Sanskrit manuscripts mentioned leprosy. Pascha believed the disease spread from India to Egypt where conditions were favorable for its development. Baron,[6] in his recent publication *Man Against Germs*, stated: "Slowly the [leprosy] germs migrated to the West. They moved along the caravan routes through Persia, crossed the rivers Tigris and Euphrates, passed through mysterious Chaldea and sinful Babylon, traveled over the sands of the Syrian desert, over the Jordan through Canaan and the Wilderness of Shur into the Land of Goshen in ancient Egypt."* Thus the signs of leprosy became a part of the Old Testament, and they were described in Egyptian papyri written 5,000 years or more before the birth of Christ.

With the development of better communications and trade and of movements of armies and immigrants, the disease was introduced in various periods of time to every continent. As may be observed on the map (Fig. 9-1), the greatest incidence of leprosy now lies between the fortieth

*From the book *Man Against Germs* by A. L. Baron. Decorations by Doris Baron. Copyright © 1957 by A. L. Baron. Published by E. P. Dutton and Co., Inc. and used with their permission.

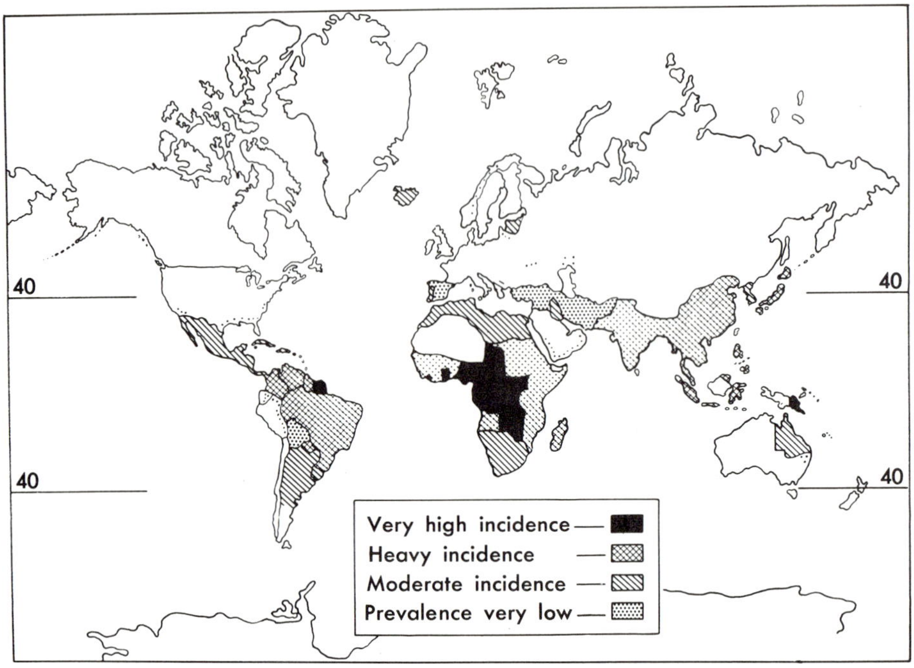

Fig. 9-1. Distribution of leprosy. (Adapted from Rogers, L., and Muir, E.: Leprosy, ed. 3, Baltimore, 1946, The Williams and Wilkins Co.)

parallels. Africa and Asia are the continents most affected. According to Holmes,[15] over 5,000,000 people are affected with leprosy throughout the world. A more accurate estimate probably is close to 10,000,000, and in the United States recent estimates vary from 5,000 to 10,000 cases. Apparently the disease was introduced into Mexico during the sixteenth century, and it is possible that leprosy spread into the part of this country that formerly belonged to Mexico. However, the earliest available reference to the disease within our present boundaries are to be found in Roman's *Concise Natural History of East and West Florida*, published in 1776. Therefore, leprosy has existed in the United States for more than 200 years, and it now is endemic in Florida, Louisiana, Texas, California, and Hawaii. It is interesting, as Cochrane[10] stated, that there is no evidence of leprosy among Eskimos.

The manner in which patients with leprosy were formerly treated, according to Badger,[4] "is not one of which we can be proud, generally they did not have a place to go. They were hounded from place to place, were isolated at the outskirts of the cities or in huts; in the woods . . . there were no institutions where they could go for treatment." Although a leprosarium had been established during the latter part of the nineteenth century in Louisiana, it was not until March 25, 1916, that Senator Ransdell presented a "Bill to Provide for the Care and Treatment of Persons afflicted with Leprosy and to Prevent the Spread of Leprosy in the United States" on a national scale. This bill became law, and in 1921 the National Leprosarium was established.

ETIOLOGY

Leprosy usually is prevalent in tropical and subtropical countries, and the incidence is greater in humid and moist countries where the rainfall is heavy. However, factors other than climate, such as poor personal hygiene and unsanitary and overcrowded living conditions, may be more important in accounting for the prevalence of the disease, for Iceland and Norway also have had large numbers of patients.

The causative agent of leprosy is an acid-fast bacillus that was originally observed by Hansen in 1873 to 1874 in Norway. Hansen[14] had seen this organism as early as 1868 in unstained scrapings of tissues from leprous lesions. When he used osmic acid for staining, he was able to demonstrate the bacilli clearly. Neisser[14] in 1879 also found the organisms in tissue sections, using fuchsin and gentian violet stains. All of Koch's postulates have not been fulfilled, since the organism cannot be cultivated nor can animals be infected;[31] however, the bacilli are considered the cause of leprosy because they are constantly present in the leprous lesions, in some instances in enormous numbers, and because there is a characteristic cellular reaction about them.

Hansen's bacillus belongs to the genus *Mycobacterium* and closely resembles the other pathogen in this genus. The organisms vary in size measuring from 1.5 to 8μ in length and 0.2 to 0.5μ in width. Some appear straight, some curved, some lie parallel to each other. Some stain solidly, but others show beading with bipolar staining. In the early stage of the disease, intracellular and extracellular masses of the bacilli adhere closely to each other, forming large bunches —"cigar bundles"—appearing in glomerations known as globi, which actually consist of tightly packed bacilli enclosed in capsular matter (Plate 3, *B*). The *Mycobacterium leprae* stains deeply and brilliantly with carbol fuchsin and resists decolorization. However, the results of the acid-fast stain may show unevenness, and some of the granular appearance may be explained on this basis. With metachromatic stains, beading and granules are observed.

Brieger[7] studied the structure *M. leprae* with the electron microscope. Since the tissues from which this material was obtained contained dead as well as living bacteria, he felt it was not possible to be certain of the appearance of a healthy organism. The bacilli rarely were found in the process of division, but some of them occasionally showed branching. In material from treated and untreated patients, the structure of the bacilli presented great variation. Many of the bacilli in sections of lepromatous nodules had lost their structure.

An interesting feature of *M. leprae* is that since its discovery, in spite of many attempts toward its propagation on a variety of artificial media, in tissue culture, in symbiosis with other organisms, and at varying temperatures, it still has never been grown. Also, attempts to transmit the disease to man[19] or to various animals with infected leprous material from human lesions have failed. However, Shepard[28] has shown that *M. leprae* can multiply to a limited extent in the footpads of mice but fail to exhibit gross signs of invasiveness. This has been confirmed by Rees[24] and by Janssens and Patlyn.[16]

The chief sites of infection are the skin, mucous membranes, and peripheral nerves. Clinical manifestations are a result of the lodgement and multiplication of the organisms in tissues and of the reaction by the host to such invasion. It is interesting that the human body has the capacity to harbor large numbers of the organisms for many years without ill effect. It may require months or years and additional stresses and strains before the resistance is broken and symptoms appear (Badger[5]). Therefore, it is not surprising that the exact incubation period in leprosy—the time interval before the infection causes disease in the individual—is not known and that it may be measured by years rather than by months or days. Rogers and Muir[26] cite Marchoux's case of a young surgeon who, while operating on a patient with leprosy, accidentally pricked his finger and developed leprosy some 8 or 9 years later. In contrast, Muir[20] observed leprosy in a 3-month-old child. Variable periods of incubation in leprosy have been recorded by a number of observers. Muir[20] considers that the average incubation period is about 3½ years and that the patient, if young, will be unable to remember the time of contact. The duration of the incubation period can be of great importance, since it may determine the source of the infection in an individual patient.

There still is uncertainty concerning transmission of Hansen's disease, although it is the oldest known infection. Prolonged and close association with someone known to have the disease is considered unwise, and yet minimal precautions seem to be adequate to prevent the infection. There is evidence that males are more susceptible than females; and in endemic areas, children and young adults are more susceptible to the infection than the middle-aged and elderly. This age difference in susceptibility is more striking in the case of leprosy than in the case of tuberculosis.

There is no simple technique to determine the presence of the leprosy bacillus in the body. The organisms may be found by acid-fast stains in large numbers in the nasal secretion even where there are no visible external signs of disease. Bacilli may be found in the dermal pulp scraped from shallow skin incisions or in material from ulcerations. Ulcerated nodules may discharge the organisms from the throat when one coughs and sneezes. The bacilli from the saliva may be swallowed and passed in the stools. Testicles, semen, and urine may contain them, as may the milk of the nursing mother. The bacilli may reach the sensory nerves from the skin or from the bloodstream. *M. leprae* may spread through the bloodstream to all parts of the body. The central nervous system is almost never invaded, perhaps because of the absence of Schwann cells, cells in which the bacilli frequently may be detected. In the severe type of leprosy the organisms are present in great numbers in the nose, mouth, skin, and upper respiratory passages and usually are increased when ulceration occurs, suggesting that the infection may occur by droplet transmission from the nose and the month as is common in tuberculosis; but, the actual pathways by which *M. leprae* enters and infects the various organs is not known.

Conclusive evidence that *M. leprae* can produce immunity is not yet available. Natural resistance to infection is present; and among different individuals exposed to infection, some acquire it, whereas others escape the disease. Rogers and Muir[26] were of the opinion that, as in tuberculosis, resistance to leprosy may be terminated by either massive or repeated infection, whereas slight infections may increase the resistance. The leprous individual may exhibit false-positive Wassermann, VDRL, and STS reactions, but reaction to the TPI test remains negative unless the patient also has syphilis. Similarly, reaction to the tuberculin test remains negative unless the individual has developed sensitivity to the tubercle bacillus. So far specific antibodies have not been demonstrated, nor is there evidence that *M. leprae* can produce general immunity. However, in 1919 Mitsuda discovered that intradermal injection of lepromin, a suspension of heat-killed *M. leprae* mixed with finely ground human skin, produced an intense inflammatory reaction, sometimes with ulceration, reaching a peak in about 3 weeks in patients with tuberculoid leprosy, but produced no visible reaction in patients with lepromatous leprosy. Later, Fernandez in 1940 reported that the late positive (Mitsuda) reaction usually was preceded by an early tuberculinlike response (Fernandez reaction) consisting of an area of erythema and edema at the site of injection 24 to 48 hours after injection. Although these tests are of no diagnostic value, experience has shown that reaction to the lepromin test is positive in individuals who have a strong biologic defense against leprosy and who develop tuberculoid or related forms of the disease.

Other immunologic studies[8] have shown that IgG, IgA, and IgM levels were significantly elevated in the sera of lepromatous patients, but serum levels in tuberculoid patients were not significantly different from control values. Serum levels of IgG and

IgA generally remained elevated even with prolonged therapy, whereas IgM levels tended to decline significantly with treatment in lepromatous leprosy. In reactional states in patients who had been treated for longer than 24 months, serum levels of IgA were unusually high.

Summary

It appears that since the discovery of *M. leprae,* the studies on leprosy have been hampered by the failure both to cultivate the bacilli and to transmit the disease to experimental animals. Very little is known about the incubation period or the mode of infection. *M. leprae* is able to live in tissues in enormous numbers for many years without producing overt disease.

CLINICAL TYPES OF LEPROSY

Clinically, leprosy may be classified as the lepromatous, tuberculoid, or dimorphous variety, although it may begin and exist for a time in an indeterminate type.

The *indeterminate* type consists of papules or pink or pale macules that have not developed sufficient character to permit differentiation into the usual varieties. The macule, usually single, commonly appears on the forehead as a circumscribed, nonelevated, hypopigmentation or hyperpigmentation exhibiting loss of sensation of cold, heat, and touch, usually in that order. The histologic appearance on biopsy of one of these lesions is that of small strands or groups of cells consisting of lymphocytes and macrophagic histiocytes with an occasional plasma cell, eosinophil, basophil, or polymorphonuclear leukocyte around the pilosebaceous system of the dermis and the nerve fibers and blood and lymph vessels, passing down toward the subcutis. Bacilli are few in number or may not be found. Reaction to the lepromin test may be weakly positive or negative. These lesions usually are precursors of tuberculoid or lepromatous leprosy but may develop into the dimorphous variety.

Lepromatous leprosy represents minimal resistance to the occurrence, multiplication, and dissemination of leprae bacilli, which are readily demonstrable individually or as

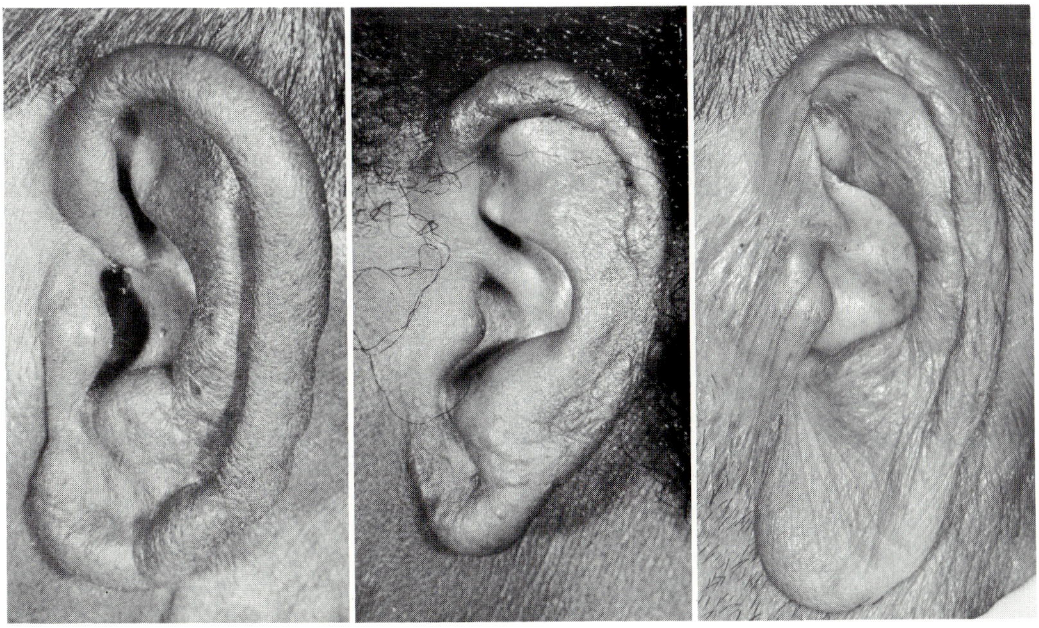

A B C

Fig. 9-2. Lepromatous nodules of the ear, **A, B,** and typical pendulous lobe, **C.**

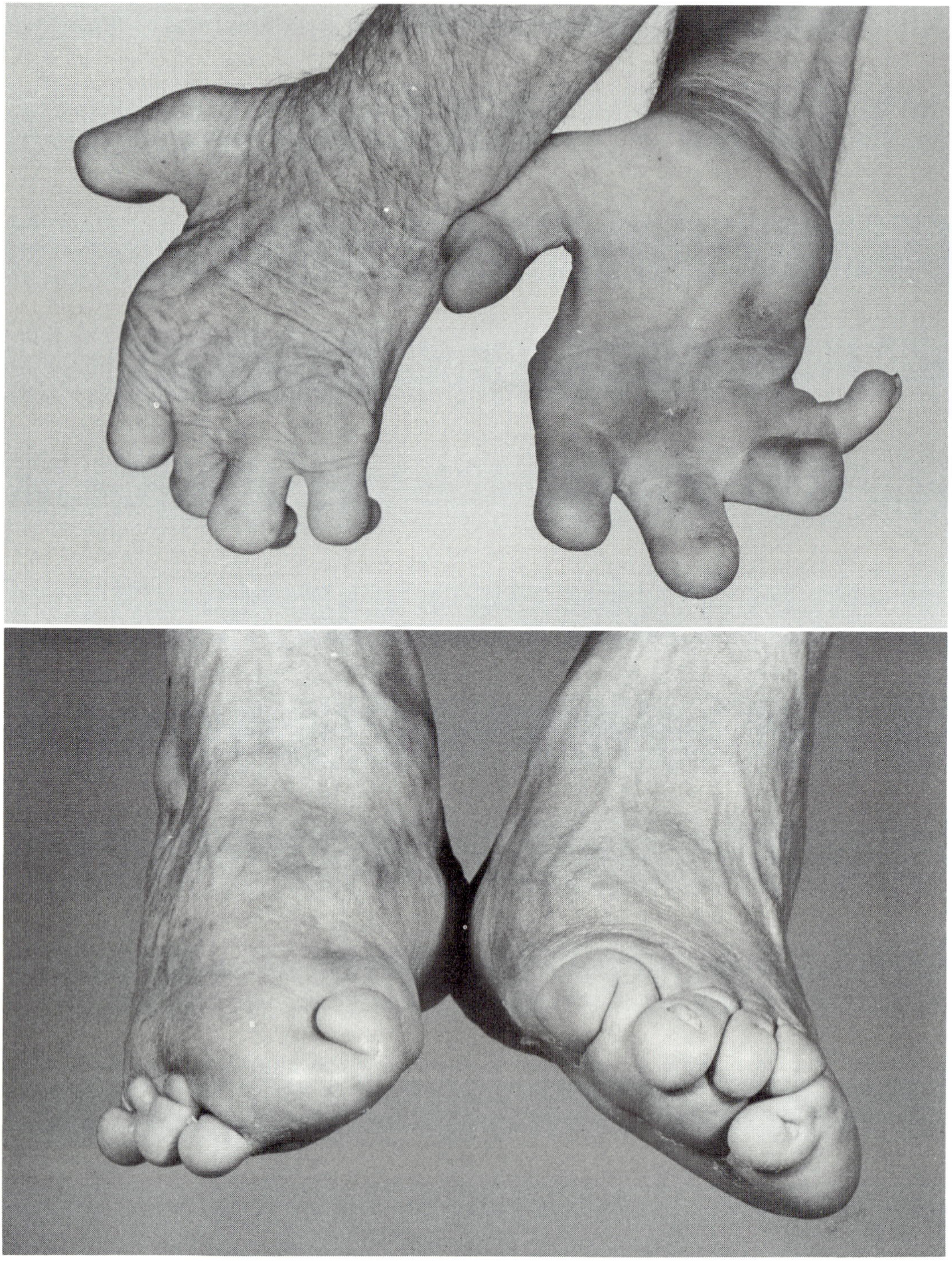

A

B

Fig. 9-3. Typical atrophic absorption of the distal phalanges.

globi in scrapings from skin and mucosa and in sections of lesions. Macular lesions of lepromatous leprosy begin as multiple, small, indistinctly marginated variations in the color of the skin and are symmetrically distributed over the whole body. They show no loss of sensation but become more erythematous and shiny in time. In some patients the macular lesions are so indefinite as to escape notice until infiltration of the skin occurs, producing either diffuse or nodular lepromatous lesions. The skin becomes infiltrated and thickened with exaggeration of the natural folds, which occasionally leads to the appearance of leonine facies. Infiltration around hair follicles causes loss of hair, especially of the brows, beginning in the outer halves. Cutaneous and subcutaneous nodules commonly develop on the brows, nose, chin, lobes and helices of the ears (Fig. 9-2), elbows buttocks, and knees. Infiltration of the mucosa also occurs especially in the nose where small punctate ulcers tend to develop especially on the septum, resulting in epistaxis. When it occurs, however, destruction of the nasal septum probably is secondary. The larynx occasionally is involved, reducing the voice to a whisper, and in some instances obstruction to respiration may require tracheotomy. Testicular involvement is followed by scarring and atrophy, but no serious involvement of the ovaries has been reported. Gynecomastia occurs frequently in older males. The bone marrow is invaded, and the distal phalanges undergo atrophic absorption (Fig. 9-3). Other lesions include rarefaction of bone and secondary osteitis, periosteitis, and osteomyelitis.

Microscopically, lepromatous leprosy is an extension of the cellular reaction described in the indeterminate stage. Groups and strands of lymphocytes and mononuclear cells are found around hair follicles, sweat glands, and sebaecous glands. The histiocytes are increased in numbers and have a foamy appearance because the pale-staining cytoplasm contains numerous vacuoles that contain leprosy bacilli. These cells are called foam cells, or lepra cells of Virchow. With time, some of these foam cells coalesce and form multinucleate giant foam cells with coalescense of the lipoid material, bacilli, and cellular debris to form globi. Later, large areas of dermis may be transformed into continuous masses of chronic inflammatory tissue with groups of mononuclear cells and plasma cells and containing large numbers of lepra cells packed with lepra bacilli. Bacilli also are found in the Schwann cells as well as in and between nerve fibers in this form of leprosy.

Tubercloid leprosy apparently represents high resistance to the occurrence, multiplication, and dissemination of the bacilli that are found with some difficulty in peripheral nerves and rarely in skin scrapings except during reactional states. The early lesion of the skin is a patch or macule with impaired sensation followed by hypopigmentation or erythema (Fig. 9-4). The typical tuberculoid lesion, or lepride, is raised and sharply marginated, with a pebbly surface, and extends by centrifugal spread, followed by atrophic changes and loss of sweating (Fig. 9-5). Basically, the lesion is an epithelioid cell reaction with giant cells of the Langhans type.

The central nervous system is not involved, but the peripheral nerves supplying the extremities are always affected in

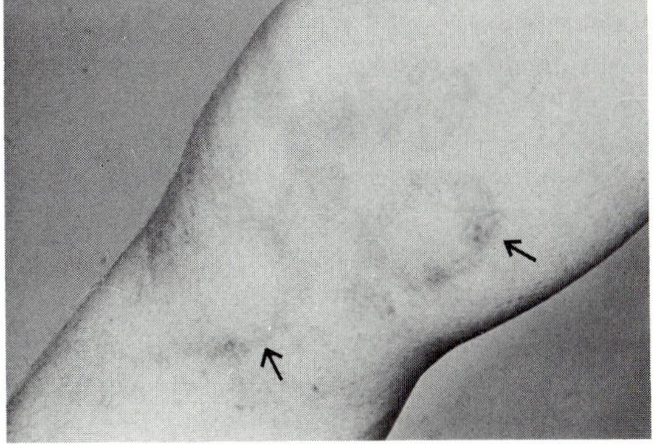

Fig. 9-4. Tuberculoid macule showing central hypopigmentation.

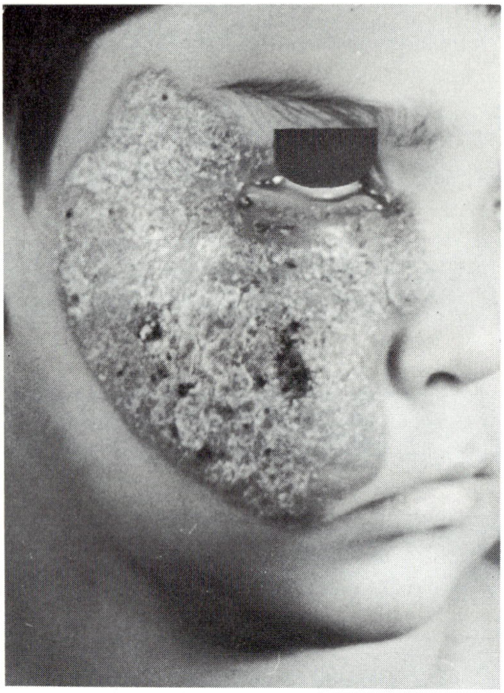

Fig. 9-5. Lepride showing typical pebbly surface and sharp margin.

advanced cases of both types of leprosy. The fine cutaneous nerves in the lesions are invaded, and ulnar, median, peroneal, posterior tibial, and auricular nerves often are enlarged and palpable. In lepromatous lesions the bacilli are in all parts of the nerve fibers and lepromas may develop within the nerve, whereas in the tuberculoid type an intense epithelioid reaction involves the nerve but bacilli are scarce. In either type, peripheral nerve involvement is followed frequently by atrophy and contracture of the muscles, associated with decalcification of the bones especially in the fingers and toes, the phalanges of which may be completely absorbed.

Polyneuritis occurs in both types but appears earlier and is more severe in tuberculoid leprosy. Numbness and tingling in an extremity may be the initial complaint.

The *dimorphous* variety has definite clinical and histologic characteristics of both lepromatous and tubercloid leprosy. The dimorphous macular lesion most commonly starts on the outer aspect of the thigh and is hypopigmented and anesthetic, with diffuse and ill-defined borders. Then for a few months or longer, multiple macular lesions appear in symmetrical distribution. Some are large with sharply defined edges; others are small with indistinct edges. The larger lesions have a dry surface with loss of light touch and thermal sensation, whereas the small ones show mixed features. The lesions are scattered symmetrically all over the body, some being larger with numerous smaller ones in between. They have indistinct borders, with the infiltration increasing centrally so that the most prominent part of the lesion is in the center, or in some cases in which healing has occurred centrally, the most prominent part is well within the spreading edge. This characteristic is in direct contrast to tuberculoid lesions, in which the edges are sharp and prominent. The superficial nerves are enlarged; the lepromin reaction is variable, being either negative or weakly positive; the histopathologic features also are mixed; however, there always is a free subepidermal zone. In some lesions the reaction, like that of tuberculoid leprosy, predominantly is lymphocytic, with epithelioid cells and giant cells frequently arranged in a focal manner. But in others that resemble lepromatous lesions the predominant cell is the macrophagic histiocyte, which develops into the foam cell of Virchow; but there are also some epithelioid cells and an occasional Langhans type of giant cell.

Reactions occur in the course of the disease and can be divided into two major categories: acute exacerbation and lepra reaction, or erythema nodosum leprosum.

Acute exacerbations may occur in any type of leprosy in preexisting lesions. The most prominent histologic feature common to all types of acute exacerbation is edema, but the other features vary with the variety of leprosy. In the lepromatous variety there often is an increase in bacterial activity along with extension of the preexisting clinical lesions. In some instances the exacerbation may represent a change from the indeterminate or dimorphous variety to the lepromatous or tuberculoid form of the disease. Generally, acute inflammatory cells are present, and even in the tuberculoid reaction there may be an increase in bacilli. But at times in pure tuberculoid leprosy there is an increase in edema in an epithelioid cell infiltrate but without bacilli. These reactions usually revert back to the more chronic phases of lepromatous or tuberculoid leprosy; however, the great danger from tuberculoid exacerbations is from increased nerve involvement and destruction.

The *lepra reaction* is an acute or subacute reaction that develops only in lepromatous leprosy and is thought by many to be a sensitization phenomenon of an allergic or anaphylactoid type to released components of the degenerated or dead bacilli. There usually are systemic symptoms as well as increased reaction in preexisting local lesions of the skin, eye, and other organs; however, the local reactions may develop where there has been no clinically apparent lesion. There are three specific lepra reactions; erythema nodosum leprosum, erythema multiforme, and erythema necroticans.

Erythema nodosum leprosum and erythema multiforme clinically resemble their like-named dermatologic counterparts but have distinctly different histologic characteristics. The reaction involves the mid and lower dermis, often extending into the subcutaneous tissue. A neutrophilic infiltrate is superimposed upon the usual lepromatous lesions. However, these cells tend to disintegrate in a few days as lymphocytes and plasma cells appear and increase in number. Vasculitis is frequent, and edema is a constant feature. Bacilli are decreased in number, especially the solid-staining forms, which may disappear completely. Fragmentation and disruption of elastic fibers and collagen in the dermis is frequent. In severe forms of the lepra reaction, a necrotizing vasculitis and fibrinoid degeneration of the vessel wall and perivascular tissues often is associated with dermal and epidermal necrosis and microabscess formation.

Erythema necroticans, or the Lucio phenomenon, is a special form of lepra reaction generally found in Central America and Mexico, in which patients without distinct skin lesions develop purple spots with scabbing on the extremities. Many of these lesions develop blisters and ulceration. In addition, they show a diffuse lepromatus infiltration of the skin, especially of the face and extremities. Histologically, there are two main types of lesions: one shows an intense neutrophilic and histiocytic infiltration into the dermis, with a severe, usually necrotizing, vasculitis of vessels larger than capillaries and of small arterioles; whereas in the more classical type there is dilation, necrosis, and thrombosis of capillaries and of larger vessels, especially in the lower dermis. In both of these severe forms the epidermis shows patches of depigmentation, degeneration, necrosis, and eventual sloughing. Numerous beaded bacilli may be found in the lesions.

OCULAR MANIFESTATIONS OF LEPROSY

Involvement of the eye rarely is a primary manifestation of leprosy; but few advanced cases, especially of lepromatous leprosy, fail to show some type of involvement of the eye or its adnexa. Ocular manifestations frequently occur in an insidious manner; therefore, Choyce[9] and Allen and Byers[2] have suggested that the eyes of patients with leprosy should be examined routinely. The relatively low incidence of ocular manifestations of infection reported in earlier studies might be because of the fact that observations of the eye were limited to gross examination. When slit-lamp biomicroscopy is applied, ocular involvements such as keratitis and iritis can be detected much earlier. Pillat[22] was the first to record the distinctive beading of corneal nerves viewed with the slit-lamp biomicroscope. Allen and Byers[2] advised the periodic repeated use of this instrument for the recognition of early corneal leprosy.

Elliot[12] made observations on the relationship of the usual time of appearance of clinical manifestations of ocular disease to the probable duration of the general disease. He concluded that the latent period varied for specific areas of the eye as much as 5 to 25 years or more. More recent observations tend to confirm the time ratio for ocular involvement in general, but these observations require modification of the details and also indicate that there may be considerable difference between the time of localization of bacilli in ocular tissues and the development of clinical signs of infection of those tissues.

Direct extension from infected adjacent skin was thought at one time to be the most common route by which the eye becomes infected with *M. leprae,* and it still may be a factor. Extension along the nerve fibers is considered the primary route of spread by some observers; however, the hematogenous route is probably the more important.

Whereas the diagnosis of the general disease is confirmed by skin biopsy, the diagnosis of ocular leprosy usually is based upon clinical findings. Skin scrapings from the lobes of the ear or from other sites, as well as biopsies of suspected lesions stained by Ziehl-Neelsen or Fite stains, may reveal acid-fast bacilli. In the lepromatous type of the disease, bacilli and globi are numerous, but in the tuberculoid type few or no bacilli may be found. Cultures and animal inoculation may be made in order to aid in differentiation from tubercle and other acid-fast bacilli. Failure to demonstrate the organisms in smears or scrapings and negative cultures do not exclude the diagnosis of leprosy. Numerous and repeated examinations are required in many patients. The techniques for the preparation of ocular scrapings and smears for examination in the laboratory are given in Chapter 4.

Large numbers of organisms may be present throughout the cornea (Plate 3, *B*). Although few changes are apparent, as Axenfeld[3] has stated, "The bacilli can spread in the lymph spaces of the cornea, where they are found in small masses between the lamellae; they often lie in small clusters cut off from the original point of entrance." Clusters of organisms under Bowman's membrane were reported by Meller.[3] Naar, as cited by Pillat,[22] was the first to make corneal scrapings. In one patient with keratitis punctata he failed to find bacilli in the corneal scrapings but did demonstrate them in smears from the conjunctiva, the nose, and the mouth. In three other patients the corneal scrapings were positive.

Pillat[22] made scrapings of corneal epithelium and of the mucous membrane of the nose, and obtained positive results. He observed that the keratitis punctata superficialis changed into uniform superficial pannus and later to interstitial keratitis. Pillat felt that "this indicates that a few leprosy bacilli in the cornea may produce miniature lepromas seen clinically as keratitis punctata superficialis. The bacilli may become destroyed, and the spot-shaped infiltrates may be absorbed. When, on the other hand, the number of bacilli is larger, the superficial infiltrates may flow together . . . and may even go into the depth of the cornea . . ." On the other hand, Pinkerton[23] has reported taking numerous scrapings from the cornea, where leukomatous spots were frequent, and found the bacilli only in a few cases. Scrapings from the lid margins frequently yield a larger number of the organisms. When diffuse infiltration of the conjunctiva occurs, a number of *M. leprae* are found. Excised nodules from the conjunctiva or lids contain a great number of *M. leprae* in the lepromatous type of the disease. Bacilli also have been found in the retina and choroid late in the course of ocular disease, but usually only in the anterior portions.

The incidence of ocular lesions of leprosy varies from approximately 15% of patients with tuberculoid leprosy to approximately 100% of patients with long-standing lepromatous leprosy. In addition, the character of the ocular involvement varies with the type of the general disease, although some ocular lesions are common to all types of leprosy.

In lepromatous leprosy one of the relatively early ocular involvements consists of thickening of the supraciliary ridges (Fig. 9-6, *A*) without evidence of inflammation. Eventually there is loss of hairs of the brows beginning laterally and progressing nasally (Fig. 9-7); but the swelling may persist for many months or even years without change or may progress with varying speed into the typical leproma of the brow (Fig. 9-8), forming a part of the so-called leonine facies of this form of leprosy. The granuloma of the brow (Fig. 9-6, *B*) may extend into the upper lid, or

A

B

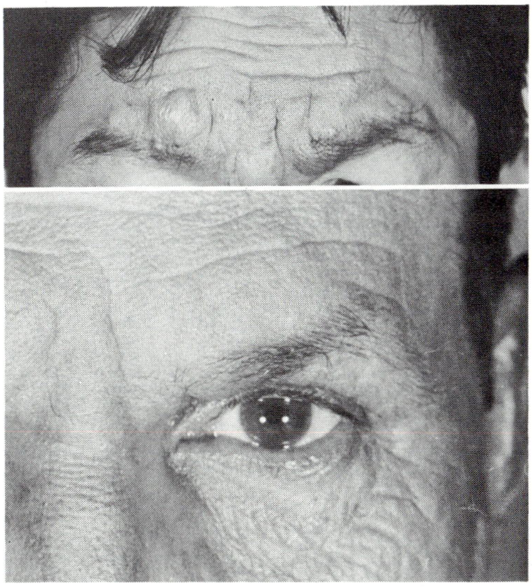

Fig. 9-6. A, Thickening of the superciliary ridges with early lepromatous nodule on the right side. **B,** Thickening of the brow with beginning loss of hairs in the outer portion.

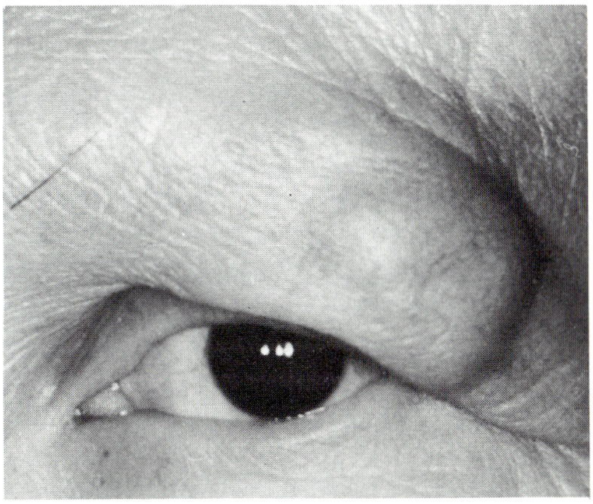

Fig. 9-8. Nodule in upper lid.

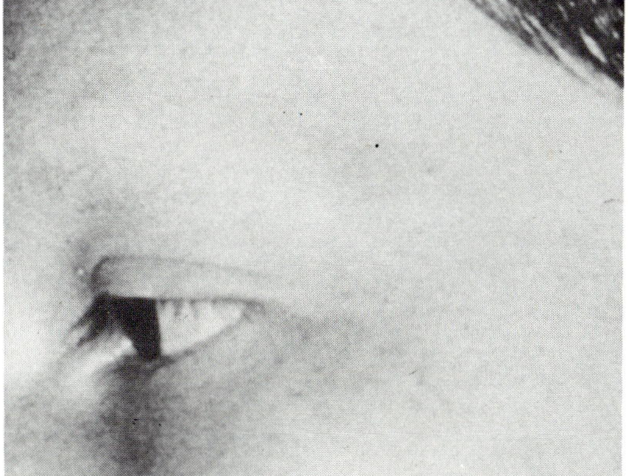

Fig. 9-7. Loss of eyebrows.

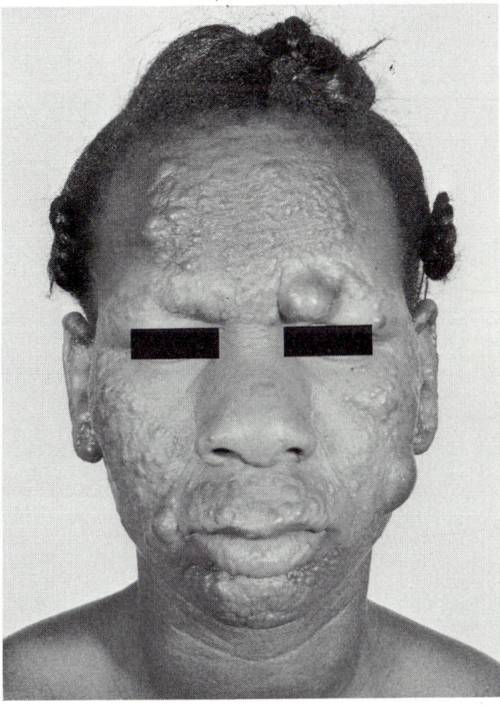

Fig. 9-9. Nodules of lid, brow, and face, showing typical distribution of early leonine character.

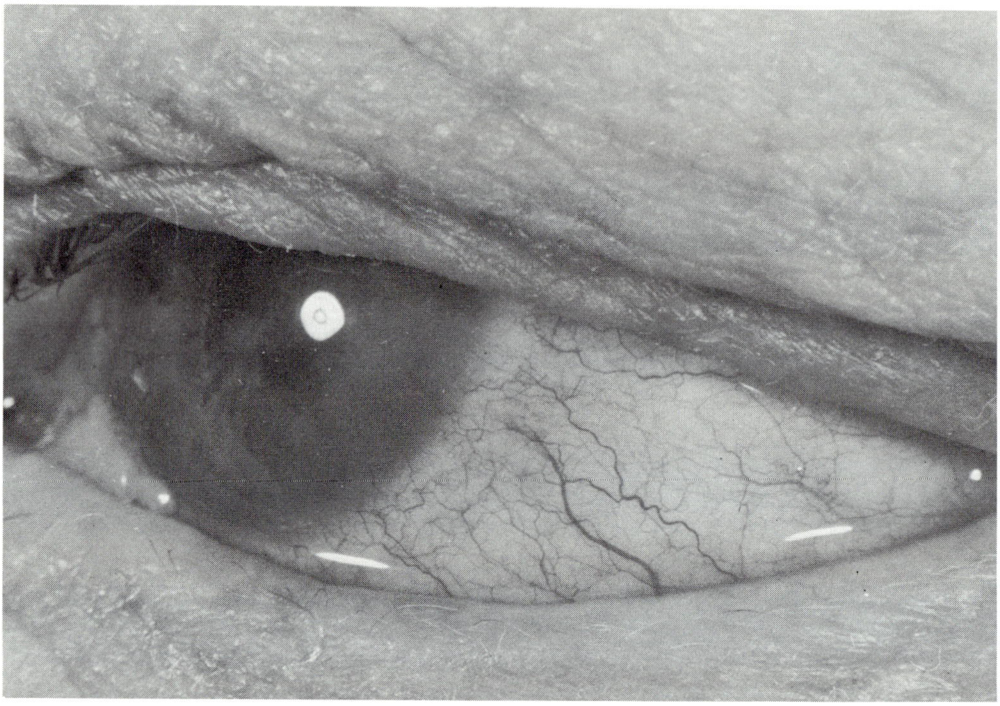

Fig. 9-10. Entropion of the upper lid.

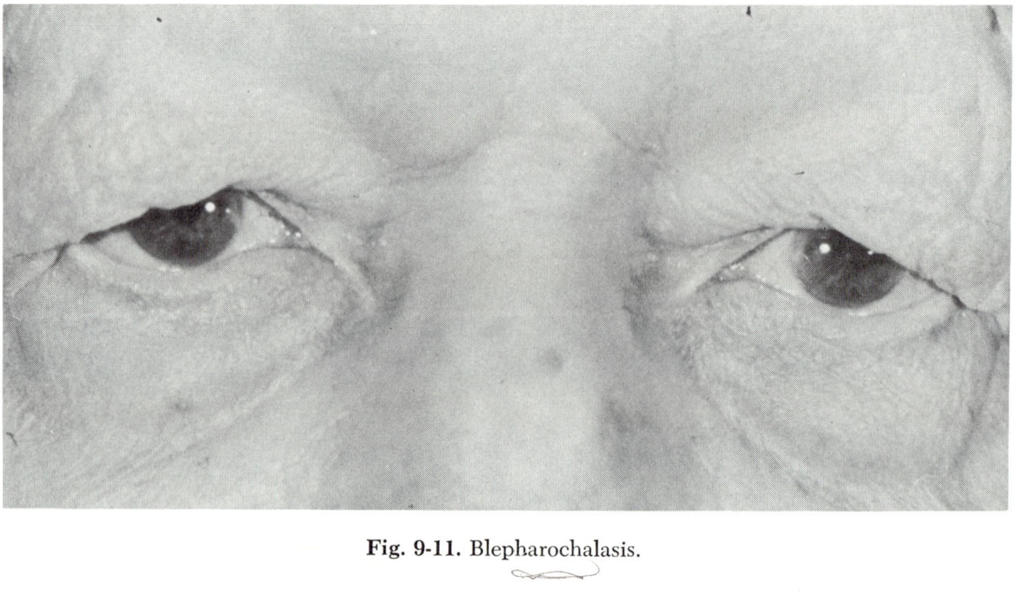

Fig. 9-11. Blepharochalasis.

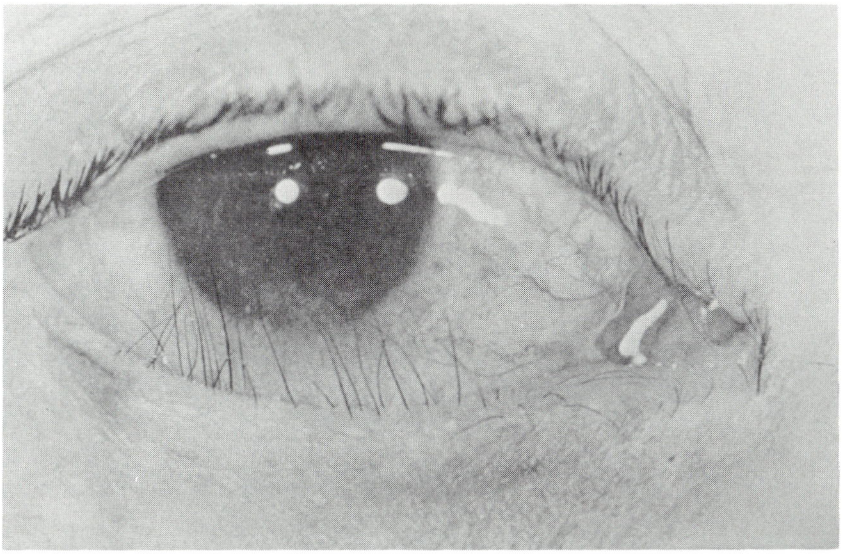

Fig. 9-12. Trichiasis. (From Holmes, W. J.: Geographic ophthalmology, Springfield, 1959, Charles C Thomas, Publisher.)

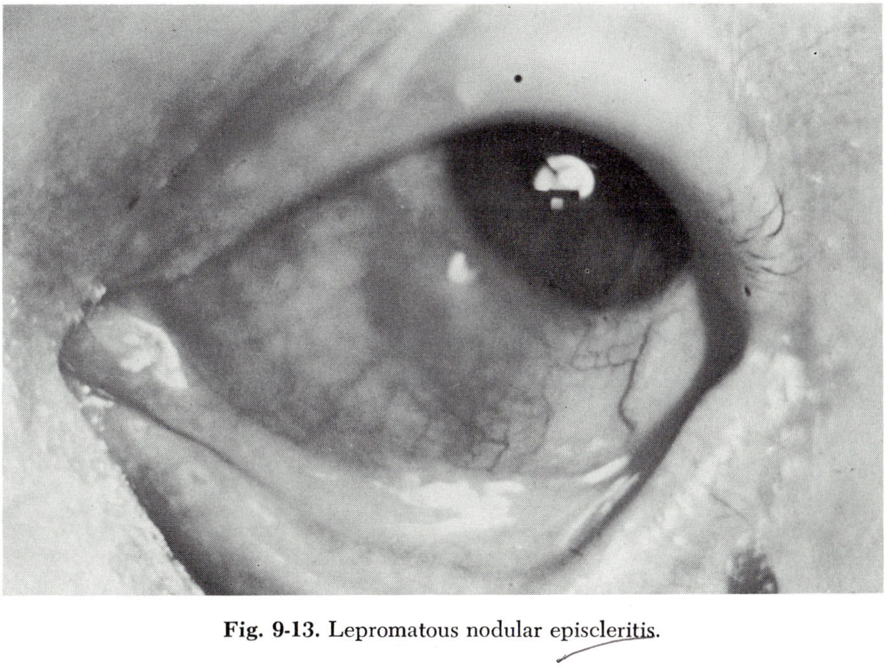

Fig. 9-13. Lepromatous nodular episcleritis.

occasionally small independent nodules may develop in the upper lid (Figs. 9-8 and 9-9), but practically never in the upper lid folds nor in the lower lid. During the inflammatory process, the tissues of the lid may become hypertrophic, or entropion may occur (Fig. 9-10). This may lead to corneal abrasions, then to spastic entropion of the lower lid. Later, after the lesions have resolved under systemic treatment, blepharochalasis may develop (Fig. 9-11). Frequently the ciliary margins of the lids are thickened, probably because of involvement of the follicles of the lashes, which results in loss of the lashes and occasionally in trichiasis (Fig. 9-12).

Conjunctivitis as such is not produced by *M. leprae*, even though the organisms may be found in variable numbers in the epithelial cells. Those patients who develop conjunctivitis do so because of coincidental superimposed infection.

Subconjunctival infiltrations occur especially around the limbus. These consist largely of chronic inflammatory cells in which miliary lepromata develop adjacent to the limbus. The infiltrations form a thin diffuse layer, which gradually disappears near the fornices. These lesions usually undergo remissions and exacerbations along with similar changes in other manifestations of the disease. However, miliary lepromata, once developed, tend to persist for long periods of time as pinpoint- to pinhead-sized, round, yellowish white, subconjunctival lesions. They are called pearls and are identical with pearls, or miliary lepromata, which occur in the cornea and other parts of the eye.

During acute reactions in the course of the disease or as the result of treatment, nodular episcleritis (Fig. 9-13) appears acutely and runs a subacute or chronic course, depending upon the course of the general reaction or the method of local treatment. Nodular episcleritis of leprosy is similar in its appearance to other types of episcleritis and in some instances, but not all, may be followed by local thinning of the sclera.

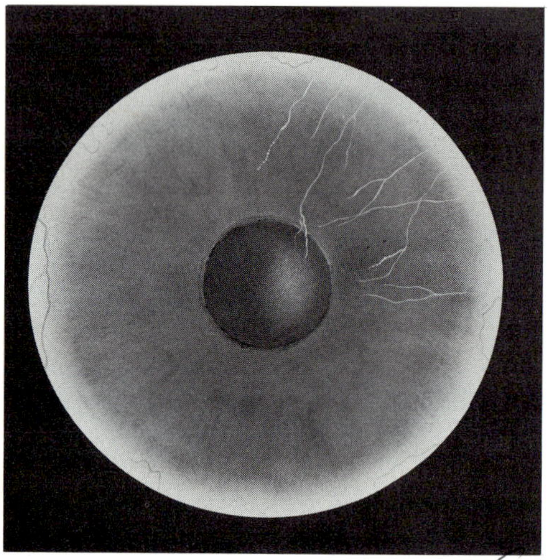

Fig. 9-14. Inflammatory opacification and beading of the corneal nerves.

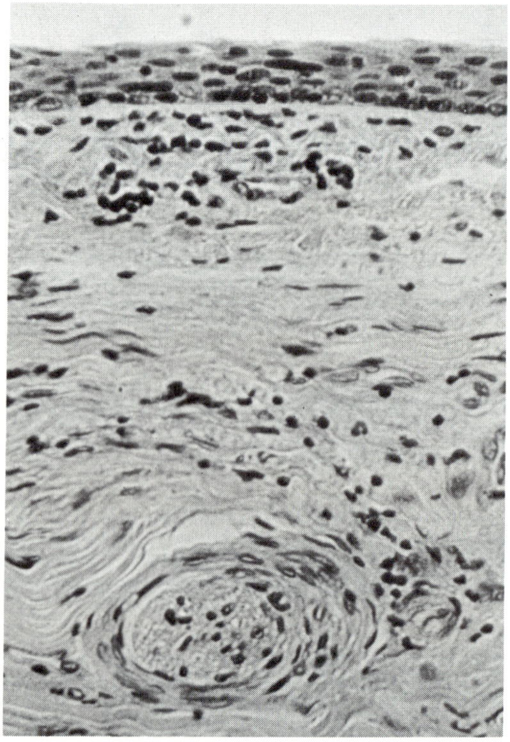

Fig. 9-15. Edema and infiltration of corneal nerve.

Japanese workers report that they have found leprosy bacilli in tears of 66% of patients without obvious ocular manifestations of the disease; however, clinical involvement of the lacrimal gland occurs only rarely.[11]

Nasolacrimal obstruction is frequent and secondary to nasal lesions of the disease, especially those producing periostitis and atrophy of the nasal bones. Secondary infection always ensues, as with any other type of nasolacrimal obstruction.

Initial clinical evidence of involvement of the globe may be inflammatory opacification of the corneal nerves (Fig. 9-14). It is a transitory asymptomatic phenomenon, visible only with magnification, which frequently first involves the nerves of the superior temporal quadrant; but several nerves in each or any quadrant may become involved atypically or in repeated episodes.

One or more, usually several, corneal nerves become swollen and opaque for a period of several weeks, then gradually become invisible again. The nerves become grayish white, fine, radial lines extending in from the limbus in the middle or deep stroma. The lines become branched after passing centrally for a short distance through the otherwise clear stroma, then gradually become finer and invisible at

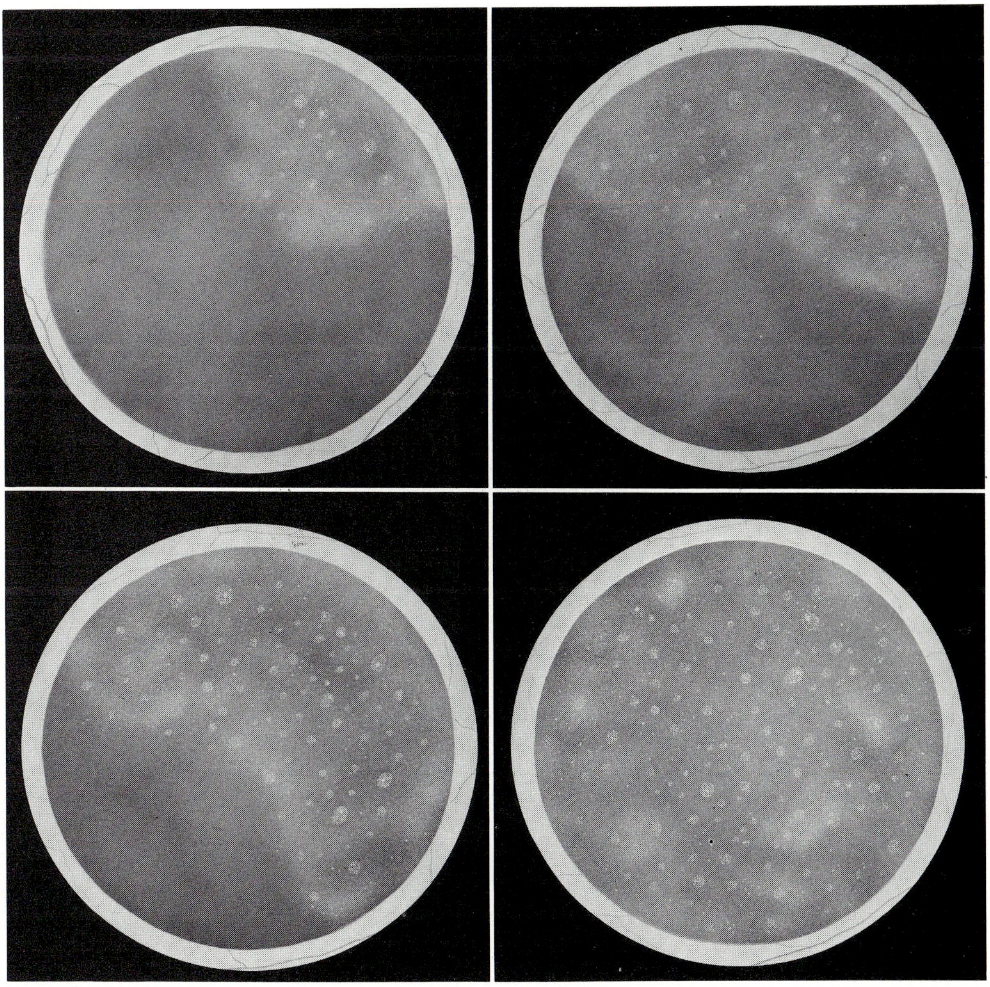

Fig. 9-16. Stages in process of typical avascular keratitis, K1, K2, K3, and K4.

variable distances from the limbus. Occasionally the opaque nerve may appear to have a sheath or sleeve around it for a short distance or have irregular swellings along its course. These changes are caused by edema of the nerves, with an accompanying infiltration of round cells and histiocytes in and around the nerves (Fig. 9-15). Bacilli may be found in Schwann cells, histiocytes, nerve fibers, and endoneural spaces.

The first corneal involvement that usually leaves permanent residual evidence of the infection is a superficial keratitis. The typical lesion begins as discrete patches of superficial thin, grayish white opacities, which appear first in the superior temporal quadrant, then coalesce, becoming confluent and diffuse, slowly spreading centrally and nasally, then downward into the inferior temporal quadrant, and finally into the inferior nasal quadrant (Fig. 9-16). Atypically, the lesion may begin in one of the other quadrants and spread to involve part or all of the cornea. Such atypical lesions may be initiated by trauma. Frequently in early stages the lesion is invisible except with magnification.

Upon biomicroscopic examination the early lesion is composed of a superficial thin, diffuse haze in Bowman's zone, with a variable number of punctate, chalky white opacities scattered through it. These chalky white opacities tend to develop in clusters composed of discrete, fine white opacities, producing a pathognomonic pattern of superficial stippling of the cornea. In some patients the keratitis, on gross examination, may seem to consist only of the diffuse haze, but even in these cases some very fine chalky deposits are seen on biomicroscopic examination. Either type of corneal involvement may spread slowly and continuously throughout the cornea, but in many patients the spread is interrupted by remissions that may occur after minimal corneal involvement or at any other stage.

As the active inflammation of the cornea subsides, the thin grayish white diffuse opacity may disappear or may persist, but the chalky white punctate opacities remain permanently. As the lesion persists, spreads, or recurs, some of the chalky white opacities increase in size and become visible, grossly measuring 0.25 to 0.5 mm in diameter. These are miliary lepromata and are sometimes called pearls. They are pathognomonic, do not disappear during remissions, and may become quite numerous.

The early and the most central opacities are superficial, lying in Bowman's zone; however, as the active lesion persists and spreads centrally and circumferentially,

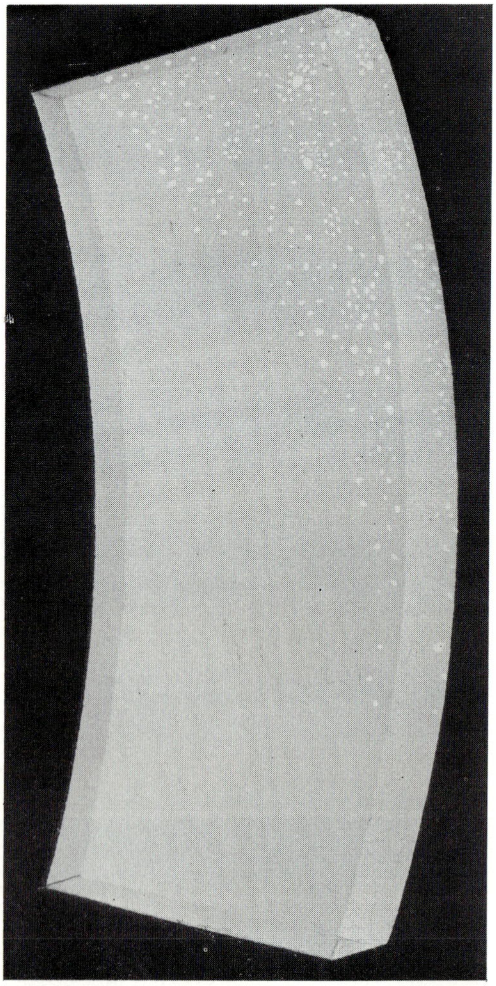

Fig. 9-17. Slit-lamp section of avascular keratitis in stage K3.

deeper layers become involved at the periphery of the cornea (Fig. 9-17). Thus the full thickness of the cornea tends to become involved first at the limbus and later centrally.

In some patients after an avascular keratitis has persisted for some time, or in some with repeated or severe attacks, superficial vessels of the limbus begin to extend into the cornea. Typically, the vascularization begins in the superior temporal quadrant, then involves the superior nasal quadrant, the inferior temporal quadrant, and finally the inferior nasal quadrant (Fig. 9-18). Vascularization of the cornea may become arrested at any stage, however. With severe or prolonged keratitis, the pannus may involve the whole surface of the cornea, but up to this point the avascular keratitis always is more extensive than the pannus. Atypical pannus formation may begin in any of the other quadrants and remain limited to that portion of the cornea or spread to involve any part or all of it.

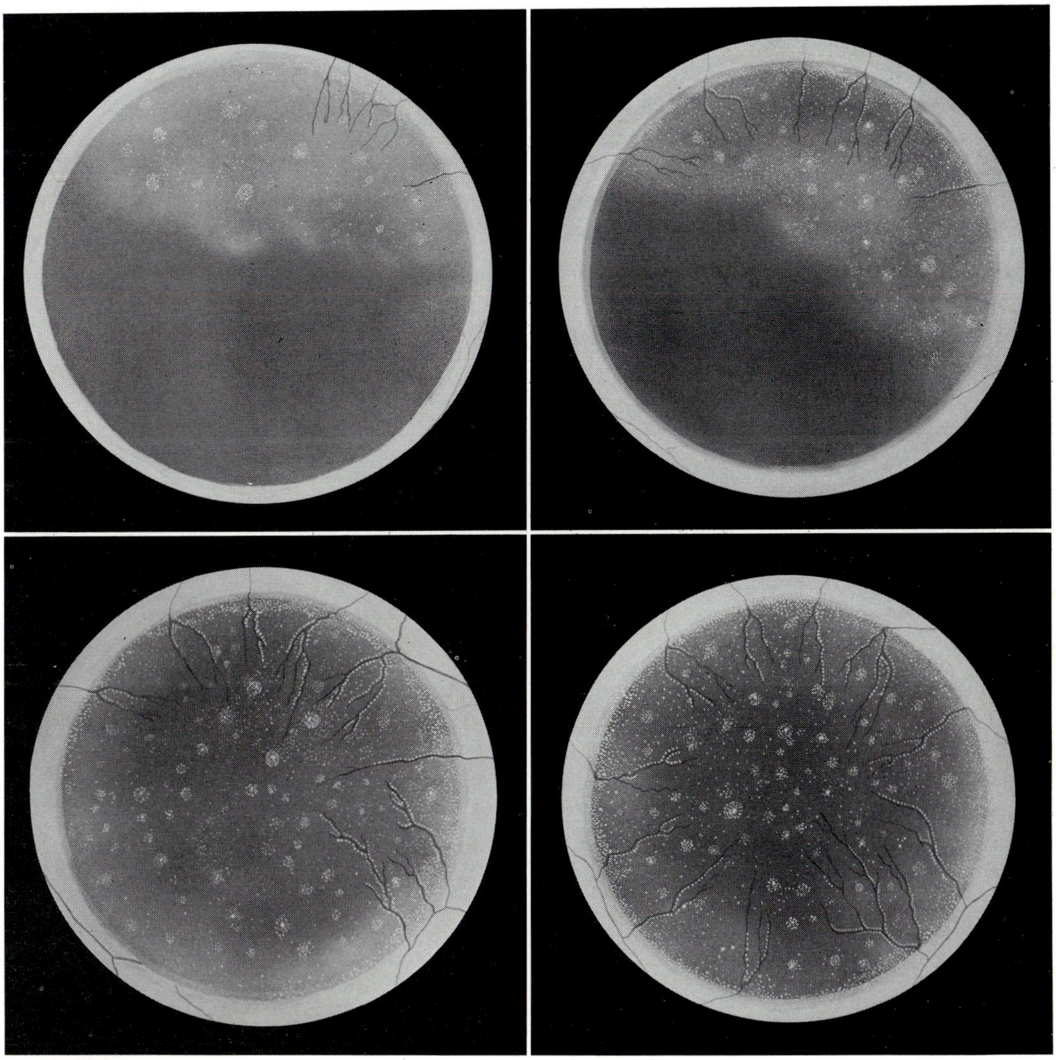

Fig. 9-18. Typical stages of pannus formation, which sometimes, but not always, follows avascular keratitis.

A

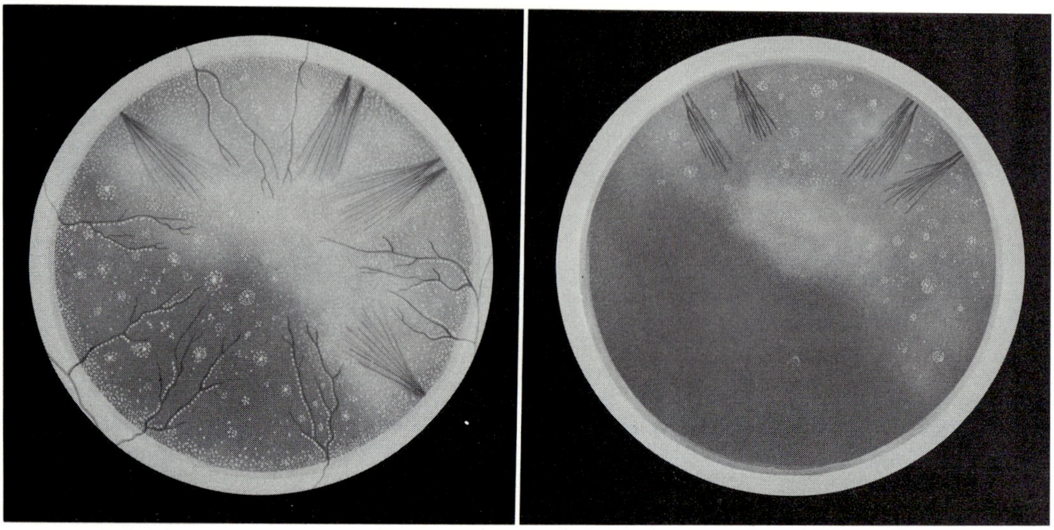

B

Fig. 9-19. A, Interstitial vascularization following pannus formation and avascular keratitis. **B,** Interstitial vascularization following avascular keratitis, but without pannus.

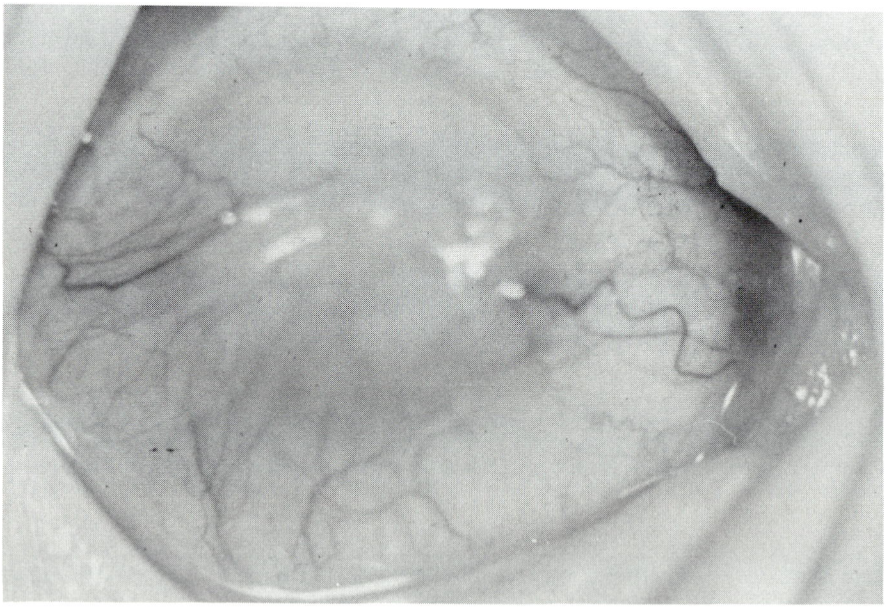

Fig. 9-20. Gross leproma of the cornea.

In a small proportion of patients deep or interstitial vascularization of the cornea may develop either in conjunction with superficial vascularization (pannus) or alone (Fig. 9-19). It develops as a consequence of prolonged deep infiltration and usually follows or accompanies pannus formation. In the typical case it begins in the superior temporal quadrant, eventually extending into other quadrants in the usual pattern of leprotic corneal spread. Not infrequently, however, deep vascularization may begin in the inferior temporal quadrant and extend from there. Interstitial

Fig. 9-21. Section of a gross leproma of the cornea.

leprotic keratitis without pannus formation occurs in some patients who have a severe granulomatous uveitis. In these patients a grayish opacity begins in the deep stroma at the limbus and extends centrally. Within the affected stroma between the newly developing vessels, there are small chalky white opacities that, once formed, usually remain, even though the inflammation may subside.

Gross lepromata occasionally develop at the limbus and involve the cornea (Fig. 9-20), or, more rarely, the involvement may be limited to the central part of the cornea. In some instances a corneal leproma resembles the fleshy type of pannus of trachoma, and in others it resembles an irregular lipoma. Pathologically, the lesions are typical lepromatous granulomas (Fig. 9-21) with both superficial and deep vessels surrounded by a heavy infiltration of lymphocytes and lepra cells, replacing practically all normal structures.

As a consequence of prolonged inflammation or repeated episodes of any of these types of corneal involvement, severe scarring ensues and under varying circumstances may undergo hyalinization or lipoid or calcium infiltration (Fig. 9-22).

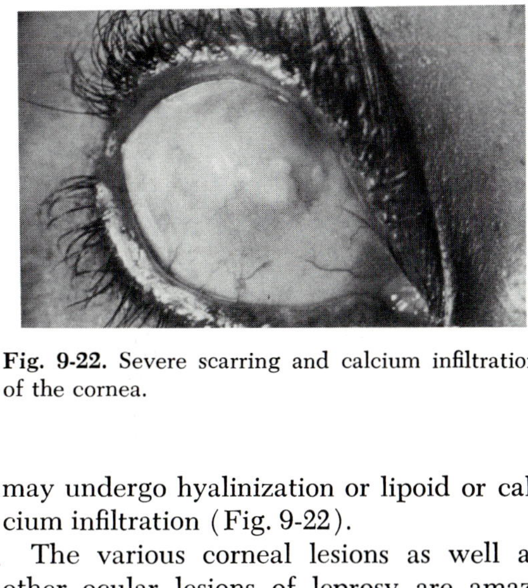

Fig. 9-22. Severe scarring and calcium infiltration of the cornea.

The various corneal lesions as well as other ocular lesions of leprosy are amazingly asymptomatic. The patient seldom is aware of corneal involvement until the lesion is well advanced, and even then he may complain only of a visual disturbance or slight circumcorneal congestion or rarely a mild photophobia. Frequently a superficial keratitis develops and subsides without the patient's knowledge. This apparently is the result of sensory nerve involvement and increases the danger of damage from relatively minor trauma.

Involvement of the iris and ciliary body occurs relatively early. Acid-fast bacilli probably are disseminated widely and localize simultaneously in the superficial or cooler tissues of the body, including the superficial tissue of the eye and adnexa as well as the anterior uvea. However, the organisms probably lie dormant or produce so little disturbance until they have increased tremendously in number that they fail to cause clinical signs for several years. Uveal involvement is limited to the iris, ciliary body, and occasionally the most anterior portion of the choroid. Organisms localize here, producing a mild subclinical inflammation consisting of an occasional lymphocyte and foam cell. After months or years, clinical signs appear in one of several forms: acute iritis, chronic iritis, nodular iritis, or chronic iritis with superimposed acute or subacute exacerbations.

An acute, subacute, or subclinical iritis always accompanies episodes of keratitis and some attacks of acute nodular episcleritis. The iritis is serous in type, characterized by an aqueous flare and fine floaters in the anterior chamber, with fine precipitates on the posterior surface of the cornea, but with relatively no symptoms. The inflammation usually has a slow onset with no gross evidence of the inflammation. In time, however, slight circumcorneal injection or mild photophobia occurs, and by this time posterior synechiae usually have begun to develop. Only by repeated periodic ocular examination with magnification may these lesions be found early. Occasionally an acute serous uveitis develops with moderate to intense circumcorneal injection, flare, and cells in the anterior chamber, but with negligible symptoms—the so-called red eye in patient terminology. The majority of these are recurrent episodes based upon an established and persistent mild or subclinical chronic anterior uveitis.

Chronic subclinical anterior uveitis probably exists in all leprosy patients from the time of the localization of organisms in the tissue and in some patients probably persists with only slight variation throughout the remainder of their lives. In others, mild, clinically apparent episodes alternate with subclinical periods; whereas in still other patients, nodular lesions (Fig. 9-23) develop in an episode of exacerbation of the iritis and persist as the acute

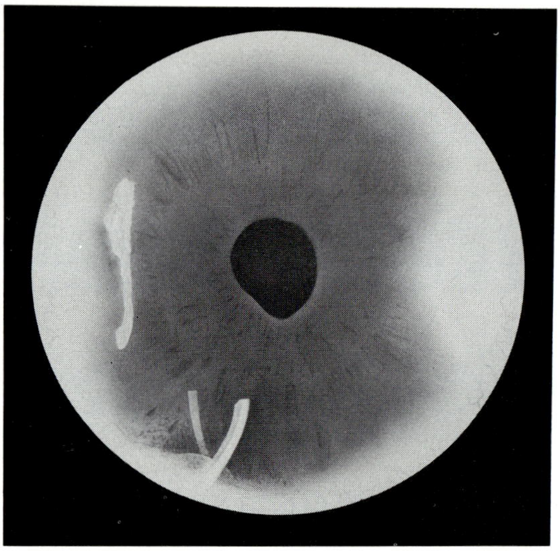

Fig. 9-23. Lepromatous nodular iritis showing nodule in the angle at 7 o'clock, as well as a pterygium and corneal scar.

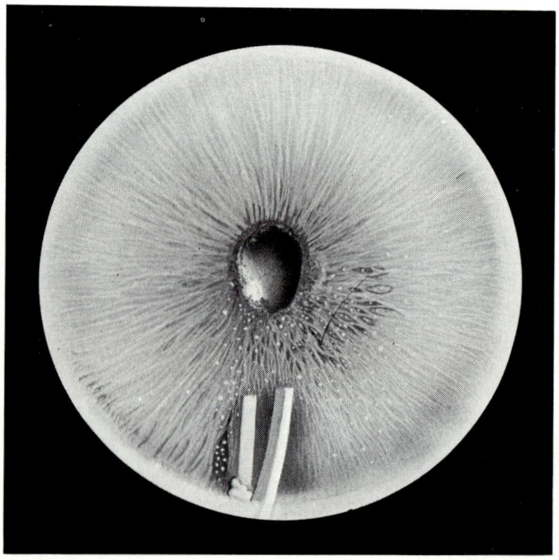

Fig. 9-24. Miliary lepromata, or pearls, of the iris, with a large conglomerate pearl in the angle at 6 o'clock.

serous lesions subside into a subclinical period. In other patients the nodules slowly and relentlessly extend, producing secondary glaucoma, cataract, and degeneration of the globe.

Miliary lepromata, or pearls of the iris (Fig. 9-24), develop as a result of the chronic inflammatory process. They develop in the stroma, usually in the pupillary portion of the iris, as the result of enlargement and fusion of lepra cells to produce giant lepra cells containing globi, which persist and eventually have calcium salts deposited in them. They continue to increase in size gradually and at other times coalesce with neighboring pearls to form a conglomerate pearl. As the pearls gradually increase in size, they appear to approach the surface of the iris, then protrude from the surface, and eventually they may break loose and migrate from place to place in the anterior chamber of the eye. Some "free" pearls have been followed for a number of years, and on the various examinations they were found in different locations on the surface of the iris or in the chamber angle. In a few instances the pearl could not be found

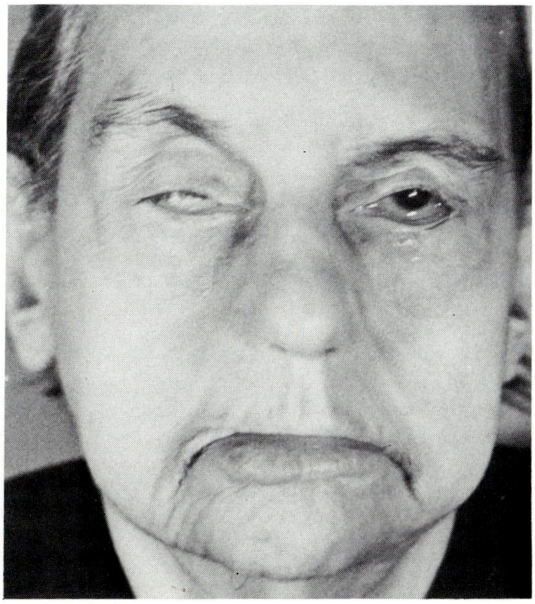

Fig. 9-25. Masklike facies in tuberculoid leprosy.

in the anterior chamber but was found there again on later examinations. Thus "free" pearls may migrate into the posterior chamber and probably even farther.[1]

Complications of anterior uveal inflammation in leprosy include peripheral anterior synechiae from nodular iris or ciliary body lesions; posterior synechiae; occlusion and seclusion of the pupil; iris bombé; secondary glaucoma, both as the result of iris bombé and peripheral anterior synechiae as well as from angle obstruction from exudate in acute episodes of anterior uveitis; cyclitic membranes; cyclitic cataract; and atrophy and degeneration of the globe.

Tuberculoid leprosy, formerly designated as anesthetic leprosy, is characterized by much less frequent clinical involvement of the eyes, but the patient may have a smooth, masklike facies (Fig. 9-25). The skin of the face and lids frequently has areas of reduced sensation or anesthesia. This reduction or loss of sensation frequently includes the conjunctiva and cornea and is manifest by diminished blinking or even loss of the blinking reflex at times. Such patients may have foreign bodies lodged on their corneas without producing symptoms, and considerable damage may result.

In approximately 10% of tuberculoid cases paralysis of the facial nerve occurs. In some instances there may be slight to moderate recovery of function after several months, but as a rule the paralysis persists and atrophy of the affected muscles ensues. As a result, inadequate closure of the lids during the acute phase is followed by ectropion of the lower lid and in some instances by atrophy and vertical shortening of the lower lid in the chronic phase. Exposure keratitis (Fig. 9-26) frequently results and is more serious when there is associated corneal anesthesia, for it may lead to corneal ulceration, perforation, and even loss of the globe. In less severe forms of exposure keratitis, healing may be followed by infiltrative lesions, which resemble those seen in lepromatous patients. More rarely lesions

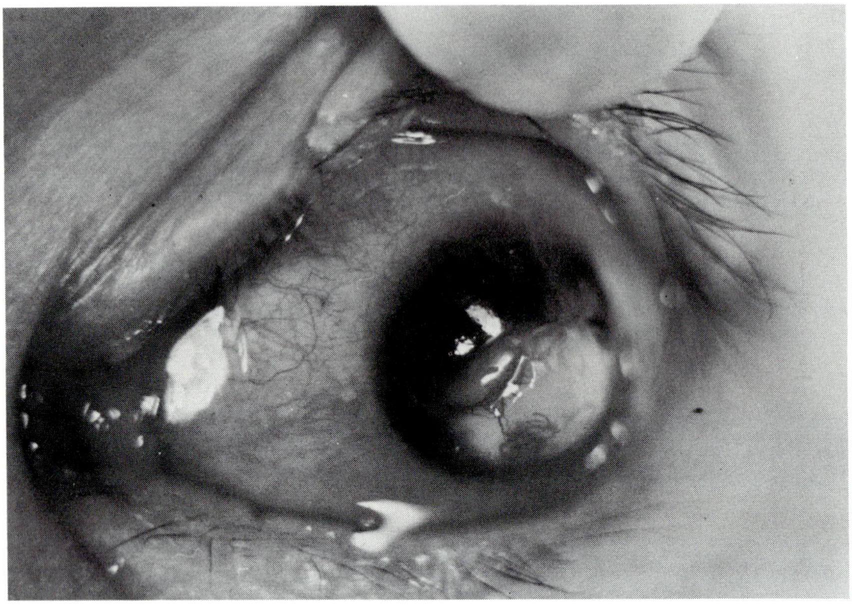

Fig. 9-26. Exposure keratitis.

identical with those seen in lepromatous leprosy also may be seen in the eyes of tuberculoid patients.

The posterior choroid probably is never involved other than by the passage of inflammatory cells in any type of leprosy until the eye begins to degenerate, and similarly there probably never is primary involvement of the retina or optic nerve. In a few instances pearls have been seen in the retina (Fig. 9-27). Although these undoubtedly are miliary lepromata, they probably represent "free" pearls that have migrated posteriorly because they always appear to be situated quite superficially in or on the retina and there is no visible associated inflammation of the retina (pearls always arise in an inflammatory infiltration in other parts of the eye). In one patient who has been observed periodically over a number of years there has been a decrease in the number of pearls visible in the retina, and in another the pearls have been observed in different positions at various times.

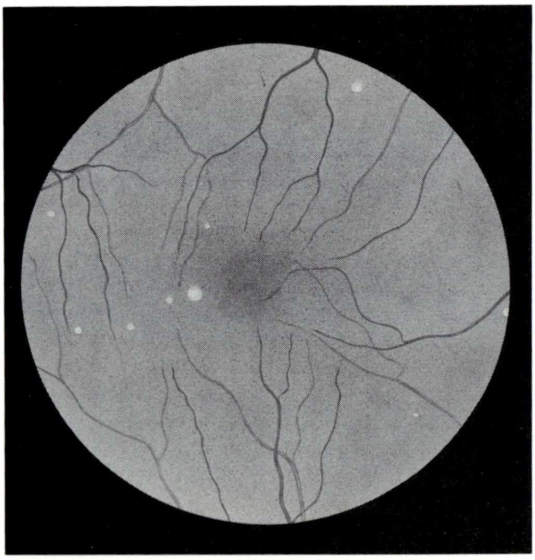

Fig. 9-27. Pearls of the retina.

The eye and adnexa participate in the several types of general reactions. The lids participate in the skin reactions of acute exacerbations and in erythema nodosum reactions. The episclera may develop diffuse infiltrative lesions, nodular episcleritis, or lepromata. Corneal lesions increase and extend, and acute exacerbations of uveitis occur. Rarely the acute iritis may be accompanied by pseudohypopyon or hyphema.

A number of lesions have been described in patients with leprosy that also have been seen in patients with or without other disease. These include such things as central chorioretinoses, night blindness, uncomplicated senile cataracts, and pterygiums. Obviously, many of these are coincidental, for even though a patient has leprosy he still is subject to all the other ills and outrages of fortune that beset man. Thus the leprologist must bear this in mind in examining his patient and diagnosing the case.

THERAPY

Since the ocular lesions of leprosy are a part of a widely disseminated disease process, the basic treatment must be general treatment of the infectious process with special therapy for its local manifestations in the eye.

As a result of the report of the usefulness of glucosulfone (Promin) in 1943 by Faget and his associates,[13] the sulfones have become the basic form of therapy. The usual daily dosage of the various drugs is: glucosulfone, 1 to 5 gm intravenously; sulfoxone (Diasone), 0.3 to 0.6 gm orally; acetosulfone (Promacetin), 0.5 to 4 gm orally; and diaminodiphenylsulfone (Avlosulfon), 0.1 gm orally. However, these dosages are "maintenance doses," and it is prudent to start with smaller and less frequent doses, building up to a full dose in approximately 6 months in order to avoid exacerbations and early treatment reactions. In addition, the patient should be followed closely for toxic or side reactions.

The sulfones are not bactericidal against the leprosy bacillus, but they apparently inhibit multiplication of the organisms and thus assist the natural defenses of the host eventually to arrest the disease in a majority of patients.

Clofazimine (B663), a new drug, apparently is especially valuable in patients with organisms resistant to sulfones. The minimal effective dosage has not yet been determined, but it has been reported that 300 mg per day reduces the incidence of reactions of erythema nodosum leprosum. B663, however, has the objectionable feature, especially for light-colored patients, of causing a reddish black pigmentation of the skin.

Rifampicin also has recently been shown to have potential usefulness.[25] It appears that it may be bactericidal; however, more extensive studies will be necessary to permit adequate evaluation.

In erythema nodosum leprosum and related reactions, immunosuppressive agents, such as the steroids, phenylbutazone, and anthiomaline, have been used extensively. More recently thalidomide, which may be immunosuppressive, has been used successfully in preliminary trials. But its extent of usefulness remains to be determined.

Ocular lesions continue to develop in patients receiving sulfone therapy, especially during acute exacerbations and episodes of erythema nodosum leprosum; but it appears that the incidence and severity of ocular lesions may be decreasing, especially in those patients who have the benefit of therapy early in the disease and before the usual time of appearance of overt ocular lesions. Close observation and institution of local palliative and supplementary therapy seems to be improving the ocular prognosis in this disease.

Local application of heat under adequate control to prevent damage to anesthetic areas is helpful in all ocular inflammatory lesions of leprosy. In addition, mydriatics are especially important in placing the iris and ciliary body at rest and in preventing posterior synechiae and secondary glaucoma. When additional control of the inflammatory lesion is necessary, topical application of steroids is valuable and important; but their use must be controlled carefully to avoid the development of steroid glaucoma and cataract.

Traumatic lesions, secondary infectious processes, or other coincidental eye lesions are treated in the same manner as in patients without leprosy.

Elective intraocular surgery, such as cataract extraction, should never be done until the eyes have been quiescent for a minimum of 6 months, but otherwise should be done after the usual preparation and in the usual manner for that particular procedure. The usual postoperative routine should be observed.

Lid surgery for the correction of trichiasis, ectropion or entropion, lid lag, and paralysis of the orbicularis may be done when deemed necessary except during reactions. Healing usually is prompt and satisfactory. Lacrimal obstruction is corrected by extirpation of the sac rather than by dacryocystorhinostomy because of the nasal bone involvement.

Glaucoma should be controlled medically until ocular inflammation has been quiet for several months; but under these conditions, iridectomy and external filtering procedures may be done without undue complications.

Keratoplasty has been reported successfully accomplished by several surgeons but should not be done until after the inflammatory lesions of the eye have been controlled for at least 6 months.

CONCLUSION

In spite of years of research and enormous numbers of references on leprosy, the disease still remains a major medical problem. However, the conception of leprosy has changed with the years, as Wayson[31] has stated: "Instead of thinking of it as a disease of foul-smelling sores and grotesque facial mutilations, we shall consider it a communicable affection of protean manifestations, whose commencement may be acute or is so insidious many times that neither the patient nor his physician is aware of the presence of the infection." Great progress has been made in the humane attitude toward patients afflicted with Hansen's disease; leprosariums have been established; and finally, the use of sulfone drugs and cortisone has resulted in an improved prognosis. Choyce[9] reported regression with delay in ocular complications, but he also stated that no patient has yet had a "complete cure" of eye lesions. Nevertheless, blindness from leprosy can be a thing of the past if an early diagnosis is made and appropriate treatment undertaken. Furthermore, recently developed drugs hold additional promise for improved results.

REFERENCES

1. Allen, J. H.: The pathology of ocular leprosy, II. Miliary lepromas of the iris, Amer. J. Ophthal. **61**:987, 1966.
2. Allen, J. H., and Byers, J. L.: The pathology of ocular leprosy, I. Cornea, Arch. Ophthal. **64**:216, 1960.
3. Axenfeld, T.: The bacteriology of the eye, translated by Angus MacNab, New York, 1908, William Wood and Co., p. 344.
4. Badger, L. F.: History of leprosy in the United States, Ann. N. Y. Acad. Sci. **54**:6, 1951.
5. Badger, L. F.: Epidemiology. In Cochrane, R. G.: Leprosy in theory and practice, Bristol, 1959, John Wright and Sons, Ltd., p. 58.
6. Baron, A. L.: Man against germs, New York, 1957, E. P. Dutton and Co., Inc., p. 80.
7. Brieger, E. M.: The submicroscopial structure of the Mycobacterium leprae. In Cochrane, R. G.: Leprosy in theory and practice, Bristol, 1959, John Wright and Sons, Ltd., pp. 22-28.
8. Bullock, W. E.: Min-Fu Ho and Mei-Jan Chen: studies of immune mechanisms in leprosy, J. Lab. Clin. Med. **75**:863, 1970.
9. Choyce, D. P.: The eyes in leprosy. In Cochrane, R. G.: Leprosy in theory and practice, Bristol, 1959, John Wright and Sons, Ltd., pp. 160-170.
10. Cochrane, R. G.: The history of leprosy and its spread throughout the world. In Cochrane, R. G.: Leprosy in theory and practice, Bristol, 1959, John Wright and Sons, Ltd., p. 1.
11. Cochrane, R. G., and Sloan, T. B. M.: Tuberculoid leprosy affecting the palpebral portion of the lachrymal gland, Leprosy in India **12**:83, 1940.
12. Elliot, D.: An interpretation of the ocular manifestations of leprosy, Ann. N. Y. Acad. Sci. **54**:84, 1951.
13. Faget, G. H., Pogge, R. C., Johansen, F. A., Prejean, B. M., and Eccles, C. G.: Pormin treatment of leprosy: a progress report, Public Health Rep. **58**:17, 1943.
14. Gay, F. P., and others: Agents of disease and host resistance, Springfield, Ill., 1935, Charles C Thomas, Publisher, p. 1027.
15. Holmes, W. J.: Geographic ophthalmology, Springfield, Ill., 1959, Charles C Thomas, Publisher, p. 42.
16. Janssens, P. G., and Patlyn, S. R.: Experience with mouse inoculation of leprosy bacilli

originating from the Congo, Int. J. Leprosy **31**:522, 1963.
17. Kellersberger, E. R.: The social stigma of leprosy, Ann. N. Y. Acad. Sci. **54**:126, 1951.
18. Kirwan, E. W. O'G.: Ocular leprosy with reference to certain cases shown, Proc. Roy. Soc. Med. **48**:108, 1955.
19. McKinley, E. B.: The bacteriology of leprosy, Int. J. Leprosy **7**:1, 1939.
20. Muir, E.: Manual of leprosy, Baltimore, 1948, The Williams and Wilkins Co., pp. 3, 10.
21. Muir, E.: Relationship of leprosy to tuberculosis, Leprosy Rev. **28**:11, 1957.
22. Pillat, A.: Leprosy bacilli in the scrapings from the diseased cornea in a leper and comments on keratitis punctata superficialis leprosy, Arch. Ophthal. **3**:306, 1930.
23. Pinkerton, F. J.: Leprosy of the eye: an analysis of the records of 512 cases of leprosy in the Hawaiian Islands, Arch. Ophthal. **56**:42, 1927.
24. Rees, R. J. W.: Limited multiplication of acid fast bacilli in footpads of mice inoculated with Mycobacterium leprae, Brit. J. Exp. Path. **45**:207, 1964.
25. Rees, R. J. W., Pearson, J. M. H., and Waters, M. F. R.: Experimental and clinical studies on Rifampicin in treatment of leprosy, Brit. Med. J. **1**:89, 1970.
26. Rogers, L., and Muir, E.: Leprosy, ed 3, Baltimore, 1946, The Williams and Wilkins Co.
27. Ruge, H. G. S., Fromm, G., Fuhner, F., and Guinto, R. S.: Serological findings in leprosy: an investigation into the specificity of various serological tests for syphilis, Bull. W.H.O. **23**:793, 1960.
28. Shepard, C. C.: The experimental disease that follows the injection of human leprosy bacilli into footpads of mice, J. Exp. Med. **112**:445, 1960.
29. Skinsnes, O. K.: Leprosy in society, I. Leprosy has appeared on the face, Leprosy Rev. **35**:21, 1964.
30. Skinsnes, O. K.: Leprosy in society, II. The pattern of concept and reaction to leprosy in oriental antiquity, Leprosy Rev. **35**:106, 1964.
31. Wayson, W. E.: Leprosy of the eye, ear, nose and throat, Laryngoscope **43**:991, 1933.

10

Allergy in relation to ophthalmology

FREDERICK H. THEODORE

In ophthalmology, perhaps more than in any other specialty, allergy has come to be accepted as a major etiologic mechanism. An increasing number of apparently unrelated ocular affections are now best explained on the basis of hypersensitivity. This is not surprising in view of the fact that the eye and its adnexa offer a unique opportunity to observe and study, both clinically and experimentally, all the fundamental allergic reactions to an extent not possible elsewhere in the body. Unfortunately, however, because ocular allergies often resemble inflammations of nonallergic origin, on many occasions allergy is not recognized when it is present, or it is erroneously diagnosed when it does not exist.

Most of the difficulties that arise in the recognition and classification of allergies, especially of the external eye, stem from the fact that allergy itself is not a single entity, but instead may assume a number of forms. If, however, the clinical manifestations of these allergies are correlated and classified in accordance with the various allergic mechanisms involved, many otherwise troublesome diagnostic problems are clarified.

GENERAL CONCEPTS OF ALLERGY

Allergy is considered a specifically altered capacity to react as judged from previous experience of the same individual or from experiences of other individuals of the same species. The term is generally used to denote acquired hypersensitivity, although tendencies to develop allergy are influenced by heredity. Atopy is the term used to designate allergy with a hereditary background, that is, with a familial history of allergy.

The concept of hypersensitivity on an immunologic basis is less than 70 years old, having first been introduced as anaphylaxis by Portier and Richet[24] in 1902. It was confirmed independently by Theobald Smith,[31] who noted the reaction with horse serum at just about the same time. The name allergy was coined by von Pirquet[45] in 1906 and has been used ever since because it is so neat an expression.

Allergens

Substances that can cause allergic states, and thereafter allergic reactions, are almost limitless. Besides foreign proteins, foods and digestive products, and nitrogenous materials as in pollens, the list includes chemicals, drugs, metals, and living or dead microorganisms (molds, yeast, fungi, bacteria, viruses, parasites, metabolites). Physical influences such as light, heat, and cold, as well as emotional factors, also cause allergic responses.

It is thus obvious that although some of these substances are true antigens, and others, such as pollens, show some degree of antigenicity, many are frankly not antigens in the usual sense. Many of these allergic excitants or allergens are believed to combine with proteins or some other substance of the host's tissues and to assume thereby the nature of an artificial antigen, thus causing allergic reactions.

The concept of allergy has until now been basically oriented on the assumption of an antigen-antibody relationship, an approach that has proved very useful but that has failed to explain many phenomena in the field of allergy. Thus even the allergist versed in immunology is often obliged to make diagnoses solely on a clinical basis. Vernal conjunctivitis, for example, the mechanism of which is still obscure, does not fit neatly into such a framework but seems to require a special chain of allergic circumstances and allergic responses for its occurrence. However, for clinical purposes it is convenient to divide allergic reactions into two types: (1) the immediate and (2) the delayed. Although the distinctions are not always clearcut and some overlapping occurs, this classification is of great value.

THE IMMEDIATE ALLERGIC RESPONSE. "Immediate" allergic responses occur shortly after application or absorption of the allergen, usually within a few minutes. Immediate reactions occur in asthma, hay fever, urticaria, unusual cases of drug sensitivity, and gastrointestinal disturbances, and are generally caused by serum or other proteins, danders, pollens, dusts, or foods. The reaction has the following characteristics:

1. Skin tests give a typical wheal in a few minutes.
2. A drop of the allergen in the eye is followed promptly by itching, tearing, redness, and chemosis.
3. Circulating antibodies, such as precipitins, are often demonstrable.
4. Passive transfer (Prausnitz-Küstner phenomenon) occurs.
5. Desensitization following the injection of minute quantities of antigen is often successful and results in the formation of "blocking antibody," which combines readily and specifically with the allergenic material and "neutralizes" it.
6. Histamine plays a dominant role in the production of the reaction.

THE DELAYED TYPE OF ALLERGIC RESPONSE. In the delayed type of allergic response, which is also known as the tuberculin or bacterial type, it is a matter of hours, often 24 to 48, before reaction occurs. Contact allergy, caused by drugs or chemicals, and bacterial allergy constitute the major clinical forms of delayed allergy.

In contradistinction to the immediate type of allergic response, the delayed type fails to show certain characteristics:

1. Circulating humoral antibodies are absent. The cells themselves are the seat of the antigen-specific reaction.
2. In most instances passive transfer cannot be produced.
3. Histamine does not appear to play a significant role in the reaction.

However, anaphylactic and immediate reactions, as well as passive transfer of serum, may occur at times with both chemical and microbial antigens, especially if the sensitization is intense.

Microbiallergic reactions

I have coined the term microbiallergic to describe reactions caused by microbial allergy.[41,42] It is now an established fact that many of the harmful effects that microorganisms can exert in the human body are manifestations of allergy. Beyond the direct toxic nature of microbial infection, allergy may convert a harmless protein product of bacterial disintegration into a violent and lethal poison. In ophthalmology such bacterial allergy, especially to staphylococcal products, is of great importance in diseases of the eyelids, conjunctiva, cornea, and sclera; hypersensitivity to products of the *Streptococcus* and the tubercle bacillus appear to play an important role in uveal diseases and in scleritis. Similar allergic reactions may result from fungal products and parasitic agents. Moreover, bacterial exotoxins may likewise incite microbiallergic reactions.

Bacterial allergy and bacterial immunity are entirely distinct and different processes.

Autogenous allergy

The concept of sensitization to antigens of one's own tissues, or autosensitization, long recognized by ophthalmologists in the causation of sympathetic ophthalmia and phacoanaphylactic uveitis, is now belatedly gaining generalized acceptance. Such diseases as acquired hemolytic anemia, thyroiditis, and certain forms of encephalitis may arise on this basis. There is evidence suggesting that the sensitization in rheumatic fever is not to the *Streptococcus* but is an autosensitivity induced by the streptococcal infection. A similar mechanism may be involved in the pathogenesis of glomerulonephritis. In ophthalmology, as will be noted later in this chapter, autogenous allergy may serve to explain many as yet incompletely elucidated reactions of apparently alergic nature.

ALLERGY OF THE CONJUNCTIVA

There could hardly be a better site in the body than the conjunctiva for the study of basic allergic reactions. The juxtaposition of readily accessible mucous membrane to the adjacent thin and delicate skin of the eyelids provides the opportunity for all types of allergic reactions to occur to an extent and with a frequency that do not exist elsewhere. Although the nasal mucosa is often the site of explosive airborne allergies, contact allergy is relatively rare. The reverse is true of the skin. Only in the conjunctiva are both of these forms of allergy, as well as reactions to microbial allergens, commonly encountered. Such allergic responses may result from local exposure to excitants or as part of a generalized hypersensitivity.

The outline below classifies the various types of conjunctival allergy that are encountered in clinical practice.

Allergy of the conjunctiva

A. Atopic conjunctivitis
   1. Atopic (familial history of allergy), rarely anaphylactic
   2. Immediate alergic response (in a few minutes) after contact
   3. Positive intradermal wheal test reaction in 15 to 30 minutes
   4. Causes: inhalants, such as pollens, dusts, fungi, animal epidermals
   5. Types
      a. Acute atopic conjunctivitis—major characteristics: chemosis, eosinophilia
      b. Chronic atopic conjunctivitis—major characteristics: itching, slight edema, eosinophilia

B. Microbiallergic conjunctivitis
   1. Atopy not essential
   2. Delayed allergic response
   3. Delayed positive intradermal test reaction (24 to 48 hours); immediate urticarial reaction may occur
   4. Causes products of microorganisms: bacteria, fungi, parasites, possibly viruses
      Major characteristics: irritation, burning, some follicles

C. Allergic dermatoconjunctivitis
   1. Atopy not essential
   2. Delayed allergic responses (24 to 48 hours)
   3. Positive patch test in 1 to 5 days
   4. Contact allergy to locally applied drugs or chemicals
      Major characteristics: eczema, eosinophilia

D. Vernal conjunctivitis

Atopic conjunctivitis

By the term atopic conjunctivitis we describe the conjunctival reaction occurring in atopy and in anaphylaxis. This is a clear-cut antigen-antibody response and should not be confused with the conjunctivitis occurring in so-called atopic dermatitis, which is by no means a clearly distinguished entity. As the outline above indicates, atopic conjunctivitis is divided into acute and chronic varieties.

ACUTE ATOPIC CONJUNCTIVITIS. This type of hypersensitivity is characterized by a sudden or immediate hyperemia and edema. Marked chemosis of a glassy appearance, associated usually with a watery, but sometimes mucopurulent, discharge, occurs. It is the conjunctival prototype of the immediate type of allergic response and usually is caused by the airborne group of allergens: pollens (Fig. 10-1), animal danders and feathers, mold spores, dusts, insects, and scents. The conjunctival reaction is similar to the general inflammatory edema that involves the upper respiratory tract.

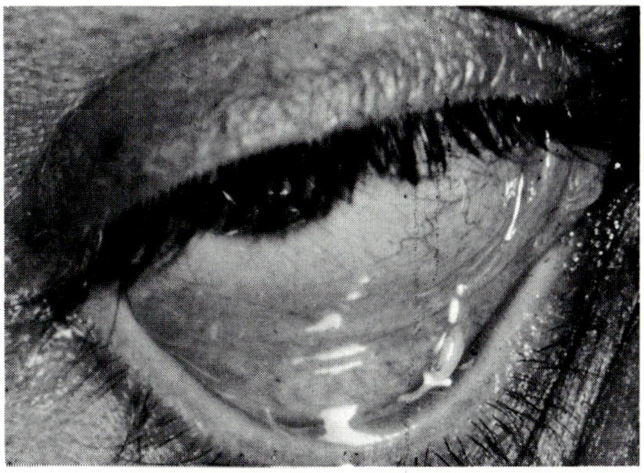

Fig. 10-1. Experimentally induced conjunctivitis in ragweed-sensitive hay fever patient. A small amount of pollen (still visible) was placed in lower conjunctival sac 7 minutes before picture was taken. (From Theodore, F. H., and Schlossman, A.: Ocular allergy, Baltimore, 1958, The Williams & Wilkins Co. © 1958, The Williams & Wilkins Co., Baltimore.)

A similar picture may occur in food allergies, such as egg and milk, but in that event the ocular reaction has only a minor role in a dramatic general systemic reaction. A familial history of allergy (atopy) is generally obtained.

As in the nasal symptoms of hay fever, the reaction may be so intense that a profuse mucopurulent discharge occurs. Itching and burning may be intolerable. Eosinophils often occur in abundance in the secretions. Vasoconstrictors, like epinephrine, phenylephrine, (Neo-Synephrine), and naphazoline (Privine), help control the reaction, which usually subsides rapidly once the excitant is removed.

Skin tests of the intradermal type give an immediate wheal response in most cases of this group. Ophthalmic tests, by means of the direct instillation of the suspected allergen in weak dilution, are generally not safe in these cases. Severe, and possibly permanent, ocular damage may result. Because of chemosis, early epidemic keratoconjunctivitis may be confused with acute allergic conjunctivitis. The virus infection, however, is differentiated by the presence of preauricular lymphadenopathy. In scrapings, mononuclear cells are found instead of eosinophils.

Chronic atopic conjunctivitis. In contradistinction to the often dramatic and generally severe reactions that occur in acute atopic conjunctivitis, the chronic form frequently shows little or no objective evidence of inflammation in comparison to the irritating subjective symptoms of itching, burning, photophobia, and dryness that are usually present. As a result, the patients often are considered psychoneurotic or else are thought to be suffering from chronic catarrhal conjunctivitis. Since the cultures from many patients with chronic conjunctivitis are found bacteriologically negative, it may be that a significant number of such so-called cases actually are overlooked forms of chronic atopic conjunctivitis. In fact, even when the ophthalmologist is alerted to the frequency of chronic atopic conjunctivitis, if he performs routine epithelial scrapings in cases of conjunctivitis, he is often surprised to find numerous eosinophils in cases that clinically did not appear allergic.

The conjunctiva often appears rather pale. A minor degree of conjunctival edema may be suggested by the somewhat juicy appearance of the conjunctiva, especially in the lower fornix in its temporal portion. Although the response is essentially papillary, follicles may appear in long-standing cases, especially where there is relatively little inflammatory reaction. In other patients, where a more subacute process occurs, there may be considerable injection, as well as slight chemosis. Corresponding to the grade of the reaction in general, there may be a slight watery discharge or a mucopurulent one. In either event, eosinophils are found in epithelial scrapings, always in diagnostic numbers, although less numerous than in acute atopic conjunctivitis.

Chronic atopic conjunctivitis is often known as simple allergic conjunctivitis.

The diagnosis of chronic atopic conjunctivitis is often difficult because the condition may resemble other forms of chronic conjunctivitis, in particular, microbiallergic conjunctivitis caused by *Staphylococcus* organisms. The differentiation of these two

Table 10-1. Differentiation of chronic types of allergic conjunctivitis

	Chronic atopic conjunctivitis	Microbiallergic conjunctivitis caused by *Staphylococcus* organisms
History	Familial allergy frequent Other associated allergies	Familial allergy infrequent No associated allergies
Allergens	Pollens, danders, inhalants, foods	Bacterial products, mainly *Staphyloccus* toxin
Allergic response	Immediate	Delayed
Symptoms	Burning, irritation, and often severe itching	Burning, irritation, and often marked dryness
Discharge	May be mucopurulent	Watery, no pus
Lid margins	Usually uninvolved	May be swollen; possible blepharitis
Conjunctiva	May appear same in both types Often little objectively noted Sometimes edematous	May appear same in both types Papillary, but scattered follicles may appear Dry, no chemosis
Cornea	Generally not involved	May have superficial punctate keratitis involving inferior portion of cornea
Epithelial scrapings	Eosinophiles almost always found sooner or later	No eosinophiles
Cultures	Normal flora	Toxin-producing staphylococci found, or normal flora
Intracutaneous *Staphylococcus* toxin or toxoid	Moderate delayed response	Marked delayed response, at least 3 × 3 cm, usually more; may have marked immediate response also; may get systemic reactions (arthritis)
Response to vasoconstrictors	Local use often affords considerable symptomatic relief	Generally no great relief
Response to antihistaminics	Oral use often affords marked symptomatic relief	Generally no relief at all
Response to cortisone	Local use usually affords marked symptomatic relief	Relief in some cases, but not the rule
Best treatment	Elimination of cause	*Staphylococcus* toxin or toxoid injections

From Theodore, F. H.: The classification and treatment of allergies of the conjunctiva, Amer. J. Ophthal. **36**:1698, 1953.

forms of chronic allergic conjunctivitis is shown in Table 10-1.

The conjunctivitis that occurs often does not permit any ready means of distinguishing between the two types, although there generally are slight differences. However, more important than the appearance of the conjunctiva itself are the history; the finding of the eosinophils; the response to certain avenues of therapy, such as local vasoconstrictors, steroids, and oral antihistamines; and, if possible, the elimination of the cause of the allergy. These clues afford the framework for differentiating chronic atopic conjunctivitis from other forms of chronic inflammation of the conjunctiva.

Microbiallergic conjunctivitis

Bacteria, fungi, and helminths appear to be capable of inciting microbiallergic conjunctivitis. The status in regard to viruses is not clear.

Bacteria. There is very little evidence that bacteria other than the *Staphylococcus* play an important role in the production of allergic conjunctivitis. Apparently a combination of two factors operates in the pathogenesis of this condition: (1) chronicity, with the opportunity for repeated reinfection, and (2) the formation of antigenic exotoxin or other sensitizing substances. Most bacteria that produce potent, but not necessarily allergic, exotoxins, such

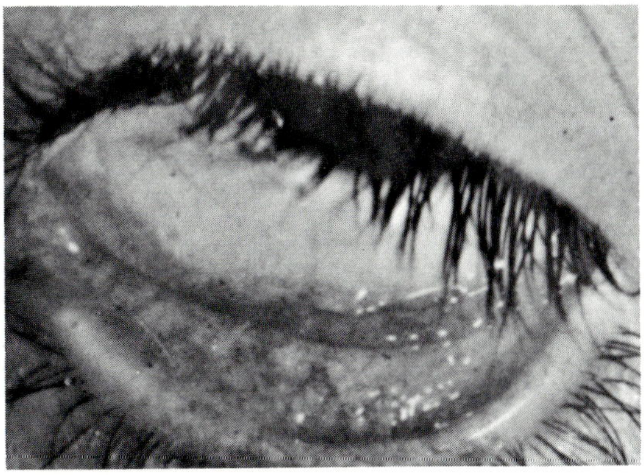

Fig. 10-2. Microbiallergic conjunctivitis caused by *Staphylococcus*. Quiescent stage, minor symptoms. Note follicles in lower lid. (From Theodore, F. H., and Schlossman, A.: Ocular allergy, Baltimore, The Williams & Wilkins Co. © 1958, The Williams & Wilkins Co., Baltimore.)

as the diphtheria and tetanus bacilli, cause acute processes. Only the *Staphylococcus* appears to fill both requirements in this respect. It must be emphasized that microbiallergic conjunctivitis may occur without the presence of or previous exposure to what we consider pathogenic organisms. Even nonpathogenic bacteria can bring about an allergic state in certain individuals. This fact would explain the occurrence of microbiallergic conjunctivitis where only so-called non-toxin-producing *Staphylococcus albus* is found. The appearance of allergic conjunctivitis caused by *Staphylococcus* is seen in Fig. 10-2.

Studies of bacteriophage typing of *Staphylococcus* organisms by Locatcher-Khorazo and Gutierrez[17] may offer, by inference, additional confirmation of the role of bacterial allergy in staphylococcal infections. Although many phage types were isolated by these authors from normal eyes, it was rare to find more than one strain in the eyes of any single individual, and the same phage type was present in the nose as well. Staphylococcal ocular infections, including unilateral postoperative infections, were caused by the same type that was present in the normal eye of the same individual. In eight postoperative infections the same phage type found before operation was responsible. Repeated ocular cultures showed that the same type of *Staphylococcus* persisted for months, even after antibiotic therapy.

Although others have reported microbiallergic conjunctivitis to streptococcal and gonococcal products, I have not encountered such instances.

In a detailed study of 60 instances of all forms of conjunctival allergy, other than those caused by contact allergy, investigated at the Fondation Rothschild in Paris, Morault[21] found that 7 cases resulted from bacterial allergy. According to Morault, the responsible organisms, which in some instances were multiple, were: *Pneumococcus,* 5 cases; *Streptococcus hemolyticus,* 5 cases; *Staphylococcus aureus,* 3 cases; *Streptococcus viridans,* 1 case; coliform bacillus, 1 case; and *Bacillus pyocyaneus,* 1 case.

Fungi. Allergic conjunctivitis caused by mold spores is not uncommon. Essentially an inhalant form of allergy, like pollen allergy, the allergic reactions these fungi arouse have many of the characteristics of the immediate atopic response. *Alternaria* and *Cladosporium,* as well as yeasts, have been responsible.

Other types of fungi, of more pathogenic character, cause allergy only after a primary infective focus has been established. Here the mechanism is of the delayed tuberculin type. The main ones of this group are the dermatophytes, such as *Trichophyton* and *Candida albicans (Monilia),* causing skin diseases. Thus, after a demonstrably infected primary dermatitis has occurred, id reactions may appear elsewhere. These id lesions reveal no fungi and disappear spontaneously when the original lesion is successfully treated. In the eye such allergic reactions occur mainly as eczematous lesions affecting the eyelids; only rarely is the conjunctiva involved.

Helminths. Conjunctival allergy to intestinal parasites has been produced experimentally. Laboratory workers in this field, especially those dissecting the parasites, may contact allergic conjunctivitis from finger-to-eye contact. Reactions similar to hay fever with conjunctivitis and asthma also occur. Slaughterhouse workers also develop similar symptoms, which disappear when such exposure no longer occurs. At least two instances of recurrent allergic conjunctivitis related to *Oxyuris* infestation have been recorded.

Allergic dermatoconjunctivitis

Contact allergy of the eyelids and conjunctiva constitutes by far the most common form of allergic reaction encountered in ophthalmic practice. It is generally caused by the local use of drugs; it can also be caused by cosmetics, articles of clothing, jewelry, animal or vegetable products, plastics, or chemicals used in industry. In those instances where the conjunctiva is the focal point of contact, as in the use of ophthalmic drugs, the allergic reaction generally begins as a conjunctivitis, soon involving the adjacent skin of the eyelids in a typical eczematous dermatitis. The particularly apt term allergic dermatoconjunctivitis has been applied to this clinical picture. When the eyelid skin is the site of the primary contact, conjunctival allergy plays little or no role in the manifestations, and an allergic eczematous dermatitis occurs without conjunctivitis. This form of reaction usually is the result of exposure to cosmetics, apparel, jewelry, metals, plants, or chemicals, and is the most frequent cause of acute ocular eczema. Although the allergic mechanism is basically the same, only allergic dematoconjunctivitis is discussed at this point; allergic eczematous dermatitis of the eyelids is discussed later in this chapter.

In contact allergy of the conjunctiva, the responsible agent almost always is an ophthalmic drug. Since the conjunctiva is usually the focal point of contact, the allergic reaction generally begins as a conjunctivitis but soon involves the adjacent skin, becoming a typical allergic dermatoconjunctivitis. The major diagnostic characteristics of this type of reaction, in the order of their appearance, are (1) severe itching of the eyes, (2) papillary conjunctivitis, (3) eczema of the skin of eyelids, and (4) conjunctival eosinophilia. If the conjunctivitis is caused by eye drops, the earliest signs of eczema may be at the canthi and on the lower lids; if ointments are the offending agents, the lid margins may first be involved. Although any ophthalmic drug may at times result in allergy in susceptible individuals, allergic dermatoconjunctivitis is usually encountered as a result of using one of the following important groups of medicaments: anesthetics, antibiotics, sulfonamides, mydriatic alkaloids, or mercurials. Moreover, even if the active drug itself is harmless, allergy may often result from other ingredients in ophthalmic vehicles, especially preservatives and ointment bases.

ALLERGENIC DRUGS. Many ophthalmic drugs are true sensitizers. On occasion probably every drug used in the treatment of the eyes has resulted in allergic dermatoconjunctivitis in some individual. In fact, there is probably no drug or chemical in existence that everyone can use with impunity. Although nonprotein substances are in themselves not primary antigens, it is believed that they form true antigens in the body by union with the host tissues. These relatively simple chemicals, known as haptens, combine with body proteins, forming conjugates that act as antigens. The hapten confers specifically on the conjugate; hence, the allergy is directed against the hapten and not against the protein. This process explains why the substitution of a drug that is chemically similar to one that previously caused sensitivity in a patient is dangerous. Thus, after allergy to one sulfonamide has developed, another sulfonamide is very likely to cause sensitivity. Benzamine and other chemicals with paralinked amino groups, as compared with ortho- and meta-attachments, are especially prone to cause contact allergies. Unfortunately, very many of the drugs we use fall into that category.

Local anesthetics. Apparently every local anesthetic that is used for instillation is a potential sensitizer and has been known to cause allergic conjunctivitis. Of the commonly used ones, those least likely to do so appear to be cocaine, phenacaine (Holocaine), and possibly piperocaine (Metycaine), tetracaine (Pontocaine), butacaine (Butyn), dibucaine (Nupercaine), and Larocaine are particularly likely to cause reactions and frequently are the causes of cases of conjunctivitis that get progressively worse despite treatment. Proparacaine (Ophthaine, Ophthetic) not only may cause a typical allergic dermatoconjunctivitis, but on occasion may incite a hyperallergenic corneal reaction of the greatest severity.

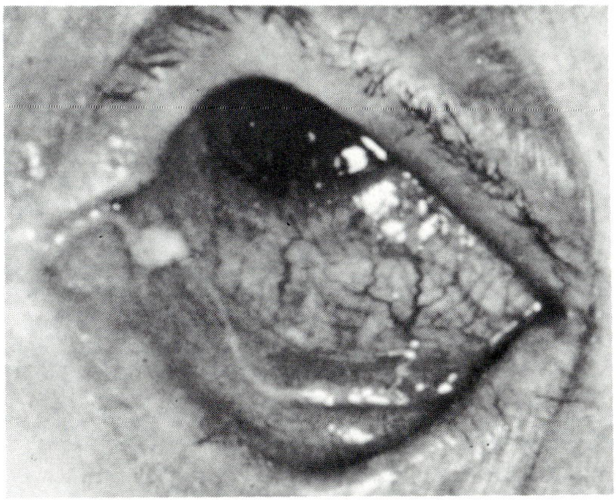

Fig. 10-3. Allergic dermatoconjunctivitis caused by hyoscine. (From Theodore, F. H., and Schlossman, A.: Ocular allergy, Baltimore, 1958, The Williams & Wilkins Co. © 1958, The Williams & Wilkins Co., Baltimore.)

Antibiotics. Penicillin is notorious for its sensitizing tendencies, especially in the ointment form. Streptomycin, bacitracin, the tetracyclines, chloramphenicol (Chloromycetin), erythromycin, and neomycin all may cause allergies. Neomycin is particularly important in this respect because it is used in combination with so many other antibiotics and corticosteroids.

Sulfonamides. All sulfonamides are prone to cause allergic reactions after prolonged topical use.

Mydriatic and miotic alkaloids. Allergy to atropine is a common occurrence, resulting in a typical allergic dermatoconjunctivitis with eczema, papillary conjunctivitis, and conjunctival eosinophilia. Both hyo-

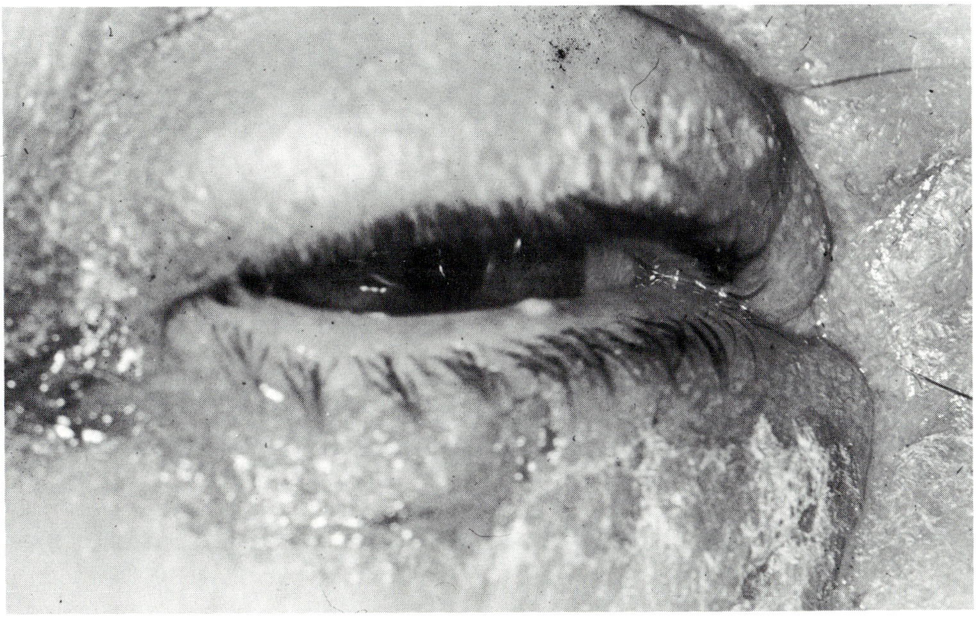

Fig. 10-4. Allergic dermatoconjunctivitis caused by eserine. Typical full-blown picture after using drops for several months. (From Theodore, F. H.: Trans. Amer. Acad. Ophthal. Otolaryng. **58**:708, 1954.)

scine and homatropine may also cause sensitivities (Fig. 10-3), but the eczematous component in these drugs is less evident. Atropine, however, sometimes causes follicular conjunctivitis without eosinophilia or dermatitis resulting from conjunctival irritation (see below).

The miotic alkaloids, on the other hand, are rarely sensitizers. The use of pilocarpine or eserine (physostigmine) only occasionally results in true allergic dermatoconjunctivitis (Fig. 10-4). Instead, the prolonged administration of these and other related synthetic miotics may produce follicular conjunctivitis, without eczema and eosinophilia, which appears to be caused by drug irritation, not by allergy. This will be elaborated upon later. Those very rare instances in which allergic dermatoconjunctivitis as well as follicles occur are explained by the fact that primary irritants may also be, or can become, sensitizers.

Other common ophthalmic drugs. Mercurials are well-recognized sensitizers, as are drugs ranging from dionin to boric acid, including zinc salts, silver protein (Argyrol), and naphazoline (Privine). Nitrofurazone (Furacin), quarternary ammonium compounds like benzalkonium (Zephiran), phenylephrine (Neo-Synephrine), and many others may also cause allergies. Even topical corticosteroids may cause an allergy. Idoxuridine may cause severe allergies, especially after prolonged use. Follicular irritation from use of this drug, involving the lower conjunctival sac, is a frequent occurrence.

Ophthalmic vehicles. Preservatives used in eye solutions may be the unsuspected cause of allergy.[43,44] This especially applies to the mercurials and to benzalkonium chloride. Ointment bases may cause allergies, usually to lanolin or petrolatum.

THE SYNDROME OF DRUG IRRITATION.[34,40] Although most cases of drug intolerance are caused by drug allergy, enough instances of drug irritation occur to make it an important differential consideration. The importance of the distinction between these two types of drug reaction lies in the fact that in drug irritations the use of the same drug may be continued if it is prepared in a manner that ensures no more irritation, whereas in allergies an entirely different drug must be substituted. Clinically, differentiation is generally simple.

Substances that come in contact with the skin and mucosa may be classified as (1) primary irritants and (2) cutaneous and mucosal sensitizers. Primary irritants cause an inflammation by direct action at the site of contact only, if allowed to act in sufficient intensity or quantity for a long enough time. Cutaneous and mucosal sensitizers usually cause no changes on first contact; but after about 5 to 7 days the entire skin, or the mucous membranes, will react specifically on further contact at the original site or elsewhere. These are the sensitizers that cause contact allergy.

Although some drugs are themselves primary irritants, and almost every drug may at times be irritating, conjunctivitis caused by irritation usually occurs following the prolonged use of preparations that are prone to deteriorate into irritating products if not prepared in a manner that will hinder or prevent this breakdown. This is particularly apt to occur with alkaloids but can also occur with the use of many chemicals, including antibiotics, especially neomycin, and, as stated above, idoxuridine.

The clinical picture of conjunctivitis resulting from drug irritation is that of a nonspecific watery inflammation of the conjunctiva in which, in contrast to allergic dermatoconjunctivitis, there is neither dermatitis of the eyelids nor conjunctival eosinophilia. In chronic cases, especially with the use of the alkaloids, follicular conjunctivitis may be the outstanding feature. The conjunctival involvement is less diffuse and less uniform than in allergy; often the area of most irritation, the lower lid and adjacent bulbar conjunctiva, bears the brunt of the involvement. Results of skin patch tests with these low-grade irritants are negative. In practice, it is important to know that such drugs as atropine usually cause allergies, but that on occasion irritation may occur because of formation of irritants like tropic acid and tropine.

The miotic alkaloids and the synthetic miotics cause irritations much more frequently than they do allergies. However, occasional allergies, especially to aromatic compounds like pilocarpine, eserine, neostigmine (Prostigmin), and Humorsol (demecarium bromide, a quaternary ammonium compound containing two neostigmine molecules), are encountered. Allergic reactions from isoflurophate (Floropryl) seem to be even rarer. The one allergic reaction I experienced with isoflurophate proved to be caused by the peanut oil vehicle. However, allergies must be expected from all miotic drugs if their use is sufficiently widespread.

The basis for the commonly encountered irritative, often follicular, conjunctivitis following the prolonged use of miotics appears to stem from their breakdown products, which are of course related to their chemical composition. All the aromatic miotics contain methyl ammonium radicals. Pilocarpine degrades to acetic acid, propionic acid, ammonia, and methylamine. Eserine forms rubeserine and methylamine through a rapid deterioration. Neostigmine appears to be more stable, but irritation occurs because of the formation of dimethylamine. Humorsol, a related drug (as noted above), appears to cause irritation by the same mechanism. Furtrethonium iodide (Furmethide) is another methyl ammonium agent with extremely marked irritating propensities, because of which the drug is no longer available.

Two aliphatic compounds, methacoline (Mecholyl) and carbachol (Doryl) also contain methyl ammonium radicals that apparently break down to methylamine or dimethylamine. There are two other aliphatic compounds that do not contain methyl ammonium radicals: isoflurophate and echothiophate (Phospholine iodide). Follicular reactions from isoflurophate are not rare. The drug degrades to hydrochloric acid and elemental fluorine. A phosphorus derivative of choline, echothiopate appears to decompose by hydrolysis to thiocholine and the diethyl ester of phosphoric acid and causes irritation possibly on this basis.

Although allergic reactions to epinephrine products used in glaucoma occur, follicular irritative reactions appear to be more common. This drug also contains a methylamine radical. Another particularly interesting phenomenon is the deposition of clumps of melanin pigment in the interior fornix after the use of epinephrine solutions, because of the "dopa" reaction. In contrast, I have encountered a number of allergic reactions to phenylephrine, another methyl ammonium compound, but no characteristic irritative follicular conjunctivitis. The absence of such irritation may lie in the fact that the drug is more stable and is used less frequently in the individual patient in such high concentrations.

Vernal conjunctivitis

Vernal conjunctivitis is a recurrent, bilateral, interstitial inflammation of the conjunctiva occurring in warm weather. Although the exact causation of the disease is unknown, it is included among allergies of the eye because the condition has so many allergic facets. The most likely hypothesis is that vernal conjunctivitis is a form of physical allergy and that it is the ocular reaction to the warm season of especially predisposed atopic individuals whose constitutions have been conditioned by a number of general organic factors.

Clinical aspects. Vernal conjunctivitis is a disease of warm climates. In the Mediterranean area an incidence of up to 2% of all patients examined in eye clinics has been reported; in New York City the probable figure is less than $\frac{1}{10}$ of 1%. The condition almost always affects boys rather than girls, in a proportion of more than 3 to 1. After puberty the incidence is about equal in both sexes. It is a disease of childhood that usually occurs in those from 6 to 20 years of age. Colored races are particularly prone to the limbic form of the disease. In most patients a familial incidence usually occurs, with an allergic background being present more than half the time. In addition, over 50% of all patients with vernal conjunctivitis suffer from other types of allergic diseases of atopic nature, such as hay fever, asthma, atopic dermatitis, and allergies to foods, dust, and molds.

SYMPTOMS. The outstanding symptom of the disease is extreme itching. Photophobia is also a frequent complaint.

OBJECTIVE FINDINGS. Vernal conjunctivitis is a bilateral condition, occurring in two main forms: the palpebral and the limbal. In a significant percentage of patients a mixed type occurs in which both the conjunctiva and the corneal limbus show changes (Fig. 10-5). The palpebral form of vernal conjunctivitis is practically always limited to the tarsal conjunctiva of the upper lid where, in long-standing cases, large cobblestone vegetations appear. Involvement of the conjunctiva of the lower lid indicates extreme severity of the condition in that individual.

The disease is essentially proliferative, characterized by a marked and continuously increasing formation of hyalinized

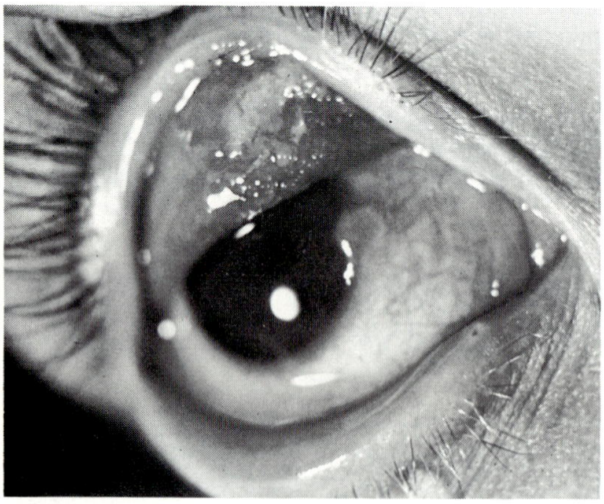

Fig. 10-5. Vernal conjunctivits. Mixed type, showing palpebral and limbal involvement with thin pseudomembrane. (Courtesy Clay-Adams Co. From Theodore, F. H., and Schlossman, A.: Ocular allergy, Baltimore, 1958, The Williams & Wilkins Co. © 1958, The Williams & Wilkins Co., Baltimore.)

connective tissue. This opaque tissue imparts a milky, bluish appearance to the conjunctiva by obscuring the capillaries, which normally give it a pink color. No matter of how many years' duration, in other forms of allergic conjunctivitis the capillary markings are never obliterated. Sometimes patients with subjectively very severe vernal conjunctivitis may show only slight conjunctival thickening and the presence of microscopic papillae containing central blood vessels ("blood points"). Usually, however, the proliferation progresses to the formation of the classical excrescences, which are hard, flat, large, polygonal papillae of varying size, with a predilection for the tarsus of the upper lid. In some countries vernal conjunctivitis, because of this predominant upper lid involvement, may be confused with trachoma. When one is in doubt, epithelial scrapings revealing conjunctival eosinophilia are helpful; however, the two conditions may coexist. Follicular conjunctivitis, despite its lower lid predilection, true follicular character, and preauricular adenopathy, is sometimes confused with vernal conjunctivitis.

The discharge that occurs in vernal conjunctivitis is of great diagnostic value. The tenacious, thick, "chewing-gum" pseudomembrane, which may be peeled off the conjunctiva without bleeding, is characteristic. Even if a membrane is not present, exposure of the everted lid, massage, or the heat of a camera light or slit lamp will result in its formation. The extremely alkaline secretion is notable for the masses of eosinophils or eosinophilic granules it contains.

The limbic proliferations of the bulbar type of vernal conjunctivitis may occur without any palpebral lesions. They appear as yellowish gray gelatinous elevations in the palpebral fissure zone. Trantas' spots are white points that sometimes cap the excrescences and are pathognomonic of vernal conjunctivitis. As differentiated from phlyctenules, vernal lesions do not stain with fluorescein. Furthermore, there is in general less inflammatory reaction than in phlyctenular keratoconjunctivitis. Conjunctival eosinophilia is a valuable differential in determining the diagnosis.

CORNEAL LESIONS. The most frequent and characteristic corneal lesion of vernal conjunctivitis is a superficial epithelial keratitis, requiring biomicroscopy for its recognition, located usually in the upper half of the cornea. The cornea looks white, as if flour had been blown over it, and stains in a punctate fashion. These lesions clear rapidly as the condition improves and thus offer an excellent index of the efficacy of treatment. Other types of corneal involvement occur much less commonly, some by direct extension of the limbal process, others being deeper and central, leaving serious opacities. Still others resemble dystrophies, such as pseudogerontoxon.

ETIOLOGY. On the basis of all the clinical characteristics of the disease and its clinical course, with exacerbations during warm weather and clearing in all but the most severe cases during the winter or in migration to a cold climate, it appears that vernal conjunctivitis is a form of physical allergy to the warm season.[46] This peculiar allergy occurs only in especially predisposed atopic individuals. The special factors that predispose these individuals may be endocrine, metabolic, allergic, and possibly infectious. In some way these factors all combine to condition atopic individuals so that the warm season acts as a trigger mechanism, setting off a sort of chain reaction precipitating the disease.

ALLERGY OF THE EYELIDS

The eyelid proper is often the site of allergic reactions. In fact, such allergies occur so commonly that they appear to constitute, at least statistically, the most important form of ocular allergy encountered in clinical practice. Because the eyelids represent a transitional structure consisting of the conjunctiva and the skin with the lid margin in between, allergies of both the conjunctiva and the eyelids often overlap and occur at the same time. The most important such combined allergic reaction, allergic dermatoconjunctivitis, has already been described. The outline below classifies all types of allergic reactions of the eyelids that are encountered.

Allergy of the eyelids

- A. Anaphylactic and atopic allergy (immediate reaction)
  - 1. Allergic edema (urticaria, angioneurotic edema, serum sickness): drugs, animal sera, insect bites
- B. Contact allergy (delayed reaction)
  - 1. Allergic dermatoconjunctivitis: drugs, chemicals
  - 2. Eczematous contact dermatitis: cosmetics, drugs, chemicals, apparel
- C. Microbial allergy (delayed reaction)
  - 1. Infectious eczematoid dermatitis
    - a. Bacterial: *Staphylococcus, Streptococcus*
    - b. Fungal: trichophytosis, moniliasis
  - 2. Infections of the lid margin (usually staphylococcal)
    - a. Blepharitis
    - b. Meibomitis
    - c. Hordeolum
    - d. Chalazion

Allergic edema of the eyelids

Anaphylactic and atopic allergic reactions of the eyelids manifest themselves basically as urticaria or angioneurotic edema. Such reactions may occur in serum sickness, from insect bites, from the ingestion of all types of foods and many drugs, and sometimes, although rarely, from inhalants and contactants. Even microbial allergens, particularly fungal products, may cause urticarial reactions about the eyelids. In addition, physical agents, such as cold or sunlight, may result in such urticaria.

Contact allergy of the eyelids

The skin of the eyelids is particularly susceptible to allergic inflammation for several special reasons.[35,36,38] First, its fine texture and extreme thinness readily permit minor traumatization and the penetration of noxious substances of all types. Second, because of their location, the eyelids are especially exposed to physical trauma (rubbing, heat, sun, wind, and ocular secretions) and to various drugs and chemicals applied elsewhere on the head and face.

Thus, although a facial cream or hair tonic is not directly applied to the eyelids, the eyelids may eventually prove to be the only site of the eczema, as even a minimal amount of the irritant, indirectly contacted by means of a pillowcase during sleep or from the fingers, is able to penetrate sufficiently here to cause the reaction, in contrast to its inability to penetrate the thicker and stronger skin where it was originally applied so generously. Similarly, the skin of the eyelids often reacts to bacterial products or ocular medicaments occurring in tears, which would be harmless elsewhere on the skin.

Contact allergy of the eyelids usually results from the use of cosmetics or ophthalmic drugs. Allergy to plants is, however, not uncommon. Less often articles of apparel, jewelry, metals, plastics, other chemicals, and animal or vegetable products, too numerous to mention, may cause a reaction. The relative importance of the various contactants causing allergic eczematous dermatitis is difficult to evaluate. However, it appears clear that cosmetics and drugs are the two major offenders. The dermatitis, which may be either unilateral or bilateral, occurs mostly in women, probably more because they use cosmetics so universally than because of any endocrine factor. There need be no previous history of familial or personal allergy. The diagnosis is based on a careful history, positive patch test reaction, and the ruling out of other major causes of eyelid eczema, namely primary ocular infection or generalized dermatoses. In addition, one must be certain that the contact dermatitis present is caused by allergy rather than by primary irritation. The highlight of therapy is to eliminate the sensitizing agent; the use of any other allergenic substance may seriously aggravate the condition.

The mechanism for the production of contact allergy has been fairly well established and has been described earlier in this chapter. The site of eczematous (contact type) allergic dermatitis appears to be strictly epidermal, not to be deeper in the

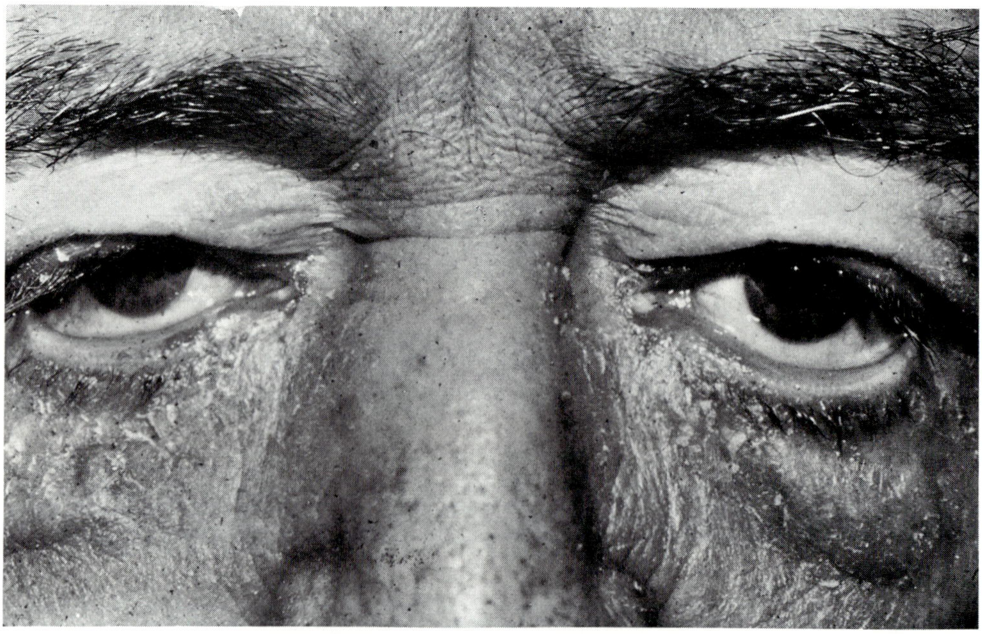

Fig. 10-6. Contact allergy to witch hazel applied to eyelids. The conjunctiva was uninvolved. (From Theodore, F. H., and Schlossman, A.: Ocular allergy, Baltimore, 1958, The Williams & Wilkins Co. © 1958, The Williams & Wilkins Co., Baltimore.)

skin. For this reason, in order to prove contact allergy, patch tests are necessary. Immediate-wheal testing by intradermal injections is not a correct diagnostic procedure here, because even the most superficial intracutaneous or scratch test carries the allergen through the epidermis into the cutis.

Ophthalmic medicaments. Most instances of this form of allergy begin in the conjunctiva, as it is generally the focal point of contact of the drug. However, certain medicaments that are directly applied to the skin of the eyelids do not cause conjunctivitis and might be briefly mentioned. These include antiseptics such as iodine, picric acid, mercurials, Zephiran, boric acid, and witch hazel (Fig. 10-6). Most of the time, however, ointments, especially those used to relieve itching or irritation of the eyelids, and particularly those containing anesthetics, antihistaminics, or even corticosteroids (Fig. 10-7), are the prime affectors. Reactions may occur from the use of all types of adhesive tapes.

Cosmetics. Cosmetics have been widely used for thousands of years, despite sporadic, unsuccessful attempts to outlaw the custom, such as the introduction of a bill in the British Parliament in the eighteenth century proposing to make the beguiling of men into matrimony by their use punishable as for witchcraft. So much is now known about the action of cosmetics on the skin, and the exclusion of harmful ingredients is so regulated by law, that the good accomplished by their use far outweighs the occasional harm that occurs. Actually, if we consider how universally cosmetics are used, the percentage incidence of allergies to them is extremely low.

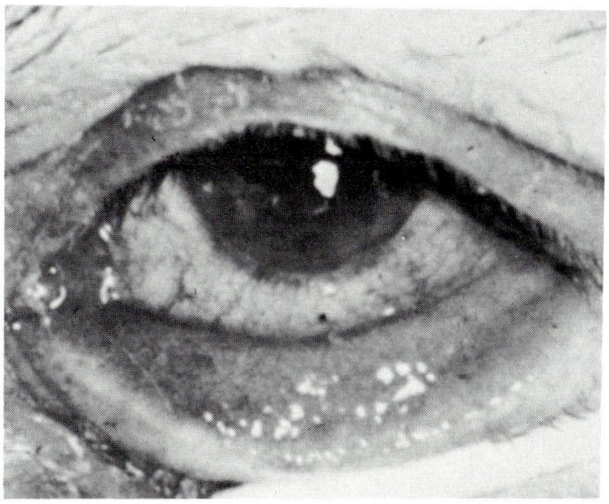

Fig. 10-7. Allergic dermatoconjunctivitis caused by cortisone. (From Theodore, F. H., and Schlossman, A.: Ocular allergy, Baltimore, 1958, The Williams & Wilkins Co. © 1958, The Williams & Wilkins Co., Baltimore.)

Unlike drug allergy, cosmetic allergy usually begins by involvement of the upper lid especially at the medial portion. Generally the conjunctiva is uninflamed and without eosinophilia. In the diagnosis of eyelid eczema caused by cosmetics it must be remembered that only the eyelids may react even if the allergenic substance was applied at a place remote to them. Careful history taking is essential; the use of leading questions often is important in opening avenues of thought otherwise dismissed by the patient as immaterial. Just because a cosmetic has been used previously without reaction for a long time does not mean it should be completely exonerated, because the manufacturer only now may have changed the formula or the method of production in some way. In diagnostic testing for cosmetic sensitivity it is better to apply the cosmetic in the way it is generally used instead of doing a patch test, because in normal use the uncovered cosmetic loses most of its substance by evaporation. Covering the cosmetic by a patch does not permit this evaporation and may give rise to false-positive reactions with cosmetics that are actually harmless when used in the ordinary manner.

Nowadays most allergies to cosmetics are caused by the perfumes used and by impurities in manufacture, as the hitherto most flagrant offenders have been eliminated. Sometimes, although the cosmetic itself is innocent, contamination of the containers may cause allergies. Allergies to a few important types of cosmetics are noted below.

Nail polish and nail lacquer. These used

to be extremely common but now occur less frequently. The offending ingredients are believed to be the synthetic resins, such as methacrylates, and possibly the dyes used.

Face powder. Face powders used to be sensitizers because of the orris root and the rice powder content. These ingredients are rarely used at present, having been replaced by titanium salts. Now any allergy to face powder is usually traced to perfumes and to impurities in manufacture and packaging.

Facial creams. Facial creams such as plain cold cream and vanishing creams are not often sensitizers. In both, the major offender is either lanolin or cocoa butter and almond oil. Emulsifying agents and perfumes should also be suspected.

Lipstick. Lipstick dermatitis is usually caused by the indelible dyes used.

Perfumes. The extremely complicated nature of perfumes makes such products especially liable to cause allergies. Their universal use, not only alone but to make other products more attractive, makes them a common cause of contact allergy.

Hair preparations. Hair dyes and rinses contain both sensitizers and toxic products. The most commonly used ingredient, paraphenylenediamine, may be both toxic and allergenic. Generally allergies occur from its use, but in susceptible persons serious toxic reactions may develop. Paraphenylenediamine is also the major sensitizer in allergy to furs. Metallic dyes may also cause allergies. Vegetable dyes are safer. Hair tonics and locations also contain sensitizing agents.

OTHER CAUSES OF CONTACT ALLERGY OF THE EYELIDS. These include wearing apparel, such as dyed fabrics, especially if the dyes tend to "bleed" or come out of the fabric easily, and laundering and dry-cleaning materials. Allergy to leather usually occurs because of the manufacturing process. Shoe polishes and shoe dyes may cause severe allergy. Furs may cause allergic dermatitis of the eyelids because of the dyes used in their manufacture. Allergy to jewelry is not uncommon, particularly to nickel, cobalt, platinum, copper, chromium, aluminum, and gold. Spectacle frames may cause allergies. Most reported cases are those caused by metallic (white gold) frames. Here the allergy is caused by the nickel content. Allergy to plastic eyeglass frames arise from sensitivities to the various chemicals used in the manufacturing process.

Infectious eczematoid dermatitis

Eczematoid dermatitis of the eyelids may arise on an allergic basis as a complication of bacterial, fungal, or other microorganismal infection present elsewhere.[37] This basically allergic response to infectants must not be confused with the numerous types of dermatitis resulting from direct invasion of the skin of the eyelids by many bacteria, viruses, and fungi. The latter are entirely distinct and not germane to the subject. Allergic reactions of the eyelids caused by bacteria generally are the result of neighboring infection of the adnexa, although rare syphilids and tuberculids need not follow this pattern. In fungal infections, however, the original focus may be far from the eye and entirely unrelated to it. In either event, the eczema appears to be in large measure a manifestation of microbial allergy. Although microbial dermatitis elsewhere on the body may result from a variety of infectants, including *Staphylococcus, Streptococcus,* and *Escherichia coli,* as far as the eyelids are concerned the important bacterium involved is the *Staphylococcus.* On rare occasions streptococci are responsible. Eyelid eczema caused by fungus sensitivity, an unusual occurrence to begin with, appears mainly as trichophytosis or moniliasis. All types of infectious eczematoid reactions appear prone to secondary contact type allergies, especially from medicaments. Since the reverse also happens—that is, the secondary infection of primarily allergic dermatitis—diagnostic problems may arise. About the eyelids, however, fewer such differential difficulties should occur, because the only type of infectious eczematoid reaction of practical importance encountered—staphylococcal eczema

—is generally associated with other ocular findings that afford valuable diagnostic clues.

Staphylococcal eczema of the eyelids

The importance of staphylococcal infections of the lid margin and conjunctiva as a major cause of eczema of the eyelids has not received the emphasis it deserves. Although dermatitis actually is a relatively infrequent complication of such common infections, recent experience indicates that the condition is the most frequent cause of chronic eyelid eczema that has been encountered; yet because its importance is not generally appreciated, it is often overlooked.

Many unrecognized, recurrent cases of

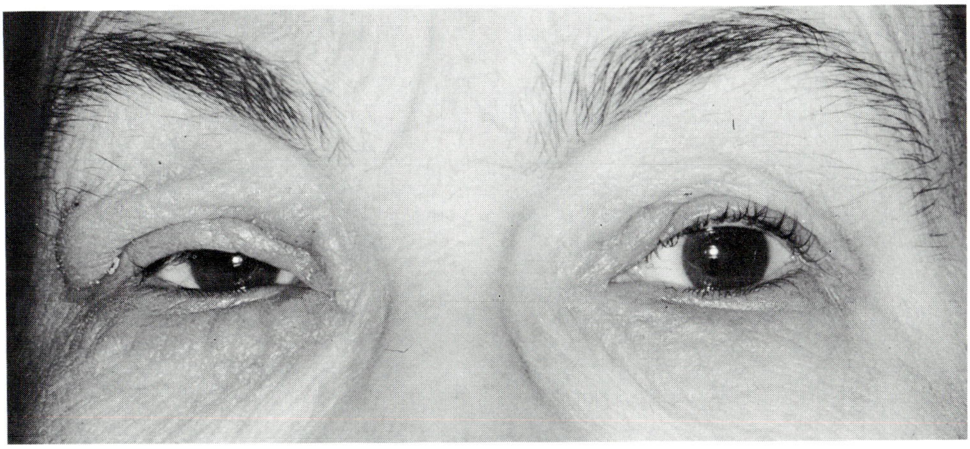

Fig. 10-8. Staphylococcal eczema of the eyelids. (From Theodore, F. H.: Trans. Amer. Acad. Ophthal. Otolaryng. **58**:708, 1954.)

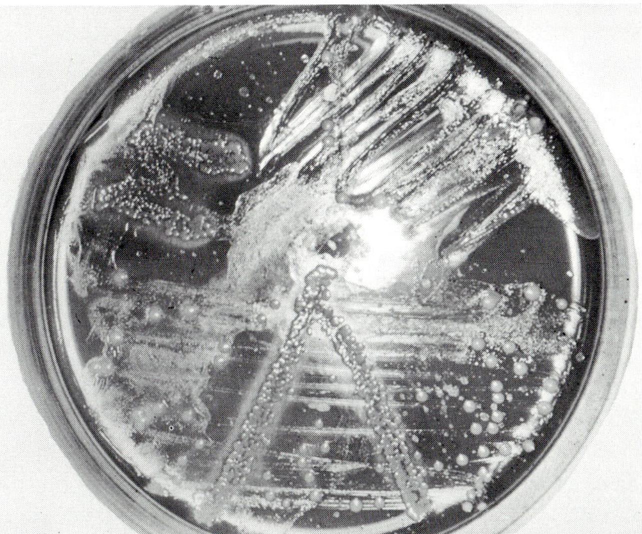

Fig. 10-9. Culture of conjunctiva and eyelids (patient seen in Fig. 10-8) revealing numerous colonies of pathogenic *Staphylococcus aureus;* growth in form of M (top left quadrant) was obtained from expressed meibomian secretion. (From Theodore, F. H.: Trans. Amer. Acad. Ophthal. Otolaryng. **58**:708, 1954.)

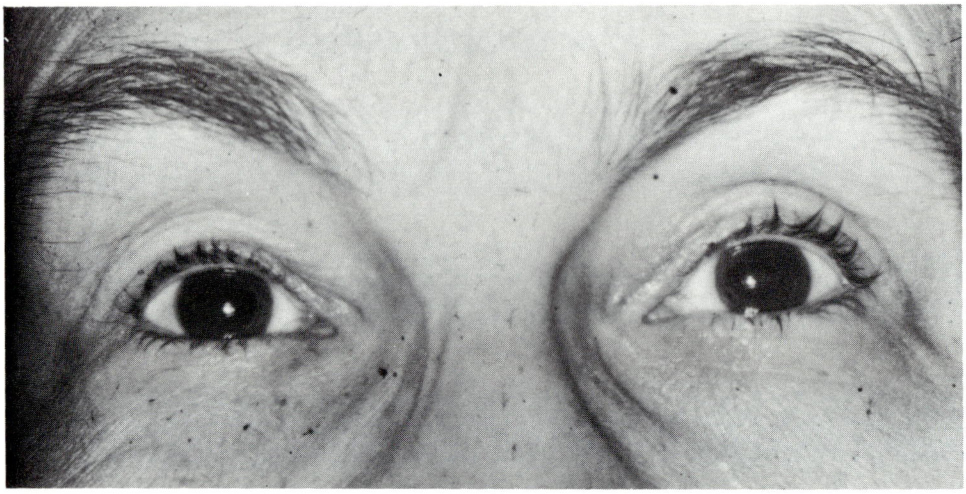

Fig. 10-10. Same patient as in Fig. 10-8. Complete cure of staphylococcal eczema. At this time bacterial cultures and skin tests were negative. (From Theodore, F. H., and Schlossman, A.: Ocular allergy, Baltimore, 1958, The Williams & Wilkins Co. © 1958, The Williams & Wilkins Co., Baltimore.)

staphylococcal eczema of the eyelids are treated unsuccessfully for years as instances of contact allergy. One cannot distinguish between the two conditions on the basis of the character of the dermatitis, as they look very much alike. What is required for the diagnosis and successful treatment of staphylococcal eczema is the demonstration that the focal point of the process is not the skin, but instead is the eye and its adnexa. When this is demonstrated and proper treatment instituted, the eczema, which is a secondary complication, gradually disappears. The key to the diagnosis is routine detailed ophthalmologic examination, both clinical and bacterial. This almost always will reveal the basis for the dermatitis, even if the primary focus is obscure, such as a minute abscess of a meibomian gland.

The following findings differentiate staphylococcal eczema of the eyelids from the allergic variety: (1) blepharitis, with scaling, and often ulcers of the eyelid margin; (2) meibomitis, either diffuse or focal; (3) superficial epithelial keratitis involving the inferior half of the cornea, readily seen on slit-lamp examination after staining with fluorescein, considered as pathognomonic of staphylococcal conjunctivitis; (4) strongly positive conjunctival and lid margin cultures, showing many toxin-producing staphylococci, often entirely out of numerical proportion to the objective clinical findings; and, (5) absence of eosinophils in epithelial scrapings, which instead usually reveal neutrophils and staphylococci, especially on the lid margins (Figs. 10-8, 10-9, and 10-10).

Lid margin affections

In long-standing blepharitis, meibomitis, recurrent hordeola, and recurrent chalazion, all conditions in which staphylococcal infection play an important role, secondary microbial allergy often supervenes. Desensitization therapy is an integral and often necessary part of the long-term treatment.

ALLERGY OF THE CORNEA

A large part of corneal disease is allergic in nature.[29] The cornea manifests all recognized forms of allergic reactions, including the atopic, microbial, and contact types. Furthermore, other diseases that occur in allergic individuals, such as atopic dermatitis and possibly rosacea, demonstrate corneal lesions.

It is interesting that some of the fundamental experimental work in ocular allergy was performed with the cornea as a shock organ. As early as 1909 Weekers[49] produced phlyctenules in rabbits sensitized to bovine tuberculosis. During the following 20 years this work was confirmed and expanded by many investigators. Wessely[51] in 1911 was a pioneer in demonstrating that the cornea is capable of reacting anaphylactically to produce parenchymatous keratitis. Thus two of the most important corneal diseases of that era, namely, interstitial keratitis and phlyctenular ophthalmia, were shown to have an allergic etiology. These pioneer investigations have been considerably extended by more modern studies, including the experimental induction of local selective tissue sensitivity of the cornea and the corneal limbus. Immune corneal rings have been produced experimentally by a number of investigators.[41]

The classification of corneal allergy

Similar to allergies of the conjunctiva, eyelids, and other parts of the eye, the various manifestations of corneal allergy can best be understood when they are considered on the basis of the allergic mechanism involved. In fact, this mechanistic approach has particular advantages in regard to the cornea, because different types of allergy may give relatively similar clinical pictures and because the type of reaction depends to a great extent on the degree of hypersensitivity, time of exposure, and other factors that alter the clinical picture. A classification of corneal allergy is presented in the following outline.

Allergy of the cornea

A. Atopic allergy (immediate reaction)
   1. Topical allergens: superficial keratitis (rarely ulceration)
   2. Generalized allergy: superficial and marginal keratitis (rarely deep)

B. Contact allergy (delayed reaction)
   1. Drugs: superficial keratitis, deep ulceration
   2. Other contactants: superficial keratitis, deep keratitis

C. Microbial allergy
   1. Lues, tubercle bacilli, other bacteria: interstitial keratitis
   2. Tubercle bacilli, staphylococci, other bacteria, coccidioides immitis, nematodes: phlyctenular ophthalmia
   3. Staphylococci, bacillary dysentery: marginal infiltrates and ring ulcers
   4. Herpes simplex virus: disciform edema of the corneal stroma
   5. Adenoviruses: subepithelial opacities
   6. Virus of trachoma: keratitis, opacities

D. Keratitis of possible allergic origin
   1. Vernal catarrh
   2. Keratitis associated with atopic dermatitis
   3. Rosacea keratitis
   4. Periarteritis nodosa

E. Allergic reactions in corneal transplants
   1. Interstitial keratitis

F. Autogenous reactions

Atopic reactions of the cornea

Atopic reactions of the cornea are most commonly caused by pollen allergy or by certain ingestants. They are chiefly superficial in nature, although occasionally deep keratitis does occur. Superficial keratitis resulting from atopy is much more common than the reports in the literature seem to indicate. Because of the mildness of atopic corneal reactions, most cases are not referred to the ophthalmologist. Others may be missed by allergists who have no experience in the use of the slit lamp. Also, many atopic reactions are self-limited or improve as a result of the treatment of the general atopy.

CORNEAL REACTIONS TO POLLENS AND DUSTS. Although the conjunctiva is commonly affected by hay fever, corneal involvement is much less frequent and occurs in more severe cases. It is generally confined to superficial keratitis with a few isolated areas that stain with fluorescein. I have also noted mild keratitis in conjunctival tests utilizing pollen extract. Such keratitis is often self-limited and responds readily to the same treatment as the conjunctivitis. This mild lesion presents no problem either diagnostically or therapeutically and should give little cause for concern. More severe reactions may occur from other pollens, such as corn and orris root. In severe generalized hypersensitivity to dusts the cornea may become involved.

CORNEAL REACTIONS TO INGESTANTS. Corneal allergies caused by foods are generally more severe than those caused by inhalants. Although they may at times be superficial, they tend to involve the corneal stroma. Foods that have been incriminated include cottonseed meal, wheat, strawberries, cucumbers, cheese, Coca-Cola, egg white, chocolate, rye, and nuts.

In addition, corneal allergy has been reported from the internal use of a mixture containing caffeine, phenacetin, and antipyrine.

Contact allergy of the cornea

The cornea is often involved in contact allergies of the eye. Usually this is part of an allergic dermatoconjunctivitis, so that the involvement of these tissues points to the correct diagnosis. Since the cornea may show only mild superficial keratitis, its involvement may be easily overlooked when other parts of the eye have a more marked reaction. In some instances the lid and conjunctiva are slightly affected, but the keratitis is extremely severe. On occasion the superficial cornea may show a similar picture to that of atopic allergic keratitis, and differential diagnosis will then depend on history and elimination of the offending allergen.

However, many times the corneal involvement may dominate the scene. Severe contact keratitis may extend into the deeper layers of the cornea and may be accompanied by an iritic condition. Occasionally the process is essentially subepithelial, with no staining of the cornea.

The responsible allergens include a variety of ophthalmic medicaments, such as local anesthetics, mydriatic alkaloids, antibiotics, sulfonamides, mercurials, quaternary ammonium compounds, lanolin, coal tar dyes, orris root, and vegetable products.

Special emphasis in regard to possible severe hyperallergic reactions from the use of some of the newer and better local anesthetics such as proparacaine, and to a lesser degree from benoxinate (Dorsacaine), must be made. I have reported a number of instances of bilateral necrotizing epithelial reactions requiring hospitalization of the patient in some cases because of severe disability from pain and poor vision after the instillation of a single drop.[42] Fortunately, recovery was complete in these acute idiosyncrasies. However, the constant home use of such anesthetics for chronic eye complaints may cause severe deeper corneal allergies that may result in permanent disabilities and visual loss.

Microbiallergic keratitis

Allergic reactions of the cornea to products of bacteria, protozoa, fungi, and helminths and to other similar protein products give a wide variety of corneal lesions. It is interesting that certain bacteria are associated with certain specific corneal changes. The classical picture of interstitial keratitis is caused by the spirochete of syphilis, and phlyctenular keratoconjunctivitis is most commonly traceable to the products of the tubercle bacillus.

INTERSTITIAL KERATITIS. An allergic mechanism appears to offer the best explanation for the occurrence of the two major types of interstitial keratitis encountered clinically; namely, luetic and tuberculous. Evidence for an allergic etiology is based on: (1) the ineffectiveness of specific (antiluetic) therapy, (2) the effectiveness of corticosteroids and foreign protein therapy, (3) the rare onset of interstitial keratitis during the course of systemic arsenic therapy (Jarisch-Herxheimer reactions), and (4) experimental production by allergic techniques of lesions identical in essential details to syphilitic parenchymatous keratitis.

PHLYCTENULAR KERATOCONJUNCTIVITIS. It is now generally accepted that phlyctenular keratoconjunctivitis is a manifestation of endogenous allergy to not only tuberculoprotein[32] but also other bacterial antigens, especially staphylococcal, and rarely to gonococcal and diplobacillary products. In-

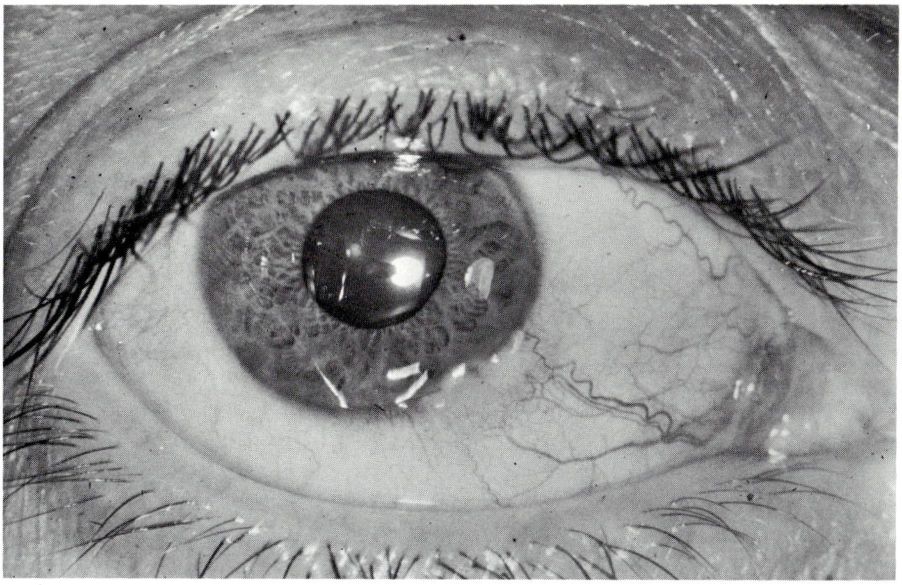

Fig. 10-11. Phlyctenular keratoconjunctivitis complicating ulcerative colitis. Excellent result with topical hydrocortisone. (From Theodore, F. H., and Schlossman, A.: Ocular allergy, Baltimore, 1958, The Williams & Wilkins Co. © 1958, The Williams & Wilkins Co., Baltimore.)

stances have been encountered in patients allergic to coccidioidin and in patients with leishmaniasis, trypanosomiasis, moniliasis, ascariasis, and helminthiasis. Phlyctenules have been produced experimentally by many allergic methods, using various antigens in appropriately sensitized animals. These include tuberculin, staphylococcal products, horse serum, and chemicals, which were instilled into the conjunctival sac. The dramatic response to corticosteroid therapy in all types of phlyctenulosis is further suggestive evidence of the allergic nature of the disease (Fig. 10-11).

MARGINAL ULCERS AND INFILTRATES. Simple marginal or catarrhal ulcers and infiltrates are probably the most common manifestations of corneal allergy. Although it has not been proved conclusively that marginal ulcers in bacterial conjunctivitis are allergic in nature, the weight of evidence favors an allergic etiology, especially as regards the *Staphylococcus;* here the antigen is probably the exotoxin of the bacterium. Marginal ulcers caused by other bacteria, such as the diplobacillus or the Koch-Weeks bacillus, may occur as a result of direct toxic action. In favor of an allergic etiology for such infiltrates and ulcers are the following findings:

1. Marginal ulcers rarely appear early in the course of conjunctivitis caused by staphylococci or diplobacilli. They occur commonly after the infection has been present for a period of time, often after repeated exacerbations.
2. Patients often show evidence of hypersensitivity to bacterial exotoxins through positive reactions to skin tests, or they may have foci of infection in sinuses and tonsils and other remote areas of the body. However, in some cases reactions to the skin tests are only slightly positive.
3. It is difficult to isolate bacteria from scrapings of marginal ulcers.
4. Thygeson[47] claims that he has never observed a catarrhal ulcer caused by the pneumococcus. This is probably because of the fact that no exotoxin has been demonstrated in association with pneumococci.
5. There are many similarities between marginal ulcers and phlyctenules,

which are universally conceded to be allergic manifestations.

6. Marginal ulcers have been observed in atopic and contact allergies, such as tetracaine sensitivity, as well as in vernal catarrh.
7. Marginal ulcers in general respond very well to corticosteroids.

RING ULCERS. Ring ulcers give the appearance of coalescing marginal ulcers but seem to have a different etiology, although they are allergic in nature. Systemic microbial allergy seems to play a role. I have encountered this condition in patients with ulcerative colitis; Thygeson has described it in bacillary dysentery.

DISCIFORM EDEMA OF THE CORNEAL STROMA. Edematous reactions of the cornea of disciform character may arise in the course of various types of keratitis, particularly herpetic, or may appear without any clearly demonstrable cause. The tendency has arisen to group all such instances of corneal edema under the term disciform keratitis. The disadvantage of considering all disciform stromal edema as one entity lies in the fact that, whereas in herpetic keratitis corticosteroids may be contraindicated, in other disciform lesions such therapy may be of value. Beneficial effects in these cases may result from the action of the corticosteroids on the possible allergic component.

Originally disciform keratitis was meant to describe a disklike opacity situated in the corneal stroma, which often assumed twice its usual thickness in the involved area and was associated with descemetitis. Usually the corneal epithelium was relatively intact, and staining with fluorescein revealed little or nothing. Most such cases were attributed to herpes simplex infection on the basis of exposure to viral infections (colds, herpes labialis), often in association with trauma, and it was hypothesized that the virus had somehow entered the stroma through the epithelium. Many times the so-called point of entrance could not be demonstrated. Such cases were of themselves very serious; resultant central opacities, after the edema of the cornea had subsidized, would reduce vision to the extent of industrial blindness. There seems no clearcut evidence, other than circumstantial, that the condition is always caused by herpes simplex. Sometimes herpes zoster and vaccinia seem causative. In still other instances the cause is not demonstrable.

The role of hypersensitivity in the causation of disciform keratitis has been accepted by many, although the evidence is partially hypothetical and inferential. Braley[4] has long been an important exponent of the allergic nature of disciform keratitis. Jones[16] has restated Braley's original concept in a graphic manner. He visualizes the corneal stroma as an immunologic blotter that absorbs antigens arising from virus proliferating in the epithelium. The subsequent arrival of antibody gives rise to a fixed tissue antigen-antibody union resulting in edema or even necrosis.

It has long been my opinion that there are two types of disciform keratitis: (1) the type without clinical evidence of herpetic keratitis and (2) the type occurring in association with demonstrable herpetic ulceration.[46] My hypothesis is that in the first (classical) variety the cause has not been demonstrated and may possibly be allergic in nature; in the second form, where a herpetic ulcer preexists, although allergy possibly operates in the production of the stromal edema, the infectious component of the process predominates in the stroma to the extent that corticosteroids should either not be used at all, or else with caution, concomitantly with idoxuridine. Fig. 10-12 illustrates an interesting example of herpes simplex keratitis in a successful graft.

ADENOVIRUS KERATITIS. A microbiallergic basis for the subepithelial capacities occurring in epidemic keratoconjunctivitis, caused by adenovirus type 8, appears to be a strong possibility. I have observed many patients in whom the opacities, which occur at a time when infectivity is over, clear dramatically with the use of topical steroids only to reappear when the medic-

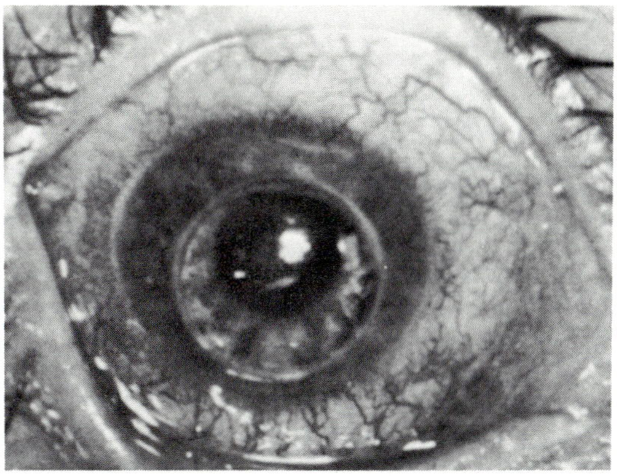

Fig. 10-12. Herpes simplex keratitis of a corneal graft. The patient had a successful keratoplasty performed 1½ years previously because of corneal opacification following disciform keratitis. The present herpetic infection is explained by the fact that the donor material assumes the susceptibilities and characteristics of recipient tissue. (From Theodore, F. H., and Schlossman, A.: Ocular allergy, Baltimore, 1958, The Williams & Wilkins Co. © 1958, The Williams & Wilkins Co., Baltimore.)

ament is discontinued.[41] Some of these patients receiving steroids show the recurrence of such opacities as much as two years after the original virus infection if the drug is stopped completely. For this reason, to avoid a form of steroid "addiction" in the patient, I do not advocate their use unless central opacities seriously bother the patient, either subjectively or visually.

Keratitis of possible allergic origin

Vernal catarrh. Except for the limbal proliferations, corneal lesions of vernal catarrh are an often overlooked feature of the disease. They may be divided into two groups: (1) those occurring as direct extension of the limbic and conjunctival disease and (2) those in which there is no such direct extension. In the first group are included crescentic, raised, fleshy processes and pannuslike formations. The lesions that do not arise from direct extension of the conjunctival process may be toxic or possibly nutritional in origin. The most important of these is superficial epithelial keratitis. Corneal ulcers and dystrophies may occur.[46]

Keratitis associated with atopic dermatitis. Corneal involvement of varying intensity may complicate chronic atopic dermatitis. According to Hogan,[15] the corneal affection may occur simultaneously with or follow repeated exacerbations of conjunctivitis. The superficial third of the periphery of the cornea is affected first. The corneal stroma near Bowman's membrane becomes clouded. The lesion gradually spreads deeper into the stroma, and after some time blood vessels push in from the limbus. The corneal epithelium covering the area of keratitis becomes edematous and shows minute punctate staining with fluorescein. In some instances the entire cornea becomes hazy and vascularized, with resultant diminished visual acuity.

Two other types of corneal lesion have been described in association with atopic dermatitis. Pillat[23] reported a case with superficial punctate keratitis. In rare instances keratoconus is associated with atopic dermatitis. Patients with this corneal condition manifest severe and extensive dermatitis with lichenification and pigmentation, severe bronchial asthma, and reactions to food and inhalant allergens. It is possible that by carefully studying such cases, ophthalmologists will gain better insight into the pathogenesis of keratoconus. Keratoconus is also a complication of vernal catarrh. Steroid therapy, both local and systemic, is the indicated form of treatment.

Rosacea keratitis. The etiology of rosacea is obscure. Although a few observers maintain that it is an allergy, there is little definitive evidence to substantiate this theory.

Keratitis in periarteritis nodosa. Keratitis has been described as a complication of periarteritis nodosa.

Allergic reactions in corneal transplants

It is probable that allergy plays a role in the clouding of corneal grafts, occurring from the tenth day to the sixth week after operation. Castroviejo[8] believes that this corneal opacification is the result of uveitis, an allergic response to foci of infection harbored in the nose or the throat. Maumenee,[18] on the other hand, showed experimentally that the recipient can become sensitized to donor material following corneal transplants, and he concluded that this allergic reaction is in large measure responsible for the clouding of grafts after the tenth day.

In a later paper Maumenee[19] stressed the importance of the homograft allergic reaction as the most important single cause of delayed opacification of the graft in an eye that has been properly selected for corneal transplantation. This reaction is characterized by progressive edema of the stroma of the graft, which eventually may become thickened and vascularized or may clear spontaneously if the reaction first occurs many months after transplantation. Evidence that this reaction is to a homograft may be summarized as follows: (1) the response is localized to the graft; (2) it occurs much more frequently in heavily vascularized than in avascular corneas; (3) a similar clouding can be experimentally produced in corneal grafts of rabbits by increasing the dose of homograft antigen by transplanting skin from the same donor to the recipient; (4) corneal stroma and endothelium are capable of eliciting the homograft reaction; and (5) penetrating autografts of the entire cornea in man do not develop a delayed opacification. It is reasonable to assume that the homograft reaction may also occur in lamellar transplants, although there is less evidence in this regard.

ALLERGY OF THE SCLERA

The widespread use of antiinfective agents and the control of syphilis and tuberculosis has resulted in the lessening importance of infection in the causation of scleritis. Thus, although scleritis in itself is a relatively infrequent disease, the percentage of scleral inflammations of probable allergic origin is very high—perhaps higher than in any other ocular structure. However, because the sclera, unlike the conjunctiva, cornea, and uvea, does not lend itself so readily to the experimental investigation of allergic phenomena, it has been largely ignored in this respect. We are therefore forced to depend chiefly on clinical observations for the evaluation of the various allergic mechanisms that appear to operate in the production of scleritis. These indicate, first, that allergy is indeed an important etiologic mechanism and, second, that many instances of scleritis are best explained on the basis of microbial allergy. The outline below offers a classification of allergic reactions of the sclera.

Allergy of the sclera

- A. Atopic allergy (immediate reaction)
  1. Inhalants: episcleritis
  2. Food allergy: episcleritis
- B. Microbial allergy (delayed reaction)
  1. *Streptococcus, Staphylococcus:* episcleritis, scleritis
  2. Tubercle bacillus: scleritis, sclerosing keratitis
  3. Other microorganisms: episcleritis, scleritis
- C. Collagen diseases (mechanism unknown)
  1. Gout, rheumatic fever, rheumatoid arthritis, periarteritis nodosa: scleritis, episcleritis

Atopic reactions

Most atopic allergic reactions causing scleritis arise from foods, generally fish, and especially seafood. Inhalant allergies are rare but may occur from ragweed.

Microbiallergic reactions

If we consider certain collagen diseases, such as rheumatoid arthritis, to result from bacterial (streptococcal) allergy, then it would be safe to say that most cases of scleritis have a microbiallergic origin. Clinical and confirmative laboratory experiences indicate that such allergy is in general evoked chiefly by three organisms: *Streptococcus, Staphylococcus,* and the tubercle bacillus. However, the well-known relationship between scleritis and what was originally considered the focus of infection, but which I prefer to call an antigenic focus, indicates that although these organisms dominate the picture, many others, such as enteric bacilli and perhaps even parasites, may be incriminated, although the evidence here is less clearcut.

Scleritis in collagen diseases

Gout, which has allergic facets, often is accompanied by scleritis. Scleritis likewise occurs in periarteritis nodosa, a collagen disease in which allergy appears to play an important role.

Tenonitis

Except for the rare instance of tenonitis arising from direct infection, such as that complicating trauma or extraocular muscle surgery, the obscure etiology of tenonitis appears best explained in certain instances on the basis of allergy. The rarity of the disease makes etiologic evaluation difficult. Its exudative character, its transient, recurrent, yet benign course, its association with collagen disease, its relationship and similarity to scleritis, and its response to corticosteroids and foreign protein therapy all suggest that an allergic component may play a role in its causation. However, we know of no known instance where tenonitis has arisen on the basis of atopy, such as food allergy. That microbial allergy may operate is suggested by its association with rheumatoid arthritis and infective antigenic foci. In this regard it is possible that the chemosis and other ocular findings in trichiniasis are caused by tenonitis arising secondary to muscle infestation, either as toxic or allergic response to products of trichinella.

ALLERGY OF THE UVEA

There is considerable evidence, based upon anatomic, experimental, and clinical data, that allergy plays an important role in the causation of endogenous uveitis.[11,41] Anatomically, the extremely vascular uvea offers a particularly favorable site for the concentration of proteins, antigens, and antibodies originating elsewhere in the body. Furthermore, it is known that local antibody formation can and does occur in the uvea itself. Since the uvea thus affords such an ideal shock organ for the study of allergic reactions, it is not surprising that investigators in the field of immunology have utilized this structure experimentally in the production of allergic reactions ever since the introduction of the concept of anaphylaxis. As a result, much of the basic knowledge of hypersensitivity in general stems from such experiments. Allergic uveitis has been induced in laboratory animals by means of horse, cat, and other foreign sera, egg albumin, various microbial products, and by tissue proteins, either alone or in combination with Freund's adjuvant (heat-killed tubercle bacilli in mineral oil).

These observations show that all forms of allergic mechanisms operate in the production of uveal inflammation. Work by Silverstein and Zimmerman[30] also confirms this. The authors indicate that endophthalmitis may be produced in the guinea pig by two distinct immunologic mechanisms: (1) delayed hypersensitivity and (2) antigen interaction with circulating antibody. By varying the method of sensitization, the route of administration, and the dose of antigen, the several immunologic mechanisms may be stimulated to produce ocular lesions. According to these authors, not only may acute nongranulomatous uveitis caused by a delayed hypersensitivity result from microbiallergic processes involving bacterial antigens, but it may also arise as a consequence of the tuberculin type of hypersensitivity to simple bland protein antigens, such as crystalline egg albumin or bovine serum albumin.

More recent contributions relating hypersensitivity to the etiology of uveitis, offering additional detailed evidence, include those by Aronson and his associates,[1] Böke,[3] Campinchi and his associates,[7] Coles,[9] Maumenee and Silverstein,[20] Remky,[27] Schlaegel,[28] Wittmer,[52] and Woods.[53]

The following classification is useful for understanding the clinical manifestations of uveal hypersensitivity.

Allergy of the uvea

A. Anaphylactic and atopic reactions: uveitis from sera, pollens, animal proteins, foods, drugs, angioneurotic edema
B. Microbiallergic reactions
   1. Nongranulomatous uveitis (allergy apparently the main factor): *Streptococcus, M. tuberculosis, Staphylococcus, Gonococcus, Pneumococcus*
   2. Granulomatous uveitis (allergy not primary factor but may alter the nature of the inflammatory response): tuberculosis, syphilis, toxoplasmosis, brucellosis, helminthiasis
   3. Viral uveitis: herpes simplex, herpes zoster, poxviruses, myxoviruses, arboviruses, others
   4. Mycotic uveitis: *Candida,* histoplasmosis
C. Autogeneous allergic reactions
   1. Sympathetic ophthalmia (allergy to autogenous uveal pigment)
   2. Endophthalmitis phacoanaphylactica (allergy to autogenous lens protein)

Atopic reactions

Anaphylactic and atopic reactions of the uvea, although rare, offer the best clinical corroboration of the fact that the uvea may participate in generalized allergic reactions. A prime example of this is uveitis complicating serum sickness, as originally described by Theodore and Lewson[45] and later described by others.[10] Other instances of atopic allergic uveitis may result from hypersensitivity to ragweed, house dust, cat dander, and other inhalants, and to foods. In this category allergy plays the primary role in the uveal inflammation.

Microbiallergic reactions

Although the terms anterior and posterior uveitis and panuveitis have much in their favor, the classification most widely used, that of nongranulomatous and granulomatous types, based on histopathologic studies, is a convenient tool. Microbiallergic reactions of the uvea may occur in both nongranulomatous and granulomatous uveitis. In the anterior nongranulomatous variety, where no direct bacterial infection can be demonstrated, allergy appears to be the primary factor. Even in those granulomatous form of uveitis where microorganisms can be demonstrated, a secondary microbial allergy often alters and aggravates the inflammatory state.

Although specific hypersensitivity to any of the bacteria listed in the outline may cause anterior nongranulomatous uveitis, allergy to *Streptococcus* organisms is most commonly implicated. However, studies have failed to establish a correlation between elevated antistreptolysin-O titers and uveitis. Furthermore, skin tests with bacterial vaccines may be nonspecific, and convincing proof of the value of desensitization in a disease as variable and unpredictable as uveitis is not yet at hand.

All types of tuberculous uveitis have allergic components. Anterior uveitis may occur as a manifestation of tubercle bacillus hypersensitivity in the absence of direct infection, just as it is known to occur in phlyctenular keratoconjunctivitis and scleritis. Even in actual tuberculous infection of the uvea, as in posterior granulomatous uveitis, a secondary allergic reaction may dominate the clinical picture. Exacerbations of uveitis following testing or therapy with tuberculin are known clinically. Moreover, it has been found that destruction of bacilli by means of chemotherapy can occasionally liberate sufficient antigen to cause a focal allergic response.

In ocular toxoplasmosis both local and systemic sensitization to the toxoplasma occurs at the time that the organism forms pseudocysts in the retina. Generalized sensitization is manifested by a positive dye titer. When years later pseudocysts rupture, the hypersensitive tissues react with the liberated organism to produce an allergic reaction clinically manifested as posterior uveitis.

Autogenous reactions

Long before the concept of autoallergy or sensitization of the host to his own tissues was utilized to explain certain systemic diseases, the ophthalmologist had recognized that two unique types of autogenous allergy occur in the eye; namely, phacoanaphylactic uveitis and sympathetic ophthalmia.

Phacoanaphylactic uveitis. Phacoanaphylactic uveitis occurs as the result of a severe allergic reaction in persons sensitized to lens protein, which has entered the anterior chamber as the result of surgery, trauma, or the altered permeability of the diseased lens capsule. Although endophthalmitis phacoanaphylactica clinically appears to be a definite allergic reaction, the fact that its experimental production in animals has required adjuvants besides lens protein, has led some investigators to believe that the mechanism is toxic rather than allergic. However, clinical experiences with the use of fish lens extract in the treatment of cataracts in human beings offers excellent experimental confirmation of two facts: (1) the allergic basis for endophthalmitis phacoanaphylactica and (2) the organ specificity, rather than species specificity, of allergy to lens protein. In three reports, uveitis was induced by repeated injections of fish lens protein and cured by intracapsular cataract extraction. [5,25,26]

Sympathetic uveitis. Analysis of what is known experimentally and clinically about sympathetic ophthalmia leads to the conclusion that the most acceptable pathogenesis of this entity is some form of hypersensitivity. The allergic reaction apparently requires special exciting factors, some of which are still unknown. However, uveal pigment appears to be the basic exciting antigen, as first shown by Elschnig in 1910. Woods[53] has demonstrated a specific antibody to uveal pigment in human beings after perforating ocular wounds, demonstrable on intradermal skin testing. Friedenwald,[14] utilizing histologic studies on the skin of persons who had received such intradermal injections of suspensions of uveal pigment, felt that the uveal pigment granules acted as a specific antigen and that the disease was primarily allergic in nature.

The concept that the allergic reaction resulting in sympathetic ophthalmia is somewhat more complicated, requiring special circumstances for its development, is suggested by recent work. Collins[12] found that in order to produce the picture in in guinea pigs it was necessary to employ Freund's adjuvant in addition to uveal pigment. Naquin,[22] however, was unable to duplicate this work. Blodi[2] recently analyzed a total of 170 eyes with sympathetic ophthalmia and found that 39, or about 23%, showed evidence of a phacoanaphylactic reaction. He felt that this association was higher than could be expected by chance coincidence and that their simultaneous occurrence lent additional weight to the concept of an allergic pathogenesis in sympathetic uveitis.

The concept of autoallergy, or allergy of the host to his own tissues, although still largely an experimental one, has other provocative applications in the field of uveitis. It has been found that in producing experimental allergic encephalitis, in which animals formed antibodies to their own central nervous system myelin following single injections of myelin extract, three quarters of the animals developed a nongranulomatous uveitis limited to the anterior segment.[6] Fernando,[13] injecting bovine albumin tagged with radioactive iodine into the vitreous of nonsensitized rabbits, found that an allergic uveitis developed on the basis of the original injection. No second, or "shocking," dose of antigen was necessary. His autoradiographs demonstrated that the antigen accumulated most intensely in the ciliary body and lens, structures rich in mucopolysaccharides. This association of antigen with tissue polysaccharide suggested to him the possible formation of a polysaccharide-antigen complex that resulted in autosensitization.

These and other experiments indicate that the mechanisms of uveal allergy are even more complex than previously presumed and that there are probably many avenues in which hypersensitivity operates in uveitis. Autoallergy may some day serve to explain many now obscure uveal reactions. Moreover, the hypothesis of autogenous hypersensitivity resulting from a microbial antigen may clarify much that is lacking for our full acceptance and understanding of microbiallergic uveitis.

ALLERGY OF THE RETINA AND OPTIC NERVE

The optic nerve and retina are frequently the site of reactions resulting from exposure to various types of noxious agents. Although most of such phenomena are obviously toxic in nature, often definite allergic manifestations appear to occur. In general, allergic reactions of the retina and optic nerve appear to follow the usual lines of allergy seen elsewhere. With the exception of contact allergy, which obviously cannot occur, anaphylactic, atopic, and microbiallergic reactions have been noted.

In evaluating reactions occurring in the retina and optic nerve, one must make a distinction between allergy and hyperreactivity, especially in regard to drugs. Everyone accepts the fact that when poisons are taken in large enough or toxic doses, tissue death will result. It is generally not realized, however, that in certain hyperreactive individuals minute amounts of the same drug may cause a similar reaction. It should be clear that this is not allergy, because the person so affected reacts in a manner only quantitatively different from that of the average person. For example, most persons will react to large doses of atropine taken internally, with the classic pharmacologic and toxic actions of the drug, such as vomiting, tachycardia, fever, and dilatation of the pupil. When, however, an individual reacts in the same manner with the same symptoms if we inject a very small amount of atropine, or reacts similarly after its local instillation into the eye, this is in a sense a toxic effect following a minute dose, or hyperreactivity. In other words, the only difference in the reaction here is a quantitative one as regards dosage. When, however, a person is allergic or hypersensitive to a drug, he responds with entirely different symptoms, which are essentially allergic in character. These are asthma, rhinitis, urticaria, skin eruptions, and other accepted allergic phenomena. These are not the pharmacologic effects of the drug. Moreover, when such reactions occur, they are the same no matter what particular drug the patient has used. Other examples of hyperreactivity that are important to the ophthalmologist occur in quinine amblyopia, reactions with minute amounts of nicotine, or following the use of atropine and cocaine. Marked ciliary spasms from extremely small amounts of pilocarpine and other miotics should also be noted.

Toxic and hyperreactive affections of the retina and optic nerve reveal little immediate objective evidence of tissue involvement, so that diagnosis depends on subjective findings, especially those of central vision and visual fields. In most instances of neuroretinal allergy, however, reactions are exudative and hemorrhagic, so that they are readily apparent on ophthalmoscopic examination. The only major exceptions appear to be those rare cases of retrobulbar neuritis arising on an allergic basis.

The outline below classifies all known instances of neuroretinal allergy.

Neuroretinal allergy

- A. Anaphylactic and atopic reactions
  1. Serum
  2. Food
  3. Pollen
  4. Drug
  5. Angioneurotic edema and urticaria
- B. Microbiallergic reactions
  1. Tuberculin
  2. Other organisms
- C. Other reactions of possible allergic origin
  1. Eye changes in periarteritis nodosa
  2. Central angiospastic retinopathy

Atopic reactions

Most instances of proved neuroretinal allergy have occurred in the course of serum sickness. Now that antisera are used less frequently, such occurrences are encountered far less often. Food allergy resulting in neuroretinitis has been reported to occur from peanuts, fish, chocolate, milk, eggs, and Coca-Cola. Pollens such as primrose and ragweed have been incriminated.

Sulfa drugs, penicillin, and procaine have been observed to give rise to allergic retinal reactions. Generalized angioedema caused by a variety of agents has resulted in retinal exudative lesions. Instances of retrobulbar neuritis occurring on the basis of atopic allergic reactions, usually caused by foods, have also been reported.

Microbiallergic reactions

Even though instances of microbiallergic neuroretinitis are rare, a number of rather convincing cases have been reported. Weizenblatt[50] has described such a reaction occurring on two occasions in a patient with marked cutaneous sensitivity to old tuberculin, 5 to 7 days after skin testing with this antigen. Other reports describe similar exudative retinitis from tuberculin injections and in trypanosomiasis.[46] Retrobulbar neuritis has resulted from prophylactic vaccination against rabies.

Other reactions of possible allergic origin

Periarteritis nodosa, a disease in which hypersensitivity mechanisms are believed to play an important role, often causes neuroretinal lesions. Central angiospastic retinopathy must also be included in this category.

Retinal detachment

Some rare instances of detachment of the retina occurring on an allergic basis have been reported. Unfortunately, in almost every instance only presumptive evidence was available to make such a diagnosis.

REFERENCES

1. Aronson, S. B., Gamble, C. N., Goodner, E. K., and O'Connor, G. R.: Clinical methods in uveitis, St. Louis, 1968, The C. V. Mosby Co.
2. Blodi, F. C.: Sympathetic uveitis as an allergic phenomenon, presented at the special program of the American Society of Ophthalmologic and Otorhinologic Allergy, October 19, 1958.
3. Böke, W.: Uveitis and allergy. In Remky, H., editor: The uveal tract and its endogenous inflammations, vol. 5, no. 3, Boston, 1965, Little, Brown and Co., pp. 755-788.
4. Braley, A. E.: Experimental herpes simplex, Amer. J. Ophthal. **35**:1737, 1952.
5. Breinin, G. M.: Danger of fish lens protein injection as treatment for cataract, J.A.M.A. **152**:698, 1953.
6. Bullington, S. J., and Waksman, B. H.: Uveitis in rabbits with experimental allergic encephalomyelitis, Arch. Ophthal. **59**:435, 1953.
7. Campinchi, R., Faure, J. P., Bloch-Michel, E., and Haut, J.: L'uvéite—phénomènes immunologiques et allergiques, Paris, 1970, Masson et Cie.
8. Castroviejo, R.: Ocular allergy, cornea and lens, Trans. Amer. Acad. Ophthal. Otolaryng. **54**:242, 1952.
9. Coles, R. S.: Uveitis. In Sorsby, A., editor: Modern ophthalmology, vol. 4, Washington, 1964, Butterworth's, p. 636.
10. Coles, R. S., and Shapiro, A.: L'uvéite biliterale anterieure compliquant la maladie du serum, Bull. Soc. Franc. Ophtal., pp. 67-70, 1971.
11. Coles, R. S., and Theodore, F. H.: Clinical aspects of uveal hypersensitivity, Arch. Ophthal. **62**:223, 1959.
12. Collins, R.: Further experimental studies on on sympathetic ophthalmia, Amer. J. Ophthal. **36**:150, 1952.
13. Fernando, A. N.: Immunological studies with I^{131} labeled antigen in experimental uveitis. Thesis presented at the New York University Post-Graduate Medical School, 1959.
14. Friedenwald, J. S.: Notes on allergic theory of sympathetic ophthalmia, Amer. J. Ophthal. **17**:1008, 1934.
15. Hogan, M. J.: Atopic keratoconjunctivitis, Amer. J. Ophthal. **36**:937, 1953.
16. Jones, R.: The management of ocular herpes, Trans. Ophthal. Soc. U.K. **79**:425, 1959.
17. Locatcher-Khorazo, D., and Gutierrez, E.: Bacteriophage typing of Staphylococcus aureus: a study of normal, infected eyes and environment, Arch. Ophthal. **63**:774, 1960.
18. Maumenee, A. E.: The influence of donor-recipient sensitization on corneal grafts, Amer. J. Ophthal. **34**:142, 1951.
19. Maumenee, A. E.: Biological responses to corneal homografts, Trans. Amer. Acad. Ophthal. Otolaryng. **64**:765, 1960.
20. Maumenee, A. E., and Silverstein, A. M.: Immunopathology of uveitis, Baltimore, 1964, The Williams and Wilkins Co.
21. Morault, S.: Enquète allergologique et traitement par desensibilisation spécifique dans les conjonctivites allergiques. Thesis presented for the degree of Doctor of Medicine to the Faculté de Médecine de Paris, July 6, 1959.
22. Naquin, H. A.: An unsuccessful attempt to

produce hypersensitivity to uveal tissue in guinea pigs, Amer. J. Ophthal. **39**:196, 1955.
23. Pillat, A.: Zur Mitteilung der Hornhaut beim Ekzem, Wien. Klin. Wschr. **50**:768, 1937.
24. Portier, P., and Richet, C.: De l'action anaphylactique de certains venins, Compt. Rend. Soc. Biol. **54**:170, 1902.
25. Posner, A.: Endophthalmitis phacoanaphylactica induced by injections of lens protein: notes on ophthalmology, Eye Ear Nose Throat Monthly **32**:42, 1953.
26. Posner, A.: Complications following injection of fish lens protein for cataract, J.A.M.A. **151**:317, 1953.
27. Remky, H.: Etiologic diagnosis of endogenous uveitis. In Remky, H., editor: The uveal tract and its endogenous inflammations, vol. 5, no. 3, Boston, 1965, Little, Brown and Co., pp. 789-822.
28. Schlaegel, T. F., Jr.: Essentials of uveitis, Boston, 1969, Little, Brown and Co.
29. Schlossman, A. X., and Theodore, F. H.: The classification of allergies of the cornea, Acta XVIII International Congress of Ophthalmology, Brussels, 1958, vol. 2, Brussels, 1959, Imprimerie Medicale et Scientifique, S. A., p. 1136.
30. Silverstein, A. M., and Zimmerman, L. E.: Immunologic endophthalmitis produced in the guinea pig by different pathogenic mechanisms, Amer. J. Ophthal. **48**:435, 1959.
31. Smith, T.: As cited by Chase, W. W.: The allergic state. In Dubos, R. J., editor: Bacterial and mycotic infections of man, Philadelphia, 1948, J. B. Lippincott Co., p. 113.
32. Sorsby, A. P.: The aetiology of phyctenular ophthalmia, Brit. J. Ophthal. **26**:159, 189, 1942.
33. Theodore, F. H.: The classification and treatment of allergies of the conjunctiva, Amer. J. Ophthal. **36**:1689, 1953.
34. Theodore, F. H.: Drug sensitivities and irritations of the conjunctiva, J.A.M.A. **151**:25, 1953.
35. Theodore, F. H.: Differentiation and treatment of eczemas of the eyelids, Trans. Amer. Acad. Ophthal. Otolaryng. **58**:708, 1954.
36. Theodore, F. H.: Ocular eczema: its classification and treatment, J. Mount Sinai Hosp. N.Y. **21**:255, 1955.
37. Theodore, F. H.: Staphylococcal eczema, Acta XVII International Congress of Ophthalmology, New York, 1954, vol. 2, Toronto, 1955, University of Toronto Press, p. 609.
38. Theodore, F. H.: Eczema of the eyelids, Ann. Allerg. **14**:484, 1956.
39. Theodore, F. H.: Bacterial allergy of the eye, Trans. Amer. Acad. Ophthal. Otolaryng. **65**:184, 1961.
40. Theodore, F. H.: Drug irritations from miotics, Eye Ear Nose Throat Monthly **42**:74, 1963.
41. Theodore, F. H.: Allergic keratitis. In King, J. H., and McTigue, J. W., editors: The Cornea World Congress, Washington, 1965, Butterworth's, p. 197.
42. Theodore, F. H.: Idiosyncratic reactions of the cornea from proparacaine, Eye Ear Nose Throat Monthly **47**:66, 1968.
43. Theodore, F. H., and Feinstein, R. R.: Practical suggestions for the preparation and maintenance of sterile ophthalmic solutions, Amer. J. Ophthal. **35**:656, 1952.
44. Theodore, F. H., and Feinstein, R. R.: Preparation and maintenance of sterile ophthalmic solutions, J.A.M.A. **152**:1631, 1953.
45. Theodore, F. H., and Lewson, A. C.: Bilateral iritis complicating serum sickness, Arch. Ophthal. **21**:82, 1939.
46. Theodore, F. H., and Schlossman, A.: Ocular allergy, Baltimore, 1958, The Williams and Wilkins Co.
47. Thygeson, P.: Marginal corneal infiltrates and ulcers, Trans. Amer. Acad. Ophthal. Otolaryng. **51**:198, 1946.
48. von Pirquet, C.: Allergie, Munchen. Med. Wschr. **53**:1457, 1906.
49. Weekers, L.: Nouvelle pathogénie des phlyctènes, Bull. Acad. Roy. Med. Belg. **23**:577, 1909.
50. Weizenblatt, S.: Allergic ocular reactions to tuberculin (bilateral cyclitis and neuroretinitis), Arch. Ophthal. **41**:436, 1949.
51. Wessely, K.: Ueber anaphylaktische Erscheinungen an der Hornhaut (experimentelle Erzeugung einer parenchymatoson Keratitis durch artfremdes Serum), Munchen. Med. Wschr. **59**:1713, 1911.
52. Witmer, R.: Klinik und Diagnostik der Uveitis in heuter Sicht, Klin. Mbl. Augenheilk. **138**:609, 1961.
53. Woods, A. C.: Endogenous inflammations of the uveal tract, Baltimore, 1961, The Williams and Wilkins Co.

11

Ocular toxoplasmosis

G. RICHARD O'CONNOR

Within the past 25 years *Toxoplasma gondii* has come to be considered one of the most important causes of retinochoroiditis. This intracellular protozoan parasite, measuring approximately 2.5 × 6.5 μ, belongs to the class Sporozoa; and as such it can be thought of as a "second cousin" to the more familiar malarial parasite.

The organism has two distinct morphological forms. In its invasive or proliferative form it is crescentic in shape (Fig. 11-1). The more sharply pointed end of this crescent presents a conoid or primitive mouth (Fig. 11-2) by means of which the organism presumably attaches itself to the host cell and invades it. *Toxoplasma* has a very distinctive type of movement in liquid media. Bringmann and Holz[4] believe that it achieves its gliding motion by periodic contraction of a network of fibrillary organelles (toxonemes) (Fig. 11-2), which originate from a ring at the base of the conoid and fan out into the cytoplasm. By electron microscopic studies Ludvik[26] has shown that the toxonemes extend only one fifth of the distance from the anterior (conoid) end of the organism to its posterior pole. There are, however, certain rodlike organelles that extend in a zig-zag fashion from one end of the organism to the other. Hogan and his co-workers[15] have confirmed the presence of rodlike contractile organelles, and they have also shown that the cell wall is made up of several discrete morphologic layers.

Until recently it was thought that the organism multiplied by simple binary (longitudinal) fission. Goldman and his associates[12] have shown, however, that the actual process of cell division is preceded by internal budding, or "endodyogeny." A pair of buds appears on the anterior surface of the parent nucleus in the earliest stages of reproduction. These buds then develop into the miniature conoids (anterior cones) of the daughter cells and seem to be the organizing influence for division of the cytoplasm (Fig. 11-3). The organism is not able to survive outside its host cell for more than a few hours and is totally unable to withstand certain elements of the external environment, such as heat, drying, or changes in osmotic concentration. For this reason the exact mode of its transmission from one host to another has always remained somewhat of a mystery.

A second, more resistant form of the organism has the appearance of a cyst. After invasion of the host cell, the organism may reproduce itself 25 to 100 times by repeated division (Fig. 11-4); however, the resulting daughter cells may remain inactive as a rounded mass within the host cell, (Fig. 11-5). A process of cyst formation may then take place in which the nucleus of the parasitized cell generally becomes pushed to one side. Until recently it was thought that the host cell membrane became thickened into a cyst wall. Because the host provided the material of the cyst

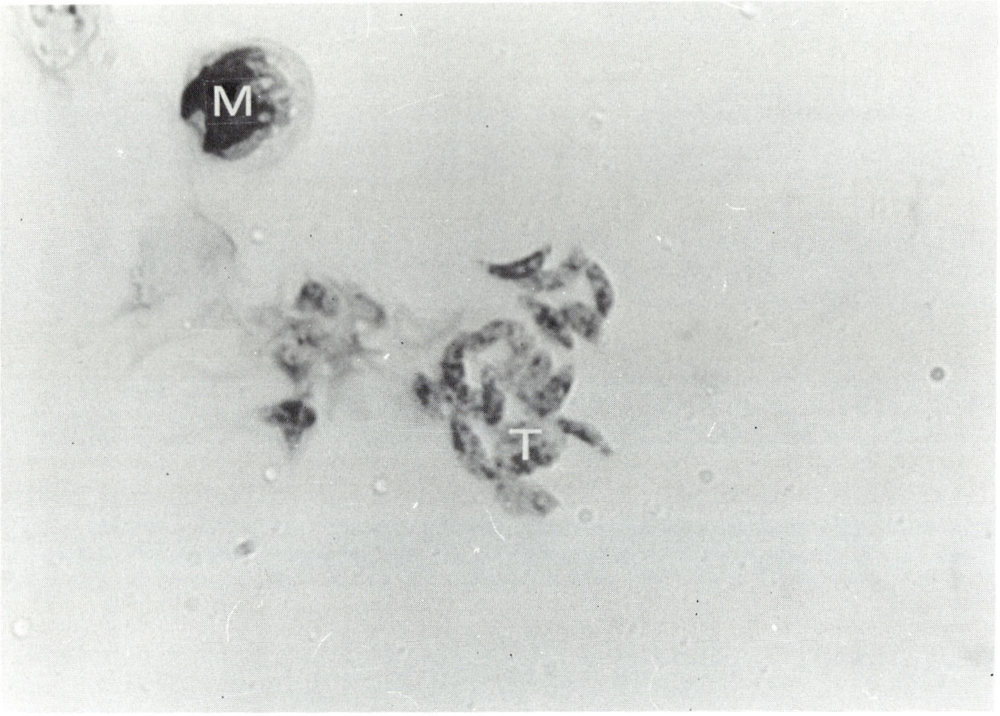

Fig. 11-1. *Toxoplasma* organisms in a Giemsa-stained smear of peritoneal exudate from an infected mouse. *M,* Macrophage; *T, Toxoplasma* organisms. (×1,000.)

wall, the cyst was generally called a pseudocyst. Hogan and his co-workers[16] believe that the preferred term is cyst, since the parasites themselves can be shown to contribute to the cyst wall. Whatever the fine points of this argument may be, the older term pseudocyst is still used by some authors; and since the work of some of these is quoted in this chapter, the reader will encounter both terms.

Although the cysts (Fig. 11-6) are nonchemotactic, they are prone to periodic rupture in the tissues. Frenkel and Jacobs[9] state that organisms liberated from the cysts are generally destroyed in the ensuing inflammatory reaction. If, however, the degree of immunity of the host has declined and a sufficient level of antibody is not available in the tissues at the site of rupture, the liberated organisms appear to be able to invade new cells and proliferate in them. It is apparently the recurrent multiplication of organisms in the retina that accounts for the chronic, progressively destructive nature of the infection.

Although only one species of *Toxoplasma* is known, many strains of widely differing virulence have been isolated. The classical RH strain, for example, has a high degree of virulence for man, mice, and rabbits. It is generally lethal to mice within 3 to 5 days after inoculation. Strain JQ, on the other hand, is relatively avirulent for mice and tends to produce chronic infections of long duration.

Much of what is now known about virulence has been provided by tissue-culture studies. The organism does not seem to be at all fastidious about the type of cell on which it grows. Chorioallantoic membranes of chick embryos have long been used as a culture medium. So, too, have HeLa cells, derived from human cervical carcinoma. Although it has long been thought that high degrees of virulence are correlated with both an increased proportion of cells in-

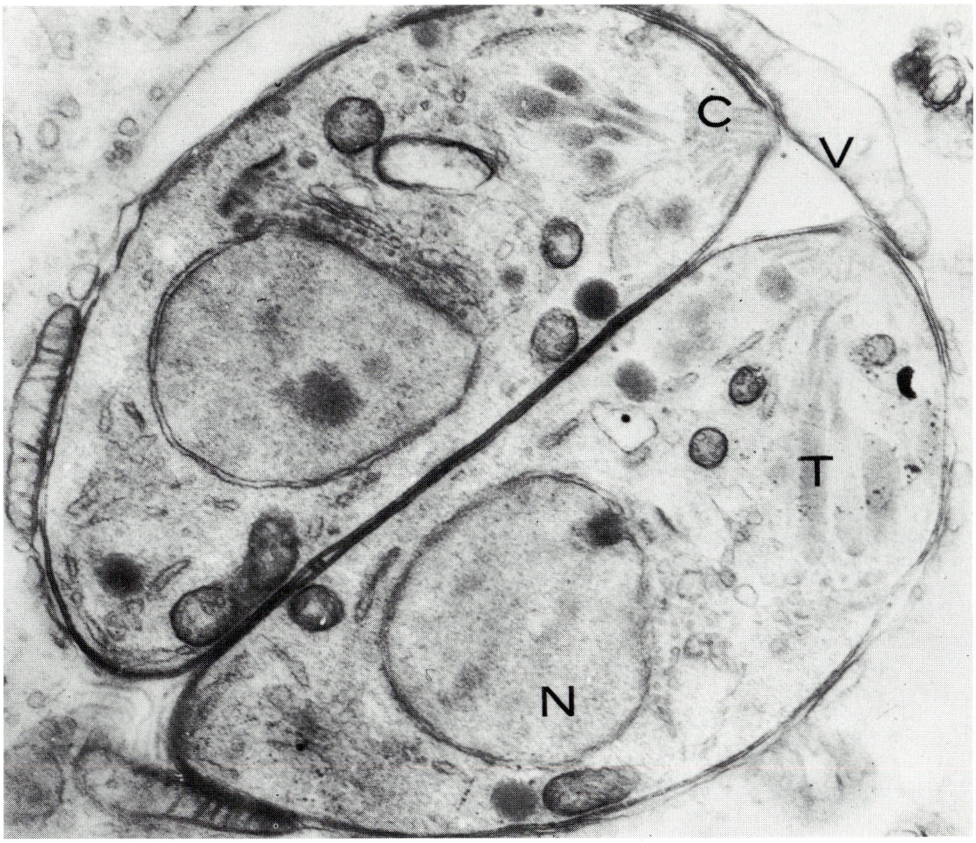

Fig. 11-2. An electron micrograph of two recently divided *Toxoplasma* organisms lying within a vacuole of host cell cystoplasm. *C*, Conoid; *V*, wall of cytoplasmic vacuole; *T*, toxoneme; *N*, nucleus. (×24,000.)

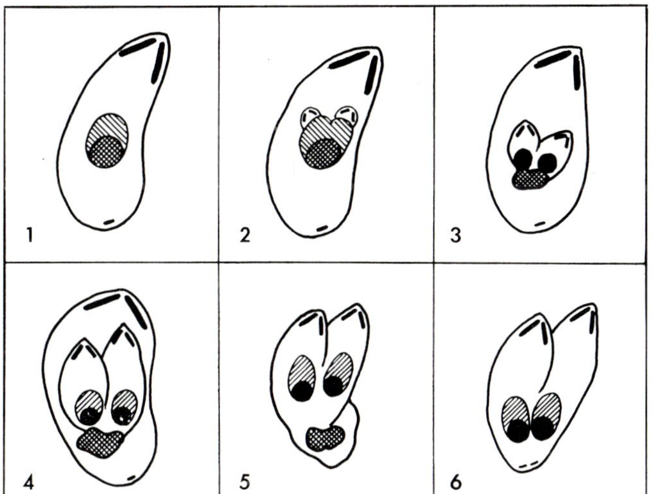

Fig. 11-3. Diagrammatic representation of division by internal budding (endodyogeny). Two polar bodies resembling the conoid of the parent cell appear at the anterior pole of the nucleus. Subsequent stages of division are apparently organized by them. (From Goldman, M., Carver, R. K., and Sulzer, A. J.: J. Parasit. **44**:161, 1958.)

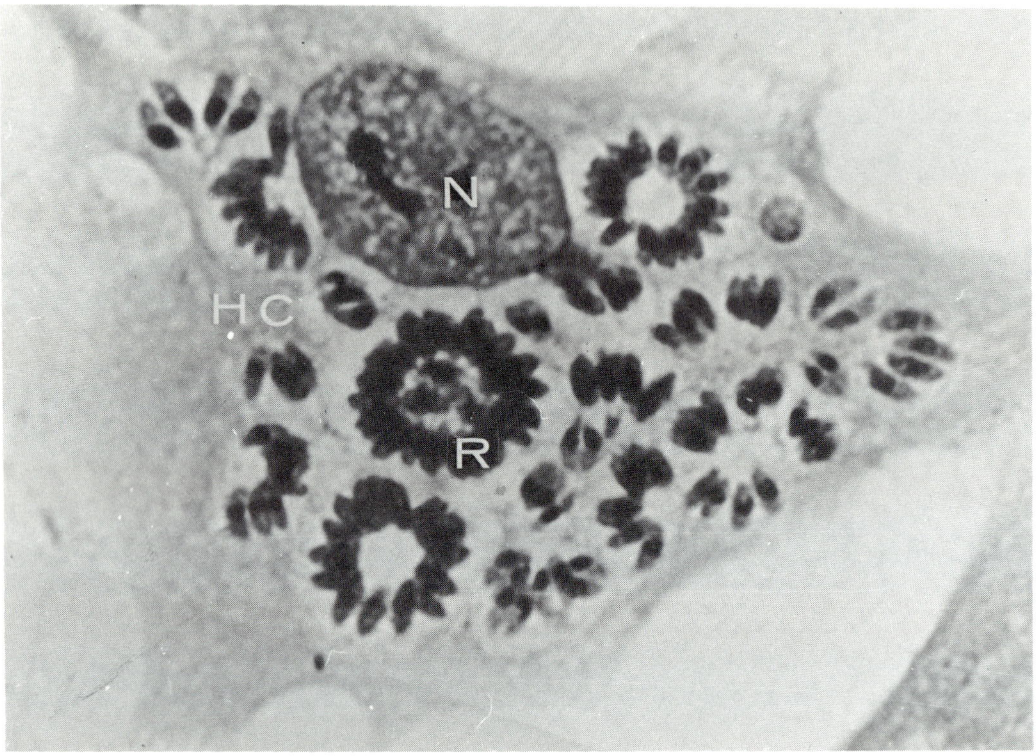

Fig. 11-4. *Toxoplasma* parasites multiplying in the cytoplasm of a HeLa cell (in tissue culture). *N*, Nucleus; *HC*, host cell; *R*, rosette. (×1,000.)

vaded and a more rapid multiplication of the organisms within the cells, the tissue-culture experiments of Hogan and his coworkers[16] and Kaufman and his associates[25] have left no doubt about this. Strains known to have a high degree of virulence in mouse infections have shown the same highly destructive effects on many types of cells in tissue culture. Strains of low virulence, on the other hand, produce much less cell destruction and have a tendency to produce terminal colonies in tissue-culture cells relatively soon after inoculation.

Toxoplasma is widely distributed among warm-blooded vertebrate hosts. Cats, dogs, pigeons, rodents, and many meat-yielding animals, such as hogs, have been shown to be infected. By and large, infected animals (including humans) do not appear to be sick, and the parasitic relationship between the organism and its host is a harmonious one. On the basis of serologic studies it appears that human populations in many areas of the world are infected. Some regions, such as Tahiti, have a very high rate of infection, whereas others, such as Iceland, have virtually none. Populations in high-altitude areas appear to be less infected than groups from tropical or sea-level areas.

Little is known about the transmission of the disease to man. Although the transplacental passage of the organisms from mother to child seems to be well established in the congenital form of the disease, the epidemiology of the adult, or acquired, form is poorly understood. It is known that cats, dogs, and rodents carry the organisms in their excreta and on their mucous membranes. Blood-sucking arthropods, such as ticks, may harbor the organism for short periods of time; and Jacobs[18] has demonstrated the persistence of viable toxoplasmas in the body louse *(Pediculus humanus)*

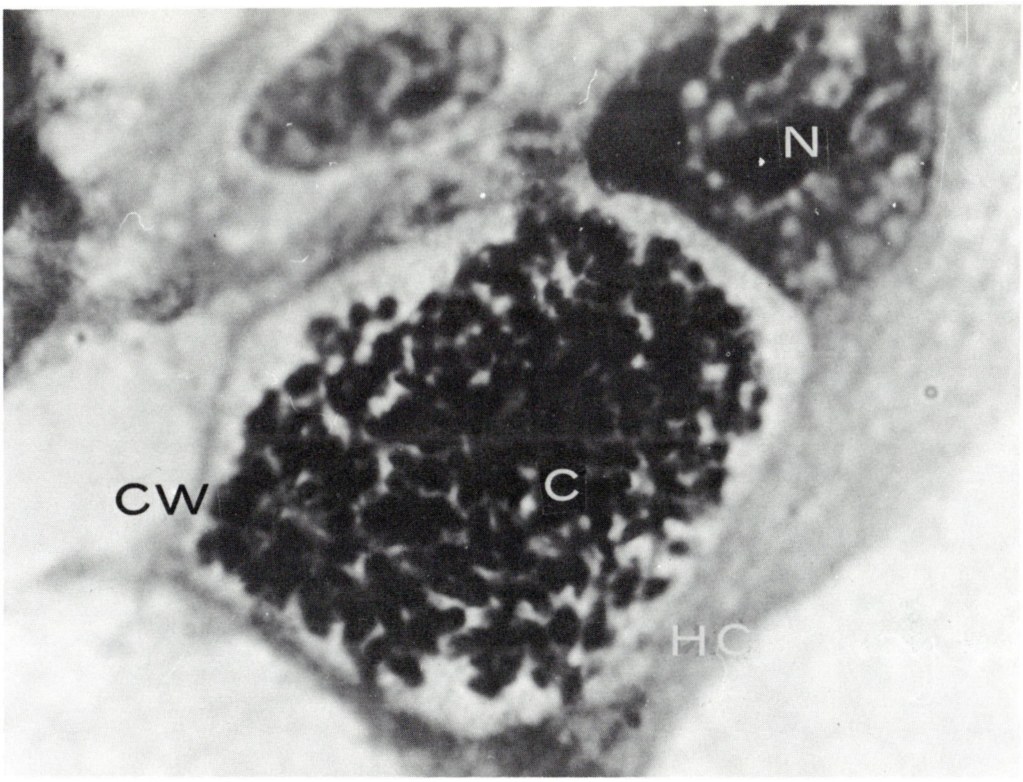

Fig. 11-5. Cyst formation in tissue culture. *N,* Nucleus; *C,* cyst; *HC,* host cell (retinoblastoma); *CW,* cyst wall. (×1,000.)

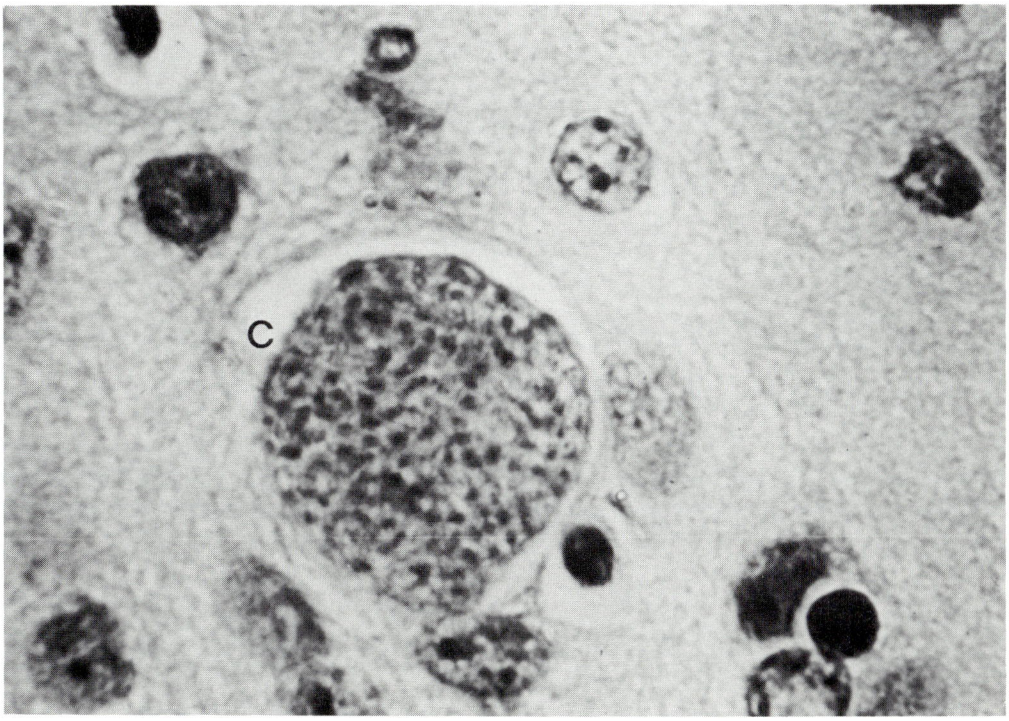

Fig. 11-6. *Toxoplasma* cyst in brain tissue. Note absence of inflammatory reaction. *C,* Cyst. Hematoxylin-eosin stain. (×1,000.)

for as long as 6 days. Jacobs, Woke, and Jones[23] also showed that the ingestion of infected flesh could cause infection in mice. It is quite possible that the same holds true for human beings, although there is still much uncertainty about the whole problem. Current work indicates that the oocysts of *Toxoplasma* may be shed in the feces of cats, and this may provide an important clue as to the transmission of the parasite to man.

Whatever the source of infection, it appears that the organisms multiply locally at the site of penetration and eventually spread by hematogenous dissemination to all parts of the body. Although the initial stages of the disease may be accompanied by remittent fevers, malaise, maculopapular rash, and myalgias, the chronic stages of the disease associated with pseudocyst formation are usually asymptomatic. Were it not for affections of the eye and central nervous system, many more cases of toxoplasmosis would undoubtedly go unnoticed.

HISTORY

Although the parasite was discovered by Nicolle and Manceaux[31] as early as 1909, its importance in human disease was not suspected until 1923, when the Czechoslovakian ophthalmologist Jankû[24] described intracellular parasites in sections of retina taken from eyes with so-called congenital coloboma of the macula. The organisms were not definitely identified as *Toxoplasma* at this time, nor were isolations of the parasite performed. In 1937 Wolf and Cowen[48] recovered *Toxoplasma* from the tissues of an infant who died of "encephalitozoic encephalomyelitis." It was recognized that this child had bilateral chorioretinitis, and the organisms were later identified in tissue sections; but living parasites were not recovered from the retinal tissues.

Wilder[46] in 1952 reported a series of 53 human eyes in which organisms with the morphologic characteristics of *T. gondii* were observed in tissue sections. Although severe necrotizing inflammatory reactions could be seen in the retina, choroid, ciliary body, iris, and sclera, the parasites themselves could be found *only* in the retina. This was a fundamental discovery, and it is the reason why the disease is more properly referred to as *retinochoroiditis* rather than chorioretinitis. As yet, however, the particular predilection of the organism for the retina itself has not been fully explained.

Conclusive proof that granulomatous retinochoroiditis could be caused by *Toxoplasma* was not obtained until the now classical work of Jacobs, Fair, and Bickerton[20] was published in 1954. These authors obtained a freshly enucleated eye removed from a 30-year-old white man who had suffered from recurrent retinochoroiditis for 8½ years. They divided the eye into two halves. One half was used for the preparation of routine histologic sections. The retina and uveal tissues from the other half were removed by dissection, ground in a mortar, and injected into mice, chick embryos, and various tissue cultures. After suitable periods of incubation, *Toxoplasma* organisms, identified by irrefutable morphologic and immunologic criteria, were obtained from all of the injected specimens. Cysts of *T. gondii* were seen on histopathologic sections of the remaining half of the eye. With the exception of the fact that the organism could not be grown on artificial media, Koch's postulates had, indeed, been fulfilled. A positive skin test for toxoplasmosis as well as a postoperative rise in the *Toxoplasma* dye test titer gave additional support to the diagnosis in this patient.

Hogan and his co-workers[17] reported a similar isolation from the eye of a 20-year-old white male in 1958. This patient had allegedly had macular choroiditis in his right eye since the age of 1½ years. The case was thought to represent congenital toxoplasmosis in which the organisms had persisted in the retina and perhaps elsewhere in the body for 20 years. Habegger[13] reported the isolation of *Toxoplasma* from the subretinal fluid of a patient who suffered an inflammatory detachment of the retina. Ansari and Minou[1] reported finding the organisms in a lesion of the conjunctiva. In the strictest sense, it is only in this type of observation, when the organisms are visualized in, and if possible, isolated from ocular tissues that an absolute diagnosis of ocular toxoplasmosis can be made.

DIAGNOSIS

Although the observations listed above leave little doubt that *Toxoplasma* is the causative agent of certain forms of ocular disease, the isolation of the organism can be achieved only rarely from tissues. From the practicing ophthalmologist's point of view, it cannot be looked upon as a practicable diagnostic method. From the patient's point of view, there is little to be gained from establishing the diagnosis on an enucleated eye. On the whole we are still forced to make the diagnosis on indirect evidence, and the greatest part of this is derived from serologic testing.

Various serologic methods are now available for the diagnosis of toxoplasmosis. Each is presented briefly from the point of view of its theoretical background, its advantages, and its disadvantages.

Neutralization test

The *Toxoplasma* neutralization test of Sabin and Ruchman[40] utilizes the principle that the serum of an immune patient will neutralize a given suspension of *Toxoplasma* organisms to the extent that the injection of a standardized mixture of serum and organisms fails to produce the usual necrotizing skin lesion in a rabbit. This test has been largely abandoned because neutralizing antibodies are heat labile and will not survive passage through the mail. They are destroyed by the usual procedures for serial dilution. In addition to these factors, they disappear relatively early in the disease and do not necessarily indicate active disease. At best, the test is cumbersome, expensive, and impractical to perform.

Complement-fixation test

A complement-fixation test for toxoplasmosis was described by Warren and Sabin[45] in 1942. The original antigen used for this purpose was a 10% suspension of infected rabbit brain in physiologic saline. Later an extract of *Toxoplasma*-infected chick chorioallantoic membrane was found to be more suitable, being less anticomplementary than the rabbit brain suspension. Complement-fixing antibodies were detected in monkeys about 2 weeks after infection, but they began to disappear at the end of 2 months. In one established case of acute toxoplasmosis in a human subject, complement-fixing antibodies did not appear until the thirty-second day of infection and disappeared within 3 months.[38] The usefulness of this test in studies of chronic disease was obviously limited; and this test, like its predecessor the neutralization test, has been largely abandoned. The test is used occasionally in determining whether the disease is in its earlier or more acute stages.

Toxoplasma dye test

The mainstay of our present diagnostic regimen is the Sabin-Feldman dye test, which was described by these two authors in 1948.[39] The test depends on the observation that the cytoplasm of living *Toxoplasma* organisms, when acted upon by a specific antibody in the presence of a complementlike "accessory factor," loses its affinity for methylene blue at pH 11. The titer refers to the highest original dilution of serum at which 50% of the extracellular *Toxoplasma* organisms lose their affinity for the dye. The reaction appears to result from minute breaks in the cell membrane of the organism and is related to the depolymerization of ribose nucleic acids within its cytoplasm. It is almost invariably accompanied by swelling of the parasite. Studies by Bringmann and Holz[3] and by Strannegård[43] on the ultramicroscopic morphology of the organism during the dye test procedure have shed some light on this subject.

As a rule dye test antibodies in titers of 1:256 to 1:4,000 develop within 10 to 20 days following infection and may remain at these high levels for 5 years or more without reinfection. It is rare indeed for a patient with an established *Toxoplasma* infection to become totally seronegative even after 5 or 10 years. Treatment with antimicrobial agents, such as pyrimethamine (Daraprim) and the sulfonamides, may be associated with a twofold to fourfold rise in titer during the first 2 weeks of therapy. It may be that this is associated with widespread necrosis of the organisms and the liberation of large amounts of antigen. In no case, however, is the patient rendered seronegative by treatment. It is assumed that the encysted forms of the organism are not reached by the antimicrobial agents. It may be that periodic breakdown of the cysts provides an antigenic stimulus adequate for the maintenance of dye test antibodies in low titer.

The dye test has been the basis for such studies as those of Hogan[14] and those of Jacobs and his associates,[22] in which attempts were made to correlate the incidence of granulomatous uveitis (chorioretinitis) with positive serum dye tests. There seems to be little doubt that patients with chorioretinitis show a higher incidence of positive dye tests than those with nongranulomatous anterior uveitis or control subjects selected at random. In evaluating these surveys, however, one must always be mindful of (1) the incidence of positive tests in the control population of the locality in which the survey is being conducted, (2) the mean age of the group being studied, and (3) the minimum titer that is thought to represent a positive test (such as 1:16 or 1:64). In the study of Jacobs, Cook, and Wilder,[19] for example, all of the 21 patients in whom toxoplasmic chorioretinitis was diagnosed histologically showed positive serum dye tests. In one case antibodies were present only in the undiluted sample, and in another case they were present at a titer of only 1:2. The ages ranged from 16 to 71 years. On the basis of Feldman's[8] population survey, 50% of the people in this age group might have been expected to show dye test antibodies anyway.

The incidence of low titers in some of the patients described by Jacobs, Cook, and Wilder[19] deserves some comment. It must be remembered that the eye is a relatively isolated organ and that retinal lesions have a relatively meager vascular supply. Therefore, as far as the production of circulating antibodies is concerned, the isolated retinal lesion must surely represent a minor antigenic stimulus. If the eye lesions were the only residue of chronic infection in a given patient, the low titer of circulating antibodies would be easy to understand. By the same token one cannot necessarily expect to find a rise in the titer of serum antibodies with each exacerbation of inflammatory disease in the retina. The point of all these observations is to emphasize the caution that must be taken in evaluating a single dye test result. The finding of dye test antibodies in the serum indicates nothing more or less than that the patient has been exposed to toxoplasmosis at some time in his life. Fourfold increases or decreases in titer (correlated with the appearance or disappearance of complement-fixing antibodies), a positive skin test, or symptomatic changes may give some indication of the activity of the disease; but single isolated blood tests are of use only when they are completely negative.

Hemagglutination test

In 1957 Jacobs and Lunde[21] described a hemagglutination test for antibodies to *Toxoplasma*. This test was based on Boyden's[2] observation that sheep erythrocytes treated with tannic acid could bind antigens to their surfaces. If these coated red cells were then suspended in a medium containing specific antibody, they could be agglutinated by the latter. The agglutination is not a gross clumping of cells as one might expect, but rather a fine network that is best demonstrated on a bowl-shaped glass surface. If antigen-coated cells are merely suspended in solution in a small glass tube to which no antibody has been added, they will sediment in such a way as to form a compact pellet or button on the bottom of the tube. If, on the other hand,

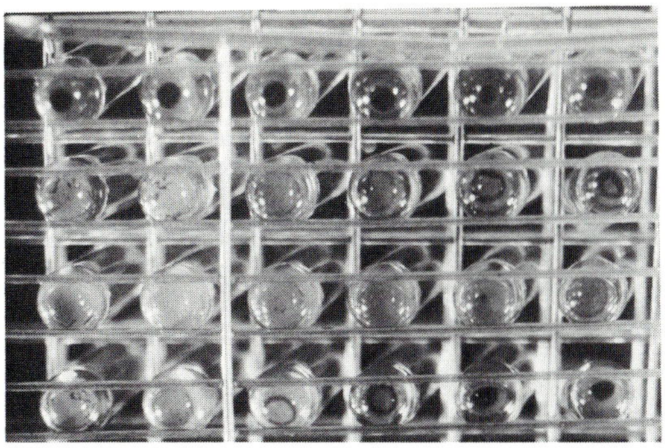

1
2
3
4

Fig. 11-7. The hemagglutination test for toxoplasmosis. Typical appearance of the tubes as seen from the botton. Fourfold dilutions of each serum are arranged in horizontal rows. Serum 1, negative; serum 2, positive, 1:1,024; serum 3, positive, 1:16,384; serum 4, positive, 1:256.

they have been mixed with specific antibody before they are allowed to settle, the sedimentation pattern is quite different. The cells will be spread out in an even, fine reticulum over the entire bottom of the tube (Fig. 11-7). As the antibody is progressively diluted, the agglutination observed is correspondingly less, and intermediate patterns represented by rings or discs are noted instead. In this way the agglutination patterns have been graded from "4+," represented by a fine network spread over the entire bottom of the tube, to "0," represented by the compact button. A "2+" reaction, represented by a diffuse ring in the bottom of the tube, is taken as the endpoint for titer determination.

The test is now performed in many laboratories throughout the United States and Europe. It is not a foolproof test, in that numerous small factors, such as the cleanliness of the glassware, the shape of the bottom of the hemagglutination tube, and the age of the normal rabbit serum used as a stabilizing medium, may all play a significant role in the readings obtained. For this reason one should take particular note of procedural details and variations from the original Jacobs-Lunde test when evaluating the results of a given laboratory.

Maloney and Kaufman[29] have developed a procedure for using formalin-fixed human type O cells in the *Toxoplasma* hemagglutination test, and Park[35] has published a method for preserving alcohol–formalin-fixed sheep cells in the lyophilized state. The advantages of these methods are obvious. If stable, antigen-coated erythrocytes can be prepared in large lots and preserved indefinitely, the normal variations in titer that are observed when a standard serum is tested against two different lots of freshly sensitized cells can be eliminated.

The use of preserved cells also makes it possible for small laboratories in outlying communities to test for hemagglutinating antibodies. Until recently, practicing ophthalmologists have been compelled to rely upon the Sabin-Feldman dye test, which is performed in relatively few laboratories and which entails the additional risks of handling and maintaining living strains of the parasite. For this reason facilities have in the past been inadequate for the testing of patients' sera, and certain laboratories have had to restrict the acceptance of serum specimens to tests on pregnant women and newborn infants.

Lunde and Jacobs[28] have shown that their hemagglutination test generally parallels the Sabin-Feldman test in titer. The difference in titer between the two tests was usually not more than one fourfold dilution. Some doubt has been cast on this issue by Mitchell and Green,[30] who found the hemagglutination tests positive in some 20 patients in whom simultaneous dye tests were negative. In general, it is felt that the hemagglutination test is somewhat more sensitive than the dye test; and if there is a difference in the titers, it is usually the hemagglutination test that gives the higher value. Some of the patients in Mitchell and Green's series may represent borderline cases in which the hemagglutination test showed the presence of antibody when the dye test did not. On the other hand, Reuss[37] was able to establish complete agreement between the hemagglutination test and the dye test as far as negative sera were concerned; and 89.9% of his dye test positive sera showed correspondingly positive hemagglutination tests, with only minor differences in titer.

Precipitin test

O'Connor[32] in 1957 first demonstrated the presence of precipitating antibodies to *Toxoplasma* in the blood and aqueous humor of certain patients suspected of having toxoplasmosis. He demonstrated these antibodies by the use of Ouchterlony's[34] double-diffusion technique in agar. According to this method, antibodies placed in a hole or reservoir in a layer of clear agar are allowed to diffuse toward a source of *Toxoplasma* antigen, located in the same layer of agar at a given distance from the antibody reservoir. As diffusion proceeds and the antibody comes into contact with the antigen, an insoluble grayish white precipitate is laid down in the agar. This takes the form of a thin straight line, which is always laid down at right angles to an imaginary line constructed between the centers of the respective antigen and antibody reservoirs.

Using an antigen solution derived from the peritoneal exudate of infected mice, O'Connor[33] was able to demonstrate multiple lines of precipitation, each corresponding to a separate antigen-antibody system. This showed that the *Toxoplasma* parasite actually elaborated at least five antigens, each of which stimulated the formation of a separate precipitating antibody in the host. The author's work has subsequently been confirmed by Uminski and Wawrzkiewicz[44] and by Strannegård,[42] who was able to show a time-relation curve correlating the appearance of precipitating antibodies, dye test antibodies, and complement-fixing antibodies in sera of rabbit and human subjects. In general, it can be said that precipitating antibodies appear in the serum early in the course of the disease, being most plentiful between 6 weeks and 2 months after onset. It is rare to find circulating precipitins later than 2 years after onset of the disease, although dye test antibodies may still be present in high titer. Thus the appearance and duration of precipitating antibodies seem to run approximately parallel with complement-fixing antibodies. In any case, a given serum might show as many as five types of precipitating antibodies or none at all, depending on the phase of the disease and the individual capacity of the patient to produce precipitins. It is likely that precipitating antibodies represent entirely different types of molecules from dye test antibodies. On the other hand, some precipitating antibodies are likely to be complement fixing.

The functions of the different antigenic components of the organism, analogous perhaps to the multiple antigens of the hemolytic *Streptococcus,* have not yet been elucidated. It may be that some of them are enzymic in nature, corresponding to the antigenic streptolysin O derived from the beta-hemolytic *Streptococcus.* It is likely, in any case, that there is a serum antibody corresponding to each of the antigenic components of *Toxoplasma* and that the individual antibodies may one day be assayable in the same way that antistreptolysin activity is currently measurable in the sera of patients with streptococcal disease.

One of the chief advantages of the precipitin method is that relatively small quantities are required for testing. For example, as little as 0.05 ml of serum, spinal fluid, or aqueous humor can be placed in the central hole of a standard Ouchterlony agar plate and tested simultaneously against five or six different antigens. In addition, specific precipitates can be differentiated from one another by means of the so-called reaction of identity. In this reaction, lines of precipitate from adjacent antigen reservoirs containing a common antigen join each other to form a smooth, rounded angle; whereas two different antigens give rise to lines that cross over each other. This principle is illustrated in Fig. 11-8. Here an unknown serum suspected of containing precipitins to either *Toxoplasma* or *Streptococcus* was in fact found to contain antibodies to both organisms.

The application of the agar-diffusion method to the study of aqueous humor ap-

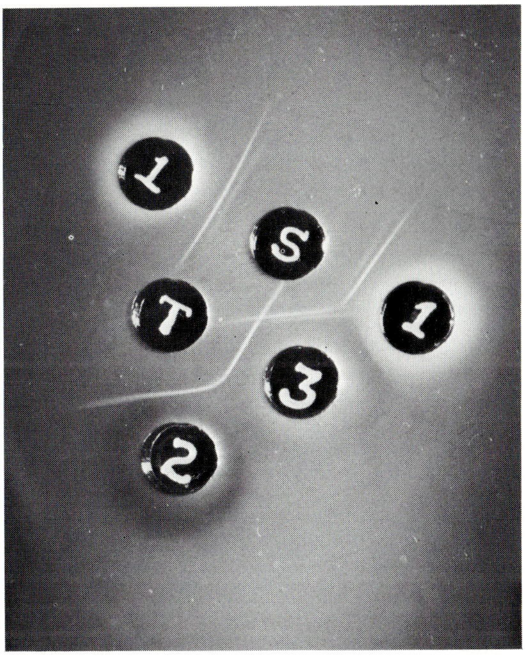

Fig. 11-8. Testing for precipitins by the agar-diffusion technique. S, Reservoir containing streptococcal antigens; *T*, *Toxoplasma* antigen; *1*, serum known to contain antibodies to *Strepotcoccus*, negative for *Toxoplasma; 2*, serum known to contain antibodies to *Toxoplasma*, negative for *Streptococcus; 3*, unknown serum, contains antibodies to both *Toxoplasma* and *Streptococcus*.

pears to be particularly promising. In 1957 O'Connor[33] demonstrated precipitating antibodies to *Toxoplasma* in 11 out of 63 human aqueous specimens tested. Of these 63 specimens, 32 came from patients seriously suspected of having ocular toxoplasmosis on the basis of high serum dye tests, positive skin tests, and "classical" or compatible lesions in the eye. Of particular interest was the fact that 7 out of 11 positive cases showed precipitating antibodies in their aqueous humor, where none could be detected in the serum. This suggested that the formation of antibody may very well take place in the eye itself; and if this is indeed the case, the testing of aqueous humor may be much more specific than the testing of blood. A positive test in the aqueous is of particularly great significance when simultaneous findings in the serum are negative. If, however, both are positive for precipitating antibodies, it should still be possible to differentiate the portion of the antibody molecules that was elaborated in the eye itself from the portion that might have seeped into the eye because of altered permeability of the blood-aqueous barrier. If analyses of the gamma globulin content of both blood and aqueous humor are performed simultaneously, it should be possible to calculate the amount of antibody present per mg of gamma globulin; and this, as Witmer[47] points out, is the really significant measurement. A quotient expressing the titer divided by the amount of gamma globulins per ml can be calculated separately for both the aqueous humor and the serum. If the value of the quotient found for the aqueous humor divided by the quotient for the serum exceeds 1.0, it may be assumed that there is a greater amount of antibody in the anterior chamber than could be accounted for on the basis of seepage through the blood-aqueous barrier alone. Witmer[47] has shown that this value often exceeds 1.0 in experimental leptospiral uveitis, and he has cited this as evidence for formation of antibody within the eye itself. These determinations have recently been performed by Desmonts[7] in cases of human toxoplasmic uveitis, where it is likely that a similar situation obtains.

The agar-diffusion method of Ouchterlony offers the possibility of testing a relatively small amount of fluid (20 to 100 μl) against multiple antigens simultaneously. Heretofore, extensive testing of the aqueous has always been hampered by the small volumes of test fluid that could be collected at any one time. However, one should not allow himself to become overly optimistic about this or other methods involving the direct examination of the aqueous humor in cases of toxoplasmic uveitis. The fact is that the majority of cases, at least in the subacute or chronic stages, involve the posterior portion of the eye exclusively. It is often true that no cells or flare can be seen in the anterior chamber despite rather fulminating disease in the retina or posterior vitreous. In the experience of the author, tests for antibody in the aqueous humor are of no avail unless there is at least some degree of flare in the anterior chamber, as seen on slit-lamp examination.

Flocculation test

Siim and Lind[41] have described a flocculation test for the detection of antibodies to *Toxoplasma.* The test involves the coating of polymethyl meta-acrylic particles with soluble antigen derived from the peritoneal exudates of infected mice. The patient's serum, suitably diluted in a solution consisting of 2% normal human serum in saline, is mixed with a 1:20 suspension of the coated particles on a glass slide and allowed to stand overnight at 4° C. The tests are read the next day, the final titer representing the reciprocal of that serum dilution which gives rise to a standard degree of flocculation.

According to Siim and Lind, 39 out of 40 sera with dye tests in the range of 1:250 to 1:6,250 showed flocculation titers in the range of 1:40 to 1:1,240. Five out of 18 sera negative for the dye test showed titers in the range of 1:10 to 1:40 for the flocculation test.

At present only preliminary reports are available on this technique, and the test is not as yet in general use.[28,41] It seems, however, that this method might be of particular use where (1) only small quantities of material are available (aqueous humor), (2) limitations of space or personnel preclude the constant maintenance of *Toxoplasma* strains, or (3) rapid screening of large segments of the population is desirable on a mass-production basis.

Fluorescein techniques

One of the most important landmarks in the field of applied immunology was passed when Coons and his co-workers[5] developed a method for labeling antibodies with fluorescein. These investigators found that they could couple fluorescein with antibodies of various types without the loss of specific immunologic activity.

The original method, involving the use of fluorescein-4-isocyanate, was both tedious and dangerous. In recent times, however, a more stable compound called fluorescein isothiocyanate has become commercially available. It is no longer difficult to prepare fluorescein-antibody conjugates by the use of this product.

Goldman[10] has labeled anti-*Toxoplasma* antibodies with fluorescein isothiocyanate and has demonstrated the fluorescence of the organisms on fixed smears after staining with the fluorescein-antibody conjugate. A layer of fluorescent antibody is actually laid down on the surface of the cell membrane, which then appears to glow as a greenish golden capsule when viewed microscopically under ultraviolet light (Fig. 11-9). Aside from its aesthetic aspects, which are considerable, the technique has great diagnostic possibilities.

Goldman[11] has also described a test for antibodies to *Toxoplasma* based on the inhibition of specific staining. If smears of the organism are first coated with the patient's serum and then treated with fluorescein-labeled antibody, the labeled antibody will not stick to the organism in those cases where the patient's serum is positive. The test is based on the assumption that there is a limited number of receptor sites on the surface of the organism. If these become occupied by specific antibodies from the patient's serum, they will no longer be able to bind the labeled antibodies that may be applied to them subsequently. Negative sera, on the other hand, will not adhere to the receptor sites; and after gentle washing with saline and application of fluorescein-labeled antibodies, the organisms will be rendered brilliantly fluorescent. In fact, Goldman has shown that the inhibition of staining is just as efficient if the patient's serum and the labeled serum are applied together in a mixture. The explanation for the efficiency of such a mixture must be that natural human antibodies are more successful in the competition for receptor sites on the organism than are the fluorescein-labeled antibodies.

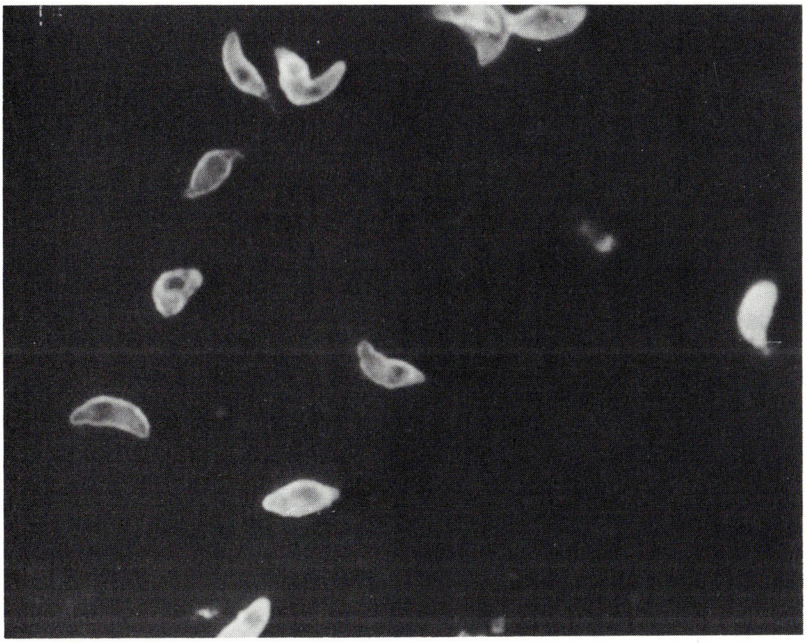

Fig. 11-9. Fixed smear of *Toxoplasma* organisms stained with fluorescein-tagged antibody. Ultraviolet light microscopy. (×1,000.)

Dallenbach and Piekarski[6] have used fluorescein-labeled antibodies to *Toxoplasma* to detect the organisms in tissue sections. The parasite has been detected in lymph node, spleen, and liver preparations from both infected mice and guinea pigs. These investigators noted, moreover, that the organisms stained much more brilliantly in acute infections and that the effectiveness of the labeled antibody preparation was directly proportional to its dye test. The method has not yet been applied to ocular tissues, but it is clear that its greatest usefulness will be in the sphere of necropsy or biopsy examinations.

Another ramification of this method has been employed in the laboratories of the Institute of Ophthalmology.* Here an indirect, or "sandwich," technique has been used to demonstrate antibodies in patients' sera. Smears of *Toxoplasma* organisms obtained from the peritoneal cavities of infected mice were first exposed to serial dilutions of sera, aqueous humor, or spinal fluid from patients thought to have toxoplasmosis. After about 30 minutes the tissues were washed and stained with fluorescein-labeled antibody to human gamma globulin. This antibody was obtained by serial intravenous injections of alum-precipitated human gamma globulin into rabbits. The smears of *Toxoplasma* that had been exposed to serum or eye fluid containing specific antibodies to *Toxoplasma* bound these antibodies and were then stained by the fluorescein-labeled antihuman gamma globulin. The organisms fluoresced in ultraviolet light, and they resembled in appearance the organisms illustrated in Fig. 11-9, which had been stained directly with fluorescent-labeled antibody to *Toxoplasma gondii*.

One hundred sera, in addition to 5 aqueous humor and 3 spinal fluid specimens,

*The author is indebted to Miss Mary-Alice Neville for most of the data presented in this study. Thanks are also due Dr. Elliot Osserman of the Francis Delafield Hospital for supplying the purified human gamma-globulin material and Drs. Beatrice Seegal and Konrad Hsu of the College of Physicians and Surgeons for their advice and help in preparing fluorescein conjugate.

Table 11-1. Results of tests for the presence of antibody to *Toxoplasma* in 100 sera, 5 aqueous humor specimens, and 3 spinal fluid specimens

Source of suspected *Toxoplasma* antibody	Precipitin test	Hemagglutination test	Indirect fluorescein test
Sera 1-7	Positive	Positive	Positive
Sera 8-56	Negative	Positive	Positive
Sera 57-65	Negative	Negative	Positive
Sera 66-100	Negative	Negative	Negative
Aqueous humor 1-3	Negative	Negative	Positive
Aqueous humor 4-5	Negative	Negative	Negative
Spinal fluid 1	Negative	Positive	Positive
Spinal fluid 2-3	Negative	Negative	Negative

have been tested simultaneously for the presence of *Toxoplasma* antibodies by the precipitin, hemagglutination, and fluorescein techniques (Table 11-1). Sera of 7 patients showed positive reactions in all 3 of these tests; sera of 48 patients were negative for precipitating antibody but positive for the other 2 tests (Table 11-1). Another 7 sera were negative for both the precipitin and hemagglutin test but positive for the fluorescein test. Finally, 35 specimens were negative for all the tests. Three of 5 aqueous humor specimens were positive for the fluorescein test, as was 1 of 3 spinal fluids. Thus it is seen that the fluorescent antibody test is sensitive and also provides a technique that can be particularly useful in testing aqueous humor. That volumes as small as 5 μl can be used to coat tiny fixed smears of *Toxoplasma* is a distinct advantage. Serial dilutions can also be carried out with relative ease. The resulting smears are clearly stained with the fluorescent material and are even easier to read than those prepared with human serum. In aqueous humor there may be less interference from viscous proteins, such as albumin, which are present in such large amounts in serum.

Radioactive methods

O'Connor[33] showed that antibodies to *Toxoplasma gondii* could be coupled with I^{131} and still retain their activity when tested subsequently on an Ouchterlony plate. It was hoped that this method might be useful in the localization of unbound antigen in tissues, such as those of the eye, where access of circulating antibody to the affected area is relatively limited. In experimental retinal lesions (in the rabbit) no significant uptake of intravenously injected radioactive antibody could be demonstrated either by scanning techniques or by radioautography of the sectioned eyes. The technique of using radioactive antibody is potentially useful, however, and a wide field is left in this area for further investigative work.

Conclusions on the diagnosis of Toxoplasma gondii infection

It is generally thought that toxoplasmosis accounts for 30% to 50% of all posterior uveitis. Under the best of circumstances the diagnosis of ocular toxoplasmosis is a difficult one to make, particularly where adult patients are concerned. There are many pitfalls, and of these the most common is that of placing too much confidence in a single blood test. It is sometimes difficult to remember that blood taken from an arm vein contains antibodies from the general circulation and that diseases such as toxoplasmosis and syphilis are *systemic* diseases. An acutely rising dye test titer, for example, is actually more likely to be associated with an active focus of toxoplasmic lymphadenitis at some remote point in the body than with an exacerbation of inflammatory changes in the retina. The ophthalmoscopic picture of the disease is often said

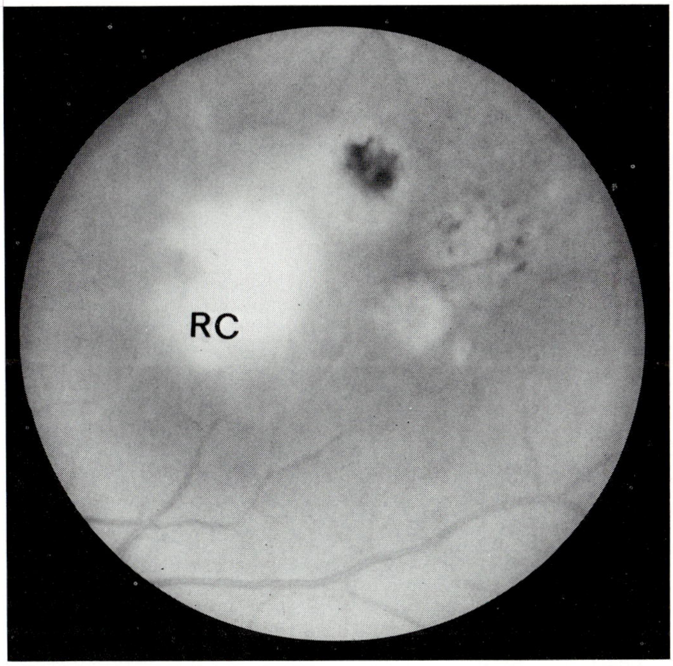

Fig. 11-10. Typical fundus picture of acute recurrent toxoplasmosis. Vitreous overlying the fresh retinochoroiditic lesion (*RC*) is slightly cloudy. Note adjacent healed, pigmented "satellite" lesions.

to be pathognomonic. It is well known, however, that certain of the granulomatous uveitises can mimic the clinical picture of other uveitises. Toxoplasmosis and tuberculosis, for example, have often been confused with each other in the past.

If we assume that the fundus and vitreous pictures (Fig. 11-10) are compatible with toxoplasmosis, what steps should then be taken to establish the diagnosis? In general, one should follow the clinical leads presented by the patient. A careful history should be taken. In many of the proved cases of acquired toxoplasmosis in adults, histories of chronic malaise, muscular aches, chronic low-grade fevers, and swollen lymph glands have been elicited. These symptoms closely resemble those of infectious mononucleosis, and indeed many cases of "infectious mononucleosis" with negative heterophile reactions have subsequently been proved to be toxoplasmosis.

The question of late, previously undetected congenital toxoplasmosis always arises. Inquiries into possible illness of the patient's mother during her pregnancy are singularly unrewarding in the author's experience. Rarely, the disease can be traced in this manner, particularly if the mother had lymphadenopathy or a skin rash during pregnancy; from this point of view, such questioning is worthwhile. If positive or suggestive information is obtained about the patient's mother, serologic tests should be performed on her as well as on the patient.

A baseline serologic test for toxoplasmosis should be performed, whether it be the dye test, the hemagglutination test, or the fluorescein-inhibition test. It should be repeated at least once for confirmation, especially if it is highly positive (over 1:1,024) or completely negative. If the test is positive, further attempts should be made to determine the stage or acuteness of the disease; the complement-fixation and precipitin tests are both helpful in this regard. In the opinion of most authors, a consistently negative dye test rules out the possibility of toxoplasmosis. A differential blood count should be performed. In proved cases of toxoplasmosis, leukopenia with a relative neutropenia has been seen, as well as large "abnormal" lymphocytes. Eosinophilia is variable.

The anterior chamber should be examined for the presence of cells and flare. If either or both is present, a diagnostic paracentesis is indicated. The specimen should be centrifuged and a smear made for routine Gram staining and Giemsa staining. Habegger[13] has found *Toxoplasma* organisms in the aqueous humor of a 2-month-old infant, but it should be emphasized that this is indeed a rare experience. In any case the small amount of effort expended in examining a Giemsa smear is certainly justified, if only to document the type of inflammatory cells present. Approximately 0.2 ml of aqueous humor usually suffices for both a dye test (initial dilution 1:4) and an agar-diffusion test for precipitating antibody. The indirect fluorescein test can be performed as an alternate procedure. In the deep-frozen state, aqueous humor can be stored in sealed glass capillary tubes almost indefinitely; serologic examination of the specimen therefore can be postponed if desired. In addition, precipitating antibodies and dye test antibodies are heat stable, and test samples can be sent through the mail.

If lymph nodes are palpable or inflamed, the patient should be urged to undergo a lymph node biopsy. Prior arrangements should be made with a well-equipped laboratory for isolation studies on the lymph node as well as routine histopathologic examination. Injection of minced lymph node tissue into the peritoneal cavity of a mouse seems, by all odds, to be the most reliable method for making an isolation. A great deal of information has been gained from lymph node studies, and parallel success in isolation has been achieved with muscle (gastrocnemius) biopsies.

Skull x-rays for intracerebral calcification may be of help, especially in suspected late congenital cases. In the few cases where this has been justified, there has usually been a long-standing history of visual impairment, strabismus, or mental deficiency.

The skin test (intradermal injection of toxoplasmin) may be employed as a confirmatory measure. Dermal hypersensitivity to *Toxoplasma* antigens usually cannot be detected until about 6 months after infection. For this reason the test is not useful in detecting early infections. It must be remembered, too, that a certain number of patients, particularly in the older age groups, will always be anergic to skin test antigens, or relatively so.

In the diagnosis of congenital toxoplasmosis among infants we stand on somewhat firmer ground. In addition to the ophthalmoscopic findings there are often other associated signs of disease, such as hydrocephalus, mental retardation, neurologic defects, and microphthalmia. Skull x-ray films sometimes show the presence of intracerebral calcifications. There may also be associated cardiovascular disease, skin rash, and lymphadenopathy. Cytomegalic inclusion disease in infants has often been observed to run a parallel course with toxoplasmosis. The reason for this association is not yet clear, although it may be linked with pathologic changes in the host's cellular defense mechanisms.

Serologic studies on both the mother and the patient are particularly important. If both react positively, attention should be directed to the type of antibody present. Remington and his associates[36] have shown that macromolecules at the 19S type will not cross the normal placenta. If an infant suspected of having toxoplasmosis shows antibody to *Toxoplasma* that is of the 19S type, it may be assumed that he himself has the disease. Smaller antibodies of the 7S type cross the placental barrier easily. If an infant's positive dye test is caused by 7S antibodies alone, it may be assumed that he has received a passive transfer of antibodies from his mother.

Finally, a word of caution should be inserted about the confirmation of the diagnosis by therapeutic trial. Although numerous studies now seem to suggest that pyrimethamine (Daraprim), particularly when given in combination with the sulfa drugs, is effective in the treatment of both the systemic and ocular forms of the disease, these effects have always been more easily demonstrable in animal experiments and tissue culture than in humans. But even if we grant that these drugs are effective, all forms of uveitis, toxoplasmosis included, undergo certain phases of spontaneous exacerbation and remission. To say that the improvement in a given patient was necessarily correlated with specific therapy is indeed spurious. Often the diagnosis itself is so uncertain that even if the retinopathy were improved by the drug therapy, it would still be difficult to ascribe the eye disease to toxoplasmosis. We must look to the future for more exact methods of diagnosis *upon the eye itself* before we can expect to make etiologic diagnosis a practical reality.

REFERENCES

1. Ansari, N., and Minou, A.: Présence de toxoplasmes dans des frottis de conjunctive palpébrale humaine, Bull. Soc. Path. Exot. **41:** 463, 1948.
2. Boyden, S. V.: The adsorption of proteins on erythrocytes treated with tannic acid and subsequent hemagglutination by antiprotein sera, J. Exp. Med. **93**:107, 1951.
3. Bringmann, G., and Holz, J.: Licht und elektronenmikroskopische Untersuchungen zum Sero-Farbtest auf Toxoplasmose nach Sabin und Feldman, Z. Ges. Hyg. **138**:151, 1953.
4. Bringmann, G., and Holz, J.: Die Bewegungsorganellen des Toxoplasma gondii, Z. Tropenmed. Parasit. **5**:54, 1954.
5. Coons, A. H., Creech, H. J., Jones, R. N., and Berliner, E.: The demonstration of pneumococcal antigen in tissues by the use of fluorescent antibody, J. Immun. **45**:159, 1942.
6. Dallenbach, F., and Piekarski, G.: Über den Nachweis von Toxoplasma gondii im Gewebe mit Hilfe markierter fluorescierender Antikörper (Methode nach Coons), Virchow Arch. Path. Anat. **333**:607, 1960.
7. Desmonts, G.: Definitive serological diagnosis of ocular toxoplasmosis, Arch. Ophthal. **76:** 839, 1966.
8. Feldman, H. A.: The clinical manifestations and laboratory diagnosis of toxoplasmosis, Amer. J. Trop. Med. **2**:420, 1953.
9. Frenkel, J. K., and Jacobs, L. J.: Ocular toxoplasmosis: pathogenesis, diagnosis and treatment, Arch. Ophthal. **59**:260, 1958.
10. Goldman, M.: Staining Toxoplasma gondii with fluorescein-labeled antibody: the reaction in smears of peritoneal exudate, J. Exp. Med. **105**:549, 1957.
11. Goldman, M.: Staining Toxoplasma gondii with fluorescein-labeled antibody: a new test for antibodies to Toxoplasma based upon inhibition of specific staining, J. Exp. Med. **105:** 557, 1957.
12. Goldman, M., Carver, R. K., and Sulzer, A. J.: Reproduction of Toxoplasma gondii by internal budding, J. Parasit. **44**:161, 1958.
13. Habegger, H.: Toxoplasmose humaine: mise en évidence des parasites dans les milieux intra-oculaires (humeur acqueuse, exsudat rétro-rétinian), Arch. Ophthal. **14**:470, 1954.
14. Hogan, M. J.: Ocular toxoplasmosis; clinical and laboratory diagnosis; evaluation of immunologic tests; treatment, Arch. Ophthal. **55**:333, 1956.
15. Hogan, M. J., Yoneda, C., Feeney, L., Zweigart, P., and Lewis, A.: Morphology and culture of Toxoplasma, Trans. Amer. Ophthal. Soc. **58**:167, 1961.
16. Hogan, M. J., Yoneda, C., and Zweigart, P.: Growth of Toxoplasma strains in tissue culture, Amer. J. Ophthal. **51**:920, 1961.
17. Hogan, M. J., Zweigart, P., and Lewis, A.: Recovery of Toxoplasma from a human eye, Arch. Ophthal. **60**:548, 1958.
18. Jacobs, L. J.: Biology of Toxoplasma, Amer. J. Trop. Med. **2**:365, 1953.
19. Jacobs, L., Cook, M. K., and Wilder, H. C.: Serologic data on adults with histologically diagnosed toxoplasmic chorioretinitis, Trans. Amer. Acad. Ophthal. Otolaryng. **58**:193, 1954.
20. Jacobs, L., Fair, J., and Bickerton, J.: Adult ocular toxoplasmosis: report of parasitologically proven case, Arch. Ophthal. **52**:63, 1954.
21. Jacobs, L., and Lunde, M. N.: A hemagglutination test for toxoplasmosis, Parasitology **43**:308, 1957.
22. Jacobs, L., Naquin, H., Hoover, R., and Woods, A. C.: A comparison of the toxoplasmin skin test, the Sabin-Feldman dye test, and the complement fixation test for toxoplasmosis in various forms of uveitis, Bull. Johns Hopkins Hosp. **99**:1, 1956.
23. Jacobs, L. J., Woke, P., and Jones, F. E.: Studies on the transmission of Toxoplasma gondii, J. Parasit. **36**:36, 1950.
24. Jankû, J.: Pathogenesis and pathologic anatomy of colomboma of macula lutea in eye of

normal dimensions and in microphthalmic eye, with parasites in retina, Cas. Lek. Cesk. **62**:1021, 1054, 1081, 1111, 1138, 1923.
25. Kaufman, H. E., Remington, J. S., and Jacobs, L.: Toxoplasmosis: the nature of virulence, Amer. J. Ophthal. **46**:255, 1958.
26. Ludvik, J.: Verschiedene elektronoptische Untersuchungen am Toxoplasma gondii und Sarcocystis nevilla, Zbl. Bakt. (Orig.) **166**:60, 1956.
27. Lunde, M. N., and Jacobs, L.: A comparison of results of hemagglutination and dye tests for toxoplasmosis in a survey of Trinidad natives, Amer. J. Trop. Med. **7**:523, 1958.
28. Lunde, M. N., and Jacobs, L.: Evaluation of a latex agglutination test for toxoplasmosis, J. Parasit. **53**:933, 1967.
29. Maloney, E. D., and Kaufman, H. E.: The rapid and convenient detection of Toxoplasma antibodies using formaldehyde-treated human erythrocytes, Amer. J. Ophthal. **50**:945, 1960.
30. Mitchell, R. G., and Green, C. A.: The hemagglutination test for Toxoplasma antibodies, J. Clin. Path. **13**:331, 1960.
31. Nicolle, C., and Manceaux, L.: Sur un protozoaire nouveau du gondi, C. R. Acad. Sci. [D] (Paris) **148**:369, 1909.
32. O'Connor, G. R.: Anti-Toxoplasma precipitins in aqueous humor: new application of the agar-diffusion technique, Arch. Ophthal. **57**: 52, 1957.
33. O'Connor, G. R.: Precipitating antibody to Toxoplasma: a follow-up study on findings in the blood and aqueous humor, Amer. J. Ophthal. **44**(4):75, 1957.
34. Ouchterlony, Ö.: Antigen-antibody reactions in gels, Acta Path. Microbiol. Scand. **26**:507, 1949.
35. Park, H. K.: Toxoplasma hemagglutination test using alcohol-formalin fixed sensitized lyophilized erythrocytes, Arch. Ophthal. **65**: 184, 1961.
36. Remington, J. S., Miller, M. J., and Brownlee, I.: IgM antibodies in acute toxoplasmosis: diagnostic significance in congenital cases and a method for their rapid demonstration, Pediatrics **41**:1082, 1968.
37. Reuss, K.: Untersuchungen zur Hemagglutinationsreaktion and Toxoplasmose nach Jacobs und Lunde, Immunitätsforschung **121**:75, 1961.
38. Sabin, A. B., Eichenwald, H., Feldman, H. A., and Jacobs, L.: Present status of clinical manifestations of toxoplasmosis in man, J.A.M.A. **150**:1063, 1952.
39. Sabin, A. B., and Feldman, H. A.: Dyes as micro-chemical indicators of a new immune phenomenon affecting a protozoan parasite (Toxoplasma), Science **108**:660, 1948.
40. Sabin, A. B., and Ruchman, I.: Characteristics of the Toxoplasma neutralizing antibody, Proc. Soc. Exp. Biol. Med. **51**:1, 1942.
41. Siim, J. C., and Lind, K.: A Toxoplasma flocculation test, Acta Path. Microbiol. Scand. **50**:445, 1960.
42. Strannegård, Ö., University of Göteborg, Sweden. Personal communication.
43. Strannegård, Ö.: An electronmicroscopic study on the immunoinactivation of Toxoplasma gondii, Acta Path. Microbiol. Scand. **71**:463, 1967.
44. Uminski, J., and Wawrzkiewicz, J.: Preliminary results of investigations concerning the application of the precipitation reaction in a gel medium for the diagnosis of toxoplasmosis, Wiadomosci Parazytologiczne Rok VII **2**:423, 1961.
45. Warren, J., and Sabin, A. B.: The complement fixation reaction in toxoplasmic infection, Proc. Soc. Exp. Biol. Med. **51**:11, 1942.
46. Wilder, H. C.: Toxoplasma chorioretinitis in adults, Arch. Ophthal. **48**:127, 1952.
47. Witmer, R. N.: Experimental leptospiral uveitis in rabbits, Arch. Ophthal. **53**:547, 1955.
48. Wolf, A., and Cowen, D.: Granulomatous encephalomyelitis due to encephalitozoan (encephalitozoic encephalomyelitis): new protozoan disease of man, Bull. Neurolog. Inst. N. Y. **6**:306, 1937.

12

Ocular toxocariasis

HAROLD W. BROWN AND DAVID H. BROWN

Ocular toxocariasis results primarily from invasion of the eye by the larval stage of *Toxocara canis*, the common roundworm of man's best friend, the dog. *Toxocara cati* of the domestic cat may also cause infection, though invasion of the eye has not been reported. The predominance of *T. canis* as opposed to that of *T. catis* in human infection may be because of the fastidious toilet habits of the cat. *T. canis* is widely distributed throughout the world, and unrecognized human infections with the larvae are probably more widespread than the few hundred reported ocular infections suggest. The adult worm lives in the small intestine of the dog, and its microscopic eggs are passed in its stools. After a short incubation period in the soil, the eggs contain infective larvae (Fig. 12-1). When eaten by a dog, they liberate their larvae in its small intestine; and after a bloodstream invasion and a pulmonary migration, they grow to maturity in the small intestine. In man, an aberrant host, the larvae hatching from the ingested embryonated *Toxocara* eggs penetrate the intestinal mucosa and by way of the bloodstream reach the liver, lungs, brain, eye, and other organs of the body. These larvae, strangers in a strange land, do not mature, but wander in the tissues for many months before dying.[4]

Ocular toxocariasis has been reported in 245 patients ranging from 2 to 31 years of age. Unfortunately, for the sake of ready diagnosis the leukocytosis and eosinophilia, which are pathognomonic of visceral larva migrans, are not usually present in uncomplicated ocular toxocariasis. The typical visceral larva migrans syndrome consists of hepatic enlargement and tenderness, cough, fever, central nervous system disturbances, marked leukocytosis, and eosinophilia. Larval overflow into the eyes may cause ocular toxocariasis. However, very few of the patients with ocular toxocariasis experienced recognizable generalized larval invasion of the body, or such involvement was not sufficient to produce symptoms of visceral larva migrans. Only 5 of the 245 patients with ocular toxocariasis had a typical visceral larva migrans syndrome before or during the ocular disease. Table 12-1 summarizes the clinical differences between ocular and visceral *T. canis* disease.[1]

PATHOLOGY

The *Toxocara* larvae probably enter the eye by way of the retinal artery. The most characteristic lesion of ocular toxocariasis is a white, elevated, circular tumor approximately one to two disk diameters, usually in the macular area or temporal to the optic disk (Plate 3, *C*). Dark areas in the lesion represent the site of the larva. The surrounding retina may be relatively normal in appearance, although tension folds may radiate from the lesion and retinal

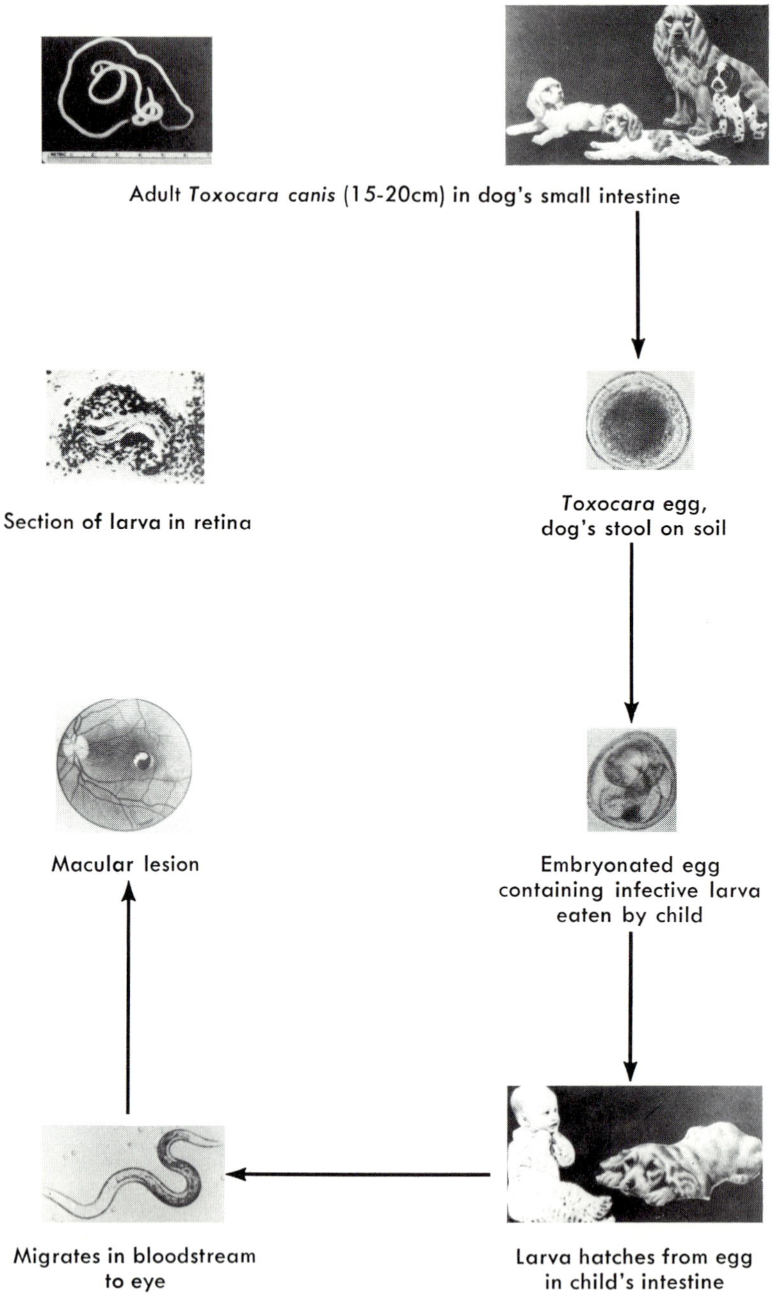

Fig. 12-1. Ocular *Toxocara canis* infection. (From Brown, D. H.: J. Pediat. Ophthal. **7**:182, 1970.)

vessels may invade and disappear into it. Complete detachment of the retina may occur. Less commonly, the larva is situated anteriorly in the fundus and is accompanied by chronic endophthalmitis, iridocyclitis, and posterior synechiae. Although vision is grossly impaired, the eye appears normal outwardly. In 6 of the 245 ocular infections there was a possibility of bilateral infections. Optic neuritis has been reported in one pathologically proved infection. *T. canis* probably accounted for 10%

Table 12-1. Clinical differences between ocular *Toxocara canis* disease and visceral larva migrans

	Ocular *Toxocara canis* disease	Visceral larva migrans
Average age of patient	7.5 years	2 years
WBC	Normal	Elevated
Eosinophilia	Normal	Elevated (>30%)
Hepatomegaly	None	Usually present
Splenomegaly	None	Usually present
Ocular findings	Posterior pole lesion, endophthalmitis, uveitis	Very rare

From Brown, D. H.: Ocular Toxocara canis, J. Pediat. Ophthal. **7**:190, 1970.

of the 150 cases of childhood uveitis reported by Perkins,[8] all of which were unilateral.

The characteristic histologic lesion includes an eosinophilic granuloma with plasma cells and lymphocytes, as well as eosinophils and necrotic fibrinoid material. Epithelioid and giant cells containing fragments of the larva are diagnostic. Table 12-2 lists the data reported by various authors on the pathologic findings in 143 of the 245 cases of *T. canis* ocular infection that have been indicated in the literature.

DIAGNOSIS

The presenting signs and symptoms of the patient, in order of frequency, are strabismus, failing vision, blind eye, and leukokoria. The clinical observations, in order of frequency, are solid retinal tumor, endophthalmitis, uveitis, chorioretinitis, retinal detachment, corneal larva, cyclitic membrane, iritis, iris granuloma, nummular keratitis, cataract, massive retinal fibrosis, and optic neuritis.

Only one out of eight patients that have been reported gave a history of contact with a dog or cat. This to a large degree represents a failure to question the patient or his parents about such contact. The majority of patients who were questioned about contact with dogs or cats gave an affirmative answer.

Ocular toxocariasis is difficult to diagnose clinically because of its resemblance to retinoblastoma, endophthalmitis, and Coats's disease. A sensitive specific serologic test would be of great value. Such tests are of some value in the diagnosis of visceral larva migrans with its extensive organ involvement.[6] However, when only the eye is invaded by larvae, the serologic tests presently available are of little value. A recent summary of serologic data on repeated sera from a patient whose eye was removed and *Toxocara* larva demonstrated showed that all tests were negative. Serologic tests with purified *Ascaris* antigen also were negative, as were passive cutaneous anaphylaxis tests. No specimens revealed an increase of IgA or IgM levels. The patient had type O blood and no significant titer to A or B erythrocytes. Thus, neither specific nor nonspecific antibodies

Table 12-2. Pathologic findings in 143 cases of ocular *Toxocara canis* infection

Slide specimens of eye	*Toxocara canis* second-stage larva	53
	Toxocara species	3
	Nematode larva	20
	No larva found, eosinophilic granuloma	60
Slide specimens of liver biopsy	*Toxocara canis* second-stage larva	1
	No larva found, eosinophilic granuloma	1
Slide specimen of corneal trephine button	*Toxocara canis* second-stage larva	1
In vivo	Nematode larva	4

From Brown, D. H.: Ocular Toxocara canis, J. Pediat. Ophthal. **7**:188, 1970.

were detected in this proved case of ocular toxocariasis.[7]

Clinical data obtained from examination of patients seen in the children's tumor clinic at the Institute of Ophthalmology, Columbia-Presbyterian Medical Center (IOCPMC) are summarized in Table 12-3. Most of the patients were referred for the purpose of ruling out retinoblastoma; thus no laboratory data was gathered, and the diagnosis depended upon the findings in the eye and on the history of the case. As is seen from the table, 10 eyes were enucleated because of continued suspicion of retinoblastoma. Microscopic study of the eye sections was done at the IOCPMC, and the pathologic findings are listed. Typical eosinophilic granulomas were found in 5 of the 10 cases, but these are only suggestive of *T. canis* since no larvae were found. Finding larvae often requires serial sections, which have not yet been made on these patients. The series indicates the difficulty in making a definitive diagnosis of *T. canis* even in enucleated eyes. Since patients with *T. canis* infection should not have the eye enucleated, it is most important to establish a helpful laboratory test. Brown[2] has shown that *T. canis* contains human blood group A antigen. As *T. canis* stimulates some production of anti-A antibodies in man, the possibility of using a test for rising antibody to group A antigen as a diagnostic tool is being further explored.

Table 12-3. Summary of 43 cases of presumed ocular *Toxocara canis* infection in the children's tumor clinic of the IOCPMC

Age	Average	4.4 years	
	Range	3 months to 11 years	
	0.4 years	23	
	5-8 years	18	
	9-12 years	2	
Sex	Male	26	
	Female	17	
Dog or cat contact		27	
Pica (geophagia)		7	
Original complaint			
Strabismus		19	
Failing vision		5	
Leukokoria		3	
Routine exam		3	
Blind eye		1	
Fixed pupil		1	
Red eye		1	
Clinical observations			
Solid retinal tumor		23	
Posterior pole		18	
Periphery		5	
Vitreous mass or vitreous haze (fundus not seen)		12	
Retinal detachment		10	
Cataract		2	
Chorioretinitis		1	
Iris heterochromia		1	
Microphthalmos		1	
Visceral larval migrans syndrome			0
Possible bilateral lesions			1
Pathology: patients 43 enucleation 10			
Toxocara canis second stage larva			0
No larva found, eosinophilic granuloma			5
Congenital anomalies			2
Coats's disease			1
Retinoblastoma			1
Toxoplasmosis			1

From Brown, D. H.: Ocular Toxocara canis, J. Pediat. Ophthal. **7**:189, 1970.

THERAPY

Therapy of ocular toxocariasis is neither specific nor satisfactory. Steroid therapy consisting of prednisone, 40 mg daily for several weeks, is of value, especially when organs other than the eye are involved. A patient harboring an adult intraretinal nematode was treated with 40 mg of prednisone daily for 3 weeks and 2 gm of thiabendazole daily for 5 days with good results.[3] The larva disappeared from the fundus within 24 hours after the initiation of therapy, and a linear retinal hemorrhage remained where the larva was last seen. Although the chemotherapy of visceral larva migrans with thiabendazole has given equivocal results, its recent 15- to 30-day use in *Capillaria* infections suggests an extended course, if necessary, in *Toxocara* infections. Light coagulation and cryotherapy are alternative treatments that might be given a trial in ocular toxocariasis.

PREVENTION

Close contact with infected dogs and cats should be avoided. Since puppies and kittens are more commonly and more heavily infected and may carry *Toxocara* eggs on their fur, small children should be protected against them. Animals under 6 months old should be dewormed with piperazine every month and older ones every 2 months. Worms passed as a result of treatment should be destroyed. Dog and cat feces passed in children's play areas should be buried; and sandboxes, which offer attractive animal defecating areas, should be covered when not in use.

REFERENCES

1. Brown, D. H.: Ocular Toxocara canis, J. Pediat. Ophthal. **7**:182, 1970.
2. Brown, D. H.: Ocular Toxocara canis: experimental immunology, Ann. Ophthal. **3**:907, 1971.
3. Brown, H. D.: Antiparasitic drugs: IV 2(4'-thiazolyl)-benzimidazole, J. Amer. Chem. Soc. **83**:1764, 1961.
4. Brown, H. W.: Basic clinical parasitology, New York, 1969, Appleton-Century-Crofts, p. 343.
5. Duquid, I. M.: Chronic endophthalmitis due to Toxocara, Brit. J. Ophthal. **45**:75, 1961.
6. Duquid, I. M.: Features of ocular infestation by Toxocara, Brit. J. Ophthal. **45**:789, 1961.
7. Kagan, I. G.: Serologic diagnosis of visceral larva migrans, Clin. Pediat. **7**:508, 1968.
8. Perkins, E. S.: Pattern of uveitis in children, Brit. J. Ophthal. **50**:169, 1966.
9. Rubin, M. L., Kaufman, H. E., and Tierney, J. P.: An intraretinal nematode, Trans. Amer. Acad. Ophthal. Otolaryng. **72**:855, 1968.

13

Fungal infections of the eye

with **ARTHUR GERARD DeVOE AND MARGARITA SILVA-HUTNER**

Interest in the possible role of fungi as ocular pathogens was evident in a few areas as early as the middle of the eighteenth century, according to Professor Theodor Leber, who in 1879 published the first report of keratomycosis in man. Leber (Fig. 13-1), an ophthalmologist, made many contributions to the field, among the more notable of which were his studies on the physiology of the circulation of the eye and on inflammation. He distinguished inflammatory response caused by physical or chemical trauma from that caused by infection. His early training in botany may have given him the interest and skill to isolate and identify the etiologic agent in a case of mycotic ocular infection. The patient was a farmer who sustained an injury to the eye from oat chaff while working with a threshing machine. During the next few days the eye became painful; there was a loss of visual acuity; and on the fifth day a corneal ulcer measuring 4 to 5 mm in diameter, with a grayish, turbid base and slightly raised edge, was found. Pus in the anterior chamber reached the level of the lower edge of the ulcer and was associated with intense chemosis. During the next 5 days the ulcer enlarged, and a membrane lifted from the upper border was found to be thickly overgrown with mycelial elements. With the assistance of Professor Reincke, a botanist, Leber identified the organism as *Aspergillus glaucus*. Leber contaminated corneal wounds of rabbits with small bits of mycelium from pure cultures of the *Aspergillus* and injected rabbit corneas with suspensions of the *Aspergillus* spores. In both instances the fungus grew with surprising swiftness and in 2 days penetrated the thickness of the cornea.[31]

Despite Leber's classic paper, only scattered accounts of mycotic infections of the eye appeared during the next 70 years. Chick and Conant[11] in their review of the literature found 148 case reports of corneal mycotic infections following Leber's observation, but 84 of these occurred between 1951 and 1962. They point out that this was an era of intensive antibiotic and steroid therapy. The fungi most frequently isolated were members of the genus *Aspergillus*. However, inspection of Table 13-1, a list of the fungi known to have caused corneal ulcers, shows that at least 95 species in 44 genera have been reported. A few, indicated by an asterisk, are known to be potential pathogens of man; but the great majority of these fungi have their normal habitat in the soil and vegetable matter as saprophytes that produce disease only when some damaging circumstance promotes their infectivity. Such fungi are referred to as opportunists.

Various conditions listed in Table 13-2 give opportunity to potential pathogenic and saprophytic fungi alike to overcome man's natural resistance to mycotic infec-

Table 13-1. Genera and species of fungi isolated in cases of keratomycosis

Genera	Species
Absidia	*A. corymbifera, A. ramosa, Absidia* species
Acremonium	*Acremonium* species
Acrostalagmus	*A. cinnabarinus*
Acrothecium	*A. hominis*
Allescheria	*A. boydii**
Alternaria	*Alternaria* species
Aspergillus	*A. flavus,* A. fumigatus,* A. glaucus, A. nidulans,* A. niger, A. niveus, A. ochraceus, A. versicolor, Aspergillus* species
Beauveria	*B. brumptii*
Blastomyces	*B. dermatitidis*
Botrytis	*B. cinerea, Botrytis* species
Candida	*C. albicans,* C. guilliermondii, C. krusei,* C. mycoderma,* C. parapsilosis,* C. pseudotropicalis,* C. tropicalis,* Candida* species
Cephalosporium	*C. arcuatum, C. curtipes, C. serrae,* Cephalosporium* species
Cryptococcus	*C. laurentii,* C. neoformans**
Curvularia	*C. geniculata, C. lunata, Curvularia* species
Fusarium	*F. conglutinans, F. oxysporum, F. nivale, F. solani, Fusarium* species
Fusidium	*F. terricola*
Gibberella	*G. fujikuroi*
Glenospora	*G. graphii*
Helminthosporium	*Helminthosporium* species
Hormodendrum	*Hormodendrum* species
Lasiodiplodia	*Lasiodiplodia* species
Microsporum	*M. canis**
Monosporium	*M. apiospermum* (imperfect state of *Allescheria boydii*)
Mucor	*M. cornealis, M. mucedo, M. racemosus, Mucor* species
Oospora	*O. floccosa, Oospora* species
Paecilomyces	*Paecilomyces* species
Penicillium	*P. citrinum, P. columnaris, P. crustaceum, P. chrysogenum, P. glacum, P. lilacidum, P. notatum, P. spinulosum, Penicillium* species
Periconia	*P. keratitidis*
Phialophora	*P. verrucosa**
Phycomyces	*Phycomyces* species
Rhizomucor	*R. parasiticus*
Rhizopus	*R. equinus, R. nigricans, R. parasiticus, Rhizopus* species
Rhodotorula	*Rhodotorula* species
Scopulariopsis	*S. brevicaulis,* S. mastigocladium, Scopulariopsis* species
Sporothrix	*S. schenkii**
Sporotrichum	*S. fonsecia, Sporotrichum* species
Sterigmatocystis	*S. nigra*
Tetraploa	*Tetraploa* species
Trichoderma	*Trichoderma* species
Trichophyton	*T. schoenleinii,* T. tonsurans,* Trichophyton* species
Trichosporon	*T. rugosi*
Tritirachium	*T. brumpti, T. roseum*
Ustilago	*U. zeae*
Verticillium	*V. graphii, V. rubrum, Verticillium* species†

Compiled from many published reports and unpublished data from 1879 to 1970. In addition to these isolates, which were classified as to genus and species, three authors reported five cases of keratomycosis by fungi belonging in the order Mycelia sterila (no reproductive structures known).
*Known to be pathogenic for man.
†The term species throughout indicates an isolate that was identified as to genus only.

Fig. 13-1. Theodor Leber (1840-1917), biologist and physician. He was introduced to ophthalmology by Hermann Knapp, and his early training was guided by Helmholz, Professor of Physiology at Heidelberg, by Ludwig in Vienna, and by Virchow in Berlin. After years spent with Liebreich in Paris, he went to Berlin with Albrecht von Graefe and later became the Director of the University Eye Clinic in Gottingen and thereafter the director in his own university at Heidelberg. (From Festschrift zur Feier des Siebzigsten Geburtstages Herrn Geheimrat Professor Dr. Th. Leber in Heidelberg, Gewidmet von Dankbaren Schülern, Graefes Arch. Ophth. **74**:1910.)

Table 13-2. Factors that may promote mycotic ocular infections

Physical injuries
Abrasions
Lacerations
Penetrating wounds
Occupational hazards
Rural living
Agricultural workers, gardeners, foresters, telephone linemen, and others have higher incidence of ocular trauma
Surgical procedures
Repair of retinal tear
Keratoplasty
Cataract extraction
Removal of intraocular foreign body
Therapeutic regimens
Broad-spectrum antibiotics used over long periods of time singly or with local or systemic steroids
Infected eye drops
Preceding nonfungal infections
Bacteria
Viruses
Trachoma
Miscellaneous causes
Incomplete removal of foreign bodies
Inadequate care after surgery
Extension to eye from infected contiguous area
Trichiasis
Exposure keratitis
Bullous keratitis

tion. First among these predisposing factors are traumatic wounds to the eye. Chick and Conant[11] stated that trauma preceded keratomycosis in 77 of the 148 patients whose case histories they reviewed. Antibiotics and steroids probably have favored fungal eye infections recently. Clinical and experimental evidence, to be described later, supports this hypothesis.[32,36,66] It has been estimated that three quarters of fungal corneal ulcers have developed in eyes treated with steroids and broad-spectrum antibiotics.[13] These figures are probably similar today in many parts of the world because diagnosis of mycotic infection is so often made late, after these therapeutic measures have been tried.

Occupational hazards exert an important influence on the incidence of fungal infection. Fig. 13-2, taken from Blake's 1969 study of ocular hazards in agriculture, presents the cause and incidence of trauma in 416 cases of eye injuries seen at the Eye and Ear Hospital in Dublin over the period 1958 to 1967.[9] Agriculture engages but one third of Ireland's workers, yet 58.5% of the injuries were those that would be expected among workers in the field and forest. It will be seen later that because ocular injuries that become infected with a fungus occur more frequently in agricultural communities, they are not unexpected and therefore may be diagnosed more rapidly; thus the added hazard of nonspecific anti-

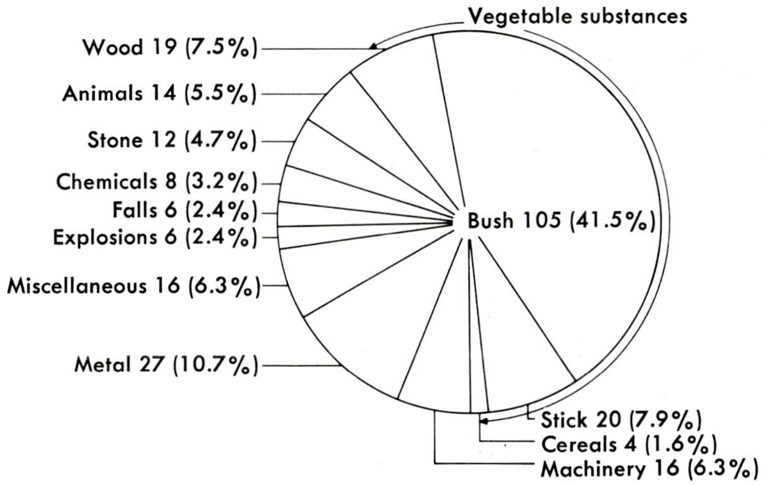

Fig. 13-2. Causes of serious injuries to the cornea. (From Blake, J.: Ophthalmologica **158**:126, 1969. S. Karger, Basel, New York.)

biotics and steroids may be reduced. Other factors that may promote fungal infections, such as preceding nonfungal infection, surgical procedures, and miscellaneous causes (Table 13-2) were encountered infrequently by Blake.

FUNGI CARRIED BY HEALTHY MAN

Information about the source of infections caused by any group of microorganisms is essential as a first step toward the control of such infections. What is the normal habitat of the organism? Does man carry it, and if so, how? Does it always cause disease when present in the tissues, or do accessory factors promote its pathogenicity?

Several studies of the distribution of fungi in the eye and body of man have been reported. A list of fungi isolated from the normal eye is presented in Table 13-3. Eighty-three species belonging to 49 genera have been grown from the normal eye by several investigators. The great majority of these fungi are considered to have their normal habitat in soil and vegetable matter. This is also true for the fungi, listed in Table 13-1, that have been recovered from patients with keratomycosis. A comparison of the findings given in Table 13-1 with those in Table 13-3 shows that 38 species from 25 genera, grown from fungal infections of the cornea, have also been grown from the normal eye. Such findings are compatible with the hypothesis that fungi found in some ocular infections may have been carried by the patient. This is, however, still speculation, since no cultures prior to mycotic ocular infection have been reported. Percebois and Raspiller[44] and Wilson and his associates[65] took repeated cultures of normal eyes at intervals of days and found either a varying mycotic flora or none on such serial examinations. They surmise that the fungal population of the normal eye is inconstant. Percebois and Raspiller report great fluctuations in the number of colonies isolated at each testing. The conclusion that might be drawn from these observations is that the condition of an eye at a moment of trauma may determine the likelihood of endogenous infection.

In association with the late Dr. Rhoda W. Benham we have studied the incidence of fungi in healthy eyes of 508 young people 10 to 18 years old and of 1,347 adults 19 to 80 years old. The results are presented in Table 13-4. This series of studies revealed the following information: (1) fungi were less frequently present in

Table 13-3. Genera and species of fungi isolated from adult normal eyes

Genera	Species
Acremonium	*Acremonium* species
Acrostalagmus	*A. cinnabarinus*
Acrothecium	*A. hominis*
Alternaria	*Alternaria* species
Aspergillus	*A. flaviceps, A. fumigatus, A. glaucus, A. niger, A. terreus, Aspergillus* species
Aureobasidium	*A. pullulans*
Botrytis	*B. cinerea, Botrytis* species
Candida	*C. albicans, C. guilliermondii, C. krusei, C. parapsilosis, C. pseudotropicalis, C. pulcherima, C. tropicalis, Candida* species
Cephalosporium	*Cephalosporium* species
Circinella	*C. minor*
Cladosporium	*C. herbarum, C. mansoni, Cladosporium* species (see also *Hormodendrum*)
Closterosporium	*Closterosporium* species
Cryptococcus	*C. albidus, C. laurentii, Cryptococcus* species
Curvularia	*C. geniculata, C. lunata, Curvularia* species
Debaryomyces	*D. hansenii*
Dematium	*Dematium* species
Eurotium	*Eurotium* species
Fusarium	*Fusarium* species
Geotrichum	*Geotrichum* species
Gliocladium	*Gliocladium* species
Haplographium	*Haplographium* species
Helminthosporium	*Helminthosporium* species
Heterobotrytis	*Heterobotrytis* species
Hormodendrum	*H. algeriensis, Hormodendrum* species (*Hormondendrum* now a synonym of *Cladosporium*)
Isaria	*Isaria* species
Madurella	*Madurella* species
Monilia	*M. sitophila*
Monosporium	*M. apiospermum, Monosporium* species
Mucor	*Mucor* species
Nigrospora	*Nigrospora* species
Oospora	*O. penicillioides, Oospora* species
Paecilomyces	*P. varioti, Paecilomyces* species
Papulaspora	*Papulaspora* species
Penicillium	*P. citrinum, P. commune, Penicillium* species
Pestalotia	*Pestalotia* species
Pullularia	*Pullularia* species
Rhizopus	*Rhizopus* species
Rhodotorula	*R. mucilaginosa, R. rubra, Rhodotorula* species
Saccharomyces	*S. cerevisiae, S. fragilis, Saccharomyces* species
Schizosaccharomyces	*S. hominis*
Scopulariopsis	*S. brevicaulis, S. mastigocladium, Scopulariopsis* species
Sporotrichum	*Sporotrichum* species
Stachybotrys	*S. atra*
Sterignatocystis	*S. nigra*
Torula	*T. chartarum, Torula* species
Torulopsis	*T. glabrata, Torulopsis* species
Trichoderma	*T. viride, Trichoderma* species
Trichophyton	*Trichophyton* species
Trichothecium	*T. roseum*

Compiled from many published reports and unpublished data. In addition to these isolates, which were classified as to genus and species, some authors reported fungi belonging in the order Mycelia sterila (no reproductive structures known).

Table 13-4. Incidence of fungi cultured from clinically healthy eyes of 508 young people 10 to 18 years of age and in 1,347 adults 19 to 80 years of age

Fungus isolated	Number of subjects positive age 10 to 18 years		Number of subjects positive age 19 to 80 years	
	Conjunctiva	Eyelid	Conjunctiva	Eyelid
Aspergillus species	—	—	10	12
Aspergillus niger	—	—	1	1
Candida species	2	4	3	8
Candida albicans	1	1	1	1
Cephalosporium species	1	2	4	7
Cryptococcus species	—	1	5	7
Fusarium species	1	1	7	10
Geotrichum species	1	1	—	3
Hormodendrum species	—	—	1	5
Penicillium species	—	1	2	2
Pullularia species	—	—	4	5
Rhizopus species	—	—	1	3
Rhodotorula species	—	—	9	11
Rhodotorula rubra	1	2	7	9
Saccharomyces species	1	2	4	7
Total number positive	8	15	59	91
Percent positive	1.5	2.9	4.4	6.7

Developed in collaboration with the late Dr. Rhoda W. Benham, Columbia-Presbyterian Medical Center.

the eyes of young people (1.5% as opposed to 4.4% in adults); (2) *Aspergillus,* the fungus most commonly cultured from adult eyes, was missing from the eyes of healthy young people; (3) the eyelids were more likely to contain fungi than were the conjunctivas; and (4) in the adult population the genera *Aspergillus* and *Rhodotorula* accounted for almost half the fungi isolated.

Some similar observations have been made in other parts of the world. Williamson and his associates[63] in Glasgow, reporting on 553 patients from newborns to those 70 years old and older, stated that children up to 9 years of age had an incidence of only 0.8% of fungi in the conjunctiva; and in those 10 to 20 years old the incidence was 1.4%, whereas the average for the older group was 9%. Wilson and his coworkers[65] in Florida found a higher incidence of fungi on the eyelids than in the conjunctiva. This latter observation indicates the need for care in culturing the conjunctiva to avoid contact with the eyelid, which might become a source of contamination for the conjunctiva. High figures (36%) for carriage rate of fungi are reported by Olson[40] in California, but examination included eyelid cultures.

The incidence of fungi in clinically normal eyes of adults reported from various areas of the world is presented in Table 13-5. The percent of cases showing positive cultures varies from 2.45% to 52%. When the figures were examined, it was found that the number of species of fungi identified by the various investigators ranged from 3 to 29. Furthermore, there was a disparity in the predominating fungus found in different geographical areas. The incidence of *Penicillium* species (27%) was twice that of any other fungus reported by Fazakas. The figures from the Columbia-Presbyterian Medical Center (CPMC) placed *Aspergillus* species at the head of the list, whereas the studies of Janke and Schwab[27] showed that 50% of the fungi were *Candida* species. Some of the differences in incidence and classification of fungi in normal eyes probably were caused by climate and occupation. On the whole, the lower carriage rates were in northern

Table 13-5. Incidence of fungi in conjunctivas of normal adult eyes in various areas of the world

Geographic area	Number people cultured	Percent positive	Reported by
Armenia, USSR: Erevan	250	2.45	Ovsepian and Osipian, 1965
Austria: Vienna	171	11.1	Janke and Schwab, 1961
England: Oxford	43	27.9	Ainley and Smith, 1965
Hungary: Debrecen	160*	24.3	Fazakas, 1935
	456*	25.65	Fazakas, 1938
	992*	25.5	Fazakas, 1953
India: Calcutta	112	13.4	Sinha and Das, 1968
Alligarh	100	22.2	Nema and associates, 1966
New Delhi	100	6.0	Agarwal and Khosla, 1963
Japan: Kumamoto	65	18.4	Mitsui and Hanabusa, 1955
Philippines: Manila	232	3.5	de Ocampo and associates, 1965
Scotland: Glasgow	553	2.9	Williamson and associates, 1968
Spain: Santiago de Compostela	42	52.0	Vasquez de Parga and Pereiro, 1965
USA: Little Rock, Arkansas	156	10.3	Hammeke and Ellis, 1960
Miami, Florida	158	22.1	Wilson and associates, 1969
New York, N.Y.	1,347	4.4	Locatcher-Khorazo and Benham, 1968

*Number of eyes cultured.

and urban areas. In the 180 individuals studied by Nema and his co-workers,[38] 92 were urban dwellers, of which 18.5% carried fungi in the eye, whereas 88 were living in the country, of which 25.4% carried fungi. Fazakas[17] found that 80% of harvest workers, when tested after returning from threshing, harbored fungi in the eye. The variety of fungi (Table 13-3) and the variability in their incidence in the eyes of healthy people around the world is not surprising when the ubiquitous nature of these organisms is considered. For example, 55 genera of airborne fungi have been identified in different parts of the world, 27 of which have been reported to dwell in the eyes and/or the body of healthy man.

All reports on ocular mycosis in relation to steroid and antibiotic therapy stress the likelihood that these agents, alone or in combination, promote fungal infection, as was first suggested in 1953 by Thygeson and his associates.[60] Mitsui and Hanabusa[36] were the first to obtain information on the effect of antibiotics and steroids on the flora of the healthy eye (Table 13-6). Hydrocortisone ointment applied for 3 weeks resulted in the emergence of fungi in the eyes of 9 out of 18 individuals who had shown no fungi before treatment. In experiments lasting for 1 week only Williamson and his co-workers found no change in the incidence of fungi in eyes treated with betamethasone; but when neomycin was added to the steroid, the carrier rate increased from 2% to 22%. Nema and his associates treated the eyes of one group of individuals previously free of fungi with hydrocortisone and another group with tetracycline for 4 weeks. The carrier rate of fungi in the first group rose to 41% and in the second group to 29%. Both Mitsui and Hanabusa and Nema and his associates present considerably more confirming data on series of individuals from whom no cultures were grown previous to the treatment but who carried substantially more organisms after treatment than those seen in healthy control groups. It may also be noted that among the fungi isolated in the experiments shown in Table 13-6, *Aspergillus, Penicillium,* and *Candida* predominated.

Summary

The data arising from the study of the normal mycotic flora in man have certain

Table 13-6. Effect of topical application of steroids alone or in combination with an antibiotic on mycotic flora of the healthy eye

Type of treatment	Number of subjects	Number positive before treatment	Number positive after treatment	Number fungi after treatment	Reported by
Hydrocortisone ointment 5-6 times daily for about 3 weeks	18	None	9	1 *Aspergillus* species 1 *Dematium* species 1 *Hyalopus* species 4 *Penicillium* species 2 *Rhodotorula* species	Mitsui and Hanabusa, 1955
Betamethasone disodium phosphate for 1 week	86	1 *Aspergillus* species 1 *Rhodotorula* species	3	1 *Aspergillus* species 1 *Penicillium* species 1 *Saccharomyces* species	Williamson and associates, 1968
Betamethasone disodium phosphate and neomycin sulfate for 1 week	52	1 *Candida albicans*	10	3 *Aspergillus* species 3 *Candida albicans* 3 *Penicillium* species 1 *Trichoderma* species	
Hydrocortisone 1% for 4 weeks	34	None	15	1 *Alternaria* species 4 *Aspergillus* species 1 *Candida* species 2 *Penicillium* species 1 *Helminthosporium* species 2 *Hormodendrum* species 1 *Monosporium* species 1 *Curvularia* species 1 *Mucor* species 1 *Trichothecium* species	Nema and associates, 1969
Tetracycline 1% for 4 weeks	37	None	18	1 *Alternaria* species 1 *Aspergillus* species 2 *Candida* species 1 *Fusarium* species 1 *Helminthosporium* species 1 *Penicillium* species 1 *Trichothecium* species	

relevance to mycotic infections of the eye, which may be summarized as follows:

1. Many species of fungi are found in the conjunctival sac of a portion of the healthy population, free of ocular disease. A larger number are carried in other parts of the body and in the air.
2. Many of the fungi found as transients or colonizers in normal man are among those known to be associated with ocular mycosis.
3. The carrier rate of fungi reported from various parts of the world varies from 2.45% to 52%. The reasons for this great spread in incidence of fungi are not certain; but a warm, moist climate, a rural agricultural environ-

ment, and advanced age may be associated with an increase in the number of people yielding positive fungal cultures.

4. It has been shown experimentally that both antibiotics and steroids increase the number of fungi isolated from the healthy eye.

INCIDENCE OF MYCOTIC OCULAR DISEASE

External eye

At the Institute of Ophthalmology, Columbia-Presbyterian Medical Center (IOC-PMC) from 1938 to 1968 there were 3,358 patients with disease of the external eye that appeared to be caused by infection but from which no bacterial pathogen was isolated. Smears and scrapings were stained, and exudates were cultured for fungi. The late Dr. Rhoda W. Benham gave early supervision and assistance in the isolation and identification of these organisms. The results of these studies are summarized in Table 13-7, where the area of infection and the fungi involved are recorded. Inspection of the table shows that a diagnosis of fungal infection was established in 114 instances. Only 10 of these infections occurred before 1950. The eyelids were involved in 37 infections, the lacrimal apparatus in 26, the conjunctiva in 13, and the cornea in 16. There also were 21 infections of the socket and one endogenous uveitis. In all infections except keratomycosis, *Candida* was the fungus most frequently recovered, with *Aspergillus* a poor second. Corneal ulcers were produced by *Candida* organisms in 8 instances, and the remainder were caused by 7 other fungi. In this group of 114 cases from which fungi were grown, smears and/or scrapings revealed the presence of fungi.

The sparsity of fungal infections of the eye is apparent in this New York series. During the same time (1938 to 1968) some 33,000 bacterial infections of the eye were seen (Chapter 5), 3,532 of which were corneal ulcers of proved bacterial origin. There are two other studies of fungal infection of the eye, covering a period of several years, which present somewhat the same picture. The first of these was made by Fazakas[17] in Hungary in the years 1934 to 1953. He found that in a total of 2,241 diseased eyes cultured for fungi there were 38 fungal infections.* Halde and Okumoto[23] report a series of fungal infections of the eye recognized during 1952 to 1966 at the Francis I. Proctor Foundation in San Franciso. In this period 13,500 cases of external eye disease were examined, in which 30 mycotic ulcers and 13 other various cases of ocular mycosis were identified.† Table 13-8 has been constructed so that the fungi found to cause external ocular disease in these three rather distantly spaced areas of the world might be compared. The fungi causing disease are listed at the left, and the number of infections caused by each fungus as found by Fazakas,[17] Locatcher-Khorazo and associates,[34] and Halde and Okumoto,[23] are indicated in the proper space. The findings for keratomycosis, mycosis of the conjunctiva, mycosis of the lacrimal apparatus, and mycosis of the eyelid are recorded. Study of the table reveals some interesting similarities and differences:

1. Keratomycosis was caused by a wider range of fungi than were found in other parts of the eye. This may be because of the almost constant association of some sort of preceding trauma. Fig. 13-3 illustrates the only case of corneal ulcer caused by *Verticillium* species (Fig. 13-20 shows a culture of the organism). The patient was diabetic.

2. *Aspergillus* accounted for 10 of the 29 cases of keratomycosis reported from

---

*Nine cases caused by *Actinomyces* organisms are omitted from the totals given by Fazakas, since these are now classified with the bacteria.

†Five cases of corneal ulcers are omitted from the totals given by Halde and Okumoto because they were caused by *Nocardia*, which is now classified with the bacteria. Thirty-four infections caused either by *Nocardia* or *Actinomyces* are deleted from the other ocular infections for the same reason (Chapter 5).

Table 13-7. Fungi causing ocular infection observed at the IOCPMC: 1938-1968

Clinical diagnosis	Number patients studied	Number fungal infections	Genera	Fungi isolated (species and number)
Infected eyelid	22	7	*Candida*	*C. albicans* (2)
				C. parapsilosis (4)
			Trichophyton	*T. mentagrophytes* (1)
Meibomitis	48	16	*Candida*	*C. albicans* (5)
				C. parapsilosis (6)
				C. tropicalis (2)
				C. pseudotropicalis (1)
			Cryptococcus	*C. neoformans* (2)
Chronic blepharitis	356	14	*Candida*	*C. albicans* (6)
				C. parapsilosis (4)
				C. mortifera (3)
			Aspergillus	*A. fumigatus* (1)
Lacrimal canaliculitis	124	15	*Candida*	*C. albicans* (6)
				C. parapsilosis (5)
				C. tropicalis (3)
				C. pseudotropicalis (1)
Lacrimal obstruction	356	11	*Candida*	*C. albicans* (3)
				C. parapsilosis (4)
				C. tropicalis (3)
			Aspergillus	*Aspergillus* species (1)
Chronic conjunctivitis	1,203	13	*Candida*	*C. albicans* (5)
				C. parapsilosis (4)
				C. tropicalis (4)
Corneal ulcer	922	16	*Candida*	*C. albicans* (6)
				C. pseudotropicalis (1)
				C. tropicalis (1)
			Cephalosporium	*Cephalosporium* species (2)
			Alternaria	*Alternaria* species (1)
			Monosporium	*M. apiospermum* (1)
			Fusarium	*Fusarium* species (1)
			Verticillium	*Verticillium* species (1)
			Cryptococcus	*C. laurentii* (1)
			Rhizopus	*Rhizopus* species (1)
Infected socket	326	21	*Candida*	*C. albicans* (5)
				C. parapsilosis (4)
				C. tropicalis (3)
				C. pseudotropicalis (1)
				C. pseudotropicalis (2)
			Aspergillus	*A. fumigatus* (4)
				A. niger (2)
Endogenous uveitis	1	1	*Aspergillus*	*A. fumigatus* (1)

Table 13-8. Summary of results from three long-term studies of mycotic infections of the eye

Species*	Keratomycosis			Mycosis of the conjunctiva			Mycosis of the lacrimal apparatus			Mycosis of the eyelids		
	(1)	(2)	(3)	(1)	(2)	(3)	(1)	(2)	(3)	(1)	(2)	(3)
Achorion schoenleinii (*Trichophyton schoenleinii*)										1		
Acrothecium hominis				1						1		
Acrostalagmus cinnabarinus	2									1		
Alternaria species		1	1									
Aspergillus flavus	2			1								
Aspergillus fumigatus	6		1								1	
Aspergillus species	2		1					1				
Candida albicans		6			5			9			13	
Candida mortifera		1									3	
Candida parapsilosis					4			9	1		14	
Candida pseudotropicalis								1			1	
Candida tropicalis		1			4			6			2	
Candida species			12						5			2
Cephalosporium species		2	1									
Coccidioides immitis						2						
Cryptococcus hominis (*C. neoformans*)	1											
Cryptococcus laurentii		1										
Cryptococcus neoformans											2	
Fusarium solani			4									
Fusarium species		1	4									
Hormodendrum algeriensis	1											
Microsporum fellineum (*M. canis*)										1		
Monosporium apiospermum		1										
Mucor cornealis	2											
Mucor species	1											
Oospora floccosa	1											
Penicillium crustaceum	3			1								
Penicillium lilacinum			1									
Penicillium species							1					
Periconia keratitidis	2											
Rhizomucor parasiticus	2											
Rhizopus species		1										
Scopulariopsis mastigocladium	1						2					
Scopulariopsis brevicaulis	2											
Sporotrichum councilmanni										1		
Sterigmatocystis niger	1											
Trichosporon rugosum							1					
Trichophyton species										2	1	
Verticillium species		1										

Sources of data are: (1) Fazakas, 1934-1953; (2) Locatcher-Khorazo, Benham, and Silva-Hutner, 1938-1968; (3) Halde and Okumoto, 1952-1966.

*Only fungi identified by culture are included. Figures to the side of names of fungi indicate number of cases.

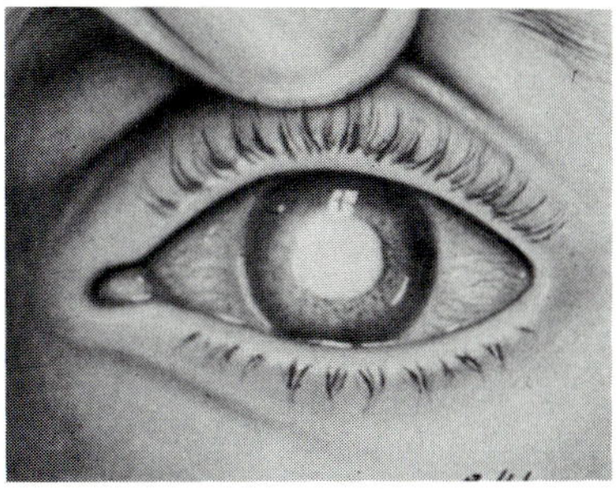

Fig. 13-3. Corneal ulcer caused by *Verticillium* species. (Courtesy Dr. C. Perera.)

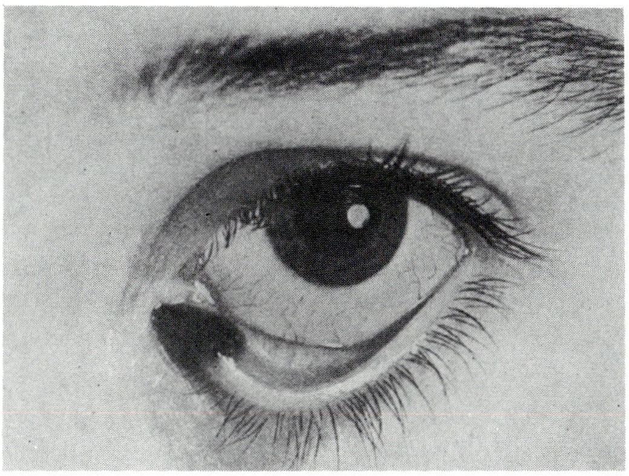

Fig. 13-4. Involvement of the lower punctum, inferior canaliculus, and palpebral conjunctiva by *Aspergillus niger*. (From Donahue, H. G.: Amer. J. Ophthal. **32**:207, 1949.)

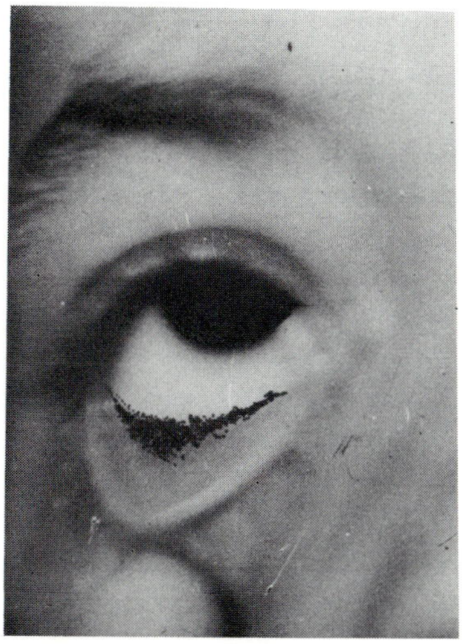

Fig. 13-5. Involvement of the conjunctiva by *Aspergillus niger*. (From Abboud, I. A., and Hanna, L. S.: Brit. J. Ophthal. **54**:480, 1970.)

Hungary but for only 2 of 40 cases from the East and West coasts of the United States. Other areas of the United States, however, have had a number of cases caused by *Aspergillus*. Figs. 13-4 and 13-5 show the eyes of two patients infected with *Aspergillus niger*, unique in that it may suggest the diagnosis on inspection.

3. *Candida* infection played the most important role in the production of keratomycosis in both the New York and San Francisco areas. Fig. 13-6 shows a hypopyon ulcer caused by *Candida albicans*. (Fig. 13-13 illustrates the primary culture from this case, and Fig. 13-15 the PAS-stained direct scraping.)

4. The number of ocular infections in the other parts of the eye as reported from Hungary and San Francisco were considerably fewer than those reported from New York, and *Candida* was the fungus most frequently found.

Other variations in the incidence of mycotic ocular disease have been sought by a survey of reported fungal corneal ulcers around the world. Table 13-9 presents data on the incidence of culturally identified keratomycosis involving 444 patients from 25 countries. The genera of fungi most frequently causing corneal ulcers proved to be *Aspergillus*, 33.3%; *Candida*, 16.4% *Fusarium*, 14.2%; *Cephalosporium*, 6.8%; and *Penicillium*, 6.7%. The remaining 105 patients (23.6% of the total) had infections

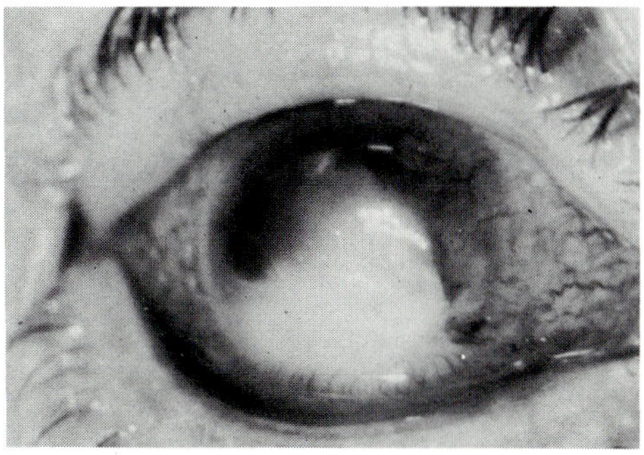

Fig. 13-6. Hypopyon ulcer caused by *Candida albicans*.

Table 13-9. Keratomycosis in various areas of the world, verified by cultures: 1879-1970

Geographic area	Number of fungi grown	Genera of fungi recovered (number)*						Total number
		Aspergillus	*Candida*	*Fusarium*	*Cephalosporium*	*Penicillium*	Others	
North America								
Canada	2							
United States	130	15	31	49	10	3	24	132
South America								
Argentina	2							
Brasil	4							
Peru	3	—	—	3	1	—	5	9
Europe								
Austria	3							
Bulgaria	8							
England	5							
Finland	1							
France	11							
Germany	21							
Greece	1							
Hungary	29							
Ireland	1							
Italy	15							
Poland	9							
Spain	9							
Switzerland	2							
USSR	26	76	17	1	3	8	36	141
Egypt	1	—	—	—	—	—	1	1
Ceylon	3							
India	91	37	7	7	4	12	27	94
China	5							
Japan	48							
Philippines	14	20	18	3	12	2	12	67
Total number of species		148 (33.3%)	73 (16.4%)	63 (14.2%)	30 (6.8%)	25 (5.7%)	105 (23.6%)	444 (100%)

Compiled from many published reports.
*Six more fungal cultures are recorded than there were patients. Nine patients had dual infections.

caused by one of the 37 other genera of fungi listed in Table 13-1. The United States, India, Japan, and Hungary had the largest series of cases. The reasons for the relatively large number of infections in these areas appeared to differ. In the United States the reports were made by 25 teams of investigators from 24 different cities spread over the country, including some in Hawaii. The largest series of cases were those from San Francisco and New York, described in Table 13-8, and a review of 39 cases reported from Miami in 1970 by Jones and his co-workers.[28] Otherwise there were 16 single cases reported, and the remaining 10 publications described 2 to 10 mycotic corneal ulcers. This variation in the apparent incidence of fungal infection in various parts of the United States may be because of concentration of interest and competence in the recognition of fungi in certain parts of the country.

Ninety-one proved cases of keratomycosis from India have been described, and it is apparent that the situation is different from that in the United States. Fungi cause a sizeable proportion of corneal ulcers in India. For example, Balakrishnan[7] of Madras states that in a 2-year period, October 1960 to November 1962, he saw 60 suspected mycotic ulcers among approximately 780 hypopyon ulcers, and in 30 of the 60 cases he identified the fungus by culture. Puttana[45] identified the fungus in 20 cases of keratomycosis seen in a 4-year period in Bangolore. Rao and Ramakrishnan,[46] from Manipal, in the period December 1965 to September 1966, saw 40 cases of corneal ulcers and 6 cases of conjunctivitis with lacrimal canaliculitis; all eyes were carefully investigated for fungi. In 13 of the ulcers (33%), fungi were isolated, and in another case a fungus was seen on the smear but not grown. Fungal infection was the cause of 4 of the 6 cases of canaliculitis. Only 8 teams of investigators made the reports on fungal infection in India. It seems, however, that fungal diseases of the eye are familiar to the physician, since diagnosis is made relatively soon and treatment is started early.[12a,13a,44a]

The 48 cases of keratomycosis from Japan were reports of single cases or small groups of not more than 5 patients. Thirty of the patients' histories were collected from the literature of 1910 to 1969 by Okuda, Matsuura, and Okabe.[39] Fazakas described 28 of the 29 cases from Hungary, which are reported in Table 13-8. They are a tribute to the energy and skill of one man.

Two other pieces of information entered in Table 13-9 are provocative. The small number of cases of mycotic infection of the cornea recorded in South America probably does not give an accurate estimate of the amount of infection caused by fungi in this area. De Almeida[4] from Brazil has given an account of 77 cases of infection following cataract extraction, of which 13 were caused by fungi. It is likely that where there are so many postoperative fungal infections there also must be keratomycosis. The single case of keratomycosis from Egypt was recognized by Lichaa in 1932.[33] He made the clinical diagnosis but sent the culture to the laboratory of Professor Brumpt in Paris where Dr. Langeron discovered that the fungus was a new species of the genus *Beauveria: Beauveria brumptii.* Other reports of keratomycosis from Egypt have not been found, although the climate and agricultural activities of the country seem to favor the disease.

The spotty distribution of the reported cases of keratomycosis in various parts of the world indicate that fungal infections of the eye probably are much more common than the published figures indicate. This point of view is supported by other data. For example, Ball,[8] Ophthalmic Surgeon of the St. Louis City Hospital, stated in 1926 that locality probably had much to do with the frequency of keratomycosis caused by *Aspergillus,* "Dr. C. F. Briegleb of St. Clair, Mo., having met with fifty cases within twenty years." The only case of the disease that Ball had seen "came from the same vicinity." Ball's case is illustrated in Fig. 13-7. Since we were not able to locate Briegleb's report, these possible 50 examples of *Aspergillus* keratitis have been omitted from Table 13-9. Kinnas[29] of Greece reported in 1965 on a corneal ulcer caused by *Acremonium* species and stated that he had treated 33 clinically similar cases, which were not cultured.

There are, furthermore, some substantial data from which one may conclude that cultures of fungi can be negative even when subsequent histologic findings show that the disease was of fungal origin. Salceda and his associates[51] in the Philippines saw 17 patients, 12 of them manual workers, all with a history of corneal trauma, who presented a picture considered typical of fungal keratitis. In 13 cases a fungus was isolated and identified, but in 4 no fungus could be grown. In these latter cases, however, fungi were identified in keratectomy

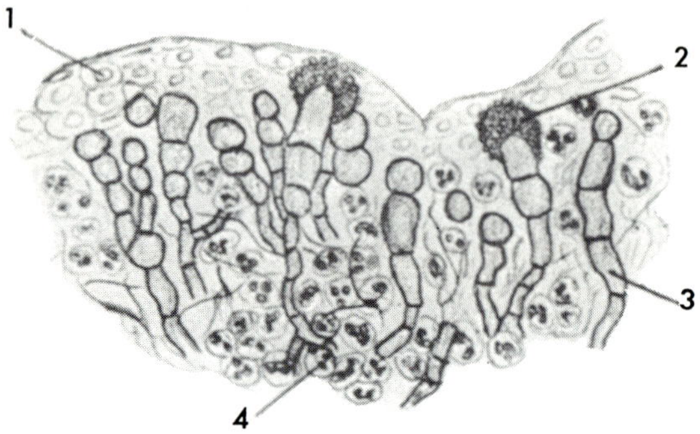

Fig. 13-7. Microscopic section of growth in keratitis caused by *Aspergillus*. *1,* Epithelium of the cornea; *2,* sporangia; *3,* mycelium; *4,* pus corpuscles. (Original drawing by Dr. Carl Fisch. From Ball, J. M.: Modern ophthalmology, vol. 1, Philadelphia, 1926, F. A. Davis Co., p. 479.)

Table 13-10. Incidence of keratomycosis in the material of the Armed Forces Institute of Pathology

Time interval	Kerato-mycosis cases	Number intact enucleated globes	Incidence
1933-1951	3	33,987	1:11,329
1952-1956	14	5,783	1:413
1957-1961	21	4,814	1:229
1962-1966	13	4,515	1:347

From Naumann, G., Green, W. R., and Zimmerman, L. E.: Mycotic keratitis: a histopathologic study of 73 cases, Amer. J. Ophthal. **64**:668, 1967.

sections. Presumably the reason for failing to grow the fungus was the difficulty of obtaining material from the active area of infection or prior death of the fungus.

There is other evidence that mycotic keratitis often is not diagnosed until histologic examination is made. Naumann and his co-workers[37] have published the results of a study of 73 cases of primary mycotic keratitis proved by means of histologic examination. Table 13-10 presents the data on 51 of these patients from whom surgically enucleated whole eyes were submitted directly to the Armed Forces Institute of Pathology. The table shows that histologic examination of 33,987 enucleated globes in the years 1933 to 1951 revealed only 3 cases of keratomycosis, or one in 11,329 cases. A combination of the figures for the remaining 15 years gives a total of 48 globes with keratomycosis among 15,112 eyes examined, or about 1 in 315 cases. This represents a 36-fold increase in incidence in recent years. The diagnosis of keratomycosis in the other 22 eyes was made from portions of eyes previously studied in other laboratories or from corneal buttons from cases treated by keratoplasty.

Naumann and his associates have supplied additional useful data from the series of 73 patients with mycotic keratitis. Ages ranged from 9 to 87 years, and one half of the patients were 51 years old or older. Men outnumbered women 51 to 16. Information was obtained in 51 instances as to the time of year in which the infection began. Particularly in the posttraumatic cases the majority occurred in the warmer months. Approximately two thirds of the specimens were contributed to the Armed Forces Institute of Ophthalmology from southern or southwestern states of the USA or from tropical countries of South America or Africa. The paper also lists the frequency of the various causes of trauma or eye diseases that preceded these cases of keratomycosis. Among the 33 cases with initial corneal trauma, 14 were caused by vegetable matter; in 6 the type of foreign body was unknown; while the remaining injuries were caused by a variety of other objects. Glaucoma with related complications preceded 10 of the infections, and 15 infections followed other miscellaneous corneal diseases. In 15 cases no preceding disease or trauma had been reported.

The histopathology of keratomycosis as described in this series of cases included ulceration of the epithelium and Bowman's membrane with migration of leukocytes into the corneal parenchyma around the organisms and into the anterior chamber. There was frequent formation of multiple abscesses separated from the main lesion. The fungi typically were found throughout the entire thickness of the cornea. Less fre-

Table 13-11. Incidence of *Pityrosporum orbiculare* in various ocular infections and in clinically healthy eyes

Clinical diagnosis	Total number of cases	Percent with seborrheic dermatitis	Percent with *Pityrosporum orbiculare* in eyelid margin scrapings
Seborrheic blepharitis	1,001	80.1	91.6
Seborrheic blepharoconjunctivitis	376	32.9	80.5
Blepharoconjunctivitis	1,201	40.4	80.1
Chronic blepharitis	1,626	25.5	74.3
Squamous blepharitis	326	0.1	27.8
Ulcerative blepharitis	216	0.6	25.4
Allergic conjunctivitis	258	—	34.8
Follicular conjunctivitis	220	—	27.2
Corneal ulcers	144	—	26.3
Chronic conjunctivitis	2,466	0.8	25.8
Keratitis	107	—	25.2
Patients awaiting ocular surgery	1,408	11.5	42.6
Normal controls	1,202	11.8	32.9

In the early days of this investigation the identification of the fungi was carried out in association with Dr. Rhoda W. Benham.

quently they were seen only in the middle or deeper layers. The corneal surface and the crater of the ulcer were usually free of fungi, although there was a profuse growth of hyphae on the surface in a few cases.

The incidence of mycotic infections in the other parts of the external eye—the eyelids, the lacrimal apparatus, and the conjunctiva—as seen at the Columbia-Presbyterian Medical Center, by Fazakas, and by Halde and Okumoto have been given in Tables 13-7 and 13-8. There are several additional reports of infection of the lacrimal canal, which add four more genera of fungi: *Rhinosporidium, Glenospora, Hyalopus,* and *Torulopsis.* According to François and his associates,[20] infection of the lacrimal sac by *Blastomyces, Sporotrichum, Cephalosporium,* and *Trichophyton* may occur as well. The series of cases of fungal conjunctivitis reported in Tables 13-7 and 13-8 are the most extensive ones that have been observed. This condition may precede or be concomitant with canaliculitis, blepharitis, meibomitis, or keratomycosis.

The series of cases of fungal infections in these parts of the eye, as reported in Table 13-7, showed equal distribution between sexes; only 6 of the patients were under 12 years of age. In 22 instances the infections were apparently acquired from neighboring areas, namely, the nose, nasopharynx, the sinuses, and adjacent skin.

Seborrheic blepharitis and blepharoconjunctivitis are associated in a high percentage of cases (Table 13-11) with the presence of a fungus, *Pityrosporum orbiculare,* which is found in 100% of cases of seborrheic dermatitis. This disease involves the eyebrows, hair, and external ears of most patients with seborrheic blepharitis and blepharoconjunctivitis (Fig. 13-8, *A*). Scrapings of the affected areas show yeastlike organisms ranging in size from 2 to 10 μ (Fig. 13-8, *B*). The organisms are most abundant when the eyelid scrapings are obtained from a site with marked inflammation. As may be seen from Table 13-11, normal eyes or eyes of people with other categories of disease had neither a high rate of seborrheic dermatitis nor as high an incidence of *Pityrosporum* on the eyelid margins. The numbers of organisms seen in the smears of scrapings was sharply decreased. Definite proof that the association of the *Pityrosporum* group of fungi with seborrheic conditions establishes an etiologic role has not been given. Gots, Thy-

A

B

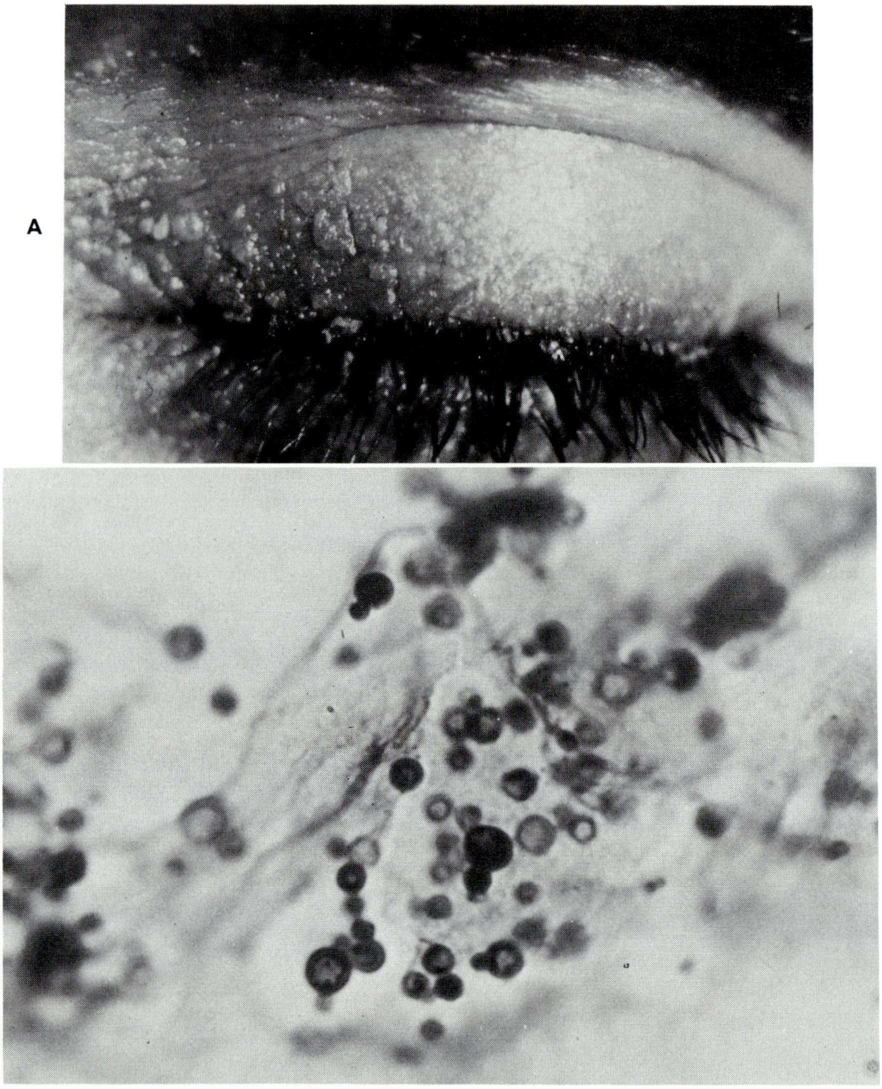

Fig. 13-8. A, Seborrheic blepharitis. (Courtesy U. S. Department of Health, Education, and Welfare.) **B,** Direct scraping from eyelid margins from patient shown in **A.** Note mostly spherical forms, some budding cells, singly or in pairs, of *Pityrosporum orbiculare.*

geson, and Waisman[21] in 1947 reported that their investigation did not result in conclusive evidence that *Pityrosporum* was pathogenic for the eye and its adnexa. However, the following results were suggestive of an etiologic relationship: (1) constant presence of yeastlike organisms in the lesions, (2) the apparent relationship between the number of the organisms found in the scrapings and the severity of the clinical disease, and (3) the absence of the organisms or their presence in only small numbers in clinically normal eyes. Thygeson and Vaughan[61] in 1955 found *Pityrosporum* in 98% of lesions in patients with seborrheic blepharitis, in 86% of lesions with mixed types of blepharitis, and in 46% of patients with staphylococcic blepharitis. In normal eyelids in patients in a hospital for the mentally ill, the organisms were present in 58% of the cases and in 20% of the cases in a series of private patients.

Table 13-12. Mycotic ocular infections following ocular surgery: summary of cases identified by cultures

Type of surgery	Day of onset first symptoms	Causative agent	Final outcome	Reported by
Cataract extraction				
OD	9 days	*Volutella cinerescens*	47th day, enucleation	Foster and associates, 1958
OS	31 days	*Volutella cinerescens*	Blind eye	
Cataract extraction		*Volutella cinerescens*	Enucleation	
Cataract extraction	15 days	*Cephalosporium* species	3½ months, enucleation	Fine and Zimmerman, 1959
Cataract extraction	2 months	*Hyalopus bogolepofii*	2½ months, enucleation	Paiva and associates, 1960
Cataract extraction	40 days	*Volutella* species	92 days, enucleation	Theodore and associates, 1961
Cataract extraction	15 days	*Cephalosporium* species	No useful vision	
Cataract extraction	14 days	*Neurospora sitophila*	Light perception	Theodore and associates, 1962
Cataract extraction	10 months	*Aspergillus fumigatus*	1 year, enucleation	Diamond and Kirk, 1962
Cataract extraction	24 days	*Sporothrix schenkii*	6 weeks, enucleation	Küper, 1962
Cataract extraction	2 weeks	*Cephalosporium* species	4 weeks, enucleation	
Cataract extraction	10 months	*Hormodendrum* species	10 months and 8 days, enucleation	François and associates, 1962
Cataract extraction	6 months	*Sporothrix schenkii*	Enucleation	
Cataract extraction	24 hours	*Cladosporium* species	Enucleation	de Azevedo, 1962
Cataract extraction	2 months	Mycelia sterila	Enucleation	
Cataract extraction	7 days	*Aspergillus niger*	5 months, enculeation	Queiroga, 1961 Cited by de Rezende, 1962
Cataract extraction	24 hours	*Cladosporium* species	8 days, enucleation	Lobo, 1961 Cited by de Rezende, 1962
Cataract extraction	5 months	*Candida lipolytica*	Enucleation	Graf, 1964
Lamellar keratoplasty (for a stromal herpes infection)	5 weeks	*Cephalosporium* species	Enucleation 5 weeks after 2nd graft	Hoffman and Naumann, 1963
Cataract extraction	2½ months	*Scopulariopsis brevicaulis*		Strampelli and Maggi, 1963
Cataract extraction	28 days	*Cephalosporium* species	Visual result poor	Green and associates, 1965
Cataract extraction		*Cephalosporium* species		Halde and Okumoto, 1966
Postoperative vitreous abscess		*Candida* species		
Keratoplasty (2 cases)		*Candida* species		
Lamellar graft for corneal dystrophy	7 days	*Aspergillus nidulans*	3 weeks, lamellar graft removed; second graft successful	White, 1969
Lamellar graft		*Pullularia pullulans*	Both cleared up with antifungal treatment	
Lamellar graft		*Trichophyton rubrum*		
Cataract extraction	35 days	*Cephalosporium* species	7 weeks, enucleation	Albesi, 1969
Cataract extraction	1 month	*Cephalosporium* species	Enucleation	

On the contrary, Parunovic and Halde[43] in 1967 reported that microscopic examination of eyelid margins demonstrated that *Pityrosporum orbiculare* was present in as many subjects with normal eyelid margins as in those with seborrheic lid margins.

Treatment that removes the *Pityrosporum* from the lid margins in cases of blepharitis and blepharoconjunctivitis results in considerable improvement in the condition of the eye. If this fungus is not acting as an etiologic agent in the disease, it is at least a very irritating factor in the eye.

Internal eye

The largest number of fungal infections that penetrate to the internal structures of the eye must undoubtedly arise from keratomycosis. In the absence of early diagnosis and treatment, hypopyon is almost inevitable, and endophthalmitis is the result in the majority of cases. This aspect of keratomycosis is considered later.

Among the remaining internal mycotic infections, those of greatest concern to the ophthalmologist follow ocular surgery. Surgical skill, careful observation of asepsis, and presurgical preparation of the eye are such that relatively few postsurgical infections occur, and of these only a small portion are caused by fungi. In their book on ocular mycosis, François and his associates[20] have included 64 examples of postoperative infection, seen from 1914 to 1965, which were proved by culture and/or histologic examination. From the literature 9 more cases, also proved by culture, have been added. Table 13-12 gives data on 29 cases in which the fungi were identified by culture. An outstanding characteristic of postoperative infection is the irregular and usually long latent period between surgery and the onset of symptoms of the infection. In the cases presented in Table 13-12 the latent period varied from 7 days to 10 months. The indolent way in which postoperative infections begin is an important diagnostic aid. The data given in column four of Table 13-12 established the fact that significant sight was the end result in only 3 cases of lamellar graft surgery. Eighteen eyes were enucleated.

The fungi involved in these postsurgical infections differed from those found in keratomycosis in the following respects: 5 cases involved *Volutella, Hyalopus, Neurospora* (Fig. 13-9), *Cladosporium,* and *Pullularia,* genera of fungi that had not been recovered from cases of keratomycosis; *Cladosporium* had been reported in the normal eye. The incidence of *Cephalosporium* was higher in postoperative infections than in keratomycosis, whereas that of *Aspergillus* and *Candida* was lower. Both *Fusarium* and *Penicillium* were absent. In such a small number of widely distributed infections these differences may not be significant, but they alert the mycologist to include a wider range of fungi among possible pathogens or opportunists.

A very few intraocular mycotic infections have followed penetration of the eye by trauma. François and his associates[20]

A

B

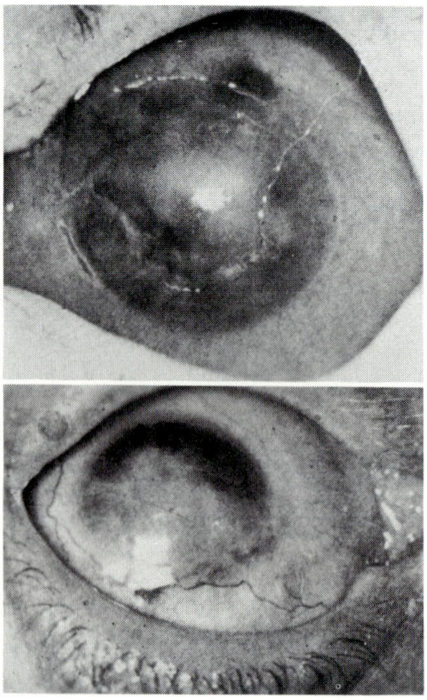

Fig. 13-9. A, Endophthalmitis, following cataract extraction, caused by *Neurospora sitophila;* 48 days after operation. **B,** Photograph of eye seen in **A,** 6 months after operation. (From Theodore, F. H., Littman, M. L., and Almeda, E.: Amer. J. Ophthal. **53**:35, 1962.)

have collected 21 examples from the years 1898 to 1968. Only 2 of these, both caused by *Aspergillus fumigatus,* were diagnosed by culture, the remainder being examined histologically. In the 3 instances in which the data were given, there was a long latent period between the trauma and the signs of infection: 11 days, 2 weeks, and 1 month.

A larger group of intraocular infections arise from a variety of endogenous sources: direct extension from the brain or opposite eye, debilitating infections, or other disease leading to generalized mycosis or focal mycosis elsewhere in the body. François has also tabulated 98 cases from the literature covering the period 1910 to 1968. Diagnosis was made by culture in 44 cases and by histologic examination in the remainder. The fungi involved were predominantly *Aspergillus, Candida,* and *Crytococcus,* the last as a sequela of cerebral or pulmonary infection. *Coccidioides immitis* was recorded 10 times; and in the 6 infections where the original site of infection was recorded, it was either generalized or in the lungs. François gives one reference to a case of retinal spots caused by *Histoplasma capsulatum* and states that more than 100 others have been described. However, the cultural identification of *Histoplasma* in the eye is recorded only by Hoefnagels and Pijpers.[26a] The mycotic infections included in this group of intraocular infections are usually seen first by the internist.

CLINICAL ASPECTS OF MYCOTIC INFECTIONS OF THE EYE

Because the distribution of fungi is so widespread and preventing contamination by them is nearly impossible, it is remarkable that serious infection of the eye or its appendages is relatively rare. Fortunately, the fungi, whose role in nature is apparently to break down dead and devitalized tissue of all kinds, have relatively little virulence when placed on normal, healthy human tissue. Numerous studies have been made concerning the incidence of fungi in normal conjunctival sacs. Variations from 2% to 80% have been noted, the higher figure being obtained from agricultural workers in the field at harvest time. Under normal circumstances the fungi do not invade, but injury with resulting devitalization of tissue can be followed rather rapidly by invasion of ocular structures. Fungal infections are much less common in an urban population with similar injuries, presumably because the size of the inoculum is so much smaller.

Mycotic infections of the conjunctiva are uncommon but may be of either superficial or granulomatous appearance. A thrushlike conjunctivitis may be caused by *Candida albicans* and by the ringworm group, such as *Microsporum* and *Trichophyton.* The role of *Pityrosporum orbiculare* in seborrheic conjunctivitis is somewhat questionable. Probably the commonest external mycotic problem seen by the opthalmologist is *Actinomyces israelii* infection of the lacrimal drainage system. This organism is classified at the present time among the higher bacteria rather than among the fungi. As such, it is more amenable to chemotherapeutic attack. The firm yellowish granules that plug the lacrimal canaliculus and sac require surgical removal. This is ordinarily done by slitting the lower punctum and removing material with a chalazion curette.

Atypical papillomatous lesions in the conjunctiva should always arouse suspicion of mycotic invasion as well as neoplasia, syphilis, tuberculosis, and leprosy. *Aspergillus, Blastomyces, Coccidioides,* and *Rhinosporidium* have been implicated in such lesions. All are rare in the northern part of North America but become increasingly common the further south one goes.

The two most common types of mycotic infection seen by the clinical ophthalmologist are those involving the cornea and those seen after intraocular surgery. Both have characteristics that tend to differentiate them from bacterial and viral infections. Keratomycosis is rarely if ever found in a healthy individual who has suffered no injury to the cornea.

The typical history of keratomycosis may be one of several types. Probably the most common is one of prolonged corneal infection, usually herpes simplex, which has been treated for months with steroids, antibiotics, and every medication available without effecting a cure. There may be a rather precipitous turn for the worse with a localized gray infiltration of the anterior stroma, totally refractory to all ordinary therapy, and slowly but relentlessly extending in depth and width. Characteristically, the inflammatory reaction of the entire globe is much less than one would anticipate with a similar bacterial infection. There may be small satellite lesions extending out from the main area of invasion. These are almost diagnostic of mycotic superinfection. If the individual has a serious systemic and debilitating disease, such as generalized carcinomatosis, diabetes, tuberculosis, or chronic alcoholism, one's suspicions should be increased.

Diagnosis is not always simple. At the first suspicion of mycotic invasion, one should immediately attempt to obtain material for histologic inspection. Superficial scraping or use of applicators will not be sufficient. We prefer to use the operating microscope and to remove first superficial necrotic tissue. This is then followed by deep and aggressive removal of tissue at the base and under the overhanging margins of the ulcer. The material is examined immediately in a saline or 10% sodium hydroxide drop so that it is frequently possible to establish the diagnosis at once. The presence of a negative result in a clinically suspicious case requires repeated and more vigorous tissue removal. On occasion this may mean the use of a trephine with lamellar keratectomy in order to obtain an adequate specimen. The material is then utilized not only for culture but for histologic section. It has been our experience that this technique is not only valuable in diagnosis but is also an effective adjuvant to therapy. Although all of the currently available antibiotics and chemotherapeutics known to have some effect upon the various fungi are of relatively little help in curing advanced disease, they may be effective when the major mass of diseased tissue is removed so that the drug can reach the deep tissues. It is our feeling, however, that antibiotics are relatively futile when used alone in advanced keratomycosis. Total excision is rarely possible, but we have repeatedly seen recovery following subtotal resection of the involved cornea. We have even salvaged the one remaining eye in a patient with a large, perforated, *Cephalosporium*-caused ulcer with extruded lens, hypopyon, and severe diabetes. The entire cornea was removed with a 10 mm trephine and replaced with a full-thickness graft. Even though it was possible to culture organisms from the vitreous face, the graft was retained and the globe salvaged without further evidence of intraocular invasion. If a major part of the involved tissue is removed, there is a reasonable chance that infection can be controlled.

Recently a technique has been devised by which specific antibody to a fungus has been conjugated with fluorescein and instilled into the eyes of animals that have been infected with *Candida albicans*. The stained antigen-antibody complex is then treated with the argon laser, destroying the fungal element. So far, this has been accomplished only in animals, but the principle offers some promise. It seems highly doubtful to us that any of the systemic antibiotics effective against fungi have much practical use, since these antibiotics are highly toxic. None has been effective enough in our experience to justify the hazard. In addition to the commonly used antibiotics (such as amphotericin B, nystatin, and pimaricin), grenz rays, ultraviolet light, cryocautery, and thermocautery have been used without dramatic improvement. The use of the thermophore unfortunately requires a temperature that destroys not only the fungus but also the corneal stroma.

Acute endophthalmitis following cataract surgery is rare, the incidence being somewhere in the order of one per thousand. Viral infections other than the presumed release of virus in needling and aspiration of rubella cataract must be extremely uncommon. The possibility exists, however, that they may be the cause of certain apparently sterile inflammatory reactions after cataract operation. Bacterial intraocular infections are characterized by acute inflammatory onset within 24 to 48 hours after surgery. If, after uneventful surgery, the eye is apparently well on the road to healing and at about the fourth or fifth week begins to develop a progressively severe low-grade inflammatory reaction, one's suspicion should turn to fungus invasion. Diagnosis here is perhaps even more difficult than in corneal infections. As soon as infection is suspected, an anterior chamber tap should be performed and suitable culture media inoculated. Direct examination of the fluid may be positive. There is no really satisfactory therapy available for mycotic endophthalmitis. If the virulence of the organism is low or the inoculum small, it is possible that with the usual supporting methods of therapy common to any ocular inflammation, the body may be able to withstand the invasion and overcome it. In general, however, the prognosis is extremely poor.

Endogenous endophthalmitis of mycotic origin is being reported with increasing frequency. Blastomycosis, actinomycosis, aspergillosis, mucormycosis, and cryptococcosis, among others, have all been described. Although cures have been reported, not only is the eye usually lost but not infrequently the disease is lethal.

Orbital cellulitis, not obviously explained by direct injury or sinusitis of bacterial origin, should alert the clinician to the possibility of fungal invasion of the orbit. The most commonly reported offender has been *Mucor*. In by far the majority of cases, approaching 90%, the patients have been either severely diabetic or overwhelmed by wasting disease. Invasion of the orbit can

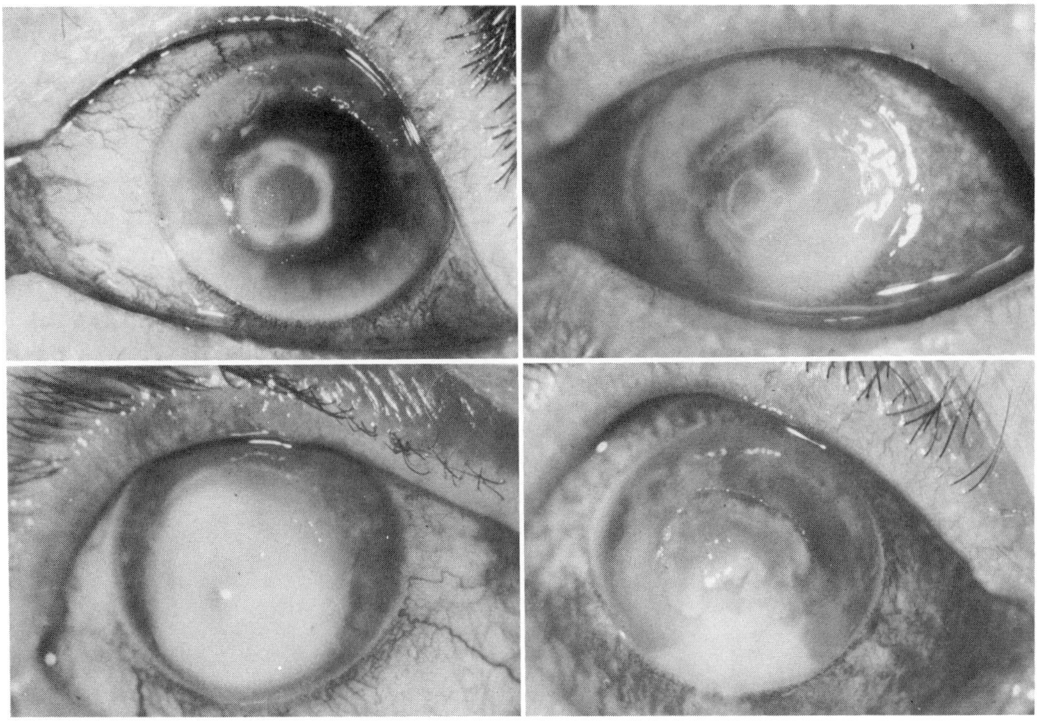

Fig. 13-10. A, Keratitis caused by *Candida tropicalis;* **B,** keratitis caused by *Alternaria;* **C,** keratitis caused by *Fusarium;* **D,** keratitis caused by *Monosporium apiospermum.*

be either by the bloodstream or more commonly by direct extension from the nose through the sinuses and then the orbit. Such orbital disease usually implies a widespread involvement, and many of these situations end fatally. Systemic antibiotics are essential in their management of the disease.

The clinical ophthalmologist should be aware that any structure of the eye can be invaded by fungi, that such infections are apt to be obscure and puzzling, and that not infrequently the diagnosis may be missed for a considerable period. Therefore, one should be on the alert to suspect such invaders in unusual clinical problems and particularly in individuals who have a systemic disorder that has resulted in severe debilitation. One must be prepared for the fact that treatment is frequently unsatisfactory.

Fig. 13-10, *A-D* illustrates keratitis produced by four genera of fungi: *Candida tropicalis, Alternaria* species, *Fusarium* species, and *Monosporium apiospermum.* Fig. 13-21 illustrates a fresh preparation from a culture of the *Alternaria* keratitis, and Fig. 13-19 shows a similar culture of *Monosporium apiospermum* from the patient shown in Fig. 13-10, *D.*

TREATMENT OF MYCOTIC KERATITIS

Various chemical or physical agents have been reported effective in the management of both clinical and experimental mycotic keratitis. They include nystatin, amphotericin B, pimaricin, potassium iodide, heat, cryotherapy, laser beam, and surgical curettage. These agents were used alone or in combination with one of the others listed.[11a]

THE LABORATORY DIAGNOSIS OF MYCOTIC KERATITIS

Collection of specimens

The fungus producing ophthalmic keratitis is confined to the base of the ulcer where it is firmly embedded in a layer rarely exceeding 0.5 mm. For this reason a cutting instrument such as an ophthalmic curette should be employed, rather than a cotton swab, in collecting a sample. The sample does not have to be larger than the head of a pin, but it should be collected aseptically and kept moist or rehydrated with sterile water at the time of examination and culture.

To avoid introducing contaminants, one should collect the sample in a draft-free room and take strict aseptic precautions, which should be maintained from the time of collection throughout the entire process of transport to and subsequent handling in the mycology laboratory.

For this purpose, sterile glass Petri dishes of standard size (10 cm in diameter and 1 cm deep) are employed at the Columbia-Presbyterian Medical Center. The dishes are wrapped individually in paper before sterilization and kept in this sterile envelope to protect the dish from air currents and the ever present aerosol of spores of contaminating bacteria and fungi. Care is taken in opening and closing the dish, which is always handled within the confines of its own loosened sterile paper wrapper.

These extra aseptic precautions are necessary because of the slow growth of pathogenic fungi. Since cultures will have to be incubated at times for several weeks, any accidental airborne contaminant will have an opportunity to develop and outgrow the pathogen.

The wet sample is deposited directly from the curette into the center of the bottom dish either by rubbing the instrument against the glass bottom of the dish or by detaching the sample with the aid of a sterile needle or another sterile curette. The lid of the dish is replaced gently, the paper wrapper pressed firmly but slowly against the surface of the closed dish, and the sample is taken to the laboratory without delay to avoid drying.

Direct microscopic examination

Immediately upon arrival at the laboratory the sample is rehydrated with a drop

Table 13-13. Guide to the recognition in tissue and isolation in cultures of fungi most frequently implicated as causes of mycotic keratitis

		Conditions for isolation	
Genus or species	**Parasitic morphology**	**Culture media**	**Temperature**
Aspergillus fumigatus	Septate branching hyphae, 2-6 μ wide, often in clusters; occasional conidial heads (Fig. 13-1)	Sabouraud's glucose agar	45° and 37° C
Candida albicans	Oval-to-elongate budding cells, 2-3 × 3-7μ, plus hyphae and pseudohyphae	Sabouraud's glucose agar, Mycosel agar*	Room temperature and 37° C
Fusarium *Cephalosporium* *Penicillium* *Monosporium* *Verticillium* *Alternaria*	Sparse septate branching hyphae, 2-4 μ, with occasional intercalary chlamydospores	Sabouraud's glucose agar	Room temperature or 23-25° C

*Sabouraud's glucose agar plus chloromycetin and cycloheximide.

of sterile distilled water. A small portion of the sample—no larger than one-fourth its total volume—is separated by means of the cutting edge of a specially flattened, stiff nichrome wire (mycologic needle) and transferred to a sterile slide. A sterile coverslip is then gently lowered over this portion of the specimen; and after the preparation is flattened with firm but gentle pressure, the slide is examined *without further additives* directly under a compound light microscope. Scanning with the low power permits locating filamentous or unicellular structures of presumably fungal nature. These can appear as hyphae or conidia or as clusters of budding cells, depending on the genus or species. Higher magnification will permit confirming the suspicion and at the same time enable preliminary identification of the fungus. Table 13-13 describes the morphology of the fungal genera and species most frequently implicated as agents of mycotic keratitis.

The successful demonstration of the invading fungus in a wet sterile preparation of a portion of the specimen will accomplish two objectives: it confirms the tentative clinical diagnosis that the lesion was indeed a mycotic infection, and it permits a preliminary identification of the particular fungus causing the problem. This information will be useful to the physician in choosing the most specific therapy available and to the mycologist in choosing the best media, temperature, and atmosphere for the cultural isolation of the disease-causing fungus.

Should microscopic examination of the first slide be negative, one or two additional fractions of the sample should be similarly mounted and examined. In the event that only one of the small fragments examined has been positive, the coverslip can be lifted and the specimen returned to the Petri dish for further subdivision before culturing. The reason for using a sterile slide and coverslip now becomes apparent, for they permit culturing the area of the specimen, which contains fungal elements.

Special stains

To help reveal the fungus in those cases where microscopic examination of wet preparations has been negative, and also to keep a permanent record of positive cases, certain microbiologic and histologic stains are useful. These are enumerated below.

GRAM STAIN. For smears or exudates, the Kopeloff and Beerman modification of the Gram stain should be employed. This is a buffered stain, recommended by the Stain Committee of the American Society of Microbiology, to reveal gram-positive microorganisms in exudates. For actinomycetes in histologic sections, the Brown and Brenn modification of the Gram stain is essential. The more widely employed Gram-Weigert stain is unsuitable. Although all fungi are gram positive, this stain should be employed only for yeast cells and actinomycetes. Filamentous fungi do not stain well with Gram stain, and other stains should be employed (see below).

PERIODIC ACID, SCHIFF REAGENT STAIN. The PAS stain and its many modifications (Gridley's, for example) are useful in helping reveal those fungi that do not stain well with H & E in histologic sections. The Gridley modification is useful for staining smears as well.

GOMORI SILVER METHENAMINE STAIN. Although a prolonged procedure, the Groccot modification of the silver methenamine stain has great sensitivity as a detector of fungi and actinomycetes in histologic sections and is the preferred procedure in screening for their presence on any suspected lesion.

FITE-FARACO STAIN. As a stain to demonstrate acid-fast organisms the Fite stain, as modified by Faraco, is unexcelled in sensitivity. Hansen's bacillus and other mycobacteria are often revealed by this stain where Ziehl-Neelsen's has failed. The Coates modification has been particularly devised for *Nocardia asteroides*.

Primary cultures

Besides the culture media listed in Table 13-13, others useful in culturing systemic fungal pathogens from the eye include Littman's oxgall agar and brain-heart infusion agar to which may be added 10% whole blood and chloromycetin or gentamicin.

As already discussed, small pieces of the sample can be deposited individually on various spots on the surface of the agar media, avoiding the edges. Streaking is seldom necessary.

Purification of cultures

As soon as a fungal colony has reached a diameter of 2 to 5 mm, it should be subcultured to fresh media to avoid overgrowth by contaminants that may be present in the primary culture. The pathogen can usually be recognized by its slower growth and by the fact that several similar colonies will appear simultaneously at various inoculated sites. If subcultures are carried out early enough, one can usually obtain a pure culture for further study. At this time, a teased mount can be made for preliminary identification of the fungus.

Identification of cultures

The criteria employed in identifying genera and species of fungi include (1) gross colony morphology; (2) microscopic morphology, with particular emphasis on the type of sporulation; (3) biochemical reactions; (4) serologic reactions; and (5) animal pathogenicity. In order for one to be able to observe all these, special media and methods are required, which are described below.

Cultural characteristics of fungi most frequently invading the eye

GENUS ASPERGILLUS. This genus has been found to be the cause of 33% of fungal eye lesions, as seen in a survey of the world literature (summarized in Table 13-9). The generic name *Aspergillus* is descriptive of its conidiophores (Fig. 13-11), whose vesicle resembles an aspergillum, a perforated globe used to sprinkle holy water in church. The most frequent pathogen in this genus is the species *fumigatus* (Figs. 13-11 and 13-12). The specific name *fumigatus* describes the smoky color of the colony, which is a grayish yellow-green when its conidia have reached maturity. This generally occurs in less than 48 hours if cultures are incubated at 45° C, the op-

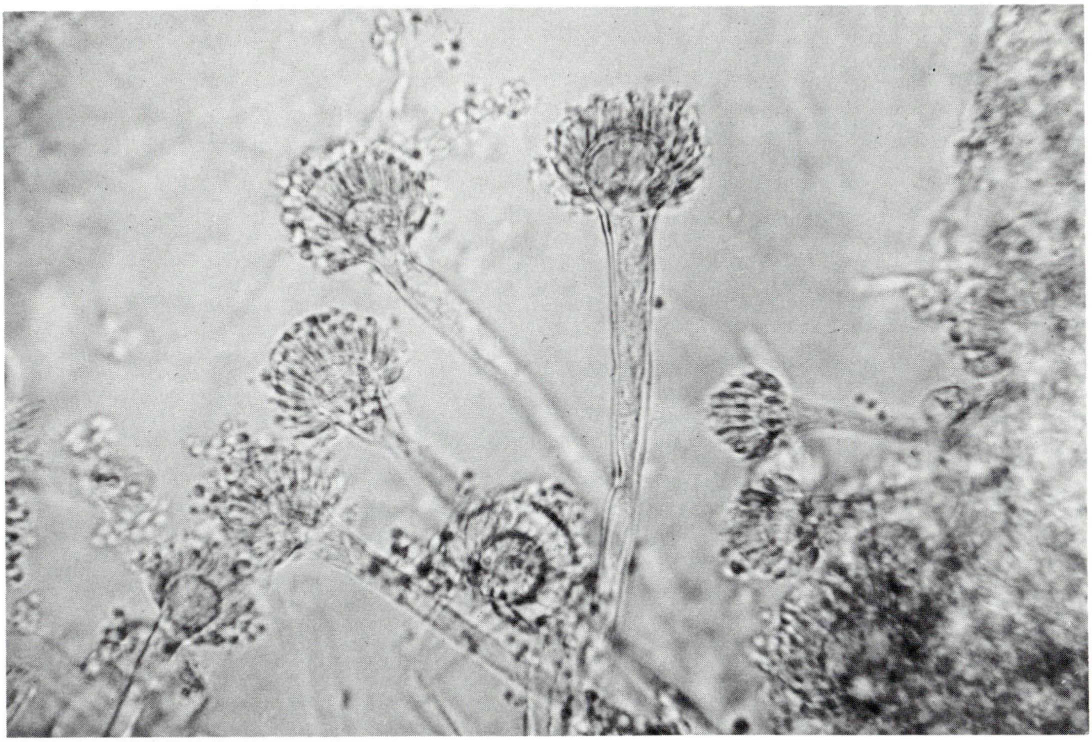

Fig. 13-11. *Aspergillus fumigatus* (sputum isolate) showing conidiophores with typical clavate vesicles and uniserate sterigmata (phialids) after 48 hours' growth on Sabouraud's glucose agar at 45° C. (approx. ×600.)

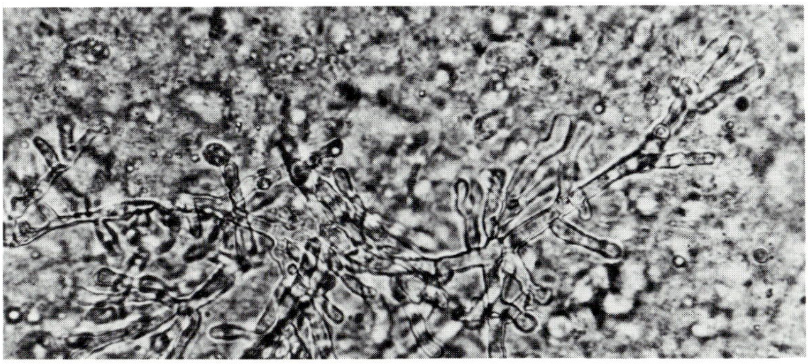

Fig. 13-12. Typical cluster of branching septate hyphae of *Aspergillus fumigatus* in a fresh preparation of sputum mounted in distilled water. (×430.)

timal temperature for growth of this thermophilic species. On Sabouraud's glucose agar, the entire surface of the slant will be covered by the fungus in a day or two. The border of a growing colony and its undersurface remain white because of the absence or immaturity of the conidia.

The species is identified by the following criteria.

The morphologic characteristics of infected tissues consist of compact aggregates of septate hyphae of uniform thickness (4 to 6 μ in diameter), exhibiting a characteristic coralloid pattern with dichotomous branching and rounded, sometimes swollen, hyphal tips (Fig. 13-12).

Microscopic examination reveals clavate vesicles with single layers of phialids on

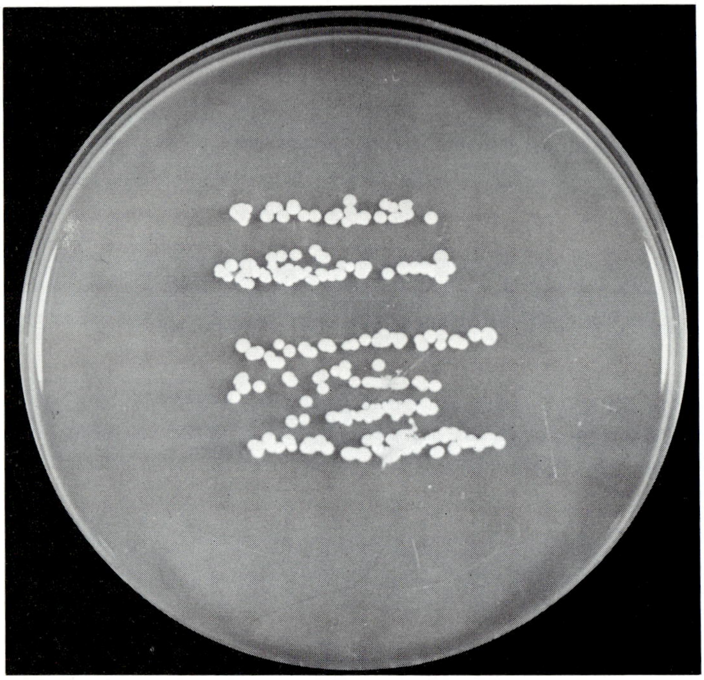

Fig. 13-13. *Candida albicans* colonies in primary culture of corneal scraping on Sabouraud's glucose agar. (×½.)

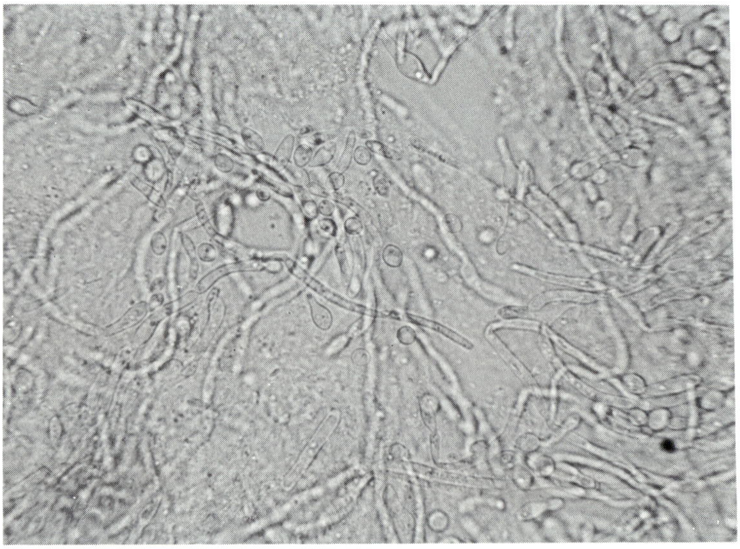

Fig. 13-14. *Candida albicans* from corneal ulcer. Budding cells, hyphae, and pseudohyphae in unstained wet preparation of direct scraping. (×400.)

their upper hemispheres from which conidia are abstricted repeatedly, forming parallel chains that adhere to each other and give the sporulating head a characteristic columnar appearance.

Aspergillus has the ability to grow at a temperature as high as 50 to 55° C; the optimal temperature is 45° C.

Besides *A. fumigatus,* the species *flavus, glaucus,* and *niger* have been reported from corneal ulcers. For a key to their specific identification, the reader is referred to Raper and Fennell's monograph *The Genus Aspergillus.*[47]

These additional specific names are descriptive of the species' conidial pigments (flavus, yellow; glaucus, white, and niger, black). All three have spherical, rather than clavate, vesicles at the sporulating end of their conidiophores; and one or two layers of phialids (uniseriate or biseriate) may extend all over the surface of the vesicle from which conidial chains radiate to give the sporulating head a spherical, rather than columnar, shape.

GENUS CANDIDA. The genus *Candida* includes a group of fermentative yeasts, which, besides reproducing vegetatively by budding, are also capable of producing pseudohyphae and true hyphae but no ascospores. The most frequent pathogen in this genus is the species *albicans,* whose description follows.

Colonies are pasty to creamy, buff colored, hemispherical in outline, and rarely exceed 2 to 3 mm in diameter during the first week of growth (Fig. 13-13). Microscopically, surface growth on Sabouraud's agar consists of oval budding cells (blastospores) of medium size (2 × 3 μ). Subsurface growth on cornmeal and Wolin-Bevis agar shows pseudohyphae and true hyphae along whose length spherical clusters of blastospores are strung at regular intervals (Fig. 13-14). Toward the ends or tips of these filaments chlamydospores are produced. These are diagnostic for the species and are spherical in shape, 8 to 10 μ in diameter, with heavy, double-contoured walls and large droplets of lipid and glycogen in their cytoplasm (Fig. 13-15).

C. albicans can also be identified by the ability of its blastospores (budding cells) to round up and germinate or extrude filamentous outgrowths when suspended in human or other mammalian blood serum and incubated at 37° C for 1 to 4 hours. Other characteristics used in its identification are: (1) resistance to cycloheximide; (2) ability to grow on various media at 37° C as well as at room temperature; (3) ability to ferment, with gas production, the sugars glucose, maltose, and sucrose, but not lactose; (4) inability to split urea or to utilize KNO_3; and (5) pathogenicity for the mouse and rabbit, causing death with miliary renal abscesses in less than 1 week.

Fig. 13-16 illustrates a smear from a culture of *C. parapsilosis.*

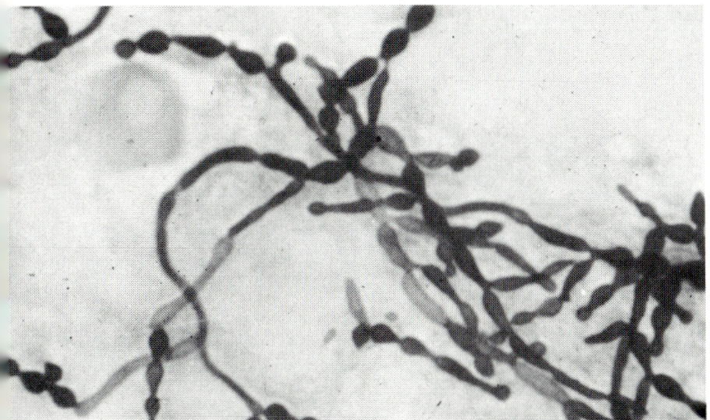

Fig. 13-15. *Candida albicans* from direct corneal scraping. Budding cells and young pseudohyphae. PAS stain. (×400.)

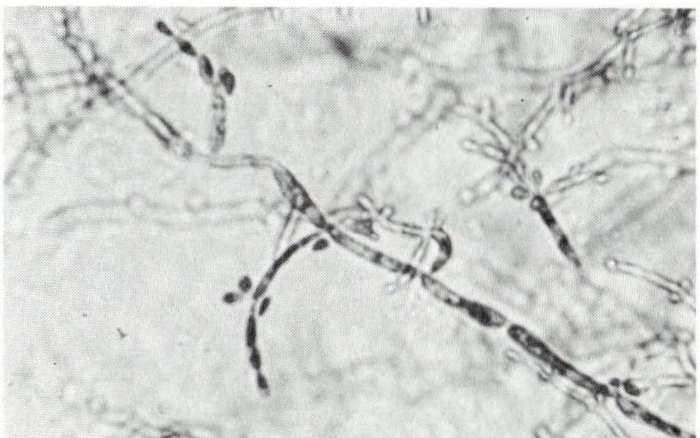

Fig. 13-16. *Candida parapsilosis.* Giant pseudohyphae in culture on cornmeal agar. From a patient with meibomitis. (×400.)

GENUS FUSARIUM. This genus produces colonies that are cottony in texture, and lavender, pink, or yellow. Their conidia are produced in clusters from phialids in palisade arrangement along the aerial (fertile) hyphae and are of two types: unicellular and multicellular. The unicellular ones, called microconidia, are oval to crescent-shaped; the multicellular ones, called macroconidia, are longer, curved into long crescents, and septate. Since all the septa are at right angles to the long axis of the spores, these macroconidia are said to resemble canoes. Clusters of conidia are often compared to bunches of bananas; they are frequently held together in a gelatinous substance by the conidiophores.

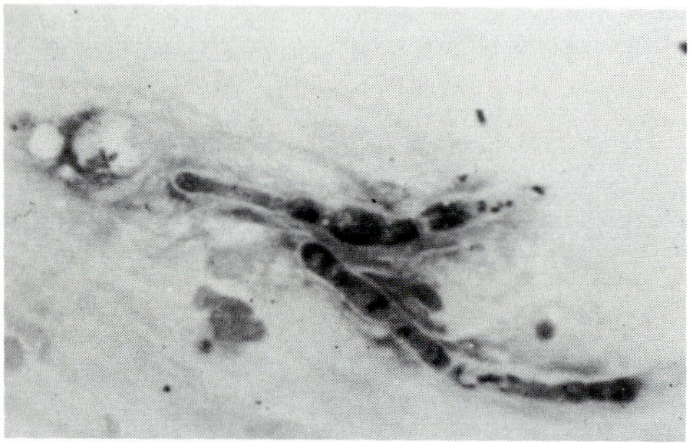

Fig. 13-17. *Cephalosporium* species. Hyphae in direct scraping from corneal ulcer, showing septa and intercalary chlamydospore. Giemsa stain. (×400.)

Fusarium is a very large genus containing numerous species, most of which are parasitic on fruit trees and other crops. Because of its importance to agriculture, the genus has been extensively studied and monographed. A good reference is Snyder and Hansen's article, published in 1940.[54]

Perhaps because of its wide distribution in nature, *Fusarium* is a frequent air contaminant. This fact plus its association with erosive vegetable particles may account for its frequent incidence in eye lesions.

GENUS CEPHALOSPORIUM. The name of this genus describes its conidiophores, borne singly as erect tapering stalks on whose apex conidia are clustered as heads. When grown slightly below room temperature and in short periods of daylight, *Cephalosporium* (Figs. 31-17 and 13-18) can also produce conidia in enclosed cushionlike structures called sporodochia. Taxonomically, this event requires transferring the species to the genus *Volutella* in the family Tuberculariaceae (other filamentous fungi mentioned thus far are in the family Moniliaceae [Saccardo's classification]). This taxonomic information is given here because several cases of *Cephalospor-*

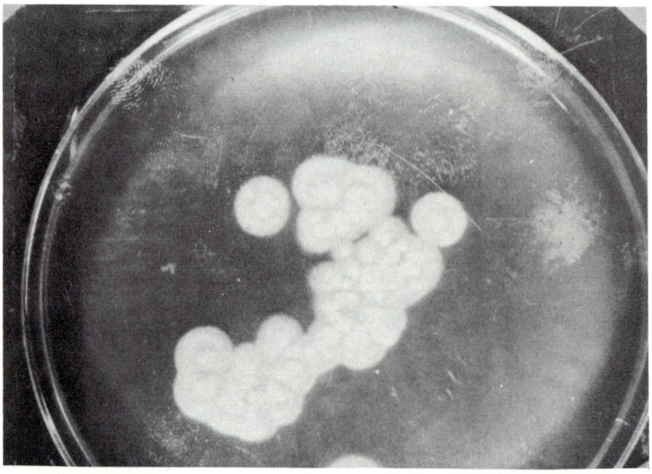

Fig. 13-18. *Cephalosporium* species. Colonies in primary culture from corneal scraping on Sabouraud's glucose agar. (×⅔.)

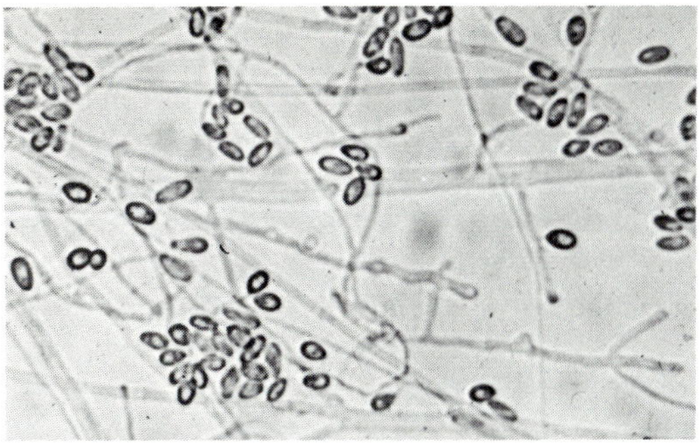

Fig. 13-19. *Monosporium apiospermum.* Conidia in culture, lactophenol mount. From a corneal ulcer. (×400.)

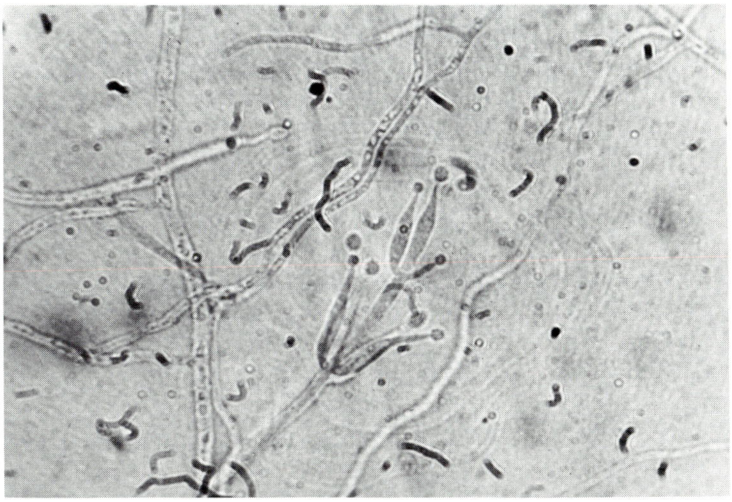

Fig. 13-20. *Verticillium* species. Slide culture showing conidia borne singly or in short chains from phialids arising in whorls at regular intervals (in verticillate arrangement) on the conidiophores. From a corneal ulcer. (×400.)

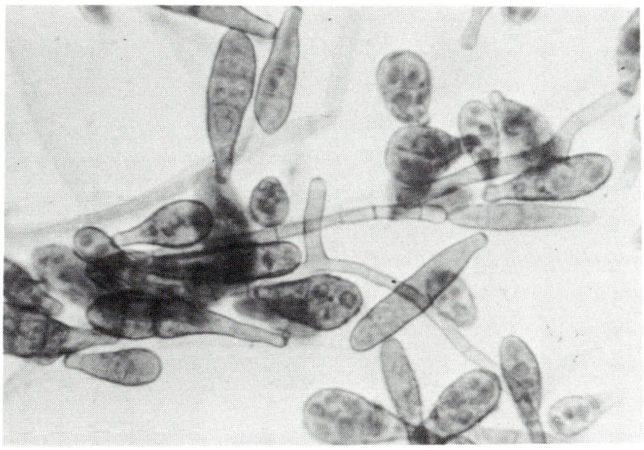

Fig. 13-21. *Alternaria* species. Typical dematiaceous (brownish) multiseptate conidia with septa in two planes (dyctiospores). From a corneal ulcer. (×1,000.)

ium and *Volutella* corneal ulcers have been described, and it is important to know that we are dealing with the same isolates grown under different environmental conditions. *Cephalosporium* (or *Volutella*) organisms have also been isolated from the cortisone eye drops used by patients.

Colonies of these two genera are velvety to fluffy, but may be membranous in species producing sparse aerial mycelium with moist conidiophores.

GENUS PENICILLIUM. This genus is closely related botanically to *Aspergillus*, but differs from it in morphologic characteristics. Its colonies are more finely powdery than those of *Aspergillus*, with shades of dark blue-green or blue-gray predominating. Colony color here, too, is the color of the mature conidia, which are also produced from phialids. These, however, are not on a vesicle but in branched conidiophores whose terminal cells (metulae) are not inflated. The conidial head therefore resembles a fine painter's brush (penicillus) rather than a dust mop or aspergillum (Plate 3, *D*). Species on this genus, of which *P. notatum* (producer of penicillin) is a well-known example, are widely distributed in nature as fruit and leather spoilers. Citrus fruits, peaches, and plums, as well as shoes, belts, and suitcases, are often seen covered with the bluish mold dust of *Penicillium* conidia.

• • •

Figs. 13-19, 13-20, and 13-21 illustrate the appearance of other fungi that have been cultured from patients with corneal ulcers.*

*For further information on identification of fungi involved in keratitis, see references 10, 25, 35, 48, 49, and 54.

REFERENCES

1. Agarwal, L. P., and Khosla, P. K.: Organismal conjunctival flora in Indians, Orient. Arch. Ophthal. **1**:145, 1963.
2. Ainley, R., and Smith, B.: Fungal flora of the conjunctival sac in healthy and diseased eyes, Brit. J. Ophthal. **49**:505, 1965.
3. Albesi, E. J.: Infeccion micotica ocular, Arch. Oftal. B. Air. **44**:101, 1969.
4. de Almeida, A. A.: Infecção pós-operatória—Bacteriana e micótica, Rev. Brasil. Oftal. **22**:53, 1963.
5. Arnold, R., and Whildin, J.: Rhinosporidiosis of the conjunctiva, Amer. J. Ophthal. **25**:1227, 1942.
6. de Azevedo, M. L.: Investigações preliminares sôbre a microflora ocular, Arq. Brasil. Oftal. **25**:41, 1962.
7. Balakrishnan, E.: Mycotic keratitis. In XIX International Ophthalmological Congress Acta 11, 1242, New Delhi, 1962.
8. Ball, J. M.: Modern ophthalmology, vol. 1, Philadelphia, 1926, F. A. Davis Co., p. 479.
9. Blake, J.: Ocular hazards in agriculture, Ophthalmologica (Basel) **158**:125, 1969.
10. Barron, G. L.: The genera of Hyphomycetes from soil, Baltimore, 1968, The Williams and Wilkins Co.
11. Chick, E. W., and Conant, N. F.: Mycotic ulcerative keratitis: a review of 148 cases from the literature, Invest. Ophthal. **1**:419, 1962.

11a. DeVoe, A. G.: Keratomycosis, Amer. J. Ophthal. **71**:406, 1971.

12. Diamond, M. A., and Kirk, H. Q.: Postoperative intraocular aspergillosis, Amer. J. Ophthal. **54**:1124, 1962.

12a. Dieckhues, B.: Die Behandlung der Keratomykose mit Nystatin und Pimaricin, Klin. Mbl. Augenheilk. **148**:895, 1966.

13. Editorial: Fungal corneal ulcers, J.A.M.A. **179**:649, 1962.

13a. Ellison, A. C., Newmark, E., and Kaufman, H. E.: Chemotherapy of experimental keratomycosis, Amer. J. Ophthal. **68**:812, 1969.

14. Fazakas, A.: Über die von Bindehaut, Hornhaut, Lidrand und Tränenwegen gezüchteten Pilze, Graefes Arch. Ophth. **133**:461, 1935.
15. Fazakas, A.: Über die Schimmelpilze des gesunden und kranken Auges, Graefes Arch. Ophth. **138**:416, 1938.
16. Fazakas, A.: Zusammenfassender Bericht über meine ophthalmologischen Pilzuntersuchungen, Ophthalmologica (Basel) **126**:91, 1953.
17. Fazakas, A.: Zusammenfassender Bericht über meine im Rahmen der Akademischen Arbeitsgemeinschaft durchgeführten ophthalmologischen Pilzuntersuchungen, Ophthalmologica (Basel) **128**:163, 1954.
18. Fine, B. S., and Zimmerman, L. E.: Exogenous intraocular fungus infections with particular reference to complications of intraocular surgery, Amer. J. Ophthal. **48**:151, 1959.
19. Foster, J. B. T., Almeda, E., Littman, M. L., and Wilson, M. E.: Some intraocular and conjunctival effects of amphotericin B in man and in the rabbit, Arch. Ophthal. **60**:555, 1958.

19a. François, J., DeVos, E., Hanssens, M., and

Elewaut-Rysselaere, M.: Mycoses intra-oculaires, Ann. Oculist. **195**:97, 1962.
20. François, J., Elewaut-Rysselaere, M., and De Vos, E.: Les mycoses oculaires, Paris, 1968, Masson et Cie.
21. Gots, J. S., Thygeson, P., and Waisman, M.: Observations on Pityrosporum ovale in seborrheic blepharitis and conjunctivitis, Amer. J. Ophthal. **30**:1485, 1947.
22. Graf, K.: Postoperative intraokulare Mykosen nach Kataraktoperation, Klin. Mbl. Augenheilk. **144**:309, 1964. (Abstract.)
22a. Green, W. R., Bennett, J. E., and Goos, R. D.: Ocular penetration of amphotericin B.: a report of laboratory studies and a case report of postsurgical Cephalosporium endophthalmitis, Arch. Ophthal. **73**:769, 1965.
23. Halde, C., and Okumoto, M.: Ocular mycosis: a study of 82 cases. In Proc. XX International Congress of Ophthalmology, part 2, 705, 1966. International Congress series no. 146. Excerpta Medica Foundation.
24. Hammeke, J. C., and Ellis, P. P.: Mycotic flora of the conjunctiva, Amer. J. Ophthal. **49**:1174, 1960.
25. Hazen, E., Gordon, M. A., and Reed, C.: Laboratory identification of pathogenic fungi simplified, ed. 3, Springfield, Ill., 1970, Charles C Thomas, Publisher.
26. Hoffmann, D. H., and Naumann, G.: Ein Beitrag zur Pilzinfektion der Hornhaut, Klin. Mbl. Augenheilk. **142**:286, 1963.
26a. Hoefnagels, K. L. J., and Pijpers, P. M.: Histoplasma capsulatum in a human eye, Amer. J. Ophthal. **63**:715, 1967.
27. Janke, R. G., and Schwab, F.: Zur Frage der Pilzflora der normalen Bindehaut des Menschen, Wien. Med. Wschr. **111**:859, 1961.
27a. Jones, B. R., Richards, A. B., and Morgan, G.: Direct fungal infection of the eye in Britain, Trans. Ophthal. Soc. U. K. **89**:727, 757, 1969.
28. Jones, D. B., Sexton, R., and Rebell, G.: Mycotic keratitis in South Florida: a review of 39 cases, Trans. Ophthal. Soc. U. K. **89**:781, 1969.
29. Kinnas, J. S.: Ophthalmic disease caused by a mycete of the giant cane, Brit. J. Ophthal. **49**:327, 1965.
30. Küper, J.: Zur Klinik postoperativer intraokularer Mykosen, Klin. Mbl. Augenheilk. **140**: 827, 1962.
31. Leber, T.: Keratomycosis aspergillina als Ursache von Hypopyonkeratitis, Graefes Arch. Ophth. **25**:285, 1879.
32. Ley, A. P., and Sanders, T. E.: Fungus keratitis: a report of three cases, Arch. Ophthal. **56**:257, 1956.
33. Lichaa, E.: Mycose oculaire primitive due au "Beauveria Brumpti–Langeron et Lichaa," Bull. Ophthal. Soc. Egypt **26**:91, 1933.
34. Locatcher-Khorazo, D., Benham, R. W., and Silva-Hutner, M.: Incidence of fungi from clinically healthy eyes of 508 young people 10-18 years of age and in 1,347 adults 19-80 years of age. Unpublished data.
35. Lodder, J., editor: The yeasts, a toxonomic study, Amsterdam, 1970, North Holland Publishing Co.
36. Mitsui, Y., and Hanabusa, J.: Corneal infections after cortisone therapy, Brit. J. Ophthal. **39**:244, 1955.
37. Naumann, G., Green, W. R., and Zimmerman, L. E.: Mycotic keratitis: a histopathologic study of 73 cases, Amer. J. Ophthal. **64**:668, 1967.
38. Nema, H. V., Ahuja, O. P., Bal, A., and Mohapatra, L. N.: Mycotic flora of the conjunctiva, Amer. J. Ophthal. **62**:968, 1966.
39. Okuda, K., Matsuura, K., and Okabe, S.: A case of keratomycosis caused by Cephalosporium sp. (light and electron microscopic observations), Folia Ophthal. Jap. **21**:285, 1970.
40. Olson, C. L.: Ocular mycosis, Gen. Practice **40**:78, August, 1969.
41. Ovsepian, T. L., and Osipian, L. L., Mycologic flora of the conjunctival sac, Zh. Eksp. Klin. Med. **5**:78, 1965.
42. Paiva, C., Chaves Batista, A., and Gomes, A.: Endoftalmite micótica pós-operatoria por Hyalopus bogolepofii, Rev. Brasil. Oftal. **19**: 193, 1960.
42a. Parfenov, L. K.: Fungal infections of the eyes in the Astrakhan region, Vestnik Oftal. **76**:17, 1963.
43. Parunovic, A., and Halde, C.: Pityrosporum orbiculare: its possible role in seborrheic blepharitis, Amer. J. Ophthal. **63**:815, 1967.
44. Percebois, G., and Raspiller, A.: Flore fongique de la conjonctive humaine, Bull. Soc. Ophtal. Franc. **48**:626, 1968.
44a. Polack, F. M., Kaufman, H. E., and Newmark, E.: Keratomycosis: medical and surgical treatment, Arch. Ophthal. **85**:410, 1971.
45. Puttanna, S. T.: Mycotic infections of the cornea, J. All India Ophthal. Soc. **15**:11, 1967.
46. Rao, P. N. S., and Ramakrishnan, T. S.: Studies on the fungus diseases affecting man in and around Manipal, Indian J. Path. Bact. **11**:53, 1968.
47. Raper, K. B., and Fennell, D. I.: The genus Aspergillus, Baltimore, 1965, The Williams and Wilkins Co.
48. Raper, K. B., and Thom, C.: Manual of the Penicillia, Baltimore, 1949, The Williams and Wilkins Co.
49. de Rezende, C.: Ocular mycosis. In XIX Concilium Ophthalmoligicum, New Delhi, 1962. Acta **1**:299, 1962.

50. Sachsenweger, R.: Klinische Befunde und experimentelle Untersuchungen üeber Augenerkrankungen durch saprophytär wachsende Pilze, Klin. Mbl. Augenheilk. **127**:721, 1955.
51. Salceda, S. R., Nievera, L.-F., and Abendaño, R.: Keratomycosis: a clinical, laboratory and therapeutic evaluation of 17 cases, Philippine J. Ophthal. **1**:31, 1969.
52. Sanders, N.: Penetrating keratoplasty in treatment of fungus keratitis, Amer. J. Ophthal **70**:24, 1970.
52a. Scholer, H. J., and Saubermann, G.: Aspergillose der Hornhaut, vol. 2, Diskussion von 90 Fällen der Literatur, Basel, 1959, S. Karger, p. 23.
53. Sinha, B. N., and Das, M. S.: Bacterial and fungal flora of the conjunctival sacs in healthy and diseased eyes, J. Indian Med. Ass. **51**:217, 1968.
54. Snyder, W. C., and Hansen, H. N.: The species concept of Fusarium, Amer. J. Bot. **27**:64, 1940.
55. Strampelli, B., and Maggi, C.: Endoftalmite micotica post-operatoria, Riv. Ital. Trachome Patol. Ocul. Esot. **15**:168, 1963.
56. Theodore, F. H., Littman, M. L., and Almeda, E.: The diagnosis and management of fungus endophthalmitis following cataract extraction, Arch. Ophthal. **66**:163, 1961.
57. Theodore, F. H., Littman, M. L., and Almeda, E.: Endophthalmitis following cataract extraction due to Neurospora sitophila, a so-called non-pathogenic fungus, Amer. J. Ophthal. **53**:35, 1962.
58. Thomas, C. I.: The cornea, Springfield, Ill., 1955, Charles C Thomas, Publisher.
59. Thygeson, P.: Etiology and treatment of blepharitis: a study in military personnel, Arch. Ophthal. **36**:445, 1946.
60. Thygeson, P., Hogan, M. J., and Kimura, S. J.: Cortisone and hydrocortisone in ocular infections, Trans. Amer. Acad. Ophthal. Otolaryng. **57**:64, 1953.
61. Thygeson, P., and Vaughan, D. G., Jr.: Seborrheic blepharitis, Trans. Amer. Ophthal. Soc. **52**:173, 1955.
62. Vasquez de Parga, A. S., and Pereiro, M.: Flora micotica de la conjunctiva, Arch. Soc. Oftal. Hisp. Amer. **25**:168, 1965.
63. White, J. H.: Fungal contamination of donor eyes, Brit. J. Ophthal. **53**:30, 1969.
64. Williamson, J., Gordon, A. M., Wood, R. Dyer, A., and Yahya, O. A.: Fungal flora of the conjunctival sac in health and disease, Brit. J. Ophthal. **52**:127, 1968.
65. Wilson, L. A., Ahern, D. G., Jones, D. B., and Sexton, R. R.: Fungi from the normal outer eye, Amer. J. Ophthal. **67**:52, 1969.
66. Zimmerman, L. E.: Keratomycosis, Survey Ophthal. **8**:1, 1963.

14

The Chlamydia: trachoma and inclusion conjunctivitis

Trachoma, a disease of the conjunctiva and cornea, is clinically defined by four stages. These start with granular conjunctivitis and epithelial keratitis. Follicular hypertrophy, particularly in the conjunctiva of the upper eyelid, is followed by pannus formation, which is most constant in the upper segment of the cornea; but when the disease is severe, the whole cornea becomes involved. Healing takes place with varying degrees of cicatrization, which determine the extent of visual impairment. Trachoma is considered a disease limited to the eye, spontaneously transmitted only from one eye of man to another. It is to be differentiated from inclusion conjunctivitis, which in the newborn is acquired from an infection of the mother's birth canal and in the adult from contact with infected material from the genitourinary tract as in swimming pools before chlorination or, rarely, from infected eyes. Initially somewhat similar in appearance to the *early* stages of stages II or IIb,[17] inclusion conjunctivitis is mild, does not produce gross pannus formation, and heals without cicatrization. But certain common features between trachoma and inclusion conjunctivitis, which are currently being investigated, make it more relevant at present to treat the two diseases in parallel in order to help clarify their relationship to each other.

EPIDEMIOLOGY

Trachoma

Trachoma is apparently as old as recorded history. According to Kugelmass,[15] the disease was known in China in the twenty-seventh century BC and was endemic in Mesopotamia in the twentieth century BC. The Ebers Papyrus, written about 1550 BC, mentions a disease whose description is consistent with that of trachoma. In the fifth century BC trachoma was common in Greece. As noted in Chapter 1, Celsus in 14 AD recorded the roughness of the eyelids, and Aetus of Amida in 550 AD described the four stages of the disease.[7] Trachoma is now endemic around the world, as may be seen in Table 14-1, which shows the distribution and prevalence of the disease in 1962.[4,5] The areas of highest concentration of trachoma lie around the Mediterranean, especially Africa, and in parts of Iran, India, and Indochina. In the United States trachoma is found in some Indian reservations in the West and among some rural groups in the Appalachian mountains. It is estimated that about 400 million people are infected with the trachoma agent and that about 20 million are blind.[6]

Trachoma is spread from eye to eye by contact with contaminated hands, towels, and other intermediate carriers. Being con-

Table 14-1. Geographic distribution of trachoma*

Majority of population affected with trachoma	Trachoma highly widespread (20%-50%)	Trachoma quite common (1%-20%)	Trachoma prevalent sporadically	Trachoma practically extinct
Egypt	Aden	Argentina	Baltic states	Austria
Eritrean highlands	Africa (equatorial East, and South—(among the native populations	Australia (among the aboriginal tribes)	Chile	Belgium
Israel (Arab population)	Borneo	Africa (West)	Finland	Canada
Kuwait	Burma	Balkan states	France (imported cases only)	Denmark
Lybia	Cambodia	Brazil	Guatamala	England
Some districts of	China (mainland)	Bulgaria	Honduras	Germany
Algeria	India (areas)	Ceylon	Ireland	Holland
China	Indonesia	Greece	Italy (central and northern)	Iceland
India (northern)	Iran	Hungary	Panama	Norway
Iran	Iraq	India (areas)	Paraguay	Scotland
Iraq	Jordan	Israel	Peru	Switzerland
Lebanon	Laos	Italy (southern)	Philippines (areas)	
Morocco	Lebanon	Japan	Salvador	
Saudi Arabia	Morocco	Korea	Sweden	
Syria	Pakistan (West)	Lebanon	Uruguay	
Tunisia	Sudan	Mediterranean Islands	United States (non-Indian populations)	
Vietnam	Syria	Mexico	Venzuela	
	Thailand	New Zealand (among the Maoris)		
	Tunisia	Pacific Islands		
	Turkey (southern)	Pakistan		
	Vietnam (South)	Poland		
		Portugal		
		Rumania		
		Spain (Mediterranean coast)		
		Turkey (central and northern)		
		USSR (southern and central)		
		United States (among the American Indians)		

*From surveys published in 1962 and 1967.[4,5]

cerned with the spread of the disease in the army, Vetch in 1807 prevented the use of common towels by a group of soldiers, some of whom had the disease, while permitting common use of towels by another similar group.[14] He observed that fresh cases of the disease were much more frequent in the latter group. Transmission of trachoma with follicular material to two human volunteers was accomplished in 1823 by Werneck, who established an incubation period of 9 days in one volunteer and 14 days in the other.[14]

Thus, more than 150 years ago information was available, which if used could reduce the spread of the disease. But this information preceded the discovery of bacteria and antisepsis, and attempt at control of trachoma was hardly exercised. The disease spread under conditions of poverty, overcrowding, and cultural patterns that opposed hygienic measures. It was not until 1907, when Halberstaedter and von Prowazek[11] (Fig. 14-1, *A* and *B*) saw viruslike inclusion bodies in the eptihelial cells from

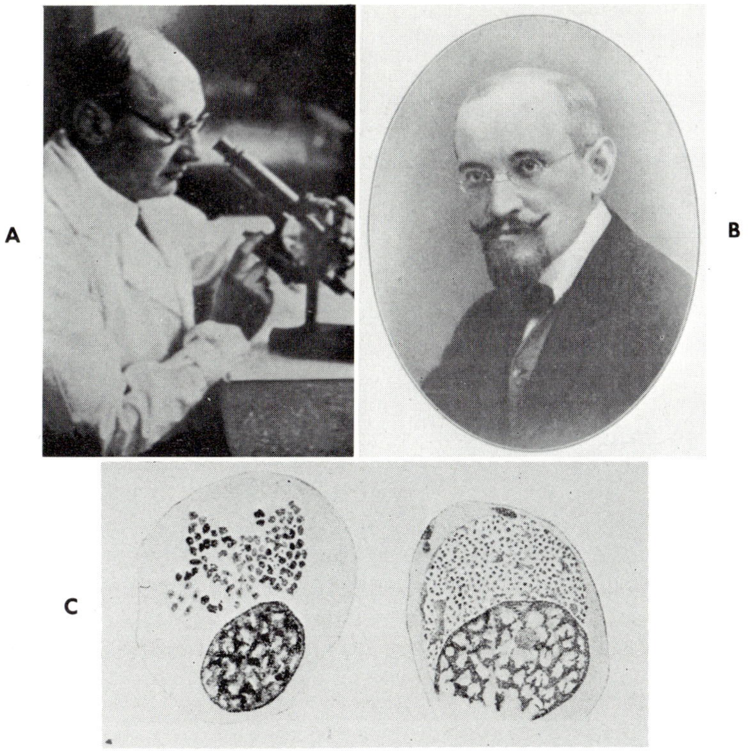

Fig. 14-1. A classic observation in trachoma was made jointly by a German, Halberstaedter (1877-1949), **A**, and an Austrian, von Prowazek (1875-1910), **B**. Both of these physicians were famous for other reasons. Halberstaedter was a pioneer radiologist who developed much of the important groundwork in this field. Von Prowazek worked in the field of rickettsial infections, and the causative organism of typhus fever is named *Rickettsia prowazekii* in his honor. Like Ricketts, he died from typhus fever. When on a joint expedition to Java to study syphilis, Halberstaedter and von Prowazek noted that severe conjunctivitis of trachoma was common. Scrapings of these conjunctivas, **C**, revealed small granular intracellular bodies that stained red with Giemsa stain and lay in a blue matrix. They believed these to be the cause of trachoma, and when the bodies were present in heavy masses, they called them inclusion bodies. They were successful in producing infection in the eyes of the orangutan with their scrapings. (**A** from Finzi, N. S.: Brit. J. Radiol. **22**:347, 1949; **B** from Olpp, G.: Charakterkopfe der Tropenmedizin, Berlin, 1936, Verlag Die Brücke Zur Heimat, p. 60; **C** from Halberstaedter, L., and von Prowazek, S.: Deutsch. Med. Wschr. **33**:1285, 1907.)

trachomatous eyes (Fig. 14-1, *C*), that the modern history of the cause and control of trachoma began.

Inclusion conjunctivitis

Awareness of the disease inclusion conjunctivitis arose in this century. In 1903 Morax[18] described a disease of the eyes of the newborn that occurred in the absence of pathogenic bacteria. Stargardt[22] and Schmeichler[21] in 1909 saw, in epithelial cells of nongonorrheal conjunctivitis of the newborn, inclusion bodies that appeared identical to those found in trachoma. Halberstaedter and von Prowazek looked for evidence of inclusion body disease in the birth canals of those mothers who had infants with inclusion conjunctivitis and found typical inclusions in the epithelial cells.[7] Lindner[16] confirmed these observations and showed that material from both the venereal disease and the conjunctivitis of the newborn proved to be infectious for the eye of the baboon. Lindner named the disease of the infants inclusion blennorrhea.

As in the case of trachoma, under natural conditions the agent of inclusion conjunctivitis is spontaneously pathogenic only for man but has been transmitted to a few primates. This disease of the newborn resembles gonococcal ophthalmitis epidemiologically in that it is acquired from the mother's birth canal. Venereal transmission probably maintains the organism as a human pathogen. The genital infection in the female is seldom recognized, since the cervix is the area of infection and there are no symptoms. However, about 40% of nongonorrheal urethritis in men is found to be of chlamydial origin.[8] The exudate is largely made up of the polymorphonuclear leukocytes that do not have inclusion bodies and of a few epithelial cells with inclusion bodies. Thygeson has commented that infection of the adult eye with the agent of inclusion conjunctivitis has become more frequent with the advent of the sexual revolution. Transmission from infants' eyes is unusual, although obstetricians and nurses may become infected.[12,30] Cases of adult eye infection that used to occur from contaminated swimming pools is now relatively rare because of chlorination.

ETIOLOGY

The organisms that cause trachoma and inclusion conjunctivitis have offered one of the most interesting problems in identification and classification. Halberstaedter and von Prowazek, who saw and first pictured the inclusion bodies of trachoma (Fig. 14-1), gave the name *Chlamydozoen* (mantle body) to these organisms. The similarity of the inclusion bodies of trachoma to those found in psittacosis was noted by Thygeson[24] in 1934. Nicolle and his associates,[19] Thygeson and Proctor,[31] and Julianelle[14] demonstrated the filterability of the infectious agents and transmission of the diseases to monkeys, apes, and man. Despite all this information and the use of the ever increasingly successful methods for cultivation of similar types of obligate intracellular parasites, the agent of trachoma was not successfully cultured until 1957. T'ang and his co-workers[23] substituted streptomycin for penicillin in the preparation of the infectious material for injection in the chick embryonic yolk sac and were successful in the cultivation of the infectious agent. They found that a mixture of streptomycin and penicillin might also be used but yielded fewer positive cultures. Now both the trachoma and inclusion conjunctivitis agents (TRIC agents) have been grown in many laboratories in various areas around the world. Their cultivation characteristics and classification are documented by Jawetz[12] in an article on the history of this milestone in microbiology.

The generic term *Chlamydia* includes the psittacosis group primarily infecting birds; the human pathogens causing lymphogranuloma venereum, trachoma, and inclusion conjunctivitis; and finally a wide variety of other mammalian agents producing such diseases as feline pneumonitis and bovine encephalitis. The various chlamydia are morphologically similar. The infectious particle, or elementary body, is 0.2 to 0.3 $m\mu$ in diameter.[32] When these particles enter a cell they change, after a lag period, to the larger (0.7 μ), much less infectious initial bodies. These in turn give rise to elementary bodies that increase in numbers and finally burst from the cell to infect fresh cells. Both DNA and RNA have been demonstrated in these bodies, and they further resemble rickettsiae and bacteria in having a cell wall, similar to that of gram-negative bacteria, which derives its rigidity from mucopeptide containing the amino sugar muramic acid.

A question still unsettled is the relation of the trachoma (TR) agent to the inclusion conjunctivitis (IC) agent. Since they are culturally identical, the effort to distinguish them has been directed toward a classification based on antigenic difference. This has been a difficult problem because of a major antigen or antigens that are common to many of the chlamydia. The most successful test has proved to be the mouse toxicity prevention test. Mice are immunized against a variety of TRIC strains, and their immunity to lethal doses of unknown strains are tested.[3] By this technique six types are currently recognized. Most of the strains (62 of 64) from eyes diagnosed ocular trachoma have been placed in types A, B, or C. Two strains from clinical trachoma and 16 strains from sources other than ocular trachoma belong to types D, E, or F.[1] This type of classification will undoubtedly be extended since, besides the theoretical importance, the whole problem of prophylactic vaccination depends on the choice of suitable antigens.

In an effort to understand the relationship of trachoma to inclusion conjunctivitis, many investigators have conducted studies of the lesions produced in man and animals by experimental infection with these agents. Thygeson and his associates[27,29] have shown that *Chlamydia* organisms isolated from trachoma produce a less severe disease in monkeys than do *Chlamydia* organisms isolated from inclusion conjunctivitis. Wang and Grayston,[33] in experimental infection carried out on 451 *Macaca cyclopis* monkeys, presented data indicating that TRIC agents derived from both trachoma and inclusion conjunctivitis may cause the clinical picture of trachoma in the eyes of these animals. The disease appeared to be more severe, and pannus formation more prominent, when the animals had a relatively high antibody titer against the organism, and it was concluded that "hypersensitivity to homologous or heterologous TRIC agents is an important part of the host response in the pathogenesis of pannus formation." Tarizzo and his coworkers,[23a] using 13 human volunteers, found that cultured inclusion conjunctivitis agents could elicit corneal as well as conjunctival changes normally considered peculiar to trachoma. The role of partial immunity in this clinical picture is also discussed by Tarizzo and his co-workers. In the natural infection in man the clinical distinction between trachoma and inclusion conjunctivitis is usually quite clear. But whether further investigations will show that the agents of these diseases are independent of one another or that one represents a mutation or that the condition of the host determines the type of infection still remains open. Immunologic data favor two distinct agents.

CLINICAL DIAGNOSIS

Trachoma

According to the classification presented by MacCallan,[17] there are four stages in the evolution of trachoma. In children the onset may be insidious with minimal symptoms.[26,28] For these reasons it is often overlooked. In the adult stage I begins 5 to 12 days after infection and is characterized by the presence of immature follicles on the upper tarsal plate. There is only slight papillary hypertrophy, little if any conjunctival exudate, and the disease is often relatively asymptomatic. The bulbar conjunctiva may not be inflamed. Epithelial keratitis is always present but usually can be seen only with the aid of some magnification and best with the biomicroscope. Fig. 14-2, *A* and *B* illustrates the developing follicles in the upper tarsal conjunctivas in the early and advanced stage I, respectively. Plate 3, *E* illustrates the eye of another patient with early lesions of the tarsal conjunctiva. This phase of the disease may last 3 months to 3 years. In Stage IIa (Fig. 14-2, *C*) the tarsal conjunctiva is swollen with sago-grainlike follicles that are mature, soft, and easily expressed, which is diagnostic of trachoma, since the follicles found in some other ocular diseases are firm and not expressible. Discharge continues to be minimal and symptoms are mild, but the keratitis has advanced and may be visible without magnification. Pannus formation (Fig. 14-3) proceeds, and limbal follicles may be present. In both stages I and II, scrapings of the affected areas will yield epithelial cells containing the characteristic inclusion bodies, which are illustrated in Fig. 14-4 and Plate 1, *E*. The discharges present in stages I and IIa are rich in neutrophils in contrast to lymphocytes found in viral eye infections. The material ex-

A

B

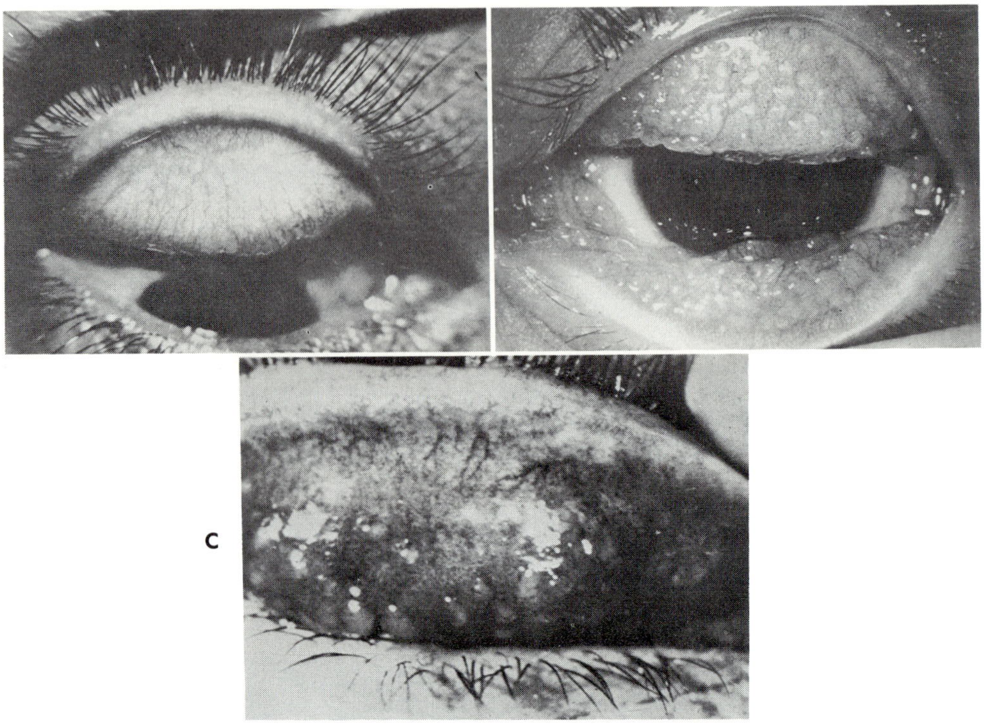

C

Fig. 14-2. A, Stage I trachoma, showing immature follicles in the upper tarsal conjunctiva. **B,** Stage I trachoma in transition to stage IIa, showing immature follicles in the upper tarsal conjunctiva and mature follicles in the fornices and lower tarsal conjunctiva. **C,** Stage IIa trachoma, showing sago-grainlike mature follicles in the upper tarsal conjunctiva. (From Thygeson, P.: Trachoma manual and atlas, Public Health Service Publication No. 541, Washington, D. C., 1960, U. S. Department of Health, Education, and Welfare.)

pressed from the follicles shows poorly staining lymphoid elements, bare nuclei, scattered cytoplasmic fragments, and numerous large mononuclear phagocytes filled with cell debris (Fig. 4-14). Stage IIa may last from 3 months to 2 to 3 years.

In some instances of highly virulent trachoma or of trachoma complicated by a bacterial infection, a stage IIb is produced in which the follicles may be masked by intense cellular infiltration and papillary hypertrophy. This condition may subside into stage IIa or pass directly into stage III.[7,25]

The time of arrival at stage III with its scar formation varies considerably. Usually progression in children is slower than in adults. Beginning scar formation is recognized with the aid of the biomicroscope. Signs of activity of the infection decrease as cicatrization progresses. With severe disease and rapid cicatrization, the scars become visible (Fig. 14-5). As the follicles undergo necrosis, they are replaced by connective tissue. Interference with blood supply may produce further necrosis and shrinking of the conjunctiva, leading to the development of trichiasis and entropion.

Pannus formation progresses through stages I and II, for the cornea is always involved early. The upper half of the cornea is involved first. The severity of the pannus usually is a measure of the severity of the disease. Corneal ulcers caused by the breakdown of trachoma pustules are often seen. Their healing may result in thick scar formation. Herbert's peripheral pits result from cicatrization of limbal follicles; and in the process, a dimple is formed, filled with transparent epithelium giving the ap-

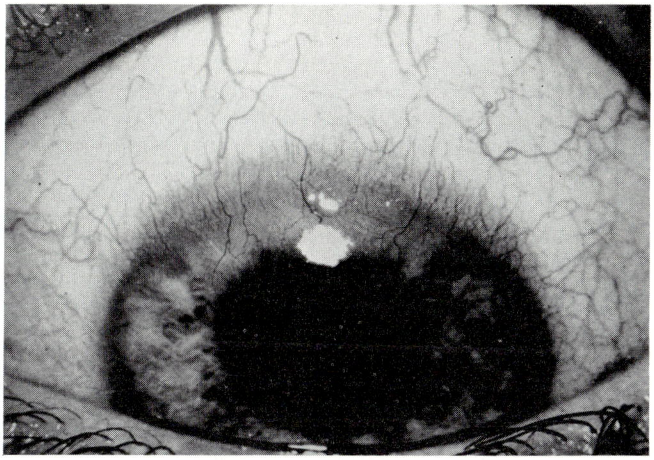

Fig. 14-3. Early pannus formation. (Courtesy Dr. P. Thygeson.)

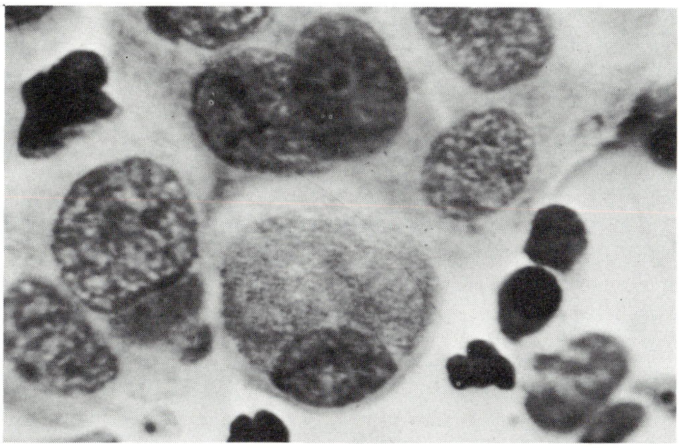

Fig. 14-4. Epithelial conjunctival scraping from a patient with acute trachoma (Plate 3, *E*), showing inclusion bodies in the cytoplasm of two epithelial cells.

pearance of a pit. These are characteristic of trachoma as seen in half the cases among American Indians.[25,30] Stage III may last for several years; but especially in children, if healing occurs spontaneously or following the use of drugs, the scars may be visible only with magnification.

Stage IV may exist indefinitely without symptoms if there are no complications such as tear deficiency and secondary bacterial infection. The biomicroscope can be used to detect the healing. If the cornea is free of subepithelial infiltrates and fluorescein will not stain the cornea, it is presumed that healing has taken place. This can be further established if treatment with topical steroids fails to exacerbate the disease. When extensive pannus formation and ulceration and cicatrization have occurred, the end result is severe impairment of the eye, which sometimes may be helped by restorative surgery.

Inclusion conjunctivitis

Inclusion conjunctivitis is a relatively mild disease that may be considered in two forms, inclusion blennorrhea (inclusion conjunctivitis) in the newborn and inclusion conjunctivitis in other age groups. The untreated disease in the newborn is char-

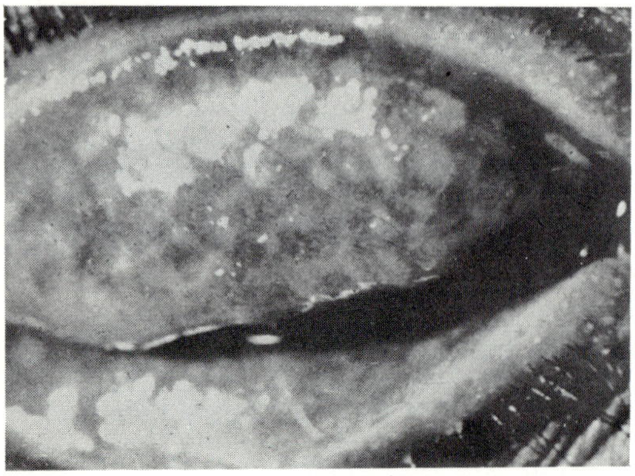

Fig. 14-5. Stage III, cicatrizing trachoma. Follicles in process of cicatrization can be seen clearly. (From Thygeson, P.: Trachoma manual and atlas, Public Health Service Publication No. 541, Washington, D. C., 1960, U. S. Department of Health, Education, and Welfare.)

acterized by a incubation period of 5 to 12 days, followed by the onset of intense inflammation of the conjunctiva and profuse discharge (Plate 3, *F*). The exudate contains many neutrophils and epithelial cells showing cytoplasmic inclusion bodies similar to those found in conjunctival epithelial cells in trachoma. In the untreated disease follicular conjunctivitis develops, but the cornea is not involved except for minor epithelial keratitis. In a matter of months the lesions heal spontaneously without residual damage. Prompt treatment with sulfonamides or antibiotics or both results in cure within a week or so, before the folliculitis occurs.

Inclusion conjunctivitis of older children and adults appears as an acute follicular conjunctivitis without major keratitis and with only moderate involvement of the conjunctiva of the upper tarsus. Untreated, it may run a course of 3 to 12 months. It too is readily treated with sulfonamides and antibiotics. The primary source for both examples of this infection is the genitourinary tract. In the female there is a subclinical cervicitis and in the male a nonspecific urethritis. The newborn becomes infected as he passes through the mother's birth canal. At one time children and those of older age groups acquired the disease most often from using swimming pools that presumably had been contaminated from the genitourinary tracts of the swimmers. This gave the name of swimming pool conjunctivitis to the disease. Today adequate chlorination of the water has almost eliminated this source of infection. Nurses or nurses aides or members of the family sometimes acquire the disease from the infant or the mother, but the source of infection is still primarily the genitourinary system.

DIFFERENTIAL DIAGNOSIS

Trachoma

Other types of follicular conjunctivitis may be mistaken for trachoma. These include pharyngoconjunctival fever, usually caused by adenovirus type 3; epidemic keratoconjunctivitis, caused by adenovirus type 8; acute primary herpetic keratoconjunctivitis; and conjunctivitis resulting from Newcastle disease caused by a paramyxovirus disease of chickens. None of these present a difficult diagnostic problem (Chapters 17, 18, and 19). The exudates from these infections lack cells containing inclusion bodies and show a predominance of lymphocytes over neutrophils. Chronic follicular conjunctivitis (Axenfeld type), a disease of unknown etiology, is the one most likely to be confused with trachoma, since this type of conjunctivitis occasionally has follicles in the upper tarsal conjunctiva. It is distinguished because of the lack of inclusion bodies; the character of the scanty exudate, which has mostly mononuclear cells; its benign course, without cicatrization or pannus, progressing to spontaneous healing in a year or two; and its failure to respond to chemotherapeutic agents.

There are three types of toxic follicular conjunctivitis to be distinguished from trachoma. The presence of molluscum nodules in the lid margins identifies the first of these: molluscum contagiosum conjunctivitis. The second is a conjunctivitis arising from the prolonged use of idoxuridine (IDU) in patients with herpetic disease and of eserine and other miotics in glaucoma patients. They may develop a follicular conjunctivitis that, if neglected, can simulate trachoma even to pannus formation and scarring. A third follicular conjunctivitis may arise in susceptible children in response to various irritants, such as ordinary bacterial conjunctivitis or Parinaud's oculoglandular conjunctivitis caused by *Leptothrix buccalis.* Finally, there is a folliculosis of unknown etiology associated with noninflammatory follicular hyperplasia of the nasopharynx, a chronic but self-limited disease distinguished from trachoma because of slight or no involvement of the upper tarsal conjunctiva. These categories of disease to be differentiated from trachoma are reviewed in the World Health Organization Technical Report Series.[9]

Inclusion conjunctivitis

The differential diagnosis of inclusion blennorrhea presents few difficulties. The incubation period of several days, the failure to detect a bacterial agent in the profuse exudate, and the presence of epithelial inclusion bodies in scrapings make the diagnosis in the newborn; and the relatively benign course and response to sulfonamides and tetracycline confirm it. In the child or adult the diagnosis may present some difficulties. A history of the possible source of infection, together with the characteristic involvement of the lower tarsal conjunctiva and the failure to compromise the cornea seriously, differentates the infection from trachoma.

EXPERIENCE WITH THE TRIC AGENTS AT THE IOCPMC

Since trachoma is not indigenous to New York City, the patients seen at the IOCPMC were for the most part people who were in transit. Diagnosis of this disease depended on the results obtained from the examination of scrapings. Table 14-2 summarizes the findings. The table shows that of 1,367 patients examined because of possible trachoma, 223 had cells containing inclusions, which were considered characteristic of the active stages of trachoma, and in 32 the findings were compatible with chronic trachoma. Inclusion blennorrhea was suspected in 1,585 newborns, 362 of which had scrapings confirming this diagnosis. In a few instances scrapings of the mother's cervix were taken, and all were positive for inclusions. Ocular scrapings were examined in 1,077 patients with a questionable diagnosis of inclusion conjunctivitis, and 201 cases showed cytoplasmic inclusions.

The predominating bacteria found in eyes of patients with acute trachoma were (1) *Staphylococcus aureus* with diphtheroids, (2) *Diplococcus pneumoniae,* and (3) *Klebsiella pneumoniae*. Fewer organisms were found in the 32 patients with chronic trachoma; but in 14 of these, *Mor-*

Table 14-2. Incidence of trachoma and inclusion conjunctivitis at the IOCPMC: 1938-1968

Number of patients examined	Diagnosis in question	Laboratory diagnosis as determined by conjunctival epithelial scrapings	Number of patients positive
1,367	Acute trachoma	Inclusions characteristic of *Chlamydia* infection; polymorphonuclear neutrophils present	223
	Chronic trachoma	Cytoplasmic inclusions very rare; presence of lymphocytes and macrophages	32
1,585	Inclusion blennorrhea	Cytoplasmic inclusions identical with those in trachoma; polymorphonuclear cells and a few lymphocytes	362
1,077	Inclusion conjunctivitis (adult)	Cytoplasmic inclusions as in trachoma; polymorphonuclear cells and lymphocytes	201

axella species were isolated in a flora that also contained S. *aureus*.

LABORATORY DIAGNOSIS

The growth of TRIC agents in the chick embryo and on cultured cells[2,8,10] is technically too complicated to be useful as a routine diagnostic tool. Two laboratory procedures, however, are of practical value. The first is the time-honored examination of conjunctival scrapings. Details for the techinque of preparing and staining these smears are given in Chapter 4. The scrapings for acute trachoma and inclusion conjunctivitis are similar (Table 14-3). The differential diagnosis of these diseases, as already stated, depends upon the character of the follicular conjunctivitis and upon the involvement of the cornea.

The immunofluorescein technique for the recognition of inclusion bodies in scrapings containing epithelial cells from trachomatous eyes is accurate, fast, and yields a higher percentage of positive tests than can be obtained by other techniques. The principal of the test is to label antibody to TRIC agents or cross-reacting lymphogranuloma antibody with fluorescein, to "stain" smears of the exudate, scrapings, or cultured cells under question with the labeled antibody, and to examine the preparation under ultraviolet light. The antibody is bound to those areas in the cell harboring the TRIC agent, and the attached fluorescein gives a bright yellow-green fluorescence that stands out brilliantly against a darker background (Plate 3, *G*).[20] In this way a rare infected cell is readily seen. The test requires the use of a source of ultraviolet light, a trained technician, and a room with greatly reduced light for the examination. Since the test is useful for the identification of viruses and often of bacteria as well, it can become a valuable part of the equipment of a microbiology laboratory in a department of ophthalmology.

The fluorescein test is only as specific as the antibody employed. As yet it has proved difficult to obtain sera that can readily be used to classify the subdivisions of the TRIC agents, and even the subdivisions of the *Chlamydia;* so as yet much remains to be done in sharpening immunologic diagnostic procedures for this group of organisms.

PROPHYLAXIS AND THERAPY

Attempts at prophylaxis for trachoma are centered about the development of a vaccine useful in preventing or ameliorating

Table 14-3. Diagnosis of stained scrapings from patients with TRIC infection

Selective affinity of TRIC agents	Sites for obtaining scrapings	Results of staining smears from scrapings	Cytoplasmic structures that may be mistaken for inclusions
		Trachoma	
Confined to epithelium	Upper tarsal conjunctiva, fornix, cornea	Intracytoplasmic inclusions present in high percentage in early phase and in acute exacerbations of late disease; polymorphonuclear cells; difficult to find inclusions in later chronic stages; in trachomatous follicular expression, macrophages contain broken-down debris, which is abundant; pale-staining large mononuclear cells; some plasma cells	Pseudoinclusion bodies Pigment granules Aggregates of nonspecific granular cytoplasm Aggregates of bacteria
		Inclusion conjunctivitis	
	Lower tarsal conjunctiva	Intracytoplasmic inclusions, polymorphonuclear cells	

the disease. The investigators at best have claimed only some reduction in the attack rate for short periods and at worst some accentuation of the disease apparently caused by acquired hypersensitivity.[13] It seems necessary to find a key antigen associated with infectivity that can stimulate a protective antibody. On theoretical grounds it is a difficult problem, since trachoma, even after years of activity, produces no lasting immunity.

The transmission of trachoma is favored by poor hygiene. According to Jawetz,[12] transmission of the disease is highest in the homes of the poor, where there is overcrowding, sharing of common towels and other possessions, and little knowledge of hygiene. In the Middle East, North Africa, and the Orient, seasonal epidemics of bacterial conjunctivitis caused by such organisms as *Hemophlius aegypticus, Neisseria gonorrhoeae, Moraxella lacunata,* and staphylococci often occur. These infections exacerbate trachoma and increase its spread by the profuse conjunctival exudate. The history of the lessening or disappearance of trachoma from most of Europe before the introduction of specific therapy must reflect the importance of such factors.

Therapy of trachoma and inclusion conjunctivitis are the same. For trachoma, sulfonamides are given orally for about 3 weeks or less. Local application of a 1% concentration of a broad-spectrum antibiotic, usually tetracycline, is given twice daily or more for 1 to 3 months. Tetracyclines taken orally make good substitutes for sulfonamides when a patient is sulfa sensitive. The period of treatment necessary for inclusion conjunctivitis is considerably less than that needed for trachoma. Though the treatment of trachoma may seem simple on paper, in practice, unless conditions of living can be improved, reinfection often occurs.

REFERENCES

1. Alexander, E. R., Wang, S. P., and Grayston, J. T.: Further classification of TRIC agents from ocular trachoma and other sources by the mouse toxicity prevention test, Amer. J. Ophthal. **63**(Suppl.):1469, 1967.
2. Al-Hussaini, M. K., Jones, B. R., and Dunlop, E. M. C.: Genital infection in association with TRIC virus infection of the eye: cytology (preliminary report), Brit. J. Vener. Dis. **40**:25, 1964.
3. Bell, S. D., Jr., and Theobald, B.: Differentiation of trachoma strains on the basis of immunization against toxic death of mice, Ann. N. Y. Acad. Sci. **98**:337, 1962.
4. Bietti, G. B., Freyche, M. J., and Vozza, R.: La diffusion actuelle du trachome dans le monde, Rev. Int. Trachome **39**:113, 1962.
5. Bietti, G., and Werner, G. H.: In Kugelmass, I. N., editor: Trachoma: prevention and treatment, Springfield, Ill., 1967, Charles C Thomas, Publisher.
6. Davis, B. D., Dulbecco, R., Eisen, H. N., Ginsberg, H. S., and Wood, W. B.: Microbiology, New York, 1967, Harper and Row, Publishers.
7. Duke-Elder, S.: System of ophthalmology, Diseases of the outer eye, vol. 8, part 1, St. Louis, 1965, The C. V. Mosby Co., pp. 260, 300.
8. Dunlop, E. M. C., Hare, M. J., Darougar, S., Jones, B. R., and Rice, N. S. C.: Detection of Chlamydia (Bedsonia) in certain infections of man: clinical study of genital tract, eye, rectum and other sites of recovery of Chlamydia, J. Infect. Dis. **120**:463, 1969.
9. Expert Committee on Trachoma: World Health Organization technical report series no. 234, third report, 1962.
10. Gordon, F. B., Harper, I. A., Quan, A. L., Treharne, J. D., Dwyer, R. St. C., and Garland, J. A.: Detection of Chlamydia (Bedsonia) in certain infections of man: laboratory procedures: comparison of yolk sac and cell culture for detection and isolation, J. Infect. Dis. **120**:451, 1969.
11. Halberstaedter, L., and von Prowazek, S.: Zur aetiologie des Trachoms, Deutsch. Med. Wschr. **33**:1285, 1907.
12. Jawetz, E.: Agents of trachoma and inclusion conjunctivitis, Ann. Rev. Microbiol. **18**:301, 1964.
13. Jawetz, E.: Summary of discussion of papers 67-73, Amer. J. Ophthal. **63**:1654, 1967.
14. Julianelle, L. A.: Etiology of trachoma, New York, 1938, The Commonwealth Fund, pp. 42, 148.
15. Kugelmass, I. N., editor: Trachoma: prevention and treatment, Springfield, Ill., 1967, Charles C Thomas, Publisher.
16. Lindner, K.: Gonoblennorrhöe, Einschlussblennorrhöe und Trachom, Graefes Arch. Ophth. **78**:345, 1911.

17. MacCallan, A. F.: Trachoma, London, 1936, Butterworth and Co.
18. Morax, V.: Sur l'etiologie des ophthalmies du nouveau-né et la déclaration obligatoire, Ann. Oculist. **129**:346, 1903.
19. Nicolle, C., Blaisot, L., and Cuénod, A.: Le magot animal réactif du trachome: filtrabilité du virus; pouvoir infectant des larmes, C. R. Acad. Sci. **155**:241, 1912.
19a. Nichols, R. L., editor: Trachoma and related disorders caused by chlamydial agents: proceeings of a symposium held in Boston, Massachusetts, 17-20 August 1970, Amsterdam, 1971, Excerpta Medica Foundation.
20. Nichols, R. L., Bobb, A. A., Haddad, N. A., and McComb, D. E.: Immunofluorescent studies of the microbiologic epidemiology of trachoma in Saudi Arabia, Amer. J. Ophthal. **63**: 1372, 1967.
21. Schmeichler, L.: Ueber Chlamydozoenbefunde bei nichtgonorrhoischer Blennorrhöe der Neugeborenen, Berl. Klin. Wschr. **46**: 2057, 1909.
22. Stargardt, K.: Über Epithelzellveränderungen beim Trachom und andern Conjunctivalerkrankungen, Graefes Arch. Ophth. **69**:525, 1909.
23. T'ang, F., Chang, H., Huang, Y., and Wang, K.: Studies of the etiology of trachoma with special reference to isolation of the virus in chick embryo, Chin. Med. J. **75**:429, 1957.
23a. Tarizzo, M. L., Nataf, R., and Nabli, B.: Experimental inoculation of thirteen volunteers with agent isolated from inclusion conjunctivitis, Amer. J. Ophthal. **63**:1120, 1967.
24. Thygeson, P.: The nature of the elementary and initial bodies of trachoma, Arch. Ophthal. **12**:307, 1934.
25. Thygeson, P.: Trachoma manual and atlas, U. S. Department of Health, Education, and Welfare, Public Health Service Publication No. 541, 1960.
26. Thygeson, P.: Corneal changes in TRIC agent infections, Amer. J. Ophthal. **63**:1278, 1967.
27. Thygeson, P., and Crocker, T. T.: Observations on experimental trachoma and inclusion conjunctivitis, Amer. J. Ophthal. **42**:76, 1956.
28. Thygeson, P., and Dawson, C. R.: Trachoma and follicular conjunctivitis in children, Arch. Ophthal. **75**:3, 1966.
29. Thygeson, P., Dawson, C., Hanna, L., Jawetz, E., and Okumoto, M.: Observations on experimental trachoma in monkeys produced by strains of virus propagated in the yolk sac, Amer. J. Ophthal. **50**:907, 1960.
30. Thygeson, P., and Hanna, L.: TRIC agents. In Lennette, E. H.,. and Schmidt, N. J., editors: Diagnostic procedures for viral and rickettsial infections, ed. 4, New York, 1969, American Public Health Association, p. 904.
31. Thygeson, P., and Proctor, F. I.: The filtrability of trachoma virus, Arch. Ophthal. **13**: 1018, 1935.
32. Thygeson, P., Proctor, F. I., and Richards, P.: Etiologic significance of the elementary body in trachoma, Amer. J. Ophthal. **18**:811, 1935.
33. Wang, S., and Grayston, J. T.: Pannus with experimental trachoma and inclusion conjunctivitis agent: infection of Taiwan monkeys, Amer. J. Ophthal. **63**:1133, 1967.

15

Rickettsiae in relation to infection of the eye

THE RICKETTSIAE

The generic term *Rickettsia* belongs to a group of agents of disease, first discovered by Dr. H. T. Ricketts[23] (Fig. 15-1) in 1906. They appear under the light microscope as coccobacillary or bacillary forms usually about 0.3 × 0.5 μ in size; they divide by binary fission. Fig. 15-2 shows massive multiplication of these organisms in a macrophage infected with rickettsiae causing Q fever. Typhus *Rickettsia* organisms growing in the yolk sac of a chick embryo are seen by electron microscopy in Fig. 15-3, where variations in form are apparent. Although these organisms seem structurally similar to bacteria, and like bacteria contain both DNA and RNA, most of them are unable to survive except in living cells. They are further remarkable because they have a wide host range and are transmitted from host to host by insect or arachnoid vectors—namely, lice and fleas or ticks and mites. Table 15-1 presents some of the characteristic features of the seven *Rickettsia* species that are important human pathogens and that also produce noteworthy infection of the eye. Many mammals are susceptible to strains pathogenic for man and act as reservoirs of the organisms. A number of possible vectors are available for most strains, and some vectors are also hosts to the organisms, passing them on to their offspring. Rickettsiae, because of their mode of division, possession of DNA and RNA, and their chemical composition and metabolic activities, are considered obligately parasitic, true bacteria.[13] In most instances, they require living cells for growth.

Rickettsial diseases of man are undoubtedly of long standing. Typhus is thought to have existed at the time of Hippocrates, and tsutsugamushi disease was described in a Chinese medical book of the sixth century AD.[30] Nevertheless, Ricketts was the first to transmit one of this group of diseases, Rocky Mountain spotted fever, to laboratory animals and to describe the organism. He injected the blood of patients, drawn during the first few days of their illness, into monkeys and guinea pigs, which proved susceptible to infection. Ricketts then demonstrated that a tick transmitted the disease to man, and in 1907[24] he reported that the "virus" was hereditary in the tick. Descriptions of the organisms of Rocky Mountain spotted fever were given by Ricketts in 1909,[24] and descriptions of the organisms of Mexican typhus were given by Ricketts and Wilder in 1910.[25] By cross-immunity experiments these investigators showed that the two diseases were distinguishable immunologically. At the age of 39, Ricketts died of typhus fever in Mexico.

The rickettsial diseases of man usually are characterized by a lesion at the site of the insect bite, which produces an eschar. The onset of symptoms occurs after 1 to 2 weeks' incubation, and the diseases are characterized by high fever, severe headache, and general malaise. In a few days a macular rash or maculopapular rash appears on the trunk and extends over much of the body. In the case of Rocky Mountain spotted fever there is a petechial rash, and in Q fever the lesion is in the lungs and only rarely is there a cutaneous rash. In this latter disease, transmission is principally by inhalation of infected droplets from expired air or contaminated dust, and here there is no eschar. The organisms in man are found chiefly in endothelial cells of capillaries, arterioles, and venules. The distribution of the lesions is widespread, but the skin and central nervous system are more severely affected. Thrombi form, followed by hemorrhage and perivascular infiltration. The organisms also may be present in smooth muscle cells of the media of some arterioles. On this basic pattern the various rickettsiae exhibit certain individual characteristics, which will be mentioned later, since they are relevant to the eye.

Fig. 15-1. Dr. Howard Taylor Ricketts (1871-1910). (From Olpp, G.: Charakterkopfe der Tropenmedizin, Berlin, 1936, Verlag Die Brücke Zur Heimat, p. 60.)

The rickettsiae, which are demonstrable in the cytoplasm and sometimes in the nuclei of cells, may be seen by staining with Giemsa stain, in which they appear bluish purple, or with Macchiavello's stain, after which they are bright red. The organism can be found in the saliva and gut of vectors and in the bloodstream and vascular endothelium of the host. Monkeys, rabbits, guinea pigs, mice, and rats are among the

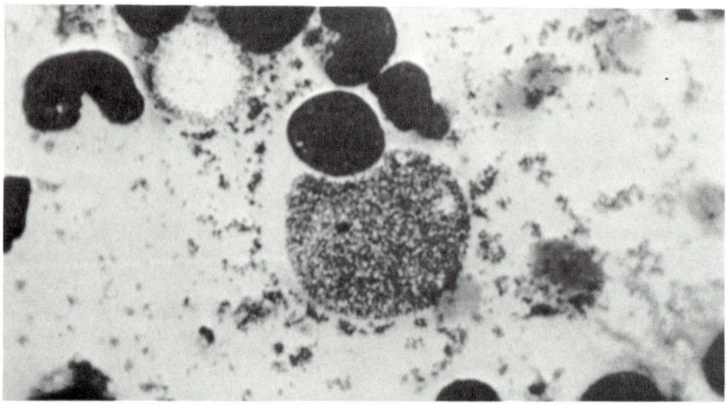

Fig. 15-2. Rickettsiae of Q fever in a smear of skin exudate. Note the macrophage packed with enormous numbers of rickettsiae. (× 1,125.) (From Burrows, W., and others: Textbook of microbiology, ed. 19, Philadelphia, 1968, W. B. Saunders Co., p. 829.)

Table 15-1. Rickettsiae producing ocular disease

Disease produced	Etiologic agent	Natural cycle: Arthropod	Natural cycle: Mammal host range	Susceptable laboratory animals	Positive Weil-Felix test	Complement-fixation test	Immunity	Geographical distribution
Typhus								
Epidemic	*Rickettsia prowazekii*	Human body louse—*Pediculus corporis*	Man only	Guinea pig—fever only	OX-19	Positive, group and type specific	For many years	Worldwide
Endemic murine	*Rickettsia mooseri*	Rat louse—*Polyplax spinulosa*; Rat flea—*Xenopsylla*	Wild rats, rodents, field mice	Guinea pig—fever, scrotal swelling	OX-19	Positive, group and type specific	For many years	Worldwide
Spotted fever								
Rocky Mountain spotted fever	*Rickettsia rickettsii*	Ticks: wood tick—*Dermacentor andersoni;* dog tick—*Dermacentor variabilis* and other species	Dogs, wild rodents	Guinea pig—fever, scrotal necrosis	OX-19 OX-2	Positive, group and type specific	High degree of immunity	Western hemisphere, Brazil, Colombia
Boutonneuse fever South African tick bite fever	*Rickettsia conori*	Ticks: *Rhipicephalus sanguineus* and a number of others	Dogs, wild rodents	Guinea pig—fever, scrotal swelling	OX-19 OX-2	Positive, group and type specific	Present	Mediterranean countries of Europe, Asia, Africa
Scrub typhus	*Rickettsia tsutsugamushi*	Several species of trombiculid mites	Wild rodents, many species	Mouse—rough fur, inactive, ascites	OX-K	Positive in about 50% of patients	Second attack may occur within a few years after initial disease	Australia, Pacific Islands, Asia
Epidemic Q fever	*Coxiella burnetti*	Ticks: number of species	Very broad; wild and domesticated animals, birds	Guinea pig—fever only	Negative	Positive, group and type specific	Second attacks have been reported	Worldwide
Trench fever	*Rickettsia quintana*	Human body louse—*Pediculus humanus*	Man	—	Negative	None available	Partial immunity	Europe, Africa, North America

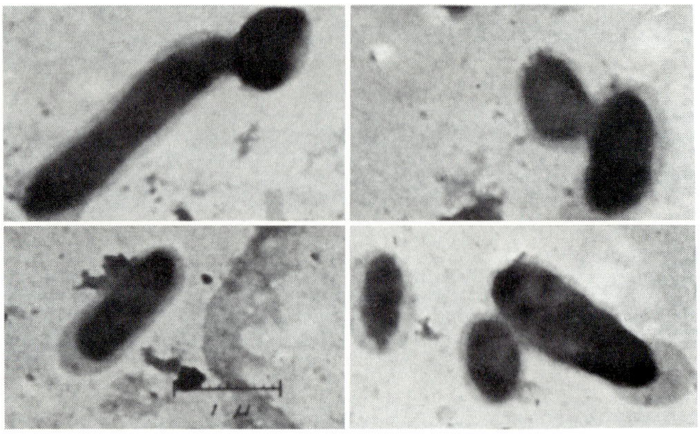

Fig. 15-3. Electron micrographs of typhus rickettsiae from egg yolk culture. Note the variability in size, ranging from cocco-bacillary to bacillary forms, and the capsular material surrounding the individual cells. (Courtesy Lilly Research Laboratories.)

mammals that can be infected experimentally with various strains of rickettsiae that are pathogenic for man. The rickettsiae may be cultivated in tissue-cultured cells or in the yolk sac of 5- to 7-day-old chick embryos. Isolation is quite difficult, however, and of only limited use in diagnosis.[9]

Immunologic classification of rickettsiae may be made by complement-fixation (Table 15-1), agglutination, and neutralization tests. Though sharing some common antigens, the rickettsiae are immunologically distinguishable, and immunity to infection is strain specific. The rickettsiae also carry antigens shared in common with certain strains of the *Proteus* group of bacilli—namely OX-19, OX-2, and OX-K. Patients suffering from one of the various diseases produced by the rickettsiae develop agglutinins to one or two of the three strains of *Proteus* involved, or to none at all. After performing an agglutination test, the Weil-Felix test, and using the patient's serum and suspensions of the *Proteus* bacilli, one should look for rising titers of antibodies during the course of the disease. This series of tests is most useful and avoids the necessity of using the rickettsiae as antigens.

Recovery from an attack of most rickettsial diseases results in lasting immunity, perhaps because of the persistence of the organisms in the human host. Brill's disease, however, is a relapse seen in some who have had epidemic typhus fever. This is a relatively mild disease, when compared with the original attack, and is thought to result from the prolonged survival of *R. prowazekii* in some tissue of the body in a latent form. It occurs many years after the initial attack.

The fact that infection with rickettsiae produces lasting immunity suggests that vaccines should be successful prophylactic agents, and this has proved to be correct in some of the diseases. In practice, however, the more practical public health measure against epidemic attacks of typhus fever is to eradicate the vector. DDT has been used with excellent results in areas where typhus is prevalent. The hairy areas of the body, which harbor lice, are sprayed with DDT, thus killing the vector and breaking the cycle.

When untreated, rickettsial infections may cause a high mortality rate; in epidemic periods the mortality for typhus has run as high as 70%. Both chloramphenicol and tetracyclines have been effective in the treatment of rickettsial disease in man.

Epidemic typhus, murine typhus, Rocky Mountain spotted fever, boutonneuse fever, African tick typhus, scrub typhus, Q fever, and trench fever may be associated with lesions of the eye. Table 15-2 presents features of the ocular lesions produced by each

Table 15-2. Ocular diseases caused by rickettsial infections

Etiologic agent	Primary disease	Incidence of ocular involvement in various parts of the eye, clinical features, and complications that may occur	References
Rickettsia prowazekii	Epidemic typhus fever	Systemic infection can be acquired through conjunctival route; conjunctivitis and conjunctival hyperemia; subconjunctival hemorrhages; photophobia; corneal ulcers; chorioretinitis and retinal hemorrhages; optic neuritis; uveitis; vitreous opacities	Brooks and Fineberg[3]* Duke-Elder[8] Lebas and Bernaerts-Lebas[15] Rouher and others[26]
Rickettsia mooseri	Endemic murine typhus	Mild conjunctivitis; systemic infection may result through conjunctiva as portal of entry; uveitis; perivascular chorioretinitis	Lebas and Bernaerts-Lebas[15] Rouher and others[26]
Rickettsia rickettsii	Rocky Mountain spotted fever	Conjunctivitis and conjunctival hyperemia; photophobia; iridocyclitis	Lebas and Bernaerts-Lebas[15] Venable and Pollack[29]
Rickettsia conori	Boutonneuse fever South African tick bite fever	Primary lesion and infection of conjunctiva results from direct contamination after squashing tick or spurting blood from crushing tick between fingers; initially, lesion is ulcerative or granulomatous; extensive involvement of preauricular and submaxillary nodes; chorioretinitis and iridocyclitis; in systemic infection, a mild preeruptive conjunctivitis	Audoueineix[2] Clavel and Hamart[5] Duke-Elder[8] Lebas and Bernaerts-Lebas[15] Rhodes and van Rooyen[22]
Rickettsia tsutsugamushi	Scrub typhus	Conjunctival hyperemia; subconjunctival hemorrhages sometimes associated with ecchymosis of eyelids; eschar on eyelid; corneal ulcers; retinal edema and hemorrhages; uveitis; vitreous opacities; engorgement of veins; nystagmus; if primary lesion appears on eyelids, a black escharotic ulcer develops with a violent conjunctival reaction and lymphadenopathy	Ahlm and Lipshutz[1] Lebas and Bernaerts-Lebas[15] Scheie[27]
Coxiella burnetti	Endemic Q fever	Infection may be transmitted through human conjunctiva; violent conjunctival hyperemia, gangrene of eyelids, uveitis, chorioretinitis, episcleritis, optic neuritis result; ocular involvement in the course of the usual disease may be associated with retroorbital pain, photophobia, and injection of conjunctiva	Audoueineix[2] Calmettes and others[4] Lebas and Bernaerts-Lebas[15]
Rickettsia quintana	Trench fever	Conjunctivitis; hyperemia of conjunctiva; retinitis	Lebas and others[15]

*Superscript numbers indicate references at end of chapter.

A

B

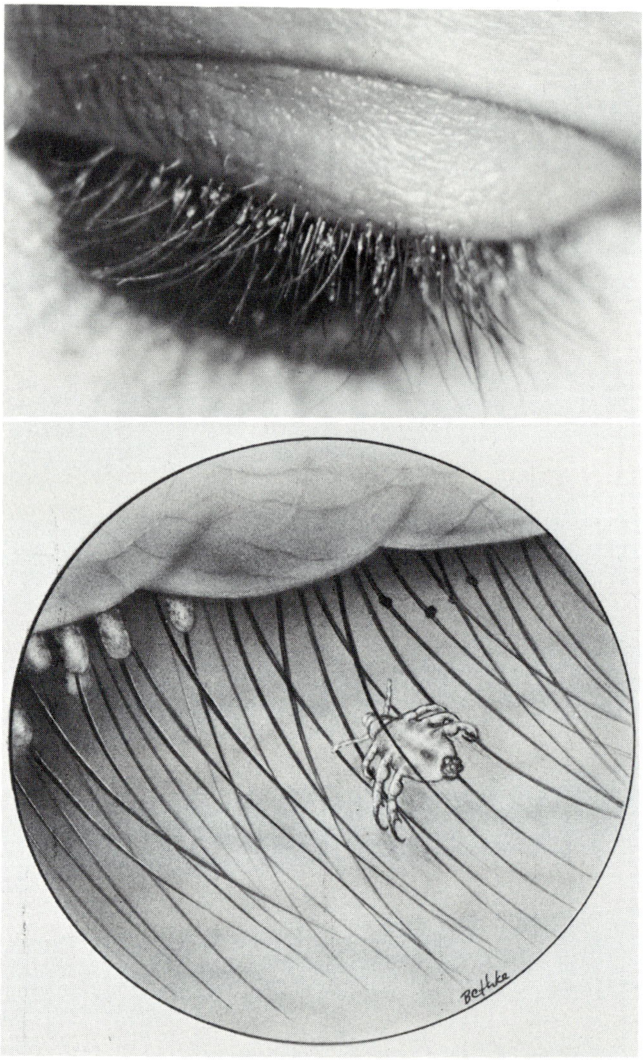

Fig. 15-4. A, Pediculosis of the eyelid. **B,** *Phthirus pubis* on eyelashes seen in **A.** (Courtesy Dr. R. Castroviejo.)

organism. Since rickettsial organisms are spread throughout the body via the bloodstream, ocular manifestations of rickettsial diseases are, in a majority of cases, an expression of the widespread distribution of the organisms in the end arterioles, capillaries, and venules. As previously stated, the inflammation produced in these vessels causes thrombus formation with leukocytic infiltration, hemorrhage, and focal necrosis. Any part of the eye may be involved, a fact that accounts for the great variety of ocular lesions reported in Table 15-2.

Infections of the eye, in rare instances, may be acquired by the primary introduction of the rickettsial organism into the eye tissue. This can come from a bite on the eyelid by the arthopod vector. Fig. 15-4, *A* illustrates the presence of lice on the lid margin and eyelashes of a patient seen at the IOCPMC. Fig. 15-4, *B*, is a drawing of an individual louse. In such circumstances the bite of an infected louse produces an initial lesion, which develops as an eschar on the lid. Infection of the eye may also occur if an individual rubs his eye after he contaminates his fingers with infected blood or feces by crushing a tick or louse or by scratching a bite on the skin. The disease spreads from the initial site to produce the generalized infection characteristic of rickettsial disease. Further involvement of the eye may then develop later in the course of the disease.

TYPHUS FEVER

Epidemic typhus fever, caused by *R. prowazekii* (named to honor von Prowazek who, like Ricketts, died of typhus fever while studying the disease), is associated with poverty, overcrowding, war, famine, and all other conditions that favor the human body louse. Between 1918 and 1922 Eastern Europe and Russia were ravaged, and some 30 million people were infected; mortality was 10%.[30] In World War II, typhus fever complicated military operations in Europe; and in the Far East, Japan, and Korea, outbreaks of typhus also developed. Deaths from typhus fever have been greater among medical personnel than among other groups, and laboratory infestions have occurred frequently.

Epidemic typhus fever is a severe febrile disease that arises spontaneously only in man. Man is both the host and the reservoir of this infection. The human body louse, *Pediculus humanus* var. *corporis*, is the most important vector, although *Pediculus capitis* may also transmit the infection. The rickettsiae fill the intestinal tract of the louse and are deposited on the person's skin when the louse draws blood. The bite causes itching, and man scratches the feces into the bite. After an incubation period of 10 to 14 days there is an abrupt onset of chills and fever with prostration. A maculopapular rash appears in 5 to 7 days first on the trunk and later on the extremities. Only in severely ill patients is the face involved. *R. prowazekii* is found in the blood of the patient during the febrile period and is infective for lice feeding on the patient during the first 10 days of the illness.

A review of ocular manifestations in typhus fever was made by Brooks and Fineberg[3] in 1951. While reviewing the literature from 1867 to 1943, they collected reports of 2,031 cases of eye involvement caused by typhus. Their paper opens with a quotation from Larionow: ". . . man solle die Augen typhöser Kranker öfter untersuchen."[14*] The great variety of lesions reported indicates that this advice should be carefully followed. Their report covers every area of the eye, and their table carries 38 types of pathologic conditions. The commonest disease reported was optic neuritis (395 examples, or 19.75%). This is followed by 274 cases of constructed visual fields (13.95%), 144 cases of retinal hyperemia (7.20%), 136 cases of retinal edema (6.80%), 110 cases of papillary hyperemia (5.10%), and 85 cases of conjunctivitis (4.20%). The cornea was involved in the disease in 31 instances; the pupils in 44; the uveal tract in 33; the media in 29; the optic nerve in addition to neuritis and papillary hyperemia in an additional 19 instances. Besides retinal hemorrhage and edema, there were 32 instances of other retinal pathologic conditions. Blood vessels of the eye were visibly implicated 26 times; and in addition to constriction of visual fields, the function of the eye was impaired in various other ways 33 times. The authors point out that, with the possible exception of constricted visual fields, most of the eye complications are those reflecting change in the circulation; and the authors conclude that in typhus one should search for changes in the eye that can result from involvement of the circulation of the eye.

In Duke-Elder's discussion of ocular manifestations in typhus fever, the role of conjunctivitis, with or without hemorrhages, is stressed.[8] He states that conjunctival hyperemia is invariable and that in some 10% of the cases, catarrhal conjunctivitis with slight secretion occurs.

The above manifestations of ocular involvement in typhus are those that take place during the acute disease and that

*"One should frequently examine the eyes of the patient with typhus fever."

fade with convalescence. P. V. Morax and his associates[18] have reported on 15 patients, 9 with uveitis and the remainder with uveitis and additional complications. The only sign pointing to the possible cause was the presence of antibodies in the patients' sera, which agglutinated one of the preparations of rickettsial antigens used for the microagglutination test (a test that is credited with a high degree of specificity).[13] Appropriate antibiotic treatment improved or cured the ocular condition and also eliminated the positive agglutination tests. Four of the patients had antibodies to *R. prowazekii.* The idea that latent rickettsial infections might produce uveitis many months later is reasonable, since it has a counterpart in Brill's disease, an exacerbation of a latent infection with *R. prowazekii,* which may occur many years after the acute manifestations have subsided.

MURINE TYPHUS

Murine typhus is an endemic disease that has been reported in many parts of the world and that clinically is similar to mild typhus fever. *R. mooseri,* the etiologic agent, resembles *R. prowazekii* in shape and staining properties. The two share a common soluble antigen, but each has immunologically distinct toxins; infection with one does not protect against the other disease. *R. mooseri* finds a reservoir in rats, and both the rat louse and the rat flea carry the infection from rat to rat. The disease is transmitted from rat to man by the rat flea and from man to man by the body louse. It is possible that food soiled with rat urine may also be a source of infection.

The incubation period of murine typhus, like that of typhus fever, is 1 to 2 weeks. An eschar is found at the site of the infecting bite, and the patient develops fever, generalized pains, and headache. The macular rash comes about 5 days after the onset of fever and spreads from the trunk to the limbs. Usually all symptoms are milder than those seen in typhus fever, and there are a smaller number of complications and a lower death rate. However, a severe case of murine typhus will simulate typhus fever. As in other rickettsial diseases, chloramphenicol or tetracycline or both are effective in treatment. Infection establishes long-term immunity.

The involvement of the eye in this disease is slight when compared with that seen in typhus fever (Table 15-2). Usually it is limited to a mild conjunctivitis. It is possible, however, for one to infect the conjunctiva directly by rubbing the eye with fingers contaminated by feces of the carrier flea. This produces a severe conjunctivitis, and the generalized infection develops from this local site. As in typhus fever the organisms are present in the blood for several days, and such blood will transmit infection to laboratory animals or chick embryos.

In this disease, as well as in typhus fever, late ocular lesions may occur—namely, uveitis with perivascular chorioretinitis. The evidence that these conditions are late sequelae of the rickettsial disease has been based on the presence of significant titers of agglutinating antibodies in the patient's serum.

ROCKY MOUNTAIN SPOTTED FEVER

Rocky Mountain spotted fever was first reported in 1873 from the Rocky Mountain areas. As has been described earlier, Ricketts found that the infection was transmitted to man by the wood tick, and he infected guinea pigs and monkeys with the blood of patients with Rocky Mountain spotted fever. He found the agent in blood smears from man, guinea pigs, and monkeys, and in the tissues of the ticks. The ticks transmit the agent to their progeny. Rocky Mountain spotted fever often is an occupational disease, seen among hunters, foresters, highway engineers, maintenance crews, and other individuals exposed to wood ticks. It is prevalent in all parts of the world. In the USA it extends to the Atlantic Coast and is found in the southeastern states. Thirteen cases from the Cape Cod area, with two deaths, were reported in 1969.[11]

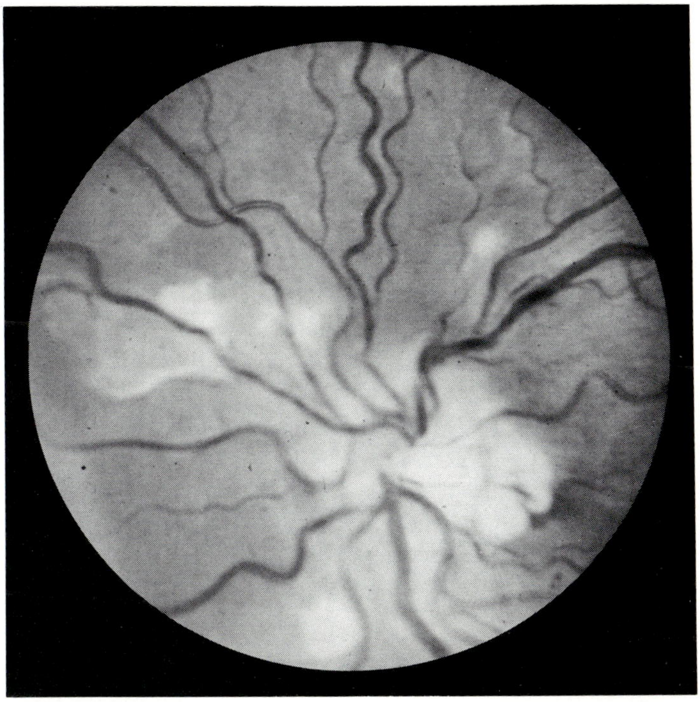

Fig. 15-5. Fundus changes in Rocky Mountain spotted fever. Eye showing papilledema, cotton-wool exudates, small, flame-shaped hemorrhages, tortuosity of the vessels, and venous engorgement. (From Presley, G. D.: Amer. J. Ophthal. **67**:263, 1969.)

The etiologic agent of Rocky Mountain spotted fever is similar in morphologic and cultural characteristics to *R. prowazekii*. It has the characteristic predilection for endothelial cells of small vessels. The incubation period of the disease is rather short, usually 4 to 8 days; the onset is abrupt, characterized by fever, chills, headache, and muscle and bone pains. In 2 to 14 days a generalized rash develops, which may be petechial or hemorrhagic. Unlike the rash of typhus fever, it usually starts on the extremities and moves centripetally; the palms of the hands and the soles of the feet are also involved. The disease seen in the western United States is carried by the wood tick, *Dermacentor andersoni*, that in the Atlantic area by the dog tick, *D. variabilis*. Immunity develops in man and in laboratory animals after recovery from infection. Effective vaccines were first employed in the prophylaxis of this disease.

The ocular manifestations of infection with *R. rickettsii* include a conjunctivitis characterized by hyperemia with photophobia and mild catarrh.[29] Iridocyclitis, occurring as a late sequela, has been described by Lebas and Bernaerts-Lebas.[15] Presley[20] in 1969 examined 6 children, 5 to 14 years old, in the acute stage of the disease at the Division of Ophthalmology, University of North Carolina. All but one showed lesions of the fundus (Fig. 15-5), which included engorgement of retinal veins, papilledema, retinal edema, cytoid bodies, flare-shaped hemorrhages, Roth spots, and an arterial occlusion. A similar case has been reported by Raab and his associates.[21]

BOUTONNEUSE FEVER AND SOUTH AFRICAN TICK BITE FEVER

Rickettsia conori was recognized and first described by Conor and Bruch in 1910 in Tunisia.[6] The disease caused by this agent is known as boutonneuse fever. The name boutonneuse refers to a local nodule, the black buttonlike necrotic primary lesion, tache noir, usually observed on the leg. *R. conori* is also the cause of South African tick bite fever. *R. conori* belongs to the Rocky Mountain spotted fever group and is morphologically and immunologically similar to but separable from *R. rickettsii*, by toxin neutralization tests and other serologic methods.

Infected tissue cells may exhibit intranuclear as well as intracytoplasmic rickettsiae, but the latter predominate. *R. conori* has been cultivated in yolk sacs of the developing chick embryo; the organisms multiply intranuclearly when growing in the chorioallantoic membrane of the fertile egg. The animal reservoirs of infection are the dog, rodents, and mice; the disease is transmitted by ticks.

Following the incubation period, which lasts 5 to 7 days, the infection is characterized by fever, chills, headache, and malaise. The lesions are similar to those seen in typhus. The generalized maculopapular rash, involving most of the body, including face, palms, and soles, appears on the fourth day. The Weil-Felix test becomes positive during the second week, and complement-fixing antibodies appear shortly thereafter. The diagnosis is established by the clinical picture, including the tache noir, positive serologic tests, and geographic location.

Ocular disease seen in boutonneuse fever (as outlined in Table 15-2) is, according to Duke-Elder,[8] probably the most interesting of the rickettsioses, from the ophthalmologic point of view, for the conjunctival infection is characteristically acquired by direct contamination. The dog tick is the vector, and infection of the eye comes about by contamination after squashing the tick.

In the case of the patient with South African tick bite fever, shown in Fig. 15-6, infected blood had squirted into the eye. Pronounced edema of the lid and soft tissues below the eye are seen.[22] Rubbing an eye with contaminated fingers will also infect the conjunctiva. The initial lesion is ulcerative or granulomatous, and the preauricular and submaxillary nodes are rapidly involved. Chorioretinitis and paramacular hemorrhages, as well as iridocyclitis, may develop. The systemic disease manifests itself in about 5 days. A mild conjunctivitis is present during the systemic infection.

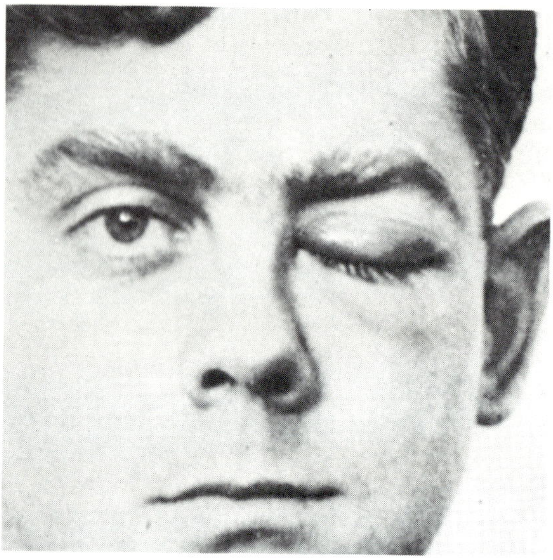

Fig. 15-6. South African tick typhus. Primary lesion on conjunctiva following spurting of blood from crushing an infected dog tick between the fingers. (Courtesy Dr. J. H. S. Gear. From Rhodes, A. J., and van Rooyen, C. E.: Textbook of virology for students and practitioners of medicine, ed. 4, Baltimore, 1962, The Williams & Wilkins Co.)

SCRUB TYPHUS

Scrub typhus, or tsutsugamushi disease, was originally described in Japan and is also found in many of the islands of Southwest and Eastern Asia and in India and Australia. The disease occurred in military units in the Asiatic-Pacific area during World War II. The animal reservoirs of the disease are the field mouse and other rodents, and the vectors are the trombiculid mites. These, like the tick vectors of the spotted fever group, transmit the rickettsiae transovarially to succeeding generations of mites.

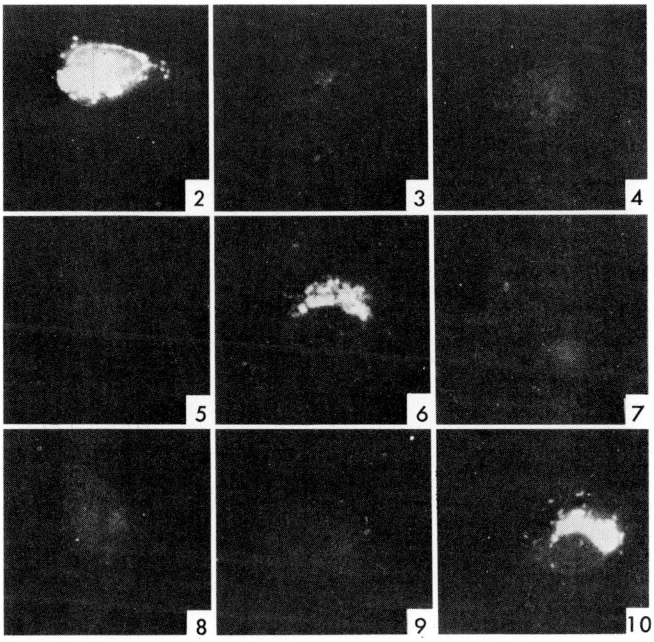

Fig. 15-7. Illustration of the use of fluorescein-labeled antibody to identify strain differences in the *R. tsutsugamushi* organism. Gilliam, Karp, and Kato strains were used to immunize guinea pigs. Immune sera obtained against all three strains were labeled with fluorescein and used to stain peritoneal exudates from infected mice. Slides 2 to 4 were prepared from mice infected with the Gilliam strain, 5 to 7 with the Karp strain, and 8 to 10 with the Kato strain. All were treated with a 1:4 dilution of the labeled antisera to Gilliam, Karp, and Kato strains, respectively. Organisms were stained only with specific antibody. (From Iida, T., Kawashima, H., and Kawamura, A.: J. Immun. **95**:1129, 1965. © 1965 The Williams & Wilkins Co., Baltimore.)

R. tsutsugamushi is characterized by variation in antigenic composition. Shishido[28] has differentiated three antigenically distinct groups: the Gilliam, Karp, and Kato strains. Fig. 15-7 illustrates the use of specific fluorescein-labeled antisera to differentiate the three groups. Slides prepared from peritoneal exudate of infected mice were stained with antibodies specific to each of the three strains; each fluorescent antibody bound only to its specific strain.[12]

The incubation period of this rickettsial disease is 6 to 21 days. A generalized maculopapular rash appears at the end of the first week. The rise of agglutinins in the patient's serum to *Proteus* OX-K strain, as determined by the Weil-Felix test, is considered the most useful diagnostic measure in scrub typhus. Immunity to this disease lasts only a few years.

Various ocular changes may occur in infection caused by *R. tsutsugamushi.* Congestion of the conjunctiva has been observed in the acute phase of the infection so constantly that it is considered a clinical feature. Sheie[27] in 1947 recorded ocular changes in the routine examination of 451 patients suffering from scrub typhus and noted the incidence of ocular pathologic characteristics as follows:

Conjunctival injection	38 %
Subconjunctival hemorrhages	6.4%
Ecchymosis of eyelids	1 %
Eschar of eyelid	0.5%
Engorgement of veins	67 %
Fixation nystagmus	0.5%
Retinal edema	36 %
Retinal hemorrhages	6.6%
Exudates	4.9%
Uveitis	1.3%
Vitreous opacities	4.6%

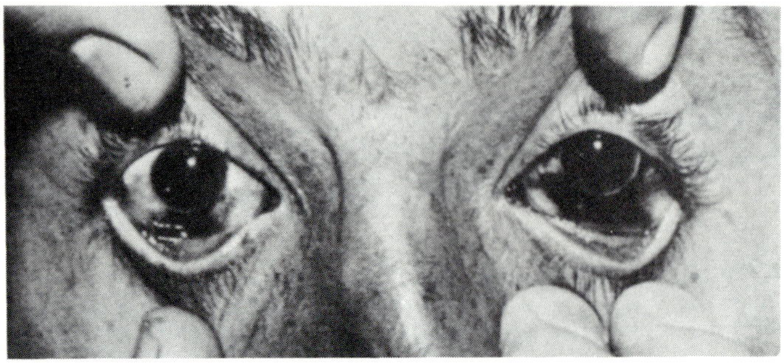

Fig. 15-8. Bilateral subconjunctival hemorrhages in patient with scrub typhus. (From Scheie, H. G.: Trans. Amer. Ophthal. Soc. **45**:637, 1947.)

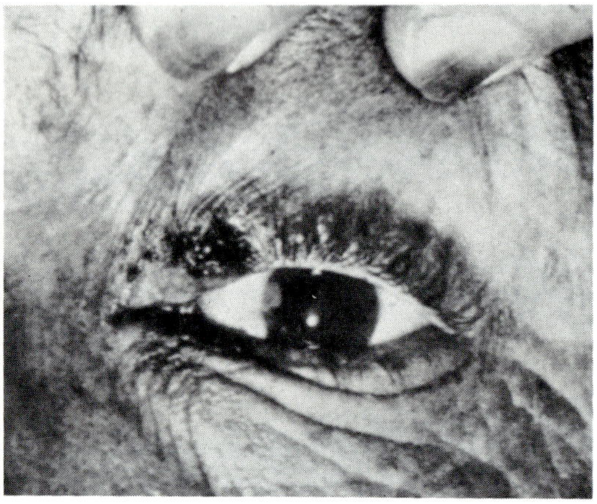

Fig. 15-9. Primary lesion of eyelid in patient with scrub typhus. (Courtesy H. G. Scheie.)

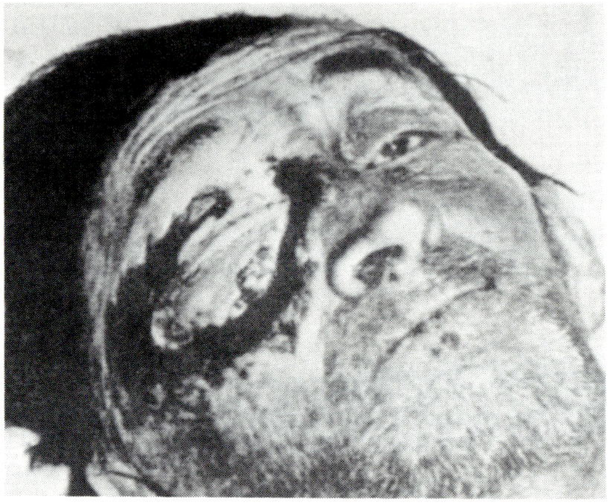

Fig. 15-10. Q fever. Shows widespread ulceration in the region of eyelids with preauricular adenopathy in a 64-year-old patient. (From Calmettes, L., Deodati, F., and Bec, P.: Bull. Mem. S. F. O. **73**:509, 1960, Masson et Cie, Paris.)

Figs. 15-8 and 15-9, from Scheie's report, illustrate bilateral subconjunctival hemorrhage and a primary lesion of the eyelid in the area where mites might like to rest. Pathologic studies of the eyes obtained at autopsy showed subacute diffuse choroiditis to be the basic lesion. As may be seen in Table 15-2, retinal hemorrhage and corneal ulcer have been added to this list by other investigators. In addition, Macaskill[17] has reported bilateral optic atrophy in one patient 6 months after the acute illness.

Q FEVER

Q fever is a self-limited infection caused by *Coxiella burnetti.* The symptoms of this disease are more characteristic of influenza or atypical pneumonia than of typhus. This is caused in part by the mode of transmission, which typically is inhalation or ingestion. Ticks transmit the agent to cattle, sheep, and goats. Cows develop a chronic infection of the udder, and rickettsiae are excreted in the milk. The placenta of cows and sheep also contain the organisms, and parturition creates infectious aerosols. Dust becomes contaminated. Thus, Q fever patients rarely give a history of tick bite. Patients and animals develop agglutinins and complement-fixing antibodies, but man does not produce agglutinins to *Proteus* antigens. The unique resistance of *C. bur-*

netti to prolonged exposure in nature contributes to the spread of the agent.

After an incubation period of 9 to 20 days following respiratory exposure, there is abrupt onset of fever, headache, muscle pain, and malaise. The temperature remains elevated with fluctuations for a period of about 3 weeks. In contrast to the other rickettsial infections, there is no rash.

General infection with *C. burnetti* may be acquired through the conjunctiva. It is characterized by violent conjunctival hyperemia and occasionally gangrene of the eyelids.[19] Uveitis, chorioretinitis, episcleritis, and optical neuritis may occur. Fig. 15-10 illustrates ulceration in the region of the eyelids with preauricular adenopathy. Harvey and his co-workers[10] in 1951 described a small outbreak of Q fever in East Kent. The ocular features in these patients consisted of retroorbital pain with photophobia and some injection of the conjunctiva.

TRENCH FEVER

Trench fever is a self-limited louse-borne rickettsial disease characterized by intermittent fever and generalized aches and pains. The agent, *R. quintana*, grows extracellularly in the louse intestine and is excreted in louse feces. The organisms can be cultivated on agar enriched with 10% fresh blood. Transmission occurs by rubbing louse feces into skin abrasions produced by the bite of the louse. The incubation period varies from 5 to 38 days, but usually is 2 to 4 weeks. History of contact with lice within this period is a useful diagnostic aid. Trench fever is associated with headache, exhaustion, fever, a roseolar rash, and relapses. This disease may show exacerbation over a period of a year.

Ocular features of infection caused by *R. quintana* include conjunctivitis, hyperemia of the conjunctiva, photophobia, and retinitis. Deduit[7] in 1963 described a case of iridocyclitis in a patient suffering from paresis. The patient was inoculated with a strain of *R. quintana* for therapeutic purposes and developed iridocyclitis and later a small focus of retinitis.

REFERENCES

1. Ahlm, C. E., and Lipshutz, J.: Tsutsugamushi fever in the Southwest Pacific Theater, J.A.M.A. **124**:1095, 1944.
2. Audoueineix, M. E.: Les complications oculaires des rickettsioses dans la Sarthe, Bull. Soc. Franc. Ophthal. **73**:477, 1960.
3. Brooks, F., and Fineberg, R.: Ocular manifestations of typhus fever: a review of the literature from 1867 to 1943, Amer. J. Ophthal. **34**:605, 1951.
4. Calmettes, L., Deodati, F., and Bec. P. Gangrène des paupières d'origine rickettsienne, Bull. Soc. Franc. Ophtal. **73**:509, 1960.
5. Clavel, A., and Hamart: Manifestations oculaires au cours d'une fièvre boutonneuse méditerranéenne autochtone, Ann. Oculit **182**:232, 1949.
6. Conor, A., and Bruch, A.: Une fièvre éruptive observée en Tunisie, Bull. Soc. Path. Exot. **3**:492, 1910.
7. Deduit, Y.: Iridocyclite et bouton de rétinite au cours d'une rickettsiose thérapeutique a Rickettsia quintana, Ann. Oculist. **196**:934, 1963.
8. Duke-Elder, S.: System of ophthalmology. Diseases of the outer eye, vol. 8, part 1, St. Louis, 1955, The C. V. Mosby Co., p. 244.
9. Elisberg, B. L., and Bozeman, F. M.: Rickettsiae. In Lennette, E. H., and Schmidt, N. J., editors: Diagnostic procedures for viral and rickettsial infections, ed. 4, New York, 1969, American Public Health Association, p. 826.
10. Harvey, M. S., Forbes, G. B., and Marmon, B. P.: An outbreak of "Q" fever in East Kent, Lancet **2**:1152, 1951.
11. Hazard, G. W., Ganz, R. N., Nevin, R. W., Nauss, A. H., Curtis, E., Bell, D. W. J., and Murray E. S.: Rocky Mountain spotted fever in the eastern United States, New Eng. J. Med. **280**:57, 1969.
12. Iida, T., Kawashima, H., and Kawamura, A.: Direct immunofluorescence for typing of tsutsugamushi disease Rickettsia, J. Immunol. **95**:1129, 1965.
13. Jawetz, E., Melnick, J. L., and Adelberg, E. A.: Review of medical microbiology, ed. 9, Los Altos, Calif., 1970, Lange Medical Publications, p. 263.
14. Larionow, N.: Erkrankungen des Sehorgans beim Typhus, Klin. Mbl. Augenheilk. **16**:487, 1878.
15. Lebas, M. P., and Bernaerts-Lebas, M. A.: Rickettsioses et affections oculaires, Bull. Soc. Belg. Ophtal. **135**:437, 1963.
16. Lennette, E. H., and Schmidt, N. J.: Diagnostic procedures for viral and rickettsial infections, ed. 4, New York, 1969, American Public Health Association.

17. Macaskill, J.: Some ocular complications of scrub typhus, Brit. J. Ophthal. **29**:537, 1945.
18. Morax, P. V., Saraut, H., Moday, S., and Coscas, G.: Les manifestastions oculaires des rickettsioses et des neorickettsioses, Ann. Oculist. **194**:929, 1961.
19. Perdriel, G., Michel, A., Guyard, M., and Hoel, J.: Les manifestations oculaires de la fièvre "Q," Ann. Oculist. **194**:957, 1961.
20. Presley, G. D.: Fundus changes in Rocky Mountain spotted fever, Amer. J. Ophthal. **67**:263, 1969.
21. Raab, E. L., Leopold, I. H., and Hodes, H. L.: Retinopathy in Rocky Mountain spotted fever, Amer. J. Ophthal. **68**:42, 1969.
22. Rhodes, A. J., and van Rooyen, C. E.: Textbook of virology, Baltimore, 1962, The Williams and Wilkins Co., p. 534.
23. Ricketts, H. T.: Further observations on Rocky Mountain spotted fever and Dermacentor occidentalis, J.A.M.A. **47**:1067, 1906.
24. Ricketts, H. T.: Further experiments with the wood-tick in relation to Rocky Mountain spotted fever, J.A.M.A. **49**:1278, 1907; and **52**:379, 1909.
25. Ricketts, H. T., and Wilder, R. M.: The etiology of the typhus fever (tabardillo) of Mexico City: a further preliminary report, J.A.M.A. **54**:1373, 1910; and **55**:309, 1910.
26. Rouher, F., Cluzel, P., Cantat, and Vaurs, M. R.: Kératite recidivante et rickettsies, Bull. Soc. Franc. Ophtal. **63**:86, 1963.
27. Scheie, H. G.: Ocular changes in scrub typhus: a study of 451 patients, Trans. Amer. Ophthal. Soc. **45**:637, 1947.
28. Shishido, A.: Identification and serological classification of the causative agent of scrub typhus in Japan, J. Med. Sci. Biol. **15**:308, 1962.
29. Venable, M. P., and Pollock, F. J.: Rickettsias in ophthalmolgoy, Arch. Ophthal. **30**:362, 1943.
30. Wolbach, S. B.: The rickettsial diseases, a general survey. In Virus and Rickettsial diseases, Cambridge, 1940, Harvard University Press.

16

Introduction to viral ocular infections

VIRAL AGENTS OF DISEASE

In 1948 Rivers' definitive textbook on viral and rickettsial infections of man brought together under one cover the current knowledge concerning these agents of human disease.[18] At that time 31 viruses and rickettsiae had been propagated in the chick embryo. Tissue-culture methods were only in the developmental stage.

An early criterion for the identification of a virus was its inability to grow in the absence of living cells. It was apparent that viruses depended heavily on the apparatus of cells for their propagation. Just how this was controlled to favor the virus was unknown. Viruses were characterized in part by their particle size, determined by filtration through materials of known pore diameter. Iwanowski[11] was the first to recognize the filterability of these agents under conditions that kept back bacteria. A tentative classification of the viruses on the basis of size was made possible through the development of methods of filtration through graded collodion membranes.[7] Passage of viruses through filters was determined by the infectivity of the filtrates.

Microscopic examination of stained preparations of lesions found in known or suspected viral diseases had shown the presence of inclusion bodies, small intracellular structures in nuclei, cytoplasm, or both. Woodruff and Goodpasture[22] demonstrated that inclusion bodies seen in fowlpox produced typical fowlpox disease when inoculated into the skin of hens.

Small bits of information were pieced together from cultural requirements, size, and inclusion body formation, but were not sufficient to establish any sort of classification comparable to that by which bacteria were recognized. Many viruses went only by the name of the disease they caused; and in many cases the virus etiology, though suspected, was not proved by culture. For example, epidemic keratoconjunctivitis (Chapter 17) was thought to be produced by a virus, since mice or rabbits injected intracerebrally with material from the early lesions developed fatal encephalitis in about a week, though no bacteria were found. Rabbits inoculated in the cornea developed keratitis. A relationship to the herpes simplex virus (Chapter 18) was suspected, as herpes simplex was known at times to produce meningoencephalitis as well as keratoconjunctivitis in man. Several years were to pass before the etiologic agent of keratoconjunctivitis was identified as a member of the adenovirus group. Trachoma and inclusion conjunctivitis were accepted as diseases of viral origin because of the presence of inclusion bodies in the cells involved in the pathogenic process and the transmissibility of the diseases by filtered material gathered from the lesions. Another 10 years went by before the infectious agents were isolated and classified in a group apart from the viruses (Chapter 14).

Today the viruses are in some respects as well known as the bacteria. Fourteen

Table 16-1. Classification of viruses into groups based on chemical and physical properties

Nucleic acid core	Capsid symmetry	Virion enveloped or naked	Ether sensitivity	Number of capsomers	Virus particle size (nm)*	Molecular weight nucleic acid virion (×10⁶)	Number of genes	Virus group
DNA	Icosahedral	Naked	Resistant	12 or 32	18–24	1.5	7	Parvovirus
				42 or 72	40–55	2–4	10	Papovavirus
				252	70–80	23	50	Adenovirus
		Enveloped	Sensitive	162	180–250 (110)†	40–84	150	Herpesvirus
	Unknown	(Complex)	Resistant		230×300	160–240	400	Poxvirus
RNA	Icosahedral	Naked	Resistant	32	18–30	2	10	Picornavirus
				92	70–75	10	50	Reovirus‡
	Unknown§	Enveloped	Sensitive	32§	50–60 (35)§	2	10	Togavirus§
					110			Tacaribe-LCM‖
					80–120			Coronavirus¶
	Helical	Enveloped	Sensitive		80–200	2–5	10	Myxovirus
					100–300	7–10	40	Paramyxovirus
					60×225	3–4	10	Rhabdovirus
	Unknown	Enveloped	Sensitive		100	10–12	50	Leukovirus#

Table prepared by J. L. Melnick. From Jawetz, E., Melnick, J. L., Adelberg, E. A.: Review of medical microbiology, ed. 9, Los Altos, Calif., 1970, Lange Medical Publications, p. 278.

*Diameter, or diameter × length.

†The naked virus, that is, the nucleocapsid, is 110 nm in diameter; however, the enveloped virion varies from 180 to 250 nm.

‡Recent evidence indicates that reovirus is one of a large group of viruses possessing double-stranded RNA (diplornaviruses).

§Included in the togavirus group are arboviruses of subgroups A and B. Rubella virus and lactic dehydrogenase (LDH) virus of mice have similar properties. One togavirus that has been studied in detail (Sindbis) has been shown to have a nucleocapsid 35 nm in diameter, containing 32 capsomers in an icosahedral surface lattice with a triangulation number of T = 3. The enveloped virion is about 50 to 60 nm in diameter.

‖Lymphocytic choriomeningitis virus recently has been shown to be identical morphologically, and to cross-react serologically, with arboviruses of the Tacaribe complex, which includes the Junin and Machupo viruses of South American hemorrhagic fevers. A new taxonomic group of these RNA-containing, enveloped viruses has been proposed: the Tacaribe-LCM group.

¶Included in this group of viruses with petal-shaped surface projections are a number of new human respiratory viruses, avian infectious bronchitis virus (IBV), and mouse hepatitis virus.

#The leukovirus group refers to leukemia viruses of avian, feline, and murine species, since a human leukemia virus has not yet been described.

major groups, which contain hundreds of immunologically distinct types (or species) pathogenic for man or animals or both, have been established (Table 16-1).[9,12,20] Other viruses are awaiting classification. In addition to these groups there is a large and important number of viruses infecting plants. It is probable that viruses are responsible for at least 50% of all infectious diseases of man and animals, as well as for some forms of cancer. The rapid growth of knowledge concerning these agents was made possible in great measure by the introduction of the monolayer growth of cultured cells on glass by Enders and his associates.[7] This provided a relatively simple method for the cultivation of large quantities of virus. Antibiotics also contributed much to viral cultivation by cutting down on bacterial contamination of both embryonated eggs and cell cultures. Initial isolation of viruses taken from contaminated areas became possible.

Structure of viruses

The physical features of animal viruses are schematically illustrated in Fig. 16-1. The core of a virus, the nucleic acid genome, is made up of either ribonucleic acid (RNA) or deoxyribonucleic acid (DNA). The capsid is a symmetric protein coat enclosing the nucleic acid genome. The capsomers are clusters of the structural units, the basic building blocks of the capsid. Some more complicated virions (complete, infective viral particles) may have one or more additional membranes. Fig. 16-2, *1* is an electron micrograph of a varicella virus particle in its enveloping membrane with a centrally placed capsid. On the right such a virus particle with ruptured membrane shows the hollow capsomers on the surface of the capsid.

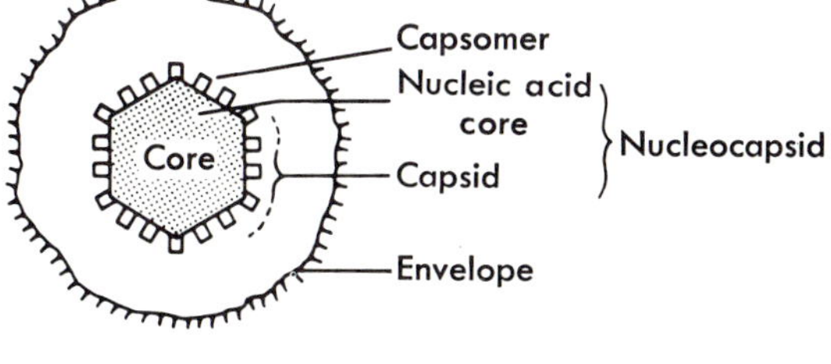

Fig. 16-1. Components of a complete virus particle, or virion. (From Horne, R. W.: Sci. Amer. **208:** 50, 1963.)

The size of viruses, still a basic criterion for classification, is now measured by electron microscopy, ultracentrifugation, or ionizing radiation, as well as by filtration. Fig. 16-3 illustrates the relative size, determined by electron microscopy, of 4 viruses that contain DNA and 4 that contain RNA in their core. These are representatives of the 14 major groups of viruses so far classified (Table 16-1). The variation in size is even more striking if it is expressed in terms of volume. The poxvirus has a mass 1,000 times greater than that of the picornavirus. By contrast, the mass of a small bacterium is 50,000 times greater than the picornavirus.

Table 16-1 presents the physical characteristics of the 14 groups of viruses listed. This table shows that specific information is available in most instances about the size, shape, and structure of the elements of the virion, including the molecular weight of the nucleic acid core and the number of genes. Ether sensitivity, included in this table, is believed to depend on the presence of lipids in the membrane.

Some of the most exciting work in micro-

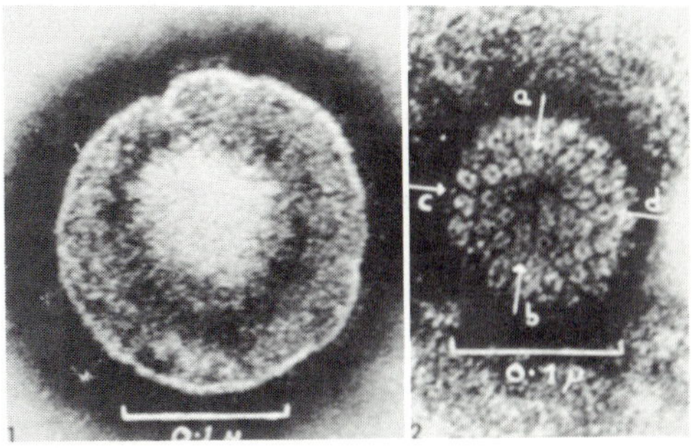

Fig. 16-2. *1*, Intact varicella virus particle showing enveloping membrane with projections and centrally placed capsid. *2*, Particle in which the envelop had ruptured, showing structure of capsid more clearly. (From Almeida, J. D., Howatson, A. F., and Williams, M. G.: Virology **16**(3): 353, 1962. Copyright © Academic Press, Inc.)

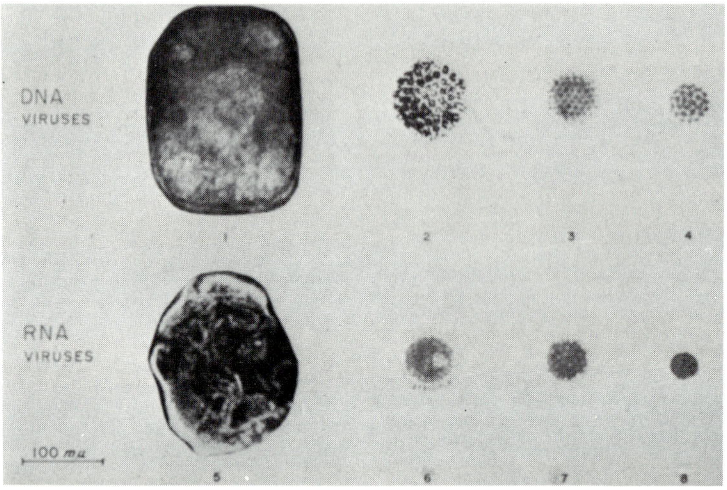

Fig. 16-3. Relative sizes of animal viruses (negative-stained virus particles). *1,* Poxvirus group (vaccinia virus); *2,* herpesvirus group (herpes simplex); *3,* adenovirus group; *4,* papovavirus group (papilloma virus); *5,* paramyxovirus group (Sendai virus); *6,* myxovirus group (influenza A virus); *7,* reovirus group; *8,* picornavirus group (polio virus). (From Hsiung, G. D.: Bact. Rev. **29**:470, 1965.)

biology deals with the studies on the manner in which viruses replicate their structure. For reviews of the subject, Erikson[7a] and Jawetz and his coauthors[12] may be consulted.

Viral antigens and antibodies

The structural proteins of the viral capsids are important for a number of reasons. They form a compact and stable, structured symmetry that protects the genome, and they play a part in the attachment of some viruses to susceptible cells. To the immunologist and physician, however, the most important characteristic of viral protein is its antigenic specificity, which permits identification of viral groups and types, thus enabling precise diagnosis of the infectious agent of a viral disease. A corollary to this is the ability to utilize preparations of living or dead viruses as vaccines to stimulate active immunity in man or to provide from animals antibodies that are useful both in therapy and in viral identification. Antibody response also may influence the type of immunopathologic reaction. The absence of immunologic response may result in enhanced susceptibility to infection on the one hand or to inapparent infection or the carrier state on the other.[12]

Antibodies are used to recognize viruses by one or more of the following procedures:

1. Myxovirus, paramyxovirus, and poxvirus groups of viruses agglutinate red blood cells of man and some animals. The source and method of operation of the hemagglutinin varies with the group, but in each case the agglutinating agent is antigenic. The antibodies produced will specifically prevent the hemagglutinating activity. Hemagglutination neutralization is a useful diagnostic test.

2. Antiviral sera fix complement in the presence of viral antigens. Some viruses produce a group-specific soluble antigen that stimulates early antibody production. The detection of the antibody by the complement-fixation test may have considerable diagnostic value.

3. Infectious virus particles induce the production of a virus-neutralizing antibody, which may protect against the transmission of the disease or prevent its develment in man. Virus-neutralizing anti-

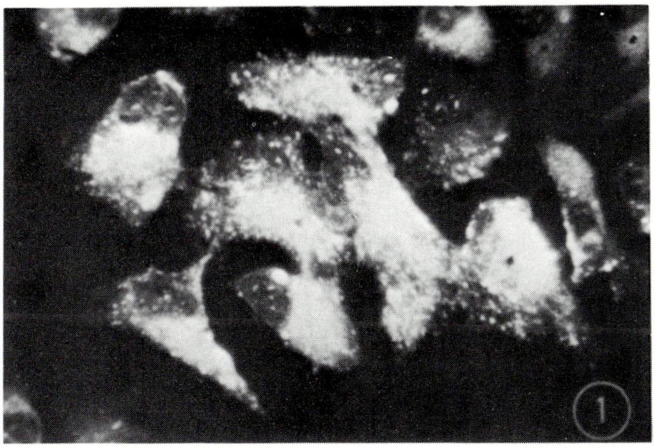

Fig. 16-4. Infection of human conjunctiva cells with cytopathogenic line of Dunai strain mumps virus. Stained with fluorescent antibody 30 hours after inoculation. Mumps virus antigen is distributed throughout the cytoplasm in small and large granules. (× 500.) (From Walker, L., and Hinze, H. C.: J. Exp. Med. **116**:750, 1962.)

body can also be used to protect tissue-cultured cells. Viruses growing in tissue culture usually produce cellular destruction, proliferation, or some other gross abnormality called cytopathogenic effect (CPE). This cellular damage will be prevented by specific neutralizing antibody. Individuals who have had viral infection develop this antibody. If it has not been possible to isolate the cause of a given infection but a virus is suspected, serum from the patient, drawn 2 or more weeks after symptoms have developed, may be used to establish a diagnosis. Such a serum is added to a series of cultured cells together with one of the several viruses that might have been the cause of the patient's disease. If the CPE of any one of the viruses tested is prevented, it is assumed that this virus was the cause of the disease.

One very important control is necessary. A sample of blood taken when the patient is first ill, the acute phase sample, must be tested simultaneously with the convalescent serum to rule out the possibility that the patient is carrying an antibody to a virus acquired during a previous exposure but unrelated to the illness under question.

4. It is possible to couple fluorescein isothiocyanate with immune globulins without impairing the immunologic activity of the globulin.[13] Such coupled virus-specific antibodies may be added to infected cultured cells, exudates, or pieces of tissue taken by biopsy from a site of infection. After a short period to permit reaction between the immunoglobulin and the specific virus in the specimen, the excess of labeled globulin is washed away. The preparation is examined with a microscope using ultraviolet light. The area where the immune globulin is bound fluoresces, emitting a yellow-green light, thus showing the presence of the virus in cells. Fig. 16-4 illustrates the appearance of mumps virus growing in cultured human conjunctival cells, examined by the fluorescein technique. The infected cells had been grown on coverslips, dried, fixed in acetone, and stained with fluorescein-labeled rabbit antibody to measles virus. The illustration shows that this virus is irregularly distributed in the cytoplasm of the cells.

A similar use of labeled antibody has been introduced into electron microscopy.[21] The antibodies are labeled with electron-

A 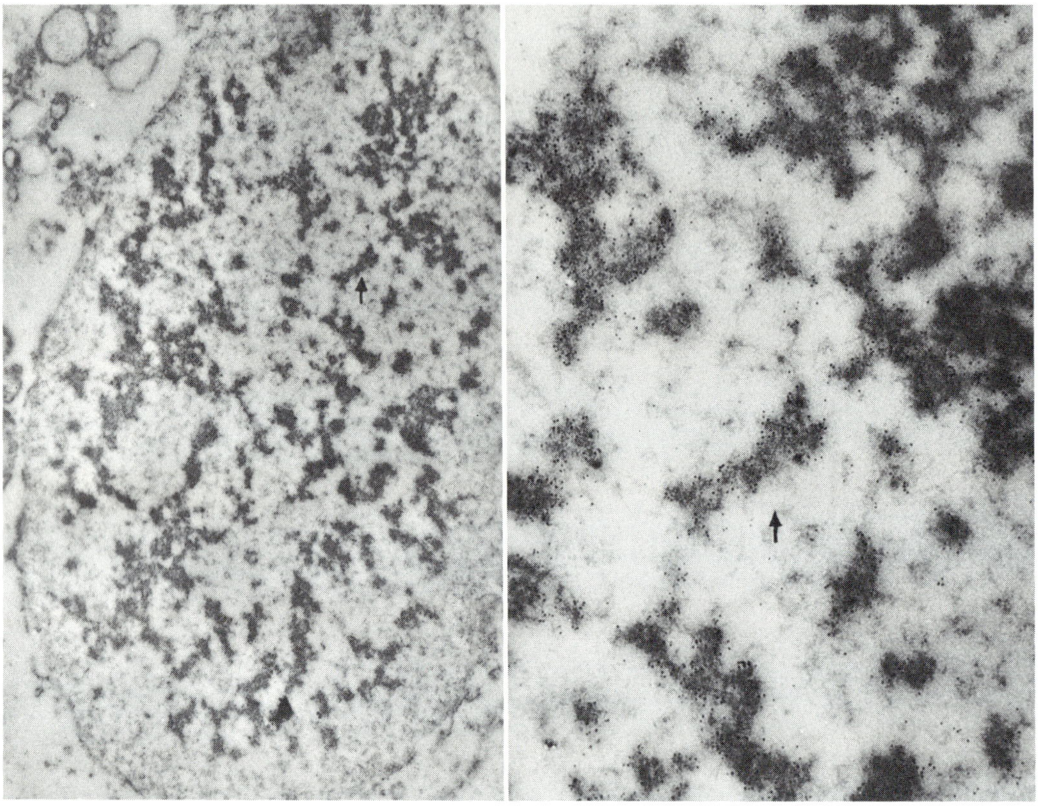 B

Fig. 16-5. **A,** Nucleus of a cell infected with influenza virus, containing aggregates of dense material (arrow). The chromatin is sparse, and the nuclear membranes have disrupted. (× 26,000.) **B,** Part of the nucleus seen in **A** (arrow), viewed at higher magnification. Ferritin-conjugated antibody to the influenza virus is present within the aggregates of dense material (arrow). The intervening nuclear matrix is nearly devoid of ferritin. (× 97,000.) (From Morgan, C., Hsu, K. C., Rifkind, R. A., Knox, A. W., and Rose, H. M.: J. Exp. Med. **114**:833, 1961.)

opaque ferritin. The tissue containing the virus is treated in a manner analogous to that used for fluorescein examination, and ultrathin sections are prepared for electron microscopic examination. The potentialities of the technique can be illustrated by influenza virus. In Fig. 16-5, *A,* the nucleus of a cell infected with this virus shows collections of amorphous electron-opaque material. Ferritin-labeled antibody to the nucleocapsid antigen of the virus was applied to the tissue prior to fixation for electron microscopic study. Fig. 16-5, *B,* a higher magnification of a portion of this nucleus, illustrates the presence of ferritin particles in the amorphous aggregates. Their presence indicates that the ferritin-labeled antibody to nucleocapsid antigen has been bound to the aggregates, which therefore must contain nucleocapsid antigen made in this area. Complete viral particles, however, are found only on or near the surface of the cell, as can be seen in Fig. 16-6. Thus the ferritin-labeled antibody identifies specific antigens of the virus. These studies and other related investigations are helping to distinguish the areas of the cell where parts of the viruses may be made or put together.

Pathogenesis of viral disease

The modes of entry of viruses into the human body are as numerous as those used by bacteria. They are: (1) inhalation

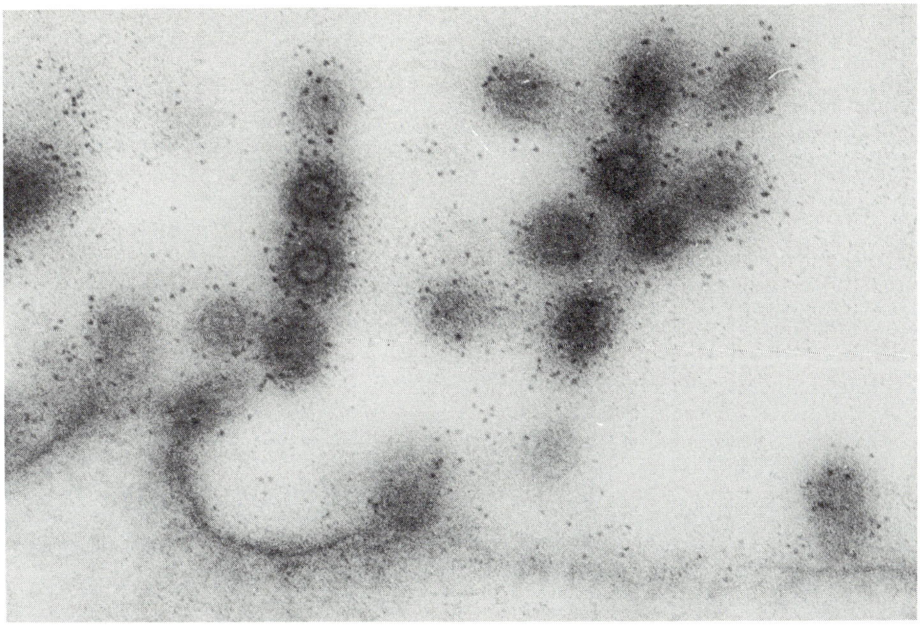

Fig. 16-6. Influenza virus at the surface of a chick chorioallantoic membrane. The surface of the cell has been cut obliquely, and cytplasm occupies the lower border of the field. It is of interest that, in addition to fully formed virus, ferritin has tagged particles in the process of budding from the cell. (× 135,000.) (Courtesy Dr. C. Morgan.)

of airborne droplets or dust, (2) ingestion, (3) direct contact, (4) bite of an infected animal, and (5) bite of an insect or arthropod. Some of the viruses inhaled and some ingested appear to remain localized either in the upper respiratory tract or the intestinal canal. The majority seem to travel rather freely via the circulation, though a predilection for a specific tissue is common.

Virus usually can be cultured from infected areas or the bloodstream only during the first few days of acute infection. Nevertheless, it has long been conjectured that a virus may persist in the host after recovery has taken place, since many viral infections are followed by lasting immunity. Evidence for the carriage of a virus in healthy cultured cells and in the tissues of some animals has been reported.[12,19] Such evidence is not as readily obtained in man. However, there is a chronic degenerative brain disorder—Dawson's inclusion body encephalitis—seen in some who have had measles a number of years previously. Measles virus has been isolated from such a case.[17] More data are needed, however, before the role of a carrier state in the maintenance of protection from infection is established.

An apparent exception to prolonged immunity afforded by viral infection is the common cold. Colds are caused by various members of the picornavirus group, including a number of enteroviruses and rhinoviruses. Since many types within the groups have been involved and immunity to some affords no protection to others, man is always potentially vulnerable to new strains.

The immune response of the body to the presence of a virus over a period of months or years may be of great importance in the pathogenesis of some viral diseases. The most familiar examples are found in the herpesvirus group. Latent infection (the carrier state) with herpes simplex can last for many years even when circulating antibodies are demonstrable. Varicella virus, a member of this group, produces chicken-

Table 16-2. Viruses causing ocular infection

Virus group and subgroups or types	Primary diseases	Ocular diseases
Papovavirus group		
Papilloma virus	Warts; tumorlike papilliform proliferations	Warts of lid margins; bulbar conjunctiva and opposite lid; verrucose conjunctivitis and keratitis
Adenovirus group		
Type 8	Epidemic keratoconjunctivitis	Epidemic keratoconjunctivitis
Type 3 (predominately)	Pharyngoconjunctival fever and other upper respiratory tract diseases	Pharyngoconjunctival fever
Types 1-11, 14-17, 20, 22, 26, 27	Fever; pharyngitis with involvement of neighboring areas	Follicular conjunctivitis; keratoconjunctivitis
Herpesvirus group		
Herpes simplex	Systemic infection at birth or herpetic stomatitis; chronic recurrent infections	Herpetic keratitis and keratouveitis
Varicella-zoster	Chickenpox in childhood; shingles in adulthood	Rare involvement of conjunctiva and cornea; conjunctivitis; keratitis; scleritis; iridocyclitis; and optic neuritis
Cytomegalo virus	Cytomegalic inclusion disease of newborn and young children; in adults 80% positive serology	Ocular involvement rare, usually includes lacrimal gland, uveal tract, and sometimes retinal changes
Poxvirus group		
Variola	Smallpox	Exanthematous, catarrhal conjunctivitis; superficial infiltration of cornea
Vaccinia	Spread of infection from vaccinated area	Ranges from uncomplicated blepharoconjunctivitis in immune person to purulent conjunctivitis and disciform keratitis
Molluscum contagiosum	Molluscum contagiosum disease characterized by small epithelial tumors	Molluscum tumors may occur on lids and palpebral conjunctiva, leading to conjunctivitis and sometimes keratitis
Picornavirus group		
Rhinovirus of cattle	Foot-and-mouth disease of cattle transmitted to man	A rare disease, in association with ulcerative condition of mouth; severe keratoconjunctivitis arises
Togavirus group		
Rubella virus	German measles	Catarrhal or follicular conjunctivitis; mild keratitis
Arbovirus B, yellow fever virus	Yellow fever, transmitted by mosquito	Intense conjunctival injection; subconjunctival hemorrhages; edema and swelling of face
Arboviruses B, dengue fever virus	Dengue fever, transmitted by mosquito	Conjunctival injection (bulbar); edema of lids; lacrimation; retrobulbar pain
Arbovirus (ungrouped), sandfly fever virus	Sandfly fever, transmitted by *Phlebotomus papatasi*	Injection of bulbar conjunctiva limited to exposed part; burning pain; lacrimation
Myxovirus group		
Influenza viruses	Influenza	Catarrhal, occasionally acute follicular conjunctivitis

Table 16-2. Viruses causing ocular infection—cont'd

Virus group and subgroups or types	Primary diseases	Ocular diseases
Paramyxovirus group		
Measles virus	Measles	Eye lesions precede exanthem; mucopurulent conjunctivitis; papules and pale spots (Koplik's spots) on mucous membrane; caruncle; punctate keratitis in severe disease
Mumps virus	Mumps	Simple conjunctivitis common; edema of bulbar conjunctiva; may develop episcleritis, keratitis
Newcastle disease virus	Primarily pathogenic for fowl; an occupational disease of man primary to the eye	Mild conjunctivitis with itching, burning, or mild gelatinous swelling with follicles, palpebral edema

pox in children; and herpes zoster, indistinguishable from varicella, produces shingles in adults. Herpes zoster can also be a source of chickenpox in children. Adults exposed to chickenpox may develop herpes zoster. The question to be answered is whether long residence in the body changes tissue response to infection by the same virus.

"Slow virus diseases," where the incubation period is a matter of several years and there is no preliminary acute phase, are under intensive investigation and offer one of the most important areas for the study of the relation of immune response to pathology. In these diseases virus-antibody complexes may circulate for long periods of time. Trapping of these complexes in affected organs may serve to initiate the pathologic sequences. Among the human diseases under investigation for such an etiologic agent are spongiform encephalopathies, kuru, amyotropic lateral sclerosis, polymyositis, and progressive leukoencephalopathy.[12]

VIRUSES INFECTING THE EYE

Eight of the 14 major groups of animal viruses described in Table 16-1 are known to contain strains that produce ocular disease. Table 16-2 lists these viruses, the primary illnesses they cause, and the nature of the lesions found in the eyes. The table shows that viruses infecting the eye come from the upper respiratory tract and the skin in most instances, though three of the viruses are first introduced into the circulation by the bite of an insect. Each of the eye infections is considered in one of the three succeeding chapters.

Diagnosis of viral infections of the eye

Although each virus requires some special handling (choice of cultured cells, length of incubation, and so on, which will be discussed in the next few chapters), there are four basic methods for diagnosis, one of which may be handled in a routine laboratory if the necessary equipment is available (see also pp. 270-272). These methods are: (1) the immunofluorescein technique, in which specific viral antibodies labeled with fluorescein isothiocyanate are used to stain smears of scrapings from the suspected lesion; (2) the hemagglutination inhibition test, in which a series of known antiviral antibodies are used in an attempt to inhibit hemagglutination produced by the unknown virus; (3) the cytopathogenic neutralization test, in which known antiviral sera are tested for their ability to neutralize or prevent the cytopathogenic effect produced by the unknown virus in tissue-cultured cells; and (4) the retrospective diagnostic test, used when the attempt to grow the unknown virus has been unsuccessful. In this test a sample of the patient's serum drawn during his convalescence is compared with an early acute phase sample for the presence of antiviral antibodies, usually determined by the hemagglutination inhibition test, cytopathogenic neutralization test, or by the complement-fixation test.[12a]

The test that can be used without having a laboratory equipped for the cultivation of viruses utilizes the immunofluorescein technique to stain smears of scrapings. Since this technique requires several control slides, it is almost impossible to apply it to the small samples available from scrapings of the eye. This technique is much more successfully used when the virus is grown in tissue-cultured cells (Fig. 16-4). However, in some instances early cases of epidemic keratoconjunctivitis, probably caused by adenovirus type 8, or corneal ulcers, probably caused by herpesvirus infection, may be confirmed by staining scrapings from the lesions with fluorescein-labeled antibodies to the suspected viral antigen. The scrapings must be obtained as early as possible in the infection.

If there are no facilities available in association with the microbiology laboratory for growing viruses, the ophthalmologist may send the material to a virology laboratory. Since viruses can be isolated from the affected tissues only in the first few days of the disease, early cases alone are suitable for culturing. If there is abundant exudate, it may be taken on a cotton swab moistened with tissue-culture fluid or broth. A scraping from the lesion or vesicle fluid is more likely to contain the virus, since it grows in living cells. The scrapings are kept moist in a tube containing a drop of tissue-culture fluid or broth. Both specimens are sent to the laboratory immediately. If they are to be forwarded to a diagnostic center, they should be packed in dry ice to preserve the virus and prevent proliferation of contaminating bacteria. They should be accompanied by the following data:

Date of onset of disease
Date of obtaining specimen
Capsule history and clinical diagnosis
Recent travel

All specimens must be accompanied by a tube of clotted blood. This blood specimen (acute phase) should be obtained within 3 to 4 days of the onset of illness. The blood is needed to attempt a retrospective diagnosis in case no viral growth is obtained. This technique requires that the blood must be drawn again 2 to 4 weeks later in order to test for a rising titer of antibodies to one of the group of stock viruses. The history and clinical diagnosis that accompany the swab and serum sample assist the laboratory in choosing which group of antisera or viruses to use in identifying the unknown virus or in looking for serum antibodies.

SUMMARY

A few points may be stressed again.

1. Viral structure, content, physiology, and antigenicity are being documented, and it is now possible to speak of viruses with some of the same familiarity formerly reserved only for bacteria.
2. There are hundreds of viruses that can be identified because of their unique antigens. Most of them have been classified in 1 of 14 groups.
3. If it has not been possible to isolate a virus, retrospective diagnosis of viral infections can be made, using the patient's newly formed antibodies.
4. The immunologic response of the host to invading viruses may establish the pathologic manifestations of the disease.
5. Viruses known to cause ocular disease are listed. They come from 8 viral groups. In only 2 instances is the ocular disease primary to the eye.

REFERENCES

1. Almeida, J. D., Howatson, A. F., and Williams, M. G.: Morphology of varicella (chickenpox) virus, Virology **16**:353, 1962.
2. Andres, G. A., Hsu, K. C., and Seegal, B. C.: Immunoferritin technique for the identification of antigens by electron microscopy. In Weir, D. M., editor: Handbook of experimental immunology, Oxford, 1967, Blackwell Scientific Publications.
3. Beale, A. J., Doane, F., and Ormsby, H. L.: Studies on adenovirus infections of the eye in Toronto, Amer. J. Ophthal. **43**:26, 1957. (Symposium on Viral Keratoconjunctivitis)
4. Davis, B. D., Dulbecco, R., Eisen, H. N., Ginsberg, H. S., and Wood, W. B.: Microbiology, New York, 1967, Harper and Row, Publishers.
5. Duke-Elder, St.: System of ophthalmology. Diseases of the outer eye, vol. 8, part I, St. Louis, 1965, The C. V. Mosby Co.
6. Elford, W. J.: A new series of graded collodion membranes suitable for general bacteriological use, especially in filterable virus studies, J. Path. Bact. **34**:505, 1931.
7. Enders, J. F., Weller, T. H., and Robbins, F. C.: Cultivation of the Lansing strain of poliomyelitis virus in cultures of various human embryonic tissues, Science **109**:85, 1949.

7a. Erikson, R. L.: Replication of RNA viruses, Ann. Rev. Microbiol. **22**:305, 1968.

8. Horne, R. W.: The structure of viruses, Sci. Amer. **208**:48, 1963.
9. Horsfall, F. L., and Tamm, I.: Viral and rickettsial infections of man, ed. 4, Philadelphia, 1965, J. B. Lippincott Co.
10. Hsiung, G. D.: Use of ultrafiltration for animal virus grouping, Bact. Rev. **29**:477, 1965.
11. Iwanowski, D.: Uber die Mosaik-Krankheit der Tabaksplanze, Bull. Acad. Imp. Sci. (St. Petersburg) **35**(N.S. 3):67, 1892.
12. Jawetz, E., Melnick, J. L., and Adelberg, E. A.: Review of medical microbiology, ed. 9, Los Altos, Calif., 1970, Lange Medical Publications.

12a. Lennette, E. H.: General principles underlying laboratory diagnosis of viral and rickettsial infections. In Lennette, E. H., and Schmidt, N. J., editors: Diagnostic procedures for viral and rickettsial infections, ed. 4, New York, 1969, American Public Health Association, p. 1.

13. Liu, C.: Fluorescent antibody technics. In Lennette, E. H., and Schmidt, N. J., editors: Diagnostic procedures for viral and rickettsial infections, ed. 4, New York, 1969, American Public Health Association, p. 179.
14. Morgan, C., Howe, C., Rose, H. M., and Moore, D. H.: Structure and development of viruses observed in the electron microscope: viruses in the RI-APC group, J. Biophys. Biochem. Cytol. **2**:351, 1956.
15. Morgan, C., Hsu, K. C., Rifkind, R. A., Knox, A. W., and Rose, H. M.: The application of ferritin-conjugated antibody to electron microscopic studies of influenza virus in infected cells: the cellular surface, J. Exp. Med. **114**:825, 1961.
16. Morgan, C., Hsu, K. C., Rifkind, R. A., Knox, A. W., and Rose, H. M.: The application of ferritin-conjugated antibody to electron microscopic studies of influenza virus in infected cells: the interior of the cell, J. Exp. Med. **114**:833, 1961.
17. Payne, F. E., Baublis, J. V., and Itabashi, H. H.: Isolation of measles virus from cell cultures of brain from a patient with subacute sclerosing panencephalitis, New Eng. J. Med. **281**:585, 1969.
18. Rivers, T. M.: Viral and rickettsial infections of man, Philadelphia, 1948, J. P. Lippincott Co.
19. Walker, D. L., and Hinze, H. C.: A carrier state of mumps virus in human conjunctiva cells: observations on intracellular transfer of virus and virus release, J. Exp. Med. **116**:751, 1962.
20. Wilner, B. I.: A classification of the major groups of human and other animal viruses, ed. 4, Minneapolis, 1969, Burgess Publishing Co.
21. Woodruff, C. E., and Goodpasture, E. W.: The infectivity of isolated inclusion bodies of fowl-pox, Amer. J. Path. **5**:1, 1929.

17

Ocular infections caused by adenovirus and papovavirus groups

ADENOVIRUS INFECTION OF THE EYE

Adenoviruses are common and widespread parasites of man. Many may be found in the absence of disease, but most if not all of the 33 types so far isolated from man can act as pathogens. The commonest disease produced by members of the adenovirus group is acute febrile pharyngitis. In civilian life young children are more susceptible than adults. Among military groups, however, the incidence of adenovirus infection of the upper respiratory tract may be high; the infection goes by the name of acute respiratory disease. The eye is often involved, in which case the disease is called pharyngoconjunctival fever. Ocular disease—either acute follicular conjunctivitis or epidemic keratoconjunctivitis—may be the primary finding.

Discovery

Rowe and his associates[82] were the first to isolate and describe a member of this group of viruses. The new agent was obtained from explanted adenoid tissues, which had been removed from children during the winter and spring of 1952 to 1953, and was cultured by the chicken plasma clot technique. Thirty-three of the 53 cultured tissues demonstrated a characteristic rounding of the peripheral epithelium after 1 to 4 weeks. When inoculated into fresh preparations of adenoids, human embryonic tissue, or HeLa cells, the culture fluids produced characteristic changes in the epithelium of the recipient cultures. When a concentrated inoculum of the culture fluid was transmitted to HeLa cells, clumping and rounding of these cells appeared in 2 or 3 days. The authors report that Dr. Henry Pinkerton examined preparations of infected embryonic tracheal epithelium and described both nuclear and cytoplasmic inclusions.

Almost simultaneously with Rowe's experiments, Hilleman and Werner[35] recovered a similar, viruslike agent from nasopharyngeal secretions of military recruits with acute respiratory disease. Huebner and his co-workers[40] reported the isolation of 6 immunologically distinct types of virus from adenoid and tonsilar tissues removed surgically from nasopharyngeal or conjunctival secretions or from feces of persons with respiratory infections. Thus, confirmation of earlier findings followed rapidly, and it was soon apparent that the isolates constituted a group of infectious agents. Following the recommendation of these early workers,[23] further investigators proposed the name adenovirus to designate this rapidly growing family or viruses, which at that time numbered 12 serologically distinct types. In the past 15 years numerous papers have appeared in many areas of the world describing isolation from man of old and new members of this group of viruses. Table 17-1 presents the original source of the 33 immunologically distinct types of adenoviruses that have been obtained from human sources. The first 14 isolates were made in the United States, while 11 of the remaining types were found in the Netherlands, Saudi Arabia, and England. The viruses have been recovered predominantly

Table 17-1. Discovery of the types of adenoviruses of human origin

Types	Geographic origins	Sources	Clinical conditions	References
1	United States	Adenoid	Hypertrophied tonsils, adenoids	Rowe and others, 1953[82]* Huebner and others, 1954[40]
2	United States	Adenoid	Hypertrophied tonsils, adenoids	Huebner and others, 1954[40] Rowe and others, 1955[83]
3	United States	Nasal washing	Common cold (volunteer)	Huebner and others, 1954[40] Rowe and others, 1955[83]
4	United States	Throat washing	Primary atypical pneumonia	Hilleman and Werner, 1954[35]
5	United States	Adenoid	Hypertrophied tonsils, adenoids	Huebner and others, 1954[40] Rowe and others, 1955[83]
6	United States	Tonsil	Hypertrophied tonsils, adenoids	Huebner and others, 1954[40] Rowe and others, 1955[83]
7	United States	Throat washing	Pharyngitis	Berge and others, 1955[7]
7a	United States	Throat washing	Undifferentiated respiratory infection	Rowe and others, 1958[80]
8	United States	Corneal, conjunctival scraping	Epidemic keratoconjunctivitis	Jawetz and others, 1955[47]
9	United States	Stool	Rheumatoid arthritis	Kibrick and others, 1957[52]
10	United States	Eye swabbing	Conjunctivitis	Rowe and others, 1956[79]
11	United States	Stool	Paralytic poliomyelitis (type 1 polio virus recovered)	Kibrick and others, 1957[52]
12	United States	Stool	Nonparalytic poliomyelitis	Kibrick and others, 1957[52]
13	United States	Stool	Healthy child	Rowe and others, 1958[80]
14	Netherlands	Throat washing	Acute respiratory disease	van der Veen and Kok, 1957[88]
15-17	Saudi Arabia	Conjunctival scraping	Conjunctivitis (early trachoma in children under 15 years of age)	Murray and others, 1957[64]
18	United States	Anal swabbing	Niemann-Pick disease	Rowe and others, 1956[79]
19-24	Saudi Arabia	Conjunctival scraping	Trachoma in children from 7 months to 3 years of age	Bell and others, 1960[6]
25-28	United States	Anal swabbing	No specific illness	Rosen and others, 1961[76]
29, 30	United States	Anal swabbing	No specific illness	Rosen and others, 1962[77]
31	England	Anal swabbing	No specific illness	Pereira and others, 1965[70]
32, 33	United States	Anal swabbing	No specific illness	Blacklow and others, 1969[10]

*Superscript numbers indicate references at end of chapter.

Table 17-2. Adenovirus associated with epidemic and sporadic ocular infections

Clinical diagnosis	Adenovirus types	References
Follicular conjunctivitis	1, 2, 3, 4, 5, 6, 7, 8, 9, 10, 11, 14, 15, 16, 17, 20, 22	Murray and others, 1957[64]* Boniuk and others, 1966[13]
Follicular conjunctivitis with or without respiratory symptoms; may be associated with corneal opacities	2, 3, 4, 6, 7, 8, 9	Jawetz and others, 1956[46] Ormsby and others, 1957[68] Huebner and Rowe, 1957[41]
Pharyngoconjunctival fever with or without corneal opacities	1, 3, 4, 5, 6, 7, 7a, 9, 14	Bell and others, 1955[4] Beale and others, 1957[2] Evans, 1957[24] Huebner and Rowe, 1957[41] Kendall and others, 1957[51] Kimura and others, 1957[54] Oker-Blom and others, 1957[66] van der Veen and van der Ploeg, 1958[89]
Chronic superficial keratitis	2, 11	Koseki, 1960[56] Boniuk and others, 1965[12]
Keratoconjunctivitis	3	Fowle and others, 1955[27]
Punctate keratitis	3, 7, 8, 11	Koseki, 1960[56]
Epidemic keratoconjunctivitis, "infantile" form, with membranous conjunctivitis or pseudomembranous conjunctivitis	8	Mitsui and others, 1959[61] Mordhorst and Kjer, 1961[63]
Epidemic keratoconjunctivitis in adult	8, 11, 9	Jawetz and others, 1955[47] Koseki, 1960[56] Profeta and others, 1963[72]

*Superscript numbers indicate references at end of chapter.

from adenoids and tonsils, nasal and throat washings, corneal or conjunctival scrapings, and anal swabbings. Many apparently were not producing disease when isolated.

Additional types of adenovirus have been cultured from apes, monkeys, dogs, cows, pigs, mice, and chickens.[75]

Associated ocular diseases

Among the viruses first studied by Huebner and his co-workers[40] (Table 17-1), type 3 was isolated from several cases of acute nasopharyngitis and conjunctivitis during a winter outbreak of respiratory infection in a hospital ward and during a summer outbreak in a country day school. Jawetz and his associates[46] recovered type 8 in a conjunctival and corneal scraping from a merchant seaman who had just arrived from the Orient with epidemic keratoconjunctivitis. The table shows that types 10, 15, 16, 17, and 19 through 24 also were first isolated from the eye. All but type 10 of these latter viruses were found in Saudi Arabia. Table 17-2 lists the 18 types of adenoviruses that have been found in patients with ocular disease. Although there is no record as yet of spontaneous ocular disease produced by the other 15 types, there is some experimental evidence to indicate that at least a few are able to infect the eye under experimental conditions.

Induced ocular infections

Table 17-3 summarizes a number of studies on the production of ocular infections in human volunteers. Bell and his associates[5] showed that swabbing the lower

Table 17-3. Production of ocular infections with various adenoviruses in human volunteers

Adenovirus types	Methods of transmission	Clinical manifestations	References
1, 3, 4, 5	Swabbing conjunctiva only Swabbing lower palpebral conjunctiva and throat	Pharyngoconjunctival fever Follicular nonpurulent conjunctivitis	Bell and others, 1956[5*]
3, 4	Swabbing lower tarsal conjunctiva and posterior pharynx	Pharyngoconjunctival fever	Ward and others, 1955[91]
3	Inoculation into lower fornix of conjunctiva	Keratoconjunctivitis with diffuse centrally placed subepithelial opacities	Fowle and others, 1955[27]
8	Scraping lightly upper fornix of conjunctiva with sharp knife	Epidemic keratoconjunctivitis	Mitsui and others, 1957[60]
	Instillation of virus into conjunctival sac 5 times at 5 minute intervals; bandage then placed over eyes, removed following day	Acute follicular conjunctivitis with preauricular adenopathy and subepithelial opacities	Bietti and Bruna, 1957[9]
16	Stroking lower conjunctival sac 10 times with cotton swab previously immersed in 10^{-3} dilution of inoculum virus	An afebrile nonpurulent conjunctivitis associated in one case with slight corneal involvement	Kasel and others, 1963[49]
16, 17, 20	Dropping 0.05 ml of culture fluid into conjunctival sac, followed by gentle rubbing of inoculum into palpebral conjunctiva with a sterile cotton swab	Follicular conjunctivitis	Feng and others, 1959[26]
26, 27	Swabbing lower conjunctiva with a cotton swab soaked in tissue culture	Acute nonpurulent conjunctivitis	Kasel and others, 1963[49]

*Superscript numbers indicate references at end of chapter.

palpebral conjunctiva of human volunteers with cultures of adenoviruses types 1, 3, 4, or 5 resulted in infection (rarely associated with fever), which occurred after a latent period of 2 to 7 days and which generally persisted from 4 to 8 days. A rise in complement-fixing antibodies was demonstrable. The incidence of infection was influenced by the immunologic state of the volunteers, because those without detectable neutralizing antibodies developed conjunctivitis, whereas only 45 of 113 volunteers with detectable antibodies became infected. A nonpurulent, follicular conjunctivitis was produced; it frequently involved both the palpebral and bulbar conjunctiva and often was accompanied by blepharospasm. The cytologic examination of conjunctival scrapings revealed no cytoplasmic inclusions and only a lymphocytic response. Histopathologic examination of a conjunctival biopsy showed a conjunctivitis characterized predominantly by migration of lymphocytes into the submucosal tissue.

Other studies of the viruses listed in Table 17-3 confirm the above observations and give additional information. Fowle and her associates[27] inoculated a culture of type 3 virus from a case of keratoconjunctivitis into the lower fornix of the conjunctiva of the left eye of a volunteer. Five days after inoculation there was a sudden onset of a foreign-body sensation with profuse tearing. The preauricular node became enlarged and tender, and a pseudomembrane was observed on the lower conjunctiva. Conjunctival scraping revealed a lymphocytic response without inclusion bodies. The right eye became involved 5 days after onset of the disease in the left eye. At that time slit-lamp examination revealed 3 diffuse paracentral subepithelial opacities in the left eye and 1 in the right eye. Blurring of vision in the left eye persisted for 10 days.

Mitsui and his associates[60] used the following technique to transmit infection with type 8 adenovirus to 5 volunteers. The upper fornix of the conjunctiva was scraped lightly with a sharp knife after a local anesthetic was administered to the eye. The virus inoculum was then instilled into the conjunctival sac 5 times at 5-minute intervals. After the last instillation a bandage was placed over the eye for 24 hours. In 3 of the volunteers an acute follicular conjunctivitis and preauricular adenopathy developed. The secretion was fibrinous and contained mononuclear cells. Subepithelial punctate keratitis typical of epidemic keratoconjunctivitis appeared about 1 week after the onset of the conjunctivitis. The results corresponded to those of epidemic keratoconjunctivitis acquired by natural infection. There was a rise in neutralizing antibodies against adenovirus type 8 in the convalescent sera of these volunteers similar to that found in spontaneous epidemic keratoconjunctivitis. An acute follicular conjunctivitis resulted in the fourth volunteer, and the symptoms were those of mild epidemic keratoconjunctivitis without keratitis. In the fifth volunteer repeated inoculations with adenovirus type 8 were negative, and serologic findings suggested a partial immunity in this person.

Bietti and Bruna[9] also tested the infectivity of type 8 virus. They found that 6 volunteers inoculated with tissue-cultured material after scarification of the cornea developed follicular conjunctivitis similar to epidemic keratoconjunctivitis. Another 8 volunteers in whom the tissue-cultured material was merely dropped into the eye did not develop lesions.

Kasel and his co-workers[49] used adenovirus type 16 to infect 3 volunteers. The viral strain used to prepare the inoculum was isolated from an anal specimen of a young child. The inoculation of volunteers was performed by stroking the lower conjunctival sac of the eye 10 times with a cotton swab previously immersed in 1.5 ml of fluid containing a 10^{-3} dilution of culture inoculum. All 3 volunteers developed conjunctivitis with adenovirus type 16; in one a tender preauricular node was noted on the third day after inoculation. The acute phase of illness was associated with excessive lacrimation, photophobia, and foreign-body sensation. All 3 volunteers shed virus from the inoculated eyes and from the throat and rectum.

Adenovirus types 16, 17, and 20 grown from conjunctival scrapings from 2 patients with trachoma[6,64] produced acute follicular conjunctivitis in 3 blind children when 0.05 ml of culture fluid was dropped into the conjunctival sac and the inoculum gently rubbed into the palpebral conjunctiva with a sterile cotton swab. Viruses were reisolated in 2 of the 3 cases, whereas immunologic response was detected in all 3.

Kasel and his co-workers[48] cultured adenovirus types 26 and 27 from rectal specimens of children and infected the eyes of volunteers by swabbing the lower conjunctival sac with 0.1 ml of culture or instilling 0.5 ml in each nostril. An acute self-limited conjunctivitis developed. Virus was isolated from the inoculated eyes during the first week of infection, and sporadically from the throat and rectum during 2 to 3 weeks of convalescence, but never from urine or serum specimens. Nasal inoculation resulted in greater throat infection than that which followed eye inoculation, but both routes produced about the same degree of enteric infection. It is noteworthy that strains 26 and 27, so far as we know, have not produced spontaneous diseases of the eyes.

Thus, the findings described present experimental data to show: (1) the protective influence of natural immunity upon induction of adenovirus ocular infection, (2) the effect on the induction of disease exerted by the method of inoculation and the quantity of virus used, (3) the capability of experimental disease to extend to contiguous areas, such as the contralateral eye, the throat, and the preauricular lymph nodes, (4) the incubation period and progression of the disease, and (5) the resulting serologic responses. In addition, the data show that viruses not known to produce spontaneous disease nevertheless can produce experimental infection in man.

Characterization of adenovirus groups

Adenoviruses cultured from man and animals are categorized structurally in Table 16-1. Those arising from animals have come from simian, bovine, canine, murine, and avian sources. Three distinct hemagglutinins separate the human viruses into subgroups. Group 1 (types 3, 7, 11, 14, 16, 20, 21, 25, and 28) agglutinates rhesus but not rat erythrocytes; group 2 (types 8, 9, 10, 13, 15, 17, 19, 22, 23, 24, 26, 27, 29, and 30) agglutinates rat but not rhesus erythrocytes; group 3 (types 1, 2, 4, 5, 6, and 12) fails to agglutinate rhesus cells and only partially agglutinates rat cells. Types 18 and 31 have no hemagglutinating activity.

The diagnostic identification of these viruses depends largely on immunologic tests. The hexon antigen is found in the major subunits of the virus capsid, the capsomers. This antigen represents the group-reactive antigen common to all adenoviruses. Antibody to it will give a positive complement-fixation test with adenoviruses irrespective of type. Another viral antigen, penton antigen, is a complex antigen consisting of capsomers from the 12 vertices of the icosahedral virion together with their attached fibers. Penton antigen has a biologic activity that causes clumping and detachment of tissue-cultured cells. The fiber part of this penton antigen repre-

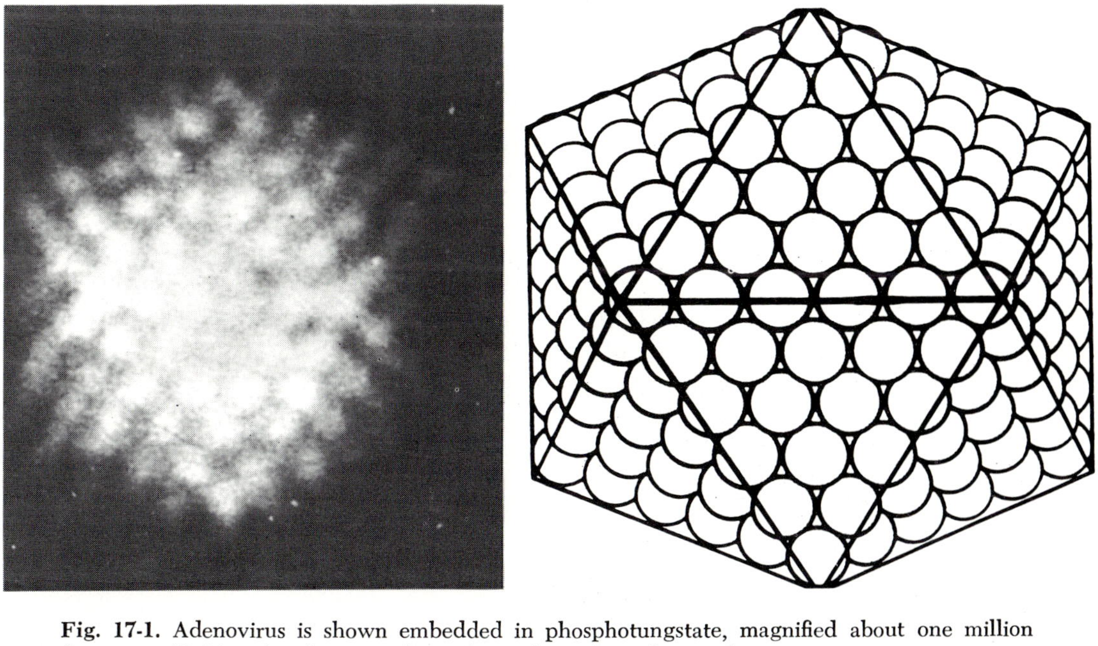

Fig. 17-1. Adenovirus is shown embedded in phosphotungstate, magnified about one million diameters (left). The drawing shows how the 252 surface subunits, or capsomers, are arranged with icosahedral symmetry. There are 12 on corners and 240 on faces or edges. (From Horne, R. W.: Sci. Amer. **208**:51, 1963.)

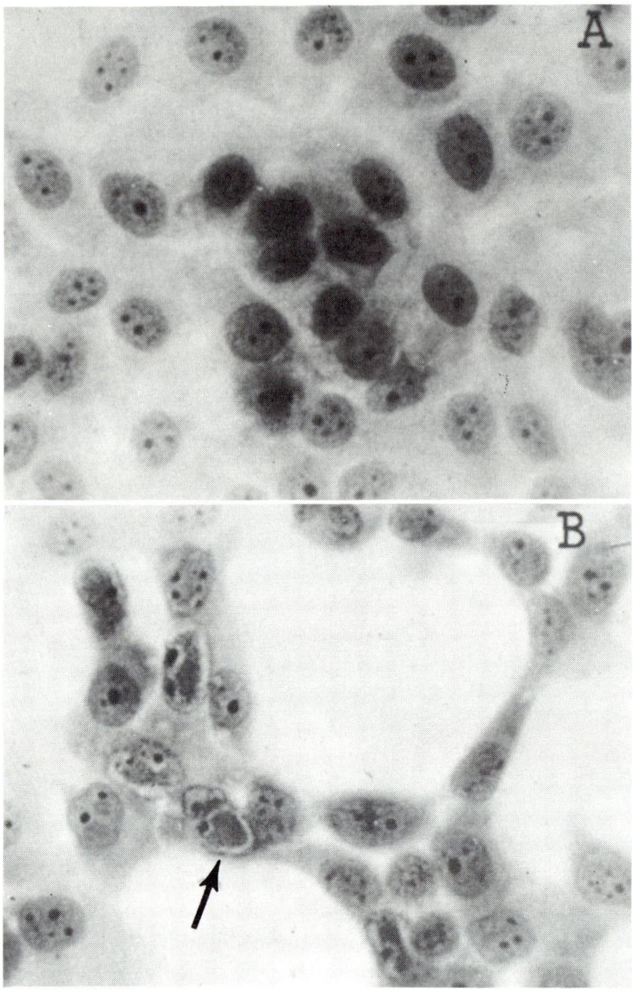

Fig. 17-2. A, Normal HeLa cells. Hematoxylin-eosin stain. (× 500.) **B,** HeLa cells 17 hours after infection with adenovirus type 3. Note aggregation of cells and nuclear inclusions (arrow). (From Beale, A. J., Doane, F., and Ormsby, H. L.: Amer. J. Ophthal. **43**:26, 1957.)

sents the *type*-specific component of the virion.[47a,75] Fig. 17-1 illustrates by electron microscopy and by schematic drawing the capsomers that compose these identifying antigens. The application of immunologic reactions to the identification and typing of adenoviruses is discussed on pp. 294-296.

The adenoviruses share certain cultural peculiarities. They have a predilection for epithelial cells, and primary isolation from human sources is readily carried out on HeLa cells derived from cervical carcinomas. Adenoviruses isolated from animals are best cultivated on epithelial cells from the specific susceptible host.[75] Characteristic cytopathogenic changes are produced in the nuclei of the infected HeLa cells by most types of these viruses. In local areas infected cells become rounded, enlarged, and show increased opacity. The cells tend to roll up and fall off the glass. Fig. 17-2 illustrates normal HeLa cells (*A*) and those infected 17 hours previously with adenovirus type 3 (*B*). The loss of some cells and the presence of inclusion bodies in surviving cells are seen in Fig. 17-2, *B*. Electron microscopic studies show that adenovirus types 3, 4, and 7 commonly

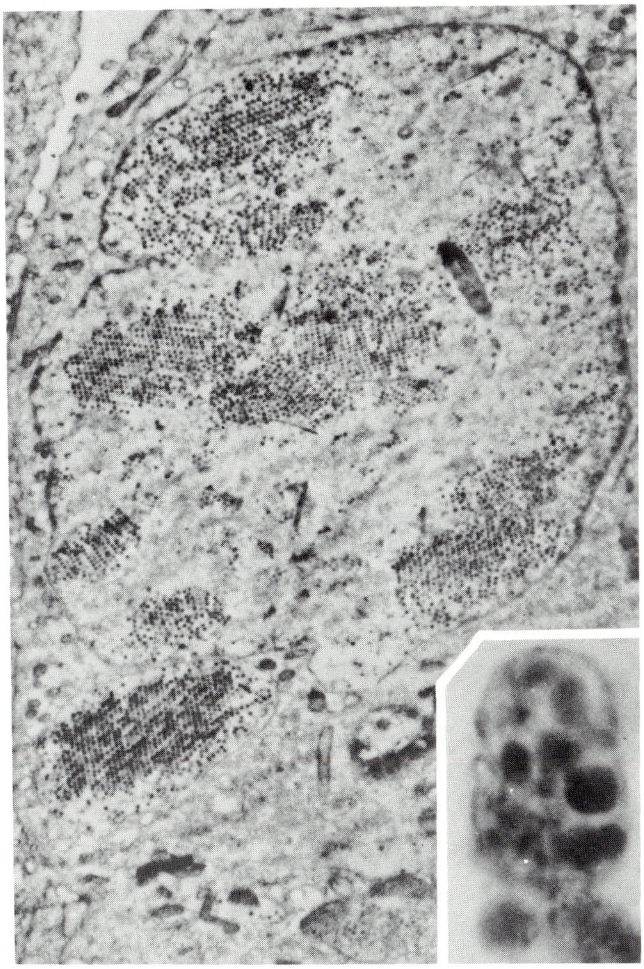

Fig. 17-3. Electron micrograph of a nucleus of a HeLa cell showing intranuclear aggregates and crystals composed of viral particles (adenovirus type 7). Each of the particles measures about 60 mμ in diameter. (× 14,000.) **Inset,** Adjacent thick section of the same nucleus after Feulgen staining, showing the aggregates seen in micrograph. Additional aggregates, which lie outside the plane of the thin section, appear in the thick section. Photomicrographs may show a slight loss of detail and quality because they have been reprinted from the original letterpress journal. (× 4,500.) (From Bloch, D. P., Morgan, C., Godman, G. C., Howe, C., and Rose, H. M.: J. Biophys. Biochem. Cytology **3**:1, 1957.)

form viral crystals as they develop in cell nuclei. Fig. 17-3 is an electron micrograph showing aggregates and crystals of viral particles of type 7 in the nucleus of a cultured HeLa cell.

Epidemiology

Laboratory diagnosis of infection by viruses can be made by testing for a rising titer of specific antibodies in the serum of the patient. The application of this technique has led to surveys on the normal distribution of antibodies to adenoviruses. The results show the presence of adenoviral antibodies in most of the population at large. In adult life antibodies to one or more of these viruses are present in most normal individuals. These antibodies pass through the placenta and are demonstrable in the serum of infants for a few weeks.

Children then develop antibodies from exposure to the viruses, and before reaching their first year of life they often have been infected by at least one of the following types: 1, 2, 3, or 5. Although widely distributed, these viruses are of low pathogenicity; and primary infection occurring at an early age, as determined by a rising antibody titer, does not necessarily manifest clinical disease. About 10% of upper respiratory tract infections of children, however, are caused by adenoviruses. Less than 10% of adolescents are susceptible to these types.

The incidence of infections with adenoviruses in various age groups has been summarized by Ginsberg[30,31] in an attempt to present the natural history of man's relation to the virus. In the population as a whole about 2% to 5% of acute upper respiratory disease is caused by adenoviruses. Following primary infection, the virus may persist in the tonsils or adenoids and perhaps in other lymphoid tissues. The chronic viral infection appears as a result of parasitism of only a rare cell but is probably responsible for the presence of the base line of antibodies found in most adults.

Adenoviruses can be important pathogens for military personnel. As early as 1954 Hilleman and Werner[35] reported type 4 in military recruits, and Berge and his associates[7] isolated types 3, 4, and 7 from soldiers suffering from respiratory illness at Fort Ord in California. Tyrrell and his co-workers[87a] at this time recorded outbreaks of adenovirus infection in England in the Army and Royal Air Force. Recent studies show that adenovirus respiratory infections in induction centers may account for 70% to 80% of acute respiratory diseases among recruits. Following primary infection, lasting immunity often results.

Since antibodies to various types of adenoviruses are so readily demonstrated in sera of young adults and the disease is transmitted to relatively few contacts, vaccination appears possible and has been tried among military personnel. Both heat- and formaldehyde-inactivated type 3 virus tested by human volunteers stimulated neutralizing antibodies in 78% of those vaccinated. These immunized individuals were subsequently challenged with live virus and proved to be immune.[5]

Epidemic keratoconjunctivitis

For many decades epidemic keratoconjunctivitis (EKC) has occurred extensively in various areas throughout the world (Table 17-4). There are two phases in the history of this disease: in the first, the clinical entity was defined but the cause of the infection was obscure; in the second period, an adenovirus was identified as the causative agent.

Clinical diagnosis. Ernest Fuchs[29] (Fig. 17-4) gave the earliest description of this disease (Table 17-4). Fuchs was an ophthalmologist in Vienna who reported a form of "superficial punctate keratitis" in a group of 36 patients. In the same article he also described similar cases seen in December of 1885, in January of 1886, in March of 1887, and in the winter of 1888. The disease described by Fuchs was characterized by an acute onset, which involved only one eye in two thirds or more of the patients. At onset there was catarrhal conjunctivitis with marked ciliary injection; but within 3 to 14 days, pain, photophobia, lacrimation, and blepharospasm were predominant. The conjunctival secretion was characteristically thin and watery. Keratitis developed as conjunctivitis subsided, and corneal lesions appeared in a few days to a few weeks. These consisted of a number of gray, sharply defined punctate dots usually situated centrally rather than peripherally. These dots, about 0.5 mm in diameter, in many cases were arranged in rows or in irregularly scattered groups (Fig. 17-5). There often were as many as 100 of these lesions, although in some patients as few as 10 were seen. Some increased in size by coalescence of several adjacent foci. Each dot consisted of a number of fine points of opacity more closely aggregated at the center. The points were located just beneath Bowman's membrane; hence, they could

Table 17-4. Early records of epidemic keratoconjunctivitis in various countries (etiologic agent not yet known)

Geographic areas	Clinical observations and remarks	References
Austria	First classical description of superficial punctate keratitis	Fuchs, 1889[29]*
England	First report of superficial punctate keratitis	Gunn, 1890[32]
	Nonrecurrent superficial punctate keratitis of the Fuchs type, more common in winter than in summer	Doggart, 1933[22]
India	Superficial punctate keratitis thought to be associated with an encapsulated bacillus	Herbert, 1901[34]
	An epidemic of macular keratitis somewhat resembling superficial punctate keratitis; evidence of the contagious nature of disease	Kilpatrick, 1920[53]
	An epidemic form of superficial punctate keratitis; attempt made to transmit the disease by direct transfer of epithelial corneal scraping to healthy cornea of man and animals; histopathologic observations showed inclusion bodies	Wright, 1930[94]
	Epidemic superficial keratitis observed in families, among students in the same hostels	Sanyal, 1933[84]
	Epidemic superficial punctate keratitis; diagnosis confirmed by the minute spots on the cornea but could be missed unless examined with the corneal microscope	Kirwan, 1933[55]
Uruguay	Epidemic superficial punctate keratitis, generally bilateral; young adults affected	Iturburu, 1931[44]
Australia Tasmania	Superficial punctate keratitis with a record of a mild epidemic; frequency of bilateral involvement; complete examination of cornea with slit-lamp or corneal microscope recommended	Hamilton, 1941[33]
Malaya	"Acute kerato-conjunctivitis of unidentified origin, occasionally becoming epidemic"; susceptibility to infection greater following trauma or surgery	Viswalingam, 1941[90]
France	Superficial punctate keratitis; respiratory infection preceded onset of ocular disease	Fradkine, 1932[28]
Germany	Superficial punctate keratitis; epidemic nature of the disease thought to be caused by an unknown virus	Schneider, 1938[85] Zur Nedden, 1938[95]
Italy	Appearance of epidemic keratoconjunctivitis in 1942 coincided with arrival of German troups; disease was prevalent in Germany at that time	Bietti and Bruna, 1957[9]
China	Superficial punctate keratitis; half of the patients were exposed to swimming pool conjunctivitis	Ling, 1936[57]
Middle East Palestine	First report of epidemic superficial punctate keratitis in patients ages 26 to 67	Feigenbaum, 1939[25]
Japan	First record of superficial punctate keratitis in Orient	Inoue, 1894[43]
Java	Reported keratitis punctata tropica	Westhoff, 1912[92]

*Superscript numbers indicate references at end of chapter.

Fig. 17-4. Ernest Fuchs (1851-1930), an Austrian ophthalmologist, is credited with the original, classical description of superficial punctate keratitis, now known as epidemic keratoconjunctivitis (EKC). Though the cause was unknown then, the clinical picture was so clearly presented that the disease acquired a defined clinical identity. (From Amer. J. Ophthal. **14**:1, 1939.)

not be removed by scraping the corneal epithelium. In 14 instances the corneal lesions remained for 6 to 10 months, and Fuchs stated that one could not be certain that the corneal lesions would disappear completely until after 1 to 2 years. Corneal sensitivity was diminished, and this also returned to normal only slowly. Fuchs observed that this punctate keratitis occurred in several members of one family, that the incidence increased in the winter time, that the onset might be associated with acute pharyngitis, and that young adults were commonly affected. Independently of Fuchs a number of Austrian ophthalmologists described the same clinical entity under different names. Apparently at that period of time epidemics existed in Austria, and the epidemic nature of this ocular disease was recognized by these early clinicians.

Since the original report by Fuchs, data have accumulated recording small, dispersed epidemics of a similar eye affection in the Orient and in the Western world. It appears that no nationality is immune from this infection and that there is no preference for sex or age. Table 17-4 lists early reports of this clinical syndrome in various countries other than the USA.

Wiener[93] in 1909 (Table 17-5) described the first case in the United States of the disease that was to be known as epidemic

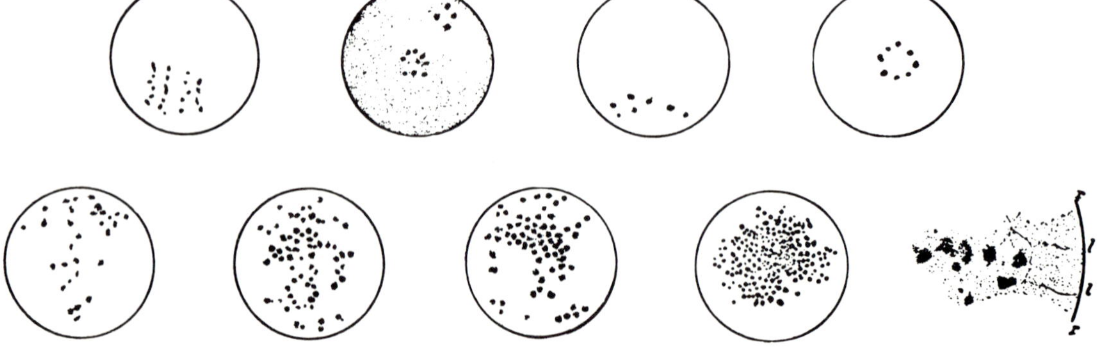

Fig. 17-5. The first pictorial representation, drawn by Fuchs, of the opacities present in EKC. The various eyes drawn illustrate the variety both in number and position of the corneal opacities. (From Fuchs, E.: Wien. Klin. Wschr. **2**:837, 1889.)

keratoconjunctivitis. His patient, a young woman aged 21, developed irritation, lacrimation, and injection of the conjunctiva in November of 1907. This was followed by about 30 corneal opacities consisting of small round grayish dots 1 mm or less in diameter, which lasted about 6 weeks.

Hobson[36] reported an acute epidemic of superficial punctate keratitis in 16 patients at the Veterans Administration Hospital, San Fernando, California. Those affected included 2 members of the kitchen help and a number of ward attendants. All were living on the hospital grounds. The disease began with a violent conjunctivitis, unilateral in only one case and associated with pronounced photophobia and lacrimation. The ocular and palpebral conjunctiva became markedly red and edematous, but pain was not severe. Twenty to 100 minute, opaque, grayish dots were seen, involving, for the most part, the second layer of the cornea. In 4 of the cases there was a persisting reduction of visual acuity in one or both eyes, because absorption of deposits in these cases was incomplete and slight haziness of the cornea remained. There were no children in this series, which was equally divided between males and females.

EKC received considerable attention during World War II, when the disease was considered a major problem in some large industrial plants. Holmes[38] in 1941 in Oahu, Hawaii, reported a rapidly spreading type of conjunctivitis referred to by the patients and doctors as "pink eye." Repeated cultures and smears made from conjunctival scrapings and secretions failed to determine a causative agent. The ex-

Table 17-5. Some early records of epidemic keratoconjunctivitis in the United States

Prominent areas	Clinical observations and remarks	References
St. Louis	First record of superficial punctate keratitis; observed numerous small round grayish dots in the cornea	Wiener, 1909[93]*
California San Fernando	The first epidemic recorded under title "acute epidemic punctate keratitis"	Hobson, 1938[36]
Hawaii	Epidemic infectious conjunctivitis in a form of "pink eye"; extremely contagious	Holmes, 1941[38]
Spokane	Called the infection keratoconjunctivitis; swelling of the preauricular gland a frequent occurrence; etiology was believed to be a specific virus; trauma to the eye a predisposing factor	deRoetth, 1942[74]
Oregon	Acute conjunctivitis with corneal involvement in shipyard workers; unilateral, contagious; startling rapid spread	Rieke, 1942[73]
California San Francisco	Abbreviated as "EKC," the name epidemic keratoconjunctivitis was suggested and quickly adopted; infection occurred in shipyard workers; popular name "shipyard, or California, conjunctivitis" was attached to this disease	Hogan and Crawford, 1942[37]
Schenectady	Outbreak of epidemic keratoconjunctivitis in an industrial plant; report based on observations in several thousand cases	Perkins and others, 1943[71]
New York	Epidemic keratoconjunctivitis observed in over 2,000 cases from the New York area; convalescent plasma or sera seemed to be the treatment of choice	Braley, 1943[15] Braley, 1945[16]

*Superscript numbers indicate references at end of chapter.

treme contagiousness of this outbreak suggested that it might assume epidemic proportions, as had occurred in large outbreaks at the Pearl Harbor Navy Yard. At the end of June, the infection was fairly widespread; and people of all walks of life, irrespective of age, sex, and race, were affected. An estimated 10,000 to 25,000 cases occurred. The clinical picture was quite uniform. The most common complication was the development of the characteristic corneal infiltrates involving the second layer of the cornea. The number of dots varied from 1 to 100, and they were arranged in groups or rows, scattered irregularly, and commonly located on the central portion of the cornea. Other features occurring after a period of 2 to 5 days were profuse tearing, photophobia, and blepharospasm. The discharge was characteristically scant or absent. Multiple subconjunctival hemorrhages appeared on the tarsal portions of the conjunctiva and were so characteristic that they were considered almost pathognomonic of the disease. The eye remained irritable for 4 to 6 weeks or longer when corneal infiltrates developed. Preauricular adenopathy was noted in about 50% of the cases. A factor thought to contribute to the spread of the infection was rubbing of the eyes with unclean fingers. This hand-to-eye type of transmission was suggested by the involvement of the second eye in three fourths of the cases and by the presence in many instances of multiple cases in one family. Holmes stated that those who were infected in Hawaii were inclined to attribute the infection to foreign material lodging in or touching the eye: those working in canneries blamed it on drops of pineapple juice reaching the affected eye; the worker at the emery wheel was certain that his attack began when a piece of dust landed in the eye; those doing much swimming were convinced that the disease was caused by the contaminated water, and so on.

In 1942 deRoetth[74] reported 12 cases of a similar disease in Spokane, Washington, and used the term epidemic keratoconjunctivitis. The main clinical features were acute conjunctivitis followed by superficial punctate keratitis. Swelling of the preauricular gland was a frequent occurrence. Trauma of the eye was a predisposing factor. The etiology was unknown but was believed to be a specific virus.

Rieke[73] described an epidemic of ocular inflammation of unknown origin in the Oregon Shipbuilding Corporation yards at Portland, Oregon; the epidemic involved 600 cases in one shipyard. The clinical descriptions of Rieke's series closely resembled those already presented. The infection involved a small percentage of exposed men and usually attacked only one eye. It appeared that factors that influenced the epidemics were heavy concentration of men from all stations of life, poor ocular hygiene, and probably the virulence of the responsible agent. The author felt that the infection came from Hawaii.

Other examples of EKC were reported during the early years of the war by Hogan and Crawford[37] in San Francisco, Perkins and his co-workers[71] in Schenectady, New York, and Braley[16] in the New York City area. Bedell[3] described the distribution of epidemic keratoconjunctivitis in the United States. He noted the failure of the infection to appear in any of the great army mobilization centers and attributed this to the health of the men in service as well as to the high standard of sanitation in the camps. There were no epidemics in the shipyards on the Gulf of Mexico or along the Atlantic Coast, a fact proving that the infection was not a shipyard disease. Thygeson[87] in his review on the epidemiology of epidemic keratoconjunctivitis stated that during the later war years (1943 to 1946) only sporadic cases were observed. Contrary to expectations, the infection did not become a military problem.

In Canada, McGavin and his associates[59] reported that no evidence of the disease was observed until the winter of 1947 to 1948 when an epidemic of keratoconjunctivitis affected four plants of the Ford

Motor Company. The economic loss caused by this epidemic was severe.

ISOLATION OF CAUSATIVE AGENT. The discovery of the etiologic agent of EKC came only about a year after the first isolations of types of the adenoviruses. Jawetz, Kimura, Hanna, Coleman, Thygeson, and Nichols[47] in a classic paper presented studies on a virus isolated from a merchant seaman traveling between the Orient and San Francisco. The patient was seen 12 days after the onset of the disease in his right eye. He showed "an intense conjunctivitis with pseudo-membrane formation and enlarged preauricular nodes. . . . The left eye became involved 4 days later." A virus, isolated on HeLa cells from each eye, proved to possess a complement-fixing antigen common to other human adenoviruses. But it was not typable with any antiserum then available and was classified by Dr. Huebner as type 8. Samples of sera were collected from patients with typical EKC in Canada, Philadelphia, Chicago, and California. All of the 25 patients with typical clinical manifestations of the disease showed neutralizing antibodies for the new type 8 after 19 days. These antibodies were demonstrable in only 1 of 5 patients with atypical keratoconjunctivitis or uveitis. Further studies by these and other workers demonstrated type 8 neutralizing antibodies, following EKC, in sera of patients from many parts of the world. An excellent correlation between clinical and laboratory diagnosis was demonstrable.

It may be seen from Table 17-2 that there are also two reports of EKC following infection with adenovirus types 9 and 11. Hence, the possibility that adenoviruses other than type 8 may sometimes produce EKC should be kept in mind.

The communicability of EKC is related to the ability of the virus to survive. Mordhorst and Kjer[63] studied an epidemic that occurred in Copenhagen in 1958. Type 8 virus was isolated from conjunctival swabbings of 5 patients taken within the first 6 days of symptoms. These swabbings were stored for 1 year before they were grown in cultured cells. Transmission of the infection has also been known to occur from contaminated tonometers, eye medicaments, and direct finger-to-eye contact. Offices, dispensaries, and hospitals may spread the infection.[8,18,18a,85a]

At the Institute of Ophthalmology, Columbia-Presbyterian Medical Center, over a thousand cases clinically diagnosed as epidemic keratoconjunctivitis have been seen in the past 30 years. Although the cases seen at the IOCPMC did not appear to be part of an epidemic, epidemics continue to occur in various areas of the country. One such epidemic, lasting from April through September of 1967, is of interest because of the hospital personnel and patients who were involved.[56a] It was the fourth largest epidemic in 18 years in the Philadelphia area. Virus isolation was accomplished in 38 of the 62 cases in which it was attempted. In an outbreak of EKC in and around Glasgow in the autumn of 1967 type 8 adenovirus was recovered from 57 of 186 patients, and the serologic reactions were positive in 21 additional infections.[3a]

Plate 3, *H* illustrates a severe follicular and membranous conjunctivitis in a 40-year-old Negro male with EKC. The patient was seen 13 days after onset of illness. In Plate 3, *I* an artist's drawing shows the appearance and position of subepithelial corneal infiltrates 3 weeks after onset of the disease in another patient.

Specific treatment for epidemic keratoconjunctivitis has not yet been discovered. The disease may impair vision for months or years, and in rare instances the corneal opacities never clear.

Pharyngoconjunctival fever

Pharyngoconjunctival fever is a highly infectious disease responsible for epidemic outbreaks in schools and military establishments. It also occurs in families as well as in sporadic form. Small epidemic outbreaks in civilian populations, private offices, dispensaries, hospitals, and among medical personnel have been recorded.

Probably the earliest report of a form of

acute follicular conjunctivitis, now called pharyngoconjunctival fever, was that made by Béal[1] in 1907. The characteristic feature of the disease was the presence of follicles, particularly on the palpebral conjunctiva and in the cul-de-sacs. The upper tarsal conjunctiva was not affected; the bulbar conjunctiva was edematous and hyperemic. Corneal changes were not seen. The conjunctival secretions were scanty, and mononuclear cells predominated. One third of Béal's patients had an associated acute upper respiratory tract infection. Enlargement of the preauricular and submaxillary nodes was noted. The infection was observed in several members of the same family; therefore, Béal believed that the disease was contagious. However, an etiologic agent was not determined. Morax[62] in 1933 described the disease under the name "acute follicular conjunctivitis, Béal's type." In this country, Thygeson[86] in 1935 reported a minor epidemic, occurring among students at the University of Iowa, of an acute follicular conjunctivitis that in all respects closely resembled the type described by Béal. Because of the absence of bacteria, Thygeson suggested that a virus was probably the etiologic agent. Derrick[20] in 1943 reported 3 cases of "swimming bath conjunctivitis" from Australia. Duke-Elder[22a] states that although the disease is characterized by the triad of fever, pharyngitis, and a nonpurulent follicular conjunctivitis often associated with regional lymphadenitis, these may occur singly or together and in all stages of severity.

In 1951 Cockburn[17] observed in Greeley, Colorado, an epidemic of a disease known as "Greeley disease," which occurred during the months of August and September. The clinical entity consisted of an unusual combination of features; acute, nonpurulent conjunctivitis with photophobia, follicular inflammation, and edema, sometimes followed by nonmembranous keratitis. Scrapings revealed mononuclear cells; no inclusions were observed. Usually one eye was involved initially, followed by infection of the other eye. Conjunctival scrapings revealed no viral inclusions, and no bacterial pathogens were isolated. Three weeks after onset, in 13 out of 54 cases slit-lamp examination revealed that the cornea had small whitish plaques, 1 mm or less in size. Other associated findings were vesicular pharyngitis, pyrexia, and severe muscle pain. The preauricular lymph nodes were not significantly enlarged, but the lymph nodes in the neck were involved. Conjunctivitis persisted from 3 days to 3 weeks. Many of the patients had all clinical features. All but 3 of the patients were under 21 years of age. One was a mother who became infected from her child. The sources of infection in the Greeley outbreak were either direct contact or association with swimming pools, since 25% to 50% of the children using the pools developed the disease. In some instances a diagnosis of poliomyelitis was made; but spinal puncture did not show any cells in the cerebrospinal fluid, and all the patients suspected of having the disease made an uneventful recovery.

Cockburn, Rowe, and Huebner[17] related the disease to adenovirus infection when they studied the sera that had been drawn from 11 patients during the Greeley epidemic in 1951. Ten of the 11 sera showed a rise in the complement-fixing antibody titer to the group antigen of adenovirus, and 8

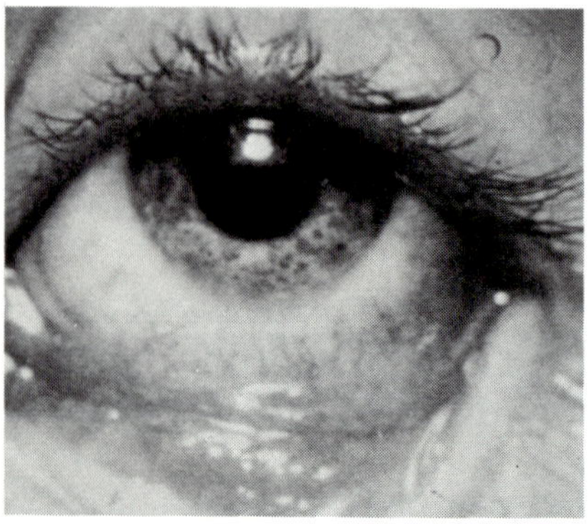

Fig. 17-6. Acute follicular conjunctivitis in pharyngoconjunctival fever caused by type 3 adenovirus. (From Thygeson, P.: Med. Clin. N. Amer. 43:1423, 1959.)

exhibited a rise in neutralizing antibody to adenovirus type 3. In 1955 blood samples were taken from several of the patients from the 1951 series to determine the persistence of neutralizing antibodies. These antibodies were again detected 41 months after the infection.

Fig. 17-6 illustrates pharyngoconjunctival conjunctivitis caused by type 3 adenovirus. Table 17-2 lists other types of adenoviruses (1, 4, 5, 6, 7, 7a, 9, and 14) that have been reported to cause the same clinical syndrome. Type 3 adenovirus has also been cultured from other ocular lesions (see Table 17-2), EKC excepted. In Table 17-6 some of the similarities and differences in EKC and pharyngoconjunctival fever are presented. Today infections of the eye with adenoviruses usually are classified in either of these two categories.

The clinical features of pharyngoconjunctival fever are varied but typically consist of acute nonmembranous follicular conjunctivitis associated with preauricular adenopathy, pharyngitis, fever, malaise, and sometimes diarrhea and muscle pains. Acute catarrhal inflammation, as seen in Fig. 17-6, occurs on the lower palpebral conjunctiva and may extend into the bulbar and upper palpebral conjunctiva. Hyperplasia of lymph follicles is most marked in the lower palpebral conjunctiva. Visual blurring may occasionally occur together with excessive lacrimation. Conjunctivitis may persist for only 3 to 4 days or last as long as 3 weeks and is followed by

Table 17-6. Epidemic keratoconjunctivitis and pharyngoconjunctival fever: similarities and differences

Epidemic keratoconjunctivitis	Pharyngoconjunctival fever
Described by Fuchs in 1889 in Vienna	Described by Béal in 1907 in France
Etiologic agent, adenovirus type 8, isolated in 1955	Etiologic agent, largely adenovirus type 3, isolated in 1955; also associated with other adenovirus infections, notably type 7
Frequently epidemic, worldwide	Frequently epidemic, worldwide
Incubation period 5 to 12 days	Incubation period 5 or 6 days
High incidence of epidemics in industrial plants, private offices, hospitals, and dispensaries; marked tendency to spread among medical personnel	Epidemics related to swimming pools and population density
Predilection for adults	Predilection for young children
Highly infectious	Highly infectious
Rarely associated with febrile systemic manifestations	Usually marked febrile and systemic manifestations
Preauricular lymphadenopathy usual	Lymphadenopathy variable
Acute follicular or membranous conjunctivitis	Acute severe follicular conjunctivitis
Corneal involvement constant; primary subepithelial infiltrates grossly visible	Corneal involvement rare and slight
Resolution 8 weeks to 2 years; permanent visual damage very rare	Resolution rapid: 1 to 3 weeks
Low incidence of antibodies to adenovirus type 8 in general population	High incidence of antibodies to various types of adenoviruses causing pharyngoconjunctival fever

complete recovery. Usually there is no corneal involvement associated with the conjunctivitis, and when keratitis is present it does not interfere with vision or last long.

Pharyngoconjunctival fever has a predilection for young children; older people appear to be less susceptible, presumably because of previous exposure to adenoviruses. Infected individuals are the common sources of spread, via the respiratory tract, from person to person. Swimming pools have been implicated as an important source of infection.[17,65,67] The incubation period is usually from 5 to 6 days; the virus, most frequently type 3, can be isolated in the first few days of the disease but is rarely found after the ninth day. Specific antibodies persist in the circulation for 3 or more years.[17] Immunity is type specific, and recovery from infection with one type does not result in resistance against infection with another type.

Acute follicular conjunctivitis

A number of adenoviruses have been isolated in acute follicular conjunctivitis (Table 17-2). The major clinical manifestations are as follows. Follicles are seen mainly in the palpebral conjunctiva and fornix of the lower lid. The bulbar conjunctiva is edematous and hyperemic; itching and burning are of moderate severity; lacrimation may be present in the absence of photophobia. As with the other adenovirus infections, conjunctival scrapings reveal a predominance of lymphocytes and no inclusions.

Laboratory identification of adenovirus in ocular infection

Two approaches are available for establishing the etiologic agent in ocular disease caused by one of the adenoviruses. The first is to isolate the virus and identify it by serologic means. The second method, a retrospective measure, is to obtain 2 or more samples of serum from the patient; the first taken during the early acute phase of the disease, and further serum to be taken during the convalescent months. A rising titer of specific antibody to one of the types of adenoviruses indicates the particular virus that caused the infection[21] (see also Chapter 16).

ISOLATION OF THE VIRUS. Conjunctival washings or conjunctival and corneal scrapings (Chapter 4) are obtained within the first few days of infection. The material recovered is placed in Earle's balanced salt solution with 500 mg per ml of penicillin and 250 mg per ml of streptomycin added. If the material is not to be on culture medium within 6 hours, it may be stored at -20° C after it has been quick-frozen in a mixture of dry ice and butanol. For prolonged storage (more than a month), a temperature of -60° C should be maintained. The HeLa cell is commonly used today for the cultivation of adenoviruses, although some laboratories use other cell lines.[75] The sample is inoculated into two or more tubes of cultured cells and incubated at 37° C. If an adenovirus is present, a typical cytopathogenic effect usually will be seen in a few days; this effect characteristically consists of enlargement and increased opacity of the cells. These round up, clump together, and eventually fall off the glass. Such changes may be observed in a few days, depending on the amount of virus present in the test specimen. Subcultures will produce cytopathogenic changes more rapidly. The adenoviruses multiply within the nuclei of the susceptible cells (Figs. 17-2 and 17-3). If these cells are stained with Giemsa stain, nuclear inclusions can be seen by the light microscope. Fluorescein-labeled specific antibody will bind to the nuclear inclusions, which contain virus antigen; and under ultraviolet light, fluorescence can be seen (Fig. 17-7). Final identification of the virus depends upon establishing the ability of a known type-specific antiserum either to interfere with hemagglutination, to neutralize cytopathogenic change in cultured cells, or to prevent infection of cultured cells. Detailed instructions for carrying out these procedures are to be found in the chapter by Rose[75] in the text edited by Lennette and Schmidt.

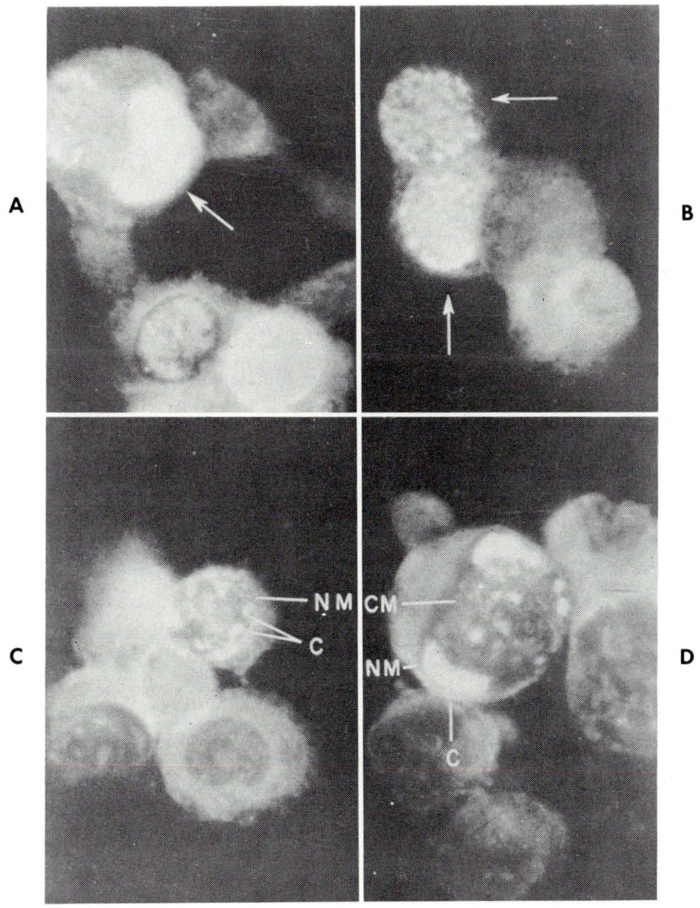

Fig. 17-7. A, HeLa cells 24 hours after infection with type 7 adenovirus stained by indirect fluorescent antibody technique demonstate intracellular virus antigen. Note intense fluorescence in infected nuclei. Cell indicated by arrow reveals diffuse nuclear staining of evenly spaced granules. **B,** HeLa cells in slightly later stage in sequence of infection. Note bright intranuclear staining (arrows) now appears in coarse aggregates. **C,** HeLa cell in still later stage, showing sharp-edged, brightly stained, crystallike masses (*C*) in nucleus. Denser central zone of nucleus can barely be distinguished. *NM* indicates site of nuclear membrane. **D,** HeLa cells late in course of infection (48 hours). Note large, sharply outlined crystalline masses (*C*) in peripheral nuclear zone and small fluorescent aggregates in central mass (*CM*). *NM* indicates site of nuclear membrane. Photomicrographs may show a slight loss of detail and quality because they have been reprinted from the original letterpress journal. (× 1,000.) (From Boyer, G. S., Denny, F. W., Jr., and Ginsberg, H. S.: J. Exp. Med. **110**:827, 1959.)

Test of rising antibody titer in serum from convalescent patients. Significant increases in complement-fixing and specific virus-neutralizing antibodies usually are found during convalescene. Initial acute phase sera must be examined to check the normal amount of antibodies present in each patient. Maximum serum levels of antibodies following infection are observed 2 or more weeks after onset of the disease. Titers then may decline, but in many individuals antibodies may be detected for more than a year. Identification of specific types is possible only in a virology laboratory that maintains many types and that is prepared to carry out extensive cultural and serologic work.[50,75]

Fluorescein-labeled antibody for identification of adenoviruses. Fluorescein-labeled antibodies to some of the commoner specific types of adenoviruses are now available commercially. They are useful in staining slides containing infected cells. The fluorescein technique for examining smears of ocular discharge or scrapings is complicated by the large number of types of adenoviruses. It is possible to make only a very limited number of smears from such material; therefore, even if the offending virus is highly suspected, as in the case of keratoconjunctivitis, it may be difficult to obtain a sufficient number of slides to set up the necessary controls. Another factor that may decrease the usefulness of direct smears or scrapings of the conjunctiva and cornea in studies with fluorescein-labeled antibody is the usual lack of demonstrable inclusion bodies in the cells of scrapings from patients with EKC and pharyngoconjunctival fever. It is these inclusions that bind the specific antibody in high concentration.[42] As is seen in Figs. 17-3 and 17-4, inclusion bodies are plentiful in tissue-cultured cells infected with adenoviruses. The usefulness of the fluorescein antibody technique in the study of infected cultured cells is well established. The fluorescein technique is presented by Liu[58] in the text edited by Lennette and Schmidt.

• • •

The story of the association of adenoviruses, especially type 8, with ocular disease has been described in more detail than is alloted to other viruses. There are several reasons for this. The first is the unique position that type 8 adenovirus occupies, because of its specific predilection for the eye. In contrast, the majority of the other types may often be associated with eye infections occurring spontaneously or produced experimentally. Perhaps the most compelling reason is that the story of the discovery of this group of viruses, their isolation from the normal and inflamed eye, and their identification has taken place through a brilliant series of studies by many investigators around the world. The first 30 of the known 33 types producing infection in man were isolated in a 10-year period, 1952 to 1962. A clear picture of the ocular disease produced and the epidemiology of the infections were largely established during this period, and much was learned of the physical structure and antigenic makeup of these agents. The immunologic responses of the host have been determined. The ophthalmologist has contributed a clinical description of the disease that, although it goes back almost a hundred years, has proved valid today. All of these facts, many of them established by our peers, provides a sense of participation in an exciting aspect of ocular disease.

PAPOVAVIRUS INFECTION OF THE EYE

The papovavirus group consists of small animal viruses that have only one representative typically infectious for man; the papilloma (wart) virus. Others of this group include the Shope rabbit papilloma, the polyoma virus of mice, and the vacuolating (SV 40) virus of monkeys. The name papovavirus is derived from the first two letters of these three viruses. They characteristically produce chronic infections in their hosts, and all are tumorigenic. Their structural characteristics are given in Table 16-1, which shows that they share several common traits with adenoviruses.

Papilloma, or wart, virus of man has been reported to produce proliferative changes in tissue-cultured cells of human embryonic skin and mucle. These are similar to tissue-culture changes produced by the rabbit papilloma virus.[5] In man the warts consist of irregular upgrowths of epidermis containing elongated papillae carrying blood vessels, and of thickened corneum. Most of the cells of the papilloma above the basal cells contain intranuclear inclusion bodies and characteristic cytoplasmic masses. Crystalline virus particles may be obtained from these warts. The extraction of elementary bodies and bits of crystals is carried out by a relatively simple procedure, which has been described by Melnick.[59a] Fig. 17-8 shows both individual elementary bodies and a three-dimensional array of

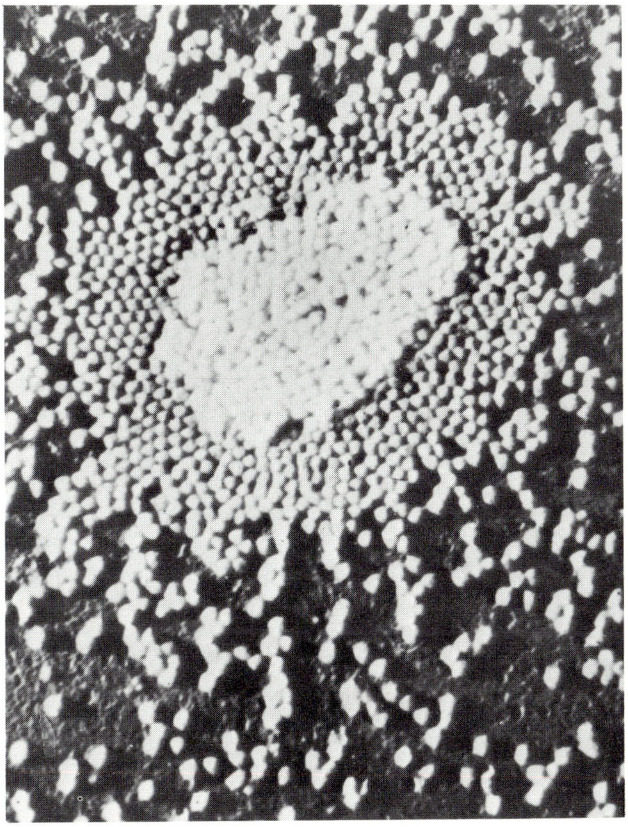

Fig. 17-8. Human wart virus, with particles showing the beginning of orderly arrangement into crystalline formation. Particles 50 mμ in diameter. (From Annals of The New York Academy of Sciences, volume 54, article 6, figure 6, p. 1216, J. L. Melnick et al. © The New York Academy of Sciences; 1952; reprinted by permission.)

these particles. This illustrates the first crystalline substance of this type obtained from any animal viral growth.

Warts, or verrucae, are spread through autoinoculation by scratching and by direct or indirect contact. Patients who are carrying warts have circulating antibody specific for the human papovavirus. This antibody is carried in the IgM fraction of the immunoglobulins.

When warts are situated on the margins of the lids between the eyelashes, a catarrhal conjunctivitis usually results, as was first described by deRoetth.[73a] It is subacute without much discharge. The scrapings fail to show inclusion bodies in the cells. The cornea also may be involved with multiple punctate erosions. Duke-Elder[22a]

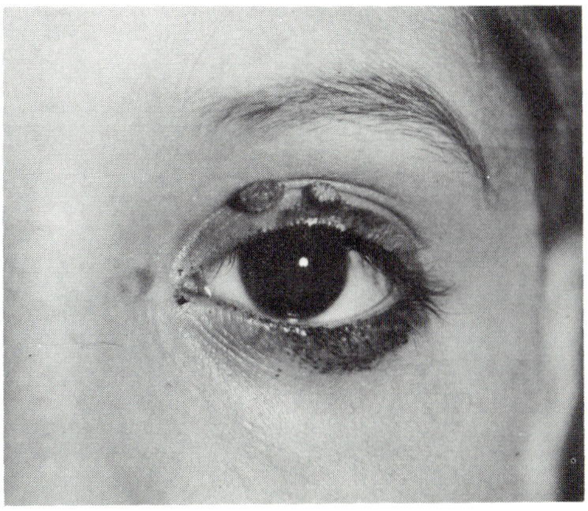

Fig. 17-9. Verrucae of upper left lid.

believes that the conjunctival and corneal lesions are probably either toxic or allergic in origin, since inclusion bodies so characteristic of the infection are not found and since removal of the warts by excision or fulguration results in prompt healing of the lesions of both the conjunctiva and cornea. Small warts may be difficult to see between the eyelashes and should be carefully searched for. Fig. 17-9 illustrates verrucae of the upper lid of a patient seen at the IOCPMC. These were associated with edema and tenderness. The verrucae were excised, and relief of the eye condition followed.

REFERENCES

1. Béal, R.: Sur une forme particulière de conjonctivite aigue avec follicules, Ann. Oculist. **87**:1, 1907.
2. Beale, A. J., Doane, F., and Ormsby, H. L.: Studies on adenovirus infections of the eye in Toronto, Amer. J. Ophthal. **43**:26, 1957.
3. Bedell, A. J.: Distribution of epidemic keratoconjunctivitis in the United States, J.A.M.A. **123**:1101, 1943.
3a. Bell, E. J., Martin, K. W., and Ross, C.A.C.: Laboratory diagnosis of epidemic keratoconjunctivitis, J. Med. Microbiol. **2**:125, 1969.
4. Bell, J. A., Rowe, W. P., Engler, J. I., Parrott, R. H., and Huebner, R. J.: Pharyngoconjunctival fever, epidemiological studies of a recently recognized disease entity, J.A.M.A. **157**:1083, 1955.
5. Bell, J. A., Ward, T. G., Huebner, R. J., Rowe, W. P., Suskind, R. G., and Paffenbarger, R. S.: Studies of adenoviruses (APC) in volunteers, Amer. J. Public Health: **46**:1130, 1956.
6. Bell, S. D., Jr., Rota, T. R., and McComb, D. E.: Adenoviruses isolated from Saudi Arabia: six new serotypes, Amer. J. Trop. Med. **9**:523, 1960.
7. Berge, T. O., England, B., Mauris, C., Shuey, H. E., and Lennette, E. H.: Etiology of acute respiratory disease among service personnel at Fort Ord, California, Amer. J. Hyg. **62**: 283, 1955.
8. Berg, R., and Ringstrom, C.-B.: Adenovirus type 8 associated follicular conjunctivitis and keratoconjunctivitis in patients subjected to tonometry, Scand. J. Infect. Dis. **1**:153, 1969.
9. Bietti, G. B., and Bruna, F.: Epidemic keratoconjunctivitis in Italy: some contributions to its clinical aspects, epidemiology and etiology (part II), Amer. J. Ophthal. **43**:50, 1957.
10. Blacklow, N. R., Hoggan, M. D., Austin, J. B., and Rowe, W. P.: Observations on two new adenovirus serotypes with unusual antigenic characteristics, Amer. J. Epidem. **90**: 501, 1969.
11. Bloch, D. P., Morgan, C., Godman, G. C., Howe, C., and Rose, H. M.: A correlated histochemical and electron microscopic study of the intranuclear crystalline aggregates of adenovirus (RI-APC virus) in HeLa cells, J. Biophys. Biochem. Cytol. **3**:1, 1957.
12. Boniuk, M., Phillips, C. A., and Friedman, J. B.: Chronic adenovirus type 2 keratitis in man, New Eng. J. Med. **273**:924, 1965.
13. Boniuk, M., Phillips, C. A., Hines, M. J., and Friedman, J. B.: Adenovirus infections of the conjunctiva and cornea, Trans. Amer. Acad. Ophthal. Otolaryng. **70**:1016, 1966.
14. Boyer, G. S., Denny, F. W., Jr., and Ginsberg, H. S.: Sequential cellular changes produced by types 5 and 7 adenoviruses in HeLa cells and in human amniotic cells: cytological studies aided by fluorescein-labelled antibody, J. Exp. Med. **110**:827, 1959.
15. Braley, A.: Epidemic keratoconjunctivitis, Med. Clin. N. Amer. **27**:1641, 1943.
16. Braley, A.: Epidemic keratoconjunctivitis, Arch. Ophthal. **33**:47, 1945.
17. Cockburn, T. A., Rowe, W. P., and Huebner, R. J.: Relationship of the 1951 Greeley, Colorado outbreak of conjunctivitis and pharyngitis to type 3 APC virus infection, Amer. J. Hyg. **63**:250, 1956.
18. Dawson, C., and Darrell, R.: Infections due to adenovirus type 8 in the United States (parts I and II), New Eng. J. Med. **268**: 1031, 1963.
18a. Dawson, C. R., Hanna, L., Wood, T. R., and Despain, R.: Adenovirus type 8 keratoconjunctivitis in the United States: epidemiologic, clinical, and microbiologic features, Amer. J. Ophthal. **69**:473, 1970.
19. Dean, A.: Superficial and punctate keratitis, Mich. State Med. Soc. **33**:295, 1934.
20. Derrick, E. H.: Swimming bath conjunctivitis, with a report of three probable cases and a note on its epidemiology, Med. J. Aust. **2**: 334, 1943.
21. Doane, F., Rhodes, A. J., and Ormsby, H. L.: Virological investigations in adenovirus infections of the conjunctiva, Toronto, 1955-56, Canad. Med. Ass. J. **77**:675, 1957.
22. Doggart, J. H.: Superficial punctate keratitis, Brit. J. Ophthal. **17**:65, 1933.
22a. Duke-Elder, S.: System of ophthalmology. Diseases of the outer eye, vol. 8, part 1, St. Louis, 1965, The C. V. Mosby Co., p. 348.
22b. Ellison, E. D., Kaufman, H. E., and Little,

J. M.: Comparison of methods for the laboratory diagnosis of ocular adenovirus type 3 infection, Invest. Ophthal. **8**:484, 1969.

23. Enders, J. F., Bell, J. A., Dingle, J. H., Francis, T., Jr., Hilleman, M. R., Huebner, R. J., and Payne, A. M. M.: "Adenoviruses," group name proposed for new respiratory-tract viruses, Science **124**:119, 1956.
24. Evans, A. S.: Acute respiratory disease in University of Wisconsin students, New Eng. J. Med. **256**:377, 1957.
25. Feigenbaum, A.: Note sur la kératite ponctuée superficielle épidémique en Palestine, Rev. Int. Trachome **16**:18, 1939.
26. Feng, M., Chang, R. S., Smith, T. R., and Snyder, J. C.: Adenoviruses isolated from Saudi Arabia: pathogenicity of certain strains for man, Amer. J. Trop. Med. **8**:501, 1959.
27. Fowle, A. M. C., Cockeram, A., and Ormsby, H. L.: Virus isolations from patients with keratoconjunctivitis, Amer. J. Ophthal. **40**:180, 1955.
28. Fradkine, J.: Trois cas de kératite ponctuée superficielle: étiologie-thérapeutique, Bull. Mem. Soc. Franc. Ophtal. **45**:135, 1932.
29. Fuchs, E.: Keratitis punctata superficialis, Wien. Klin. Wschr. **2**:837, 1889.
30. Ginsberg, H. S.: Newer aspects of adenovirus infections, Amer. J. Public Health **49**:1480, 1959.
31. Ginsberg, H. S., and Dingle, J. H.: The adenovirus group. In Horsfall, F. L., Jr., and Tamm, I., editors: Viral and rickettsial infections of man, ed. 4, Philadelphia, 1965, J. P. Lippincott Co., pp. 860-891.
32. Gunn, R. M.: Two cases of superficial punctate keratitis (Fuchs), Trans. Ophthal. Soc. U. K. **10**:79, 1890.
33. Hamilton, J. B.: A survey of superficial punctate keratitis in Tasmania with the record of a mild epidemic, Brit. J. Ophthal. **25**:1, 1941.
34. Herbert, H.: Superficial keratitis associated with an encapsulated bacillus, Ophthal. Rev. **20**:339, 1901.
35. Hilleman, M. R., and Werner, J. H.: Recovery of new agent from patients with acute respiratory illness, Proc. Soc. Exp. Biol. Med. **85**:183, 1954.
36. Hobson, L. C.: Acute epidemic superficial punctate keratitis, Amer. J. Ophthal. **21**:1153, 1938.
37. Hogan, M. J., and Crawford, J. W.: Epidemic kerato-conjunctivitis, Amer. J. Ophthal. **25**: 1059, 1942.
38. Holmes, W. J.: Epidemic infectious conjunctivitis, Hawaii Med. J. **1**:11, 1941.

38a. Horne, R. W.: The structure of viruses, Sci. Amer. **208**:48, 1963.

39. Horsfall, F. L., Jr., and Tamm, I., editors: Viral and rickettsial infections of man, ed. 4, Philadelphia, 1965, J. B. Lippincott Co., p. 872.
40. Huebner, R. J., Rowe, W. P., Ward, T. G., Parrott, R. H., and Bell, J. A.: Adenoidal pharyngeal-conjunctival agents: a newly recognized group of common viruses of the respiratory system, New Eng. J. Med. **251**: 1077, 1954.
41. Huebner, R. J., and Rowe, W. P.: Adenoviruses as etiologic agents in conjunctivitis and keratoconjunctivitis, Amer. J. Ophthal. **43**:20, 1957.
42. Imre, G., Korchmáros, I., Geck, P., Nasz, I. and Dán, P.: Antigenic specificity of inclusion bodies in epidemic keratoconjunctivitis, Ophthalmologica **148**:7, 1964.
43. Inouse, M.: Superficial punctate keratitis, Ganka Zasshi **1**:273, 1894.
44. Iturburu, J. C.: Queratitis punctata superficial de carácter epidémico, Rev. Otoneuro-oftal. (S. Amer.) **6**:500, 1931.
45. Jawetz, E., Hanna, L., Kimura, S. J., and Thygeson, P.: Clinical diseases of adults associated with sporadic infections of APC viruses, Arch. Intern. Med. **98**:71, 1956.
46. Jawetz, E., Hanna, L., Kimura, S. J., and Thygeson, P.: A new type of APC virus from follicular conjunctivitis, Amer. J. Ophthal. **41**: 231, 1956.
47. Jawetz, E., Kimura, S. J., Hanna, L., Coleman, V. R., Thygeson, P., and Nicholas, A.: Studies on the etiology of epidemic keratoconjunctivitis, Amer. J. Ophthal. **40** (5, part 2):200, 1955.

47a. Jawetz, E., Melnick, J. L., and Adelberg, E. A.: Review of medical microbiology, ed. 9, Los Altos, Calif., 1970, Lange Medical Publications.

48. Kasel, J. A., Evans, H. E., Spickard, A., and Knight, V.: Conjunctivitis and enteric infection with adenovirus types 26 and 27: responses to primary, secondary and reciprocal cross-challenges, Amer. J. Hyg. **77**:265, 1963.
49. Kasel, J. A., Loda, F., and Knight, V.: Infection of volunteers with adenovirus type 16, Proc. Soc. Exp. Biol. Med. **114**:621, 1963.
50. Katz, S., Jordan, W. S., Jr., Badger, G. F., and Dingle, J. H.: Studies of complement-fixing and neutralizing antibodies against certain adenoviruses, J. Immun. **73**:118, 1957.
51. Kendall, E. J. C., Riddle, R. W., Tuck, H. A., Rodan, K. S., Andrews, B. E., and McDonald, J. C.: Pharyngo-conjunctival fever school outbreak in England during the summer of 1955 associated with adenovirus types 3, 7, 14, Brit. Med. J. **2**:131, 1957.
52. Kibrick, S., Melendez, L., and Enders, J. F.: Clinical associations of enteric viruses with particular reference to agents exhibiting prop-

erties of the echo group, Ann. N. Y. Acad. Sci. **67**:311, 1957.

53. Kilpatrick, H.: An epidemic of macular keratitis, Brit. J. Ophthal. **4**:16, 1920.

54. Kimura, S. J., Hanna, L., Nicholas, A., Thygeson, P., and Jawetz, E.: Sporadic cases of pharyngoconjunctival fever, Amer. J. Ophthal. **43**:14, 1957.

55. Kirwan, E. O'G.: Epidemic superficial punctate keratitis in Bengal, Proc. All-India Ophthal. Soc. **3**:1, 1933.

56. Koseki, S.: Studies on adenovirus type 11, infection of the eye, Jap. J. Ophthal. **4**:92, 1960.

56a. Laibson, P. R., Ortolan, G., and Dupre-Strachan, S.: Community and hospital outbreak of epidemic keratoconjunctivitis, Arch. Ophthal. **80**:467, 1968.

57. Ling, W. P.: Keratitis punctata superficialis and swimming pool conjunctivitis, Chin. Med. J. **50**:1381, 1936.

58. Liu, C.: Fluorescent antibody techniques. In Lennette, E. H., and Schmidt, N. J., editors: Diagnostic procedures for viral and rickettsial infections, ed. 4, New York, 1969, American Public Health Association.

59. McGavin, M. F., Boley, J. P., and Ormsby, H. L.: Epidemic keratoconjunctivitis, Canad. Med. Ass. J. **69**:137, 1953.

59a. Melnick, J. L., Bunting, H., Banfield, W. G., Strauss, M. J., and Gaylord, W. H.: Electron microscopy of viruses of human papilloma, molluscum contagiosum, and vaccinia, including observations on the formation of virus within the cell, Ann. N. Y. Acad. Sci. **54**: 1214, 1952.

60. Mitsui, Y., Hanabusa, J., Minoda, R., and Ogata, S.: Effect of inoculating adenovirus (APC virus) type 8 into human volunteers, Amer. J. Ophthal. **43**:84, 1957. (Symposium on viral keratoconjunctivitis)

61. Mitsui, Y., Hanna, L., Hanabusa, J., Minoda, R., Ogata, S., Kurihara, H., Okamura, R., and Miura, M.: Association of adenovirus type 8 with epidemic keratoconjunctivitis, Arch. Ophthal. **61**:891, 1959.

62. Morax, V.: Les conjonctivites folliculaires, Paris, 1933, Maison et Cie, p. 108.

63. Mordhorst, C. H., and Kjer, P.: Studies on an epidemic of keratoconjunctivitis caused by adenovirus type 8: virus isolation in human amniotic cells and serological observations, Acta Ophthal. (Copenhagen) **39**:974, 1961.

64. Murray, E. S., Chang, R. S., Bell, S. D., Tarizzo, M. L., and Snyder, J. C.: Agents recovered from acute conjunctivitis cases in Saudi Arabia, Amer. J. Ophthal. **43**:32, 1957.

65. Okamura, R.: A Study of pharyngoconjunctival fever, Acta Soc. Ophthal. Jap. **64**:96, 1960.

66. Oker-Blom, N., Wager, O., Strandström, H., Makela, P., and Jansson, E.: Adenoviruses associated with pharyngoconjunctival fever, Ann. Med. Exp. Biol. Fenn. **35**:342, 1957.

67. Ormsby, H. L., and Aitchison, W. S.: The role of the swimming pool in the transmission of pharyngoconjunctival fever, Canad. Med. Ass. J. **73**:864, 1955.

68. Ormsby, H. L., Fowle, M. C., and Doane, F.: Canadian cases of adenovirus infections, 1951-1956, Amer. J. Ophthal. **43**:17, 1957.

69. Pereira, H. G., Allison, A. C., Niven, J. S. F.: Fatal infection of new-born hamsters by an adenovirus of human origin, Nature **196**:244, 1962.

70. Pereira, M. S., Pereira, H. G., Clarke, S. K. R.: Human adenovirus type 31, a new serotype with oncogenic properties, Lancet **1**:21, 1965.

71. Perkins, J. E., Korns, R. F., and Westphal, R. S.: Epidemiology of epidemic keratoconjunctivitis, Amer. J. Public Health **33**:1187, 1943.

72. Profeta, M. L., Verdi, G. P., and Orzalesi, N.: Epidemia di cheratocongiuntivite epidemica (CCE) da adenovirus tipo 9, Ann. Sclavo **5**:454, 1963.

73. Rieke, F. E.: Epidemic conjunctivitis of presumed virus causation, report of an estimated six hundred cases in one shipyard, J.A.M.A. **119**:942, 1942.

73a. de Roetth, A. F.: Common wart as an etiological factor in certain cases of conjunctivtis and keratitis Arch. Ophthal. **21**:409, 1939.

74. de Roetth, A. F.: Epidemic keratoconjunctivitis, superficial punctate keratitis, Northwest Med. **41**:246, 1942.

75. Rose, H. M.: The adenoviruses. In Lennette, E. H., and Schmidt, N. J., editors: Diagnostic procedures for viral and rickettsial Infections, ed. 4, New York, 1969, American Public Health Association.

76. Rosen, L., Baron, S., and Bell, J. A.: Four newly recognized adenoviruses, Proc. Soc. Exp. Biol. Med. **107**:434, 1961.

77. Rosen, L., Hovis, J. F., and Bell, J. A.: Further observation on typing adenoviruses and a description of two possible additional serotypes, Proc. Soc. Exp. Biol. Med. **110**:710, 1962.

78. Rowe, W. P., and Hartley, J. W.: A general review of the adenoviruses, comparative virology (Ar. 2), Ann. N. Y. Acad. Sci. **110**: 466, 1962.

79. Rowe, W. P., Hartley, J. W., and Huebner, R. J.: Additional serotypes of the APC virus group, Proc. Soc. Exp. Biol. Med. **91**:260, 1956.

80. Rowe, W. P., Hartley, J. W., and Huebner, R. J.: Serotype composition of the adenovirus

group, Proc. Soc. Exp. Biol. Med. **97**:465, 1958.

81. Rowe, W. P., Huebner, R. J., and Bell, J. A.: Definition and outline of contemporary information on the adenovirus group, Ann. N. Y. Acad. Sci. **67**:255, 1956-1957.
82. Rowe, W. P., Huebner, R. J., Gilmore, L. K., Parrott, R. H., and Ward, T. G.: Isolation of cytopathogenic agent from human adenoids undergoing spontaneous degeneration in tissue culture, Proc. Soc. Exp. Biol. Med. **84**: 570, 1953.
83. Rowe, W. P., Huebner, R. J., Hartley, J. W., Ward, T. G., and Parrott, R. H.: Studies of the adenoidal-pharyngeal conjunctival (APC) group of viruses, Amer. J. Hyg. **61**:197, 1955.
84. Sanyal, S.: Epidemic superficial keratitis, Amer. J. Ophthal. **16**:390, 1933.
85. Schneider, R.: Ueber eine epidemische keratokonjunktivitis in Munchen and Umgebung, Munchen. Med. Wschr. **85**:1981, 1938.

85a. Svartz-Malmberg, G., and Germanis, M.: Adenovirus type 8 associated keratoconjunctivitis: hospital infections and secondary spread in Stockholm, 1967, Scand. J. Infect. Dis. **1**:161, 1969.

86. Thygeson, P.: Acute follicular conjunctivitis, Beal's type, Arch. Ophthal. **13**:853, 1935.
87. Thygeson, P.: The epidemiology of epidemic keratoconjunctivitis, Amer. J. Ophthal. **32**: 951, 1949.

87a. Tyrrell, D. A. S., Balducci, D., and Zaiman, T. E.: Acute infections of the respiratory tract and the adenoviruses, Lancet **2**:1326, 1956.

88. van der Veen, J., and Kok, G.: Isolation and typing of adenoviruses recovered from recruits with acute respiratory disease in the Netherlands, Amer. J. Hyg. **65**:119, 1957.
89. van der Veen, J., and van der Ploeg, G.: An outbreak of pharyngoconjunctival fever caused by types 3 and 4 adenovirus at Waalwijk, the Netherlands, Amer. J. Hyg. **68**:95, 1958.
90. Viswalingam, A.: Epidemic superficial punctate keratitis in Malaya, Brit. J. Ophthtal. **25**:313, 1941.
91. Ward, T. G., Huebner, R. J., Rowe, W. P., Ryan, R. W., and Bell, J. A.: Production of pharyngoconjunctival fever in human volunteers inoculated with APC viruses, Science **122**:1086, 1955.
92. Westhoff, C. H. A.: Keratitis punctata tropica, Zbl. Pract. Augenhed. **36**:289, 1912.
93. Wiener, M.: Keratitis punctata superficialis with report of a case, Arch. Ophthal. **38**:120, 1909.
94. Wright, R. E.: Superficial punctate keratitis, Brit. J. Ophthal. **14**:257, 1930.
95. Zur Nedden: Üeber eine Epidemie von Keratitis superficialis punktata, Klin. Mbl. Augenheilk. **101**:567, 1938.

18

Ocular infections caused by the herpesvirus group

RICHARD W. DARRELL

Four of the more than twenty viruses within the herpesvirus group are known to affect man. Their major common denominator is their similar structure revealed by electron microscopy. They all have a central core of DNA enveloped in a protein coat. This coat exhibits an icosahedral symmetry and is covered in turn by an envelope containing lipids. The infectivity of herpesvirus can be destroyed by lipid solvents, such as ether, because of this lipid envelope. In addition, human herpesviruses have in common the ability to cause different types of infection in the nonimmune and immune host. Although the primary infection in a nonimmune host is usually subclinical, it may have acute manifestations. Thereafter the virus remains in the body in a latent state, from which occasional acute infections reappear years later.

The four human herpesviruses are:

1. Herpes simplex virus types 1 and 2
2. Varicella-zoster virus
3. Cytomegalovirus
4. EB virus

HERPES SIMPLEX VIRUS INFECTIONS

Herpes simplex virus is responsible for the majority of severe eye infections, at least in temporate climates, and demands our first consideration. Although acute herpes simplex keratitis can be suppressed by chemotherapy, the chronic forms of keratitis still present great problems in management.

Primary

Herpes simplex virus is widespread.[36] Adults periodically shed virus into saliva and tears, usually in the absence of clinical illness.[38,39,48] The spread of infection to susceptible children usually occurs within the family. Within closed populations, as in an orphanage, the spread of virus is relatively rapid. For example, in one group of 51 children under 3 years of age, 20 developed a primary stomatitis within a year's observation, at an average age of 14 months.[1] The period in which primary infection occurs outside institutions is usually early childhood or late adolescence. Jones[35] has suggested that kissing is important in spreading the infection because of this peak frequency of primary infections in early childhood, presumably caused by parental kisses, followed by a peak in late adolescence, coinciding with burgeoning interest in the opposite sex. The primary attack is subclinical in about 85% of the patients and is recognized only in the 15% who developed a variety of illnesses, ranging from mild stomatitis to severe gingivostomatitis and keratoconjunctivitis.

In newborn infants, herpes simplex virus can present a spectrum of disease. Microcephaly, seizures, and death can follow neonatal infection; type 2 herpes simplex virus (genital strain) is usually isolated.[70] Bleeding disorders, often with a fatal outcome, have been reported.[51] However, herpes simplex infection in the neonate can be confined to the cornea, as in a 24-day-old infant who developed a large epithelial dendrite without other signs.[7] The mother's cervix was negative for virus on culture, so the infection was probably acquired after birth. On the other hand, another neonate with vesicles on the lids developed keratitis and subsequently died of an overwhelming infection.[5] In addition to the cornea, the lens, retina, and choroid can be involved with subsequent congenital cataract and pigmentary changes in the peripheral retina.[15,16,29]

Infants are protected by antibodies received from their mothers during the first 6 months of life. The peak incidence of primary infection occurs after 6 months up to 5 years of age. Primary infection takes the form of gingivostomatitis, regional lymphadenitis, and follicular conjunctivitis, but fatalities are rare. Dendritic and disciform keratitis is occasionally seen. Diagnosis is made primarily by slit-lamp examination of the cornea, with help from frequent presence of fever blisters on the lip or eyelid, and by a history of recent blisters in a parent or sibling. Laboratory diagnosis is made by isolation of herpes simplex virus in tissue culture, by fluorescent antibody identification, or by demonstration of a rise in antibody titer compared with the serum taken in the acute stage of disease. Although primary herpes simplex infection usually occurs in childhood, in one series an adult of 48 developed a primary infection.[55]

Treatment of primary ocular herpetic infection in children depends in part on the age of the patient. Dendritic keratitis can be treated either by topical 5-iodo-2′-deoxyuridine (IDU), or by curettage of the entire corneal epithelium followed by ether scrub. With very young children slit-lamp examination is difficult, and administration of eye drops is uncertain. The medicine may not go into the eye or may be washed out by crying. A general anesthesia may be preferable to permit careful examination of the cornea under the operating microscope, followed by definitive treatment by epithelial curettage and ether scrub, should treatment be indicated. Appropriate material for laboratory diagnosis can also be taken.

In older children, who present no obstacles to careful examination and treatment, a trial of IDU is indicated. The relative advantages of curettage and of IDU depend in part on the location and character of the ulcer and are discussed later in this chapter.

Recurrent

Herpes simplex virus remains in some form within cells of the body following primary infection, in spite of the presence of circulating antibodies. Recurrent herpetic infection occurs when the balance between latent virus and host protective mechanisms is disturbed. Among the protective mechanisms, the level of serum antibody seems to be unrelated to timing of recurrences.[14] The amount of antibody in saliva shows no correlation with serum antibody level or recurrence of infection in the oral mucous membranes.[20] Similarly, there is little correlation between secretory antibodies in tears and recurrence of corneal herpetic infection. Virus proliferation periodically occurs within mucous membranes and secretory glands leading to the appearance of infectious virus in saliva and tears without any evidence of clinical illness.[38,39,45,48] Recurrences become manifest when acute fever blisters appear on the lids or lips or when the cornea becomes involved. An attack can be precipitated by a febrile illness, by menses, or by exposure to sunlight. An unusual trigger is surgical section of the fifth nerve ganglion for tic douloureux, which may be followed by crops of fever blisters on the skin supplied by that nerve.[24] Reactivation of virus shedding into tears has also been reported to follow injections of epinephrine in the rabbit.[44a] A clue to the mechanism by which epinephrine acts to stimulate virus shedding is provided by recent experiments in epinephrine-induced skin lesions.[69a] A characteristic purple-colored hemorrhage was produced in rabbits at the site of intradermal injection of 100 μg of epinephrine when herpes simplex virus had been injected on the same site 3 to 4 hours previously. Histologic examination showed that the purple-colored hemorrhagic spots were caused by marked capillary dilation with extravasation of erythrocytes. Systemic administration of epinephrine in rabbits may have a similar vascular effect at sites of chronic virus multiplication.

Although the dendritic figure is the one classical sign of ocular herpes simplex infection (Plate 4, *A*), corneal lesions may be atypical or even absent. A chronic follicular conjunctivitis alone may be the only sign of herpetic disease. Serial viral culture in a series of such cases established the diagnosis in spite of a normal cornea.[9] An acute florid conjunctivitis, often follicular, may precede the onset of herpetic keratitis by several days. Such a conjunctivitis should not be immediately treated with corticosteroids until it becomes clear from the clinical course that herpes simplex virus is not the cause.

If a dendritic lesion enlarges and develops rounded, rather than serpiginous edges, it is termed a geographic ulcer. In one study there was a relationship between prior use of corticosteroids and subsequent development of geographic ulcer.[55]

At a later stage an epithelial ulcer may persist for weeks or months in the absence of demonstrable virus proliferation in the epithelium; these ulcers have been called metaherpetic keratitis (Plate 4, *B*). The failure of the epithelium to re-cover the corneal stroma is thought to lie in a damaged basement membrane. The desmosomes of the basement membrane can be destroyed by infection, trauma, or caustic chemicals. Destruction of the desmosomes has been shown to occur in herpes simplex virus infection of the rabbit cornea.[18b,72] An analogous situation has been shown in a case of recurrent erosion, in which the corneal epithelium and superficial stroma were removed at therapeutic lamellar keratoplasty and studied by electron microscopy.[27] To protect the basement membrane, we prefer to use ether following epithelial curettage because it probably causes less damage to the desmosomes as compared with the use of iodine and because ether directly destroys the infectivity of the lipid-containing herpes simplex virus.

The stroma underlying the epithelial ulcer rapidly develops pathologic changes, which often spread in spite of healing of the epithelial lesion. The entire thickness of the corneal stroma is invaded by inflammatory cells, with development of corneal edema, decompensation of the endothelium, and stromal scarring.[32] This chronic form of ocular herpetic infection is often termed disciform keratitis because of the frequent disc-shaped area of stromal edema (Plate 4, *C*). At the present time the most important mechanism involved in disciform keratitis thought to be delayed hypersensitivity, which can be suppressed in part by corticosteroids.[2,9,42,68]

Although herpes simplex virus particles have been identified within corneal epithelium and stroma in disciform keratitis,[19,22,66,67] they rarely cause stromal disease by active multiplication, as is the case in epithelial disease. Rather, the presence of virus within the keratocytes of the diseased stroma causes a new antigen to appear on the infected cell's surface membrane. It is thought that these neoantigens provoke the delayed hypersensitivity mechanism.[2] In addition, the endothelium is probably directly damaged by herpes simplex virus infection, and infectious virus has been cultured from it.[33]

Recurrences of dendritic ulcers often occur immediately above or adjacent to an underlying stromal herpetic scar. Does the

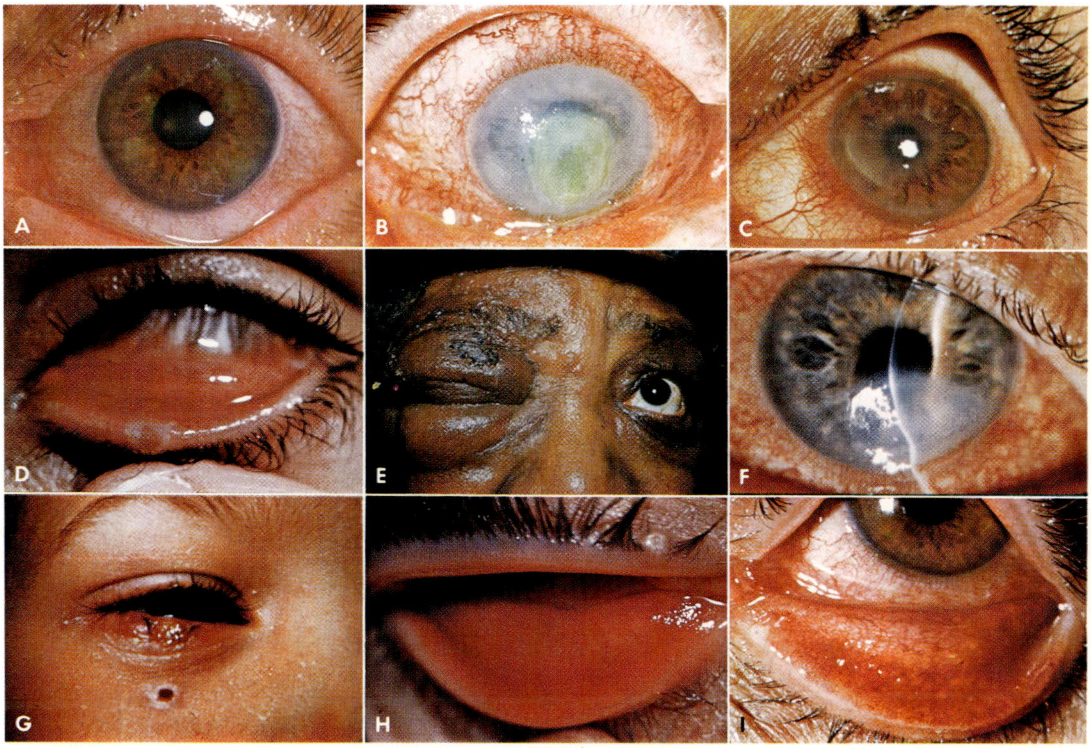

Plate 4. A, Acute herpetic keratitis with seven dendritic figures. **B,** Metaherpetic keratitis with persistent failure of epithelization. **C,** Disciform keratitis with semicircular Wessely ring. **D,** Varicella ulcer of lid margin. **E,** Herpes zoster ophthalmicus. **F,** Herpes zoster with corneal ulceration. **G,** Vaccinia of the eyelid. **H,** Molluscum contagiosum on lid, with secondary follicular conjunctivitis. **I,** Newcastle disease conjunctivitis in a laboratory worker.

virus responsible for a recurrence come from chronically infected superficial keratocytes, from inside out, or is the virus brought to the epithelium by the tear film? On clinical grounds the closeness of the old stromal and new epithelial disease often makes it logical to assume that virus from superficial keratocytes is responsible. There is ample evidence from electron microscopic studies of human stroma removed at keratoplasty at the convalescent stage of herpetic infection, that virus is present within keratocytes of chronically infected stroma.[19] To test the hypothesis that virus within keratocytes is responsible for recurrent epithelial disease, herpes-convalescent rabbit stroma was placed by penetrating keratoplasty into normal rabbit eyes and observed in vivo for up to 15 months. These grafts did not shed virus into tears and did not cause seroconversion in recipient rabbits.[18a] Thus, although virus can be found within keratocytes by electron microscopy, attempts at isolation of virus from chronically infected keratocytes have been unsuccessful both in vitro and in vivo.

Outside the cornea, other sites of latent virus infection have been found adjacent to the eye. Within the lacrimal and adnexal glands, virus has been identified by fluorescent antibody techniques; these glands must be the source of infectious virus shed into tears when the cornea has no ulceration[39,52] or is absent because of enucleation.[8]

Recent experiments have pointed out a third site of latent virus infection, within the sensory ganglia. Fifteen years ago recurrence of facial herpes was reported to follow surgical section of the trigeminal nerve for tic douloureux; 18 out of 26 patients treated in this way developed facial herpes over the denervated area.[24] The introduction of fluorescent antibody techniques and refinements in tissue culture have provided the tools to show latent herpes simplex virus infection in spinal ganglia. Following footpad injection of virus in mice, infection can be traced up the sciatic nerve to the sciatic ganglia, where a chronic virus infection is instituted.[64a] Similarly, even after herpetic keratitis resolves in rabbits, virus can still be isolated for up to 8 months by organ culture of the trigeminal ganglion, but not from the tear film.[64b] Chronic infection of the trigeminal ganglion is probably the dominant factor in recurrent ocular herpes.

Treatment of ocular disease

Dendritic keratitis. If there is only minmal stromal involvement accompanying acute dendritic keratitis, and particularly if corticosteroids have not been used, therapy either with IDU or by epithelial curettage and ether scrub yields equally good results. Following the initial report of the success of IDU in acute dendritic keratitis,[37] several double-blind trials confirmed its value.[12,46,54] There were some dissenting voices pointing out that up to 75% of the placebo group had spontaneous remissions.[34,49] Nevertheless, IDU has been established as an excellent treatment for acute dendritic keratitis. When IDU treatment is compared with curettage of the infected epithelium rather than with placebo drops, however, both IDU and curettage yield equally good results. In a recent double-blind study of 77 such cases, there was no statistical difference in results of treatment with IDU or by curettage.[55]

One major advantage of IDU over curettage is that treatment with IDU is painless. However, should the virus be resistant and treatment fail, the stromal disease may well become more extensive. The advantage of curettage is that the infected epithelium is immediately removed before irreversible stromal disease can be established. Unfortunately, after curettage there may be some pain, and a patch must be worn for a week. Visual acuity may not return to normal for a week or more, until the epithelium reestablishes a normal anterior smooth surface. Nevertheless, if the dendritic ulcer is located in the pupillary axis, immediate curettage is the treatment of choice, because it eliminates the risk of further stromal disease. In the pupillary axis, stromal disease cannot be risked. However, if the epithelial ulcer is not in the pupillary axis, a trial of IDU is certainly worthwhile. If the ulcer becomes smaller in 3 to 5 days, treatment should be continued; if not, curettage is indicated. Young children with recurrent herpetic infection may have to be examined under general anesthesia for accurate slit-lamp diagnosis, and at that time curettage of the corneal epithelium should be carried out.

Progression of a dendritic ulcer to a geographic ulcer can be caused by injudicious use of steroids. IDU is superior to curettage in treatment of these steroid-related geographic ulcers.[55] Dendritic ulcers that are unrelated to use of steroids, however, respond equally well either to IDU or to curettage.

METAHERPETIC KERATITIS. The corneal epithelium may fail to cover the cornea not because of active virus proliferation but because of severe damage to the basement membrane. The essential problem in such a case is lack of an epithelial barrier on the cornea against bacterial and fungal infection. Patching of the eye sometimes enhances epithelization, but more definitive steps are usually required. The epithelium can be replaced by a conjunctival flap, by lamellar keratoplasty, or by penetrating keratoplasty. We have had good results with lamellar keratoplasty to reestablish an anterior epithelium and basement membrane. Twelve months later, if the eye is quiet, penetrating keratoplasty through the previous lamellar graft often restores vision.

HERPETIC IRITIS. Herpes simplex virus has been identified in the iris stroma and cultured from the anterior chamber; clinically it can produce a very severe iritis.[56] The iritis may be the predominant problem, leading to endothelial decompensation, stromal edema, epithelial disease, and glaucoma. Although it is said that steroids should never be used in the presence of epithelial ulceration, exception must be made when the iritis is thought to be the major factor. Nonspecific epithelial ulceration with rounded edges may be secondary to an edematous stroma. When the iritis is controlled by mydriatics and steroids, the epithelial ulceration often heals. A cover of IDU is, of course, also used.

STROMAL KERATITIS. The most challenging problem for therapy is stromal herpesvirus infection, which presents in a variety of forms from limited anterior stromal disease, to larger, but delineated disciform opacity, and finally to total corneal edema, inflammatory reaction, and scarring. Shortly after the introduction of the corticosteroids 20 years ago, the initial clinical enthusiasm was sharply tempered by a growing feeling that steroids make herpetic keratitis worse.[40,43,68,69] They were then contraindicated in herpetic infection until IDU was introduced and the various stages of herpetic keratitis were more carefully delineated. Steroids are still inadvisable in simple dendritic keratitis associated at the most with minimal stromal disease. Beyond that, steroids are often essential in treatment, particularly if combined with IDU.[2,41,42] The risk of recurrent epithelial disease that has been associated with use of steroids is clearly reduced by concomitant use of IDU, as shown definitely by a recent double-blind trial. Patients receiving both IDU and steroids had 66% fewer recurrences of epithelial keratitis as compared with patients treated with steroids alone.[55] One way IDU works is by reducing the amount of infectious virus in the tear film. Although infectious virus can be isolated by a conjunctival swab from 75% of untreated patients who have dendritic keratitis, positive isolations number less than 4% in patients previously treated with IDU.[17] Steroids are indicated whenever the primary problem is inflammatory stromal disease and iritis in the absence of dendritic ulceration; small nonspecific epithelial ulcers are not a contraindication if steroid treatment is judged essential. Unfortunately, curettage and ether scrub are often unsuccessful in the presence of stromal keratitis.

Stromal herpetic infection may be difficult to diagnose in the absence of prior history or documentation of dendritic ulceration. For example, one patient presented with vessels on Descemet's membrane and deep stromal keratitis in association with a moderate iritis. Epithelial disease was completely absent. For 7 years she was maintained in moderate comfort by the sparing application of steroid drops once every other day during flare-ups. The specific diagnosis became apparent only after isolation of herpes simplex virus from a conjunctival swab on a routine clinic visit; 1 week later the same eye developed a dendritic ulcer. Curettage of the ulcer was done immediately, and from the epithelial specimen virus was again isolated.

Treatment of complications. The most common complication is permanent stromal scarring with loss of visual acuity. When indicated, penetrating keratoplasty is the only way to restore vision. A severe iritis may lead to acute glaucoma and later to chronic glaucoma secondary to anterior and posterior synechiae. Miotics, as well as carbonic anhydrase inhibitors, usually control the situation. Glaucoma surgery is contraindicated unless there is reasonable hope of restoring vision. Bacterial and fungal infections are treated as outlined in other chapters. Descemetocoele and corneal perforation require emergency keratoplasty or application of tissue adhesives. Uncommon complications include hyphema and occlusion of the nasolacrimal canaliculi.[61]

Recent methods of treatment. Cryotherapy has been greeted with enthusiasm in Europe but is less widely used in the United States. Freezing undoubtedly destroys some virus, but certainly not all, and it is a method used in the laboratory to release virus from infected tissue-culture cells when increased titer of infectious virus is desired. In our hands, cryotherapy has not proved superior to other methods of therapy.

Stimulation of endogenous interferon production by administration of polyinosinic-cytidylic complex (poly IC) has been studied in rabbits and in man, and its use has been suggested as a new therapy.[53] At the present these compounds are not licensed for general human use, and final judgment will have to await laboratory and clinical trials.

VARICELLA-ZOSTER VIRUS INFECTIONS

The varicella-zoster virus (V-Z virus) causes in the nonimmune host a common acute exantheum of childhood, and in the immune host, usually an adult, the herpes zoster syndrome. The close relationship between chickenpox and herpes zoster is shown by epidemiology, by the identity of the two viruses in tissue culture, and by similar immune responses to both infections. V-Z virus can be isolated from skin vesicles in both chickenpox and herpes zoster; and patients with one form of the disease can cause the appearance of the other form to appear in close contacts. For example, a 3-year-old girl with herpes zoster visited a nursery, and 4 other children developed chickenpox.[26] In another report, a nurse with herpes zoster was attended by an intern who developed chickenpox 16 days later. The intern was also taking care of a patient receiving immunosuppressive drugs following kidney transplant. This patient developed herpes zoster.[6]

In children and nonimmune adults the V-Z virus causes chickenpox, a mild but highly contagious disease of childhood characterized by vesicular eruptions in successive crops on the skin and mucous membranes. The peak incidence is in children from 2 to 6 years of age, and transmission of infection is thought to occur by droplet spread followed by infection of the respiratory tract. After an incubation period of 14 to 17 days, malaise, fever, and then the rash make their appearance. The exanthem is diagnosed by (1) rapid evolution of lesions even within a few hours, through states of macular, pustular, and crusting forms; (2) successive crops at 2 to 5 days; and (3) distribution of lesions primarily on the trunk, neck, and face, and proximal but not distal extremities. Although the majority of children recover promptly without sequela, rare but severe complications of chickenpox, such as pneumonia and encephalitis, have been known to occur. Mortality from varicella may be high in neonates.[10]

Corneal and conjunctival manifestations of chickenpox occur in about 4% of the cases.[65] Vesicles may be found on the conjunctiva, particularly on the semilunar fold. Vesicles may occur on the eyelids (Plate 4, *D*) and may lead to eventual scarring of the lid margin and to distortion of the cilia. Limbal phlyctenules and disciform keratitis are occasionally seen on the cornea. Bilateral papilledema has been reported to follow chickenpox.[47] In addition, encephalitis, optic neuritis, oculomotor paralysis, and tonic pupil are occasionally encountered.[28] Although IDU is effective against varicella in vitro, there is little information available that is of benefit in the occasional varicella eye infections. Zoster immune globulin has been recommended to treat severe vaccinia in the newborn or in very ill children.[11]

After primary infection the V-Z virus probably remains in a latent form within the dorsal root ganglions.[14] Reactivation usually occurs without any specific exciting factor, but local pressure on a sensory ganglion from a tumor or a defect in systemic immunity may be responsible for a recurrence. After reactivation the virus spreads peripherally from the dorsal root ganglion or extramedullary cranial nerve ganglion along the sensory nerve; hence, sensory alterations occur before the appearance of vesicles. Occasionally, the virus spreads centrally to cause meningitis, myelitis, and encephalitis, with concomitant paralysis. The skin eruption is limited to the dermatomes corresponding to the involved sensory ganglion and consists of successive crops of maculopapular plaques that eventually crust in 1 to 3 weeks.

Infection of the ophthalmic division of the trigeminal nerve produces the characteristic eruption of skin lesions on the forehead, lids, and nose, as well as damage to the cornea (Plate 4, *E*). The corneal changes range from mild, with punctate staining of the corneal epithelium and no stromal alteration, to severe, with epithelial and stromal ulceration and opacity (Plate 4, *F*). Most of these changes are secondary to the temporary sensory denervation of the cornea, changes also observed when the trigeminal nerve ganglion is sectioned for tic douloureux. In addition, optic neuritis, oculomotor paralysis, levator paralysis, iritis, paralytic mydriasis, and bilateral hemorrhagic retinopathy have been reported.[3,59,62]

Atropine and local steroids are the best treatment for the corneal lesions and iritis. Local treatment with cortisone was shown to be clearly superior in a recent study of 70 patients, one third of whom served as controls.[5] Systemic steroids for a 3-week period during the acute stage of infection were used in one double-blind study; the postherpetic neuralgia was reported to be shortened.[21] In addition, calamine lotion, with 0.25% menthol and 0.25% phenol can be used to dry the skin eruptions. Antibiotics are not usually necessary unless the skin lesions or corneal ulceration becomes infected. The duration of the postherpetic neuralgia may be very long, up to several years, and may require mild analgesics and constant reassurance of the patient.

CYTOMEGALOVIRUS INFECTIONS

Cytomegaloviruses are classified within the herpesvirus group because their morphologic characteristics are identical to those of herpes simplex virus when observed by electron microscopy.[23,30] They also have in common the ability to cause latent or asymptomatic infections; and in both, excretion of infectious virus into urine and saliva may occur for long periods of time in spite of the presence of circulating antibodies. Serologic studies show that 4% of children under age 5, 15% at age 10, and 54% of adults over age 35 have circulating antibodies against the cytomegaloviruses. These data indicate widespread infection of the population, although there is usually no history of a primary illness.[60] However, neonates may develop overt signs of primary infection.

In a prospective study of 185 pregnant women, 6 were found to be excretors of cytomegalovirus.[25] All pregnancies were full term and normal, and only 1 of the 6 neonates yielded virus in urine, throat, and placenta cultures. The only sign of infection was a few petechiae. A study of all babies born during one year in an urban hospital showed that 1.2% were shedding virus.[64] These prospective studies show the relative frequency of mild cytomegalovirus infection at birth without sequelae. However, congenital infection of unborn children by transplacental transfer of cytomegalovirus from their mothers, who may be asymptomatic, can produce a disastrous neonatal illness. The clinical manifestations of cytomegalic inclusion disease are hepatomegaly, splenomegaly, mental retardation, microcephaly, jaundice, petechiae, chorioretinitis, and cerebral calcifications.[23,30,50] The congenital form of infection may also be caused by exchange transfusion immediately after birth when blood carrying cytomegalovirus is inadvertently used. Specific diagnosis requires virologic and serologic data in addition to clinical findings. The presence of infection can be established by (1) isolation of cytomegalovirus, (2) demonstration of specific intranuclear inclusions within markedly enlarged cells, (3) a rising anticytomegalovirus antibody titer, and (4) detection of cytomegalovirus macroglobulin. Diagnosis based on clinical manifestations alone is subject to error because similar illnesses can be produced by congenital toxoplasmosis and congenital rubella; these can be excluded by the toxoplasmosis dye test and the rubella hemagglutination inhibition test. When present, periventricular calcification is strongly suggestive of cytomegalic inclusion disease; however, it is present only in the minority of cases. Treatment is unsatisfactory, although IDU has been used in a few cases. In one 14-week-old girl there was a clinical improvement and marked drop in virus excretion following systemic administration of IDU.[18]

Within the retina of eyes obtained at autopsy of infants who have died of cytomegalic inclusion disease, typical large mononuclear cells with both intranuclear and cytoplasmic inclusions are seen; on electron microscopy, viruses within these inclusions have morphologic characteristics identical to those of herpes simplex virus.[63,71]

Infection acquired after infancy is usually asymptomatic, although a transient hepatitis can often be detected by liver function tests in subclinical cases. Children may remain virus excretors up to 4 years of age. Infection of the adult is rarely diagnosed but may present a clinical picture similar to that of infectious mononucleosis, with predominance of fever and liver disease, and relatively less pharyngitis, lymph node enlargement, and hematologic changes than are usually found in classical infectious mononucleosis.[44] In addition, the heterophil agglutination test is negative for cytomegalic virus–related infectious mononucleosis but positive in most patients with classical infectious mononucleosis. Adults may acquire infection through blood transfusions, particularly after the multiple transfusions required for open heart surgery.[23] The eye is not affected in adult forms of the disease.

INFECTIOUS MONONUCLEOSIS

Infectious mononucleosis is characterized by protracted fever (2 to 3 weeks), enlargement of the lymph glands and spleen, and increased numbers of monocytes and large lymphocytes displaying morphologic abnormalities in the circulating blood.[13,31,58] In addition, 50% to 80% of patients with infectious mononucleosis develop heterophil antibodies. The ocular signs of the disease include periorbital edema, particularly of the upper lid; ocular muscle pain; photophobia; and occasionally a mild conjunctivitis. The relative frequency of the various signs and symptoms are:[13]

	Percent
Adenopathy	100
Malaise and fatigue	90 - 100
Fever	80 - 95
Sweating	80 - 95
Sore throat	80 - 85
Splenomegaly	50 - 60
Periorbital edema	25 - 40
Ocular muscle pain	10 - 20
Photophobia	5 - 10
Conjunctivitis	less than 5

Infectious mononucleosis is included in this chapter because there is considerable evidence that it is caused by the EB virus, a virus indistinguishable under the electron microscope from other human herpesviruses.[13] Antibodies to EB virus are found in populations in many parts of the world and appear early in life. Although the majority of EB virus infections are clinically inapparent, during infectious mononucleosis antibodies to EB virus rise; they also rise in patients with Burkitt's lymphoma. However, the role of EB virus in infectious mononucleosis and Burkitt's lymphoma is still unsettled because of failure to transmit the disease to laboratory animals or human volunteers.

REFERENCES

1. Anderson, S. G., and Hamilton, J.: The epidemiology of primary herpes simplex infection, Med. J. Aust. **1**:308, 1949.
2. Aronson, S. B., and Moore, T. E., Jr.. Corticosteroid therapy in central stromal keratitis, Amer. J. Ophthal. **67**:873, 1969.
3. Bartlett, R. E., Mumma, C. S., and Irvine, A. R.: Herpes zoster ophthalmicus with bilateral hemorrhagic retinopathy, Amer. J. Ophthal. **34**:45, 1951.
4. Bergaust, B., and Westby, R. K.: Zoster ophthalmicus: local treatment with cortisone, Acta Ophthal. **45**:787, 1967.
5. Berkovich, S., and Ressel, M.: Neonatal herpes keratitis, J. Pediat. **69**:652, 1966.
6. Berlin, B. S., and Campbell, T.: Hospital-acquired herpes zoster following exposure to chicken pox, J.A.M.A. **211**:1831, 1970.
7. Bobo, C. B., Antine, B., and Manos, J. P.: Neonatal herpes simplex infection limited to the cornea, Arch. Ophthal. **84**:697, 1970.
8. Brown, D. C., and Kaufman, H. E.: Chronic herpes simplex infection of the ocular adnexa, Arch. Ophthal. **81**:837, 1969.
9. Brown, D. C., Nesburn, A. B., Nauheim, J. S., Pavan-Langston, D., and Kaufman, H. E.: Recurrent herpes simplex conjunctivitis, Arch. Ophthal. **79**:733, 1968.
10. Brunell, P. A.: Varicella-zoster infections in pregnancy, J.A.M.A. **199**:315, 1967.
11. Brunell, P. A., Ross, A., Miller, L. H., and Kuo, B.: Prevention of varicella by zoster immune globulin, New Eng. J. Med. **280**:1191, 1969.
12. Burns, R. P.: A double-blind study of IDU in human herpes simplex keratitis, Arch. Ophthal. **70**:381, 1963.
13. Carter, R. L., and Penman, H. G.: Infectious mononucleosis, Oxford, 1969, Blackwell Scientific Publications.
14. Chang, T.: Recurrent viral infection (reinfection), New Eng. J. Med. **284**:765, 1971.
15. Cibis, A., and Burde, R. M.: Herpes simplex virus-induced congenital cataracts, Arch. Ophthal. **85**:220, 1961.
16. Cogan, D. G., Kuwabara, T., Young, G. F., and Kuck, D. L.: Herpes simplex retinopathy in an infant, Arch. Ophthal. **72**:641, 1964.
17. Coleman, V. R., Thygeson, P., Dawson, C. R., and Jawetz, E.: Isolation of virus from herpetic keratitis, Arch. Ophthal. **81**:22, 1969.
18. Conchie, A. F., Barton, B. W., and Tobin, J. O'H.: Congenital cytomegalovirus infection treated with idoxuridine, Brit. Med. J. **4**:162, 1968.

18a. Darrell, R. W.: Keratitis following penetrating grafts of herpes-convalescent cornea in the rabbit, Arch. Ophthal. In press.

18b. Darrell, R. W., and Vrabec, F.: The corneal stroma in experimental herpetic keratitis in rabbits, Docum. Ophthal. **29**:243, 1971.

19. Dawson, C., Togni, B., and Moore, T. E., Jr.: Structural changes in chronic herpetic keratitis, Arch. Ophthal. **79**:740, 1968.
20. Douglas, R. G., Jr., and Couch, R. B.: A prospective study of chronic herpes simplex infection and recurrent herpes labialis in humans, J. Immun. **104**:289, 1970.
21. Eaglstein, W. H., Katz, R., and Brown, J. A.: The effects of early corticosteroid therapy on the skin eruption and pain of herpes zoster, J.A.M.A. **211**:1681, 1970.
22. Edelhauser, H. F., Schultz, R. O., and Van Horn, D. L.: Experimental herpes simplex keratitis, Amer. J. Ophthal. **68**:458, 1969.
23. Editorial: Diseases caused by cytomegaloviruses, Brit. Med. J. **1**:72, 1968.
24. Ellison, S. A., Carton, C. A., and Rose, H. M.: Studies of recurrent herpes simplex infections following section of the trigeminal nerve, J. Infect. Dis. **105**:161, 1959.
25. Feldman, R. A.: Cytomegalovirus infection during pregnancy, Amer. J. Dis. Child. **117**: 517, 1969.
26. Garrett, F. E.: Herpes zoster ophthalmicus, Amer. J. Ophthal. **46**:741, 1958.

27. Goldman, J. N., Dohlman, C. H., and Kravitt, B. A.: The basement membrane of the human cornea in recurrent epithelial erosion syndrome, Trans. Amer. Acad. Ophthal. Otolaryng. **73**:471, 1969.
28. Goldsmith, M. O.: Tonic pupil following varicella, Amer. J. Ophthal. **66**:551, 1968.
29. Hagler, W. S., Walters, P. V., and Nahmias, A. J.: Ocular involvement in neonatal herpes simplex virus infection, Arch. Ophthal. **82**: 169, 1969.
30. Hanshaw, J. B.: Cytomegaloviruses. In Virology monographs No. 3, New York, 1968, Springer-Verlag New York, Inc., pp. 1-23.
31. Hoagland, R. J.: Infectious mononucleosis, New York, 1967, Grune and Stratton.
32. Hogan, M. J., Kimura, S. J., and Thygeson, P.: Pathology of herpes simplex kerato-iritis, Amer. J. Ophthal. **57**:551, 1964.
33. Irvine, A. R., and Kimura, S. J.: Experimental stromal herpes simplex keratitis in rabbits, Arch. Ophthal. **78**:654, 1967.
34. Jepson, C. N.: Treatment of herpes simplex of the cornea with IDU: a double blind study, Amer. J. Ophthal. **57**:213, 1964.
35. Jones, B.: The management of ocular herpes, Trans. Ophthal. Soc. U. K. **79**:425, 1959.
36. Kaplan, A. S.: Herpes simplex and pseudorabies viruses. In Virology monographs No. 5, New York, 1969, Springer-Verlag New York, Inc.
37. Kaufman, H. E.: Clinical cure of herpes simplex keratitis by 5-iodo-2′-deoxyuridine, Proc. Soc. Exp. Biol. Med. **109**:250, 1962.
38. Kaufman, H. E., Brown, D. C., and Ellison, E. D.: Herpesvirus in the lacrimal gland, conjunctiva and cornea of man—a chronic infection, Amer. J. Ophthal. **65**:32, 1968.
39. Kaufman, H. E., Brown, D. C., and Ellison, E. M.: Recurrent herpes in rabbit and man, Science **156**:1628, 1967.
40. Kaufman, H. E., and Maloney, E. D.: Experimental herpes simplex keratitis, Arch. Ophthal. **66**:99, 1961.
41. Kaufman, H. E., and Maloney, E. D.: IDU and hydrocortisone in experimental herpes simplex keratitis, Arch. Ophthal. **68**:396, 1962.
42. Kaufman, H. E., Martola, E., and Dohlman, C. H.: Herpes simplex treatment with IDU and corticosteroids, Trans. Amer. Acad. Ophthal. Otolaryng. **67**:695, 1963.
43. Kimura, S. J., Diaz-Bonnet, V., Okumoto, M., and Hogan, M. J.: The effect of corticosteroid hormones on experimental herpes simplex keratitis, Amer. J. Ophthal. **75**:945, 1961.
44. Klemola, E., Kaariainen, L., von Essen, R., Haltia, K., Koivuniemi, A., and von Bonsdorff, C. H.: Further studies on cytomegalovirus mononucleosis in previously healthy individuals, Acta Med. Scand. **182**:311, 1967.
44a. Laibson, P. R., and Kibrick, S.: Reactivation of herpetic keratitis in the rabbit, Arch. Ophthal. **77**:244, 1967.
45. Laibson, P. R., and Kibrick, S.: Recurrence of herpes simplex virus in rabbit eyes: results of a three-year study, Invest. Ophthal. **8**:346, 1969.
46. Laibson, P. R., and Leopold, I. H.: An evaluation of double-blind IDU therapy in 100 cases of herpetic keratitis, Trans. Amer. Acad. Ophthal. Otolaryng. **68**:22, 1964.
47. Liioi, J. A., and Aiello, M. V.: Bilateral papilledema following chicken pox, J. Pediat. Ophthal. **7**:155, 1970.
48. Lindgren, K. M., Douglas, R. G., Jr., and Couch, R. B.: Significance of herpesvirus hominis in respiratory secretions of man, New Eng. J. Med. **278**:517, 1968.
49. Luntz, M. H., and MacCallum, F. O.: Treatment of herpes simplex keratitis with 5-iodo-2′-deoxyuridine, Brit. J. Ophthal. **47**:449, 1963.
50. McCracken, G. H., Schinefield, H. R., Cobb, K., Rausen, A. R., Dische, M. R., and Eichenwald, H. F.: Congenital cytomegalic inclusion disease, Amer. J. Dis. Child. **117**:522, 1969.
51. Miller, D. R., Hanshaw, J. B., O'Leary, D. S., and Hnilicka, J. V.: Fatal disseminated herpes simplex virus infection and hemorrhage in the neonate, J. Pediat. **76**:409, 1970.
52. Nesburn, A. B., Elliott, J. H., and Leibowitz, H. M.: Spontaneous reactivation of experimental herpes simplex keratitis in rabbits, Arch. Ophthal. **78**:523, 1967.
53. Park, J. H., Galin, M., Billau, A., and Baron, S.: Prophylaxis of herpetic keratoconjunctivitis with interferon inducers, Arch. Ophthal. **81**:840, 1969.
54. Patterson, A., Fox, A. D., and Davies, G., and others: Controlled studies of IDU in the treatment of herpetic keratitis, Trans. Ophthal. Soc. U. K. **83**:583, 1963.
55. Patterson, A., and Jones, B. R.: The management of ocular herpes, Trans. Ophthal. Soc. U. K. **87**:59, 1967.
56. Pavan-Langston, D., and Brockhurst, R. J.: Herpes simplex panuveitis, Arch. Ophthal. **81**: 783, 1969.
57. Pavan-Langston, D., and Nesburn, A. B.: The chronology of primary herpes simplex infection of the eye and adnexal glands, Arch. Ophthal. **80**:258, 1968.
58. Pejme, J.: Infectious mononucleosis, Acta Med. Scand. Suppl. **413**:1, 1964.
59. Ramsell, T. G.: Complications of herpes zoster ophthalmicus, Amer. J. Ophthal. **63**: 1796, 1967.
60. Rifkind, D.: Cytomegalovirus mononucleosis, Ann. Inst. Med. **69**:842, 1968.

61. Sandford-Smith, J. H.: Herpes simplex canalicular obstruction, Brit. J. Ophthal. **54:** 456, 1970.
62. Singha, S. S.: and Nirankari, M. S.: Herpes zoster ophthalmicus, Orient. Arch. Ophthal. **8**:21, 1970.
63. Smith, M. E., Zimmerman, L. E., and Harley, R. D.: Ocular involvement in congenital cytomegalic inclusion disease, Arch. Ophthal. **76**:696, 1966.
64. Starr, J. G., Bart, R. D., and Gold, E.: Inapparent congenital cytomegalovirus infection, New Eng. J. Med. **282**:1075, 1970.
64a. Stevens, J. G., and Cook, M. L.: Latent herpes simplex virus in spinal ganglia of mice, Science **173**:843, 1971.
64b. Stevens, J. G., Nesburn, A. B., and Cook, M. L.: Recovery of latent Herpes simplex virus from trigeminal ganglia of rabbits with recurrent eye infection, Nature. In press.
65. Stucchi, C. A., and Bianchi, G.: Complications oculaires graves post-varicelleuses chez l'adulte, Ophthalmologica **161**:108, 1970.
66. Tanaka, N., and Kimura, S. J.: Localization of herpes simplex antigen and virus, Arch. Ophthal. **78**:68, 1962.
67. Tanaka, N., and Togni, B.: Further investigations by electron microscopy of herpes simplex virus in the corneal stroma of rabbits with experimental keratitis, Exp. Eye Res. **7**:142, 1968.
68. Thygeson, P.: Chronic herpetic keratouveitis, Trans. Amer. Ophthal. Soc. **65**:211, 1967.
69. Thygeson, P., Hogan, M. J., and Kimura, S. J.: The unfavorable effect of topical steroid therapy on herpetic keratitis, Trans. Amer. Ophthal. Soc. **58**:245, 1960.
69a. Tokumaru, T.: Production of epinephrine-induced skin lesions and fever following local and systemic injections of herpes simplex and vesicular stomatitis viruses in the rabbit, Arch. Ges. Virusforsch. **34**:179, 1971.
70. Torphy, D. E., Ray, C. G., McAlister, R., and Du, J. N. H.: Herpes simplex virus infection in infants: a spectrum of disease, J. Pediat. **76**:405, 1970.
71. Tsukahara, I., Veno, I., and Kawanishi, H.: Retinal changes in human cytomegalovirus infection, Amer. J. Ophthal. **62**:1153, 1966.
72. Vrabec, F., and Darrell, R. W.: The corneal epithelium in experimental herpetic keratitis in rabbits, Docum. Ophthal. **29**:212, 1971.
73. Witmer, R., and Iwamoto, T.: Electron microscope observation of herpes-like particles in the iris, Arch. Ophthal. **79**:331, 1968.

19

Ocular infections caused by the poxvirus, togavirus, myxovirus, paramyxovirus, and picornavirus groups

RICHARD W. DARRELL

Viruses can be grouped together on the basis of similar morphologic characteristics as seen on electron microscopic observation and on the basis of antigenic characteristics shared in common. Although two viruses may be in the same group, the clinical illnesses they produce are often very different from each other. This chapter contains a discussion of ocular infections caused by the following virus groups:

1. Poxvirus group
   a. Variola (smallpox)
   b. Vaccinia
   c. Orf (contagious ecthyma)
   d. Molluscum contagiosum
2. Togavirus group
   a. Rubella (German measles)
   b. Arbovirus disease
3. Myxovirus group (influenza)
4. Paramyxovirus group
   a. Measles (rubeola)
   b. Mumps (epidemic parotitis)
   c. Newcastle disease
5. Picornavirus group

POXVIRUS GROUP

When members of the poxvirus group are observed under the electron microscope in a dry condition, they are found to have a brick-shaped structure; but when observed in thin sections of infected cells, they have an elliptical shape. They all contain DNA and measure 200 to 300 nm. They are dermotropes and develop within the cytoplasm of the infected cell.

Variola (smallpox)

Classical smallpox, or variola major, is characterized by an incubation period of 12 days before the onset of fever and malaise, which last for 1 to 5 days and are followed by the appearance of an exanthem.[14,81] The skin lesions appear all in one crop and are distributed primarily on the face, chest, distal extremities, and palms and soles, all in contrast to the eruption associated with varicella virus infection (chickenpox). The lesions go through papular, vesicular, pustular, and crusting stages before healing, often with residual scarring. There are several types of infection, depending on the virulence of the virus and degree of resistance or immunity of the host. Most severe is hemorrhagic smallpox, in which bleeding appears 2 to 3 days after onset, and death may precede the exanthem. Subconjunctival hemorrhages and bleeding from the mouth and gums are seen. In malignant smallpox at about the tenth day the entire cornea may lose its luster, appear necrotic, and slough, releasing the intraocular contents; death usually follows.[14]

The pocks of classical smallpox may occur on the palpebral or ocular conjunctiva, but they do not develop on the cornea unless the cornea is vascularized. In countries where smallpox is endemic, trachoma is often a common problem and accounts for many vascularized corneal leukomas that subsequently develop pocks during smallpox infection. In nonvascularized corneas, about 2 weeks after onset of smallpox infection a marginal corneal ulcer may develop and proceed to spread across the cornea to the opposite limbus. If the process is slow enough, healing takes place with vascularization following the advancing border of the ulcer; otherwise, perforation of the cornea occurs.[14]

Conjunctivitis may appear 5 days after the skin lesions. In a study of 84 smallpox patients who concurrently developed conjunctivitis, variola virus was isolated from conjunctival swabs in three fourths of the patients.[40] However, if the conjunctivitis persisted longer than 2 weeks, variola virus was not recovered—a finding that suggests that this persistent conjunctivitis was caused by a different infectious agent.

When the virulence of the virus is low or resistance of the patient is high, a mild form of smallpox occurs, with only a limited skin eruption and often without eruption of any kind. A mild conjunctivitis may be the only sign. A study was made of conjunctivitis occurring in relatives or friends of smallpox patients, who brought the patient to the hospital and stayed there during the illness.[40] Of 55 of these family contacts, 21 developed conjunctivitis but no other illness; from over half of these cases of conjunctivitis, variola virus was isolated. These family contacts were presumably protected from severe illness because of previous smallpox infection or vaccination.

Vaccinia

Vaccinia virus is a laboratory virus used for vaccination against smallpox and was originally derived from cowpox or smallpox virus strains. The overwhelming majority of persons vaccinated with vaccinia virus develop no complications. However, since viremia does develop after vaccination, the open skin lesions of children with eczema can become infected and cause severe generalized vaccinia. Postvaccinial encephalitis carries a mortality of up to 40% but occurs rarely—in about 1.9 per million vaccinations.[57] Vaccinia necrosum, or progressive vaccinia, is seen in children undergoing immunosuppressive therapy or otherwise weakened by a chronic illness; this occurs in 1.1 per million vaccinations and is usually fatal.[57] Vaccinia immune globulin is protective when administered promptly.[39] In 1963 there were 7 deaths out of 6 million persons receiving primary vaccinations and 7 million receiving revaccinations in the United States.[57]

Ocular infection occurs when infectious virus is inadvertently transferred from the inoculation site to the eye (autoinoculation), from the vaccination site of a parent or sibling to another person's eye, or rarely, by transfer by unusual means, such as a swimming pool.[20] In 1919 Bedell[5] collected the reports of 91 cases of ocular vaccinia and found one third to have corneal complications; 20 years later Atkinson[2] brought the collection of case reports up to date. Since then several case reports have appeared describing the most common ocular complication, vaccinia of the eyelid (Plate 4, *G*).[9,20,27,55]

Usually the vaccinia vesicle appears on the lid margin, and the conjunctiva becomes markedly swollen; yet there is little pain.[27] Vaccinia of both the eyelids and the vulva occurred in a 3-year-old girl who inoculated herself from her sister's vaccination site.[62] Conjunctivitis alone without lid lesions has also been reported; diagnosis in this case was made by virus isolation.[12] Stenosis of the lacrimal sac can follow ocular vaccinia.[6] Corneal vaccinia takes the form of superficial punctate keratitis, epithelial ulcers, and disciform keratitis. These epithelial ulcers may develop a dumbbell shape superficially resembling herpetic keratitis.[13]

Diagnosis of ocular vaccinia is made by a history of recent vaccination, by isolation of vaccinia virus, and by cytology of infected epithelial cells. In 1892 Guarnieri[28] first described the cytoplasmic inclusions within epithelial cells; these inclusions are still known as Guarnieri bodies. They can be easily seen in cultures of corneal epithelial cells infected with vaccinia virus.[7]

Ocular vaccinia has been treated by idoxuridine,[21,34,37] vaccinia hyperimmune globuline (VIG),[21,23,39] cytosine arabinoside,[26] and monkey interferon.[35] One report suggested that VIG treatment of corneal vaccinia in rabbits caused a larger scar than did treatment with IDU.[24] However, at the present time the accepted treatment is systemic administration of VIG[69] as well as local application of IDU.

Through the assistance of the Red Cross, which provides VIG upon request, Ruben and Lane[69] conducted a retrospective study of ocular vaccinia treated by VIG. Of the 348 cases, only 6% had acute corneal complications, and only 1.2% had permanent corneal scarring. Most of the patients had lid and conjunctival vaccinia; only 2% of them were left with loss of lashes or scarring of the lid margin. Three fourths of the 348 patients were children under 4 years of age who developed ocular vaccinia 7 to 10 days after receiving a primary vaccination.

Orf (contagious ecthyma)

Orf, a viral disease of sheep, is caused by a member of the poxvirus group. Young lambs are most susceptible and develop lesions around their lips, nostrils, and eyes. The infection is occasionally transmitted to persons in contact with infected animals, such as shepherds, veterinarians, butchers, and hunters of wild goats.[10,43,45,60]

In man, the infectious agent probably enters through scratches in the skin. After an incubation period of a few days up to a week, macular lesions appear, the most likely sites being the hands, fingers, wrists, face, and legs. The macular lesions progress through a target stage, an acute vesicular stage, and a regenerative stage, and then regress during a period of 35 to 40 days before healing without scars. Infection of the eyelids is uncommon; the forehead and temple are more usual sites.

Diagnosis is made by history of contact with sheep, by biopsy of a skin lesion, by finding typical intracytoplasmic inclusions on histologic sections, and by noting the 5-week course of the disease.

Molluscum contagiosum

Typical skin lesions are small, waxy, pink tumors with an umbilicated center full of a cheesy material composed of epithelial cell husks. On histologic examination of infected basal epithelial cells, cytoplasmic inclusion bodies are seen, which on electron microscopy can be identified as masses of oval or brick-shaped virus particles typical of the poxvirus group. The virus is spread from person to person by close physical contact; for example, from a baby's face to mother's breast, from sibling to sibling, surgeon to patients, husband to wife, masseuse to clients, prostitutes to soldiers, or in the setting of a swimming pool or Turkish bath.[17] In one family group of 46 people, ranging in age from 10 months to 13 years of age, 25% became infected with molluscum contagiosum; and the tumors lasted from 2 weeks to 1½ years.[59] Transmission of molluscum contagiosum virus by sexual intercourse can be followed by outbreaks of tumors on the skin of the genitalia. This is particularly common in certain military groups.[49] Although the virus has never been cultivated in vitro, transmission is possible by direct skin inoculation of volunteers, with lesions appearing after an incubation period of 16 to 50 days; antibody against molluscum contagiosum virus also appears.[17]

Ocular disease is caused primarily by

the toxic effect of the cheesy material in the center of the molluscum. If a tumor is located on the upper eyelid, this cheesy material may easily fall into the conjunctival fornix and cause a severe follicular conjunctivitis (Plate 4, *H*). If the cause of the conjunctivitis is not recognized, the follicular conjunctivitis can progress to total corneal vascularization.[3,50,78] Treatment by electrocautery, chemical cautery, or excision of the tumor all yield equally good results.

TOGAVIRUS GROUP

Members of this group include rubella virus and many arboviruses. The enveloped virion containing RNA is 50 to 60 nm in diameter, enclosed by a lipid envelope.

Rubella (German measles)

Rubella is a common disease of childhood characterized by low-grade fever, malaise, and morbilliform rash on the face, trunk, and extremities; it is of short duration (3 to 4 days). There are no significant ocular signs.

However, rubella infection during pregnancy can cause persistent virus infection of the fetus and placenta, with devastating effects on the fetus if infection occurs early enough in gestation. During the 1964 to 1965 rubella epidemic, the rate of fetal abnormalities was 50% if infection occurred in the first month of pregnancy, 22% if in the second, and 6% if in the third.[71] Abnormalities include ocular malformation, heart defects, deafness, microcephaly, mental retardation, and chronic infection with virus excretion. Among children who reached their tenth birthday in spite of the fact that their mothers had had rubella during the first 4 months of pregnancy, the frequency of malformation was lower: 15% of these children had permanent abnormalities of the eye, ear, and heart; and 16% were left with minor abnormalities of these organs.[72]

The triad of microphthalmos, iris hypoplasia, and cataract is characteristic of ocular rubella;[68] whereas retinopathy, glaucoma, and strabismus are less common. Rubella retinopathy is not progressive and can help to label other congenital malformations, such as heart disease or cataract, as being of rubella origin.[74] Rubella virus is excreted for as long as 2 years following birth and on many occasions has been isolated from cataractous lenses at surgery; it has also been isolated as long as 3 years after the birth of the child.[52,56,65,70]

Rubella vaccine became possible following the first isolation of rubella virus in tissue culture.[79] In years past, young girls were often exposed to others with German measles in the hope that they might contract the disease and thereafter be protected by complete immunity. Now that rubella vaccine is available, large-scale immunization may prevent rubella in the future. However, rubella will probably not be eradicated, because not everyone will be vaccinated. The question remains whether immunity provided by vaccination is as adequate as natural infection and whether women vaccinated 20 years before will continue to be immune to rubella during pregnancy. Vaccination of women of childbearing age can be hazardous; one mother vaccinated inadvertently at 3 weeks' gestation was later aborted for therapeutic reasons, and rubella virus was recovered from the distorted conceptus.[61] However, given a successful program of vaccination throughout the population, congenital rubella infections should be much decreased.

Arbovirus disease

Over 241 types of viruses have been included in the arbovirus group, so named because arthropods are the principal vectors of infection, and hence the viruses are arthropod borne.[11] In the northeastern United States several varieties of arbovirus encephalitis are recognized: eastern equine, western equine, St. Louis, and Californian encephalitides.[16] Diagnosis is usually made in the context of an epidemic situation, which helps to differentiate the syndrome from aseptic meningitis following mumps, measles, vaccinia, or more rarely, herpes simplex infection. A rash on the eyelids and conjunctivitis occasionally occur. Ocular symptoms are more prominent in dengue and yellow fever, both more prevalent in tropical areas. In particular, dengue fever is characterized by severe deep pain behind the eyes and an erythematous rash affecting the skin of the eyelids as well as other parts of the body.[4,25,66] The illness starts with an acute fever of sudden onset and severe pain in the limbs and trunk, lasting 6 days. The erythema has been compared with that of measles.[66]

Although arbovirus infection during pregnancy has never been associated with congenital malformations, cataracts can be produced in rats by intracerebral inoculation of St. Louis encephalitis virus.[30]

MYXOVIRUS GROUP (INFLUENZA)

The influenza viruses, which comprise the myxovirus group, contain RNA in helical form, enclosed by an ether-sensitive envelope. They vary in shape from 80 to 250 nm in diameter and are found both in spherical and filamentous forms. The presence of 3 distinct antigens serves to classify the viruses into types A, B, and C.

Influenza infection is characterized by a 1- to 4-day incubation period, sudden onset of headache, lassitude, sweating, muscular pain, coryza, cough, congestion of the conjunctiva, and recovery usually within 1 week.[1,32,75] The infection occurs in epidemics, as in 1889, 1918, 1957, and 1968. Ocular symptoms were not significant in the 1918 epidemic, which was noted, however, for causing the deaths of millions of people, including many previously healthy young men. In contrast, during the Asian flu epidemic of 1957, mortality was lower, but ocular symptoms were the rule. In both England and the United States during the 1957 epidemic, conjunctivitis was a frequent sign: 48% of the Royal Air Force personnel contracting influenza during 1957 developed conjunctivitis, as did 60% of the patients studied in Cleveland.[31,36]

Corneal pathology is not usually associated with influenza infection, but there has been one report from the Soviet Union of 87 patients who suffered from a variety of corneal lesions during the Asian flu epidemic; the lesions included superficial punctate keratitis, deep interstitial keratitis, and dendritic keratitis associated with iridocyclitis.[41] Dendritic keratitis may be secondary to the high fever of influenza, a factor known to provoke recurrent herpes simplex virus infection. In Iran during the Asian flu epidemic, follicular conjunctivitis associated with painful preauricular lymphadenopathy was common. Dendritic keratitis, neuroretinitis, vitreous hemorrhage, and ocular muscle paralysis were also reported.[82] Superficial punctate keratitis was seen in 3 cases of influenza in France in 1932.[19]

PARAMYXOVIRUS GROUP

Within this group, viruses causing ocular disease are measles, mumps, and Newcastle disease virus; infections without ocular signs are caused by the parainfluenza and respiratory syncytial viruses. These viruses have an inner helix of RNA 18 nm in diameter and an ether-sensitive envelope 100 to 300 nm in diameter.

Measles (rubeola)

Measles is an acute infectious disease of childhood characterized by a maculopapular, blotchy rash. After an incubation period of about 10 days, fever develops, with the rash appearing 4 days later. During the prodrome, Koplik's spots can be found on the buccal mucosa; and fever, coryza, cough, and red eyes are present.

Photophobia is a characteristic symptom of measles, and children prefer to be in a darkened room. An ophthalmologic examination is rarely done; but when it is, a mild conjunctivitis and epithelial keratitis are nearly always found. Koplik's spots, tiny bluish white spots on a bright red base, may be found on the semilunar fold. Measles virus can be identified within these Koplik's spots as viral microtubular aggregates within nuclei and cytoplasm.[76] The conjunctivitis and keratitis almost always resolve without complication. But in parts of the world where malnutrition is a problem, measles conjunctivitis and keratitis can lead to secondary bacterial infection, perforation of the globe, and blindness. In the Philippines a history of measles preceding or accompanying the corneal lesion has been found in one fifth of the patients with corneal blindness at a general hospital and in one third of the students at a school for the blind. However, none of the patients treated for active cases of measles at the hospital had any ocular complication.[58] The role of malnutrition and poor resistance is pointed out by a report from Zambia, in which measles (50%), chickenpox (25%), or smallpox (25%) shared responsibility for corneal blindness; and three fourths of 686 totally blind persons were associated with these 3 infectious diseases.[51] The epidemic of measles in Haiti during 1966 to 1967 caused severe ocular complications. In 25 patients with corneal ulceration, over half the globes perforated.[21] Reports from different areas in Africa confirm this association between measles and corneal blindness, usually in the context of malnutrition.[42,64]

Even in normally nourished children rare complications of measles can occur. In one case unilateral optic neuritis was the presenting symptom before the rash appeared.[73] Measles virus can pass the placental barrier early in gestation, and the resulting infection may terminate in spontaneous abortion, or the child may be born with congenital measles. Anomalies associated with congenital measles are cataract, dacryostenosis, pigmentary retinopathy, and cardiopathy. Pigmentary retinopathy rarely follows adult measles but is common in congenital measles and must be differentiated from a similar retinopathy caused by chickenpox and influenza viruses.[53] Encephalitis is a very rare complication of measles infection in childhood.

Following administration of measles vaccine, a live attenuated virus strain, clinical measles may develop in about 6 days. Cough, conjunctivitis, rhinorrhea, fever, pneumonia, and thrombocytopenic purpura were noted in one such case.[80] Many people hoped large-scale measles vaccination of the population would eradicate the disease. However, not everyone could be vaccinated; and in 1971 a full-scale measles epidemic was again in progress in the United States.

Mumps (epidemic parotitis)

Mumps is an acute contagious disease characterized by enlargement of one or both parotid and other salivary glands, which become painful. The testicles and ovaries may be affected in adolescents and adults. The major complications of mumps affect the central nervous system and include encephalitis, aseptic meningitis, hemiplegia, quadriplegia, blindness, cranial nerve palsies, deafness, and seizures.[47] Estimates of the frequency of central nervous system complications vary from less than 1% to 30%.[47] Other rare complications of mumps are polyarthritis, pancreatitis, nephritis, and arthritis.

A dramatic keratitis occasionally develops 5 days after onset of parotitis. It is unilateral, causes abrupt loss of visual acuity, and is associated with tearing and photophobia but with little pain; on slit-lamp examination, the keratitis is found to be deep. The entire cornea becomes hazy and gray, with folds in Descemet's membrane. In spite of the dramatic appearance of the keratitis, recovery within 3 weeks is usual.[54,67]

In addition to keratitis, dacryoadenitis occurs in as many as 20% in epidemic situations.[46,47] It is characterized by sudden onset of orbital pain, lid edema, and a mass in the lacrimal fossa. Other rare ocular complications include iritis, conjunctivitis, scleritis, and central retinal occlusion.[67]

Newcastle disease

Newcastle disease, an often fatal pneumoencephalitis of fowl, was first identified at Newcastle-on-Tyne, England.[15] In spite of the neurotropic behavior of the virus in infection of fowls, the virus produces only a self-limited conjunctivitis in man. After the first human case was reported by Burnet[8] in 1943, occasional cases have been reported, usually among poultry workers and laboratory technicians.[23,29,33,38,44,48,63,77] The conjunctivitis is typically unilateral, follicular, and associated with preauricular adenopathy, but without corneal changes (Plate 4, *I*). Recovery is complete in 10 to 14 days.

Diagnosis is made by history of contact with infected birds, isolation of the virus, and by a rise in Newcastle antibodies, which are detected by hemagglutination inhibition, complement-fixation, or neutralization tests.

PICORNAVIRUS GROUP

This group is made up of small ether-resistant RNA viruses. Those of human origin are: (1) enteroviruses (polioviruses, coxsackieviruses, echoviruses, and numbered enteroviruses), and (2) rhinoviruses. Several of these viruses can cause aseptic meningitis and transient or permanent paralysis of the cranial nerves, including the oculomotor nerves. Others are responsible for a variety of upper respiratory infections with predominant involvement of the oropharynx and cervical lymph nodes. Conjunctivitis is not usual, although watery eyes are a frequent occurrence in the common cold. Some of the echoviruses cause aseptic meningitis, rash, and conjunctivitis. However, ocular signs are uncommon in infections caused by the picornavirus group.

Coxsackievirus causes a common infection of cloven-footed animals, the foot-and-mouth disease. Occasionally, infection is transmitted by contact between animal and man, and infection of the human cornea and conjunctiva can occur.

REFERENCES

1. Andrewes, C. H.: Epidemiology of influenza. In Influenza: a review of current research, Geneva, 1954, World Health Organization.
2. Atkinson, W. S., and Scullard, G.: Vaccinia with ocular involvement, Arch. Ophthal. **23**: 584, 1940.
3. Balakrishnan, E.: Molluscum contagiosum conjunctivitis, J. All-India Ophthal. Soc. **12**: 173, 1964.
4. Barkan, H.: The ocular complications of dengue fever, Amer. J. Ophthal. **2**:650, 1919.
5. Bedell, A. J.: Multiple vaccination of the eyelids, Trans. Amer. Ophthal. Soc. **17**:273, 1919.
6. Bill, I.: A case of stenosis in the lacrimal sac caused by vaccinia, Amer. J. Ophthal. **37**:33, 1959.
7. Bland, J. O. W., and Robinow, C. F.: The inclusion bodies of vaccinia and their relationship to the elementary bodies studied in cultures of the rabbit's cornea, J. Path. Bact. **48**:381, 1939.
8. Burnet, F. M.: Human infection with the virus of Newcastle disease of fowls, Med. J. Aust. **2**:313, 1943.
9. Bybee, J. D., Phillips, C. A., Ory, E. M., and Brunschwig, J. P.: Vaccinia of the eyelid, J.A.M.A. **199**:126, 1967.
10. Carr, R. W.: A case of orf (ecthyma contagiosum, contagious pustular dermatitis) contracted by a human from a wild Alaskan goat, Alaska Med. **10**:75, 1968.
11. Catalogue of arthropod-borne viruses of the world, Amer. J. Trop. Med. **19**:1081, 1970.
12. Croffead, G. W., and Harrison, S. W.: Vaccinia conjunctivitis, Amer. J. Ophthal. **53**: 531, 1962.
13. Darrell, R. W., and Vrabec, F.: Vaccinia virus infection of the rabbit cornea, Arch. Ophthal. **86**:568, 1971.
14. Dixon, C. W.: Smallpox, London, 1962, J. & A. Churchill, pp. 94-96.
15. Doyle, T. M.: A hitherto unrecorded disease of fowls due to a filter-passing virus, J. Comp. Path. **40**:144, 1927.
16. Editorial: Arboviruses in the northeastern United States, New Eng. J. Med. **279**:380, 1968.
17. Editorial: Molluscum contagiosum, Brit. Med. J. **1**:459, 1968.
18. Editorial: Prophylaxis against rubella and mumps, Brit. Med. J. **2**:282, 1970.
19. Fradkine, J.: Trois cas de keratite ponctuee superficielle, Soc. Franc. Ophtal. **45**:135, 1932.
20. Frampton, G., and Smith, C.: Primary vaccinia of the eyelid, Brit. J. Ophthal. **36**:214, 1952.
21. Francois, J., De Molder, E., and Gildemyn, H.: Ocular vaccinia, Acta Ophthal. **45**:25, 1967.
22. Frederique, G., Howard, R. O., and Boniuk, V.: Corneal ulcers in rubeola, Amer. J. Ophthal. **68**:996, 1969.
23. Freymann, M. W., and Bang, F. B.: Human

conjunctivitis due to Newcastle virus in the U. S. A., Bull. Johns Hopkins Hosp. **84:** 409, 1949.

24. Fulginiti, V. A., Winograd, L. A., Jackson, M., and Ellis, P.: Therapy of experimental vaccinial keratitis, Arch. Ophthal. **74**:539, 1965.
25. Gill, W. D.: Ocular symptomatology in dengue, Arch. Ophthal. **57**:628, 1928.
26. Gordon, D. M., and Advocate, S.: Vaccinial blepharokeratitis, Amer. J. Ophthal. **59**:480, 1965.
27. Gould, E. L., Havener, W. H., and Andrew, J. M.: Vaccinia lid infection, Amer. J. Ophthal. **56**:830, 1963.
28. Guarnieri, G.: Ricerche sulla patogenesi ed etiologia dell'infezione vaccinica e vaiolosa, Arch. Sci. Med. (Torino) **16**:403, 1892.
29. Gustafson, D. P., and Moses, H. E.: Isolation of Newcastle disease virus from the eye of a human being, J. Amer. Vet. Med. Ass. **118:** 1, 1951.
30. Hanna, C., Jarman, R. V., Keatts, J. G., and Duffy, C. E.: Virus-induced cataracts, Arch. Ophthal. **79**:59, 1968.
31. Holland, W. W.: A clinical study of influenza in the Royal Air Force, Lancet **2**:840, 1957.
32. Hoyle, L.: The influenza viruses. In Virology monographs no. 4, New York, 1968, Springer-Verlag New York, Inc.
33. Ingalls, W. L., and Mahoney, A.: Isolation of virus of Newcastle disease from human beings, Amer. J. Public Health **39**:737, 1949.
34. Jack, M. K., and Sorenson, R. W.: Vaccinial keratitis treated with IDU, Arch. Ophthal. **69**:730, 1963.
35. Jones, B. R., Galbraith, J. E. K., and Al-Hussaini, M. K.: Vaccinial keratitis treated with interferon, Lancet **1**:875, 1962.
36. Jordan, W. S., Jr., Denny, F. W., Jr., Badger, G. F., Curtiss, C., Dingle, J. H., Oseasohn, R., and Stevens, D. A.: A study of illness in a group of Cleveland families, Amer. J. Hyg. **68**:190, 1958.
37. Kaufman, H. E., Nesburn, A. B., and Maloney, E. D.: Cure of vaccinia infection by 5-iodo-2′-deoxyuridine, Virology **18**:567, 1962.
38. Keeney, A. H., and Hunter, M. L.: Human infection with the Newcastle virus of fowls, Arch. Ophthal. **44**:573, 1950.
39. Kempe, C. H., Berge, T. O., and England, B.: Hyperimmune vaccinal gamma globulin, Pediatrics **18**:177, 1956.
40. Kempe, C. H., Dekking, F., St. Vincent, L., Rao, A. R., and Downie, A. W.: Conjunctivitis and subclinical infection in smallpox, J. Hyg. (Camb.) **67**:631, 1969.
41. Kulikova, L. A.: Lesions of the cornea during influenza epidemics, Vestn. Oftal. **79**:72, 1966.
42. Lagraulet, J., and Bard, J.: Lesions oculaires de la rougeole dans les milieux ruraux en Afrique noire, Bull. Soc. Path. Exot. **60:** 203, 1967.
43. Lambie, J. A., and Gustafson, A. A.: Contagious ecthyma in man, J. Lancet **87**:400, 1967.
44. Lamer, L.: Sur un cas conjunctivite de la maladie de Newcastle, Canad. J. Ophthal. **4**:390, 1969.
45. Leavell, V. W., McNamara, M. J., Muelling, R., Talbert, W. M., Rucker, R. C., and Dalton, A. J.: Orf: report of 19 human cases with clinical and pathological observations, J.A.M.A. **204**:657, 1968.
46. Lemke, L., Jütte, A., Oehring, H., and Stelzner, I.: Mumpsinfektion als Ursache der Konjunktivitis Parinaud, Med. Klin. **61**:1107, 1966.
47. Levitt, L. P., Rich, T. A., Kinde, S. W., Lewis, A. L., Gates, E. H., and Bond, J. O.: Central nervous system mumps—a review of 64 cases, Neurology **20**:829, 1970.
48. Lippmann, O.: Human conjunctivitis due to the Newcastle-disease virus of fowls, Amer. J. Ophthal. **35**:1021, 1952.
49. Lynch, P. J., and Minkin, W.: Molluscum contagiosum of the adult: probable venereal transmission, Arch. Derm. (Chicago) **98**:141, 1968.
50. Mathur, S. P.: Ocular complications in molluscum contagiosum, Brit. J. Ophthal. **44:** 572, 1960.
51. McGlashan, W. D.: Measles malnutrition and blindness in Luapula Province, Zambia, Trop. Geogr. Med. **21**:157, 1969.
52. Menser, M. A., Harley, J. D., Hertzberg, R., Dorman, D. C., and Murphy, A. M.: Persistence of virus in lens for three years after prenatal rubella, Lancet **2**:387, 1967.
53. Metz, H. S., and Harkey, M. E.: Pigmentary retinopathy following maternal measles (morbilli) infection, Amer. J. Ophthal. **66**:1107, 1968.
54. Mickatavage, R., and Amdur, J.: A case report of mumps keratitis, Arch. Ophthal. **69:** 758, 1963.
55. Moses, L.: Vaccinia palpebrorum, Amer. J. Ophthal. **60**:331, 1965.
56. Murphy, A. M., Reid, R. R., Pollard, I., Gillespie, A. M., Dorman, D. C., Menser, M. A., Harley, J. D., and Hertzberg, R.: Rubella cataracts, Amer. J. Ophthal. **64**:1109, 1967.
57. Neff, J. M., Levine, R. H., Lane, J. M., and others: Complications of smallpox vaccination in the United States, 1963, Pediatrics **39:** 916, 1967.
58. Ocompo, G., and Olivar, E.: Corneal blind-

ness associated with measles, Philippine J. Ophthal. **1**:5, 1969.
59. Overfield, T. M., and Brody, J. A.: An epidemiologic study of molluscum contagiosum in Anchorage, Alaska, J. Pediat. **69**:640, 1966.
60. Parnell, A. G.: Ecthyma contagiosum (orf), Brit. J. Oral Surg. **3**:128, 1965.
61. Phillips, C. A., Maeck, J. V. S., Rogers, R. A., and Savel, H.: Intrauterine rubella infection following immunization with rubella vaccine, J.A.M.A. **213**:624, 1970.
62. Piazza, M., Piccinino, F., Picciotto, L., and Visconti, M.: Infezione da virus vaccinico a localizzazione palpebrale e vulvare insorta in sequito a contatto indiretto con un soggetto vaccinato di recente, Riv. Ist. Sieroter. Ital. **44**:59, 1969.
63. Pilsworth, R., and Wall, B. J.: The isolation of Newcastle disease virus from human infections, Great Britain Ministry of Health—Monthly Bulletin **23**:122, 1964.
64. Quere, M. A.: Les complications oculaires de la rougeole, cause majeure de cecite chez l'enfant en pays tropical, Ophthalmologica **148**:107, 1964.
65. Rawls, W. E., Phillips, C. A., Melnick, J. L., and Desmond, M. M.: Persistent virus infection in congenital rubella, Arch. Ophthal. **77**:430, 1967.
66. Riad, M.: Some eye complications of dengue fever, Bull. Ophthal. Soc. Egypt **22**:60, 1929.
67. Riffenburgh, R. S.: Ocular manifestations of mumps, Arch. Ophthal. **66**:739, 1961.
68. Roy, F. H., Hiatt, R. L., Korones, S. B., and Roane, J.: Ocular manifestations of congenital rubella syndrome, Arch. Ophthal. **75**:601, 1966.
69. Ruben, F. L., and Lane, J. M.: Ocular vaccinia: an epidemiologic analysis of 348 cases, Arch. Ophthal. **84**:45, 1970.
70. Scheie, H. G., Schaffer, D. B., Plotkin, S. A., and Kertesz, E. D.: Congenital rubella cataract, Arch. Ophthal. **77**:440, 1967.
71. Sever, J. L.: The epidemiology of rubella, Arch. Ophthal. **77**:427, 1967.
72. Sheridan, M. D.: Final report of a prospective study of children whose mothers had rubella in early pregnancy, Brit. Med. J. **2**:536, 1964.
73. Srivastave, S. P., and Nema, H. V.: Optic neuritis in measles, Brit. J. Ophthal. **47**:180, 1963.
74. Stark, G.: Rubella retinopathy, Arch. Dis. Child. **41**:420, 1966.
75. Stuart-Harris, C. H., and Andrewes, C.: Influenza, London, 1965, Edward Arnold (Publishers), Ltd.
76. Suringa, D. W. R., Bank, L., and Ackerman, A. B.: Role of measles virus in skin lesions and Koplik's spots, New Eng. J. Med. **283**:1139, 1971.
77. Trott, D. G.: Outbreak of conjunctivitis due to Newcastle disease virus among workers of chicken broiler factory, Brit. Med. J. **5477**:1514, 1965.
78. Vannas, S., and Lapinleima, K.: Molluscum contagiosum in the skin, caruncle and conjunctiva, Acta Ophthal. **45**:314, 1967.
79. Weller, T. H., and Neva, F. A.: Propagation in tissue culture of cytopathogenic agents from patients with rubella-like illness, Proc. Soc. Exp. Biol. Med. **111**:215, 1962.
80. Wilhelm, D. J., and Peagle, R. D.: Thrombocytopenic purpura and pneumonia following measles vaccination, Amer. J. Dis. Child. **113**:534, 1967.
81. World Health Organization Scientific Group: Smallpox eradication, W. H. O. Techn. Rep. Ser. **393**:1, 1968.
82. Zarrabi: Quelques complications oculaires en Iran pendant la grippe asiatique, Arch. Ophthal. **18**:59, 1958.

20

Ocular syphilis

RANDOLPH WHITFIELD, Jr. AND EMIL WIROSTKO

Following Mahoney's[34] 1943 report on penicillin's great effectiveness for treating syphilis, members of the medical community generally felt that another great affliction of man had finally been conquered by modern medical science. This attitude was strongly reinforced by the dramatic decrease of infectious syphilis following the increase in availability and use of penicillin in the late 1940's. From a high of 106,000 cases of primary and secondary syphilis reported in 1947, the incidence fell to only 6,251 cases by 1957.[57] The results of this apparent success were readily seen in the hospitals and medical schools across the country. The insistence on routine serologic evaluations on all inpatients became so relaxed that these tests often became the exception rather than the rule. In the medical schools the study of syphilis developed such a low priority that the students soon received merely a few hours of instruction in the diagnosis and management of the disease. At present there is a whole generation of medical school graduates who are but vaguely aware of the clinical manifestations and natural course of syphilis. The decrease in routine serologic testing and in physicians' interest in and knowledge of the disease must have had profound, if unmeasurable, effects on the reported incidence of the disease and must have significantly contributed to the sharp decrease in reported cases. In addition, the early indiscriminate use of penicillin for treating many infectious disorders contributed to the apparent decrease in the incidence of clinically recognizable early syphilis. It is a well-known fact that minute amounts of penicillin can alter the natural course of the disease by causing it to evolve directly into the latent stage, aborting the clinically evident early manifestations, with the appearance later of symptomatic late gummatous, cardiovascular, or central nervous system syphilis.

Despite these many factors operating to decrease the reported incidence, the downward trend has reversed itself; and in recent years there has been a marked increase in the number of reported cases of infectious syphilis. In 1970 the figure was 20,186, an increase of 320% over the 1957 level.[57] The significance of the present problem is pointed out by the results of a 1968 report sponsored by the American Public Health Association. This study revealed that private physicians, treating more than 75% of all cases of infectious syphilis in this country, reported only one eighth of the cases to public health authorities.[14] The study concludes that the true incidence of infectious syphilis is actually at least four times the reported incidence, totaling over 80,000 cases in 1970. Syphilis is now far from being under control in this country, and a greater awareness of the problem, especially on the part of physicians, is urgently needed.

Ophthalmologists find themselves in a

unique position concerning the diagnosis and treatment of syphilis. Not only are ocular complaints often the reason for an undiagnosed syphilitic patient to seek medical attention,[62] but in some forms of the disease, such as neurorecurrence and tardive heredosyphilis, ocular lesions are the most common and often the most serious clinical manifestations. In addition, the recent controversial reports of intraocular treponemes in seronegative patients, with the persistence of intraocular treponemes following supposedly adequate therapy, have stimulated a resurgence of basic and clinical research into the nature of the disease, much of which is being conducted by ophthalmologists.

This chapter includes, first, a brief review of the various tests available for the diagnosis of syphilis; second, a presentation of the natural course of the disease and the different ways syphilis may be clinically manifest in the eye; and third, a discussion of some of the more recent research in ocular syphilis and the implications of this research for the clinician as regards diagnosis and treatment of ocular syphilis.

DIAGNOSTIC PROCEDURES

More than 200 different types of serologic tests for syphilis have been developed since August von Wasserman in 1905 described his classical complement-fixing reaction using a syphilitic patient's serum and liver extract from a stillborn syphilitic fetus. All the tests have in common the ability to demonstrate the presence of one or more of the antibodies elicited in the immunologic response of an individual to infection with *Treponema pallidum.* Today there are 2 general types of serologic tests for syphilis: nonspecific reagin tests and specific treponemal tests.

Reagin tests

The Wasserman test and its progeny all detect the presence of the same reactive substance (reagin) found in syphilitic serum. Although reagin antibodies are undetectable in the serum of uninfected individuals, they may occur transiently in acute infections and following inoculations; and, less commonly, they may be found chronically in the serum of some patients suffering with diseases such as systemic lupus erythematosus, lepromatous leprosy, malaria, and other conditions characterized by serum immunoglobulin elevations. The level of reagin in the serum of an infected person varies in a predictable way, corresponding to the clinical stage of the disease. It also falls in a typical manner following adequate therapy, especially in the early stages of the disease (Fig. 20-1). These quantifiable tests, when performed serially, are indispensable in determining the stage of the disease, in detecting reinfection in a patient having had previously adequate therapy, and also in ascertaining the effectiveness of treatment. Because they are inexpensive and simple to perform, they are well suited for routine testing and mass screening of sera. A final use of the reagin test is to examine the cerebrospinal fluid for possible neurologic involvement of the disease. False-positive reactions in the cerebrospinal fluid are quite rare, although false-negative reactions may occur. Thus reagin tests, though not specific for infection with *T. pallidum,* have great value in the diagnosis and management of syphilis. The quantitative reagin tests in common use today are the Venereal Disease Research Laboratory (VDRL) test, the Kahn, Kline, and Mazzini tests, all of which are flocculation tests, and the Kolmer, a complement-fixation test. Of these, the VDRL is the test of choice in the United States because of its reliability, simplicity, and widespread availability.

Treponemal tests

The first serologic test to measure specific antitreponemal antibodies was the *Treponema pallidum* immobilization test (TPI) developed by Nelson[42] in 1949. Although expensive, technically difficult to perform, and lacking sensitivity in early acquired syphilis, it soon became the serologic test against which all others were

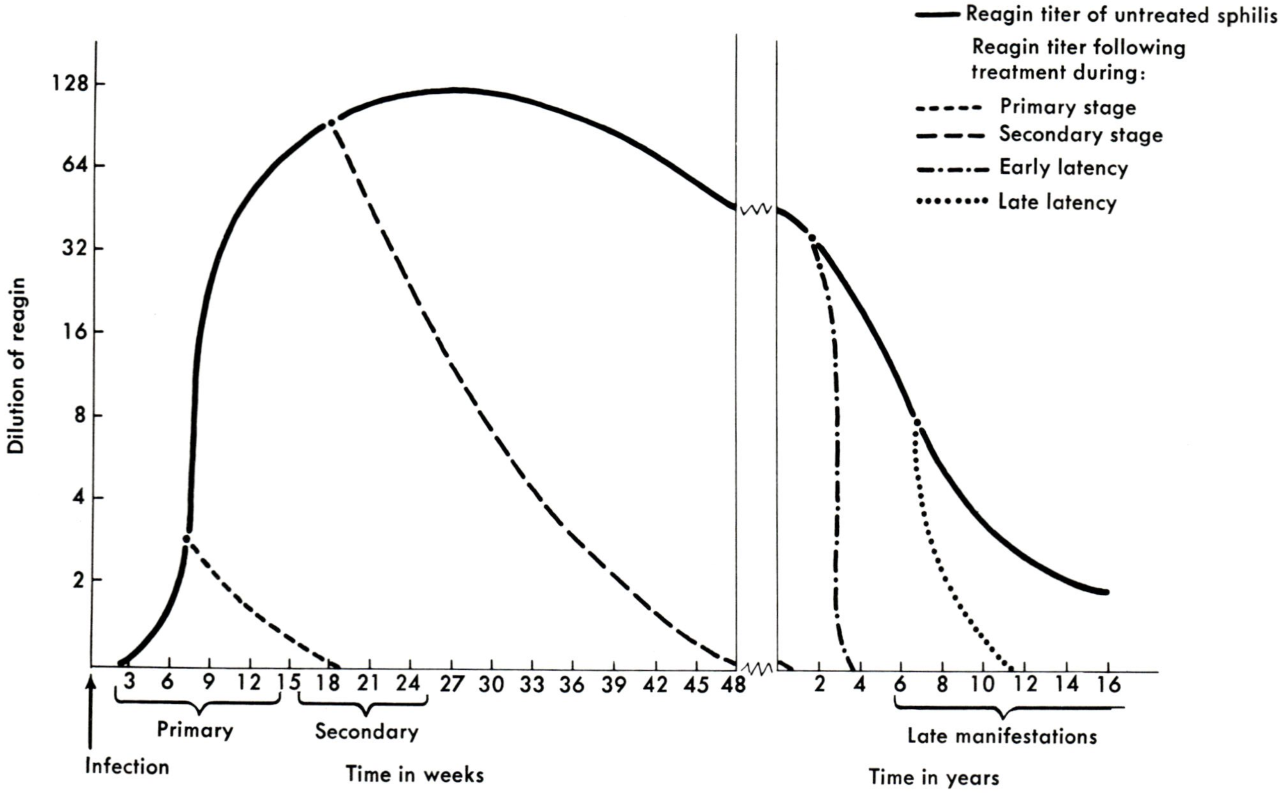

Fig. 20-1. Typical serum reagin titer changes during the course of untreated syphilis and following treatment during the primary, secondary, early latent, and late latent stages. (Adapted from Syphilis—modern diagnosis and management, U. S. Public Health Service Publication No. 743, Washington, D. C., 1961, pp. 17-21.)

measured because of its great specificity and excellent sensitivity in late acquired syphilis. Many treponemal tests quickly followed the TPI, but none of them were able to replace it. However, in 1957 Deacon and his associates[9] at the Venereal Disease Research Laboratories developed an immunofluorescent technique for identifying antitreponemal antibodies. It held great promise as an inexpensive test possessing the specificity and sensitivity of the TPI. It was evaluated and improved over the next several years, and in 1964 the fluorescent treponemal antibody absorption test (FTA-ABS) was presented as a modification of the original FTA test.[21] In its present form the FTA-ABS test is widely available and with its increased sensitivity and specificity is the confirmatory test of choice.[55]

The limitations of the treponemal tests, such as the TPI and the FTA-ABS, should be recognized, however. They are qualitative not quantitative tests, and they do not usually revert to negative following adequate therapy. They have no value in serial use to determine either the stage of disease or the adequacy of treatment. Also, the treponemal tests on the cerebrospinal fluid (CSF) have not been proved reliable.[49]

Table 20-1 is a comparison of the relative sensitivities of the VDRL, the TPI, and the FTA-ABS tests in various stages of syphilis.[33] Because of the greater sensitivity of the FTA-ABS test, it is the test of choice for the ophthalmologist in determining whether syphilis is a possible cause of ocular disease.

CSF examination

The parameters measured in a CSF examination for evidence of neurosyphilitic activity must include cell count, total protein level, and a titered VDRL test. Any cell count above 4 round cells, total protein level above approximately 40 mg/100 ml (depending upon the laboratory), and any reactivity of the VDRL test is considered abnormal. As increased cells and protein in the CSF are nonspecific indications of meningeal inflammation, a reactive VDRL test is the only certain indicator of past or present luetic involvement of the central nervous system (CNS). Soon after infection there is CNS invasion with treponemes in virtually all untreated cases of syphilis. Although the associated involvement of the meninges, vessels, or parenchyma of the CNS may be so slight as to cause no clinical or spinal fluid manifestations, it is known that those who do develop even transient spinal fluid abnormalities at this time have an increased tendency to develop one of the clinical forms of neurosyphilis months or years later. Spinal fluid examination is an essential step in the diagnosis and management of every case of syphilis.

The indications for performing a CSF examination are:

1. Before discharge, following adequate treatment of early acquired or congenital syphilis
2. Before initiating treatment in any patient with late latent, late acquired, or late congenital syphilis
3. Whenever there is clinical or sero-

Table 20-1. Sensitivities of three tests in the various stages of syphilis

	Percent reactivity			
Test	**Primary syphilis**	**Secondary syphilis**	**Latent syphilis**	**Late congenital and symptomatic syphilis**
VDRL	77	97	75	75
FTA-ABS	85	99	95	96
TPI	56	94	82	93

Adapted from Lucas, J. B.: New developments in venereal disease research, Med. Ann. D. C. **35**:375, 1966.

logic evidence of syphilitic activity or reinfection in a patient having had previous adequate therapy for syphilis

4. In any patient suspected of being a chronic biologic false-positive reactor
5. In any patient being treated for neurosyphilis (Repeated CSF examinations must be performed.)

ACQUIRED OCULAR SYPHILIS

Primary lesions of the eye

Primary syphilis is contracted following prolonged, moist, warm intimate contact with an infectious lesion or with material containing motile treponemes. Following successful inoculation, a generalized spirochetemia ensues within 24 hours. The incubation period preceding the appearance of the chancre is usually 2 to 3 weeks but ranges from 10 to 90 days. The chancre, which is evident in approximately two thirds of all cases, is the first clinical sign of syphilis and appears at the site of original inoculation. It is typically an elevated, eroded, painless papule, associated with rubbery, nontender enlargement of the lymph nodes draining the area of the chancre. Extragenital chancres are not uncommon. The lesion lasts 1 to 5 weeks and then disappears spontaneously. It is infectious, and a properly performed darkfield examination usually reveals a large number of motile treponemes. At the first appearance of the chancre the serologic tests may or may not be reactive, though they rapidly become so during the following month.

Although the primary chancre is usually genital, it may occur almost anywhere on the body; and conjunctival primary lesions have been reported repeatedly since the first written description by Desmarres in 1854.[11] The conjunctival chancre, typically presenting as a nontender, firm, eroded papule, associated with enlarged, painless, preauricular and submandibular lymph nodes, is not different from one in any other location. It may arise from the bulbar or palpebral conjunctiva, with a predilection for the lower fornix. Stokes makes special mention of the conjunctiva as a common site for nonvenereally acquired chancres in physicians: "Two of four such lesions . . . were in physicians, and in none did the ophthalmologist consulted make or even suggest the diagnosis."[54] History and darkfield and serologic examination, combined with physical findings, usually make the diagnosis a simple one. The problem of diagnosis is that one must be aware of the possibility of syphilis in the first place.

Secondary lesions of the eye

The secondary, or mucocutaneous, stage of the disease has its onset within 6 months of the inoculation, occurring usually within a few weeks of and occasionally concomitant with the regression of the chancre. The signs and symptoms of the secondary disease may be pronounced or may be so slight as to pass unnoticed by the patient. Fever, headache, alopecia, iritis, and malaise may be prominent heralding symptoms; but the hallmarks of secondary syphilis are generalized lymphadenopathy and papulosquamous lesions of the skin and mucous membranes. These lesions are infectious, darkfield positive for treponemes, and in early secondary syphilis are usually discrete, generalized, and bilaterally symmetrical. Antibody production is at its height, and the serologic tests are all strongly reactive. The signs and symptoms of this stage regress slowly over a period of several weeks, and the patient thereafter becomes completely asymptomatic.

Ocular complications occur in 4.5% of all patients with secondary syphilis and in 9.3% of those with relapsing secondary syphilis.[61]

LIDS AND CONJUNCTIVA. The skin of the lids is often involved in the generalized mucocutaneous eruption, which is so typical of secondary syphilis with macular, papular, squamous, and pustular eruptions, or any combination of these. The lesions are usually dry, nonpruritic, and bilaterally symmetrical. They are darkfield positive and infectious. The alopecia of secondary syphilis may be manifest by the temporary loss of the eyebrows, especially a bilaterally symmetrical loss of the outer portions.

A transient nonspecific conjunctivitis often appears concomitantly with the development of systemic symptoms and skin rash. It may vary from mild to severe and usually lasts only a few days.

Specific luetic conjunctivitis is much less common, occurs later in the secondary stage, and has three forms. The most common form is that of diffuse papillary conjunctivitis, with the palpebral conjunctiva involved in a hypertrophic papillary inflammatory process. Its frequent association with a pannus and its granular configuration bring trachoma into the differential diagnosis. However, scrapings reveal motile treponemes on darkfield examination; results of serologic tests are strongly positive; and no follicles are present on examination of the conjunctiva.

Syphilitic mucous patches, commonly found on other mucous membranes in this stage of the disease, form the second, much less common, form of specific syphilitic involvement of the conjunctiva. The patches vary from 2 to 10 mm in diameter, are slightly elevated with an erythematous base, and have an overlying pearly gray membrane. These may occasionally develop into papules or, rarely, into frankly flowering condyloma lata. The third form of specific luetic conjunctivitis, called syphilitic scleroconjunctivitis, is extremely rare.[11] It is a perilimbal inflammation, with edema, injection, and elevation of the perilimbal conjunctiva, and with involvement of the episclera as well. It may be associated with a mild iritis or iridocyclitis.

A final form of mucous membrane inflammation that may affect the ocular adnexa is the rare extension of secondary rhinitis to involve the tear sac, causing an acute obstructive dacryocystitis. All these forms of early inflammation respond dramatically to specific antisyphilitic therapy.

CORNEA AND SCLERA. Simple scleritis occurs rarely, and episcleritis occurs only slightly less rarely in secondary syphilis. Interstitial keratitis is a rare complication of acquired early syphilis, accounting for less than 3% of all cases of the disorder. It is often associated with anterior uveitis, and but for its tendency to be unilateral, its rapid evolution, and relatively benign course, is quite similar to the interstitial keratitis of congenital syphilis.

IRIS AND CILIARY BODY. According to Hogan and Zimmerman,[20] the most common syphilitic eye lesion is the iritis of secondary syphilis, and it comprises about 75% of all the eye lesions of the early acquired disease. Approximately 4% of those in the secondary stage develop iritis. It is slightly more common in males than in females and occurs twice as frequently in Negroes as in Caucasians.[40] In 1943 Woods[61] reported that between 30% and 40% of all cases of iritis were syphilitic in origin, but with the decrease in the incidence of syphilis its importance as a cause of iritis also diminished. However, serologic determinations are certainly indicated in any case of iritis of unknown or doubtful cause. Usually the diagnosis is not difficult, as the iritis is associated with florid mucocutaneous lesions in 75% of the cases, and in every case there usually is other evidence of early syphilis.[40]

There are two special types of iritis that deserve mention. The first is recurrent iridocyclitis, called iridorelapse, which follows inadequate therapy for early syphilis. Although morphologically nonspecific and usually quite responsive to specific antiluetic therapy, iridorelapse is important because it is associated with concomitant neurorecurrence in about 25% of the cases and is thought to indicate a predisposition to the development of symptomatic late neurosyphilis.[39] It is usually unilateral and occurs within 4 months of antiluetic therapy.

Iritis may also develop as a complication following antisyphilitic therapy in the unusual systemic reaction called the Jarisch-Herxheimer reaction. This generalized reaction occurs within 24 hours after treatment of a syphilitic patient with even small amounts of penicillin and is thought to be a reaction to the systemic release of toxins following the rapid destruction of large numbers of treponemes throughout the body. It is characterized by low fever, chills, and generalized malaise lasting 24 to 36 hours, and occasionally by sudden inflammatory activation of previously quiescent syphilitic lesions. Sudden loss of vision following initial treatment of syphilitic optic neuritis has been reported, and even more calamitous reactions following initial treatment of cardiovascular or central nervous system syphilis are on record. However, the iritis of the Jarisch-Herxheimer reaction, though occasionally intense, is typically self-limited, responds well to the usual therapeutic measures, and rarely results in permanent ocular damage.

In addition to iridorelapse and the iritis of the Jarish-Herxheimer reaction, there are two general types of iritis in early acquired syphilis. The first is roseola of the iris, often a symptom heralding secondary syphilis and occurring within the first few months after infection. It is acute in onset, bilateral in about one half of the cases, is manifested by engorgement of the superficial iris vessels, and is accompanied by photophobia and circumcorneal injection. It is usually mild, responds well to specific antiluetic therapy and does not recur.

The second type is the plastic iritis of secondary syphilis. It is often bilateral, but in contrast to roseola occurs later in the secondary stage, and is often quite severe, producing marked decrease in visual acuity, intense pain, and photophobia. Slit-lamp examination reveals a violent anterior chamber reaction with a copious cellular, fibrinous, and occasionally hemorrhagic exudate. Ocular hypotony with inflammation in the anterior vitreous is commonly associated with severe iritis, indicating a significant cyclitis. The process may remain simple and plastic, or may, in a more typically syphilitic reaction, include the development of nodules on the surface of the iris. The iris lesions in this papular iritis vary in size from 0.5 to 3 or 4 mm and range in appearance from multiple, slightly elevated, pink nests of newly formed capillaries to isolated, large, fleshy, elevated yellowish papules. If untreated, the eye may develop extensive posterior synechiae with iris bombé and secondary glaucoma as frequent complications. If the patient receives prompt specific antiluetic therapy, along with topical mydriatics and steroids, the prognosis is usually good, with permanent severe sequelae uncommon. Asymptomatic isolated grayish patches of atrophic iris at the site of previous iris nodules are occasionally found as the sole sign of previous papular iritis.

Other than the extremely rare occurrence of a ciliary syphiloma in late secondary syphilis, the ciliary body is rarely involved alone but is frequently associated with any intense and extensive iritis described above.

Choroid and retina. Although the choroid or retina may occasionally be independently affected as a complication of secondary syphilis, they are typically involved together in various forms of chorioretinitis, which depending upon the investigator, are reported to account for 8% to 16% of all cases of chorioretinitis arising in adults.[3] There are no typical morphologic characteristics of syphilitic chorioretinitis, but when it occurs as a complication of acquired syphilis it is usually restricted to the posterior pole and the midperiphery, whereas the more frequent chorioretinitis of congenital syphilis involves the midperiphery and far periphery. It characteristically has its onset in the late secondary stage, though it may appear as late as 10 years after the initial infection.

The diffuse chorioretinitis of Förster is by far the most common form of acquired syphilitic chorioretinitis. Bilateral in about 50% of the cases, this diffuse syphilitic chorioretinitis is usually restricted to the posterior segment, involving the macula and peripapillary areas most frequently and the far periphery only rarely. The initial symptoms include diminished visual acuity, flashes of light, and metamorphopsia. There is often an associated ring scotoma, which is so typical of acute syphilitic chorioretinitis that, when present, it should suggest a syphilitic cause of the disorder.[12] Although there is no diagnostic ophthalmoscopic appearance, there is typically a dense vitreous haze through which slight blurring of the disc margins may be seen. In the posterior pole a grayish edematous retina overlies multiple areas of yellowish, deep inflammatory infiltrates of varying sizes. Abundant superficial perivascular exudates and scattered flame-shaped hemorrhages in the posterior pole are also common findings. The course of the untreated disease is prolonged, remittent, unpredictable, and is characterized by progression, with enlargement and coalescence of the inflammatory patches and the almost invariable exuberant reaction of the retinal pigment epithelium with peripheral pigment aggregates. Occasionally, a secondary anterior uveitis may complicate the picture, resulting in a chronic degenerative panuveitis that is refractory to all therapy. One may see perivascular and midzone diffuse pigmentary deposits, patchy areas of chorioretinal atrophy, with variable subretinal gliosis and occasionally varying degrees of optic atrophy in the end stage of the disease. The ring scotoma is rarely present in the late stages, having either progressed to marked field contracture by breaking through to the periphery if the disease was severe and generalized or regressed to irregular scotomas corresponding to the areas of scarring and atrophy. The severe atrophic form with marked field contracture is easily confused with primary retinal pigmentary degeneration, although the optic atrophy and vascular atresia in the syphilitic disease is not as pronounced as in the hereditary disorder. The history of a chancre and a highly reactive serologic test, with the presence of other signs and symptoms of secondary syphilis, make the diagnosis relatively straightforward. Visual loss is highly variable, rarely complete, and often significant.

Several other forms of chorioretinitis have been described as occurring within 2 years of infection with the spirochete. All of these forms probably represent clinical variations of Förster's chorioretinitis and are not important as distinct clinical entities. However, they do have value in indicating the range of host response to infection by *T. pallidum*. Many of these various types of chorioretinitis may be grouped together conveniently under the term focal disseminated chorioretinitis. They differ from Förster's chorioretinitis in that only a few circumscribed areas become inflamed in these forms. All of these forms are rare. Included in this category are chorioretinitis juxtapapillaris, central chorioretinitis, and anterior chorioretinitis; the names indicate the location of the majority of the lesions. Anterior chorioretinitis, certainly atypical in acquired syphilis, is interesting in that it is occasionally associated with iridocyclitis and interstitial keratitis.

A rare but distinct syphilitic entity is the neuritis papulosa of Fuchs. It occurs within 2 years of infection and is unilateral. It is an intense exudative papillitis, with peripapillary chorioretinitis occurring with a marked vasculitis and dense vitreous reaction. The exudate arises from the disc and blooms up into the vitreous, obscuring the disc in the acute phase. Early visual loss is the rule; but if the patient receives prompt penicillin therapy, the inflammation recedes, with fair return of vision, leaving an atrophic band of scar tissue extending from the disc to the area where chorioretinitis had occurred.

The retina is occasionally involved in a discrete inflammatory process in which the choroid remains essentially free of disease. The inflammation begins as a retinal vasculitis at the posterior pole, with associated papillitis and retinitis. It may remain confined to the area around the disc (retinitis circumpapillaris),[26] or the retinitis may extend diffusely to the equator and beyond. Marked retinal edema, macular star formation, perivascular exudates with sheathing, and superficial, intraretinal, and vitreous hemorrhages are prominent features. A secondary iridocyclitis may complicate the process. Visual loss is quite variable. If the patient receives adequate specific antisyphilitic therapy, recurrences are rare. The response of the retinal pigment epithelium varies from the formation of occasional spicule-shaped pigment clumps following the milder forms of inflammation associated with little retinal or optic atrophy to widespread disorganization of the pigment epithelium, with extensive diffuse pigment mobilization and clumping found with significant retinal and optic atrophy in the severe inflammations.

Several other forms of syphilitic retinal involvement have been described. Syphilitic retinal perivasculitis is simply a mild form of inflammation in which the process does not spread from the vessels to cause a significant retinitis or retinochoroiditis. The arteries are more often involved than the veins, and the process includes vascular sheathing, perivascular exudates, occasionally intraretinal hemorrhages, and rarely vitreal hemorrhage with the development of retinitis proliferans. A rare complication of syphilitic retinal vascular inflammation is thrombosis of the central retinal artery or vein, though branch occlusions of either the artery or vein are not rare.

A final form is recurrent central syphilitic retinitis, also called toxic exudative syphilitic retinitis. It is characterized by recurrent, often bilateral, macular edema, indistinguishable clinically from central serous retinopathy. Indeed, Woods[61] doubts that the cases reported in the literature are syphilitic in origin at all and believes that they represent idiopathic central serous retinopathy. Specific antiluetic therapy, always indicated to prevent later syphilitic systemic complications, may be curative in the above forms of chorioretinal complications, although these complications may be refractory to even the most intensive specific and nonspecific measures. Förster's chorioretinitis and simple retinitis usually respond in the most satisfactory manner.

OPTIC NERVE. Mention should be made here of neurorecurrence and early neurosyphilis, as they occur within the first 2 years after infection. As a result of the generalized spirochetemia following infection, treponemes are distributed widely throughout the body; and the central nervous system is no exception. Stokes[54] states that involvement of the CNS occurs in an "overwhelming proportion of early cases of syphilis." The involvement is usually transitory and asymptomatic and is associated with similarly transient abnormalities in the CSF. Woods indicates that abnormal results of spinal fluid examinations occur in 28% to 56% of all cases of early syphilis.[59] Indeed, treponemes may be found in the spinal fluid sometime during the course of early syphilis in over 40% of all cases.[54] In about 25% of all cases of early syphilis (meaning primary, secondary, and early latent stages), the involvement of the CNS is sufficient to cause some clinical signs and symptoms, usually those of transient mild meningeal irritation. In a much smaller percentage gross signs of CNS disease develop and almost invariably indicate "serious damage, not infrequently irreparable."[54]

The only other form of early neurosyphilis is called neurorecurrence. Although uncommon, it is important because it occurs following inadequate therapy for primary or secondary syphilis. In fact, those receiving inadequate therapy for early syphilis have a much greater chance of developing neurosyphilis than those not treated at all.[39] Neurorecurrence is typically meningovascular in type, often presenting as a sudden cranial nerve palsy or hemiplegia within the first 2 years of the disease. The prognosis is usually good with prompt therapy because the signs and symptoms are caused by acute, potentially reversible, lesions rather than by old scarred changes of late neurosyphilis. Neurorecurrence definitely predisposes the patient to the later development of symptomatic late neurosyphilis. Thus, it is imperative that a normal result of a CSF examination be obtained before a patient is discharged as having been adequately treated for early syphilis.

Changes occurring in the optic nerve in early acquired syphilis include papilledema, optic neuritis (called neuroretinitis when the retina is extensively involved, as described above), and retrobulbar neuritis.

Papilledema is only occasionally seen in early acquired syphilis and is usually a complication of acute extensive syphilitic meningitis. Occurring as a result of increased intracranial pressure, it is typically associated with other obvious signs of acute meningitis, such as severe frontal headache, meningismus, and sudden cranial nerve palsies, which are often manifest as acute diplopia, ptosis, or unilateral central deafness. Choked discs are usually bilateral and often cause little loss of vision unless present for a long time. The prognosis for vision is excellent, if therapy is promptly instituted.

Inflammation of the optic nerve, occurring anywhere from the optic disc to the chiasm, is common in late secondary syphilis and is one of the most frequent manifestations of early neurosyphilis and of neurorecurrence. Minimal degrees of optic neuritis may be the only manifestation of early involvement of the central nervous system by the treponeme. Stokes[54] states that 13% of all those with secondary syphilis develop some degree of optic neuritis, though it may be transient and slight, and he also reports that involvement of the optic nerve in early syphilis accounts for 11.1% of all syphilitic eye complications. Moore emphasizes the importance of ruling out optic neuritis in early syphilis because of its prognostic significance. When present, optic neuritis indicates greatly increased chances of developing late symptomatic neurosyphilis.[39] It is interesting that minimal changes in the optic nerve are the most common signs of CNS involvement in early syphilis. Therefore, careful ophthalmologic examination is mandatory in every case.

Optic neuritis, or inflammation of the intraocular portion of the optic nerve, may range in severity from a common evanescent hyperemia of the disc to a rare severe neuroretinitis. It usually occurs in the late secondary stage, often is bilateral, and may occur as an isolated ocular lesion in conjunction with syphilitic meningitis or as a manifestation—occasionally the sole manifestation—of neurorecurrence. Clinical symptoms may be minimal, but diminished visual acuity of varying degree is commonly reported. The ophthalmoscopic appearance is typical for optic neuritis with blurred disc margins, hyperemia, and edematous elevation of the disc, which is occasionally obscured by an overlying vitreous reaction. In the severe cases there is edema of the surrounding retina, perivascular sheathing at the nerve head, and superficial hemorrhages in the retina and disc. The mild forms display no visual field changes. In moderately severe cases there is usually enlargement of the blind spot, and with severe inflammation there is often a dense central or paracentral scotoma with peripheral contraction of the fields.

Syphilis causes 4% to 7% of all cases of retrobulbar neuritis.[6,58] Retrobulbar neuritis is rare in congenital syphilis and only slightly more common in late acquired neurosyphilis, occurring usually in the secondary stage of the disease. Symptoms include pain on movement of the eye and visual loss proportional to the severity of the disease. There are usually no ophthalmoscopic signs of the neuritis, though other syphilitic ocular lesions may be found. Central, paracentral, and sector scotomas, with peripheral loss of vision, also may occur. All forms of optic nerve inflammation in early acquired syphilis respond rapidly to early antiluetic therapy.

LATENT SYPHILIS

The period from the disappearance of the secondary signs and symptoms to 4 years after original infection is arbitrarily called early latent syphilis. During this time the patient remains potentially infectious. This stage is characterized by the recurrence of secondary lesions in approximately 25% of the cases. As time passes, recurrences become less symmetrical, more sporadic and isolated, and begin to assume the arciform configuration typical of the skin lesions of late gummatous syphilis, though they usually remain darkfield positive. Between secondary relapses, which become uncommon after the first year, the patient is completely asymptomatic. The only indication that he is infected with *T. pallidum* is a reactive serologic test, with both reagin and treponemal tests being uniformly strongly reactive.

Four years after infection the patient is said to enter the late latent stage, having a reactive serologic test as the only evidence of infection. The period of late latency may last a lifetime or be followed 1 to 20 years later by the appearance of late symptomatic syphilis. Of those persons remaining untreated, approximately one third will slowly revert to a negative serologic reaction over the years, with no other signs or symptoms of syphilis. One third will retain a positive serologic reaction but remain asymptomatic. The final third will go on to develop one or a combination of the forms of late symptomatic syphilis, including gummatous (15%), cardiovascular (10%), and symptomatic central nervous system syphilis (8%).

ACQUIRED LATE SYMPTOMATIC SYPHILIS

Gummatous syphilis is also known as late benign cutaneous syphilis because one who develops late lesions in the skin often does not develop the other two potentially more serious forms of late syphilis. The gumma may involve practically any tissue in the body. Skin, bone, and the liver are affected most frequently. Gummas of the central nervous system, heart, and lungs occur uncommonly but are potentially more dangerous. The gumma, thought to be a manifestation of hypersensitivity to the spirochete, is essentially a granuloma composed of round cells and giant cells with central necrosis. It may occur as a caseating tumor or as a destructive infiltrative lesion. Treponemes are present, though scarce, and the serologic tests are strongly reactive.

Gummatous lesions of the eye

ORBIT. Gummatous disease of the orbit is rare, accounting for less than 1% of all ocular syphilis.[58] It may occur as early as the late secondary stage but usually has its onset several years following infection. Gummatous periostitis or osteitis in the bones of the orbit is the most frequent form, producing local pain, especially at night, and tenderness of the orbital rim to palpation. Exophthalmos with diffuse gummatous infiltration of the retrobulbar tissues and extraocular muscle palsies is not uncommon. If the infiltrative process is pronounced, any of the intraorbital structure may be invaded, including the muscles, nerves, and the globe itself. When the process is located in the apex of the orbit, painful ophthalmoplegia, ptosis, and paresis of the trochlear and abducens nerves along with corneal hypesthesia may occur, forming the clinical entity known as the superior orbital fissure, or Tolosa-Hunt syndrome.[28]

Other causes of this syndrome include other inflammatory disease processes, tumor, hemorrhage, and fracture. The diagnosis of syphilis in any form of orbital gumma is made by the presence of a reactive serologic test and prompt regression of signs and symptoms with luetic therapy. Conjunctival gummas are also rare, occurring predominantly on the perilimbal bulbar conjunctiva. The initial lesion is a small, nontender, pink limbal nodule associated with painless preauricular and submandibular lymphadenopathy. It responds well to penicillin; untreated it may continue to enlarge, breaking down to form a chronic, painless ulcer penetrating the globe. Gummas of the lids have been reported and may involve the skin alone or begin as an infiltrative tarsitis. The skin lesions are typical of skin gummas elsewhere, being asymmetrical, arciform, destructive lesions, which spread peripherally, leaving either indolent painless ulcers or atrophic "cigarette paper" scarring. Gummatous tarsitis is quite rare, presenting initially as a painless thickening of the lid, and without treatment it progresses to ulceration and extensive destruction of the lid structures.

CORNEA AND SCLERA. The scleritis of late

acquired syphilis may take many forms, all of which are felt to be essentially gummatous in nature. Isolated gummas of the sclera do occur, though they are usually extensions through the sclera of a ciliary gumma.[11] Also, diffuse or localized scleritis or episcleritis may occur on the basis of gummatous infiltration. The response to penicillin is rapid.

The several corneal complications of late acquired syphilis may all be described as gummatous lesions, though the pathogenesis is obscure in some complications. All have in common a frequent association with anterior uveitis, and all are felt to be late allergic reactions to the treponeme. The first is the isolated corneal gumma, which is quite rare and is similar to isolated gummas of the conjunctiva and sclera, described above, appearing as a slowly enlarging nodular opacity anywhere on the cornea. Prompt diagnosis and antibiotic therapy are needed because of the tendency of the gumma to perforate the cornea.

Interstitial keratitis of late acquired syphilis is similar to the rare interstitial inflammation occurring in secondary infection mentioned above, although it is somewhat more frequent. The time from infection to onset is usually 8 years or longer. Unilateral and segmental involvement limited to the periphery is the rule, though occasionally, when associated with an intense iritis, the lesion may be fulminant, with severe residual scarring. An unusual complication of this latter form is the development of a posterior corneal abscess, which ruptures through Descemet's membrane into the anterior chamber. This is keratitis pustuliformis profunda, which is associated with a severe gummatous iridocyclitis, and has single or multiple peripheral infiltrates in the posterior layers of the stroma.[56] The uveitis is severe and is accompanied by marked pain, photophobia, and diminution of vision. The response to antisyphilitic therapy is prompt, the amount of disability depending on how long the process was allowed to continue before treatment.

A final form of syphilitic keratitis is the keratitis punctata profunda of Mauthner. Occurring in association with syphilitic iritis, it is characterized by multiple small, round, deep stromal opacities. There is little associated corneal inflammatory reaction, and with control of the iritis the opacities usually regress rapidly, leaving little or no scarring.[56]

Iris and ciliary body. Gummas of the ciliary body, though uncommon, have been reported in the literature regularly since the first description by von Arlt in 1853.[12] They may occur as an isolated phenomenon or, as in the usual case, in association with a severe iridocyclitis or anterior uveitis. They are usually first seen as they protrude into the anterior chamber at the angle, appearing as a pale gray or pinkish globular tumor. Untreated they go on to necrosis, greatly exacerbating the uveitis, with penetration of the globe and endophthalmitis as late sequelae. Primary iris gummas are a great rarity. The clinical appearance and course is similar to those of ciliary gummas.

The most common ocular inflammatory lesion of late acquired syphilis is chronic gummatous iritis, or more properly, uveitis, with an incidence of 3.1%.[40] It occurs as an isolated lesion only 25% of the time, complicated most often by an accompanying keratitis as mentioned above and slightly less frequently with an extensive posterior uveitis or chorioretinitis. The average time between infection and onset is 9 years. As in early luetic iritis, uveitis is more common in males than in females and is twice as prevalent among Negroes as it is among Caucasians. It is associated with extraocular manifestations of late syphilis in about one half the cases. Its onset is typically abrupt, with pain and redness of the eye and sudden decrease in vision. The slit lamp reveals a violent anterior chamber reaction, a copious fibrinous exudate, extensive mutton-fat precipitates on the corneal endothelium and anterior lens capsule, and occasionally hypopyon or hyphema. There is an early tendency to extensive posterior synechiae formation, though the cyclitis is often so severe as to cause a prolonged secondary ocular hypotension. Unlike the iritis of early syphilis, the tendency in late syphilis is for remission and relapse with progressive damage to the eye. Cataract, secondary glucoma, retinal detachment, and occasionally phthisis bulbi are the serious sequelae that occur in approximately 25% of all cases. Significant permanent visual loss occurs in over 50% of the patients.[40] The diagnosis may be difficult because of the nonspecific nature of the inflammation. The VDRL test is reactive in approximately 40% to 50% of the cases, though the FTA-ABS test should be positive in over 90%. The therapeutic response to antibiotic treatment is variable as in many other forms of ocular syphilis. Therapy may hasten the resolution of an acute episode and decrease the chances for recurrence, but late luetic uveitis is often refractory to penicillin and all other therapy, including atropine and steroids. Primary isolated gummas of the retina, choroid, and optic nerve have been reported but are so rare as to warrant only passing mention.

LENS. Although primary involvement of the lens in a luetic process is not known, secondary cataracts are common. In addition, lenticular dislocation, whether or not associated with trauma, is associated with a reactive FTA-ABS test in over 50% of the cases.[49] The pathologic mechanism is not known.[23]

Neurosyphilitic lesions of the eye

In all forms of neurosyphilis, ocular examination is a most valuable clinical tool for evaluating neurologic involvement. The following list* indicates the great frequency of eye signs in neurosyphilis.

Some syphilitic ocular complications	63%
Argyll Robertson pupils	35%
Irregular pupils	21%
Anisocoria	19%
Fundus changes	15%
Muscular paralysis	12%

*Modified from Stokes, J. H., Beerman, H., and Ingraham, N. R.: Modern clinical syphilology, Philadelphia, 1945, W. B. Saunders Co., p. 980.

The ocular lesions of neurosyphilis are usually limited to pupillary changes, extraocular muscle palsies, and retinal, optic nerve, and optic radiation involvement. Other ocular structures may be involved concomitantly in distinct luetic processes unrelated to the neurosyphilitic disease.

The pathologic process common to all types of neurosyphilis and indeed basic to the production of almost all syphilitic lesions is perivascular inflammation, with round cell cuffing of small vessels and a concomitant obliterative endarteritis. If this process is so slight as to cause changes in the CSF only, with no clinical manifestations of the disease, asymptomatic neurosyphilis is said to be present. If the involvement is chiefly limited to the meninges and the vessels of the central nervous system, the disease is called meningovascular syphilis; and if the pathologic changes involve primary destruction of the neural tissue of the brain or spinal cord, the process is labelled parenchymatous neurosyphilis. Often the types are not easy to distinguish, but as a general classification they have both clinical and pathologic validity.

Asymptomatic neurosyphilis, comprising about one third of all cases of neurosyphilis, differs from latent syphilis only in that the CSF is abnormal. There are no signs or symptoms by definition. The disease is important because it predisposes the patient to later serious CNS involvement. When the CSF is abnormal, it should be repeatedly studied after adequate therapy until it reverts to normal.

MENINGOVASCULAR NEUROSYPHILIS. The signs and symptoms of meningovascular neurosyphilis are referable to meningeal inflammation and vascular occlusion. As indicated above, they may have their onset as early as the late primary stage but usually occur several years after infection. Three major types recognized are cerebral meningeal, cerebral vascular, and meningovascular syphilis of the spinal cord.

Cerebral meningeal neurosyphilis, which may appear within the first 2 years of infection, is not common. The meningeal involvement may be diffuse or restricted to the convexity or basilar areas of the brain. The signs and symptoms relate to meningeal inflammation above the foramen magnum, secondary hydrocephalus, cranial nerve damage, and focal cortical injury. They include stiff neck, papilledema, diplopia, facial palsy, confusion, delirium, and seizures.

Cerebral vascular neurosyphilis comprises about 10% of all cases of neurosyphilis and appears, on the average, 7 years after infection. The common pathologic process is syphilitic arteritis. The signs and symptoms depend upon the particular vessel or vessels involved and are indistinguishable from cerebral thrombosis from other causes.

Spinal meningovascular syphilis is simply a grouping of those patients having signs and symptoms referable to vascular occlusion and meningeal involvement of the cord. Symptoms include peripheral motor weakness and sensory loss. In all forms of meningovascular syphilis, CSF changes and reactive serologic tests are the rule.

Cranial nerve palsies are so frequent in meningovascular syphilis as to be its most characteristic feature. Table 20-2 shows the relative frequency of individual cranial nerve involvement in the meningeal form of the disease.

Table 20-2. Cranial nerve involvement in meningeal neurosyphilis

Cranial nerve	Percent involvement
I	1.5
II	26
III	23
IV	2.5
V	11
VI	21
VII	40
VIII	41
IX and X	6
XI	0.5
XII	3.5

Adapted from Stokes, J. H., Beerman, H., and Ingraham, N. R.: Modern clinical syphilology, Philadelphia, 1945, W. B. Saunders Co., p. 998.

Of the nerves serving the extraocular muscles, the third nerve is most frequently involved, followed closely by the sixth, with the trochlear nerve being much less commonly affected. Acute ptosis is said to be the most frequent symptom.[58] The more important associated symptoms include severe headache, diplopia, and often sudden unilateral or bilateral deafness. Cranial nerve palsies may be isolated or multiple, depending on the location and extent of involvement. An interesting collection of signs and symptoms called Roger's syndrome is occasionally seen in which there is pain and hypesthesia over the distribution of the first two divisions of the fifth cranial nerve, associated with homolateral abducens paresis and Horner's syndrome. The causative lesion is an area of syphilitic meningeal inflammation localized to the tip of the petrous bone.[58]

Pupillary changes are common in meningovascular syphilis, with an incidence of approximately 30%. Argyll Robertson pupils are a rarity in the meningeal form and are only slightly more common in the vascular form. The typical pupillary changes are anisocoria and occasionally bilaterally dilated unresponsive pupils. This is easily understandable in view of the high frequency of third nerve involvement and the fact that the pupillomotor fibers run peripherally in the nerve and are thus subject to early involvement in meningeal disease.

As is seen in Table 20-2, the optic nerve is the most frequently involved cranial nerve, except for the facial and acoustic nerves. Primary optic neuritis probably does not occur in neurovascular syphilis. The changes in the optic nerve, chiasm, optic tract, and optic radiation are secondary to local inflammatory or infiltrative meningeal disease and vascular occlusion of small or large vessels and occur less commonly from increased intracranial pressure. In the rare acute meningeal form of this disease papilledema is quite common. Visual field changes occur frequently in both the meningeal and vascular forms. When the meningeal involvement is predominantly basilar, chiasmal and optic tract damage occurs, frequently leading to variable central scotomas and peripheral field changes and occasionally to loss of vision in one eye and peripheral field defects in the other. Hemianopsia occurs much less frequently and is usually homonymous, with bitemporal defects occurring in less than 20% of the cases.[58] In the vascular form field defects are common and are enormously variable. Merritt[35] reports that the middle cerebral artery or one of its branches is occluded in the great majority of cases. Thus, total homonymous hemianopsia is the predominant defect, although partial or quadrantic defects frequently occur, their congruity increasing with the proximity of the involved area to the occipital pole. Optic atrophy is unusual in meningovascular syphilis but occurs occasionally secondary to prolonged papilledema or to involvement of the intracranial portion of the optic nerve or chiasm, with scarring from extensive meningeal inflammation. The typical appearance of secondary optic atrophy includes blurring of disc margins and obliteration of the normal optic cup with proliferation of gray glial tissue. The response of the meningeal form to penicillin is usually rapid, with prompt remission of symptoms. The progress of the vascular form will be arrested with treatment, though the neurologic deficit produced by the occlusive disease is often permanent.

PARENCHYMATOUS NEUROSYPHILIS. In parenchymatous neurosyphilis the major changes are within the neural substance of the brain and spinal cord. The two major types are tabes dorsalis and general paresis.

Tabes dorsalis. Tabes dorsalis accounts for about one third of all cases of neurosyphilis. The major pathologic changes in tabes consist of degeneration in the optic nerves, posterior columns, and dorsal roots of the spinal cord, with signs and symptoms of optic atrophy, ataxia, lancinating pains, areflexia, bladder dysfunction, and Argyll Robertson pupils. The pathogenesis of tabes remains obscure. The period between infection and onset is much greater than for any other form of late syphilis, with an average of 15 years and a range of 5 to 50 years. Unlike paresis, in which many treponemes are found in the brain substance surrounded by profuse inflammation, in tabes treponemes are not present in the involved nerves and tracts, and extensive degeneration often occurs in the absence of any signs of inflammation. In addition, serum and CSF reagin titers are typically low. The CSF shows some abnormalities in 90% of the cases, with the VDRL test reactive in 75%. The progress of the disorder can usually be halted with the use of large amounts of penicillin.

Ocular complications of tabes are frequent and often severe. Extraocular muscle paresis occurs in 12% to 25% of the cases,[35,54] with the third nerve most often involved and less commonly the sixth and the fourth. Bilateral permanent ptosis is a frequent presenting sign, and when associated with pupillary signs, syphilis is the most probable diagnosis. Transient diplopia may be an early symptom and precedes the development of frank tabes by many years. Whether cranial nerve involvement in tabes is secondary to nuclear damage or to basilar meningitis is not known.

Pupillary abnormalities occur in over 90% of tabetic patients, with the incidence of Argyll Robertson pupils reported to be 48% to 80%.[35,54] Pupillary changes other than light-near dissociation include atrophy, fixed miotic pupils, anisocoria, irregular pupils, and dilated pupils unresponsive to light or accommodation. Pupillary atrophy in tabes may be secondary to previous luetic uveitis or may be neurogenic in origin. Neurogenic tabetic pupillary atrophy may occur concomitantly with pupillomotor paresis or light-near dissociation or may occur as an independent phenomenon. The iris atrophy may be generalized or segmental. There is usually depigmentation and stromal atrophy with marked thinning and loss of normal surface details, all resulting in a grayish, smooth iris often with a small irregular pupil.

The Argyll Robertson pupil has long been regarded as a pathognomonic sign of late CNS syphilis, and indeed in over 90% syphilis is the cause of this abnormality.

Four criteria must be found for the complete Argyll Robertson pupil syndrome to be present.

1. No reaction to light, either by direct or consensual stimulation
2. Intact near synkinesis
3. Miosis
4. Enough light perception to assure that loss of the light reflex is not on the basis of amaurosis

Poor response to mydriatics has often been added as a fifth criterion because Argyll Robertson[45] in his original description of four patients noted that there was poor dilatation from atropine in all cases. Loewenfeld in her recent elegant and extensive review of the subject has shown that poor response to mydriatics is a feature that, though often present, is not related to the neurologic defect but is secondary to the frequently associated iris atrophy.[32] Several other observations that have importance in the Argyll Robertson syndrome as a diagnostic sign of neurosyphilis were made by Loewenfeld. Perhaps the most important suggestion, supported by other authors,[49] is that if the criterion of absolutely no response to light is strictly adhered to, many cases of neurosyphilis will be missed. The major feature of the phenomenon is light-near dissociation, and even in the presence of a "fairly ample" response to light a partial Argyll-Robertson syndrome exists if a "clear-cut dissociation" can be demonstrated.[32] The partial syndrome is just as important a sign of neurosyphilis as is the fully developed form in which there is absolutely no movement of the pupil to light. It must be kept in mind, however, that an intense light source must be used before light-near dissociation can be said to exist. The phenomenon is bilateral in the majority of cases but may exist as a strictly unilateral phenomenon. Loewenfeld also mentions the fact that though small pupils are the rule, the intense meiosis is not a strict criterion. In bright light the Argyll Robertson pupil may actually be larger than a normal one; however, in dim illumination the normal pupil promptly dilates, while the Argyll Robertson pupil remains stationary.[32]

Loewenfeld suggests that the site of the lesion responsible for the Argyll Robertson syndrome is a small area immediately rostral to the Edinger-Westphal nucleus where the intercalated light reflex fibers from the pretectum and supranuclear inhibitory pathways from the cortex are in close proximity.[32] Interruption of the intercalated light reflex fibers would abolish the homolateral light reflex, and interruption of the inhibitory pathways from above would account for the meiosis. A lesion confined to this site would spare the near reflex fibers and the oculomotor outflow, which are situated ventrally.

Several other conditions have been reported to cause the Argyll Robertson syndrome; these include diabetes, multiple sclerosis, chronic alcoholism, encephalitis, mesencephalic tumor, and trauma. However, as neurosyphilis causes over 90% of the cases, great effort must be made to exclude that diagnosis before another process is accepted as the cause. An entity that may easily be confused with the Argyll Robertson syndrome is the tonic pupil of Adie. It reacts so slowly to light that a mistaken diagnosis of Argyll Robertson syndrome may be made on brief examination. Adie's pupil is unilateral in 80% of the cases, is usually dilated rather than meiotic, contracts slowly to light, and constricts promptly with a dilute (2.5%) solution of methacholine (Mecholyl).[46]

Optic atrophy may be the only manifestation of CNS syphilis; it is the most common ocular sign of tabes after pupillary changes. It occurs in 15% to 20% of all tabetic patients[22] and is the most serious ocular complication of neurosyphilis, resulting in total blindness in the great majority of untreated cases.[58] It is more common in men than in women and in Caucasians than in Negroes. Woods[61] proposed a luetic cause in 23.3% of all optic neuropathies and in 51% of all cases of primary optic atrophy. Although this may be a bit higher than is acceptable today, syphilis remains the most common cause of optic atrophy, and serologic studies are indicated in every case.[5] Tabetic optic atrophy is primary in type and is called "interstitial neuritis" by Merritt.[35] The time from primary infection to development of optic atrophy is quite long, averaging 24 years.[41] It does not result from syphilitic arachnoiditis or basilar meningeal inflammation but is a primary process occurring in the intracranial portion of the nerve anterior to the chiasm. Loss of myelin sheaths occurs early, followed by degeneration of nerve fibers. Glial scarring occurs late in the disease.

There are three general patterns of degeneration of the optic nerves resulting in distinctive visual field changes. The first is progressive concentric atrophy, which arising at the periphery of the nerve results in slow, progressive contraction of peripheral fields and terminates in extremely small central fields often with surprisingly good acuity. The second form is degeneration of wedge-shaped sectors of the nerve, giving rise to localized peripheral scotomas, often with normal central vision unless the papillomacular bundle is involved. The third form is diffuse degeneration of the entire thickness of the nerve, causing central, paracentral, and centrocentral scotomas, associated with peripheral contracture. Altitudinal, binasal, bitemporal and homonymous defects may rarely be produced. The clinical appearance includes a slowly developing bluish pallor of the discs, with no associated fundus lesions or other evidence of inflammatory disease. The margins are quite sharp, and the cup may have a slight atrophic accentuation. Attenuation of the vessels occurs in the late stages, perhaps from glial proliferation and constriction of the vessels centrally. Prognosis must be guarded in every case, for despite antisyphilitic therapy, significant visual loss is common. Treatment must be prompt and vigorous. In a series of 249 untreated cases reported by Moore and Woods[41] all patients lost their vision within 9 years after onset of symptoms.

Paresis. General paralysis of the insane, or paresis, is chronic syphilitic meningoencephalitis, resulting in diffuse extensive parenchymal damage to the brain. The clinical manifestations are enormously variable. Common signs include facial and tongue tremors, dysarthria, hyperactive deep tendon reflexes, and pupillary changes. Early symptoms are usually quite nonspecific, often including minor personality changes with progress within months to psychosis and dementia. It is a fatal disease if untreated, the late stages made manifest by dementia and paralysis. The spinal fluid is almost invariably indicative of infection, and the VDRL test is reactive in the blood and spinal fluid in approximately 90% of all cases.

Extraocular muscle palsies and optic atrophy are rare in the pure form of paresis, though in the mixed disorder, known as taboparesis, both occur not infrequently. Pupillary changes are the most common ocular manifestation of paresis. The list below indicates the frequency of various pupillary changes.*

Altered light reflex	78%
Anisocoria	68%
Lost light reflex	52%
Irregular pupils	39%
Diminished accommodation reflex	17%
Mydriasis	13%

As can be seen in the list, complete loss of light reflexes is present in 52%, but some diminution of light reflex can be detected in 78%, thus emphasizing the importance of the partial Argyll Robertson syndrome, or light-near dissociation, in making the diagnosis of CNS syphilis.

CONGENITAL SYPHILIS

Heredosyphilis is in the main quite similar to acquired syphilis, although there are

*Adapted from Stokes, J. H., Beerman, H., and Ingraham, N. R.: Modern clinical syphilology, Philadelphia, 1945, W. B. Saunders Co., p. 1002.

enough differences to warrant a separate description. The fetus is infected transplacentally, and as in early acquired syphilis a generalized spirochetemia rapidly ensues, with all tissues of the body invaded by treponemes. Symptoms may be present at birth, or it may be several years before the infection becomes manifest.

There are two clinical forms of congenital syphilis: infantile and tardive.

Early congenital syphilis

Infantile, or early congenital, syphilis makes its appearance within the first 2 years of life. The skin lesions are usually infectious and correspond to secondary lesions in the acquired form. The signs and symptoms of early congenital syphilis include skin and mucous membrane lesions, widespread osteoperiostitis, anemia, and hepatosplenomegaly. The earlier the infantile disease becomes manifest, the poorer are the infant's chances for survival. If symptomatic at birth, the infant is usually desperately ill and often does not survive despite intensive therapy. As both reagin and treponemal antibodies cross the placenta, the newborn will have a reactive serologic test if the mother does. A diphasic VDRL titer over the first 3 months of life is diagnostic of infection in the newborn as the infant clears the passively acquired maternal antibodies and begins to produce his own. After the first few months of life the serologic tests are typically strongly reactive in early congenital syphilis. As in early acquired syphilis the CSF must be evaluated, for a significant percentage will develop asymptomatic neurosyphilis.

Ocular lesions of early congenital syphilis are uncommon and are even less commonly observed or reported because those that do occur are predominantly intraocular and therefore rarely visualized and recognized in the infant. The lids and conjunctiva may occasionally be involved in the generalized cutaneous eruptions. These lesions are infectious and are similar to the well-known rash of secondary syphilis, except for the occurrence of bullae, which are peculiar to congenital syphilis and are not seen in the acquired disease.

The most common cause of acute iritis in infancy is reported to be early congenital syphilis.[12] Reliable estimates of its incidence are not available, and it is probable that the majority of cases escape diagnosis unless the iritis is quite severe. Typically, the iritis begins late in the first year of life, with a variable outcome; it responds well to antisyphilitic therapy.

Interstitial keratitis and anterior uveitis may also occur within the first 2 years of life and have even been reported to be present at birth,[11] but these are distinct rarities. Extraocular muscle palsies and secondary optic atrophy are each thought to occur in approximately 1% of the cases.[54] The two important ocular lesions of infantile heredosyphilis are optic neuritis and chorioretinitis.

Carroll reported that in a series of 41 cases of optic neuritis in children 2 were caused by congenital syphilis; the incidence of optic neuritis among congenial syphilitic patients has been reported to be approximately 5%.[29] The clinical appearance is in no way different from the optic neuritis of the acquired disease and if untreated may progress to secondary optic atrophy in a small percentage of cases.

Although the chorioretinitis of congenital syphilis is usually not discovered until after the first several years of life, it is thought to begin during infancy.[25] Its incidence among congenital syphilitic patients is approximately 14%, and it accounts for over 80% of all cases of chorioretinitis in children.[58] There is a much higher incidence of chorioretinitis, especially the peripheral form, in congenital syphilitic patients with interstitial keratitis. The course of the acute disease is probably similar to that of the chorioretinitis of acquired syphilis; but since chorioretinitis is only rarely observed in its active stages, this hypothesis is speculative. When discovered, the chorioretinitis is usually old and inactive; and though it may take many different forms, there are some morphologic characteristics that are helpful in establishing the cause. The typical form of heredosyphilitic chorioretinitis is the salt-and-pepper fundus in which the changes are predominantly confined to the peripheral retina. Examination of the fundus reveals granular areas of diffusely distributed fine dots of pigment interspersed among similarly sized pale yellow lesions on a reddish yellow background. The process, which is invariably bilateral, may be segmental or may involve the full circumference of the periphery and is associated with perivascular sheathing and narrowing of both arteries and veins over and around the disc in a high percentage of cases.[25] There may occasionally be an accompanying pallor or frank atrophy of the disc. An important associated finding is the frequently occurring fine punctate haze in the anterior vitreous. Variations on the typical configuration occur frequently. Isolated, sparse pigment clumps of varying size scattered from the midzone to the periphery may be the only manifestation of a previous chorioretinitis. Scattered peripheral yellowish red areas of chorioretinal atrophy with sparse central or peripheral pigment depositions are seen less frequently; and the typical retinitis pigmentosalike fundus with midzone pigmentary degeneration, prominent bone spicule pigment forms, vascular atresia, and beeswax pallor of the disc is now felt to be a rare manifestation.[25] A mixture of the various forms is seen in the majority of cases. The diagnosis of syphilis is suggested by the predominantly peripheral location of the involvement, the occurrence of pepper-and-salt changes alone or in association with other types of lesions, and finally by the presence of perivascular sheathing, especially at the posterior pole.[25] The syphilitic cause is best determined by a history of syphilis in the parents, corneal changes of interstitial keratitis, or other stigmata of heredosyphilis rather than by any specific fundus abnormality. The VDRL test may be nonreactive in 25%, though the FTA-ABS test will be positive over 90% of the time. Vision is often unimpaired, but optic atrophy or macular involvement may cause severe visual loss in a small percentage of cases.

Tardive heredosyphilis

Late, or tardive, congenital syphilis is that which has persisted past the first 2 years of life. In over half the cases it is asymptomatic, detectable only serologically. It roughly corresponds to the late forms of acquired syphilis and is noninfectious. A large percentage of late congenital syphilitic patients have one or more stigmata of the disease without ever developing active late destructive lesions. Typical stigmata are caused by osteoperiostitis, the hallmark of congenital syphilis. These lesions may be hypertrophic and sclerotic with frontal bossing, saber shins, and clavicular swellings, or destructive, as in the saddle nose deformity and palate perforations. Various ectodermal defects are common, including dental dystrophies, hypertrichosis, and perioral scarring. Approximately 26% of untreated late congenital syphilitic patients will develop neurosyphilis,[54] usually paretic in type. In addition, isolated neural deafness is seen in 10%, and interstitial keratitis is seen in 52%.[54] Gummas, especially of bone and nasopharynx, are common serious late sequelae. Late congenital cardiovascular syphilis is a great rarity. Serologic tests in untreated late congenital syphilis are unusual, with great fluctuations in reagin titer when the VDRL test is reactive. In 25% of the cases the VDRL test is nonreactive. The FTA-ABS test is reactive 96% of the time in late congenital syphilis. Again, a CSF evaluation is imperative.

Ocular complications occur in over 75% of patients with late congenital syphilis.[54] Gummatous involvement of the lids, conjunctiva, and orbit occurs in a small percentage of patients with late heredosyphilis and is similar to the acquired disease in all respects, except that obstructive gummatous dacryocystitis is much more common in congenital syphilis, with an incidence of 2%.[54] Chronic late uveitis occurs in 7% to 10% of the cases and corresponds to late acquired luetic uveitis. The ocular signs and complications of the various forms of congenital neurosyphilis are no different from those occurring in the corresponding acquired forms. Paresis accounts for one fifth of all cases of congenital neurosyphilis. Almost all congenital paretic patients have abnormal pupils, and light-near dissociation is frequent. Optic atrophy, the most serious ocular complication of syphilis, occurs in 6% of all cases of congenital syphilis.[29] Pure tabes dorsalis is an uncommon complication of heredosyphilis, accounting for about 10% of all cases of hereditary neurosyphilis. It is typically mild in its manifestations, except for its ocular complications. Essentially all have abnormal pupils; extraocular muscle palsies and progressive primary optic atrophy each occur in 50% of the cases.[35] Treatment of congenital neurosyphilis is the same as for the acquired disease, though the results are often less satisfactory, especially in juvenile tabes.

Interstitial keratitis is the most common lesion of late congenital syphilis, occurring in approximately one half of all untreated patients. Although it may not appear until the patient is 30 years old, and although it has also been reported to be present at birth, interstitial keratitis typically occurs in the early teens, with the vast majority of cases occurring in patients between the ages of 5 and 20. It is more common in females than in males, with a ratio of 3:2. Approximately 95% of all cases of interstitial keratitis is caused by syphilis, the congenital form accounting for over 90%, whereas only 3% is seen in the acquired disease. Mycobacteria and other bacteria rarely cause a similar keratitis. Syphilis is the cause of such an overwhelming majority of cases that it must be carefully excluded before another cause can be considered. The pathogenesis of the disease is not understood, but it is felt to be a manifestation of allergy to the treponeme. Supporting the theory of allergic reaction are the facts that it is quite rare in the early stages of both congenital and acquired syphilis where there are great numbers of treponemes throughout the tissues and in the general circulation, that it occurs approximately 10 years after infection in both the congenital and acquired disease at a time when treponemes are rarely found in the blood or tissues, and that it is refractory to any antiluetic treatment. However, it responds well to the corticosteroids. It is felt that the cornea is sensitized to treponemes during the initial spirochetemia following infection and becomes the site of an acute anamnestic allergic response years later with the release of treponemes into the general circulation. Why there should be a release of treponemes years later has never been adequately explained. It is a bilateral disease in most untreated cases; initially one eye may be involved with an interval of weeks to months before the second eye becomes affected.

The disease begins typically with an initial infiltrative stage, characterized by decreased acuity, redness, pain, lacrimation, photophobia, and blepharospasm. Slit-lamp examination reveals a rapidly progressing stromal haze, which may begin peripherally, centrally, or in scattered patchy areas. Within a few days the stroma is involved in all of its layers, with a diffuse infiltrate resulting in a generalized ground-glass haze and visual acuity often reduced to hand movements or light perception. In practically every case a severe iritis or anterior chorioretinitis develops concomitantly with the initial stages. It is severe, with turbid aqueous, extensive plastic keratic precipitates on the endothelium, and often with ocular hypotony indicative of ciliary body involvement. Widespread necrosis of the corneal lamellae may occur, with stromal edema and folds in Descemet's membrane occurring as prominent secondary changes. New blood vessels make their appearance in all levels of the stroma at the periphery, heralding the stage of vascularization. The new vessels grow slowly inward from the periphery over a period of months; during this time the eye remains acutely inflamed, red, painful, and functionally blind. This invasion of the cornea, with inward progression of straight, radially oriented vessels at all stromal levels, forming rich anastomoses in the areas of denser stromal haze and necrosis, is characteristic of the disease. The central ends of the vessels branch out to form fine brushlike vascular structures. When the vessels finally meet axially, there is a dramatic regression of the disease; and the patient enters the final stage of the process, in which the corneal infiltrates are resorbed and the vessels are obliterated. This final process may take many months, the entire course lasting 1½ to 2 years. There is often a remarkable repair and clearing of the cornea. The vessels never disappear completely but become atretic and persist as fine lines or ghost vessels. Patches of deep stromal opacities may remain and if central can significantly reduce vision. The folds in Descemet's membrane may persist to form typical hyaline ridges, which are most often vertical or radial in orientation. Permanent linear or concentric splits in the membrane occur, as do localized areas of thickening where extensive keratic precipitates have become organized into plaques, often with small flecks of inspissated pigment. Rarely are the changes so severe as to lead to eventual degeneration of the cornea, with loss of substance, band keratopathy, persistent discomfort, and eventual perforation as the final outcome. Intractable glaucoma occasionally may also occur, secondary to severe iritis.[31] In general, however, the outlook is surprisingly good for vision, with 70% of these patients having 20/20 to 20/100 acuity and only 10% having poorer than 20/200 acuity.[11]

Without treatment the disease recurs in 27% of such cases, and eventual bilateral involvement is the rule.[54] Although specific antisyphilitic therapy does not greatly affect the course of the acute inflammation, it does reduce the chances for recurrence in the same eye to 3.6%; and when instituted during the involvement of the first eye, therapy significantly decreases the occurrence of bilateral disease.[54] Topical steroids have a definite place in the management of the acute process. They only suppress the inflammatory response, which will reappear upon cessation of therapy. If steroids are used continuously for 1½ to 2 years, the disease may run its course suppressed; and the severe sequelae will be avoided.[11] The most effective treatment, of course, is prevention by treating the infected mother during her pregnancy. If this is accomplished, the later development of active interstitial keratitis in the child is a rare phenomenon.

• • •

Before some of the recent research in ocular syphilis is discussed, a summary of the more important ocular lesions of syphilis and an outline of antibiotic therapy is presented.

The following list indicates the reactivity of the FTA-ABS test in patients with various types of ocular disease seen in a large municiple eye clinic.[47]

Any pupillary abnormality	46%
Pupillary light-near dissociation	67%
Interstitial keratitis	100%
Chronic uveitis	30%
Primary optic atrophy	67%
Dislocated lens	53%

The importance of any reactivity of the FTA-ABS test, including a borderline positive reaction, has recently been emphasized by Seeley and his associates[46a] as a significant indicator of luetic involvement. Although the FTA-ABS test is the most sensitive and specific serologic test that is available for syphilis, great efforts should be made to rule out syphilis in a patient having any one of the above findings, even in the face of a negative FTA-ABS test. Despite reports to the contrary, Smith[49] estimates that the FTA-ABS test is nonreactive in 40% to 50% of all cases of tardive heredosyphilis. In addition, the possibility of syphilis should be carefully evaluated in a patient with any of the following ocular findings that are not obviously caused by another process.

1. Pigmentary chorioretinal changes, especially of the salt-and-pepper type or if predominantly anterior in location
2. Acute iritis
3. Marked visual field changes, particularly extensive contracture of field
4. Ptosis or other extraocular muscle palsy
5. Significant active perivasculitis or old perivascular sheathing

Penicillin is the drug of choice in the treatment of all stages of syphilis. Tetracycline and erythromycin have also been extensively evaluated in early syphilis and have been found to be effective substitutes when penicillin cannot be used. The following dosage schedules are recommended by the Venereal Disease Program of the National Center for Disease Control.[55]

Primary, secondary, and latent syphilis

Benzathine penicillin G—2.4 million units IM at one session

Procaine penicillin G with 2% Aluminum Monostereate in Oil (PAM)—4.8 million units IM total dose; give 2.4 million units initially, followed by 1.2 mμ given twice at 3-day intervals

Aqueous procaine penicillin G (APPG)—4.8 million units IM total dose, given in eight 600,000 unit daily injections

Tetracycline HCl—total oral dose of 30 to 40 gm, given in 4 equally divided doses daily over a period of 10 to 15 days

Erythromycin—same as for tetracycline

Congenital syphilis

If patient is an adult, then treatment is the same as for late syphilis. If patient is a child, then use:

Benzathine penicillin G—50,000 μ per kg of body weight in single treatment

APPG—100,000 μ per kg of body weight, given in 10 equally divided daily doses

Late syphilis (includes asymptomatic and symptomatic neurosyphilis and cardiovascular and gummatous syphilis)

Benzathine penicillin G—6 to 9 million units IM total dose, given as 2 or 3 injections of 3.0 million units each given 7 days apart

PAM—6 to 9 million units IM total dose, given as repeated injections of 1.2 million units at 3-day intervals

APPG—6.0 to 9.0 million units total IM dose, given in daily injections of 600,000 units

Tetracycline and erythromycin—total oral dose of 60 to 80 gm, given in equally divided 4 doses daily over a period of 20 to 30 days

If the above treatment schedules are adhered to, success in treating syphilis in the infectious stages is reported to be effective in over 95% of the cases,[48] with rapid disappearance of signs and symptoms and a return of the serum to a nonreactive state when tested by reagin and treponemal tests. Treatment of late symptomatic syphilis has proved to be somewhat less effective. Adequate treatment does eradicate all signs of active clinical infection and arrests the progress of the disease in the great majority of cases. However, it has long been recognized that clinical progression of ocular and central nervous system syphilis occasionally occurs despite intensive antibiotic therapy. In addition, permanent sequelae of late symptomatic syphilis secondary to the irreversible damage of tissues before treatment are not ususual. Finally, measurable levels of specific antitreponemal antibodies typically persist indefinitely, and detectable levels of reagin may sometimes remain.

Despite the occasional exceptions, penicillin has over the years continued to have undiminished treponemicidal effects; no strain of *T. pallidum* has yet been reported to be resistant to penicillin, and its clinical efficacy in early syphilis has remained unchallenged. What, then, is the significance of the recent work indicating persistence of treponemes in human beings and in laboratory animals that have received supposedly adequate penicillin therapy?

It should first be emphasized that this is not a new finding. In 1945 Levaditi and Vaisman[30] reported the persistence of treponemes in rabbits, following both penicillin and heavy metal therapy. This work did not stimulate much interest, because effective therapy was determined in the main on clinical grounds alone and because sensitive techniques for detecting the presence of low concentrations of treponemes in tissues were lacking. The decreased effectiveness of penicillin in curing previously untreated cases of long-standing syphilis and the persistence of reactive specific treponemal tests in these patients following supposedly adequate penicillin therapy prompted a French worker, Pierre Collart, to search for treponemes in syphilitic patients and in rabbits treated with penicillin. In 1962 he reported finding viable treponemes in rabbits and human beings treated with massive doses of penicillin (one rabbit had received 9 million units of penicillin per kg of body weight, and a syphilitic patient had been treated with 151,000,000 units of the drug).[8] He was also able to demonstrate the infectivity of these persistent organisms by passive transfer to rabbits. He concluded that:

1. Therapy administered long after contraction of syphilis cannot destroy the Treponema that have been longtime residents in the host body.
2. The presence of spirochetes is the determining factor in the persistence of antibodies in the blood serum.
3. The Treponema discovered were still viable, but had lost all or part of their virulence in the tissues in the state of commensal organisms.
4. Under biological conditions still not ascertained, these organisms can revert and become pathogenic, at least to the host that shelters them.

   Hence the behavior of syphilis can be likened to that of other long term infectious diseases.[8]

Because of penicillin's great clinical value and of the fact that Collart's technique for locating the sparse organisms entailed significant technical difficulty in recognizing persisting organisms with altered morphologic characteristics, his work was doubted by other workers in the field, especially in this country where penicillin's effectiveness had been discovered. His results were confirmed by the work of Del Carpio[10] in 1963 and Boncinelli[4] in 1966, both in Italy. At the Venereal Disease Research Laboratories Yobs and her associates[63] in 1968 reported finding treponemes in lymph nodes from 11 of 45 TPI-reactive male volunteers who had been treated for syphilis in the past. By passage to animals these organisms proved to be infectious treponemes. After retreatment 3 out of 5 patients continued to harbor treponemes.

In 1967 Smith and Israel[51] reported finding motile treponemes in aqueous humor, spinal fluid, and liver biopsy specimens from 4 patients with signs and symptoms of late syphilis and negative reagin serologic tests. Three of the four patients had reactive or weakly reactive FTA-ABS tests but completely normal spinal fluids. The fourth patient's serum and spinal fluid remained nonreactive to VDRL, TPI, and FTA-ABS tests on repeated occasions, though his spinal fluid did exhibit an elevated protein level. The organisms from all of the patients took up the specific antitreponemal fluorescent antibody stain and fluoresced on ultraviolet darkfield microscopy (FADF). In another paper published a month later Smith and Israel[51] further described the findings in one of the four original patients. The patient had bilateral anterior uveitis and neuroretinitis. She gave no history of syphilis and had repeatedly negative TPI tests but had weakly reactive VDRL and FTA-ABS tests on several occasions. The organisms were motile, morphologically identical to *T. pallidum,* took up the specific fluorescent antibody stain, and were found in the aqueous both before and after treatment with 9,200,000 units of benzathine penicillin G. The two reports not only confirmed the earlier work of Collart but reaffirmed the position of many of the earlier syphilologists that active and progressive late syphilis, especially ocular syphilis and neurosyphilis, could exist in the presence of negative reagin tests and normal spinal fluid examinations.

The presence of intraocular treponemes was not a new finding,[1,2] but the use of the highly specific FADF technique for aqueous examination and the intraocular presence of treponemes after therapy had not been reported previously. Confirmatory reports soon appeared. Goldman and

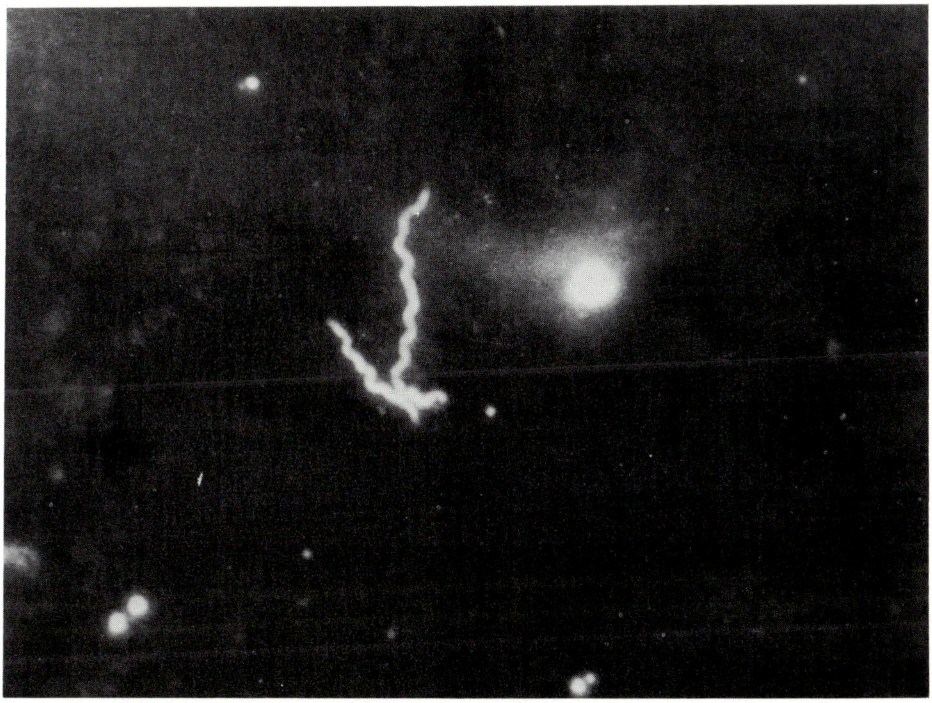

Fig. 20-2. Darkfield preparation of aqueous fluid from patient with uveitis. These spirochetes reacted specifically with fluorescein-labeled rabbit antibody to *T. pallidum*.

Girard[17] reported finding motile, FADF-positive treponemes in the aqueous of 2 congenital syphilitic patients who had been repeatedly treated with penicillin. Christman and his co-workers[7] found similar organisms in 12 of 36 patients thought to have ocular syphilis. Six of the 12 had uveitis. Intensive therapy of these 6 patients, using ampicillin and probenecid, a regime reported to produce significantly elevated intraocular levels of the treponemicidal antibiotic,[16] resulted in the temporary resolution of the uveitis, with concomitant disappearance of the intraocular treponemes in 5 of the patients; 2 of the 5 patients had subsequent recurrence of uveitis and treponemes. Dunlop and his associates[13] and Rice and his associates[44] found treponemes in the aqueous humor of 5 of 9 patients having late syphilis and ocular inflammatory disease. More recently, Whitfield and Wirostko,[59] in a controlled study of 50 patients with uveitis of unknown cause, found 6 to be harboring organisms morphologically indistinguishable from *T. pallidum*. Fig. 20-2 is a microphotograph of organisms found in the aqueous of one of these patients. The organisms from 5 of the 6 reacted specifically with fluorescein-labeled anti–*T. pallidum* rabbit globulin. No organisms resembling *T. pallidum* were found in aqueous specimens obtained at cataract surgery from a control group of 50 patients having senile cataracts, negative serologic tests, and no other clinically evident ocular disorder. Three of the 6 patients harboring treponemes had weakly reactive VDRL or FTA-ABS tests. Serologic tests on the remaining 3 were repeatedly nonreactive. Subsequent treatment of 3 of the 6 patients with a full course of ampicillin and probenecid had no clinically evident effect on the course of their disease.[60] Rice and his co-workers[43] found treponemelike forms in the aqueous or spinal fluid in 33 of 75 patients. Eight of the 33 had previously been treated for syphilis.

The above reports have not been accepted without controversy. In 1969 Montenegro and his associates[38] presented an interesting report of the various artefacts that may closely mimic treponemes on darkfield examination. Problems with artefacts on darkfield examination are certainly not new, Stokes wrote in 1944: "Certain artefacts in the darkfield have caused inexperienced observers a good deal of difficulty at times."[54] For an experienced darkfield microscopist artefacts are a nuisance but present no major problem. In the FADF test, use of identically processed known *T. pallidum* controls for comparison, scrupluously cleaned slides, and frequent alternation of white light with an ultraviolet source ensure that the microscopist will have no difficulty distinguishing organisms from artefacts.

It has also been suggested that the treponemelike forms are not *T. pallidum* but are saprophytic organisms, perhaps from the mouth.[15] There are some data to support the contention that in a small percentage of cases some gingival treponemes may fluoresce on FADF testing.[18] However, the specificity of the fluorescent antibody staining technique has been extensively investigated, and it has been confirmed that when properly used, the FADF test has a very high degree of specificity.[37] It is entirely possible that organisms that have been indistinguishable morphologically from *T. pallidum* and that take up the specific anti–*T. pallidum* fluorescent stain but that are other than *T. pallidum* have on occasion been found in the eye and elsewhere, but it is highly unlikely that this has occurred in more than a small percentage of cases. Typical morphologic characteristics, specific antibody staining, and on occasion typical motility on wet darkfield examination have been repeatedly found; and all support the thesis that *T. pallidum* does persist in the aqueous humor, spinal fluid, and other tissues after treatment.

Perhaps the strongest evidence that these organisms are *T. pallidum* has come from successful passive transfer to laboratory animals. Collart, Boncinelli and his associates, and Yobs and her associates all successfully infected rabbits, using the persisting treponemes they found in the lymph nodes of previously treated rabbits and human beings. Hardy and his co-workers[19] found pathogenic *T. pallidum* in the ocular tissues of a newborn who had been unsuccessfully treated for congenital syphilis with huge doses of penicillin, and they were able to infect a rabbit with the material. In his extensive monograph on the subject of seronegative late syphilis and in other publications Smith has reported successfully infecting rabbits and primates with syphilis on multiple occasions, using specimens of aqueous humor and spinal fluid from patients with treated syphilis.[49,52,53]

There appears to be substantial evidence that *T. pallidum* may persist in the eye and other tissues following even massive antibiotic therapy, especially in late ocular and central nervous system syphilis. This probably has not occurred as the result of the development of penicillin-resistant strains, but rather by using new and more sensitive techniques we have become aware of a process that is integral to the disease. It is likely, as Collart suggests, that a persistent reactive serologic test following treatment indicates the continued presence of treponemes stimulating antibody production. There is no reason to think that syphilis is uniquely different from other infectious diseases nor that treatment with penicillin can completely eradicate all treponemes. All that is needed is antibiotic assistance to eradicate enough organisms to allow the host's protective apparatus to take over. The relative isolation of the eye and central nervous system from the general circulation is likely quite important in the occasional persistence of treponemes in the aqueous humor and spinal fluid following treatment.

We also have good evidence that treponemes may persist and that active ocular syphilis or neurosyphilis may be present despite negative reagin and treponemal tests of the blood and spinal fluid.[52]

Yobs and her associates[63] and Collart[8] point out that the persisting organisms often have lowered virulence and modified antigenic structure as a result of their long and intimate association with the host tissues, perhaps as a direct effect of the host's antibodies developed during that association. It is certainly a possibility that the antigenic structure of the treponemes might become so altered that it no longer stimulates measurable treponemal or reagin antibody production in the host nor strongly takes up specific antibody stains itself. This is a possible explanation for both the phenomenon of late seronegative syphilis and for the presence of variably staining organisms occasionally found in late ocular syphilis, especially in late congenital syphilis.[18]

The major import of this recent finding is to underscore strongly the old dictum that the diagnosis of syphilis is a clinical one. Although the myriad new and complex laboratory tests are often of great assistance and their intelligent use is important, the disease must first be considered in the differential diagnosis, and the appropriate tests must be decided upon. There is no substitute for careful history and physical examination. The FTA-ABS test is the most sensitive and specific serologic procedure available for the diagnosis of syphilis and should be used whenever the diagnosis is considered; yet it has its occasional false-negative results, and false-positive results may also, although rarely, occur. Fresh darkfield and FADF studies of the aqueous can be quite helpful in difficult cases, though the procedures are technically demanding.

The treatment of early syphilis, using the currently accepted modes of therapy, achieves excellent clinical results; and there are no indications to modify it. The treatment of late syphilis continues to be unsatisfactory, though the above guidelines have proved to be effective in the great majority of cases. There appears to be good evidence for the persistence of pathogenic *T. pallidum* in a small percentage of patients thought to have received adequate treatment. Efforts to determine the effects of the persistent organism on the eye and other tissues and to develop appropriate effective therapy must continue. In such cases a trial of intensive retreatment is certainly indicated. Meyer, Boudin, and Collart[36] recently suggested using 1,000,000 units of aqueous penicillin G daily for 30 days. Other regimens, including those utilizing antibiotics alternated with intensive steroid therapy, are being tried; but these remain in the experimental stage. The best treatment of the complications of late syphilis continues to be prevention, in the form of epidemiologic tracing of contacts of infectious individuals, early diagnosis, and adequate therapy of the disease in its early stages.

REFERENCES

1. Agarwal, L. P.: Aqueous humor examination in granulomatous iridocyclitis, Docum. Ophthal. **14**:97, 1960.
2. Amsler, M., and others: L'Humeur aqueuse et ses fonctions, Paris, 1955, Masson et Cie.
3. Balentyne, H. J., and Michaelson, I. C.: The fundus of the eye, Baltimore, 1963, The Williams and Wilkins Co., p. 564.
4. Boncinelli, U., and others: Richerche sulla persistenza del Treponema pallidum, nelle linfoghiandole di luetici trattati, G. Ital. Derm. **107**:1, 1966.
5. Carrol, F. D.: Optic atrophy, Trans. Pacif. Coast Otoophthal. Soc. **31**:36, 1950.
6. Carrol, F. D.: Retrobulbar neuritis, Arch. Ophthal. **24**:44, 1970.
7. Christman, E. H., Hamilton, R. W., Heaton, C. L., and Hoffmeyer, I. M.: Intraocular treponemes, Arch. Ophthal. **80**:303, 1968.
8. Collart, P.: Persistence of T. pallidum in late syphilis in rabbits and humans, notwithstanding treatment: 1962 proceedings of the World Forum on Syphilis and Other Treponematoses, U. S. Public Health Service Publication No. 997, Washington, D. C., 1964, U. S. Government Printing Office, pp. 289, 291.
9. Deacon, W. E., Falcone, V. H., and Harris, A.: A fluorescent test for treponemal antibodies, Proc. Soc. Exp. Biol. Med. **96**:477, 1957.
10. Del Carpio, C.: Effetti di un trattamento massiccio con penicillina nella sifilide sperimentale de coniglio a diversa distanza dall'

infezione, Riv. Ist. Sieroter. Ital. **38**:166, 1963.

11. Duke-Elder, S.: System of ophthalmology. Diseases of the outer eye, vol. 8, St. Louis, 1965, The C. V. Mosby Co., pp. 237, 239, 539, 828, 829, 1032.
12. Duke-Elder, S.: System of ophthalmology. Diseases of the uveal tract, vol. 9, St. Louis, 1966, The C. V. Mosby Co., pp. 300, 306, 307.
13. Dunlop, E. M. C., King, A. J., and Wilkinson, R. E.: Study of late ocular syphilis: demonstrations of treponemes in aqueous humor and cerebrospinal fluid. III. General and serological findings, Trans. Ophthal. Soc. U. K. **88**:275, 1968.
14. Fleming, W. L., Brown, W. J., Donohue, F., and others: National survey of venereal disease treated by physicians in 1968, J.A.M.A. **211**:1827, 1970.
15. Golden, B., and Thompson, H. S.: Implications of spiral forms in the eye, Survey Ophthal. **14**:179, 1968.
16. Goldman, E. E., McClean, J. H., and Smith, J. L.: Penicillins and aqueous humor, Amer. J. Ophthal. **65**:717, 1968.
17. Goldman, J. N., and Girard, K. F.: Intraocular treponemes in treated congenital syphilis, Arch. Ophthal. **78**:47, 1967.
18. Goldman, J. N., and Girard, K. F.: Fluorescent staining of treponemes in uveitis, Arch. Ophthal. **79**:716, 1968.
19. Hardy, J. B., Hardy, P. H., Oppenheimer, E. H., Ryan, S. J., and Sheff, R. N.: Failure of penicillin in a newborn with congenital syphilis, J.A.M.A. **212**:1345, 1970.
20. Hogan, M. J., and Zimmerman, L. E.: Ophthalmic pathology, Philadelphia, 1968, W. B. Saunders Co., p. 397.
21. Hunter, E. F., Deacon, W. E., and Meyer, P. E.: An improved FTA test for syphilis, the absorption procedure (FTA-ABS), Public Health Rep. **79**:410, 1964.
22. Igersheimer, J.: Optic atrophy in tabes, Arch. Ophthal. **42**:170, 1949.
23. Jarrett, W. H.: Dislocation of the lens, Arch. Ophthal. **78**:290, 1967.
24. Kennedy, C., and Carrol, F. D.: Optic neuritis in children, Arch. Ophthal. **63**:748, 1960.
25. Klauder, J. V., and Meyer, G. P.: Chorioretinitis of congenital syphilis, Arch. Ophthal. **48**:147, 1953.
26. Knapp, A.: Bilateral circumpapillary chorioretinitis with detachment of the retina in syphilis, Trans. Amer. Acad. Ophthal. Otolaryng. **25**:132, 1920.
27. Knox, J. M., and Glicksman, J. M.: The FTA-ABS test for syphilis: performance in 1033 patients, Brit. J. Vener. Dis. **42**:18, 1966.
28. Lakke, J. P. W. F.: Orbital fissure syndrome, Arch. Neurol. (Chicago) **7**:289, 1962.
29. Lennarson, V. E., and Jeans, P. C.: Congenital syphilis of the eye, Amer. J. Syph. Gon. Vener. Dis. **21**:90, 1937.
30. Levaditi, C., and Vaisman, A.: Sort des treponemes syphilomateux chez les lapins traites par la penicilline, Bull. Acad. Nat. Med. (Paris) **129**:394, 1945.
31. Licher, P. R., and Shaffer, R. N.: Interstitial keratitis and glaucoma, Amer. J. Ophthal. **68**: 241, 1969.
32. Loewenfeld, I. E.: The Argyll Robertson pupil 1869-1969: a critical survey of the literature, Survey Ophthal. **14**:199, 1969.
33. Lucas, J. B.: New developments in venereal disease research, Med. Ann. D. C. **35**:375, 1966.
34. Mahoney, J. F., and others: Penicillin treatment of early syphilis: a preliminary report, Amer. J. Public Health **33**:1387, 1943.
35. Merritt, H. H.: Textbook of neurology, ed. 3, Philadelphia, 1963, Lea and Febiger, pp. 128, 134, 137, 153.
36. Meyer, J., Boudin, G., Collart, P., and others: Latent syphilis, Concours. Med. **91**:4651, 1969.
37. Meyer, P. E., and Hunter, E. F.: Antigenic relationships of 14 treponemes by immunofluorescence, J. Bact. **93**:784, 1967.
38. Montenegro, E. N. R., Nicol, W. G., and Smith, J. L.: Treponemalike forms and artifacts, Amer. J. Ophthal. **68**:197, 1969.
39. Moore, J. E.: The relation of neurorecurrences to late syphilis, Arch. Neurol. Psych. **21**:117, 1929.
40. Moore, J. E.: Syphilitic iritis, Amer. J. Ophthal. **14**:121, 1931.
41. Moore, J. E., and Woods, A. C.: The pathology and pathogenesis of syphilitic primary optic atrophy, Amer. J. Ophthal. **23**:4, 1940.
42. Nelson, R. A., and Mayer, M. M.: Immobilization of T. pallidum in vitro by antibody produced in syphilitic infection, J. Exp. Med. **89**:369, 1949.
43. Rice, N. S. C., Dunlop, E. M. C., Jones, B., Hare, M. J., and others: Demonstration of treponema-like forms in cases of treated and untreated late syphilis, and of treated early syphilis, Brit. J. Vener. Dis. **46**:1, 1970.
44. Rice, N. S. C., Jones, B., and Wilkinson, A. E.: Study of late ocular syphilis, Trans. Ophthal. Soc. U. K. **88**:259, 1968.
45. Robertson, D. A.: Four cases of spinal myosis: with remarks on the action of light on the pupil, Edin. Med. J. **15**:487, 1869.
46. Scheie, H. G.: Site of disturbance of Adie's syndrome, Arch. Ophthal. **24**:225, 1940.

46a. Seeley, R. L., Sarker, M., and Smith, J. L.: The borderline fluorescent treponemal anti-

body absorption test reactor, Arch. Ophthal. **87**:16, 1972.
47. Slight, J. R.: Personal communication, 1967.
48. Smith, C. A., and others: Benzathine penicillin G in the treatment of syphilis, Bull. W.H.O. **15**:1087, 1956.
49. Smith, J. L.: Spirochetes in late seronegative syphilis, penicillin notwithstanding, Springfield, Ill., 1969, Charles C Thomas, Publisher, pp. 30, 90, 124, 255, 261, 305.
50. Smith, J. L., and Israel, C. W.: Spirochetes in the aqueous humor in seronegative ocular syphilis: persistence after penicillin therapy, Arch. Ophthal. **77**:474, 1967.
51. Smith, J. L., and Israel, C. W.: The presence of spirochetes in late seronegative syphilis, J.A.M.A. **199**:980, 1967.
52. Smith, J. L., and Israel, C. W.: Recovery of spirochetes in the monkey by passive transfer from human late seronegative syphilis, Brit. J. Vener. Dis. **44**:109, 1968.
53. Smith, J. L., Israel, C. W., and McCrary, J. A.: Recovery of T. pallidum from aqueous humor removed at cataract surgery in man by passive transfer to rabbit testes, Amer. J. Ophthal. **65**:242, 1968.
54. Stokes, J. H., Beerman, H., and Ingraham, N. R.: Modern clinical syphilology, Philadelphia, 1945, W. B. Saunders Co., pp. 59, 504, 604, 607, 609, 1010, 1011, 1113, 1114, 1115, 1117.
55. Syphilis—a synopsis, U. S. Public Health Service Publication No. 1660, Washington, D. C., 1967, U. S. Government Printing Office.
56. Thomas, C. I.: The cornea, Springfield, Ill., 1955, Charles C Thomas, Publisher, p. 581.
57. Venereal disease fact sheet, U. S. Public Health Service Publication, Washington, D. C., 1970, U. S. Government Printing Office.
58. Walsh, F. W., and Hoyt, W. F.: Clinical neuro-ophthalmology, vol. 2, ed. 3, Baltimore, 1969, The Williams and Wilkins Co., pp. 1578, 1583, 1592, 1593, 1605, 1959.
59. Whitfield, R., and Wirostko, E.: Uveitis and intraocular treponemes, Arch. Ophthal. **84**:12, 1970.
60. Whitfield, R., and Wirostko, E.: Unpublished data.
61. Woods, A. C.: Syphilis of the eye, Amer. J. Syph. Gon. Vener. Dis. **27**:133, 1943.
62. Wuepper, K. D.: Red eye as the presenting sign of syphilis d'emblee, Calif. Med. **107**:518, 1967.
63. Yobs, A. R., Clark, J. W., Jr., Mothershed, S. E., and others: Further observations on the persistence of T. pallidum after treatment in rabbits and humans, Brit. J. Vener. Dis. **44**:116, 1968.

Index